Table of Measurement Abbreviations

U.S. Customary System

Length

in.	inch
ft	foot
yd	yard
mi	mile

Capacity

oz	ounce
c	cup
qt	quart
gal	gallon

Weight

oz	ounce
lb	pound

Area

in^2	square inch
ft^2	square foot

Metric System

Length

mm	millimeter (0.001 m)
cm	centimeter (0.01 m)
dm	decimeter (0.1 m)
m	meter
dam	decameter (10 m)
hm	hectometer (100 m)
km	kilometer (1000 m)

Capacity

ml	milliliter (0.001 L)
cl	centiliter (0.01 L)
dl	deciliter (0.1 L)
L	liter
dal	decaliter (10 L)
hl	hectoliter (100 L)
kl	kiloliter (1000 L)

Weight/Mass

mg	milligram (0.001 g)
cg	centigram (0.01 g)
dg	decigram (0.1 g)
g	gram
dag	decagram (10 g)
hg	hectogram (100 g)
kg	kilogram (1000 g)

Area

cm^2	square centimeter
m^2	square meter

Time

h	hour
min	minute
s	second

Table of Symbols

+	add
−	subtract
·, (a)(b)	multiply
$\frac{a}{b}$, ÷	divide
( )	parentheses, a grouping symbol
[ ]	brackets, a grouping symbol
π	pi, a number approximately equal to $\frac{22}{7}$ or 3.14
−a	the opposite, or additive inverse, of a
$\frac{1}{a}$	the reciprocal, or multiplicative inverse, of a
=	is equal to
≈	is approximately equal to
≠	is not equal to
<	is less than
≤	is less than or equal to
>	is greater than
≥	is greater than or equal to

(a, b)	an ordered pair whose first component is a and whose second component is b
°	degree (for angles and temperature)
$\sqrt{a}$	the principal square root of a
|a|	the absolute value of a
HM^3	reference to the HM^3 Tutorial
	reference to the Videotapes Library
	indicates graphing calculator topics
	indicates writing exercises
	indicates data analysis exercises
	indicates exercises referencing the Internet

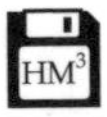

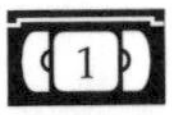

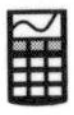

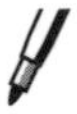

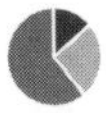

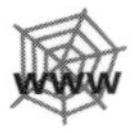

INSTRUCTOR'S ANNOTATED EDITION

Intermediate Algebra

with Applications

Fifth Edition

Richard N. Aufmann
Palomar College, California

Vernon C. Barker
Palomar College, California

Joanne S. Lockwood
Plymouth State College, New Hampshire

HOUGHTON MIFFLIN COMPANY Boston New York

Senior Sponsoring Editor: Maureen O'Connor
Senior Associate Editor: Dawn M. Nuttall
Editorial Assistant: Amanda Bafaro
Senior Project Editor: Nancy Blodget
Editorial Assistant: Joy Park
Senior Production/Design Coordinator: Carol Merrigan
Senior Manufacturing Coordinator: Sally Culler
Marketing Manager: Ros Kane

Cover design by Diana Coe/ko Design Studio
Cover photograph © Shigeru Tanaka/Photonica

Photo credits: *Chapter 1,* page 1: David Frazier/Photo Researchers, Inc. (#7J7829); *Chapter 2,* page 53: Peter Poulides/TSI; *Chapter 3,* page 117: Ed Lallo/Liaison International; *Chapter 4,* page 193: Bruce Ayers/TSI; *Chapter 5,* page 251: Brownie Harris/The Stock Market; *Chapter 6,* page 325: Michael Newman/PhotoEdit; *Chapter 7,* page 377: David Frazier/Photo Researchers, Inc.; *Chapter 8,* page 429: Bruce Hands/TSI; *Chapter 9,* page 495: Michael Rosenfeld/TSI; *Chapter 10,* page 539: Lee Snyder/Photo Researchers, Inc. (#7T0321); *Chapter 11,* page 597: Charles Thatcher/TSI; *Chapter 12,* page 637: Kathy Ferguson/PhotoEdit.

Printed in the U.S.A.

Library of Congress Catalog Card Number: 99-72029

ISBNs:
Text: 0-395-96961-1
Instructor's Annotated Edition: 0-395-96962-X

123456789-WC 03 02 01 00 99

CONTENTS

PREFACE

The fifth edition of *Intermediate Algebra with Applications* provides a mathematically sound and comprehensive coverage of the topics considered essential in an intermediate algebra course. The text has been designed not only to meet the needs of the traditional college student but also to serve the needs of returning students whose mathematical proficiency may have declined during their years away from formal education.

In this new edition of *Intermediate Algebra with Applications,* careful attention has been given to implementing the standards suggested by NCTM and AMATYC. Each chapter begins with a mathematical vignette consisting of an application, a historical note, or a curiosity related to mathematics as well as a career note. At the end of each section there are Applying Concepts exercises that include writing, synthesis, critical thinking, and challenge problems. The chapter ends with Focus on Problem Solving and Projects and Group Activities. The Focus on Problem Solving feature demonstrates various proven problem-solving strategies and then asks the student to use the strategies to solve problems. The Projects and Group Activities feature extensions or applications of a concept that was covered in the chapter. These projects can be used for cooperative learning activities or extra credit.

Instructional Features

Interactive Approach

Intermediate Algebra with Applications uses an interactive style that gives the student an opportunity to try a skill as it is presented. Each section is divided into objectives, and every objective contains one or more sets of matched-pair examples. The first example in each pair is worked out; the second example, labeled Problem, is for the student to work. By solving this problem, the student practices using concepts as they are presented in the text. There are *complete* worked-out solutions to these problems in a Solutions section at the end of the book. By comparing their solutions to model solutions, students can get immediate feedback on and reinforcement of the concepts.

Emphasis on Problem-Solving Strategies

Intermediate Algebra with Applications features a carefully sequenced approach to application problems that emphasizes using proven strategies to solve problems. Students are encouraged to develop their own strategies, to draw diagrams, and to write their strategies as part of the solution to each application problem. In each case, model strategies are presented as guides for students to follow as they attempt the matched-pair Problem.

Emphasis on Applications

The traditional approach to teaching algebra covers only the straightforward manipulation of numbers and variables and thereby fails to teach students the practical value of algebra. By contrast, *Intermediate Algebra with Applications* contains an extensive collection of contemporary application problems. Wherever appropriate, the last objective of a section presents applications that require the student to use the skills covered in that section to solve practical problems. This carefully integrated, applied approach generates awareness on the student's part of the value of algebra as a real-life tool.

Integrated Learning System Organized by Objectives

Each chapter begins with a list of the learning objectives included within that chapter. Each objective is then restated in the chapter to remind the student of the current topic of discussion. The same objectives that organize the text are reflected in the structure for exercises, for the testing programs, and for the HM³ Tutorial. Associated with every objective in the text is a corresponding computer tutorial and a corresponding set of test questions.

The Interactive Approach

Instructors have long recognized the need for a text that requires the student to use a skill as it is being taught. *Intermediate Algebra with Applications* uses an interactive technique that meets this need. Every objective, including the one shown below, contains at least one pair of examples. One of the examples is worked.

SECTION 6.1

Introduction to Rational Functions

1 Find the domain of a rational function

An explanatory passage begins each skill objective.

An expression in which the numerator and denominator are polynomials is called a **rational expression.** Examples of rational expressions are shown at the right.

$$\frac{9}{z} \qquad \frac{3x+4}{2x^2+1} \qquad \frac{x^3-x+1}{x^2-3x-5}$$

The expression $\frac{\sqrt{x}+3}{x}$ is not a rational expression because $\sqrt{x}+3$ is not a polynomial.

A function that is written in terms of a rational expression is a **rational function.** Each of the following equations represents a rational function.

$$f(x)=\frac{x^2+3}{2x-1} \qquad g(t)=\frac{3}{t^2-4} \qquad R(z)=\frac{z^2+3z-1}{z^2+z-12}$$

To evaluate a rational function, replace the variable by its value. Then simplify.

Paired examples follow the explanatory passage.

Example 1 Given $f(x)=\frac{3x-4}{x^2-2x+1}$, find $f(-3)$.

Solution

$f(x)=\frac{3x-4}{x^2-2x+1}$

$f(-3)=\frac{3(-3)-4}{(-3)^2-2(-3)+1}$ ▸ Substitute −3 for x.

$f(-3)=\frac{-9-4}{9+6+1}$

$f(-3)=\frac{-13}{16}$

$f(-3)=-\frac{13}{16}$

The Problem is the key to the interactive approach. It has not been worked so that the student may practice the skill, referring to the worked example above if necessary.

Problem 1 Given $f(x)=\frac{3-5x}{x^2+5x+6}$, find $f(2)$.

Reference to the page where the problem is solved in the Solutions section allows the student to check solutions immediately.

Solution See page S16.

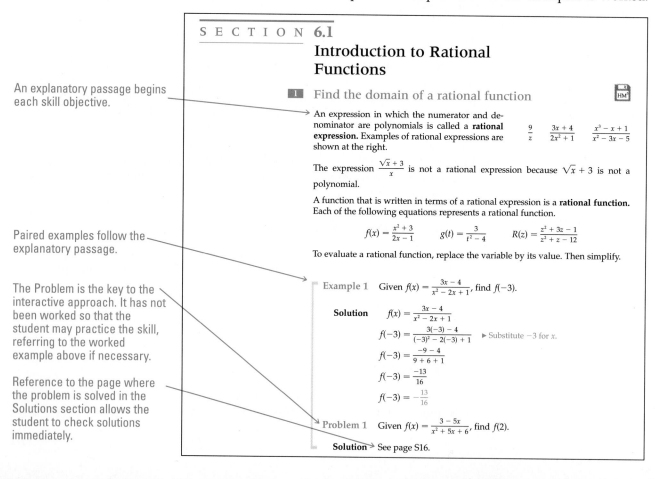

The second example in the pair (the Problem) is not worked so that the student may "interact" with the text by solving it. In order to provide immediate feedback, a complete solution to this Problem is provided in the Solutions section. The benefit of this interactive style is that students can check whether they have learned the new skill before they attempt a homework assignment.

Emphasis on Applications

The solution of an application problem in *Intermediate Algebra with Applications* comprises two parts: **Strategy** and **Solution**. The strategy is a written description of the steps that are necessary to solve the problem; the solution is the implementation of the strategy. Using this format provides students with a structure for problem solving. It also encourages students to write strategies for solving problems, which in turn fosters their organizing problem-solving strategies in a logical way. Having students write strategies is a natural way to incorporate writing into the math curriculum.

A strategy that the student may use in solving an application problem is stated.

This strategy is used in the solution of the worked example.

When students compare their solutions to those in the Solutions section, they will see a complete solution of the problem along with a written strategy for solving the problem.

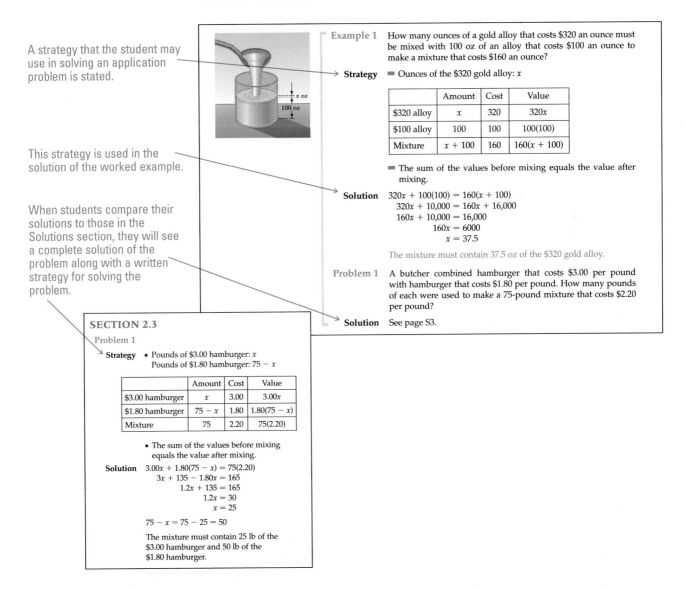

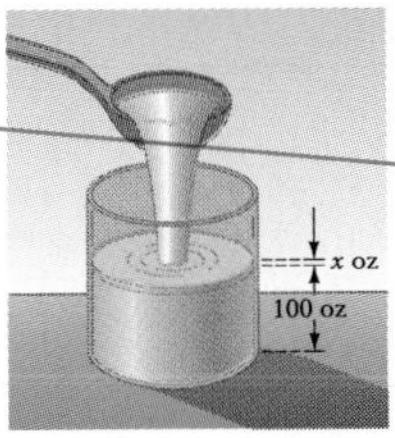

Example 1   How many ounces of a gold alloy that costs $320 an ounce must be mixed with 100 oz of an alloy that costs $100 an ounce to make a mixture that costs $160 an ounce?

Strategy   ■ Ounces of the $320 gold alloy: x

	Amount	Cost	Value
$320 alloy	x	320	$320x$
$100 alloy	100	100	$100(100)$
Mixture	$x + 100$	160	$160(x + 100)$

■ The sum of the values before mixing equals the value after mixing.

Solution
$$320x + 100(100) = 160(x + 100)$$
$$320x + 10,000 = 160x + 16,000$$
$$160x + 10,000 = 16,000$$
$$160x = 6000$$
$$x = 37.5$$

The mixture must contain 37.5 oz of the $320 gold alloy.

Problem 1   A butcher combined hamburger that costs $3.00 per pound with hamburger that costs $1.80 per pound. How many pounds of each were used to make a 75-pound mixture that costs $2.20 per pound?

Solution   See page S3.

SECTION 2.3

Problem 1

Strategy   ■ Pounds of $3.00 hamburger: x
Pounds of $1.80 hamburger: $75 - x$

	Amount	Cost	Value
$3.00 hamburger	x	3.00	$3.00x$
$1.80 hamburger	$75 - x$	1.80	$1.80(75 - x)$
Mixture	75	2.20	$75(2.20)$

■ The sum of the values before mixing equals the value after mixing.

Solution
$$3.00x + 1.80(75 - x) = 75(2.20)$$
$$3x + 135 - 1.80x = 165$$
$$1.2x + 135 = 165$$
$$1.2x = 30$$
$$x = 25$$

$$75 - x = 75 - 25 = 50$$

The mixture must contain 25 lb of the $3.00 hamburger and 50 lb of the $1.80 hamburger.

The Objective-Specific Approach

Many texts in mathematics are not organized in a manner that facilitates management of learning. Typically, students are left to wander through a maze of apparently unrelated lessons, exercise sets, and tests. *Intermediate Algebra with Applications* solves this problem by organizing all lessons, exercises sets, computer tutorials, and tests around a carefully constructed hierarchy of objectives. The advantage of this objective-by-objective organization is that it enables the student who is uncertain at any step in the learning process to refer easily to the original presentation and review that material.

The objective-specific approach also gives the instructor greater control over the management of student progress. The computerized testing program and the

An objective statement names the topic of each lesson.

The exercise sets correspond to the objectives in the text.

The answers to the odd-numbered exercises are provided in the Answer section.

The answers to the Chapter Review Exercises, the Chapter Test, and the Cumulative Review Exercises show the objective to study if the student incorrectly answers the exercise.

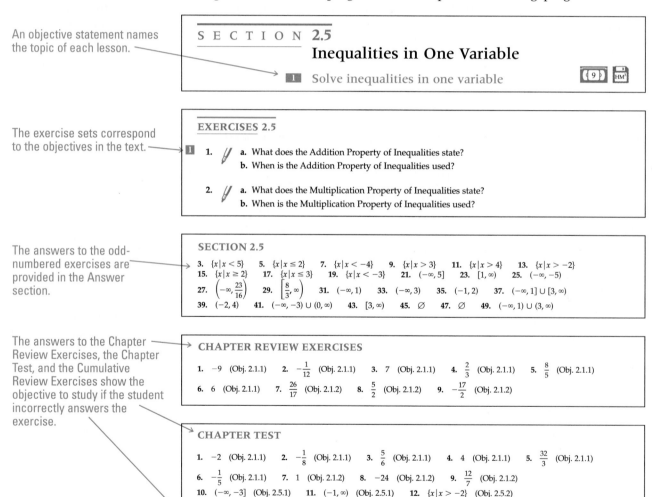

SECTION **2.5**

Inequalities in One Variable

1 Solve inequalities in one variable

EXERCISES 2.5

1. **a.** What does the Addition Property of Inequalities state?
   b. When is the Addition Property of Inequalities used?

2. **a.** What does the Multiplication Property of Inequalities state?
   b. When is the Multiplication Property of Inequalities used?

SECTION 2.5

3. $\{x | x < 5\}$ **5.** $\{x | x \le 2\}$ **7.** $\{x | x < -4\}$ **9.** $\{x | x > 3\}$ **11.** $\{x | x > 4\}$ **13.** $\{x | x > -2\}$
15. $\{x | x \ge 2\}$ **17.** $\{x | x \le 3\}$ **19.** $\{x | x < -3\}$ **21.** $(-\infty, 5]$ **23.** $[1, \infty)$ **25.** $(-\infty, -5)$
27. $\left(-\infty, \frac{23}{16}\right)$ **29.** $\left[\frac{8}{3}, \infty\right)$ **31.** $(-\infty, 1)$ **33.** $(-\infty, 3)$ **35.** $(-1, 2)$ **37.** $(-\infty, 1] \cup [3, \infty)$
39. $(-2, 4)$ **41.** $(-\infty, -3) \cup (0, \infty)$ **43.** $[3, \infty)$ **45.** $\varnothing$ **47.** $\varnothing$ **49.** $(-\infty, 1) \cup (3, \infty)$

CHAPTER REVIEW EXERCISES

1. -9 (Obj. 2.1.1)   **2.** $-\frac{1}{12}$ (Obj. 2.1.1)   **3.** 7 (Obj. 2.1.1)   **4.** $\frac{2}{3}$ (Obj. 2.1.1)   **5.** $\frac{8}{5}$ (Obj. 2.1.1)

6. 6 (Obj. 2.1.1)   **7.** $\frac{26}{17}$ (Obj. 2.1.2)   **8.** $\frac{5}{2}$ (Obj. 2.1.2)   **9.** $-\frac{17}{2}$ (Obj. 2.1.2)

CHAPTER TEST

1. -2 (Obj. 2.1.1)   **2.** $-\frac{1}{8}$ (Obj. 2.1.1)   **3.** $\frac{5}{6}$ (Obj. 2.1.1)   **4.** 4 (Obj. 2.1.1)   **5.** $\frac{32}{3}$ (Obj. 2.1.1)

6. $-\frac{1}{5}$ (Obj. 2.1.1)   **7.** 1 (Obj. 2.1.2)   **8.** -24 (Obj. 2.1.2)   **9.** $\frac{12}{7}$ (Obj. 2.1.2)
10. $(-\infty, -3]$ (Obj. 2.5.1)   **11.** $(-1, \infty)$ (Obj. 2.5.1)   **12.** $\{x | x > -2\}$ (Obj. 2.5.2)

CUMULATIVE REVIEW EXERCISES

1. -108 (Obj. 1.2.3)   **2.** 3 (Obj. 1.2.4)   **3.** -64 (Obj. 1.2.4)   **4.** -8 (Obj. 1.3.2)   **5.** The Commutative
Property of Addition (Obj. 1.3.1)   **6.** $\{3, 9\}$ (Obj. 1.1.2)   **7.** $-17x + 2$ (Obj. 1.3.3)   **8.** $25y$ (Obj. 1.3.3)
9. 2 (Obj. 2.1.1)   **10.** $\frac{1}{2}$ (Obj. 2.1.1)   **11.** 1 (Obj. 2.1.1)   **12.** 24 (Obj. 2.1.1)   **13.** 2 (Obj. 2.1.1)

14. 2 (Obj. 2.1.2)   **15.** $-\frac{13}{5}$ (Obj. 2.1.2)   **16.** $\{x | x \le -3\}$ (Obj. 2.5.1)   **17.** $\varnothing$ (Obj. 2.5.2)

printed testing program are organized in terms of the same objectives as the text. These references are provided with the answers to the test items. This allows the instructor to identify quickly those objectives for which a student may need additional instruction.

The HM^3 Tutorial is also organized around the objectives of the text. As a result, supplemental instruction is available for any objectives that are troublesome for a student.

Features of This Edition

Topical Coverage

Intermediate Algebra with Applications carefully integrates a balance of applications and skill development to help students understand the connection between mathematics and its application. The concept of function is developed early in the text and is used in a variety of situations as a model of how one quantity depends on another. Examples of polynomial, exponential, and logarithmic functions are presented at appropriate times in the text.

The application problems in the text are diverse and contemporary. We have included applications from business, sports, economics, and medicine to name but a few. Many of these applications are new and include topics not included in earlier editions. A complete list of the applications can be found in the Index of Applications.

Margin Notes

There are two types of margin notes in the student text. *Point of Interest* notes feature interesting sidelights of the topic being discussed. The *Look Closely* feature warns students that a procedure may be particularly involved or reminds students that there are certain checks of their work that should be performed. In addition, there are *Instructor Notes* that are printed only in the Instructor's Annotated Edition. These notes provide suggestions for presenting the material or related material that can be used in class.

Study Tips

Effective study skills are an important factor in achieving success in any discipline. This is especially true in mathematics. To help students acquire these skills, we offer Study Tips boxes throughout the first few chapters of the text. Each box contains a suggestion that will lead to improved study habits.

Focus on Problem Solving

Although successful problem solvers use a variety of techniques, it has been well established that there are basic recurring strategies. Among these are trying to solve a related problem, finding a counterexample, trying to solve an easier problem, working backwards, and trial and error. The Focus on Problem Solving features present some of these methods in the context of a problem to be solved. Students are encouraged to apply these strategies in solving similar problems.

Projects and Group Activities

The Projects and Group Activities appear near the end of each chapter. These projects can be used for extra credit or as cooperative learning activities. Through these projects, some of the strategies suggested by AMATYC can be implemented. These projects offer an opportunity for students to explore topics relating to functions, geometry, statistics, science, and business.

HM^3 Tutorial

This state-of-the-art tutorial software is a networkable, interactive, algorithmically driven software package that supports *every* objective in the text. Written by the authors, the HM^3 Tutorial and the text are in the same voice. Features include full-color graphics, a glossary, extensive hints, animated examples, and a comprehensive classroom management system.

The algorithmic feature essentially provides an infinite number of practice problems for students to attempt. The algorithms have been carefully crafted to present a variety of problem types from easy to difficult. Helpful hints and a complete worked-out solution are available for every problem. When a student completes a problem, there is an option to repeat a similar problem or to move on to another type of problem. A quiz feature is now also part of the package.

The interactive feature asks students to respond to questions about the topic in the current lesson. In this way, students can assess their understanding of concepts as they are presented. These interactive questions are also algorithmically driven and offer hints as well as a full solution.

The user-friendly classroom management system allows instructors to create a syllabus, post notes to individual students or a class, enter their own assessment items (i.e., homework, class participation, etc.), and choose the mastery level. There are a variety of printable reports offered; these can be called up by student, by class, by objective, etc., and many reports will tell at a glance which students are not achieving the mastery level the instructor specified.

Index of Applications

The Index of Applications that follows the Preface provides a quick reference for application problems from a wide variety of fields.

Graphing Calculators

Graphing calculators are incorporated as an optional feature at appropriate places throughout the text. Graphing calculator material is designated by the graphing calculator icon shown at the beginning of this paragraph. The graphing calculator may help some students as they struggle with new concepts. However, all graphing calculator material can be omitted without destroying the integrity of the course. Students can consult the appendix, Guidelines for Using Graphing Calculators, for help with keystroking procedures for several models of graphing calculators.

Chapter Summaries

The Chapter Summaries have been written to be a useful guide for students as they review for a test. The Chapter Summary includes the Key Words and the Essential Rules and Procedures that were covered in the chapter. Each key word and essential rule is accompanied by an example of that concept.

Glossary

There is a Glossary that includes definitions of terms used in the text.

Exercises

End-of-Section Exercises

Intermediate Algebra with Applications contains more than 6000 exercises. At the end of each section there are exercise sets that are keyed to the corresponding learning objectives. The exercises have been carefully developed to ensure that students can apply the concepts in the section to a variety of problem situations.

Concept Review Exercises

These "Always true, Sometimes true, or Never true" exercises precede every exercise set and are designed to test a student's understanding of the material. Using Concept Review exercises as oral exercises at the end of a class session can lead to interesting class discussions.

Applying Concepts Exercises

The end-of-section exercises are followed by Applying Concepts exercises. These contain a variety of exercise types:

- Challenge problems (designated by [C] in the Instructor's Annotated Edition only)
- Problems that ask students to interpret and work with real-world data
- Writing exercises
- Problems that ask students to determine incorrect procedures
- Problems that require a more in-depth analysis

Data Exercises

These exercises, designated by 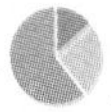, ask students to analyze and solve problems taken from actual situations. Students are often required to work with tables, graphs, and charts drawn from a variety of disciplines.

Writing Exercises

Writing exercises, denoted by 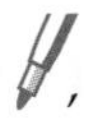, occur at the beginning of many exercise sets and within Applying Concepts exercises. These exercises ask students to write about a topic in the section or to research and report on a related topic. There are also writing exercises in some of the application problems. These exercises ask students to write a sentence that describes the meaning of their answers in the context of the problem.

Chapter Review Exercises and Chapter Tests

Chapter Review Exercises and a Chapter Test are found at the end of each chapter. These exercises are selected to help the student integrate all the topics presented in the chapter. The answers to all Chapter Review and Chapter Test exercises are given in the Answer section. Along with the answer, there is a reference to the objective that pertains to each exercise.

Cumulative Review Exercises and Final Exam

Cumulative Review Exercises, which appear at the ends of Chapters 2–11, help the student maintain skills learned in previous chapters. In addition, a Final Exam appears at the end of Chapter 12. The answers to all Cumulative Review and Final Exam exercises are given in the Answer section. Along with the answer, there is a reference to the objective that pertains to each exercise.

New to This Edition

The sequence of chapters has been slightly changed so as to allow a better pedagogical organization. The chapter on *Rational Expressions,* which requires a lot of factoring, is now Chapter 6, and immediately follows Chapter 5, *Polynomials and Exponents,* where factoring is introduced. The chapter on *Rational Exponents and Radicals* is now Chapter 7, which lends support to Chapter 8, *Quadratic Equations and Inequalities.*

Functions are still introduced in Chapter 3, *Linear Functions and Inequalities in Two Variables,* but the material, which many students find difficult to grasp, has been expanded to include more detailed explanations. In particular, a more detailed and illustrated definition of domain and range has been added.

The calculator material within the exposition has been expanded. We have given more prominence to the in-text calculator commentary by boxing this material and identifying it with an icon.

We have more than doubled the number of projects found at the end of each chapter. In addition, those projects involving calculators and internet sites have been marked for easy identification with the appropriate icon. Graph interpretation exercises have also been added to the projects.

Career notes and photos have been added to the chapter openers to illustrate to students the diverse ways in which mathematics is used in the workplace.

In response to suggestions by users, the writing and data icons appear in the student text as well as in the instructor's edition. We have also integrated writing exercises within the exercise sets rather than limit them to the Applying Concepts section of the exercises.

All the real sourced data exercises and examples have been updated, and many new data problems have been added. We have also titled application exercises to help motivate student interest.

To be consistent with our objective-specific approach, we have expanded the Table of Contents to include objective statements.

Supplements for the Instructor

Instructor's Annotated Edition

The Instructor's Annotated Edition is an exact replica of the student text except that answers to all exercises are given in the text. Also, there are Instructor Notes in the margins that offer suggestions for presenting the material in that objective.

Instructor's Resource Manual with Solutions Manual

The Instructor's Resource Manual includes suggestions for course sequencing and gives sample answers for the writing exercises. The Solutions Manual contains worked-out solutions for all end-of-section exercises, Concept Review exercises, Focus on Problem Solving exercises, Projects and Group Activities, Chapter Review exercises, Chapter Test exercises, Cumulative Review exercises, and Final Exam exercises.

Computerized Test Generator

The Computerized Test Generator is the first of three sources of testing material. The database contains more than 2000 test items. These questions are unique to the Test Generator. The Test Generator is designed to provide an unlimited number of chapter tests, cumulative chapter tests, and final exams. It is available for Microsoft Windows® and the Macintosh. Both versions provide **algorithms, online testing,** and **gradebook** functions.

Printed Test Bank with Chapter Tests

The Printed Test Bank, the second component of the testing material, is a printout of all items in the Computerized Test Generator. Instructors who do not have access to a computer can use the test bank to select items to include on tests prepared by hand. The Chapter Tests comprise the printed testing program, which is the third source of testing material. Eight printed tests, four free-response and four multiple-choice, are provided for each chapter. In addition, there are cumulative tests for use after Chapters 3, 6, 9, and 12, and a final exam.

Supplements for the Student

Student Solutions Manual

The *Student Solutions Manual* contains complete solutions to all odd-numbered exercises in the text.

HM^3 Tutorial

The content of this new interactive state-of-the-art tutorial software was written by the authors and is in the same voice as the text. The HM^3 Tutorial supports every objective in the text. Problems are algorithmically generated; lessons provide animated examples; lessons and problems are presented in a colorful, lively manner; and an integrated classroom management system tracks and reports student performance. Features include:

- assessment
- free-response problems
- ability to repeat same problem type
- printing capability
- glossary
- resizable screen
- pedagogical animations
- interactivity within lessons
- algorithms
- extensive classroom management with syllabus customization and a variety of reports

The HM^3 Tutorial can be used in several ways: (1) to cover material the student missed because of absence; (2) to reinforce instruction on a concept that the student has not yet mastered; (3) to review material in preparation for exams; and (4) for self-instruction.

This networkable HM^3 Tutorial is available on CD-ROM for Windows. There is also an alternate version offered for the Macintosh, available only on floppy disk. The HM^3 Tutorial is free to any school upon adoption of this text; however, students can purchase a non-networkable copy of the HM^3 Tutorial for home use.

Within each section of the book, a computer disk icon HM^3 appears next to each objective. The icon serves as a reminder that there is an HM^3 Tutorial lesson corresponding to that objective.

Videotapes

The videotape series contains lessons that accompany *Intermediate Algebra with Applications.* These lessons follow the format and style of the text and are closely tied to specific sections of the text. Each videotape begins with an application, and then the mathematics needed to solve that application is presented. The tape closes with the solution of the application problem.

1 Within each section of the book, a videotape icon appears next to each objective that is covered by a videotape lesson. The icon contains the reference number of the appropriate video.

Acknowledgments

The authors would like to thank the people who have reviewed this manuscript and provided many valuable suggestions:

Ron Harris, Bowling Green State University, OH
Carol Hughes, Westmoreland County Community College, PA
Cheryl Keeton, College of the Albemarle, NC

N. Elaine Kenzie, St. Clair County Community College, MI
LuAnn Malik, Community College of Aurora, CO
Linda J. Murphy, Northern Essex Community College, MA
Nancy K. Nickerson, Northern Essex Community College, MA
Lauri Semarne
William A. Steed, Jr., Saddleback College, CA
Harriet Thompson, Albany State University, GA
Grace Nicklas Warne

INDEX of Applications

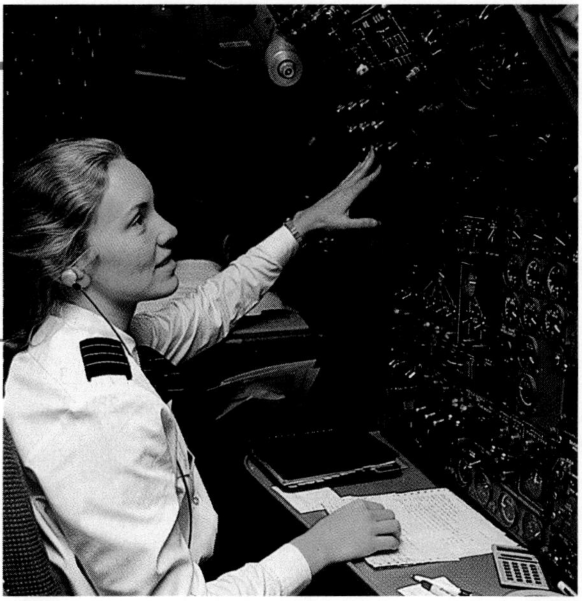

The job of a flight engineer is to assist the pilot and copilot of a large aircraft in checking the equipment on a plane before takeoff, to operate certain navigational devices during flight, to record performance levels, and to report any mechanical problems to the pilot. Airlines look for candidates who have completed at least two years of college, including courses in mathematics and science.

1

Review of Real Numbers

OBJECTIVES

Early Egyptian Number System

The early Egyptian type of picture writing shown at the left below is known as hieroglyphics.

The Egyptian hieroglyphic method of representing numbers differs from our modern version in an important way. For the hieroglyphic number, the symbol indicated the value. For example, the symbol ∩ meant 10. The symbol was repeated to get larger values.

The markings at the right represent the number 743. Each vertical stroke represents 1, each ∩ represents 10, and each 𓍢 represents 100.

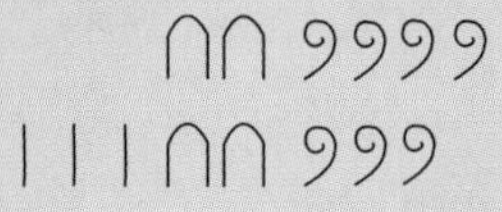

There are 3 ones, 4 tens, and 7 hundreds representing the number 743.

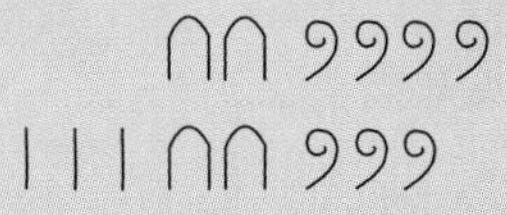

3 + 40 + 700

743

In our system, the position of a number is important. The number 5 in 356 means 5 tens, but the number 5 in 3517 means 5 hundreds. Our system is called a positional number system.

Consider the hieroglyphic number at the right. Note that the ∩ is missing. For the early Egyptians, when a certain group of ten was not needed, it was just omitted.

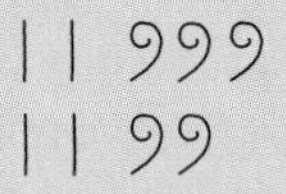

There are 4 ones and 5 hundreds. Thus the number 504 is represented by this group of markings.

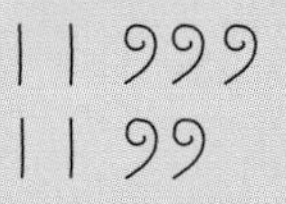

4 + 500

504

In a positional system of notation like ours, a zero is used to show that a certain group of ten is not needed. This may seem like a fairly simple idea, but it was not until near the end of the 7th century that a zero was introduced into a number system.

SECTION 1.1
Introduction to Real Numbers

1 Inequality and absolute value

POINT OF INTEREST

The Big Dipper, known to the Greeks as *Ursa Major,* the great bear, is a constellation that can be seen from northern latitudes. The stars of the Big Dipper are Alkaid, Mizar, Alioth, Megrez, Phecda, Merak, and Dubhe. The star at the bend of the handle, Mizar, is actually two stars, Mizar and Alcor. An imaginary line from Merak through Dubhe passes through Polaris, the north star.

It seems to be a human characteristic to put similar items into the same place. For instance, an astronomer places stars in *constellations* and a geologist divides the history of the earth into *eras.*

Mathematicians likewise place objects with similar properties in *sets.* A **set** is a collection of objects. The objects are called **elements** of the set. Sets are denoted by placing braces around the elements in the set.

The numbers that we use to count things, such as the number of books in a library or the number of CDs sold by a record store, have similar characteristics. These numbers are called the *natural numbers.*

$$\textbf{Natural numbers} = \{1, 2, 3, 4, 5, 6, 7, 8, 9, 10, 11, \ldots\}$$

Each natural number other than 1 is either a *prime* number or a *composite* number. A **prime number** is a number greater than 1 that is divisible (evenly) only by itself and 1. For example, 2, 3, 5, 7, 11, and 13 are the first six prime numbers. A natural number other than 1 that is not a prime number is a **composite number.** The numbers 4, 6, 8, and 9 are the first four composite numbers.

POINT OF INTEREST

The concept of zero developed very gradually over many centuries. It has been variously denoted by a blank space, by a dot, and finally as 0. Negative numbers, although evident in Chinese manuscripts dating from 200 B.C., were not fully integrated into mathematics until late in the 14th century.

The natural numbers do not have a symbol to denote the concept of none—for instance, the number of trees taller than 1000 feet. The *whole numbers* include zero and the natural numbers.

$$\textbf{Whole numbers} = \{0, 1, 2, 3, 4, 5, 6, 7, 8, \ldots\}$$

The whole numbers alone do not provide all the numbers that are useful in applications. For instance, a meteorologist needs numbers below zero and above zero.

$$\textbf{Integers} = \{\ldots, -5, -4, -3, -2, -1, 0, 1, 2, 3, 4, 5, \ldots\}$$

The integers $\ldots, -5, -4, -3, -2, -1$ are **negative integers.** The integers 1, 2, 3, 4, 5, ... are **positive integers.** Note that the natural numbers and the positive integers are the same set of numbers. The integer zero is neither a positive nor a negative integer.

Still other numbers are necessary to solve the variety of application problems that exist. For instance, a landscape architect may need to purchase irrigation pipe that has a diameter of $\frac{5}{8}$ in. Numbers that can be written in the form of a fraction, $\frac{p}{q}$, where p and q are integers and $q \neq 0$, are called *rational numbers.*

$$\textbf{Rational numbers} = \left\{\frac{p}{q}, \text{ where } p \text{ and } q \text{ are integers and } q \neq 0\right\}$$

Examples of rational numbers are $\frac{2}{3}$, $-\frac{9}{2}$, and $\frac{5}{1}$. Note that $\frac{5}{1} = 5$, so all integers are rational numbers. The number $\frac{4}{\pi}$ is not a rational number because π is not an integer.

A rational number written as a fraction can be written in decimal notation by dividing the numerator by the denominator.

Write $\frac{3}{8}$ as a decimal.

Divide 3 by 8.

$$\begin{array}{r} 0.375 \\ 8\overline{)3.000} \\ -2\,4 \\ \hline 60 \\ -56 \\ \hline 40 \\ -40 \\ \hline 0 \end{array}$$

0.375 ← This is a terminating decimal.

0 ← The remainder is zero.

$\frac{3}{8} = 0.375$

Write $\frac{2}{15}$ as a decimal.

Divide 2 by 15.

$$\begin{array}{r} 0.133 \\ 15\overline{)2.000} \\ -1\,5 \\ \hline 50 \\ -45 \\ \hline 50 \\ -45 \\ \hline 5 \end{array}$$

0.133 ← This is a repeating decimal.

5 ← The remainder is never zero.

$\frac{2}{15} = 0.1\overline{3}$ ← The bar over 3 indicates that this digit repeats.

INSTRUCTOR NOTE

The number 0.01001... shown at the right has a pattern but not a block of digits that repeat. This lack of a *repeating* block of digits distinguishes the number as an irrational number. Many students overlook this fact.

Some numbers cannot be written as terminating or repeating decimals. Such numbers include 0.01001000100001..., $\sqrt{7} \approx 2.6457513$, and $\pi \approx 3.1415927$. These numbers have decimal representations that neither terminate nor repeat. They are called **irrational numbers.** The rational numbers and the irrational numbers taken together are the *real numbers.*

Real numbers = {rational numbers and irrational numbers}

The relationship among the various sets of numbers is shown in the following figure.

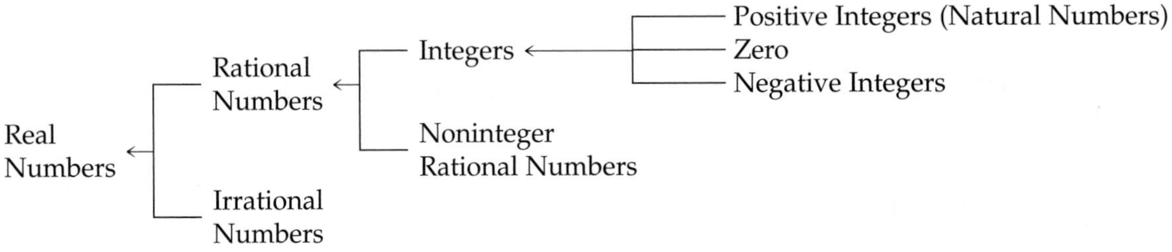

The **graph of a real number** is made by placing a heavy dot on a number line directly above the number. The graphs of some real numbers are shown below.

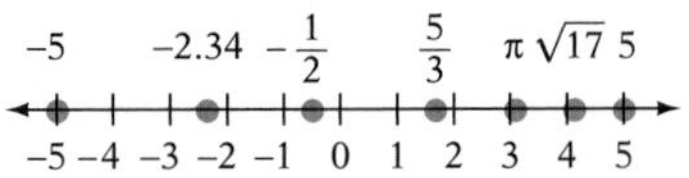

Consider these sentences:

A restaurant's chef prepared a dinner and served *it* to the customers.
A maple tree was planted, and *it* grew two feet in one year.

In the first sentence, "it" means dinner; in the second sentence, "it" means tree. In language, this word can stand for many different objects. Similarly, in mathematics, a letter of the alphabet can be used to stand for some number. A letter used in this way is called a **variable.**

It is convenient to use a variable to represent, or stand for, any one of the elements of a set. For instance, the statement "x is an element of the set $\{0, 2, 4, 6\}$" means that x can be replaced by 0, 2, 4, or 6. The set $\{0, 2, 4, 6\}$ is called the **domain** of the variable.

The symbol for "is an element of" is $\in$; the symbol for "is not an element of" is $\notin$. For example,

$$2 \in \{1, 2, 4, 6\} \qquad 6 \in \{0, 2, 4, 6\} \qquad 7 \notin \{0, 2, 4, 6\}$$

Variables are used in the next definition.

Definition of Inequality Symbols

If a and b are two real numbers and a is to the left of b on the number line, then a **is less than** b. This is written $a < b$.

If a and b are two real numbers and a is to the right of b on the number line, then a **is greater than** b. This is written $a > b$.

Here are some examples.

$$5 < 9 \qquad -4 > -10 \qquad \pi < \sqrt{17} \qquad 0 > -\frac{2}{3}$$

The inequality symbols $\leq$ (is less than or equal to) and $\geq$ (is greater than or equal to) are also important. Note the following examples.

$4 \leq 5$ is a true statement because $4 < 5$.

$5 \leq 5$ is a true statement because $5 = 5$.

The numbers 5 and -5 are the same distance from zero on the number line but on opposite sides of zero. The numbers 5 and -5 are called **additive inverses,** or **opposites,** of each other.

The additive inverse (or opposite) of 5 is -5. The additive inverse of -5 is 5. The symbol for additive inverse is $-$.

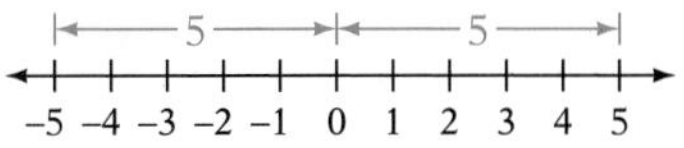

$-(4)$ means the opposite of *positive* 4. $\qquad -(4) = -4$

$-(-6)$ means the opposite of *negative* 6. $\qquad -(-6) = 6$

The **absolute value** of a number is a measure of its distance from zero on the number line. The symbol for absolute value is $|\ |$.

Note from the figure above that the distance from 0 to 5 is 5. Thus $|5| = 5$. The figure also shows that the distance from 0 to -5 is 5. Thus $|-5| = 5$.

Absolute Value

The absolute value of a positive number is the number itself. The absolute value of a negative number is the opposite of the negative number. The absolute value of zero is zero.

Evaluate: $-|-12|$

From the definition of absolute value, $|-12| = 12$. The absolute value sign does not affect the negative sign in front of the absolute value sign. $-|-12| = -12$

Example 1 Let $y \in \{-7, 0, 6\}$. For which values of y is the inequality $y < 4$ a true statement?

Solution

$y < 4$

$-7 < 4$ True ► Replace y by each of the elements of the set, and determine whether the inequality is true.

$0 < 4$ True

$6 < 4$ False

The inequality is true for -7 and 0.

Problem 1 Let $z \in \{-10, -5, 6\}$. For which values of z is the inequality $z > -5$ a true statement?

Solution See page S1. 6

Example 2 Let $a \in \{-12, 0, 4\}$. Determine $-a$, the opposite of a, for each element of the set.

Solution

$-a$

$-(-12) = 12$ ► Replace a by each element of the set, and determine the value of the expression.

$-(0) = 0$ ► 0 is neither positive nor negative.

$-(4) = -4$

Problem 2 Let $z \in \{-11, 0, 8\}$. Evaluate $|z|$ for each element of the set.

Solution See page S1. 11, 0, 8

2 Set operations and interval notation

HM³

The **roster method** of writing a set encloses the list of the elements of the set in braces. The set of whole numbers, written as $\{0, 1, 2, 3, 4, \ldots\}$, and the set of natural numbers, written as $\{1, 2, 3, 4, \ldots\}$, are **infinite sets.** The pattern of numbers continues without end. It is impossible to list all the elements of an infinite set.

The set of even natural numbers less than 10 is written $\{2, 4, 6, 8\}$. This is an example of a **finite set;** all the elements of the set can be listed.

The **empty set,** or **null set,** is the set that contains no elements. The symbol $\varnothing$ or $\{\,\}$ is used to represent the empty set.

INSTRUCTOR NOTE

Students will need a few examples of set-builder notation before they are comfortable with the idea.

A second method of representing a set is **set-builder notation.** Set-builder notation can be used to describe almost any set, but it is especially useful when writing infinite sets. In set-builder notation, the set of integers greater than -3 is written

$$\{x \mid x > -3, x \in \text{integers}\}$$

and is read "the set of all x such that x is greater than -3 and x is an element of the integers." This is an infinite set. It is impossible to list all the elements of the set, but the set can be described by using set-builder notation.

The set of real numbers less than 5 is written

$$\{x \mid x < 5, x \in \text{real numbers}\}$$

and is read "the set of all x such that x is less than 5 and x is an element of the real numbers."

Example 3 Use the roster method to write the set of natural numbers less than 10.

Solution $\{1, 2, 3, 4, 5, 6, 7, 8, 9\}$

Problem 3 Use the roster method to write the set of positive odd integers less than 12.

Solution See page S1. $\{1, 3, 5, 7, 9, 11\}$

Example 4 Use set-builder notation to write the set of integers greater than -8.

Solution $\{x \mid x > -8, x \in \text{integers}\}$

Problem 4 Use set-builder notation to write the set of real numbers less than 7.

Solution See page S1. $\{x \mid x < 7, x \in \text{real numbers}\}$

Just as operations such as addition and multiplication are performed on real numbers, operations are performed on sets. Two operations performed on sets are *union* and *intersection.*

POINT OF INTEREST

The symbols $\in$, $\cup$, and $\cap$ were first used by Giuseppe Peano in *Arithmetices Principia, Nova Exposita* (The Principle of Mathematics, a New Method of Exposition), published in 1889. The purpose of this book was to deduce the principles of mathematics from pure logic.

The **union** of two sets, written $A \cup B$, is the set of all elements that belong to either A **or** B. In set-builder notation, this is written

$$A \cup B = \{x \mid x \in A \quad \text{or} \quad x \in B\}$$

Given $A = \{2, 3, 4\}$ and $B = \{0, 1, 2, 3\}$, the union of A and B contains all the elements that belong to either A or B. The elements that belong to both sets are listed only once.

$$A \cup B = \{0, 1, 2, 3, 4\}$$

The **intersection** of two sets, written $A \cap B$, is the set of all elements that are common to both A **and** B. In set-builder notation, this is written

$$A \cap B = \{x \mid x \in A \quad \text{and} \quad x \in B\}$$

Given $A = \{2, 3, 4\}$ and $B = \{0, 1, 2, 3\}$, the intersection of A and B contains all the elements that are common to both A and B.

$$A \cap B = \{2, 3\}$$

Example 5   Find $C \cup D$ given $C = \{1, 5, 9, 13, 17\}$ and $D = \{3, 5, 7, 9, 11\}$.

Solution $C \cup D = \{1, 3, 5, 7, 9, 11, 13, 17\}$

Problem 5   Find $A \cup C$ given $A = \{-2, -1, 0, 1, 2\}$ and $C = \{-5, -1, 0, 1, 5\}$.

Solution   See page S1. $\{-5, -2, -1, 0, 1, 2, 5\}$

Example 6   Find $A \cap B$ given $A = \{x \mid x \in$ natural numbers$\}$ and $B = \{x \mid x \in$ negative integers$\}$.

Solution   There are no natural numbers that are also negative integers.
$A \cap B = \varnothing$

Problem 6   Find $E \cap F$ given $E = \{x \mid x \in$ odd integers$\}$ and $F = \{x \mid x \in$ even integers$\}$.

Solution   See page S1. $\varnothing$

Set-builder notation and the inequality symbols $>$, $<$, $\geq$, and $\leq$ are used to describe infinite sets of real numbers. These sets can be graphed on the real number line.

The graph of $\{x \mid x > -2, x \in$ real numbers$\}$ is shown below. The set is the real numbers greater than -2. The parenthesis on the graph indicates that -2 is not included in the set.

The graph of $\{x \mid x \geq -2, x \in$ real numbers$\}$ is shown below. The set is the real numbers greater than or equal to -2. The bracket at -2 indicates that -2 is included in the set.

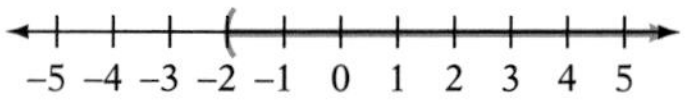

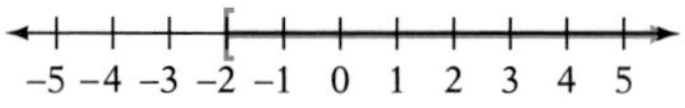

For the remainder of this section, all variables will represent real numbers. Using this convention, the above set is written $\{x \mid x \geq -2\}$.

INSTRUCTOR NOTE

There is a tendency for students to think that $x \geq -2$ means that $x = -2, -1, 0, 1, 2, \ldots$. These students forget that every real number greater than or equal to -2 satisfies the inequality. One useful oral exercise is to ask students the difference between various sets, such as $\{x \mid x > 2\}$ and $\{x \mid x > 2, x \in$ integers$\}$.

Example 7   Graph: $\{x \mid x \leq 3\}$

Solution

+++++++++++ number line -5 -4 -3 -2 -1 0 1 2 3 4 5

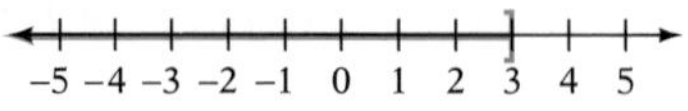

▶ The set is the real numbers less than or equal to 3. Draw a right bracket at 3, and darken the number line to the left of 3.

Problem 7 Graph: $\{x \mid x > -3\}$

Solution See page S1.

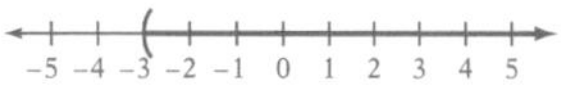

The union of two sets is the set of all elements belonging to either one or the other of the two sets. The set $\{x \mid x \leq -1\} \cup \{x \mid x > 3\}$ is the set of real numbers that are either less than or equal to -1 or greater than 3.

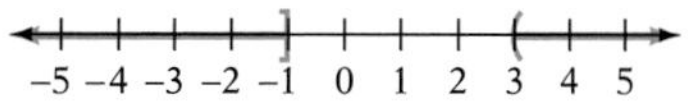

The set is written $\{x \mid x \leq -1 \text{ or } x > 3\}$.

The set $\{x \mid x > 2\} \cup \{x \mid x > 4\}$ is the set of real numbers that are either greater than 2 or greater than 4.

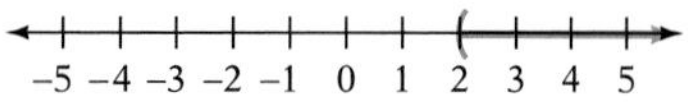

The set is written $\{x \mid x > 2\}$.

The intersection of two sets is the set that contains the elements common to both sets. The set $\{x \mid x > -2\} \cap \{x \mid x < 5\}$ is the set of real numbers that are greater than -2 and less than 5.

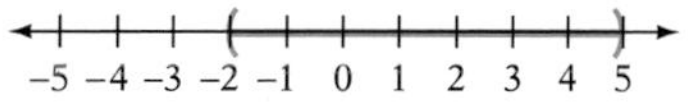

The set can be written $\{x \mid x > -2 \text{ and } x < 5\}$. However, it is more commonly written $\{x \mid -2 < x < 5\}$. This set is read "the set of all x such that x is greater than -2 and less than 5."

The set $\{x \mid x < 4\} \cap \{x \mid x < 5\}$ is the set of real numbers that are less than 4 and less than 5.

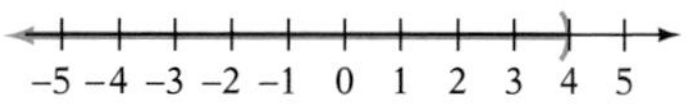

The set is written $\{x \mid x < 4\}$.

Example 8 Graph: $\{x \mid x < 0\} \cap \{x \mid x > -3\}$

Solution The set is $\{x \mid -3 < x < 0\}$.

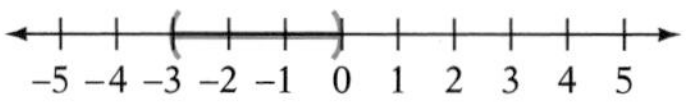

Problem 8 Graph: $\{x \mid x \geq 1\} \cup \{x \mid x \leq -3\}$

Solution See page S1.

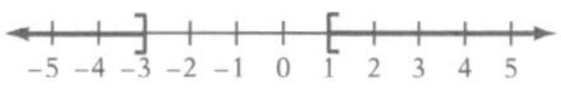

Some sets can also be expressed using **interval notation.** For example, the interval notation $(-3, 2]$ indicates the interval of all real numbers greater than -3 and less than or equal to 2. As on the graph of a set, the left parenthesis indicates that -3 is not included in the set. The right bracket indicates that 2 is included in the set.

An interval is said to be **closed** if it includes both endpoints; it is **open** if it does not include either endpoint. An interval is **half-open** if one endpoint is included and the other is not. In each example given below, -3 and 2 are the **endpoints** of the interval. In each case, the set notation, the interval notation, and the graph of the set are shown.

$\{x \mid -3 < x < 2\}$ — $(-3, 2)$ Open interval

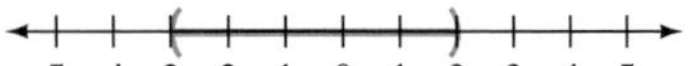

$\{x \mid -3 \le x \le 2\}$ — $[-3, 2]$ Closed interval

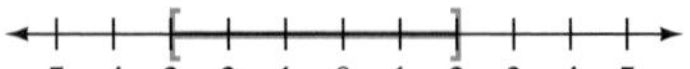

$\{x \mid -3 \le x < 2\}$ — $[-3, 2)$ Half-open interval

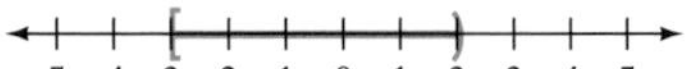

$\{x \mid -3 < x \le 2\}$ — $(-3, 2]$ Half-open interval

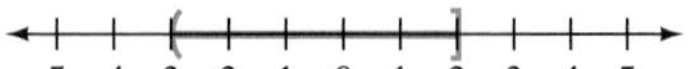

To indicate an interval that extends forever in one or both directions using interval notation, we use the **infinity symbol** ∞ or the **negative infinity symbol** $-\infty$. The infinity symbol is not a number; it is simply used as a notation to indicate that the interval is unlimited. In interval notation, a parenthesis is always used to the right of an infinity symbol or to the left of a negative infinity symbol, as shown in the following examples.

$\{x \mid x > 1\}$ — $(1, \infty)$

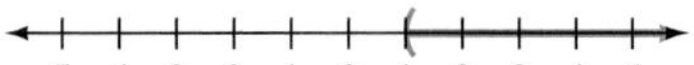

$\{x \mid x \ge 1\}$ — $[1, \infty)$

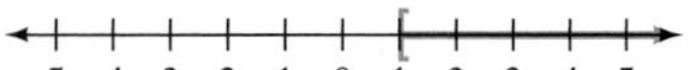

$\{x \mid x < 1\}$ — $(-\infty, 1)$

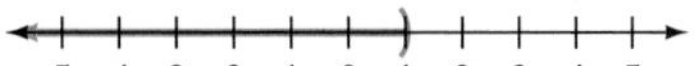

$\{x \mid x \le 1\}$ — $(-\infty, 1]$

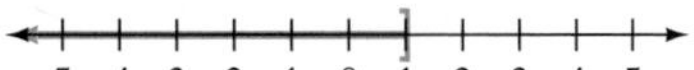

$\{x \mid -\infty < x < \infty\}$ — $(-\infty, \infty)$

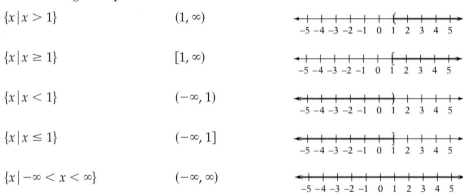

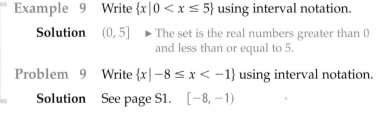

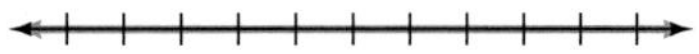

Example 9 Write $\{x \mid 0 < x \le 5\}$ using interval notation.

Solution $(0, 5]$ ▸ The set is the real numbers greater than 0 and less than or equal to 5.

Problem 9 Write $\{x \mid -8 \le x < -1\}$ using interval notation.

Solution See page S1. $[-8, -1)$

Example 10 Write $(-\infty, 9]$ using set-builder notation.

Solution $\{x \mid x \le 9\}$ ▸ The set is the real numbers less than or equal to 9.

Problem 10 Write $(-12, \infty)$ using set-builder notation.

Solution See page S1. $\{x \mid x > -12\}$

Example 11 Graph: $(-\infty, 3) \cap [-1, \infty)$

Solution

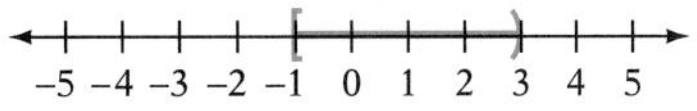

▶ $(-\infty, 3) \cap [-1, \infty)$ is the set of real numbers greater than or equal to -1 and less than 3.

Problem 11 Graph: $(-\infty, -2) \cup (-1, \infty)$

Solution See page S1.

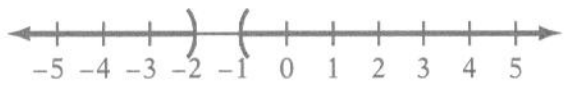

STUDY TIPS

Know Your Instructor's Requirements

To do your best in this course, you must know exactly what your instructor requires. If you don't, you probably will not meet his or her expectations and are not likely to earn a good grade in the course.

Instructors ordinarily explain course requirements during the first few days of class. Course requirements may be stated in a *syllabus*, which is a printed outline of the main topics of the course, or they may be presented orally. When they are listed in a syllabus or on other printed pages, keep them in a safe place. When they are presented orally, be sure to take complete notes. In either case, understand them completely and follow them exactly.

INSTRUCTOR NOTE

The Concept Review feature of this text appears before each exercise set. After presenting the material in the section, you may elect to use these exercises as oral classroom exercises or as a basis for class discussion on the topics of the lesson.

CONCEPT REVIEW 1.1

Determine whether the following statements are always true, sometimes true, or never true.

1. The absolute value of a number is positive. Sometimes true

2. Given $|x| = 3$, then x is a positive number. Sometimes true

3. Given x is an integer, then $-x$ is a negative integer. Sometimes true

4. If a, b, c, and d are real numbers, then $a < b$ and $c < d$ ensures that $a + c < b + d$. Always true

5. If a and b are real numbers, then $a < b < 0$ ensures that $a^2 < b^2$. Never true

6. If a, b, and c are nonzero real numbers and $a < b < c$, then $\frac{1}{a} < \frac{1}{b} < \frac{1}{c}$. Never true

7. If x is less than zero, then the absolute value of x is $-x$. Always true

EXERCISES 1.1

1 Determine which of the numbers are **a.** natural numbers, **b.** whole numbers, **c.** integers, **d.** positive integers, **e.** negative integers. List all that apply.

1. $-14, 9, 0, 53, 7.8, -626$
a. 9, 53 **b.** 0, 9, 53
c. $-14, 9, 0, 53, -626$
d. 9, 53 **e.** $-14, -626$

2. $31, -45, -2, 9.7, 8600, \frac{1}{2}$
a. 31, 8600 **b.** 31, 8600
c. $31, -45, -2, 8600$
d. 31, 8600 **e.** $-45, -2$

Determine which of the numbers are **a.** integers, **b.** rational numbers, **c.** irrational numbers, **d.** real numbers. List all that apply.

3. $-\frac{15}{2}, 0, -3, \pi, 2.\overline{33}, 4.232232223\ldots, \frac{\sqrt{5}}{4}, \sqrt{7}$
a. $0, -3$ **b.** $-\frac{15}{2}, 0, -3, 2.\overline{33}$
c. $\pi, 4.232232223\ldots, \frac{\sqrt{5}}{4}, \sqrt{7}$ **d.** all

4. $-17, 0.3412, \frac{3}{\pi}, -1.010010001\ldots, \frac{27}{91}, 6.1\overline{2}$
a. -17 **b.** $-17, 0.3412, \frac{27}{91}, 6.1\overline{2}$
c. $\frac{3}{\pi}, -1.010010001\ldots$ **d.** all

5. What is a terminating decimal? Provide an example.

6. What is a repeating decimal? Provide an example.

7. What is the additive inverse of a number?

8. What is the absolute value of a number?

Find the additive inverse of each of the following.

9. 27 -27 **10.** -3 3 **11.** $\frac{3}{4}$ $-\frac{3}{4}$ **12.** $\sqrt{17}$ $-\sqrt{17}$ **13.** 0 0

14. $-\pi$ π **15.** $-\sqrt{33}$ $\sqrt{33}$ **16.** -1.23 1.23 **17.** -91 91 **18.** $-\frac{2}{3}$ $\frac{2}{3}$

Solve.

19. Let $x \in \{-3, 0, 7\}$. For which values of x is $x < 5$ true? $-3, 0$

20. Let $z \in \{-4, -1, 4\}$. For which values of z is $z > -2$ true? $-1, 4$

21. Let $y \in \{-6, -4, 7\}$. For which values of y is $y > -4$ true? 7

22. Let $x \in \{-6, -3, 3\}$. For which values of x is $x < -3$ true? -6

23. Let $w \in \{-2, -1, 0, 1\}$. For which values of w is $w \leq -1$ true? $-2, -1$

24. Let $p \in \{-10, -5, 0, 5\}$. For which values of p is $p \geq 0$ true? 0, 5

25. Let $b \in \{-9, 0, 9\}$. Evaluate $-b$ for each element of the set. $9, 0, -9$

26. Let $a \in \{-3, -2, 0\}$. Evaluate $-a$ for each element of the set. 3, 2, 0

27. Let $c \in \{-4, 0, 4\}$. Evaluate $|c|$ for each element of the set. 4, 0, 4

28. Let $q \in \{-3, 0, 7\}$. Evaluate $|q|$ for each element of the set. 3, 0, 7

29. Let $m \in \{-6, -2, 0, 1, 4\}$. Evaluate $-|m|$ for each element of the set. $-6, -2, 0, -1, -4$

30. Let $x \in \{-5, -3, 0, 2, 5\}$. Evaluate $-|x|$ for each element of the set. $-5, -3, 0, -2, -5$

2 **31.** Explain the difference between the union of two sets and the intersection of two sets.

32. Explain the difference between $\{x \mid x < 5\}$ and $\{x \mid x \leq 5\}$.

Use the roster method to write the set.

33. the integers between -3 and 5 $\{-2, -1, 0, 1, 2, 3, 4\}$

34. the integers between -4 and 0 $\{-3, -2, -1\}$

35. the even natural numbers less than 14 $\{2, 4, 6, 8, 10, 12\}$

36. the odd natural numbers less than 14 $\{1, 3, 5, 7, 9, 11, 13\}$

37. the positive-integer multiples of 3 that are less than or equal to 30 $\{3, 6, 9, 12, 15, 18, 21, 24, 27, 30\}$

38. the negative-integer multiples of 4 that are greater than or equal to -20 $\{-20, -16, -12, -8, -4\}$

39. the negative-integer multiples of 5 that are greater than or equal to -35 $\{-35, -30, -25, -20, -15, -10, -5\}$

40. the positive-integer multiples of 6 that are less than or equal to 36 $\{6, 12, 18, 24, 30, 36\}$

Use set-builder notation to write the set.

41. the integers greater than 4 $\{x \mid x > 4, x \in \text{integers}\}$

42. the integers less than -2 $\{x \mid x < -2, x \in \text{integers}\}$

43. the real numbers greater than or equal to -2 $\{x \mid x \geq -2\}$

44. the real numbers less than or equal to 2 $\{x \mid x \leq 2\}$

45. the real numbers between 0 and 1 $\{x \mid 0 < x < 1\}$

46. the real numbers between -2 and 5 $\{x \mid -2 < x < 5\}$

47. the real numbers between 1 and 4, inclusive $\{x \mid 1 \leq x \leq 4\}$

48. the real numbers between 0 and 2, inclusive $\{x \mid 0 \leq x \leq 2\}$

Find $A \cup B$.

49. $A = \{1, 4, 9\}, B = \{2, 4, 6\}$ $\{1, 2, 4, 6, 9\}$

50. $A = \{-1, 0, 1\}, B = \{0, 1, 2\}$ $\{-1, 0, 1, 2\}$

51. $A = \{2, 3, 5, 8\}, B = \{9, 10\}$ $\{2, 3, 5, 8, 9, 10\}$

52. $A = \{1, 3, 5, 7\}, B = \{2, 4, 6, 8\}$ $\{1, 2, 3, 4, 5, 6, 7, 8\}$

53. $A = \{-4, -2, 0, 2, 4\}, B = \{0, 4, 8\}$ $\{-4, -2, 0, 2, 4, 8\}$

54. $A = \{-3, -2, -1\}, B = \{-2, -1, 0, 1\}$ $\{-3, -2, -1, 0, 1\}$

55. $A = \{1, 2, 3, 4, 5\}, B = \{3, 4, 5\}$ $\{1, 2, 3, 4, 5\}$

56. $A = \{2, 4\}, B = \{0, 1, 2, 3, 4, 5\}$ $\{0, 1, 2, 3, 4, 5\}$

Find $A \cap B$.

57. $A = \{6, 12, 18\}, B = \{3, 6, 9\}$ $\{6\}$

58. $A = \{-4, 0, 4\}, B = \{-2, 0, 2\}$ $\{0\}$

59. $A = \{1, 5, 10, 20\}, B = \{5, 10, 15, 20\}$ $\{5, 10, 20\}$

60. $A = \{1, 3, 5, 7, 9\}, B = \{1, 9\}$ $\{1, 9\}$

61. $A = \{1, 2, 4, 8\}, B = \{3, 5, 6, 7\}$ $\varnothing$

62. $A = \{-3, -2, -1, 0\}, B = \{1, 2, 3, 4\}$ $\varnothing$

63. $A = \{2, 4, 6, 8, 10\}, B = \{4, 6\}$ $\{4, 6\}$

64. $A = \{-9, -5, 0, 7\}, B = \{-7, -5, 0, 5, 7\}$ $\{-5, 0, 7\}$

Graph.

65. $\{x \mid -1 < x < 5\}$

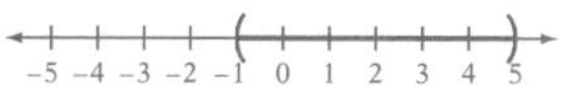

66. $\{x \mid 1 < x < 3\}$

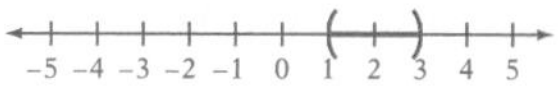

67. $\{x \mid 0 \le x \le 3\}$ 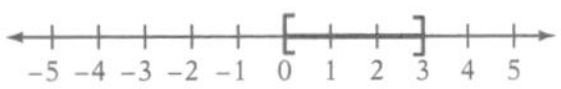

68. $\{x \mid -1 \le x \le 1\}$ 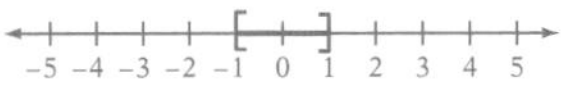

69. $\{x \mid x < 2\}$ 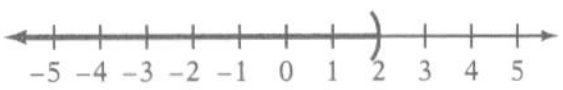

70. $\{x \mid x < -1\}$ 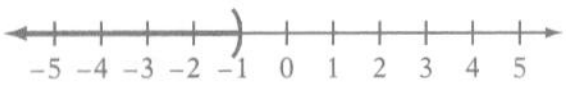

71. $\{x \mid x \ge 1\}$ 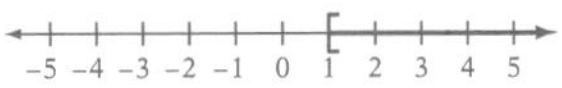

72. $\{x \mid x \le -2\}$ 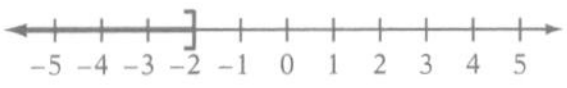

73. $\{x \mid x > 1\} \cup \{x \mid x < -1\}$ 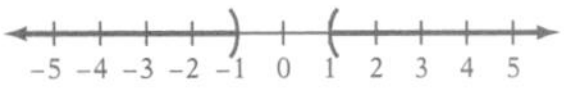

74. $\{x \mid x \le 2\} \cup \{x \mid x > 4\}$ 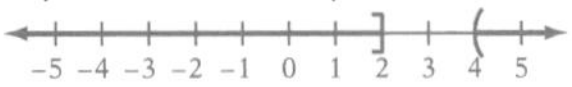

75. $\{x \mid x \le 2\} \cap \{x \mid x \ge 0\}$ 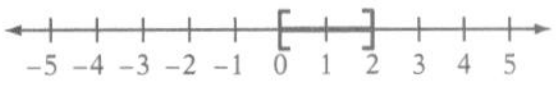

76. $\{x \mid x > -1\} \cap \{x \mid x \le 4\}$ 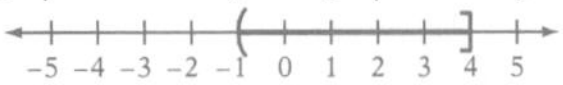

77. $\{x \mid x > 1\} \cap \{x \mid x \ge -2\}$ 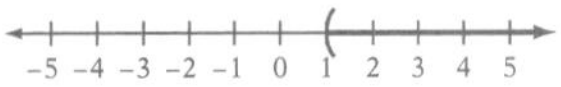

78. $\{x \mid x < 4\} \cap \{x \mid x \le 0\}$ 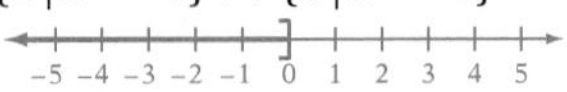

79. $\{x \mid x > 2\} \cup \{x \mid x > 1\}$ 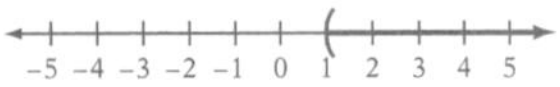

80. $\{x \mid x < -2\} \cup \{x \mid x < -4\}$ 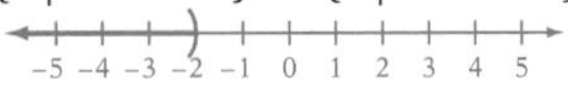

Write each interval in set-builder notation.

81. $(0, 8)$ $\{x \mid 0 < x < 8\}$
82. $(-2, 4)$ $\{x \mid -2 < x < 4\}$
83. $[-5, 7]$ $\{x \mid -5 \le x \le 7\}$
84. $[3, 4]$ $\{x \mid 3 \le x \le 4\}$
85. $[-3, 6)$ $\{x \mid -3 \le x < 6\}$
86. $(4, 5]$ $\{x \mid 4 < x \le 5\}$
87. $(-\infty, 4]$ $\{x \mid x \le 4\}$
88. $(-\infty, -2)$ $\{x \mid x < -2\}$
89. $(5, \infty)$ $\{x \mid x > 5\}$
90. $[-2, \infty)$ $\{x \mid x \ge -2\}$

Write each set of real numbers in interval notation.

91. $\{x \mid -2 < x < 4\}$ $(-2, 4)$
92. $\{x \mid 0 < x < 3\}$ $(0, 3)$
93. $\{x \mid -1 \le x \le 5\}$ $[-1, 5]$
94. $\{x \mid 0 \le x \le 3\}$ $[0, 3]$
95. $\{x \mid x < 1\}$ $(-\infty, 1)$
96. $\{x \mid x \le 6\}$ $(-\infty, 6]$
97. $\{x \mid -2 \le x < 6\}$ $[-2, 6)$
98. $\{x \mid x \ge 3\}$ $[3, \infty)$
99. $\{x \mid x \in \text{real numbers}\}$ $(-\infty, \infty)$
100. $\{x \mid x > -1\}$ $(-1, \infty)$

Graph.

101. $(-2, 5)$ 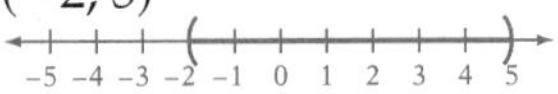

102. $(0, 3)$ 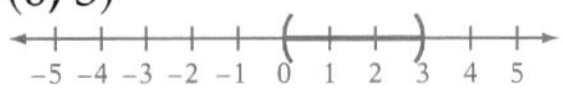

103. $[-1, 2]$ 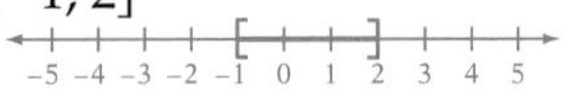

104. $[-3, 2]$ 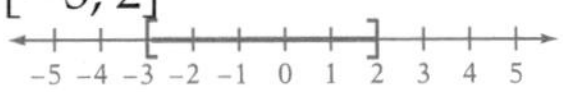

105. $(-\infty, 3]$ 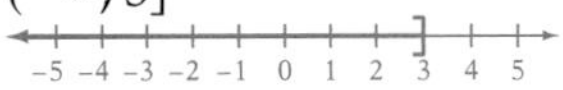

106. $(-\infty, -1)$ 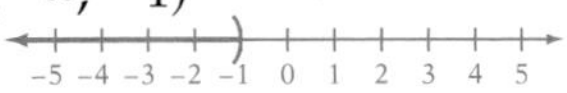

107. $[3, \infty)$ 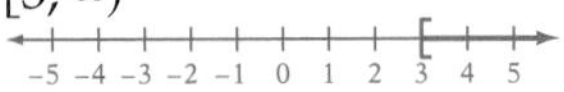

108. $[-2, \infty)$ 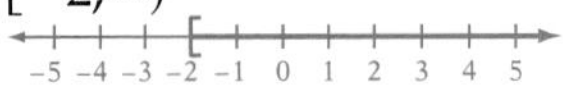

109. $(-\infty, 2] \cup [4, \infty)$ 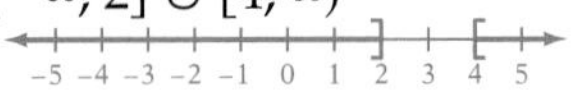

110. $(-3, 4] \cup [-1, 5)$ 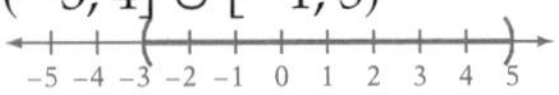

111. $[-1, 2] \cap [0, 4]$ 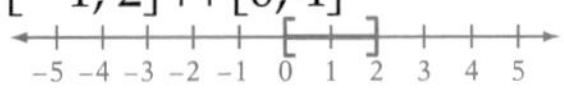

112. $[-5, 4) \cap (-2, \infty)$ 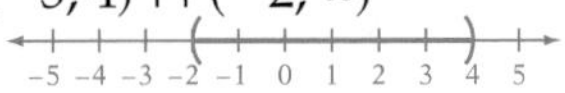

113. $(2, \infty) \cup (-2, 4]$ 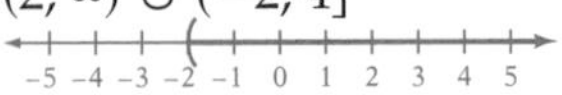

114. $(-\infty, 2] \cup (4, \infty)$ 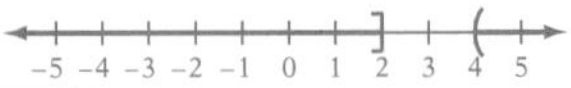

INSTRUCTOR NOTE

Each group of Applying Concepts exercises contains one or more challenge, data, or writing exercises, indicated by a [C], , or next to the exercise number. Sample answers for all writing exercises are given in the Instructor's Resource Manual.

APPLYING CONCEPTS 1.1

Let $R = \{\text{real numbers}\}$, $A = \{x \mid -1 \le x \le 1\}$, $B = \{x \mid 0 \le x \le 1\}$, $C = \{x \mid -1 \le x \le 0\}$, and $\varnothing$ = empty set. Answer the following using R, A, B, C, or $\varnothing$.

115. $A \cup B$ A
116. $A \cup A$ A
117. $B \cap B$ B
118. $A \cup C$ A
119. $A \cap R$ A

120. $C \cap R$ C **121.** $B \cup R$ R **122.** $A \cup R$ R **123.** $R \cup R$ R **124.** $R \cap \emptyset$ $\emptyset$

125. The set $B \cap C$ cannot be expressed using R, A, B, C, or $\emptyset$. What real number is represented by $B \cap C$? 0

[C] **126.** A student wrote $-3 > x > 5$ as the inequality that represents the real numbers less than -3 or greater than 5. Explain why this is incorrect.
$-3 > x > 5$ means a number less than -3 and greater than 5. This is not possible.

Graph the set.

127. $\left\{x \mid x > \frac{3}{2}\right\} \cup \left\{x \mid x < -\frac{1}{2}\right\}$

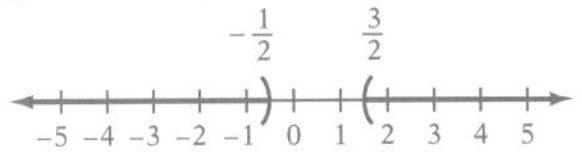

128. $\{x \mid x \le -2.5\} \cup \{x \mid x > 1.5\}$

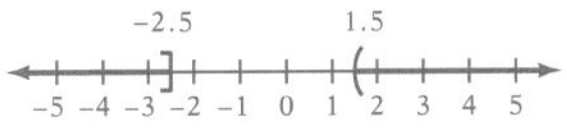

129. $\left\{x \mid x > -\frac{5}{2}\right\} \cap \left\{x \mid x \le \frac{7}{3}\right\}$

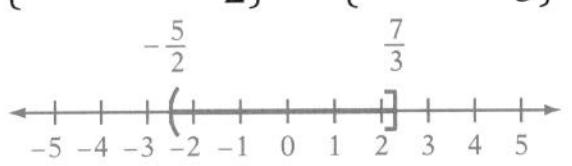

130. $\{x \mid x > -0.5\} \cap \{x \mid x \ge 3.5\}$

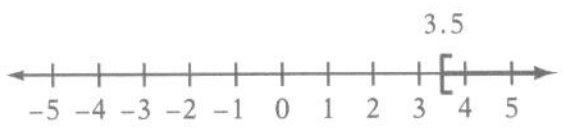

Graph the solution set.

131. $|x| < 2$

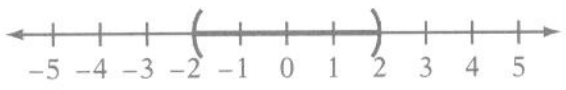

132. $|x| < 5$

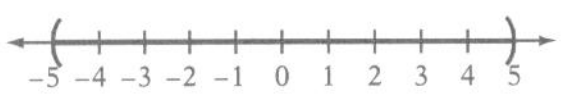

133. $|x| > 3$

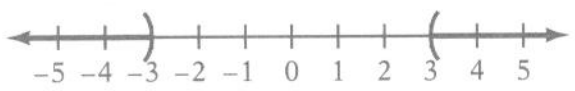

134. $|x| > 4$

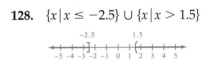

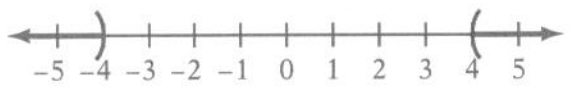

Use set-builder notation to write $A \cup B$.

135. $A = \{1, 3, 5, 7, \ldots\}$
$B = \{2, 4, 6, 8, \ldots\}$
$\{x \mid x > 0, x \text{ is an integer}\}$

136. $A = \{\ldots, -6, -4, -2\}$
$B = \{\ldots, -5, -3, -1\}$
$\{x \mid x < 0, x \text{ is an integer}\}$

Use set-builder notation to write $A \cap B$.

137. $A = \{15, 17, 19, 21, \ldots\}$
$B = \{11, 13, 15, 17, \ldots\}$
$\{x \mid x \ge 15, x \text{ is an odd integer}\}$

138. $A = \{-12, -10, -8, -6, \ldots\}$
$B = \{-4, -2, 0, 2, \ldots\}$
$\{x \mid x \ge -4, x \text{ is an even integer}\}$

[C] **139.** Given that a, b, c, and d are positive real numbers, which of the following will ensure that $\frac{a-b}{c-d} \le 0$? b and c

a. $a \ge b$ and $c > d$ **b.** $a \le b$ and $c > d$ **c.** $a \ge b$ and $c < d$ **d.** $a \le b$ and $c < d$

SECTION 1.2

Operations on Rational Numbers

1 Operations on integers

An understanding of the operations on all numbers is necessary to succeed in algebra. Let's review those properties, beginning with the rules for addition.

POINT OF INTEREST

Rules for operating with positive and negative numbers have existed for a long time. Although there are older records of these rules (from the 3rd century), one of the most thorough is contained in *The Correct Astronomical System of Brahma,* written by the Indian mathematician Brahmagupta around A.D. 600.

Rules for Addition

Numbers that have the same sign
To add numbers with the same sign, add the absolute values of the numbers. Then attach the sign of the addends.

Numbers that have different signs
To add numbers with different signs, find the absolute value of each number. Subtract the smaller of these absolute values from the larger. Then attach the sign of the number with the larger absolute value.

Add. A. $-65 + (-48)$ B. $27 + (-53)$

A. The signs are the same. Add the absolute values of the numbers. Then attach the sign of the addends. $-65 + (-48) = -113$

B. The signs are different. $27 + (-53)$

Find the absolute value of each number. $|27| = 27$ $|-53| = 53$

Subtract the smaller absolute value from the larger. $53 - 27 = 26$

Because $|-53| > |27|$, attach the sign of -53. $27 + (-53) = -26$

Subtraction is defined as addition of the additive inverse.

Rule for Subtraction

If a and b are real numbers, then $a - b = a + (-b)$.

Subtract. A. $48 - (-22)$ B. $-31 - 18$

A. Write the subtraction as the addition of the opposite number. $48 - (-22) = 48 + 22$

Add the numbers. $= 70$

B. Write the subtraction as the addition of the opposite number. $-31 - 18 = -31 + (-18)$

Then add. $= -49$

Sign Rules for Multiplication

Numbers that have the same sign
The product of two numbers with the same sign is positive.

Numbers that have different signs
The product of two numbers with different signs is negative.

⇒ Multiply. A. $-4(-9)$ B. $84(-4)$

A. The product of two numbers with the same sign is positive.

$$-4(-9) = 36$$

B. The product of two numbers with different signs is negative.

$$84(-4) = -336$$

The **multiplicative inverse** of a nonzero number a is $\frac{1}{a}$. This number is also called the **reciprocal** of a. For instance, the reciprocal of 2 is $\frac{1}{2}$, and the reciprocal of $-\frac{3}{4}$ is $-\frac{4}{3}$. Division of real numbers is defined in terms of multiplication by the multiplicative inverse.

Division of Real Numbers

> If a and b are real numbers, and $b \neq 0$, then $a \div b = a \cdot \frac{1}{b}$.

Because division is defined in terms of multiplication, the sign rules for dividing are the same as the sign rules for multiplying.

⇒ Divide. A. $\dfrac{-54}{9}$ B. $(-21) \div (-7)$

A. The quotient of two numbers with different signs is negative.

$$\frac{-54}{9} = -6$$

B. The quotient of two numbers with the same sign is positive.

$$(-21) \div (-7) = 3$$

Note that $\dfrac{-12}{3} = -4$, $\dfrac{12}{-3} = -4$, and $-\dfrac{12}{3} = -4$. These results suggest the following rule.

> If a and b are real numbers, and $b \neq 0$, then $\dfrac{-a}{b} = \dfrac{a}{-b} = -\dfrac{a}{b}$.

Properties of Zero and One in Division

Zero divided by any number other than zero is zero.

$$\frac{0}{a} = 0, a \neq 0$$

Division by zero is not defined.

INSTRUCTOR NOTE

Another way to convince students that division by 0 is not allowed is to use the model of division as the operation that separates items into equal groups. For instance, if equal shares of $6 are given to 3 people, each person receives $2; $\frac{6}{3} = 2$. Now ask, "If equal shares of $6 are given to 0 people, each person receives how many dollars?"

To understand that division by zero is not permitted, suppose $\dfrac{4}{0} = n$, where n is some number. Because each division problem has a related multiplication problem, $\dfrac{4}{0} = n$ means $n \cdot 0 = 4$. But $n \cdot 0 = 4$ is impossible because any number times 0 is 0. Therefore, division by 0 is not defined.

Any number other than zero divided by itself is 1.

$$\frac{a}{a} = 1, a \neq 0$$

Any number divided by one is the number.

$$\frac{a}{1} = a$$

Example 1 Subtract: $-3 - (-5) - 9$

Solution $-3 - (-5) - 9 = -3 + 5 + (-9)$ ▶ Write subtraction as the addition of the opposite.

$= 2 + (-9)$ ▶ Do the additions from left to right.

$= -7$

Problem 1 Subtract: $6 - (-8) - 10$

Solution See page S1. 4

Example 2 Simplify. A. $-6|-5|(-15)$ B. $-\frac{42}{-3}$

Solution A. $-6|-5|(-15) = -6(5)(-15)$ ▶ $|-5| = 5$

$= -30(-15)$ ▶ Do multiplication from left to right.

$= 450$

B. $-\frac{42}{-3} = -\left(\frac{42}{-3}\right) = -(-14) = 14$

Problem 2 Simplify. A. $2|-7|(-8)$ B. $-\frac{-36}{-3}$

Solution See page S1. A. -112 B. -12

2 Operations on rational numbers

Recall that a rational number is one that can be written in the form $\frac{p}{q}$, where p and q are integers and $q \neq 0$. Examples of rational numbers are $-\frac{5}{9}$ and $\frac{12}{5}$. The number $\frac{9}{\sqrt{7}}$ is not a rational number because $\sqrt{7}$ is not an integer. All integers are rational numbers. Terminating and repeating decimals are also rational numbers.

The sign rules are the same for rational numbers as for integers.

➡ Add: $-15.23 + (-18.1)$

To add two decimals with the same sign, add the absolute values of the decimals. Then attach the sign of the addends.

$-15.23 + (-18.1) = -33.33$

Subtract: $-18.42 - (-9.354)$

To subtract decimals, write the subtraction as the addition of the opposite, and then add.

$$-18.42 - (-9.354) = -18.42 + 9.354 = -9.066$$

Multiply: $(-0.23)(0.04)$

The signs of the factors are different, so the product is negative.

$$(-0.23)(0.04) = -0.0092$$

Divide: $(-2.835) \div (-1.35)$

The signs of the dividend and divisor are negative, so the quotient is positive.

$$(-2.835) \div (-1.35) = 2.1$$

Addition of Fractions

The sum of two fractions with the same denominator is the sum of the numerators over the common denominator.

$$\frac{a}{c} + \frac{b}{c} = \frac{a+b}{c}$$

To add or subtract rational numbers written as fractions, first rewrite the fractions as equivalent fractions with a common denominator. The common denominator is the **least common multiple (LCM)** of the denominators.

Add: $\frac{5}{6} + \left(-\frac{7}{8}\right)$

The common denominator is 24. Write each fraction in terms of the common denominator.

$$\frac{5}{6} + \left(-\frac{7}{8}\right) = \frac{5}{6} \cdot \frac{4}{4} + \left(\frac{-7}{8} \cdot \frac{3}{3}\right) = \frac{20}{24} + \left(\frac{-21}{24}\right)$$

Add the numerators, and place the sum over the common denominator.

$$= \frac{20 + (-21)}{24} = \frac{-1}{24} = -\frac{1}{24}$$

Multiplication of Fractions

The product of two fractions is the product of the numerators over the product of the denominators.

$$\frac{a}{b} \cdot \frac{c}{d} = \frac{ac}{bd}$$

Multiply: $-\frac{5}{12} \cdot \frac{8}{15}$

The signs are different, so the product is negative. Multiply the numerators, and multiply the denominators.

$$-\frac{5}{12} \cdot \frac{8}{15} = -\frac{5 \cdot 8}{12 \cdot 15}$$
$$= -\frac{40}{180}$$
$$= -\frac{20 \cdot 2}{20 \cdot 9}$$

Write the answer in simplest form.

$$= -\frac{2}{9}$$

In the last problem, the fraction was written in simplest form by dividing the numerator and denominator by 20, which is the largest integer that divides both 40 and 180. The number 20 is called the **greatest common factor (GCF)** of 40 and 180. To write a fraction in simplest form, divide the numerator and denominator by the GCF. If you have difficulty finding the GCF, try finding the prime factorization of the numerator and denominator, and then divide by the common prime factors. For instance,

$$-\frac{5}{12} \cdot \frac{8}{15} = -\frac{\overset{1}{\cancel{5}} \cdot (\overset{1}{\cancel{2}} \cdot \overset{1}{\cancel{2}} \cdot 2)}{(\underset{1}{\cancel{2}} \cdot \underset{1}{\cancel{2}} \cdot 3) \cdot (3 \cdot \underset{1}{\cancel{5}})} = -\frac{2}{9}$$

Division of Fractions

To divide two fractions, multiply the first fraction by the reciprocal of the divisor.

$$\frac{a}{b} \div \frac{c}{d} = \frac{a}{b} \cdot \frac{d}{c}$$

Divide: $\frac{3}{8} \div \frac{9}{16}$

Multiply the first fraction by the reciprocal of the divisor.

$$\frac{3}{8} \div \frac{9}{16} = \frac{3}{8} \cdot \frac{16}{9}$$

Write the answer in simplest form.

$$= \frac{2}{3}$$

Example 3 Multiply: $\left(-\frac{5}{12}\right)\left(\frac{4}{25}\right)$

Solution $\left(-\frac{5}{12}\right)\left(\frac{4}{25}\right) = -\frac{5 \cdot 4}{12 \cdot 25} = -\frac{1}{15}$

Problem 3 Divide: $\frac{5}{8} \div \left(-\frac{15}{16}\right)$

Solution See page S1. $-\frac{2}{3}$

Example 4   Simplify.

A. $\frac{3}{8} + \frac{5}{12} - \frac{9}{16}$ B. $6.329 - 12.49$ C. $14 - |3 - 18|$

Solution   A. $\frac{3}{8} + \frac{5}{12} - \frac{9}{16} = \frac{18}{48} + \frac{20}{48} + \left(-\frac{27}{48}\right) = \frac{18 + 20 + (-27)}{48} = \frac{11}{48}$

B. $6.329 - 12.49 = 6.329 + (-12.49) = -6.161$

C. $14 - |3 - 18|$
$= 14 - |-15|$ ▶ Perform the indicated operation inside the absolute value symbols.

$= 14 - 15$ ▶ Rewrite $|-15|$ as 15.
$= -1$

Problem 4   Simplify.

A. $-\frac{5}{12} \div \frac{10}{27}$ B. $-8.729 + 12.094$ C. $-8 - |5 - 12|$

Solution   See page S1.   A. $-\frac{9}{8}$ B. 3.365   C. -15

Example 5   Divide $0.0527 \div (-0.27)$. Round to the nearest hundredth.

Solution $0.0527 \div (-0.27) \approx -0.20$ ▶ The symbol $\approx$ is used to indicate that the quotient is an approximate value after being rounded off.

Problem 5   Multiply $-4.027(0.49)$. Round to the nearest hundredth.

Solution   See page S1. -1.97

---

3   # Exponential expressions

Repeated multiplication of the same factor can be written using an exponent.

$2 \cdot 2 \cdot 2 \cdot 2 \cdot 2 \cdot 2 = 2^6 \longleftarrow$ exponent $b \cdot b \cdot b \cdot b \cdot b = b^5 \longleftarrow$ exponent
$\uparrow$ ———— base $\uparrow$ ———— base

The **exponent** indicates how many times the factor, called the **base,** occurs in the multiplication. The multiplication $2 \cdot 2 \cdot 2 \cdot 2 \cdot 2 \cdot 2$ is in **factored form.** The exponential expression 2^6 is in **exponential form.**

2^1 is read "the first power of two" or just "two." $\longrightarrow$ Usually the exponent 1 is not written.

2^2 is read "the second power of two" or "two squared."

2^3 is read "the third power of two" or "two cubed."

2^4 is read "the fourth power of two."

2^5 is read "the fifth power of two."

b^5 is read "the fifth power of b."

nth Power of a

If a is a real number and n is a positive integer, then the nth power of a is the product of n factors of a.

$$a^n = \underbrace{a \cdot a \cdot a \cdot \cdots \cdot a}_{a \text{ as a factor } n \text{ times}}$$

LOOK CLOSELY

Examine the results of $(-3)^4$ and -3^4 very carefully. As another example, $(-2)^6 = 64$ but $-2^6 = -64$.

$$5^3 = 5 \cdot 5 \cdot 5 = 125$$

$$(-3)^4 = (-3)(-3)(-3)(-3) = 81$$

$$-3^4 = -(3)^4 = -(3 \cdot 3 \cdot 3 \cdot 3) = -81$$

Note the difference between $(-3)^4$ and -3^4. The placement of the parentheses is very important.

Example 6 Evaluate $(-6)^4$ and -6^4.

Solution $(-6)^4 = (-6)(-6)(-6)(-6) = 1296$

$$-6^4 = -(6 \cdot 6 \cdot 6 \cdot 6)$$
$$= -1296$$

▶ The negative of a number is taken to a power only when the negative sign is *inside* the parentheses.

Problem 6 Evaluate $(-4)^2$ and -4^2.

Solution See page S1. 16 and -16

Example 7 Evaluate $\left(-\frac{2}{3}\right)^2 \cdot 3^3$.

Solution $\left(-\frac{2}{3}\right)^2 \cdot 3^3 = \left(-\frac{2}{3}\right)\left(-\frac{2}{3}\right) \cdot (3)(3)(3) = 12$

Problem 7 Evaluate $-\left(\frac{2}{5}\right)^3 \cdot 5^2$.

Solution See page S1. $-\frac{8}{5}$

4 The Order of Operations Agreement

To prevent there being more than one value for the same numerical expression, an **Order of Operations Agreement** is followed.

The Order of Operations Agreement

Step 1 Perform operations inside grouping symbols. Grouping symbols include parentheses ( ), brackets [ ], the absolute value symbol | |, and the fraction bar.

continued

Step 2 Simplify exponential expressions.

Step 3 Do multiplication and division as they occur from left to right.

Step 4 Do addition and subtraction as they occur from left to right.

Simplify: $8 - \frac{12-2}{4+1} \div 2^2$

$$8 - \frac{12-2}{4+1} \div 2^2$$

Perform operations above and below the fraction bar. $= 8 - \frac{10}{5} \div 2^2$

Simplify exponential expressions. $= 8 - \frac{10}{5} \div 4$

Do multiplication and division as they occur from left to right. Note that a fraction bar can be read "÷." $= 8 - 2 \div 4$

$= 8 - \frac{1}{2}$

Do addition and subtraction as they occur from left to right. $= \frac{15}{2}$

One or more of the steps shown in the last example may not be needed to simplify an expression. In that case, proceed to the next step in the Order of Operations Agreement.

When an expression has grouping symbols inside grouping symbols, perform the operations inside the inner grouping symbols first.

Simplify: $4.3 - [(25 - 9) \div 2]^2$

$$4.3 - [(25 - 9) \div 2]^2$$

Perform operations inside grouping symbols. $= 4.3 - [16 \div 2]^2$

$= 4.3 - [8]^2$

Simplify exponential expressions. $= 4.3 - 64$

Perform addition and subtraction as they occur from left to right. $= -59.7$

Example 8 Simplify: $(-1.2)^3 - 8.4 \div 2.1$

Solution $(-1.2)^3 - 8.4 \div 2.1$

$= -1.728 - 8.4 \div 2.1$ ▶ Simplify exponential expressions.

$= -1.728 - 4$ ▶ Do multiplication and division.

$= -5.728$ ▶ Do addition and subtraction.

Problem 8 Simplify: $(3.81 - 1.41)^2 \div 0.036 - 1.89$

Solution See page S1. 158.11

Example 9 Simplify: $\left(\frac{1}{2}\right)^3 - \left[\left(\frac{2}{3} + \frac{1}{4}\right) \div \frac{5}{6}\right]$

Solution

$$\left(\frac{1}{2}\right)^3 - \left[\left(\frac{2}{3} + \frac{1}{4}\right) \div \frac{5}{6}\right]$$

$$= \left(\frac{1}{2}\right)^3 - \left[\frac{11}{12} \div \frac{5}{6}\right]$$
► Perform operations inside the inner grouping symbols.

$$= \left(\frac{1}{2}\right)^3 - \frac{11}{10}$$
► Perform operations inside the grouping symbols.

$$= \frac{1}{8} - \frac{11}{10}$$
► Simplify exponential expressions.

$$= -\frac{39}{40}$$
► Do addition and subtraction.

Problem 9 Simplify: $\frac{1}{3} + \frac{5}{8} \div \frac{15}{16} - \frac{7}{12}$

Solution See page S1. $\frac{5}{12}$

A **complex fraction** is a fraction whose numerator or denominator (or both) contains one or more fractions. Examples of complex fractions are shown below.

$$\frac{\frac{2}{3}}{\frac{1}{3}} \qquad \frac{\frac{2}{5}+1}{\frac{7}{8}} \longleftarrow \text{main fraction bar}$$

To simplify a complex fraction, perform operations above and below the main fraction bar as the first step in the Order of Operations Agreement.

⇒ Simplify: $\dfrac{\frac{3}{4} - \frac{1}{2}}{\frac{2}{3} + \frac{1}{4}}$

Perform operations above and below the main fraction bar.

$$\frac{\frac{3}{4} - \frac{1}{2}}{\frac{2}{3} + \frac{1}{4}} = \frac{\frac{3}{4} - \frac{2}{4}}{\frac{8}{12} + \frac{3}{12}}$$

Multiply the numerator of the complex fraction by the reciprocal of the denominator of the complex fraction.

$$= \frac{\frac{1}{4}}{\frac{11}{12}} = \frac{1}{4} \cdot \frac{12}{11}$$

$$= \frac{1 \cdot 12}{4 \cdot 11}$$

$$= \frac{3}{11}$$

Note that $\dfrac{\frac{1}{4}}{\frac{11}{12}} = \frac{1}{4} \div \frac{11}{12} = \frac{1}{4} \cdot \frac{12}{11}$.

Example 10 Simplify: $9 \cdot \dfrac{\frac{5}{6} - 2}{\frac{3}{8}} \div \frac{7}{6}$

Solution

$$9 \cdot \frac{\frac{5}{6} - 2}{\frac{3}{8}} \div \frac{7}{6}$$

$$= 9 \cdot \frac{-\frac{7}{6}}{\frac{3}{8}} \div \frac{7}{6}$$

▶ Do operations above the main fraction bar.

$$= 9 \cdot \left(-\frac{7}{6} \cdot \frac{8}{3}\right) \div \frac{7}{6}$$

▶ Multiply the numerator of the complex fraction by the reciprocal of the denominator of the complex fraction.

$$= 9 \cdot \left(-\frac{28}{9}\right) \div \frac{7}{6}$$

$$= -28 \div \frac{7}{6}$$

▶ Do multiplication and division as they occur from left to right.

$$= -28 \cdot \frac{6}{7}$$

$$= -24$$

Problem 10 Simplify: $\frac{11}{12} - \frac{\frac{5}{4}}{2 - \frac{7}{2}} \cdot \frac{3}{4}$

Solution See page S2. $\frac{37}{24}$

STUDY TIPS

SURVEY THE CHAPTER

Before you begin reading a chapter, take a few minutes to survey it. Glancing through the chapter will give you an overview of its content and help you see how the pieces fit together as you read.

Begin by reading the chapter title. The title summarizes what the chapter is about. Next read the section headings. The section headings summarize the major topics presented in the chapter. Then read the objectives under each section heading. The objective headings describe the learning goals for that section. Keep these headings in mind as you work through the material. They provide direction as you study.

CONCEPT REVIEW 1.2

Determine whether the following statements are always true, sometimes true, or never true.

1. The sum of two numbers of opposite signs is negative. Sometimes true

2. The product of two numbers with the same sign is positive. Always true

3. The sum of two fractions is the sum of the numerators over the sum of the denominators. Never true

4. To divide two fractions, multiply the first fraction by the reciprocal of the divisor. Always true

5. The sum of a number and its additive inverse is zero. Always true

6. The rule for multiplying two fractions is to multiply the numerators and place the product over the LCD of the denominators. Never true

7. The Order of Operations Agreement says to simplify exponential expressions before performing operations inside grouping symbols. Never true

8. The Order of Operations Agreement says to do multiplication before division. Sometimes true

9. If $-3x = 0$, then $x = 0$. Always true

EXERCISES 1.2

1 1. Explain how to add **a.** two integers with the same sign and **b.** two integers with different signs.

2. Explain the meaning of the words *minus* and *negative*.

3. Explain how to rewrite $8 - (-12)$ as addition of the opposite.

4. Explain how to multiply **a.** two integers with the same sign and **b.** two integers with different signs.

Simplify.

5. $-18 + (-12)$ -30
6. $-18 - 7$ -25
7. $5 - 22$ -17
8. $16 \cdot (-60)$ -960
9. $3 \cdot 4(-8)$ -96
10. $18 \cdot 0(-7)$ 0
11. $18 \div (-3)$ -6
12. $25 \div (-5)$ -5
13. $-60 \div (-12)$ 5
14. $(-9)(-2)(-3)(10)$ -540
15. $-20(35)(-16)$ $11{,}200$
16. $54(19)(-82)$ $-84{,}132$
17. $-271(-365)$ $98{,}915$
18. $|(-16)(10)|$ 160
19. $|12(-8)|$ 96
20. $|7 - 18|$ 11
21. $|15 - (-8)|$ 23
22. $|-16 - (-20)|$ 4
23. $|-56 \div 8|$ 7
24. $|81 \div (-9)|$ 9
25. $|-153 \div (-9)|$ 17
26. $|-4| - |-2|$ 2
27. $-|-8| + |-4|$ -4
28. $-|-16| - |24|$ -40
29. $-30 + (-16) - 14 - 2$ -62
30. $3 - (-2) + (-8) - 11$ -14
31. $-2 + (-19) - 16 + 12$ -25
32. $-6 + (-9) - 18 + 32$ -1
33. $13 - |6 - 12|$ 7
34. $-9 - |-7 - (-15)|$ -17
35. $738 - 46 + (-105)$ 587
36. $-871 - (-387) - 132$ -616
37. $-442 \div (-17)$ 26
38. $621 \div (-23)$ -27
39. $-4897 \div 59$ -83
40. $-17(-5)$ 85

2 41. Describe **a.** the least common multiple of two numbers and **b.** the greatest common factor of two numbers.

42. Explain how to divide two rational numbers.

Simplify.

43. $\frac{7}{12} + \frac{5}{16}$ $\frac{43}{48}$

44. $\frac{3}{8} - \frac{5}{12}$ $-\frac{1}{24}$

45. $-\frac{5}{9} - \frac{14}{15}$ $-\frac{67}{45}$

46. $\frac{1}{2} + \frac{1}{7} - \frac{5}{8}$ $\frac{1}{56}$

47. $-\frac{1}{3} + \frac{5}{9} - \frac{7}{12}$ $-\frac{13}{36}$

48. $\frac{1}{3} + \frac{19}{24} - \frac{7}{8}$ $\frac{1}{4}$

49. $\frac{2}{3} - \frac{5}{12} + \frac{5}{24}$ $\frac{11}{24}$

50. $-\frac{7}{10} + \frac{4}{5} + \frac{5}{6}$ $\frac{14}{15}$

51. $\frac{5}{8} - \frac{7}{12} + \frac{1}{2}$ $\frac{13}{24}$

52. $-\frac{1}{3} \cdot \frac{5}{8}$ $-\frac{5}{24}$

53. $\left(\frac{6}{35}\right)\left(-\frac{5}{16}\right)$ $-\frac{3}{56}$

54. $\frac{2}{3}\left(-\frac{9}{20}\right) \cdot \frac{5}{12}$ $-\frac{1}{8}$

55. $-\frac{8}{15} \div \frac{4}{5}$ $-\frac{2}{3}$

56. $-\frac{2}{3} \div \left(-\frac{6}{7}\right)$ $\frac{7}{9}$

57. $-\frac{11}{24} \div \frac{7}{12}$ $-\frac{11}{14}$

58. $\frac{7}{9} \div \left(-\frac{14}{27}\right)$ $-\frac{3}{2}$

59. $\left(-\frac{5}{12}\right)\left(\frac{4}{35}\right)\left(\frac{7}{8}\right)$ $-\frac{1}{24}$

60. $\frac{6}{35}\left(-\frac{7}{40}\right)\left(-\frac{8}{21}\right)$ $\frac{2}{175}$

61. $-14.27 + 1.296$ -12.974

62. $-0.4355 + 172.5$ 172.0645

63. $1.832 - 7.84$ -6.008

64. $(3.52)(4.7)$ 16.544

65. $(0.03)(10.5)(6.1)$ 1.9215

66. $(1.2)(3.1)(-6.4)$ -23.808

67. $5.418 \div (-0.9)$ -6.02

68. $-0.2645 \div (-0.023)$ 11.5

69. $-0.4355 \div 0.065$ -6.7

70. $-6.58 - 3.97 + 0.875$ -9.675

Simplify. Round to the nearest hundredth.

71. $38.241 \div [-(-6.027)] - 7.453$ -1.11

72. $-9.0508 - (-3.177) + 24.77$ 18.90

73. $-287.3069 \div 0.1415$ -2030.44

74. $6472.3018 \div (-3.59)$ -1802.87

3 Simplify.

75. 5^3 125

76. 3^4 81

77. -2^3 -8

78. -4^3 -64

79. $(-5)^3$ -125

80. $(-8)^2$ 64

81. $2^2 \cdot 3^4$ 324

82. $4^2 \cdot 3^3$ 432

83. $-2^2 \cdot 3^2$ -36

84. $-3^2 \cdot 5^3$ -1125

85. $(-2)^3(-3)^2$ -72

86. $(-4)^3(-2)^3$ 512

87. $4 \cdot 2^3 \cdot 3^3$ 864

88. $-4(-3)^2(4^2)$ -576

89. $2^2 \cdot (-10)(-2)^2$ -160

90. $-3(-2)^2(-5)$ 60

91. $(-3)^3 \cdot 15 \cdot (-2)^4$ -6480

92. $-4 \cdot 2^2 \cdot 5^5$ $-50{,}000$

93. $2^5 \cdot (-3)^4 \cdot 4^5$ $2{,}654{,}208$

94. $4^4(-3)^5(6)^2$ $-2{,}239{,}488$

4 95. Why do we need an Order of Operations Agreement?

96. Describe each step in the Order of Operations Agreement.

Simplify.

97. $5 - 3(8 \div 4)^2$ -7

98. $4^2 - (5 - 2)^2 \cdot 3$ -11

99. $16 - \frac{2^2 - 5}{3^2 + 2}$ $\frac{177}{11}$

100. $\frac{4(5 - 2)}{4^2 - 2^2} \div 4$ $\frac{1}{4}$

101. $\frac{3 + \frac{2}{3}}{\frac{11}{16}}$ $\frac{16}{3}$

102. $\frac{\frac{11}{14}}{4 - \frac{6}{7}}$ $\frac{1}{4}$

103. $5[(2-4)\cdot 3-2]$ -40

104. $2[(16\div 8)-(-2)]+4$ 12

105. $16-4\left(\frac{8-2}{3-6}\right)\div\frac{1}{2}$ 32

106. $25\div 5\left(\frac{16+8}{-2^2+8}\right)-5$ 25

107. $6[3-(-4+2)\div 2]$ 24

108. $12-4[2-(-3+5)-8]$ 44

109. $\frac{1}{2}-\left(\frac{2}{3}\div\frac{5}{9}\right)+\frac{5}{6}$ $\frac{2}{15}$

110. $\left(-\frac{3}{5}\right)^2-\frac{3}{5}\cdot\frac{5}{9}+\frac{7}{10}$ $\frac{109}{150}$

111. $\frac{1}{2}-\frac{\frac{17}{25}}{4-\frac{3}{5}}+\frac{1}{5}$ $\frac{1}{2}$

112. $\frac{3}{4}+\frac{3-\frac{7}{9}}{\frac{5}{6}}\cdot\frac{2}{3}$ $\frac{91}{36}$

113. $\frac{2}{3}-\left[\frac{3}{8}+\frac{5}{6}\right]\div\frac{3}{5}$ $-\frac{97}{72}$

114. $\frac{3}{4}\div\left[\frac{5}{8}-\frac{5}{12}\right]+2$ $\frac{28}{5}$

115. $0.4(1.2-2.3)^2+5.8$ 6.284

116. $5.4-(0.3)^2\div 0.09$ 4.4

117. $1.75\div 0.25-(1.25)^2$ 5.4375

118. $(3.5-4.2)^2-3.50\div 2.5$ -0.91

119. $25.76\div(6.96-3.27)^2$ 1.891878

120. $(3.09-4.77)^2-4.07\cdot 3.66$ -12.0738

APPLYING CONCEPTS 1.2

121. A number that is its own additive inverse is __0__.

122. Which two numbers are their own multiplicative inverse? -1 and 1

123. Do all real numbers have a multiplicative inverse? If not, which ones do not have a multiplicative inverse? No; 0

[C] **124.** What is the tens digit of 11^{22}? 2

[C] **125.** What is the ones digit of 7^{18}? 9

[C] **126.** What are the last two digits of 5^{33}? 25

[C] **127.** What are the last three digits of 5^{234}? 625

128. Does $(2^3)^4 = 2^{(3^4)}$? If not, which expression is larger? No; $2^{(3^4)}$

129. What is the Order of Operations Agreement for a^{b^c}? (*Note:* Even calculators that normally follow the Order of Operations Agreement may not do so for this expression.) Find b^c; then find $a^{(b^c)}$.

SECTION 1.3 Variable Expressions

1 The Properties of the Real Numbers

The Properties of the Real Numbers describe the way operations on numbers can be performed. Following is a list of some of the Properties of the Real Numbers and an example of each property.

The Commutative Property of Addition

$$a + b = b + a$$

$3 + 2 = 2 + 3$
$5 = 5$

The Commutative Property of Multiplication

$$a \cdot b = b \cdot a$$

$(3)(-2) = (-2)(3)$
$-6 = -6$

The Associative Property of Addition

$$(a + b) + c = a + (b + c)$$

$(3 + 4) + 5 = 3 + (4 + 5)$
$7 + 5 = 3 + 9$
$12 = 12$

The Associative Property of Multiplication

$$(a \cdot b) \cdot c = a \cdot (b \cdot c)$$

$(3 \cdot 4) \cdot 5 = 3 \cdot (4 \cdot 5)$
$12 \cdot 5 = 3 \cdot 20$
$60 = 60$

The Addition Property of Zero

$$a + 0 = 0 + a = a$$

$3 + 0 = 0 + 3 = 3$

The Multiplication Property of Zero

$$a \cdot 0 = 0 \cdot a = 0$$

$8 \cdot 0 = 0 \cdot 8 = 0$

The Multiplication Property of One

$$a \cdot 1 = 1 \cdot a = a$$

$5 \cdot 1 = 1 \cdot 5 = 5$

The Inverse Property of Addition

$$a + (-a) = (-a) + a = 0$$

$4 + (-4) = (-4) + 4 = 0$

$-a$ is called the **additive inverse** of a. Also, a is the additive inverse of $-a$. The sum of a number and its additive inverse is 0.

The Inverse Property of Multiplication

$$a \cdot \frac{1}{a} = \frac{1}{a} \cdot a = 1, \quad a \neq 0$$

$$(4)\left(\frac{1}{4}\right) = \left(\frac{1}{4}\right)(4) = 1$$

$\frac{1}{a}$ is called the **multiplicative inverse** of a. It is also called the reciprocal of a. The product of a number and its multiplicative inverse is 1.

The Distributive Property

$$a(b + c) = ab + ac$$

$$3(4 + 5) = 3 \cdot 4 + 3 \cdot 5$$
$$3 \cdot 9 = 12 + 15$$
$$27 = 27$$

Example 1 Complete the statement by using the Inverse Property of Addition.

$3x + ? = 0$

Solution $3x + (-3x) = 0$

Problem 1 Complete the statement by using the Commutative Property of Multiplication.

$(x)\left(\frac{1}{4}\right) = (?)(x)$

Solution See page S2. $(x)\left(\frac{1}{4}\right) = \left(\frac{1}{4}\right)(x)$

Example 2 Identify the property that justifies the statement.

$3(x + 4) = 3x + 12$

Solution The Distributive Property

Problem 2 Identify the property that justifies the statement.

$(a + 3b) + c = a + (3b + c)$

Solution See page S2. The Associative Property of Addition

2 Evaluate variable expressions

An expression that contains one or more variables is called a **variable expression.**

A variable expression is shown at the right. The expression has four addends that are called **terms** of the expression. The variable expression has three variable terms and one constant term.

4 terms

$$4x^2y \quad - \quad 2z \quad - \quad x \quad + \quad 4$$

Variable terms — Constant term

Each variable term is composed of a **numerical coefficient** and a **variable part.** When the numerical coefficient is 1 or -1, the 1 is usually not written.

Numerical coefficient

$$4x^2y - 2z - 1x + 4$$

Variable part

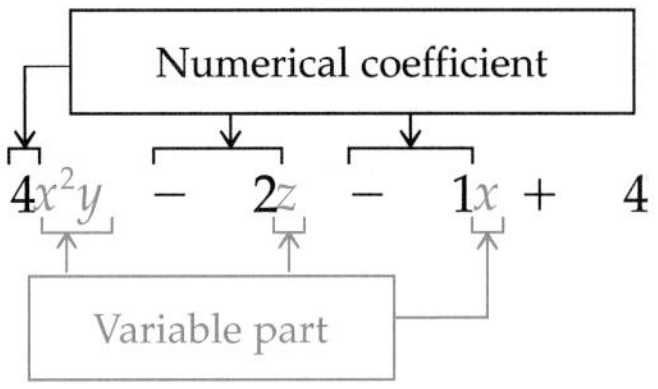

Replacing the variable in a variable expression by a numerical value and then simplifying the resulting expression is called **evaluating the variable expression.**

➡ Evaluate $3 - 2|3x - 2y^2|$ when $x = -1$ and $y = 2$.

Replace each variable with its value.

$$3 - 2|3x - 2y^2|$$
$$3 - 2|3(-1) - 2(2)^2|$$

Use the Order of Operations Agreement to simplify the resulting numerical expression.

$$= 3 - 2|3(-1) - 2(4)|$$
$$= 3 - 2|-3 - 8|$$
$$= 3 - 2|-11|$$
$$= 3 - 2(11)$$
$$= 3 - 22$$
$$= -19$$

Example 3   Evaluate $a^2 - (ab - c)$ when $a = -2$, $b = 3$, and $c = -4$.

Solution $a^2 - (ab - c)$
$(-2)^2 - [(-2)(3) - (-4)]$ ▶ Replace each variable in the expression with its value.

$= (-2)^2 - [-6 - (-4)]$ ▶ Use the Order of Operations Agreement to simplify the resulting numerical expression.
$= (-2)^2 - [-2]$
$= 4 - [-2]$
$= 6$

Problem 3   Evaluate $(b - c)^2 \div ab$ when $a = -3$, $b = 2$, and $c = -4$.

Solution   See page S2. -24

Example 4   The radius of the base of a cylinder is 3 in. and the height is 6 in. Find the volume of the cylinder. Round to the nearest hundredth.

Solution $V = \pi r^2 h$ ▶ Use the formula for the volume of a cylinder.
$V = \pi(3)^2 6$ ▶ Substitute 3 for r and 6 for h.
$V = 54\pi$ ▶ The exact volume of the cylinder is 54π in³. An approximate measure can be found by using the π key on a calculator.

$V \approx 169.65$

The approximate volume of the cylinder is 169.65 in³.

Problem 4   Find the surface area of a right circular cone with a radius of 5 cm and a slant height of 12 cm. Give both the exact area and an approximation to the nearest hundredth.

Solution   See page S2. 85π cm², 267.04 cm²

STUDENT NOTE

For your reference, geometric formulas are printed inside the front cover of this textbook.

INSTRUCTOR NOTE

The calculator logo signals that what follows is instruction on the use of a graphing calculator or computer graphing utility or on its use in solving problems or illustrating a point. Coverage of this material is optional.

A graphing calculator can be used to evaluate variable expressions. When the value of each variable is stored in the calculator's memory and a variable expression is then entered on the calculator, the calculator evaluates that variable expression for the values of the variables stored in its memory. See the Appendix on page 679 for a description of keystroking procedures for different models of graphing calculators.

3 Simplify variable expressions

INSTRUCTOR NOTE

A common mistake for students is to combine x^2 terms with x terms. Citing a physical example, such as trying to combine square feet and feet, may help these students.

Like terms of a variable expression are terms with the same variable part.

Constant terms are like terms.

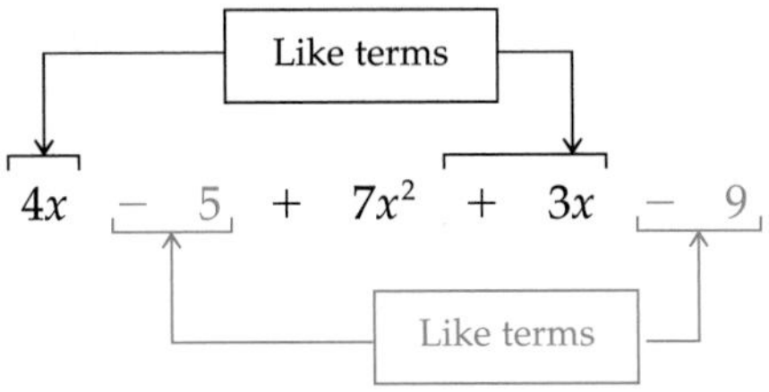

To **combine like terms,** use the Distributive Property $ba + ca = (b + c)a$ to add the coefficients.

$3x + 2x = (3 + 2)x = 5x$

Example 5 Simplify: $2(x + y) + 3(y - 3x)$

Solution $2(x + y) + 3(y - 3x)$

$= 2x + 2y + 3y - 9x$ ▶ Use the Distributive Property to remove parentheses.

$= (2x - 9x) + (2y + 3y)$ ▶ Use the Commutative and Associative Properties of Addition to rearrange and group like terms.

$= -7x + 5y$ ▶ Combine like terms.

Problem 5 Simplify: $(2x + xy - y) - (5x - 7xy + y)$

Solution See page S2. $-3x + 8xy - 2y$

Example 6 Simplify: $4y - 2[x - 3(x + y) - 5y]$

Solution $4y - 2[x - 3(x + y) - 5y]$

$= 4y - 2[x - 3x - 3y - 5y]$ ▶ Use the Distributive Property to remove parentheses.

$= 4y - 2[-2x - 8y]$ ▶ Combine like terms.

$= 4y + 4x + 16y$ ▶ Use the Distributive Property to remove brackets.

$= 4x + 20y$ ▶ Combine like terms.

Problem 6 Simplify: $2x - 3[y - 3(x - 2y + 4)]$

Solution See page S2. $11x - 21y + 36$

STUDY TIPS

Use the Textbook to Learn the Material

For each objective studied, read very carefully all of the material from the objective statement to the examples provided for that objective. As you read, note carefully the formulas and words printed in **boldface** type. It is important for you to know these formulas and the definitions of these words.

You will note that each example is paired with a problem. The example is worked out for you; the problem is left for you to do. After studying the example, do the problem. Immediately look up the answer to this problem in the Solutions section at the back of the text. The page number on which the solution appears is printed on the solution line below the problem statement. If your answer is correct, continue. If your answer is incorrect, check your solution against the one given in the Solutions section. It may be helpful to review the worked-out example also. Determine where you made your mistakes.

Next, do the problems in the exercise set that correspond to the objective you just studied. The answers to all the odd-numbered exercises appear in the Answers section in the back of the book. Check your answers to the exercises against these.

If you have difficulty solving problems in the exercise set, review the material in the text. Many examples are solved within the text material. Review the solutions to these problems. Reread the boxed examples provided for the objective. If, after checking these sources and trying to find your mistakes, you are still unable to solve a problem correctly, make a note of the exercise number so that you can ask someone for help with that problem.

CONCEPT REVIEW 1.3

Determine whether the following statements are always true, sometimes true, or never true.

1. The reciprocal of a whole number is a whole number. Sometimes true

2. The Distributive Property is used to combine like terms. Always true

3. Like terms are terms with the same variables. Sometimes true

4. The sum of a number and its multiplicative inverse is -1. Never true

5. The product of a number and its multiplicative inverse is 1. Always true

6. A real number has a multiplicative inverse. Sometimes true

EXERCISES 1.3

1 Use the given Property of the Real Numbers to complete the statement.

1. The Commutative Property of Multiplication
$3 \cdot 4 = 4 \cdot ?$ 3

2. The Commutative Property of Addition
$7 + 15 = ? + 7$ 15

3. The Associative Property of Addition
$(3 + 4) + 5 = ? + (4 + 5)$ 3

4. The Associative Property of Multiplication
$(3 \cdot 4) \cdot 5 = 3 \cdot (? \cdot 5)$ 4

5. A Division Property of Zero
$\frac{5}{?}$ is undefined. 0

6. The Addition Property of Zero
$4 + ? = 4$ 0

7. The Distributive Property
$3(x + 2) = 3x + ?$ 6

8. The Distributive Property
$5(y + 4) = ? \cdot y + 20$ 5

9. A Division Property of Zero
$\frac{?}{-6} = 0$ 0

10. The Inverse Property of Addition
$(x + y) + ? = 0$ $[-(x + y)]$

11. The Inverse Property of Multiplication
$\frac{1}{mn}(mn) = ?$ 1

12. The Multiplication Property of One
$? \cdot 1 = x$ x

13. The Associative Property of Multiplication
$2(3x) = ? \cdot x$ $(2 \cdot 3)$

14. The Commutative Property of Addition
$ab + bc = bc + ?$ ab

Identify the property that justifies the statement.

15. $\frac{0}{-5} = 0$
A Division Property of Zero

16. $-8 + 8 = 0$
The Inverse Property of Addition

17. $(-12)\left(-\frac{1}{12}\right) = 1$
The Inverse Property of Multiplication

18. $(3 \cdot 4) \cdot 2 = 2 \cdot (3 \cdot 4)$
The Commutative Property of Multiplication

19. $y + 0 = y$
The Addition Property of Zero

20. $2x + (5y + 8) = (2x + 5y) + 8$
The Associative Property of Addition

21. $\frac{-9}{0}$ is undefined.
A Division Property of Zero

22. $(x + y)z = xz + yz$
The Distributive Property

23. $6(x + y) = 6x + 6y$
The Distributive Property

24. $0 + 2 = 2$
The Addition Property of Zero

25. $(ab)c = a(bc)$
The Associative Property of Multiplication

26. $(x + y) + z = (y + x) + z$
The Commutative Property of Addition

2 **27.** Explain the meaning of the phrase "evaluate a variable expression."

28. What is the difference between the meaning of "the value of the variable" and the meaning of "the value of the variable expression"?

Evaluate the variable expression when $a = 2$, $b = 3$, $c = -1$, and $d = -4$.

29. $ab + dc$ 10

30. $2ab - 3dc$ 0

31. $4cd \div a^2$ 4

32. $b^2 - (d - c)^2$ 0

33. $(b - 2a)^2 + c$ 0

34. $(b - d)^2 \div (b - d)$ 7

35. $(bc + a)^2 \div (d - b)$ $-\frac{1}{7}$

36. $\frac{1}{3}b^3 - \frac{1}{4}d^3$ 25

37. $\frac{1}{4}a^4 - \frac{1}{6}bc$ $\frac{9}{2}$

38. $2b^2 \div \dfrac{ad}{2}$ $-\dfrac{9}{2}$

39. $\dfrac{3ac}{-4} - c^2$ $\dfrac{1}{2}$

40. $\dfrac{2d - 2a}{2bc}$ 2

41. $\dfrac{3b - 5c}{3a - c}$ 2

42. $\dfrac{2d - a}{b - 2c}$ -2

43. $\dfrac{a - d}{b + c}$ 3

44. $|a^2 + d|$ 0

45. $-a|a + 2d|$ -12

46. $d|b - 2d|$ -44

47. $\dfrac{2a - 4d}{3b - c}$ 2

48. $\dfrac{3d - b}{b - 2c}$ -3

49. $-3d \div \left|\dfrac{ab - 4c}{2b + c}\right|$ 6

50. $-2bc + \left|\dfrac{bc + d}{ab - c}\right|$ 7

51. $2(d - b) \div (3a - c)$ -2

52. $(d - 4a)^2 \div c^3$ -144

53. $-d^2 - c^3 a$ -14

54. $a^2 c - d^3$ 60

55. $-d^3 + 4ac$ 56

56. b^a 9

57. $4^{(a)^2}$ 256

58. a^b 8

Geometry Find the volume of each figure. For calculations involving π, give both the exact value and an approximation to the nearest hundredth.

59.
6 in.
14 in. 10 in.
840 in³

60.
14 ft
12 ft
168π ft³, 527.79 ft³

61.
5 ft
3 ft
3 ft
15 ft³

62.
7.5 m
7.5 m 7.5 m
421.875 m³

63.
3 cm
4.5π cm³, 14.14 cm³

64.
8 cm
8 cm
128π cm³, 402.12 cm³

Geometry Find the surface area of each figure. For calculations involving π, give both the exact value and an approximation to the nearest hundredth.

65.
3 m
4 m 5 m
94 m²

66.
14 ft
14 ft 14 ft
1176 ft²

67.
5 m
4 m 4 m
56 m²

68.
2 cm
4π cm², 12.57 cm²

69.
2 in.
6 in.
96π in², 301.59 in²

70.
9 ft
3 ft
15.75π ft², 49.48 ft²

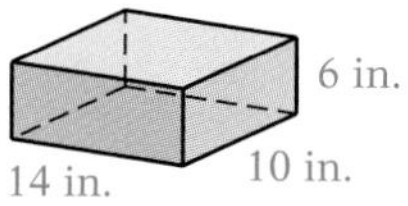

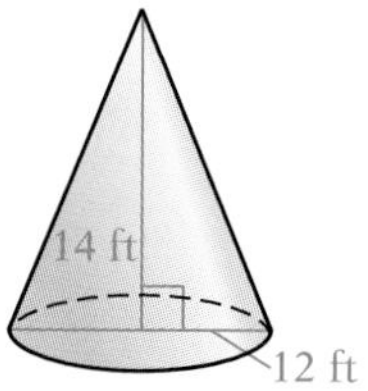

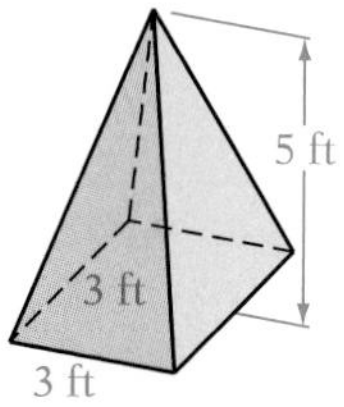

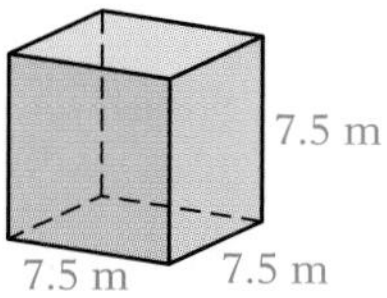

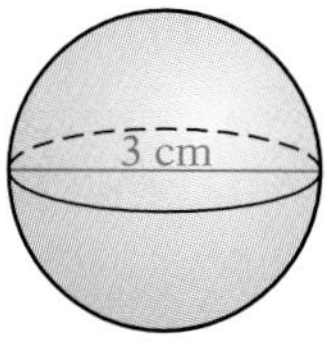

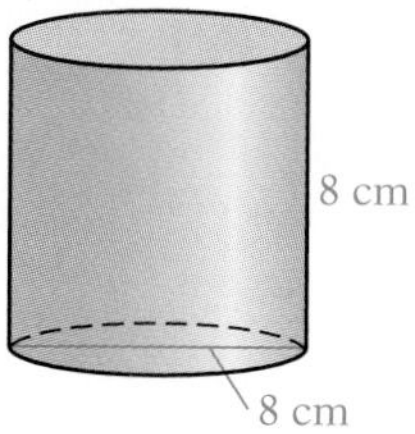

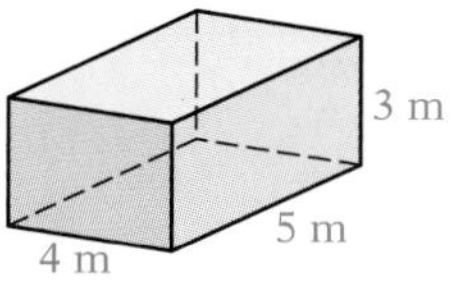

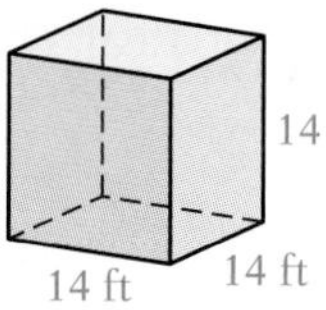

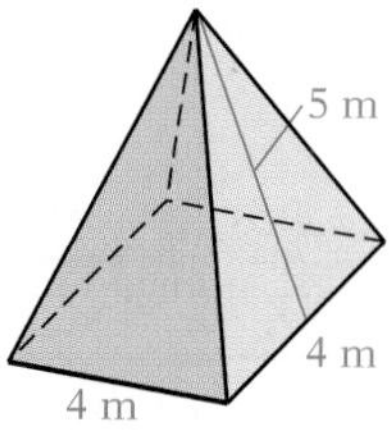

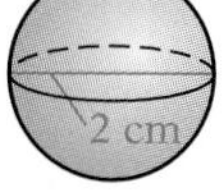

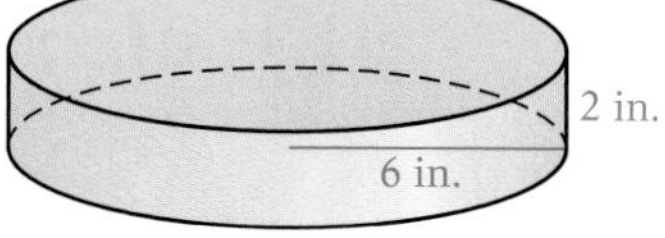

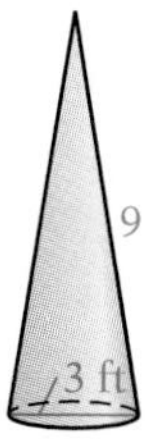

3 71. Explain how the Distributive Property is used to combine like terms.

Simplify.

72. $5x + 7x$ $12x$

73. $3x + 10x$ $13x$

74. $-8ab - 5ab$ $-13ab$

75. $-2x + 5x - 7x$ $-4x$

76. $3x - 5x + 9x$ $7x$

77. $-2a + 7b + 9a$ $7a + 7b$

78. $5b - 8a - 12b$ $-8a - 7b$

79. $12\left(\frac{1}{12}x\right)$ x

80. $\frac{1}{3}(3y)$ y

81. $-3(x - 2)$ $-3x + 6$

82. $-5(x - 9)$ $-5x + 45$

83. $(x + 2)5$ $5x + 10$

84. $-(x + y)$ $-x - y$

85. $-(-x - y)$ $x + y$

86. $3(a - 5)$ $3a - 15$

87. $3(x - 2y) - 5$ $3x - 6y - 5$

88. $4x - 3(2y - 5)$ $4x - 6y + 15$

89. $-2a - 3(3a - 7)$ $-11a + 21$

90. $3x - 2(5x - 7)$ $-7x + 14$

91. $2x - 3(x - 2y)$ $-x + 6y$

92. $3[a - 5(5 - 3a)]$ $48a - 75$

93. $5[-2 - 6(a - 5)]$ $140 - 30a$

94. $3[x - 2(x + 2y)]$ $-3x - 12y$

95. $5[y - 3(y - 2x)]$ $-10y + 30x$

96. $-2(x - 3y) + 2(3y - 5x)$ $-12x + 12y$

97. $4(-a - 2b) - 2(3a - 5b)$ $-10a + 2b$

98. $5(3a - 2b) - 3(-6a + 5b)$ $33a - 25b$

99. $-7(2a - b) + 2(-3b + a)$ $-12a + b$

100. $3x - 2[y - 2(x + 3[2x + 3y])]$ $31x + 34y$

101. $2x - 4[x - 4(y - 2[5y + 3])]$ $-2x - 144y - 96$

102. $4 - 2(7x - 2y) - 3(-2x + 3y)$ $4 - 8x - 5y$

103. $3x + 8(x - 4) - 3(2x - y)$ $5x - 32 + 3y$

104. $\frac{1}{3}[8x - 2(x - 12) + 3]$ $2x + 9$

105. $\frac{1}{4}[14x - 3(x - 8) - 7x]$ $x + 6$

APPLYING CONCEPTS 1.3

In each of the following, it is possible that at least one of the Properties of Real Numbers has been applied incorrectly. If the statement is incorrect, state the incorrect application of the Properties of Real Numbers and correct the answer. If the statement is correct, state the Property of Real Numbers that is being used.

106. $-4(5x - y) = -20x + 4y$
Distributive Property

107. $4(3y + 1) = 12y + 4$
Distributive Property

108. $6 - 6x = 0x = 0$
Incorrect use of the Distributive Property
$6 - 6x = 6 - 6x$

109. $2 + 3x = (2 + 3)x = 5x$
Incorrect use of the Distributive Property
$2 + 3x = 2 + 3x$

110. $3a - 4b = 4b - 3a$
Incorrect use of the Commutative Property of Addition
$3a - 4b = -4b + 3a$

111. $2(3y) = (2 \cdot 3)(2y) = 12y$
Incorrect use of the Associative Property of Multiplication $2(3y) = (2 \cdot 3)y = 6y$

112. $x^4 \cdot \frac{1}{x^4} = 1, x \neq 0$
Inverse Property of Multiplication

113. $-x^2 + y^2 = y^2 - x^2$
Commutative Property of Addition

Name the property that justifies each lettered step used in simplifying the expression.

[C] **114.** $3(x + y) + 2x$

a. $(3x + 3y) + 2x$
The Distributive Property

b. $(3y + 3x) + 2x$
The Commutative Property of Addition

c. $3y + (3x + 2x)$
The Associative Property of Addition

d. $3y + (3 + 2)x$
$3y + 5x$ The Distributive Property

[C] **115.** $3a + 4(b + a)$

a. $3a + (4b + 4a)$
The Distributive Property

b. $3a + (4a + 4b)$
The Commutative Property of Addition

c. $(3a + 4a) + 4b$
The Associative Property of Addition

d. $(3 + 4)a + 4b$
$7a + 4b$ The Distributive Property

[C] **116.** $y + (3 + y)$

a. $y + (y + 3)$
The Commutative Property of Addition

b. $(y + y) + 3$
The Associative Property of Addition

c. $(1y + 1y) + 3$
The Multiplication Property of One

d. $(1 + 1)y + 3$
$2y + 3$ The Distributive Property

[C] **117.** $5(3a + 1)$

a. $5(3a) + 5(1)$
The Distributive Property

b. $(5 \cdot 3)a + 5(1)$
$15a + 5(1)$
The Associative Property of Multiplication

c. $15a + 5$
The Multiplication Property of One

SECTION 1.4
Verbal Expressions and Variable Expressions

1 Translate a verbal expression into a variable expression

5 HM[3]

One of the major skills required in applied mathematics is translating a verbal expression into a mathematical expression. Doing so requires recognizing the verbal phrases that translate into mathematical operations. Following is a partial list of the verbal phrases used to indicate the different mathematical operations.

Addition	more than	8 more than w	$w + 8$
	added to	x added to 9	$9 + x$
	the sum of	the sum of z and 9	$z + 9$
	the total of	the total of r and s	$r + s$
	increased by	x increased by 7	$x + 7$
Subtraction	less than	12 less than b	$b - 12$
	the difference between	the difference between x and 1	$x - 1$
	minus	z minus 7	$z - 7$
	decreased by	17 decreased by a	$17 - a$

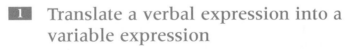

POINT OF INTEREST

Mathematical symbolism, as shown on pages 37 and 38, has advanced through various stages: rhetorical, syncoptical, and modern. In the rhetorical stage, all mathematical description was through words. In the syncoptical stage, there was a combination of words and symbols. For instance, *x* plano 4 in *y* meant 4*xy*. The modern stage, which is used today, began in the 17th century. Modern symbolism is also changing. For example, there are advocates of a system of symbolism that would place all operations last. Using this notation, 4 plus 7 would be written 4 7 + and 6 divided by 4 would be written 6 4 ÷.

INSTRUCTOR NOTE

Translating the words *sum, difference, product,* and *quotient* is difficult for students. Suggest to students that these words imply the use of parentheses. As an example, 7 less than twice x is $2x - 7$, whereas twice the difference between x and 7 is $2(x - 7)$.

Multiplication	times	negative 2 times c	$-2c$
	the product of	the product of x and y	xy
	multiplied by	3 multiplied by n	$3n$
	of	three-fourths of m	$\frac{3}{4}m$
	twice	twice d	$2d$
Division	divided by	v divided by 15	$\frac{v}{15}$
	the quotient of	the quotient of y and 3	$\frac{y}{3}$
	ratio	the ratio of x to 7	$\frac{x}{7}$
Power	the square of or the second power of	the square of x	x^2
	the cube of or the third power of	the cube of r	r^3
	the fifth power of	the fifth power of a	a^5

Translating a phrase that contains the word *sum*, *difference*, *product*, or *quotient* can sometimes cause a problem. In the examples at the right, note where the operation symbol is placed.

the *sum* of x and y (+)	$x + y$
the *difference* between x and y (−)	$x - y$
the *product* of x and y (·)	$x \cdot y$
the *quotient* of x and y (÷)	$\frac{x}{y}$

The sum of two numbers is 37. If x represents the smaller number, translate "twice the larger number" into a variable expression.

Write an expression for the larger number by subtracting the smaller number, x, from 37. — larger number: $37 - x$

Identify the words that indicate the mathematical operations. — twice the larger number

Use the identified words to write a variable expression. — $2(37 - x)$

Translate "five less than twice the difference between a number and seven" into a variable expression. Then simplify.

Assign a variable to the unknown number. — the unknown number: x

Identify words that indicate the mathematical operations. — 5 less than twice the difference between x and 7

Use the identified words to write the variable expression. — $2(x - 7) - 5$

Simplify the expression. — $= 2x - 14 - 5 = 2x - 19$

Example 1   Translate and simplify "the total of five times a number and twice the difference between the number and three."

Solution   the unknown number: n ▶ Assign a variable to one of the unknown quantities.

five times the number: $5n$ ▶ Use the assigned variable to write
the difference between the   an expression for any other unnumber and three: $n - 3$ known quantity.
twice the difference between
the number and three:
$2(n - 3)$

$5n + 2(n - 3)$ ▶ Use the assigned variable to write
the variable expression.

$= 5n + 2n - 6$ ▶ Simplify the variable expression.
$= 7n - 6$

Problem 1   Translate and simplify "a number decreased by the difference between eight and twice the number."

Solution   See page S2. $n - (8 - 2n); 3n - 8$

Example 2   Translate and simplify "fifteen minus one-half the sum of a number and ten."

Solution   the unknown number: n ▶ Assign a variable to one of the unknown quantities.

the sum of the number and   ▶ Use the assigned variable to
ten: $n + 10$ write an expression for any other
one-half the sum of the   unknown quantity.

number and ten: $\frac{1}{2}(n + 10)$

$15 - \frac{1}{2}(n + 10)$ ▶ Use the assigned variable to write
the variable expression.

$= 15 - \frac{1}{2}n - 5$ ▶ Simplify the variable expression.

$= -\frac{1}{2}n + 10$

Problem 2   Translate and simplify "the sum of three-eighths of a number and five-twelfths of the number."

Solution   See page S2. $\frac{3}{8}n + \frac{5}{12}n; \frac{19}{24}n$

2   Application problems

Many of the applications of mathematics require that you identify the unknown quantity, assign a variable to that quantity, and then attempt to express other unknowns in terms of that quantity.

Thirty gallons of paint were poured into two containers of different sizes. Express the amount of paint poured into the smaller container in terms of the amount poured into the larger container.

Assign a variable to the amount of paint poured into the larger container.

the number of gallons of paint poured into the larger container: g

Express the amount of paint poured into the smaller container in terms of g (g gallons of paint were poured into the larger container).

the number of gallons of paint poured into the smaller container: $30 - g$

Example 3 A cyclist is riding at a rate that is twice the speed of a runner. Express the speed of the cyclist in terms of the speed of the runner.

Solution the speed of the runner: r
the speed of the cyclist is twice r: $2r$

Problem 3 A mixture of candy contains 3 lb more of milk chocolate than of caramel. Express the amount of milk chocolate in the mixture in terms of the amount of caramel in the mixture.

Solution See page S2. $x + 3$

Example 4 The length of a rectangle is 2 ft more than 3 times the width. Express the length of the rectangle in terms of the width.

Solution the width of the rectangle: W
the length is 2 more than 3 times W: $3W + 2$

Problem 4 The depth of the deep end of a swimming pool is 2 ft more than twice the depth of the shallow end. Express the depth of the deep end in terms of the depth of the shallow end.

Solution See page S2. $2x + 2$

STUDY TIPS

USE THE END-OF-CHAPTER MATERIAL

To help you review the material presented within a chapter, a Chapter Summary appears at the end of each chapter. In the Chapter Summary, definitions of the important terms and concepts introduced in the chapter are provided under "Key Words." Listed under "Essential Rules and Procedures" are the formulas and procedures presented in the chapter. After completing a chapter, be sure to read the Chapter Summary. Use it to check your understanding of the material presented and to determine what concepts you need to review.

Each chapter ends with Chapter Review Exercises and a Chapter Test. The problems these contain summarize what you should have learned when you have finished the chapter. Do these exercises as you prepare

continued

for an examination. Check your answers against those in the back of the book. Answers to all Chapter Review and Chapter Test exercises are provided there. The objective being reviewed by any particular problem is written in parentheses following the answer. For any problem you answer incorrectly, review the material corresponding to that objective in the textbook. Determine *why* your answer was wrong.

CONCEPT REVIEW 1.4

Determine whether the following statements are always true, sometimes true, or never true.

1. The sum of two numbers is 12. If x represents the larger number, then the smaller number is represented by $x - 12$. Never true

2. A rope L feet long is cut into two pieces. If x is the length of one piece of the rope, the length of the other piece is $x - L$. Never true

3. The expression $2(x + 4)$ can be described by the statement "the sum of twice x and 4." Never true

4. Four times the difference between x and 3 is given by $4x - 3$. Never true

5. The square of $-x$ is $-x^2$. Sometimes true

6. The expressions $\frac{2x}{3}$ and $\frac{2}{3}x$ represent the same number. Always true

EXERCISES 1.4

1 Translate into a variable expression. Then simplify.

1. a number minus the sum of the number and two $n - (n + 2); -2$

2. a number decreased by the difference between five and the number $n - (5 - n); 2n - 5$

3. the sum of one-third of a number and four-fifths of the number $\frac{1}{3}n + \frac{4}{5}n; \frac{17}{15}n$

4. the difference between three-eighths of a number and one-sixth of the number $\frac{3}{8}n - \frac{1}{6}n; \frac{5}{24}n$

5. five times the product of eight and a number $5(8n); 40n$

6. a number increased by two-thirds of the number $n + \frac{2}{3}n; \frac{5}{3}n$

7. the difference between the product of seventeen and a number and twice the number $17n - 2n; 15n$

8. one-half of the total of six times a number and twenty-two $\frac{1}{2}(6n + 22); 3n + 11$

9. the difference between the square of a number and the total of twelve and the square of the number $n^2 - (12 + n^2); -12$

10. eleven more than the square of a number added to the difference between the number and seventeen $(n - 17) + (n^2 + 11); n^2 + n - 6$

11. the sum of five times a number and twelve added to the product of fifteen and the number
$15n + (5n + 12); 20n + 12$

12. four less than twice the sum of a number and eleven $2(n + 11) - 4; 2n + 18$

13. The sum of two numbers is 15. Using x to represent the smaller of the two numbers, translate "the sum of twice the smaller number and two more than the larger number" into a variable expression. Then simplify.
$2x + (15 - x + 2); x + 17$

14. The sum of two numbers is 20. Using x to represent the smaller of the two numbers, translate "the difference between five times the larger number and three less than the smaller number" into a variable expression.
$5(20 - x) - (x - 3); -6x + 103$

15. The sum of two numbers is 34. Using x to represent the larger of the two numbers, translate "the difference between two more than the smaller number and twice the larger number" into a variable expression. Then simplify.
$(34 - x + 2) - 2x; -3x + 36$

16. The sum of two numbers is 33. Using x to represent the larger of the two numbers, translate "the difference between six more than twice the smaller number and three more than the larger number" into a variable expression. Then simplify. $[2(33 - x) + 6] - (x + 3); -3x + 69$

2 17. *Demographics* The population of San Paolo, Brazil is four times the population of Milan, Italy. Express the population of San Paolo in terms of the population of Milan, Italy. (Source: *Information Please Almanac*)
$4P$

18. *Health* There are seven times the number of deaths each year by heart disease as there are deaths by accidents. Express the number of deaths each year by heart disease in terms of the number of deaths by accidents. (Source: *Wall Street Journal Almanac*) $7d$

19. *Sports Figures* The amount that Dennis Rodman earned in 1997 from endorsements was $\frac{2}{3}$ the amount that Arnold Palmer earned from endorsements the same year. Express the amount earned by Dennis Rodman in terms of the amount that Arnold Palmer earned. $\frac{2}{3}A$

20. *Construction* The longest rail tunnel (from Hanshu to Hokkaido, Japan) is 23.36 mi longer than the longest road tunnel (from Goschenen to Airo, Switzerland). Express the length of the longest rail tunnel in terms of the length of the longest road tunnel. (Source: *Wall Street Journal Almanac*)
$L + 23.36$

21. *Geometry* The measure of angle A of a triangle is twice the measure of angle B. The measure of angle C is twice the measure of angle A. Write expressions for angle A and angle C in terms of angle B.
Measure of angle A: $2x$; measure of angle C: $4x$

22. *Geometry* The length of a rectangle is three more than twice the width. Express the length of the rectangle in terms of the width. $2W + 3$

23. *Travel* The total flying time for a round trip between New York and Los Angeles is 12 h. Because of the jet stream, the time going is not equal to the time returning. Express the flying time between New York and Los Angeles in terms of the flying time between Los Angeles and New York. $12 - t$

24. *Business* The retail sales for DVD video disc players in December of 1997 were 24,557 greater than the retail sales for these players in November of the same year. Express the number of sales of DVD video discs in December 1997 in terms of the number sold in November 1997.
$S + 24{,}557$

APPLYING CONCEPTS 1.4

For each of the following, write a phrase that would translate into the given expression.

25. $2x + 3$ **26.** $5y - 4$ **27.** $2(x + 3)$ **28.** $5(y - 4)$

25. the sum of twice a number and three
26. four less than five times a number
27. twice the sum of a number and three
28. the product of five and four less than a number

Some English phrases require more than one variable in order to be translated into a variable expression.

29. *Physics* Translate "the product of one-half the acceleration due to gravity (g) and the time (t) squared" into a variable expression. (This expression gives the distance a dropped object will fall during a certain time interval.)
$\frac{1}{2}gt^2$

30. *Physics* Translate "the product of mass (m) and acceleration (a)" into a variable expression. (This expression is used to calculate the force exerted on an accelerating object.) ma

31. *Physics* Translate "the product of the area (A) and the square of the velocity (v)" into a variable expression. (This expression is used to compute the force a wind exerts on a sail.) Av^2

[C] **32.** *Physics* Translate "the square root of the quotient of the spring constant (k) and the mass (m)" into a variable expression. (This is part of the expression that is used to compute the frequency of oscillation of a mass on the end of a spring.) $\sqrt{\frac{k}{m}}$

Focus on Problem Solving

Polya's Four-Step Process

Your success in mathematics and your success in the workplace are heavily dependent on your ability to solve problems. One of the foremost mathematicians to study problem solving was George Polya (1887–1985). The basic structure that Polya advocated for problem solving has four steps, as outlined below. (There are many sites on the internet dealing with problem solving. Two such sites are www.askdrmath.com and www.mcallister.org. You can also conduct a search for "problem solving.")

POINT OF INTEREST

George Polya was born in Hungary and moved to the United States in 1940. He lived in Providence, Rhode Island, where he taught at Brown University until 1942, when he moved to California. There he taught at Stanford University until his retirement. While at Stanford, he published 10 books and a number of articles for mathematics journals. Of the books Polya published, *How To Solve It* (1945) is one of his best known. In this book, Polya outlines a strategy for solving problems. This strategy, although frequently applied to mathematics, can be used to solve problems from virtually any discipline.

1. Understand the Problem

You must have a clear understanding of the problem. To help you focus on understanding the problem, here are some questions to think about.

- Can you restate the problem in your own words?
- Can you determine what is known about these types of problems?
- Is there missing information that you need in order to solve the problem?
- Is there information given that is not needed?
- What is the goal?

2. Devise a Plan

Successful problem solvers use a variety of techniques when they attempt to solve a problem. Here are some frequently used strategies.

- Make a list of the known information.
- Make a list of information that is needed to solve the problem.
- Make a table or draw a diagram.
- Work backwards.
- Try to solve a similar but simpler problem.
- Research the problem to determine whether there are known techniques for solving problems of its kind.
- Try to determine whether some pattern exists.
- Write an equation.

INSTRUCTOR NOTE

The feature entitled Focus on Problem Solving appears at the end of every chapter of the text. It provides optional material that can be used to enhance your students' problem-solving skills.

3. Carry Out the Plan

Once you have devised a plan, you must carry it out.

- Work carefully.
- Keep an accurate and neat record of all your attempts.
- Realize that some of your initial plans will not work and that you may have to return to Step 2 and devise another plan or modify your existing plan.

4. Review Your Solution

Once you have found a solution, check the solution against the known facts.

- Ensure that the solution is consistent with the facts of the problem.
- Interpret the solution in the context of the problem.
- Ask yourself whether there are generalizations of the solution that could apply to other problems.
- Determine the strengths and weaknesses of your solution. For instance, is your solution only an approximation to the actual solution?
- Consider the possibility of alternative solutions.

We will use Polya's four-step process to solve the following problem.

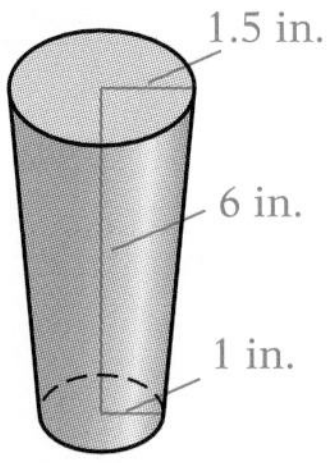

A large soft drink costs \$1.25 at a college cafeteria. The dimensions of the cup are shown at the left. Suppose you don't put any ice in the cup. Determine the cost per ounce for the soft drink.

1. *Understand the problem.* We must determine the cost per ounce for the soda. To do this, we need the dimensions of the cup (which are given), the cost of the drink (given), and a formula for the volume of the cup (unknown). Also, because the dimensions are given in inches, the volume will be in cubic inches; we need a conversion factor that will convert cubic inches to fluid ounces.

2. *Devise a plan.* Consult a resource book that gives the volume of the figure, which is called a **frustrum.** The formula for the volume is

$$V = \frac{\pi h}{3}(r^2 + rR + R^2)$$

where h is the height, r is the radius of the base, and R is the radius of the top. Also from a reference book, $1 \text{ in}^3 \approx 0.55$ fl oz. The general plan is to calculate the volume, convert the answer to fluid ounces, and then divide the cost by the number of fluid ounces.

3. *Carry out the plan.* Using the information from the drawing, evaluate the formula for the volume.

$$V = \frac{6\pi}{3}[1^2 + 1(1.5) + 1.5^2] = 9.5\pi \approx 29.8451 \text{ in}^3$$

$$V \approx 29.8451(0.55) \approx 16.4148 \text{ fl oz}$$ ▶ Convert to fluid ounces.

$$\text{Cost per ounce} \approx \frac{1.25}{16.4148} \approx 0.07615$$ ▶ Divide the cost by the volume.

The cost of the soft drink is approximately 7.62 cents per ounce.

4. *Review the solution.* The cost of a 12-ounce can of soda from a vending machine is generally about 75¢. Therefore, the cost of canned soda is 75¢ ÷ 12 = 6.25¢ per ounce. This is consistent with our solution. This does not mean our solution is correct, but it does indicate that it is at least reasonable. Why might soda from a cafeteria be more expensive per ounce than soda from a vending machine?

Is there an alternative way to obtain the solution? There are probably many, but one possibility is to get a measuring cup, pour the soft drink into it, and read the number of ounces. Name an advantage and a disadvantage of this method.

Use the four-step solution process for the following problems.

1. A cup dispenser next to a water cooler holds cups that have the shape of a right circular cone. The height of the cone is 4 in. and the radius of the circular top is 1.5 in. How many ounces of water can the cup hold?

2. Soft drink manufacturers do research into the preferences of consumers with regard to the look and feel and size of a soft drink can. Suppose that a manufacturer has determined that people want to have their hand reach around approximately 75% of the can. If this preference is to be achieved, how tall should the can be if it contains 12 oz of fluid? Assume the can is a right circular cylinder.

Projects and Group Activities

Water Displacement

When an object is placed in water, the object displaces an amount of water that is equal to the volume of the object.

➡ A sphere with a diameter of 4 in. is placed in a rectangular tank of water that is 6 in. long and 5 in. wide. How much does the water level rise? Round to the nearest hundredth.

$$V = \frac{4}{3}\pi r^3 \qquad \blacktriangleright \text{ Use the formula for the volume of a sphere.}$$

$$V = \frac{4}{3}\pi(2^3) = \frac{32}{3}\pi \qquad \blacktriangleright r = \frac{1}{2}d = \frac{1}{2}(4) = 2$$

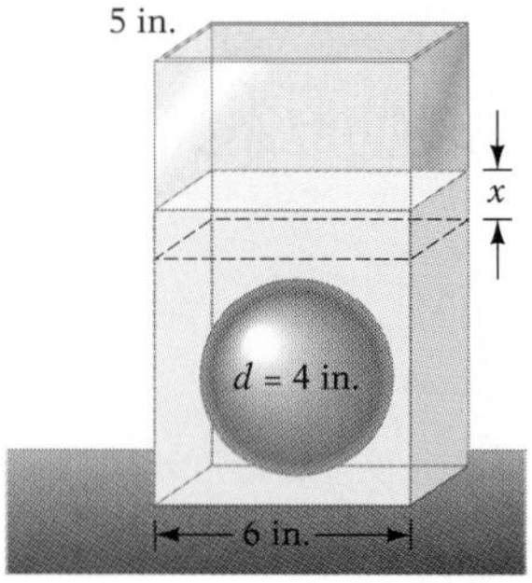

Let x represent the amount of the rise in water level. The volume of the sphere will equal the volume displaced by the water. As shown at the left, this volume is the rectangular solid with width 5 in., length 6 in., and height x in.

$$V = LWH \qquad \blacktriangleright \text{ Use the formula for the volume of a rectangular solid.}$$

$$\frac{32}{3}\pi = (6)(5)x \qquad \blacktriangleright \text{ Substitute } \frac{32}{3}\pi \text{ for } V, 5 \text{ for } W, \text{ and } 6 \text{ for } L.$$

$$\frac{32}{90}\pi = x \qquad \blacktriangleright \text{ The exact height that the water will fill is } \frac{32}{90}\pi.$$

$$1.12 \approx x \qquad \blacktriangleright \text{ Use a calculator to find an approximation.}$$

The water will rise approximately 1.12 in.

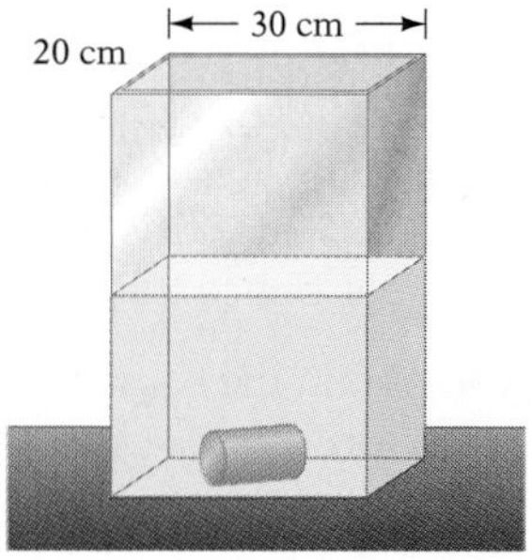

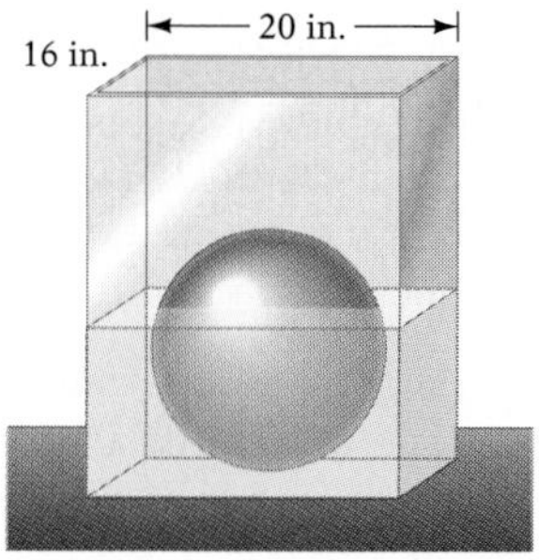

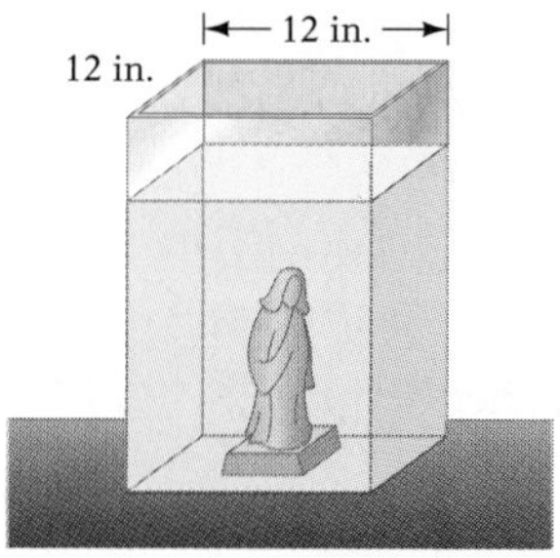

1. A cylinder with a 2-centimeter radius and a height of 10 cm is submerged in a tank of water that is 20 cm wide and 30 cm long (see Figure 1). How much does the water level rise? Round to the nearest hundredth.

2. A sphere with a radius of 6 in. is placed in a rectangular tank of water that is 16 in. wide and 20 in. long (see Figure 2). The sphere displaces water until two-thirds of the sphere is submerged. How much does the water level rise? Round to the nearest hundredth.

3. A chemist wants to know the density of a statue that weighs 15 lb. The statue is placed in a rectangular tank of water that is 12 in. long and 12 in. wide (see Figure 3). The water level rises 0.42 in. Find the density of the statue. Round to the nearest hundredth. (*Hint:* density = weight ÷ volume)

Chapter Summary

Key Words

The *integers* are ..., −4, −3, −2, −1, 0, 1, 2, 3, ... (Objective 1.1.1)

The *negative integers* are the integers ..., −4, −3, −2, −1. The *positive integers* are the integers 1, 2, 3, 4, ... The positive integers are also called the *natural numbers*. (Objective 1.1.1)

The positive integers and zero are called the *whole numbers*. (Objective 1.1.1)

0, 1, 2, 3, 4, ...

A *rational number* is a number of the form $\frac{a}{b}$, where a and b are integers and b is not equal to zero. (Objective 1.1.1)

$\frac{5}{6}, \frac{-3}{4}$ and $\frac{6}{1}$ are rational numbers.

An *irrational number* is a number whose decimal representation never terminates or repeats. (Objective 1.1.1)

$\sqrt{3}$, π, and 0.21211211121111... are irrational numbers.

The rational numbers and the irrational numbers taken together are called the *real numbers*. (Objective 1.1.1)

$-3, \sqrt{5}, \pi, \frac{-6}{7}$, and 0.232332333... are real numbers.

The *additive inverse*, or *opposite*, of a number is the same distance from zero on the number line, but on the opposite side. (Objective 1.1.1)

The additive inverse of −3 is 3.

The *absolute value* of a number is the measure of its distance from zero on the number line. (Objective 1.1.1)

$|4| = 4 \qquad |-3| = 3 \qquad -|-6| = -6$

A *set* is a collection of objects. The objects in the set are called the *elements* of the set. The *roster method* of writing a set encloses a list of the elements of the set in braces. (Objective 1.1.2)

The set $A = \{4, 5, 6, 7\}$ contains the elements 4, 5, 6, and 7.

The *empty set*, or *null set*, is the set that contains no elements. (Objective 1.1.2)

The empty set is written as $\emptyset$ or { }.

A *finite set* is a set in which the elements can be counted. (Objective 1.1.2)

$A = \{1, 2, 3, 4\}$

An *infinite set* is a set in which it is impossible to list all the elements. (Objective 1.1.2)

$A = \{1, 2, 3, 4, \ldots\}$

The *union* of two sets, written $A \cup B$, is the set that contains all the elements of A and all the elements of B. The elements that are in both set A and set B are listed only once. (Objective 1.1.2)

If $A = \{2, 3, 4\}$ and $B = \{4, 5, 6\}$, then $A \cup B = \{2, 3, 4, 5, 6\}$.

The *intersection* of two sets, written $A \cap B$, is the set that contains the elements that are common to both A and B. (Objective 1.1.2)

If $A = \{2, 3, 4\}$ and $B = \{4, 5, 6\}$, then $A \cap B = \{4\}$.

Interval notation is an alternative method of representing a set. An interval is *closed* if it includes both *endpoints*. An interval is *open* if it does not include either endpoint. An interval is *half-open* if one endpoint is included and the other endpoint is not included. A bracket is used to indicate a closed interval, and a parenthesis is used to indicate an open interval. (Objective 1.1.2)

Closed interval

Open interval

Half-open interval

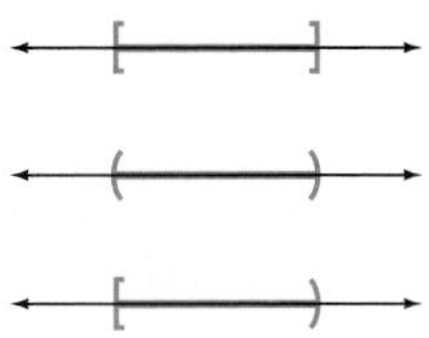

The expression a^n is in *exponential form*, where a is the *base* and n is the *exponent*. (Objective 1.2.3)

5^3 is in exponential form.

A *complex fraction* is a fraction whose numerator or denominator (or both) contains one or more fractions. (Objective 1.2.4)

$\dfrac{\frac{3}{5} - 2}{\frac{4}{5}}$ is a complex fraction.

The *multiplicative inverse* of a number is the reciprocal of the number. (Objective 1.3.1)

The multiplicative inverse of $\frac{3}{5}$ is $\frac{5}{3}$.

A *variable expression* is an expression that contains one or more variables. The *terms* of a variable expression are the addends of the expression. (Objective 1.3.2)

$3xy + x^2 - 4$ is a variable expression. $3xy$, x^2, and -4 are the terms of the expression.

A *variable term* is composed of a *numerical coefficient* and a *variable part*. (Objective 1.3.2)

The numerical coefficient of $3xy$ is 3, and the variable part is xy.

Essential Rules and Procedures

To add two numbers with the same sign, add the absolute values of the two numbers and attach the sign of the numbers. (Objective 1.2.1)

$-5 + (-6) = -11$
$4 + 2 = 6$

To add two numbers with opposite signs, find the difference between the absolute values of the two numbers and attach the sign of the number with the larger absolute value. (Objective 1.2.1)

$-4 + 8 = 4$
$6 + (-11) = -5$

To subtract b from a, add the opposite of b to a. (Objective 1.2.1)

$-4 - (-5) = -4 + 5 = 1$
$5 - 12 = 5 + (-12) = -7$

The product of two numbers with the same sign is positive.
The product of two numbers with different signs is negative. (Objective 1.2.1)

$(4)(5) = 20$ $(-3)(-7) = 21$
$(-4)(2) = -8$ $(5)(-4) = -20$

The quotient of two numbers with the same sign is positive.
The quotient of two numbers with different signs is negative. (Objective 1.2.1)

$8 \div 2 = 4$ $(-20) \div (-5) = 4$
$(-10) \div 5 = -2$ $15 \div (-3) = -5$

Order of Operations Agreement (Objective 1.2.4)
Step 1 Perform operations inside grouping symbols.
Step 2 Simplify exponential expressions.
Step 3 Do multiplication and division as they occur from left to right.
Step 4 Do addition and subtraction as they occur from left to right.

$$\begin{aligned}(5 - 2) + 9^2 \div 3 &\\ = 3 + 9^2 \div 3 &\\ = 3 + 81 \div 3 &\\ = 3 + 27 &\\ = 30 &\end{aligned}$$

Properties of the Real Numbers for Addition
(Objective 1.3.1)
Commutative: $a + b = b + a$
Associative: $(a + b) + c = a + (b + c)$
Property of Zero: $a + 0 = 0 + a = 0$
Inverse Property: $a + (-a) = (-a) + a = 0$

$3 + 2 = 2 + 3$
$(4 + 5) + 9 = 4 + (5 + 9)$
$0 + 4 = 4$
$8 + (-8) = 0$

Properties of the Real Numbers for Multiplication
(Objective 1.3.1)
Commutative: $a \cdot b = b \cdot a$
Associative: $(a \cdot b) \cdot c = a \cdot (b \cdot c)$
Property of One: $a \cdot 1 = 1 \cdot a = a$
Inverse Property: $a \cdot \dfrac{1}{a} = \dfrac{1}{a} \cdot a = 1$

$3 \cdot 5 = 5 \cdot 3$
$(2 \cdot 3) \cdot 5 = 2 \cdot (3 \cdot 5)$
$6 \cdot 1 = 6$
$3\left(\dfrac{1}{3}\right) = 1$

The Distributive Property $a(b + c) = ab + ac$
(Objective 1.3.1)

$3(x + 5) = 3x + 3(5) = 3x + 15$

Chapter Review Exercises

1. Find the additive inverse of $-\dfrac{3}{4}$. $\dfrac{3}{4}$

2. Let $x \in \{-4, -2, 0, 2\}$. For what values of x is $x > -1$ true? $0, 2$

3. Let $p \in \{-4, 0, 7\}$. Evaluate $-|p|$ for each element of the set. $-4, 0, -7$

4. Use the roster method to write the set of integers between -3 and 4. $\{-2, -1, 0, 1, 2, 3\}$

5. Use set-builder notation to write the set of real numbers less than -3. $\{x \mid x < -3\}$

6. Write $[-2, 3]$ in set-builder notation. $\{x \mid -2 \leq x \leq 3\}$

7. Find $A \cup B$ given $A = \{1, 3, 5, 7\}$ and $B = \{2, 4, 6, 8\}$. $\{1, 2, 3, 4, 5, 6, 7, 8\}$

8. Find $A \cap B$ given $A = \{0, 1, 2, 3\}$ and $B = \{2, 3, 4, 5\}$. $\{2, 3\}$

9. Graph: $[-3, \infty)$

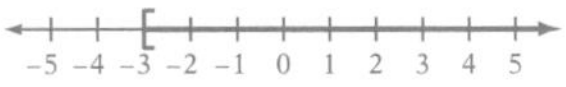

10. Graph: $\{x \mid x < 1\}$

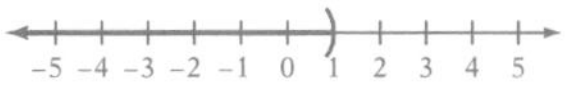

11. Graph: $\{x \mid x \leq -3\} \cup \{x \mid x > 0\}$

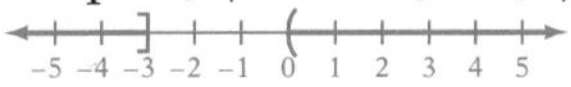

12. Graph: $(-2, 4]$

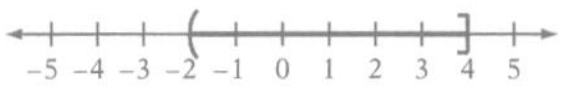

13. Subtract: $-10 - (-3) - 8$ -15

14. Divide: $-204 \div (-17)$ 12

15. Simplify: $18 - |-12 + 8|$ 14

16. Simplify: $-2 \cdot (4^2) \cdot (-3)^2$ -288

17. Simplify: $-\dfrac{3}{8} + \dfrac{3}{5} - \dfrac{1}{6}$ $\dfrac{7}{120}$

18. Multiply: $\dfrac{3}{5}\left(-\dfrac{10}{21}\right)\left(-\dfrac{7}{15}\right)$ $\dfrac{2}{15}$

19. Divide: $-\dfrac{3}{8} \div \dfrac{3}{5}$ $-\dfrac{5}{8}$

20. Simplify: $-4.07 + 2.3 - 1.07$ -2.84

21. Divide: $-3.286 \div (-1.06)$ 3.1

22. Simplify: $20 \div \frac{3^2 - 2^2}{3^2 + 2^2}$ 52

23. Evaluate $2a^2 - \frac{3b}{a}$ when $a = -3$ and $b = 2$. 20

24. Evaluate $(a - 2b^2) \div (ab)$ when $a = 4$ and $b = -3$. $\frac{7}{6}$

25. Use the Distributive Property to complete the statement.
$6x - 21y = ?(2x - 7y)$ 3

26. Use the Commutative Property of Addition to complete the statement.
$3(x + y) = 3(? + x)$ y

27. Use the Commutative Property of Multiplication to complete the statement.
$(ab)14 = 14?$ (ab)

28. Use the Associative Property of Addition to complete the statement.
$3 + (4 + y) = (3 + ?) + y$ 4

29. Identify the property that justifies the statement.
$(-4) + 4 = 0$ The Inverse Property of Addition

30. Identify the property that justifies the statement.
$2(3x) = (2 \cdot 3)x$ The Associative Property of Multiplication

31. Simplify: $-2(x - 3) + 4(2 - x)$ $-6x + 14$

32. Simplify: $4y - 3[x - 2(3 - 2x) - 4y]$
$16y - 15x + 18$

33. Translate and simplify "four times the sum of a number and four." $4(x + 4); 4x + 16$

34. Translate and simplify "eight more than twice the difference between a number and two."
$2(x - 2) + 8; 2x + 4$

35. The sum of two numbers is 40. Using x to represent the smaller of the two numbers, translate "the sum of twice the smaller number and five more than the larger number" into a variable expression. Then simplify.
$2x + (40 - x) + 5; x + 45$

36. The sum of two numbers is 9. Using x to represent the larger of the two numbers, translate "the difference between three more than twice the smaller number and one more than the larger number" into a variable expression. Then simplify. $[2(9 - x) + 3] - (x + 1); -3x + 20$

37. The length of a rectangle is 3 ft less than three times the width. Express the length of the rectangle in terms of the width. $3W - 3$

38. A second integer is 5 more than four times the first integer. Express the second integer in terms of the first integer. $4x + 5$

Chapter Test

1. Find the additive inverse of -12. 12

2. Let $x \in \{-5, 3, 7\}$. For which values of x is $-1 > x$ true? -5

3. Simplify: $2 - (-12) + 3 - 5$ 12

4. Multiply: $(-2)(-3)(-5)$ -30

5. Divide: $-180 \div 12$ -15

6. Simplify: $|-3 - (-5)|$ 2

7. Simplify: $-5^2 \cdot 4$ -100

8. Simplify: $(-2)^3(-3)^2$ -72

9. Simplify: $\frac{2}{3} - \frac{5}{12} + \frac{4}{9}$ $\frac{25}{36}$

10. Multiply: $\left(-\frac{2}{3}\right)\left(\frac{9}{15}\right)\left(\frac{10}{27}\right)$ $-\frac{4}{27}$

11. Simplify: $4.27 - 6.98 + 1.3$ -1.41

12. Divide: $-15.092 \div 3.08$ -4.9

13. Simplify: $12 - 4\left(\frac{5^2 - 1}{3}\right) \div 16$ 10

14. Simplify: $8 - 4(2 - 3)^2 \div 2$ 6

15. Evaluate $(a - b)^2 \div (2b + 1)$ when $a = 2$ and $b = -3$. -5

16. Evaluate $\frac{b^2 - c^2}{a - 2c}$ when $a = 2$, $b = 3$, and $c = -1$. 2

17. Use the Commutative Property of Addition to complete the statement.
$(3 + 4) + 2 = (? + 3) + 2$ 4

18. Identify the property that justifies the statement.
$-2(x + y) = -2x - 2y$
The Distributive Property

19. Simplify: $3x - 2(x - y) - 3(y - 4x)$ $13x - y$

20. Simplify: $2x - 4[2 - 3(x + 4y) - 2]$
$14x + 48y$

21. Translate and simplify "thirteen decreased by the product of three less than a number and nine." $13 - (n - 3)(9)$; $40 - 9n$

22. Translate and simplify "one-third of the total of twelve times a number and twenty-seven."
$\frac{1}{3}(12n + 27)$; $4n + 9$

23. Find $A \cup B$ given $A = \{1, 3, 5, 7\}$ and $B = \{2, 3, 4, 5\}$. $\{1, 2, 3, 4, 5, 7\}$

24. Find $A \cup B$ given $A = \{-2, -1, 0, 1, 2, 3\}$ and $B = \{-1, 0, 1\}$. $\{-2, -1, 0, 1, 2, 3\}$

25. Find $A \cap B$ given $A = \{1, 3, 5, 7\}$ and $B = \{5, 7, 9, 11\}$. $\{5, 7\}$

26. Find $A \cap B$ given $A = \{-3, -2, -1, 0, 1, 2, 3\}$ and $B = \{-1, 0, 1\}$. $\{-1, 0, 1\}$

27. Graph: $(-\infty, -1]$

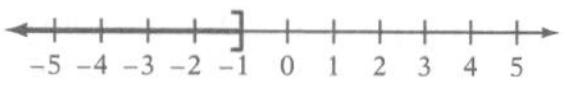

28. Graph: $(3, \infty)$

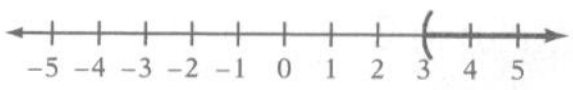

29. Graph: $\{x \mid x \le 3\} \cup \{x \mid x < -2\}$

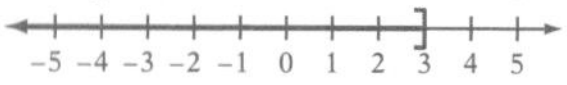

30. Graph: $\{x \mid x < 3\} \cap \{x \mid x > -2\}$

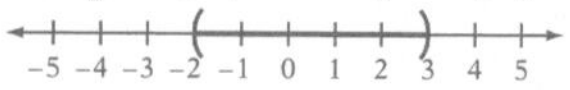

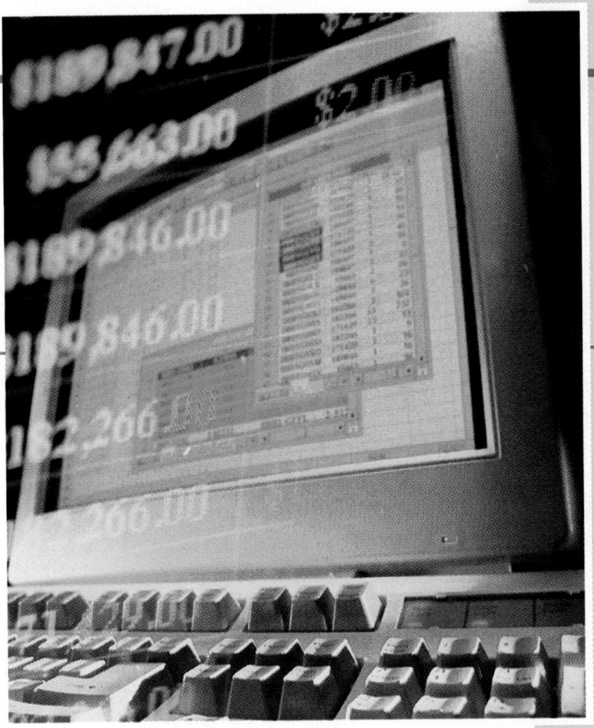

Accountants are responsible for preparing and analyzing financial reports. Many public accountants *work within an accounting firm.* Management accountants *generally work for large companies and handle their financial records.* Government accountants *examine records of either government agencies or private firms whose business is regulated by the government. All accountants work with equations. The fundamental accounting equation is*
Assets = Liabilities + Owner's Equity.

2

First-Degree Equations and Inequalities

OBJECTIVES

2.1.1	Solve equations using the Addition and Multiplication Properties of Equations
2.1.2	Solve equations using the Distributive Property
2.1.3	Application problems
2.2.1	Coin and stamp problems
2.2.2	Integer problems
2.3.1	Value mixture problems
2.3.2	Uniform motion problems
2.4.1	Investment problems
2.4.2	Percent mixture problems
2.5.1	Solve inequalities in one variable
2.5.2	Solve compound inequalities
2.5.3	Application problems
2.6.1	Absolute value equations
2.6.2	Absolute value inequalities
2.6.3	Application problems

Moscow Papyrus

Most of the early history of mathematics can be traced to the Egyptians and Babylonians. This early work began around 3000 B.C. The Babylonians used clay tablets and a type of writing called *cuneiform* (wedge-shape) to record their thoughts and discoveries.

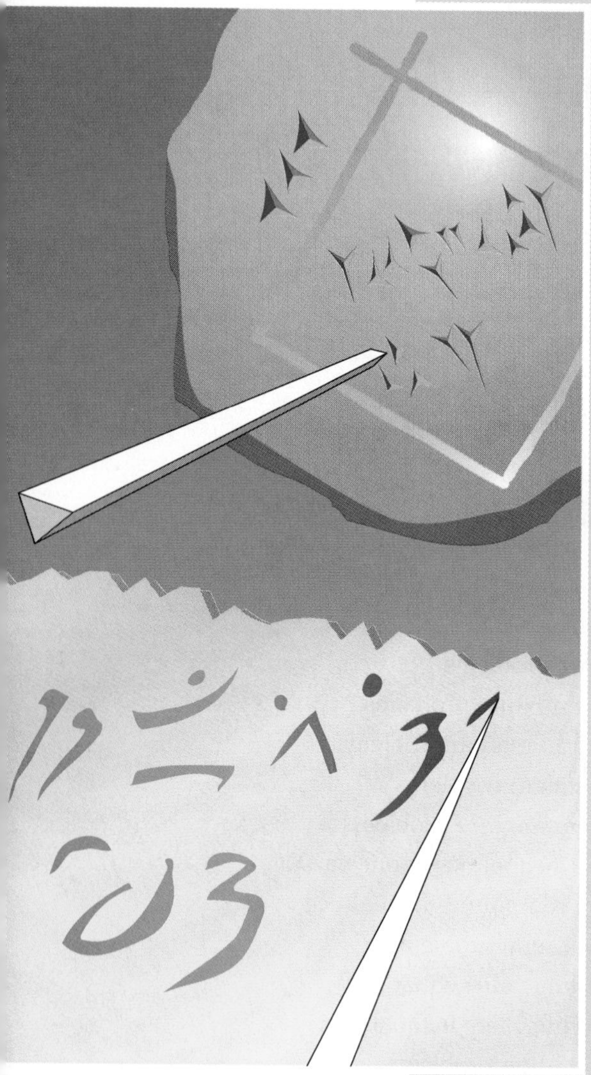

Here are the symbols for 10, 1, and subtraction.

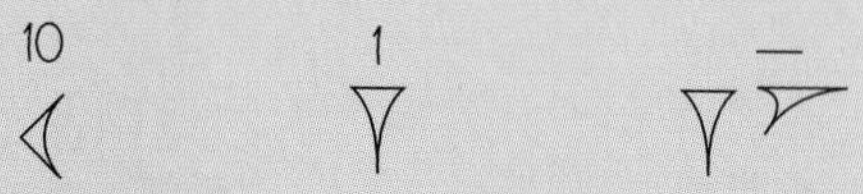

Simple groupings of these symbols were used to represent numbers less than 60. The numbers 32 and 28 are shown below.

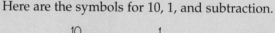

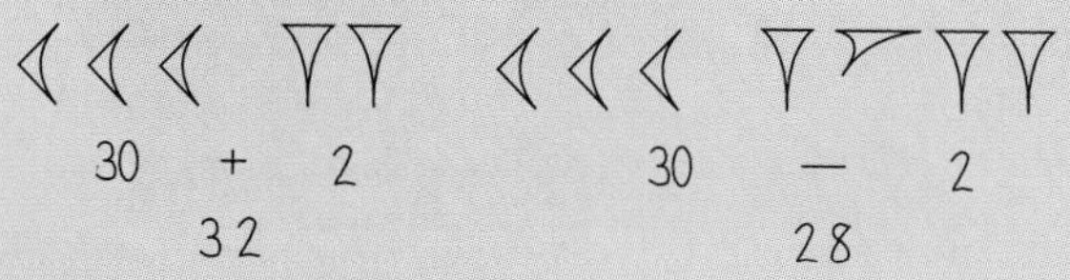

The Egyptians used papyrus, a plant reed dried and pounded thin, and hieratic writing, which was derived from hieroglyphics, to make records. A number of papyrus documents have been discovered over the years. One particularly famous one called the Moscow Papyrus dates from 1850 B.C. It is approximately 18 ft long and 3 in. wide and contains 25 problems.

The Moscow Papyrus dealt with solving practical problems related to geometry, food preparation, and grain allotments. Here is one of the problems:

The width of a rectangle is $\frac{3}{4}$ the length and the area is 12.

Find the dimensions of the rectangle.

You might recognize this as a type of problem you have solved before. Word problems are very old indeed. This one is around 3900 years old.

SECTION 2.1 Equations in One Variable

1 Solve equations using the Addition and Multiplication Properties of Equations

An **equation** expresses the equality of two mathematical expressions. The expressions can be either numerical or variable expressions.

$2 + 8 = 10$ $x + 8 = 11$ $x^2 + 2y = 7$	Equations

The equation at the right is a **conditional equation.** The equation is *true* if the variable is replaced by 3. The equation is *false* if the variable is replaced by 4.

$x + 2 = 5$	Conditional equation
$3 + 2 = 5$	A true equation
$4 + 2 = 5$	A false equation

The replacement values of the variable that will make an equation true are called the **roots,** or **solutions,** of the equation.

The solution of the equation $x + 2 = 5$ is 3.

The equation at the right is an **identity.** Any replacement for x will result in a true equation.

$x + 2 = x + 2$	Identity

The equation at the right has *no solution* because there is no number that equals itself plus 1. Any replacement value for x will result in a false equation.

$x = x + 1$	No solution

INSTRUCTOR NOTE

Students have a tendency to think of equations and expressions as the same thing. Emphasize that equations have equals signs. Also, we *solve* equations and *simplify* expressions.

Each of the equations at the right is a **first-degree equation in one variable.** All variables have an exponent of 1.

$x + 2 = 12$ $3y - 2 = 5y$ $3(a + 2) = 14a$	First-degree equations

Solving an equation means finding a solution of the equation. The simplest equation to solve is an equation of the form **variable = constant,** because the constant is the solution.

If $x = 3$, then 3 is the solution of the equation because $3 = 3$ is a true equation.

In solving an equation, the goal is to rewrite the given equation in the form *variable = constant.* The Addition Property of Equations can be used to rewrite an equation in this form.

The Addition Property of Equations

If a, b, and c are algebraic expressions, then the equation $a = b$ has the same solutions as the equation $a + c = b + c$.

INSTRUCTOR NOTE

The standard model of an equation as a balance scale still applies.

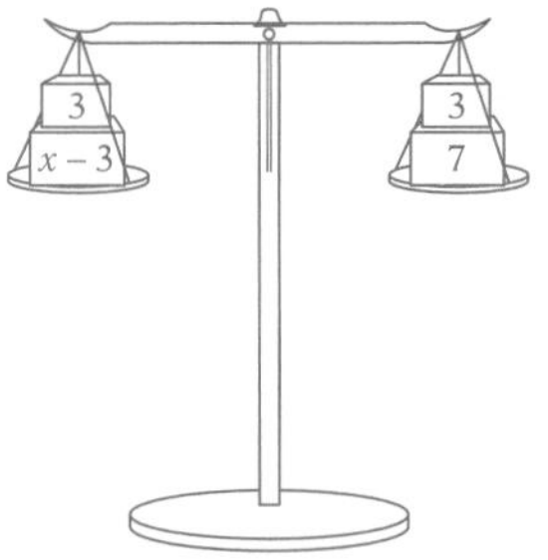

Adding a weight to one side of the equation requires adding the same weight to the other side of the equation so that the pans remain in balance.

The Addition Property of Equations states that the same quantity can be added to each side of an equation without changing the solution of the equation. This property is used to remove a term from one side of the equation by adding the opposite of that term to each side of the equation.

Solve: $x - 3 = 7$

Add the opposite of the constant term -3 to each side of the equation and simplify. After simplifying, the equation is in the form *variable* = *constant*.

$$x - 3 = 7$$
$$x - 3 + 3 = 7 + 3$$
$$x + 0 = 10$$
$$x = 10$$

To check the solution, replace the variable with 10. Simplify the left side of the equation. Because $7 = 7$ is a true equation, 10 is a solution.

Check: $x - 3 = 7$

$$\begin{array}{c|c} 10 - 3 & 7 \\ 7 = 7 & \end{array}$$

The solution is 10.

Because subtraction is defined in terms of addition, the Addition Property of Equations enables us to subtract the same number from each side of the equation without changing the solution of the equation.

LOOK CLOSELY

Remember to check the solution.

$$x + \frac{7}{12} = \frac{1}{2}$$

$$\begin{array}{r|l} -\frac{1}{12} + \frac{7}{12} & \frac{1}{2} \\ \frac{6}{12} & \frac{1}{2} \\ \frac{1}{2} = \frac{1}{2} & \end{array}$$

Solve: $x + \frac{7}{12} = \frac{1}{2}$

Add the opposite of the constant term $\frac{7}{12}$ to each side of the equation. This is equivalent to subtracting $\frac{7}{12}$ from each side of the equation.

$$x + \frac{7}{12} = \frac{1}{2}$$
$$x + \frac{7}{12} - \frac{7}{12} = \frac{1}{2} - \frac{7}{12}$$
$$x + 0 = \frac{6}{12} - \frac{7}{12}$$
$$x = -\frac{1}{12}$$

The solution is $-\frac{1}{12}$.

The Multiplication Property of Equations can also be used to rewrite an equation in the form *variable* = *constant*.

The Multiplication Property of Equations

If a, b, and c are algebraic expressions, and $c \neq 0$, then the equation $a = b$ has the same solution as the equation $ac = bc$.

The Multiplication Property of Equations states that we can multiply each side of an equation by the same nonzero number without changing the solution of the equation. This property is used to remove a coefficient from a variable term in an equation by multiplying each side of the equation by the reciprocal of the coefficient.

Solve: $-\dfrac{3}{4}x = 12$

Multiply each side of the equation by $-\dfrac{4}{3}$, the reciprocal of $-\dfrac{3}{4}$. After simplifying, the equation is in the form *variable* = *constant*.

$$-\frac{3}{4}x = 12$$
$$\left(-\frac{4}{3}\right)\left(-\frac{3}{4}\right)x = \left(-\frac{4}{3}\right)12$$
$$1x = -16$$
$$x = -16$$

Check: $\qquad -\dfrac{3}{4}x = 12$

$$\begin{array}{c|c} -\dfrac{3}{4}(-16) & 12 \\ \hline & \\ 12 & = 12 \end{array}$$

The solution is -16.

Because division is defined in terms of multiplication, the Multiplication Property of Equations enables us to divide each side of an equation by the same nonzero number without changing the solution of the equation.

Solve: $-5x = 9$

Multiply each side of the equation by the reciprocal of -5. This is equivalent to dividing each side of the equation by -5.

$$-5x = 9$$
$$\frac{-5x}{-5} = \frac{9}{-5}$$
$$1x = -\frac{9}{5}$$
$$x = -\frac{9}{5}$$

You should check the solution.

The solution is $-\dfrac{9}{5}$.

When using the Multiplication Property of Equations, it is usually easier to multiply each side of the equation by the reciprocal of the coefficient when the coefficient is a fraction. Divide each side of the equation by the coefficient when the coefficient is an integer or a decimal.

In solving an equation, it is often necessary to apply both the Addition and the Multiplication Properties of Equations.

Solve: $4x - 3 + 2x = 8 + 9x - 12$

Simplify each side of the equation by combining like terms.

$$4x - 3 + 2x = 8 + 9x - 12$$
$$6x - 3 = 9x - 4$$

Subtract $9x$ from each side of the equation. Then simplify.

$$6x - 9x - 3 = 9x - 9x - 4$$
$$-3x - 3 = -4$$

Add 3 to each side of the equation. Then simplify.

$$-3x - 3 + 3 = -4 + 3$$
$$-3x = -1$$

Divide each side of the equation by −3. $\frac{-3x}{-3} = \frac{-1}{-3}$

Then simplify. $x = \frac{1}{3}$

You should check the solution. The solution is $\frac{1}{3}$.

Example 1 Solve: $-\frac{2}{7}x = \frac{4}{21}$

Solution

$$-\frac{2}{7}x = \frac{4}{21}$$

$$\left(-\frac{7}{2}\right)\left(-\frac{2}{7}x\right) = \left(-\frac{7}{2}\right)\left(\frac{4}{21}\right)$$

▶ Multiply both sides of the equation by the reciprocal of $-\frac{2}{7}$. Then simplify.

$$x = -\frac{2}{3}$$

The solution is $-\frac{2}{3}$.

Problem 1 Solve: $-3x = 18$

Solution See page S2. -6

Example 2 Solve: $3x - 5 = x + 2 - 7x$

Solution

$$3x - 5 = x + 2 - 7x$$

$$3x - 5 = -6x + 2$$

▶ Simplify the right side of the equation by combining like terms.

$$3x + 6x - 5 = -6x + 6x + 2$$

▶ Add $6x$ to each side of the equation.

$$9x - 5 = 2$$

$$9x - 5 + 5 = 2 + 5$$

▶ Add 5 to each side of the equation.

$$9x = 7$$

$$\frac{9x}{9} = \frac{7}{9}$$

▶ Divide each side of the equation by the coefficient 9.

$$x = \frac{7}{9}$$

The solution is $\frac{7}{9}$.

Problem 2 Solve: $6x - 5 - 3x = 14 - 5x$

Solution See page S2. $\frac{19}{8}$

2 Solve equations using the Distributive Property

When an equation contains parentheses, one of the steps in solving the equation requires the use of the Distributive Property.

➡ Solve: $3(x - 2) + 3 = 2(6 - x)$

$$3(x - 2) + 3 = 2(6 - x)$$

Use the Distributive Property to remove parentheses. Simplify.

$$3x - 6 + 3 = 12 - 2x$$

$$3x - 3 = 12 - 2x$$

Add $2x$ to each side of the equation. $5x - 3 = 12$

Add 3 to each side of the equation. $5x = 15$

Divide each side of the equation by the coefficient 5. $x = 3$

You should check the solution. The solution is 3.

Example 3 Solve: $5(2x - 7) + 2 = 3(4 - x) - 12$

Solution

$5(2x - 7) + 2 = 3(4 - x) - 12$

$10x - 35 + 2 = 12 - 3x - 12$ ▶ Use the Distributive Property.

$10x - 33 = -3x$ ▶ Simplify.

$-33 = -13x$ ▶ Subtract $10x$ from each side of the equation.

$\frac{33}{13} = x$ ▶ Divide each side of the equation by -13.

The solution is $\frac{33}{13}$.

Problem 3 Solve: $6(5 - x) - 12 = 2x - 3(4 + x)$

Solution See page S3. 6

To solve an equation that contains fractions, first **clear denominators** by multiplying each side of the equation by the least common multiple (LCM) of the denominators.

Solve: $\frac{x}{2} - \frac{7}{9} = \frac{x}{6} + \frac{2}{3}$

$$\frac{x}{2} - \frac{7}{9} = \frac{x}{6} + \frac{2}{3}$$

Multiply each side of the equation by 18, the LCM of 2, 9, 6, and 3.

$$18\left(\frac{x}{2} - \frac{7}{9}\right) = 18\left(\frac{x}{6} + \frac{2}{3}\right)$$

Use the Distributive Property to remove parentheses.

$$\frac{18x}{2} - \frac{18 \cdot 7}{9} = \frac{18x}{6} + \frac{18 \cdot 2}{3}$$

$$9x - 14 = 3x + 12$$

Subtract $3x$ from each side of the equation. $6x - 14 = 12$

Add 14 to each side of the equation. $6x = 26$

Divide each side of the equation by the coefficient 6. $x = \frac{13}{3}$

You should check the solution. The solution is $\frac{13}{3}$.

Example 4 Solve: $\frac{3x - 2}{12} - \frac{x}{9} = \frac{x}{2}$

Solution

$\frac{3x - 2}{12} - \frac{x}{9} = \frac{x}{2}$ ▶ The LCM of 12, 9, and 2 is 36.

$36\left(\frac{3x - 2}{12} - \frac{x}{9}\right) = 36\left(\frac{x}{2}\right)$ ▶ Multiply each side of the equation by the LCM.

$\frac{36(3x - 2)}{12} - \frac{36x}{9} = \frac{36x}{2}$ ▶ Use the Distributive Property.

$$\begin{aligned} 3(3x-2)-4x &= 18x \quad \text{▶ Simplify.} \\ 9x-6-4x &= 18x \\ 5x-6 &= 18x \\ -6 &= 13x \\ -\frac{6}{13} &= x \end{aligned}$$

The solution is $-\frac{6}{13}$.

Problem 4 Solve: $\frac{2x-7}{3}-\frac{5x+4}{5}=\frac{-x-4}{30}$

Solution See page S3. −10

3 Application problems

6 HM³

Solving application problems is primarily a skill in translating sentences into equations and then solving the equations.

An equation states that two mathematical expressions are equal. Therefore, to translate a sentence into an equation requires recognizing the words or phrases that mean *equals*. These phrases include "is," "is equal to," "amounts to," and "represents."

Once the sentence is translated into an equation, solve the equation by rewriting the equation in the form *variable* = *constant*.

An electrician works 34 h and uses \$642 worth of materials repairing the wiring in a home. The electrician receives an hourly wage of \$22. Find the total cost of repairing the wiring.

Strategy
To find the total cost of repairing the wiring, write and solve an equation using C to represent the total cost. The total cost is the sum of the cost of labor and the cost of materials.

Solution
$C = (34)(22) + 642$
$C = 748 + 642$
$C = 1390$

The total cost of repairing the wiring is \$1390.

Example 5 The charges for a long-distance telephone call are \$2.14 for the first three minutes and \$.47 for each additional minute or fraction of a minute. If the charges for a long-distance call were \$17.65, how many minutes did the phone call last?

Strategy To find the length of the phone call in minutes, write and solve an equation using n to represent the total number of minutes of the call. Then $n-3$ is the number of additional minutes after the first three minutes of the phone call. The fixed charge for the three minutes plus the charge for the additional minutes is the total cost of the phone call.

Solution

$$\begin{aligned} 2.14 + 0.47(n - 3) &= 17.65 \\ 2.14 + 0.47n - 1.41 &= 17.65 \\ 0.47n + 0.73 &= 17.65 \\ 0.47n &= 16.92 \\ n &= 36 \end{aligned}$$

The phone call lasted 36 min.

Problem 5 You are making a salary of \$14,500 and receive an 8% raise for next year. Find next year's salary.

Solution See page S3. \$15,660

STUDY TIPS

KEEP UP TO DATE WITH COURSE WORK

College terms start out slowly. Then they gradually get busier and busier, reaching a peak of activity at final examination time. If you fall behind in the work for a course, you will find yourself trying to catch up at a time when you are very busy with all of your other courses. Don't fall behind—keep up to date with course work.

Keeping up with course work is doubly important for a course in which information and skills learned early in the course are needed to learn information and skills later in the course. Skills must be learned immediately and reviewed often.

Your instructor gives assignments to help you acquire a skill or understand a concept. Do each assignment as it is assigned or you may well fall behind and have great difficulty catching up. Keeping up with course work also makes it easier to prepare for each exam.

CONCEPT REVIEW 2.1

Determine whether the following statements are always true, sometimes true, or never true.

1. $6x + 5$ is an equation. Never true
2. A first-degree equation has no solution. Sometimes true
3. An equation expresses the equality between two variable expressions. Sometimes true
4. The equation $x - 3 = x - 3$ has no solution. Never true
5. The equation $x = x - 4$ is an identity. Never true
6. Zero is a solution of $5x = 2x$. Always true
7. Suppose a, b, and c are real numbers and $a = b$. Then $a + c = b + c$. Always true

8. Suppose a, b, and c are real numbers and $a = b$. Then $\frac{a}{c} = \frac{b}{c}$. Sometimes true

EXERCISES 2.1

1

1. What is the difference between an equation and an expression?

2. Explain why the goal in solving an equation is to write the equation in the form *variable* = *constant*.

3. **a.** When is the Addition Property of Equations used to solve an equation?
   b. When is the Multiplication Property of Equations used to solve an equation?

4. Why, when the Multiplication Property of Equations is used, must the number that multiplies each side of the equation be nonzero?

Solve and check.

5. $x - 2 = 7$ 9

6. $x - 8 = 4$ 12

7. $a + 3 = -7$ -10

8. $-12 = x - 3$ -9

9. $3x = 12$ 4

10. $8x = 4$ $\frac{1}{2}$

11. $\frac{2}{7} + x = \frac{17}{21}$ $\frac{11}{21}$

12. $x + \frac{2}{3} = \frac{5}{6}$ $\frac{1}{6}$

13. $\frac{5}{8} - y = \frac{3}{4}$ $-\frac{1}{8}$

14. $\frac{2}{3}y = 5$ $\frac{15}{2}$

15. $\frac{3}{5}y = 12$ 20

16. $\frac{3t}{8} = -15$ -40

17. $\frac{3a}{7} = -21$ -49

18. $-\frac{5}{8}x = \frac{4}{5}$ $-\frac{32}{25}$

19. $-\frac{5}{12}y = \frac{7}{16}$ $-\frac{21}{20}$

20. $-\frac{3}{4}x = -\frac{4}{7}$ $\frac{16}{21}$

21. $b - 14.72 = -18.45$ -3.73

22. $b + 3.87 = -2.19$ -6.06

23. $3x + 5x = 12$ $\frac{3}{2}$

24. $2x - 7x = 15$ -3

25. $2x - 4 = 12$ 8

26. $3x - 12 = 5x$ -6

27. $4x + 2 = 4x$ No solution

28. $3m - 7 = 3m$ No solution

29. $2x + 2 = 3x + 5$ -3

30. $7x - 9 = 3 - 4x$ $\frac{12}{11}$

31. $2 - 3t = 3t - 4$ 1

32. $7 - 5t = 2t - 9$ $\frac{16}{7}$

33. $2a - 3a = 7 - 5a$ $\frac{7}{4}$

34. $3a - 5a = 8a + 4$ $-\frac{2}{5}$

35. $\frac{5}{8}b - 3 = 12$ 24

36. $\frac{1}{3} - 2b = 3$ $-\frac{4}{3}$

37. $b + \frac{1}{5}b = 2$ $\frac{5}{3}$

38. $3x - 2x + 7 = 12 - 4x$ 1

39. $2x - 9x + 3 = 6 - 5x$ $-\frac{3}{2}$

40. $7 + 8y - 12 = 3y - 8 + 5y$ No solution

41. $2y - 4 + 8y = 7y - 8 + 3y$ No solution

42. $2x - 5 + 7x = 11 - 3x + 4x$ 2

43. $9 + 4x - 12 = -3x + 5x + 8$ $\frac{11}{2}$

44. $3.24a + 7.14 = 5.34a$ 3.4

45. $5.3y + 0.35 = 5.02y$ -1.25

2 Solve and check.

46. $2x + 2(x + 1) = 10$ 2

47. $2x + 3(x - 5) = 15$ 6

48. $2(a - 3) = 2(4 - 2a)$ $\frac{7}{3}$

49. $5(2 - b) = -3(b - 3)$ $\frac{1}{2}$

50. $3 - 2(y - 3) = 4y - 7$ $\frac{8}{3}$

51. $3(y - 5) - 5y = 2y + 9$ -6

52. $4(x - 2) + 2 = 4x - 2(2 - x)$ -1

53. $2x - 3(x - 4) = 2(3 - 2x) + 2$ $-\frac{4}{3}$

54. $2(2d + 1) - 3d = 5(3d - 2) + 4d$ $\frac{2}{3}$

55. $-4(7y - 1) + 5y = -2(3y + 4) - 3y$ $\frac{6}{7}$

56. $4[3 + 5(3 - x) + 2x] = 6 - 2x$ $\frac{33}{5}$

57. $2[4 + 2(5 - x) - 2x] = 4x - 7$ $\frac{35}{12}$

58. $2[b - (4b - 5)] = 3b + 4$ $\frac{2}{3}$

59. $-3[x + 4(x + 1)] = x + 4$ -1

60. $4[a - (3a - 5)] = a - 7$ 3

61. $5 - 6[2t - 2(t + 3)] = 8 - t$ -33

62. $-3(x - 2) = 2[x - 4(x - 2) + x]$ 10

63. $3[x - (2 - x) - 2x] = 3(4 - x)$ 6

64. $\frac{2}{9}t - \frac{5}{6} = \frac{1}{12}t$ 6

65. $\frac{3}{4}t - \frac{7}{12}t = 1$ 6

66. $\frac{2}{3}x - \frac{5}{6}x - 3 = \frac{1}{2}x - 5$ 3

67. $\frac{1}{2}x - \frac{3}{4}x + \frac{5}{8} = \frac{3}{2}x - \frac{5}{2}$ $\frac{25}{14}$

68. $\frac{3x - 2}{4} - 3x = 12$ $-\frac{50}{9}$

69. $\frac{2a - 9}{5} + 3 = 2a$ $\frac{3}{4}$

70. $\frac{x - 2}{4} - \frac{x + 5}{6} = \frac{5x - 2}{9}$ $-\frac{40}{17}$

71. $\frac{2x - 1}{4} + \frac{3x + 4}{8} = \frac{1 - 4x}{12}$ $-\frac{4}{29}$

72. $\frac{2}{3}(15 - 6a) = \frac{5}{6}(12a + 18)$ $-\frac{5}{14}$

73. $\frac{1}{5}(20x + 30) = \frac{1}{3}(6x + 36)$ 3

74. $\frac{1}{3}(x - 7) + 5 = 6x + 4$ $-\frac{4}{17}$

75. $2(y - 4) + 8 = \frac{1}{2}(6y + 20)$ -10

76. $\frac{1}{4}(2b + 50) = \frac{5}{2}\left(15 - \frac{1}{5}b\right)$ 25

77. $\frac{1}{4}(7 - x) = \frac{2}{3}(x + 2)$ $\frac{5}{11}$

78. $-4.2(p + 3.4) = 11.13$ -6.05

79. $-1.6(b - 2.35) = -11.28$ 9.4

3 80. *Temperature* The Fahrenheit temperature is 59°. This is 32° more than $\frac{9}{5}$ of the Celsius temperature. Find the Celsius temperature. 15°C

81. *Consumerism* A local feed store sells a 100-pound bag of feed for $10.90. If a customer buys more than one bag, each additional bag costs $10.50. A customer bought $84.40 worth of feed. How many 100-pound bags of feed did this customer purchase? 8 bags

82. *Consumerism* A Chinese restaurant charges $11.95 for the adult buffet and $7.25 for the children's buffet. One family's bill came to $79.35. If there were three adults in the family, how many children were there? 6 children

83. *Labor* The repair bill on your car was $316.55. The charge for parts was $148.55. A mechanic worked on your car for 4 h. What was the charge per hour for labor? $42

84. *Paychecks* An elementary school teacher has 26.8% of his gross biweekly paycheck deducted to cover union dues, insurance, and taxes. The biweekly pay is $1620.00. Find the teacher's take-home pay. $1185.84

85. *Wages* A machine shop employee earned $374.50 last week. She worked 40 h on the line at her standard rate and 9 h at a time-and-a-half rate. Find her regular hourly rate. $7

86. *Consumerism* The Showcase Cinema of Lawrence charges $7.75 for an adult ticket and $4.75 for a child's ticket for all shows after 6:00 P.M. If a family of six pays $34.50 to get into an evening show, how many adult tickets and how many children's tickets did the family purchase? 2 adult tickets, 4 children's tickets

87. *Consumerism* A group of 12 fraternity brothers decided to go to a Boston Red Sox game. They were unable to get seats together. Some of them sat in the $20.00 outfield grandstand seats. The others sat in the $27.00 upper field box seats. How many of each type of ticket were purchased if the total price of the tickets was $275.00? 7 grandstand tickets and 5 upper field box tickets

88. *Consumerism* The admission charge for a family at a city zoo is $7.50 for the first person and $4.25 for each additional member of the family. How many people are in a family that is charged $28.75 for admission? 6 people

89. *Consumerism* A family of seven bought tickets for the afternoon show of a play. Not all the members of the family could sit together. Some were in the mezzanine ($50.00) and some were in the balcony ($35.00). The total cost of the tickets was $275.00. Find the number of mezzanine tickets that were purchased. 2 mezzanine tickets

APPLYING CONCEPTS 2.1

Solve.

90. $8 \div \frac{1}{x} = -3$ $-\frac{3}{8}$

91. $\frac{1}{\frac{1}{y}} = -9$ -9

92. $\frac{6}{\frac{7}{a}} = -18$ -21

93. $\frac{10}{\frac{3}{x}} - 5 = 4x$ $-\frac{15}{2}$

Solve. If the equation has no solution, write "no solution."

94. $3[4(y + 2) - (y + 5)] = 3(3y + 1)$ No solution

95. $2[3(x + 4) - 2(x + 1)] = 5x + 3(1 - x)$ No solution

96. $\frac{4(x - 5) - (x + 1)}{3} = x - 7$ All real numbers

97. $\frac{4[(x - 3) + 2(1 - x)]}{5} = x + 1$ -1

[C] 98. $2584 \div x = 54\frac{46}{x}$ 47

[C] 99. $3(2x + 2) - 4(x - 3) = 2(x + 9)$ All real numbers

100. *Consumerism* Consumers in the United States are using more mail-order catalogs for their shopping. The graph at the right shows actual and projected catalog sales. An equation that models this data is $C = 5.34Y + 52.9$, where C is the catalog sales in billions of dollars and Y is the year. 1992 corresponds to $Y = 0$.

a. What catalog sales does this model predict for 1997? $79.6 billion

b. What is the absolute value of the difference between the amount predicted for 1997 by the model and the amount shown in the graph? $1 billion

c. What is the percent error? Use the formula

$$\text{Percent error} = \left(\frac{|\text{predicted} - \text{actual}|}{\text{actual}}\right)100 \quad 1.3\%$$

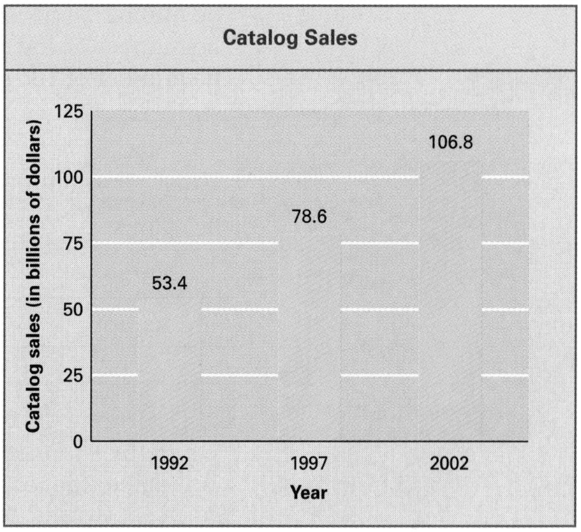

Source: Direct Marketing Association

101. Solve the equation $4c + 1 = 9$, showing every step. State the Property of Real Numbers or the Property of Equations that is used in each step.

SECTION 2.2 Coin, Stamp, and Integer Problems

1 Coin and stamp problems

POINT OF INTEREST

The Rhind papyrus, purchased by A. Henry Rhind in 1858, contains much of the historical evidence of ancient Egyptian mathematics. The papyrus was written in demotic script (a kind of written hieroglyphics). The key to deciphering this script is contained in the *Rosetta Stone*, which was discovered by an expedition of Napoleon's soldiers along the Rosetta branch of the Nile in 1799. This stone contains a passage written in hieroglyphics, in demotic script, and in Greek. By translating the Greek passage, archeologists were able to determine how to translate demotic script, which they applied to translating the Rhind papyrus.

In solving problems dealing with coins or stamps of different values, it is necessary to represent the value of the coins or stamps in the same unit of money. The unit of money is frequently cents. For example,

The value of five 8¢ stamps is $5 \cdot 8$, or 40, cents.
The value of four 20¢ stamps is $4 \cdot 20$, or 80, cents.
The value of n 10¢ stamps is $n \cdot 10$, or $10n$, cents.

Solve: A collection of stamps consists of 5¢, 13¢, and 18¢ stamps. The number of 13¢ stamps is two more than three times the number of 5¢ stamps. The number of 18¢ stamps is five less than the number of 13¢ stamps. The total value of all the stamps is $1.68. Find the number of 18¢ stamps.

STRATEGY for solving a stamp problem

- For each denomination of stamp, write a numerical or variable expression for the number of stamps, the value of the stamp, and the total value of the stamps in cents. The results can be recorded in a table.

The number of 5¢ stamps: x
The number of 13¢ stamps: $3x + 2$
The number of 18¢ stamps: $(3x + 2) - 5 = 3x - 3$

	Number of stamps	·	Value of stamp in cents	=	Total value in cents
5¢ stamp	x	·	5	=	$5x$
13¢ stamp	$3x + 2$	·	13	=	$13(3x + 2)$
18¢ stamp	$3x - 3$	·	18	=	$18(3x - 3)$

■ Determine the relationship among the total values of the stamps. Use the fact that the sum of the total values of each denomination of stamp is equal to the total value of all the stamps.

The sum of the total values of each denomination of stamp is equal to the total value of all the stamps (168 cents).

$$\begin{aligned} 5x + 13(3x + 2) + 18(3x - 3) &= 168 \\ 5x + 39x + 26 + 54x - 54 &= 168 \\ 98x - 28 &= 168 \\ 98x &= 196 \\ x &= 2 \end{aligned}$$

The number of 18¢ stamps is $3x - 3$. Replace x by 2 and evaluate.

$$3x - 3 = 3(2) - 3 = 3$$

There are three 18¢ stamps in the collection.

Example 1 A coin bank contains $1.80 in nickels and dimes. In all, there are twenty-two coins in the bank. Find the number of nickels and the number of dimes in the bank.

Strategy ■ Number of nickels: x
Number of dimes: $22 - x$

	Number	Value	Total value
Nickel	x	5	$5x$
Dime	$22 - x$	10	$10(22 - x)$

■ The sum of the total values of each denomination of coin equals the total value of all the coins (180 cents).

Solution

$$\begin{aligned} 5x + 10(22 - x) &= 180 \\ 5x + 220 - 10x &= 180 \\ -5x + 220 &= 180 \\ -5x &= -40 \\ x &= 8 \end{aligned}$$

▶ There are 8 nickels in the bank.

$22 - x = 22 - 8 = 14$ ▶ Substitute the value of x into the variable expression for the number of dimes.

The bank contains 8 nickels and 14 dimes.

Problem 1 A collection of stamps contains 3¢, 10¢, and 15¢ stamps. The number of 10¢ stamps is two more than twice the number of 3¢ stamps. There are three times as many 15¢ stamps as there are 3¢ stamps. The total value of the stamps is $1.56. Find the number of 15¢ stamps.

Solution See page S3. 6 stamps

2 Integer problems

Recall that an **even integer** is an integer that is divisible by 2. An **odd integer** is an integer that is not divisible by 2.

Consecutive integers are integers that follow one another in order. Examples of consecutive integers are shown at the right.

8, 9, 10
$-3, -2, -1$
$n, n + 1, n + 2$, where n is an integer

Examples of **consecutive even integers** are shown at the right.

16, 18, 20
$-6, -4, -2$
$n, n + 2, n + 4$, where n is an even integer

Examples of **consecutive odd integers** are shown at the right.

11, 13, 15
$-23, -21, -19$
$n, n + 2, n + 4$, where n is an odd integer

Solve: The sum of three consecutive even integers is seventy-eight. Find the integers.

STRATEGY for solving an integer problem

- Let a variable represent one of the integers. Express each of the other integers in terms of that variable. Remember that for consecutive integer problems, consecutive integers will differ by 1. Consecutive even or consecutive odd integers will differ by 2.

Represent three consecutive even integers.

First even integer: n
Second even integer: $n + 2$
Third even integer: $n + 4$

- Determine the relationship among the integers.

The sum of the three even integers is 78.

$$n + (n + 2) + (n + 4) = 78$$
$$3n + 6 = 78$$
$$3n = 72$$
$$n = 24$$

$$n + 2 = 24 + 2 = 26$$
$$n + 4 = 24 + 4 = 28$$

The three consecutive even integers are 24, 26, and 28.

Example 2 Five times the first of three consecutive even integers is five more than the product of four and the third integer. Find the integers.

Strategy ■ First even integer: n
Second even integer: $n + 2$
Third even integer: $n + 4$
■ Five times the first integer equals five more than the product of four and the third integer.

Solution $5n = 4(n + 4) + 5$
$5n = 4n + 16 + 5$
$5n = 4n + 21$
$n = 21$

Because 21 is not an even integer, there is no solution.

Problem 2 The sum of three numbers is eighty-one. The second number is twice the first number, and the third number is three less than four times the first number. Find the numbers.

Solution See page S3.   12, 24, 45

STUDY TIPS

ATTENDANCE

Attending class is vital if you are to succeed in your algebra course. Your instructor will provide not only information but also practice in the skills you are learning. Be sure to arrive on time. You are responsible for *everything* that happens in class, even if you are absent. If you must be absent from a class session:

1. Deliver due assignments to the instructor as soon as possible.
2. Contact a classmate to learn about assignments or tests announced in your absence.
3. Hand copy or photocopy notes taken by a classmate while you were absent.

CONCEPT REVIEW 2.2

Determine whether the following statements are always true, sometimes true, or never true.

1. In solving coin problems, we find the total value of all the coins of one type by multiplying the value of one coin by the number of coins. Always true

2. The expressions n and $n + 2$ represent two consecutive odd integers. Sometimes true

3. The expression $2n + 1$, where n is an integer, represents an odd integer. Always true

4. If the sum of two numbers is n and one of the numbers is x, the other number is $x - n$. Never true

5. Three consecutive odd integers are given by $n + 1$, $n + 3$, and $n + 5$. Sometimes true

EXERCISES 2.2

1 Coin and Stamp Problems

1. Explain how to represent the total value of x quarters using the equation

   Number of coins · Value of the coin in cents = Total value in cents.

2. Suppose a coin purse contains only nickels and dimes. In the context of this situation, explain the meaning of the statement "The sum of the total values of each denomination of coin is equal to the total value of all the coins."

3. A collection of 56 coins has a value of \$4.00. The collection contains only nickels and dimes. Find the number of dimes in the collection. 24 dimes

4. A collection of 22 coins has a value of \$4.45. The collection contains dimes and quarters. Find the number of quarters in the collection. 15 quarters

5. A cashier has \$730 in twenty-dollar bills and five-dollar bills. In all, the cashier has 68 bills. How many twenty-dollar bills does the cashier have? 26 twenty-dollar bills

6. In conducting its daily business, a department store uses twice as many five-dollar bills as ten-dollar bills. \$2500 was obtained in five- and ten-dollar bills for the day's business. How many five-dollar bills were obtained? 250 five-dollar bills

7. A stamp collector has some 15¢ stamps and some 20¢ stamps. The number of 15¢ stamps is eight less than three times the number of 20¢ stamps. The total value is \$4. Find the number of each type of stamp in the collection. 20¢ stamps: 8; 15¢ stamps: 16

8. An office has some 20¢ stamps and some 28¢ stamps. Altogether the office has 140 stamps for a total value of $31.20. How many of each type of stamp does the office have? 20¢ stamps: 100; 28¢ stamps: 40

9. A coin bank contains 25 coins in nickels, dimes, and quarters. There are four times as many dimes as quarters. The value of the coins is $2.05. How many dimes are in the bank? 8 dimes

10. A coin collection contains nickels, dimes, and quarters. There are twice as many dimes as quarters and seven more nickels than dimes. The total value of all the coins is $3.10. How many quarters are in the collection? 5 quarters

11. A stamp collection consists of 3¢, 8¢, and 13¢ stamps. The number of 8¢ stamps is three less than twice the number of 3¢ stamps. The number of 13¢ stamps is twice the number of 8¢ stamps. The total value of all the stamps is $2.53. Find the number of 3¢ stamps in the collection. 5 stamps

12. An account executive bought 300 stamps for $73.80. The purchase included 15¢ stamps, 20¢ stamps, and 40¢ stamps. The number of 20¢ stamps is four times the number of 15¢ stamps. How many 40¢ stamps were purchased? 80 stamps

13. A stamp collector has 8¢, 11¢, and 18¢ stamps. The collector has twice as many 8¢ stamps as 18¢ stamps. There are three more 11¢ than 18¢ stamps. The total value of the stamps in the collection is $3.48. Find the number of 18¢ stamps in the collection. 7 stamps

14. A stamp collection consists of 3¢, 12¢, and 15¢ stamps. The number of 3¢ stamps is five times the number of 12¢ stamps. The number of 15¢ stamps is four less than the number of 12¢ stamps. The total value of the stamps in the collection is $3.18. Find the number of 15¢ stamps in the collection. 5 stamps

2 Integer Problems

15. Explain how to represent three consecutive integers using only one variable.

16. Explain why both consecutive even integers and consecutive odd integers can be represented algebraically as $n, n + 2, n + 4, \ldots$.

17. The sum of two integers is 10. Three times the larger integer is three less than eight times the smaller integer. Find the integers. 3 and 7

18. The sum of two integers is thirty. Eight times the smaller integer is six more than five times the larger integer. Find the integers. 12 and 18

19. One integer is eight less than another integer. The sum of the two integers is fifty. Find the integers. 21 and 29

20. One integer is four more than another integer. The sum of the integers is twenty-six. Find the integers. 11 and 15

21. The sum of three numbers is one hundred twenty-three. The second number is two more than twice the first number. The third number is five less than the product of three and the first number. Find the three numbers.
21, 44, and 58

22. The sum of three numbers is forty-two. The second number is twice the first number, and the third number is three less than the second number. Find the three numbers. 9, 18, and 15

23. The sum of three consecutive integers is negative fifty-seven. Find the integers. -20, -19, and -18

24. The sum of three consecutive integers is one hundred twenty-nine. Find the integers. 42, 43, and 44

25. Five times the smallest of three consecutive odd integers is ten more than twice the largest. Find the integers. No solution

26. Find three consecutive even integers such that twice the sum of the first and third integers is twenty-one more than the second integer. No solution

27. Find three consecutive odd integers such that three times the middle integer is seven more than the sum of the first and third integers. 5, 7, and 9

28. Find three consecutive even integers such that four times the sum of the first and third integers is twenty less than six times the middle integer.
-12, -10, and -8

APPLYING CONCEPTS 2.2

[C] 29. *Coin Problem* A coin bank contains only nickels and dimes. The number of dimes in the bank is two less than twice the number of nickels. There are 52 coins in the bank. How much money is in the bank? $4.30

[C] 30. *Integer Problem* Four times the first of three consecutive odd integers is five less than the product of three and the third integer. Find the integers.
7, 9, and 11

[C] 31. *Stamp Problem* A collection of stamps consists of 3¢ stamps, 5¢ stamps, and 7¢ stamps. There are six more 3¢ stamps than 5¢ stamps and two more 7¢ stamps than 3¢ stamps. The total value of the stamps is $1.94. How many 3¢ stamps are in the collection? 14 stamps

[C] 32. *Integer Problem* The sum of four consecutive even integers is 100. What is the sum of the smallest and the largest of the four integers? 50

[C] 33. *Integer Problem* The sum of four consecutive odd integers is -64. What is the sum of the smallest and the largest of the four integers? -32

34. *Integer Problem* Find three consecutive odd integers such that the product of the second and third minus the product of the first and second is 42.
No solution

[C] **35.** *Integer Problem* The sum of the digits of a three-digit number is 6. The tens digit is one less than the units digit, and the number is 12 more than 100 times the hundreds digit. Find the number. 312

SECTION 2.3 Value Mixture and Motion Problems

1 Value mixture problems

A **value mixture problem** involves combining two ingredients that have different prices into a single blend. For example, a coffee manufacturer may blend two types of coffee into a single blend.

The solution of a value mixture problem is based on the equation $AC = V$, where A is the amount of the ingredient, C is the cost per unit of the ingredient, and V is the value of the ingredient.

The value of 12 lb of coffee costing $5.25 per pound is

$$V = AC$$
$$V = 12(\$5.25)$$
$$V = \$63$$

Solve: How many pounds of peanuts that cost $2.25 per pound must be mixed with 40 lb of cashews that cost $6.00 per pound to make a mixture that costs $3.50 per pound?

STRATEGY for solving a value mixture problem

- For each ingredient in the mixture, write a numerical or variable expression for the amount of the ingredient used, the unit cost of the ingredient, and the value of the amount used. For the mixture, write a numerical or variable expression for the amount, the unit cost of the mixture, and the value of the amount. The results can be recorded in a table.

INSTRUCTOR NOTE

For value mixture problems, the sum of the values of the ingredients equals the value of the mixture. If students organize these problems in a table such as the one at the right, the last column can be used to write the equation. The sum of the values of the ingredients, $2.25x + 6.00(40)$, equals the value of the mixture, $3.50(40 + x)$.

Amount of peanuts: x

	Amount, A	·	Unit cost, C	=	Value, V
Peanuts	x	·	2.25	=	$2.25x$
Cashews	40	·	6.00	=	$6.00(40)$
Mixture	$x + 40$	·	3.50	=	$3.50(x + 40)$

■ Determine how the values of the individual ingredients are related. Use the fact that the sum of the values of these ingredients is equal to the value of the mixture.

The sum of the values of the peanuts and the cashews is equal to the value of the mixture.

$$2.25x + 6.00(40) = 3.50(x + 40)$$
$$2.25x + 240 = 3.50x + 140$$
$$-1.25x + 240 = 140$$
$$-1.25x = -100$$
$$x = 80$$

The mixture must contain 80 lb of peanuts.

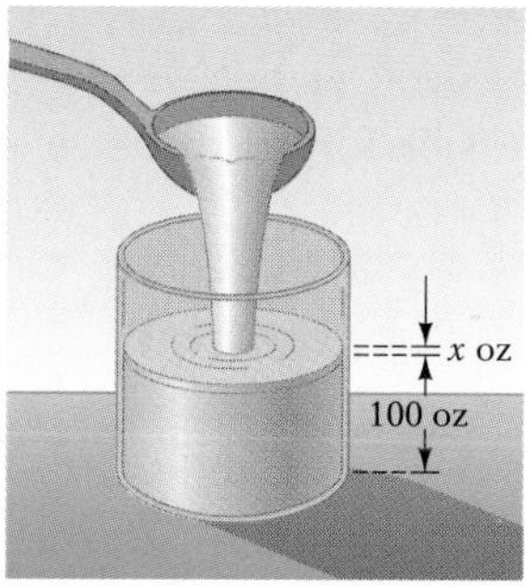

Example 1   How many ounces of a gold alloy that costs $320 an ounce must be mixed with 100 oz of an alloy that costs $100 an ounce to make a mixture that costs $160 an ounce?

Strategy   ■ Ounces of the $320 gold alloy: x

	Amount	Cost	Value
$320 alloy	x	320	$320x$
$100 alloy	100	100	100(100)
Mixture	$x + 100$	160	$160(x + 100)$

■ The sum of the values before mixing equals the value after mixing.

Solution $320x + 100(100) = 160(x + 100)$
$320x + 10,000 = 160x + 16,000$
$160x + 10,000 = 16,000$
$160x = 6000$
$x = 37.5$

The mixture must contain 37.5 oz of the $320 gold alloy.

Problem 1   A butcher combined hamburger that costs $3.00 per pound with hamburger that costs $1.80 per pound. How many pounds of each were used to make a 75-pound mixture that costs $2.20 per pound?

Solution   See page S3. $3/lb hamburger: 25 lb; $1.80/lb hamburger: 50 lb

2   Uniform motion problems

A car that travels constantly in a straight line at 55 mph is in uniform motion. **Uniform motion** means that the speed of an object does not change.

The solution of a uniform motion problem is based on the equation $rt = d$, where r is the rate of travel, t is the time spent traveling, and d is the distance traveled.

A car travels 55 mph for 3 h.

$$d = rt$$
$$d = 55(3)$$
$$d = 165$$

The car travels a distance of 165 mi.

Solve: An executive has an appointment 785 mi from the office. The executive takes a helicopter from the office to the airport and a plane from the airport to the business appointment. The helicopter averages 70 mph, and the plane averages 500 mph. The total time spent traveling is 2 h. Find the distance from the executive's office to the airport.

STRATEGY for solving a uniform motion problem

- For each object, write a numerical or variable expression for the distance, rate, and time. The results can be recorded in a table.

INSTRUCTOR NOTE

One of the complications of distance-rate problems is that the variable may not directly represent the unknown. The example at the right illustrates this. The unknown is distance, but the variable is time.

The total time of travel is 2 h.

Unknown time in the helicopter: t
Time in the plane: $2 - t$

	Rate, r	·	Time, t	=	Distance, d
Helicopter	70	·	t	=	$70t$
Plane	500	·	$2 - t$	=	$500(2 - t)$

- Determine how the distances traveled by the individual objects are related. For example, the total distance traveled by both objects may be known, or it may be known that the two objects traveled the same distance.

The total distance traveled is 785 mi.

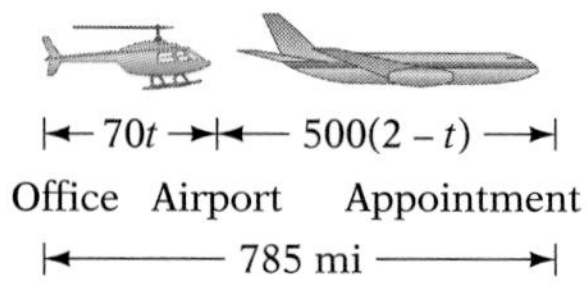

$$70t + 500(2 - t) = 785$$
$$70t + 1000 - 500t = 785$$
$$-430t + 1000 = 785$$
$$-430t = -215$$
$$t = 0.5$$

The time spent traveling from the office to the airport in the helicopter is 0.5 h. To find the distance between these two points, substitute the values of r and t into the equation $rt = d$.

$$rt = d$$
$$70(0.5) = d$$
$$35 = d$$

The distance from the office to the airport is 35 mi.

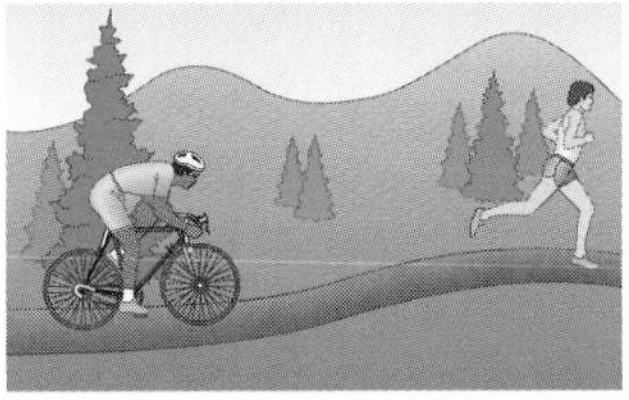

Example 2   A long-distance runner started a course running at an average speed of 6 mph. One and one-half hours later, a cyclist traveled the same course at an average speed of 12 mph. How long after the runner started did the cyclist overtake the runner?

Strategy   ■ Unknown time for the cyclist: t
Time for the runner: $t + 1.5$

	Rate	Time	Distance
Runner	6	$t + 1.5$	$6(t + 1.5)$
Cyclist	12	t	$12t$

■ The runner and the cyclist traveled the same distance.

Solution $6(t + 1.5) = 12t$
$6t + 9 = 12t$
$9 = 6t$
$\dfrac{3}{2} = t$ ▶ The cyclist traveled for 1.5 h.

$t + 1.5 = 1.5 + 1.5$ ▶ Substitute the value of t into the variable
$= 3$ expression for the runner's time.

The cyclist overtook the runner 3 h after the runner started.

Problem 2   Two small planes start from the same point and fly in opposite directions. The first plane is flying 30 mph faster than the second plane. In 4 h the planes are 1160 mi apart. Find the rate of each plane.

Solution   See page S4.   First plane: 160 mph; second plane: 130 mph

---

STUDY TIPS

DETERMINING WHEN TO STUDY

Spaced practice is generally superior to massed practice. For example, four half-hour study periods will produce more learning than one two-hour study session. The following suggestions may help you decide when you will study.

1. A free period immediately before class is the best time to study about the lecture topic for the class.

2. A free period immediately after class is the best time to review notes taken during the class.

3. A brief period of time is good for reciting or reviewing information.

4. A long period of an hour or more is good for doing challenging activities such as learning to solve a new type of problem.

5. Free periods just before you go to sleep are good times for learning information. (There is evidence that information learned just before sleep is remembered longer than information learned at other times.)

CONCEPT REVIEW 2.3

Determine whether the following statements are always true, sometimes true, or never true.

1. A gold alloy costing \$300 per ounce was mixed with a gold alloy costing \$200 per ounce. The cost of the mixture was \$250 per ounce. Sometimes true

2. Two planes are flying in opposite directions. One plane is flying at a rate of x mph, and the other plane is flying at a rate of y mph. The distance between the planes is increasing at a rate of $(x + y)$ mph. Always true

3. A runner starts a course running at a rate of x mph. One-half hour later, a cyclist travels the same course at a rate of y mph and overtakes the runner. If t is the total time for the runner, $t + \frac{1}{2}$ is the total time for the cyclist. Never true

4. A car travels 2 h at a rate of 40 mph and another 2 h at a rate of 60 mph. The average speed is 50 mph. Always true

5. A car travels 120 mi at a rate of 40 mph. On the return trip, the car travels at a rate of 60 mph. The average speed is 50 mph. Never true

EXERCISES 2.3

1 Value Mixture Problems

1. Use the equation Amount · Unit cost = Value to explain how to represent the value of x quarts of juice that cost \$.85 per quart.

2. Suppose you are mixing peanuts that cost \$4 per pound with raisins that cost \$2 per pound. In the context of this situation, explain the meaning of the statement "The sum of the values of the individual ingredients is equal to the value of the mixture."

3. A restaurant chef mixes 20 lb of snow peas costing \$1.99 a pound with 14 lb of petite onions costing \$1.19 a pound to make a vegetable medley for the evening meal. Find the cost per pound of the mixture. \$1.66

4. A coffee merchant combines coffee costing \$5.50 per pound with coffee costing \$3.00 per pound. How many pounds of each should be used to make 40 lb of a blend costing \$4.00 per pound? 16 lb of \$5.50 coffee, 24 lb of \$3.00 coffee

5. Adult tickets for a play cost \$5.00 and children's tickets cost \$2.00. For one performance, 460 tickets were sold. Receipts for the performance were \$1880. Find the number of adult tickets sold. 320 adult tickets

6. The tickets for a local production of Gilbert and Sullivan's *H.M.S. Pinafore* cost \$6.00 for adults and \$3.00 for children and seniors. The total receipts for 505 tickets were \$1977. Find the number of adult tickets sold. 154 adult tickets

7. Fifty liters of pure maple syrup that costs \$9.50 per liter are mixed with imitation maple syrup that costs \$4.00 per liter. How much imitation maple syrup is needed to make a mixture that costs \$5.00 per liter? 225 L

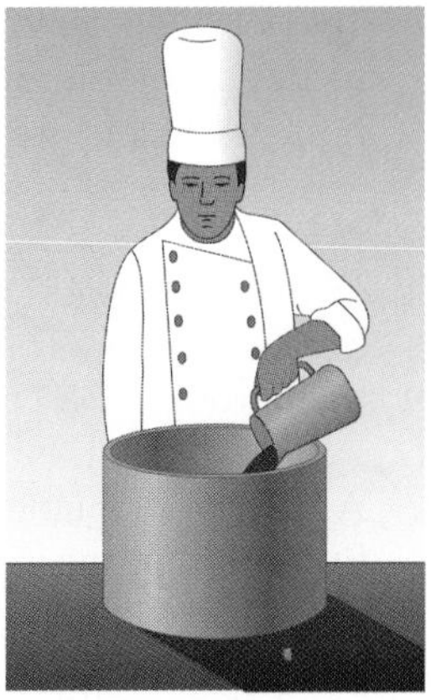

8. To make a flour mix, a miller combined soybeans that cost \$8.50 per bushel with wheat that costs \$4.50 per bushel. How many bushels of each were used to make a mixture of 800 bushels that costs \$5.50 per bushel?
200 bushels of soybeans, 600 bushels of wheat

9. A hiking instructor is preparing a trail mix for a group of young hikers. She mixes nuts that cost \$3.99 a pound with pretzels that cost \$1.29 a pound to make a 20-pound mixture that costs \$2.37 a pound. Find the number of pounds of nuts used. 8 lb

10. A silversmith combined pure silver that costs \$5.20 an ounce with 50 oz of a silver alloy that costs \$2.80 an ounce. How many ounces of the pure silver were used to make an alloy of silver that costs \$4.40 an ounce? 100 oz

11. A tea mixture was made from 30 lb of tea that costs \$6.00 per pound and 70 lb of tea that costs \$3.20 per pound. Find the cost per pound of the tea mixture. \$4.04/lb

12. Find the cost per ounce of a salad dressing mixture made from 64 oz of olive oil that costs \$8.29 and 20 oz of vinegar that costs \$1.99. Round to the nearest cent. \$.12

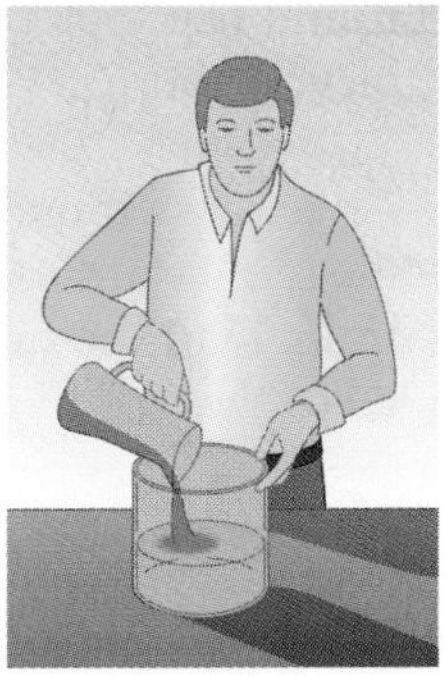

13. A fruitstand owner combined cranberry juice that costs \$4.20 per gallon with 50 gal of apple juice that costs \$2.10 per gallon. How much cranberry juice was used to make cranapple juice that costs \$3.00 per gallon? 37.5 gal

14. Walnuts that cost \$4.05 per kilogram were mixed with cashews that cost \$7.25 per kilogram. How many kilograms of each were used to make a 50-kilogram mixture that costs \$6.25 per kilogram? Round to the nearest tenth.
15.6 kg of walnuts, 34.4 kg of cashews

2 Uniform Motion Problems

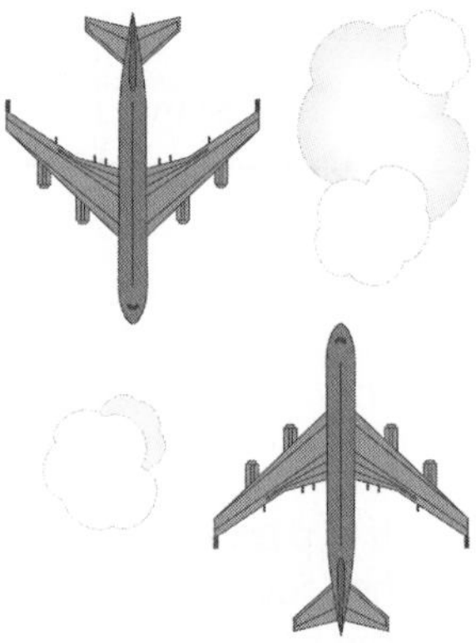

15. A bicyclist traveling at 18 mph overtakes an in-line skater who is traveling at 10 mph and had a 0.5-hour head start. How far from the starting point does the bicyclist overtake the in-line skater? 11.25 mi

16. A helicopter traveling 120 mph overtakes a speeding car traveling at 90 mph. The car had a 0.5-hour head start. How far from the starting point did the helicopter overtake the car? 180 mi

17. Two planes are 1380 mi apart and traveling toward each other. One plane is traveling 80 mph faster than the other plane. The planes meet in 1.5 h. Find the speed of each plane. 1st plane: 420 mph; 2nd plane: 500 mph

18. Two jet skiers leave the same dock at the same time and travel in opposite directions. One skier is traveling 14 mph slower than the other skier. In half an hour the skiers are 48 mi apart. Find the rate of the slower skier. 41 mph

19. A ferry leaves a harbor and travels to a resort island at an average speed of 18 mph. On the return trip, the ferry travels at an average speed of 12 mph because of fog. The total time for the trip is 6 h. How far is the island from the harbor? 43.2 mi

20. A commuter plane provides transportation from an international airport to the surrounding cities. One commuter plane averaged 210 mph flying to a city and 140 mph returning to the international airport. The total flying time was 4 h. Find the distance between the two airports. 336 mi

21. Two planes start from the same point and fly in opposite directions. The first plane is flying 50 mph slower than the second plane. In 2.5 h, the planes are 1400 mi apart. Find the rate of each plane. 1st plane: 255 mph; 2nd plane: 305 mph

22. Two hikers start from the same point and hike in opposite directions around a lake whose shoreline is 13 mi long. One hiker walks 0.5 mph faster than the other hiker. How fast did each hiker walk if they meet in 2 h? 3 mph, 3.5 mph

23. A student rode a bicycle to the repair shop and then walked home. The student averaged 14 mph riding to the shop and 3.5 mph walking home. The round trip took one hour. How far is it between the student's home and the bicycle shop? 2.8 mi

24. An express train leaves Grand Central Station 1 h after a freight train leaves the same station. The express train is traveling 15 mph faster than the freight train. Find the rate at which each train is traveling if the express train overtakes the freight train in 3 h. freight train: 45 mph; express train: 60 mph

25. At noon a train leaves Washington, D.C., headed for Pittsburgh, Pennsylvania, a distance of 260 mi. The train travels at a speed of 60 mph. At 1 P.M. a second train leaves Pittsburgh headed for Washington, D.C., traveling at 40 mph. How long after the train leaves Pittsburgh will the two trains pass each other? 2 h

26. A plane leaves an airport at 3 P.M. At 4 P.M. another plane leaves the same airport traveling in the same direction at a speed 150 mph faster than that of the first plane. Four hours after the first plane takes off, the second plane is 250 mi ahead of the first plane. How far did the second plane travel? 1050 mi

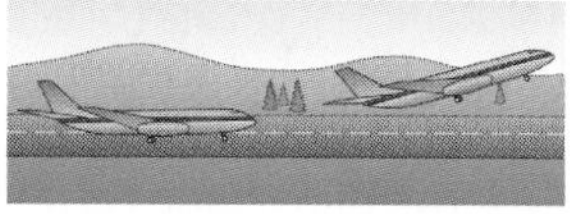

APPLYING CONCEPTS 2.3

Uniform Motion Problems

[C] 27. If a parade 2 mi long is proceeding at 3 mph, how long will it take a runner, jogging at 6 mph, to travel from the front of the parade to the end of the parade? $\frac{2}{9}$ h

[C] **28.** If a parade 2 mi long is proceeding at 3 mph, how long will it take a runner, jogging at 6 mph, to travel from the end of the parade to the start of the parade? $\frac{2}{3}$ h

29. Two cars are headed directly toward each other at rates of 40 mph and 60 mph. How many miles apart are they 2 min before impact? $3\frac{1}{3}$ mi

[C] **30.** The following problem appears in a math text written around A.D. 1200. Two birds start flying from the tops of two towers 50 ft apart at the same time and at the same rate. One tower is 30 ft high, and the other tower is 40 ft high. The birds reach a grass seed on the ground at exactly the same time. How far is the grass seed from the 40-foot tower? 18 ft

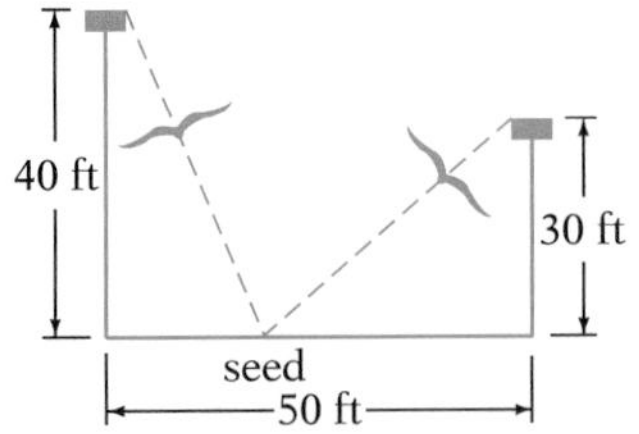

[C] **31.** A car travels at an average speed of 30 mph for 1 mi. Is it possible for it to increase its speed during the next mile so that its average speed for the 2 mi is 60 mph? No

32. In December of 1986, pilots Dick Rutan and Jeana Yeager flew the *Voyager* in the first nonstop, nonrefueled flight around the world. They flew east from Edwards Air Force Base in California on December 14, traveled 24,986.727 mi around the world, and returned to Edwards 216 h 3 min 44 s after their departure.

a. On what date did Rutan and Yeager land at Edwards Air Force Base after their flight? December 23

b. What was their average speed in miles per hour? Round to the nearest whole number. 116 mph

c. Find the circumference of Earth in a reference almanac, and then calculate the approximate distance above Earth that the flight was flown. Round to the nearest whole number. 14 mi

33. A student jogs 100 yd at a rate of 8 mph and jogs back at a rate of 6 mph. Does it seem reasonable that the average rate is 7 mph? Why or why not? Support your answer.

SECTION 2.4 Applications: Problems Involving Percent

1 Investment problems

The annual simple interest that an investment earns is given by the equation $Pr = I$, where P is the principal, or the amount invested, r is the simple interest rate, and I is the simple interest. The solution of an investment problem is based on this equation.

The annual interest rate on a $3000 investment is 9%. The annual simple interest earned on the investment is

$$I = Pr$$
$$I = \$3000(0.09)$$
$$I = \$270$$

Solve: You have a total of $8000 invested in two simple interest accounts. On one account, a money market fund, the annual simple interest rate is 11.5%. On the second amount, a bond fund, the annual simple interest rate is 9.75%. The total annual interest earned by the two accounts is $823.75. How much do you have invested in each account?

STRATEGY for solving a problem involving money deposited in two simple interest accounts

- For each amount invested, use the equation $Pr = I$. Write a numerical or variable expression for the principal, the interest rate, and the interest earned. The results can be recorded in a table.

The total amount invested is $8000.

Amount invested at 11.5%: x
Amount invested at 9.75%: $8000 - x$

INSTRUCTOR NOTE

These problems are similar to value mixture problems. The last column of the table can be used to write the equation.

	Principal, P	·	Interest rate, r	=	Interest earned, I
Amount at 11.5%	x	·	0.115	=	$0.115x$
Amount at 9.75%	$8000 - x$	·	0.0975	=	$0.0975(8000 - x)$

- Determine how the amounts of interest earned on the individual investments are related. For example, the total interest earned by both accounts may be known, or it may be known that the interest earned on one account is equal to the interest earned on the other account.

The total annual interest earned is $823.75.

$$0.115x + 0.0975(8000 - x) = 823.75$$
$$0.115x + 780 - 0.0975x = 823.75$$
$$0.0175x + 780 = 823.75$$
$$0.0175x = 43.75$$
$$x = 2500$$

The amount invested at 9.75% is $8000 - x$.
Replace x by 2500 and evaluate.

$$8000 - x = 8000 - 2500 = 5500$$

The amount invested at 11.5% is $2500.
The amount invested at 9.75% is $5500.

Example 1 An investment of $4000 is made at an annual simple interest rate of 4.9%. How much additional money must be invested at an annual simple interest rate of 7.4% so that the total interest earned is 6.4% of the total investment?

Strategy ■ Additional amount to be invested at 7.4%: x

	Principal	Rate	Interest
Amount at 4.9%	4000	0.049	0.049(4000)
Amount at 7.4%	x	0.074	0.074x
Amount at 6.4%	4000 + x	0.064	0.064(4000 + x)

■ The sum of the interest earned by the two investments equals the interest earned by the total investment.

Solution $0.049(4000) + 0.074x = 0.064(4000 + x)$
$$196 + 0.074x = 256 + 0.064x$$
$$196 + 0.01x = 256$$
$$0.01x = 60$$
$$x = 6000$$

$6000 must be invested at an annual simple interest rate of 7.4%.

Problem 1 An investment of $3500 is made at an annual simple interest rate of 5.2%. How much additional money must be invested at an annual simple interest rate of 7.5% so that the total interest earned is $575?

Solution See page S4. $5240

---

2 Percent mixture problems

The amount of a substance in a solution or alloy can be given as a percent of the total solution or alloy. For example, in a 10% hydrogen peroxide solution, 10% of the total solution is hydrogen peroxide. The remaining 90% is water.

The solution of a percent mixture problem is based on the equation $Ar = Q$, where A is the amount of solution or alloy, r is the percent of concentration, and Q is the quantity of a substance in the solution or alloy.

The number of grams of silver in 50 g of a 40% silver alloy is

$$Q = Ar$$
$$Q = 50(0.40)$$
$$Q = 20$$

Solve: A chemist mixes an 11% acid solution with a 4% acid solution. How many milliliters of each solution should the chemist use to make a 700-milliliter solution that is 6% acid?

STRATEGY *for solving a percent mixture problem*

■ For each solution, use the equation $Ar = Q$. Write a numerical or variable expression for the amount of solution, the percent of concentration, and the quantity of the substance in the solution. The results can be recorded in a table.

The total amount of solution is 700 ml.

Amount of 11% solution: x
Amount of 4% solution: $700 - x$

	Amount of solution, A	⋅	Percent of concentration, r	=	Quantity of substance, Q
11% solution	x	⋅	0.11	=	$0.11x$
4% solution	$700 - x$	⋅	0.04	=	$0.04(700 - x)$
6% solution	700	⋅	0.06	=	$0.06(700)$

■ Determine how the quantities of the substance in the individual solutions are related. Use the fact that the sum of the quantities of the substances being mixed is equal to the quantity of the substance after mixing.

The sum of the amounts of acid in the 11% solution and the 4% solution is equal to the amount of acid in the 6% solution.

$$0.11x + 0.04(700 - x) = 0.06(700)$$
$$0.11x + 28 - 0.04x = 42$$
$$0.07x + 28 = 42$$
$$0.07x = 14$$
$$x = 200$$

The amount of 4% solution is $700 - x$. Replace x by 200 and evaluate.

$$700 - x = 700 - 200 = 500$$

The chemist should use 200 ml of the 11% solution and 500 ml of the 4% solution.

Example 2   How many grams of pure acid must be added to 60 g of an 8% acid solution to make a 20% acid solution?

Strategy   ■ Grams of pure acid: x

	Amount	Percent	Quantity
Pure acid (100%)	x	1.00	x
8%	60	0.08	0.08(60)
20%	$x + 60$	0.20	$0.20(x + 60)$

- The sum of the quantities before mixing equals the quantity after mixing.

Solution

$$x + 0.08(60) = 0.20(x + 60)$$
$$x + 4.8 = 0.20x + 12$$
$$0.8x + 4.8 = 12$$
$$0.8x = 7.2$$
$$x = 9$$

To make the 20% acid solution, 9 g of pure acid must be used.

Problem 2 A butcher has some hamburger that is 22% fat and some that is 12% fat. How many pounds of each should be mixed to make 80 lb of hamburger that is 18% fat?

Solution See page S4. 48 lb of the hamburger that is 22% fat; 32 lb of the hamburger that is 12% fat

STUDY TIPS

DETERMINING HOW MUCH TO STUDY

Instructors often advise students to spend twice as much time studying outside of class as they spend in the classroom. For example, if a course meets for three hours each week, instructors customarily advise students to study for six hours each week outside of class.

It is often necessary to practice a skill more than a teacher requires. For example, this textbook may provide 50 practice problems on a specific objective and the instructor may assign only 25 of them. However, some students may need to do 30, 40, or all 50 problems.

If you are an accomplished athlete, musician, or dancer, you know that long hours of practice are necessary to acquire a skill. Do not cheat yourself of the practice you need to develop the abilities taught in this course.

Study followed by reward is usually productive. Schedule something enjoyable to do following study sessions. If you know that you have only two hours to study because you have scheduled a pleasant activity for yourself, you may be inspired to make the best use of the two hours that you have set aside for studying.

CONCEPT REVIEW 2.4

Determine whether the following statements are always true, sometimes true, or never true.

1. $10,000 is invested in two accounts. If x is invested in one account, then $x - 10{,}000$ is invested in the other account. Never true

2. The equation $Q = Ar$, where A is the amount of solution or alloy, r is the percent concentration, and Q is the quantity of a substance in the solution or alloy, is used in solving percent mixture problems. Always true

3. If one solution has a 5% concentration and another solution has a 20% concentration, the mixture will have a 25% concentration. Never true

4. An investment of \$8000 is divided equally between two accounts with one account paying 6% and the other account paying 8%. The interest received is equivalent to 7% of the total investment. Always true

EXERCISES 2.4

1 Investment Problems

1. Explain the meaning of each variable in the equation $I = Pr$. Give an example of how this equation is used.

2. Suppose \$5000 is invested in two simple interest accounts. On one account the annual simple interest rate is 6%, and on the other the annual simple interest rate is 7%. The total annual interest earned on the two accounts is \$330. In the context of this situation, explain each term of the equation

$$0.06x + 0.07(5000 - x) = 330$$

3. Two investments earn a total annual income of \$2825. One investment is in a 6.75% annual simple interest certificate of deposit. The other is in a 7.25% tax-free annual simple interest account. The total amount invested is \$40,000. Find the amount invested in the certificate of deposit. \$15,000

4. Two investments earn an annual income of \$765. One investment earns an annual simple interest rate of 8.5%, and the other investment earns an annual simple interest rate of 10.2%. The total amount invested is \$8000. How much is invested in each account? \$3000 at 8.5%, \$5000 at 10.2%

5. An investment club invested \$5000 at an annual simple interest rate of 8.4%. How much additional money must be invested at an annual simple interest rate of 10.5% so that the total interest earned will be 9% of the total investment? \$2000

6. Two investments earn an annual income of \$465. One investment is a 5.5% tax-free annual simple interest account, and the other is a 4.5% annual simple interest certificate of deposit. The total amount invested is \$9600. How much is invested in each account? \$3300 at 5.5%, \$6300 at 4.5%

7. Two investments earn an annual income of \$575. One investment earns an annual simple interest rate of 8.5%, and the other investment earns an annual simple interest rate of 6.4%. The total amount invested is \$8000. How much is invested in each account? \$3000 at 8.5%, \$5000 at 6.4%

8. An investment club invested \$6000 at an annual simple interest rate of 4.0%. How much additional money must be invested at an annual simple interest rate of 6.5% so that the total annual interest earned will be 5% of the total investment? \$4000

9. Dee Pinckney made an investment of \$6000 at an annual simple interest rate of 5.5%. How much additional money must she invest at an annual simple interest rate of 10% in order for the total annual interest earned to be 7% of the investment? \$3000

10. An account executive deposited \$42,000 into two simple interest accounts. On the tax-free account the annual simple interest rate is 3.5%, and on the money market fund the annual simple interest rate is 4.5%. How much should be invested in each account so that both accounts earn the same annual interest? \$18,375 at 4.5%, \$23,625 at 3.5%

11. An investment club invested \$13,600 into two simple interest accounts. On one account, the annual simple interest rate is 4.2%. On the other, the annual simple interest rate is 6%. How much should be invested in each account so that both accounts earn the same annual interest? \$8000 at 4.2%, \$5600 at 6%

12. Orlando Salavarrio, a financial planner, recommended that 25% of a client's investment be placed in a 4% annual simple interest tax-free account, that 40% be placed in 6% certificates of deposit, and that the remainder be placed in a 9% high-risk investment. The total interest earned from the investments would be \$6550. Find the total amount to be invested. \$100,000

2 Percent Mixture Problems

13. Suppose you are mixing a 6% acid solution with a 12% acid solution. In the context of this situation, explain the meaning of the statement "The sum of the quantities of the substances being mixed is equal to the quantity of the substance after mixing."

14. Suppose a goldsmith mixes 25 g of a 75% gold alloy with 40 g of a 50% gold alloy. In the context of this situation, explain each term of the equation

$$25(0.75) + 40(0.50) = 65x$$

15. A jeweler mixed 15 g of a 60% silver alloy with 45 g of a 20% silver alloy. What is the percent concentration of silver in the resulting alloy? 30%

16. A goldsmith mixed 10 g of a 50% gold alloy with 40 g of a 15% gold alloy. What is the percent concentration of the resulting alloy? 22%

17. A silversmith mixed 25 g of a 70% silver alloy with 50 g of a 15% silver alloy. What is the percent concentration of the resulting alloy? $33\frac{1}{3}\%$

18. A chemist mixed 100 ml of an 8% saline solution with 60 ml of a 5% saline solution. Find the percent concentration of the resulting mixture. 6.875%

19. How many pounds of a 12% aluminum alloy must be mixed with 400 lb of a 30% aluminum alloy to make a 20% aluminum alloy? 500 lb

20. How many pounds of a 20% copper alloy must be mixed with 600 lb of a 30% copper alloy to make a 27.5% copper alloy? 200 lb

21. A hospital staff member mixed a 65% disinfectant solution with a 15% disinfectant solution. How many liters of each were used to make 50 L of a 40% disinfectant solution? 25 L of 65% solution, 25 L of 15% solution

22. A butcher has some hamburger that is 20% fat and some hamburger that is 12% fat. How many pounds of each should be mixed to make 80 lb of hamburger that is 17% fat? 50 lb of 20% fat, 30 lb of 12% fat

23. How many quarts of water must be added to 5 qt of an 80% antifreeze solution to make a 50% antifreeze solution? 3 qt

24. Rubbing alcohol is typically diluted with water to 70% strength. If you need 3.5 oz of 45% rubbing alcohol, how many ounces of 70% rubbing alcohol and how much water should you combine?
2.25 oz of 70% rubbing alcohol, 1.25 oz of water

25. How many ounces of pure water must be added to 60 oz of a 7.5% salt solution to make a 5% salt solution? 30 oz

26. How much water must be evaporated from 10 gal of a 12% sugar solution in order to obtain a 15% sugar solution? 2 gal

27. Many fruit drinks are actually only 5% real fruit juice. If you let 2 oz of water evaporate from 12 oz of a drink that is 5% fruit juice, what is the percent concentration of the result? 6%

28. A student mixed 50 ml of a 3% hydrogen peroxide solution with 30 ml of a 12% hydrogen peroxide solution. Find the percent concentration of the resulting mixture. Round to the nearest tenth of a percent. 6.4%

29. Eighty pounds of a 54% copper alloy is mixed with 200 lb of a 22% copper alloy. Find the percent concentration of the resulting mixture. Round to the nearest tenth of a percent. 31.1%

30. A druggist mixed 100 cc (cubic centimeters) of a 15% alcohol solution with 50 cc of pure alcohol. Find the percent concentration of the resulting mixture. Round to the nearest tenth of a percent. 43.3%

APPLYING CONCEPTS 2.4

31. *Investment Problem* A financial manager invested 25% of a client's money in bonds paying 9% annual simple interest, 30% in an 8% annual simple

interest account, and the remainder in 9.5% corporate bonds. Find the amount invested in each if the total annual interest earned is \$1785.
\$5000 at 9%, \$6000 at 8%, \$9000 at 9.5%

32. *Percent Mixture Problem* A silversmith mixed 90 g of a 40% silver alloy with 120 g of a 60% silver alloy. Find the percent concentration of the resulting alloy. Round to the nearest tenth of a percent. 51.4%

33. *Value Mixture Problem* Find the cost per pound of a tea mixture made from 50 lb of tea that costs \$5.50 per pound and 75 lb of tea that costs \$4.40 per pound. \$4.84/lb

[C] **34.** *Percent Mixture Problem* How many kilograms of water must be evaporated from 75 kg of a 15% salt solution to produce a 20% salt solution?
18.75 kg

[C] **35.** *Percent Mixture Problem* A chemist added 20 g of pure acid to a beaker containing an unknown number of grams of pure water. The resulting solution was 25% acid. How many grams of water were in the beaker before the acid was added? 60 g

[C] **36.** *Percent Mixture Problem* A radiator contains 6 L of a 25% antifreeze solution. How much should be drained and replaced with pure antifreeze to produce a 50% antifreeze solution? 2 L

37. *Consumer Price Index* The Consumer Price Index (CPI), or the "cost-of-living index," measures the change in the price of consumer goods and services. The graph at the right shows the percent increase in the CPI for the years 1991 through 1997.

a. During which year shown was the percent increase in the CPI the greatest? 1996

b. During which year was the cost of consumer goods and services highest? 1997

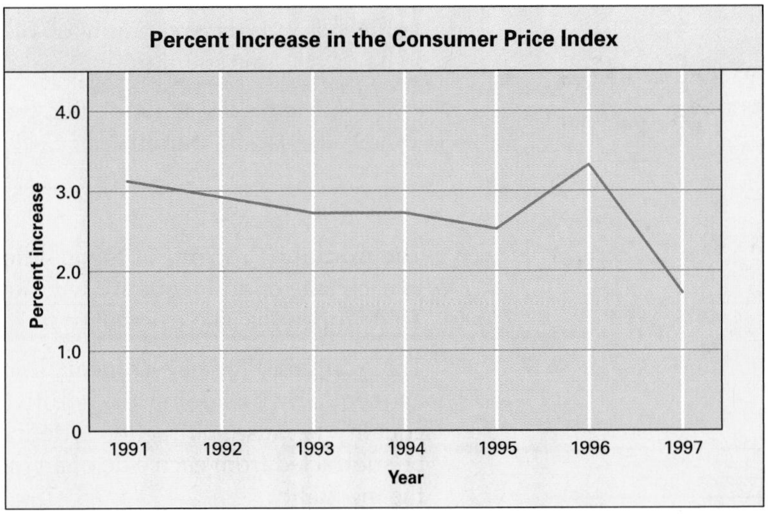

Source: Bureau of Labor Statistics

38. Write an essay on the topic of annual percentage rates.

39. Write a report on series trade discounts. Explain how to convert a series discount to a single-discount equivalent.

S E C T I O N   **2.5**

Inequalities in One Variable

1   Solve inequalities in one variable

The **solution set of an inequality** is a set of numbers, each element of which, when substituted for the variable, results in a true inequality.

The inequality at the right is true if the variable is replaced by 3, -1.98, or $\frac{2}{3}$.

$$x - 1 < 4$$
$$3 - 1 < 4$$
$$-1.98 - 1 < 4$$
$$\frac{2}{3} - 1 < 4$$

There are many values of the variable x that will make the inequality $x - 1 < 4$ true. The solution set of the inequality is any number less than 5. The solution set can be written in set-builder notation as $\{x \mid x < 5\}$.

The graph of the solution set of $x - 1 < 4$ is shown at the right.

$$\xleftarrow{\quad\quad} \overset{\displaystyle +\ +\ +\ +\ +\ +\ +\ +\ +\ )}{\underset{-5\ -4\ -3\ -2\ -1\ \ 0\ \ 1\ \ 2\ \ 3\ \ 4\ \ 5}{\quad}} \rightarrow$$

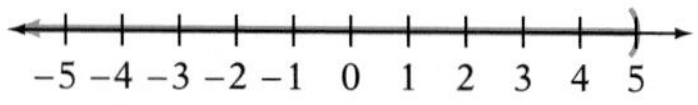

In solving an inequality, use the Addition and Multiplication Properties of Inequalities to rewrite the inequality in the form *variable* $<$ *constant* or *variable* $>$ *constant*.

The Addition Property of Inequalities

> If $a > b$ and c is a real number, then the inequalities $a > b$ and $a + c > b + c$ have the same solution set.
>
> If $a < b$ and c is a real number, then the inequalities $a < b$ and $a + c < b + c$ have the same solution set.

The Addition Property of Inequalities states that the same number can be added to each side of an inequality without changing the solution set of the inequality. This property is also true for an inequality that contains the symbol $\le$ or $\ge$.

The Addition Property of Inequalities is used to remove a term from one side of an inequality by adding the additive inverse of that term to each side of the inequality. Because subtraction is defined in terms of addition, the same number can be subtracted from each side of an inequality without changing the solution set of the inequality.

⇒ Solve: $3x - 4 < 2x - 1$

	$3x - 4 < 2x - 1$
Subtract $2x$ from each side of the inequality.	$x - 4 < -1$
Add 4 to each side of the inequality.	$x < 3$
Write the solution set. The solution set can be written in either set-builder notation or interval notation.	$\{x \mid x < 3\}$
	$(-\infty, 3)$ ⬅

INSTRUCTOR NOTE

Students are confused by the need to change the inequality symbol when multiplying or dividing by a negative number. Emphasize that what matters is what you multiply or divide *by*. For instance,

$$2x < -8$$
$$\frac{2x}{2} < \frac{-8}{2}$$
$$x < -4$$

Here, we divide *by* 2. It does not matter that we're dividing into a negative number.

$$-2x < 8$$
$$\frac{-2x}{-2} > \frac{8}{-2}$$
$$x > -4$$

Now we have divided *by* a negative number; the inequality symbol must be reversed.

Another source of confusion is problems of the following type: $-4x > 0$ and $3x < 0$.

The Multiplication Property of Inequalities is used to remove a coefficient from one side of an inequality so that the inequality can be written in the form *variable* < *constant* or *variable* > *constant*.

The Multiplication Property of Inequalities

Rule 1

If $a > b$ and $c > 0$, then the inequalities $a > b$ and $ac > bc$ have the same solution set.

If $a < b$ and $c > 0$, then the inequalities $a < b$ and $ac < bc$ have the same solution set.

Rule 2

If $a > b$ and $c < 0$, then the inequalities $a > b$ and $ac < bc$ have the same solution set.

If $a < b$ and $c < 0$, then the inequalities $a < b$ and $ac > bc$ have the same solution set.

Rule 1 states that when each side of an inequality is multiplied by a positive number, the inequality symbol remains the same. However, Rule 2 states that when each side of an inequality is multiplied by a negative number, the inequality symbol must be reversed.

Here are some examples of this property.

Rule 1		Rule 2	
$3 > 2$	$2 < 5$	$3 > 2$	$2 < 5$
$3(4) > 2(4)$	$2(4) < 5(4)$	$3(-4) < 2(-4)$	$2(-4) > 5(-4)$
$12 > 8$	$8 < 20$	$-12 < -8$	$-8 > -20$

Because division is defined in terms of multiplication, when each side of an inequality is divided by a positive number, the inequality symbol remains the same. When each side of an inequality is divided by a negative number, the inequality symbol must be reversed.

The Multiplication Property of Inequalities is also true for the symbols ≤ and ≥.

LOOK CLOSELY

Each side of the inequality is divided *by* a negative number; the inequality symbol must be reversed.

Solve: $-3x > 9$
Write the solution set in interval notation.

$$-3x > 9$$

Divide each side of the inequality by the coefficient -3, and reverse the inequality symbol. Simplify.

$$\frac{-3x}{-3} < \frac{9}{-3}$$
$$x < -3$$

Write the solution set. $(-\infty, -3)$

Example 1 Solve: $x + 3 > 4x + 6$
Write the solution set in set-builder notation.

Solution

$$x + 3 > 4x + 6$$
$$-3x + 3 > 6$$ ▸ Subtract $4x$ from each side of the inequality.

$$-3x > 3$$ ▸ Subtract 3 from each side of the inequality.
$$x < -1$$ ▸ Divide each side of the inequality by -3, and reverse the inequality symbol.

$$\{x \mid x < -1\}$$ ▸ Write the solution set.

Problem 1   Solve: $2x - 1 < 6x + 7$
Write the solution set in set-builder notation.

Solution   See page S4. $\{x \mid x > -2\}$

When an inequality contains parentheses, the first step in solving the inequality is to use the Distributive Property to remove the parentheses.

Example 2   Solve: $5(x - 2) \geq 9x - 3(2x - 4)$
Write the solution set in set-builder notation.

Solution $5(x - 2) \geq 9x - 3(2x - 4)$
$5x - 10 \geq 9x - 6x + 12$ ▸ Use the Distributive Property to remove parentheses.

$5x - 10 \geq 3x + 12$ ▸ Simplify.
$2x - 10 \geq 12$ ▸ Subtract $3x$ from each side of the inequality.

$2x \geq 22$ ▸ Add 10 to each side of the inequality.
$x \geq 11$ ▸ Divide each side of the inequality by 2.

$\{x \mid x \geq 11\}$

Problem 2   Solve: $5x - 2 \leq 4 - 3(x - 2)$
Write the solution set in interval notation.

Solution   See page S4. $\left(-\infty, \dfrac{3}{2} \right]$

2   ## Solve compound inequalities

A **compound inequality** is formed by joining two inequalities with a connective word such as "and" or "or." The inequalities shown below are compound inequalities.

$$2x < 4 \quad \text{and} \quad 3x - 2 > -8$$
$$2x + 3 > 5 \quad \text{or} \quad x + 2 < 5$$

The solution set of a compound inequality with the connective word *and* is the set of all elements common to the solution sets of both inequalities. Therefore, it is the *intersection* of the solution sets of the two inequalities.

▸ Solve: $2x < 6$ and $3x + 2 > -4$
Write the solution set in interval notation.

$$2x < 6 \quad \text{and} \quad 3x + 2 > -4$$
Solve each inequality. $$x < 3 \qquad\qquad 3x > -6$$
$$x > -2$$

$\{x \mid x < 3\} \qquad \{x \mid x > -2\}$

Find the intersection of the solution sets. $\{x \mid x < 3\} \cap \{x \mid x > -2\} = (-2, 3)$

Solve: $-3 < 2x + 1 < 5$
Write the solution set in set-builder notation.

This inequality is equivalent to the compound inequality $-3 < 2x + 1$ and $2x + 1 < 5$.

$$-3 < 2x + 1 < 5$$

$$\begin{aligned} -3 &< 2x + 1 \quad \text{and} \quad & 2x + 1 &< 5 \\ -4 &< 2x & 2x &< 4 \\ -2 &< x & x &< 2 \end{aligned}$$

$\{x \mid x > -2\}$ $\quad$ $\{x \mid x < 2\}$

Find the intersection of the solution sets. $\{x \mid x > -2\} \cap \{x \mid x < 2\} = \{x \mid -2 < x < 2\}$

There is an alternative method for solving the inequality in the last example.

$$-3 < 2x + 1 < 5$$

Subtract 1 from each of the three parts of the inequality.

$$-3 - 1 < 2x + 1 - 1 < 5 - 1$$
$$-4 < 2x < 4$$

Divide each of the three parts of the inequality by the coefficient 2.

$$\frac{-4}{2} < \frac{2x}{2} < \frac{4}{2}$$
$$-2 < x < 2$$

$\{x \mid -2 < x < 2\}$

The solution set of a compound inequality with the connective word *or* is the *union* of the solution sets of the two inequalities.

Solve: $2x + 3 > 7$ or $4x - 1 < 3$
Write the solution set in set-builder notation.

Solve each inequality.

$$\begin{aligned} 2x + 3 &> 7 \quad \text{or} \quad & 4x - 1 &< 3 \\ 2x &> 4 & 4x &< 4 \\ x &> 2 & x &< 1 \end{aligned}$$

$\{x \mid x > 2\}$ $\quad$ $\{x \mid x < 1\}$

Find the union of the solution sets. $\{x \mid x > 2\} \cup \{x \mid x < 1\} = \{x \mid x > 2 \text{ or } x < 1\}$

Example 3 Solve: $1 < 3x - 5 < 4$
Write the solution set in interval notation.

Solution

$$1 < 3x - 5 < 4$$

$$1 + 5 < 3x - 5 + 5 < 4 + 5$$ ▶ Add 5 to each of the three parts of the inequality.

$$6 < 3x < 9$$ ▶ Simplify.

$$\frac{6}{3} < \frac{3x}{3} < \frac{9}{3}$$ ▶ Divide each of the three parts of the inequality by 3.

$$2 < x < 3$$

$(2, 3)$ ▶ Write the solution set.

Problem 3 Solve: $-2 \le 5x + 3 \le 13$
Write the solution set in interval notation.

Solution See page S4. $[-1, 2]$

Example 4 Solve: $11 - 2x > -3$ and $7 - 3x < 4$
Write the solution set in set-builder notation.

Solution

$$\begin{aligned} 11 - 2x &> -3 \\ -2x &> -14 \\ x &< 7 \end{aligned} \qquad \text{and} \qquad \begin{aligned} 7 - 3x &< 4 \\ -3x &< -3 \\ x &> 1 \end{aligned}$$

▶ Solve each inequality.

$\{x \mid x < 7\}$ $\qquad$ $\{x \mid x > 1\}$

$\{x \mid x < 7\} \cap \{x \mid x > 1\} = \{x \mid 1 < x < 7\}$

▶ Find the intersection of the solution sets.

Problem 4 Solve: $5 - 4x > 1$ and $6 - 5x < 11$
Write the solution set in interval notation.

Solution See page S4. $(-1, 1)$

Example 5 Solve: $3 - 4x > 7$ or $4x + 5 < 9$
Write the solution set in set-builder notation.

Solution

$$\begin{aligned} 3 - 4x &> 7 \\ -4x &> 4 \\ x &< -1 \end{aligned} \qquad \text{or} \qquad \begin{aligned} 4x + 5 &< 9 \\ 4x &< 4 \\ x &< 1 \end{aligned}$$

▶ Solve each inequality.

$\{x \mid x < -1\}$ $\qquad$ $\{x \mid x < 1\}$

$\{x \mid x < -1\} \cup \{x \mid x < 1\} = \{x \mid x < 1\}$

▶ Find the union of the solution sets.

Problem 5 Solve: $2 - 3x > 11$ or $5 + 2x > 7$
Write the solution set in set-builder notation.

Solution See page S4. $\{x \mid x < -3 \text{ or } x > 1\}$

3 Application problems

Example 6 Company A rents cars for \$6 a day and 14¢ for every mile driven. Company B rents cars for \$12 a day and 8¢ for every mile driven. You want to rent a car for 5 days. How many miles can you drive a Company A car during the 5 days if it is to cost less than a Company B car?

Strategy To find the number of miles, write and solve an inequality using N to represent the number of miles.

Solution Cost of Company A car < cost of Company B car

$$\begin{aligned} 6(5) + 0.14N &< 12(5) + 0.08N \\ 30 + 0.14N &< 60 + 0.08N \\ 30 + 0.06N &< 60 \end{aligned}$$

$$0.06N < 30$$
$$N < 500$$

It is less expensive to rent from Company A if the car is driven less than 500 mi.

Problem 6 The base of a triangle is 12 in., and the height is $(x + 2)$ in. Express as an integer the maximum height of the triangle when the area is less than 50 in^2.

Solution See page S5. 8 in.

Example 7 Find three consecutive odd integers whose sum is between 27 and 51.

Strategy To find the three consecutive odd integers, write and solve a compound inequality using x to represent the first odd integer.

Solution $\text{Lower limit of the sum} < \text{sum} < \text{upper limit of the sum}$

$$27 < x + (x + 2) + (x + 4) < 51$$
$$27 < 3x + 6 < 51$$
$$27 - 6 < 3x + 6 - 6 < 51 - 6$$
$$21 < 3x < 45$$
$$\frac{21}{3} < \frac{3x}{3} < \frac{45}{3}$$
$$7 < x < 15$$

The three integers are 9, 11, and 13; or 11, 13, and 15; or 13, 15, and 17.

Problem 7 An average score of 80 to 89 in a history course receives a B grade. A student has grades of 72, 94, 83, and 70 on four exams. Find the range of scores on the fifth exam that will give the student a B for the course.

Solution See page S5. $81 \le N \le 100$

STUDY TIPS

TAKE CAREFUL NOTES IN CLASS

You need a notebook in which to keep class notes and records about assignments and tests. Make sure to take complete and well-organized notes. Your instructor will explain text material that may be difficult for you to understand on your own and may provide important information that is not provided in the textbook. Be sure to include in your notes everything that is written on the chalkboard.

continued

Information recorded in your notes about assignments should explain exactly what they are, how they are to be done, and when they are due. Information about tests should include exactly what text material and topics will be covered on each test and the dates on which the tests will be given.

CONCEPT REVIEW 2.5

Determine whether the following statements are always true, sometimes true, or never true.

1. If $a > b$, then $-a < -b$. Always true

2. If $a < b$ and $a \neq 0$, $b \neq 0$, then $\frac{1}{a} < \frac{1}{b}$. Never true

3. When dividing both sides of an inequality by an integer, we must reverse the inequality symbol. Sometimes true

4. When subtracting the same real number from both sides of an inequality, we must reverse the inequality symbol. Never true

5. Suppose $a < 1$. Then $a^2 < a$. Sometimes true

6. Suppose $a < b < 0$ and $c < d < 0$. Then $ac > bd$. Always true

EXERCISES 2.5

1 **1.** **a.** What does the Addition Property of Inequalities state?
b. When is the Addition Property of Inequalities used?

2. **a.** What does the Multiplication Property of Inequalities state?
b. When is the Multiplication Property of Inequalities used?

Solve. Write the solution set in set-builder notation.

3. $x - 3 < 2$ $\{x | x < 5\}$

4. $x + 4 \geq 2$ $\{x | x \geq -2\}$

5. $4x \leq 8$ $\{x | x \leq 2\}$

6. $6x > 12$ $\{x | x > 2\}$

7. $-2x > 8$ $\{x | x < -4\}$

8. $-3x \leq -9$ $\{x | x \geq 3\}$

9. $3x - 1 > 2x + 2$ $\{x | x > 3\}$

10. $5x + 2 \geq 4x - 1$ $\{x | x \geq -3\}$

11. $2x - 1 > 7$ $\{x | x > 4\}$

12. $4x + 3 \leq -1$ $\{x | x \leq -1\}$

13. $6x + 3 > 4x - 1$ $\{x | x > -2\}$

14. $7x + 4 < 2x - 6$ $\{x | x < -2\}$

15. $8x + 1 \geq 2x + 13$ $\{x \mid x \geq 2\}$

16. $5x - 4 < 2x + 5$ $\{x \mid x < 3\}$

17. $7 - 2x \geq 1$ $\{x \mid x \leq 3\}$

18. $3 - 5x \leq 18$ $\{x \mid x \geq -3\}$

19. $4x - 2 < x - 11$ $\{x \mid x < -3\}$

20. $6x + 5 \geq x - 10$ $\{x \mid x \geq -3\}$

Solve. Write the solution set in interval notation.

21. $x + 7 \geq 4x - 8$ $(-\infty, 5]$

22. $3x + 1 \leq 7x - 15$ $[4, \infty)$

23. $6 - 2(x - 4) \leq 2x + 10$ $[1, \infty)$

24. $4(2x - 1) > 3x - 2(3x - 5)$ $\left(\frac{14}{11}, \infty\right)$

25. $2(1 - 3x) - 4 > 10 + 3(1 - x)$ $(-\infty, -5)$

26. $2 - 5(x + 1) \geq 3(x - 1) - 8$ $(-\infty, 1]$

27. $\frac{3}{5}x - 2 < \frac{3}{10} - x$ $\left(-\infty, \frac{23}{16}\right)$

28. $\frac{5}{6}x - \frac{1}{6} \leq x - 4$ $[23, \infty)$

29. $\frac{1}{3}x - \frac{3}{2} \geq \frac{7}{6} - \frac{2}{3}x$ $\left[\frac{8}{3}, \infty\right)$

30. $\frac{7}{12}x - \frac{3}{2} < \frac{2}{3}x + \frac{5}{6}$ $(-28, \infty)$

31. $\frac{1}{2}x - \frac{3}{4} > \frac{7}{4}x - 2$ $(-\infty, 1)$

32. $\frac{2 - x}{4} - \frac{3}{8} \geq \frac{2}{5}x$ $\left(-\infty, \frac{5}{26}\right]$

33. $2 - 2(7 - 2x) < 3(3 - x)$ $(-\infty, 3)$

34. $3 + 2(x + 5) \geq x + 5(x + 1) + 1$ $\left(-\infty, \frac{7}{4}\right]$

2 Solve. Write the solution set in interval notation.

35. $3x < 6$ and $x + 2 > 1$ $(-1, 2)$

36. $x - 3 \leq 1$ and $2x \geq -4$ $[-2, 4]$

37. $x + 2 \geq 5$ or $3x \leq 3$ $(-\infty, 1] \cup [3, \infty)$

38. $2x < 6$ or $x - 4 > 1$ $(-\infty, 3) \cup (5, \infty)$

39. $-2x > -8$ and $-3x < 6$ $(-2, 4)$

40. $\frac{1}{2}x > -2$ and $5x < 10$ $(-4, 2)$

41. $\frac{1}{3}x < -1$ or $2x > 0$ $(-\infty, -3) \cup (0, \infty)$

42. $\frac{2}{3}x > 4$ or $2x < -8$ $(-\infty, -4) \cup (6, \infty)$

43. $x + 4 \geq 5$ and $2x \geq 6$ $[3, \infty)$

44. $3x < -9$ and $x - 2 < 2$ $(-\infty, -3)$

45. $-5x > 10$ and $x + 1 > 6$ $\varnothing$

46. $7x < 14$ and $1 - x < 4$ $(-3, 2)$

47. $2x - 3 > 1$ and $3x - 1 < 2$ $\varnothing$

48. $4x + 1 < 5$ and $4x + 7 > -1$ $(-2, 1)$

49. $3x + 7 < 10$ or $2x - 1 > 5$ $(-\infty, 1) \cup (3, \infty)$

50. $6x - 2 < -14$ or $5x + 1 > 11$
$(-\infty, -2) \cup (2, \infty)$

Solve. Write the solution set in set-builder notation.

51. $-5 < 3x + 4 < 16$ $\{x \mid -3 < x < 4\}$

52. $5 < 4x - 3 < 21$ $\{x \mid 2 < x < 6\}$

53. $0 < 2x - 6 < 4$ $\{x \mid 3 < x < 5\}$

54. $-2 < 3x + 7 < 1$ $\{x \mid -3 < x < -2\}$

55. $4x - 1 > 11$ or $4x - 1 \leq -11$
$\left\{x \mid x > 3 \text{ or } x \leq -\frac{5}{2}\right\}$

56. $3x - 5 > 10$ or $3x - 5 < -10$
$\left\{x \mid x > 5 \text{ or } x < -\frac{5}{3}\right\}$

57. $2x + 3 \geq 5$ and $3x - 1 > 11$ $\{x \mid x > 4\}$

58. $6x - 2 < 5$ or $7x - 5 < 16$ $\{x \mid x < 3\}$

59. $9x - 2 < 7$ and $3x - 5 > 10$ $\varnothing$

60. $8x + 2 \leq -14$ and $4x - 2 > 10$ $\varnothing$

61. $3x - 11 < 4$ or $4x + 9 \geq 1$ $\{x \mid x \in \text{real numbers}\}$

62. $5x + 12 \geq 2$ or $7x - 1 \leq 13$ $\{x \mid x \in \text{real numbers}\}$

63. $3 - 2x > 7$ and $5x + 2 > -18$ $\{x \mid -4 < x < -2\}$

64. $1 - 3x < 16$ and $1 - 3x > -16$ $\left\{x \mid -5 < x < \frac{17}{3}\right\}$

65. $5 - 4x > 21$ or $7x - 2 > 19$ $\{x \mid x < -4 \text{ or } x > 3\}$

66. $6x + 5 < -1$ or $1 - 2x < 7$ $\{x \mid x \in \text{real numbers}\}$

67. $3 - 7x \leq 31$ and $5 - 4x > 1$ $\{x \mid -4 \leq x < 1\}$

68. $9 - x \geq 7$ and $9 - 2x < 3$ $\varnothing$

69. $\frac{2}{3}x - 4 > 5$ or $x + \frac{1}{2} < 3$ $\left\{x \mid x > \frac{27}{2} \text{ or } x < \frac{5}{2}\right\}$

70. $\frac{5}{8}x + 2 < -3$ or $2 - \frac{3}{5}x < -7$ $\{x \mid x < -8 \text{ or } x > 15\}$

71. $-\frac{3}{8} \leq 1 - \frac{1}{4}x \leq \frac{7}{2}$ $\left\{x \mid -10 \leq x \leq \frac{11}{2}\right\}$

72. $-2 \leq \frac{2}{3}x - 1 \leq 3$ $\left\{x \mid -\frac{3}{2} \leq x \leq 6\right\}$

3

73. *Integer Problem* Five times the difference between a number and two is greater than the quotient of two times the number and three. Find the smallest integer that will satisfy the inequality. 3

74. *Integer Problem* Two times the difference between a number and eight is less than or equal to five times the sum of the number and four. Find the smallest number that will satisfy the inequality. −12

75. *Geometry* The length of a rectangle is 2 ft more than four times the width. Express as an integer the maximum width of the rectangle when the perimeter is less than 34 ft. 2 ft

76. *Geometry* The length of a rectangle is 5 cm less than twice the width. Express as an integer the maximum width of the rectangle when the perimeter is less than 60 cm. 11 cm

77. *Integer Problem* Find four consecutive integers whose sum is between 62 and 78. 15, 16, 17, 18; 16, 17, 18, 19; or 17, 18, 19, 20

78. *Integer Problem* Find three consecutive even integers whose sum is between 30 and 52. 10, 12, 14; 12, 14, 16; or 14, 16, 18

79. *Geometry* One side of a triangle is 1 in. longer than the second side. The third side is 2 in. longer than the second side. Find, to the nearest whole number, the length of the second side of the triangle if the perimeter is more than 15 in. and less than 25 in. 5, 6, or 7 in.

80. *Geometry* The length of a rectangle is 4 ft more than twice the width. Find, to the nearest whole number, the width of the rectangle if the perimeter is more than 28 ft and less than 40 ft. 4 or 5 ft

81. *Consumerism* A cellular phone company offers its customers a rate of $99 for up to 200 min per month of cellular phone time, or a rate of $35 per month plus $.40 for each minute of cellular phone time. For how many minutes per month can a customer who chooses the second option use a cellular phone before the charges exceed those of the first option? 160 min

82. *Consumerism* In 1997, the computer service America Online offered its customers a rate of $19.95 per month for unlimited use or $4.95 per month with 3 free hours plus $2.50 for each hour thereafter. How many hours can you use this service per month if the second plan is to cost you less than the first? less than 9 h

83. *Consumerism* AirTouch advertises local paging service for $6.95 per month for up to 400 pages, and $.10 per page thereafter. A competitor advertises paging service for $3.95 per month for up to 400 pages and $.15 per page thereafter. For what number of pages per month is the AirTouch plan less expensive? more than 460 pages

84. *Consumerism* Suppose PayRite Rental Cars rents compact cars for $32 per day with unlimited mileage, and Otto Rentals offers compact cars for $19.99 per day but charges $.19 for each mile beyond 100 mi driven per day. You want to rent a car for one week. How many miles can you drive during the week if Otto Rental is to be less expensive than PayRite? less than 1143 mi

85. *Consumerism* During a weekday, to call a city 40 mi away from a certain pay phone costs $.70 for the first 3 min and $.15 for each additional minute. If you use a calling card, there is a $.35 fee and then the rates are $.196 for the first minute and $.126 for each additional minute. How many minutes must a call last if it is to be cheaper to pay with coins rather than a calling card? 7 min or less

86. *Consumerism* Heritage National Bank offers two different checking accounts. The first charges $3 per month, and $.50 per check after the first 10 checks. The second account charges $8 per month with unlimited check writing. How many checks can be written per month if the first account is to be less expensive than the second account? less than 20 checks

87. *Consumerism* Glendale Federal Bank offers a checking account to small businesses. The charge is $8 per month plus $.12 per check after the first 100 checks. A competitor is offering an account for $5 per month plus $.15 per check after the first 100 checks. If a business chooses the Glendale Federal Bank account, how many checks does the business write monthly if it is assumed that this account will cost less than the competitor's account? more than 200 checks

88. *Commissions* George Stoia earns $1000 per month plus 5% commission on the amount of sales. George's goal is to earn a minimum of $3200 per month. What amount of sales will enable George to earn $3200 or more per month? $44,000 or more

89. *Automobiles* A new car will average at least 22 mpg for city driving and at most 27.5 mpg for highway driving. Find the range of miles that the car can travel on a full tank (19.5 gal) of gasoline. Between 429 mi and 536.25 mi

90. *Education* An average score of 90 or above in an English course receives an A grade. A student has grades of 85, 88, 90, and 98 on four tests. Find the range of scores on the fifth test that will give the student an A grade. $89 \le N \le 100$

APPLYING CONCEPTS 2.5

Use the roster method to list the set of positive integers that are solutions of the inequality.

91. $8x - 7 < 2x + 9$ $\{1, 2\}$

92. $2x + 9 \ge 5x - 4$ $\{1, 2, 3, 4\}$

93. $5 + 3(2 + x) > 8 + 4(x - 1)$ $\{1, 2, 3, 4, 5, 6\}$

94. $6 + 4(2 - x) > 7 + 3(x + 5)$ $\varnothing$

95. $-3x < 15$ and $x + 2 < 7$ $\{1, 2, 3, 4\}$

96. $3x - 2 > 1$ and $2x - 3 < 5$ $\{2, 3\}$

97. $-4 \le 3x + 8 < 16$ $\{1, 2\}$

98. $5 < 7x - 3 \le 24$ $\{2, 3\}$

99. *Temperature* The relationship between Celsius temperature and Fahrenheit temperature is given by the formula $F = \frac{9}{5}C + 32$. If the temperature is between 77°F and 86°F, what is the temperature range in degrees Celsius? Between 25°C and 30°C

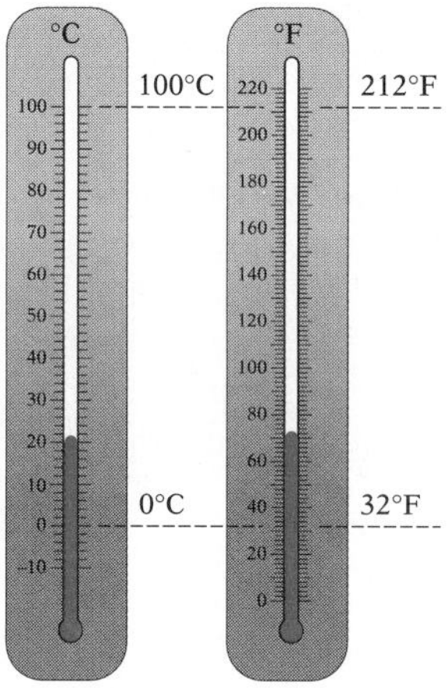

[C] **100.** *Integer Problem* The average of two negative integers is less than or equal to −15. The smaller integer is 7 less than the larger integer. Find the greatest possible value for the smaller integer. −19

[C] **101.** *Consumerism* The charges for a long-distance telephone call are $1.56 for the first three minutes and $.52 for each additional minute or fraction of a minute. What is the largest whole number of minutes a call can last if it is to cost you less than $5.40? 10 min

SECTION 2.6 Absolute Value Equations and Inequalities

1 Absolute value equations

The **absolute value** of a number is its distance from zero on the number line. Distance is always a positive number or zero. Therefore, the absolute value of a number is always a positive number or zero.

The distance from 0 to 3 or from 0 to -3 is 3 units.

$|3| = 3$ $|-3| = 3$

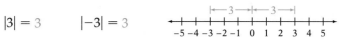

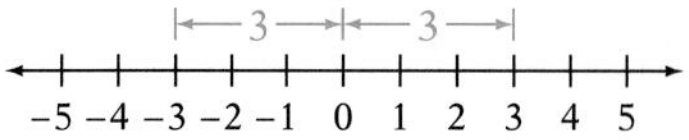

An equation containing an absolute value symbol is called an **absolute value equation**.

$$\left.\begin{array}{l} |x| = 3 \\ |x + 2| = 8 \\ |3x - 4| = 5x - 9 \end{array}\right\} \begin{array}{l} \text{Absolute} \\ \text{value} \\ \text{equations} \end{array}$$

Absolute Value Equations

If $a \geq 0$ and $|x| = a$, then $x = a$ or $x = -a$.

Given $|x| = 3$, then $x = 3$ or $x = -3$.

➡ Solve: $|x + 2| = 8$

$$|x + 2| = 8$$

Remove the absolute value sign, and rewrite as two equations.

$x + 2 = 8 \qquad x + 2 = -8$

Solve each equation.

$x = 6 \qquad\qquad x = -10$

Check:
$$\begin{array}{c|c} |x + 2| = 8 & |x + 2| = 8 \\ \hline |6 + 2| \ \ \big| \ \ 8 & |-10 + 2| \ \ \big| \ \ 8 \\ |8| \ \ \big| & |-8| \ \ \big| \\ 8 = 8 & 8 = 8 \end{array}$$

Write the solution.

The solutions are 6 and -10.

Example 1   Solve.   A. $|x| = 15$ B. $|2 - x| = 12$

C. $|2x| = -4$ D. $3 - |2x - 4| = -5$

Solution   A. $|x| = 15$

$x = 15 \qquad x = -15$

▶ Remove the absolute value sign, and rewrite as two equations.

The solutions are 15 and -15.

B. $|2 - x| = 12$

$2 - x = 12 \qquad 2 - x = -12$ ▶ Remove the absolute value sign, and rewrite as two equations.

$-x = 10 \qquad\quad -x = -14$

$x = -10 \qquad\quad x = 14$ ▶ Solve each equation.

The solutions are -10 and 14.

C. $|2x| = -4$ ▶ The absolute value of a number must be nonnegative.

There is no solution to the equation.

LOOK CLOSELY

$3 - \|2x - 4\| = -5$	
$3 - \|2(6) - 4\|$	-5
$3 - \|12 - 4\|$	-5
$3 - \|8\|$	-5
$3 - 8$	-5
$-5 = -5$	

$3 - \|2x - 4\| = -5$	
$3 - \|2(-2) - 4\|$	-5
$3 - \|-4 - 4\|$	-5
$3 - \|-8\|$	-5
$3 - 8$	-5
$-5 = -5$	

D. $3 - |2x - 4| = -5$

$-|2x - 4| = -8$ ▶ Solve for the absolute value.

$|2x - 4| = 8$ ▶ Multiply each side of the equation by -1.

$2x - 4 = 8 \qquad 2x - 4 = -8$ ▶ Remove the absolute value sign, and rewrite as two equations.

$2x = 12 \qquad 2x = -4$

$x = 6 \qquad x = -2$ ▶ Solve each equation.

The solutions are 6 and -2.

Problem 1 Solve. A. $|x| = 25$ B. $|2x - 3| = 5$
C. $|x - 3| = -2$ D. $5 - |3x + 5| = 3$

Solution See page S5.
A. $25, -25$ B. $4, -1$ C. No solution D. $-1, -\frac{7}{3}$

2 Absolute value inequalities

Recall that absolute value represents the distance between two points. For example, the solutions of the absolute value equation $|x - 1| = 3$ are the numbers whose distance from 1 is 3. Therefore, the solutions are -2 and 4.

The solutions of the **absolute value inequality** $|x - 1| < 3$ are the numbers whose distance from 1 is *less than* 3. Therefore, the solutions are the numbers greater than -2 and less than 4. The solution set is $\{x \mid -2 < x < 4\}$. *Note:* In this text, solutions to absolute value inequalities will always be written in set-builder notation.

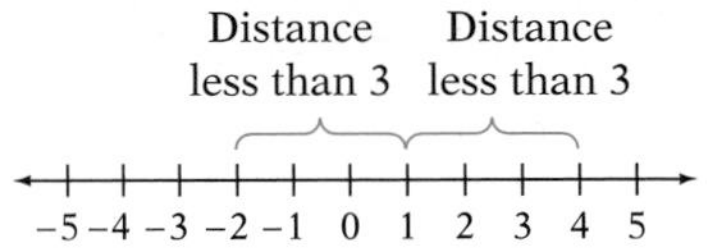

Absolute Value Inequalities of the Form $|ax + b| < c$

To solve an absolute value inequality of the form $|ax + b| < c$, solve the equivalent compound inequality $-c < ax + b < c$.

The compound inequality $-c < ax + b < c$ is equivalent to the intersection of the two inequalities $-c < ax + b$ and $ax + b < c$. The intersection of these two inequalities is shown on the following number line.

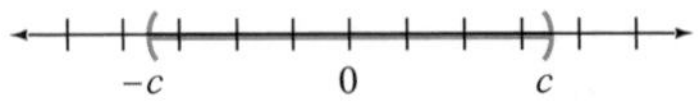

From this number line, we see that the solution of $|ax + b| < c$ is all the numbers for which $ax + b$ is greater than $-c$ and less than c.

Example 2 Solve: $|4x - 3| < 5$

Solution $|4x - 3| < 5$

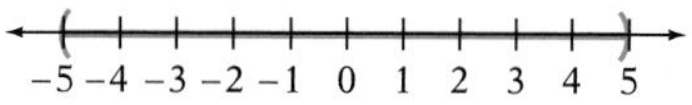

▸ The distance from zero is less than 5 units. $4x - 3$ is between -5 and 5.

$$-5 < 4x - 3 < 5$$

▸ Solve the equivalent compound inequality.

$$-5 + 3 < 4x - 3 + 3 < 5 + 3$$

▸ Add 3 to each of the three parts of the inequality.

$$-2 < 4x < 8$$

$$\frac{-2}{4} < \frac{4x}{4} < \frac{8}{4}$$

▸ Divide each of three parts of the inequality by 4.

$$-\frac{1}{2} < x < 2$$

$$\left\{x \mid -\frac{1}{2} < x < 2\right\}$$

▸ Write the solution set.

Problem 2 Solve: $|3x + 2| < 8$

Solution See page S5. $\left\{x \mid -\frac{10}{3} < x < 2\right\}$

Example 3 Solve: $|x - 3| < 0$

Solution $|x - 3| < 0$ ▸ The absolute value of a number must be nonnegative.

$\varnothing$ ▸ The solution set is the empty set.

Problem 3 Solve: $|3x - 7| < 0$

Solution See page S5. $\varnothing$

The solutions of the absolute value inequality $|x + 1| > 2$ are the numbers whose distance from -1 is *greater than* 2. Therefore, the solutions are the numbers less than -3 or greater than 1. The solution set is $\{x \mid x < -3 \text{ or } x > 1\}$.

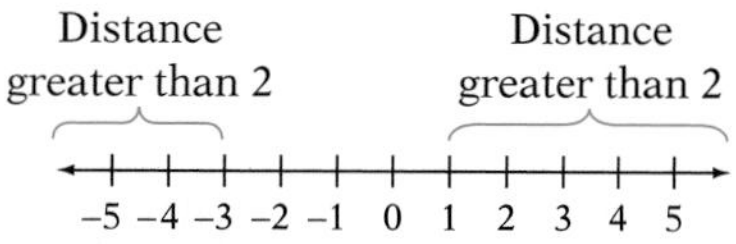

Absolute Value Inequalities of the Form $|ax + b| > c$

> To solve an absolute value inequality of the form $|ax + b| > c$, solve the equivalent compound inequality $ax + b < -c$ or $ax + b > c$.

The solution of the compound inequality $ax + b < -c$ or $ax + b > c$ is shown on the following number line.

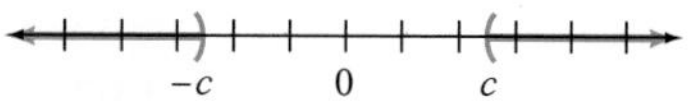

From this number line we see that the solution of $|ax + b| > c$ is the union of the numbers less than $-c$ and greater than c.

Example 4   Solve: $|2x - 1| > 7$

Solution $|2x - 1| > 7$

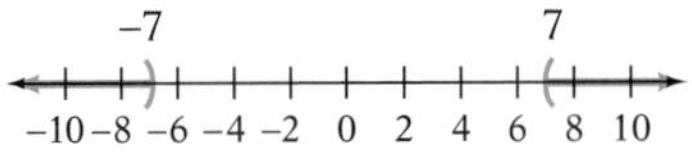

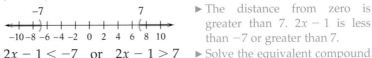

▸ The distance from zero is greater than 7. $2x - 1$ is less than -7 or greater than 7.

$2x - 1 < -7$ or $2x - 1 > 7$
$\qquad 2x < -6 \qquad\qquad 2x > 8$
$\qquad\quad x < -3 \qquad\qquad\quad x > 4$

▸ Solve the equivalent compound inequality.

$\{x \,|\, x < -3\} \qquad\qquad \{x \,|\, x > 4\}$

$\{x \,|\, x < -3\} \cup \{x \,|\, x > 4\}$
$\quad = \{x \,|\, x < -3 \text{ or } x > 4\}$

▸ Find the union of the solution sets.

Problem 4   Solve: $|5x + 3| > 8$

Solution   See page S5. $\left\{x \,\middle|\, x < -\dfrac{11}{5} \text{ or } x > 1\right\}$

3   Application problems

The **tolerance** of a component, or part, is the acceptable amount by which the component may vary from a given measurement. For example, the diameter of a piston may vary from the given measurement of 9 cm by 0.001 cm. This is written as $9 \text{ cm} \pm 0.001 \text{ cm}$, which is read "9 centimeters plus or minus 0.001 centimeter." The maximum diameter, or **upper limit,** of the piston is $9 \text{ cm} + 0.001 \text{ cm} = 9.001 \text{ cm}$. The minimum diameter, or **lower limit,** of the piston is $9 \text{ cm} - 0.001 \text{ cm} = 8.999 \text{ cm}$.

The lower and upper limits of the diameter could also be found by solving the absolute value inequality $|d - 9| \le 0.001$, where d is the diameter of the piston.

$$|d - 9| \le 0.001$$

$$-0.001 \le d - 9 \le 0.001$$
$$-0.001 + 9 \le d - 9 + 9 \le 0.001 + 9$$
$$8.999 \le d \le 9.001$$

The lower and upper limits of the diameter are 8.999 cm and 9.001 cm.

Example 5   A doctor has prescribed 2 cc of medication for a patient. The tolerance is 0.03 cc. Find the lower and upper limits of the amount of medication to be given.

Strategy   Let p represent the prescribed amount of medication, T the tolerance, and m the given amount of medication. Solve the absolute value inequality $|m - p| \le T$ for m.

Solution $|m - p| \le T$
$|m - 2| \le 0.03$

▸ Substitute the values of p and T into the inequality.

$-0.03 \le m - 2 \le 0.03$ ▸ Solve the equivalent compound inequality.
$1.97 \le m \le 2.03$

The lower and upper limits of the amount of medication to be given to the patient are 1.97 cc and 2.03 cc.

Problem 5 A machinist must make a bushing that has a tolerance of 0.003 in. The diameter of the bushing is 2.55 in. Find the lower and upper limits of the diameter of the bushing.

Solution See page S5. 2.547 in. and 2.553 in.

CONCEPT REVIEW 2.6

Determine whether the following statements are always true, sometimes true, or never true.

1. If $a > 0$ and $|x| = a$, then $x = a$ or $x = -a$. Always true
2. The absolute value inequality $|x| < a$, $a > 0$, is equivalent to the compound inequality $x > a$ or $x < -a$. Never true
3. If $|x + 3| = 12$, then $|x + 3| = 12$ or $|x + 3| = -12$. Never true
4. An absolute value equation has two solutions. Sometimes true
5. If $|x + b| < c$, then $x + b < -c$ or $x + b > c$. Never true

EXERCISES 2.6

1

1. Explain why, if $a \geq 0$ and $|x| = a$, $x = a$ or $x = -a$.
2. Explain how to solve an absolute value equation.

Solve.

3. $|x| = 7$ $-7, 7$
4. $|a| = 2$ $-2, 2$
5. $|-t| = 3$ $-3, 3$
6. $|-a| = 7$ $-7, 7$
7. $|-t| = -3$ No solution
8. $|-y| = -2$ No solution
9. $|x + 2| = 3$ $-5, 1$
10. $|x + 5| = 2$ $-7, -3$
11. $|y - 5| = 3$ $8, 2$
12. $|y - 8| = 4$ $4, 12$
13. $|a - 2| = 0$ 2
14. $|a + 7| = 0$ -7
15. $|x - 2| = -4$ No solution
16. $|x + 8| = -2$ No solution
17. $|2x - 5| = 4$ $\frac{1}{2}, \frac{9}{2}$
18. $|4 - 3x| = 4$ $0, \frac{8}{3}$
19. $|2 - 5x| = 2$ $0, \frac{4}{5}$
20. $|2x - 3| = 0$ $\frac{3}{2}$
21. $|5x + 5| = 0$ -1
22. $|3x - 2| = -4$ No solution
23. $|2x + 5| = -2$ No solution
24. $|x - 2| - 2 = 3$ $-3, 7$
25. $|x - 9| - 3 = 2$ $4, 14$
26. $|3a + 2| - 4 = 4$ $-\frac{10}{3}, 2$
27. $|8 - y| - 3 = 1$ $4, 12$
28. $|2x - 3| + 3 = 3$ $\frac{3}{2}$
29. $|4x - 7| - 5 = -5$ $\frac{7}{4}$
30. $|2x - 3| + 4 = -4$ No solution
31. $|3x - 2| + 1 = -1$ No solution
32. $|6x - 5| - 2 = 4$ $\frac{11}{6}, -\frac{1}{6}$

33. $|4b + 3| - 2 = 7$ $-3, \frac{3}{2}$

34. $|3t + 2| + 3 = 4$ $-\frac{1}{3}, -1$

35. $|5x - 2| + 5 = 7$ $\frac{4}{5}, 0$

36. $3 - |x - 4| = 5$ No solution

37. $2 - |x - 5| = 4$ No solution

38. $|2x - 8| + 12 = 2$ No solution

39. $|3x - 4| + 8 = 3$ No solution

40. $2 + |3x - 4| = 5$ $\frac{7}{3}, \frac{1}{3}$

41. $5 + |2x + 1| = 8$ $-2, 1$

42. $5 - |2x + 1| = 5$ $-\frac{1}{2}$

43. $3 - |5x + 3| = 3$ $-\frac{3}{5}$

44. $8 - |1 - 3x| = -1$ $-\frac{8}{3}, \frac{10}{3}$

2 45. Explain how to solve an absolute value inequality of the form $|ax + b| < c$.

46. Explain how to solve an absolute value inequality of the form $|ax + b| > c$.

Solve.

47. $|x| > 3$ $\{x \mid x > 3 \text{ or } x < -3\}$

48. $|x| < 5$ $\{x \mid -5 < x < 5\}$

49. $|x + 1| > 2$ $\{x \mid x > 1 \text{ or } x < -3\}$

50. $|x - 2| > 1$ $\{x \mid x < 1 \text{ or } x > 3\}$

51. $|x - 5| \le 1$ $\{x \mid 4 \le x \le 6\}$

52. $|x - 4| \le 3$ $\{x \mid 1 \le x \le 7\}$

53. $|2 - x| \ge 3$ $\{x \mid x \le -1 \text{ or } x \ge 5\}$

54. $|3 - x| \ge 2$ $\{x \mid x \le 1 \text{ or } x \ge 5\}$

55. $|2x + 1| < 5$ $\{x \mid -3 < x < 2\}$

56. $|3x - 2| < 4$ $\left\{x \mid -\frac{2}{3} < x < 2\right\}$

57. $|5x + 2| > 12$ $\left\{x \mid x > 2 \text{ or } x < -\frac{14}{5}\right\}$

58. $|7x - 1| > 13$ $\left\{x \mid x < -\frac{12}{7} \text{ or } x > 2\right\}$

59. $|4x - 3| \le -2$ $\varnothing$

60. $|5x + 1| \le -4$ $\varnothing$

61. $|2x + 7| > -5$ $\{x \mid x \in \text{real numbers}\}$

62. $|3x - 1| > -4$ $\{x \mid x \in \text{real numbers}\}$

63. $|4 - 3x| \ge 5$ $\left\{x \mid x \le -\frac{1}{3} \text{ or } x \ge 3\right\}$

64. $|7 - 2x| > 9$ $\{x \mid x < -1 \text{ or } x > 8\}$

65. $|5 - 4x| \le 13$ $\left\{x \mid -2 \le x \le \frac{9}{2}\right\}$

66. $|3 - 7x| < 17$ $\left\{x \mid -2 < x < \frac{20}{7}\right\}$

67. $|6 - 3x| \le 0$ $\{x \mid x = 2\}$

68. $|10 - 5x| \ge 0$ $\{x \mid x \in \text{real numbers}\}$

69. $|2 - 9x| > 20$ $\left\{x \mid x < -2 \text{ or } x > \frac{22}{9}\right\}$

70. $|5x - 1| < 16$ $\left\{x \mid -3 < x < \frac{17}{5}\right\}$

3 71. *Mechanics* The diameter of a bushing is 1.75 in. The bushing has a tolerance of 0.008 in. Find the lower and upper limits of the diameter of the bushing. 1.742 in., 1.758 in.

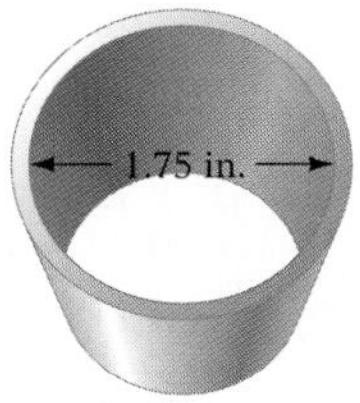

72. *Mechanics* A machinist must make a bushing that has a tolerance of 0.004 in. The diameter of the bushing is 3.48 in. Find the lower and upper limits of the diameter of the bushing. 3.476 in., 3.484 in.

73. *Medicine* A doctor has prescribed 2.5 cc of medication for a patient. The tolerance is 0.2 cc. Find the lower and upper limits of the amount of medication to be given. 2.3 cc, 2.7 cc

74. *Computers* A power strip is utilized on a computer to prevent the loss of programming by electrical surges. The power strip is designed to allow 110 volts plus or minus 16.5 volts. Find the lower and upper limits of voltage to the computer. 93.5 volts, 126.5 volts

75. *Appliances*   An electric motor is designed to run on 220 volts plus or minus 25 volts. Find the lower and upper limits of voltage on which the motor will run.   195 volts, 245 volts

76. *Automobiles*   A piston rod for an automobile is $10\frac{3}{8}$ in. with a tolerance of $\frac{1}{32}$ in. Find the lower and upper limits of the length of the piston rod.

$$10\frac{11}{32} \text{ in., } 10\frac{13}{32} \text{ in.}$$

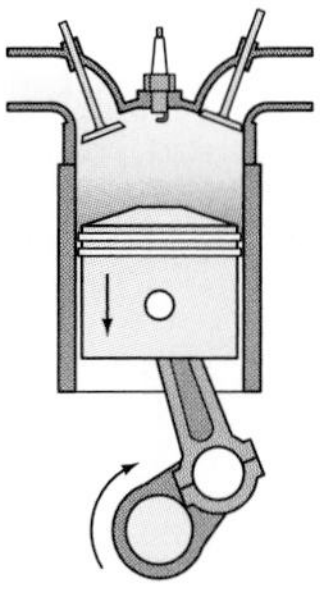

77. *Automobiles*   The diameter of a piston for an automobile is $3\frac{5}{16}$ in. with a tolerance of $\frac{1}{64}$ in. Find the lower and upper limits of the diameter of the piston. $3\frac{19}{64}$ in., $3\frac{21}{64}$ in.

The tolerance of the resistors used in electronics is given as a percent.

78. *Electronics*   Find the lower and upper limits of a 29,000-ohm resistor with a 2% tolerance.   28,420 ohms, 29,580 ohms

79. *Electronics*   Find the lower and upper limits of a 15,000-ohm resistor with a 10% tolerance.   13,500 ohms, 16,500 ohms

80. *Electronics*   Find the lower and upper limits of a 25,000-ohm resistor with a 5% tolerance.   23,750 ohms, 26,250 ohms

81. *Electronics*   Find the lower and upper limits of a 56-ohm resistor with a 5% tolerance.   53.2 ohms, 58.8 ohms

APPLYING CONCEPTS 2.6

Solve.

82. $\left|\dfrac{2x-5}{3}\right| = 7$ 83. $\left|\dfrac{3x-2}{4}\right| + 5 = 6$ 84. $\left|\dfrac{4x-2}{3}\right| > 6$ 85. $\left|\dfrac{2x-1}{5}\right| \le 3$

13, −8 $2, -\dfrac{2}{3}$ $\{x \mid x > 5 \text{ or } x < -4\}$ $\{x \mid -7 \le x \le 8\}$

For what values of the variable is the equation true? Write the solution set in set-builder notation.

[C] 86. $|x + 3| = x + 3$ [C] 87. $|y + 6| = y + 6$ [C] 88. $|a - 4| = 4 - a$ [C] 89. $|b - 7| = 7 - b$
$\{x \mid x \ge -3\}$ $\{y \mid y \ge -6\}$ $\{a \mid a \le 4\}$ $\{b \mid b \le 7\}$

[C] 90. For real numbers x and y, which of the following is always true?

  a. $|x + y| = |x| + |y|$ **b.** $|x + y| \le |x| + |y|$ **c.** $|x + y| \ge |x| + |y|$ b

91. Write an absolute value inequality to represent all real numbers within 5 units of 2. $|x - 2| < 5$

92. Write an absolute value inequality to represent all real numbers within k units of j. (Assume $k > 0$). $|x - j| < k$

93. Replace the question mark with $\leq$, $\geq$, or $=$.

a. $|x + y| \,?\, |x| + |y|$ $\leq$ **b.** $|x - y| \,?\, |x| - |y|$ $\geq$ **c.** $||x| - |y|| \,?\, |x| - |y|$ $\geq$

d. $\left|\frac{x}{y}\right| \,?\, \frac{|x|}{|y|}, y \neq 0$ $=$ **e.** $|xy| \,?\, |x||y|$ $=$

Focus on Problem Solving

Understand the Problem

The first of the four steps that Polya advocated to solve problems is to *understand the problem.* This aspect of problem solving is frequently not given enough attention. There are various exercises that you can try to achieve a good understanding of a problem. Some of these are stated in the Focus on Problem Solving in the chapter entitled "Review of Real Numbers." They are also listed here:

- Try to restate the problem in your own words.
- Determine what is known about this type of problem.
- Determine what information is given.
- Determine what information is unknown.
- Determine whether any of the information given is unnecessary.
- Determine the goal.

To illustrate this aspect of problem solving, consider the following famous ancient riddle.

As I was going to St. Ives,
I met a man with seven wives;
Each wife had seven sacks,
Each sack had seven cats,
Each cat had seven kits:
Kits, cats, sacks, and wives,
How many were going to St. Ives?

To answer the question in the poem, we will ask and answer some of the questions listed above.

POINT OF INTEREST

Try this brain teaser: You have two U.S. coins that add up to $.55. One is not a nickel. What are the two coins?

1. What is the goal?

The goal is to determine how many were going to St. Ives. (We know this from reading the last line of the poem.)

2. What information is necessary and what information is unnecessary?

The first line indicates that the poet was going to St. Ives. The next five lines describe a man the poet *met* on the way. This information is irrelevant.

The answer to the question, then, is 1. Only the poet was going to St. Ives.

INSTRUCTOR NOTE

The answer to the brain teaser above is that the two coins are a half dollar and a nickel. One of these coins is not a nickel; it is a half dollar.

There are many other examples of the importance, in problem solving, of recognizing *irrelevant information.* One that frequently makes the college circuit is posed in the form of a test. The first line of a 100-question test states, "Read the entire test before you begin." The last line of the test reads, "Choose any *one* question to answer." Many people ignore the information given in the first line and just begin the test, only to find out much later that they did a lot more work than necessary.

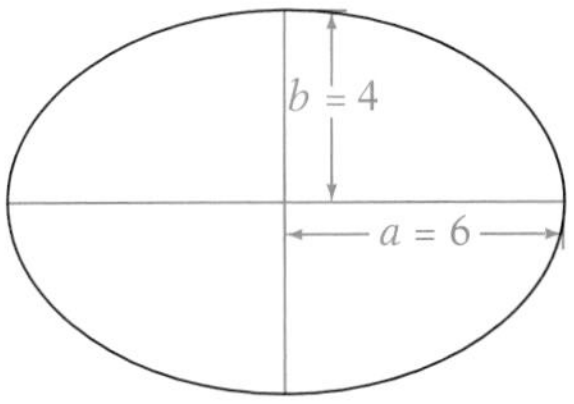

To illustrate another aspect of Polya's first step in the problem-solving process, consider the problem of finding the area of the oval-shaped region (called an *ellipse*) shown in the diagram at the left. This problem can be solved by doing some research to determine *what information is known about this type of problem.* Mathematicians have found a formula for the area of an ellipse. That formula is $A = \pi ab$, where a and b are as shown in the diagram. Therefore, $A = \pi(6)(4) = 24\pi \approx 75.40$ square units. Without the formula, this problem is difficult to solve. With the formula, it is fairly easy.

For each of the following problems, examine the problem in terms of the first step in Polya's problem-solving method. *Do not solve the problem.*

1. Johanna spent one-third of her allowance on a book. She then spent \$5.00 for a sandwich and iced tea. The cost of the iced tea was one-fifth the cost of the sandwich. Find the cost of the iced tea.
2. A flight from Los Angeles to Boston took 6 h. What was the average speed of the plane?
3. A major league baseball is approximately 5 in. in diameter and is covered with cowhide. Approximately how much cowhide is used to cover 10 baseballs?
4. How many donuts are in seven baker's dozen?
5. The smallest prime number is 2. Twice the difference between the eighth and the seventh prime numbers is two more than the smallest prime number. How large is the smallest prime number?

Projects and Group Activities

Venn Diagrams

Diagrams can be very useful when we are working with sets. The set diagrams shown below are called **Venn diagrams.**

In the Venn diagram at the right, the rectangle represents the set U, and the circles A and B represent subsets of set U. The common area of the two circles represents the intersection of A and B.

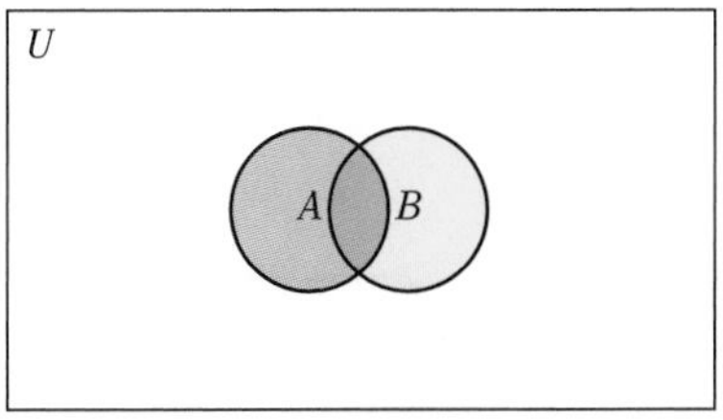

This Venn diagram represents the set $U = \{1, 2, 3, 4, 5, 6, 7, 8\}$. The sets $A = \{2, 3, 4, 5\}$ and $B = \{4, 5, 6, 7\}$ are subsets of U. Note that 1 and 8 are in set U but not in A or B. The numbers 4 and 5 are in both set A and set B.

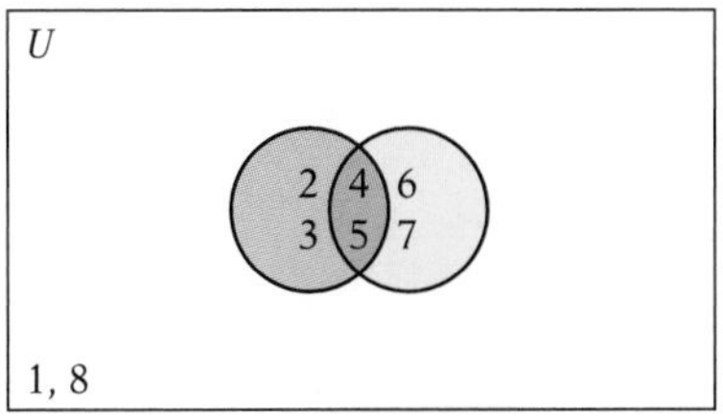

Sixty-five students at a small college enrolled in the following courses.

39 enrolled in English.
26 enrolled in mathematics.
35 enrolled in history.
19 enrolled in English and history.
11 enrolled in English and mathematics.
9 enrolled in mathematics and history.
2 enrolled in all three courses.

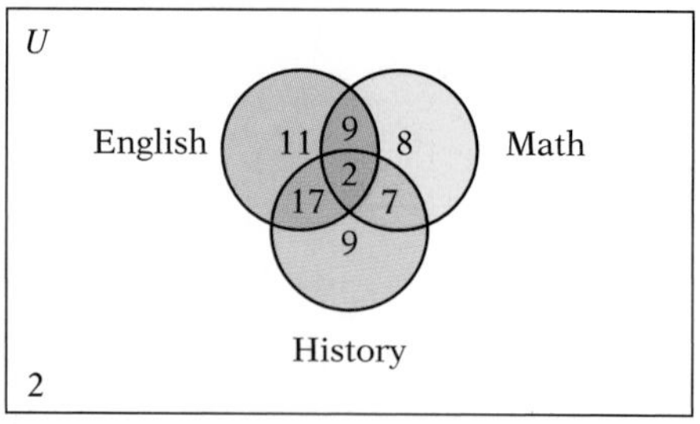

The Venn diagram is shown above. The diagram was drawn by placing the 2 students in the intersection of all three courses. Because 9 students took history and mathematics and 2 students are already in the intersection of history and mathematics, 7 more students are placed in the intersection of history and mathematics. Continue in this manner until all information is used.

Complete the following.

1. Use the Venn diagram above that shows course enrollments.
   a. How many students enrolled only in English and mathematics?
   b. How many students enrolled in English but did not enroll in mathematics or history?
   c. How many students did not enroll in any of the three courses?

2. Sketch the Venn diagram for $U = \{1, 2, 3, 4, 5, 6, 7, 8\}$, $A = \{2, 3, 4, 5, 6\}$, $B = \{4, 5, 6, 7, 8\}$, and $C = \{4, 5\}$.

3. A busload of 50 scouts stopped at a fast-food restaurant and ordered hamburgers. The scouts could have pickles, and/or tomatoes, and/or lettuce on their hamburgers.

   31 ordered pickles.
   36 ordered tomatoes.
   31 ordered lettuce.
   21 ordered pickles and tomatoes.
   24 ordered tomatoes and lettuce.
   22 ordered pickles and lettuce.
   17 ordered all three.

   a. How many scouts ordered a hamburger with pickles only?
   b. How many scouts ordered a hamburger with lettuce and tomatoes without the pickles?
   c. How many scouts ordered a hamburger without pickles, tomatoes, or lettuce?

Absolute Value Equations and Inequalities

The absolute value equation $|ax + b| = |cx + d|$ can be solved by using the definition of absolute value.

If both $ax + b$ and $cx + d$ are positive, then: $\quad ax + b = cx + d$

If both $ax + b$ and $cx + d$ are negative, then: $-(ax + b) = -(cx + d)$ or $ax + b = cx + d$

If either $ax + b$ or $cx + d$ is negative, we obtain: $-(ax + b) = cx + d$ or $ax + b = -(cx + d)$

From these three equations, we see that if $|ax + b| = |cx + d|$ is a true equation, then either

$$ax + b = cx + d \quad \text{or} \quad ax + b = -(cx + d)$$

Solve: $|2x - 5| = |3 - 5x|$

The expressions are equal or the expressions are opposite.

$$\begin{aligned} 2x - 5 &= 3 - 5x \\ 7x &= 8 \\ x &= \frac{8}{7} \end{aligned} \qquad \begin{aligned} 2x - 5 &= -(3 - 5x) \\ 2x - 5 &= -3 + 5x \\ -3x &= 2 \\ x &= -\frac{2}{3} \end{aligned}$$

The solutions are $-\frac{2}{3}$ and $\frac{8}{7}$.

Solve: $|x - 4| > |3 - x|$

This problem can be solved by squaring each side of the absolute value inequality. This ensures that each side of the absolute value inequality will be positive.

$$\begin{aligned} |x - 4| &> |3 - x| \\ (x - 4)^2 &> (3 - x)^2 \\ x^2 - 8x + 16 &> 9 - 6x + x^2 \\ -2x &> -7 \\ x &< \frac{7}{2} \end{aligned}$$

The solution is $\left\{x \mid x < \frac{7}{2}\right\}$.

Solve.

1. $|3x - 4| = |5x - 8|$
2. $|5x - 3| = |3x - 3|$
3. $|2x - 1| = |x|$
4. $|x + 1| = |x|$
5. $\left|\frac{2x - 3}{5}\right| = |2 - x|$
6. $\left|\frac{3 - 2x}{3}\right| = \left|\frac{x - 4}{2}\right|$
7. $|2x - 3| > |2x - 5|$
8. $|x + 4| < |3 - x|$
9. $|3x - 2| < |3x + 2|$

Chapter Summary

Key Words

An *equation* expresses the equality of two mathematical expressions. (Objective 2.1.1) $3 + 2 = 5$ $\quad 2x - 5 = 4$

The simplest equation to solve is an equation of the form *variable* = *constant*. (Objective 2.1.1)

$x = 5 \qquad -2 = x$

A *solution,* or *root, of an equation* is a replacement value for the variables that will make the equation true. (Objective 2.1.1)

The root of $x + 5 = 2$ is -3.

Solving an equation means finding a solution of the equation. The goal is to rewrite the equation in the form *variable* = *constant*, because the constant is the solution. (Objective 2.1.1)

The equation $x = 12$ is in the form *variable* = *constant*. The constant 12 is the solution of the equation.

An equation of the form $ax + b = c$, $a \neq 0$, is called a *first-degree equation.* (Objective 2.1.1)

$3x - 2 = 5$

The *solution set of an inequality* is a set of numbers, each element of which, when substituted in the inequality, results in a true inequality. (Objective 2.5.1)

Any number greater than 4 is a solution of the inequality $x > 4$.

A *compound inequality* is formed by joining two inequalities with a connective word such as "and" or "or." (Objective 2.5.2)

$3x > 6$ and $2x + 5 < 7$

$2x + 2 < 3$ or $x + 2 > 4$

An *absolute value equation* is an equation that contains an absolute value symbol. (Objective 2.6.1)

$|x - 2| = 3$

An *absolute value inequality* is an inequality that contains an absolute value symbol. (Objective 2.6.2)

$|x - 4| < 5 \qquad |2x - 3| > 6$

Essential Rules and Procedures

The Addition Property of Equations
If $a = b$, then $a + c = b + c$. (Objective 2.1.1)

$$\begin{aligned} x + 5 &= -3 \\ x + 5 - 5 &= -3 - 5 \\ x &= -8 \end{aligned}$$

The Multiplication Property of Equations
If $a = b$ and $c \neq 0$, then $ac = bc$. (Objective 2.1.1)

$$\frac{2}{3}x = 4$$

$$\left(\frac{3}{2}\right)\left(\frac{2}{3}x\right) = \left(\frac{3}{2}\right)4$$

$$x = 6$$

Coin and Stamp Equation
$$\begin{matrix}\text{Number} \\ \text{of items}\end{matrix} \cdot \begin{matrix}\text{Value of} \\ \text{each item}\end{matrix} = \begin{matrix}\text{Total value} \\ \text{of items}\end{matrix}$$ (Objective 2.2.1)

A collection of stamps consists of 17¢ and 27¢ stamps. In all there are 15 stamps with a value of \$3.55. How many 17¢ stamps are in the collection?

$$17n + 27(15 - n) = 355$$

Consecutive Integers
$n, n + 1, n + 2, \ldots$ (Objective 2.2.2)

The sum of three consecutive integers is 57.

$$n + (n + 1) + (n + 2) = 57$$

Consecutive Even or Odd Integers
$n, n + 2, n + 4, \ldots$ (Objective 2.2.2)

The sum of three consecutive even integers is 132.

$$n + (n + 2) + (n + 4) = 132$$

Value Mixture Equation
$V = AC$ (Objective 2.3.1)

A merchant combines coffee that costs \$6 per pound with coffee that costs \$3.20 per pound. How many pounds of each should be used to make 60 lb of a blend that costs \$4.50 per pound?

$$6x + 3.20(60 - x) = 4.50(60)$$

Uniform Motion Equation
$d = rt$ (Objective 2.3.2)

Two planes are 1640 mi apart and traveling toward each other. One plane is traveling 60 mph faster than the other plane. The planes meet in 2 h. Find the speed of each plane.

$$2r + 2(r + 60) = 1640$$

Annual Simple Interest Equation
$I = Pr$ (Objective 2.4.1)

An investment of \$4000 is made at an annual simple interest rate of 5%. How much additional money must be invested at an annual simple interest rate of 6.5% so that the total interest earned is \$720?

$$0.05(4000) + 0.065x = 720$$

Percent Mixture Problems
$Q = Ar$ (Objective 2.4.2)

A silversmith mixed 120 oz of an 80% silver alloy with 240 oz of a 30% silver alloy. Find the percent concentration of the resulting silver alloy.

$$0.80(120) + 0.30(240) = x(360)$$

The Addition Property of Inequalities
If $a > b$, then $a + c > b + c$.
The property is also true for $a < b$. (Objective 2.5.1)

$$\begin{aligned} x + 3 &> -2 \\ x + 3 - 3 &> -2 - 3 \\ x &> -5 \end{aligned}$$

The Multiplication Property of Inequalities
(Objective 2.5.1)

Rule 1 If $a > b$ and $c > 0$, then $ac > bc$.
If $a < b$ and $c > 0$, then $ac < bc$.

$$\begin{aligned} 3x &> 2 \\ \left(\frac{1}{3}\right)(3x) &> \left(\frac{1}{3}\right)2 \\ x &> \frac{2}{3} \end{aligned}$$

Rule 2 If $a > b$ and $c < 0$, then $ac < bc$.
If $a < b$ and $c < 0$, then $ac > bc$.

$$\begin{aligned} -2x &< 5 \\ \frac{-2x}{-2} &> \frac{5}{-2} \\ x &> -\frac{5}{2} \end{aligned}$$

Chapter Review Exercises

1. Solve: $x + 4 = -5$ -9

2. Solve: $\frac{2}{3} = x + \frac{3}{4}$ $-\frac{1}{12}$

3. Solve: $-3x = -21$ 7

4. Solve: $\frac{2}{3}x = \frac{4}{9}$ $\frac{2}{3}$

5. Solve: $3y - 5 = 3 - 2y$ $\frac{8}{5}$

6. Solve: $3x - 3 + 2x = 7x - 15$ 6

7. Solve: $2(x - 3) = 5(4 - 3x)$ $\frac{26}{17}$

8. Solve: $2x - (3 - 2x) = 4 - 3(4 - 2x)$ $\frac{5}{2}$

9. Solve: $\frac{1}{2}x - \frac{5}{8} = \frac{3}{4}x + \frac{3}{2}$ $-\frac{17}{2}$

10. Solve: $\frac{2x - 3}{3} + 2 = \frac{2 - 3x}{5}$ $-\frac{9}{19}$

11. Solve: $3x - 7 > -2$
Write the solution set in interval notation.
$\left(\frac{5}{3}, \infty\right)$

12. Solve: $2x - 9 < 8x + 15$
Write the solution set in interval notation.
$(-4, \infty)$

13. Solve: $\frac{2}{3}x - \frac{5}{8} \geq \frac{5}{4}x + 3$
Write the solution set in set-builder notation.
$\left\{x \mid x \leq -\frac{87}{14}\right\}$

14. Solve: $2 - 3(x - 4) \leq 4x - 2(1 - 3x)$
Write the solution set in set-builder notation.
$\left\{x \mid x \geq \frac{16}{13}\right\}$

15. Solve: $-5 < 4x - 1 < 7$
Write the solution set in interval notation.
$(-1, 2)$

16. Solve: $5x - 2 > 8$ or $3x + 2 < -4$
Write the solution set in interval notation.
$(-\infty, -2) \cup (2, \infty)$

17. Solve: $3x < 4$ and $x + 2 > -1$
Write the solution set in set-builder notation.
$\left\{x \mid -3 < x < \frac{4}{3}\right\}$

18. Solve: $3x - 2 > -4$ or $7x - 5 < 3x + 3$
Write the solution set in set-builder notation.
$\{x \mid x \in \text{real numbers}\}$

19. Solve: $|2x - 3| = 8$ $-\frac{5}{2}, \frac{11}{2}$

20. Solve: $|5x + 8| = 0$ $-\frac{8}{5}$

21. Solve: $6 + |3x - 3| = 2$ No solution

22. Solve: $|2x - 5| \leq 3$ $\{x \mid 1 \leq x \leq 4\}$

23. Solve: $|4x - 5| \geq 3$ $\left\{x \mid x \leq \frac{1}{2} \text{ or } x \geq 2\right\}$

24. Solve: $|5x - 4| < -2$ $\varnothing$

25. The diameter of a bushing is 2.75 in. The bushing has a tolerance of 0.003 in. Find the lower and upper limits of the diameter of the bushing.
2.747 in., 2.753 in.

26. A doctor has prescribed 2 cc of medication for a patient. The tolerance is 0.25 cc. Find the lower and upper limits of the amount of medication to be given. 1.75 cc, 2.25 cc

27. The sum of two integers is twenty. Five times the smaller integer is two more than twice the larger integer. Find the integers. 6 and 14

28. Find three consecutive integers such that five times the middle integer is twice the sum of the other two integers. −1, 0, 1

29. A coin collection contains 30 coins in nickels, dimes, and quarters. There are three more dimes than nickels. The value of the coins is \$3.55. Find the number of quarters in the collection. 7 quarters

30. A silversmith combines 40 oz of pure silver that costs \$8.00 per ounce with 200 oz of a silver alloy that costs \$3.50 per ounce. Find the cost per ounce of the mixture. \$4.25/oz

31. A grocer mixed apple juice that costs \$3.20 per gallon with 40 gal of cranberry juice that costs \$5.50 per gallon. How much apple juice was used to make cranapple juice that costs \$4.20 per gallon? 52 gal

32. Two planes are 1680 mi apart and traveling toward each other. One plane is traveling 80 mph faster than the other plane. The planes meet in 1.75 h. Find the speed of each plane. 440 mph, 520 mph

33. Two investments earn an annual income of \$635. One investment is earning 10.5% annual simple interest, and the other investment is earning 6.4% annual simple interest. The total investment is \$8000. Find the amount invested in each account. \$3000 invested at 10.5%, \$5000 invested at 6.4%

34. An alloy containing 30% tin is mixed with an alloy containing 70% tin. How many pounds of each were used to make 500 lb of an alloy containing 40% tin? 375 lb of 30% tin, 125 lb of 70% tin

35. A sales executive earns \$800 per month plus a 4% commission on the amount of sales. The executive's goal is to earn at least \$3000 per month. What amount of sales will enable the executive to earn \$3000 or more per month? \$55,000 or more

36. An average score of 80 to 90 in a psychology class receives a B grade. A student has grades of 92, 66, 72, and 88 on four tests. Find the range of scores on the fifth test that will give the student a B for the course. $82 \le N \le 100$

Chapter Test

1. Solve: $x - 2 = -4$ $\quad -2$

2. Solve: $x + \frac{3}{4} = \frac{5}{8}$ $\quad -\frac{1}{8}$

3. Solve: $-\frac{3}{4}y = -\frac{5}{8}$ $\quad \frac{5}{6}$

4. Solve: $3x - 5 = 7$ $\quad 4$

5. Solve: $\frac{3}{4}y - 2 = 6$ $\quad \frac{32}{3}$

6. Solve: $2x - 3 - 5x = 8 + 2x - 10$ $\quad -\frac{1}{5}$

7. Solve: $2[x - (2 - 3x) - 4] = x - 5$ $\quad 1$

8. Solve: $\frac{2}{3}x - \frac{5}{6}x = 4$ $\quad -24$

9. Solve: $\frac{2x+1}{3} - \frac{3x+4}{6} = \frac{5x-9}{9}$ $\frac{12}{7}$

10. Solve: $2x - 5 \geq 5x + 4$
Write the solution set in interval notation.
$(-\infty, -3]$

11. Solve: $4 - 3(x + 2) < 2(2x + 3) - 1$
Write the solution set in interval notation.
$(-1, \infty)$

12. Solve: $3x - 2 > 4$ or $4 - 5x < 14$
Write the solution set in set-builder notation.
$\{x \mid x > -2\}$

13. Solve: $4 - 3x \geq 7$ and $2x + 3 \geq 7$
Write the solution set in set-builder notation.
$\varnothing$

14. Solve: $|3 - 5x| = 12$ $3, -\frac{9}{5}$

15. Solve: $2 - |2x - 5| = -7$ $7, -2$

16. Solve: $|3x - 1| \leq 2$ $\left\{x \mid -\frac{1}{3} \leq x \leq 1\right\}$

17. Solve: $|2x - 1| > 3$ $\{x \mid x > 2 \text{ or } x < -1\}$

18. Solve: $4 + |2x - 3| = 1$ No solution

19. Agency A rents cars for \$12 per day and 10¢ for every mile driven. Agency B rents cars for \$24 per day with unlimited mileage. How many miles per day can you drive an Agency A car if it is to cost you less than an Agency B car? Less than 120 mi

20. A doctor prescribed 3 cc of medication for a patient. The tolerance is 0.1 cc. Find the lower and upper limits of the amount of medication to be given. 2.9 cc, 3.1 cc

21. A stamp collection contains 11¢, 15¢, and 24¢ stamps. There are twice as many 11¢ stamps as 15¢ stamps. There are 30 stamps in all with a value of \$4.40. How many 24¢ stamps are in the collection? 6 stamps

22. A butcher combines 100 lb of hamburger that costs \$1.60 per pound with 60 lb of hamburger that costs \$3.20 per pound. Find the cost of the hamburger mixture. \$2.20/lb

23. A jogger runs a distance at a speed of 8 mph and returns the same distance running at a speed of 6 mph. Find the total distance that the jogger runs if the total time running is 1 h and 45 min. 12 mi

24. An investment of \$12,000 is deposited into two simple interest accounts. On one account, the annual simple interest rate is 7.8%. On the other, the annual simple interest rate is 9%. The total interest earned for one year is \$1020. How much was invested in each account? \$5000 at 7.8%, \$7000 at 9%

25. How many ounces of pure water must be added to 60 oz of an 8% salt solution to make a 3% salt solution? 100 oz

Cumulative Review Exercises

1. Simplify: $-2^2 \cdot 3^3$ -108

2. Simplify: $4 - (2 - 5)^2 \div 3 + 2$ 3

3. Simplify: $4 \div \frac{\frac{3}{8} - 1}{5} \cdot 2$ -64

4. Evaluate $2a^2 - (b - c)^2$ when $a = 2$, $b = 3$, and $c = -1$. -8

5. Identify the property that justifies the statement.

$(2x + 3y) + 2 = (3y + 2x) + 2$

The Commutative Property of Addition

6. Find $A \cap B$ given $A = \{3, 5, 7, 9\}$ and $B = \{3, 6, 9\}$. $\{3, 9\}$

7. Simplify: $3x - 2[x - 3(2 - 3x) + 5]$ $-17x + 2$

8. Simplify: $5[y - 2(3 - 2y) + 6]$ $25y$

9. Solve: $4 - 3x = -2$ 2

10. Solve: $-\frac{5}{6}b = -\frac{5}{12}$ $\frac{1}{2}$

11. Solve: $2x + 5 = 5x + 2$ 1

12. Solve: $\frac{5}{12}x - 3 = 7$ 24

13. Solve: $2[3 - 2(3 - 2x)] = 2(3 + x)$ 2

14. Solve: $3[2x - 3(4 - x)] = 2(1 - 2x)$ 2

15. Solve: $\frac{3x - 1}{4} - \frac{4x - 1}{12} = \frac{3 + 5x}{8}$ $-\frac{13}{5}$

16. Solve: $3x - 2 \geq 6x + 7$
Write the solution set in set-builder notation.
$\{x \mid x \leq -3\}$

17. Solve: $5 - 2x \geq 6$ and $3x + 2 \geq 5$
Write the solution set in set-builder notation.
$\varnothing$

18. Solve: $4x - 1 > 5$ or $2 - 3x < 8$
Write the solution set in set-builder notation.
$\{x \mid x > -2\}$

19. Solve: $|3 - 2x| = 5$ $-1, 4$

20. Solve: $3 - |2x - 3| = -8$ -4 and 7

21. Solve: $|3x - 5| \leq 4$ $\left\{x \middle| \frac{1}{3} \leq x \leq 3\right\}$

22. Solve: $|4x - 3| > 5$ $\left\{x \middle| x > 2 \text{ or } x < -\frac{1}{2}\right\}$

23. Graph: $\{x \mid x \geq -2\}$

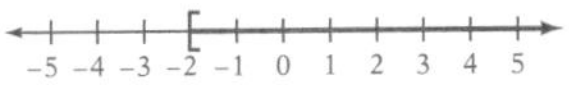

24. Graph: $\{x \mid x \geq 1\} \cup \{x \mid x < -2\}$

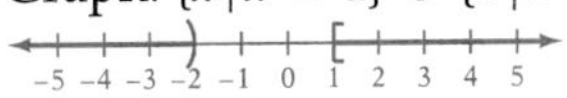

25. Translate and simplify "the sum of three times a number and six added to the product of three and the number." $(3n + 6) + 3n$; $6n + 6$

26. Three times the sum of the first and third of three consecutive odd integers is fifteen more than the second integer. Find the first integer. 1

27. A stamp collection consists of 9¢ and 11¢ stamps. The number of 9¢ stamps is five less than twice the number of 11¢ stamps. The total value of the stamps is $1.87. Find the number of 9¢ stamps. 11 stamps

28. Tickets for a school play sold for $2.25 for each adult and $.75 for each child. The total receipts for 75 tickets were $128.25. Find the number of adult tickets sold. 48 adult tickets

29. Two planes are 1400 mi apart and traveling toward each other. One plane is traveling 120 mph faster than the other plane. The planes meet in 2.5 h. Find the speed of the faster plane. 340 mph

30. How many liters of a 12% acid solution must be mixed with 4 L of a 5% acid solution to make an 8% acid solution? 3 L

31. An investment advisor invested $10,000 in two accounts. One investment earned 9.8% annual simple interest, and the other investment earned 12.8% annual simple interest. The amount of interest earned in one year was $1085. How much was invested in the 9.8% account? $6500

3

Linear Functions and Inequalities in Two Variables

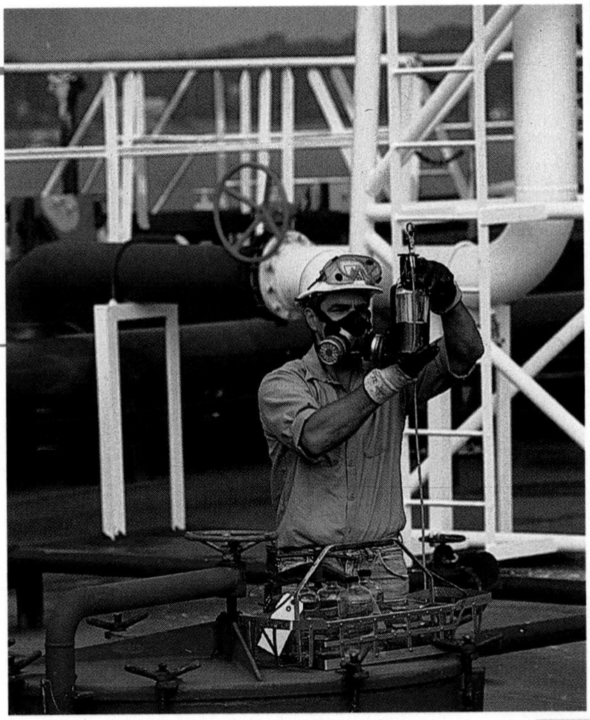

Petroleum technicians measure and record physical and geologic conditions in oil or gas wells from instruments lowered into wells. They also perform a chemical analysis of the mud from wells to determine petroleum and mineral content. By measuring the rate at which oil or gas is being pumped from a well, the technician can estimate the useful life of the well. Rates are one of the topics of this chapter.

OBJECTIVES

Brachistochrone Problem

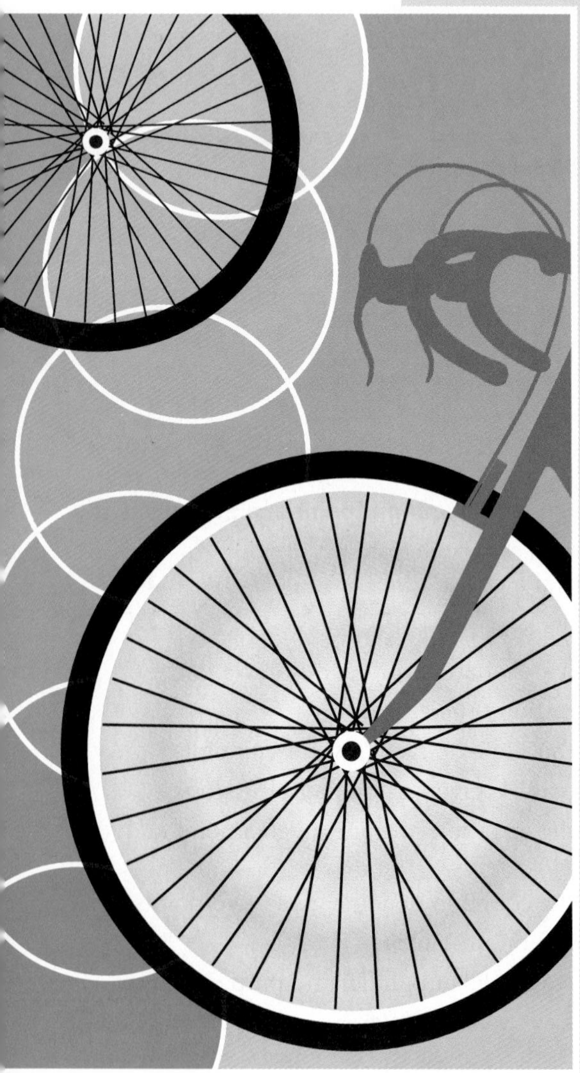

Consider the diagram at the right. What curve should be drawn so that a ball allowed to roll along the curve will travel from A to B in the shortest time?

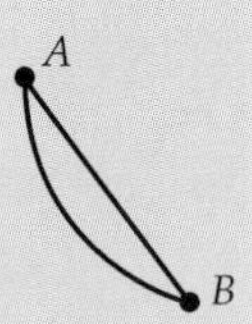

At first thought, one might conjecture that a straight line should connect the two points, because that shape is the shortest *distance* between the two points. Actually, however, the answer is half of one arch of an inverted cycloid.

A cycloid is shown below as the graph in bold. One way to draw this curve is to think of a wheel rolling along a straight line without slipping. Then a point on the rim of the wheel traces a cycloid.

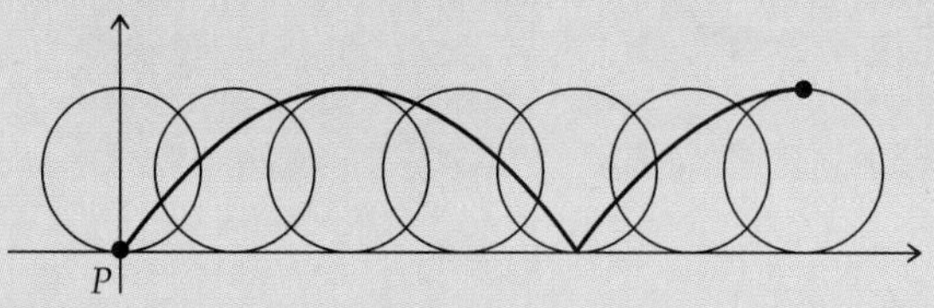

A cycloid

There are many applications of the idea of finding the shortest time between two points. As the above problem illustrates, the path of shortest time is not necessarily the path of shortest distance. Problems involving paths of shortest time are called *brachistochrone* problems.

SECTION 3.1 The Rectangular Coordinate System

1 Points on a rectangular coordinate system

POINT OF INTEREST

A rectangular coordinate system is also called a **Cartesian** coordinate system in honor of Descartes.

Before the 15th century, geometry and algebra were considered separate branches of mathematics. That changed when René Descartes, a French mathematician who lived from 1596 to 1650, founded **analytic geometry.** In this geometry, a *coordinate system* is used to study relationships between variables.

INSTRUCTOR NOTE

Although Descartes is normally given credit for the coordinate system, Pierre de Fermat played a prominent role in its development. Fermat's discussion of these ideas was never formally published but rather was circulated as manuscript.

The coordinate system as we think of it today did not appear in any of Descartes's writings. The essence of his contribution was an oblique coordinate system similar to the following.

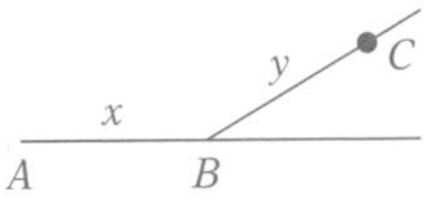

The significance was that all other quantities were related to the line segments *AB* and *BC*.

All Descartes's work was in the first quadrant. Negative numbers were still not universally accepted.

A **rectangular coordinate system** is formed by two number lines, one horizontal and one vertical, that intersect at the zero point of each line. The point of intersection is called the **origin.** The two lines are called **coordinate axes,** or simply **axes.**

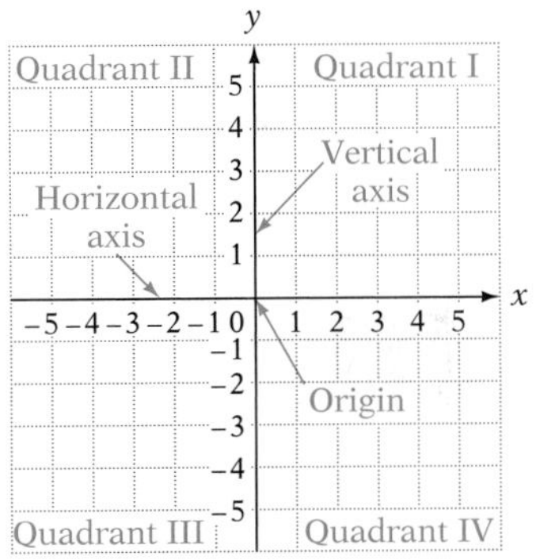

The axes determine a **plane,** which can be thought of as a large, flat sheet of paper. The two axes divide the plane into four regions called **quadrants.** The quadrants are numbered counterclockwise from I to IV.

Each point in the plane can be identified by a pair of numbers called an **ordered pair.** The first number of the pair measures a horizontal distance and is called the **abscissa.** The second number of the pair measures a vertical distance and is called the **ordinate.** The **coordinates** of the point are the numbers in the ordered pair associated with the point. The abscissa is also called the **first coordinate,** or ***x*-coordinate,** of the ordered pair, and the ordinate is also called the **second coordinate,** or ***y*-coordinate,** of the ordered pair.

Horizontal distance → Vertical distance

Ordered pair → (2, 3)

x-coordinate → *y*-coordinate

POINT OF INTEREST

Gottfried Leibnitz introduced the words *abscissa* and *ordinate. Abscissa* is from Latin meaning "to cut off." Originally, Leibnitz used *abscissa linea,* "cut off a line." The root of *ordinate* is a Latin word used to suggest the sense of order.

When drawing a rectangular coordinate system, we often label the horizontal axis x and the vertical axis y. In this case, the coordinate system is called an ***xy*-coordinate system.** To graph or plot a point in the xy-coordinate system, place a dot at the location given by the ordered pair. The **graph of an ordered pair** is the dot drawn at the coordinates of the point in the xy-coordinate system. The points whose coordinates are (3, 4) and (−2.5, −3) are graphed in the figure at the right.

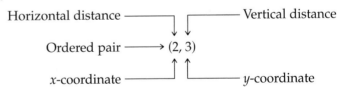

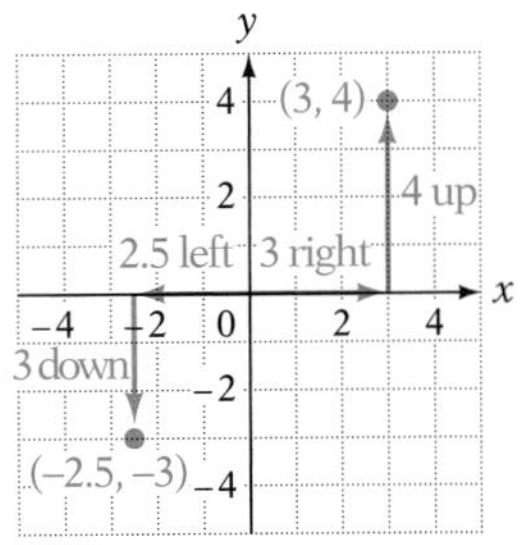

The points whose coordinates are $(3, -1)$ and $(-1, 3)$ are graphed at the right. Note that the graphs are in different locations. The *order* of the coordinates of an ordered pair is important.

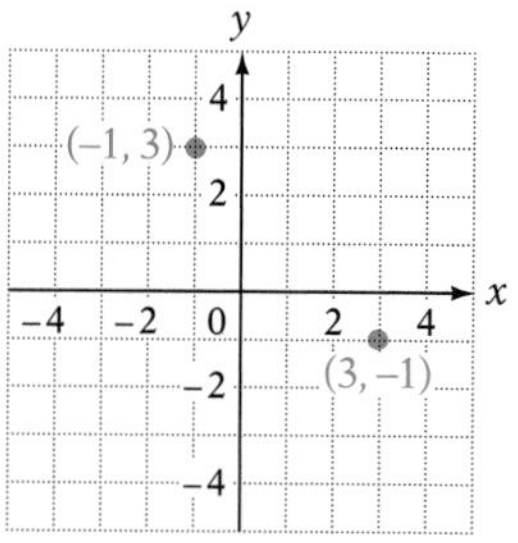

The *xy*-coordinate system is used to graph equations in *two variables*. Examples of equations in two variables are shown at the right.

$$y = 3x + 7$$
$$y = x^2 - 4x + 3$$
$$x^2 + y^2 = 25$$
$$x = \frac{y}{y^2 + 4}$$

A **solution of an equation in two variables** is an ordered pair (x, y) whose coordinates make the equation a true statement.

▶ Is the ordered pair $(-3, 7)$ a solution of the equation $y = -2x + 1$?

Replace x by -3 and replace y by 7. Then simplify.

$$\begin{array}{c|c} \multicolumn{2}{l}{y = -2x + 1} \\ \hline 7 & -2(-3) + 1 \\ 7 & 6 + 1 \\ \multicolumn{2}{l}{7 = 7} \end{array}$$

Compare the results. If the resulting equation is a true statement, the ordered pair is a solution of the equation. If it is not a true statement, the ordered pair is not a solution of the equation.

Yes, the ordered pair $(-3, 7)$ is a solution of the equation. ◀

Besides the ordered pair $(-3, 7)$, there are many other ordered-pair solutions of the equation $y = -2x + 1$. For example, $(-5, 11)$, $(0, 1)$, $\left(-\frac{3}{2}, 4\right)$, and $(4, -7)$ are also solutions of the equation.

In general, an equation in two variables has an infinite number of solutions. By choosing any value of x and substituting that value into the equation, we can calculate a corresponding value of y. The resulting ordered-pair solution (x, y) of the equation can be graphed in a rectangular coordinate system.

▶ Graph the solutions (x, y) of $y = x^2 - 1$ when x equals $-2, -1, 0, 1,$ and 2.

Substitute each value of x into the equation and solve for y. Then graph the resulting ordered pairs. It is convenient to record the ordered-pair solutions in a table similar to the one shown below.

x	$y = x^2 - 1$	y	(x, y)
-2	$y = (-2)^2 - 1$	3	$(-2, 3)$
-1	$y = (-1)^2 - 1$	0	$(-1, 0)$
0	$y = 0^2 - 1$	-1	$(0, -1)$
1	$y = 1^2 - 1$	0	$(1, 0)$
2	$y = 2^2 - 1$	3	$(2, 3)$

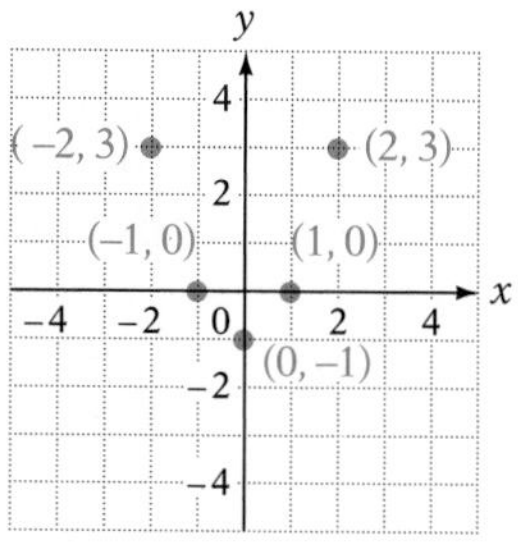

Example 1 Determine the ordered-pair solution of $y = \frac{x}{x-2}$ that corresponds to $x = 4$.

Solution

$y = \frac{x}{x-2}$

$y = \frac{4}{4-2}$ ▶ Replace x by 4 and solve for y.

$y = 2$

The ordered-pair solution is (4, 2).

Problem 1 Determine the ordered-pair solution of $y = \frac{3x}{x+1}$ that corresponds to $x = -2$.

Solution See page S6. (−2, 6)

Example 2 Graph the ordered-pair solutions of $y = x^2 - x$ when $x = -1, 0, 1$, and 2.

Solution

x	y
−1	2
0	0
1	0
2	2

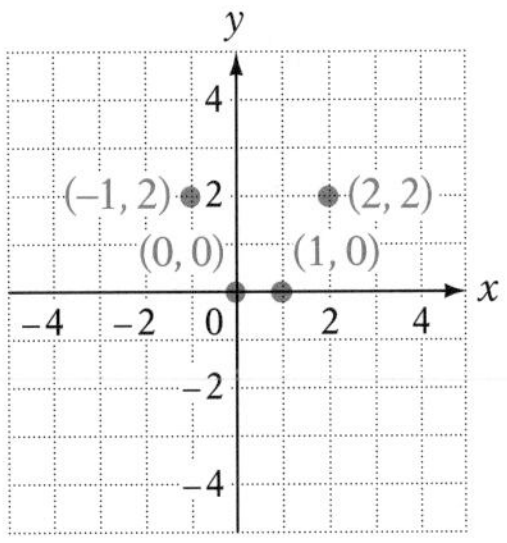

Problem 2 Graph the ordered-pair solutions of $y = |x + 1|$ when $x = -3, -2, -1, 0$, and 1.

Solution See page S6.

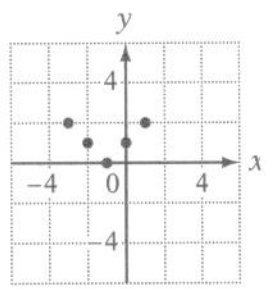

2 Find the length and midpoint of a line segment

POINT OF INTEREST

Although Pythagoras (c. 550 B.C.) is given credit for this theorem, a clay tablet called Plimpton 322 shows that the Babylonians knew the theorem 1000 years before Pythagoras!

The distance between two points in an xy-coordinate system can be calculated by using the Pythagorean Theorem.

Pythagorean Theorem

If a and b are the lengths of the legs of a right triangle and c is the length of the hypotenuse, then $a^2 + b^2 = c^2$.

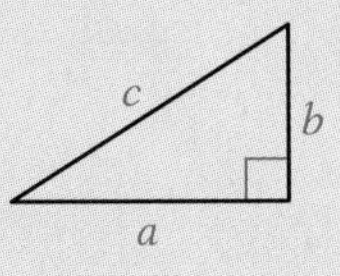

Consider the two points and the right triangle shown at the right. The vertical distance between $P_1(x_1, y_1)$ and $P_2(x_2, y_2)$ is $|y_2 - y_1|$. The horizontal distance between the points $P_1(x_1, y_1)$ and $P_2(x_2, y_2)$ is $|x_2 - x_1|$.

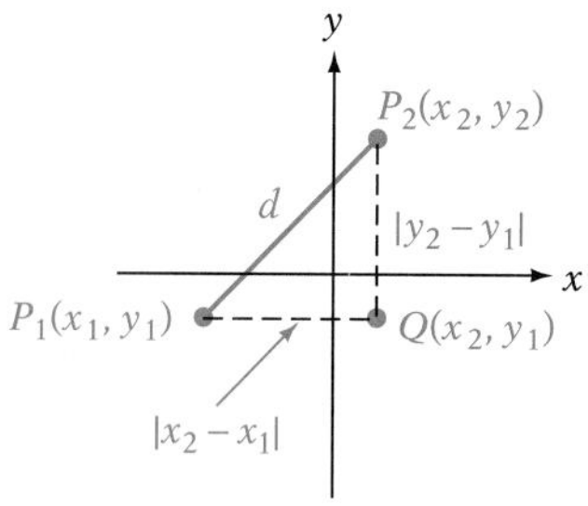

The quantity d^2 is calculated by applying the Pythagorean Theorem to the right triangle. $d^2 = |x_2 - x_1|^2 + |y_2 - y_1|^2$

Because the square of a number is always nonnegative, the absolute value signs are not necessary. $d^2 = (x_2 - x_1)^2 + (y_2 - y_1)^2$

The distance d is the square root of d^2. $d = \sqrt{(x_2 - x_1)^2 + (y_2 - y_1)^2}$

The Distance Formula

If $P_1(x_1, y_1)$ and $P_2(x_2, y_2)$ are two points in the plane, then the distance d between the two points is given by

$$d = \sqrt{(x_2 - x_1)^2 + (y_2 - y_1)^2}$$

Example 3 Find the exact distance between the points whose coordinates are (−3, 2) and (4, −1).

Solution

$$\begin{aligned} d &= \sqrt{(x_2 - x_1)^2 + (y_2 - y_1)^2} && \text{▶ Use the distance formula.} \\ &= \sqrt{[4 - (-3)]^2 + (-1 - 2)^2} && \text{▶ } (x_1, y_1) = (-3, 2) \text{ and } (x_2, y_2) = (4, -1). \\ &= \sqrt{7^2 + (-3)^2} = \sqrt{49 + 9} \\ &= \sqrt{58} \end{aligned}$$

The distance between the points is $\sqrt{58}$.

Problem 3 Find the exact distance between the points whose coordinates are (5, −2) and (−4, 3).

Solution See page S6. $\sqrt{106}$

INSTRUCTOR NOTE

Before you begin the discussion of the midpoint formula, review with students the following property of similar triangles.

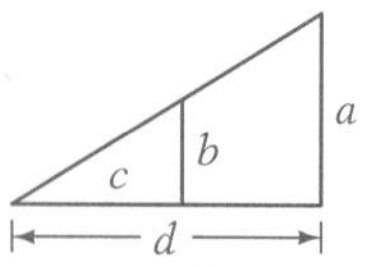

$$\frac{a}{d} = \frac{b}{c}$$

The midpoint of a line segment is equidistant from its endpoints. The coordinates of the midpoint of the line segment P_1P_2 are (x_m, y_m). The intersection of the horizontal line segment through P_1 and the vertical line segment through P_2 is Q, with coordinates (x_2, y_1).

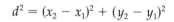

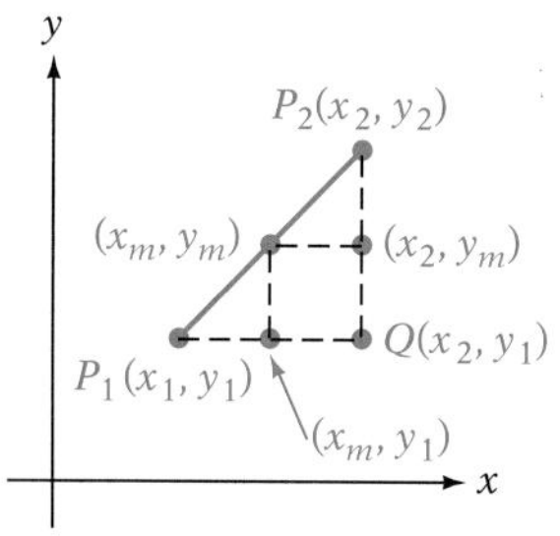

The x-coordinate x_m of the midpoint of the line segment P_1P_2 is the same as the x-coordinate of the midpoint of the line segment P_1Q. It is the average of the x-coordinates of the points P_1 and P_2.

$$x_m = \frac{x_1 + x_2}{2}$$

Similarly, the y-coordinate y_m of the midpoint of the line segment P_1P_2 is the same as the y-coordinate of the midpoint of the line segment P_2Q. It is the average of the y-coordinates of the points P_1 and P_2.

$$y_m = \frac{y_1 + y_2}{2}$$

The Midpoint Formula

If $P_1(x_1, y_1)$ and $P_2(x_2, y_2)$ are the endpoints of a line segment, then the coordinates of the midpoint (x_m, y_m) of the line segment are given by

$$x_m = \frac{x_1 + x_2}{2} \quad \text{and} \quad y_m = \frac{y_1 + y_2}{2}$$

Example 4 Find the coordinates of the midpoint of the line segment with endpoints $(-5, 4)$ and $(-3, 7)$.

Solution

$x_m = \frac{x_1 + x_2}{2}$ $\qquad$ $y_m = \frac{y_1 + y_2}{2}$ $\qquad$ ▶ Use the midpoint formula.

$= \frac{-5 + (-3)}{2}$ $\qquad$ $= \frac{4 + 7}{2}$ $\qquad$ ▶ $(x_1, y_1) = (-5, 4)$ and $(x_2, y_2) = (-3, 7)$.

$= -4$ $\qquad$ $= \frac{11}{2}$

The midpoint is $\left(-4, \frac{11}{2}\right)$.

Problem 4 Find the coordinates of the midpoint of the line segment with endpoints $(-3, -5)$ and $(-2, 3)$.

Solution See page S6. $\left(-\frac{5}{2}, -1\right)$

3 Graph a scatter diagram

Discovering a relationship between two variables is an important task in the study of mathematics. These relationships occur in many forms and in a wide variety of applications. Here are some examples.

- A botanist wants to know the relationship between the number of bushels of wheat yielded per acre and the amount of watering per acre.
- An environmental scientist wants to know the relationship between the incidence of skin cancer and the amount of ozone in the atmosphere.
- A business analyst wants to know the relationship between the price of a product and the number of products that are sold at that price.

A researcher may investigate the relationship between two variables by means of *regression analysis,* which is a branch of statistics. The study of the relationship between two variables may begin with a **scatter diagram,** which is a graph of the ordered pairs of the known data.

The following table shows data collected by a university registrar comparing the grade-point average (GPAs) of graduating high school seniors and their scores on a national test.

GPA, x	3.25	3.00	3.00	3.50	3.50	2.75	2.50	2.50	2.00	2.00	1.50
Score, y	1200	1200	1000	1500	1100	1000	1000	900	800	900	700

The scatter diagram for these data is shown at the right. Each ordered pair represents the GPA and test score for a student. For example, the ordered pair (2.75, 1000) indicates that one student with a GPA of 2.75 had a test score of 1000.

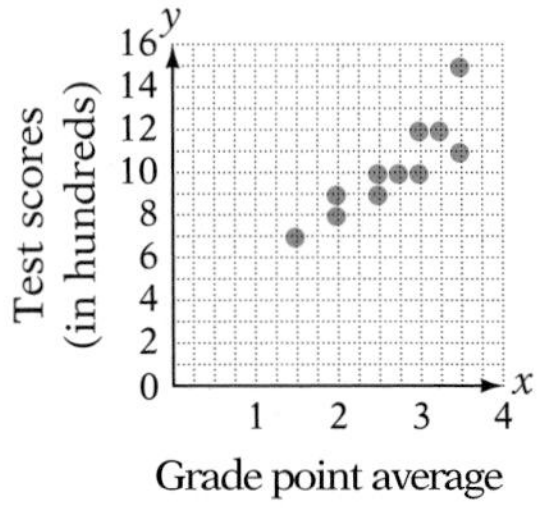

Example 5 As the sales of high-capacity external disk drives have increased, the sales of floppy disks have declined. The scatter diagram below shows the sales of floppy disks, in billions, between 1994 and 1998.

a. How many floppy disks were sold in 1995?

b. In which year did floppy disk sales first fall below 4 billion floppy disks?

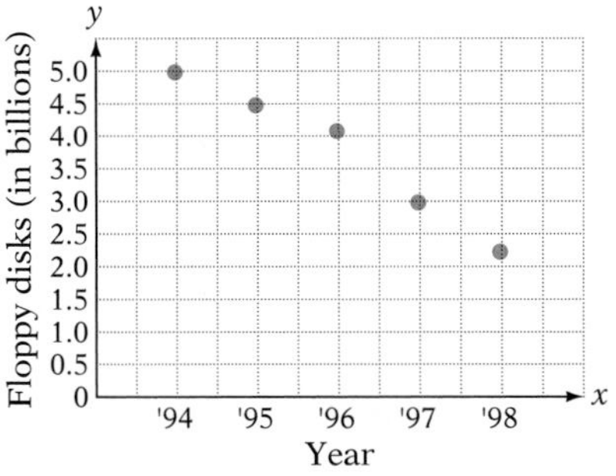

Source: San Diego Union-Tribune 8/25/98

Strategy **a.** Read the graph. Find 1995 on the horizontal axis. Follow that mark up vertically to determine the number of floppy disks sold that year.

b. Find the horizontal line that represents 4 billion. Determine the year for which the number of floppy disks sold is below that line.

Solution **a.** There were approximately 4.5 billion floppy disks sold in 1995.

b. In 1997 the sales of floppy disks first fell below 4 billion floppy disks.

Problem 5 High-capacity external disk drives, such as a Zip drive, have become increasingly popular. The table below shows the number of these disk drives, in millions, that have been shipped for the years 1995 through 2001. Make a scatter diagram of these data. (*Source: San Diego Union-Tribune,* 25 August 1998)

Year	1995	1996	1997	1998	1999*	2000*	2001*
External drives (in millions)	1	4	8	15	21	27	33

*Projected shipments

Solution See page S6.

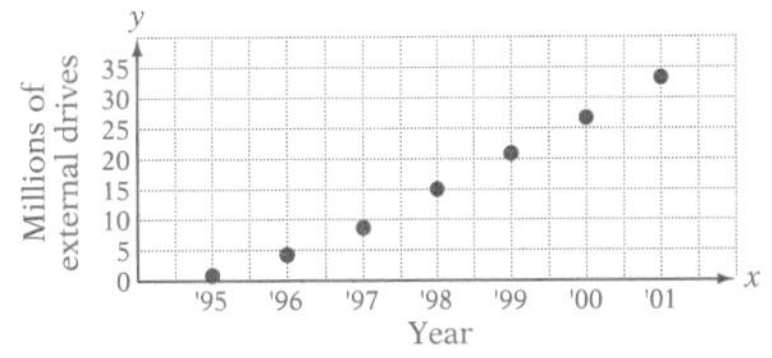

STUDY TIPS

GET HELP FOR ACADEMIC DIFFICULTIES

If you do have trouble in this course, teachers, counselors, and advisers can help. They usually know of study groups, tutors, or other sources of help that are available. They may suggest visiting an office of academic skills, a learning center, a tutorial service, or some other department or service on campus.

Students who have already taken the course and who have done well in it may be a source of assistance. If they have a good understanding of the material, they may be able to help by explaining it to you.

CONCEPT REVIEW 3.1

Determine whether the following statements are always true, sometimes true, or never true.

1. An equation in two variables has an infinite number of solutions.
   Always true
2. A rectangular coordinate system has two real number axes. Always true
3. In the graph of an ordered pair, the first number is the y-coordinate and the second number is the x-coordinate. Never true
4. The solution of an equation with two variables is an ordered pair.
   Always true
5. The point $(-2, -4)$ is in the fourth quadrant. Never true
6. In quadrant III, the y-coordinate of a point is negative. Always true

EXERCISES 3.1

1

1. Describe a rectangular coordinate system.

2. Discuss the sign of a and the sign of b as they pertain to the quadrant in which the ordered pair (a, b) is graphed.

3. Graph the ordered pairs $(0, -1)$, $(2, 0)$, $(3, 2)$, and $(-1, 4)$.

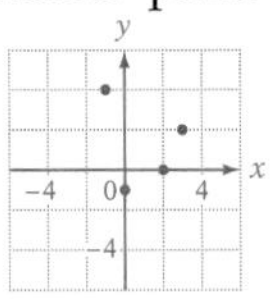

4. Graph the ordered pairs $(-1, -3)$, $(0, -4)$, $(0, 4)$, and $(3, -2)$.

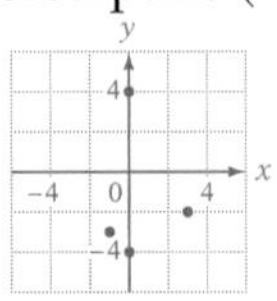

5. Find the coordinates of each point.

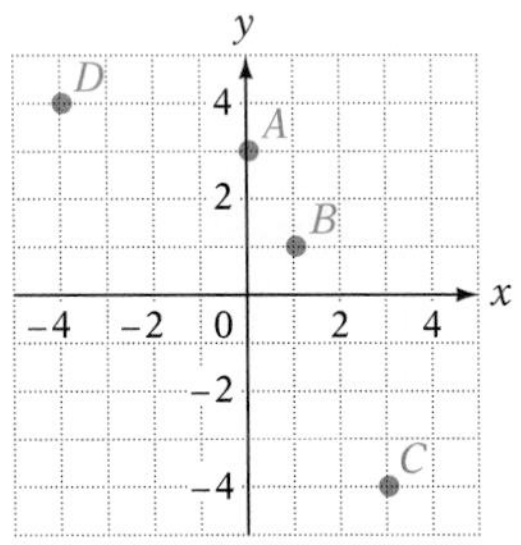

$A(0, 3)$
$B(1, 1)$
$C(3, -4)$
$D(-4, 4)$

6. Find the coordinates of each point.

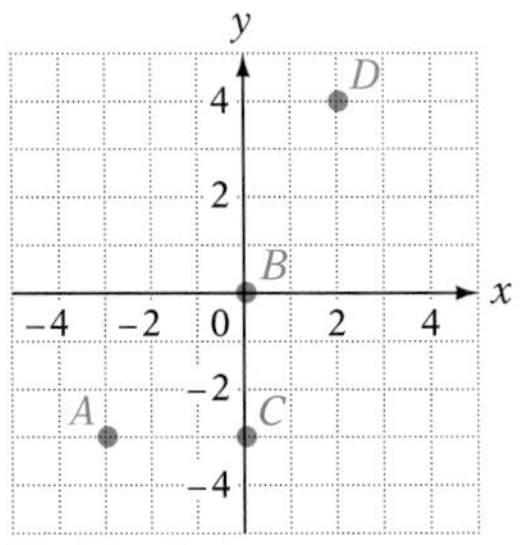

$A(-3, -3)$
$B(0, 0)$
$C(0, -3)$
$D(2, 4)$

7. Draw a line through all points with an abscissa of 2.

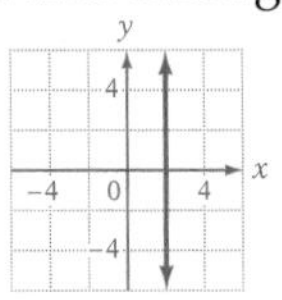

8. Draw a line through all points with an abscissa of -3.

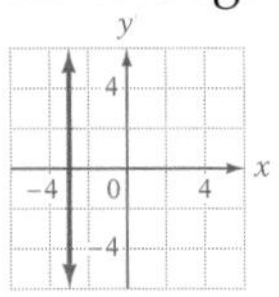

9. Draw a line through all points with an ordinate of -3.

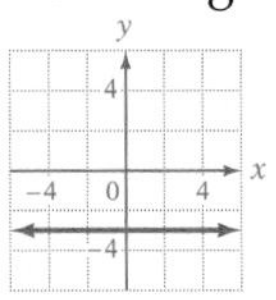

10. Draw a line through all points with an ordinate of 4.

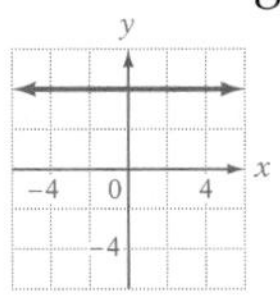

11. Graph the ordered-pair solutions of $y = x^2$ when $x = -2, -1, 0, 1$, and 2.

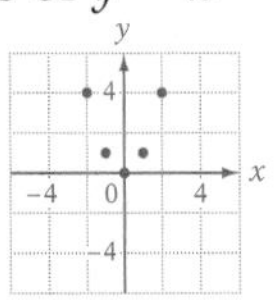

12. Graph the ordered-pair solutions of $y = -x^2 + 1$ when $x = -2, -1, 0, 1$, and 2.

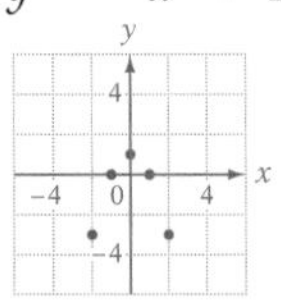

13. Graph the ordered-pair solutions of $y = |x + 1|$ when $x = -5, -3, 0, 3$, and 5.

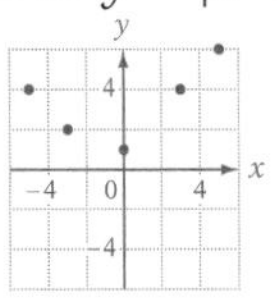

14. Graph the ordered-pair solutions of $y = -2|x|$ when $x = -3, -1, 0, 1$, and 3.

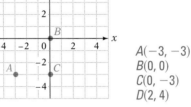

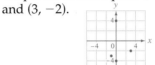

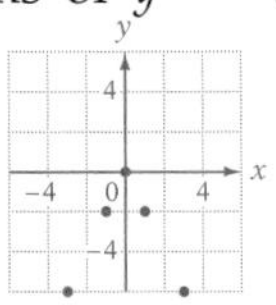

15. Graph the ordered-pair solutions of $y = -x^2 + 2$ when $x = -2, -1, 0, 1,$ and 2.

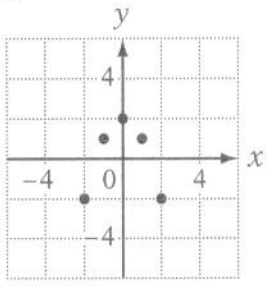

16. Graph the ordered-pair solutions of $y = -x^2 + 4$ when $x = -3, -1, 0, 1,$ and 3.

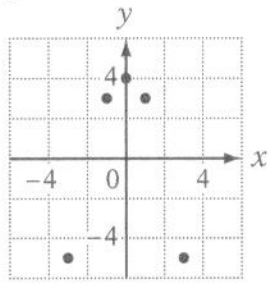

17. Graph the ordered-pair solutions of $y = x^3 - 2$ when $x = -1, 0, 1,$ and 2.

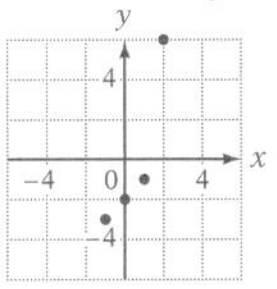

18. Graph the ordered-pair solutions of $y = -x^3 + 1$ when $x = -1, 0, 1,$ and $\frac{3}{2}$.

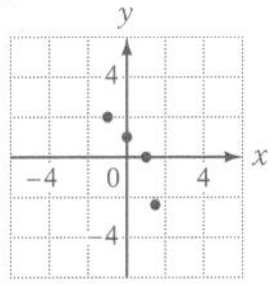

2 19. Explain how to find the distance between two points in the plane.

20. What is the midpoint of a line segment and how is it calculated?

Find the length and midpoint of the line segment between the given points.

21. $P_1(3, 5)$ and $P_2(4, 1)$
$\sqrt{17}; \left(\frac{7}{2}, 3\right)$

22. $P_1(-2, 3)$ and $P_2(5, -1)$
$\sqrt{65}; \left(\frac{3}{2}, 1\right)$

23. $P_1(0, 3)$ and $P_2(-2, 4)$
$\sqrt{5}; \left(-1, \frac{7}{2}\right)$

24. $P_1(6, -1)$ and $P_2(-3, -2)$
$\sqrt{82}; \left(\frac{3}{2}, -\frac{3}{2}\right)$

25. $P_1(-3, -5)$ and $P_2(2, -4)$
$\sqrt{26}; \left(-\frac{1}{2}, -\frac{9}{2}\right)$

26. $P_1(-7, -5)$ and $P_2(-2, -1)$
$\sqrt{41}; \left(-\frac{9}{2}, -3\right)$

27. $P_1(5, -2)$ and $P_2(-1, 5)$
$\sqrt{85}; \left(2, \frac{3}{2}\right)$

28. $P_1(3, -5)$ and $P_2(6, 0)$
$\sqrt{34}; \left(\frac{9}{2}, -\frac{5}{2}\right)$

29. $P_1(5, -5)$ and $P_2(2, -5)$
$3; \left(\frac{7}{2}, -5\right)$

3 Solve.

30. *Chemistry* The amount of a substance that can be dissolved in a fixed amount of water usually increases as the temperature of the water increases. Cerium selenate, however, does not behave in this manner. The graph at the right shows the number of grams of cerium selenate that will dissolve in 100 mg of water for various temperatures, in degrees Celsius.

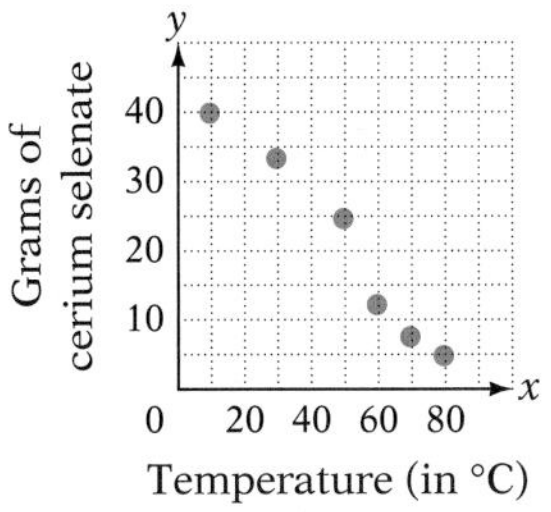

a. Determine the temperature at which 25 g of cerium selenate will dissolve. 50°C

b. Determine the number of grams of cerium selenate that will dissolve when the temperature is 80°C. 5 g

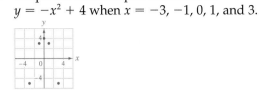

31. *Nutrition* Sodium content in foods is a measure of the salt content in the food. The scatter diagram at the right shows the approximate amount of sodium and the number of calories in various sandwiches offered at a McDonald's restaurant. (*Source:* Gourmet Connection website/Convence Labs)

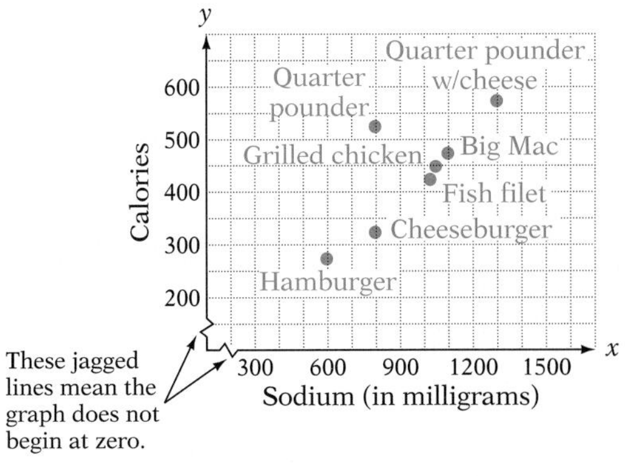

a. How many calories are in a hamburger?

b. How many milligrams of sodium are in a Big Mac?

a. 275 calories b. 1100 mg

32. *Automotive Technology* When water freezes, it expands. This expansion is a potential problem for cars, because the freezing water can cause the engine block of a car to crack. Ethylene glycol is mixed with water to produce an antifreeze that helps prevent the water in the car from freezing. The scatter diagram at the right shows the changes in the freezing point of a solution of water and ethylene glycol that occur as the percent of ethylene glycol is increased.

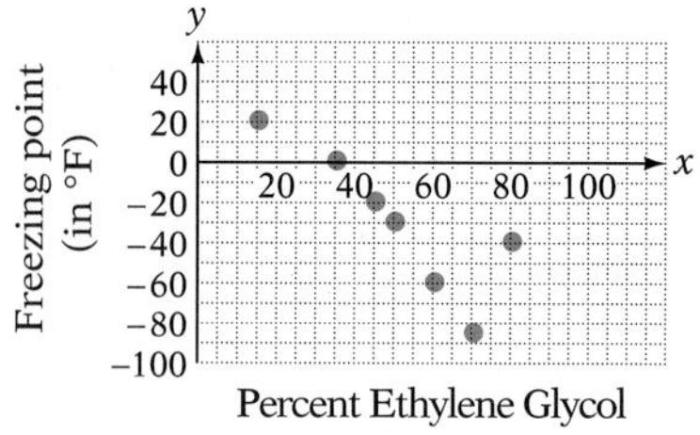

a. At what temperature will a 45% mixture of ethylene glycol and water freeze?

b. What percent of ethylene glycol is required to protect the car to temperatures as low as −30°F?

a. −20°F b. 50%

33. *Telecommunications* The number of people using cell phones has soared over the past few years. The scatter diagram at the right shows the number of cell phones in use for a 6-year period. (*Source: Wall Street Journal Almanac,* 1998)

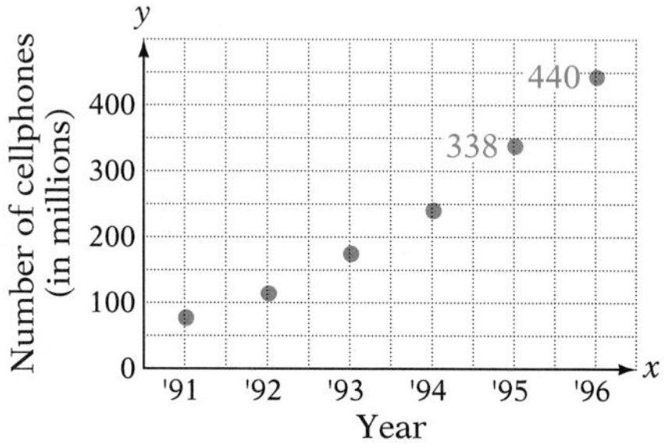

a. In what year did the number of cell phones in use first exceed 200 million?

b. If the number of people using cell phones increased between 1996 and 1997 by the same percent increase as between 1995 and 1996, how many people were using cell phones in 1997? Round to the nearest million.

a. 1994 b. 573 million people

34. *Marketing* As shown in the table below, the number of manufacturers' coupons redeemed by consumers has begun to decline, causing companies to rethink how best to promote their products. Draw a scatter diagram for these data. (*Source:* HCH Promotional Services)

Year, x	1985	1988	1990	1992	1994	1996
Coupons redeemed (in billions), y	6.5	7.1	7.1	7.7	6.2	5.3

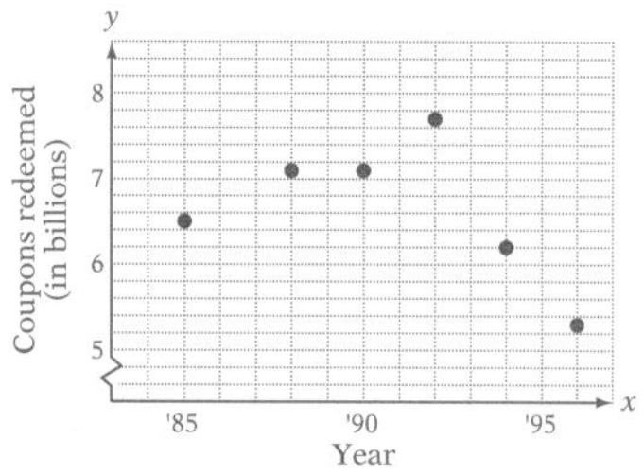

35. *Telecommunications* As shown in the table below, the percent of households with telephones has increased over the years. Draw a scatter diagram for these data. (*Source:* U.S. Census Bureau/Federal Communications Commission)

Year, x	'20	'30	'40	'50	'60	'70	'80	'90	'00*
Percent, y	35	41	37	62	78	90	93	95	96

*Projected

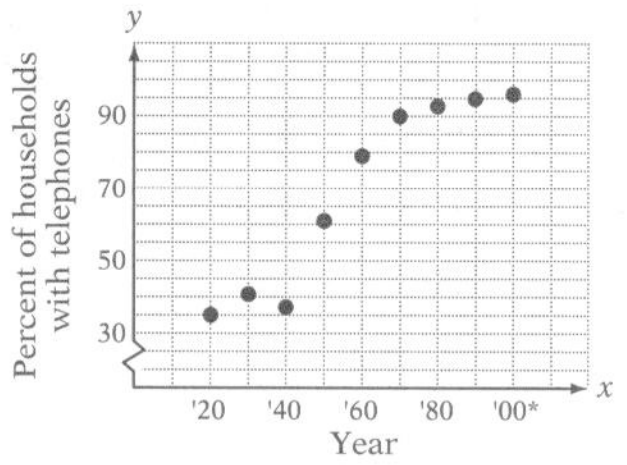

*Projected

APPLYING CONCEPTS 3.1

36. Graph the ordered pairs $\left(x, \frac{1}{x}\right)$, where $x \in \left\{-2, -1, -\frac{1}{2}, -\frac{1}{3}, \frac{1}{3}, \frac{1}{2}, 1, 2\right\}$.

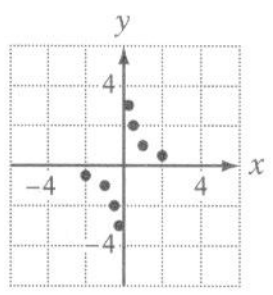

37. Graph the ordered pairs (x, x^2), where $x \in \{-2, -1, 0, 1, 2\}$.

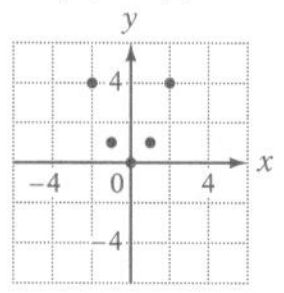

38. Draw a line passing through every point whose ordinate is the additive inverse of its abscissa.

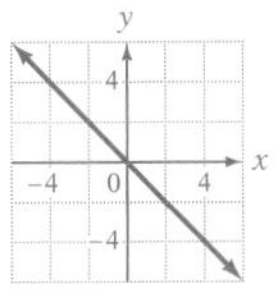

39. Draw a line passing through every point whose abscissa equals its ordinate.

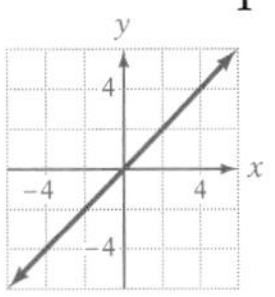

40. Describe the graph of all the ordered pairs (x, y) that are 5 units from the origin.

41. Consider two distinct fixed points in the plane. Describe the graph of all the points (x, y) that are equidistant from these fixed points.

SECTION 3.2 Introduction to Functions

1 Evaluate a function

In mathematics and its application, there are many times when it is necessary to investigate a relationship between two quantities. Here is a financial application: Consider a person who is planning to finance the purchase of a car. If the current interest rate for a five-year loan is 9%, the equation that describes the relationship between the amount that is borrowed B and the monthly payment P is $P = 0.020758B$.

INSTRUCTOR NOTE

The examples on this page show various functions. As a classroom activity, have students suggest relationships between two quantities. Record these on the board. Then, when you have finished defining the term *function,* return to the suggestions and determine which are functions. Some possible suggestions:

- The distance a rock falls in a given time
- The points scored by a football team and whether the team won or lost

For each amount B the purchaser may borrow, there is a certain monthly payment P. The relationship between the amount borrowed and the payment can be recorded as ordered pairs, where the first coordinate is the amount borrowed and the second coordinate is the monthly payment. Some of these ordered pairs are shown at the right.

$0.020758B = P$

(5000, 103.79)
(6000, 124.55)
(7000, 145.31)
(8000, 166.07)

A relationship between two quantities is not always given by an equation. The table at the right describes a grading scale that defines a relationship between a score on a test and a letter grade. For each score, the table assigns only one letter grade. The ordered pair (84, B) indicates that a score of 84 receives a letter grade of B.

Score	Grade
90–100	A
80–89	B
70–79	C
60–69	D
0–59	F

The graph at the right also shows a relationship between two quantities. It is a graph of the viscosity V of SAE 40 motor oil at various temperatures T. Ordered pairs can be approximated from the graph. The ordered pair (120, 250) indicates that the viscosity of the oil at 120°F is 250 units.

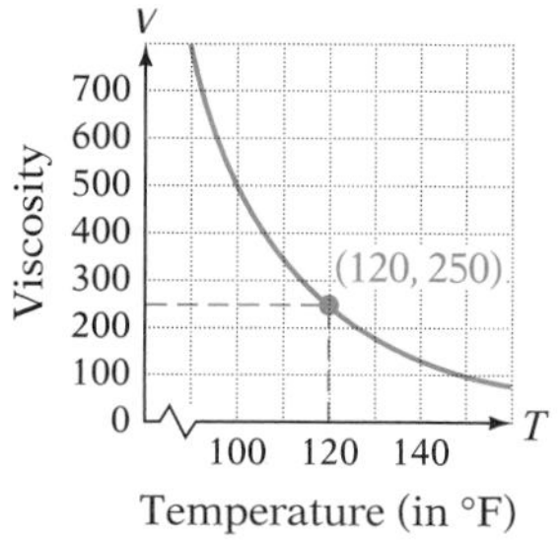

In each of these examples, there is a rule (an equation, a table, or a graph) that determines a certain set of ordered pairs.

Definition of Relation

A **relation** is a set of ordered pairs.

Here are some of the ordered pairs for the relations just described.

Relation	Some of the Ordered Pairs of the Relation
Car payment	(2500, 51.90), (3750, 77.85), (4396, 91.26)
Grading scale	(78, C), (98, A), (70, C), (81, B), (94, A)
Oil viscosity	(100, 500), (120, 250), (130, 175), (150, 100)

These three relations actually represent a special type of relation called a function. Functions play an important role in mathematics and its application.

Definition of Function

A **function** is a relation in which no two ordered pairs have the same first coordinate and different second coordinates.

The **domain** of a function is the set of the first coordinates of all the ordered pairs of the function. The **range** is the set of the second coordinates of all the ordered pairs of the function.

For the function defined by the ordered pairs

$$\{(2, 3), (4, 5), (6, 7), (8, 9)\}$$

the domain is {2, 4, 6, 8} and the range is {3, 5, 7, 9}.

Example 1   Find the domain and range of the function {(2, 3), (4, 6), (6, 8), (10, 6)}.

Solution   The domain is {2, 4, 6, 10}.   ▶ The domain of the function is the set of the first components in the ordered pairs.

The range is {3, 6, 8}.   ▶ The range of the function is the set of the second components in the ordered pairs.

> **Problem 1** Find the domain and range of the function
> $\{(-1, 5), (3, 5), (4, 5), (6, 5)\}$.
>
> **Solution** See page S6.   D: $\{-1, 3, 4, 6\}$; R: $\{5\}$

For each element of the domain of a function there is a corresponding element in the range of the function. A possible diagram for the function of Example 1 is

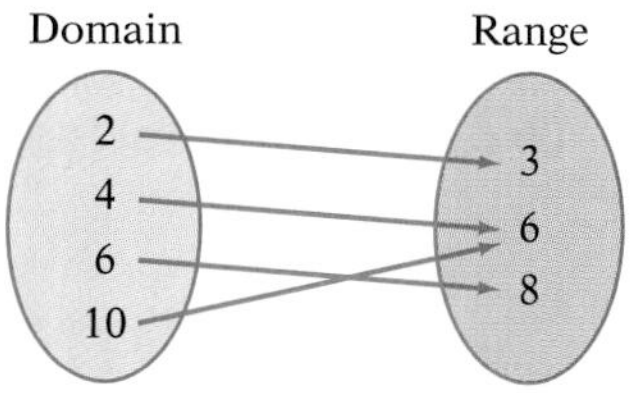

$\{(2, 3), (4, 6), (6, 8), (10, 6)\}$

Functions defined by tables or graphs, such as those described at the beginning of this section, have important applications. However, a major focus of this text is functions defined by equations in two variables.

The "square" function, which pairs each real number with its square, can be defined by the equation

$$y = x^2$$

This equation states that for a given value of x in the domain, the value of y in the range is the square of x. For instance, if $x = 6$, then $y = 36$ and if $x = -7$, then $y = 49$. Because the value of y *depends* on the value of x, y is called the **dependent variable** and x is called the **independent variable.**

A pictorial representation of this function is shown below for selected elements from the domain of the function. The function acts as a machine that turns a number from the domain into the square of the number.

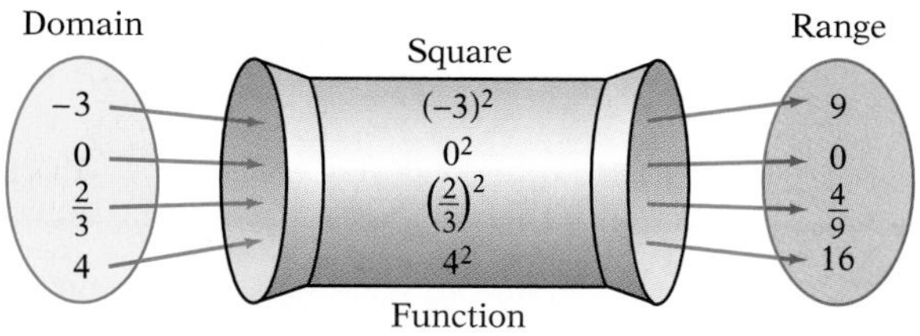

The ordered pairs for the values shown above are $(-3, 9)$, $(0, 0)$, $\left(\frac{2}{3}, \frac{4}{9}\right)$, and $(4, 16)$.

For this function, the second coordinate is the square of the first coordinate. If we let x represent the first coordinate, then the second coordinate is x^2 and we have the ordered pair (x, x^2).

A function cannot have two ordered pairs with *different* second coordinates and the same first coordinate. However, a function may contain ordered pairs with the *same* second coordinate. For instance, the square function given above has the ordered pairs $(-3, 9)$ and $(3, 9)$: the same second coordinate but different first coordinates.

Not every equation in two variables defines a function. For instance, consider

$$y^2 = x^2 + 9$$

This equation does not define a function. Because

$$5^2 = 4^2 + 9 \qquad \text{and} \qquad (-5)^2 = 4^2 + 9$$

the ordered pairs (4, 5) and (4, −5) are both solutions of the equation. Consequently, there are two ordered pairs that have the same first coordinate, 4, but *different* second coordinates, 5 and −5. Therefore, the equation does not define a function. Other ordered pairs for this equation are (0, −3), (0, 3), ($\sqrt{7}$, −4), and ($\sqrt{7}$, 4). A graphical representation of these ordered pairs is shown below.

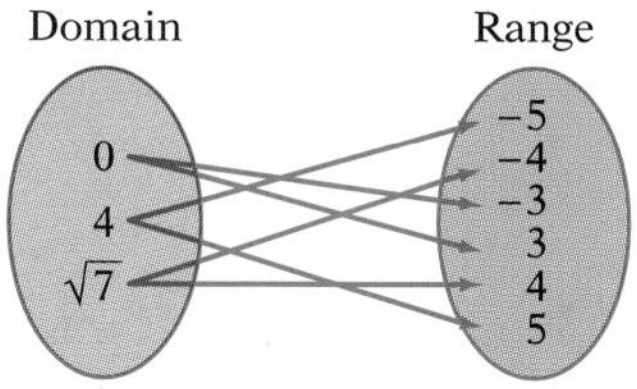

Note from this graphical representation that an element from the domain has two arrows to different elements in the range. Any time this occurs, the situation does not represent a function. However, this diagram does represent a relation. The relation for the values shown is {(0, −3), (0, 3), (4, −5), (4, 5), ($\sqrt{7}$, −4), ($\sqrt{7}$, 4)}.

The phrase "y is a function of x," or a similar phrase with different variables, is used to describe those equations in two variables that define functions. To emphasize that the equation represents a function, **functional notation** is used.

Just as x is commonly used as a variable to represent a number, the letter f is commonly used to name a function. The square function is written in functional notation as follows:

LOOK CLOSELY

The dependent variable y and $f(x)$ can be used interchangeably.

This is the value of the function.
It is the number that is paired with x.

$$f(x) = x^2$$

The name of the function is f.

This is an algebraic expression that defines the relationship between the dependent and independent variables.

The symbol $f(x)$ is read "the *value* of f at x" or "f of x."

It is important to note that $f(x)$ does not mean f times x. The symbol $f(x)$ is the **value of the function** and represents the value of the dependent variable for a given value of the independent variable. We often write $y = f(x)$ to emphasize the relationship between the independent variable, x, and the dependent variable, y. Remember that y and $f(x)$ are different symbols for the same number.

The letters used to represent a function are somewhat arbitrary. All of the following equations represent the same function.

$$\left.\begin{aligned} f(x) &= x^2 \\ s(t) &= t^2 \\ P(v) &= v^2 \end{aligned}\right\} \text{ These equations represent the square function.}$$

The process of determining $f(x)$ for a given value of x is called **evaluating the function.** For instance, to evaluate $f(x) = x^2$ when $x = 4$, replace x by 4 and simplify.

$$f(x) = x^2$$
$$f(4) = 4^2 = 16$$

The *value* of the function is 16 when $x = 4$. An ordered pair of the function is (4, 16).

INSTRUCTOR NOTE

One way of helping students evaluate functions is by writing the equation of the function with open parentheses.

$$g(t) = 3(t)^2 - 5(t) + 1$$
$$g(\ \ ) = 3(\ \ )^2 - 5(\ \ ) + 1$$

To evaluate the function, fill each of the parentheses with the same number. Then simplify.

Evaluate $g(t) = 3t^2 - 5t + 1$ when $t = -2$.

Replace t by -2 and then simplify.

$$\begin{aligned} g(t) &= 3t^2 - 5t + 1 \\ g(-2) &= 3(-2)^2 - 5(-2) + 1 \\ &= 3(4) - 5(-2) + 1 \\ &= 12 + 10 + 1 = 23 \end{aligned}$$

When t is -2, the value of the function is 23.
Therefore an ordered pair of the function is $(-2, 23)$.

It is possible to evaluate a function for a variable expression.

Evaluate $P(z) = 3z - 7$ when $z = 3 + h$.

Replace z by $3 + h$ and then simplify.

$$\begin{aligned} P(z) &= 3z - 7 \\ P(3 + h) &= 3(3 + h) - 7 \\ &= 9 + 3h - 7 \\ &= 3h + 2 \end{aligned}$$

When z is $3 + h$, the value of the function is $3h + 2$.
Therefore, an ordered pair of the function is $(3 + h, 3h + 2)$.

Example 2 Given $s(t) = 3t^2 - 4t + 5$, find $s(-4)$.

Solution

$$\begin{aligned} s(t) &= 3t^2 - 4t + 5 \\ s(-4) &= 3(-4)^2 - 4(-4) + 5 && \blacktriangleright \text{Replace } t \text{ by } -4. \\ &= 3(16) - 4(-4) + 5 && \blacktriangleright \text{Use the Order of Operations Agreement to simplify the numerical expression.} \\ &= 48 + 16 + 5 \\ &= 69 \end{aligned}$$

Problem 2 Evaluate $G(x) = \dfrac{2x}{x + 4}$ when $x = -2$.

Solution See page S6. -2

There are several ways to use a graphing calculator to evaluate a function. The screens at the right show one way that will work on many calculators. See the Graphing Calculator Appendix for other methods of evaluating a function.

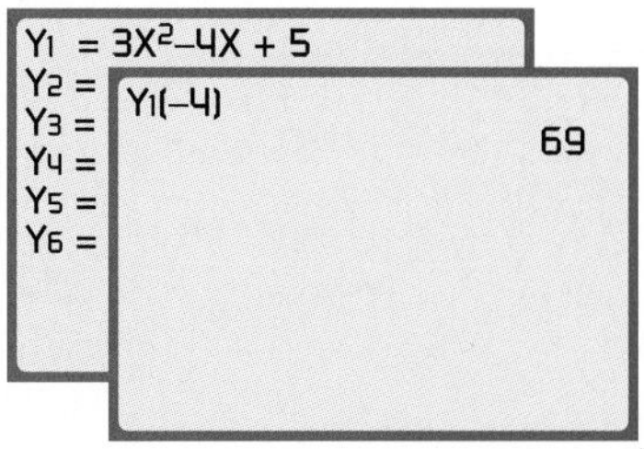

The range of a function contains all the elements that result from applying the function to each element of the domain. If the domain contains an infinite number of elements, then it may be difficult to find the range. However, if the domain has only a finite number of elements, then the range can be found by evaluating the function for each element in the domain.

Example 3 Find the range of $f(x) = x^3 + x$ if the domain is $\{-2, -1, 0, 1, 2\}$.

Solution

$$f(x) = x^3 + x$$
$$f(-2) = (-2)^3 + (-2) = -10$$
$$f(-1) = (-1)^3 + (-1) = -2$$
$$f(0) = 0^3 + 0 = 0$$
$$f(1) = 1^3 + 1 = 2$$
$$f(2) = 2^3 + 2 = 10$$

▶ Replace x by each member of the domain. The range includes the values of $f(-2)$, $f(-1)$, $f(0)$, $f(1)$, and $f(2)$.

The range is $\{-10, -2, 0, 2, 10\}$.

Problem 3 Find the range of $f(x) = x^2 - x - 2$ if the domain is $\{-3, -2, -1, 0, 1, 2\}$.

Solution See page S6. $\{-2, 0, 4, 10\}$

The TABLE feature of a calculator can be used to determine range values of a function for selected values in the domain.

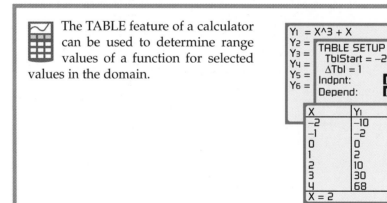

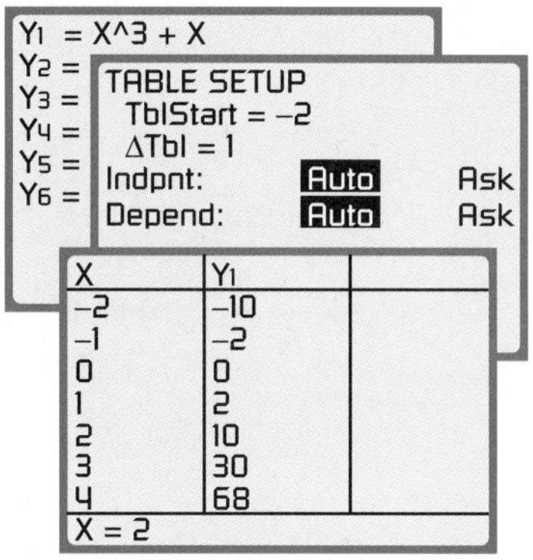

When a function is represented by an equation, the domain of the function is all real numbers for which the value of the function is a real number. For instance:

- The domain of $f(x) = x^2$ is all real numbers, because the square of every real number is a real number.
- The domain of $g(x) = \frac{1}{x - 2}$ is all real numbers except 2, because when $x = 2$, $g(2) = \frac{1}{2 - 2} = \frac{1}{0}$, which is not a real number.

The domain of the grading-scale function is the set of whole numbers from 0 to 100. In set-builder notation, this is written $\{x \mid 0 \le x \le 100, x \in \text{whole numbers}\}$. The range is {A, B, C, D, F}.

Score	Grade
90–100	A
80–89	B
70–79	C
60–69	D
0–59	F

Example 4 What values, if any, are excluded from the domain of $f(x) = 2x^2 - 7x + 1$?

Solution Because the value of $2x^2 - 7x + 1$ is a real number for any value of x, the domain of the function is all real numbers. No values are excluded from the domain of $f(x) = 2x^2 - 7x + 1$.

Problem 4 What values, if any, are excluded from the domain of $f(x) = \frac{2}{x-5}$?

Solution See page S6. 5

STUDY TIPS

REVIEWING MATERIAL

Reviewing material is the repetition that is essential for learning. Much of what we learn is soon forgotten unless we review it. If you find that you do not remember information that you studied previously, you probably have not reviewed it sufficiently. *You will remember best what you review most.*

One method of reviewing material is to begin a study session by reviewing a concept you have studied previously. For example, before trying to solve a new type of problem, spend a few minutes solving a kind of problem you already know how to solve. Not only will you provide yourself with the review practice you need, but you are also likely to put yourself in the right frame of mind for learning how to solve the new type of problem.

CONCEPT REVIEW 3.2

Determine whether the following statements are always true, sometimes true, or never true.

1. A set of ordered pairs is called a function. Sometimes true
2. A function is a relation. Always true
3. The domain of a function is the set of real numbers. Sometimes true
4. Zero is excluded from the domain of a function. Sometimes true
5. $f(a)$ represents a value in the range of a function when a is a value in the domain of the function. Always true
6. Two ordered pairs of a function can have the same first coordinate. Never true

EXERCISES 3.2

1 **1.** In your own words, explain what a function is.

2. What is the domain of a function? What is the range of a function?

3. Does the diagram below represent a function? Explain your answer. Yes

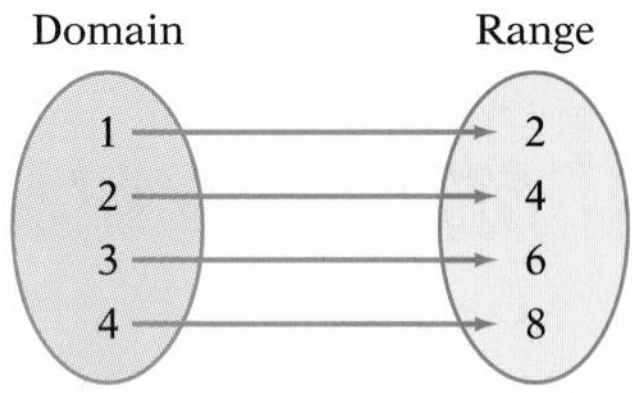

4. Does the diagram below represent a function? Explain your answer. Yes

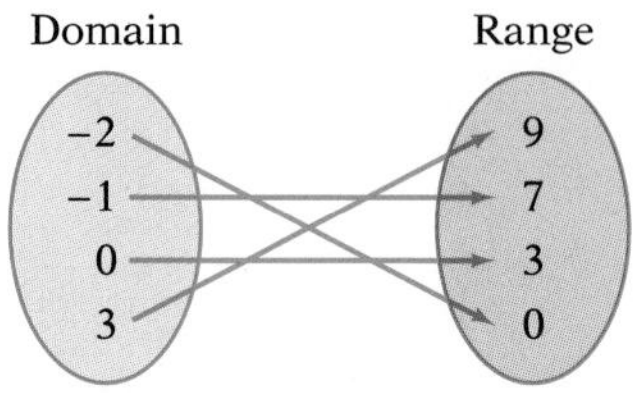

5. Does the diagram below represent a function? Explain your answer. Yes

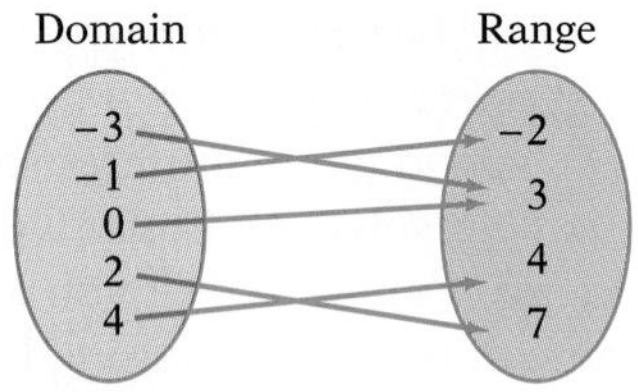

6. Does the diagram below represent a function? Explain your answer. Yes

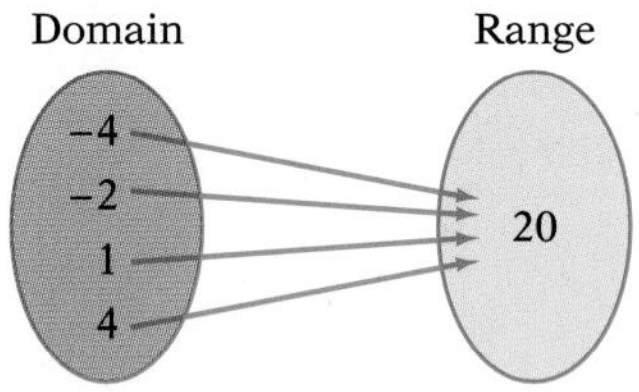

7. Does the diagram below represent a function? Explain your answer. No

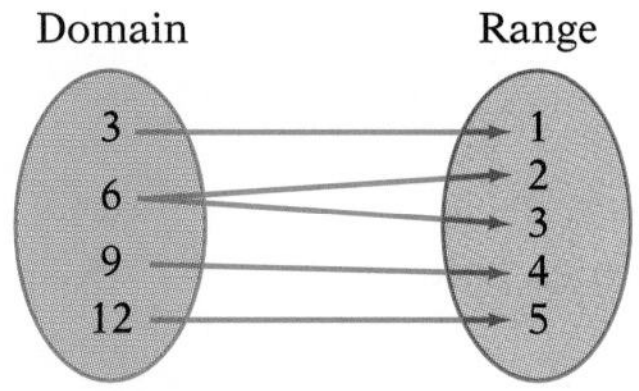

8. Does the diagram below represent a function? Explain your answer. No

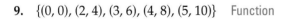

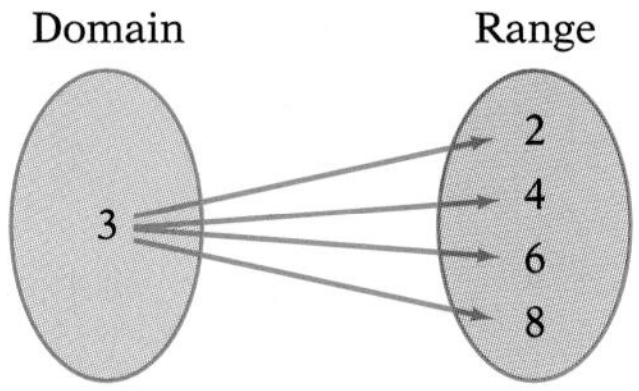

State whether each of the following relations is a function.

9. $\{(0, 0), (2, 4), (3, 6), (4, 8), (5, 10)\}$ Function

10. $\{(1, 3), (3, 5), (5, 7), (7, 9)\}$ Function

11. $\{(-2, -1), (-4, -5), (0, -1), (3, 5)\}$ Function

12. $\{(-3, -1), (-1, -1), (0, 1), (2, 6)\}$ Function

13. $\{(-2, 3), (-1, 3), (0, -3), (1, 3), (2, 3)\}$ Function

14. $\{(0, 0), (1, 0), (2, 0), (3, 0), (4, 0)\}$ Function

15. $\{(1, 1), (4, 2), (9, 3), (1, -1), (4, -2)\}$ Not a function

16. $\{(3, 1), (3, 2), (3, 3), (3, 4)\}$ Not a function

Solve.

17. *Commerce* The table below shows the cost in 1998 to send an overnight package using United Parcel Service.

Weight (in pounds)	Cost
$0 < x \leq 1$	\$21.00
$1 < x \leq 2$	\$23.50
$2 < x \leq 3$	\$25.75
$3 < x \leq 4$	\$28.00
$4 < x \leq 5$	\$30.25

Does this table define a function? If not, explain why it is not a function. Yes

18. *Commerce* The table below shows the cost in 1998 to send an overnight package using the U.S. Postal Service.

Weight (in pounds)	Cost
$0 < x < 2$	\$15.00
$2 \leq x < 3$	\$17.25
$3 \leq x < 4$	\$19.40
$4 \leq x < 5$	\$21.55

Does this table define a function? If not, explain why it is not a function. Yes

19. What does it mean to evaluate a function? Explain how to evaluate $f(x) = 3x$ when $x = 2$.

20. What is the value of a function?

Given $f(x) = 5x - 4$, evaluate:

21. $f(3)$ 11

22. $f(-2)$ −14

23. $f(0)$ −4

24. $f(-1)$ −9

Given $G(t) = 4 - 3t$, evaluate:

25. $G(0)$ 4

26. $G(-3)$ 13

27. $G(-2)$ 10

28. $G(4)$ −8

Given $q(r) = r^2 - 4$, evaluate:

29. $q(3)$ 5

30. $q(4)$ 12

31. $q(-2)$ 0

32. $q(-5)$ 21

Given $F(x) = x^2 + 3x - 4$, evaluate:

33. $F(4)$ 24

34. $F(-4)$ 0

35. $F(-3)$ −4

36. $F(-6)$ 14

Given $H(p) = \frac{3p}{p+2}$, evaluate:

37. $H(1)$ 1

38. $H(-3)$ 9

39. $H(t)$ $\frac{3t}{t+2}$

40. $H(v)$ $\frac{3v}{v+2}$

Given $s(t) = t^3 - 3t + 4$, evaluate:

41. $s(-1)$ 6

42. $s(2)$ 6

43. $s(a)$ $a^3 - 3a + 4$

44. $s(w)$ $w^3 - 3w + 4$

45. *Accounting* The amount of state income tax, I, in Massachusetts for 1997 for individuals earning at least \$40,000 but less than \$40,250 is shown in the table below.

Income	Tax
$40{,}000 \le I < 40{,}050$	\$2381
$40{,}050 \le I < 40{,}100$	\$2384
$40{,}100 \le I < 40{,}150$	\$2387
$40{,}150 \le I < 40{,}200$	\$2390
$40{,}200 \le I < 40{,}250$	\$2393

Find the tax for an income (I) of:

a. \$40,135 **b.** \$40,249

(*Source:* Massachusetts state income tax forms)

a. \$2387 b. \$2393

46. *Insurance* The semiannual cost, C, for comprehensive auto insurance depends on the value, V, of the car. The costs of car insurance in Massachusetts for 1998, for cars valued between \$11,200 and \$15,000, are shown in the table below.

Car Value	Cost of Insurance
$11{,}200 \le V < 12{,}500$	\$649
$12{,}500 \le V < 13{,}000$	\$659
$13{,}000 \le V < 13{,}500$	\$669
$13{,}500 \le V < 14{,}000$	\$679
$14{,}000 \le V < 14{,}500$	\$689
$14{,}500 \le V < 15{,}000$	\$699

Find the cost of insurance if the value of a car (V) is:

a. \$13,227 **b.** \$14,500

(*Source:* Registry of Motor Vehicles of Massachusetts)

a. \$669 b. \$699

47. *Real Estate* A real estate appraiser charges a fee, F, that depends on the estimated value, V, of the property. A table of the fees charged for various estimated values of the real estate is given below.

Value of Property (in dollars)	Appraisal Fee
$V < 100{,}000$	\$350
$100{,}000 \le V < 500{,}000$	\$525
$500{,}000 \le V < 1{,}000{,}000$	\$950
$1{,}000{,}000 \le V < 5{,}000{,}000$	\$2500
$5{,}000{,}000 \le V < 10{,}000{,}000$	\$3000

Evaluate this function when:

a. $V = \$5{,}000{,}000$ **b.** $V = \$767{,}000$

\$3000 \$950

48. *Commerce* The cost, C, to mail a priority overnight package by Federal Express depends on the weight, w, of the package. A table of the costs for selected weights is given below.

Weight (in pounds)	Cost
$0 < w < 1$	\$22.75
$1 \le w < 2$	\$25.50
$2 \le w < 3$	\$28.00
$3 \le w < 4$	\$30.50
$4 \le w < 5$	\$32.50

Evaluate this function when:

a. $w = 2$ lb 3 oz **b.** $w = 1.9$ lb

\$28.00 \$25.50

Find the domain and range of the function.

49. $\{(1, 1), (2, 4), (3, 7), (4, 10), (5, 13)\}$
D: $\{1, 2, 3, 4, 5\}$; R: $\{1, 4, 7, 10, 13\}$

50. $\{(2, 6), (4, 18), (6, 38), (8, 66), (10, 102)\}$
D: $\{2, 4, 6, 8, 10\}$; R: $\{6, 18, 38, 66, 102\}$

51. $\{(0, 1), (2, 2), (4, 3), (6, 4)\}$
D: $\{0, 2, 4, 6\}$; R: $\{1, 2, 3, 4\}$

52. $\{(0, 1), (1, 2), (4, 3), (9, 4)\}$
D: $\{0, 1, 4, 9\}$; R: $\{1, 2, 3, 4\}$

53. $\{(1, 0), (3, 0), (5, 0), (7, 0), (9, 0)\}$
D: $\{1, 3, 5, 7, 9\}$; R: $\{0\}$

54. $\{(-2, -4), (2, 4), (-1, 1), (1, 1), (-3, 9), (3, 9)\}$
D: $\{-3, -2, -1, 1, 2, 3\}$; R: $\{-4, 1, 4, 9\}$

55. $\{(0,0), (1, 1), (-1, 1), (2, 2), (-2, 2)\}$
D: $\{-2, -1, 0, 1, 2\}$; R: $\{0, 1, 2\}$

56. $\{(0, -5), (5, 0), (10, 5), (15, 10)\}$
D: $\{0, 5, 10, 15\}$; R: $\{-5, 0, 5, 10\}$

What values, if any, are excluded from the domain of the function?

57. $f(x) = \frac{1}{x-1}$ 1

58. $g(x) = \frac{1}{x+4}$ −4

59. $h(x) = \frac{x+3}{x+8}$ −8

60. $F(x) = \frac{2x-5}{x-4}$ 4

61. $f(x) = 3x + 2$ None

62. $g(x) = 4 - 2x$ None

63. $G(x) = x^2 + 1$ None

64. $H(x) = \frac{1}{2}x^2$ None

65. $f(x) = \frac{x-1}{x}$ 0

66. $g(x) = \frac{2x+5}{7}$ None

67. $H(x) = x^2 - x + 1$ None

68. $f(x) = 3x^2 + x + 4$ None

69. $f(x) = \frac{2x-5}{3}$ None

70. $g(x) = \frac{3-5x}{5}$ None

71. $H(x) = \frac{x-2}{x+2}$ −2

72. $h(x) = \frac{3-x}{6-x}$ 6

73. $f(x) = \frac{x-2}{2}$ None

74. $G(x) = \frac{2}{x-2}$ 2

Find the range of the function defined by each equation.

75. $f(x) = 4x - 3$; domain = $\{0, 1, 2, 3\}$
R: $\{-3, 1, 5, 9\}$

76. $G(x) = 3 - 5x$; domain = $\{-2, -1, 0, 1, 2\}$
R: $\{-7, -2, 3, 8, 13\}$

77. $g(x) = 5x - 8$; domain = $\{-3, -1, 0, 1, 3\}$
R: $\{-23, -13, -8, -3, 7\}$

78. $h(x) = 3x - 7$; domain = $\{-4, -2, 0, 2, 4\}$
R: $\{-19, -13, -7, -1, 5\}$

79. $h(x) = x^2$; domain = $\{-2, -1, 0, 1, 2\}$
R: $\{0, 1, 4\}$

80. $H(x) = 1 - x^2$; domain = $\{-2, -1, 0, 1, 2\}$
R: $\{-3, 0, 1\}$

81. $f(x) = 2x^2 - 2x + 2$; domain = $\{-4, -2, 0, 4\}$
R: $\{2, 14, 26, 42\}$

82. $G(x) = -2x^2 + 5x - 2$;
domain = $\{-3, -1, 0, 1, 3\}$
R: $\{-35, -9, -5, -2, 1\}$

83. $H(x) = \frac{5}{1-x}$; domain = $\{-2, 0, 2\}$
R: $\left\{-5, \frac{5}{3}, 5\right\}$

84. $g(x) = \frac{4}{4-x}$; domain = $\{-5, 0, 3\}$
R: $\left\{\frac{4}{9}, 1, 4\right\}$

85. $f(x) = \frac{2}{x-4}$; domain = $\{-2, 0, 2, 6\}$
R: $\left\{-1, -\frac{1}{2}, -\frac{1}{3}, 1\right\}$

86. $g(x) = \frac{x}{3-x}$; domain = $\{-2, -1, 0, 1, 2\}$
R: $\left\{-\frac{2}{5}, -\frac{1}{4}, 0, \frac{1}{2}, 2\right\}$

87. $H(x) = 2 - 3x - x^2$; domain = $\{-5, 0, 5\}$
R: $\{-38, -8, 2\}$

88. $G(x) = 4 - 3x - x^3$; domain = $\{-3, 0, 3\}$
R: $\{-32, 4, 40\}$

APPLYING CONCEPTS 3.2

89. Given $P(x) = 4x + 7$, write $P(-2 + h) - P(-2)$ in simplest form. $4h$

90. Given $G(t) = 9 - 2t$, write $G(-3 + h) - G(-3)$ in simplest form. $-2h$

91. Given $f(z) = 8 - 3z$, write $f(-3 + h) - f(-3)$ in simplest form. $-3h$

92. Given $M(n) = 5 - 2n$, write $M(-6 + h) - M(-6)$ in simplest form. $-2h$

93. *Automotive Technology* The distance, s (in feet), a car will skid on a certain road surface after the brakes are applied is a function of the car's velocity, v (in miles per hour). The function can be approximated by $s = f(v) = 0.017v^2$. How far will a car skid after its brakes are applied if it is traveling 60 mph? 61.2 ft

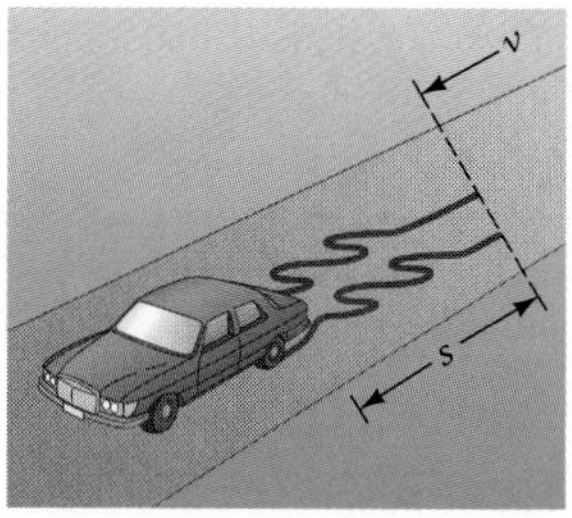

94. *Energy* The power a windmill can generate is a function of the velocity of the wind. The function can be approximated by $P = f(v) = 0.015v^3$, where P is the power in watts and v is the velocity of the wind in meters per second. How much power will be produced by a windmill when the velocity of the wind is 15 m/s? 50.625 watts

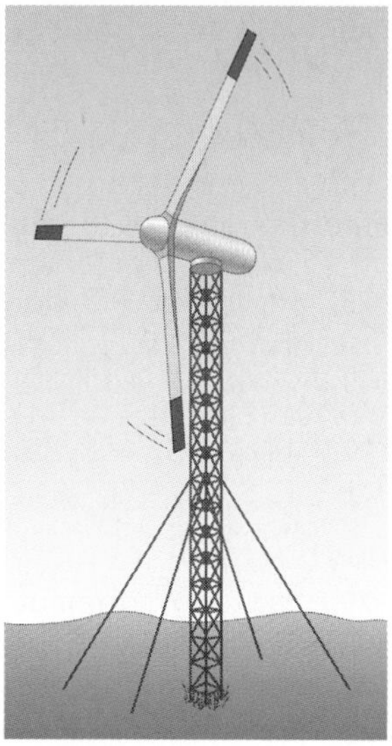

Each of the following graphs defines a function. Evaluate the function by estimating the ordinate (which is the value of the function) for the given value of t.

95. *Physics* The graph in Figure 1 shows the speed, v (in feet per second), at which a parachutist is falling during the first 20 s after jumping out of a plane. Estimate the speed at which the parachutist is falling when:

a. $t = 5$ s 20 ft/s **b.** $t = 15$ s 28 ft/s

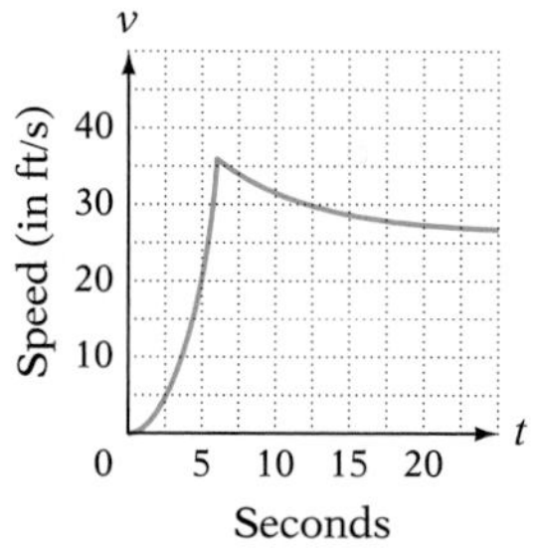

Figure 1

96. *Psychology* The graph in Figure 2 shows what an industrial psychologist has determined to be the average percent score, P, for an employee taking a performance test t weeks after training begins. Estimate the score an employee would receive on this test when:

a. $t = 4$ weeks 90% **b.** $t = 10$ weeks 100%

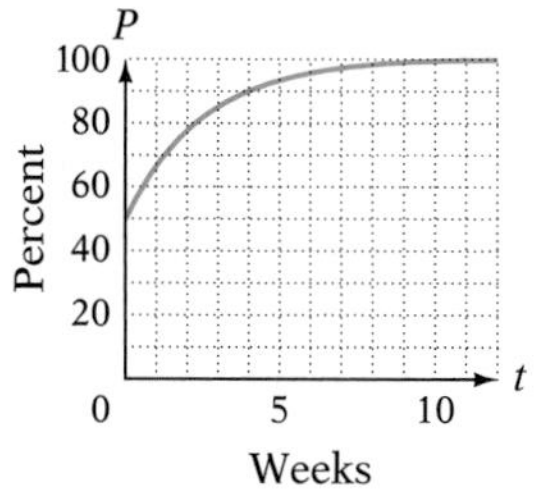

Figure 2

97. *Physics* The graph in Figure 3 shows the temperature, T (in degrees Fahrenheit), of a can of cola t hours after it is placed in a refrigerator. Use the graph to estimate the temperature of the cola when:

a. $t = 10$ h 60°F **b.** $t = 20$ h 50°F

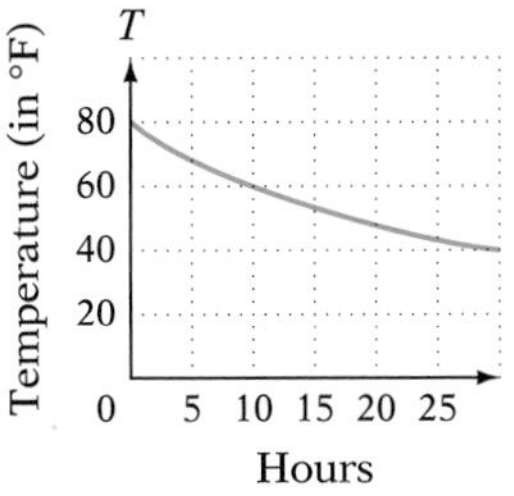

Figure 3

98. *Health Science* The graph in Figure 4 shows the decrease in the heart rate, r (in beats per minute), of a runner t minutes after the completion of a race. Use the graph to estimate the heart rate of a runner when:

a. $t = 5$ min 110 beats/min **b.** $t = 20$ min 75 beats/min

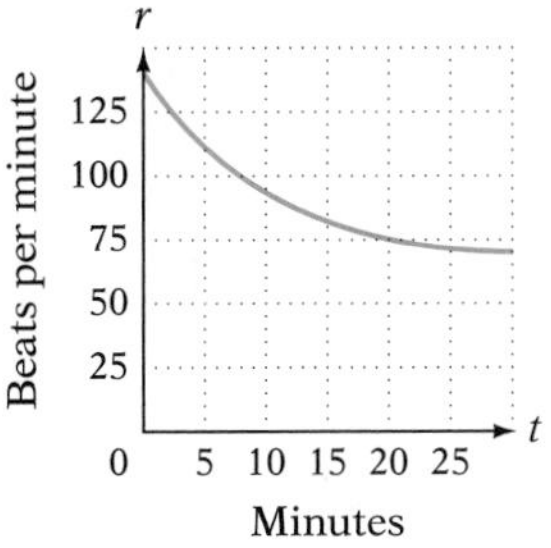

Figure 4

99. *Social Science* The list that follows gives the fifty states in the United States of America in terms of the Post Office two-letter abbreviations. The first number following the name of each state is the figure provided by the Bureau of the Census for the population of the state in 1990; these figures are in millions and are rounded to the nearest hundred thousand. The second number following the name of each state is the number of representatives that state has in the House of Representatives.

AL: 4.0; 7	HI: 1.1; 2	MA: 6.0; 10	NM: 1.5; 3	SD: 0.7; 1
AK: 0.6; 1	ID: 1.0; 2	MI: 9.3; 16	NY: 18.0; 31	TN: 4.9; 9
AZ: 3.7; 6	IL: 11.4; 20	MN: 4.4; 8	NC: 6.6; 12	TX: 17.0; 30
AR: 2.4; 4	IN: 5.5; 10	MS: 2.6; 5	ND: 0.6; 1	UT: 1.7; 3
CA: 29.8; 52	IA: 2.8; 5	MO: 5.1; 9	OH: 10.8; 19	VT: 0.6; 1
CO: 3.3; 6	KS: 2.5; 4	MT: 0.8; 1	OK: 3.1; 6	VA: 6.2; 11
CT: 3.3; 6	KY: 3.7; 6	NE: 1.6; 3	OR: 2.8; 5	WA: 4.9; 9
DE: 0.7; 1	LA: 4.2; 7	NV: 1.2; 2	PA: 11.9; 21	WV: 1.8; 3
FL: 12.9; 23	ME: 1.2; 2	NH: 1.1; 2	RI: 1.0; 2	WI: 4.9; 9
GA: 6.5; 11	MD: 4.8; 8	NJ: 7.7; 13	SC: 3.5; 6	WY: 0.5; 1

a. Draw a scatter diagram for the data. Label the horizontal axis "State population (in millions)" and the vertical axis "Number of representatives."

b. Explain the relationship between a state's population and the number of representatives that state has in the House of Representatives.

a.

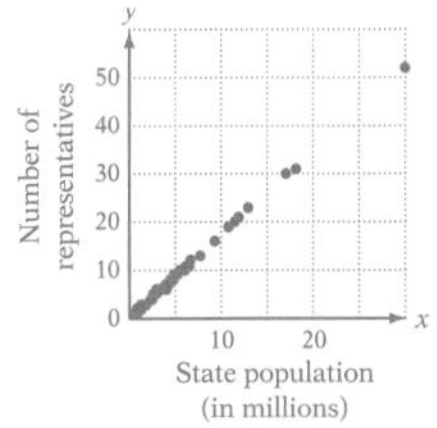

b. The 435 representatives in the House of Representatives are apportioned among the states according to their populations.

100. Write a report on the longitude-latitude coordinate system used to locate a position on Earth. Include in your report the definition of *nautical mile* and the coordinates of the city in which your school is located.

101. Investigate the relationship between two variables by collecting data, graphing the scatter diagram for these data, and discussing the results. For example, for each person in your classes who commutes to class by car, find out how many miles he or she commutes and the total number of miles on the odometer of the car he or she drives to class. Discuss whether you think there is a relationship between the two variables.

102. Give a real-world example of a relation that is not a function. Is it possible to give an example of a function that is not a relation? If so, give one. If not, explain why it is not possible.

SECTION 3.3

Linear Functions

1 Graph a linear function

INSTRUCTOR NOTE

It is important for students to make a connection between a function and its graph. A linear function has a characteristic graph, which is different from the graph of a quadratic function, which is different from the graph of a cubic function, and so on. Before you even graph a straight line, you might ask students whether

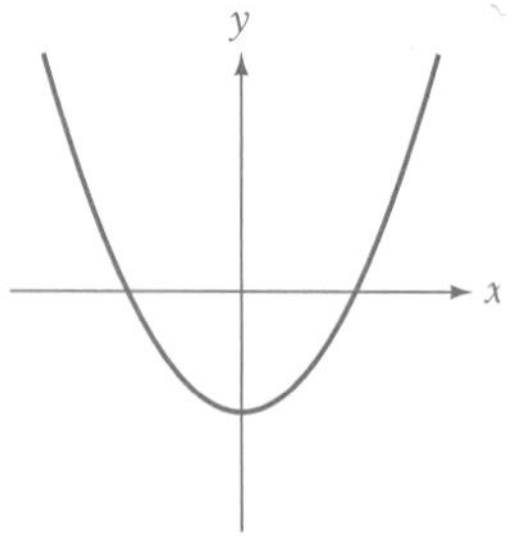

(or any other nonlinear graph) is the graph of a linear function.

Recall that the ordered pairs of a function can be written as $(x, f(x))$ or (x, y). The **graph of a function** is a graph of the ordered pairs (x, y) that belong to the function. Certain functions have characteristic graphs. A function that can be written in the form $f(x) = mx + b$ (or $y = mx + b$) is called a **linear function** because its graph is a straight line.

Examples of linear functions are shown at the right. Note that the exponent on each variable is 1.

$$f(x) = 2x + 5 \qquad (m = 2, b = 5)$$
$$P(t) = 3t - 2 \qquad (m = 3, b = -2)$$
$$y = -2x \qquad (m = -2, b = 0)$$
$$y = -\frac{2}{3}x + 1 \qquad \left(m = -\frac{2}{3}, b = 1\right)$$
$$g(z) = z - 2 \qquad (m = 1, b = -2)$$

The equation $y = x^2 + 4x + 3$ is not a linear function because it includes a term with a variable squared. The equation $f(x) = \frac{3}{x - 2}$ is not a linear function because a variable occurs in the denominator. Another example of an equation that is not a linear function is $y = \sqrt{x} + 4$; this equation contains a variable within a radical and so is not a linear function.

Consider $f(x) = 2x + 1$. Evaluating the linear function when $x = -3, -2, -1, 0, 1,$ and 2 produces some of the ordered pairs of the function. It is convenient to record the results in a table similar to the one at the right. The graph of the ordered pairs is shown in Figure 1 below.

x	$f(x) = 2x + 1$	y	(x, y)
-3	$2(-3) + 1$	-5	$(-3, -5)$
-2	$2(-2) + 1$	-3	$(-2, -3)$
-1	$2(-1) + 1$	-1	$(-1, -1)$
0	$2(0) + 1$	1	$(0, 1)$
1	$2(1) + 1$	3	$(1, 3)$
2	$2(2) + 1$	5	$(2, 5)$

Evaluating the function when x is not an integer produces more ordered pairs to graph, such as $\left(-\frac{5}{2}, -4\right)$ and $\left(\frac{3}{2}, 4\right)$, as shown in Figure 2 below. Evaluating the function for still other values of x would result in more and more ordered pairs to graph. The result would be so many dots that the graph would look like the straight line shown in Figure 3, which is the graph of $f(x) = 2x + 1$.

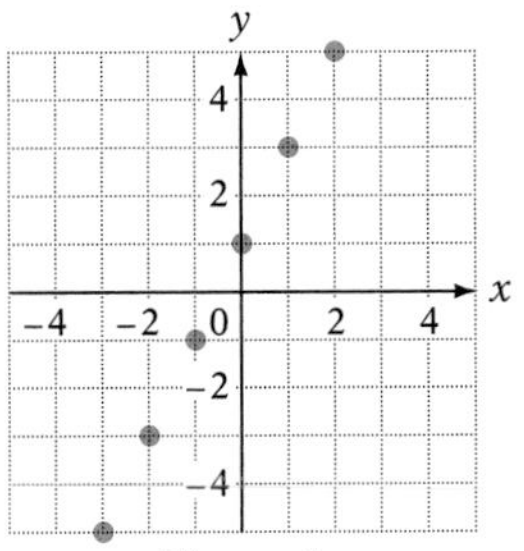

Figure 1

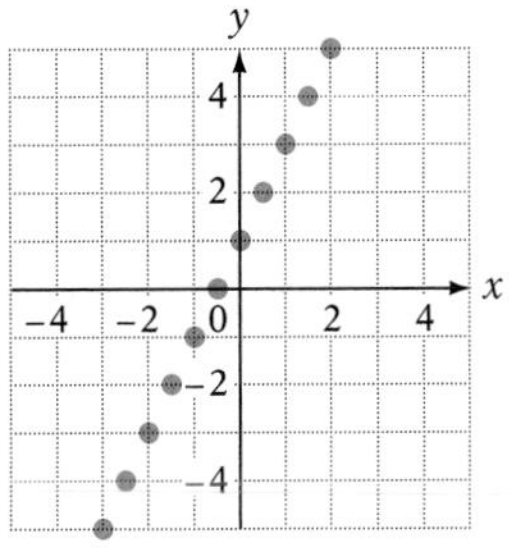

Figure 2

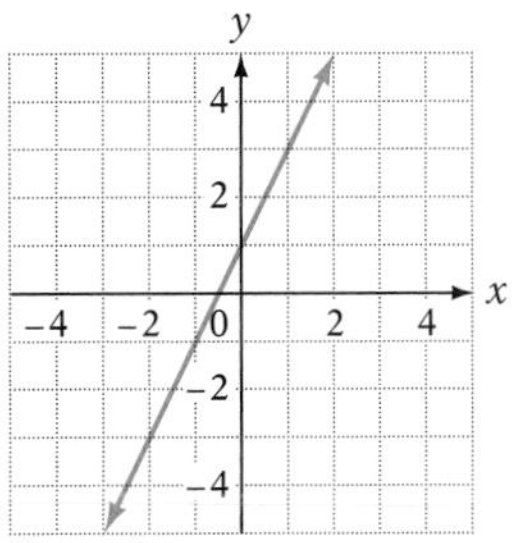

Figure 3

No matter what value of x is chosen, $2x + 1$ is a real number. This means the domain of $f(x) = 2x + 1$ is all real numbers. Therefore, we can use any real number when evaluating the function. Normally, however, values such as π or $\sqrt{5}$ are not used because it is difficult to graph the resulting ordered pairs.

INSTRUCTOR NOTE

It is important for the student to understand this. For each point whose coordinates are on the graph, the ordered pair belongs to the function, and any ordered pair that belongs to the function is the coordinates of a point on the graph.

Note from the graph of $f(x) = 2x + 1$ shown at the right that $(-1.5, -2)$ and $(3, 7)$ are the coordinates of points on the graph and that $f(-1.5) = -2$ and $f(3) = 7$. Note also that the point whose coordinates are $(2, 1)$ is not a point on the graph and that $f(2) \neq 1$. Every point on the graph is an ordered pair that belongs to the function, and every ordered pair that belongs to the function corresponds to a point on the graph.

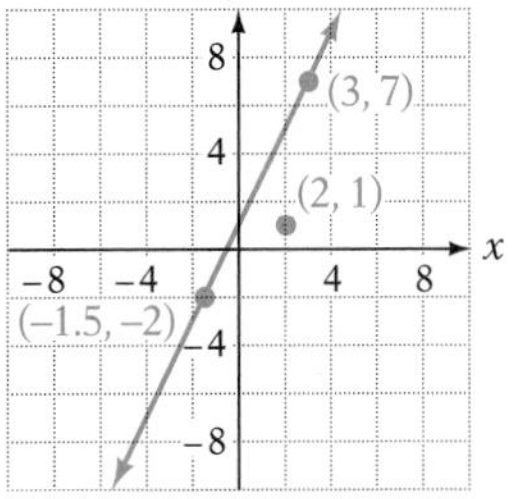

Whether an equation is written as $f(x) = mx + b$ or as $y = mx + b$, the equation represents a linear function, and the graph of the equation is a straight line.

Because the graph of a linear function is a straight line, and a straight line is determined by two points, the graph of a linear function can be drawn by finding only two of the ordered pairs of the function. However, it is recommended that you find at least *three* ordered pairs to ensure accuracy.

INSTRUCTOR NOTE

Students need to be constantly reminded that $f(x)$ and y have the same value for a given x.

Example 1 Graph: $f(x) = -\frac{3}{2}x - 3$

Solution

x	$y = f(x)$
0	−3
−2	0
−4	3

▶ Find at least three ordered pairs. When the coefficient of x is a fraction, choose values of x that will simplify the evaluations. The ordered pairs can be displayed in a table.

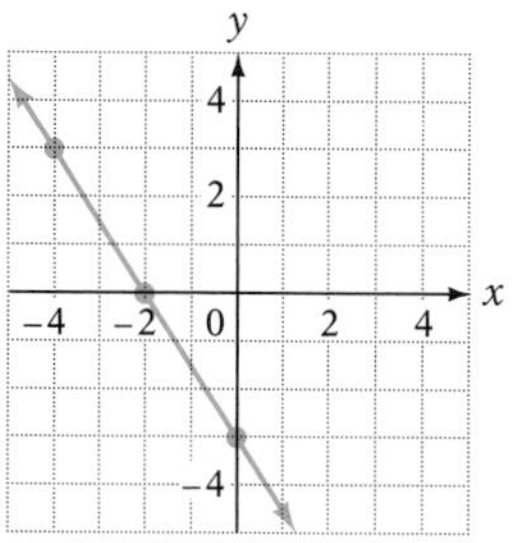

▶ Graph the ordered pairs and draw a line through the points.

Problem 1 Graph: $y = \frac{3}{5}x - 4$

Solution See page S6.

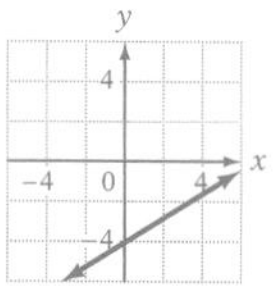

Graphing utilities create graphs by plotting points and then connecting the points to form a curve. Using a graphing utility, enter the equation $y = -\frac{3}{2}x - 3$ and verify the graph drawn in Example 1. (Refer to the Appendix on page 679 for suggestions on what domain and range to use.) Trace along the graph and verify that (0, −3), (−2, 0), and (−4, 3) are the coordinates of points on the graph. Now enter the equation $y = \frac{3}{5}x - 4$ given in Problem 1. Verify that the ordered pairs you found for this function are the coordinates of points on the graph.

2 Graph an equation of the form $Ax + By = C$

INSTRUCTOR NOTE

There are many times when an equation in two variables must be solved for y before it is graphed. For instance, most graphing calculators require that an equation be in that form. Rewriting linear equations in this form is good practice for a student. Later, we develop the graph of the equation $Ax + By = C$ by using its intercepts.

A **literal equation** is an equation with more than one variable. Examples of literal equations are $P = 2L + 2W$, $V = LWH$, $d = rt$, and $3x + 2y = 6$.

Linear equations of the form $y = mx + b$ are literal equations. In some cases, a linear equation has the form $Ax + By = C$. In such a case, it may be convenient to solve the equation for y to get the equation in the form $y = mx + b$. To solve for y, we use the same rules and procedures that we use to solve equations with numerical values.

➡ Write $4x - 3y = 6$ in the form $y = mx + b$.

$$4x - 3y = 6$$

Subtract $4x$ from each side of the equation.

$$4x - 4x - 3y = 6 - 4x$$

Simplify. $-3y = 6 - 4x$

Divide both sides of the equation by -3. $\frac{-3y}{-3} = \frac{6 - 4x}{-3}$

Simplify. $y = \frac{6 - 4x}{-3}$

Divide each term in the numerator by the denominator. $y = \frac{6}{-3} - \frac{4x}{-3}$

Simplify. $y = -2 + \frac{4}{3}x$

Write the equation in the form $y = mx + b$. $y = \frac{4}{3}x - 2$

We will show two methods of graphing an equation of the form $Ax + By = C$. In the first method, we solve the equation for y and then follow the same procedure used for graphing an equation of the form $y = mx + b$.

Example 2 Graph: $3x + 2y = 6$

Solution $3x + 2y = 6$

$2y = -3x + 6$ ▶ Solve the equation for y.

$y = -\frac{3}{2}x + 3$

▶ Find at least three solutions.

x	y
0	3
2	0
4	−3

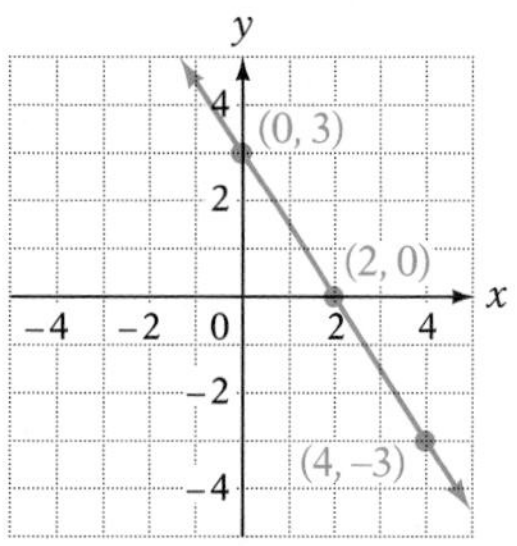

▶ Graph the ordered pairs in a rectangular coordinate system. Draw a straight line through the points.

Problem 2 Graph: $-3x + 2y = 4$

Solution See page S6.

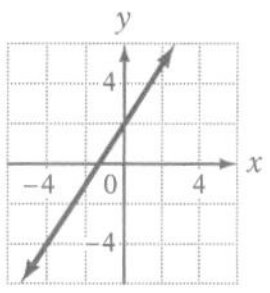

Use a graphing utility to graph the equation $3x + 2y = 6$. First solve the equation for y and then enter the equation $y = -\frac{3}{2}x + 3$. Now trace along the graph and verify that (0, 3), (2, 0), and (4, −3) are the coordinates of points on the graph. Follow the same procedure for Problem 2.

An equation in which one of the variables is missing has a graph that is either a horizontal or a vertical line.

The equation $y = -2$ can be written

$$0 \cdot x + y = -2$$

Because $0 \cdot x = 0$ for any value of x, y is -2 for every value of x.

Some of the possible ordered-pair solutions of $y = -2$ are given in the following table. The graph is shown at the right.

x	y
-2	-2
0	-2
3	-2

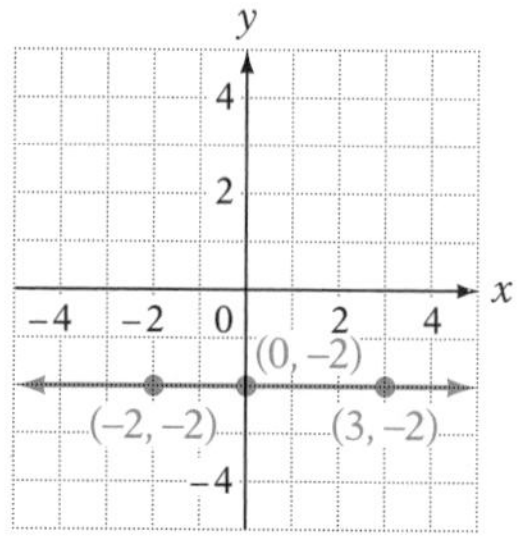

The equation $y = -2$ represents a function. Some of the ordered pairs of this function are $(-2, -2)$, $(0, -2)$, and $(3, -2)$. In functional notation we would write $f(x) = -2$. This function is an example of a constant function. No matter what value of x is selected, $f(x) = -2$.

Definition of the Constant Function

A function given by $f(x) = b$, where b is a constant, is a **constant function.** The graph of a constant function is a horizontal line passing through $(0, b)$.

Graph: $y + 4 = 0$

Solve for y.

$$y + 4 = 0$$
$$y = -4$$

The graph of $y = -4$ is a line passing through $(0, -4)$.

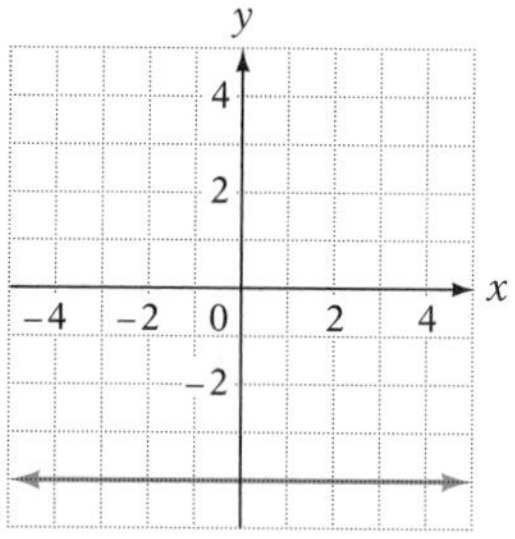

INSTRUCTOR NOTE

For students who have difficulty evaluating a constant function, you might try the evaluation model we used earlier.

$f(x) = 4$

$f(\quad) = 4$

Because there are no parentheses on the righthand side, there is no way to change the value of the function.

For each value in the domain of a constant function, the value of the function is the same (that is, it is constant). For instance, if $f(x) = 4$, then $f(2) = 4$, $f(3) = 4$, $f(\sqrt{3}) = 4$, $f(\pi) = 4$, and so on. The value of $f(x)$ is 4 for all values of x.

Evaluate $P(t) = -7$ when $t = 6$.

The value of the constant function is the same for all values of the variable.

$$P(t) = -7$$
$$P(6) = -7$$

For the equation $y = -2$, the coefficient of x is zero. For the equation $x = 2$, the coefficient of y is zero. For instance, the equation $x = 2$ can be written

$$x + 0 \cdot y = 2$$

No matter what value of y is chosen, $0 \cdot y = 0$ and therefore x is always 2.

Some of the possible ordered-pair solutions are given in the following table. The graph is shown at the right.

x	y
2	6
2	1
2	−4

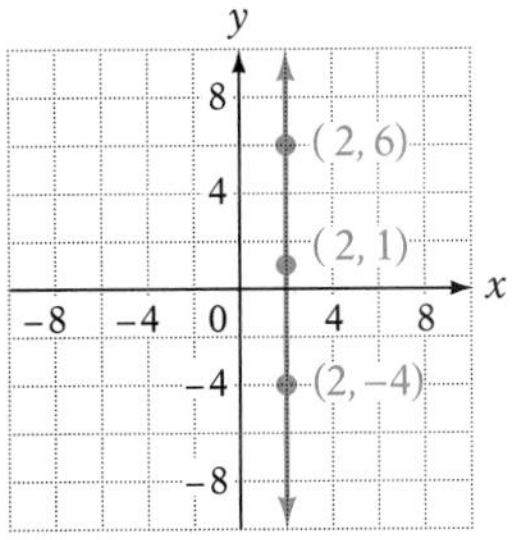

Graph of $x = a$

The graph of $x = a$ is a vertical line passing through the point $(a, 0)$.

LOOK CLOSELY

The equation $y = b$ represents a function. The equation $x = a$ does *not* represent a function. Remember, not all equations represent functions.

Recall that a function is a set of ordered pairs in which no two ordered pairs have the same first coordinate and different second coordinates. Because (2, 6), (2, 1), and (2, −4) are ordered pairs belonging to the equation $x = 2$, this equation does not represent a function, and the graph is not the graph of a function.

Example 3 Graph: $x = -4$

Solution

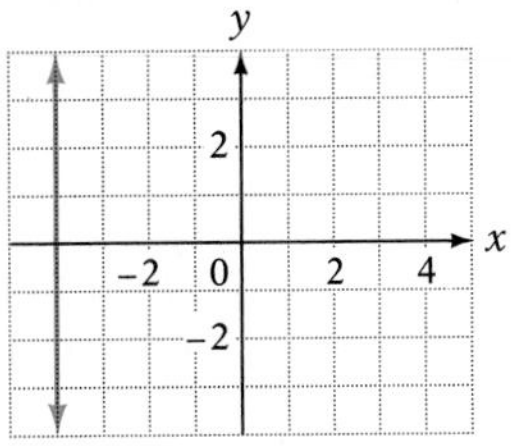

▸ The graph of an equation of the form $x = a$ is a vertical line passing through the point whose coordinates are $(a, 0)$.

Problem 3 Graph: $y - 3 = 0$

Solution See page S6.

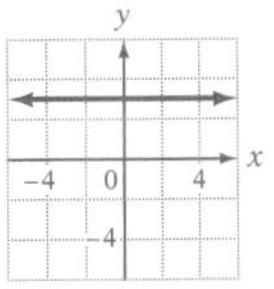

A second method of graphing straight lines uses the *intercepts* of the graph.

The graph of the equation $x - 2y = 4$ is shown at the right. The graph crosses the x-axis at the point (4, 0). This point is called the ***x*-intercept.** The graph also crosses the y-axis at the point (0, −2). This point is called the ***y*-intercept.**

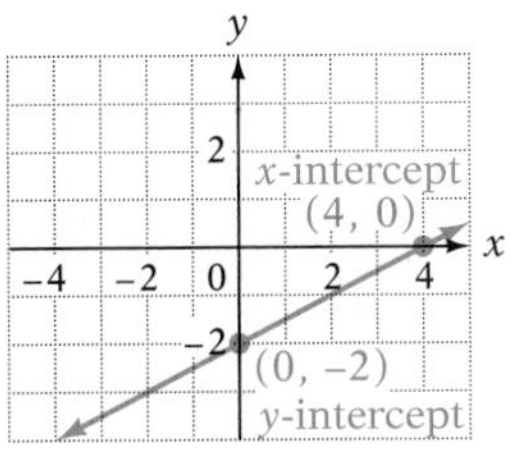

➡ Find the x- and y-intercepts of the graph of the equation $3x + 4y = -12$.

LOOK CLOSELY

The x-intercept occurs when $y = 0$. The y-intercept occurs when $x = 0$.

To find the x-intercept, let $y = 0$.
(Any point on the x-axis has y-coordinate 0.)

$$3x + 4y = -12$$
$$3x + 4(0) = -12$$
$$3x = -12$$
$$x = -4$$

The x-intercept is (−4, 0).

To find the y-intercept, let $x = 0$.
(Any point on the y-axis has x-coordinate 0.)

$$\begin{aligned} 3x + 4y &= -12 \\ 3(0) + 4y &= -12 \\ 4y &= -12 \\ y &= -3 \end{aligned}$$

The y-intercept is $(0, -3)$.

Find the y-intercept of the graph of the equation $y = \frac{2}{3}x + 7$.

To find the y-intercept, let $x = 0$.

$$\begin{aligned} y &= \frac{2}{3}x + 7 \\ y &= \frac{2}{3}(0) + 7 \\ y &= 7 \end{aligned}$$

The y-intercept is $(0, 7)$.

Note that the y-coordinate of the y-intercept $(0, 7)$ has the same value as the constant term of $y = \frac{2}{3}x + 7$. As shown below, this is always true.

To find the y-intercept, let $x = 0$.

$$\begin{aligned} y &= mx + b \\ y &= m(0) + b \\ y &= b \end{aligned}$$

Thus the y-intercept is $(0, b)$.

y-Intercept of a Straight Line

For any equation of the form $y = mx + b$, the y-intercept is $(0, b)$.

A linear equation can be graphed by finding the x- and y-intercepts and then drawing a line through the two points.

Example 4 Graph $4x - y = 4$ by using the x- and y-intercepts.

Solution x-intercept:

$$\begin{aligned} 4x - y &= 4 \\ 4x - 0 &= 4 \\ 4x &= 4 \\ x &= 1 \end{aligned}$$

▸ To find the x-intercept, let $y = 0$.

The x-intercept is $(1, 0)$.

y-intercept:

$$\begin{aligned} 4x - y &= 4 \\ 4(0) - y &= 4 \\ -y &= 4 \\ y &= -4 \end{aligned}$$

▸ To find the y-intercept, let $x = 0$.

The y-intercept is $(0, -4)$.

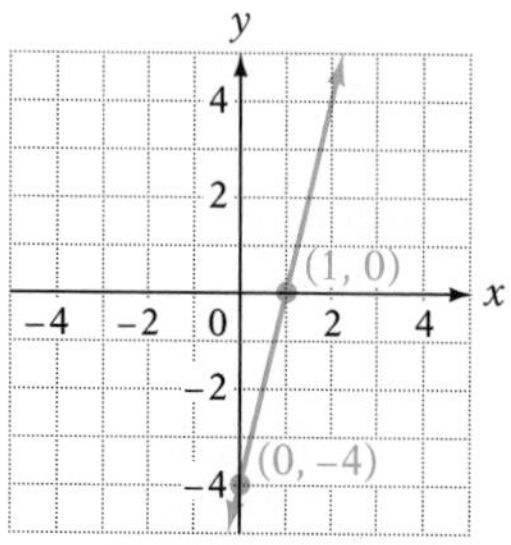

▸ Graph the points $(1, 0)$ and $(0, -4)$. Draw a line through the two points.

Problem 4 Graph $3x - y = 2$ by using the x- and y-intercepts.

Solution See page S7. x-intercept: $\left(\frac{2}{3}, 0\right)$
y-intercept: $(0, -2)$

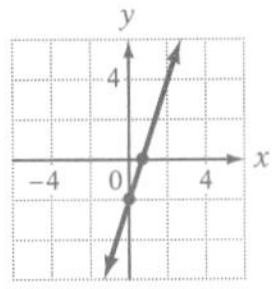

3 Application problems

10 12 HM³

INSTRUCTOR NOTE

Part b of this example asks the student to explain, in the context of the application, the meaning of an ordered pair. This important exercise requires students to make a connection between an ordered pair and its meaning.

There are a variety of applications of linear functions.

On the basis of data from *The Joy of Cooking*, the daily caloric allowance for a woman can be approximated by the equation $C = -7.5A + 2187.5$, where C is the caloric intake and A is the age of the woman.

a. Graph this equation for $25 \le A \le 75$.

b. The point whose coordinates are (45, 1850) is on the graph. Write a sentence that describes the meaning of this ordered pair.

a.

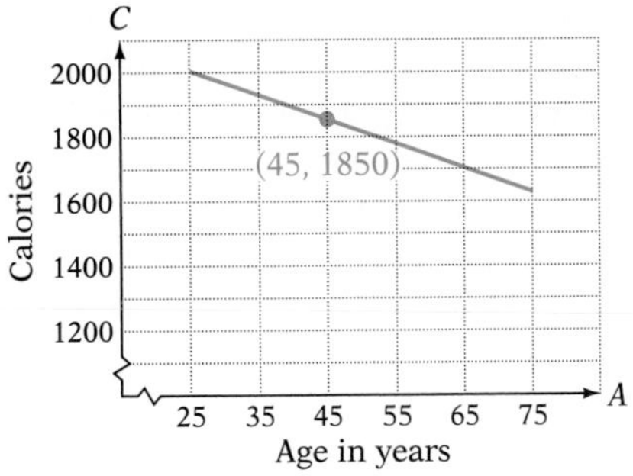

b. The ordered pair (45, 1850) means that the caloric allowance of a 45-year-old woman is 1850 calories per day.

Example 5 An electronics technician charges \$45 plus \$1 per minute to repair defective wiring in a home or apartment. The equation that describes the total cost, C (in dollars), to have defective wiring repaired is given by the equation $C = t + 45$, where t is the number of minutes the technician works. Graph this equation for $0 \le t \le 60$. The point whose coordinates are (15, 60) is on the graph. Write a sentence that describes the meaning of this ordered pair.

Solution

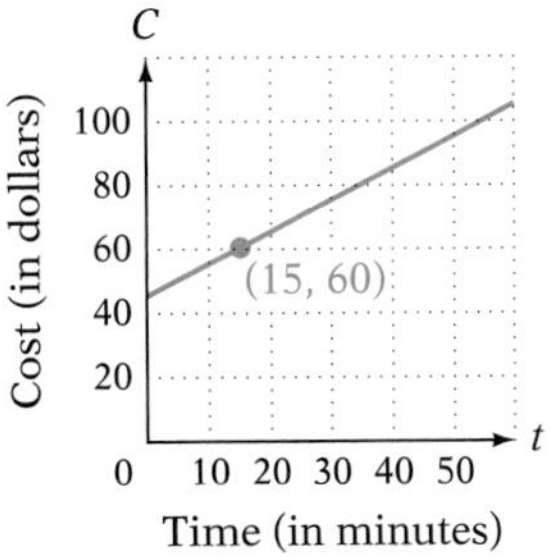

The ordered pair (15, 60) means that it costs \$60 for the technician to work 15 min.

Problem 5 The height, h (in inches), of a person and the length, L (in inches), of that person's stride while walking are related. The equation $h = \frac{3}{4}L + 50$ approximates the relationship. Graph this equation for $15 \le L \le 40$. The point whose coordinates are $(32, 74)$ is on the graph. Write a sentence that describes the meaning of this ordered pair.

Solution See page S7.

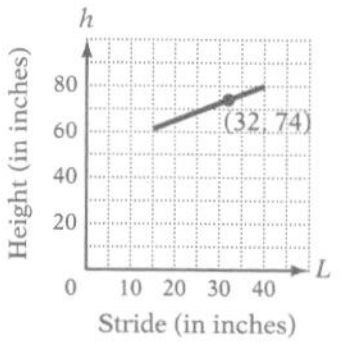

The ordered pair (32, 74) means that a person with a stride of 32 in. is 74 in. tall.

STUDY TIPS

FIND GOOD STUDY AREAS

Find a place to study where you are comfortable and can concentrate well. Many students find the campus library to be a good place. You might select two or three places at the college library where you like to study. Or there may be a small, quiet lounge on the third floor of a building where you find you can study well. Take the time to find places that promote good study habits.

CONCEPT REVIEW 3.3

Determine whether the following statements are always true, sometimes true, or never true.

1. The graph of $y = \frac{1}{x} + 2$ is the graph of a linear function. Never true
2. The graph of $y - 3 = 0$ is a vertical line. Never true
3. The graph of a straight line crosses the x-axis when $y = 0$. Always true
4. The graph of a linear equation is a straight line. Always true
5. The point at which a line crosses the y-axis is the x-intercept. Sometimes true
6. $xy + 2 = 0$ is an example of a linear equation. Never true

EXERCISES 3.3

1 1. Explain how to graph a linear function by plotting points.

2. Give one example of a linear function and one example of a nonlinear function.

Graph.

3. $y = 3x - 4$

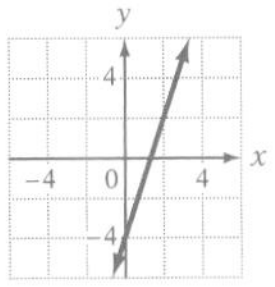

4. $y = -2x + 3$

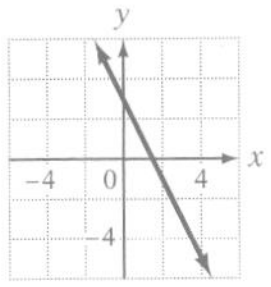

5. $y = -\frac{2}{3}x$

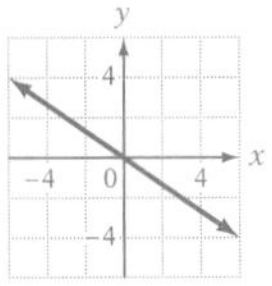

6. $y = \frac{3}{2}x$

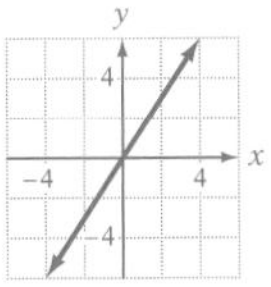

7. $y = \frac{2}{3}x - 4$

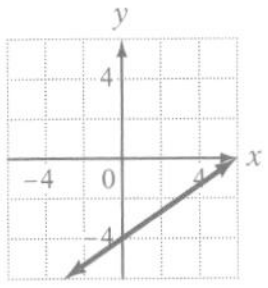

8. $y = \frac{3}{4}x + 2$

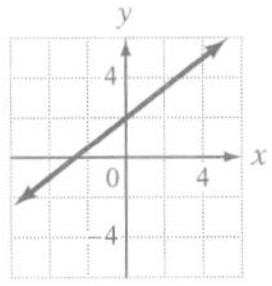

9. $y = -\frac{1}{3}x + 2$

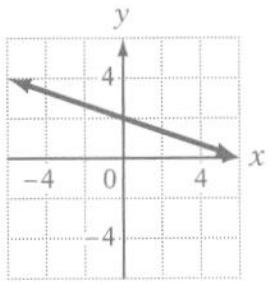

10. $y = -\frac{3}{2}x - 3$

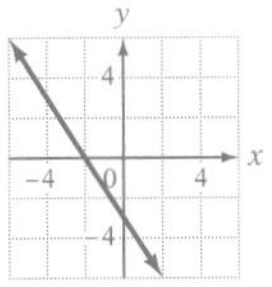

11. $y = \frac{3}{5}x - 1$

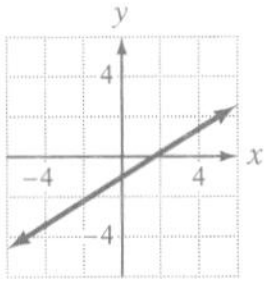

2 **12.** Explain how to find the x-intercept of the graph of an equation.

13. Explain how to find the y-intercept of the graph of an equation.

Graph.

14. $2x - y = 3$

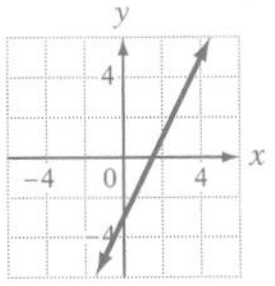

15. $2x + y = -3$

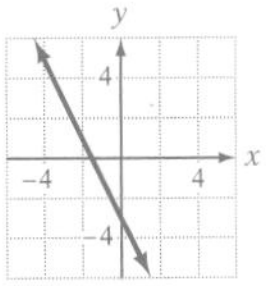

16. $2x + 5y = 10$

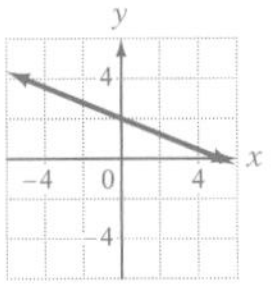

17. $x - 4y = 8$

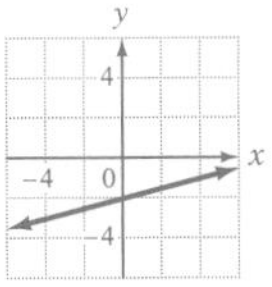

18. $y = -2$

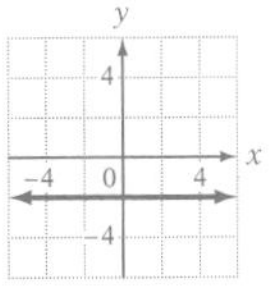

19. $x - 3y = 0$

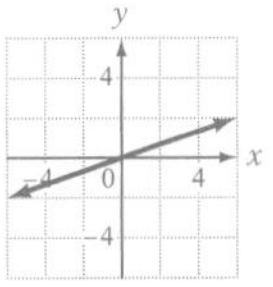

20. $2x - 3y = 12$

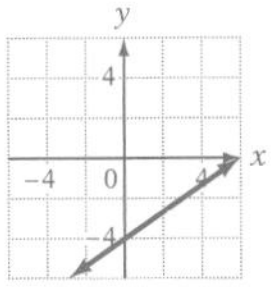

21. $3x - y = -2$

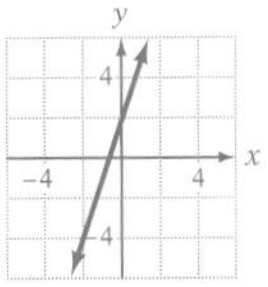

22. $3x - 2y = 8$

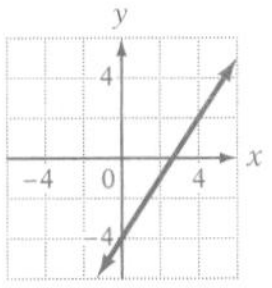

Find the x- and y-intercepts and graph.

23. $x - 2y = -4$

$(-4, 0)$; $(0, 2)$

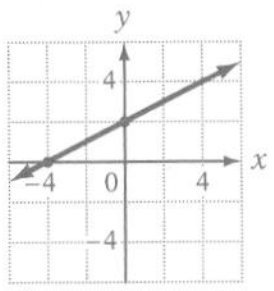

24. $3x + y = 3$

$(1, 0)$; $(0, 3)$

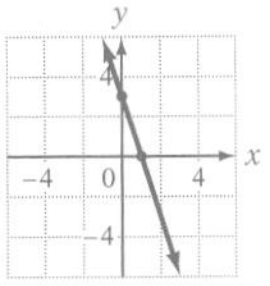

25. $2x - 3y = 9$

$\left(\frac{9}{2}, 0\right)$; $(0, -3)$

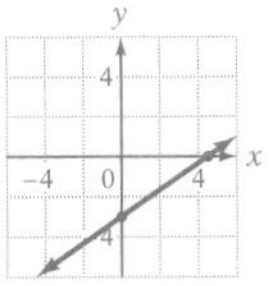

26. $4x - 2y = 5$

$\left(\frac{5}{4}, 0\right); \left(0, -\frac{5}{2}\right)$

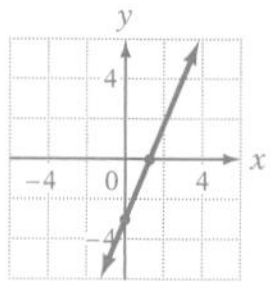

27. $2x - y = 4$

$(2, 0); (0, -4)$

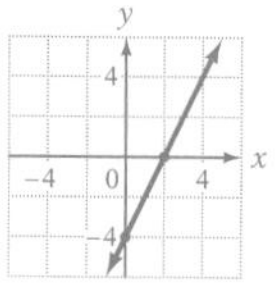

28. $2x + y = 3$

$\left(\frac{3}{2}, 0\right); (0, 3)$

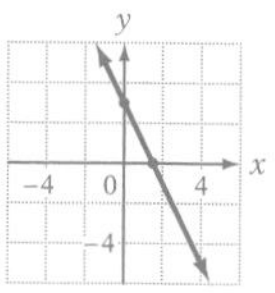

29. $3x + 2y = 5$

$\left(\frac{5}{3}, 0\right); \left(0, \frac{5}{2}\right)$

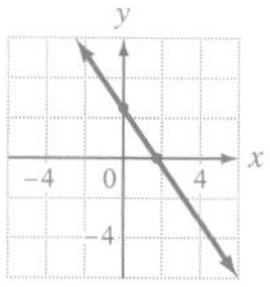

30. $4x - 3y = 8$

$(2, 0); \left(0, -\frac{8}{3}\right)$

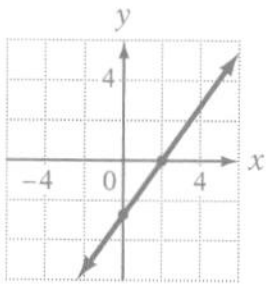

31. $3x + 2y = 4$

$\left(\frac{4}{3}, 0\right); (0, 2)$

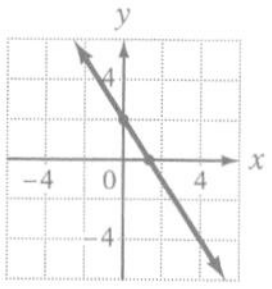

32. $2x - 3y = 4$

$(2, 0); \left(0, -\frac{4}{3}\right)$

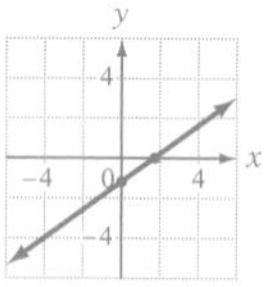

33. $3x - 5y = 9$

$(3, 0); \left(0, -\frac{9}{5}\right)$

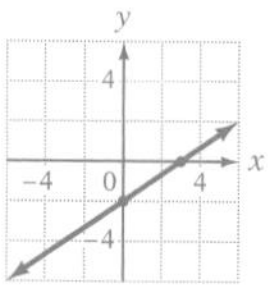

34. $4x - 3y = 6$

$\left(\frac{3}{2}, 0\right); (0, -2)$

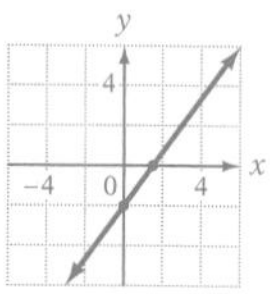

3 Solve.

35. *Human Resources* An unskilled laborer receives \$5.65 (minimum wage) to pick apples at the local orchard. The equation that describes the wages, W (in dollars), for the laborer is $W = 5.65t$, where t is the number of hours worked. Use the coordinate axes at the right to graph this equation for $0 \le t \le 40$. The point (30, 169.50) is on the graph. Write a sentence that describes the meaning of this ordered pair.

The laborer will earn \$169.50 for working 30 h.

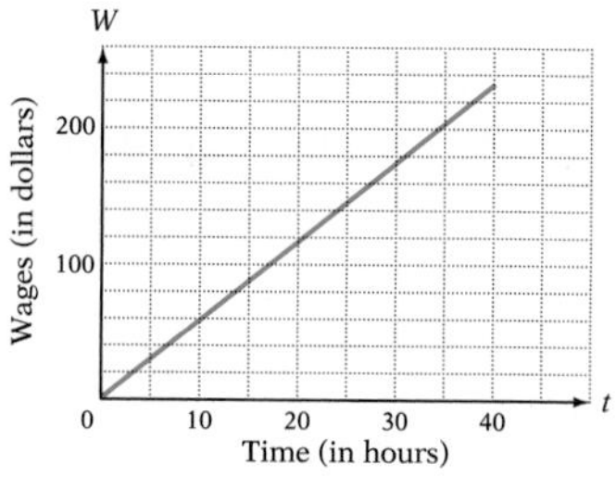

36. *Sports* A tennis pro is paid \$40 an hour for a private lesson by members of a condominium group. The equation that describes the total earnings, E (in dollars), received by the pro is $E = 40t$, where t is the total number of hours worked. Use the coordinate axes at the right to graph this equation for $0 \le t \le 30$. The point (25, 1000) is on the graph. Write a sentence that describes the meaning of this ordered pair.

The tennis pro will earn \$1000 for working 25 h.

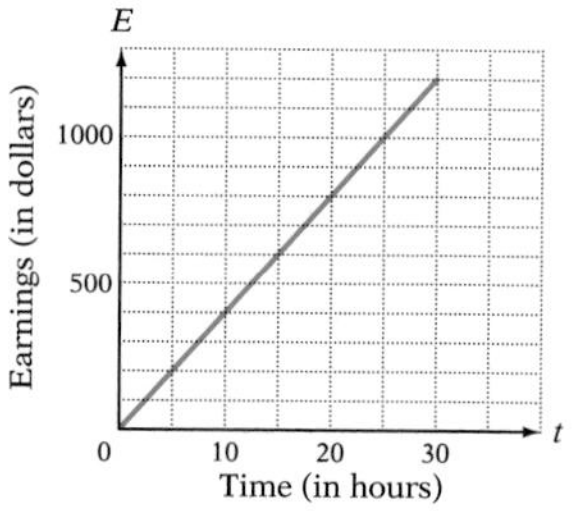

37. *Real Estate* A realtor receives \$400 per month plus 6% commission on sales. The equation that describes the total monthly income, I (in dollars), of the realtor is $I = 0.06s + 400$, where s is the amount of sales. Use the coordinate axes at the right to graph this equation for $0 \le s \le 150{,}000$. The point (60,000, 4000) is on the graph. Write a sentence that describes the meaning of this ordered pair.

The realtor will earn \$4000 for selling \$60,000 worth of property.

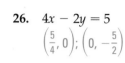

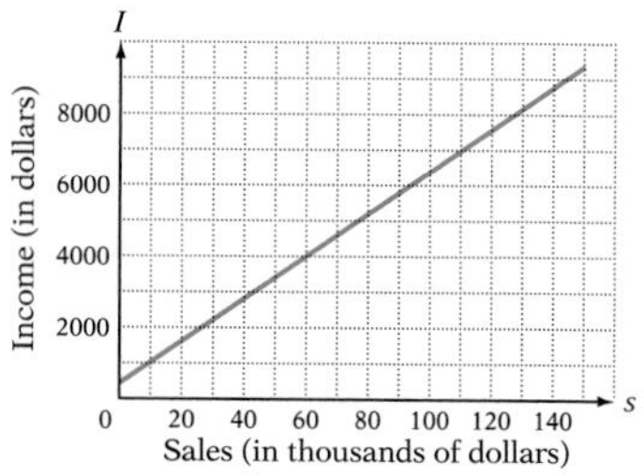

38. *Human Resources* An electronics company pays line workers \$9 per hour plus \$.05 for each transistor produced. The equation that describes the hourly wage, W (in dollars), is $W = 0.05n + 9$, where n is the number of transistors produced during any given hour. Use the coordinate axes at the right to graph this equation for $0 \le n \le 20$. The point (16, 9.80) is on the graph. Write a sentence that describes the meaning of this ordered pair.
The line worker will earn \$9.80 during an hour in which 16 transistors are produced.

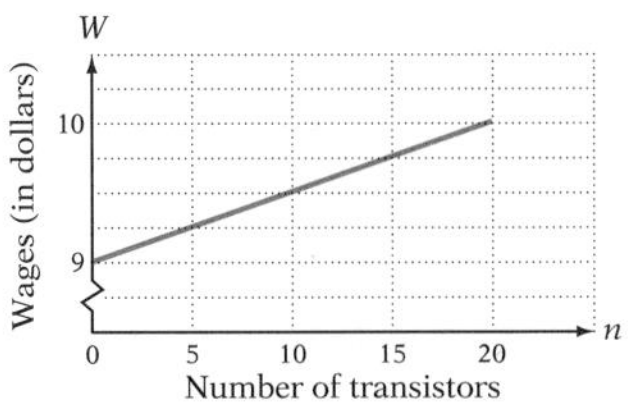

39. *Food Science* A caterer charges a flat rate of \$500 plus \$.95 for each hot appetizer. The equation that describes the cost, C (in dollars), of catering a dinner party is $C = 0.95n + 500$, where n is the number of hot appetizers. Use the coordinate axes at the right to graph this equation for $0 \le n \le 300$. The point (120, 614) is on the graph. Write a sentence that describes the meaning of this ordered pair.
The caterer will charge \$614 for 120 hot appetizers.

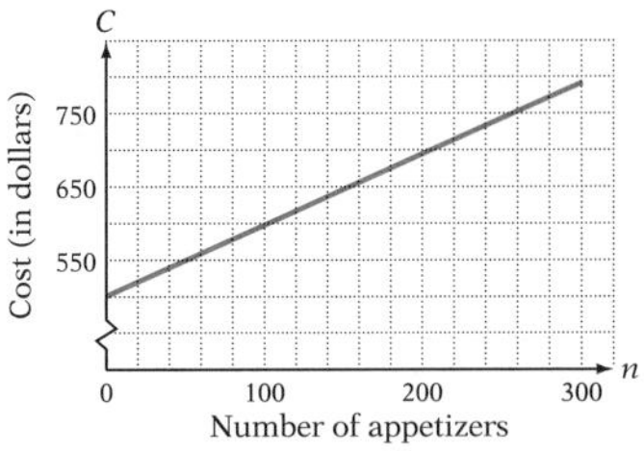

40. *Environmental Science* A tree service charges \$60 plus \$35 for each tree removed. The equation that describes the cost, C (in dollars), is $C = 35n + 60$, where n is the number of trees removed. Use the coordinate axes at the right to graph this equation for $0 \le n \le 100$. The point (50, 1810) is on the graph. Write a sentence that describes the meaning of this ordered pair.
The tree service will charge \$1810 to remove 50 trees.

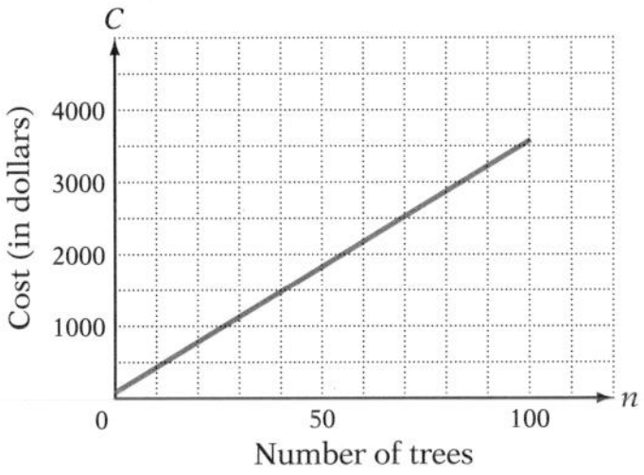

APPLYING CONCEPTS 3.3

41. *Business* The sale price of an item is a function, s, of the original price, p, where $s(p) = 0.80p$. If an item's original price is \$200, what is the sale price of the item? \$160

42. *Business* The markup of an item is a function, m, of its cost, c, where $m(c) = 0.25c$. If the cost of an item is \$150, what is the markup on the item?
\$37.50

Use a graphing utility to draw the graph of each of the following equations.

43. $f(x) = 1.2x + 2.3$

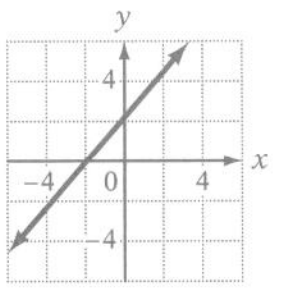

44. $f(x) = 2.4x + 0.5$

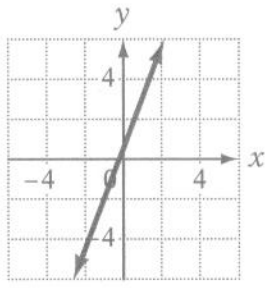

45. $f(x) = \frac{2x}{3} - \frac{5}{3}$

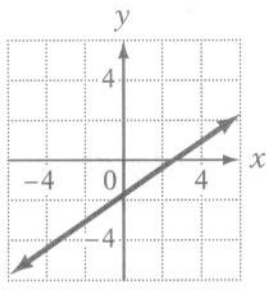

46. $y = -\frac{3x}{4} - \frac{5}{2}$

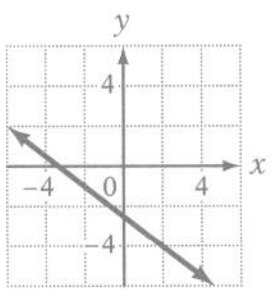

47. $3x - y = 4$

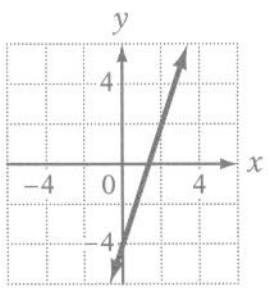

48. $2x + y = 3$

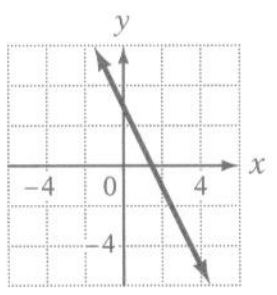

49. *Meteorology* The wind-chill factor is the temperature of still air that would have the same effect on exposed human skin as a given combination of wind speed and air temperature. For example, given a wind speed of 10 mph and a temperature reading of 20°F, the wind-chill factor is 3°F. In the following set of ordered pairs, the abscissa is the air temperature in degrees Fahrenheit, and the ordinate is the wind-chill factor when the wind speed is 10 mph. (*Source: Information Please Almanac*)

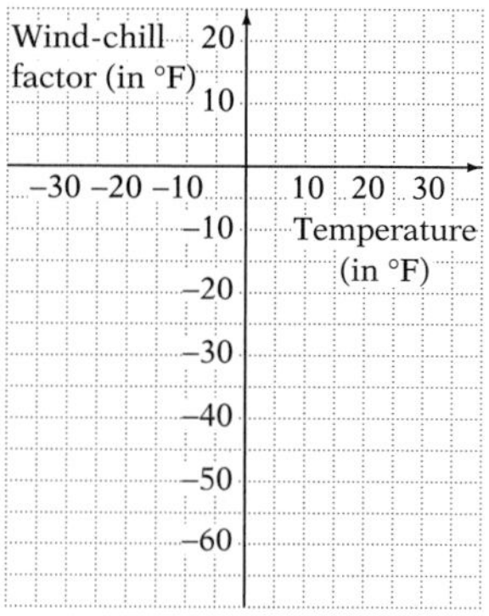

$$\{(35, 22), (30, 16), (25, 10), (20, 3), (15, -3), (10, -9), (5, -15), (0, -22), (-5, -27), (-10, -34), (-15, -40), (-20, -46), (-25, -52), (-30, -58), (-35, -64)\}$$

a. Use the coordinate axes at the right to graph the ordered pairs.

b. Is the set of ordered pairs a function? Yes

c. List the domain of the relation. List the range of the relation.

d. The equation of the line that approximately models the wind-chill factor when the wind speed is 10 mph is $y = 1.2321x - 21.2667$, where x is the air temperature in degrees Fahrenheit and y is the wind-chill factor. Evaluate the function $f(x) = 1.2321x - 21.2667$ to determine which of the ordered pairs listed above do not satisfy this function. Round to the nearest integer. $(0, -22)$ does not satisfy the function.

e. Find the x-intercept of the graph of $f(x) = 1.2321x - 21.2667$. Round to the nearest integer. What does the x-intercept represent? Find the y-intercept of the graph of the function. Round to the nearest tenth. What does the y-intercept represent?
The x-intercept is $(17, 0)$. Given a wind speed of 10 mph, the wind-chill factor is 0°F when the air temperature is 17°F. The y-intercept is $(0, -21.3)$. Given a wind speed of 10 mph, the wind-chill factor is −21.3°F when the air temperature is 0°F.

a.

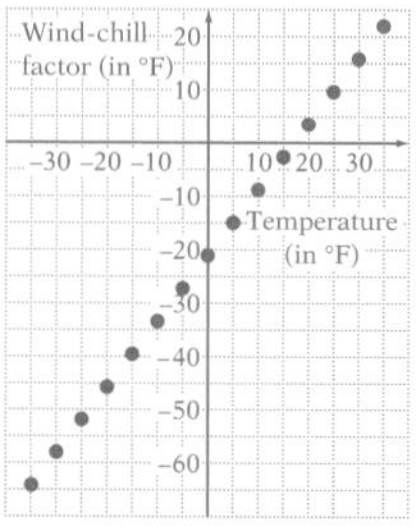

c. D: $\{-35, -30, -25, -20, -15, -10, -5, 0, 5, 10, 15, 20, 25, 30, 35\}$
R: $\{-64, -58, -52, -46, -40, -34, -27, -22, -15, -9, -3, 3, 10, 16, 22\}$

50. Explain the relationship between the zero of a linear function and the x-intercept of the graph of that function.

51. There is a relationship between the number of times a cricket chirps and the air temperature. The function $f(C) = 7C - 30$, where C is the temperature in degrees Celsius, can be used to approximate the number of times per minute that a cricket chirps. Discuss the domain and range of this function, and explain the significance of its x-intercept.

52. Explain what the graph of an equation represents.

53. Explain why you cannot graph the equation $4x + 3y = 0$ by using just its intercepts.

SECTION 3.4 Slope of a Straight Line

1 Find the slope of a line given two points

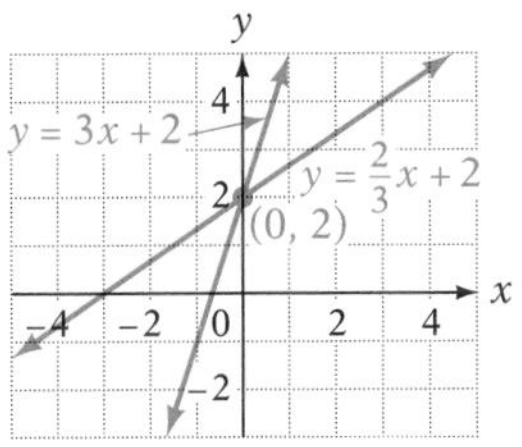

The graphs of $y = 3x + 2$ and $y = \frac{2}{3}x + 2$ are shown at the left. Each graph crosses the y-axis at the point (0, 2), but the graphs have different slants. The **slope** of a line is a measure of the slant of the line. The symbol for slope is m.

The slope of a line containing two points is the ratio of the change in the y values between the two points to the change in the x values. The line containing the points whose coordinates are $(-1, -3)$ and $(5, 2)$ is shown below.

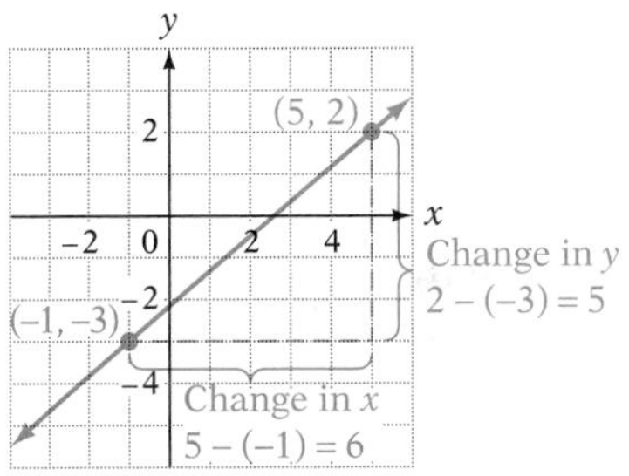

The change in the y values is the difference between the y-coordinates of the two points.

$$\text{Change in } y = 2 - (-3) = 5$$

The change in the x values is the difference between the x-coordinates of the two points.

$$\text{Change in } x = 5 - (-1) = 6$$

The slope of the line between the two points is the ratio of the change in y to the change in x.

$$\text{Slope} = m = \frac{\text{change in } y}{\text{change in } x} = \frac{5}{6} \qquad m = \frac{2-(-3)}{5-(-1)} = \frac{5}{6}$$

INSTRUCTOR NOTE

When calculating the slope of a line, show students that the choice of P_1 and P_2 does not alter the value of the slope.

In general, if $P_1(x_1, y_1)$ and $P_2(x_2, y_2)$ are two points on a line, then

$$\text{Change in } y = y_2 - y_1 \qquad \text{Change in } x = x_2 - x_1$$

Using these ideas, we can state a formula for slope.

Slope Formula

The slope of the line containing the two points $P_1(x_1, y_1)$ and $P_2(x_2, y_2)$ is given by

$$m = \frac{y_2 - y_1}{x_2 - x_1}, \; x_1 \neq x_2$$

Frequently, the Greek letter Δ is used to designate the change in a variable. Using this notation, we can write the equations for the change in y and the change in x as follows:

$$\text{Change in } y = \Delta y = y_2 - y_1 \qquad \text{Change in } x = \Delta x = x_2 - x_1$$

With this notation, the slope formula is written $m = \dfrac{\Delta y}{\Delta x}$.

INSTRUCTOR NOTE

When you have finished discussing positive and negative slope, draw some random lines and ask students whether the slope is positive or negative.

Now draw a line and ask students how they could approximate the slope of the line. Do this twice, once with a line that has positive slope and once with a line that has negative slope.

➡ Find the slope of the line containing the points whose coordinates are (−2, 0) and (4, 5).

Let $P_1 = (-2, 0)$ and $P_2 = (4, 5)$. (It does not matter which point is named P_1 and which P_2; the slope will be the same.)

$$m = \frac{y_2 - y_1}{x_2 - x_1} = \frac{5 - 0}{4 - (-2)} = \frac{5}{6}$$

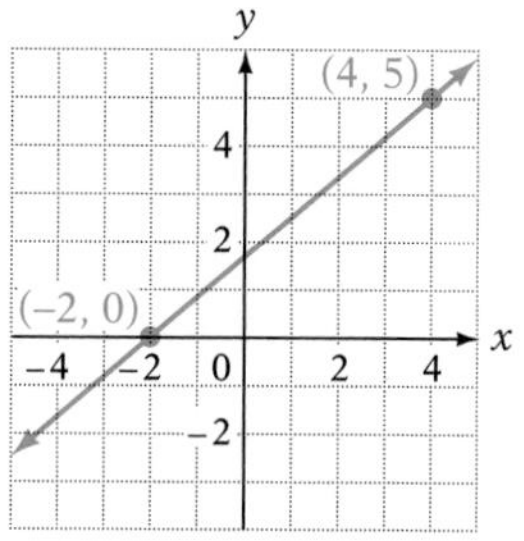

Positive slope

A line that slants upward to the right always has a **positive slope.**

➡ Find the slope of the line containing the points whose coordinates are (−3, 4) and (4, 2).

Let $P_1 = (-3, 4)$ and $P_2 = (4, 2)$.

$$m = \frac{y_2 - y_1}{x_2 - x_1} = \frac{2 - 4}{4 - (-3)} = \frac{-2}{7} = -\frac{2}{7}$$

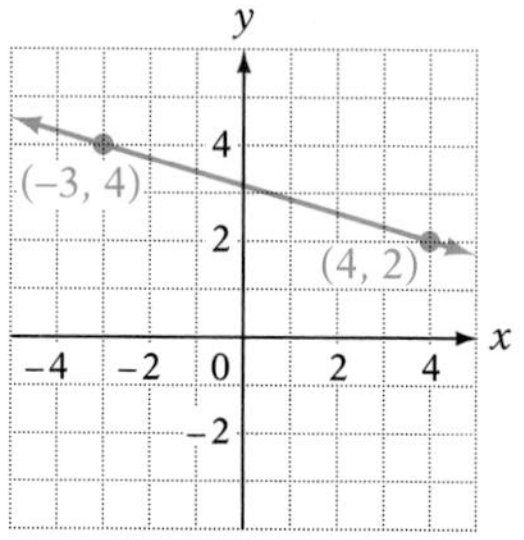

Negative slope

A line that slants downward to the right always has a **negative slope.**

INSTRUCTOR NOTE

Students have a tendency to think that zero slope means *no slope.* Emphasize that a slope of zero refers to a line that is parallel to the x-axis; a line whose slope is undefined is parallel to the y-axis.

➡ Find the slope of the line containing the points whose coordinates are (−2, 2) and (4, 2).

Let $P_1 = (-2, 2)$ and $P_2 = (4, 2)$.

$$m = \frac{y_2 - y_1}{x_2 - x_1} = \frac{2 - 2}{4 - (-2)} = \frac{0}{6} = 0$$

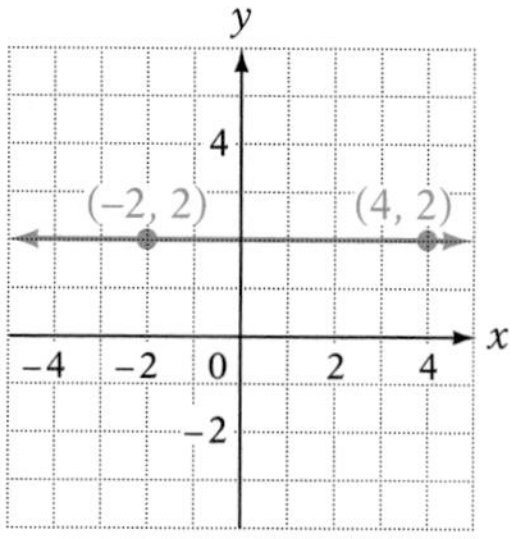

Zero slope

A horizontal line has **zero slope.**

➡ Find the slope of the line containing the points whose coordinates are (1, −2) and (1, 3).

Let $P_1 = (1, -2)$ and $P_2 = (1, 3)$.

$$m = \frac{y_2 - y_1}{x_2 - x_1} = \frac{3 - (-2)}{1 - 1} = \frac{5}{0} \quad \text{Not a real number}$$

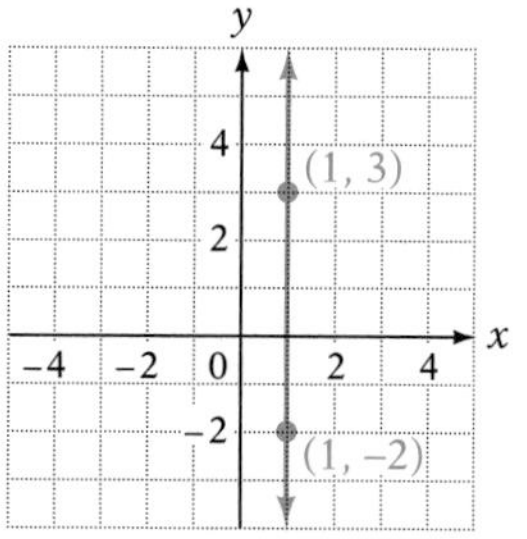

Undefined slope

The slope of a vertical line is **undefined.**

Example 1 Find the slope of the line containing the points whose coordinates are (2, −5) and (−4, 2).

Solution $m = \dfrac{y_2 - y_1}{x_2 - x_1} = \dfrac{2 - (-5)}{-4 - 2} = \dfrac{7}{-6} = -\dfrac{7}{6}$ ▶ Let $P_1 = (2, -5)$ and $P_2 = (-4, 2)$.

The slope is $-\dfrac{7}{6}$.

Problem 1 Find the slope of the line containing the points whose coordinates are (4, −3) and (2, 7).

Solution See page S7. −5

POINT OF INTEREST

One of the motivations for the discovery of calculus was to solve a more complicated version of the distance-rate problem at the right.

You may be familiar with twirling a ball on the end of a string. If you release the string, the ball flies off in a path as shown below.

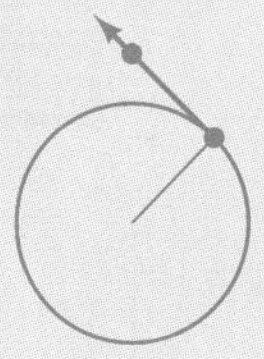

The question that mathematicians tried to answer was essentially "What is the slope of the line represented by the arrow?"

Answering similar questions led to the development of one aspect of calculus.

INSTRUCTOR NOTE

These two examples illustrate that the concept of slope has real applications. You might have students make a list of some quantities that we use every day that are treated mathematically as slope; for instance, miles per gallon, miles per hour, and cost per unit.

In a sense, just as we translate the phrase "the sum of" as addition, a phrase that contains "per" is translated as slope.

There are many applications of slope. Here are two possibilities.

The first record for the one-mile run was recorded in 1865 in England. Richard Webster ran the mile in 4 min 36.5 s. His average speed was approximately 19 ft/s. The graph at the right shows the distance he ran during that run. From the graph, note that after 60 s (1 min) he had traveled 1140 ft and that after 180 s (3 min) he had traveled 3420 ft. The slope of the line between these two points is

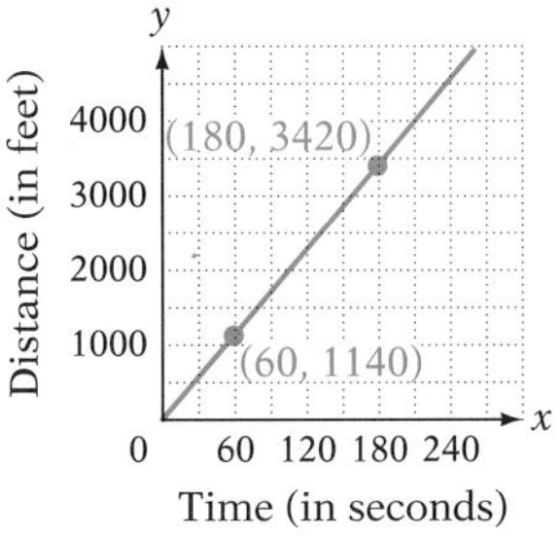

$$m = \frac{3420 - 1140}{180 - 60} = \frac{2280}{120} = 19$$

Note that the slope of the line is the same as Webster's average speed, 19 ft/s. Average speed is related to slope.

Here is another example, this one related to economics.

During the 1980s the value of a 1957 Chevrolet increased dramatically, reaching about \$26,000 by 1990. Since then, however, the value has been decreasing at a rate of approximately \$3500 per year. The graph at the right shows the value of the car for the years 1990 through 1994. From the graph, note that the value of the car in 1991 was \$22,500 and that its value in 1993 was \$15,500. The slope of the line between these two points is

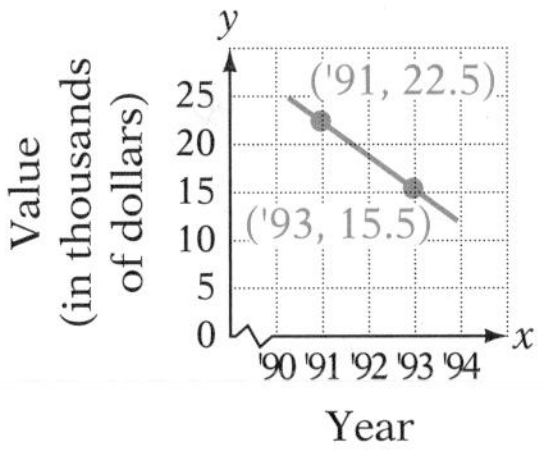

$$m = \frac{15{,}500 - 22{,}500}{1993 - 1991} = \frac{-7000}{2} = -3500$$

Note that if we interpret a negative slope as a decrease, then the slope of the line is the same as the rate at which the value of the car is decreasing, \$3500 per year.

In general, any quantity that is expressed by using the word *per* is represented mathematically as slope. In the first example, slope was 19 feet *per* second. In the second example, slope was −\$3500 *per* year.

Example 2 The graph shows the relationship between the cost of an item and the sales tax. Find the slope of the line between the two points shown on the graph. Write a sentence that states the meaning of the slope.

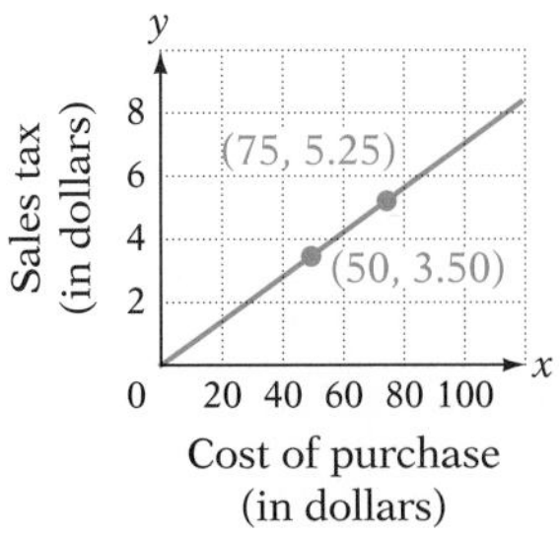

Solution $m = \frac{5.25 - 3.50}{75 - 50} = \frac{1.75}{25} = 0.07$

A slope of 0.07 means that the sales tax is \$.07 per dollar.

Problem 2 The graph shows the decrease in the value of a printing press for a period of six years. Find the slope of the line between the two points shown on the graph. Write a sentence that states the meaning of the slope.

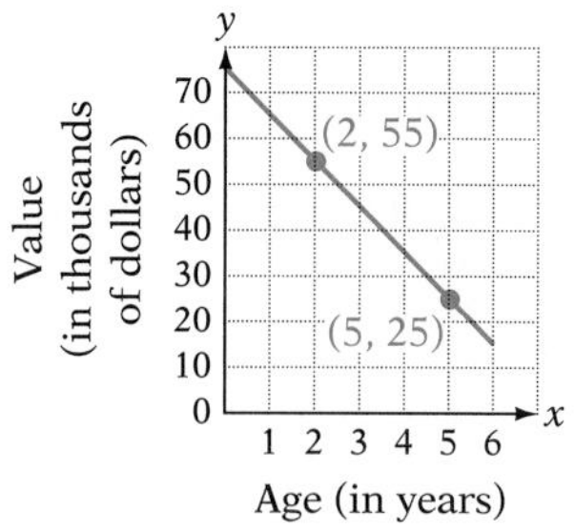

Solution See page S7. $m = -10{,}000$; a slope of $-10{,}000$ means that the value of the printing press is decreasing by $10,000 per year.

2 Graph a line given a point and the slope

The graph of the equation $y = -\frac{3}{4}x + 4$ is shown at the right. The points $(-4, 7)$ and $(4, 1)$ are on the graph. The slope of the line is

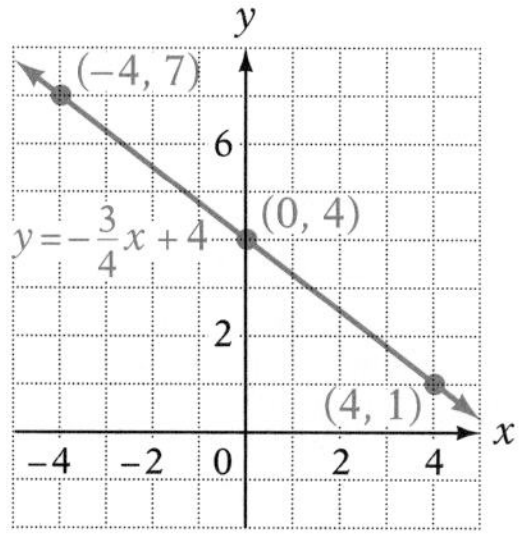

$$m = \frac{7-1}{-4-4} = \frac{6}{-8} = -\frac{3}{4}$$

Note that the slope of the line has the same value as the coefficient of x. The y-intercept is $(0, 4)$.

Slope-Intercept Form of a Straight Line

The equation $y = mx + b$ is called the **slope-intercept** form of a straight line. The slope of the line is m, the coefficient of x. The y-intercept is $(0, b)$.

When the equation of a straight line is in the form $y = mx + b$, the graph can be drawn by using the slope and the y-intercept. First locate the y-intercept. Use the slope to find a second point on the line. Then draw a line through the two points.

Graph $y = \frac{5}{3}x - 4$ by using the slope and the y-intercept.

The slope is the coefficient of x.

$$m = \frac{5}{3} = \frac{\text{change in } y}{\text{change in } x}$$

The y-intercept is $(0, -4)$.

Beginning at the y-intercept $(0, -4)$, move right 3 units (change in x) and then up 5 units (change in y).

The point whose coordinates are $(3, 1)$ is a second point on the graph. Draw a line through the points $(0, -4)$ and $(3, 1)$.

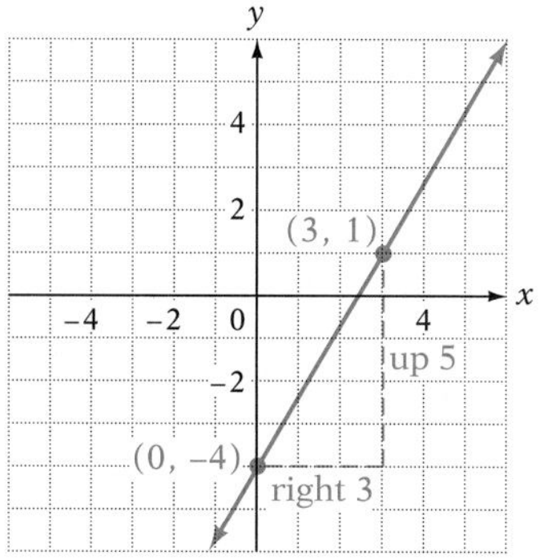

LOOK CLOSELY

When graphing a line by using its slope and y-intercept, *always* start at the y-intercept.

INSTRUCTOR NOTE

For consistency, we always graph a slope by moving to the right and then up or down, depending on whether the slope is positive or negative. An interesting class activity is to graph a slope in other ways. For instance, a slope of $-\frac{1}{2}$ can be graphed by moving left 2 and up 1. It can also be graphed by moving right 1 and down $\frac{1}{2}$, right 4 and down 2, or several other ways. Doing this will help confirm for students that slope measures the ratio of the change in y to the change in x.

Graph $x + 2y = 4$ by using the slope and the y-intercept.

Solve the equation for y.

$$x + 2y = 4$$
$$2y = -x + 4$$
$$y = -\frac{1}{2}x + 2$$

The slope is $m = -\frac{1}{2}$ and the y-intercept is (0, 2).

Beginning at the y-intercept (0, 2), move right 2 units (change in x) and then down 1 unit (change in y).

The point whose coordinates are (2, 1) is a second point on the graph. Draw a line through the points (0, 2) and (2, 1).

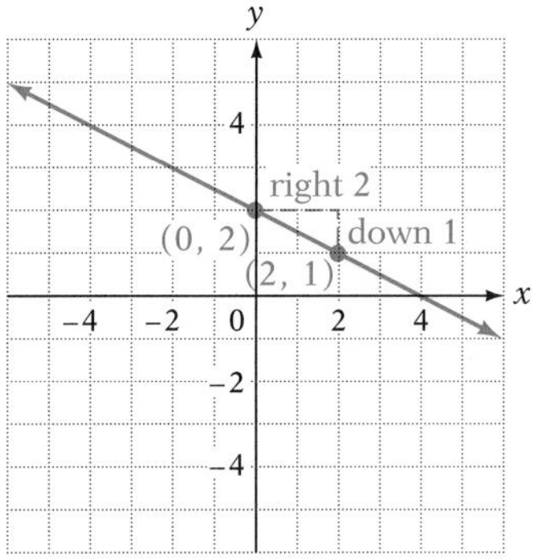

Example 3 Graph $y = -\frac{3}{2}x + 4$ by using the slope and the y-intercept.

Solution y-intercept $= (0, 4)$ ▸ Locate the y-intercept.

$m = -\frac{3}{2} = \frac{-3}{2}$ ▸ $m = \frac{\text{change in } y}{\text{change in } x}$

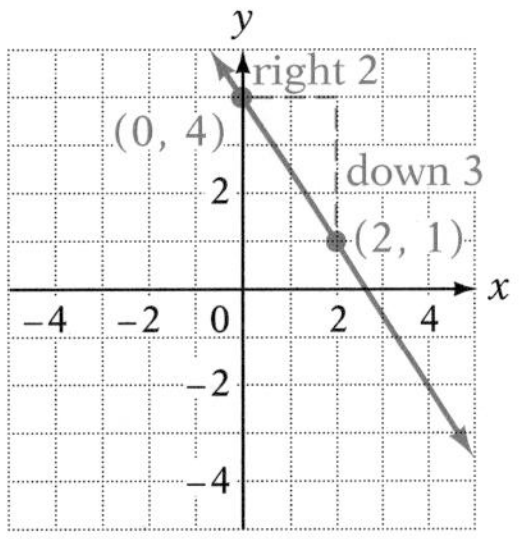

▸ Beginning at the y-intercept, (0, 4), move right 2 units and then down 3 units. (2, 1) is a second point on the graph.

▸ Draw a line through the points (0, 4) and (2, 1).

Problem 3 Graph $2x + 3y = 6$ by using the slope and the y-intercept.

Solution See page S7.

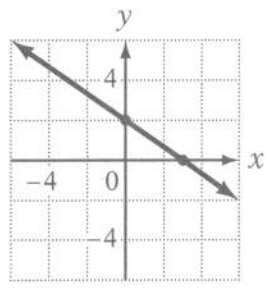

The graph of a line can be drawn when any point on the line and the slope of the line are given.

Graph the line that passes through the point $(-4, -4)$ and has slope 2.

When the slope is an integer, write it as a fraction with denominator 1.

$$m = 2 = \frac{2}{1} = \frac{\text{change in } y}{\text{change in } x}$$

LOOK CLOSELY

This example differs from the others in that a point other than the y-intercept is used. In this case, start at the given point.

Locate $(-4, -4)$ in the coordinate plane. Beginning at that point, move right 1 unit (change in x) and then up 2 units (change in y).

The point whose coordinates are $(-3, -2)$ is a second point on the graph. Draw a line through the points $(-4, -4)$ and $(-3, -2)$.

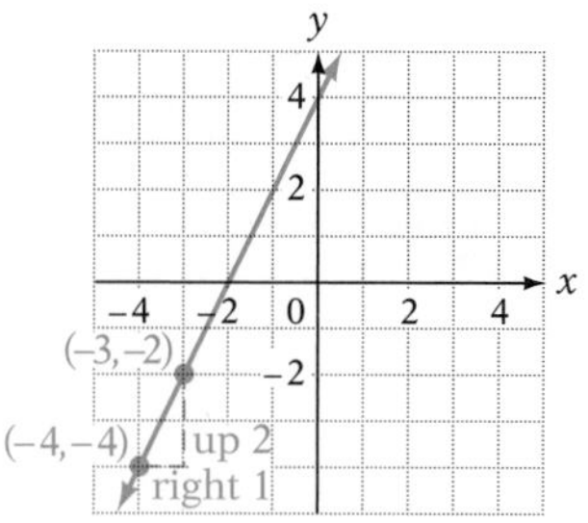

Example 4 Graph the line that passes through the point $(-2, 3)$ and has slope $= -\frac{4}{3}$.

Solution

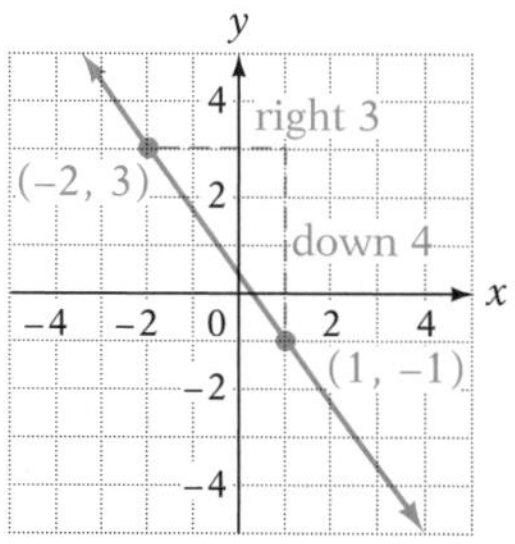

- $(x_1, y_1) = (-2, 3)$; $m = -\frac{4}{3} = \frac{-4}{3} = \frac{\text{change in } y}{\text{change in } x}$
- Beginning at the point $(-2, 3)$, move right 3 units and then down 4 units. $(1, -1)$ is a second point on the line.
- Draw a line through the points $(-2, 3)$ and $(1, -1)$.

Problem 4 Graph the line that passes through the point $(-3, -2)$ and has slope 3.

Solution See page S7.

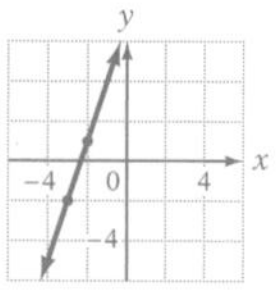

CONCEPT REVIEW 3.4

Determine whether the following statements are always true, sometimes true, or never true.

1. The slope of a line that slants downward to the left is positive. Always true

2. The y-intercept of a line is the point at which the line crosses the y-axis. Always true

3. The slope of a vertical line is zero. Never true

4. A line whose slope is undefined is parallel to the y-axis. Always true

EXERCISES 3.4

1 1. Explain in your own words how to find the slope of a line given two points on the line.

2. What is the difference between a line that has positive slope and one that has negative slope? Can a line have zero slope? If so, explain what its graph would look like.

Find the slope of the line containing the points P_1 and P_2.

3. $P_1(1, 3), P_2(3, 1)$ -1

4. $P_1(2, 3), P_2(5, 1)$ $-\frac{2}{3}$

5. $P_1(-1, 4), P_2(2, 5)$ $\frac{1}{3}$

6. $P_1(3, -2), P_2(1, 4)$ -3

7. $P_1(-1, 3), P_2(-4, 5)$ $-\frac{2}{3}$

8. $P_1(-1, -2), P_2(-3, 2)$ -2

9. $P_1(0, 3), P_2(4, 0)$ $-\frac{3}{4}$

10. $P_1(-2, 0), P_2(0, 3)$ $\frac{3}{2}$

11. $P_1(2, 4), P_2(2, -2)$ Undefined

12. $P_1(4, 1), P_2(4, -3)$ Undefined

13. $P_1(2, 5), P_2(-3, -2)$ $\frac{7}{5}$

14. $P_1(4, 1), P_2(-1, -2)$ $\frac{3}{5}$

15. $P_1(2, 3), P_2(-1, 3)$ 0

16. $P_1(3, 4), P_2(0, 4)$ 0

17. $P_1(0, 4), P_2(-2, 5)$ $-\frac{1}{2}$

18. $P_1(-2, 3), P_2(-2, 5)$ Undefined

19. $P_1(-3, -1), P_2(-3, 4)$ Undefined

20. $P_1(-2, -5), P_2(-4, -1)$ -2

21. *Business* The graph below shows the relationship between the March 1998 price of the "Flutter" Beanie Baby and the August 1998 price. Find the slope of the line between the two points shown on the graph. (Note that we let 1 represent January, 2 represent February, and so on.) Write a sentence that explains the meaning of the slope. (*Source: Beanie World Monthly*, August 1998)

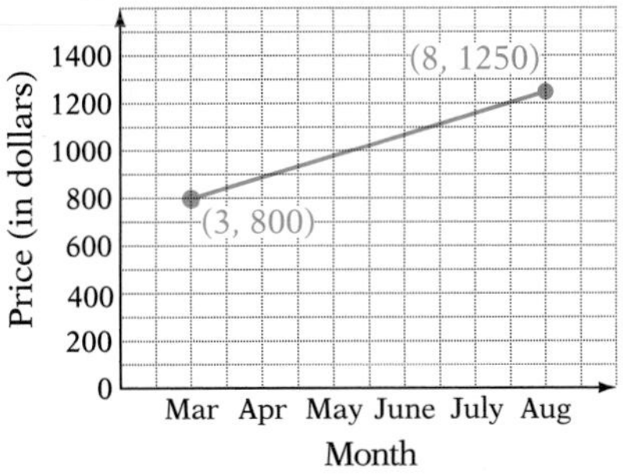

$m = 90$
The price of the "Flutter" Beanie Baby was increasing at a rate of \$90 per month.

22. *Aeronautics* The graph below shows how the altitude of an airplane above the runway changes after take-off. Find the slope of the line. Write a sentence that states the meaning of the slope.

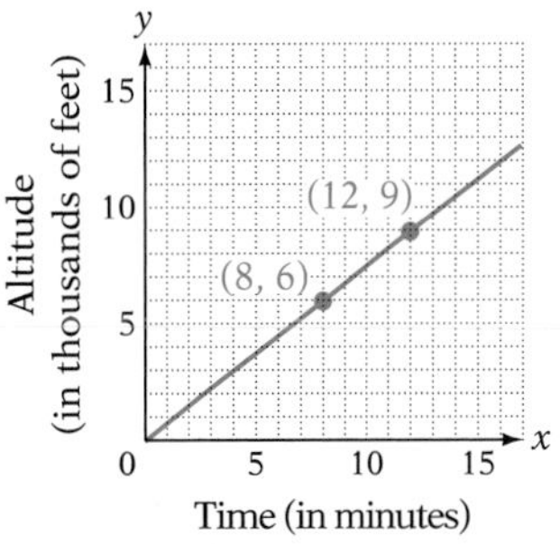

$m = 750$
The altitude of the plane increases at 750 ft/min.

23. *Automotive Technology* The graph below shows how the amount of gas in the tank of a car decreases as the car is driven. Find the slope of the line. Write a sentence that states the meaning of the slope.

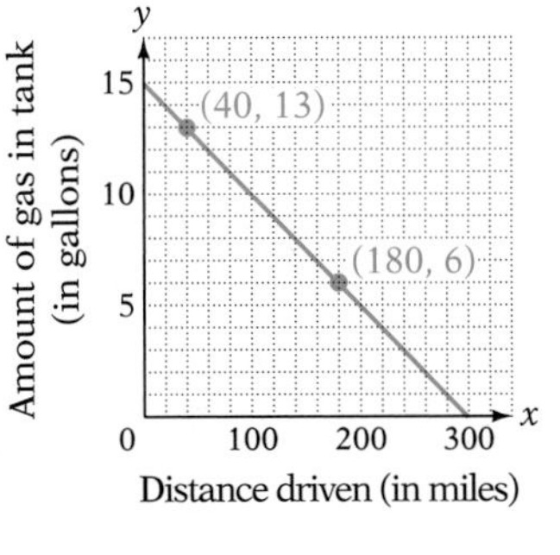

$m = -0.05$
For each mile the car is driven, approximately 0.05 gal of fuel is used.

24. *Meteorology* The troposphere extends from the surface of Earth to an elevation of approximately 11 km. The graph below shows the decrease in temperature of the troposphere as altitude increases. Find the slope of the line. Write a sentence that states the meaning of the slope.

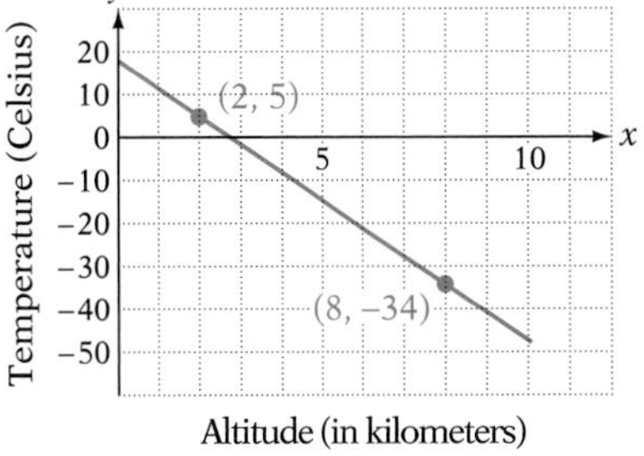

$m = -6.5$
The temperature of the troposphere decreases 6.5°C/km.

25. *Sports* Lois and Tanya start from the same place on a jogging course. Lois is jogging at 9 km/h, and Tanya is jogging at 6 km/h. The graphs below show the total distance traveled by each jogger and the total distance between Lois and Tanya. Which lines represent which distances?

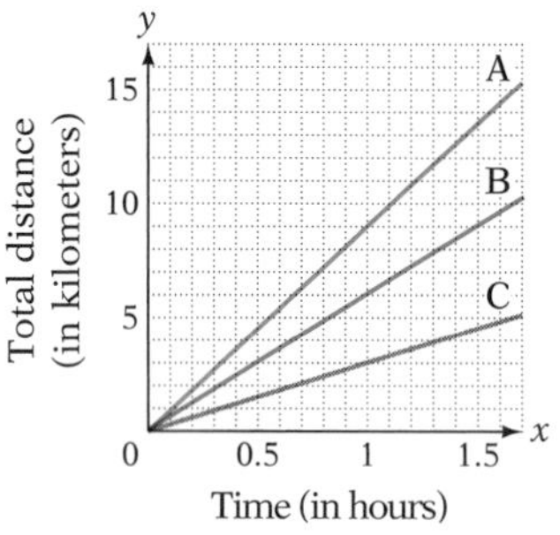

A - Lois
B - Tanya
C - Distance between

26. *Chemistry* A chemist is filling two cans from a faucet that releases water at a constant rate. Can 1 has a diameter of 20 mm, and can 2 has a diameter of 30 mm. The depth of the water in each can is measured at 5-second intervals. The graph of the results is shown below. On the graph, which line represents the depth of the water for which can?

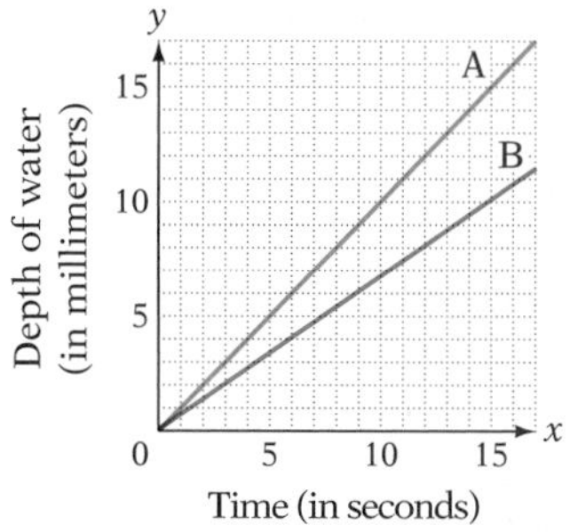

Can 1 - A
Can 2 - B

27. *Health Science* The American National Standards Institute (ANSI) states that the slope for a wheelchair ramp must not exceed $\frac{1}{12}$.

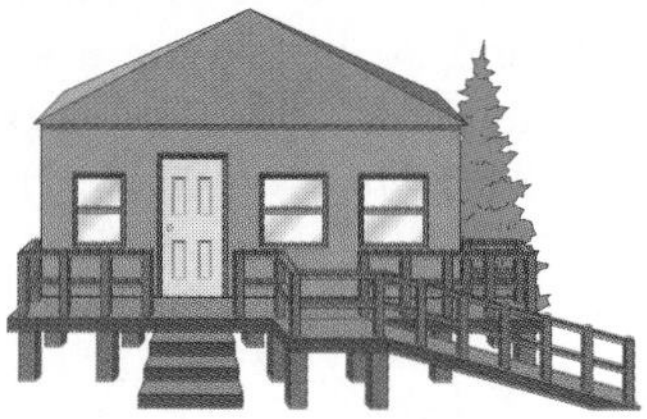

a. Does a ramp that is 6 in. high and 5 ft long meet the requirements of ANSI?

b. Does a ramp that is 12 in. high and 170 in. long meet the requirements of ANSI?

a. No **b.** Yes

28. *Health Science* A ramp for a wheelchair must be 14 in. high. What minimum length must this ramp be so that it meets the ANSI requirements given in Exercise 27? 169 in.

2 Graph by using the slope and the y-intercept.

29. $y = \frac{1}{2}x + 2$

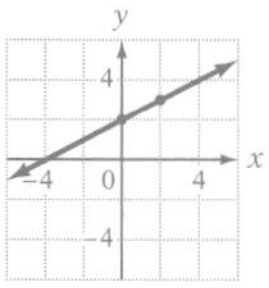

30. $y = \frac{2}{3}x - 3$

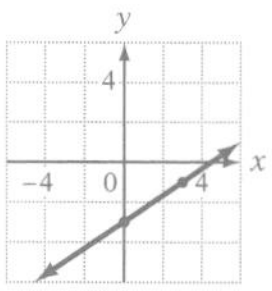

31. $y = -\frac{3}{2}x$

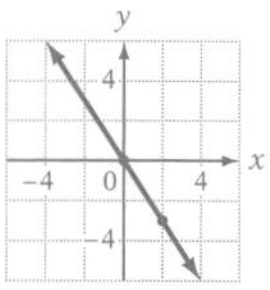

32. $y = \frac{3}{4}x$

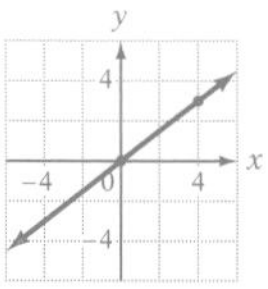

33. $y = -\frac{1}{2}x + 2$

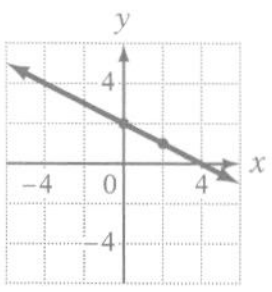

34. $y = \frac{2}{3}x - 1$

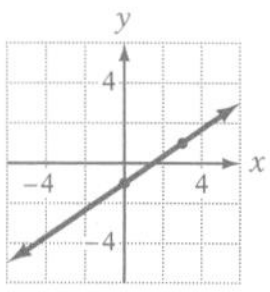

35. $x - 3y = 3$

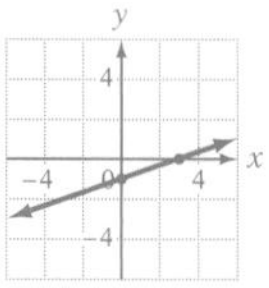

36. $3x + 2y = 8$

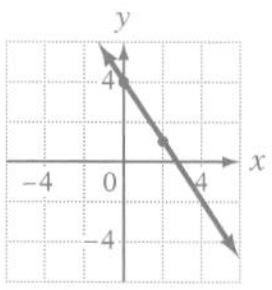

37. $4x + y = 2$

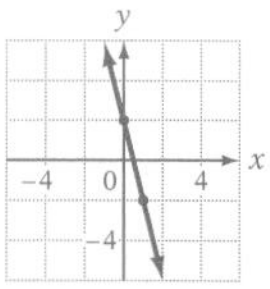

38. Graph the line that passes through the point $(-1, -3)$ and has slope $\frac{4}{3}$.

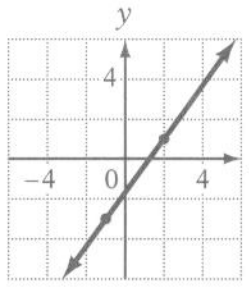

39. Graph the line that passes through the point $(-2, -3)$ and has slope $\frac{5}{4}$.

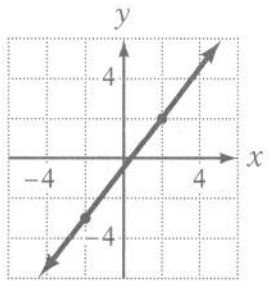

40. Graph the line that passes through the point $(-3, 0)$ and has slope -3.

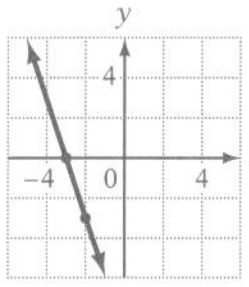

41. Graph the line that passes through the point $(2, 0)$ and has slope -1.

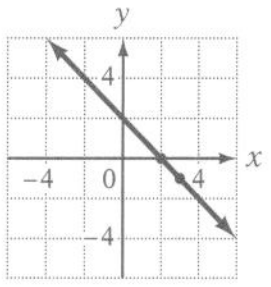

APPLYING CONCEPTS 3.4

42. Let $f(x)$ be the digit in the nth decimal place of the repeating digit $0.\overline{387}$. For example, $f(3) = 7$ because 7 is the digit in the third decimal place. Find $f(14)$.
8

Complete the sentences.

43. If a line has a slope of 2, then the value of y increases/decreases by ______ as the value of x increases by 1. increases by 2

44. If a line has a slope of -3, then the value of y increases/decreases by ______ as the value of x increases by 1. decreases by 3

45. If a line has a slope of $\frac{1}{2}$, then the value of y increases/decreases by ______ as the value of x increases by 1. increases by $\frac{1}{2}$

46. If a line has a slope of $-\frac{2}{3}$, then the value of y increases/decreases by ______ as the value of x increases by 1. decreases by $\frac{2}{3}$

Determine the value of k such that the points whose coordinates are given below lie on the same line.

47. $(3, 2), (4, 6), (5, k)$ 10

48. $(k, 1), (0, -1), (2, -2)$ -4

49. *Computer Science* According to Jupiter Communications, the world on-line market is growing. A study was conducted to find the total number of on-line households between 1995 and 1998 with projections through 1999. The findings are given in the following table.

Year	Number of Households (in millions)
1995	15
1996	23
1997	34
1998	45
1999	56

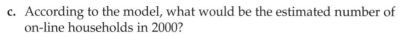

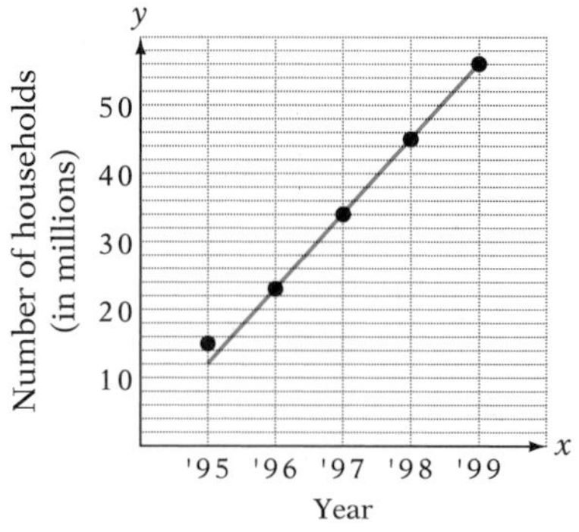

a. Use the coordinate axes at the right to graph the data in the table.

b. The equation of the line that approximately models the data in the table is $y = 11x - 21933$, where x is the year and y is the number of on-line households. Graph this line on the same coordinate axes.

c. According to the model, what would be the estimated number of on-line households in 2000?

d. According to the model, what is the increase in the number of on-line households between 1997 and 1998?

e. Write a sentence that gives an interpretation of the slope of the straight-line model.

f. If the model were used to estimate the total number of on-line households prior to 1995, what would be the estimated number for 1994?

g. What are some of the problems in using this model to predict the total number of on-line households beyond the year 2000?

c. 67 million households **d.** 11 million households
e. The number of on-line households is increasing at a rate of 11 million per year.
f. 1 million households

50. A warning sign for drivers on a mountain road might read, "Caution: 8% downgrade next 2 miles." Explain this statement.

51. Graph $y = 2x + 3$ and $y = 2x - 1$ on the same coordinate system. Explain how the graphs are different and how they are the same. If b is any real number, how is the graph of $y = 2x + b$ related to the two graphs you have drawn?

52. Explain how you can use the slope of a line to determine whether three given points lie on the same line. Then use your procedure to determine whether all of the following points lie on the same line.

a. (2, 5), (−1, −1), (3, 7) **b.** (−1, 5), (0, 3), (−3, 4)

SECTION 3.5 Finding Equations of Lines

1 Find the equation of a line given a point and the slope

13 HM³

When the slope of a line and a point on the line are known, the equation of the line can be determined. If the particular point is the y-intercept, use the slope-intercept form, $y = mx + b$, to find the equation.

⇨ Find the equation of the line that contains the point $(0, 3)$ and has slope $\frac{1}{2}$.

The known point is the y-intercept, $(0, 3)$.

Use the slope-intercept form. $y = mx + b$

Replace m with $\frac{1}{2}$, the given slope, and replace $y = \frac{1}{2}x + 3$

b with 3, the y-coordinate of the y-intercept.

The equation of the line is $y = \frac{1}{2}x + 3$.

One method of finding the equation of a line when the slope and any point on the line are known involves using the *point-slope formula*. This formula is derived from the formula for the slope of a line.

Let (x_1, y_1) be the given point on the line, and let (x, y) be any other point on the line.

Use the formula for the slope of a line. $\frac{y - y_1}{x - x_1} = m$

Multiply both sides of the equation by $(x - x_1)$. $\frac{y - y_1}{x - x_1}(x - x_1) = m(x - x_1)$

Then simplify. $y - y_1 = m(x - x_1)$

Point-Slope Formula

Let m be the slope of a line, and let (x_1, y_1) be the coordinates of a point on the line. The equation of the line can be found using the **point-slope formula:**

$$y - y_1 = m(x - x_1)$$

⇨ Find the equation of the line that contains the point $(4, -1)$ and has slope $-\frac{3}{4}$.

Use the point-slope formula. $y - y_1 = m(x - x_1)$

Substitute the slope, $-\frac{3}{4}$, and the coordinates of the $y - (-1) = -\frac{3}{4}(x - 4)$

given point, $(4, -1)$, into the point-slope formula. Then simplify. $y + 1 = -\frac{3}{4}x + 3$

                                                        $y = -\frac{3}{4}x + 2$

The equation of the line is $y = -\frac{3}{4}x + 2$.

Find the equation of the line that passes through the point whose coordinates are (4, 3) and whose slope is undefined.

Because the slope is undefined, the point-slope formula cannot be used to find the equation. Instead, recall that when the slope is undefined, the line is vertical and that the equation of a vertical line is $x = a$, where a is the x-coordinate of the x-intercept. Because the line is vertical and passes through (4, 3), the x-intercept is (4, 0).

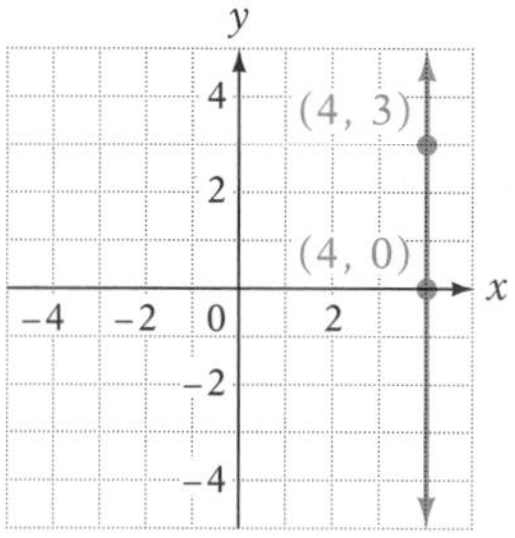

The equation of the line is $x = 4$.

Example 1 Find the equation of the line that contains the point (−2, 4) and has slope 2.

Solution

$y - y_1 = m(x - x_1)$ ▸ Use the point-slope formula.

$y - 4 = 2[x - (-2)]$ ▸ Substitute the slope, 2, and the coordinates of the given point, (−2, 4), into the point-slope formula.

$y - 4 = 2(x + 2)$

$y - 4 = 2x + 4$

$y = 2x + 8$

The equation of the line is $y = 2x + 8$.

Problem 1 Find the equation of the line that contains the point (4, −3) and has slope −3.

Solution See page S7. $y = -3x + 9$

2 Find the equation of a line given two points

INSTRUCTOR NOTE

Students need frequent reminders of the following fact. If the point-slope formula is used to determine the equation of a line, it is *always* necessary to know a point on the line and the slope of the line. Finding the equation of a line given two points is a good example. First the slope formula must be used to determine m.

INSTRUCTOR NOTE

A good classroom exercise is to have students use (x_2, y_2) in this problem and verify that the same equation results.

The point-slope formula and the formula for slope are used to find the equation of a line when two points are known.

Find the equation of the line containing the points (3, 2) and (−5, 6).

To use the point-slope formula, we must know the slope. Use the formula for slope to determine the slope of the line between the two given points.

Let $(x_1, y_1) = (3, 2)$ and $(x_2, y_2) = (-5, 6)$.

$$m = \frac{y_2 - y_1}{x_2 - x_1} = \frac{6 - 2}{-5 - 3} = \frac{4}{-8} = -\frac{1}{2}$$

Now use the point-slope formula with $m = -\frac{1}{2}$ and $(x_1, y_1) = (3, 2)$.

$$y - y_1 = m(x - x_1)$$
$$y - 2 = -\frac{1}{2}(x - 3)$$
$$y - 2 = -\frac{1}{2}x + \frac{3}{2}$$
$$y = -\frac{1}{2}x + \frac{7}{2}$$

The equation of the line is $y = -\frac{1}{2}x + \frac{7}{2}$.

Example 2 Find the equation of the line containing the points (2, 3) and (4, 1).

Solution

$$m = \frac{y_2 - y_1}{x_2 - x_1} = \frac{1 - 3}{4 - 2} = \frac{-2}{2} = -1$$

▶ Find the slope. Let $(x_1, y_1) = (2, 3)$ and $(x_2, y_2) = (4, 1)$.

$$y - y_1 = m(x - x_1)$$
$$y - 3 = -1(x - 2)$$
$$y - 3 = -x + 2$$
$$y = -x + 5$$

▶ Substitute the slope and the coordinates of either one of the known points into the point-slope formula.

The equation of the line is $y = -x + 5$.

Problem 2 Find the equation of the line containing the points (2, 0) and (5, 3).

Solution See page S7. $y = x - 2$

3 Application problems

Linear functions can be used to model a variety of applications in science and business. For each application, data are collected and the independent and dependent variables are selected. Then a linear function is determined that models the data.

Example 3 In 1950, there were 13 million adults 65 years old or older in the United States. Data from the Census Bureau shows that the population of these adults has been increasing at a constant rate of approximately 0.5 million per year. This rate of increase is expected to continue through the year 2010. Find the equation of the line that approximates the population of adults 65 years old or older in terms of the year. Use your equation to approximate the population of these adults in 2005.

Strategy Select the independent and dependent variables. Because we want to determine the population of adults 65 years or older, that quantity is the *dependent* variable, y. The year is the *independent* variable.

From the data, the ordered pair (1950, 13), where the y-coordinate is in millions of people, is a point on the line. The slope of the line is the *rate* of increase, 0.5 million per year.

Solution Use the point-slope formula.

$$y - y_1 = m(x - x_1)$$

$$y - 13 = 0.5(x - 1950)$$

▶ $m = 0.5, (x_1, y_1) = (1950, 13)$

$$y - 13 = 0.5x - 975$$
$$y = 0.5x - 962$$

The linear function is $f(x) = 0.5x - 962$.

$$\begin{aligned} f(x) &= 0.5x - 962 \\ f(2005) &= 0.5(2005) - 962 \\ &= 1002.5 - 962 \\ &= 40.5 \end{aligned}$$

▶ Evaluate the function at 2005 to predict the population in 2005.

There will be approximately 40.5 million people 65 years or older in 2005.

Problem 3 Gabriel Daniel Fahrenheit invented the mercury thermometer in 1717. In terms of readings on this thermometer, water freezes at 32°F and boils at 212°F. In 1742 Anders Celsius invented the Celsius temperature scale. On this scale, water freezes at 0° and boils at 100°. Determine a linear function that can be used to predict the Celsius temperature when the Fahrenheit temperature is known.

Solution See page S7. $f(F) = \frac{5}{9}(F - 32)$

CONCEPT REVIEW 3.5

Determine whether the following statements are always true, sometimes true, or never true.

1. Increasing the value of m in the equation $y = mx + b$ increases the slope of the line that is the graph of the equation. Always true
2. Decreasing the value of b in the equation $y = mx + b$ decreases the slope of the line that is the graph of the equation. Never true
3. The point-slope formula for the equation of a line is $y = m(x - x_1)$. Never true
4. $x = a$ is the equation for a vertical line. Always true
5. The line represented by the equation $y = 2x - \frac{1}{2}$ has slope $-\frac{1}{2}$ and x-intercept 2. Never true
6. If $y = mx + b$, then m represents the rate of change of y with respect to x. Always true
7. A horizontal line has no slope. Never true

EXERCISES 3.5

1 1. What is the point-slope formula and how is it used?

2. Explain one way to find the equation of a line given its slope and its y-intercept.

Find the equation of the line that contains the given point and has the given slope.

3. Point $(0, 5)$, $m = 2$
$y = 2x + 5$

4. Point $(0, 3)$, $m = 1$
$y = x + 3$

5. Point $(2, 3)$, $m = \frac{1}{2}$
$y = \frac{1}{2}x + 2$

6. Point $(5, 1)$, $m = \frac{2}{3}$
$y = \frac{2}{3}x - \frac{7}{3}$

7. Point $(3, 0)$, $m = -\frac{5}{3}$
$y = -\frac{5}{3}x + 5$

8. Point $(-2, 0)$, $m = \frac{3}{2}$
$y = \frac{3}{2}x + 3$

9. Point $(-1, 7)$, $m = -3$
$y = -3x + 4$

10. Point $(-2, 4)$, $m = -4$
$y = -4x - 4$

11. Point $(0, 0)$, $m = \frac{1}{2}$
$y = \frac{1}{2}x$

12. Point $(0, 0)$, $m = \frac{3}{4}$
$y = \frac{3}{4}x$

13. Point $(2, -3)$, $m = 3$
$y = 3x - 9$

14. Point $(4, -5)$, $m = 2$
$y = 2x - 13$

15. Point $(3, 5)$, $m = -\frac{2}{3}$
$y = -\frac{2}{3}x + 7$

16. Point $(5, 1)$, $m = -\frac{4}{5}$
$y = -\frac{4}{5}x + 5$

17. Point $(0, -3)$, $m = -1$
$y = -x - 3$

18. Point $(2, 0)$, $m = \frac{5}{6}$
$y = \frac{5}{6}x - \frac{5}{3}$

19. Point $(3, -4)$, slope is undefined $x = 3$

20. Point $(-2, 5)$, slope is undefined $x = -2$

21. Point $(-2, -3)$, $m = 0$
$y = -3$

22. Point $(-3, -2)$, $m = 0$
$y = -2$

23. Point $(4, -5)$, $m = -2$
$y = -2x + 3$

24. Point $(-3, 5)$, $m = 3$
$y = 3x + 14$

25. Point $(-5, -1)$, slope is undefined $x = -5$

26. Point $(0, 4)$, slope is undefined $x = 0$

2 Find the equation of the line that contains the given points.

27. $P_1(0, 2)$, $P_2(3, 5)$
$y = x + 2$

28. $P_1(0, 4)$, $P_2(1, 5)$
$y = x + 4$

29. $P_1(0, -3)$, $P_2(-4, 5)$
$y = -2x - 3$

30. $P_1(0, -2)$, $P_2(-3, 4)$
$y = -2x - 2$

31. $P_1(-1, 3)$, $P_2(2, 4)$
$y = \frac{1}{3}x + \frac{10}{3}$

32. $P_1(-1, 1)$, $P_2(4, 4)$
$y = \frac{3}{5}x + \frac{8}{5}$

33. $P_1(0, 3)$, $P_2(2, 0)$
$y = -\frac{3}{2}x + 3$

34. $P_1(0, 4)$, $P_2(2, 0)$
$y = -2x + 4$

35. $P_1(-2, -3)$, $P_2(-1, -2)$
$y = x - 1$

36. $P_1(4, 1)$, $P_2(3, -2)$
$y = 3x - 11$

37. $P_1(2, 3)$, $P_2(5, 5)$
$y = \frac{2}{3}x + \frac{5}{3}$

38. $P_1(7, 2)$, $P_2(4, 4)$
$y = -\frac{2}{3}x + \frac{20}{3}$

39. $P_1(2, 0)$, $P_2(0, -1)$
$y = \frac{1}{2}x - 1$

40. $P_1(0, 4)$, $P_2(-2, 0)$
$y = 2x + 4$

41. $P_1(3, -4)$, $P_2(-2, -4)$
$y = -4$

42. $P_1(-3, 3)$, $P_2(-2, 3)$
$y = 3$

43. $P_1(0, 0)$, $P_2(4, 3)$
$y = \frac{3}{4}x$

44. $P_1(2, -5)$, $P_2(0, 0)$
$y = -\frac{5}{2}x$

45. $P_1(-2, 5)$, $P_2(-2, -5)$
$x = -2$

46. $P_1(3, 2)$, $P_2(3, -4)$
$x = 3$

47. $P_1(2, 1)$, $P_2(-2, -3)$
$y = x - 1$

48. $P_1(-3, -2)$, $P_2(1, -4)$
$y = -\frac{1}{2}x - \frac{7}{2}$

49. $P_1(0, 3)$, $P_2(3, 0)$
$y = -x + 3$

50. $P_1(1, -3)$, $P_2(-2, 4)$
$y = -\frac{7}{3}x - \frac{2}{3}$

3 51. *Aeronautics* The pilot of a Boeing 757 jet takes off from Boston's Logan Airport, which is at sea level, and climbs to a cruising altitude of 32,000 ft at a constant rate of 1200 ft/min. Write a linear equation for the height of the plane in terms of the time after take-off. Use your equation to find the height of the plane 11 min after take-off. $y = 1200x$; 13,200 ft

52. *Physical Fitness* A jogger running at 9 mph burns approximately 14 calories per minute. Write a linear equation for the number of calories burned by the jogger in terms of the number of minutes run. Use your equation to find the number of the calories burned after jogging for 32 min. $y = 14x$; 448 calories

53. *Telecommunications* A cellular phone company offers several different options for using a cellular telephone. One option, for people who plan on using the phone only in emergencies, costs the user \$4.95 per month plus \$.59 per minute for each minute the phone is used. Write a linear equation for the monthly cost of the phone in terms of the number of minutes the phone is used. Use your equation to find the monthly cost of using the cellular phone for 13 min in one month. $y = 0.59x + 4.95$; \$12.62

54. *Aeronautics* An Airbus 320 plane takes off from Denver International Airport, which is 5200 ft above sea level, and climbs to 30,000 ft at a constant rate of 1000 ft/min. Write a linear equation for the height of the plane in terms of the time after take-off. Use your equation to find the height of the plane 8 min after take-off. $y = 1000x + 5200$; 13,200 ft

55. *Nutrition* There are approximately 126 calories in a 2-ounce serving of lean hamburger and approximately 189 calories in a 3-ounce serving. Write a linear equation for the number of calories in lean hamburger in terms of the size of the serving. Use your equation to estimate the number of calories in a 5-ounce serving of lean hamburger. $y = 63x$; 315 calories

56. *Chemistry* At sea level, the boiling point of water is 100°C. At an altitude of 2 km, the boiling point of water is 93°C. Write a linear equation for the boiling point of water in terms of the altitude above sea level. Use your equation to predict the boiling point of water on the top of Mount Everest, which is approximately 8.85 km above sea level. Round to the nearest degree. $y = -3.5x + 100$; 69°C

57. *History* In 1927, Charles Lindbergh made history by making the first transatlantic flight from New York to Paris. It took Lindbergh approximately 33.5 h to make the trip. In 1997, the Concorde could make the trip in approximately 3.3 h. Write a linear equation of the time, in hours, it takes to cross the Atlantic in terms of the year. Use your equation to predict how long a flight between the two cities would have taken in 1967. Round your answer to the nearest tenth. On the basis of your answer, do you think a linear model accurately predicts how the flying time between New York and Paris has changed? $y = -0.431x + 864$; 16.2 h

58. *Computers* On the basis of data from the U.S. Department of Commerce, there were 24 million homes with computers in 1991. The average rate of growth in computers in homes is expected to increase by 2.4 million homes per year through 2005. Write a linear equation for the number of computers in homes in terms of the year. Use your equation to find the number of computers that will be in homes in 2001.
$y = 2.4x - 4754.4$; 48 million

59. *Automotive Technology* The gas tank of a certain car contains 16 gal when the driver of the car begins a trip. Each mile driven by the driver decreases the amount of gas in the tank by 0.032 gal. Write a linear equation for the number of gallons of gas in the tank in terms of the number of miles driven. Use your equation to find the number of gallons in the tank after driving 150 mi. $y = -0.032x + 16$; 11.2 gal

60. *Oceanography* Whales, dolphins, and porpoises communicate using high-pitched sounds that travel through the water. The speed at which the sound travels depends on many factors, one of which is the depth of the water. At approximately 1000 m below sea level, the speed of sound is 1480 m/s. Below 1000 m, the speed of sound increases at a constant rate of 0.017 m/s for each additional meter below 1000 m. Write a linear equation for the speed of sound in terms of the number of meters below sea level. Use your equation to approximate the speed of sound 2500 m below sea level. Round to the nearest meter per second. $y = 0.017x + 1463$; 1506 m/s

APPLYING CONCEPTS 3.5

61. What is the x-intercept of the graph of $y = mx + b$? $\left(-\frac{b}{m}, 0\right)$

62. A line contains the points $(4, -1)$ and $(2, 1)$. Find the coordinates of three other points that are on this line. Possible answers are $(0, 3)$, $(1, 2)$, and $(3, 0)$.

63. Given that f is a linear function for which $f(1) = 3$ and $f(-1) = 5$, determine $f(4)$. 0

64. Find the equation of the line that passes through the midpoint of the line segment between $P_1(2, 5)$ and $P_2(-4, 1)$ and has slope -2. $y = -2x + 1$

65. If $y = mx + b$, where m is a given constant, how does the graph of this equation change as the value of b changes?
Changing b moves the graph of the line up or down.

66. *Meteorology* In Exercise 49 of Section 3.3, the wind-chill factor for different air temperatures and a wind speed of 10 mph was given. In the set of ordered pairs that follows, the abscissa is the air temperature in degrees Fahrenheit, and the ordinate is the wind-chill factor when the wind speed is 20 mph. Note that the abscissa in each case is the same as in the earlier exercise. However, the wind speed has increased to 20 mph. Therefore, the wind-chill factor is lower. (*Source: Information Please Almanac*)

$\{(35, 12), (30, 4), (25, -3), (20, -10), (15, -17), (10, -24), (5, -31), (0, -39), (-5, -46), (-10, -53), (-15, -60), (-20, -67), (-25, -74), (-30, -81), (-35, -88)\}$

a. Graph these ordered pairs on the same coordinate axes used for the data in Exercise 49 of Section 3.3, and draw a straight line that as closely as possible goes through all these points.

b. Is the graph of the line that connects these ordered pairs parallel to the line that connects the ordered pairs given in Exercise 49 of Section 3.3? Why?

a.

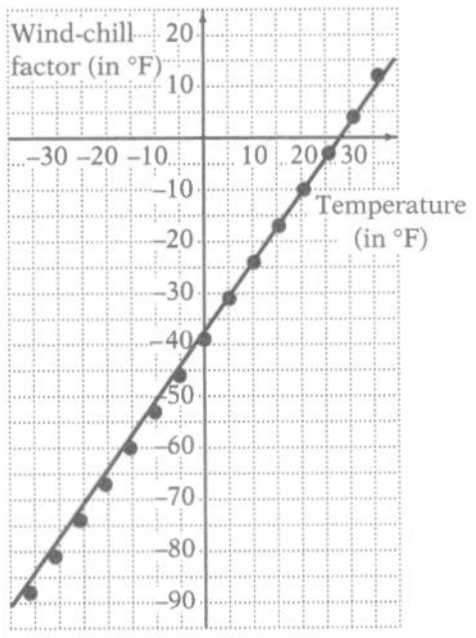

b. No. At higher temperatures the wind speed has less of an effect on the wind-chill factor. At lower temperatures the wind has more of an effect on the wind-chill factor.

67. Explain the similarities and differences between the point-slope formula and the slope-intercept form of a straight line.

68. Explain why the point-slope formula cannot be used to find the equation of a line that is parallel to the y-axis.

SECTION 3.6 Parallel and Perpendicular Lines

1 Find equations of parallel and perpendicular lines

Two lines that have the same slope do not intersect and are called **parallel lines.**

The slope of each of the lines at the right is $\frac{2}{3}$.

The lines are parallel.

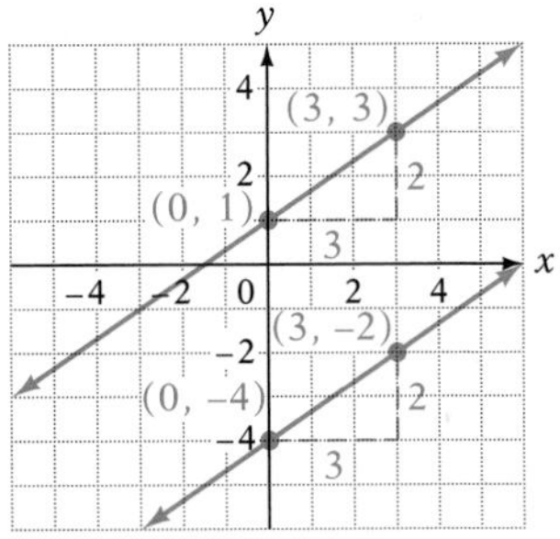

Slopes of Parallel Lines

Two nonvertical lines with slopes of m_1 and m_2 are parallel if and only if $m_1 = m_2$. Any two vertical lines are parallel.

Is the line that contains the points $(-2, 1)$ and $(-5, -1)$ parallel to the line that contains the points $(1, 0)$ and $(4, 2)$?

Find the slope of the line through $(-2, 1)$ and $(-5, -1)$.

$$m_1 = \frac{-1 - 1}{-5 - (-2)} = \frac{-2}{-3} = \frac{2}{3}$$

Find the slope of the line through $(1, 0)$ and $(4, 2)$.

$$m_2 = \frac{2 - 0}{4 - 1} = \frac{2}{3}$$

Because $m_1 = m_2$, the lines are parallel.

INSTRUCTOR NOTE

When students are attempting to find the equations of parallel or perpendicular lines, it may help to use the model given in the last section.

$$y - y_1 = m(x - x_1)$$
$$\downarrow \quad \downarrow \quad \downarrow$$
$$y - (\ ) = (\ )[x - (\ )]$$

If the lines are parallel, m has the same value as the slope of the line to which it is parallel.

If the lines are perpendicular, m is the negative reciprocal of the slope of the line to which it is perpendicular.

Find the equation of the line that contains the point $(2, 3)$ and is parallel to the line $y = \frac{1}{2}x - 4$.

The slope of the given line is $\frac{1}{2}$. Because parallel lines have the same slope, the slope of the unknown line is also $\frac{1}{2}$.

Use the point-slope formula.

$$y - y_1 = m(x - x_1)$$

Substitute $m = \frac{1}{2}$ and $(x_1, y_1) = (2, 3)$.

$$y - 3 = \frac{1}{2}(x - 2)$$

Solve for y.

$$y - 3 = \frac{1}{2}x - 1$$
$$y = \frac{1}{2}x + 2$$

The equation of the line is $y = \frac{1}{2}x + 2$.

Find the equation of the line that contains the point $(-1, 4)$ and is parallel to the line $2x - 3y = 5$.

Because the lines are parallel, the slope of the unknown line is the same as the slope of the given line. Solve $2x - 3y = 5$ for y and determine its slope.

$$2x - 3y = 5$$
$$-3y = -2x + 5$$
$$y = \frac{2}{3}x - \frac{5}{3}$$

The slope of the given line is $\frac{2}{3}$. Because the lines are parallel, the slope of the unknown line is also $\frac{2}{3}$.

Use the point-slope formula.

$$y - y_1 = m(x - x_1)$$

Substitute $m = \frac{2}{3}$ and $(x_1, y_1) = (-1, 4)$.

$$y - 4 = \frac{2}{3}[x - (-1)]$$

Solve for y.

$$y - 4 = \frac{2}{3}(x + 1)$$
$$y - 4 = \frac{2}{3}x + \frac{2}{3}$$
$$y = \frac{2}{3}x + \frac{14}{3}$$

The equation of the line is $y = \frac{2}{3}x + \frac{14}{3}$.

INSTRUCTOR NOTE

The concept of *perpendicular* is very important in mathematics, but students often do not see its significance. This is in part because of the difficulty of the applications. However, one simple example is the coordinate system itself.

Another example that is more complicated but can be understood by most students is a ball being rotated at the end of a string. The direction of the force is perpendicular to the tangent line.

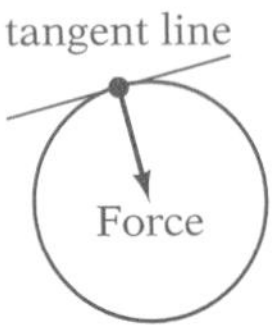

Two lines that intersect at right angles are **perpendicular lines.**

Any horizontal line is perpendicular to any vertical line. For example, $x = 3$ is perpendicular to $y = -2$ as shown in the graph at the right.

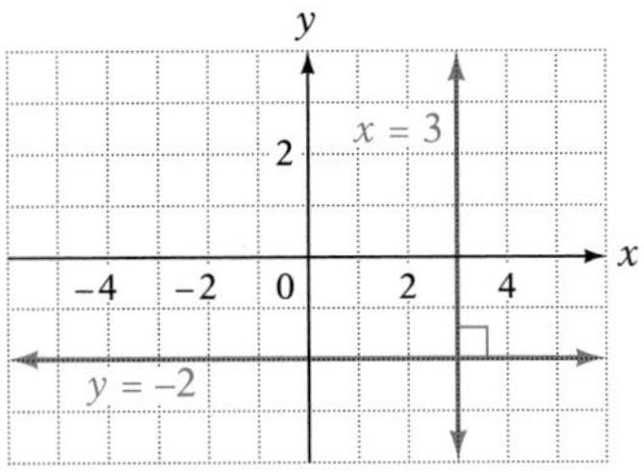

Slopes of Perpendicular Lines

If m_1 and m_2 are the slopes of two lines, neither of which is vertical, then the lines are perpendicular if and only if $m_1 \cdot m_2 = -1$.

A vertical line is perpendicular to a horizontal line.

Solving $m_1 \cdot m_2 = -1$ for m_1 gives $m_1 = -\frac{1}{m_2}$. This last equation states that the slopes of perpendicular lines are *negative reciprocals* of each other.

Is the line that contains the points (4, 2) and (−2, 5) perpendicular to the line that contains the points (−4, 3) and (−3, 5)?

Find the slope of the line that passes through the points (4, 2) and (−2, 5). $m_1 = \frac{5-2}{-2-4} = \frac{3}{-6} = -\frac{1}{2}$

Find the slope of the line that passes through the points (−4, 3) and (−3, 5). $m_2 = \frac{5-3}{-3-(-4)} = \frac{2}{1} = 2$

Find the product of the two slopes. $m_1 \cdot m_2 = -\frac{1}{2}(2) = -1$

Because $m_1 \cdot m_2 = -1$, the lines are perpendicular.

Are the graphs of the lines whose equations are $3x + 4y = 8$ and $8x + 6y = 5$ perpendicular?

To determine whether the lines are perpendicular, solve each equation for y and find the slope of each line.

$$3x + 4y = 8 \qquad 8x + 6y = 5$$
$$4y = -3x + 8 \qquad 6y = -8x + 5$$
$$y = -\frac{3}{4}x + 2 \qquad y = -\frac{4}{3}x + \frac{5}{6}$$
$$m_1 = -\frac{3}{4} \qquad m_2 = -\frac{4}{3}$$

Then check whether the equation $m_1 \cdot m_2 = -1$ is true.

$$m_1 \cdot m_2 = \left(-\frac{3}{4}\right)\left(-\frac{4}{3}\right) = 1$$

Because $m_1 \cdot m_2 = 1 \neq -1$, the lines are not perpendicular.

Find the equation of the line that contains the point (3, −4) and is perpendicular to the line $2x - y = -3$.

Determine the slope of the given line by solving the equation for y.

$$2x - y = -3$$
$$-y = -2x - 3$$
$$y = 2x + 3$$

The slope is 2.

The slope of the line perpendicular to the given line is $-\frac{1}{2}$, the negative reciprocal of 2.

Use the point-slope formula. $y - y_1 = m(x - x_1)$

Substitute $m = -\frac{1}{2}$ and $(x_1, y_1) = (3, -4)$. $y - (-4) = -\frac{1}{2}(x - 3)$

Solve for y. $y + 4 = -\frac{1}{2}x + \frac{3}{2}$

$$y = -\frac{1}{2}x - \frac{5}{2}$$

The equation of the perpendicular line is $y = -\frac{1}{2}x - \frac{5}{2}$.

Example 1 Is the line that contains the points $(-4, 2)$ and $(1, 6)$ parallel to the line that contains the points $(2, -4)$ and $(7, 0)$?

Solution $(x_1, y_1) = (-4, 2), (x_2, y_2) = (1, 6)$

$m_1 = \frac{6 - 2}{1 - (-4)} = \frac{4}{5}$ ▶ Find the slope of each line.

$(x_1, y_1) = (2, -4), (x_2, y_2) = (7, 0)$

$m_2 = \frac{0 - (-4)}{7 - 2} = \frac{4}{5}$

$m_1 = m_2 = \frac{4}{5}$ ▶ Check whether the slopes are equal.

Yes, the lines are parallel.

Problem 1 Is the line that contains the points $(-2, -3)$ and $(7, 1)$ perpendicular to the line that contains the points $(4, 1)$ and $(6, -5)$?

Solution See page S7. No

Example 2 Are the lines $4x - y = -2$ and $x + 4y = -12$ perpendicular?

Solution

$4x - y = -2$ ▶ Solve each equation for y to find the slopes.

$-y = -4x - 2$

$y = 4x + 2$ $m_1 = 4$

$x + 4y = -12$

$4y = -x - 12$

$y = -\frac{1}{4}x - 3$ $m_2 = -\frac{1}{4}$

$m_1 \cdot m_2 = 4\left(-\frac{1}{4}\right) = -1$ ▶ Find the product of the slopes.

Yes, the lines are perpendicular.

Problem 2 Are the lines $5x + 2y = 2$ and $5x + 2y = -6$ parallel?

Solution See page S8. Yes

Example 3 Find the equation of the line that contains the point $(3, -1)$ and is parallel to the line $3x - 2y = 4$.

Solution

$$3x - 2y = 4$$ ▶ Solve the equation for y to find the slope.

$$-2y = -3x + 4$$

$$y = \frac{3}{2}x - 2$$

$$m = \frac{3}{2}$$

$$y - y_1 = m(x - x_1)$$ ▶ Use the point-slope formula.

$$y - (-1) = \frac{3}{2}(x - 3)$$ ▶ $m = \frac{3}{2}, (x_1, y_1) = (3, -1)$

$$y + 1 = \frac{3}{2}x - \frac{9}{2}$$

$$y = \frac{3}{2}x - \frac{11}{2}$$

Problem 3 Find the equation of the line that contains the point $(-2, 3)$ and is perpendicular to the line $x - 4y = 3$.

Solution See page S8. $y = -4x - 5$

STUDY TIPS

BE PREPARED FOR TESTS

The Chapter Test at the end of a chapter should be used to prepare for an examination. We suggest that you try the Chapter Test a few days before your actual exam. Do these exercises in a quiet place and try to complete the exercises in the same amount of time as you will be allowed for your exam. When completing the exercises, practice the strategies of successful test takers:

- Look over the entire test before you begin to solve any problem.
- Write down any rules or formulas you may need so they are readily available.
- Read the directions carefully.
- Work the problems that are easiest for you first.
- Check your work, looking particularly for careless errors.

When you have completed the exercises in the Chapter Test, check your answers. If you missed a question, review the material in that objective and rework some of the exercises from that objective. This will strengthen your ability to perform the skills in that objective.

CONCEPT REVIEW 3.6

Determine whether the following statements are always true, sometimes true, or never true.

1. Perpendicular lines have the same y-intercept. Sometimes true

2. Parallel lines have the same slope. Always true

3. The lines given by the equations $y = \frac{3}{2}x + 2$ and $y = -\frac{3}{2}x + 2$ are perpendicular. Never true

4. Two lines are perpendicular if $m_1 \cdot m_2 = 1$. Never true

5. A vertical line is perpendicular to a horizontal line. Always true

6. A line parallel to the y-axis has zero slope. Never true

7. The lines $y = \frac{1}{2}x - 2$ and $y = \frac{1}{2}x - 4$ are parallel. Always true

EXERCISES 3.6

1 1. Given the slopes of two lines, explain how to determine whether the two lines are parallel.

2. Given the slopes of two lines, how can you determine whether the two lines are perpendicular?

3. Is the line $x = -2$ perpendicular to the line $y = 3$? Yes

4. Is the line $y = \frac{1}{2}$ perpendicular to the line $y = -4$? No

5. Is the line $x = -3$ parallel to the line $y = \frac{1}{3}$? No

6. Is the line $x = 4$ parallel to the line $x = -4$? Yes

7. Is the line $y = \frac{2}{3}x - 4$ parallel to the line $y = -\frac{3}{2}x - 4$? No

8. Is the line $y = -2x + \frac{2}{3}$ parallel to the line $y = -2x + 3$? Yes

9. Is the line $y = \frac{4}{3}x - 2$ perpendicular to the line $y = -\frac{3}{4}x + 2$? Yes

10. Is the line $y = \frac{1}{2}x + \frac{3}{2}$ perpendicular to the line $y = -\frac{1}{2}x + \frac{3}{2}$? No

11. Are the lines $2x + 3y = 2$ and $2x + 3y = -4$ parallel? Yes

12. Are the lines $2x - 4y = 3$ and $2x + 4y = -3$ parallel? No

13. Are the lines $x - 4y = 2$ and $4x + y = 8$ perpendicular? Yes

14. Are the lines $4x - 3y = 2$ and $4x + 3y = -7$ perpendicular? No

15. Is the line that contains the points $(3, 2)$ and $(1, 6)$ parallel to the line that contains the points $(-1, 3)$ and $(-1, -1)$? No

16. Is the line that contains the points $(4, -3)$ and $(2, 5)$ parallel to the line that contains the points $(-2, -3)$ and $(-4, 1)$? No

17. Is the line that contains the points $(-3, 2)$ and $(4, -1)$ perpendicular to the line that contains the points $(1, 3)$ and $(-2, -4)$? Yes

18. Is the line that contains the points $(-1, 2)$ and $(3, 4)$ perpendicular to the line that contains the points $(-1, 3)$ and $(-4, 1)$? No

19. Is the line that contains the points $(-5, 0)$ and $(0, 2)$ parallel to the line that contains the points $(5, 1)$ and $(0, -1)$? Yes

20. Is the line that contains the points $(3, 5)$ and $(-3, 3)$ perpendicular to the line that contains the points $(2, -5)$ and $(-4, 4)$? No

21. Find the equation of the line that contains the point $(-2, -4)$ and is parallel to the line $2x - 3y = 2$. $y = \frac{2}{3}x - \frac{8}{3}$

22. Find the equation of the line that contains the point $(3, 2)$ and is parallel to the line $3x + y = -3$. $y = -3x + 11$

23. Find the equation of the line that contains the point $(4, 1)$ and is perpendicular to the line $y = -3x + 4$. $y = \frac{1}{3}x - \frac{1}{3}$

24. Find the equation of the line that contains the point $(2, -5)$ and is perpendicular to the line $y = \frac{5}{2}x - 4$. $y = -\frac{2}{5}x - \frac{21}{5}$

25. Find the equation of the line that contains the point $(-1, -3)$ and is perpendicular to the line $3x - 5y = 2$. $y = -\frac{5}{3}x - \frac{14}{3}$

26. Find the equation of the line that contains the point $(-1, 3)$ and is perpendicular to the line $2x + 4y = -1$. $y = 2x + 5$

APPLYING CONCEPTS 3.6

27. If the graphs of $A_1x + B_1y = C_1$ and $A_2x + B_2y = C_2$ are perpendicular, express $\frac{A_1}{B_1}$ in terms of A_2 and B_2. $\frac{A_1}{B_1} = -\frac{B_2}{A_2}$

28. If the graphs of $A_1x + B_1y = C_1$ and $A_2x + B_2y = C_2$ are parallel, express $\frac{A_1}{B_1}$ in terms of A_2 and B_2. $\frac{A_1}{B_1} = \frac{A_2}{B_2}$

29. The graphs of $y = -\frac{1}{2}x + 2$ and $y = \frac{2}{3}x - 5$ intersect at the point whose coordinates are $(6, -1)$. Find the equation of a line whose graph intersects the graphs of the given lines to form a right triangle. (*Hint:* There is more than one answer to this question.)

Any equation of the form $y = 2x + b$, where $b \neq -13$, or of the form $y = -\frac{3}{2}x + c$, where $c \neq 8$.

30. A theorem from geometry states that a line passing through the center of a circle and through a point P on the circle is perpendicular to the tangent line at P. (See the figure at the right.) If the coordinates of P are $(5, 4)$ and the coordinates of C are $(3, 2)$, what is the equation of the tangent line?
$y = -x + 9$

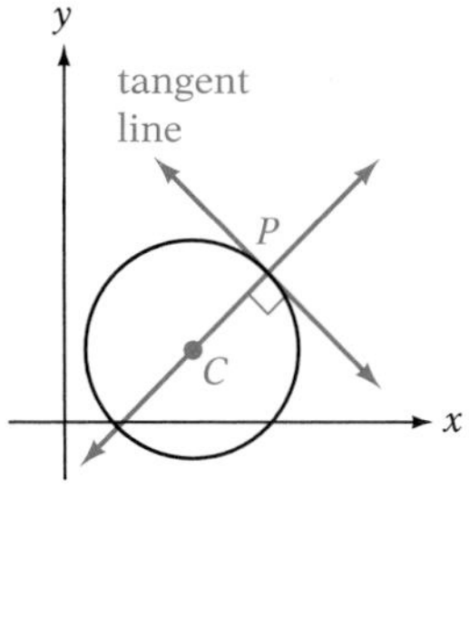

31. Find the x- and y-intercepts of the tangent line in Exercise 30.
x-intercept: $(9, 0)$; y-intercept: $(0, 9)$

32. Explain how to determine whether the graphs of two lines are:
a. parallel **b.** perpendicular

SECTION 3.7

Inequalities in Two Variables

1 Graph the solution set of an inequality in two variables

The graph of the linear equation $y = x - 1$ separates the plane into three sets: the set of points on the line, the set of points above the line, and the set of points below the line.

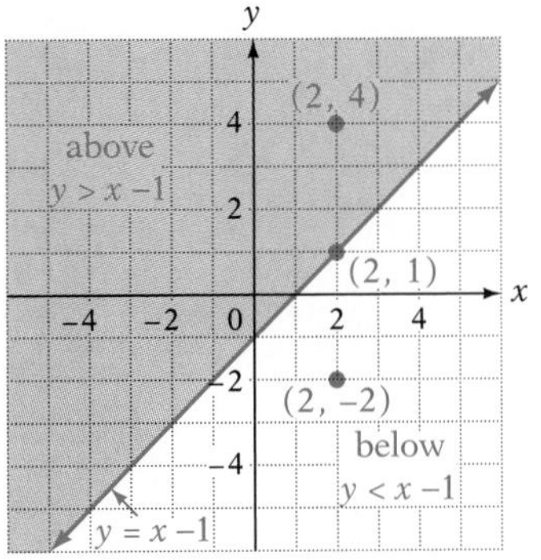

The point whose coordinates are (2, 1) is a solution of $y = x - 1$ and is a point on the line.

The point whose coordinates are (2, 4) is a solution of $y > x - 1$ and is a point above the line.

The point whose coordinates are (2, −2) is a solution of $y < x - 1$ and is a point below the line.

The set of points on the line are the solutions of the equation $y = x - 1$. The set of points above the line are the solutions of the inequality $y > x - 1$. These points form a **half-plane.** The set of points below the line are solutions of the inequality $y < x - 1$. These points also form a half-plane.

An inequality of the form $y > mx + b$ or $Ax + By > C$ is a **linear inequality in two variables.** (The inequality symbol could be replaced by $\geq$, $<$, or $\leq$.) The solution set of a linear inequality in two variables is a half-plane.

The following illustrates the procedure for graphing the solution set of a linear inequality in two variables.

Graph the solution set of $3x - 4y < 12$.

Solve the inequality for y.

$$3x - 4y < 12$$
$$-4y < -3x + 12$$
$$y > \frac{3}{4}x - 3$$

Change the inequality $y > \frac{3}{4}x - 3$ to the equality $y = \frac{3}{4}x - 3$, and graph the line.

If the inequality contains $\leq$ or $\geq$, the line belongs to the solution set and is shown by a *solid line.* If the inequality contains $<$ or $>$, the line is not part of the solution set and is shown by a *dashed line.*

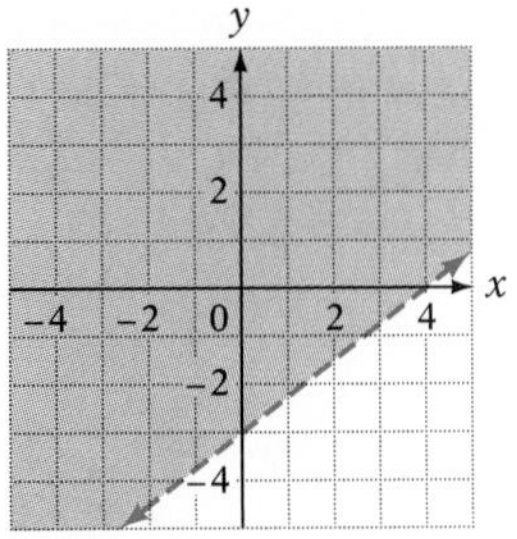

If the inequality contains $>$ or $\geq$, shade the upper half-plane. If the inequality contains $<$ or $\leq$, shade the lower half-plane.

As a check, use the ordered pair (0, 0) to determine whether the correct region of the plane has been shaded. If (0, 0) is a solution of the inequality, then (0, 0) should be in the shaded region. If (0, 0) is not a solution of the inequality, then (0, 0) should not be in the shaded region.

As shown below, (0, 0) is a solution of the inequality in the example above.

$$y > \frac{3}{4}x - 3$$
$$0 > \frac{3}{4}(0) - 3$$
$$0 > 0 - 3$$
$$0 > -3 \qquad \text{True}$$

Because (0, 0) is a solution of the inequality, (0, 0) should be in the shaded region. The solution set as graphed above is correct.

If the line passes through the point (0, 0), another point, such as (0, 1), must be used as a check.

From the graph of $y > \frac{3}{4}x - 3$, note that for a given value of x, more than one value of y can be paired with the value of x. For instance, (4, 1), (4, 3), (5, 1), and $\left(5, \frac{9}{4}\right)$ are all ordered pairs that belong to the graph. Because there are ordered pairs with the same first component and different second components, the inequality does not represent a function. The inequality is a relation but not a function.

Example 1 Graph the solution set of $x + 2y \le 4$.

Solution

$$x + 2y \le 4 \qquad \blacktriangleright \text{ Solve the inequality for } y.$$
$$2y \le -x + 4$$
$$y \le -\frac{1}{2}x + 2$$

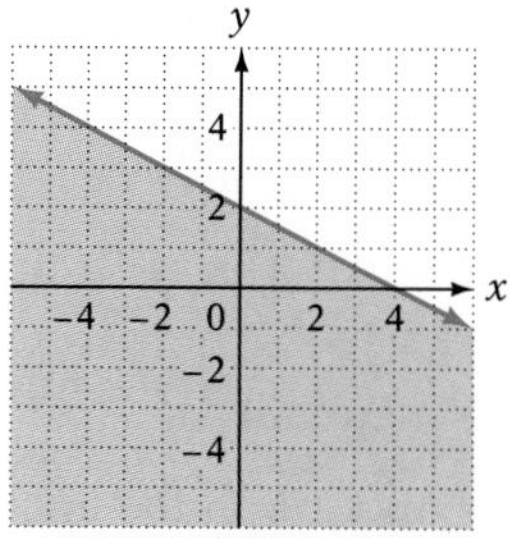

▸ Graph $y = -\frac{1}{2}x + 2$ as a solid line. Shade the lower half-plane.

Problem 1 Graph the solution set of $x + 3y > 6$.

Solution See page S8.

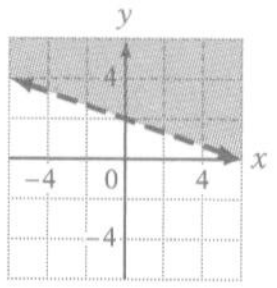

Example 2 Graph the solution set of $x \geq -1$.

Solution

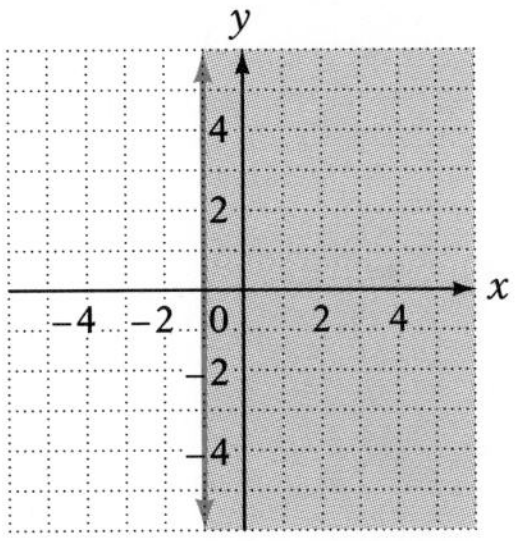

▶ Graph $x = -1$ as a solid line.
▶ The point $(0, 0)$ satisfies the inequality. Shade the half-plane to the right of the line.

Problem 2 Graph the solution set of $y < 2$.

Solution See page S8.

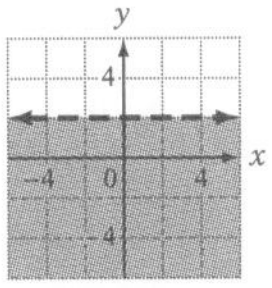

CONCEPT REVIEW 3.7

Determine whether the following statements are always true, sometimes true, or never true.

1. It is possible to write a linear inequality in two variables that has no solutions. Never true
2. The exponents on the variables in a linear inequality in two variables are 1. Always true
3. The graph of a linear inequality is a half-plane. Always true
4. The graph of a linear inequality in two variables represents a function. Never true
5. The solution of a linear inequality in two variables containing $\leq$ or $\geq$ includes the line separating the half-planes. Always true
6. The solution of the inequality $y > x + 2$ is all the points above the line $y = x + 2$. Always true

EXERCISES 3.7

1 1. What is a half-plane?

2. Explain a method that you can use to check that the graph of a linear inequality in two variables has been shaded correctly.

Graph the solution set.

3. $y \leq \frac{3}{2}x - 3$

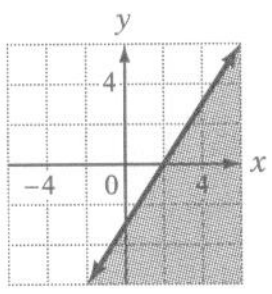

4. $y \geq \frac{4}{3}x - 4$

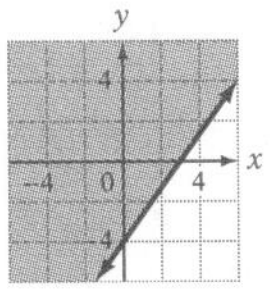

5. $y < \frac{4}{5}x - 2$

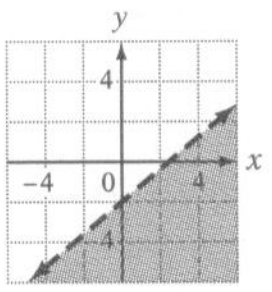

6. $y < \frac{3}{5}x - 3$

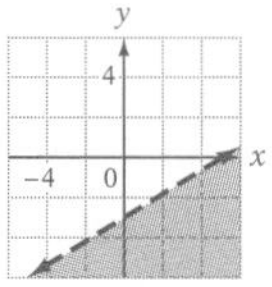

7. $y < -\frac{1}{3}x + 2$

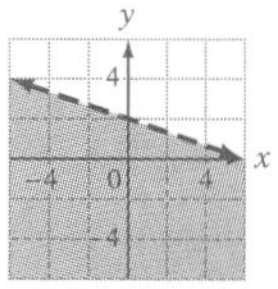

8. $y < -\frac{4}{3}x + 3$

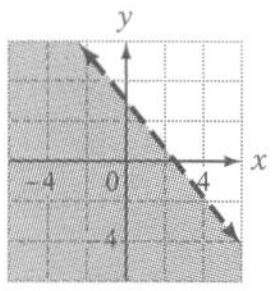

9. $x + 3y < 4$

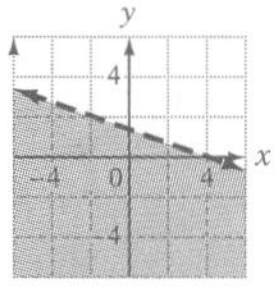

10. $2x - 5y \leq 10$

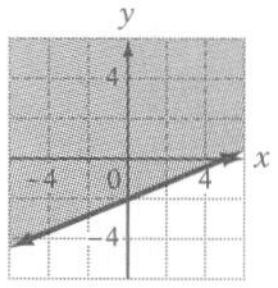

11. $2x + 3y \geq 6$

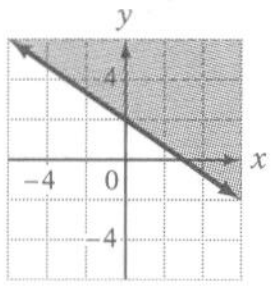

12. $3x + 2y < 4$

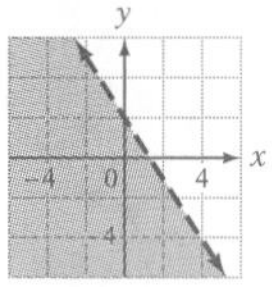

13. $-x + 2y > -8$

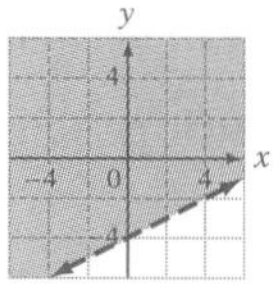

14. $-3x + 2y > 2$

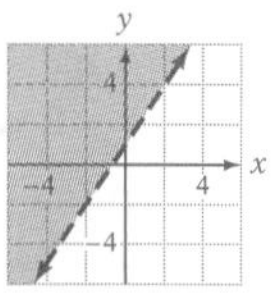

15. $y - 4 < 0$

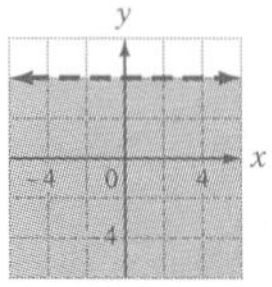

16. $x + 2 \geq 0$

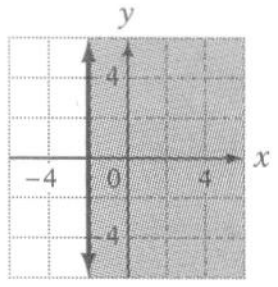

17. $6x + 5y < 15$

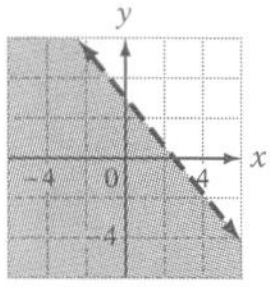

18. $3x - 5y < 10$

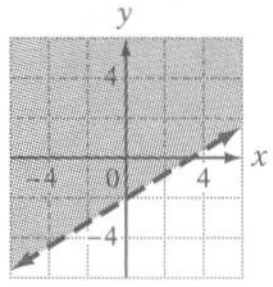

19. $-5x + 3y \geq -12$

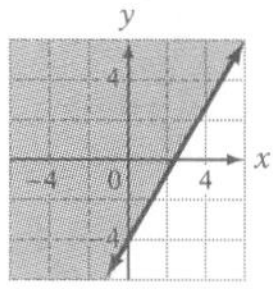

20. $3x + 4y \geq 12$

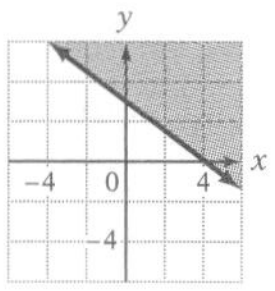

APPLYING CONCEPTS 3.7

21. Does the inequality $y < 3x - 1$ represent a function? Explain your answer.

22. Are there any points whose coordinates satisfy both $y \leq x + 3$ and $y \geq -\frac{1}{2}x + 1$? If so, give the coordinates of three such points. If not, explain why not.

23. Are there any points whose coordinates satisfy both $y \leq x - 1$ and $y \geq x + 2$? If so, give the coordinates of three such points. If not, explain why not.

Focus on Problem Solving

Find a Pattern

Polya's four recommended problem-solving steps are stated below.

1. Understand the problem.
2. Devise a plan.
3. Carry out the plan.
4. Review your solution.

POINT OF INTEREST

Karl Friedrich Gauss (1777–1855) has been called the "Prince of Mathematicians" by some historians. He applied his genius to many areas of mathematics and science. A unit of magnetism, the *gauss*, is named in his honor. Some electronic equipment (televisions, for instance) contain a *degausser* that controls magnetic fields.

One of the several ways of devising a plan is first to try to find a pattern. Karl Friedrich Gauss supposedly used this method to solve a problem that was given to his math class when he was in elementary school. As the story goes, his teacher wanted to grade some papers while the class worked on a math problem. The problem given to the class was to find the sum

$$1 + 2 + 3 + 4 + \cdots + 100$$

Gauss quickly solved the problem by seeing a pattern. Here is what he saw.

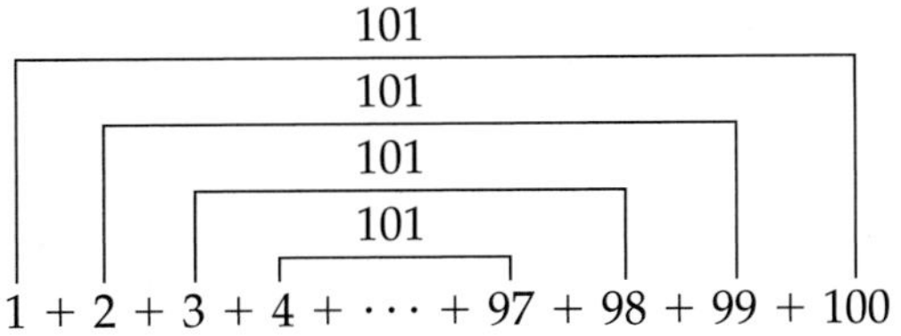

Note that

$$1 + 100 = 101$$
$$2 + 99 = 101$$
$$3 + 98 = 101$$
$$4 + 97 = 101$$

Gauss noted that there were 50 sums of 101. Therefore, the sum of the first 100 natural numbers is

$$1 + 2 + 3 + 4 + \cdots + 97 + 98 + 99 + 100 = 50(101) = 5050$$

Try to solve the following problems by finding a pattern.

1. Find the sum $2 + 4 + 6 + \cdots + 96 + 98 + 100$.

2. Find the sum $1 + 3 + 5 + \cdots + 97 + 99 + 101$.

3. Find another method of finding the sum $1 + 3 + 5 + \cdots + 97 + 99 + 101$ given in the previous exercise.

4. Find the sum $\frac{1}{1 \cdot 2} + \frac{1}{2 \cdot 3} + \frac{1}{3 \cdot 4} + \cdots + \frac{1}{49 \cdot 50}$.

Hint: $\frac{1}{1 \cdot 2} = \frac{1}{2}, \frac{1}{1 \cdot 2} + \frac{1}{2 \cdot 3} = \frac{2}{3}, \frac{1}{1 \cdot 2} + \frac{1}{2 \cdot 3} + \frac{1}{3 \cdot 4} = \frac{3}{4}$

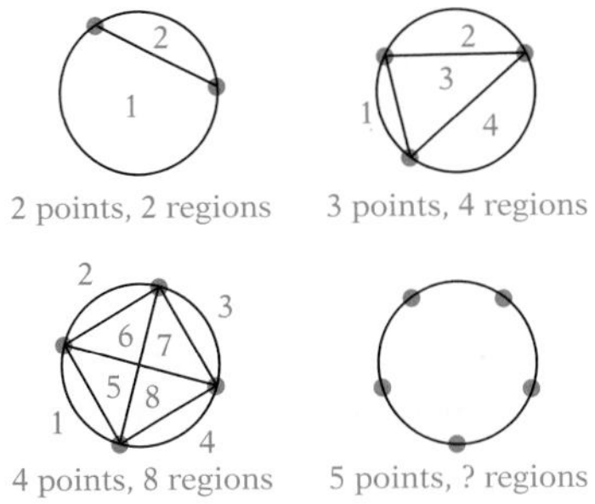

2 points, 2 regions 3 points, 4 regions 4 points, 8 regions 5 points, ? regions

5. The following problem shows that checking a few cases does not always result in a conjecture that is true for *all* cases. Select any two points on a circle and draw a *chord* (see the drawing in the left margin), the line connecting the points. The chord divides the circle into two regions. Now select three different points and draw chords connecting each of the three points with every other point. The chords divide the circle into four regions. Now select four points and connect each of the points with every other point. Make a conjecture about the relationship between the number of regions and the number of points on the circle. Does your conjecture work for five points? six points?

6. A *polygonal number* is a number that can be represented by arranging that number of dots in rows to form a geometric figure such as a triangle, square, pentagon, or hexagon. For instance, the first four *triangular* numbers, 3, 6, 10, and 15, are given (and shown) below. What are the next two triangular numbers?

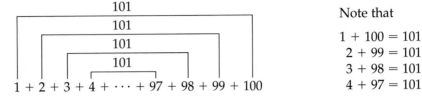

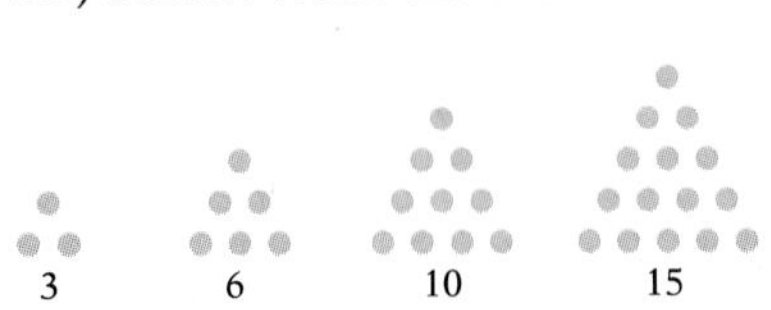

Projects and Group Activities

Evaluating a Function with a Graphing Calculator

You can use a graphing calculator to evaluate some functions. We will give the keystrokes to evaluate $f(x) = 3x^2 + 2x - 1$ when $x = -2$ for three different calculators.[1] If you try these keystrokes, you should get 7 as the value of the function.

TI-83	*SHARP EL-9600*	*Casio CFX-9850G*
[Y=] [CLEAR] 3 [X,T,θ,n] [∧] 2 [+] 2 [X,T,θ,n] [−] 1 [2nd] QUIT [(−)] 2 [STO] [▶] [X,T,θ,n] [ENTER] [VARS] [▶] 1 1 [ENTER]	[Y=] [CL] 3 [X/θ/T/n] [a^b] [▶] 2 [+] 2 [X/θ/T/n] [−] 1 [+ − × ÷] [(−)] 2 [STO] [X/θ/T/n] [ENTER] [VARS] [ENTER] [ENTER] 1 [ENTER]	[MENU] 5 [F2] [F1] 3 [X,θ,T] [∧] 2 [+] 2 [X,θ,T] [−] 1 [EXE] [MENU] 1 [(−)] 2 [→] [X,θ,T] [EXE] [VARS] [F4] [F1] 1 [EXE]

Introduction to Graphing Calculators

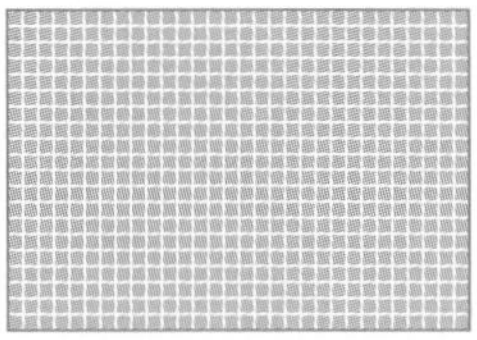

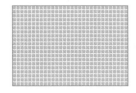

Calculator Screen

There are a variety of computer programs and calculators that can graph an equation. A computer or graphing calculator screen is divided into pixels. Depending on the computer or calculator, there are approximately 6000 to 790,000 pixels available on the screen. The greater the number of pixels, the smoother the graph will appear. A portion of a screen is shown at the left. Each little rectangle represents one pixel.

A graphing calculator draws a graph in a manner similar to the way we have shown in this chapter. Values of x are chosen and ordered pairs calculated. Then a graph is drawn through those points by illuminating pixels (an abbreviation for "picture element") on the screen.

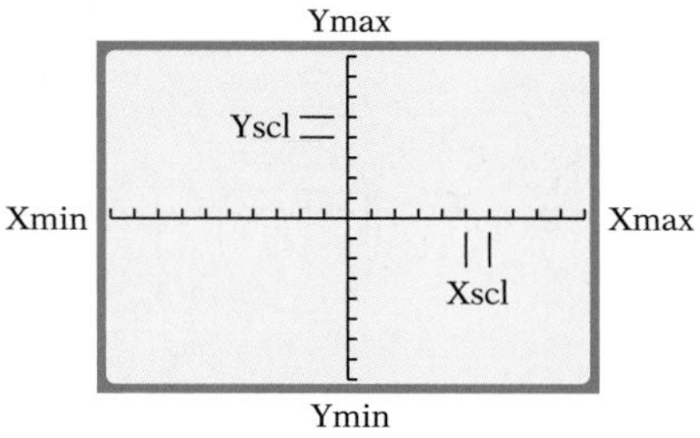

Graphing utilities can display only a portion of the xy-plane, called a window. The window [Xmin, Xmax] by [Ymin, Ymax] consists of those points (x, y) that satisfy both of the following:

$$\text{Xmin} \le x \le \text{Xmax} \quad \text{and} \quad \text{Ymin} \le y \le \text{Ymax}$$

The user sets these values before a graph is drawn.

The numbers Xscl and Yscl are the distances between the tick marks that are drawn on the x- and y-axes. If you do not want tick marks on the axes, set Xscl = 0 and Yscl = 0.

[1]There are other methods of evaluating functions. This is just one option.

The graph at the right is a portion of the graph of $y = \frac{1}{2}x + 1$, as it was drawn with a graphing calculator. The window is

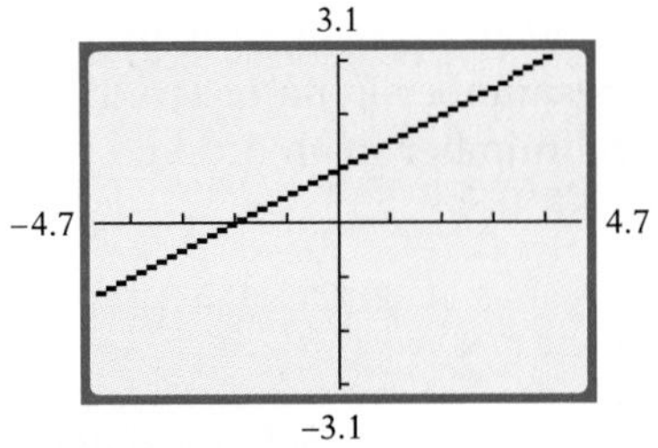

Xmin = −4.7, Xmax = 4.7
Ymin = −3.1, Ymax = 3.1
Xscl = 1 and Yscl = 1

Using interval notation, this is shown as $[-4.7, 4.7]$ by $[-3.1, 3.1]$.

The window $[-4.7, 4.7]$ by $[-3.1, 3.1]$ gives "nice" coordinates in the sense that each time the ▶ or the ◀ is pressed, the change in x is 0.1. The reason for this is that the horizontal distance from the middle of the first pixel to the middle of the last pixel is 94 units. By using Xmin = −4.7 and Xmax = 4.7, we have[1]

$$\text{Change in } x = \frac{\text{Xmax} - \text{Xmin}}{94} = \frac{4.7 - (-4.7)}{94} = \frac{9.4}{94} = 0.1$$

Similarly, the vertical distance from the middle of the first pixel to the last pixel is 62 units. Therefore, using Ymin = −3.1 and Ymax = 3.1 will give nice coordinates in the vertical direction.

Graph each of the following by using a graphing calculator.

LOOK CLOSELY

The Appendix "Guidelines for Using Graphing Calculators" contains keystroking suggestions to use with this project.

1. $y = 2x + 1$ — For $2x$, you may enter $2 \times x$ or just $2x$. The times sign $\times$ is not necessary on many graphing calculators.

2. $y = -x + 2$ — Many calculators use the (−) key to enter a negative sign.

3. $3x + 2y = 6$ — Solve for y. Then enter the equation.

4. $y = 50x$ — You must adjust the viewing window. Try the window given by $[-4.7, 4.7]$ by $[-250, 250]$ with Yscl = 50.

5. $y = \frac{2}{3}x - 3$ — You may enter $\frac{2}{3}x$ as $2x/3$ or $(2/3)x$. Although entering $2/3x$ works on some calculators, it is not recommended.

6. $4x + 3y = 75$ — You must adjust the viewing window.

Chapter Summary

Key Words

A *rectangular coordinate system* is formed by two number lines, one horizontal and one vertical, that intersect at the zero point of each line. The number lines that make up a coordinate system are called the *coordinate axes* or simply the *axes.* The *origin* is the point of intersection of the two coordinate axes. Generally, the horizontal axis is labeled the *x-axis*, and the vertical axis is labeled the *y-axis*. A rectangular coordinate system divides the *plane* determined by the axes into four regions called *quadrants.* (Objective 3.1.1)

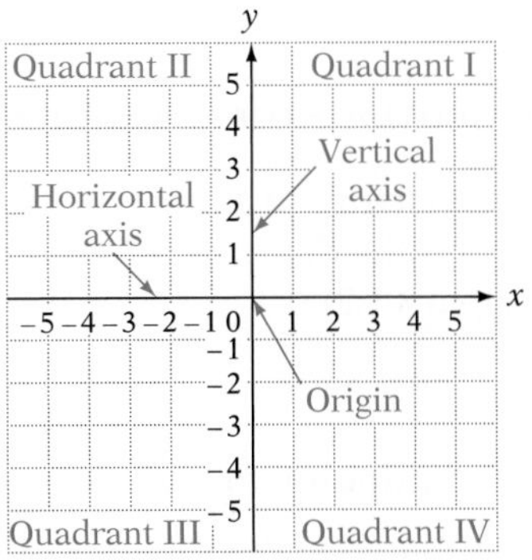

[1]Some calculators have screen widths of 126 pixels. For those calculators, use Xmin = −6.3 and Xmax = 6.3 to obtain nice coordinates.

An *ordered pair* (x, y) is used to locate a point in a plane. The first number in an ordered pair is called the *abscissa.* The second number in an ordered pair is called the *ordinate.* (Objective 3.1.1)

(3, 4) is an ordered pair.

3 is the abscissa.

4 is the ordinate.

A *scatter diagram* is a graph of ordered-pair data. (Objective 3.1.3)

A *relation* is any set of ordered pairs. (Objective 3.2.1)

$\{(2, 3), (2, 4), (3, 4), (5, 7)\}$

A *function* is a set of ordered pairs in which no two ordered pairs have the same first coordinate and different second coordinates. (Objective 3.2.1)

$\{(2, 3), (3, 5), (5, 7), (6, 9)\}$

Functional notation is used for those equations that represent functions. For the equation at the right, x is the *independent variable* and y is the *dependent variable.* (Objective 3.2.1)

In functional notation, $y = 3x + 7$ is written as $f(x) = 3x + 7$.

The *domain of a function* is the set of first coordinates of all the ordered pairs of the function. The *range of a function* is the set of second coordinates of all the ordered pairs of the function. (Objective 3.2.1)

$\{(1, 2), (2, 4), (5, 7), (8, 3)\}$

Domain: $\{1, 2, 5, 8\}$

Range: $\{2, 3, 4, 7\}$

A *linear function* is one that can be expressed in the form $f(x) = mx + b$. The graph of a linear function is a straight line. The point at which a graph crosses the x-axis is called the *x-intercept,* and the point at which a graph crosses the y-axis is called the *y-intercept.* (Objectives 3.3.1 and 3.3.2)

$f(x) = -\frac{3}{2}x + 3$

The *slope* of a line is a measure of the slant, or tilt, of a line. The symbol for slope is m. A line that slants upward to the right has a *positive slope,* and a line that slants downward to the right has a *negative slope.* (Objective 3.4.1)

The line $y = 2x - 3$ slants upward to the right.

The line $y = --5x + 2$ slants downward to the right.

A horizontal line has *zero slope.* The slope of a vertical line is *undefined.* (Objective 3.4.1)

$y = 3$ is an equation of a line with zero slope.
The slope of the line $x = -3$ is undefined.

An inequality of the form $y > mx + b$ or of the form $Ax + By > C$ is a *linear inequality in two variables.* (The symbol $>$ could be replaced by $\geq$, $<$, or $\leq$.) The solution set of an inequality in two variables is a *half-plane.* (Objective 3.7.1)

Essential Rules and Procedures

The **distance formula** $d = \sqrt{(x_2 - x_1)^2 + (y_2 - y_1)^2}$ can be used to find the distance between any two points in a plane. (Objective 3.1.2)

$(x_1, y_1) = (2, 4), (x_2, y_2) = (-3, 5)$

$$d = \sqrt{(-3 - 2)^2 + (5 - 4)^2} = \sqrt{25 + 1} = \sqrt{26}$$

The **midpoint formula** $(x_m, y_m) = \left(\frac{x_1 + x_2}{2}, \frac{y_1 + y_2}{2}\right)$ is used to find the midpoint of a line segment. (Objective 3.1.2)

$(x_1, y_1) = (2, 3), (x_2, y_2) = (-6, -1)$

$x_m = \frac{2 + (-6)}{2} = -2, y_m = \frac{3 + (-1)}{2} = 1$

$(x_m, y_m) = (-2, 1)$

To find the x-intercept of $Ax + By = C$, let $y = 0$. **To find the y-intercept,** let $x = 0$. (Objective 3.3.2)

$3x + 4y = 12$

Let $y = 0$:
$3x + 4(0) = 12$
$x = 4$

The x-intercept is $(4, 0)$.

Let $x = 0$:
$3(0) + 4y = 12$
$y = 3$

The y-intercept is $(0, 3)$.

The **slope formula** $m = \frac{y_2 - y_1}{x_2 - x_1}$, $x_1 \neq x_2$, is used to find the slope of a line when two points on the line are known. (Objective 3.4.1)

$(x_1, y_1) = (-3, 2), (x_2, y_2) = (1, 4)$

$$m = \frac{4 - 2}{1 - (-3)} = \frac{1}{2}$$

The equation $y = mx + b$ is called the **slope-intercept form of a straight line.** (Objective 3.4.2)

$y = -3x + 2$
The slope is -3; the y-intercept is $(0, 2)$.

The equation $y - y_1 = m(x - x_1)$ is called the **point-slope formula.** (Objective 3.5.1)

The equation of the line that passes through the point $(4, 2)$ and has slope -3 is

$$\begin{aligned} y - y_1 &= m(x - x_1) \\ y - 2 &= -3(x - 4) \\ y - 2 &= -3x + 12 \\ y &= -3x + 14 \end{aligned}$$

Two lines that have the same slope do not intersect and are called **parallel lines.** (Objective 3.6.1)

$y = 3x - 4, \quad m_1 = 3$
$y = 3x + 2, \quad m_2 = 3$

Because $m_1 = m_2$, the lines are parallel.

Two lines that intersect at right angles are called **perpendicular lines.** The product of the slopes of two nonvertical perpendicular lines is -1. (Objective 3.6.1)

$y = \frac{1}{2}x - 1, \quad m_1 = \frac{1}{2}$
$y = -2x + 2, \quad m_2 = -2$

Because $m_1 \cdot m_2 = -1$, the lines are perpendicular.

Chapter Review Exercises

1. Determine the ordered-pair solution of $y = \frac{x}{x - 2}$ that corresponds to $x = 4$. $(4, 2)$

2. Find the midpoint and length of the line segment with endpoints $(-2, 4)$ and $(3, 5)$. $\left(\frac{1}{2}, \frac{9}{2}\right)$; $\sqrt{26}$

3. Graph the ordered-pair solutions of $y = x^2 - 2$ when $x = -2, -1, 0, 1$, and 2.

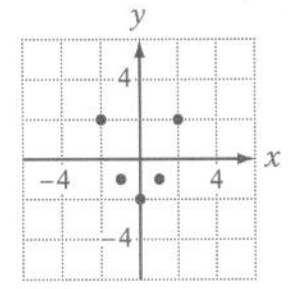

4. Draw a line through all points with an ordinate of -2.

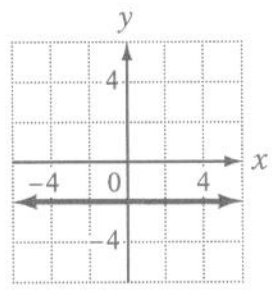

5. Given $P(x) = 3x + 4$, evaluate $P(-2)$ and $P(a)$.
$P(-2) = -2$; $P(a) = 3a + 4$

6. Find the domain and range of the function $\{(-1, 0), (0, 2), (1, 2), (2, 0), (5, 3)\}$.
D: $\{-1, 0, 1, 2, 5\}$; R: $\{0, 2, 3\}$

7. Find the range of $f(x) = x^2 - 2$ if the domain is $\{-2, -1, 0, 1, 2\}$. R: $\{-2, -1, 2\}$

8. What value of x is excluded from the domain of $f(x) = \frac{x}{x + 4}$? -4

9. Find the x- and y-intercepts and graph $4x - 6y = 12$. $(3, 0)$; $(0, -2)$

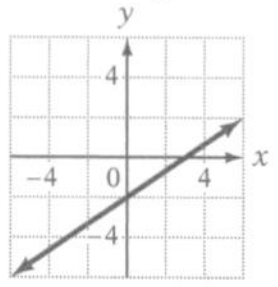

10. Graph: $y = -2x + 2$

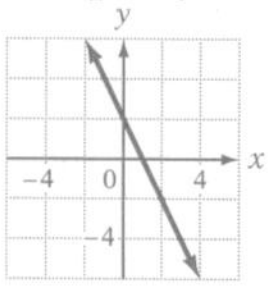

11. Graph: $2x - 3y = -6$

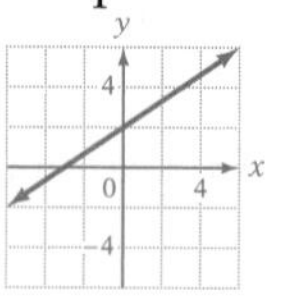

12. Find the slope of the line that contains the points $(3, -2)$ and $(-1, 2)$. -1

13. Graph $3x + 2y = -4$ by using the x- and y-intercepts.

x-intercept: $\left(-\frac{4}{3}, 0\right)$

y-intercept: $(0, -2)$

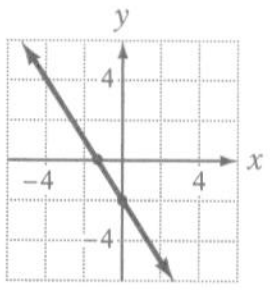

14. Graph the line that passes through the point $(-1, 4)$ and has slope $-\frac{1}{3}$.

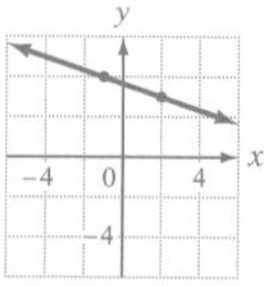

15. Find the equation of the line that contains the point $(-3, 4)$ and has slope $\frac{5}{2}$. $y = \frac{5}{2}x + \frac{23}{2}$

16. Find the equation of the line that contains the points $(-2, 4)$ and $(4, -3)$. $y = -\frac{7}{6}x + \frac{5}{3}$

17. Find the equation of the line that contains the point $(3, -2)$ and is parallel to the line $y = -3x + 4$. $y = -3x + 7$

18. Find the equation of the line that contains the point $(-2, -4)$ and is parallel to the line $2x - 3y = 4$. $y = \frac{2}{3}x - \frac{8}{3}$

19. Find the equation of the line that contains the point (2, 5) and is perpendicular to the line $y = -\frac{2}{3}x + 6$. $y = \frac{3}{2}x + 2$

20. Find the equation of the line that contains the point $(-3, -1)$ and is perpendicular to the line $4x - 2y = 7$. $y = -\frac{1}{2}x - \frac{5}{2}$

21. Graph the solution set of $y \geq 2x - 3$.

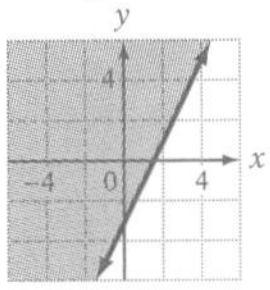

22. Graph the solution set of $3x - 2y < 6$.

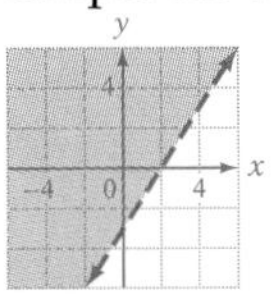

23. A car is traveling at 55 mph. The equation that describes the distance, d (in miles), traveled is $d = 55t$, where t is the number of hours driven. Use the coordinate axes below to graph this equation for $0 \leq t \leq 6$. The point (4, 220) is on the graph. Write a sentence that describes the meaning of this ordered pair.

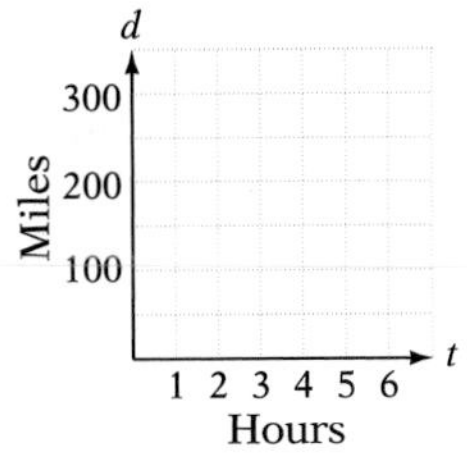

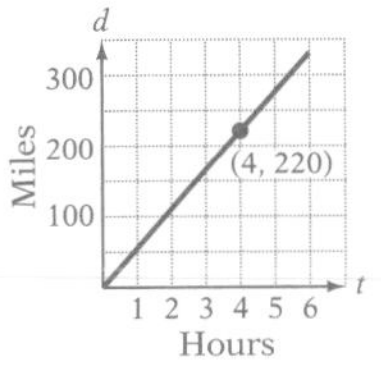

After 4 h the car has traveled 220 mi.

24. The graph below shows the relationship between the cost of manufacturing calculators and the number of calculators manufactured. Find the slope of the line between the two points shown on the graph. Write a sentence that states the meaning of the slope.

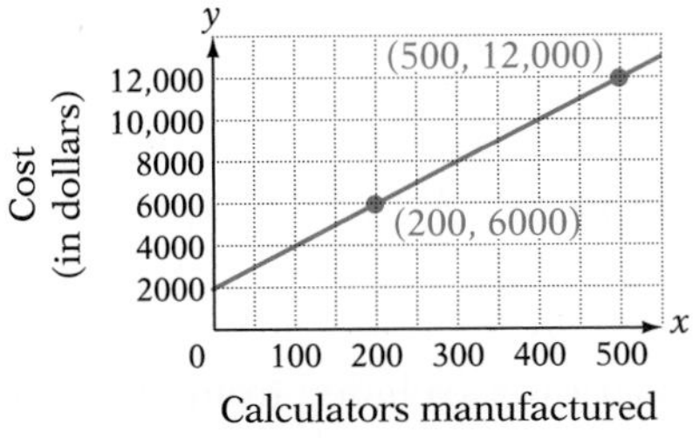

$m = 20$
The cost of manufacturing one calculator is \$20.

25. A building contractor estimates that the cost to build a new home is \$25,000 plus \$80 for each square foot of floor space in the house. Determine a linear function that will give the cost to build a house that contains a given number of square feet. Use this model to determine the cost of building a house that contains 2000 ft^2. $y = 80x + 25{,}000$; \$185,000

Chapter Test

1. Graph the ordered-pair solutions of $P(x) = 2 - x^2$ when $x = -2, -1, 0, 1$, and 2.

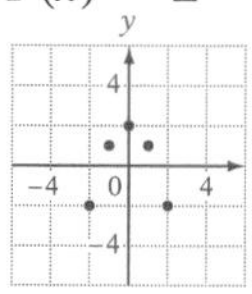

2. Find the ordered-pair solution of $y = 2x + 6$ that corresponds to $x = -3$. $(-3, 0)$

3. Graph: $y = \frac{2}{3}x - 4$

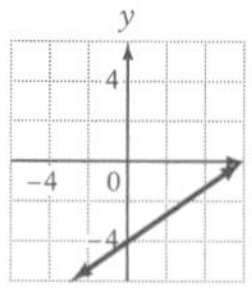

4. Graph: $2x + 3y = -3$

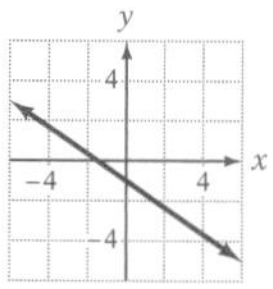

5. Find the equation of the vertical line that contains the point $(-2, 3)$. $x = -2$

6. Find the midpoint and length of the line segment with endpoints $(4, 2)$ and $(-5, 8)$. $\left(-\frac{1}{2}, 5\right); \sqrt{117}$

7. Find the slope of the line that contains the points $(-2, 3)$ and $(4, 2)$. $-\frac{1}{6}$

8. Given $P(x) = 3x^2 - 2x + 1$, evaluate $P(2)$. 9

9. Graph $2x - 3y = 6$ by using the x- and y-intercepts.

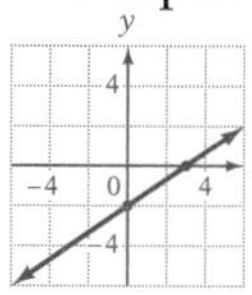

10. Graph the line that passes through the point $(-2, 3)$ and has slope $-\frac{3}{2}$.

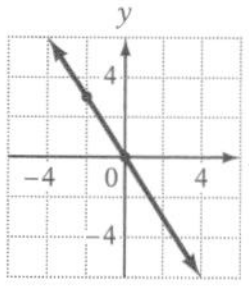

11. Find the equation of the line that contains the point $(-5, 2)$ and has slope $\frac{2}{5}$. $y = \frac{2}{5}x + 4$

12. What value of x is excluded from the domain of $f(x) = \frac{2x + 1}{x}$? 0

13. Find the equation of the line that contains the points $(3, -4)$ and $(-2, 3)$. $y = -\frac{7}{5}x + \frac{1}{5}$

14. Find the equation of the horizontal line that contains the point $(4, -3)$. $y = -3$

15. Find the domain and range of the function $\{(-4, 2), (-2, 2), (0, 0), (3, 5)\}$. D: $\{-4, -2, 0, 3\}$; R: $\{0, 2, 5\}$

16. Find the equation of the line that contains the point (1, 2) and is parallel to the line $y = -\frac{3}{2}x - 6$. $y = -\frac{3}{2}x + \frac{7}{2}$

17. Find the equation of the line that contains the point $(-2, -3)$ and is perpendicular to the line $y = -\frac{1}{2}x - 3$. $y = 2x + 1$

18. Graph the solution set of $3x - 4y > 8$.

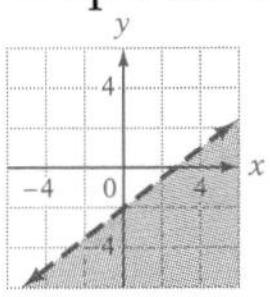

19. The director of a baseball camp estimates that 100 students will enroll if the tuition is \$250. For each \$20 increase in tuition, 6 fewer students will enroll. Determine a linear function that will predict the number of students who will enroll at a given tuition. Use this model to predict enrollment when the tuition is \$300. $y = -\frac{3}{10}x + 175$; 85 students

20. The graph at the right shows the relationship between the value of a rental house and the depreciation allowed for income tax purposes. Find the slope between the two points shown on the graph. Write a sentence that states the meaning of the slope.

$m = -\frac{10{,}000}{3}$

The value of the house decreases by \$3333.33 per year.

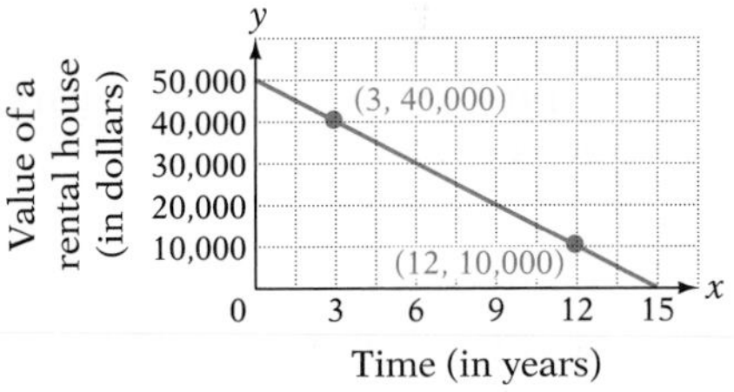

Cumulative Review Exercises

1. Identify the property that justifies the statement. $(x + y) \cdot 2 = 2 \cdot (x + y)$
Commutative Property of Multiplication

2. Solve: $3 - \frac{x}{2} = \frac{3}{4}$ $\frac{9}{2}$

3. Solve: $2[y - 2(3 - y) + 4] = 4 - 3y$ $\frac{8}{9}$

4. Solve: $\frac{1 - 3x}{2} + \frac{7x - 2}{6} = \frac{4x + 2}{9}$ $-\frac{1}{14}$

5. Solve: $x - 3 < -4$ or $2x + 2 > 3$
$\left\{x \mid x < -1 \text{ or } x > \frac{1}{2}\right\}$

6. Solve: $8 - |2x - 1| = 4$ $\frac{5}{2}, -\frac{3}{2}$

7. Solve: $|3x - 5| < 5$ $\left\{x \mid 0 < x < \frac{10}{3}\right\}$

8. Simplify: $4 - 2(4 - 5)^3 + 2$ 8

9. Evaluate $(a - b)^2 \div ab$ when $a = 4$ and $b = -2$.
-18

10. Graph: $\{x \mid x < -2\} \cup \{x \mid x > 0\}$

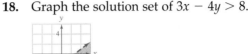

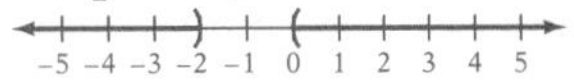

11. Solve: $3x - 1 < 4$ and $x - 2 > 2$ $\varnothing$

12. Given $P(x) = x^2 + 5$, evaluate $P(-3)$. 14

13. Find the ordered-pair solution of $y = -\frac{5}{4}x + 3$ that corresponds to $x = -8$. $(-8, 13)$

14. Find the slope of the line that contains the points $(-1, 3)$ and $(3, -4)$. $-\frac{7}{4}$

15. Find the equation of the line that contains the point $(-1, 5)$ and has slope $\frac{3}{2}$. $y = \frac{3}{2}x + \frac{13}{2}$

16. Find the equation of the line that contains the points $(4, -2)$ and $(0, 3)$. $y = -\frac{5}{4}x + 3$

17. Find the equation of the line that contains the point $(2, 4)$ and is parallel to the line $y = -\frac{3}{2}x + 2$. $y = -\frac{3}{2}x + 7$

18. Find the equation of the line that contains the point $(4, 0)$ and is perpendicular to the line $3x - 2y = 5$. $y = -\frac{2}{3}x + \frac{8}{3}$

19. Graph $3x - 5y = 15$ by using the x- and y-intercepts.

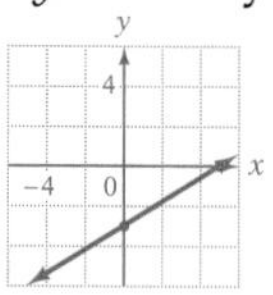

20. Graph the line that passes through the point $(-3, 1)$ and has slope $-\frac{3}{2}$.

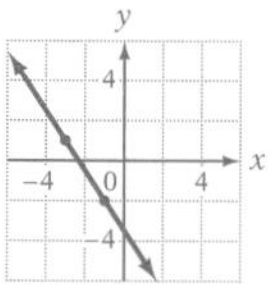

21. Graph the solution set of $3x - 2y \geq 6$.

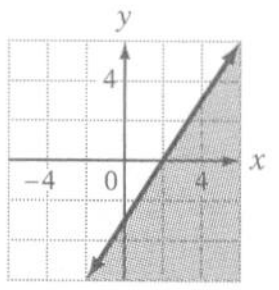

22. A coin purse contains coins with a value of \$1.60. The purse contains nickels and quarters. There are three times as many nickels as quarters. Find the number of nickels in the purse. 12 nickels

23. Two planes are 1800 mi apart and are traveling toward each other. One plane is traveling twice as fast as the other plane. The planes will meet in 3 h. Find the speed of each plane. First plane: 200 mph; second plane: 400 mph

24. A grocer combines coffee that costs \$8.00 per pound with coffee that costs \$3.00 per pound. How many pounds of each should be used to make 80 lb of a blend that costs \$5.00 per pound? 32 lb of \$8 coffee; 48 lb of \$3 coffee

25. The relationship between the depreciated value of a truck for income tax purposes and its age in years is shown in the graph at the right. Write the equation for the line that represents the depreciated value of the truck. Write a sentence that explains the meaning of the slope in the context of this exercise.
$y = -2500x + 15{,}000$; the value of the truck decreases by \$2500 each year.

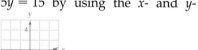

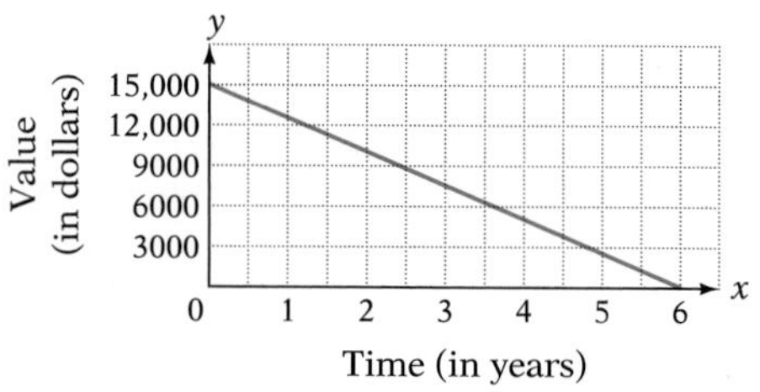

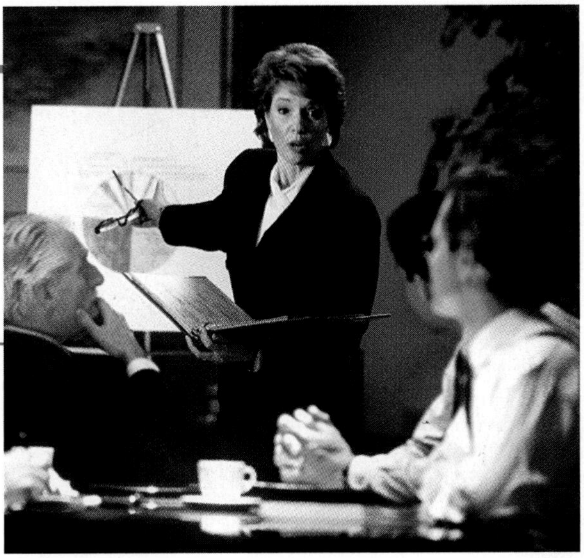

A budget analyst provides advice and technical assistance in the preparation of annual budgets. Managers and department heads submit proposed operating and financial plans to the budget analyst for review. These plans outline proposed program increases or new products, estimated costs and expenses, and capital expenditures needed to finance these programs. To analyze the different needs of each department or division requires solving large systems of equations.

4

Systems of Equations and Inequalities

OBJECTIVES

- 4.1.1 Solve a system of linear equations by graphing
- 4.1.2 Solve a system of linear equations by the substitution method
- 4.2.1 Solve a system of two linear equations in two variables by the addition method
- 4.2.2 Solve a system of three linear equations in three variables by the addition method
- 4.3.1 Evaluate determinants
- 4.3.2 Solve systems of linear equations by using Cramer's Rule
- 4.3.3 Solve systems of linear equations by using matrices
- 4.4.1 Rate-of-wind and water-current problems
- 4.4.2 Application problems
- 4.5.1 Graph the solution set of a system of linear inequalities

Analytic Geometry

Euclid's geometry (plane geometry is still taught in high school) was developed around 300 B.C. Euclid depends on a synthetic proof—a proof based on pure logic.

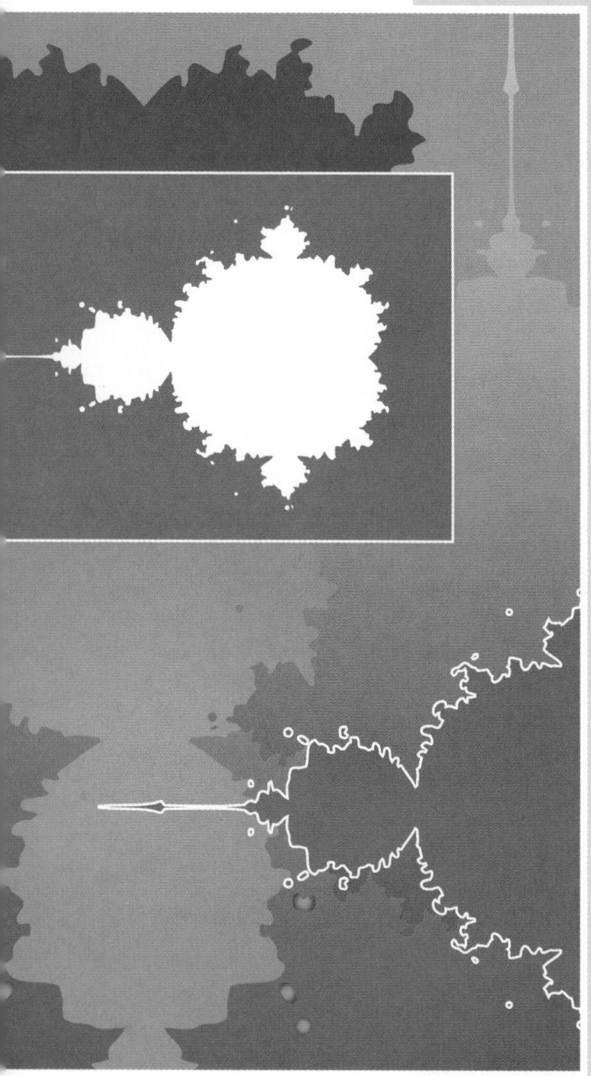

Algebra is the science of solving equations such as $ax^2 + bx + c = 0$. As early as the 3rd century, equations were solved by trial and error. However, the appropriate terminology and symbols of operation were not fully developed until the 17th century.

Analytic geometry is the combining of algebraic methods with geometric concepts. The development of analytic geometry in the 17th century is attributed to Descartes.

The development of analytic geometry depended on the invention of a system of coordinates in which any point in the plane could be represented by an ordered pair of numbers. This invention allowed lines and curves (such as the conic sections) to be represented and described by algebraic equations.

Some equations are quite simple and have a simple graph. For example, the graph of the equation $y = 2x - 3$ is a line.

Other graphs are quite complicated and have complicated equations. The graph of the Mandelbrot set, shown at the left, is very complicated. This graph is called a fractal.

SECTION 4.1 Solving Systems of Linear Equations by Graphing and by the Substitution Method

1 Solve a system of linear equations by graphing

INSTRUCTOR NOTE

The idea of a system of equations is fairly difficult for students. You might try to motivate the discussion by first asking, "Find two numbers whose sum is 20." It will not take students too long to realize that there are an infinite number of solutions to this question. By guiding your students a little further, you can show that the solution set is the set of points whose coordinates satisfy $x + y = 20$.

Now ask, "Find two numbers whose difference is 4." This will provide another line.

Finally, ask, "Find two numbers whose sum is 20 and whose difference is 4." You might have students guess and check until they find the solution and then show them that the solution is the coordinates of the point of intersection of the two lines.

You can extend this number problem to illustrate inconsistent and dependent systems of equations:

"Find two numbers whose sum is 5 and whose sum is 8." This is obviously not possible, and the graphs of the lines are parallel.

"Find two numbers whose sum is 10 such that the larger number is the difference between 10 and the smaller number." This is always true, and the graphs are the same line.

A **system of equations** is two or more equations considered together. The system at the right is a system of two linear equations in two variables. The graphs of the equations are straight lines.

$$3x + 4y = 7$$
$$2x - 3y = 6$$

A **solution of a system of equations in two variables** is an ordered pair that is a solution of each equation of the system.

Is $(3, -2)$ a solution of the system of equations shown at the right?

$$2x - 3y = 12$$
$$5x + 2y = 11$$

Replace x by 3 and y by -2.

$2x - 3y = 12$		$5x + 2y = 11$	
$2(3) - 3(-2)$	12	$5(3) + 2(-2)$	11
$6 - (-6)$		$15 + (-4)$	
$12 = 12$		$11 = 11$	

Yes, because $(3, -2)$ is a solution of each equation, it is a solution of the system of equations.

A solution of a system of linear equations can be found by graphing the lines of the system on the same coordinate axes. The point of intersection of the lines is the ordered pair that lies on both lines. It is the solution of the system of equations.

Solve by graphing: $x + 2y = 4$
$\qquad 2x + y = -1$

Graph each line.

Find the point of intersection.

The ordered pair $(-2, 3)$ lies on each line.

The solution is $(-2, 3)$.

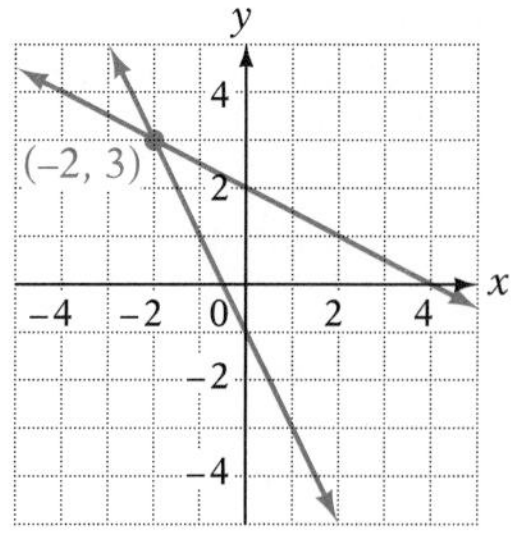

When the graphs of a system of equations intersect at only one point, the system of equations is called an **independent system of equations.**

A system of equations can be solved by using a graphing utility. First (if necessary), solve each equation for y. Then enter the two equations and graph them. The point at which the graphs of the two equations intersect is the solution of the system of equations.

If the INTERSECT feature on a graphing calculator is used to find the solution of the system of equations given above, the result will be $(-2, 3)$, the solution of the equation. However, if the ZOOM feature is used to estimate the solution, the estimated solution might be $(-2.05, 3.03)$ or some other ordered pair that contains decimals. When these values are rounded to the nearest integer, the ordered pair becomes $(-2, 3)$. This solution can be verified by replacing x by -2 and y by 3 in the system of equations.

⮞ Solve by graphing: $2x + 3y = 6$
$4x + 6y = -12$

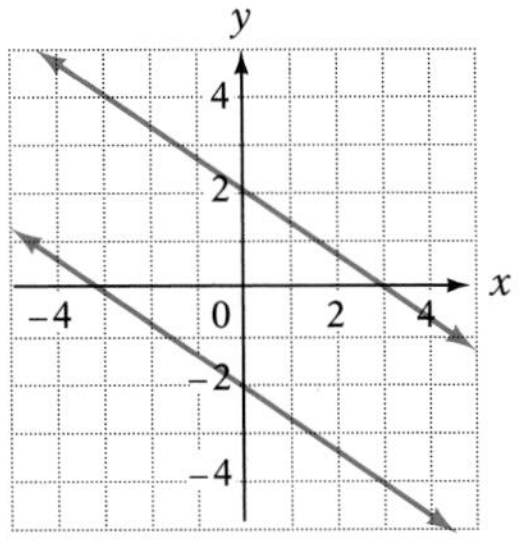

Graph each line.

The lines are parallel and therefore do not intersect. The system of equations has no solution. ⬅

When a system of equations has no solution, it is called an **inconsistent system of equations.**

⮞ Solve by graphing: $x - 2y = 4$
$2x - 4y = 8$

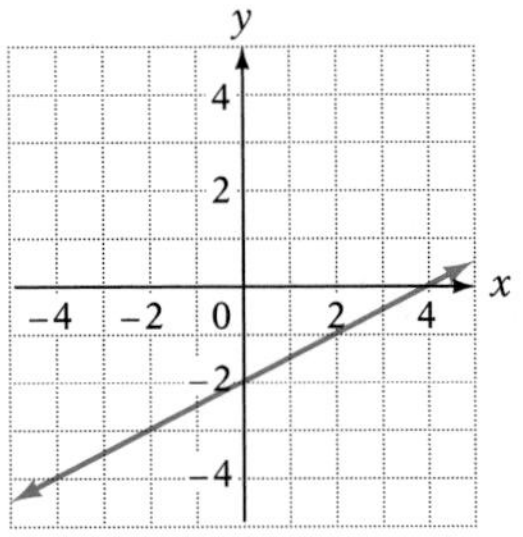

Graph each line.

The two equations represent the same line. This is a **dependent system of equations.** When a system of two equations is dependent, solve (if necessary) one of the equations of the system for y. The solutions of the dependent system are $(x, mx + b)$, where $y = mx + b$. For this system,

$$x - 2y = 4$$
$$-2y = -x + 4$$
$$y = \frac{1}{2}x - 2$$

The solutions are the ordered pairs $\left(x, \frac{1}{2}x - 2\right)$. ⬅

⌐ **Example 1**   Solve by graphing.

A. $2x - y = 3$ B. $2x + 3y = 6$
$3x + y = 2$ $y = -\frac{2}{3}x + 1$

Solution A.

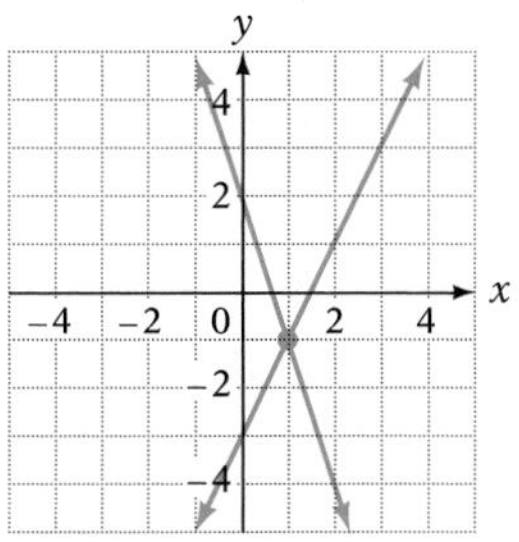

The solution is (1, −1).

B.

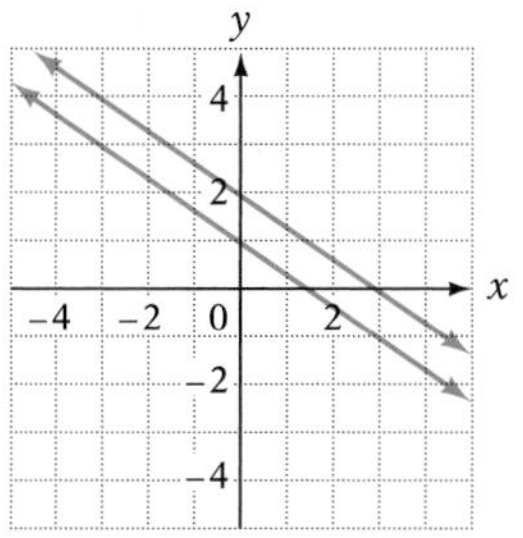

The lines are parallel and therefore do not intersect. The system of equations has no solution. It is inconsistent.

Problem 1 Solve by graphing. A. $x + y = 1$ B. $3x - 4y = 12$

$$2x + y = 0 \qquad y = \frac{3}{4}x - 3$$

Solution See page S8. A. $(-1, 2)$ B. $\left(x, \frac{3}{4}x - 3\right)$

2 Solve a system of linear equations by the substitution method

INSTRUCTOR NOTE

When students evaluate a variable expression, they replace a variable by a constant. Here the student is replacing a variable with a variable expression. Mentioning this connection will help some students grasp the substitution method of solving a system of equations. Also, have students imagine parentheses around the variable. This will help ensure that they apply the Distributive Property correctly.

A graphical solution of a system of equations is based on approximating the coordinates of a point of intersection. An algebraic method called the **substitution method** can be used to find an exact solution of a system of equations. To use the substitution method, we must write one of the equations of the system in terms of x or in terms of y.

Solve by the substitution method: $3x + y = 5$
$4x + 5y = 3$

(1) $3x + y = 5$
(2) $4x + 5y = 3$

Solve equation (1) for y. The result is equation (3).

$3x + y = 5$
(3) $y = -3x + 5$

Use equation (2). Substitute $-3x + 5$ for y and solve for x.

$$\begin{aligned} 4x + 5y &= 3 \\ 4x + 5(-3x + 5) &= 3 \\ 4x - 15x + 25 &= 3 \\ -11x &= -22 \\ x &= 2 \end{aligned}$$

Now substitute the value of x into equation (3) and solve for y.

$$\begin{aligned} y &= -3x + 5 \\ y &= -3(2) + 5 \\ y &= -6 + 5 \\ y &= -1 \end{aligned}$$

The solution is (2, −1).

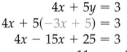

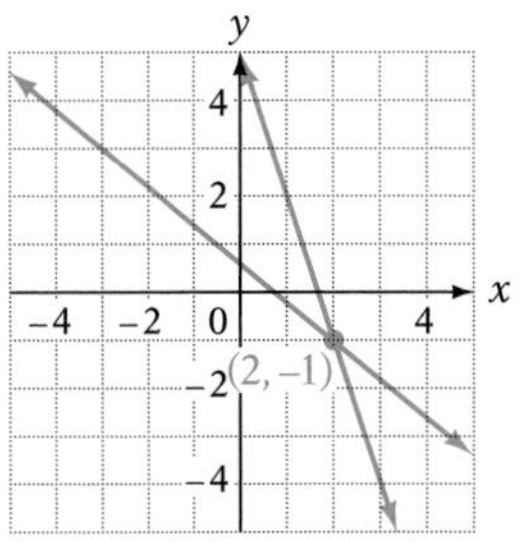

The graph of the system is shown at the left. Note that the graphs intersect at the point whose coordinates are (2, −1), the solution of the system of equations.

Example 2 Solve by substitution.

A. $3x - 3y = 2$
$y = x + 2$

B. $9x + 3y = 12$
$y = -3x + 4$

Solution A. (1) $3x - 3y = 2$
(2) $y = x + 2$ ▸ Equation (2) states that $y = x + 2$.

$3x - 3(x + 2) = 2$ ▸ Substitute $x + 2$ for y in equation (1).
$3x - 3x - 6 = 2$
$-6 = 2$

$-6 = 2$ is not a true equation. The system of equations is inconsistent. The system does not have a solution.

B. (1) $9x + 3y = 12$
(2) $y = -3x + 4$ ▸ Equation (2) states that $y = -3x + 4$.

$9x + 3(-3x + 4) = 12$ ▸ Substitute $-3x + 4$ for y in equation (1).
$9x - 9x + 12 = 12$
$12 = 12$

$12 = 12$ is a true equation. The system of equations is dependent. The solutions are the ordered pairs $(x, -3x + 4)$.

Problem 2 Solve by substitution. A. $3x - y = 3$, $6x + 3y = -4$ B. $6x - 3y = 6$, $2x - y = 2$

Solution See page S8–S9. A. $\left(\frac{1}{3}, -2\right)$ B. $(x, 2x - 2)$

CONCEPT REVIEW 4.1

Determine whether the following statements are always true, sometimes true, or never true.

1. A solution of a system of equations in two variables is an ordered pair that is a solution of either equation in the system. Always true

2. An inconsistent system of equations has no solution. Always true

3. A system of two equations with two unknowns has an infinite number of solutions. Sometimes true

4. A system of two equations with the same slope and different y-intercepts has an infinite number of solutions. Never true

5. When the graphs of a system of equations intersect at only one point, the system of equations is called an independent system of equations. Always true

6. The system $y = \frac{3}{2}x + 2$ and $y = -4x + 2$ has exactly one solution. Always true

EXERCISES 4.1

1 **1.** Describe the solution of an independent system of linear equations in two variables.

2. How does a dependent system of equations differ from an independent system of equations?

Solve by graphing.

3. $x + y = 2$
$x - y = 4$
(3, −1)

4. $x + y = 1$
$3x - y = -5$
(−1, 2)

5. $x - y = -2$
$x + 2y = 10$
(2, 4)

6. $2x - y = 5$
$3x + y = 5$
(2, −1)

7. $3x - 2y = 6$
$y = 3$
(4, 3)

8. $x = 4$
$3x - 2y = 4$
(4, 4)

9. $x = 4$
$y = -1$
(4, −1)

10. $x + 2 = 0$
$y - 1 = 0$
(−2, 1)

11. $y = x - 5$
$2x + y = 4$
(3, −2)

12. $2x - 5y = 4$
$y = x + 1$
(−3, −2)

13. $y = \frac{1}{2}x - 2$
$x - 2y = 8$
Inconsistent

14. $2x + 3y = 6$
$y = -\frac{2}{3}x + 1$
Inconsistent

15. $2x - 5y = 10$
$y = \frac{2}{5}x - 2$
$\left(x, \frac{2}{5}x - 2\right)$

16. $3x - 2y = 6$
$y = \frac{3}{2}x - 3$
$\left(x, \frac{3}{2}x - 3\right)$

17. $3x - 4y = 12$
$5x + 4y = -12$
(0, −3)

18. $2x - 3y = 6$
$2x - 5y = 10$
(0, −2)

2 **19.** When you solve a system of equations using the substitution method, how can you tell whether the system of equations is inconsistent?

20. When you solve a system of equations using the substitution method, how can you tell whether the system of equations is dependent?

Solve by substitution.

21. $3x - 2y = 4$
$x = 2$
(2, 1)

22. $y = -2$
$2x + 3y = 4$
(5, −2)

23. $y = 2x - 1$
$x + 2y = 3$
(1, 1)

24. $y = -x + 1$
$2x - y = 5$
(2, −1)

25. $4x - 3y = 5$
$y = 2x - 3$
(2, 1)

26. $3x + 5y = -1$
$y = 2x - 8$
(3, −2)

27. $x = 2y + 4$
$4x + 3y = -17$
(−2, −3)

28. $3x - 2y = -11$
$x = 2y - 9$
(−1, 4)

29. $5x + 4y = -1$
$y = 2 - 2x$
(3, −4)

30. $3x + 2y = 4$
$y = 1 - 2x$
(−2, 5)

31. $7x - 3y = 3$
$x = 2y + 2$
(0, −1)

32. $3x - 4y = 6$
$x = 3y + 2$
(2, 0)

33. $2x + 2y = 7$ $\left(\frac{1}{2}, 3\right)$
$y = 4x + 1$

34. $3x + 7y = -5$ $\left(\frac{2}{3}, -1\right)$
$y = 6x - 5$

35. $3x + y = 5$ $(1, 2)$
$2x + 3y = 8$

36. $4x + y = 9$ $(2, 1)$
$3x - 4y = 2$

37. $x + 3y = 5$ $(-1, 2)$
$2x + 3y = 4$

38. $x - 4y = 2$ $(-2, -1)$
$2x - 5y = 1$

39. $3x + 4y = 14$ $(-2, 5)$
$2x + y = 1$

40. $5x + 3y = 8$ $(4, -4)$
$3x + y = 8$

41. $3x + 5y = 0$ $(0, 0)$
$x - 4y = 0$

42. $2x - 7y = 0$ $(0, 0)$
$3x + y = 0$

43. $5x - 3y = -2$ $(-4, -6)$
$-x + 2y = -8$

44. $2x + 7y = 1$ $(-3, 1)$
$-x + 4y = 7$

45. $y = 3x + 2$ $(1, 5)$
$y = 2x + 3$

46. $y = 3x - 7$ $(2, -1)$
$y = 2x - 5$

47. $x = 2y + 1$ $(5, 2)$
$x = 3y - 1$

48. $x = 4y + 1$ $(-3, -1)$
$x = -2y - 5$

49. $y = 5x - 1$ $(1, 4)$
$y = 5 - x$

50. $y = 3 - 2x$ $(-1, 5)$
$y = 2 - 3x$

APPLYING CONCEPTS 4.1

For what values of k will the system of equations be inconsistent?

51. $2x - 2y = 5$ 2
$kx - 2y = 3$

52. $6x - 3y = 4$ $\frac{3}{2}$
$3x - ky = 1$

53. $x = 6y + 6$ $\frac{1}{2}$
$kx - 3y = 6$

54. $x = 2y + 2$ 4
$kx - 8y = 2$

Solve using two variables.

55. *Number Problem*   The sum of two numbers is 44. One number is 8 less than the other number. Find the numbers.   26 and 18

56. *Number Problem*   The sum of two numbers is 76. One number is 12 less than the other number. Find the numbers.   44 and 32

57. *Number Problem*   The sum of two numbers is 19. Five less than twice the first number equals the second number. Find the numbers.   8 and 11

58. *Number Problem*   The sum of two numbers is 22. Three times the first plus 2 equals the second number. Find the numbers.   5 and 17

Solve. (*Hint:* These equations are not linear equations. First rewrite the equations as linear equations by substituting x for $\frac{1}{a}$ and y for $\frac{1}{b}$.)

[C] **59.** $\frac{2}{a} + \frac{3}{b} = 4$
$\frac{4}{a} + \frac{1}{b} = 3$ $(2, 1)$

[C] **60.** $\frac{2}{a} + \frac{1}{b} = 1$
$\frac{8}{a} - \frac{2}{b} = 0$ $\left(6, \frac{3}{2}\right)$

[C] **61.** $\frac{1}{a} + \frac{3}{b} = 2$
$\frac{4}{a} - \frac{1}{b} = 3$ $\left(\frac{13}{11}, \frac{13}{5}\right)$

[C] **62.** $\frac{3}{a} + \frac{4}{b} = -1$
$\frac{1}{a} + \frac{6}{b} = 2$ $(-1, 2)$

Use a graphing utility to solve each of the following systems of equations. Round answers to the nearest hundredth.

63. $y = -\frac{1}{2}x + 2$
$y = 2x - 1$
(1.20, 1.40)

64. $y = 1.2x + 2$
$y = -1.3x - 3$
(−2.00, −0.40)

65. $y = \sqrt{2}x - 1$
$y = -\sqrt{3}x + 1$
(0.64, −0.10)

66. $y = \pi x - \frac{2}{3}$
$y = -x + \frac{\pi}{2}$
(0.54, 1.03)

67. *Demography* Using data from the latest report from the Census Bureau, a model of the population decline since 1980 of Detroit, Michigan, can be given by $y_1 = -0.0175x + 35.853$, where y_1 is the population (in millions) of Detroit in year x. According to the data from the same report, Phoenix, Arizona, has experienced an increase in population that can be modeled by $y_2 = 0.0146x - 28.658$, where y_2 is the population (in millions) of Phoenix in year x.

a. Assuming this model is accurate for future years, will the population of Phoenix exceed the population of Detroit in the year 2000? No

b. In what year will the population of Phoenix first exceed the population of Detroit? 2009

c. What does the slope of the line that models Detroit's population mean in the context of this problem? The population is decreasing at a rate of 17,500 people per year.

d. What does the slope of the line that models Phoenix's population mean in the context of this problem? The population is increasing at a rate of 14,600 people per year.

68. Suppose $\begin{array}{l} a_1x + b_1y = c_1 \\ a_2x + b_2y = c_2 \end{array}$ is a system of equations and $\frac{a_1}{b_1} = \frac{a_2}{b_2}$. Can the system of equations be independent? Explain your answer.

69. The Substitution Property of Equality states that if $a = b$, then a may be replaced by b in any expression that involves a. Explain how this property applies to solving a system of equations by using the substitution method.

70. Solving a system of equations by using a graphing utility requires a mathematical understanding of systems of equations. Using the viewing window Xmin = −5, Xmax = 5, Ymin = −5, Ymax = 5, try to determine graphically the solution of each of the following three systems of equations.

$$\begin{array}{lll} 2x - y = 1 & y = x + 2 & y = 1.4x - 2 \\ 4x - 2y = 2 & y = x + 1 & y = 1.5x - 7 \end{array}$$

In the first case, the system is dependent, so there are an infinite number of solutions. In the second case, the system of equations is inconsistent. And in the third case, there is exactly one solution. Explain how to determine, without solving or graphing the equations, whether a system of equations is **a.** dependent, **b.** independent, and **c.** inconsistent.

SECTION 4.2 Solving Systems of Linear Equations by the Addition Method

1 Solve a system of two linear equations in two variables by the addition method

INSTRUCTOR NOTE

The addition method is not clear to many students. Try showing them that this method is really a variation of the substitution method.

Using the system of equations at the right, add $2x + 3y$ to each side of $5x - 3y = 14$.

$$5x - 3y + (2x + 3y) = 14 + (2x + 3y)$$

Simplify the left side; substitute -7 for $2x + 3y$ on the right side. The result is $7x = 7$.

The **addition method** is an alternative method for solving a system of equations. This method is based on the Addition Property of Equations. Use the addition method when it is not convenient to solve one equation for one variable in terms of another variable.

Note, for the system of equations at the right, the effect of adding equation (2) to equation (1). Because $-3y$ and $3y$ are additive inverses, adding the equations results in an equation with only one variable.

$$\begin{aligned} (1)\quad 5x - 3y &= 14 \\ (2)\quad 2x + 3y &= -7 \\ 7x + 0y &= 7 \\ 7x &= 7 \end{aligned}$$

The solution of the resulting equation is the first component of the ordered-pair solution of the system.

$$\begin{aligned} 7x &= 7 \\ x &= 1 \end{aligned}$$

The second component is found by substituting the value of x into equation (1) or (2) and then solving for y. Equation (1) is used here.

$$\begin{aligned} (1)\quad 5x - 3y &= 14 \\ 5(1) - 3y &= 14 \\ 5 - 3y &= 14 \\ -3y &= 9 \\ y &= -3 \end{aligned}$$

The solution is $(1, -3)$.

POINT OF INTEREST

There are records of Babylonian mathematicians solving systems of equations 3600 years ago. Here is a system of equations from that time (in our modern notation):

$$\frac{2}{3}x = \frac{1}{2}y + 500$$
$$x + y = 1800$$

We say *modern notation* for many reasons. Foremost is the fact that using variables did not become widespread until the 17th century. There are many other reasons, however. The equals sign had not been invented, 2 and 3 did not look like they do today, and zero had not even been considered as a possible number.

Sometimes adding the two equations does not eliminate one of the variables. In this case, use the Multiplication Property of Equations to rewrite one or both of the equations so that when the equations are added, one of the variables is eliminated. To do this, first choose which variable to eliminate. The coefficients of that variable must be additive inverses. Multiply each equation by a constant that will produce coefficients that are additive inverses.

➡ Solve by the addition method: $3x + 4y = 2$, $2x + 5y = -1$

$$\begin{aligned} (1)\quad 3x + 4y &= 2 \\ (2)\quad 2x + 5y &= -1 \end{aligned}$$

Eliminate x. Multiply equation (1) by 2 and equation (2) by -3. Note how the constants are selected. The negative sign is used so that the coefficients will be additive inverses.

$$\begin{aligned} 2(3x + 4y) &= 2(2) \\ -3(2x + 5y) &= -3(-1) \end{aligned}$$

The coefficients of the x-terms are additive inverses.

$$\begin{aligned} 6x + 8y &= 4 \\ -6x - 15y &= 3 \end{aligned}$$

Add the equations.
Solve for y.

$$\begin{aligned} -7y &= 7 \\ y &= -1 \end{aligned}$$

Substitute the value of y into one of the equations, and solve for x. Equation (1) is used here.

$$(1)\quad 3x + 4y = 2$$
$$3x + 4(-1) = 2$$
$$3x - 4 = 2$$
$$3x = 6$$
$$x = 2$$

The solution is (2, −1).

Solve by the addition method: $\frac{2}{3}x + \frac{1}{2}y = 4$
$\frac{1}{4}x - \frac{3}{8}y = -\frac{3}{4}$

$$(1)\quad \frac{2}{3}x + \frac{1}{2}y = 4$$
$$(2)\quad \frac{1}{4}x - \frac{3}{8}y = -\frac{3}{4}$$

Clear the fractions. Multiply each equation by the LCM of the denominators.

$$6\left(\frac{2}{3}x + \frac{1}{2}y\right) = 6(4)$$
$$8\left(\frac{1}{4}x - \frac{3}{8}y\right) = 8\left(-\frac{3}{4}\right)$$
$$4x + 3y = 24$$
$$2x - 3y = -6$$

Eliminate y. Add the equations.
Solve for x.

$$6x = 18$$
$$x = 3$$

Substitute the value of x into equation (1), and solve for y.

$$\frac{2}{3}x + \frac{1}{2}y = 4$$
$$\frac{2}{3}(3) + \frac{1}{2}y = 4$$
$$2 + \frac{1}{2}y = 4$$
$$\frac{1}{2}y = 2$$
$$y = 4$$

The solution is (3, 4).

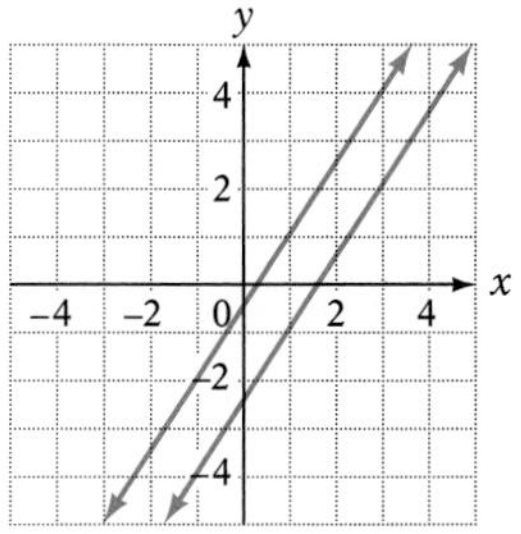

To solve the system of equations at the right, eliminate x. Multiply equation (1) by −2 and add to equation (2).

$$(1)\quad 3x - 2y = 5$$
$$(2)\quad 6x - 4y = 1$$

$$-6x + 4y = -10$$
$$6x - 4y = 1$$
$$0 = -9$$

$0 = -9$ is not a true equation. The system of equations is inconsistent. The system does not have a solution.

The graph of the equations of the system is shown at the left. Note that the lines are parallel and therefore do not intersect.

Example 1 Solve by the addition method.

A. $3x - 2y = 2x + 5$
$2x + 3y = -4$

B. $4x - 8y = 36$
$3x - 6y = 27$

Solution A. (1) $3x - 2y = 2x + 5$
(2) $2x + 3y = -4$

$x - 2y = 5$
$2x + 3y = -4$
▶ Write equation (1) in the form $Ax + By = C$.

$-2(x - 2y) = -2(5)$
$2x + 3y = -4$
▶ To eliminate x, multiply each side of equation (1) by -2.

$-2x + 4y = -10$
$2x + 3y = -4$
$7y = -14$ ▶ Add the equations.
$y = -2$

$2x + 3y = -4$ ▶ Replace y in equation (2) by its value.
$2x + 3(-2) = -4$
$2x - 6 = -4$
$2x = 2$
$x = 1$

The solution is $(1, -2)$.

B. (1) $4x - 8y = 36$
(2) $3x - 6y = 27$

$3(4x - 8y) = 3(36)$
$-4(3x - 6y) = -4(27)$
▶ To eliminate x, multiply each side of equation (1) by 3 and each side of equation (2) by -4.

$12x - 24y = 108$
$-12x + 24y = -108$
$0 = 0$ ▶ Add the equations.

$0 = 0$ is a true equation. The system of equations is dependent. The solutions are the ordered pairs $\left(x, \frac{1}{2}x - \frac{9}{2}\right)$.

Problem 1 Solve by the addition method.

A. $2x + 5y = 6$
$3x - 2y = 6x + 2$

B. $2x + y = 5$
$4x + 2y = 6$

Solution See page S9. A. $(-2, 2)$ B. Inconsistent

2 Solve a system of three linear equations in three variables by the addition method

16 HM³

An equation of the form $Ax + By + Cz = D$, where A, B, and C are coefficients and D is a constant, is a **linear equation in three variables.** Examples of these equations are shown at the right. The graph of a linear equation in three variables is a plane.

$2x + 4y - 3z = 7$
$x - 6y + z = -3$

INSTRUCTOR NOTE

A good model of a three-dimensional coordinate system is the corner of the floor in a room. The *xy*-plane is the floor, the *xz*-plane one wall, and the *yz*-plane the other wall. Have students indicate where points are for which some of the coordinates are negative numbers.

Graphing an equation in three variables requires a third coordinate axis perpendicular to the *xy*-plane. The third axis is commonly called the *z*-axis. The result is a three-dimensional coordinate system called the *xyz*-coordinate system. To help visualize a three-dimensional coordinate system, think of a corner of a room: the floor is the *xy*-plane, one wall is the *yz*-plane, and the other wall is the *xz*-plane. A three-dimensional coordinate system is shown at the right.

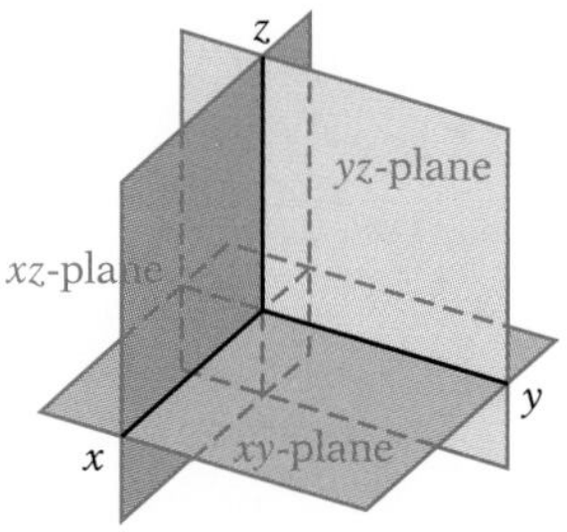

The graph of a point in an *xyz*-coordinate system is an **ordered triple** (x, y, z). Graphing an ordered triple requires three moves, the first along the *x*-axis, the second parallel to the *y*-axis, and the third parallel to the *z*-axis. The graphs of the points $(-4, 2, 3)$ and $(3, 4, -2)$ are shown at the right.

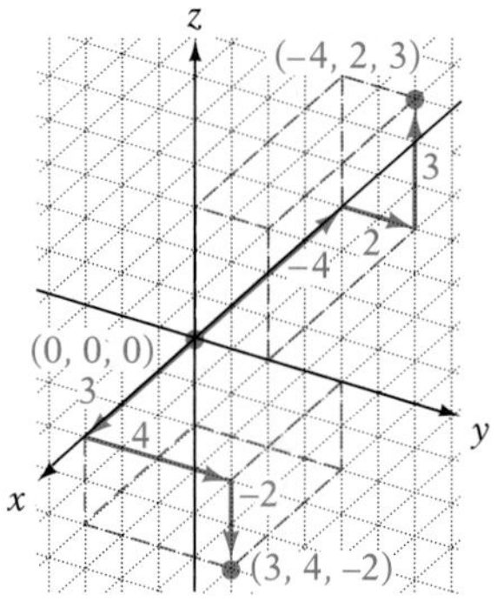

The graph of a linear equation in three variables is a plane. That is, if all the solutions of a linear equation in three variables were plotted in an *xyz*-coordinate system, the graph would look like a large piece of paper extending infinitely. The graph of $x + y + z = 3$ is shown at the right.

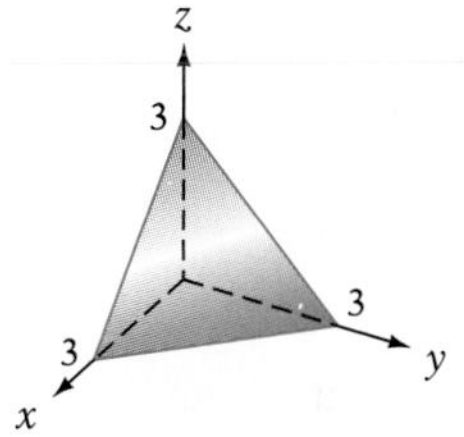

INSTRUCTOR NOTE

Your book and a desk or table top can be used to model various ways in which planes can intersect.

There are different ways in which three planes can be oriented in an *xyz*-coordinate system. The systems of equations represented by the planes below are inconsistent.

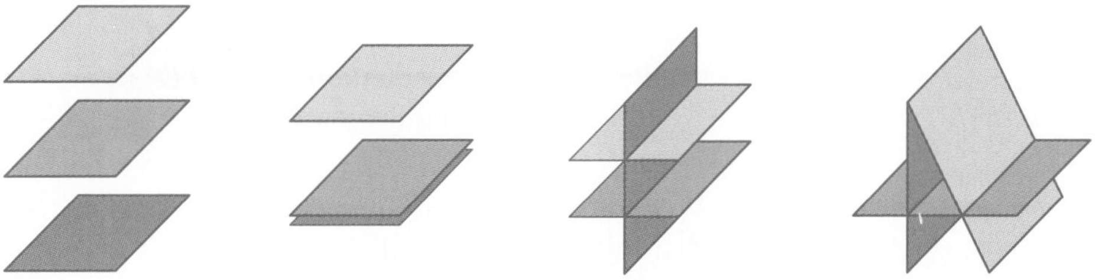

A **B** **C** **D**

Graphs of Inconsistent Systems of Equations

For a system of three equations in three variables to have a solution, the graphs of the planes must intersect at a single point, they must intersect along a common line, or all equations must have a graph that is the same plane. These situations are shown in the figures that follow.

The three planes shown in Figure E below intersect at a point. A system of equations represented by planes that intersect at a point is independent.

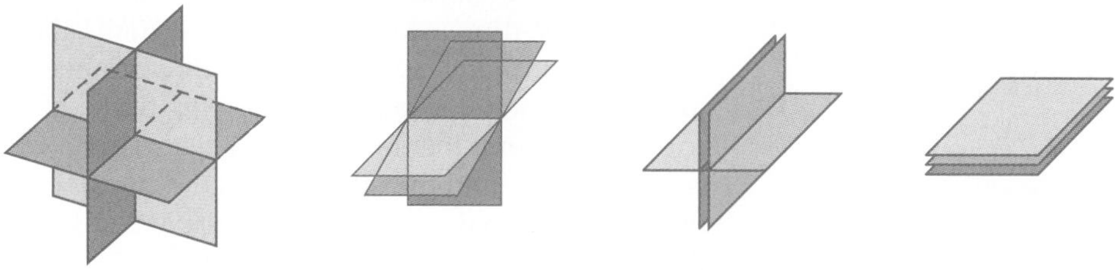

E F G H

An Independent System of Equations

Dependent Systems of Equations

The planes shown in Figures F and G above intersect along a common line. The system of equations represented by the planes in Figure H has a graph that is the same plane. The systems of equations represented by these three graphs are dependent.

Just as a solution of an equation in two variables is an ordered pair (x, y), a **solution of an equation in three variables** is an ordered triple (x, y, z). For example, $(2, 1, -3)$ is a solution of the equation $2x - y - 2z = 9$. The ordered triple $(1, 3, 2)$ is not a solution.

POINT OF INTEREST

In the early 1980s, Stephen Hoppe became interested in winning Monopoly strategies. Finding these strategies required solving a system that contained 123 equations with 123 variables!

A **system of linear equations in three variables** is shown at the right. A **solution of a system of equations in three variables** is an ordered triple that is a solution of each equation of the system.

$$\begin{aligned} x - 2y + z &= 6 \\ 3x + y - 2z &= 2 \\ 2x - 3y + 5z &= 1 \end{aligned}$$

A system of linear equations in three variables can be solved by using the addition method. First, eliminate one variable from any two of the given equations. Then eliminate the same variable from any other two equations. The result will be a system of two equations in two variables. Solve this system by the addition method.

INSTRUCTOR NOTE

Show students that the objective of solving a system of three equations with three variables is to replace it with an equivalent system of two equations in two variables that can be solved by using the techniques of the previous section.

Solve: $x + 4y - z = 10$, $3x + 2y + z = 4$, $2x - 3y + 2z = -7$

$$\begin{aligned} x + 4y - z &= 10 \quad (1) \\ 3x + 2y + z &= 4 \quad (2) \\ 2x - 3y + 2z &= -7 \quad (3) \end{aligned}$$

Eliminate z from equations (1) and (2) by adding the two equations.

$$\begin{aligned} x + 4y - z &= 10 \\ 3x + 2y + z &= 4 \\ 4x + 6y &= 14 \quad (4) \end{aligned}$$

Eliminate z from equations (1) and (3). Multiply equation (1) by 2 and add it to equation (3).

$$\begin{aligned} 2x + 8y - 2z &= 20 \\ 2x - 3y + 2z &= -7 \\ 4x + 5y &= 13 \quad (5) \end{aligned}$$

Solve the system of two equations in two variables.

$$\begin{aligned} 4x + 6y &= 14 \quad (4) \\ 4x + 5y &= 13 \quad (5) \end{aligned}$$

Eliminate x. Multiply equation (5) by -1 and add it to equation (4).

$$\begin{aligned} 4x + 6y &= 14 \\ -4x - 5y &= -13 \\ y &= 1 \end{aligned}$$

Substitute the value of y into equation (4) or (5), and solve for x. Equation (4) is used here.

$$\begin{aligned} 4x + 6y &= 14 \\ 4x + 6(1) &= 14 \\ 4x + 6 &= 14 \\ 4x &= 8 \\ x &= 2 \end{aligned}$$

Substitute the value of y and the value of x into one of the equations in the original system. Equation (2) is used here.

$$\begin{aligned} 3x + 2y + z &= 4 \\ 3(2) + 2(1) + z &= 4 \\ 6 + 2 + z &= 4 \\ 8 + z &= 4 \\ z &= -4 \end{aligned}$$

The solution is (2, 1, −4).

Solve: $2x - 3y - z = 1$
$x + 4y + 3z = 2$
$4x - 6y - 2z = 5$

$$\begin{aligned} (1) \quad 2x - 3y - z &= 1 \\ (2) \quad x + 4y + 3z &= 2 \\ (3) \quad 4x - 6y - 2z &= 5 \end{aligned}$$

Eliminate x from equations (1) and (2). Multiply equation (2) by −2 and add to equation (1).

$$\begin{aligned} 2x - 3y - z &= 1 \\ -2x - 8y - 6z &= -4 \\ -11y - 7z &= -3 \end{aligned}$$

Eliminate x from equations (1) and (3). Multiply equation (1) by −2 and add to equation (3). This is not a true equation.

$$\begin{aligned} -4x + 6y + 2z &= -2 \\ 4x - 6y - 2z &= 5 \\ 0 &= 3 \end{aligned}$$

The system is inconsistent.

Example 2 Solve: $3x - y + 2z = 1$
$2x + 3y + 3z = 4$
$x + y - 4z = -9$

Solution

$$\begin{aligned} (1) \quad 3x - y + 2z &= 1 \\ (2) \quad 2x + 3y + 3z &= 4 \\ (3) \quad x + y - 4z &= -9 \end{aligned}$$

$$\begin{aligned} 3x - y + 2z &= 1 \\ x + y - 4z &= -9 \\ 4x - 2z &= -8 \end{aligned}$$

▶ Eliminate y. Add equations (1) and (3).

$$(4) \quad 2x - z = -4$$

▶ Multiply each side of the equation by $\frac{1}{2}$.

$$\begin{aligned} 9x - 3y + 6z &= 3 \\ 2x + 3y + 3z &= 4 \\ (5) \quad 11x + 9z &= 7 \end{aligned}$$

▶ Multiply equation (1) by 3 and add to equation (2).

$$\begin{aligned} (4) \quad 2x - z &= -4 \\ (5) \quad 11x + 9z &= 7 \end{aligned}$$

▶ Solve the system of two equations.

$$\begin{aligned} 18x - 9z &= -36 \\ 11x + 9z &= 7 \\ 29x &= -29 \\ x &= -1 \end{aligned}$$

▶ Multiply equation (4) by 9 and add to equation (5).

$$\begin{aligned} 2x - z &= -4 \\ 2(-1) - z &= -4 \\ -2 - z &= -4 \\ -z &= -2 \\ z &= 2 \end{aligned}$$

▶ Replace x by -1 in equation (4).

$$\begin{aligned} x + y - 4z &= -9 \\ -1 + y - 4(2) &= -9 \\ -1 + y - 8 &= -9 \\ -9 + y &= -9 \\ y &= 0 \end{aligned}$$

▶ Replace x by -1 and z by 2 in equation (3).

The solution is $(-1, 0, 2)$.

Problem 2 Solve:
$$\begin{aligned} x - y + z &= 6 \\ 2x + 3y - z &= 1 \\ x + 2y + 2z &= 5 \end{aligned}$$

Solution See page S9. $(3, -1, 2)$

CONCEPT REVIEW 4.2

Determine whether the following statements are always true, sometimes true, or never true.

1. The graph of a linear equation in three variables is a plane. Always true
2. A solution of an equation in three variables is an ordered triple (x, y, z). Always true
3. The solution of a system of linear equations in three variables is a straight line. Sometimes true
4. A system of linear equations in three variables that has an infinite number of solutions is an independent system. Never true
5. When you are solving a system of equations by the addition method and the final result is $0 = 5$, the system is inconsistent and has an infinite number of solutions. Never true

EXERCISES 4.2

1 1. If you begin to solve the system of equations $\begin{aligned} x + y &= 6 \\ x - y &= 4 \end{aligned}$ by adding the two equations, the result is $2x = 10$, which means $x = 5$. Does this mean that 5 is a solution of the system of equations? Why or why not?

2. Suppose you must solve the system of equation $\begin{aligned} 2x - 5y &= 13 \\ 3x + 2y &= 10 \end{aligned}$.
   a. Explain how you would eliminate x from the system of equations.
   b. Explain how you would eliminate y from the system of equations.

Solve by the addition method.

3. $x - y = 5$ (6, 1)
$x + y = 7$

4. $x + y = 1$ (2, −1)
$2x - y = 5$

5. $3x + y = 4$ (1, 1)
$x + y = 2$

6. $x - 3y = 4$ (1, −1)
$x + 5y = -4$

7. $3x + y = 7$ (2, 1)
$x + 2y = 4$

8. $x - 2y = 7$ (1, −3)
$3x - 2y = 9$

9. $3x - y = 4$ $(x, 3x - 4)$
$6x - 2y = 8$

10. $x - 2y = -3$ $\left(x, \frac{1}{2}x + \frac{3}{2}\right)$
$-2x + 4y = 6$

11. $2x + 5y = 9$ $\left(-\frac{1}{2}, 2\right)$
$4x - 7y = -16$

12. $8x - 3y = 21$ $\left(\frac{3}{2}, -3\right)$
$4x + 5y = -9$

13. $4x - 6y = 5$ Inconsistent
$2x - 3y = 7$

14. $3x + 6y = 7$ Inconsistent
$2x + 4y = 5$

15. $3x - 5y = 7$ (−1, −2)
$x - 2y = 3$

16. $3x + 4y = 25$ (3, 4)
$2x + y = 10$

17. $3x + 2y = 16$ (2, 5)
$2x - 3y = -11$

18. $2x - 5y = 13$ (4, −1)
$5x + 3y = 17$

19. $4x + 4y = 5$ $\left(\frac{1}{2}, \frac{3}{4}\right)$
$2x - 8y = -5$

20. $3x + 7y = 16$ (3, 1)
$4x - 3y = 9$

21. $5x + 4y = 0$ (0, 0)
$3x + 7y = 0$

22. $3x - 4y = 0$ (0, 0)
$4x - 7y = 0$

23. $3x - 6y = 6$ $\left(\frac{2}{3}, -\frac{2}{3}\right)$
$9x - 3y = 8$

24. $4x - 8y = 5$ $\left(\frac{1}{4}, -\frac{1}{2}\right)$
$8x + 2y = 1$

25. $5x + 2y = 2x + 1$ (1, −1)
$2x - 3y = 3x + 2$

26. $3x + 3y = y + 1$ (−3, 5)
$x + 3y = 9 - x$

27. $\frac{2}{3}x - \frac{1}{2}y = 3$ $\left(x, \frac{4}{3}x - 6\right)$
$\frac{1}{3}x - \frac{1}{4}y = \frac{3}{2}$

28. $\frac{3}{4}x + \frac{1}{3}y = -\frac{1}{2}$ (−2, 3)
$\frac{1}{2}x - \frac{5}{6}y = -\frac{7}{2}$

29. $\frac{2}{5}x - \frac{1}{3}y = 1$ (5, 3)
$\frac{3}{5}x + \frac{2}{3}y = 5$

30. $\frac{5}{6}x + \frac{1}{3}y = \frac{4}{3}$ (2, −1)
$\frac{2}{3}x - \frac{1}{2}y = \frac{11}{6}$

31. $\frac{3}{4}x + \frac{2}{5}y = -\frac{3}{20}$ $\left(\frac{1}{3}, -1\right)$
$\frac{3}{2}x - \frac{1}{4}y = \frac{3}{4}$

32. $\frac{2}{5}x - \frac{1}{2}y = \frac{13}{2}$ (10, −5)
$\frac{3}{4}x - \frac{1}{5}y = \frac{17}{2}$

33. $4x - 5y = 3y + 4$ $\left(\frac{5}{3}, \frac{1}{3}\right)$
$2x + 3y = 2x + 1$

34. $5x - 2y = 8x - 1$ (−3, 5)
$2x + 7y = 4y + 9$

35. $2x + 5y = 5x + 1$ Inconsistent
$3x - 2y = 3y + 3$

36. When you solve a system of equations using the addition method, how can you tell whether the system of equations is inconsistent?

37. When you solve a system of equations using the addition method, how can you tell whether the system of equations is dependent?

38. Describe the graph of a linear equation in three variables.

39. Describe the solution of an independent system of linear equations in three variables.

Solve by the addition method.

40. $x + 2y - z = 1$, $2x - y + z = 6$, $x + 3y - z = 2$ (2, 1, 3)

41. $x + 3y + z = 6$, $3x + y - z = -2$, $2x + 2y - z = 1$ (−1, 2, 1)

42. $2x - y + 2z = 7$, $x + y + z = 2$, $3x - y + z = 6$ (1, −1, 2)

43. $x - 2y + z = 6$, $x + 3y + z = 16$, $3x - y - z = 12$ (6, 2, 4)

44. $3x + y = 5$, $3y - z = 2$, $x + z = 5$ (1, 2, 4)

45. $2y + z = 7$, $2x - z = 3$, $x - y = 3$ (4, 1, 5)

46. $x - y + z = 1$, $2x + 3y - z = 3$, $-x + 2y - 4z = 4$ (2, −1, −2)

47. $2x + y - 3z = 7$, $x - 2y + 3z = 1$, $3x + 4y - 3z = 13$ (3, 1, 0)

48. $2x + 3z = 5$, $3y + 2z = 3$, $3x + 4y = -10$ (−2, −1, 3)

49. $3x + 4z = 5$, $2y + 3z = 2$, $2x - 5y = 8$ (−1, −2, 2)

50. $2x + 4y - 2z = 3$, $x + 3y + 4z = 1$, $x + 2y - z = 4$ Inconsistent

51. $x - 3y + 2z = 1$, $x - 2y + 3z = 5$, $2x - 6y + 4z = 3$ Inconsistent

52. $2x + y - z = 5$, $x + 3y + z = 14$, $3x - y + 2z = 1$ (1, 4, 1)

53. $3x - y - 2z = 11$, $2x + y - 2z = 11$, $x + 3y - z = 8$ (2, 1, −3)

54. $3x + y - 2z = 2$, $x + 2y + 3z = 13$, $2x - 2y + 5z = 6$ (1, 3, 2)

55. $4x + 5y + z = 6$, $2x - y + 2z = 11$, $x + 2y + 2z = 6$ (2, −1, 3)

56. $2x - y + z = 6$, $3x + 2y + z = 4$, $x - 2y + 3z = 12$ (1, −1, 3)

57. $3x + 2y - 3z = 8$, $2x + 3y + 2z = 10$, $x + y - z = 2$ (6, −2, 2)

58. $3x - 2y + 3z = -4$, $2x + y - 3z = 2$, $3x + 4y + 5z = 8$ (0, 2, 0)

59. $3x - 3y + 4z = 6$, $4x - 5y + 2z = 10$, $x - 2y + 3z = 4$ (0, −2, 0)

60. $3x - y + 2z = 2$, $4x + 2y - 7z = 0$, $2x + 3y - 5z = 7$ (1, 5, 2)

61. $2x + 2y + 3z = 13$, $-3x + 4y - z = 5$, $5x - 3y + z = 2$ (2, 3, 1)

62. $2x - 3y + 7z = 0$, $x + 4y - 4z = -2$, $3x + 2y + 5z = 1$ (−2, 1, 1)

63. $5x + 3y - z = 5$, $3x - 2y + 4z = 13$, $4x + 3y + 5z = 22$ (1, 1, 3)

APPLYING CONCEPTS 4.2

Solve. (*Hint:* First multiply both sides of each equation in the system by a multiple of 10 so that the coefficients and constants are integers.)

64. $0.2x - 0.3y = 0.5$, $0.3x - 0.2y = 0.5$ (1, −1)

65. $0.4x - 0.9y = -0.1$, $0.3x + 0.2y = 0.8$ (2, 1)

66. $1.25x - 0.25y = -1.5$, $1.5x + 2.5y = 1$ (−1, 1)

67. $2.25x + 1.5y = 3$, $1.75x + 2.25y = 1.25$ (2, −1)

68. $1.5x + 2.5y + 1.5z = 8$, $0.5x - 2y - 1.5z = -1$, $2.5x - 1.5y + 2z = 2.5$ (3, 2, −1)

69. $1.6x - 0.9y + 0.3z = 2.9$, $1.6x + 0.5y - 0.1z = 3.3$, $0.8x - 0.7y + 0.1z = 1.5$ (2, 0, −1)

Solve.

70. The point of intersection of the graphs of the equations $Ax + 3y = 6$ and $2x + By = -4$ is $(3, -2)$. Find A and B. $A = 4, B = 5$

71. The point of intersection of the graphs of the equations $Ax + 3y + 2z = 8$, $2x + By - 3z = -12$, and $3x - 2y + Cz = 1$ is $(3, -2, 4)$. Find A, B, and C.
$A = 2, B = 3, C = -3$

[C] **72.** The distance between a point and a line is the perpendicular distance from the point to the line. Find the distance between the point $(3, 1)$ and the line $y = x$. $\sqrt{2}$

[C] **73.** *Coin Problem* A coin bank contains only nickels, dimes, and quarters. There is a total of 30 coins in the bank. The value of all the coins is \$3.25. Find the number of nickels, dimes, and quarters in the bank. (*Hint:* There is more than one solution.)
(nickels, dimes, quarters) $\rightarrow (3z - 5, -4z + 35, z)$, where $z = 2, 3, 4, 5, 6, 7$, or 8

The following are not linear systems of equations. However, they can be solved by using a modification of the addition method. Solve each system of equations.

[C] **74.** $\frac{1}{x} - \frac{2}{y} = 3$
$\frac{2}{x} + \frac{3}{y} = -1$ $(1, -1)$

[C] **75.** $\frac{1}{x} + \frac{2}{y} = 3$
$\frac{1}{x} - \frac{3}{y} = -2$ $(1, 1)$

[C] **76.** $\frac{3}{x} + \frac{2}{y} = 1$
$\frac{2}{x} + \frac{4}{y} = -2$ $(1, -1)$

[C] **77.** $\frac{3}{x} - \frac{5}{y} = -\frac{3}{2}$
$\frac{1}{x} - \frac{2}{y} = -\frac{2}{3}$ $(3, 2)$

78. For Exercise 74, solve each equation of the system for y. Then use a graphing utility to verify the solution you determined algebraically. Note that the graphs are not straight lines.

79. When is it possible to solve a system of three linear equations in two variables and when is it not possible?

80. Describe the graph of each of the following equations in an xyz-coordinate system.

a. $x = 3$ **b.** $y = 4$ **c.** $z = 2$ **d.** $y = x$

81. Explain, graphically, the following situations when they are related to a system of three linear equations in three variables.

a. The system of equations has no solution.

b. The system of equations has exactly one solution.

c. The system of equations has infinitely many solutions.

SECTION 4.3

Solving Systems of Equations by Using Determinants and by Using Matrices

1 Evaluate determinants

POINT OF INTEREST

The word *matrix* was first used (in a mathematical context) in 1850. The root of this word is the Latin *mater.* A matrix was thought of as being an object from which something else originates. The idea was that a determinant, discussed below, originated (was born) from a matrix. Today, matrices are one of the most widely used tools of applied mathematics.

A **matrix** is a rectangular array of numbers. Each number in the matrix is called an **element** of the matrix. The matrix at the right, with three rows and four columns, is called a 3 × 4 (read "3 by 4") matrix.

$$A = \begin{bmatrix} 1 & -3 & 2 & 4 \\ 0 & 4 & -3 & 2 \\ 6 & -5 & 4 & -1 \end{bmatrix}$$

A matrix of m rows and n columns is said to be of **order $m \times n$.** Matrix A above has order 3 × 4. The notation a_{ij} refers to the element of a matrix in the ith row and the jth column. For matrix A, $a_{23} = -3$, $a_{31} = 6$, and $a_{13} = 2$.

A **square matrix** is one that has the same number of rows as columns. A 2 × 2 matrix and a 3 × 3 matrix are shown at the right.

$$\begin{bmatrix} -1 & 3 \\ 5 & 2 \end{bmatrix} \qquad \begin{bmatrix} 4 & 0 & 1 \\ 5 & -3 & 7 \\ 2 & 1 & 4 \end{bmatrix}$$

Associated with every square matrix is a number called its **determinant.**

Determinant of a 2 × 2 Matrix

The determinant of a 2 × 2 matrix $\begin{bmatrix} a_{11} & a_{12} \\ a_{21} & a_{22} \end{bmatrix}$ is written $\begin{vmatrix} a_{11} & a_{12} \\ a_{21} & a_{22} \end{vmatrix}$. The value of this determinant is given by the formula

$$\begin{vmatrix} a_{11} & a_{12} \\ a_{21} & a_{22} \end{vmatrix} = a_{11}a_{22} - a_{21}a_{12}$$

Note that vertical bars are used to represent the determinant and that brackets are used to represent the matrix.

Find the value of the determinant $\begin{vmatrix} 3 & 4 \\ -1 & 2 \end{vmatrix}$.

$$\begin{vmatrix} 3 & 4 \\ -1 & 2 \end{vmatrix} = 3 \cdot 2 - (-1)(4) = 6 - (-4) = 10$$

The value of the determinant is 10.

For a square matrix whose order is 3 × 3 or greater, the value of the determinant of that matrix is found by using 2 × 2 determinants.

The **minor of an element** in a 3 × 3 determinant is the 2 × 2 determinant that is obtained by eliminating the row and column that contain that element.

➡ Find the minor of -3 for the determinant $\begin{vmatrix} 2 & -3 & 4 \\ 0 & 4 & 8 \\ -1 & 3 & 6 \end{vmatrix}$.

The minor of -3 is the 2×2 determinant created by eliminating the row and column that contain -3.

Eliminate the row and column as shown: $\begin{vmatrix} 2 & -3 & 4 \\ 0 & 4 & 8 \\ -1 & 3 & 6 \end{vmatrix}$

The minor of -3 is $\begin{vmatrix} 0 & 8 \\ -1 & 6 \end{vmatrix}$.

Definition of a Cofactor

The **cofactor** of an element of a determinant is $(-1)^{i+j}$ times the minor of that element, where i is the row number of the element and j is the column number of the element.

➡ For the determinant $\begin{vmatrix} 3 & -2 & 1 \\ 2 & -5 & -4 \\ 0 & 3 & 1 \end{vmatrix}$, find the cofactor of -2 and of -5.

Because -2 is in the first row and the second column, $i = 1$ and $j = 2$. Therefore, $i + j = 1 + 2 = 3$, and $(-1)^{i+j} = (-1)^3 = -1$. The cofactor of -2 is $(-1)\begin{vmatrix} 2 & -4 \\ 0 & 1 \end{vmatrix}$.

Because -5 is in the second row and the second column, $i = 2$ and $j = 2$. Therefore, $i + j = 2 + 2 = 4$, and $(-1)^{i+j} = (-1)^4 = 1$. The cofactor of -5 is $1 \cdot \begin{vmatrix} 3 & 1 \\ 0 & 1 \end{vmatrix}$.

Note from this example that the cofactor of an element is -1 times the minor of that element or 1 times the minor of that element, depending on whether the sum $i + j$ is an odd or an even integer.

The value of a 3×3 or larger determinant can be found by **expanding by cofactors** of *any* row or *any* column. The result of expanding by cofactors using the first row of a 3×3 determinant is shown here.

$$\begin{vmatrix} a_{11} & a_{12} & a_{13} \\ a_{21} & a_{22} & a_{23} \\ a_{31} & a_{32} & a_{33} \end{vmatrix} = a_{11}(-1)^{1+1}\begin{vmatrix} a_{22} & a_{23} \\ a_{32} & a_{33} \end{vmatrix} + a_{12}(-1)^{1+2}\begin{vmatrix} a_{21} & a_{23} \\ a_{31} & a_{33} \end{vmatrix} + a_{13}(-1)^{1+3}\begin{vmatrix} a_{21} & a_{22} \\ a_{31} & a_{32} \end{vmatrix}$$

$$= a_{11}\begin{vmatrix} a_{22} & a_{23} \\ a_{32} & a_{33} \end{vmatrix} - a_{12}\begin{vmatrix} a_{21} & a_{23} \\ a_{31} & a_{33} \end{vmatrix} + a_{13}\begin{vmatrix} a_{21} & a_{22} \\ a_{31} & a_{32} \end{vmatrix}$$

➡ Find the value of the determinant $\begin{vmatrix} 2 & -3 & 2 \\ 1 & 3 & -1 \\ 0 & -2 & 2 \end{vmatrix}$.

Expand by cofactors of the first row.

$$\begin{vmatrix} 2 & -3 & 2 \\ 1 & 3 & -1 \\ 0 & -2 & 2 \end{vmatrix} = 2\begin{vmatrix} 3 & -1 \\ -2 & 2 \end{vmatrix} - (-3)\begin{vmatrix} 1 & -1 \\ 0 & 2 \end{vmatrix} + 2\begin{vmatrix} 1 & 3 \\ 0 & -2 \end{vmatrix}$$

$$= 2(6 - 2) - (-3)(2 - 0) + 2(-2 - 0)$$
$$= 2(4) - (-3)(2) + 2(-2) = 8 - (-6) + (-4)$$
$$= 10$$

To illustrate a statement made earlier, the value of this determinant will now be found by expanding by cofactors of the second column.

$$\begin{vmatrix} 2 & -3 & 2 \\ 1 & 3 & -1 \\ 0 & -2 & 2 \end{vmatrix} = -3(-1)^{1+2}\begin{vmatrix} 1 & -1 \\ 0 & 2 \end{vmatrix} + 3(-1)^{2+2}\begin{vmatrix} 2 & 2 \\ 0 & 2 \end{vmatrix} + (-2)(-1)^{3+2}\begin{vmatrix} 2 & 2 \\ 1 & -1 \end{vmatrix}$$

$$= -3(-1)\begin{vmatrix} 1 & -1 \\ 0 & 2 \end{vmatrix} + 3 \cdot 1\begin{vmatrix} 2 & 2 \\ 0 & 2 \end{vmatrix} + (-2)(-1)\begin{vmatrix} 2 & 2 \\ 1 & -1 \end{vmatrix}$$

$$= 3(2 - 0) + 3(4 - 0) + 2(-2 - 2)$$
$$= 3(2) + 3(4) + 2(-4) = 6 + 12 + (-8)$$
$$= 10$$

Example 1 Evaluate the determinant. A. $\begin{vmatrix} 3 & -2 \\ 6 & -4 \end{vmatrix}$ B. $\begin{vmatrix} -2 & 3 & 1 \\ 4 & -2 & 0 \\ 1 & -2 & 3 \end{vmatrix}$

Solution A. $\begin{vmatrix} 3 & -2 \\ 6 & -4 \end{vmatrix} = 3(-4) - (6)(-2) = -12 + 12 = 0$

The value of the determinant is 0.

B. $\begin{vmatrix} -2 & 3 & 1 \\ 4 & -2 & 0 \\ 1 & -2 & 3 \end{vmatrix} = -2\begin{vmatrix} -2 & 0 \\ -2 & 3 \end{vmatrix} - 3\begin{vmatrix} 4 & 0 \\ 1 & 3 \end{vmatrix} + 1\begin{vmatrix} 4 & -2 \\ 1 & -2 \end{vmatrix}$

$$= -2(-6 + 0) - 3(12 - 0) + 1(-8 + 2)$$
$$= -2(-6) - 3(12) + 1(-6)$$
$$= 12 - 36 - 6$$
$$= -30$$

The value of the determinant is -30.

Problem 1 Evaluate the determinant.

A. $\begin{vmatrix} -1 & -4 \\ 3 & -5 \end{vmatrix}$ B. $\begin{vmatrix} 1 & 4 & -2 \\ 3 & 1 & 1 \\ 0 & -2 & 2 \end{vmatrix}$

Solution See page S10. A. 17 B. -8

2 Solve systems of linear equations by using Cramer's Rule

The connection between determinants and systems of equations can be understood by solving a general system of linear equations.

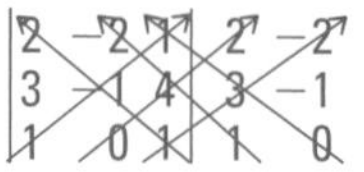

Solve: $\begin{aligned} a_{11}x + a_{12}y &= b_1 \\ a_{21}x + a_{22}y &= b_2 \end{aligned}$

(1) $a_{11}x + a_{12}y = b_1$
(2) $a_{21}x + a_{22}y = b_2$

Eliminate y. Multiply equation (1) by a_{22} and equation (2) by $-a_{12}$.

$a_{11}a_{22}x + a_{12}a_{22}y = b_1a_{22}$
$-a_{21}a_{12}x - a_{12}a_{22}y = -b_2a_{12}$

Add the equations.

$(a_{11}a_{22} - a_{21}a_{12})x = b_1a_{22} - b_2a_{12}$

Assuming $a_{11}a_{22} - a_{21}a_{12} \neq 0$, solve for x.

$$x = \frac{b_1a_{22} - b_2a_{12}}{a_{11}a_{22} - a_{21}a_{12}}$$

The denominator $a_{11}a_{22} - a_{21}a_{12}$ is the determinant of the coefficients of x and y. This is called the **coefficient determinant.**

$$a_{11}a_{22} - a_{21}a_{12} = \begin{vmatrix} a_{11} & a_{12} \\ a_{21} & a_{22} \end{vmatrix}$$

Coefficients of x ————————↑
Coefficients of y ————————↑

The numerator for x, $b_1a_{22} - b_2a_{12}$, is the determinant obtained by replacing the first column in the coefficient determinant by the constants b_1 and b_2. This is called the **numerator determinant.**

$$b_1a_{22} - b_2a_{12} = \begin{vmatrix} b_1 & a_{12} \\ b_2 & a_{22} \end{vmatrix}$$

Constants of ————————↑
the equations

Following a similar procedure and eliminating x, we can also express the y-component of the solution in determinant form. These results are summarized in Cramer's Rule.

Cramer's Rule

The solution of the system of equations $\begin{aligned} a_{11}x + a_{12}y &= b_1 \\ a_{21}x + a_{22}y &= b_2 \end{aligned}$ is given by $x = \dfrac{D_x}{D}$ and $y = \dfrac{D_y}{D}$, where

$$D = \begin{vmatrix} a_{11} & a_{12} \\ a_{21} & a_{22} \end{vmatrix},\ D_x = \begin{vmatrix} b_1 & a_{12} \\ b_2 & a_{22} \end{vmatrix},\ D_y = \begin{vmatrix} a_{11} & b_1 \\ a_{21} & b_2 \end{vmatrix},\ \text{and } D \neq 0.$$

Example 2   Solve by using Cramer's Rule: $\begin{aligned} 2x - 3y &= 8 \\ 5x + 6y &= 11 \end{aligned}$

Solution $D = \begin{vmatrix} 2 & -3 \\ 5 & 6 \end{vmatrix} = 27$ ▶ Find the value of the coefficient determinant.

$D_x = \begin{vmatrix} 8 & -3 \\ 11 & 6 \end{vmatrix} = 81$ ▶ Find the value of each of the numerator determinants.

$D_y = \begin{vmatrix} 2 & 8 \\ 5 & 11 \end{vmatrix} = -18$

$x = \dfrac{D_x}{D} = \dfrac{81}{27} = 3$ ▶ Use Cramer's Rule to write the solutions.

$y = \dfrac{D_y}{D} = \dfrac{-18}{27} = -\dfrac{2}{3}$

The solution is $\left(3, -\dfrac{2}{3}\right)$.

Problem 2 Solve by using Cramer's Rule: $6x - 6y = 5$
$2x - 10y = -1$

Solution See page S10. $\left(\frac{7}{6}, \frac{1}{3}\right)$

For the system shown at the right, $D = 0$. Therefore, $\frac{D_x}{D}$ and $\frac{D_y}{D}$ are undefined.

$$6x - 9y = 5$$
$$4x - 6y = 4$$

When $D = 0$, the system of equations is dependent if both D_x and D_y are zero. The system of equations is inconsistent if $D = 0$ and either D_x or D_y is not zero.

$$D = \begin{vmatrix} 6 & -9 \\ 4 & -6 \end{vmatrix} = 0$$

A procedure similar to that followed for two equations in two variables can be used to extend Cramer's Rule to three equations in three variables.

Cramer's Rule for Three Equations in Three Variables

The solution of the system of equations
$$a_{11}x + a_{12}y + a_{13}z = b_1$$
$$a_{21}x + a_{22}y + a_{23}z = b_2$$
$$a_{31}x + a_{32}y + a_{33}z = b_3$$
is given by $x = \frac{D_x}{D}$, $y = \frac{D_y}{D}$, and $z = \frac{D_z}{D}$, where

$$D = \begin{vmatrix} a_{11} & a_{12} & a_{13} \\ a_{21} & a_{22} & a_{23} \\ a_{31} & a_{32} & a_{33} \end{vmatrix}, D_x = \begin{vmatrix} b_1 & a_{12} & a_{13} \\ b_2 & a_{22} & a_{23} \\ b_3 & a_{32} & a_{33} \end{vmatrix}, D_y = \begin{vmatrix} a_{11} & b_1 & a_{13} \\ a_{21} & b_2 & a_{23} \\ a_{31} & b_3 & a_{33} \end{vmatrix},$$

$$D_z = \begin{vmatrix} a_{11} & a_{12} & b_1 \\ a_{21} & a_{22} & b_2 \\ a_{31} & a_{32} & b_3 \end{vmatrix}, \text{ and } D \neq 0.$$

Example 3 Solve by using Cramer's Rule: $3x - y + z = 5$
$x + 2y - 2z = -3$
$2x + 3y + z = 4$

Solution

$$D = \begin{vmatrix} 3 & -1 & 1 \\ 1 & 2 & -2 \\ 2 & 3 & 1 \end{vmatrix} = 28$$

▶ Find the value of the coefficient determinant.

$$D_x = \begin{vmatrix} 5 & -1 & 1 \\ -3 & 2 & -2 \\ 4 & 3 & 1 \end{vmatrix} = 28$$

▶ Find the value of each of the numerator determinants.

$$D_y = \begin{vmatrix} 3 & 5 & 1 \\ 1 & -3 & -2 \\ 2 & 4 & 1 \end{vmatrix} = 0$$

$$D_z = \begin{vmatrix} 3 & -1 & 5 \\ 1 & 2 & -3 \\ 2 & 3 & 4 \end{vmatrix} = 56$$

$x = \frac{D_x}{D} = \frac{28}{28} = 1$ ▸ Use Cramer's Rule to write the solution.

$y = \frac{D_y}{D} = \frac{0}{28} = 0$

$z = \frac{D_z}{D} = \frac{56}{28} = 2$

The solution is (1, 0, 2).

Problem 3 Solve by using Cramer's Rule:

$$\begin{aligned} 2x - y + z &= -1 \\ 3x + 2y - z &= 3 \\ x + 3y + z &= -2 \end{aligned}$$

Solution See page S10. $\left(\frac{3}{7}, -\frac{1}{7}, -2\right)$

3 Solve systems of linear equations by using matrices

HM[3]

As previously stated in this section, a **matrix** is a rectangular array of numbers. Each number in the matrix is called an **element** of the matrix. The matrix shown below, with 3 rows and 4 columns, is a 3 × 4 matrix.

$$\begin{bmatrix} 1 & 4 & -3 & 6 \\ -2 & 5 & 2 & 0 \\ -1 & 3 & 7 & -4 \end{bmatrix}$$

The elements $a_{11}, a_{22}, a_{33}, \ldots, a_{nn}$ form the **main diagonal** of a matrix. The elements 1, 5, and 7 form the main diagonal of the matrix above.

By considering only the coefficients and constants for the following system of equations, we can form the corresponding 3 × 4 **augmented matrix.**

System of Equations

$$\begin{aligned} 3x - 2y + z &= 2 \\ x \qquad - 3z &= -2 \\ 2x - y + 4z &= 5 \end{aligned}$$

Augmented Matrix

$$\begin{bmatrix} 3 & -2 & 1 & 2 \\ 1 & 0 & -3 & -2 \\ 2 & -1 & 4 & 5 \end{bmatrix}$$

Note that when a term is missing from one of the equations of the system, the coefficient of that term is 0, and a 0 is entered in the matrix.

A system of equations can be written from an augmented matrix.

$$\begin{bmatrix} 2 & -1 & 4 & 1 \\ 1 & 1 & 0 & 3 \\ 3 & -2 & -1 & 5 \end{bmatrix} \qquad \begin{aligned} 2x - y + 4z &= 1 \\ x + y \qquad &= 3 \\ 3x - 2y - z &= 5 \end{aligned}$$

A system of equations can be solved by writing the system in matrix form and then performing operations on the matrix similar to those performed on the equations of the system. These operations are called **elementary row operations.**

Elementary Row Operations

1. Interchange two rows.
2. Multiply all the elements in a row by the same nonzero number.
3. Replace a row by the sum of that row and a multiple of any other row.

The goal is to use the elementary row operations to rewrite the matrix with 1′s down the main diagonal and 0′s to the left of the 1′s in all rows except the first. This is called the **echelon form** of the matrix. Examples of echelon form are shown below.

$$\begin{bmatrix} 1 & 2 & 2 \\ 0 & 1 & -1 \end{bmatrix} \quad \begin{bmatrix} 1 & -3 & -1 & -6 \\ 0 & 1 & -2 & 7 \\ 0 & 0 & 1 & -2 \end{bmatrix} \quad \begin{bmatrix} 1 & \frac{1}{2} & -\frac{1}{3} & 2 \\ 0 & 1 & -\frac{2}{3} & -2 \\ 0 & 0 & 1 & 6 \end{bmatrix}$$

A system of equations can be solved by using elementary row operations to rewrite the augmented matrix of a system of equations in echelon form.

Solve by using elementary row operations: $2x + 5y = 8$
$3x + 4y = 5$

First write the system in matrix form. Then use elementary row operations to write the matrix in echelon form.

$$\begin{bmatrix} 2 & 5 & 8 \\ 3 & 4 & 5 \end{bmatrix}$$

Element a_{11} must be a 1. Multiply row 1 by $\frac{1}{2}$. (Elementary row operation 2)

$$\begin{bmatrix} 1 & \frac{5}{2} & 4 \\ 3 & 4 & 5 \end{bmatrix}$$

Element a_{21} must be a 0. Multiply row 1 by -3, and add it to row 2. Replace row 2 by the sum. (Elementary row operation 3)

$$\begin{bmatrix} 1 & \frac{5}{2} & 4 \\ 0 & -\frac{7}{2} & -7 \end{bmatrix}$$

Element a_{22} must be a 1. Multiply row 2 by $-\frac{2}{7}$.

(Elementary row operation 2)

$$\begin{bmatrix} 1 & \frac{5}{2} & 4 \\ 0 & 1 & 2 \end{bmatrix}$$

The matrix is now in echelon form.

Write the system of equations represented by the matrix.

(1) $x + \frac{5}{2}y = 4$

(2) $y = 2$

Substitute the value of y into equation (1), and solve for x.

$$x + \frac{5}{2}(2) = 4$$
$$x + 5 = 4$$
$$x = -1$$

The solution is $(-1, 2)$.

The order in which the elements in the 2 × 3 matrix are changed is important.

1. Change a_{11} to a 1.
2. Change a_{21} to a 0.
3. Change a_{22} to a 1.

$$\begin{bmatrix} a_{11} & a_{12} & a_{13} \\ a_{21} & a_{22} & a_{23} \end{bmatrix}$$

Example 4 Solve by using a matrix: $3x + 2y = 3$
$2x - 3y = 15$

Solution

$$\begin{bmatrix} 3 & 2 & 3 \\ 2 & -3 & 15 \end{bmatrix}$$ ▸ Write the system in matrix form.

$$\begin{bmatrix} 1 & \frac{2}{3} & 1 \\ 2 & -3 & 15 \end{bmatrix}$$ ▸ Multiply row 1 by $\frac{1}{3}$.

$$\begin{bmatrix} 1 & \frac{2}{3} & 1 \\ 0 & -\frac{13}{3} & 13 \end{bmatrix}$$ ▸ Multiply row 1 by −2, and add it to row 2.

$$\begin{bmatrix} 1 & \frac{2}{3} & 1 \\ 0 & 1 & -3 \end{bmatrix}$$ ▸ Multiply row 2 by $-\frac{3}{13}$.

(1) $x + \frac{2}{3}y = 1$
(2) $y = -3$

▸ Write the system of equations represented by the matrix.

$$x + \frac{2}{3}(-3) = 1$$
$$x - 2 = 1$$
$$x = 3$$

▸ Substitute the value of y into equation (1), and solve for x.

The solution is (3, −3).

Problem 4 Solve by using a matrix: $3x - 5y = -12$
$4x - 3y = -5$

Solution See page S10. (1, 3)

The matrix method of solving systems of equations can be extended to larger systems of equations. A system of three equations in three unknowns is written as a 3 × 4 augmented matrix.

The order in which the elements in a 3 × 4 matrix are changed is as follows:

1. Change a_{11} to a 1.
2. Change a_{21} and a_{31} to 0's.
3. Change a_{22} to a 1.
4. Change a_{32} to a 0.
5. Change a_{33} to a 1.

$$\begin{bmatrix} a_{11} & a_{12} & a_{13} & a_{14} \\ a_{21} & a_{22} & a_{23} & a_{24} \\ a_{31} & a_{32} & a_{33} & a_{34} \end{bmatrix}$$

To solve the system shown at the right, write the system in matrix form.

$$\begin{aligned} 2x + 3y + 3z &= -2 \\ x + 2y - 3z &= 9 \\ 3x - 2y - 4z &= 1 \end{aligned} \qquad \begin{bmatrix} 2 & 3 & 3 & -2 \\ 1 & 2 & -3 & 9 \\ 3 & -2 & -4 & 1 \end{bmatrix}$$

Element a_{11} must be a 1. Interchange rows 1 and 2.

$$\begin{bmatrix} 1 & 2 & -3 & 9 \\ 2 & 3 & 3 & -2 \\ 3 & -2 & -4 & 1 \end{bmatrix}$$

Element a_{21} must be a 0. Multiply row 1 by -2, and add it to row 2. Replace row 2 by the sum.
Element a_{31} must be a 0. Multiply row 1 by -3, and add it to row 3. Replace row 3 by the sum.

$$\begin{bmatrix} 1 & 2 & -3 & 9 \\ 0 & -1 & 9 & -20 \\ 0 & -8 & 5 & -26 \end{bmatrix}$$

Element a_{22} must be a 1. Multiply row 2 by -1.

$$\begin{bmatrix} 1 & 2 & -3 & 9 \\ 0 & 1 & -9 & 20 \\ 0 & -8 & 5 & -26 \end{bmatrix}$$

Element a_{32} must be a 0. Multiply row 2 by 8, and add it to row 3. Replace row 3 by the sum.

$$\begin{bmatrix} 1 & 2 & -3 & 9 \\ 0 & 1 & -9 & 20 \\ 0 & 0 & -67 & 134 \end{bmatrix}$$

Element a_{33} must be a 1. Multiply row 3 by $-\frac{1}{67}$.

$$\begin{bmatrix} 1 & 2 & -3 & 9 \\ 0 & 1 & -9 & 20 \\ 0 & 0 & 1 & -2 \end{bmatrix}$$

Write the system represented by the matrix.

$$\begin{aligned} (1)\quad x + 2y - 3z &= 9 \\ (2)\quad y - 9z &= 20 \\ (3)\quad z &= -2 \end{aligned}$$

Substitute the value of z into equation (2), and solve for y.

$$\begin{aligned} y - 9z &= 20 \\ y - 9(-2) &= 20 \\ y + 18 &= 20 \\ y &= 2 \end{aligned}$$

Substitute the values of y and z into equation (1), and solve for x.

$$\begin{aligned} x + 2y - 3z &= 9 \\ x + 2(2) - 3(-2) &= 9 \\ x + 4 + 6 &= 9 \\ x &= -1 \end{aligned}$$

The solution is $(-1, 2, -2)$.

Solve by using matrices:

$$\begin{aligned} x - y + z &= 2 \\ x + 2y - z &= 3 \\ 3x + 3y - z &= 6 \end{aligned}$$

Write the system in matrix form.

$$\begin{bmatrix} 1 & -1 & 1 & 2 \\ 1 & 2 & -1 & 3 \\ 3 & 3 & -1 & 6 \end{bmatrix}$$

Element a_{11} is a 1. Element a_{21} must be a 0. Multiply row 1 by -1, and add it to row 2. Replace row 2 by the sum.
Element a_{31} must be a 0. Multiply row 1 by -3, and add it to row 3. Replace row 3 by the sum.

$$\begin{bmatrix} 1 & -1 & 1 & 2 \\ 0 & 3 & -2 & 1 \\ 0 & 6 & -4 & 0 \end{bmatrix}$$

Element a_{22} must be a 1. Multiply row 2 by $\frac{1}{3}$.

$$\begin{bmatrix} 1 & -1 & 1 & 2 \\ 0 & 1 & -\frac{2}{3} & \frac{1}{3} \\ 0 & 6 & -4 & 0 \end{bmatrix}$$

Element a_{32} must be a 0. Multiply row 2 by -6, and add it to row 3. Replace row 3 by the sum.

$$\begin{bmatrix} 1 & -1 & 1 & 2 \\ 0 & 1 & -\frac{2}{3} & \frac{1}{3} \\ 0 & 0 & 0 & -2 \end{bmatrix}$$

Write the system represented by the matrix. The equation $0 = -2$ is not a true equation. The system of equations is inconsistent.

$$\begin{aligned} x - y + z &= 2 \\ y - \tfrac{2}{3}z &= \tfrac{1}{3} \\ 0 &= -2 \end{aligned}$$

The system of equations has no solution.

Example 5 Solve by using a matrix:

$$\begin{aligned} 3x + 2y + 3z &= 2 \\ 2x - 3y + 4z &= 5 \\ x + 4y + 2z &= 8 \end{aligned}$$

Solution

$$\begin{bmatrix} 3 & 2 & 3 & 2 \\ 2 & -3 & 4 & 5 \\ 1 & 4 & 2 & 8 \end{bmatrix}$$

▶ Write the system in matrix form.

$$\begin{bmatrix} 1 & 4 & 2 & 8 \\ 2 & -3 & 4 & 5 \\ 3 & 2 & 3 & 2 \end{bmatrix}$$

▶ Interchange rows 1 and 3.

$$\begin{bmatrix} 1 & 4 & 2 & 8 \\ 0 & -11 & 0 & -11 \\ 0 & -10 & -3 & -22 \end{bmatrix}$$

▶ Multiply row 1 by -2, and add it to row 2.

▶ Multiply row 1 by -3, and add it to row 3.

$$\begin{bmatrix} 1 & 4 & 2 & 8 \\ 0 & 1 & 0 & 1 \\ 0 & -10 & -3 & -22 \end{bmatrix}$$

▶ Multiply row 2 by $-\frac{1}{11}$.

$$\begin{bmatrix} 1 & 4 & 2 & 8 \\ 0 & 1 & 0 & 1 \\ 0 & 0 & -3 & -12 \end{bmatrix}$$

▶ Multiply row 2 by 10, and add it to row 3.

$$\begin{bmatrix} 1 & 4 & 2 & 8 \\ 0 & 1 & 0 & 1 \\ 0 & 0 & 1 & 4 \end{bmatrix}$$

▶ Multiply row 3 by $-\frac{1}{3}$.

$$\begin{aligned} (1)\quad x + 4y + 2z &= 8 \\ (2)\quad y &= 1 \\ (3)\quad z &= 4 \end{aligned}$$

▶ Write the system of equations represented by the matrix.

$$\begin{aligned} x + 4y + 2z &= 8 \\ x + 4(1) + 2(4) &= 8 \\ x + 4 + 8 &= 8 \\ x &= -4 \end{aligned}$$

▶ Substitute the values of y and z into equation (1), and solve for x.

The solution is $(-4, 1, 4)$.

Problem 5 Solve by using a matrix:
$$\begin{aligned} 3x - 2y - 3z &= 5 \\ x + 3y - 2z &= -4 \\ 2x + 6y + 3z &= 6 \end{aligned}$$

Solution See page S10. (3, −1, 2)

CONCEPT REVIEW 4.3

Determine whether the following statements are always true, sometimes true, or never true.

1. A 3×3 determinant can be evaluated by expanding about any row or column of the matrix. Always true

2. Cramer's Rule can be used to solve a system of linear equations in three variables. Sometimes true

3. A matrix has the same number of rows and columns. Sometimes true

4. Associated with a matrix is a number called its determinant. Sometimes true

5. A system of two linear equations in two variables has no solution, exactly one solution, or infinitely many solutions. Always true

EXERCISES 4.3

1

1. How do you find the value of the determinant associated with a 2×2 matrix?

2. What is the cofactor of a given element in a matrix?

Evaluate the determinant.

3. $\begin{vmatrix} 2 & -1 \\ 3 & 4 \end{vmatrix}$ 11

4. $\begin{vmatrix} 5 & 1 \\ -1 & 2 \end{vmatrix}$ 11

5. $\begin{vmatrix} 6 & -2 \\ -3 & 4 \end{vmatrix}$ 18

6. $\begin{vmatrix} -3 & 5 \\ 1 & 7 \end{vmatrix}$ −26

7. $\begin{vmatrix} 3 & 6 \\ 2 & 4 \end{vmatrix}$ 0

8. $\begin{vmatrix} 5 & -10 \\ 1 & -2 \end{vmatrix}$ 0

9. $\begin{vmatrix} 1 & -1 & 2 \\ 3 & 2 & 1 \\ 1 & 0 & 4 \end{vmatrix}$ 15

10. $\begin{vmatrix} 4 & 1 & 3 \\ 2 & -2 & 1 \\ 3 & 1 & 2 \end{vmatrix}$ 3

11. $\begin{vmatrix} 3 & -1 & 2 \\ 0 & 1 & 2 \\ 3 & 2 & -2 \end{vmatrix}$ −30

12. $\begin{vmatrix} 4 & 5 & -2 \\ 3 & -1 & 5 \\ 2 & 1 & 4 \end{vmatrix}$ −56

13. $\begin{vmatrix} 4 & 2 & 6 \\ -2 & 1 & 1 \\ 2 & 1 & 3 \end{vmatrix}$ 0

14. $\begin{vmatrix} 3 & 6 & -3 \\ 4 & -1 & 6 \\ -1 & -2 & 3 \end{vmatrix}$ −54

2 **15.** Explain how Cramer's Rule is used to solve a system of two equations in two unknowns.

16. Explain why Cramer's Rule cannot be used to solve a system of equations for which the determinant of the coefficient matrix is 0.

Solve by using Cramer's Rule, if possible.

17. $2x - 5y = 26$ $(3, -4)$
$5x + 3y = 3$

18. $3x + 7y = 15$ $(-2, 3)$
$2x + 5y = 11$

19. $x - 4y = 8$ $(4, -1)$
$3x + 7y = 5$

20. $5x + 2y = -5$ $(-3, 5)$
$3x + 4y = 11$

21. $2x + 3y = 4$ $\left(\frac{11}{14}, \frac{17}{21}\right)$
$6x - 12y = -5$

22. $5x + 4y = 3$ $\left(-\frac{3}{5}, \frac{3}{2}\right)$
$15x - 8y = -21$

23. $2x + 5y = 6$ $\left(\frac{1}{2}, 1\right)$
$6x - 2y = 1$

24. $7x + 3y = 4$ $(1, -1)$
$5x - 4y = 9$

25. $-2x + 3y = 7$ Not possible
$4x - 6y = 9$ by Cramer's Rule

26. $9x + 6y = 7$ Not possible
$3x + 2y = 4$ by Cramer's Rule

27. $2x - 5y = -2$ $(-1, 0)$
$3x - 7y = -3$

28. $8x + 7y = -3$ $\left(-\frac{41}{2}, 23\right)$
$2x + 2y = 5$

29. $2x - y + 3z = 9$ $(1, -1, 2)$
$x + 4y + 4z = 5$
$3x + 2y + 2z = 5$

30. $3x - 2y + z = 2$ $(-1, -2, 1)$
$2x + 3y + 2z = -6$
$3x - y + z = 0$

31. $3x - y + z = 11$ $(2, -2, 3)$
$x + 4y - 2z = -12$
$2x + 2y - z = -3$

32. $x + 2y + 3z = 8$
$2x - 3y + z = 5$
$3x - 4y + 2z = 9$
$(-3, -2, 5)$

33. $4x - 2y + 6z = 1$
$3x + 4y + 2z = 1$
$2x - y + 3z = 2$
Not possible by Cramer's Rule

34. $x - 3y + 2z = 1$
$2x + y - 2z = 3$
$3x - 9y + 6z = -3$
Not possible by Cramer's Rule

35. $5x - 4y + 2z = 4$
$3x - 5y + 3z = -4$ $\left(\frac{68}{25}, \frac{56}{25}, -\frac{8}{25}\right)$
$3x + y - 5z = 12$

36. $2x + 4y + z = 7$
$x + 3y - z = 1$ $\left(\frac{53}{19}, -\frac{1}{19}, \frac{31}{19}\right)$
$3x + 2y - 2z = 5$

3 **37.** Explain how to write the augmented matrix of a system of equations.

38. Is the matrix $\begin{bmatrix} 1 & -2 & 3 & 1 \\ 0 & 1 & 2 & 8 \\ 0 & 1 & 0 & 1 \end{bmatrix}$ in echelon form? Why or why not?

Solve by using matrices.

39. $3x + y = 6$ $(1, 3)$
$2x - y = -1$

40. $2x + y = 3$ $(2, -1)$
$x - 4y = 6$

41. $x - 3y = 8$ $(-1, -3)$
$3x - y = 0$

42. $2x + 3y = 16$ $(2, 4)$
$x - 4y = -14$

43. $y = 4x - 10$ $(3, 2)$
$2y = 5x - 11$

44. $2y = 4 - 3x$ $(-2, 5)$
$y = 1 - 2x$

45. $2x - y = -4$ Inconsistent
$y = 2x - 8$

46. $3x - 2y = -8$ Inconsistent
$y = \frac{3}{2}x - 2$

47. $4x - 3y = -14$ $(-2, 2)$
$3x + 4y = 2$

48. $\begin{aligned} 5x + 2y &= 3 \\ 3x + 4y &= 13 \end{aligned}$ (−1, 4)

49. $\begin{aligned} 5x + 4y + 3z &= -9 \\ x - 2y + 2z &= -6 \\ x - y - z &= 3 \end{aligned}$ (0, 0, −3)

50. $\begin{aligned} x - y - z &= 0 \\ 3x - y + 5z &= -10 \\ x + y - 4z &= 12 \end{aligned}$ (1, 3, −2)

51. $\begin{aligned} 5x - 5y + 2z &= 8 \\ 2x + 3y - z &= 0 \\ x + 2y - z &= 0 \end{aligned}$ (1, −1, −1)

52. $\begin{aligned} 2x + y - 5z &= 3 \\ 3x + 2y + z &= 15 \\ 5x - y - z &= 5 \end{aligned}$ (2, 4, 1)

53. $\begin{aligned} 2x + 3y + z &= 5 \\ 3x + 3y + 3z &= 10 \\ 4x + 6y + 2z &= 5 \end{aligned}$ Inconsistent

54. $\begin{aligned} x - 2y + 3z &= 2 \\ 2x + y + 2z &= 5 \\ 2x - 4y + 6z &= -4 \end{aligned}$ Inconsistent

55. $\begin{aligned} 3x + 2y + 3z &= 2 \\ 6x - 2y + z &= 1 \\ 3x + 4y + 2z &= 3 \end{aligned}$ $\left(\frac{1}{3}, \frac{1}{2}, 0\right)$

56. $\begin{aligned} 2x + 3y - 3z &= -1 \\ 2x + 3y + 3z &= 3 \\ 4x - 4y + 3z &= 4 \end{aligned}$ $\left(\frac{1}{2}, 0, \frac{2}{3}\right)$

57. $\begin{aligned} 5x - 5y - 5z &= 2 \\ 5x + 5y - 5z &= 6 \\ 10x + 10y + 5z &= 3 \end{aligned}$ $\left(\frac{1}{5}, \frac{2}{5}, -\frac{3}{5}\right)$

58. $\begin{aligned} 3x - 2y + 2z &= 5 \\ 6x + 3y - 4z &= -1 \\ 3x - y + 2z &= 4 \end{aligned}$ $\left(\frac{2}{3}, -1, \frac{1}{2}\right)$

59. $\begin{aligned} 4x + 4y - 3z &= 3 \\ 8x + 2y + 3z &= 0 \\ 4x - 4y + 6z &= -3 \end{aligned}$ $\left(\frac{1}{4}, 0, -\frac{2}{3}\right)$

APPLYING CONCEPTS 4.3

Solve for x.

60. $\begin{vmatrix} 3 & 2 \\ 4 & x \end{vmatrix} = -11$ −1

61. $\begin{vmatrix} 1 & 0 & 2 \\ 4 & 3 & -1 \\ 0 & 2 & x \end{vmatrix} = -24$ −14

62. $\begin{vmatrix} -2 & 1 & 3 \\ 0 & x & 4 \\ -1 & 2 & -3 \end{vmatrix} = -24$ −4

Complete.

[C] **63.** If all the elements in one row or one column of a 2×2 matrix are zeros, the value of the determinant of the matrix is ___0___.

[C] **64.** If all the elements in one row or one column of a 3×3 matrix are zeros, the value of the determinant of the matrix is ___0___.

[C] **65.** **a.** The value of the determinant $\begin{vmatrix} x & x & a \\ y & y & b \\ z & z & c \end{vmatrix}$ is ___0___.

b. If two columns of a 3×3 matrix contain identical elements, the value of the determinant is ___0___.

66. Show that $\begin{vmatrix} a & b \\ c & d \end{vmatrix} = -\begin{vmatrix} c & d \\ a & b \end{vmatrix}$. Complete solution is available in the Solutions Manual.

67. *Surveying* Surveyors use a formula to find the area of a plot of land. The *surveyor's area formula* states that if the vertices (x_1, y_1), (x_2, y_2), (x_3, y_3), ..., (x_n, y_n) of a simple polygon are listed counterclockwise around the perimeter, then the area of the polygon is

$$A = \frac{1}{2}\left\{\begin{vmatrix} x_1 & x_2 \\ y_1 & y_2 \end{vmatrix} + \begin{vmatrix} x_2 & x_3 \\ y_2 & y_3 \end{vmatrix} + \begin{vmatrix} x_3 & x_4 \\ y_3 & y_4 \end{vmatrix} + \cdots + \begin{vmatrix} x_n & x_1 \\ y_n & y_1 \end{vmatrix}\right\}$$

Use the surveyor's area formula to find the area of the polygon with vertices $(9, -3)$, $(26, 6)$, $(18, 21)$, $(16, 10)$, and $(1, 11)$. Measurements given are in feet.
239 ft^2

68. If the determinant of the denominator is zero when using Cramer's Rule, the system of equations is either dependent or inconsistent. Explain how you can determine which it is.

69. What is echelon form? Explain the steps that are necessary to write a 3×3 matrix in echelon form.

70. Consider the system of equations $\begin{aligned} 3x - 2y &= 12 \\ 2x + 5y &= -11 \end{aligned}$. Graph the equations. Now multiply the second equation by 3, add it to the first equation, and replace the second equation by the new equation. What is the new system of equations? Graph the equations of this system on the same coordinate system. The lines intersect at the same point, so the new system of equations is equivalent to the original system. Explain how this is related to the third elementary row operation.

71. Suppose that applying elementary row operations to an augmented matrix produces the matrix $\begin{bmatrix} 1 & 2 & -3 & 5 \\ 0 & 1 & -2 & -2 \\ 0 & 0 & 0 & -3 \end{bmatrix}$. Explain why the original system of equations has no solution.

SECTION 4.4 Application Problems

1 Rate-of-wind and water-current problems

Motion problems that involve an object moving with or against a wind or current normally require two variables to solve.

Solve: A motorboat traveling with the current can go 24 mi in 2 h. Against the current, it takes 3 h to go the same distance. Find the rate of the motorboat in calm water and the rate of the current.

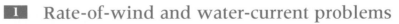

STRATEGY for solving rate-of-wind and water-current problems

▪ Choose one variable to represent the rate of the object in calm conditions and a second variable to represent the rate of the wind or current. Using these variables, express the rate of the object with and against the wind or current. Use the equation $rt = d$ to write expressions for the distance traveled by the object. The results can be recorded in a table.

Rate of the boat in calm water: x
Rate of the current: y

	Rate	·	Time	=	Distance
With current	$x + y$	·	2	=	$2(x + y)$
Against current	$x - y$	·	3	=	$3(x - y)$

▪ Determine how the expressions for the distance are related.

The distance traveled with the current is 24 mi. $2(x + y) = 24$
The distance traveled against the current is 24 mi. $3(x - y) = 24$

Solve the system of equations.

$2(x + y) = 24$ $\dfrac{1}{2} \cdot 2(x + y) = \dfrac{1}{2} \cdot 24$ $x + y = 12$

$3(x - y) = 24$ $\dfrac{1}{3} \cdot 3(x - y) = \dfrac{1}{3} \cdot 24$ $x - y = 8$

$$2x = 20$$
$$x = 10$$

Replace x by 10 in the equation $x + y = 12$. Solve for y. $x + y = 12$
$$10 + y = 12$$
$$y = 2$$

The rate of the boat in calm water is 10 mph.
The rate of the current is 2 mph.

Example 1 Flying with the wind, a plane flew 1000 mi in 5 h. Flying against the wind, the plane could fly only 500 mi in the same amount of time. Find the rate of the plane in calm air and the rate of the wind.

Strategy ▪ Rate of the plane in still air: p
Rate of the wind: w

	Rate	Time	Distance
With wind	$p + w$	5	$5(p + w)$
Against wind	$p - w$	5	$5(p - w)$

The distance traveled with the wind is 1000 mi.
The distance traveled against the wind is 500 mi.

Solution $5(p + w) = 1000$
$5(p - w) = 500$

$p + w = 200$ ▶ Multiply each side of the equations by $\frac{1}{5}$.
$p - w = 100$
$2p = 300$ ▶ Add the equations.
$p = 150$

$p + w = 200$ ▶ Substitute the value of p into one of the equations.
$150 + w = 200$
$w = 50$

The rate of the plane in calm air is 150 mph.
The rate of the wind is 50 mph.

Problem 1 A rowing team rowing with the current traveled 18 mi in 2 h. Against the current, the team rowed 10 mi in 2 h. Find the rate of the rowing team in calm water and the rate of the current.

Solution See page S11. Rowing team: 7 mph; current: 2 mph

2 Application problems

The application problems in this section are varieties of those problems solved earlier in the text. Each of the strategies for the problems in this section will result in a system of equations.

Solve: A store owner purchased twenty 60-watt light bulbs and thirty fluorescent lights for a total cost of \$40. A second purchase, at the same prices, included thirty 60-watt bulbs and ten fluorescent lights for a total cost of \$25. Find the cost of a 60-watt bulb and of a fluorescent light.

STRATEGY for solving an application problem in two variables

- Choose one variable to represent one of the unknown quantities and a second variable to represent the other unknown quantity. Write numerical or variable expressions for all the remaining quantities. These results can be recorded in two tables, one for each of the conditions.

Cost of 60-watt bulb: b
Cost of a fluorescent light: f

First purchase

	Amount	·	Unit cost	=	Value
60-watt	20	·	b	=	$20b$
Fluorescent	30	·	f	=	$30f$

Second purchase

	Amount	·	Unit cost	=	Value
60-watt	30	·	b	=	$30b$
Fluorescent	10	·	f	=	$10f$

▪ Determine a system of equations. The strategies presented in Chapter 2 can be used to determine the relationships between the expressions in the tables. Each table will give one equation of the system.

The total of the first purchase was $40.　　$20b + 30f = 40$

The total of the second purchase was $25.　　$30b + 10f = 25$

Solve the system of equations.

$$20b + 30f = 40$$
$$30b + 10f = 25$$

→

$$3(20b + 30f) = 3 \cdot 40$$
$$-2(30b + 10f) = -2 \cdot 25$$

→

$$60b + 90f = 120$$
$$-60b - 20f = -50$$
$$70f = 70$$
$$f = 1$$

Replace f by 1 in the equation $20b + 30f = 40$. Solve for b.

$$20b + 30f = 40$$
$$20b + 30(1) = 40$$
$$20b + 30 = 40$$
$$20b = 10$$
$$b = 0.5$$

The cost of a 60-watt bulb was $.50.

The cost of a fluorescent light was $1.00.

Example 2　The total value of the nickels and dimes in a coin bank is $2.50. If the nickels were dimes and the dimes were nickels, the total value of the coins would be $3.05. Find the number of dimes and the number of nickels in the bank.

Strategy　▪ Number of nickels in the bank: n

　　　　Number of dimes in the bank: d

Coins in the bank now:

Coin	Number	Value	Total value
Nickels	n	5	$5n$
Dimes	d	10	$10d$

Coins in the bank if the nickels were dimes and the dimes were nickels:

Coin	Number	Value	Total value
Nickels	d	5	$5d$
Dimes	n	10	$10n$

■ The value of the nickels and dimes in the bank is $2.50.
The value of the nickels and dimes in the bank would be $3.05.

Solution

(1) $5n + 10d = 250$
(2) $10n + 5d = 305$

$$10n + 20d = 500$$
$$-10n - 5d = -305$$
$$15d = 195$$
$$d = 13$$

▶ Multiply equation (1) by 2 and equation (2) by -1.

▶ Add the equations.

$$5n + 10d = 250$$
$$5n + 10(13) = 250$$
$$5n + 130 = 250$$
$$5n = 120$$
$$n = 24$$

▶ Substitute the value of d into one of the equations.

There are 24 nickels and 13 dimes in the bank.

Problem 2   A citrus fruit grower purchased 25 orange trees and 20 grapefruit trees for $290. The next week, at the same prices, the grower bought 20 orange trees and 30 grapefruit trees for $330. Find the cost of an orange tree and the cost of a grapefruit tree.

Solution   See page S11.   Orange tree: $6; grapefruit tree: $7

Example 3 is an application problem that requires more than two variables. The solution of this problem also illustrates how the substitution method can be used to solve a system of linear equations in three variables.

d_1 d_3
d_2
$w_3 = 4$
$w_2 = 3$
$w_1 = 2$

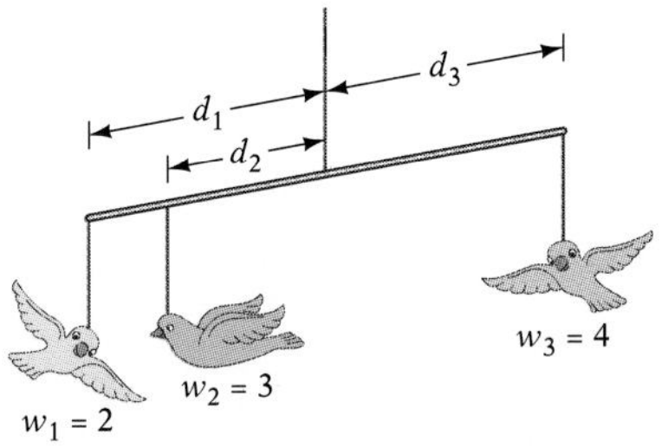

Example 3   An artist is creating a mobile from which three objects will be suspended from a light rod that is 18 in. long as shown at the left. The weight, in ounces, of each object is shown in the diagram. For the mobile to balance, the objects must be positioned so that $w_1d_1 + w_2d_2 = w_3d_3$. The artist wants d_1 to be 1.5 times d_2. Find the distances d_1, d_2 and d_3 so that the mobile will balance.

Strategy   There are three unknowns for this problem. Use the information in the problem to write three equations with d_1, d_2, and d_3 as the variables.

The length of the rod is 18 in. Therefore, $d_1 + d_3 = 18$. Using $w_1d_1 + w_2d_2 = w_3d_3$, we have $2d_1 + 3d_2 = 4d_3$. The artist wants d_1 to be 1.5 times d_2. Thus $d_1 = 1.5d_2$.

Solution

(1) $d_1 + d_3 = 18$
(2) $2d_1 + 3d_2 = 4d_3$
(3) $d_1 = 1.5d_2$

We will solve the system of equations by substitution. Use equation (3) to replace d_1 in equation (1) and in equation (2).

$$1.5d_2 + d_3 = 18$$
$$2(1.5d_2) + 3d_2 = 4d_3$$

(4) $1.5d_2 + d_3 = 18$
(5) $6d_2 = 4d_3$
(6) $1.5d_2 = d_3$ ► Divide each side of equation (5) by 4.

Replace d_3 by $1.5d_2$ in equation (4) and solve for d_2.

$$1.5d_2 + 1.5d_2 = 18$$
$$3d_2 = 18$$
$$d_2 = 6$$

From equation (6), $d_3 = 1.5d_2 = 1.5(6) = 9$.

Substituting the value of d_3 into equation (1), we have $d_1 = 9$.

The values are $d_1 = 9$ in., $d_2 = 6$ in., and $d_3 = 9$ in.

Problem 3 A science museum charges $10 for an admission ticket, but members receive a discount of $3, and students are admitted for half price. Last Saturday, 750 tickets were sold for a total of $5400. If 20 more student tickets than full-price tickets were sold, how many of each type of ticket were sold?

Solution See page S11–S12.
General admission, 190; members, 350; students, 210

CONCEPT REVIEW 4.4

Determine whether the following statements are always true, sometimes true, or never true.

1. The rate of a current in a river is x mph, and the rate of a boat in still water is y mph. The rate of a boat going down the river is $x + y$, and the rate going up the river is $x - y$. Never true

2. The speed of a plane is 500 mph. There is a headwind of 50 mph. The speed of the plane relative to an observer on the ground is 550 mph. Never true

3. A contractor bought 50 yd of nylon carpet for x dollars per yard and bought 100 yd of wool carpet for y dollars per yard. The total cost of the carpet is $50x + 100y$. Always true

4. A coin bank contains nickels and dimes. The total value of the coins is $.70. The bank had 5 dimes and 4 nickels. Sometimes true

EXERCISES 4.4

1 Rate-of-wind or current problems

1. Flying with the wind, a small plane flew 320 mi in 2 h. Against the wind, the plane could fly only 280 mi in the same amount of time. Find the rate of the plane in calm air and the rate of the wind. Plane: 150 mph; wind: 10 mph

2. A jet plane flying with the wind went 2100 mi in 4 h. Against the wind, the plane could fly only 1760 mi in the same amount of time. Find the rate of the plane in calm air and the rate of the wind.
Plane: 482.5 mph; wind: 42.5 mph

3. A cabin cruiser traveling with the current went 48 mi in 3 h. Against the current, it took 4 h to travel the same distance. Find the rate of the cabin cruiser in calm water and the rate of the current. Cabin cruiser: 14 mph; current: 2 mph

4. A motorboat traveling with the current went 48 mi in 2 h. Against the current, it took 3 h to travel the same distance. Find the rate of the boat in calm water and the rate of the current. Boat: 20 mph; current: 4 mph

5. Flying with the wind, a pilot flew 450 mi between two cities in 2.5 h. The return trip against the wind took 3 h. Find the rate of the plane in calm air and the rate of the wind. Plane: 165 mph; wind: 15 mph

6. A turbo-prop plane flying with the wind flew 600 mi in 2 h. Flying against the wind, the plane required 3 h to travel the same distance. Find the rate of the wind and the rate of the plane in calm air. Plane: 250 mph; wind: 50 mph

7. A motorboat traveling with the current went 88 km in 4 h. Against the current, the boat could go only 64 km in the same amount of time. Find the rate of the boat in calm water and the rate of the current. Boat: 19 km/h; current: 3 km/h

8. A rowing team rowing with the current traveled 18 km in 2 h. Rowing against the current, the team rowed 12 km in the same amount of time. Find the rate of the rowing team in calm water and the rate of the current.
Rowing team: 7.5 km/h; current: 1.5 km/h

9. A plane flying with a tailwind flew 360 mi in 3 h. Against the wind, the plane required 4 h to fly the same distance. Find the rate of the plane in calm air and the rate of the wind. Plane: 105 mph; wind: 15 mph

10. Flying with the wind, a plane flew 1000 mi in 4 h. Against the wind, the plane required 5 h to fly the same distance. Find the rate of the plane in calm air and the rate of the wind. Plane: 225 mph; wind: 25 mph

11. A motorboat traveling with the current went 54 mi in 3 h. Against the current, it took 3.6 h to travel the same distance. Find the rate of the boat in calm water and the rate of the current. Boat: 16.5 mph; current: 1.5 mph

12. A plane traveling with the wind flew 3625 mi in 6.25 h. Against the wind, the plane required 7.25 h to fly the same distance. Find the rate of the plane in calm air and the rate of the wind. Plane: 540 mph; wind: 40 mph

Solve.

13. *Construction* A carpenter purchased 50 ft of redwood and 90 ft of pine for a total cost of $31.20. A second purchase, at the same prices, included 200 ft of redwood and 100 ft of pine for a total cost of $78. Find the cost per foot of redwood and of pine. Pine: $.18/ft; redwood: $.30/ft

14. *Business* A merchant mixed 10 lb of a cinnamon tea with 5 lb of spice tea. The 15-pound mixture cost $40. A second mixture included 12 lb of the cinnamon tea and 8 lb of the spice tea. The 20-pound mixture cost $54. Find the cost per pound of the cinnamon tea and of the spice tea.
Cinnamon tea: $2.50/lb; spice tea: $3.00/lb

15. *Home Economics* During one month, a homeowner used 400 units of electricity and 120 units of gas for a total cost of $73.60. The next month, 350 units of electricity and 200 units of gas were used for a total cost of $72. Find the cost per unit of gas. $.08

16. *Construction* A contractor buys 20 yd^2 of nylon carpet and 28 yd^2 of wool carpet for $1360. A second purchase, at the same prices, includes 15 yd^2 of nylon carpet and 20 yd^2 of wool carpet for $990. Find the cost per square yard of the wool carpet. $30

17. *Coin Problem* The total value of the quarters and dimes in a coin bank is $6.90. If the quarters were dimes and the dimes were quarters, the total value of the coins would be $7.80. Find the number of quarters in the bank.
18 quarters

18. *Coin Problem* A coin bank contains only nickels and dimes. The total value of the coins in the bank is $2.50. If the nickels were dimes and the dimes were nickels, the total value of the coins would be $3.20. Find the number of nickels in the bank. 26 nickels

19. *Manufacturing* A company manufactures both color and black-and-white television sets. The cost of materials for a black-and-white TV is $25, whereas the cost of materials for a color TV is $75. The cost of labor to manufacture a black-and-white TV is $40, whereas the cost of labor to manufacture a color TV is $65. During a week when the company has budgeted $4800 for materials and $4380 for labor, how many color TVs does the company plan to manufacture? 60 color TVs

20. *Manufacturing* A company manufactures both 10-speed and standard model bicycles. The cost of materials for a 10-speed bicycle is $40, whereas the cost of materials for a standard bicycle is $30. The cost of labor to manufacture a 10-speed bicycle is $50, whereas the cost of labor to manufacture a standard bicycle is $25. During a week when the company has budgeted $1740 for materials and $1950 for labor, how many 10-speed bicycles does the company plan to manufacture? 30 10-speed bicycles

21. *Health Science* A pharmacist has two vitamin-supplement powders. The first powder is 25% vitamin B_1 and 15% vitamin B_2. The second is 15% vitamin B_1 and 20% vitamin B_2. How many milligrams of each of the two powders should the pharmacist use to make a mixture that contains 117.5 mg of vitamin B_1 and 120 mg of vitamin B_2?
First powder: 200 mg; second powder: 450 mg

22. *Chemistry* A chemist has two alloys, one of which is 10% gold and 15% lead and the other of which is 30% gold and 40% lead. How many grams of each of the two alloys should be used to make an alloy that contains 60 g of gold and 88 g of lead? First alloy: 480 g; second alloy: 40 g

23. *Business* On Monday, a computer manufacturing company sent out three shipments. The first order, which contained a bill for \$114,000, was for 4 Model II, 6 Model VI, and 10 Model IX computers. The second shipment, which contained a bill for \$72,000, was for 8 Model II, 3 Model VI, and 5 Model IX computers. The third shipment, which contained a bill for \$81,000, was for 2 Model II, 9 Model VI, and 5 Model IX computers. What does the manufacturer charge for a Model VI computer? \$4000

24. *Health Care* A relief organization supplies blankets, cots, and lanterns to victims of fires, floods, and other natural disasters. One week the organization purchased 15 blankets, 5 cots, and 10 lanterns for a total cost of \$1250. The next week, at the same prices, the organization purchased 20 blankets, 10 cots, and 15 lanterns for a total cost of \$2000. The next week, at the same prices, the organization purchased 10 blankets, 15 cots, and 5 lanterns for a total cost of \$1625. Find the cost of one blanket, the cost of one cot, and the cost of one lantern. Blanket, \$25; cot, \$75; lantern, \$50

25. *Investments* An investor has a total of \$18,000 deposited in three different accounts, which earn annual interest of 9%, 7%, and 5%. The amount deposited in the 9% account is twice the amount in the 5% account. If the three accounts earn total annual interest of \$1340, how much money is deposited in each account? \$8000 at 9%, \$6000 at 7%, \$4000 at 5%

26. *Investments* An investor has a total of \$15,000 deposited in three different accounts, which earn annual interest of 9%, 6%, and 4%. The amount deposited in the 6% account is \$2000 more than the amount in the 4% account. If the three accounts earn total annual interest of \$980, how much money is deposited in each account? \$5250 at 9%, \$5875 at 6%, \$3875 at 4%

27. *Art* A sculptor is creating a mobile from which three objects will be suspended from a light rod that is 15 in. long as shown at the right. The weight, in ounces, of each object is shown in the diagram. For the mobile to balance, the objects must be positioned so that $w_1d_1 = w_2d_2 + w_3d_3$. The artist wants d_3 to be three times d_2. Find the distances d_1, d_2, and d_3 so that the mobile will balance.
$d_1 = 6$ in., $d_2 = 3$ in., $d_3 = 9$ in.

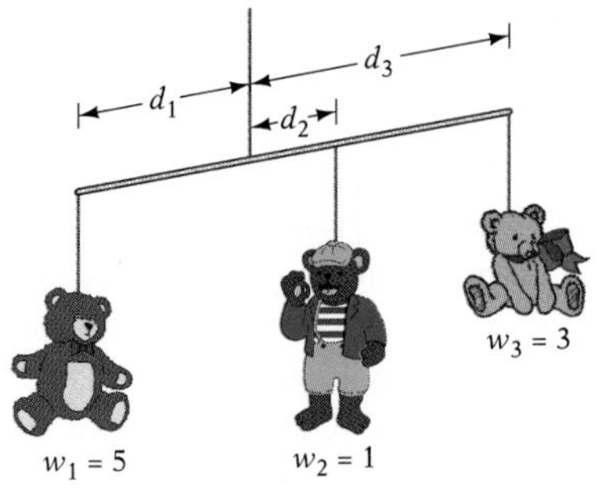

28. *Art* A mobile is made by suspending three objects from a light rod that is 20 in. long as shown at the right. The weight, in ounces, of each object is shown in the diagram. For the mobile to balance, the objects must be positioned so that $w_1d_1 + w_2d_2 = w_3d_3$. The artist wants d_3 to be twice d_2. Find the distances d_1, d_2, and d_3 so that the mobile will balance.
$d_1 = 8$ in., $d_2 = 6$ in., $d_3 = 12$ in.

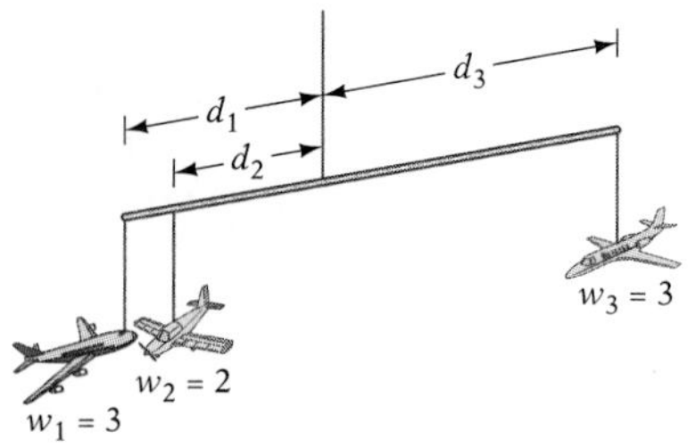

29. *Coin Problem* A coin bank contains only nickels, dimes, and quarters. The value of the 19 coins in the bank is $2. There are twice as many nickels as dimes. Find the number of each type of coin in the bank.
10 nickels, 5 dimes, 4 quarters

30. *Coin Problem* The total value of the nickels, dimes, and quarters in a coin bank is $4. If the nickels were dimes and the dimes were nickels, the total value of the coins would be $3.75. If the quarters were dimes and the dimes were quarters, the total value of the coins would be $6.25. Find the number of nickels, dimes, and quarters in the bank. 15 nickels, 20 dimes, 5 quarters

31. *Investment* A financial planner invested $33,000 of a client's money, part at 9%, part at 12%, and the remainder at 8%. The total annual income from these three investments was $3290. The amount invested at 12% was $5000 less than the combined amount invested at 9% and 8%. Find the amount invested at each rate. $14,000 at 12%, $10,000 at 8%, $9000 at 9%

32. *Travel Industry* Smith Travel Research listed the top three hotel and motel chains in the United States according to the total number of rooms. Holiday Inn, Best Western, and Days Inn had a combined total of 558,963 rooms. The Best Western and Days Inn had a combined total of 108,279 more rooms than the Holiday Inn. The Days Inn had 30,469 fewer rooms than the Best Western. Find the total number of rooms in each of the three chains.
Holiday Inn, 225,342 rooms; Best Western, 182,045 rooms; Days Inn, 151,576 rooms

33. *Sports* The Sports Marketing Letter of Westport, Connecticut, reports that the top three sports stars with the largest estimated annual income from endorsements are Michael Jordan, Shaquille O'Neal, and Arnold Palmer. In 1997, the combined income from endorsements for the three stars was $77 million. Michael earned $1 million less than the combined incomes of Shaquille and Arnold. Shaquille earned $7 million more than Arnold. Find the amount of earnings, from endorsements, for each player.
Michael Jordan, $38 million; Shaquille O'Neal, $23 million; Arnold Palmer, $16 million

34. *Labor Market* The U.S. Bureau of Labor Statistics has projected that the top three occupations with the largest numerical job decline between 1994 and 2005 are farmers, typists/word processors, and clerks (bookkeeping, accounting, and auditing). These three occupations are expected to decline by a total of 663,000 jobs. The decline in the number of jobs in farming will be 117,000 less than the combined loss in jobs for typists/word processors and clerks. The job decline in clerk positions will be 95,000 less than that for farmers. Find the job decline for each of the three occupations.
farmers: 273,000 jobs; typists/word processors: 212,000 jobs; clerks: 178,000 jobs

APPLYING CONCEPTS 4.4

35. *Geometry* Two angles are complementary. The larger angle is 9° more than eight times the measure of the smaller angle. Find the measure of the two angles. (Complementary angles are two angles whose sum is 90°.)
9° and 81°

36. *Geometry* Two angles are supplementary. The larger angle is 40° more than three times the measure of the smaller angle. Find the measure of the two angles. (Supplementary angles are two angles whose sum is 180°.)
35° and 145°

[C] **37.** *Coin Problem* The total value of the nickels, dimes, and quarters in a coin bank is $3.50. If the nickels were dimes and the dimes were nickels, the total value of the coins would be $4.25. If the quarters were dimes and the dimes were quarters, the total value of the coins would be $4.25. Find the number of nickels, dimes, and quarters in the bank.
25 nickels, 10 dimes, and 5 quarters

38. *Health Care* The following table shows the per capita spending on health care, by country, for the years 1970, 1980, and 1990. (*Source:* The Organization for Economic Cooperation and Development.) In the right-hand column is the linear equation that approximately models the data for each country. In each equation, x is the last two digits of the year, and y is the per capita spending on health care.

Country	1970	1980	1990	Model Equation
Australia	204	595	1151	$y = 47.35x - 3138$
Canada	274	806	1795	$y = 76.05x - 5125.67$
Denmark	209	571	963	$y = 37.70x - 2435$
France	192	656	1379	$y = 59.35x - 4005.67$
Greece	62	196	406	$y = 17.20x - 1154.67$
Italy	147	541	1113	$y = 48.30x - 3263.67$
Japan	126	515	1113	$y = 49.35x - 3363.33$
Norway	153	624	1281	$y = 56.40x - 3826$
Spain	82	322	730	$y = 32.40x - 2214$
Sweden	274	859	1421	$y = 57.35x - 3736.67$
United Kingdom	144	445	909	$y = 38.25x - 2560.67$
United States	346	1063	2566	$y = 111x - 7555$

a. Use the model equations to determine in which country the per capita spending is increasing at the most rapid rate and in which country the per capita spending is increasing least rapidly. Most rapidly: United States; least rapidly: Greece

b. The graph of the model equation for Sweden contains the point (83, 1023.38). Write a sentence that gives an interpretation of this ordered pair. In 1983, Sweden's per capita spending on health care was $1023.38.

c. For each pair of countries given below, determine during what year the model equations predict that the per capita spending on health care was the same. To the nearest ten dollars, what was the per capita spending on health care during that year?

    (1) Norway and Australia   1976; $460

    (2) Denmark and Italy   1978; $510

    (3) Japan and the United Kingdom   1972; $190

39.   Read the article "Matrix Mathematics: How to Win at Monopoly" by Dr. Crypton in the September 1985 issue of *Science Digest,* and prepare a report. Include in your report the relationship between systems of equations and Monopoly.

S E C T I O N   **4.5**

Solving Systems of Linear Inequalities

1   Graph the solution set of a system of linear inequalities

Two or more inequalities considered together are called a **system of inequalities.** The **solution set of a system of inequalities** is the intersection of the solution sets of the individual inequalities. To graph the solution set of a system of inequalities, first graph the solution set of each inequality. The solution set of the system of inequalities is the region of the plane represented by the intersection of the two shaded areas.

➡ Graph the solution set: $2x - y \le 3$
                               $3x + 2y > 8$

Solve each equation for y.

$$2x - y \le 3 \qquad\qquad 3x + 2y > 8$$
$$-y \le -2x + 3 \qquad\qquad 2y > -3x + 8$$
$$y \ge 2x - 3 \qquad\qquad y > -\frac{3}{2}x + 4$$

Graph $y = 2x - 3$ as a solid line. Because the inequality is $\ge$, shade above the line.

Graph $y = -\frac{3}{2}x + 4$ as a dashed line. Because the inequality is $>$, shade above the line.

The solution set of the system is the region of the plane represented by the intersection of the solution sets of the individual inequalities.

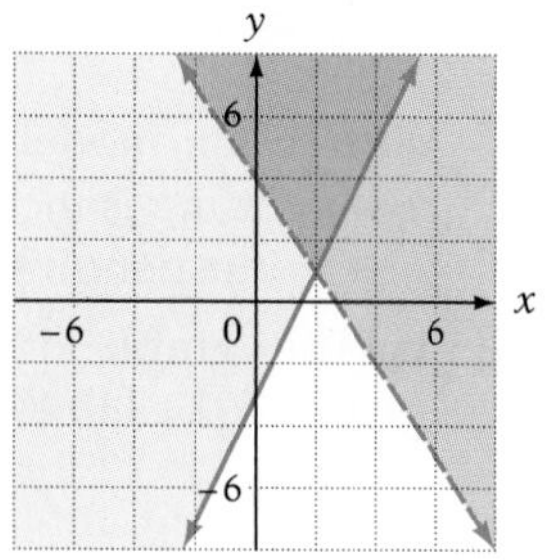

POINT OF INTEREST

Large systems of linear inequalities containing over 100 inequalities have been used to solve application problems in such diverse areas as providing health care and hardening a nuclear missile silo.

Graph the solution set: $-x + 2y \geq 4$
$x - 2y \geq 6$

Solve each equation for y.

$$-x + 2y \geq 4 \qquad\qquad x - 2y \geq 6$$
$$2y \geq x + 4 \qquad\qquad -2y \geq -x + 6$$
$$y \geq \frac{1}{2}x + 2 \qquad\qquad y \leq \frac{1}{2}x - 3$$

Shade above the solid line $y = \frac{1}{2}x + 2$.

Shade below the solid line $y = \frac{1}{2}x - 3$.

Because the solution sets of the two inequalities do not intersect, the solution set of the system is the empty set.

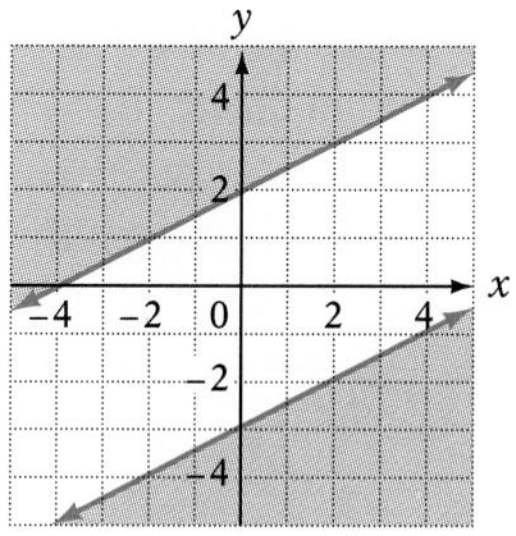

Example 1 Graph the solution set.

A. $y \geq x - 1$
$y < -2x$

B. $2x + 3y > 9$
$y < -\frac{2}{3}x + 1$

Solution A. Shade above the solid line $y = x - 1$.

Shade below the dashed line $y = -2x$.

The solution set of the system is the intersection of the solution sets of the individual inequalities.

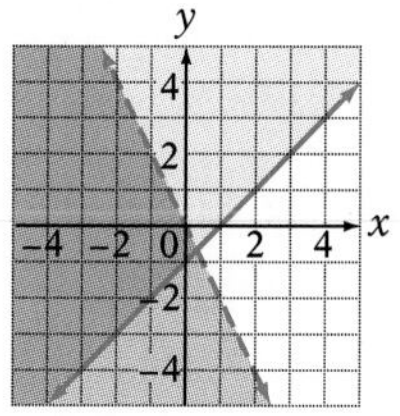

B. Solve $2x + 3y > 9$ for y.

$$2x + 3y > 9$$
$$3y > -2x + 9$$
$$y > -\frac{2}{3}x + 3$$

Shade above the dashed line $y = -\frac{2}{3}x + 3$.

Shade below the dashed line $y = -\frac{2}{3}x + 1$.

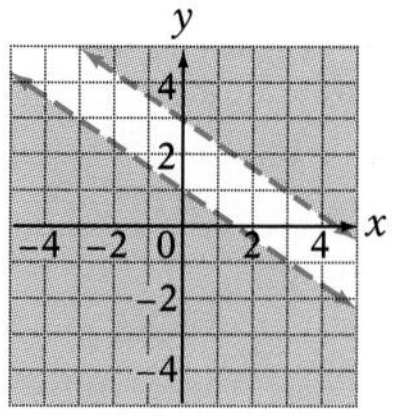

The intersection of the system is the empty set because the solution sets of the two inequalities do not intersect.

Problem 1 Graph the solution set.

A. $y \geq 2x - 3$
$y > -3x$

B. $3x + 4y > 12$
$y < \frac{3}{4}x - 1$

Solution See page S11. A.

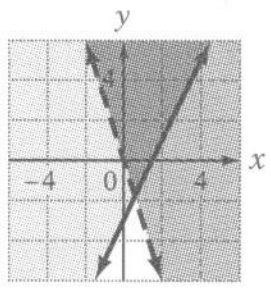

B.

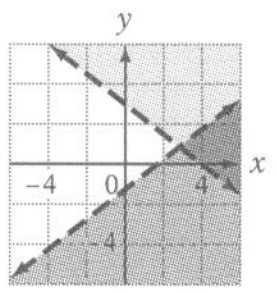

CONCEPT REVIEW 4.5

Determine whether the following statements are always true, sometimes true, or never true.

1. The solution set of a system of linear inequalities is an infinite set. Sometimes true
2. The solution set of a system of linear inequalities is the empty set. Sometimes true
3. The solution set of a system of linear inequalities is the intersection of the solution sets of the individual inequalities. Always true

EXERCISES 4.5

1 1. What is the solution set of a system of linear inequalities in two variables?

2. If the solution set of a system of linear inequalities in two variables is the empty set, what can be said about the slopes of the lines of the linear equations?

Graph the solution set.

3. $y \le x - 3$; $y \le -x + 5$

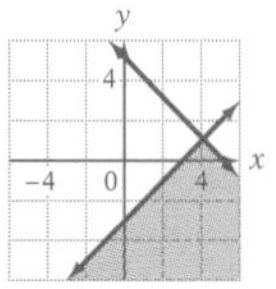

4. $y > 2x - 4$; $y < -x + 5$

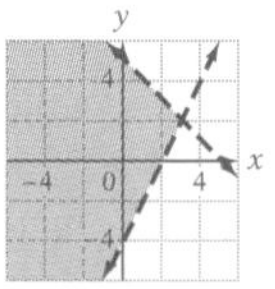

5. $y > 3x - 3$; $y \ge -2x + 2$

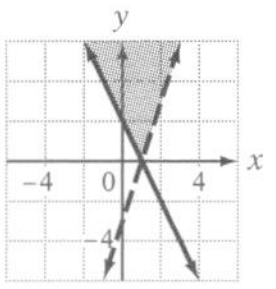

6. $y \le -\frac{1}{2}x + 3$; $y \ge x - 3$

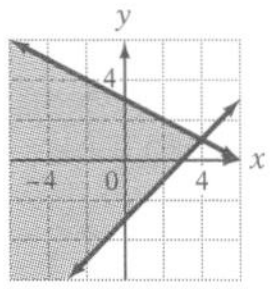

7. $2x + y \ge -2$; $6x + 3y \le 6$

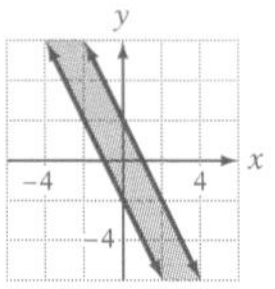

8. $x + y \ge 5$; $3x + 3y \le 6$
No solution

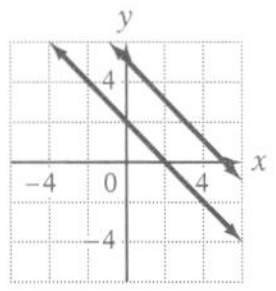

9. $3x - 2y < 6$; $y \le 3$

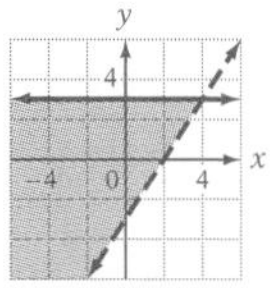

10. $x \le 2$; $3x + 2y > 4$

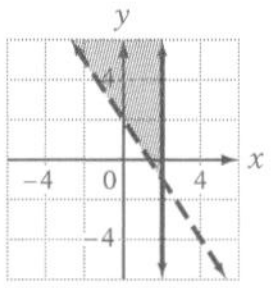

11. $y > 2x - 6$; $x + y < 0$

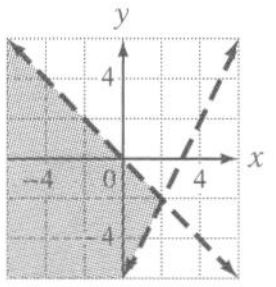

12. $x < 3$; $y < -2$

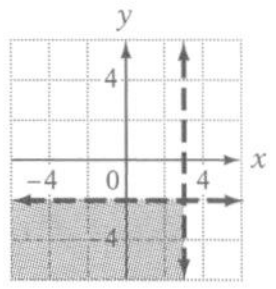

13. $x + 1 \ge 0$; $y - 3 \le 0$

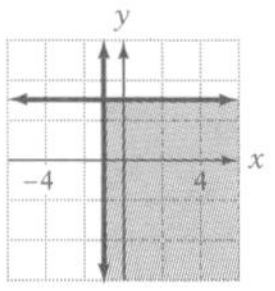

14. $5x - 2y \ge 10$; $3x + 2y \ge 6$

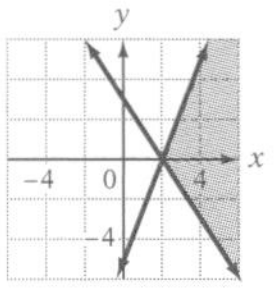

15. $2x + y \ge 4$; $3x - 2y < 6$

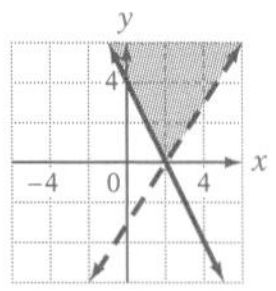

16. $3x - 4y < 12$; $x + 2y < 6$

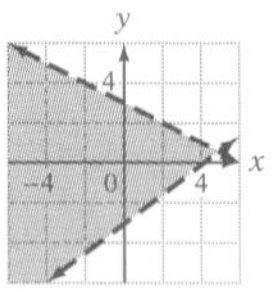

17. $x - 2y \le 6$; $2x + 3y \le 6$

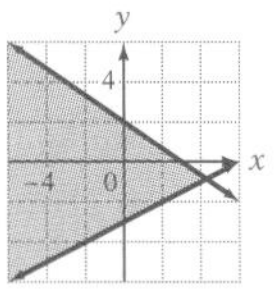

18. $x - 3y > 6$; $2x + y > 5$

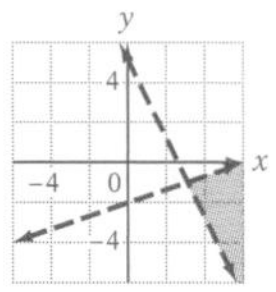

19. $x - 2y \le 4$; $3x + 2y \le 8$; $x > -1$

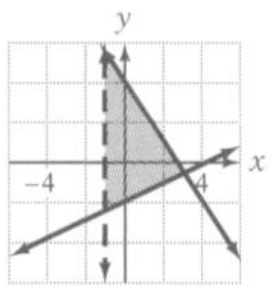

20. $3x - 2y < 0$; $5x + 3y > 9$; $y < 4$

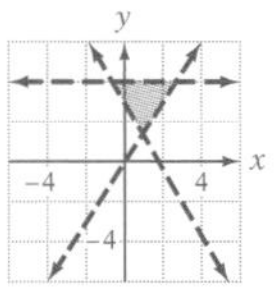

APPLYING CONCEPTS 4.5

Graph the solution set.

21. $2x + 3y \le 15$
$3x - y \le 6$
$y \ge 0$

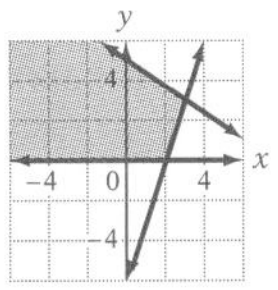

22. $x + y \le 6$
$x - y \le 2$
$x \ge 0$

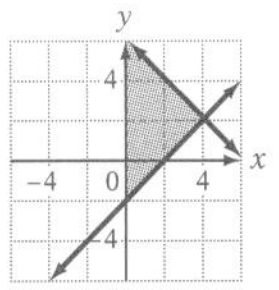

23. $x - y \le 5$
$2x - y \ge 6$
$y \ge 0$

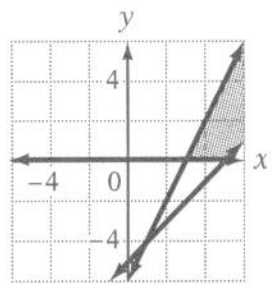

24. $x - 3y \le 6$
$5x - 2y \ge 4$
$y \ge 0$

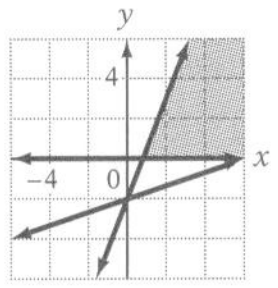

25. $2x - y \le 4$
$3x + y < 1$
$y \le 0$

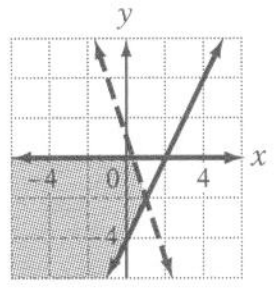

26. $x - y \le 4$
$2x + 3y > 6$
$x \ge 0$

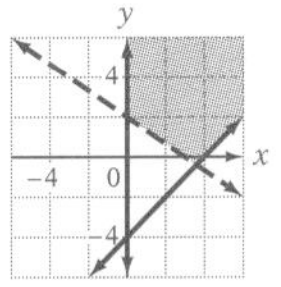

Focus on Problem Solving

Solve an Easier Problem

We begin this problem-solving feature with a restatement of the four steps of Polya's recommended problem-solving model.

1. Understand the problem.
2. Devise a plan.
3. Carry out the plan.
4. Review your solution.

One of the several methods of devising a plan is to try to solve an easier problem. Suppose you are in charge of your softball league, which consists of 15 teams. You must devise a schedule in which each team plays every other team once. How many games must be scheduled?

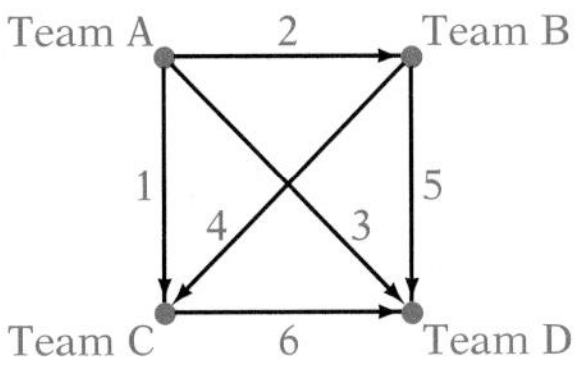

To solve this problem, we will attempt an easier problem first. Suppose that your league contains only a small number of teams. For instance, if there were only 1 team, you would schedule 0 games. If there were 2 teams, you would schedule 1 game. If there were 3 teams, you would schedule 3 games. The diagram at the left shows that 6 games must be scheduled when there are 4 teams in the league.

Here is a table of our results so far. (Remember that making a table is another strategy to be used in problem solving.)

Number of Teams	*Number of Games*	*Possible Pattern*
1	0	0
2	1	1
3	3	1 + 2
4	6	1 + 2 + 3

1. Draw a diagram with 5 dots to represent the teams. Draw lines from each dot to a second dot, and determine the number of games required.

2. What is the apparent pattern for the number of games required?
3. Assuming that the pattern continues, how many games must be scheduled for the 15 teams of the original problem?

After solving a problem, good problem solvers ask whether it is possible to solve the problem in a different manner. Here is a possible alternative method of solving the scheduling problem.

Begin with one of the 15 teams (say team A) and ask, "How many games must this team play?" Because there are 14 teams left to play, you must schedule 14 games. Now move to team B. It is already scheduled to play team A, and it does not play itself, so there are 13 teams left for it to play. Consequently, you must schedule 14 + 13 games.

4. Continue this reasoning for the remaining teams and determine the number of games that must be scheduled. Does this answer correspond to the answer you obtained with the first method?
5. Visit www. sports.com to get a listing of the professional sports teams and how they are separated into leagues and divisions. To appreciate how complicated the game schedules can be, try to create a partial schedule where each team in one of the conferences in the National Football League—for instance, the NFC East—plays each other team in that conference twice. An actual schedule would require scheduling an additional eight games for each team in the conference with selected teams in the other conferences.
6. Making connections to other problems you have solved is an important step toward becoming an excellent problem solver. How does the answer to the scheduling of teams relate to the triangular numbers discussed in the Focus on Problem Solving in the chapter titled Linear Functions and Inequalities in Two Variables?

Projects and Group Activities

Using a Graphing Calculator to Solve a System of Equations

A graphing calculator can be used to solve a system of equations. For this procedure to work on most calculators, it is necessary that the point of intersection be on the screen. This means that you may have to experiment with Xmin, Xmax, Ymin, and Ymax values until the graphs intersect on the screen.

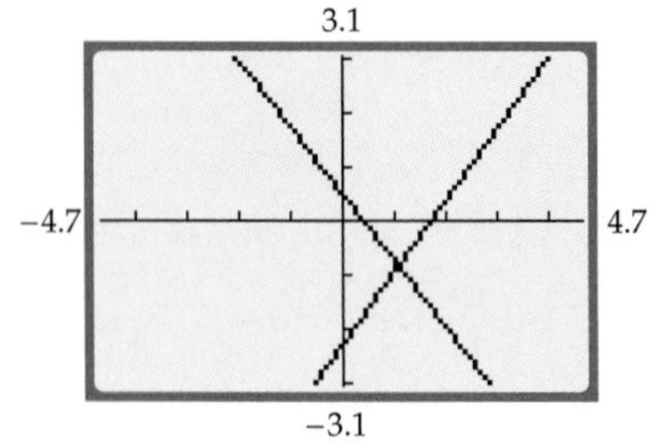

To solve a system of equations graphically, solve each equation for y. Then graph the equations of the system. Their point of intersection is the solution.

For instance, to solve the system of equations

$$4x - 3y = 7$$
$$5x + 4y = 2$$

first solve each equation for y.

$$4x - 3y = 7 \Rightarrow y = \frac{4}{3}x - \frac{7}{3}$$
$$5x + 4y = 2 \Rightarrow y = -\frac{5}{4}x + \frac{1}{2}$$

The keystrokes to solve this system are given below. We are using a viewing window $[-4.7, 4.7]$ by $[-3.1, 3.1]$.

TI-83	*SHARP EL-9600*	*Casio CFX-9850G*
[Y=] [CLEAR] 4 [X,T,θ,n] [÷] 3 [−] 7 [÷] 3 [ENTER] [CLEAR] [(−)] 5 [X,T,θ,n] [÷] 4 [+] 1 [÷] 2 [GRAPH]	[Y=] [CL] 4 [X/θ/T/n] [÷] 3 [−] 7 [÷] 3 [ENTER] [CL] [(−)] 5 [X/θ/T/n] [÷] 4 [+] 1 [÷] 2 [GRAPH]	[MENU] 5 [F2] [F1] 4 [X,θ,T] [÷] 3 [−] 7 [÷] 3 [EXE] [F2] [F1] [CLEAR] [(−)] 5 [X,θ,T] [÷] 4 [+] 1 [÷] 2 [EXE] [GRAPH]

Once the calculator has drawn the graphs, use the TRACE feature and move the cursor to the approximate point of intersection. This will give you an approximate solution of the system of equations. The approximate solution is (1.096774, −0.870968). The method by which a more accurate solution can be determined depends on the type of calculator you use. Many calculators have an INTERSECT feature that can be used to solve a system of equations. Consult the user's manual under "systems of equations."

Some of the exercises in the first section of this chapter asked you to solve a system of equations by graphing. Try those exercises again, this time using your graphing calculator.

Current models of calculators do not allow you to solve graphically a system of equations in three variables. However, these calculators do have matrix and determinant operations that can be used to solve these systems.

Solving a First-Degree Equation with a Graphing Calculator

A first-degree equation in one variable can be solved graphically[1] by using a graphing calculator. The idea is to rewrite the equation as a system of equations and then solve the system of equations as we illustrated on the previous page. For instance, to solve the equation

$$2x + 1 = 5x - 5$$

write the equation as the following system of equations.

$$y = 2x + 1$$
$$y = 5x - 5$$

Note that we have used the left and right sides of the original equation to form the system of equations.

[1]Some graphing calculators offer nongraphical techniques for solving an equation.

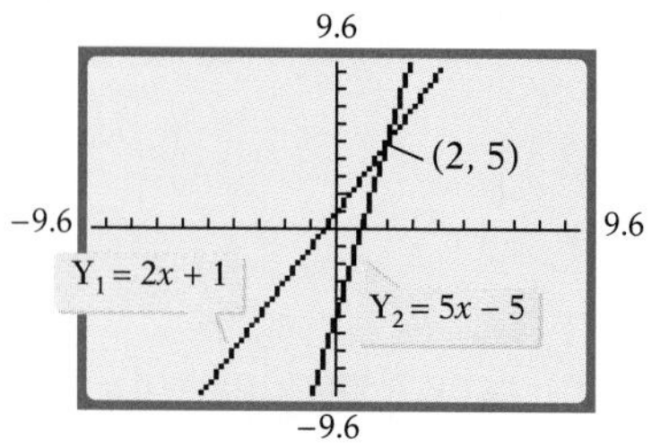

Now graph the equations and find the point of intersection. This is shown at the left using a viewing window of [−9.6, 9.6] by [−9.6, 9.6]. The solution of the original equation, $2x + 1 = 5x - 5$, is the x-coordinate of the point of intersection. Thus the solution of the equation is 2.

Recall that not all equations have a solution. Consider the equation

$$x - 3(x + 2) = 2x - 2(2x - 2)$$

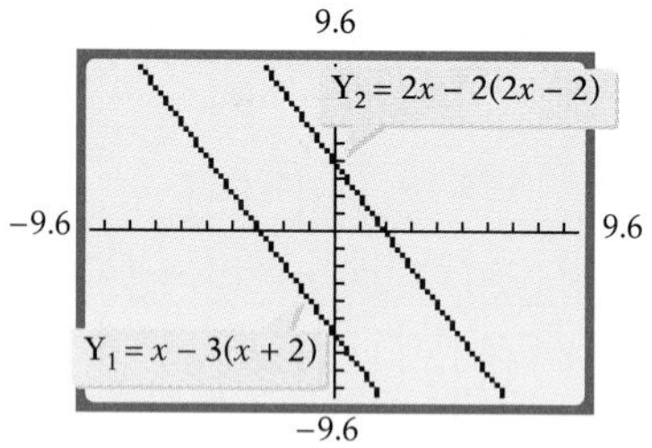

The graphs of the left and right sides of the equation are shown at the left. Note that the lines appear to be parallel and therefore do not intersect. Because the lines do not intersect, there is no solution of the system of equations and thus no solution of the original equation. We algebraically verify this result below.

$$x - 3(x + 2) = 2x - 2(2x - 2)$$
$$x - 3x - 6 = 2x - 4x + 4 \quad \blacktriangleright \text{ Use the Distributive Property.}$$
$$-2x - 6 = -2x + 4 \quad \blacktriangleright \text{ Simplify.}$$
$$-6 = 4 \quad \blacktriangleright \text{ Add } 2x \text{ to each side of the equation.}$$

Because $-6 = 4$ is not a true equation, there is no solution.

One final point about the last equation. Note that in the third line of the solution we obtain the expressions $-2x - 6$ and $-2x + 4$. If we were to graph $y = -2x - 6$ and $y = -2x + 4$, the slopes of the lines would be equal (both are -2) and therefore the graphs of the lines are parallel.

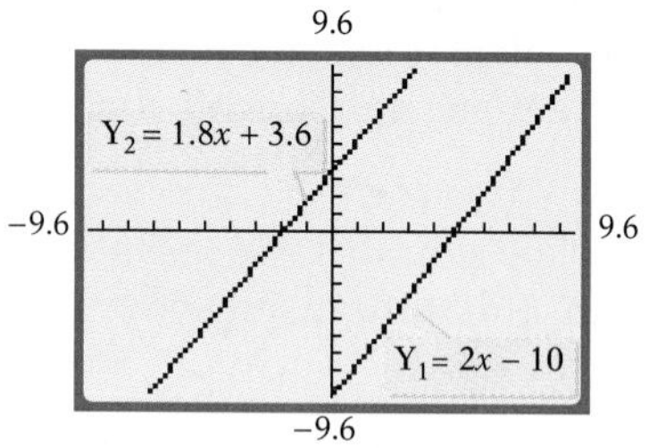

Before we leave graphical solutions, a few words of caution. Graphs can be deceiving and appear not to intersect when they do, or appear to intersect when they do not. For instance, an attempt to graphically solve

$$2x - 10 = 1.8x + 3.6$$

is shown at the left.

If you use the viewing window [−9.6, 9.6] by [−9.6, 9.6], the graphs will appear to be parallel. However, the graphs of $y = 2x - 10$ and $y = 1.8x + 3.6$ do not have the same slope and therefore must intersect. For this equation, you need a larger viewing window to see the point of intersection.

Solve the following equations by using a graphing calculator.

1. $3x - 1 = 5x + 1$

2. $3x + 2 = 4$

3. $3 + 2(2x - 4) = 5(x - 3)$

4. $2x - 4 = 5x - 3(x + 2) + 2$

5. Explain how Problem 4 relates to an *identity* as explained in Section 2.1.

6. Find an appropriate viewing window so that you can determine the solution of the equation $2x - 10 = 1.8x + 3.6$ given above.

Chapter Summary

Key Words

A *system of equations* is two or more equations considered together. A *solution of a system of equations in two variables* is an ordered pair that is a solution of each equation of the system. (Objective 4.1.1)

The solution of the system

$$\begin{aligned} x + y &= 2 \\ x - y &= 4 \end{aligned}$$

is the ordered pair (3, −1). (3, −1) is the only ordered pair that is a solution of both equations.

When the graphs of a system of equations intersect at only one point, the system is called an *independent system of equations.* (Objective 4.1.1)

$$\begin{aligned} 3x - 2y &= 4 \\ 5x + 2y &= 6 \\ 8x \quad\;\; &= 10 \end{aligned}$$

When the graphs of a system of equations coincide, the system is called a *dependent system of equations.* (Objective 4.1.1)

$$\begin{aligned} -3x + 2y &= 4 \\ 3x - 2y &= -4 \\ 0 &= 0 \end{aligned}$$

When the graphs of a system of equations do not intersect, the system has no solution and is called an *inconsistent system of equations.* (Objective 4.1.1)

$$\begin{aligned} 2x - 3y &= -2 \\ -2x + 3y &= 5 \\ 0 &= 3 \end{aligned}$$

An equation of the form $Ax + By + Cz = D$ is called a *linear equation in three variables.* (Objective 4.2.2)

$3x + 2y - 5z = 12$ is a linear equation in three variables.

A *solution of a system of equations in three variables* is an ordered triple that is a solution of each equation of the system. (Objective 4.2.2)

The ordered triple (1, 2, 1) is a solution of the system of equations

$$\begin{aligned} 3x + y - 3z &= 2 \\ -x + 2y + 3z &= 6 \\ 2x + 2y - 2z &= 4 \end{aligned}$$

A *matrix* is a rectangular array of numbers. (Objective 4.3.1)

$\begin{bmatrix} 2 & 3 & 6 \\ -1 & 2 & 4 \end{bmatrix}$ is a 2×3 matrix.

A *square matrix* has the same number of rows as columns. (Objective 4.3.1)

$\begin{bmatrix} 2 & 3 \\ 4 & -1 \end{bmatrix}$ is a square matrix.

A *determinant* is a number associated with a square matrix. (Objective 4.3.1)

$$\begin{aligned} \begin{vmatrix} 2 & -3 \\ -4 & 5 \end{vmatrix} &= 2(5) - (-4)(-3) \\ &= 10 - 12 \\ &= -2 \end{aligned}$$

The *minor of an element* in a 3 × 3 determinant is the 2 × 2 determinant obtained by eliminating the row and column that contain the element. (Objective 4.3.1)

$\begin{vmatrix} 2 & 1 & 3 \\ 4 & 6 & 2 \\ 1 & 8 & 3 \end{vmatrix}$ The minor of 4 is $\begin{vmatrix} 1 & 3 \\ 8 & 3 \end{vmatrix}$.

The *cofactor of an element of a determinant* is $(-1)^{i+j}$ times the minor of the element, where i is the row number of the element and j is its column number. (Objective 4.3.1)

In the determinant above, 4 is from the second row and the first column. The cofactor of 4 is

$$(-1)^{2+1}\begin{vmatrix} 1 & 3 \\ 8 & 3 \end{vmatrix} = -\begin{vmatrix} 1 & 3 \\ 8 & 3 \end{vmatrix}$$

The evaluation of the determinant of a 3 × 3 or larger matrix is accomplished by *expanding by cofactors.* (Objective 4.3.1)

Expand by cofactors of the first column.

$$\begin{vmatrix} 2 & 1 & 3 \\ 4 & 6 & 2 \\ 1 & 8 & 3 \end{vmatrix} = 2\begin{vmatrix} 6 & 2 \\ 8 & 3 \end{vmatrix} - 4\begin{vmatrix} 1 & 3 \\ 8 & 3 \end{vmatrix} + 1\begin{vmatrix} 1 & 3 \\ 6 & 2 \end{vmatrix}$$

Inequalities considered together are called a *system of inequalities.* The *solution set of a system of inequalities* is the intersection of the solution sets of the individual inequalities. (Objective 4.5.1)

$x + y > 3$
$x - y > -2$

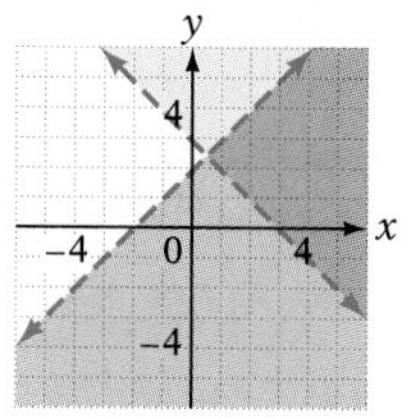

Essential Rules and Procedures

A system of equations can be solved by:

a. Graphing (Objective 4.1.1)

$y = \frac{1}{2}x + 2$

$y = \frac{5}{2}x - 2$

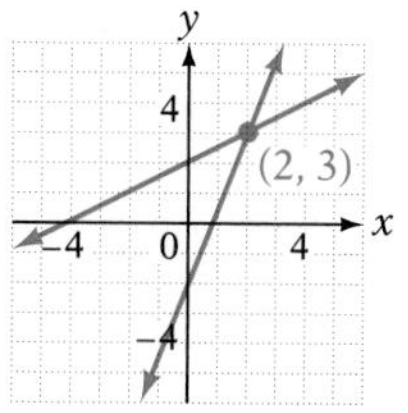

b. The substitution method (Objective 4.1.2)

(1) $2x - 3y = 4$
(2) $y = -x + 2$

Substitute the value of y into equation (1).

$$2x - 3(-x + 2) = 4$$

c. The addition method (Objective 4.2.1)

$$\begin{aligned} -2x + 3y &= 7 \\ 2x - 5y &= 2 \\ -2y &= 9 \end{aligned}$$

Cramer's Rule (Objective 4.3.2)

For two variables: $a_{11}x + a_{12}y = b_1$
$a_{21}x + a_{22}y = b_2$

$$x = \frac{D_x}{D},\ y = \frac{D_y}{D}, \text{ where } D = \begin{vmatrix} a_{11} & a_{12} \\ a_{21} & a_{22} \end{vmatrix},\ D_x = \begin{vmatrix} b_1 & a_{12} \\ b_2 & a_{22} \end{vmatrix}$$

$$D_y = \begin{vmatrix} a_{11} & b_1 \\ a_{21} & b_2 \end{vmatrix}, \text{ and } D \neq 0.$$

$2x - y = 6$
$x + 3y = 4$

$$D = \begin{vmatrix} 2 & -1 \\ 1 & 3 \end{vmatrix} = 7,\ D_x = \begin{vmatrix} 6 & -1 \\ 4 & 3 \end{vmatrix} = 22,$$

$$D_y = \begin{vmatrix} 2 & 6 \\ 1 & 4 \end{vmatrix} = 2$$

$$x = \frac{D_x}{D} = \frac{22}{7},\ y = \frac{D_y}{D} = \frac{2}{7}$$

For three variables: $a_{11}x + a_{12}y + a_{13}z = b_1$
$a_{21}x + a_{22}y + a_{23}z = b_2$
$a_{31}x + a_{32}y + a_{33}z = b_3$

$$x = \frac{D_x}{D},\ y = \frac{D_y}{D}, \text{ and } z = \frac{D_z}{D}, \text{ where } D = \begin{vmatrix} a_{11} & a_{12} & a_{13} \\ a_{21} & a_{22} & a_{23} \\ a_{31} & a_{32} & a_{33} \end{vmatrix},$$

$$D_x = \begin{vmatrix} b_1 & a_{12} & a_{13} \\ b_2 & a_{22} & a_{23} \\ b_3 & a_{32} & a_{33} \end{vmatrix},\ D_y = \begin{vmatrix} a_{11} & b_1 & a_{13} \\ a_{21} & b_2 & a_{23} \\ a_{31} & b_3 & a_{33} \end{vmatrix},$$

$$D_z = \begin{vmatrix} a_{11} & a_{12} & b_1 \\ a_{21} & a_{22} & b_2 \\ a_{31} & a_{32} & b_3 \end{vmatrix}, \text{ and } D \neq 0$$

$x - y + z = 2$
$2x + y - 2z = -2$
$x - 2y + 3z = 6$

$$D = \begin{vmatrix} 1 & -1 & 1 \\ 2 & 1 & -2 \\ 1 & -2 & 3 \end{vmatrix} = 2,$$

$$D_x = \begin{vmatrix} 2 & -1 & 1 \\ -2 & 1 & -2 \\ 6 & -2 & 3 \end{vmatrix} = 2,$$

$$D_y = \begin{vmatrix} 1 & 2 & 1 \\ 2 & -2 & -2 \\ 1 & 6 & 3 \end{vmatrix} = 4,$$

$$D_z = \begin{vmatrix} 1 & -1 & 2 \\ 2 & 1 & -2 \\ 1 & -2 & 6 \end{vmatrix} = 6$$

$$x = \frac{D_x}{D} = \frac{2}{2} = 1,\ y = \frac{D_y}{D} = \frac{4}{2} = 2,$$

$$z = \frac{D_z}{D} = \frac{6}{2} = 3$$

Elementary row operations to rewrite a matrix in *echelon form:*

1. Interchange two rows.
2. Multiply all the elements in a row by the same nonzero number.
3. Replace a row by the sum of that row and a multiple of any other row.
(Objective 4.3.3)

$$\begin{bmatrix} 1 & 4 & -2 & 3 \\ 0 & 1 & 3 & -1 \\ 0 & 0 & 1 & 6 \end{bmatrix} \quad \text{Echelon form}$$

Chapter Review Exercises

1. Solve by substitution: $2x - 6y = 15$
$x = 3y + 8$
Inconsistent

2. Solve by substitution: $3x + 12y = 18$
$x + 4y = 6$
$\left(x, -\frac{1}{4}x + \frac{3}{2}\right)$

3. Solve by the addition method: $3x + 2y = 2$
$x + y = 3$
$(-4, 7)$

4. Solve by the addition method: $5x - 15y = 30$
$x - 3y = 6$
$\left(x, \frac{1}{3}x - 2\right)$

5. Solve by the addition method: $3x + y = 13$
$2y + 3z = 5$
$x + 2z = 11$
$(5, -2, 3)$

6. Solve by the addition method:
$3x - 4y - 2z = 17$
$4x - 3y + 5z = 5$
$5x - 5y + 3z = 14$ $(3, -1, -2)$

7. Evaluate the determinant: $\begin{vmatrix} 6 & 1 \\ 2 & 5 \end{vmatrix}$ 28

8. Evaluate the determinant: $\begin{vmatrix} 1 & 5 & -2 \\ -2 & 1 & 4 \\ 4 & 3 & -8 \end{vmatrix}$ 0

9. Solve by using Cramer's Rule: $2x - y = 7$
$3x + 2y = 7$
$(3, -1)$

10. Solve by using Cramer's Rule: $3x - 4y = 10$
$2x + 5y = 15$
$\left(\frac{110}{23}, \frac{25}{23}\right)$

11. Solve by using Cramer's Rule:
$x + y + z = 0$
$x + 2y + 3z = 5$
$2x + y + 2z = 3$ $(-1, -3, 4)$

12. Solve by using Cramer's Rule:
$x + 3y + z = 6$
$2x + y - z = 12$
$x + 2y - z = 13$ $(2, 3, -5)$

13. Solve by the addition method:
$x - 2y + z = 7$
$3x - z = -1$
$3y + z = 1$ $(1, -1, 4)$

14. Solve by using Cramer's Rule:
$3x - 2y = 2$
$-2x + 3y = 1$ $\left(\frac{8}{5}, \frac{7}{5}\right)$

15. Solve by using a matrix: $2x - 2y - 6z = 1$
$4x + 2y + 3z = 1$
$2x - 3y - 3z = 3$
$\left(\frac{1}{2}, -1, \frac{1}{3}\right)$

16. Evaluate the determinant: $\begin{vmatrix} 3 & -2 & 5 \\ 4 & 6 & 3 \\ 1 & 2 & 1 \end{vmatrix}$ 12

17. Solve by using Cramer's Rule: $4x - 3y = 17$
$3x - 2y = 12$
$(2, -3)$

18. Solve by using a matrix: $3x + 2y - z = -1$
$x + 2y + 3z = -1$
$3x + 4y + 6z = 0$
$(2, -3, 1)$

19. Solve by graphing: $x + y = 3$
$3x - 2y = -6$
$(0, 3)$

20. Solve by graphing: $2x - y = 4$
$y = 2x - 4$
$(x, 2x - 4)$

21. Graph the solution set: $x + 3y \le 6$
$2x - y \ge 4$

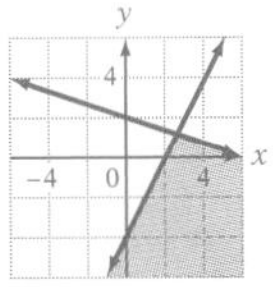

22. Graph the solution set: $2x + 4y \ge 8$
$x + y \le 3$

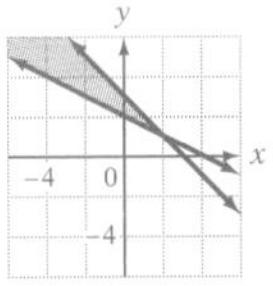

23. A cabin cruiser traveling with the current went 60 mi in 3 h. Against the current, it took 5 h to travel the same distance. Find the rate of the cabin cruiser in calm water and the rate of the current. Cabin cruiser: 16 mph; current: 4 mph

24. A plane flying with the wind flew 600 mi in 3 h. Flying against the wind, the plane required 4 h to travel the same distance. Find the rate of the plane in calm air and the rate of the wind. Plane: 175 mph; wind: 25 mph

25. At a movie theater, admission tickets are \$5 for children and \$8 for adults. The receipts for one Friday evening were \$2500. The next day there were three times as many children as the preceding evening and only half as many adults as the night before, yet the receipts were still \$2500. Find the number of children who attended the movie Friday evening. 100 children

26. A chef wants to prepare a lowfat, low sodium meal using lean meat, roasted potatoes, and green beans. A 1-ounce serving of meat contains 50 Cal, 20 g of protein, and 16 mg of sodium. A 1-ounce serving of potatoes contains 9 Cal, 1 g of protein, and 3 mg of sodium. A 1-ounce serving of green beans contains 12 Cal, 75 g of protein, and 17 mg of sodium. If the chef wants to prepare the meal so that it contains 243 Cal, 365 g of protein, and 131 mg of sodium, how many ounces of each ingredient should be prepared?
meat, 3 oz; potatoes, 5 oz; green beans, 4 oz

Chapter Test

1. Solve by substitution: $3x + 2y = 4$
$x = 2y - 1$
$\left(\frac{3}{4}, \frac{7}{8}\right)$

2. Solve by substitution: $5x + 2y = -23$
$2x + y = -10$
$(-3, -4)$

3. Solve by substitution: $y = 3x - 7$
$y = -2x + 3$
$(2, -1)$

4. Solve by using a matrix: $3x + 4y = -2$
$2x + 5y = 1$
$(-2, 1)$

5. Solve by the addition method: $4x - 6y = 5$
$6x - 9y = 4$
Inconsistent

6. Solve by the addition method:
$3x - y = 2x + y - 1$
$5x + 2y = y + 6$
$(1, 1)$

7. Solve by the addition method:

$$\begin{aligned} 2x + 4y - z &= 3 \\ x + 2y + z &= 5 \\ 4x + 8y - 2z &= 7 \end{aligned}$$

Inconsistent

8. Solve by using a matrix:

$$\begin{aligned} x - y - z &= 5 \\ 2x + z &= 2 \\ 3y - 2z &= 1 \end{aligned}$$

$(2, -1, -2)$

9. Evaluate the determinant: $\begin{vmatrix} 3 & -1 \\ -2 & 4 \end{vmatrix}$ 10

10. Evaluate the determinant: $\begin{vmatrix} 1 & -2 & 3 \\ 3 & 1 & 1 \\ 2 & -1 & -2 \end{vmatrix}$ -32

11. Solve by using Cramer's Rule:

$$\begin{aligned} x - y &= 3 \\ 2x + y &= -4 \end{aligned}$$

$\left(-\frac{1}{3}, -\frac{10}{3}\right)$

12. Solve by using Cramer's Rule:

$$\begin{aligned} x - y + z &= 2 \\ 2x - y - z &= 1 \\ x + 2y - 3z &= -4 \end{aligned}$$

$\left(\frac{1}{5}, -\frac{6}{5}, \frac{3}{5}\right)$

13. Solve by using Cramer's Rule:

$$\begin{aligned} 3x + 2y + 2z &= 2 \\ x - 2y - z &= 1 \\ 2x - 3y - 3z &= -3 \end{aligned}$$

$(0, -2, 3)$

14. Solve by graphing:

$$\begin{aligned} 2x - 3y &= -6 \\ 2x - y &= 2 \end{aligned}$$

$(3, 4)$

15. Solve by graphing:

$$\begin{aligned} x - 2y &= -5 \\ 3x + 4y &= -15 \end{aligned}$$

$(-5, 0)$

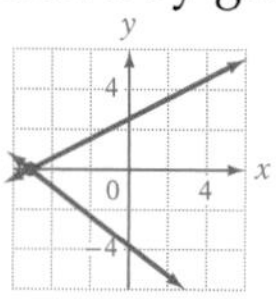

16. Graph the solution set:

$$\begin{aligned} 2x - y &< 3 \\ 4x + 3y &< 11 \end{aligned}$$

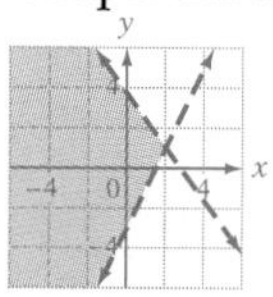

17. Graph the solution set:

$$\begin{aligned} x + y &> 2 \\ 2x - y &< -1 \end{aligned}$$

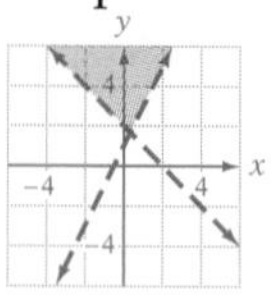

18. Use a graphing utility to solve the system to the nearest hundredth.

$$\begin{aligned} 2x + 3y &= 7 \\ x - 2y &= -5 \end{aligned}$$

$(-0.14, 2.43)$

19. A plane flying with the wind went 350 mi in 2 h. The return trip, flying against the wind, took 2.8 h. Find the rate of the plane in calm air and the rate of the wind. Plane: 150 mph; wind: 25 mph

20. A clothing manufacturer purchased 60 yd of cotton and 90 yd of wool for a total cost of \$1800. Another purchase, at the same prices, included 80 yd of cotton and 20 yd of wool for a total cost of \$1000. Find the cost per yard of the cotton and of the wool. Cotton: \$9; wool: \$14

Cumulative Review Exercises

1. Solve: $\frac{3}{2}x - \frac{3}{8} + \frac{1}{4}x = \frac{7}{12}x - \frac{5}{6}$ $-\frac{11}{28}$

2. Find the equation of the line that contains the points (2, −1) and (3, 4). $y = 5x - 11$

3. Simplify: $3[x - 2(5 - 2x) - 4x] + 6$ $3x - 24$

4. Evaluate $a + bc \div 2$ when $a = 4$, $b = 8$, and $c = -2$. -4

5. Solve: $2x - 3 < 9$ or $5x - 1 < 4$ $\{x \mid x < 6\}$

6. Solve: $|x - 2| - 4 < 2$ $\{x \mid -4 < x < 8\}$

7. Solve: $|2x - 3| > 5$ $\{x \mid x > 4 \text{ or } x < -1\}$

8. Given $f(x) = 3x^3 - 2x^2 + 1$, evaluate $f(-3)$. -98

9. Find the range of $f(x) = 3x^2 - 2x$ if the domain is $\{-2, -1, 0, 1, 2\}$. $\{0, 1, 5, 8, 16\}$

10. Given $F(x) = x^2 - 3$, find $F(2)$. 1

11. Given $f(x) = 3x - 4$, write $f(2 + h) - f(2)$ in simplest form. $3h$

12. Graph: $\{x \mid x \le 2\} \cap \{x \mid x > -3\}$

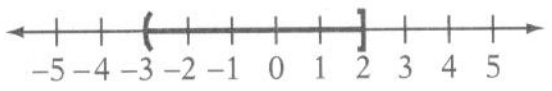

13. Find the equation of the line that contains the point (−2, 3) and has slope $-\frac{2}{3}$. $y = -\frac{2}{3}x + \frac{5}{3}$

14. Find the equation of the line that contains the point (−1, 2) and is perpendicular to the line $2x - 3y = 7$. $y = -\frac{3}{2}x + \frac{1}{2}$

15. Find the distance between the points (−4, 2) and (2, 0). $2\sqrt{10}$

16. Find the midpoint of the line connecting the points (−4, 3) and (3, 5). $\left(-\frac{1}{2}, 4\right)$

17. Graph $2x - 5y = 10$ by using the slope and y-intercept.

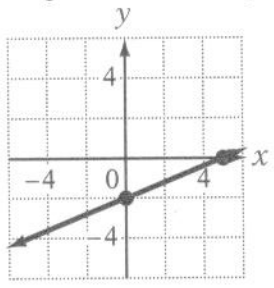

18. Graph the solution set of the inequality $3x - 4y \ge 8$.

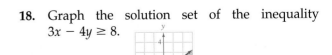

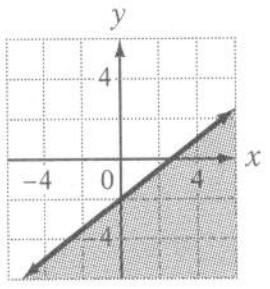

19. Solve by substitution: $3x - 2y = 7$
$y = 2x - 1$
$(-5, -11)$

20. Solve by the addition method: $3x + 2z = 1$
$2y - z = 1$
$x + 2y = 1$
$(1, 0, -1)$

21. Evaluate the determinant: $\begin{vmatrix} 2 & -5 & 1 \\ 3 & 1 & 2 \\ 6 & -1 & 4 \end{vmatrix}$ 3

22. Solve by graphing: $5x - 2y = 10$
$3x + 2y = 6$
$(2, 0)$

23. Solve by using Cramer's Rule: $4x - 3y = 17$
$3x - 2y = 12$
$(2, -3)$

24. Graph the solution set: $3x - 2y \geq 4$
$x + y < 3$

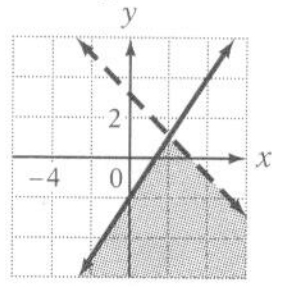

25. A coin purse contains 40 coins in nickels, dimes, and quarters. There are three times as many dimes as quarters. The total value of the coins is $4.10. How many nickels are in the coin purse? 16 nickels

26. How many milliliters of pure water must be added to 100 ml of a 4% salt solution to make a 2.5% salt solution? 60 ml

27. Flying with the wind, a small plane required 2 h to fly 150 mi. Against the wind, it took 3 h to fly the same distance. Find the rate of the wind. 12.5 mph

28. A restaurant manager buys 100 lb of hamburger and 50 lb of steak for a total cost of $490. A second purchase, at the same prices, includes 150 lb of hamburger and 100 lb of steak. The total cost is $860. Find the price of one pound of steak. $5

29. Find the lower and upper limits of a 12,000-ohm resistor with a 15% tolerance. Lower limit: 10,200 ohms; upper limit: 13,800 ohms

30. The graph below shows the relationship between the monthly income, in dollars, and sales, in thousands of dollars, of an account executive. Find the slope of the line between the two points shown on the graph. Write a sentence that states the meaning of the slope. $m = 40$; The account executive earns $40 for each $1000 of sales.

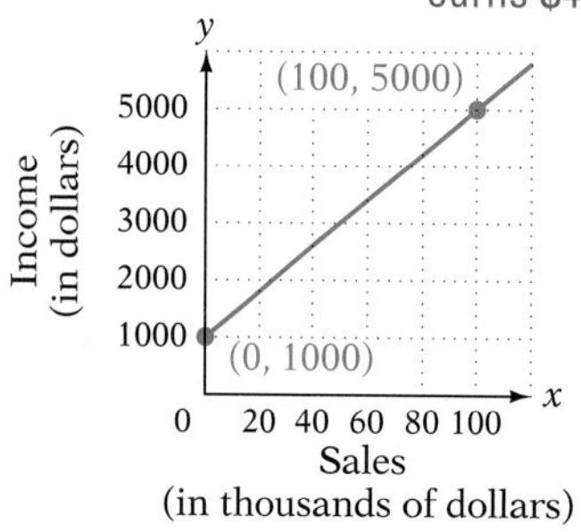

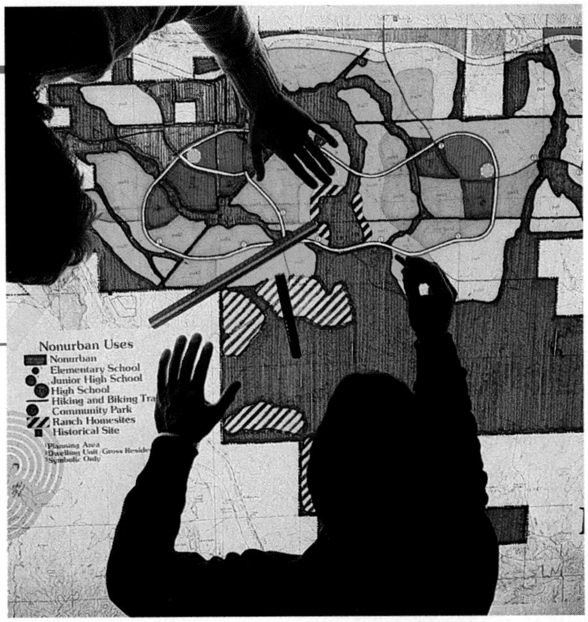

It is the job of urban planners to work toward growth and revitalization of urban areas. They must deal not only with such current problems as deterioration and pollution, but also with projections of long-term needs and plans for the future. Making estimations of changes in population, health care requirements, housing needs, and transportation systems requires working with variables.

5

Polynomials and Exponents

OBJECTIVES

5.1.1 Multiply monomials
5.1.2 Divide monomials and simplify expressions with negative exponents
5.1.3 Scientific notation
5.1.4 Application problems
5.2.1 Evaluate polynomial functions
5.2.2 Add or subtract polynomials
5.3.1 Multiply a polynomial by a monomial
5.3.2 Multiply two polynomials
5.3.3 Multiply polynomials that have special products
5.3.4 Application problems
5.4.1 Divide polynomials
5.4.2 Synthetic division
5.4.3 Evaluate a polynomial using synthetic division
5.5.1 Factor a monomial from a polynomial
5.5.2 Factor by grouping
5.5.3 Factor trinomials of the form $x^2 + bx + c$
5.5.4 Factor trinomials of the form $ax^2 + bx + c$
5.6.1 Factor the difference of two perfect squares and factor perfect-square trinomials
5.6.2 Factor the sum or the difference of two cubes
5.6.3 Factor trinomials that are quadratic in form
5.6.4 Factor completely
5.7.1 Solve equations by factoring
5.7.2 Application problems

Origins of the Word Algebra

The word *algebra* has its origins in an Arabic book written around A.D. 825 entitled *Hisab al-jabr w' almuqa-balah,* by al-Kwarizmi. The word *al-jabr,* which literally translated means "reunion," was written as the word *algebra* in Latin translations of al-Kwarizmi's work and became synonymous with equations and the solutions of equations. It is interesting to note that an early meaning of the Spanish word *algebrista* was "bonesetter" or "reuniter of broken bones."

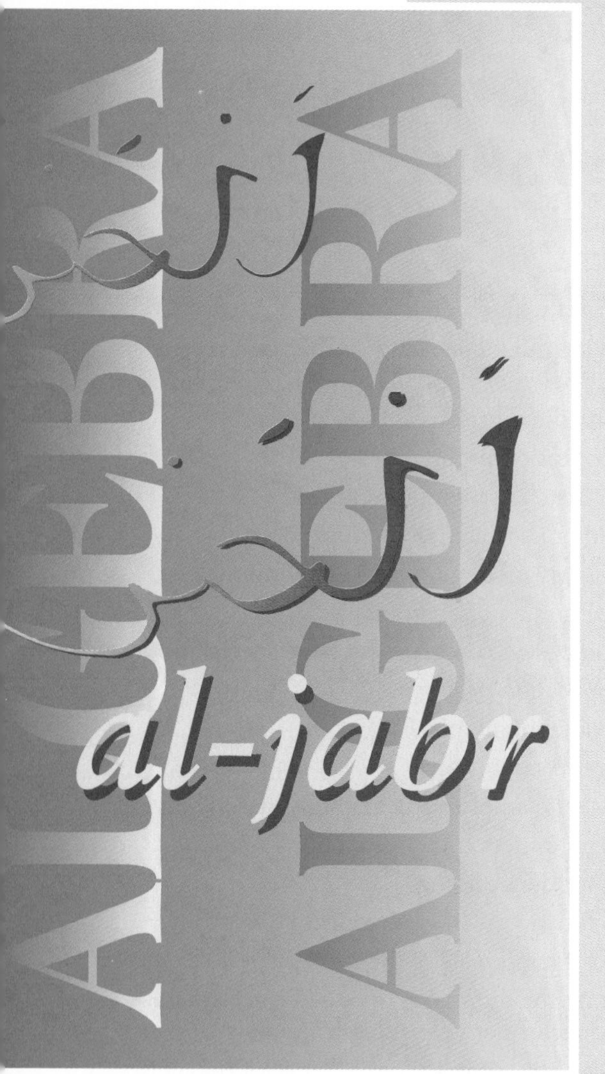

Our language of mathematics also owes another term to al-Kwarizmi. One of the translations of his work into Latin shortened his name to *Algoritmi.* A further modification of this word gives us our present word *algorithm.* An algorithm is a procedure or set of instructions that is used to solve different types of problems. Computer scientists use algorithms when writing computer programs.

A further historical note is not about the word *algebra* but about Omar Khayyam, a Persian who probably read al-Kwarizmi's work. Omar Khayyam is especially noted as a poet and the author of the *Rubiat.* However, he was also an excellent mathematician and astronomer and made many contributions to mathematics.

Exponential Expressions

1   Multiply monomials

A **monomial** is a number, a variable, or a product of a number and variables.

The examples at the right are monomials. The **degree of a monomial** is the sum of the exponents of the variables.

$$x \qquad \text{degree 1 } (x = x^1)$$
$$3x^2 \qquad \text{degree 2}$$
$$4x^2y \qquad \text{degree 3}$$
$$6x^3y^4z^2 \qquad \text{degree 9}$$
$$x^n \qquad \text{degree } n$$

The degree of a nonzero constant term is zero. $6 \qquad \text{degree 0}$

The expression $5\sqrt{x}$ is not a monomial because $\sqrt{x}$ cannot be written as a product of variables. The expression $\dfrac{x}{y}$ is not a monomial because it is a quotient of variables.

The expression x^4 is an exponential expression. The exponent, 4, indicates the number of times the base, x, occurs as a factor.

The product of exponential expressions with the *same* base can be simplified by writing each expression in factored form and writing the result with an exponent.

$$x^3 \cdot x^4 = \overbrace{(x \cdot x \cdot x)}^{3 \text{ factors}} \cdot \overbrace{(x \cdot x \cdot x \cdot x)}^{4 \text{ factors}}$$
$$\underbrace{\qquad\qquad\qquad\qquad}_{7 \text{ factors}}$$
$$= x^7$$

Note that adding the exponents results in the same product.

$$x^3 \cdot x^4 = x^{3+4} = x^7$$

Rule for Multiplying Exponential Expressions

If m and n are positive integers, then $x^m \cdot x^n = x^{m+n}$.

Example 1   Simplify: $(5a^2b^4)(2ab^5)$

Solution $(5a^2b^4)(2ab^5)$
$= (5 \cdot 2)(a^2 \cdot a)(b^4 \cdot b^5)$ ▶ Use the Commutative and Associative Properties to rearrange and group factors.

$= 10a^{2+1}b^{4+5}$ ▶ Multiply variables with like bases by
$= 10a^3b^9$ adding the exponents.

Problem 1   Simplify: $(7xy^3)(-5x^2y^2)(-xy^2)$

Solution   See page S12. $35x^4y^7$

As shown below, the power of a monomial can be simplified by writing the power in factored form and then using the Rule for Multiplying Exponential Expressions. It can also be simplified by multiplying each exponent inside the parentheses by the exponent outside the parentheses.

Write in factored form.	$(a^2)^3 = a^2 \cdot a^2 \cdot a^2$	$(x^3y^4)^2 = (x^3y^4)(x^3y^4)$
Use the Rule for Multiplying Exponential Expressions.	$= a^{2+2+2}$ $= a^6$	$= x^{3+3}y^{4+4}$ $= x^6y^8$
Multiply each exponent inside the parentheses by the exponent outside the parentheses.	$(a^2)^3 = a^{2\cdot 3} = a^6$	$(x^3y^4)^2 = x^{3\cdot 2}y^{4\cdot 2} = x^6y^8$

Rule for Simplifying a Power of an Exponential Expression

If m and n are positive integers, then $(x^m)^n = x^{mn}$.

Rule for Simplifying Powers of Products

If m, n, and p are positive integers, then $(x^my^n)^p = x^{mp}y^{np}$.

Example 2 Simplify.

A. $(x^4)^5$ B. $(x^2)^n$ C. $(2a^3b^4)^3$ D. $(2ab)(3a)^2 + 5a(2a^2b)$

Solution

A. $(x^4)^5 = x^{4\cdot 5} = x^{20}$ ▶ Multiply the exponents.

B. $(x^2)^n = x^{2n}$ ▶ Multiply the exponents.

C. $(2a^3b^4)^3 = 2^{1\cdot 3}a^{3\cdot 3}b^{4\cdot 3}$ ▶ Use the Rule for Simplifying Powers of Products.

$= 2^3a^9b^{12}$

$= 8a^9b^{12}$

D. $(2ab)(3a)^2 + 5a(2a^2b) = (2ab)(3^2a^2) + 10a^3b$

$= (2ab)(9a^2) + 10a^3b$

$= 18a^3b + 10a^3b = 28a^3b$

Problem 2 Simplify.

A. $(y^3)^6$ B. $(x^n)^3$ C. $(-2ab^3)^4$ D. $6a(2a)^2 + 3a(2a^2)$

Solution See page S12. A. y^{18} B. x^{3n} C. $16a^4b^{12}$ D. $30a^3$

2 Divide monomials and simplify expressions with negative exponents

18 HM³

The quotient of two exponential expressions with the same base can be simplified by writing each expression in factored form, dividing by common factors, and writing the result with an exponent.

$$\frac{x^5}{x^2} = \frac{\overset{1}{\cancel{x}} \cdot \overset{1}{\cancel{x}} \cdot x \cdot x \cdot x}{\underset{1}{\cancel{x}} \cdot \underset{1}{\cancel{x}}} = x^3$$

Subtracting the exponents gives the same result.

$$\frac{x^5}{x^2} = x^{5-2} = x^3$$

INSTRUCTOR NOTE

Here we are just verbalizing the rule for division of monomials. The theorem comes after we define negative exponents.

Have students copy this rule onto a piece of paper and then practice a few exercises such as $\frac{a^9}{a^2}$ and $\frac{y^8}{y}$.

It may also help to give $\frac{a^9}{b^5}$ to emphasize that the bases must be the same.

To divide two monomials with the same base, subtract the exponents of the like bases.

➡ Simplify: $\frac{z^8}{z^2}$

The bases are the same. Subtract the exponents. $\frac{z^8}{z^2} = z^{8-2} = z^6$ ⬅

➡ Simplify: $\frac{a^5b^9}{a^4b}$

Subtract the exponents of the like bases. $\frac{a^5b^9}{a^4b} = a^{5-4}b^{9-1} = ab^8$ ⬅

Consider the expression $\frac{x^4}{x^4}$, $x \neq 0$. This expression can be simplified by subtracting exponents or by dividing by common factors.

$$\frac{x^4}{x^4} = x^{4-4} = x^0 \qquad \frac{x^4}{x^4} = \frac{\overset{1}{\cancel{x}} \cdot \overset{1}{\cancel{x}} \cdot \overset{1}{\cancel{x}} \cdot \overset{1}{\cancel{x}}}{\underset{1}{\cancel{x}} \cdot \underset{1}{\cancel{x}} \cdot \underset{1}{\cancel{x}} \cdot \underset{1}{\cancel{x}}} = 1$$

The equations $\frac{x^4}{x^4} = x^0$ and $\frac{x^4}{x^4} = 1$ suggest the following definition of x^0.

Definition of Zero as an Exponent

If $x \neq 0$, then $x^0 = 1$. The expression 0^0 is not defined.

➡ Simplify: $(16z^5)^0$, $z \neq 0$

Any nonzero expression to the zero power is 1. $(16z^5)^0 = 1$ ⬅

➡ Simplify: $-(7x^4y^3)^0$

The negative outside the parentheses is not affected by the exponent. $-(7x^4y^3)^0 = -(1) = -1$ ⬅

Consider the expression $\frac{x^4}{x^6}$, $x \neq 0$. This expression can be simplified by subtracting exponents or by dividing by common factors.

$$\frac{x^4}{x^6} = x^{4-6} = x^{-2} \qquad \frac{x^4}{x^6} = \frac{\overset{1}{\cancel{x}} \cdot \overset{1}{\cancel{x}} \cdot \overset{1}{\cancel{x}} \cdot \overset{1}{\cancel{x}}}{\underset{1}{\cancel{x}} \cdot \underset{1}{\cancel{x}} \cdot \underset{1}{\cancel{x}} \cdot \underset{1}{\cancel{x}} \cdot x \cdot x} = \frac{1}{x^2}$$

The equations $\frac{x^4}{x^6} = x^{-2}$ and $\frac{x^4}{x^6} = \frac{1}{x^2}$ suggest that $x^{-2} = \frac{1}{x^2}$.

Definition of a Negative Exponent

If $x \neq 0$ and n is a positive integer, then

$$x^{-n} = \frac{1}{x^n} \quad \text{and} \quad \frac{1}{x^{-n}} = x^n$$

POINT OF INTEREST

In the 15th century, the expression $12^{\overline{2m}}$ was used to mean $12x^{-2}$. The use of $\overline{m}$ reflects an Italian influence, where m was used for minus and p was used for plus. It was understood that $2\overline{m}$ referred to an unnamed variable. Isaac Newton, in the 17th century, advocated the current use of a negative exponent.

➡ Evaluate: 2^{-4}

Use the Definition of a Negative Exponent to write the expression with a positive exponent. Then simplify.

$$2^{-4} = \frac{1}{2^4} = \frac{1}{16}$$

Consider the expression $\left(\dfrac{x^3}{y^4}\right)^2$, $y \neq 0$. This expression can be simplified by squaring $\dfrac{x^3}{y^4}$ or by multiplying each exponent in the quotient by the exponent outside the parentheses.

$$\left(\frac{x^3}{y^4}\right)^2 = \left(\frac{x^3}{y^4}\right)\left(\frac{x^3}{y^4}\right) = \frac{x^3 \cdot x^3}{y^4 \cdot y^4} = \frac{x^{3+3}}{y^{4+4}} = \frac{x^6}{y^8} \qquad \left(\frac{x^3}{y^4}\right)^2 = \frac{x^{3\cdot 2}}{y^{4\cdot 2}} = \frac{x^6}{y^8}$$

Rule for Simplifying Powers of Quotients

If m, n, and p are integers and $y \neq 0$, then $\left(\dfrac{x^m}{y^n}\right)^p = \dfrac{x^{mp}}{y^{np}}$.

➡ Simplify: $\left(\dfrac{a^2}{b^3}\right)^{-2}$

Use the Rule for Simplifying Powers of Quotients.

$$\left(\frac{a^2}{b^3}\right)^{-2} = \frac{a^{2(-2)}}{b^{3(-2)}}$$
$$= \frac{a^{-4}}{b^{-6}}$$

Use the Definition of a Negative Exponent to write the expression with positive exponents.

$$= \frac{b^6}{a^4}$$

An exponential expression is in simplest form when it contains only positive exponents.

➡ Simplify: $\dfrac{2}{5a^{-4}}$

Use the Definition of a Negative Exponent to rewrite the expression with a positive exponent.

$$\left(\frac{2}{5a^{-4}}\right) = \frac{2}{5} \cdot \frac{1}{a^{-4}} = \frac{2}{5} \cdot a^4 = \frac{2a^4}{5}$$

Now that zero and negative exponents have been defined, a rule for dividing exponential expressions can be stated.

Rule for Dividing Exponential Expressions

If m and n are integers and $x \neq 0$, then $\dfrac{x^m}{x^n} = x^{m-n}$.

Simplify: $\frac{x^4}{x^9}$

Use the Rule for Dividing Exponential Expressions. Subtract the exponents, and then use the Definition of a Negative Exponent to rewrite the expression with a positive exponent.

$$\frac{x^4}{x^9} = x^{4-9} = x^{-5} = \frac{1}{x^5}$$

The rules for simplifying exponential expressions and powers of exponential expressions are true for all integers. These rules are restated here for convenience.

Rules of Exponents

If m, n, and p are integers, then

$x^m \cdot x^n = x^{m+n}$ $\quad$ $(x^m)^n = x^{mn}$ $\quad$ $(x^m y^n)^p = x^{mp}y^{np}$

$\frac{x^m}{x^n} = x^{m-n}, x \neq 0$ $\quad$ $\left(\frac{x^m}{y^n}\right)^p = \frac{x^{mp}}{y^{np}}, y \neq 0$ $\quad$ $x^{-n} = \frac{1}{x^n}, x \neq 0$

$x^0 = 1, x \neq 0$

Example 3 Simplify. A. $(3x^2y^{-3})(6x^{-4}y^5)$ B. $\frac{x^2y^{-4}}{x^{-5}y^{-2}}$ C. $\left(\frac{3a^2b^{-2}c^{-1}}{27a^{-1}b^2c^{-4}}\right)^{-2}$

Solution A. $(3x^2y^{-3})(6x^{-4}y^5)$

$= 18x^{2+(-4)}y^{-3+5}$ ▸ Use the Rule for Multiplying Exponential Expressions.

$= 18x^{-2}y^2$

$= \frac{18y^2}{x^2}$ ▸ Use the Definition of a Negative Exponent to rewrite the expression without negative exponents.

B. $\frac{x^2y^{-4}}{x^{-5}y^{-2}} = x^{2-(-5)}y^{-4-(-2)}$ ▸ Use the Rule for Dividing Exponential Expressions.

$= x^7y^{-2}$

$= \frac{x^7}{y^2}$ ▸ Use the Definition of a Negative Exponent to rewrite the expression without negative exponents.

C. $\left(\frac{3a^2b^{-2}c^{-1}}{27a^{-1}b^2c^{-4}}\right)^{-2} = \left(\frac{a^3b^{-4}c^3}{9}\right)^{-2}$ ▸ Simplify inside the parentheses by using the Rule for Dividing Exponential Expressions.

$= \frac{a^{-6}b^8c^{-6}}{9^{-2}}$ ▸ Multiply each exponent inside the parentheses by the exponent outside the parentheses.

$= \frac{9^2b^8}{a^6c^6}$ ▸ Use the Definition of a Negative Exponent to rewrite the expression without negative exponents.

$= \frac{81b^8}{a^6c^6}$ ▸ Simplify.

Problem 3 Simplify. A. $(2x^{-5}y)(5x^4y^{-3})$ B. $\frac{a^{-1}b^4}{a^{-2}b^{-2}}$ C. $\left(\frac{2^{-1}x^2y^{-3}}{4x^{-2}y^{-5}}\right)^{-2}$

Solution See page S12. A. $\frac{10}{xy^2}$ B. ab^6 C. $\frac{64}{x^8y^4}$

3 Scientific notation

POINT OF INTEREST

Astronomers measure the distance of some stars by using the *parsec.* One parsec is approximately 1.92×10^{13} mi.

Very large and very small numbers are encountered in the fields of science and engineering. For example, the mass of the electron is 0.000000000000000000000000000911 g. Numbers such as this one are difficult to read and write, so a more convenient system for writing them has been developed. It is called **scientific notation.**

To express a number in scientific notation, write the number as the product of a number between 1 and 10 and a power of 10. The form for scientific notation is $a \times 10^n$, where $1 \le a < 10$.

For numbers greater than 10, move the decimal point to the right of the first digit. The exponent n is positive and equal to the number of places the decimal point has been moved.

$965{,}000 = 9.65 \times 10^5$

$3{,}600{,}000 = 3.6 \times 10^6$

$92{,}000{,}000{,}000 = 9.2 \times 10^{10}$

For numbers less than 1, move the decimal point to the right of the first nonzero digit. The exponent n is negative. The absolute value of the exponent is equal to the number of places the decimal point has been moved.

$0.0002 = 2 \times 10^{-4}$

$0.0000000974 = 9.74 \times 10^{-8}$

$0.000000000086 = 8.6 \times 10^{-11}$

Example 4 Write 0.000041 in scientific notation.

Solution $0.000041 = 4.1 \times 10^{-5}$ ▶ The decimal point must be moved 5 places to the right. The exponent is negative.

Problem 4 Write 942,000,000 in scientific notation.

Solution See page S12. 9.42×10^8

Converting a number written in scientific notation to decimal notation requires moving the decimal point.

When the exponent is positive, move the decimal point to the right the same number of places as the exponent.

$1.32 \times 10^4 = 13{,}200$

$1.4 \times 10^8 = 140{,}000{,}000$

When the exponent is negative, move the decimal point to the left the same number of places as the absolute value of the exponent.

$1.32 \times 10^{-2} = 0.0132$

$1.4 \times 10^{-4} = 0.00014$

Example 5 Write 3.3×10^7 in decimal notation.

Solution $3.3 \times 10^7 = 33{,}000{,}000$ ▶ Move the decimal point 7 places to the right.

Problem 5 Write 2.7×10^{-5} in decimal notation.

Solution See page S12. 0.000027

Numerical calculations involving numbers that have more digits than a hand-held calculator is able to handle can be performed using scientific notation.

Example 6 Simplify: $\dfrac{2{,}400{,}000{,}000 \times 0.0000063}{0.00009 \times 480}$

Solution

$$\frac{2{,}400{,}000{,}000 \times 0.0000063}{0.00009 \times 480}$$

$$= \frac{2.4 \times 10^{9} \times 6.3 \times 10^{-6}}{9 \times 10^{-5} \times 4.8 \times 10^{2}}$$ ► Write the numbers in scientific notation.

$$= \frac{(2.4)(6.3) \times 10^{9+(-6)-(-5)-2}}{(9)(4.8)}$$ ► Simplify.

$$= 0.35 \times 10^{6}$$

$$= 3.5 \times 10^{5}$$ ► Write in scientific notation.

Problem 6 Simplify: $\dfrac{5{,}600{,}000 \times 0.000000081}{900 \times 0.000000028}$

Solution See page S12. 1.8×10^{4}

4 Application problems

Example 7 How many miles does light travel in 1 day? The speed of light is 186,000 mi/s. Write the answer in scientific notation.

Strategy To find the distance traveled:

- Write the speed of light in scientific notation.
- Write the number of seconds in 1 day in scientific notation.
- Use the equation $d = rt$, where r is the speed of light and t is the number of seconds in 1 day.

Solution $186{,}000 = 1.86 \times 10^{5}$

$24 \cdot 60 \cdot 60 = 86{,}400 = 8.64 \times 10^{4}$

$$d = rt$$
$$d = (1.86 \times 10^{5})(8.64 \times 10^{4})$$
$$d = 1.86 \times 8.64 \times 10^{9}$$
$$d = 16.0704 \times 10^{9}$$
$$d = 1.60704 \times 10^{10}$$

Light travels 1.60704×10^{10} mi in one day.

Problem 7 A computer can do an arithmetic operation in 1×10^{-7} s. How many arithmetic operations can the computer perform in 1 min? Write the answer in scientific notation.

Solution See page S12. 6×10^{8} operations

CONCEPT REVIEW 5.1

Determine whether the following statements are always true, sometimes true, or never true.

1. $2^{-4} = 16$ Never true

2. If x is negative, x^{-1} is positive. Never true

3. $(2 + 3)^{-1} = \frac{1}{2} + \frac{1}{3}$. Never true

4. When multiplying two numbers with the same base, the rule is to multiply the exponents. Never true

5. When adding two numbers with the same base, the rule is to add the exponents. Never true

EXERCISES 5.1

1

1. Explain how to multiply two exponential expressions with the same base.

2. Suppose a friend is having difficulty simplifying $(2a^3b^2)(3a^2b)$. Explain to your friend how to simplify the expression.

Simplify.

3. $(ab^3)(a^3b)$ a^4b^4
4. $(-2ab^4)(-3a^2b^4)$ $6a^3b^8$
5. $(9xy^2)(-2x^2y^2)$ $-18x^3y^4$
6. $(x^2y)^2$ x^4y^2
7. $(x^2y^4)^4$ x^8y^{16}
8. $(-2ab^2)^3$ $-8a^3b^6$
9. $(-3x^2y^3)^4$ $81x^8y^{12}$
10. $(2^2a^2b^3)^3$ $64a^6b^9$
11. $(3^3a^5b^3)^2$ $729a^{10}b^6$
12. $(xy)(x^2y)^4$ x^9y^5
13. $(x^2y^2)(xy^3)^3$ x^5y^{11}
14. $[(2x)^4]^2$ $256x^8$
15. $[(3x)^3]^2$ $729x^6$
16. $[(x^2y)^4]^5$ $x^{40}y^{20}$
17. $[(ab)^3]^6$ $a^{18}b^{18}$
18. $[(2ab)^3]^2$ $64a^6b^6$
19. $[(2xy)^3]^4$ $4096x^{12}y^{12}$
20. $[(3x^2y^3)^2]^2$ $81x^8y^{12}$
21. $[(2a^4b^3)^3]^2$ $64a^{24}b^{18}$
22. $y^n \cdot y^{2n}$ y^{3n}
23. $x^n \cdot x^{n+1}$ x^{2n+1}
24. $y^{2n} \cdot y^{4n+1}$ y^{6n+1}
25. $y^{3n} \cdot y^{3n-2}$ y^{6n-2}
26. $(a^n)^{2n}$ a^{2n^2}
27. $(a^{n-3})^{2n}$ a^{2n^2-6n}
28. $(y^{2n-1})^3$ y^{6n-3}
29. $(x^{3n+2})^5$ x^{15n+10}
30. $(b^{2n-1})^n$ b^{2n^2-n}
31. $(2xy)(-3x^2yz)(x^2y^3z^3)$ $-6x^5y^5z^4$
32. $(x^2z^4)(2xyz^4)(-3x^3y^2)$ $-6x^6y^3z^8$
33. $(3b^5)(2ab^2)(-2ab^2c^2)$ $-12a^2b^9c^2$
34. $(-c^3)(-2a^2bc)(3a^2b)$ $6a^4b^2c^4$
35. $(-2x^2y^3z)(3x^2yz^4)$ $-6x^4y^4z^5$
36. $(2a^2b)^3(-3ab^4)^2$ $72a^8b^{11}$
37. $(-3ab^3)^3(-2^2a^2b)^2$ $-432a^7b^{11}$
38. $(4ab)^2(-2ab^2c^3)^3$ $-128a^5b^8c^9$
39. $(-2ab^2)(-3a^4b^5)^3$ $54a^{13}b^{17}$

2 **40.** Explain how to divide two exponential expressions with the same base.

41. If a variable has a negative exponent, how can you rewrite it with a positive exponent?

Simplify.

42. $\frac{x^3}{x^{12}}$ $\frac{1}{x^9}$

43. $\frac{a^8}{a^5}$ a^3

44. $\frac{x^3y^6}{x^3y^3}$ y^3

45. $\frac{a^7b}{a^2b^4}$ $\frac{a^5}{b^3}$

46. 2^{-3} $\frac{1}{8}$

47. $\frac{1}{3^{-5}}$ 243

48. $\frac{1}{x^{-4}}$ x^4

49. $\frac{1}{y^{-3}}$ y^3

50. $\frac{2x^{-2}}{y^4}$ $\frac{2}{x^2y^4}$

51. $\frac{a^3}{4b^{-2}}$ $\frac{a^3b^2}{4}$

52. $x^{-4}x^4$ 1

53. $x^{-3}x^{-5}$ $\frac{1}{x^8}$

54. $(3x^{-2})^2$ $\frac{9}{x^4}$

55. $(5x^2)^{-3}$ $\frac{1}{125x^6}$

56. $\frac{x^{-3}}{x^2}$ $\frac{1}{x^5}$

57. $\frac{x^4}{x^{-5}}$ x^9

58. $a^{-2} \cdot a^4$ a^2

59. $a^{-5} \cdot a^7$ a^2

60. $(x^2y^{-4})^2$ $\frac{x^4}{y^8}$

61. $(x^3y^5)^{-2}$ $\frac{1}{x^6y^{10}}$

62. $(2a^{-1})^{-2}(2a^{-1})^4$ $\frac{4}{a^2}$

63. $(3a)^{-3}(9a^{-1})^{-2}$ $\frac{1}{2187a}$

64. $(x^{-2}y)^2(xy)^{-2}$ $\frac{1}{x^6}$

65. $(x^{-1}y^2)^{-3}(x^2y^{-4})^{-3}$ $\frac{y^6}{x^3}$

66. $\frac{6^2a^{-2}b^3}{3ab^4}$ $\frac{12}{a^3b}$

67. $\left(\frac{x^2y^{-1}}{xy}\right)^{-4}$ $\frac{y^8}{x^4}$

68. $\frac{-48ab^{10}}{32a^4b^3}$ $-\frac{3b^7}{2a^3}$

69. $\frac{a^2b^3c^7}{a^6bc^5}$ $\frac{b^2c^2}{a^4}$

70. $\frac{(-4x^2y^3)^2}{(2xy^2)^3}$ $2x$

71. $\frac{(-3a^2b^3)^2}{(-2ab^4)^3}$ $-\frac{9a}{8b^6}$

72. $\left(\frac{x^{-3}y^{-4}}{x^{-2}y}\right)^{-2}$ x^2y^{10}

73. $\left(\frac{a^{-2}b}{a^3b^{-4}}\right)^2$ $\frac{b^{10}}{a^{10}}$

74. $\frac{-x^{5n}}{x^{2n}}$ $-x^{3n}$

75. $\frac{y^{2n}}{-y^{8n}}$ $-\frac{1}{y^{6n}}$

76. $\frac{a^{3n-2}b^{n+1}}{a^{2n+1}b^{2n+2}}$ $\frac{a^{n-3}}{b^{n+1}}$

77. $\frac{x^{2n-1}y^{n-3}}{x^{n+4}y^{n+3}}$ $\frac{x^{n-5}}{y^6}$

78. $\frac{(2a^{-3}b^{-2})^3}{(a^{-4}b^{-1})^{-2}}$ $\frac{8}{a^{17}b^8}$

79. $\frac{(3x^{-2}y)^{-2}}{(4xy^{-2})^{-1}}$ $\frac{4x^5}{9y^4}$

80. $\left(\frac{4^{-2}xy^{-3}}{x^{-3}y}\right)^3\left(\frac{8^{-1}x^{-2}y}{x^4y^{-1}}\right)^{-2}$ $\frac{x^{24}}{64y^{16}}$

81. $\left(\frac{9ab^{-2}}{8a^{-2}b}\right)^{-2}\left(\frac{3a^{-2}b}{2a^2b^{-2}}\right)^3$ $\frac{8b^{15}}{3a^{18}}$

82. $[(xy^{-2})^3]^{-2}$ $\frac{y^{12}}{x^6}$

83. $[(x^{-2}y^{-1})^2]^{-3}$ $x^{12}y^6$

84. $\left[\left(\frac{x}{y^2}\right)^{-2}\right]^3$ $\frac{y^{12}}{x^6}$

85. $\left[\left(\frac{a^2}{b}\right)^{-1}\right]^2$ $\frac{b^2}{a^4}$

3 Write in scientific notation.

86. 0.00000467 4.67×10^{-6}

87. 0.00000005 5×10^{-8}

88. 0.00000000017 1.7×10^{-10}

89. 4,300,000 4.3×10^6

90. 200,000,000,000 2×10^{11}

91. 9,800,000,000 9.8×10^9

Write in decimal notation.

92. 1.23×10^{-7} 0.000000123

93. 6.2×10^{-12} 0.0000000000062

94. 8.2×10^{15} 8,200,000,000,000,000

95. 6.34×10^5 634,000

96. 3.9×10^{-2} 0.039

97. 4.35×10^9 4,350,000,000

Simplify. Write the answer in scientific notation.

98. $(3 \times 10^{-12})(5 \times 10^{16})$ 1.5×10^5

99. $(8.9 \times 10^{-5})(3.2 \times 10^{-6})$ 2.848×10^{-10}

100. (0.0000065)(3,200,000,000,000) 2.08×10^7

101. (480,000)(0.0000000096) 4.608×10^{-3}

102. $\dfrac{9 \times 10^{-3}}{6 \times 10^{5}}$ 1.5×10^{-8}

103. $\dfrac{2.7 \times 10^{4}}{3 \times 10^{-6}}$ 9×10^{9}

104. $\dfrac{0.0089}{500{,}000{,}000}$ 1.78×10^{-11}

105. $\dfrac{4{,}800}{0.00000024}$ 2×10^{10}

106. $\dfrac{0.00056}{0.000000000004}$ 1.4×10^{8}

107. $\dfrac{0.000000346}{0.0000005}$ 6.92×10^{-1}

108. $\dfrac{(3.2 \times 10^{-11})(2.9 \times 10^{15})}{8.1 \times 10^{-3}}$ 1.14567901×10^{7}

109. $\dfrac{(6.9 \times 10^{27})(8.2 \times 10^{-13})}{4.1 \times 10^{15}}$ 1.38

110. $\dfrac{(0.00000004)(84{,}000)}{(0.0003)(1{,}400{,}000)}$ 8×10^{-6}

111. $\dfrac{(720)(0.0000000039)}{(26{,}000{,}000{,}000)(0.018)}$ 6×10^{-15}

4 Solve. Write the answer in scientific notation.

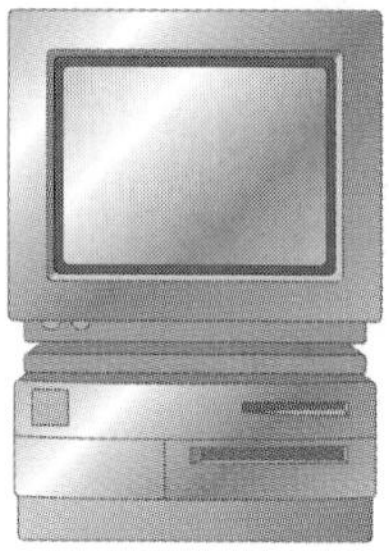

112. *Computer Science* A computer can do an arithmetic operation in 5×10^{-7} s. How many arithmetic operations can the computer perform in 1 h? 7.2×10^{9} operations

113. *Computer Science* A computer can do an arithmetic operation in 2×10^{-9} s. How many arithmetic operations can the computer perform in 1 min? 3×10^{10} operations

114. *Physics* How many meters does light travel in 8 h? The speed of light is 3×10^{8} m/s. 8.64×10^{12} m

115. *Physics* How many meters does light travel in 1 day? The speed of light is 3×10^{8} m/s. 2.592×10^{13} m

116. *Physics* A high-speed centrifuge makes 4×10^{8} revolutions each minute. Find the time in seconds for the centrifuge to make one revolution. 1.5×10^{-7} s

117. *Physics* The mass of an electron is 9.109×10^{-31} kg. The mass of a proton is 1.673×10^{-27} kg. How many times heavier is a proton than an electron? 1.83664508×10^{3} times

118. *Geology* The mass of Earth is 5.9×10^{24} kg. The mass of the sun is 2×10^{30} kg. How many times heavier is the sun than Earth? 3.3898305×10^{5} times

119. *Astronomy* One light-year, an astronomical unit of distance, is the distance that light will travel in 1 year. Light travels 1.86×10^{5} mi/s. Find the measure of 1 light-year in miles. Use a 365-day year. 5.865696×10^{12} mi

120. *Astronomy* The sun is 3.67×10^{9} mi from Pluto. How long does it take light to travel to Pluto from the sun? The speed of light is 1.86×10^{5} mi/s. 1.97311828×10^{4} s

121. *Measurement* The weight of 31 million orchid seeds is 1 oz. Find the weight of one orchid seed. 3.2258065×10^{-8} oz

122. *Astronomy* The light from the star Alpha Centauri takes 4.3 years to reach Earth. Light travels 1.86×10^{5} mi/s. How far is Alpha Centauri from Earth? Use a 365-day year. $2.52224928 \times 10^{13}$ mi

123. *Astronomy* The distance to Saturn is 8.86×10^{8} mi. A satellite leaves Earth traveling at a constant rate of 1×10^{5} mph. How long does it take for the satellite to reach Saturn? 8.86×10^{3} h

124. *Astronomy* The diameter of Neptune is 3×10^{4} mi. Use the formula $SA = 4\pi r^2$ to find the surface area of Neptune in square miles. 2.82743339×10^{9} mi^2

125. *Biology* The radius of a cell is 1.5×10^{-4} mm. Use the formula $V = \frac{4}{3}\pi r^3$ to find the volume of the cell. $1.41371669 \times 10^{-11}$ mm^3

126. *Chemistry* One gram of hydrogen contains 6.023×10^{23} atoms. Find the weight of one atom of hydrogen. $1.66030218 \times 10^{-24}$ g

127. *Astronomy* Our galaxy is estimated to be 6×10^{17} mi across. How long (in hours) would it take a space ship to cross the galaxy traveling at 25,000 mph? 2.4×10^{13} h

APPLYING CONCEPTS 5.1

State whether or not the expression is a polynomial.

128. $\frac{1}{3}x - 1$ Yes

129. $\frac{3}{x} - 1$ No

130. $5\sqrt{x} + 2$ No

131. $\sqrt{5}x + 2$ Yes

132. $\frac{1}{4y^2} + \frac{1}{3y}$ No

133. $x + \sqrt{3}$ Yes

For what value of k is the given equation an identity?

134. $(2x^3 + 3x^2 + kx + 5) - (x^3 + x^2 - 5x - 2) = x^3 + 2x^2 + 3x + 7$ -2

135. $(6x^3 + kx^2 - 2x - 1) - (4x^3 - 3x^2 + 1) = 2x^3 - x^2 - 2x - 2$ -4

Solve.

136. *Geometry* The width of a rectangle is x^n. The length of the rectangle is $3x^n$. Find the perimeter of the rectangle in terms of x^n. $8x^n$

137. *Geometry* The base of an isosceles triangle is $4x^n$. The length of each of the equal sides is $3x^n$. Find the perimeter of the triangle in terms of x^n. $10x^n$

138. *Geometry* The width of a rectangle is $4ab$. The length is $6ab$. Find the area of the rectangle in terms of a and b. $24a^2b^2$

139. *Geometry* The height of a triangle is $5xy$. The length of the base of the triangle is $8xy$. Find the area of the triangle in terms of x and y. $20x^2y^2$

Simplify.

140. $\frac{4m^4}{n^{-2}} + \left(\frac{n^{-1}}{m^2}\right)^{-2}$ $5m^4n^2$

141. $\frac{5x^3}{y^{-6}} + \left(\frac{x^{-1}}{y^2}\right)^{-3}$ $6x^3y^6$

142. $\left(\frac{3a^{-2}b}{a^{-4}b^{-1}}\right)^2 \div \left(\frac{a^{-1}b}{9a^2b^3}\right)^{-1}$ ab^2

143. $\left(\frac{2m^3n^{-2}}{4m^4n}\right)^{-2} \div \left(\frac{mn^5}{m^{-1}n^3}\right)^3$ $\frac{4}{m^4}$

Solve.

[C] **144.** Use the expressions $(2 + 3)^{-2}$ and $2^{-2} + 3^{-2}$ to show that $(x + y)^{-2} \neq x^{-2} + y^{-2}$. $\frac{1}{25} \neq \frac{13}{36}$

[C] **145.** If a and b are real nonzero numbers and $a < b$, is $a^{-1} < b^{-1}$ always a false statement? No; let $a = -1$ and $b = 1$.

[C] **146.** Correct the error in the following expressions. Explain which rule or property was used incorrectly.

a. $2x + 3x = 5x^2$	$= 5x$	The Distributive Property
b. $a - (b - c) = a - b - c$	$= a - b + c$	The Distributive Property
c. $x^0 = 0$	$= 1$	Definition of Zero as an Exponent
d. $(x^4)^5 = x^9$	$= x^{20}$	Rule for Simplifying the Power of an Exponential Expression
e. $x^2 \cdot x^3 = x^6$	$= x^5$	Rule for Multiplying Exponential Expressions
f. $b^{m+n} = b^m + b^n$	$= b^m \cdot b^n$	Rule for Multiplying Exponential Expressions

SECTION 5.2

Introduction to Polynomials

1 Evaluate polynomial functions

A **polynomial** is a variable expression in which the terms are monomials.

A polynomial of one term is a **monomial.** $5x$

A polynomial of two terms is a **binomial.** $5x^2y + 6x$

A polynomial of three terms is a **trinomial.** $3x^2 + 9xy - 5y$

Polynomials with more than three terms do not have special names.

The **degree of a polynomial** is the greatest of the degrees of any of its terms.

$3x + 2$ degree 1
$3x^2 + 2x - 4$ degree 2
$4x^3y^2 + 6x^4$ degree 5
$3x^{2n} - 5x^n - 2$ degree $2n$

The terms of a polynomial in one variable are usually arranged so that the exponents of the variable decrease from left to right. This is called **descending order.**

$2x^2 - x + 8$

$3y^3 - 3y^2 + y - 12$

For a polynomial in more than one variable, descending order may refer to any one of the variables.

The polynomial at the right is shown first in descending order of the x variable and then in descending order of the y variable.

$2x^2 + 3xy + 5y^2$

$5y^2 + 3xy + 2x^2$

The linear function given by $f(x) = mx + b$ is an example of a polynomial function. It is a polynomial function of degree one. A second-degree polynomial function, called a **quadratic function,** is given by the equation $f(x) = ax^2 + bx + c$, $a \neq 0$. A third-degree polynomial function is called a **cubic function.** In general, a polynomial function is an expression whose terms are monomials.

To evaluate a polynomial function, replace the variable by its value and simplify.

Given $P(x) = x^3 - 3x^2 + 4$, evaluate $P(-3)$.

Substitute -3 for x and simplify.

$$P(x) = x^3 - 3x^2 + 4$$
$$P(-3) = (-3)^3 - 3(-3)^2 + 4$$
$$P(-3) = -27 - 27 + 4$$
$$P(-3) = -50$$

Example 1 To overcome the resistance of the tires on the road and the wind, the horsepower (hp), P, required by a cyclist to keep a certain bicycle moving at v mph is given by $P(v) = 0.00003v^3 + 0.00211v$. How much horsepower must the cyclist supply to keep this bicycle moving at 20 mph?

Strategy To find the horsepower, evaluate the function when $v = 20$.

Solution

$$P(v) = 0.00003v^3 + 0.00211v$$
$$P(20) = 0.00003(20)^3 + 0.00211(20) \quad \text{▸ Replace } v \text{ by 20.}$$
$$= 0.2822 \quad \text{▸ Simplify.}$$

The cyclist must supply 0.2822 hp.

Problem 1 The velocity of air that is expelled during a cough can be modeled by $v(t) = 600r^2 - 1000r^3$, where v is the velocity of the air in centimeters per second and r is the radius of the trachea in centimeters. What is the velocity of expelled air during a cough when the radius of the trachea is 0.4 cm?

Solution See page S13. 32 cm/s

The **leading coefficient** of a polynomial function is the coefficient of the variable with the largest exponent. The constant term is the term without a variable.

Find the leading coefficient, the constant term, and the degree of the polynomial function $P(x) = 7x^4 - 3x^2 + 2x - 4$.

The leading coefficient is 7, the constant term is -4, and the degree is 4.

Example 2 Find the leading coefficient, the constant term, and the degree of the polynomial $P(x) = 5x^6 - 4x^5 - 3x^2 + 7$.

Solution The leading coefficient is 5, the constant term is 7, and the degree is 6.

Problem 2 Find the leading coefficient, the constant term, and the degree of the polynomial $R(x) = -3x^4 + 3x^3 + 3x^2 - 2x - 12$.

Solution See page S13. The leading coefficient is -3, the constant term is -12, and the degree is 4.

The following three functions do not represent polynomial functions.

$f(x) = 3x^2 + 2x^{-1}$ A polynomial function does not have a variable raised to a negative power.

$g(x) = 2\sqrt{x} - 3$ — A polynomial function does not have a variable expression within a radical.

$h(x) = \dfrac{x}{x - 1}$ — A polynomial function does not have a variable in the denominator of a fraction.

Example 3 State whether the function is a polynomial function.

A. $P(x) = 3x^{1/2} + 2x^2 - 3$ B. $T(x) = 3\sqrt{x} - 2x^2 - 3x + 2$

C. $R(x) = 14x^3 - \pi x^2 + 3x + 2$

Solution A. This is not a polynomial function. A polynomial function does not have a variable raised to a fractional power.

B. This is not a polynomial function. A polynomial function does not have a variable expression within a radical.

C. This is a polynomial function.

Problem 3 State whether the function is a polynomial function.

A. $R(x) = 5x^{14} - 5$ B. $V(x) = -x^{-1} + 2x - 7$

C. $P(x) = 2x - 2\sqrt{x} - 3$

Solution See page S13. A. Yes B. No C. No

The graph of a linear function is a straight line and can be found by plotting just two points. The graph of a polynomial function of degree greater than one is a curve. Consequently, many points may have to be found before an accurate graph can be drawn.

Evaluating the quadratic function given by the equation $f(x) = x^2 - x - 6$ when $x = -3, -2, -1, 0, 1, 2, 3$, and 4 gives the points shown in Figure 1. For instance, $f(-3) = 6$, so $(-3, 6)$ is graphed; $f(2) = -4$, so $(2, -4)$ is graphed; and $f(4) = 6$, so $(4, 6)$ is graphed. Evaluating the function when x is not an integer, such as $x = -\frac{3}{2}$ and $x = \frac{5}{2}$, produces more points to graph, as shown in Figure 2. Connecting the points with a smooth curve results in Figure 3, which is the graph of f.

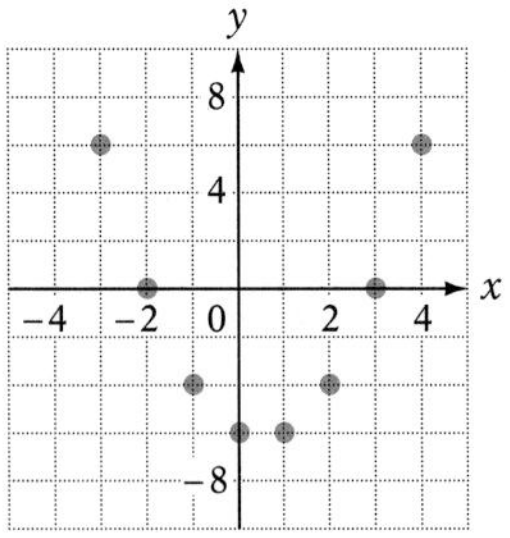

Figure 1

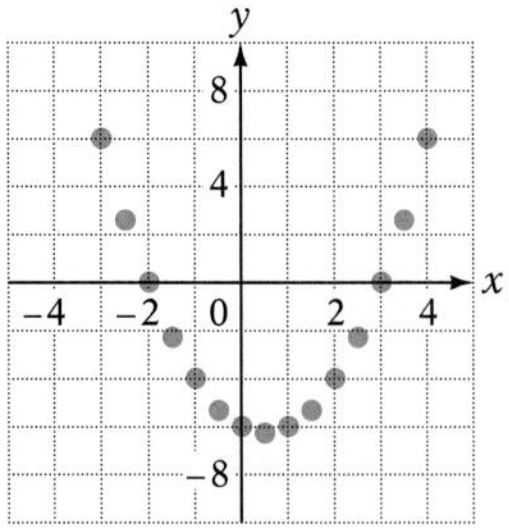

Figure 2

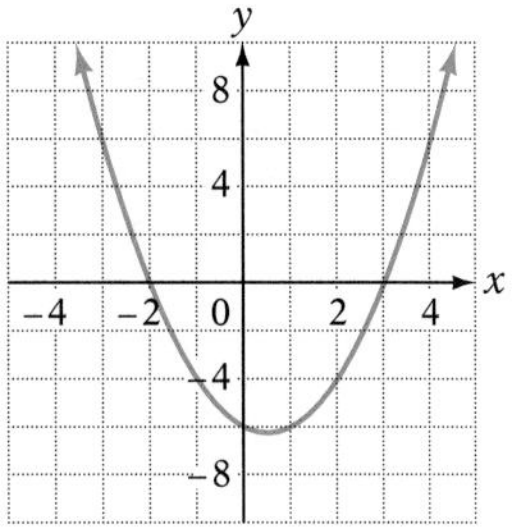

Figure 3

Here is an example of graphing a cubic function, $P(x) = x^3 - 2x^2 - 5x + 6$. Evaluating the function when $x = -2, -1, 0, 1, 2, 3$, and 4 gives the graph in Figure 4. Evaluating at some noninteger values gives the graph in Figure 5. Finally, connecting the dots with a smooth curve gives the graph in Figure 6.

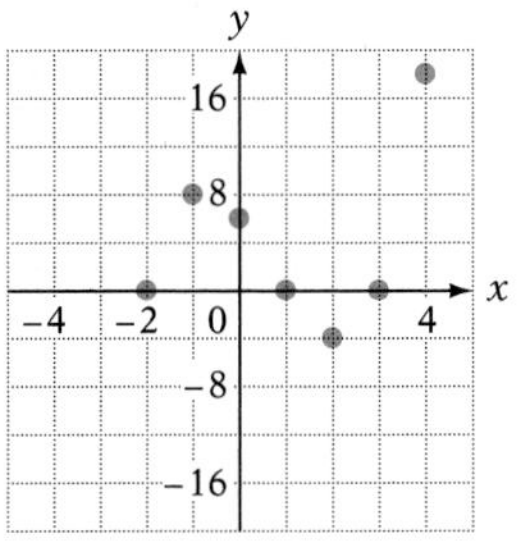

Figure 4

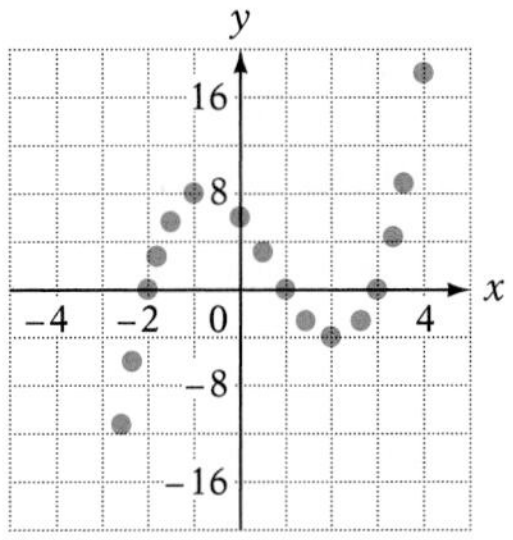

Figure 5

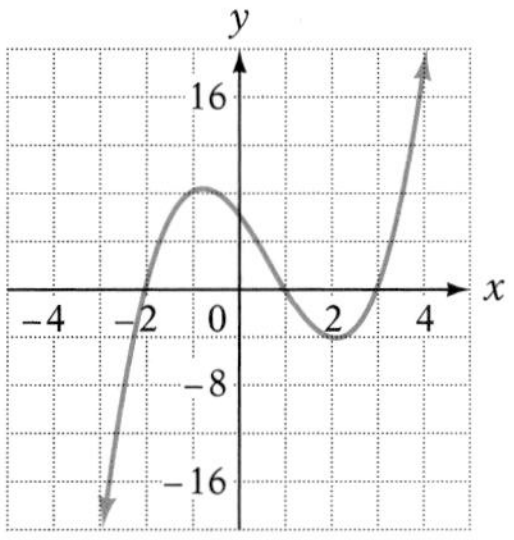

Figure 6

Example 4 Graph: $f(x) = x^2 - 2$

Solution

x	$y = f(x)$
−3	7
−2	2
−1	−1
0	−2
1	−1
2	2
3	7

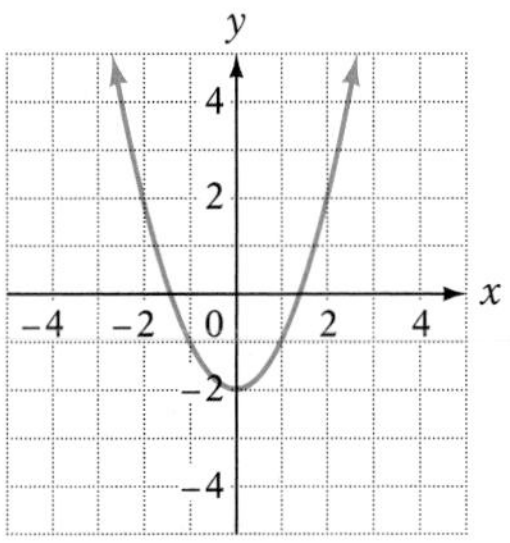

Problem 4 Graph: $f(x) = x^2 - 2x$

Solution See page S13.

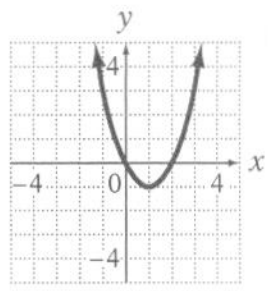

Example 5 Graph: $f(x) = x^3 - 1$

Solution

x	$y = f(x)$
−2	−9
−1	−2
0	−1
1	0
2	7

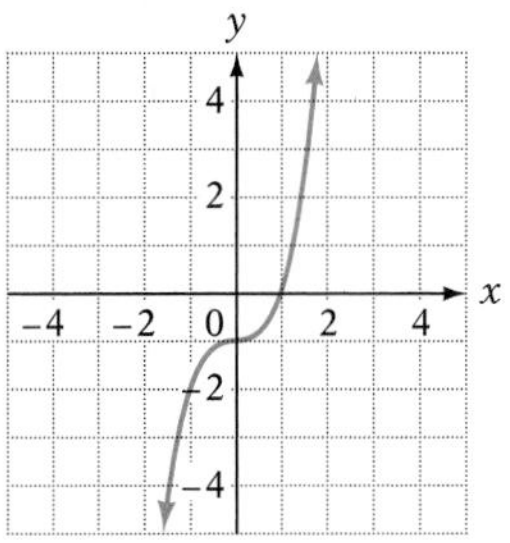

Problem 5 Graph: $F(x) = -x^3 + 1$

Solution See page S13.

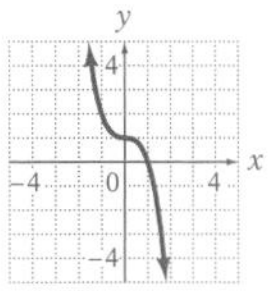

It may be necessary to plot a large number of points before drawing the graph in Example 5. Graphing utilities also create graphs by plotting a large number of points and then connecting the points to form a curve. Using a graphing utility, enter the equation $y = x^3 - 1$ and verify the graph drawn in Example 5. Now trace along the graph and verify that $(-2, -9)$, $(-1, -2)$, $(0, -1)$, $(1, 0)$, and $(2, 7)$ are the coordinates of points on the graph. Follow the same procedure for Problem 5.

Example 6 Graph $f(x) = x^2 - 2x - 3$. Find two values of x, to the nearest tenth, for which $f(x) = 2$.

Solution To find the values of x for which $f(x) = 2$, use a graphing utility to graph f.

Recall that $f(x)$ is the y-coordinate of a point on the graph. Thus the values of x for which $f(x) = 2$ are those for which y is 2. The x-coordinates where the line $y = 2$ crosses the graph of f are the desired values.

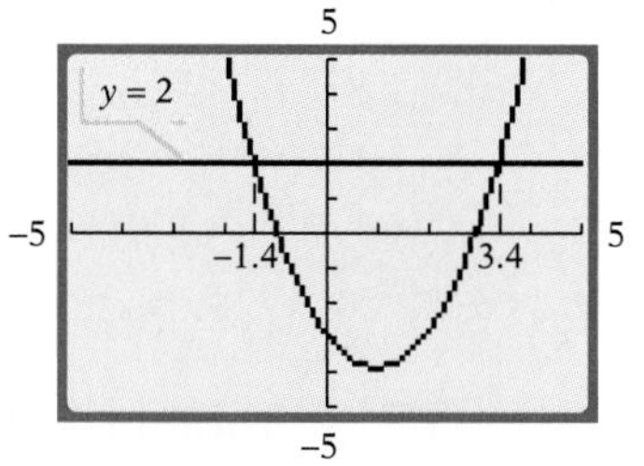

Using the TRACE feature of the graphing utility, move the cursor along the graph of f until $y = 2$. The values of x to the nearest tenth are -1.4 and 3.4.

Problem 6 Graph $S(x) = x^3 - x - 2$. Find a value of x, to the nearest tenth, for which $S(x) = 2$.

Solution See page S13. 1.8

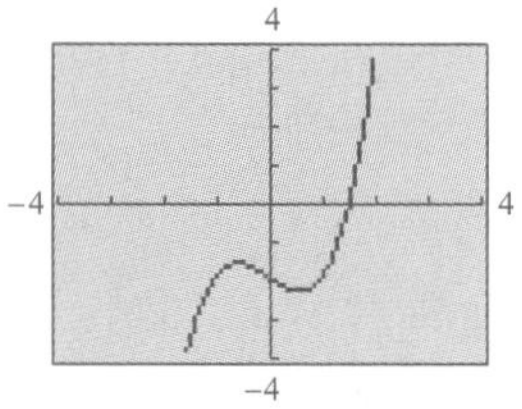

2 Add or subtract polynomials

18 HM³

Polynomials can be added by combining like terms. Either a vertical or a horizontal format can be used.

Add $(3x^2 + 2x - 7) + (7x^3 - 3 + 4x^2)$. Use a horizontal format.

Use the Commutative and Associative Properties of Addition to rearrange and group like terms.

$(3x^2 + 2x - 7) + (7x^3 - 3 + 4x^2)$
$= 7x^3 + (3x^2 + 4x^2) + 2x + (-7 - 3)$

Combine like terms.

$= 7x^3 + 7x^2 + 2x - 10$

INSTRUCTOR NOTE

It may help to show students that adding polynomials is related to adding whole numbers. Adding the coefficients of like terms is similar to arranging whole numbers to be added in columns with the unit digits aligned.

Add $(4x^2 + 5x - 3) + (7x^3 - 7x + 1) + (2x - 3x^2 + 4x^3 + 1)$. Use a vertical format.

Arrange the terms of each polynomial in descending order with like terms in the same column.

$$\begin{array}{rrrr} & 4x^2 & + 5x & - 3 \\ 7x^3 & & - 7x & + 1 \\ 4x^3 & - 3x^2 & + 2x & + 1 \\ \hline \end{array}$$

Add the terms in each column.

$$11x^3 + x^2 \quad - 1$$

The additive inverse of the polynomial $x^2 + 5x - 4$ is $-(x^2 + 5x - 4)$.

To simplify the additive inverse of a polynomial, change the sign of every term inside the parentheses.

$-(x^2 + 5x - 4) = -x^2 - 5x + 4$

LOOK CLOSELY

This is the same definition used for subtraction of integers: Subtraction is addition of the opposite.

To subtract two polynomials, add the additive inverse of the second polynomial to the first.

Subtract $(3x^2 - 7xy + y^2) - (-4x^2 + 7xy - 3y^2)$. Use a horizontal format.

Rewrite the subtraction as the addition of the additive inverse.

$(3x^2 - 7xy + y^2) - (-4x^2 + 7xy - 3y^2)$
$= (3x^2 - 7xy + y^2) + (4x^2 - 7xy + 3y^2)$

Combine like terms.

$= 7x^2 - 14xy + 4y^2$

Subtract $(6x^3 - 3x + 7) - (3x^2 - 5x + 12)$. Use a vertical format.

Rewrite the subtraction as the addition of the additive inverse.

$(6x^3 - 3x + 7) - (3x^2 - 5x + 12)$
$= (6x^3 - 3x + 7) + (-3x^2 + 5x - 12)$

Arrange the terms of each polynomial in descending order with like terms in the same column.

$$\begin{array}{rrrr} 6x^3 & & - 3x & + 7 \\ & - 3x^2 & + 5x & - 12 \\ \hline \end{array}$$

Add the terms in each column.

$$6x^3 - 3x^2 + 2x - 5$$

Example 7 Add: $(3x^2 - 2x + 5) + (2 - 3x + 4x^3) + (x - 3x^2 - x^3 - 4)$
Use a vertical format.

Solution

$$\begin{array}{rrrr} & 3x^2 & - 2x & + 5 \\ 4x^3 & & - 3x & + 2 \\ -x^3 & - 3x^2 & + x & - 4 \\ \hline 3x^3 & & - 4x & + 3 \end{array}$$

▶ Arrange the terms of each polynomial in descending order with like terms in the same column.

▶ Combine like terms in each column.

Problem 7 Add: $(5x^2 + 3x - 1) + (2x^2 + 4x - 6) + (8 - 7x + x^2)$
Use a vertical format.

Solution See page S13. $8x^2 + 1$

Example 8 Subtract: $(3x^2 - 2x + 4) - (7x^2 + 3x - 12)$
Use a vertical format.

Solution $(3x^2 - 2x + 4) + (-7x^2 - 3x + 12)$ ▶ Rewrite the subtraction as the addition of the additive inverse.

$$\begin{array}{r} 3x^2 - 2x + 4 \\ \underline{-7x^2 - 3x + 12} \end{array}$$ ▶ Arrange the polynomials in descending order in a vertical format.

$$-4x^2 - 5x + 16$$ ▶ Combine like terms in each column.

Problem 8 Subtract: $(-5x^2 + 2x - 3) - (6x^2 + 3x - 7)$
Use a vertical format.

Solution See page S13. $-11x^2 - x + 4$

Functional notation may be used when adding or subtracting polynomials. For instance, if $P(x) = 3x^2 - 2x + 4$ and $R(x) = -5x^3 + 4x + 7$, then

$$\begin{aligned} P(x) + R(x) &= (3x^2 - 2x + 4) + (-5x^3 + 4x + 7) \\ &= -5x^3 + 3x^2 + 2x + 11 \end{aligned}$$

$$\begin{aligned} P(x) - R(x) &= (3x^2 - 2x + 4) - (-5x^3 + 4x + 7) \\ &= (3x^2 - 2x + 4) + (5x^3 - 4x - 7) \\ &= 5x^3 + 3x^2 - 6x - 3 \end{aligned}$$

Example 9 Given $P(x) = -3x^2 + 2x - 6$ and $R(x) = 4x^3 - 3x + 4$, find $S(x) = P(x) + R(x)$.

Solution
$$\begin{aligned} S(x) &= (-3x^2 + 2x - 6) + (4x^3 - 3x + 4) \\ &= 4x^3 - 3x^2 - x - 2 \end{aligned}$$

Problem 9 Given $P(x) = 4x^3 - 3x^2 + 2$ and $R(x) = -2x^2 + 2x - 3$, find $D(x) = P(x) - R(x)$.

Solution See page S13. $4x^3 - x^2 - 2x + 5$

CONCEPT REVIEW 5.2

Determine whether the following statements are always true, sometimes true, or never true.

1. The degree of a polynomial is the greatest of the degrees of its individual terms. Always true

2. The leading coefficient of $4 - 2x - 3x^2$ is 4. Never true

3. Terms with the same variable parts can be added. Always true

4. The additive inverse of a polynomial is a polynomial with every sign changed to the opposite sign. Always true

5. The sum of two trinomials is a trinomial. Sometimes true

6. $f(x) = 3x^2 - 2x + x^{-1}$ is a polynomial function. Never true

EXERCISES 5.2

1 1. Explain how to determine the degree of a polynomial in one variable. Give two examples of a third-degree polynomial.

2. Explain how to evaluate a polynomial.

3. Given $P(x) = 3x^2 - 2x - 8$, evaluate $P(3)$. 13

4. Given $P(x) = -3x^2 - 5x + 8$, evaluate $P(-5)$. -42

5. Given $R(x) = 2x^3 - 3x^2 + 4x - 2$, evaluate $R(2)$. 10

6. Given $R(x) = -x^3 + 2x^2 - 3x + 4$, evaluate $R(-1)$. 10

7. Given $f(x) = x^4 - 2x^2 - 10$, evaluate $f(-1)$. -11

8. Given $f(x) = x^5 - 2x^3 + 4x$, evaluate $f(2)$. 24

9. *Oceanography* The length of a water wave close to land depends on several factors, one of which is the depth of the water. However, the length, L (in meters), of deep-water waves (those far away from land) can be approximated by $L(s) = 0.641s^2$, where s is the speed of the wave in meters per second. Find the length of a deep-water wave that has a speed of 6 m/s. Round to the nearest tenth of a meter. 23.1 m

10. *Physics* The distance, s (in feet), that an object on the moon will fall in t seconds is given by $s(t) = 2.735t^2$. How far will an object fall on the moon in 3 s? 24.615 ft

11. *Sports* The total number of softball games, T, that must be scheduled in a league that has n teams such that each team plays every other team twice is given by $T(n) = n^2 - n$. What is the total number of games that must be scheduled for a league that has 8 teams? 56

12. *Geometry* A diagonal of a polygon is a line from a vertex to a nonadjacent vertex. Seven of the possible diagonals for a decagon (10-sided figure) are shown at the right. The total number, T, of diagonals for an n-sided polygon is given by $T(n) = \frac{1}{2}n^2 - \frac{3}{2}n$. Find the total number of diagonals for a decagon. 35

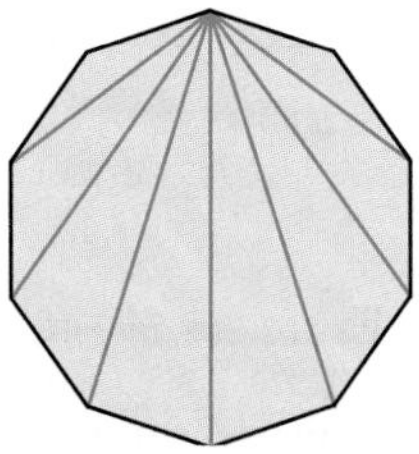

13. *Food Science* Baked Alaska is a dessert that is made by putting a 1-inch meringue coating around a hemisphere of ice cream. The amount of meringue that is needed depends on the radius of the hemisphere of ice cream and is given by $M(r) = 6.14r^2 + 6.14r + 2.094$, where $M(r)$ is the volume of meringue needed (in cubic inches) and r is the radius of the ice cream in inches. Find the amount of meringue that is needed for a baked Alaska that has a mound of ice cream with a 6-inch radius. Round to the nearest whole number. 260 in^3

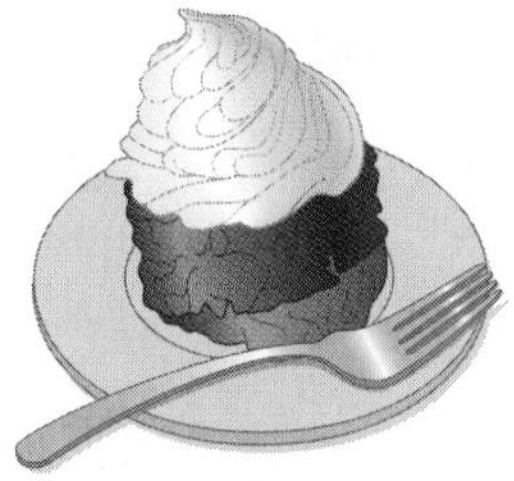

14. Suppose $f(x) = x^2 + 1$ and $g(x) = x^3 + x$. Is $f(x) > g(x)$ or is $f(x) < g(x)$ when $0 < x < 1$? Explain your answer. $f(x) > g(x)$

State whether the function is a polynomial function. For those that are polynomial functions, identify **a.** the leading coefficient, **b.** the constant term, and **c.** the degree.

15. $P(x) = -x^2 + 3x + 8$ $-1, 8, 2$

16. $P(x) = 3x^4 - 3x - 7$ $3, -7, 4$

17. $R(x) = \dfrac{x}{x+1}$ Not a polynomial

18. $R(x) = \dfrac{3x^2 - 2x + 1}{x}$ Not a polynomial

19. $f(x) = \sqrt{x} - x^2 + 2$ Not a polynomial

20. $f(x) = x^2 - \sqrt{x+2} - 8$ Not a polynomial

21. $g(x) = 3x^5 - 2x^2 + \pi$ $3, \pi, 5$

22. $g(x) = -4x^5 - 3x^2 + x - \sqrt{7}$ $-4, -\sqrt{7}, 5$

23. $P(x) = 3x^2 - 5x^3 + 2$ $-5, 2, 3$

24. $P(x) = x^2 - 5x^4 - x^6$ $-1, 0, 6$

25. $R(x) = 14$ $14, 14, 0$

26. $R(x) = \dfrac{1}{x} + 2$ Not a polynomial

Graph.

27. $P(x) = x^2 - 1$

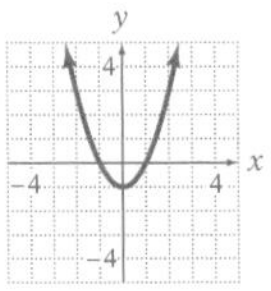

28. $P(x) = 2x^2 + 3$

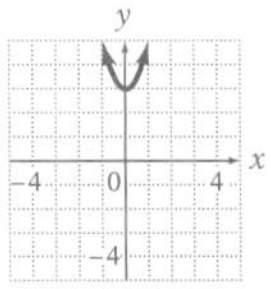

29. $R(x) = x^3 + 2$

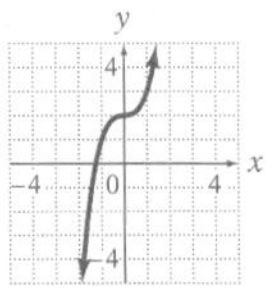

30. $R(x) = x^4 + 1$

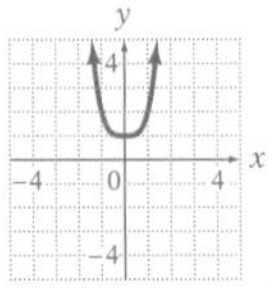

31. $f(x) = x^3 - 2x$

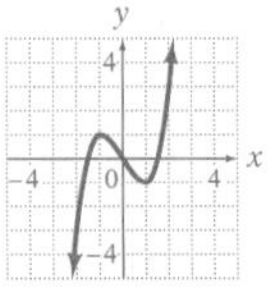

32. $f(x) = x^2 - x - 2$

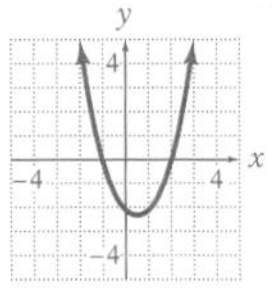

2 **33.** What is the additive inverse of a polynomial?

34. Explain **a.** how to add and **b.** how to subtract two polynomials.

Simplify. Use a vertical format.

35. $(5x^2 + 2x - 7) + (x^2 - 8x + 12)$ $6x^2 - 6x + 5$

36. $(3x^2 - 2x + 7) + (-3x^2 + 2x - 12)$ -5

37. $(x^2 - 3x + 8) - (2x^2 - 3x + 7)$ $-x^2 + 1$

38. $(2x^2 + 3x - 7) - (5x^2 - 8x - 1)$ $-3x^2 + 11x - 6$

For Exercises 39–42, simplify using a horizontal format.

39. $(3y^2 - 7y) + (2y^2 - 8y + 2)$ $5y^2 - 15y + 2$

40. $(-2y^2 - 4y - 12) + (5y^2 - 5y)$ $3y^2 - 9y - 12$

41. $(2a^2 - 3a - 7) - (-5a^2 - 2a - 9)$ $7a^2 - a + 2$

42. $(3a^2 - 9a) - (-5a^2 + 7a - 6)$ $8a^2 - 16a + 6$

43. Given $P(x) = 3x^3 - 4x^2 - x + 1$ and $R(x) = 2x^3 + 5x - 8$, find $P(x) + R(x)$. $5x^3 - 4x^2 + 4x - 7$

44. Given $P(x) = 5x^3 - 3x - 7$ and $R(x) = 2x^3 - 3x^2 + 8$, find $P(x) - R(x)$. $3x^3 + 3x^2 - 3x - 15$

45. Given $P(x) = x^{2n} + 7x^n - 3$ and $R(x) = -x^{2n} + 2x^n + 8$, find $P(x) + R(x)$. $9x^n + 5$

46. Given $P(x) = 2x^{2n} - x^n - 1$ and $R(x) = 5x^{2n} + 7x^n + 1$, find $P(x) - R(x)$. $-3x^{2n} - 8x^n - 2$

47. Given $P(x) = 3x^4 - 3x^3 - x^2$ and $R(x) = 3x^3 - 7x^2 + 2x$, find $S(x) = P(x) + R(x)$. $3x^4 - 8x^2 + 2x$

48. Given $P(x) = 3x^4 - 2x + 1$ and $R(x) = 3x^5 - 5x - 8$, find $S(x) = P(x) + R(x)$. $3x^5 + 3x^4 - 7x - 7$

49. Given $P(x) = x^2 + 2x + 1$ and $R(x) = 2x^3 - 3x^2 + 2x - 7$, find $D(x) = P(x) - R(x)$. $-2x^3 + 4x^2 + 8$

50. Given $P(x) = 2x^4 - 2x^2 + 1$ and $R(x) = 3x^3 - 2x^2 + 3x + 8$, find $D(x) = P(x) - R(x)$. $2x^4 - 3x^3 - 3x - 7$

APPLYING CONCEPTS 5.2

The value of the x-coordinates of the x-intercepts of a function are called the zeros of the function. Find, to the nearest tenth, the zeros of the following functions.

51. $f(x) = 2x - 3$ 1.5

52. $f(x) = 1.4x - 2.6$ 1.9

53. $f(x) = x^2 - 4x + 2$ 3.4, 0.6

54. $f(x) = 2x^2 + 3x - 1$ 0.3, −1.8

55. $f(x) = x^3 - 3x + 1$ −1.9, 0.35, 1.5

56. $f(x) = x^3 + x^2 - 1$ 0.8

57. Given $f(x) = x^2 - x + 3$, find, to the nearest tenth, the values of x for which $f(x) = 4$. 1.6, −0.6

58. Given $f(x) = x^3 - 2x^2 + x - 1$, find, to the nearest tenth, the value of x for which $f(x) = -2$. −0.5

Two polynomials are equal if the coefficients of like powers are equal. For example,

$$6x^3 - 7x^2 + 2x - 8 = 2x - 7x^2 - 8 + 6x^3$$

In Exercises 59 and 60, use the definition of equality of polynomials to find the value of k that makes the equation an identity.

59. $(2x^3 + 3x^2 + kx + 5) - (x^3 + x^2 - 5x - 2) = x^3 + 2x^2 + 3x + 7$ −2

60. $(6x^3 + kx^2 - 2x - 1) - (4x^3 - 3x^2 + 1) = 2x^3 - x^2 - 2x - 2$ −4

61. If $P(x)$ is a third-degree polynomial and $Q(x)$ is a fourth-degree polynomial, what can be said about the degree of $P(x) + Q(x)$? Give some examples of polynomials that support your answer.
$P(x) + Q(x)$ is a fourth-degree polynomial.

62. If $P(x)$ is a fifth-degree polynomial and $Q(x)$ is a fourth-degree polynomial, what can be said about the degree of $P(x) - Q(x)$? Give some examples of polynomials that support your answer.
$P(x) - Q(x)$ is a fifth-degree polynomial.

63. *Sports* The deflection, D (in inches), of a beam that is uniformly loaded is given by the polynomial function $D(x) = 0.005x^4 - 0.1x^3 + 0.5x^2$, where x is the distance in feet from one end of the beam. See the figure at the right. The maximum deflection occurs when x is the midpoint of the beam. Determine the maximum deflection for the beam in the diagram. 3.125 in.

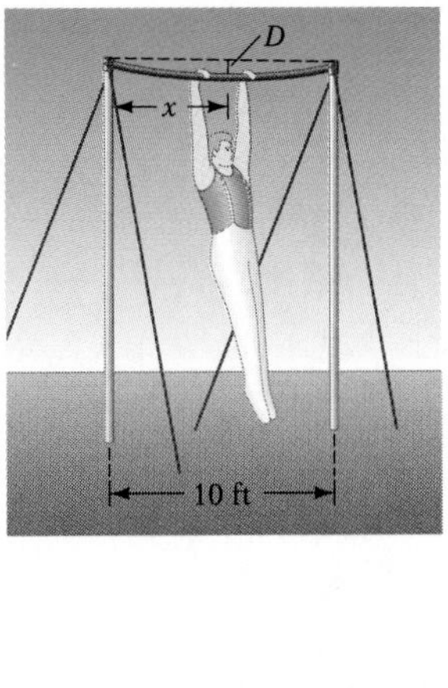

64. If $P(2) = 3$ and $P(x) = 2x^3 - 4x^2 - 2x + c$, find the value of c. 7

65. If $P(-1) = -3$ and $P(x) = 4x^4 - 3x^2 + 6x + c$, find the value of c. 2

66. Graph $f(x) = x^2$, $g(x) = (x - 3)^2$, and $h(x) = (x + 4)^2$ on the same coordinate grid. From the graphs, make a conjecture about the shape and location of $k(x) = (x - 2)^2$. Test your conjecture by graphing k.
The graph of k is the graph of f moved 2 units to the right.

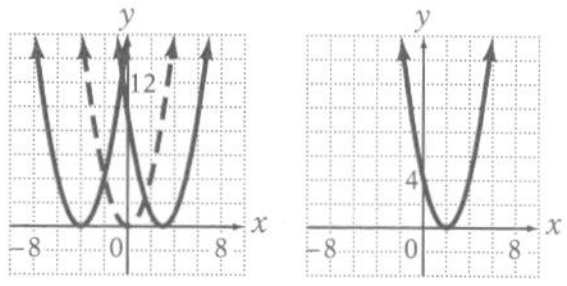

67. Graph $f(x) = x^2$, $g(x) = x^2 - 3$, and $h(x) = x^2 + 4$ on the same coordinate grid. From the graphs, make a conjecture about the shape and location of $k(x) = x^2 - 2$. Test your conjecture by graphing k.
The graph of k is the graph of f moved 2 units down.

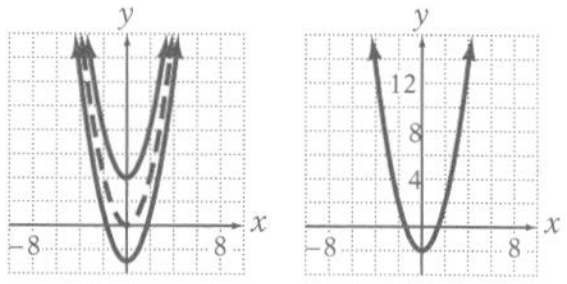

68. *Engineering* Construction of the Golden Gate Bridge, which is a suspension bridge, was completed in 1937. The length of the main span of the bridge is 4200 ft. The height, in feet, of the suspension cables above the roadway varies from 0 ft at the center of the bridge to 525 ft at the towers that support the cables.

Distance from the center of the bridge	Height of the cables above the roadway
0 ft	0 ft
1050 ft	150 ft
2100 ft	525 ft

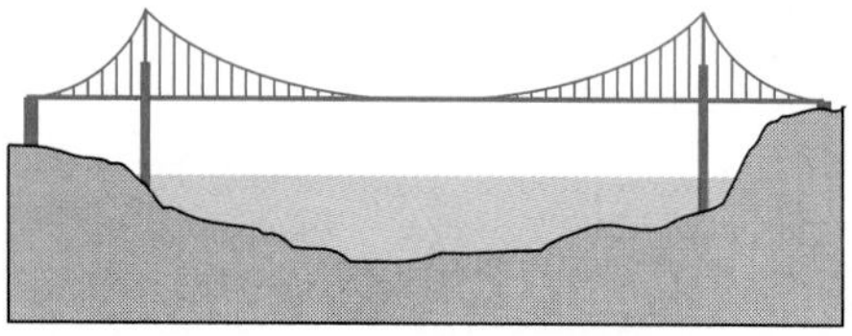

The function that approximately models the data is $f(x) = \frac{1}{8820}x^2 + 25$, where x is the distance from the center of the bridge and $f(x)$ is the height of the cables above the roadway. Use this model to approximate the height of the cables at a distance of **a.** 1000 ft from the center of the bridge and **b.** 1500 ft from the center of the bridge. Round to the nearest tenth.
a. 138.4 ft **b.** 280.1 ft

69. Write a report on *cubic splines.*

70. Write a report on *polygonal numbers.*

71. Explain the similarities and differences among the graphs of $f(x) = x^2$, $g(x) = (x - 3)^2$, and $h(x) = x^2 - 3$.

SECTION 5.3 Multiplication of Polynomials

1 Multiply a polynomial by a monomial

To multiply a polynomial by a monomial, use the Distributive Property and the Rule for Multiplying Exponential Expressions.

Example 1 Multiply.
A. $-5x(x^2 - 2x + 3)$ B. $x^2 - x[3 - x(x - 2) + 3]$
C. $x^n(x^n - x^2 + 1)$

Solution   A. $-5x(x^2 - 2x + 3)$
$$= -5x(x^2) - (-5x)(2x) + (-5x)(3)$$ ▶ Use the Distributive Property.

$$= -5x^3 + 10x^2 - 15x$$ ▶ Use the Rule for Multiplying Exponential Expressions.

B. $x^2 - x[3 - x(x - 2) + 3]$
$$= x^2 - x[3 - x^2 + 2x + 3]$$ ▶ Use the Distributive Property to remove the inner grouping symbols.

$$= x^2 - x[6 - x^2 + 2x]$$ ▶ Combine like terms.
$$= x^2 - 6x + x^3 - 2x^2$$ ▶ Use the Distributive Property to remove the brackets.

$$= x^3 - x^2 - 6x$$ ▶ Combine like terms, and write the polynomial in descending order.

C. $x^n(x^n - x^2 + 1)$
$$= x^{2n} - x^{n+2} + x^n$$ ▶ Use the Distributive Property and the Rule for Multiplying Exponential Expressions.

Problem 1   Multiply.
A. $-4y(y^2 - 3y + 2)$ B. $x^2 - 2x[x - x(4x - 5) + x^2]$
C. $y^{n+3}(y^{n-2} - 3y^2 + 2)$

Solution   See page S13.   A. $-4y^3 + 12y^2 - 8y$ B. $6x^3 - 11x^2$
C. $y^{2n+1} - 3y^{n+5} + 2y^{n+3}$

---

2   Multiply two polynomials

The product of two polynomials is the polynomial obtained by multiplying each term of one polynomial by each term of the other polynomial and then combining like terms.

▶ Multiply: $(2x^2 - 2x + 1)(3x + 2)$

Use the Distributive Property to multiply the trinomial by each term of the binomial.

$(2x^2 - 2x + 1)(3x + 2)$
$$= (2x^2 - 2x + 1)(3x) + (2x^2 - 2x + 1)(2)$$

Use the Distributive Property.

$$= (6x^3 - 6x^2 + 3x) + (4x^2 - 4x + 2)$$

Combine like terms.

$$= 6x^3 - 2x^2 - x + 2$$

A more convenient method of multiplying two polynomials is to use a vertical format similar to that used for multiplication of whole numbers.

$$\begin{array}{r} 2x^2 - 2x + 1 \\ 3x + 2 \\ \hline \end{array}$$

Like terms are written in the same column.

$$\begin{array}{r} 4x^2 - 4x + 2 = 2(2x^2 - 2x + 1) \\ 6x^3 - 6x^2 + 3x = 3x(2x^2 - 2x + 1) \\ \hline \end{array}$$

Combine like terms.

$$6x^3 - 2x^2 - x + 2$$

Example 2 Multiply: $(4a^3 - 3a + 7)(a - 5)$

Solution

$$\begin{array}{rrrrr} & & 4a^3 & -3a & +7 \\ & & & a & -5 \\ \hline & -20a^3 & & +15a & -35 \\ 4a^4 & & -3a^2 & +7a & \\ \hline 4a^4 & -20a^3 & -3a^2 & +22a & -35 \end{array}$$

Problem 2 Multiply: $(-2b^2 + 5b - 4)(-3b + 2)$

Solution See page S13. $6b^3 - 19b^2 + 22b - 8$

LOOK CLOSELY

FOIL is not really a different way of multiplying. It is based on the Distributive Property.

$(3x - 2)(2x + 5)$
$= 3x(2x + 5) - 2(2x + 5)$
$= 6x^2 + 15x - 4x - 10$
$= 6x^2 + 11x - 10$

FOIL is an efficient way to remember how to do binomial multiplication.

It is frequently necessary to find the product of two binomials. The product can be found by using a method called **FOIL**, which is based on the Distributive Property. The letters of FOIL stand for **F**irst, **O**uter, **I**nner, and **L**ast.

Simplify: $(3x - 2)(2x + 5)$

Multiply the First terms.	$(3x - 2)(2x + 5)$	$3x \cdot 2x = 6x^2$
Multiply the Outer terms.	$(3x - 2)(2x + 5)$	$3x \cdot 5 = 15x$
Multiply the Inner terms.	$(3x - 2)(2x + 5)$	$-2 \cdot 2x = -4x$
Multiply the Last terms.	$(3x - 2)(2x + 5)$	$-2 \cdot 5 = -10$

F O I L

Add the products. $(3x - 2)(2x + 5) = 6x^2 + 15x - 4x - 10$

Combine like terms. $= 6x^2 + 11x - 10$

Example 3 Multiply. A. $(6x - 5)(3x - 4)$ B. $(a^n - 2b^n)(3a^n - b^n)$

Solution A. $(6x - 5)(3x - 4) = 6x(3x) + 6x(-4) + (-5)(3x) + (-5)(-4)$
$= 18x^2 - 24x - 15x + 20$
$= 18x^2 - 39x + 20$

B. $(a^n - 2b^n)(3a^n - b^n)$
$= a^n(3a^n) + a^n(-b^n) + (-2b^n)(3a^n) + (-2b^n)(-b^n)$
$= 3a^{2n} - a^nb^n - 6a^nb^n + 2b^{2n}$
$= 3a^{2n} - 7a^nb^n + 2b^{2n}$

Problem 3 Multiply. A. $(5a - 3b)(2a + 7b)$ B. $(2x^n + y^n)(x^n - 4y^n)$

Solution See page S13. A. $10a^2 + 29ab - 21b^2$ B. $2x^{2n} - 7x^ny^n - 4y^{2n}$

3 Multiply polynomials that have special products

18 HM³

Using FOIL, a pattern can be found for the product of the sum and difference of two terms and for the square of a binomial.

The Sum and Difference of Two Terms

$$(a + b)(a - b) = a^2 - ab + ab - b^2$$
$$= a^2 - b^2$$

Square of the first term (a^2)
Square of the second term (b^2)

The Square of a Binomial

$$(a + b)^2 = (a + b)(a + b) = a^2 + ab + ab + b^2$$
$$= a^2 + 2ab + b^2$$

Square of the first term (a^2)
Twice the product of the two terms ($2ab$)
Square of the second term (b^2)

Example 4 Simplify. A. $(4x + 3)(4x - 3)$ B. $(2x - 3y)^2$ C. $(x^n + 5)(x^n - 5)$ D. $(x^{2n} - 2)^2$

Solution A. $(4x + 3)(4x - 3) = (4x)^2 - 3^2 = 16x^2 - 9$

▶ This is the sum and difference of two terms.

B. $(2x - 3y)^2 = (2x)^2 + 2(2x)(-3y) + (-3y)^2 = 4x^2 - 12xy + 9y^2$

▶ This is the square of a binomial.

C. $(x^n + 5)(x^n - 5) = x^{2n} - 25$

D. $(x^{2n} - 2)^2 = x^{4n} - 4x^{2n} + 4$

Problem 4 Simplify. A. $(3x - 7)(3x + 7)$ B. $(3x - 4y)^2$ C. $(2x^n + 3)(2x^n - 3)$ D. $(2x^n - 8)^2$

Solution See page S13. A. $9x^2 - 49$ B. $9x^2 - 24xy + 16y^2$ C. $4x^{2n} - 9$ D. $4x^{2n} - 32x^n + 64$

4 Application problems

18 HM³

Example 5 The length of a rectangle is $(2x + 3)$ ft. The width is $(x - 5)$ ft. Find the area of the rectangle in terms of the variable x.

Strategy To find the area, replace the variables L and W in the equation $A = LW$ by the given values, and solve for A.

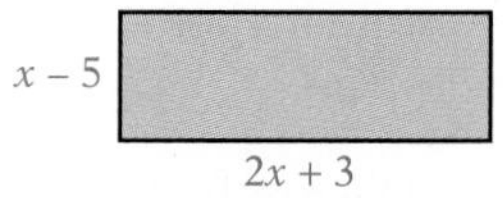

Solution
$$A = LW$$
$$A = (2x + 3)(x - 5)$$
$$A = 2x^2 - 10x + 3x - 15$$
$$A = 2x^2 - 7x - 15$$

The area is $(2x^2 - 7x - 15)$ ft^2.

Problem 5 The base of a triangle is $(2x + 6)$ ft. The height is $(x - 4)$ ft. Find the area of the triangle in terms of the variable x.

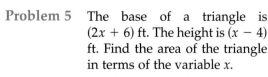

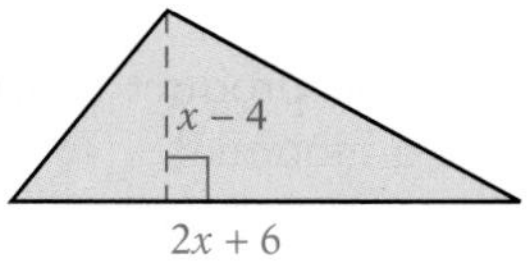

Solution See page S13. $(x^2 - x - 12)$ ft^2

Example 6 The corners are cut from a rectangular piece of cardboard measuring 8 in. by 12 in. The sides are folded up to make a box. Find the volume of the box in terms of the variable x, where x is the length of the side of the square cut from each corner of the rectangle.

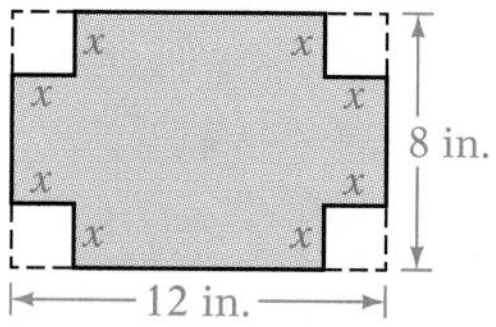

Strategy Length of the box: $12 - 2x$
Width of the box: $8 - 2x$
Height of the box: x
To find the volume, replace the variables, L, W, and H in the equation $V = LWH$, and solve for V.

Solution
$$V = LWH$$
$$V = (12 - 2x)(8 - 2x)x$$
$$V = (96 - 24x - 16x + 4x^2)x$$
$$V = (96 - 40x + 4x^2)x$$
$$V = 96x - 40x^2 + 4x^3$$
$$V = 4x^3 - 40x^2 + 96x$$

The volume is $(4x^3 - 40x^2 + 96x)$ in^3.

Problem 6 Find the volume of the rectangular solid shown in the diagram. All dimensions are given in feet.

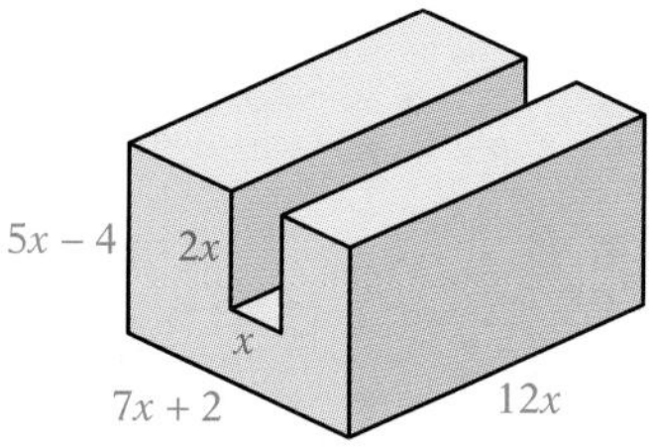

Solution See page S14.
$(396x^3 - 216x^2 - 96x)$ ft^3

CONCEPT REVIEW 5.3

Determine whether the following statements are always true, sometimes true, or never true.

1. The product of two polynomials is a polynomial. Always true

2. A cubic polynomial is a polynomial that contains three terms.
Sometimes true

3. The degree of the product of two polynomials is the sum of the degrees of the polynomials. Always true

4. The product of two binomials is a trinomial. Sometimes true

5. The product of the sum and difference of two different variable terms is a binomial. Always true

6. The Distributive Property is used in multiplying polynomials so that each term of one polynomial multiplies each term of the other polynomial.
Always true

EXERCISES 5.3

1

1. When is the Distributive Property used?

2. Explain how to multiply a monomial and a polynomial by using the Distributive Property.

Simplify.

3. $2x(x - 3)$ $2x^2 - 6x$

4. $2a(2a + 4)$ $4a^2 + 8a$

5. $3x^2(2x^2 - x)$ $6x^4 - 3x^3$

6. $-4y^2(4y - 6y^2)$ $-16y^3 + 24y^4$

7. $3xy(2x - 3y)$ $6x^2y - 9xy^2$

8. $-4ab(5a - 3b)$ $-20a^2b + 12ab^2$

9. $x^n(x + 1)$ $x^{n+1} + x^n$

10. $y^n(y^{2n} - 3)$ $y^{3n} - 3y^n$

11. $x^n(x^n + y^n)$ $x^{2n} + x^ny^n$

12. $x - 2x(x - 2)$ $-2x^2 + 5x$

13. $2b + 4b(2 - b)$ $-4b^2 + 10b$

14. $-2y(3 - y) + 2y^2$ $4y^2 - 6y$

15. $-2a^2(3a^2 - 2a + 3)$ $-6a^4 + 4a^3 - 6a^2$

16. $4b(3b^3 - 12b^2 - 6)$ $12b^4 - 48b^3 - 24b$

17. $3b(3b^4 - 3b^2 + 8)$ $9b^5 - 9b^3 + 24b$

18. $(2x^2 - 3x - 7)(-2x^2)$ $-4x^4 + 6x^3 + 14x^2$

19. $-5x^2(4 - 3x + 3x^2 + 4x^3)$ $-20x^5 - 15x^4 + 15x^3 - 20x^2$

20. $-2y^2(3 - 2y - 3y^2 + 2y^3)$ $-4y^5 + 6y^4 + 4y^3 - 6y^2$

21. $-2x^2y(x^2 - 3xy + 2y^2)$ $-2x^4y + 6x^3y^2 - 4x^2y^3$

22. $3ab^2(3a^2 - 2ab + 4b^2)$ $9a^3b^2 - 6a^2b^3 + 12ab^4$

23. $x^n(x^{2n} + x^n + x)$ $x^{3n} + x^{2n} + x^{n+1}$

24. $x^{2n}(x^{2n-2} + x^{2n} + x)$ $x^{4n-2} + x^{4n} + x^{2n+1}$

25. $a^{n+1}(a^n - 3a + 2)$ $a^{2n+1} - 3a^{n+2} + 2a^{n+1}$

26. $a^{n+4}(a^{n-2} + 5a^2 - 3)$ $a^{2n+2} + 5a^{n+6} - 3a^{n+4}$

27. $2y^2 - y[3 - 2(y - 4) - y]$ $5y^2 - 11y$

28. $3x^2 - x[x - 2(3x - 4)]$ $8x^2 - 8x$

29. $2y - 3[y - 2y(y - 3) + 4y]$ $6y^2 - 31y$

30. $4a^2 - 2a[3 - a(2 - a + a^2)]$ $2a^4 - 2a^3 + 8a^2 - 6a$

2 Simplify.

31. $(5x - 7)(3x - 8)$ $15x^2 - 61x + 56$

32. $(2x - 3y)(2x + 5y)$ $4x^2 + 4xy - 15y^2$

33. $(7x - 3y)(2x - 9y)$ $14x^2 - 69xy + 27y^2$

34. $(2a - 3b)(5a + 4b)$ $10a^2 - 7ab - 12b^2$

35. $(3a - 5b)(a + 7b)$ $3a^2 + 16ab - 35b^2$

36. $(5a + 2b)(3a + 7b)$ $15a^2 + 41ab + 14b^2$

37. $(5x + 9y)(3x + 2y)$ $15x^2 + 37xy + 18y^2$

38. $(3x - 7y)(7x + 2y)$ $21x^2 - 43xy - 14y^2$

39. $(5x - 9y)(6x - 5y)$ $30x^2 - 79xy + 45y^2$

40. $(xy + 4)(xy - 3)$ $x^2y^2 + xy - 12$

41. $(xy - 5)(2xy + 7)$ $2x^2y^2 - 3xy - 35$

42. $(2x^2 - 5)(x^2 - 5)$ $2x^4 - 15x^2 + 25$

43. $(x^2 - 4)(x^2 - 6)$ $x^4 - 10x^2 + 24$

44. $(5x^2 - 5y)(2x^2 - y)$ $10x^4 - 15x^2y + 5y^2$

45. $(x^2 - 2y^2)(x^2 + 4y^2)$ $x^4 + 2x^2y^2 - 8y^4$

46. $(x^n + 2)(x^n - 3)$ $x^{2n} - x^n - 6$

47. $(x^n - 4)(x^n - 5)$ $x^{2n} - 9x^n + 20$

48. $(2a^n - 3)(3a^n + 5)$ $6a^{2n} + a^n - 15$

49. $(5b^n - 1)(2b^n + 4)$ $10b^{2n} + 18b^n - 4$

50. $(2a^n - b^n)(3a^n + 2b^n)$ $6a^{2n} + a^nb^n - 2b^{2n}$

51. $(3x^n + b^n)(x^n + 2b^n)$ $3x^{2n} + 7x^nb^n + 2b^{2n}$

52. $(x - 2)(x^2 - 3x + 7)$
$x^3 - 5x^2 + 13x - 14$

53. $(x + 3)(x^2 + 5x - 8)$
$x^3 + 8x^2 + 7x - 24$

54. $(x + 5)(x^3 - 3x + 4)$
$x^4 + 5x^3 - 3x^2 - 11x + 20$

55. $(a + 2)(a^3 - 3a^2 + 7)$
$a^4 - a^3 - 6a^2 + 7a + 14$

56. $(2a - 3b)(5a^2 - 6ab + 4b^2)$
$10a^3 - 27a^2b + 26ab^2 - 12b^3$

57. $(3a + b)(2a^2 - 5ab - 3b^2)$
$6a^3 - 13a^2b - 14ab^2 - 3b^3$

58. $(2y^2 - 1)(y^3 - 5y^2 - 3)$
$2y^5 - 10y^4 - y^3 - y^2 + 3$

59. $(2b^2 - 3)(3b^2 - 3b + 6)$
$6b^4 - 6b^3 + 3b^2 + 9b - 18$

60. $(2x - 5)(2x^4 - 3x^3 - 2x + 9)$
$4x^5 - 16x^4 + 15x^3 - 4x^2 + 28x - 45$

61. $(2a - 5)(3a^4 - 3a^2 + 2a - 5)$
$6a^5 - 15a^4 - 6a^3 + 19a^2 - 20a + 25$

62. $(x^2 + 2x - 3)(x^2 - 5x + 7)$
$x^4 - 3x^3 - 6x^2 + 29x - 21$

63. $(x^2 - 3x + 1)(x^2 - 2x + 7)$
$x^4 - 5x^3 + 14x^2 - 23x + 7$

64. $(a - 2)(2a - 3)(a + 7)$
$2a^3 + 7a^2 - 43a + 42$

65. $(b - 3)(3b - 2)(b - 1)$
$3b^3 - 14b^2 + 17b - 6$

66. $(x^n + y^n)(x^n - 2x^ny^n + 3y^n)$
$x^{2n} - 2x^{2n}y^n + 4x^ny^n - 2x^ny^{2n} + 3y^{2n}$

67. $(x^n - y^n)(x^{2n} - 3x^ny^n - y^{2n})$
$x^{3n} - 4x^{2n}y^n + 2x^ny^{2n} + y^{3n}$

3 Simplify.

68. $(a - 4)(a + 4)$ $a^2 - 16$

69. $(b - 7)(b + 7)$ $b^2 - 49$

70. $(3x - 2)(3x + 2)$
$9x^2 - 4$

71. $(b - 11)(b + 11)$ $b^2 - 121$

72. $(3a + 5b)^2$ $9a^2 + 30ab + 25b^2$

73. $(5x - 4y)^2$
$25x^2 - 40xy + 16y^2$

74. $(x^2 - 3)^2$ $x^4 - 6x^2 + 9$

75. $(x^2 + y^2)^2$ $x^4 + 2x^2y^2 + y^4$

76. $(10 + b)(10 - b)$
$100 - b^2$

77. $(2a - 3b)(2a + 3b)$ $4a^2 - 9b^2$

78. $(5x - 7y)(5x + 7y)$ $25x^2 - 49y^2$

79. $(x^2 + 1)(x^2 - 1)$
$x^4 - 1$

80. $(x^2 + y^2)(x^2 - y^2)$ $x^4 - y^4$

81. $(2x^n + y^n)^2$ $4x^{2n} + 4x^ny^n + y^{2n}$

82. $(a^n + 5b^n)^2$
$a^{2n} + 10a^nb^n + 25b^{2n}$

83. $(5a - 9b)(5a + 9b)$
$25a^2 - 81b^2$

84. $(3x + 7y)(3x - 7y)$ $9x^2 - 49y^2$

85. $(2x^n - 5)(2x^n + 5)$
$4x^{2n} - 25$

86. $(4y + 1)(4y - 1)$ $16y^2 - 1$

87. $(6 - x)(6 + x)$ $36 - x^2$

88. $(2x^2 - 3y^2)^2$
$4x^4 - 12x^2y^2 + 9y^4$

89. $(3a - 4b)^2$ $9a^2 - 24ab + 16b^2$

90. $(2x^2 + 5)^2$ $4x^4 + 20x^2 + 25$

91. $(3x^n + 2)^2$
$9x^{2n} + 12x^n + 4$

92. $(4b^n - 3)^2$ $16b^{2n} - 24b^n + 9$

93. $(x^n + 3)(x^n - 3)$ $x^{2n} - 9$

94. $(x^n + y^n)(x^n - y^n)$
$x^{2n} - y^{2n}$

95. $(x^n - 1)^2$ $x^{2n} - 2x^n + 1$

96. $(a^n - b^n)^2$ $a^{2n} - 2a^nb^n + b^{2n}$

97. $(2x^n + 5y^n)^2$
$4x^{2n} + 20x^ny^n + 25y^{2n}$

4 **98.** *Geometry* The length of a rectangle is $(3x + 3)$ ft. The width is $(x - 4)$ ft. Find the area of the rectangle in terms of the variable x.
$(3x^2 - 9x - 12)$ ft²

99. *Geometry* The base of a triangle is $(x + 2)$ ft. The height is $(2x - 3)$ ft. Find the area of the triangle in terms of the variable x.
$\left(x^2 + \frac{x}{2} - 3\right)$ ft²

100. *Geometry* Find the area of the figure shown below. All dimensions given are in meters.

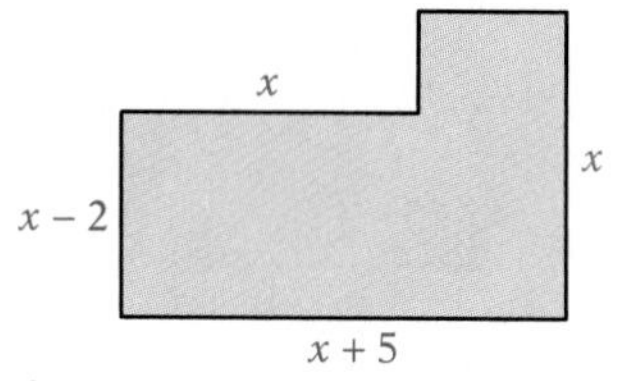

$(x^2 + 3x)$ m²

101. *Geometry* Find the area of the figure shown below. All dimensions given are in feet.

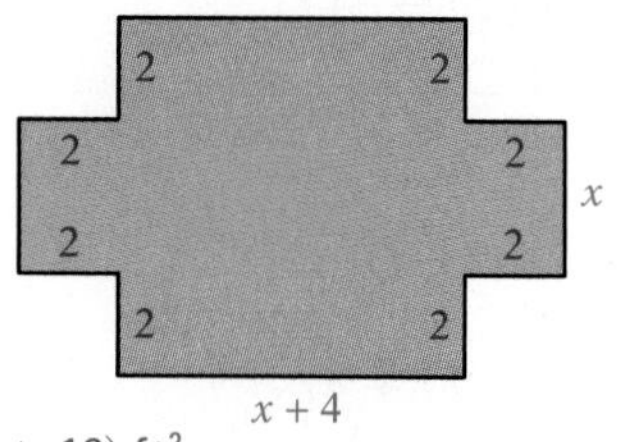

$(x^2 + 12x + 16)$ ft²

102. *Construction* A trough is made from a rectangular piece of metal by folding up the sides as shown in the figure below. What is the volume of the trough in terms of the side folded up?

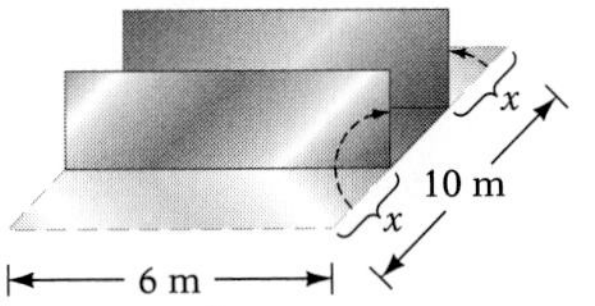

$(-12x^2 + 60x)\ \text{m}^3$

103. *Construction* A square sheet of cardboard measuring 18 in. by 18 in. is used to make an open box by cutting squares of equal size from the corners and folding up the sides. Find the volume of the box in terms of the variable x, where x is the length of the side of the square cut from the cardboard.

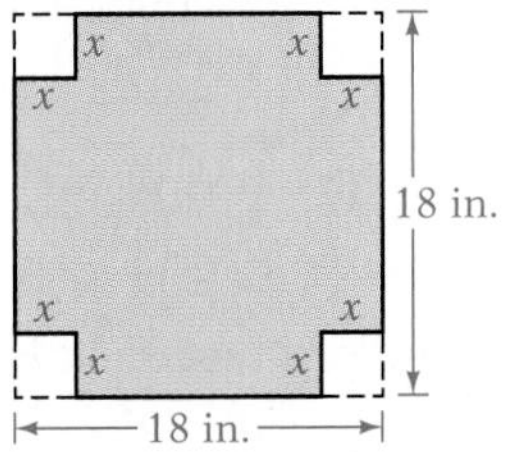

$(4x^3 - 72x^2 + 324x)\ \text{in}^3$

104. *Geometry* The length of the side of a cube is $(x - 2)$ cm. Find the volume of the cube in terms of the variable x.
$(x^3 - 6x^2 + 12x - 8)\ \text{cm}^3$

105. *Geometry* The length of a box is $(2x + 3)$ cm, the width is $(x - 5)$ cm, and the height is x cm. Find the volume of the box in terms of the variable x. $(2x^3 - 7x^2 - 15x)\ \text{cm}^3$

106. *Geometry* Find the volume of the figure shown below. All dimensions given are in inches.

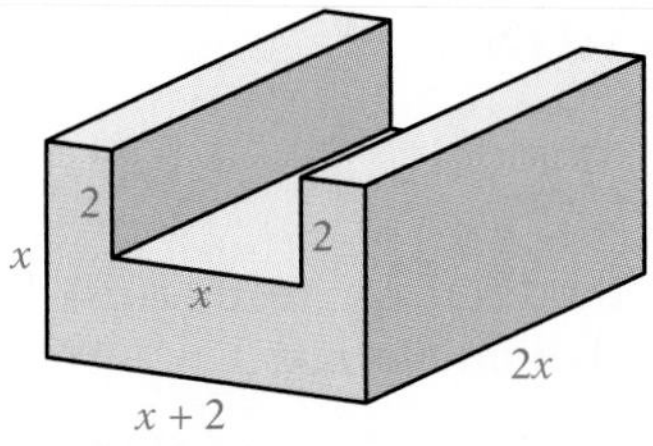

$2x^3\ \text{in}^3$

107. *Geometry* Find the volume of the figure shown below. All dimensions given are in centimeters.

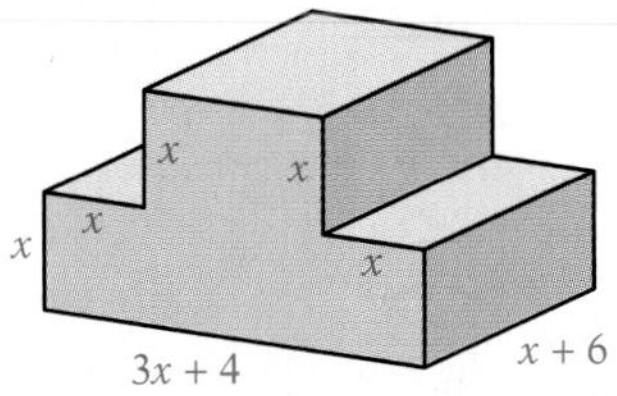

$(4x^3 + 32x^2 + 48x)\ \text{cm}^3$

APPLYING CONCEPTS 5.3

Simplify.

108. $\dfrac{(2x + 1)^5}{(2x + 1)^3}$ $4x^2 + 4x + 1$

109. $\dfrac{(3x - 5)^6}{(3x - 5)^4}$ $9x^2 - 30x + 25$

110. $(a - b)^2 - (a + b)^2$ $-4ab$

111. $(x + 2y)^2 + (x + 2y)(x - 2y)$ $2x^2 + 4xy$

112. $(4y + 3)^2 - (3y + 4)^2$ $7y^2 - 7$

113. $2x^2(3x^3 + 4x - 1) - 5x^2(x^2 - 3)$
$6x^5 - 5x^4 + 8x^3 + 13x^2$

114. $(2b + 3)(b - 4) + (3 + b)(3 - 2b)$ $-8b - 3$

115. $(3x - 2y)^2 - (2x - 3y)^2$ $5x^2 - 5y^2$

116. $[x + (y + 1)][x - (y + 1)]$ $x^2 - y^2 - 2y - 1$

117. $[x^2(2y - 1)]^2$ $4x^4y^2 - 4x^4y + x^4$

For what value of k is the given equation an identity?

118. $(5x - k)(3x + k) = 15x^2 + 4x - k^2$ 2

119. $(kx - 7)(kx + 2) = k^2x^2 + 5x - 14$ -1

Complete.

120. If $m = n + 1$, then $\frac{a^m}{a^n} = $ ___a___.

121. If $m = n + 2$, then $\frac{a^m}{a^n} = $ ___a^2___.

[C] **122.** What polynomial when divided by $x - 4$ has a quotient of $2x + 3$? $2x^2 - 5x - 12$

[C] **123.** What polynomial when divided by $2x - 3$ has a quotient of $x + 7$? $2x^2 + 11x - 21$

[C] **124.** Subtract the product of $4a + b$ and $2a - b$ from $9a^2 - 2ab$. $a^2 + b^2$

[C] **125.** Subtract the product of $5x - y$ and $x + 3y$ from $6x^2 + 12xy - 2y^2$. $x^2 - 2xy + y^2$

[C] **126.** Find $(3n^4)^3$ if $5(n - 1) = 2(3n - 2)$. 27

[C] **127.** Find $(-2n^3)^2$ if $3(2n - 1) = 5(n - 1)$. 256

SECTION 5.4
Division of Polynomials

1 Divide polynomials

18 HM³

INSTRUCTOR NOTE

Begin this section by showing an example of long division of whole numbers.

To divide two polynomials, use a method similar to that used for division of whole numbers. To check division of polynomials, use

Dividend = (quotient × divisor) + remainder

Divide: $(x^2 + 5x - 7) \div (x + 3)$

Step 1

$$\begin{array}{r} x \phantom{{}+5x-7} \\ x + 3\overline{)x^2 + 5x - 7} \\ \underline{x^2 + 3x} \phantom{{}-7}\downarrow \\ 2x - 7 \end{array}$$

Think: $x\overline{)x^2} = \frac{x^2}{x} = x$

Multiply: $x(x + 3) = x^2 + 3x$

Subtract: $(x^2 + 5x) - (x^2 + 3x) = 2x$

Bring down -7.

Step 2

$$\begin{array}{r} x + 2 \\ x + 3\overline{)x^2 + 5x - 7} \\ \underline{x^2 + 3x} \phantom{{}-7} \\ 2x - 7 \\ \underline{2x + 6} \\ -13 \end{array}$$

Think: $x\overline{)2x} = \frac{2x}{x} = 2$

Multiply: $2(x + 3) = 2x + 6$

Subtract: $(2x - 7) - (2x + 6) = -13$

The remainder is -13.

Check: $(x + 2)(x + 3) + (-13) = x^2 + 3x + 2x + 6 - 13 = x^2 + 5x - 7$

$(x^2 + 5x - 7) \div (x + 3) = x + 2 - \frac{13}{x + 3}$

INSTRUCTOR NOTE

After you have completed this objective, a possible extra-credit problem is

$2x - 4\overline{)3x^3 - 5x^2 - 6x + 8}$

➡ Divide: $\dfrac{6 - 6x^2 + 4x^3}{2x + 3}$

Arrange the terms in descending order. Note that there is no term of x in $4x^3 - 6x^2 + 6$. Insert a zero as $0x$ for the missing term so that like terms will be in the same columns.

$$
\begin{array}{r}
2x^2 - 6x + 9 \\
2x + 3\overline{)4x^3 - 6x^2 + 0x + 6} \\
\underline{4x^3 + 6x^2} \phantom{{}+0x+6} \\
-12x^2 + 0x \phantom{{}+6} \\
\underline{-12x^2 - 18x} \phantom{{}+6} \\
18x + 6 \\
\underline{18x + 27} \\
-21
\end{array}
$$

$$\frac{4x^3 - 6x^2 + 6}{2x + 3} = 2x^2 - 6x + 9 - \frac{21}{2x + 3}$$

Example 1 Divide. A. $\dfrac{12x^2 - 11x + 10}{4x - 5}$ B. $\dfrac{x^3 + 1}{x + 1}$

Solution A.

$$
\begin{array}{r}
3x + 1 \\
4x - 5\overline{)12x^2 - 11x + 10} \\
\underline{12x^2 - 15x} \phantom{{}+10} \\
4x + 10 \\
\underline{4x - 5} \\
15
\end{array}
$$

$$\frac{12x^2 - 11x + 10}{4x - 5} = 3x + 1 + \frac{15}{4x - 5}$$

B.

$$
\begin{array}{r}
x^2 - x + 1 \\
x + 1\overline{)x^3 + 0x^2 + 0x + 1} \\
\underline{x^3 + x^2} \phantom{{}+0x+1} \\
-x^2 + 0x \phantom{{}+1} \\
\underline{-x^2 - x} \phantom{{}+1} \\
x + 1 \\
\underline{x + 1} \\
0
\end{array}
$$

$$\frac{x^3 + 1}{x + 1} = x^2 - x + 1$$

Problem 1 Divide. A. $\dfrac{15x^2 + 17x - 20}{3x + 4}$ B. $\dfrac{3x^3 + 8x^2 - 6x + 2}{3x - 1}$

Solution See page S14. A. $5x - 1 - \dfrac{16}{3x + 4}$ B. $x^2 + 3x - 1 + \dfrac{1}{3x - 1}$

2 Synthetic division

Synthetic division is a shorter method of dividing a polynomial by a binomial of the form $x - a$. This method of dividing uses only the coefficients of the variable terms.

Both long division and synthetic division are used below to simplify the expression $(3x^2 - 4x + 6) \div (x - 2)$.

LONG DIVISION

Compare the coefficients in this problem worked by long division with the coefficients in the same problem worked by synthetic division.

$$
\begin{array}{r}
3x + 2 \\
x - 2\overline{)3x^2 - 4x + 6} \\
\underline{3x^2 - 6x} \phantom{{}+6} \\
2x + 6 \\
\underline{2x - 4} \\
10
\end{array}
$$

$$(3x^2 - 4x + 6) \div (x - 2) = 3x + 2 + \frac{10}{x - 2}$$

SYNTHETIC DIVISION

$x - a = x - 2; a = 2$

Value of a Coefficients of the dividend

$$2 \,\vert\, \begin{array}{ccc} 3 & -4 & 6 \end{array}$$

Bring down the 3.

$$\begin{array}{c|ccc} 2 & 3 & -4 & 6 \\ \hline & 3 \end{array}$$

Multiply $2 \cdot 3$ and add the product (6) to -4.

$$\begin{array}{c|ccc} 2 & 3 & -4 & 6 \\ & & 6 & \\ \hline & 3 & 2 \end{array}$$

Multiply $2 \cdot 2$ and add the product (4) to 6.

$$\begin{array}{c|ccc} 2 & 3 & -4 & 6 \\ & & 6 & 4 \\ \hline & 3 & 2 & 10 \end{array}$$

$$\underbrace{}_{\substack{\text{Coefficients of} \\ \text{the quotient}}} \quad \underbrace{}_{\text{Remainder}}$$

INSTRUCTOR NOTE

Remind students that they can check an answer to a synthetic-division problem in the same way that they check an answer to a long-division problem.

The degree of the first term of the quotient is one degree less than the degree of the first term of the dividend.

$(3x^2 - 4x + 6) \div (x - 2) = 3x + 2 + \dfrac{10}{x - 2}$

Check: $(3x + 2)(x - 2) + 10$
$ = 3x^2 - 6x + 2x - 4 + 10$
$ = 3x^2 - 4x + 6$

➡️ Divide: $(2x^3 + 3x^2 - 4x + 8) \div (x + 3)$

Write down the value of a and the coefficients of the dividend.

$x - a = x + 3 = x - (-3); a = -3$

Bring down the 2. Multiply $-3 \cdot 2$ and add the product (-6) to 3. Continue until all the coefficients have been used.

$$\begin{array}{c|cccc} -3 & 2 & 3 & -4 & 8 \\ & & -6 & 9 & -15 \\ \hline & 2 & -3 & 5 & -7 \end{array}$$

$$\underbrace{}_{\substack{\text{Coefficients of} \\ \text{the quotient}}} \quad \underbrace{}_{\text{Remainder}}$$

Write the quotient. The degree of the quotient is one less than the degree of the dividend.

$(2x^3 + 3x^2 - 4x + 8) \div (x + 3)$
$= 2x^2 - 3x + 5 - \dfrac{7}{x + 3}$

Example 2   Divide.   A. $(5x^2 - 3x + 7) \div (x - 1)$

B. $(3x^4 - 8x^2 + 2x + 1) \div (x + 2)$

Solution   A.
$$\begin{array}{c|ccc} 1 & 5 & -3 & 7 \\ & & 5 & 2 \\ \hline & 5 & 2 & 9 \end{array}$$

▶ $x - a = x - 1; a = 1$

$(5x^2 - 3x + 7) \div (x - 1)$
$= 5x + 2 + \dfrac{9}{x - 1}$

B.
$$\begin{array}{c|ccccc} -2 & 3 & 0 & -8 & 2 & 1 \\ & & -6 & 12 & -8 & 12 \\ \hline & 3 & -6 & 4 & -6 & 13 \end{array}$$

▶ Insert a zero for the missing term.

$x - a = x + 2; a = -2$

$(3x^4 - 8x^2 + 2x + 1) \div (x + 2)$
$= 3x^3 - 6x^2 + 4x - 6 + \dfrac{13}{x + 2}$

Problem 2 Divide. A. $(6x^2 + 8x - 5) \div (x + 2)$
B. $(2x^4 - 3x^3 - 8x^2 - 2) \div (x - 3)$

Solution See page S14.

A. $6x - 4 + \frac{3}{x + 2}$ B. $2x^3 + 3x^2 + x + 3 + \frac{7}{x - 3}$

3 Evaluate a polynomial using synthetic division

HM[3]

A polynomial can be evaluated by using synthetic division. Consider the polynomial $P(x) = 2x^4 - 3x^3 + 4x^2 - 5x + 1$. One way to evaluate the polynomial when $x = 2$ is to replace x by 2 and then simplify the numerical expression.

$$\begin{aligned} P(x) &= 2x^4 - 3x^3 + 4x^2 - 5x + 1 \\ P(2) &= 2(2)^4 - 3(2)^3 + 4(2)^2 - 5(2) + 1 \\ P(2) &= 2(16) - 3(8) + 4(4) - 5(2) + 1 \\ P(2) &= 32 - 24 + 16 - 10 + 1 \\ P(2) &= 15 \end{aligned}$$

Now use synthetic division to divide $(2x^4 - 3x^3 + 4x^2 - 5x + 1) \div (x - 2)$.

$$\begin{array}{r|rrrrr} 2 & 2 & -3 & 4 & -5 & 1 \\ & & 4 & 2 & 12 & 14 \\ \hline & 2 & 1 & 6 & 7 & 15 \end{array}$$

Coefficients of the quotient (2, 1, 6, 7) — Remainder (15)

Note that the remainder is 15, which is the same value as $P(2)$. This is not a coincidence. The following theorem states that this situation is always true.

Remainder Theorem

If the polynomial $P(x)$ is divided by $x - a$, the remainder is $P(a)$.

Evaluate $P(x) = x^4 - 3x^2 + 4x - 5$ when $x = -2$ by using the Remainder Theorem.

The value at which the polynomial is evaluated: -2

$$\begin{array}{r|rrrrr} -2 & 1 & 0 & -3 & 4 & -5 \\ & & -2 & 4 & -2 & -4 \\ \hline & 1 & -2 & 1 & 2 & -9 \end{array}$$

A 0 is inserted for the x^3 term.

-9 ← The remainder

$P(-2) = -9$

Example 3 Use the Remainder Theorem to evaluate $P(-2)$ when $P(x) = x^3 - 3x^2 + x + 3$.

Solution Use synthetic division with $a = -2$.

$$\begin{array}{r|rrrr} -2 & 1 & -3 & 1 & 3 \\ & & -2 & 10 & -22 \\ \hline & 1 & -5 & 11 & -19 \end{array}$$

By the Remainder Theorem, $P(-2) = -19$.

Problem 3 Use the Remainder Theorem to evaluate $P(3)$ when $P(x) = 2x^3 - 4x - 5$.

Solution See page S14. 37

CONCEPT REVIEW 5.4

Determine whether the following statements are always true, sometimes true, or never true.

1. The quotient of two monomials is a monomial. Sometimes true
2. Synthetic division is a shorter method of dividing a polynomial by a binomial. Sometimes true
3. In synthetic division, the degree of the quotient is always one less than the degree of the divisor. Always true
4. If a polynomial $P(x)$ is divided by $x - a$, the remainder is $P(a)$. Always true

EXERCISES 5.4

1 1. Explain how the degree of the quotient of two polynomials is related to the degrees of the dividend and the divisor.

2. Explain how you can check the result of dividing two polynomials.

Divide by using long division.

3. $(x^2 + 3x - 40) \div (x - 5)$ $x + 8$
4. $(x^2 - 14x + 24) \div (x - 2)$ $x - 12$
5. $(x^3 - 3x^2 + 2) \div (x - 3)$ $x^2 + \frac{2}{x - 3}$
6. $(x^3 + 4x^2 - 8) \div (x + 4)$ $x^2 - \frac{8}{x + 4}$
7. $(6x^2 + 13x + 8) \div (2x + 1)$ $3x + 5 + \frac{3}{2x + 1}$
8. $(12x^2 + 13x - 14) \div (3x - 2)$ $4x + 7$
9. $(10x^2 + 9x - 5) \div (2x - 1)$ $5x + 7 + \frac{2}{2x - 1}$
10. $(18x^2 - 3x + 2) \div (3x + 2)$ $6x - 5 + \frac{12}{3x + 2}$
11. $(8x^3 - 9) \div (2x - 3)$ $4x^2 + 6x + 9 + \frac{18}{2x - 3}$
12. $(64x^3 + 4) \div (4x + 2)$ $16x^2 - 8x + 4 - \frac{2}{2x + 1}$

13. $(6x^4 - 13x^2 - 4) \div (2x^2 - 5)$ $3x^2 + 1 + \frac{1}{2x^2 - 5}$

14. $(12x^4 - 11x^2 + 10) \div (3x^2 + 1)$ $4x^2 - 5 + \frac{15}{3x^2 + 1}$

15. $\frac{3x^3 - 8x^2 - 33x - 10}{3x + 1}$ $x^2 - 3x - 10$

16. $\frac{8x^3 - 38x^2 + 49x - 10}{4x - 1}$ $2x^2 - 9x + 10$

17. $\frac{4 - 7x + 5x^2 - x^3}{x - 3}$ $-x^2 + 2x - 1 + \frac{1}{x - 3}$

18. $\frac{4 + 6x - 3x^2 + 2x^3}{2x + 1}$ $x^2 - 2x + 4$

19. $\frac{16x^2 - 13x^3 + 2x^4 - 9x + 20}{x - 5}$ $2x^3 - 3x^2 + x - 4$

20. $\frac{x + 3x^4 - x^2 + 5x^3 - 2}{x + 2}$ $3x^3 - x^2 + x - 1$

21. $\frac{x^3 - 4x^2 + 2x - 1}{x^2 + 1}$ $x - 4 + \frac{x + 3}{x^2 + 1}$

22. $\frac{3x^3 - 2x^2 - 8}{x^2 + 5}$ $3x - 2 - \frac{15x - 2}{x^2 + 5}$

23. $\frac{2x^3 - x + 4 - 3x^2}{x^2 - 1}$ $2x - 3 + \frac{1}{x - 1}$

24. $\frac{2 - 3x^2 + 5x^3}{x^2 + 3}$ $5x - 3 - \frac{15x - 11}{x^2 + 3}$

25. $\frac{6x^3 + 2x^2 + x + 4}{2x^2 - 3}$ $3x + 1 + \frac{10x + 7}{2x^2 - 3}$

26. $\frac{9x^3 + 6x^2 + 2x + 1}{3x^2 + 2}$ $3x + 2 - \frac{4x + 3}{3x^2 + 2}$

2 Divide by using synthetic division.

27. $(2x^2 - 6x - 8) \div (x + 1)$ $2x - 8$

28. $(3x^2 + 19x + 20) \div (x + 5)$ $3x + 4$

29. $(3x^2 - 4) \div (x - 1)$ $3x + 3 - \frac{1}{x - 1}$

30. $(4x^2 - 8) \div (x - 2)$ $4x + 8 + \frac{8}{x - 2}$

31. $(x^2 - 9) \div (x + 4)$ $x - 4 + \frac{7}{x + 4}$

32. $(x^2 - 49) \div (x + 5)$ $x - 5 - \frac{24}{x + 5}$

33. $(2x^2 + 24) \div (2x + 4)$ $x - 2 + \frac{16}{x + 2}$

34. $(3x^2 - 15) \div (x + 3)$ $3x - 9 + \frac{12}{x + 3}$

35. $(2x^3 - x^2 + 6x + 9) \div (x + 1)$ $2x^2 - 3x + 9$

36. $(3x^3 + 10x^2 + 6x - 4) \div (x + 2)$ $3x^2 + 4x - 2$

37. $(x^3 - 6x^2 + 11x - 6) \div (x - 3)$ $x^2 - 3x + 2$

38. $(x^3 - 4x^2 + x + 6) \div (x + 1)$ $x^2 - 5x + 6$

39. $(6x - 3x^2 + x^3 - 9) \div (x + 2)$ $x^2 - 5x + 16 - \frac{41}{x + 2}$

40. $(5 - 5x + 4x^2 + x^3) \div (x - 3)$ $x^2 + 7x + 16 + \frac{53}{x - 3}$

41. $(x^3 + x - 2) \div (x + 1)$ $x^2 - x + 2 - \frac{4}{x + 1}$

42. $(x^3 + 2x + 5) \div (x - 2)$ $x^2 + 2x + 6 + \frac{17}{x - 2}$

43. $(18 + x - 4x^3) \div (2 - x)$ $4x^2 + 8x + 15 + \frac{12}{x - 2}$

44. $(12 - 3x^2 + x^3) \div (3 + x)$ $x^2 - 6x + 18 - \frac{42}{x + 3}$

45. $\frac{16x^2 - 13x^3 + 2x^4 - 9x + 20}{x - 5}$ $2x^3 - 3x^2 + x - 4$

46. $\frac{2x^3 - x^2 - 10x + 15 + x^4}{x - 2}$ $x^3 + 4x^2 + 7x + 4 + \frac{23}{x - 2}$

47. $\frac{5 + 5x - 8x^2 + 4x^3 - 3x^4}{2 - x}$ $3x^3 + 2x^2 + 12x + 19 + \frac{33}{x - 2}$

48. $\frac{3 - 13x - 5x^2 + 9x^3 - 2x^4}{3 - x}$ $2x^3 - 3x^2 - 4x + 1$

49. $\frac{3x^4 + 3x^3 - x^2 + 3x + 2}{x + 1}$ $3x^3 - x + 4 - \frac{2}{x + 1}$

50. $\frac{4x^4 + 12x^3 - x^2 - x + 2}{x + 3}$ $4x^3 - x + 2 - \frac{4}{x + 3}$

51. $\frac{2x^4 - x^2 + 2}{x - 3}$ $2x^3 + 6x^2 + 17x + 51 + \frac{155}{x - 3}$

52. $\frac{x^4 - 3x^3 - 30}{x + 2}$ $x^3 - 5x^2 + 10x - 20 + \frac{10}{x + 2}$

53. $\frac{x^3 + 125}{x + 5}$ $x^2 - 5x + 25$

54. $\frac{x^3 + 343}{x + 7}$ $x^2 - 7x + 49$

3 Use the Remainder Theorem to evaluate each of the polynomials.

55. $P(x) = 2x^2 - 3x - 1; P(3)$ 8

56. $Q(x) = 3x^2 - 5x - 1; Q(2)$ 1

57. $R(x) = x^3 - 2x^2 + 3x - 1; R(4)$ 43

58. $F(x) = x^3 + 4x^2 - 3x + 2; F(3)$ 56

59. $P(z) = 2z^3 - 4z^2 + 3z - 1; P(-2)$ −39

60. $R(t) = 3t^3 + t^2 - 4t + 2; R(-3)$ −58

61. $Q(x) = x^4 + 3x^3 - 2x^2 + 4x - 9; Q(2)$ 31

62. $Y(z) = z^4 - 2z^3 - 3z^2 - z + 7; Y(3)$ 4

63. $F(x) = 2x^4 - x^3 - 2x - 5; F(-3)$ 190

64. $Q(x) = x^4 - 2x^3 + 4x - 2; Q(-2)$ 22

65. $P(x) = x^3 - 3; P(5)$ 122

66. $S(t) = 4t^3 + 5; S(-4)$ −251

67. $R(t) = 4t^4 - 3t^2 + 5; R(-3)$ 302

68. $P(z) = 2z^4 + z^2 - 3; P(-4)$ 525

69. $Q(x) = x^5 - 4x^3 - 2x^2 + 5x - 2; Q(2)$ 0

70. $T(x) = 2x^5 + 4x^4 - x^2 + 4; T(3)$ 805

APPLYING CONCEPTS 5.4

Divide by using long division.

71. $\frac{3x^2 - xy - 2y^2}{3x + 2y}$ $x - y$

72. $\frac{12x^2 + 11xy + 2y^2}{4x + y}$ $3x + 2y$

73. $\frac{a^3 - b^3}{a - b}$ $a^2 + ab + b^2$

74. $\frac{a^4 + b^4}{a + b}$ $a^3 - a^2b + ab^2 - b^3 + \frac{2b^4}{a + b}$

75. $\frac{x^5 + y^5}{x + y}$ $x^4 - x^3y + x^2y^2 - xy^3 + y^4$

76. $\frac{x^6 - y^6}{x - y}$ $x^5 + x^4y + x^3y^2 + x^2y^3 + xy^4 + y^5$

For what value of k will the remainder be zero?

[C] **77.** $(x^3 - 3x^2 - x + k) \div (x - 3)$ 3

[C] **78.** $(x^3 - 2x^2 + x + k) \div (x - 2)$ -2

[C] **79.** $(x^2 + kx - 6) \div (x - 3)$ -1

[C] **80.** $(x^3 + kx + k - 1) \div (x - 1)$ 0

81. When $x^2 + x + 2$ is divided by a polynomial, the quotient is $x + 4$, and the remainder is 14. Find the polynomial. $x - 3$

SECTION 5.5 Factoring Polynomials

1 Factor a monomial from a polynomial

The GCF of two or more exponential expressions with the same base is the exponential expression with the smallest exponent.

$$2^5 \qquad x^5$$
$$2^2 \qquad x^7$$
$$2^9 \qquad x$$
$$\text{GCF} = 2^2 = 4 \qquad \text{GCF} = x$$

The GCF of two or more monomials is the product of each common factor with the smallest exponent.

$$16a^4b = 2^4 \cdot a^4 \cdot b$$
$$40a^2b^5 = 2^3 \cdot 5 \cdot a^2 \cdot b^5$$
$$\text{GCF} = 2^3 \cdot a^2 \cdot b = 8a^2b$$

To **factor a polynomial** means to write the polynomial as a product of other polynomials.

LOOK CLOSELY

$3x^2 + 6x$ is a sum. $3x(x + 2)$ is a product. When we factor a polynomial, we write it as a product.

In the example at the right, $3x$ is the GCF of the terms $3x^2$ and $6x$. $3x$ is a **monomial factor** of the terms of the binomial. $x + 2$ is a **binomial factor** of $3x^2 + 6x$.

Multiply

Polynomial **Factors**

$$3x^2 + 6x = 3x(x + 2)$$

Factor

Example 1 Factor.

A. $4x^3y^2 + 12x^3y + 20xy^2$ B. $x^{2n} + x^{n+1} + x^n, (n > 0)$

Solution A. Find the GCF of $4x^3y^2$, $12x^3y$, and $20xy^2$.

$$4x^3y^2 = 2^2 \cdot x^3 \cdot y^2$$
$$12x^3y = 2^2 \cdot 3 \cdot x^3 \cdot y$$
$$20xy^2 = 2^2 \cdot 5 \cdot x \cdot y^2$$
$$\text{GCF} = 2^2 \cdot x \cdot y = 4xy$$

Rewrite each term of the polynomial as a product with the GCF as one of the factors.

$$4x^3y^2 + 12x^3y + 20xy^2$$
$$= 4xy(x^2y) + 4xy(3x^2) + 4xy(5y)$$
$$= 4xy(x^2y + 3x^2 + 5y)$$

INSTRUCTOR NOTE

In part B, it may help some students to show that when factoring x^n from an expression, we use the term with the least exponent. For instance,

$$x^8 + 2x^5 - 3x^4 = x^4(x^4 + 2x - 3)$$

B. Because $n > 0$, $n < 2n$, and $n < n + 1$, the GCF of x^{2n}, x^{n+1}, and x^n is x^n.

$$x^{2n} + x^{n+1} + x^n = x^n(x^n) + x^n(x) + x^n(1) = x^n(x^n + x + 1)$$

Problem 1 Factor. A. $3x^3y - 6x^2y^2 - 3xy^3$ B. $6t^{2n} - 9t^n$

Solution See page S14. A. $3xy(x^2 - 2xy - y^2)$ B. $3t^n(2t^n - 3)$

2 Factor by grouping

19 HM³

In the examples at the right, the binomials in parentheses are called binomial factors.

$$4x^4(2x - 3)$$
$$-2r^2s(5r + 2s)$$

The Distributive Property is used to factor a common binomial factor from an expression.

⇨ Factor: $4a(2b + 3) - 5(2b + 3)$

The common binomial factor is $(2b + 3)$. Use the Distributive Property to write the expression as a product of factors.

$$4a(2b + 3) - 5(2b + 3) = (2b + 3)(4a - 5)$$

Consider the binomial $y - x$. Factoring -1 from this binomial gives

$$y - x = -(x - y)$$

This equation is sometimes used to factor a common binomial from an expression.

LOOK CLOSELY

For the simplification at the right,

$$6r(r - s) - 7(s - r) = 6r(r - s) - 7[(-1)(r - s)] = 6r(r - s) + 7(r - s)$$

⇨ Factor: $6r(r - s) - 7(s - r)$

$$6r(r - s) - 7(s - r)$$

Rewrite the expression as a sum of terms that have a common binomial factor. Use $s - r = -(r - s)$.

$$= 6r(r - s) + 7(r - s)$$

Write the expression as a product of factors.

$$= (r - s)(6r + 7)$$

Some polynomials can be factored by grouping terms so that a common binomial factor is found.

⇨ Factor: $8y^2 + 4y - 6ay - 3a$

$$8y^2 + 4y - 6ay - 3a$$

Group the first two terms and the last two terms. Note that $-6ay - 3a = -(6ay + 3a)$.

$$= (8y^2 + 4y) - (6ay + 3a)$$

Factor the GCF from each group.

$$= 4y(2y + 1) - 3a(2y + 1)$$

Write the expression as a product of factors.

$$= (2y + 1)(4y - 3a)$$

Example 2 Factor. A. $3x(y - 4) - 2(4 - y)$ B. $xy - 4x - 2y + 8$

Solution A. $3x(y - 4) - 2(4 - y)$

$= 3x(y - 4) + 2(y - 4)$ ▶ Write the expression as a sum of terms that have a common factor. Note that $4 - y = -(y - 4)$.

$= (y - 4)(3x + 2)$ ▶ Write the expression as a product of factors.

B. $xy - 4x - 2y + 8$
$\quad = (xy - 4x) - (2y - 8)$ ▶ Group the first two terms and the
$\qquad\qquad\qquad\qquad\qquad\qquad$ last two terms.

$\quad = x(y - 4) - 2(y - 4)$ ▶ Factor out the GCF from each
$\quad = (y - 4)(x - 2)$ group.

Problem 2   Factor.   A. $6a(2b - 5) + 7(5 - 2b)$ B. $3rs - 2r - 3s + 2$

Solution   See page S14.   A. $(2b - 5)(6a - 7)$ B. $(3s - 2)(r - 1)$

3   Factor trinomials of the form $x^2 + bx + c$

A **quadratic trinomial** is a trinomial of the form $ax^2 + bx + c$, where a and b are nonzero coefficients and c is a nonzero constant. The degree of a quadratic trinomial is 2. Here are examples of quadratic trinomials:

$$4x^2 - 3x - 7 \qquad\qquad z^2 + z + 10 \qquad\qquad 2y^2 + 4y - 9$$
$$(a = 4, b = -3, c = -7) \quad (a = 1, b = 1, c = 10) \quad (a = 2, b = 4, c = -9)$$

To **factor a quadratic trinomial** means to express the trinomial as the product of two binomials. For example,

Trinomial		Factored Form
$2x^2 - x - 1$	$=$	$(2x + 1)(x - 1)$
$y^2 - 3y + 2$	$=$	$(y - 1)(y - 2)$

In this objective, trinomials of the form $x^2 + bx + c$ $(a = 1)$ will be factored. The next objective deals with trinomials where $a \neq 1$.

The method by which factors of a trinomial are found is based on FOIL. Consider the following binomial products, noting the relationship between the constant terms of the binomials and the terms of the trinomial.

Sum of binomial constants

Product of binomial constants

$$\quad\quad\quad F \quad O \quad I \quad L$$
$$(x + 4)(x + 5) = x \cdot x + 5x + 4x + 4 \cdot 5 \qquad = x^2 + 9x + 20$$
$$(x - 6)(x + 8) = x \cdot x + 8x - 6x + (-6) \cdot 8 \ = x^2 + 2x - 48$$
$$(x - 3)(x - 2) = x \cdot x - 2x - 3x + (-3)(-2) = x^2 - 5x + 6$$

Observe two important points from these examples:

1. The constant term of the trinomial is the *product* of the constant terms of the binomial. The coefficient of x in the trinomial is the *sum* of the constant terms of the binomial.

2. When the constant term of the trinomial is positive, the constant terms of the binomials have the *same* sign. When the constant term of the trinomial is negative, the constant terms of the binomials have *opposite* signs.

Use the points listed above to factor a trinomial. For example, to factor

$$x^2 - 5x - 24$$

find two numbers whose sum is -5 and whose product is -24. Because the constant term of the trinomial is negative (-24), the numbers will have opposite signs.

A systematic method of finding these numbers involves listing the factors of the constant term of the trinomial and the sum of those factors.

Factors of -24	Sum of the Factors
$1, -24$	$1 + (-24) = -23$
$-1, 24$	$-1 + 24 = 23$
$2, -12$	$2 + (-12) = -10$
$-2, 12$	$-2 + 12 = 10$
$3, -8$	**$3 + (-8) = -5$**
$-3, 8$	$-3 + 8 = 5$
$4, -6$	$4 + (-6) = -2$
$-4, 6$	$-4 + 6 = 2$

3 and -8 are two numbers whose sum is -5 and whose product is -24. Write the binomial factors of the trinomial.

$$x^2 - 5x - 24 = (x + 3)(x - 8)$$

LOOK CLOSELY

You can always check a proposed factorization by multiplying all the factors.

Check: $(x + 3)(x - 8) = x^2 - 8x + 3x - 24 = x^2 - 5x - 24$

By the Commutative Property of Multiplication, the binomial factors can also be written as

$$x^2 - 5x - 24 = (x - 8)(x + 3)$$

Example 3 Factor.

A. $x^2 + 8x + 12$ B. $x^2 + 5x - 84$ C. $x^2 + 7xy + 12y^2$

Solution A. $x^2 + 8x + 12$

Factors of 12	Sum
1, 12	13
2, 6	8
3, 4	7

► Try only positive factors of 12.

► Once the correct pair is found, the other factors need not be tried.

$x^2 + 8x + 12 = (x + 2)(x + 6)$ ► Write the factors of the trinomial.

Check: $(x + 2)(x + 6) = x^2 + 6x + 2x + 12$
$= x^2 + 8x + 12$

B. $x^2 + 5x - 84$

$(-7)(12) = -84$
$-7 + 12 = 5$

► The factors must be of opposite signs. Find two factors of -84 whose sum is 5.

$x^2 + 5x - 84 = (x + 12)(x - 7)$

Check: $(x + 12)(x - 7) = x^2 - 7x + 12x - 84$
$= x^2 + 5x - 84$

C. $x^2 + 7xy + 12y^2 = (x + 3y)(x + 4y)$

Check: $(x + 3y)(x + 4y) = x^2 + 4xy + 3xy + 12y^2$
$= x^2 + 7xy + 12y^2$

Problem 3 Factor.

A. $x^2 + 13x + 42$ B. $x^2 - x - 20$ C. $x^2 + 5xy + 6y^2$

Solution See page S14. A. $(x + 6)(x + 7)$ B. $(x + 4)(x - 5)$ C. $(x + 2y)(x + 3y)$

Not all trinomials can be factored using only integers. Consider $x^2 - 3x - 6$.

Factors of -6	Sum
$1, -6$	-5
$-1, 6$	5
$2, -3$	-1
$-2, 3$	1

Because none of the pairs of factors of -6 have a sum of -3, the trinomial is not factorable. The trinomial is said to be **nonfactorable over the integers.**

Example 4 Factor: $x^2 + 2x - 4$

Solution The trinomial is nonfactorable over the integers. ▸ There are no factors of -4 whose sum is 2.

Problem 4 Factor: $x^2 + 5x - 1$

Solution See page S14. Nonfactorable over the integers

4 Factor trinomials of the form $ax^2 + bx + c$

INSTRUCTOR NOTE

Factoring $ax^2 + bx + c$ by using trial factors and by using the *ac* method is discussed in this objective.

There are various methods of factoring trinomials of the form $ax^2 + bx + c$, where $a \neq 1$. Factoring by using trial factors and factoring by grouping will be discussed in this objective. Factoring by using trial factors is illustrated first.

To use the trial factor method, use the factors of a and the factors of c to write all the possible binomial factors of the trinomial. Then use FOIL to determine the correct factorization. To reduce the number of trial factors that must be considered, remember the following:

1. Use the signs of the constant term and the coefficient of x in the trinomial to determine the signs of the binomial factors. If the constant term is positive, the signs of the binomial factors will be the same as the sign of the coefficient of x in the trinomial. If the constant term is negative, the constant terms in the binomials will have opposite signs.
2. If the terms of the trinomial do not have a common factor, then the terms in either one of the binomial factors will not have a common factor.

➡ Factor: $4x^2 + 31x - 8$

The terms of the trinomial do not have a common factor; therefore, the binomial factors will not have a common factor.

Because the constant term, c, of the trinomial is negative (-8), the constant terms of the binomial factors will have opposite signs.

LOOK CLOSELY

The binomial factor $4x - 8$ has a common factor of 4: $4x - 8 = 4(x - 2)$. Because $4x^2 + 31x + 8$ does not have a common factor, one of its binomial factors cannot have a common factor.

Find the factors of a (4) and the factors of c (-8).

Factors of 4	Factors of -8
1, 4	1, -8
2, 2	-1, 8
	2, -4
	-2, 4

Using these factors, write trial factors, and use FOIL to check the middle term of the trinomial.

Remember that if the terms of the trinomial do not have a common factor, then a binomial factor cannot have a common factor. Such trial factors need not be checked.

The correct factors have been found.

Trial Factors	Middle Term
$(x + 1)(4x - 8)$	Common factor
$(x - 1)(4x + 8)$	Common factor
$(x + 2)(4x - 4)$	Common factor
$(x - 2)(4x + 4)$	Common factor
$(2x + 1)(2x - 8)$	Common factor
$(2x - 1)(2x + 8)$	Common factor
$(2x + 2)(2x - 4)$	Common factor
$(2x - 2)(2x + 4)$	Common factor
$(4x + 1)(x - 8)$	$-32x + x = -31x$
$\mathbf{(4x - 1)(x + 8)}$	$\mathbf{32x - x = 31x}$

$$4x^2 + 31x - 8 = (4x - 1)(x + 8)$$

Other trial factors need not be checked.

The last example illustrates that many of the trial factors may have common factors and thus need not be tried. For the remainder of this chapter, the trial factors with a common factor will not be listed.

Example 5 Factor. A. $2x^2 - 21x + 10$ B. $6x^2 + 17x - 10$

Solution A. $2x^2 - 21x + 10$

Factors of 2	Factors of 10
1, 2	$-1, -10$
	$-2, -5$

▶ Use negative factors of 10.

Trial Factors	Middle Term
$(x - 2)(2x - 5)$	$-5x - 4x = -9x$
$(2x - 1)(x - 10)$	$-20x - x = -21x$

▶ Write trial factors. Use FOIL to check the middle term.

$$2x^2 - 21x + 10 = (2x - 1)(x - 10)$$

B. $6x^2 + 17x - 10$

Factors of 6	Factors of -10
1, 6	1, -10
2, 3	-1, 10
	2, -5
	-2, 5

▶ Find the factors of a (6) and the factors of c (-10).

Trial Factors	Middle Term
$(x + 2)(6x - 5)$	$-5x + 12x = 7x$
$(x - 2)(6x + 5)$	$5x - 12x = -7x$
$(2x + 1)(3x - 10)$	$-20x + 3x = -17x$
$(2x - 1)(3x + 10)$	$20x - 3x = 17x$

▶ Write trial factors. Use FOIL to check the middle term.

$$6x^2 + 17x - 10 = (2x - 1)(3x + 10)$$

Problem 5 Factor. A. $4x^2 + 15x - 4$ B. $10x^2 + 39x + 14$

Solution See page S14. A. $(4x - 1)(x + 4)$ B. $(5x + 2)(2x + 7)$

Trinomials of the form $ax^2 + bx + c$ can also be factored by grouping. This method is an extension of the method discussed in Objective 5.5.2.

To factor $ax^2 + bx + c$, first find the factors of $a \cdot c$ whose sum is b. Use the two factors to rewrite the middle term of the trinomial as the sum of two terms. Then factor by grouping to write the factorization of the trinomial.

Factor: $3x^2 + 11x + 8$

Find two positive factors of 24 $(a \cdot c = 3 \cdot 8)$ whose sum is 11, the coefficient of x.

Positive Factors of 24	Sum
1, 24	25
2, 12	14
3, 8	11

The required sum has been found. The remaining factors need not be checked.

Use the factors of 24 whose sum is 11 to write $11x$ as $3x + 8x$. Factor by grouping.

$$\begin{aligned} 3x^2 + 11x + 8 &= 3x^2 + 3x + 8x + 8 \\ &= (3x^2 + 3x) + (8x + 8) \\ &= 3x(x + 1) + 8(x + 1) \\ &= (x + 1)(3x + 8) \end{aligned}$$

Check: $(x + 1)(3x + 8) = 3x^2 + 8x + 3x + 8 = 3x^2 + 11x + 8$

Factor: $4z^2 - 17z - 21$

Find two factors of -84 $[a \cdot c = 4 \cdot (-21)]$ whose sum is -17, the coefficient of z.

When the required sum is found, the remaining factors need not be checked.

Factors of −84	Sum
1, −84	−83
−1, 84	83
2, −42	−40
−2, 42	40
3, −28	−25
−3, 28	25
4, −21	−17

Use the factors of -84 whose sum is -17 to write $-17z$ as $4z - 21z$. Factor by grouping. Recall that $-21z - 21 = -(21z + 21)$.

$$\begin{aligned} 4z^2 - 17z - 21 &= 4z^2 + 4z - 21z - 21 \\ &= (4z^2 + 4z) - (21z + 21) \\ &= 4z(z + 1) - 21(z + 1) \\ &= (z + 1)(4z - 21) \end{aligned}$$

Factor: $3x^2 - 11x + 4$

Find two negative factors of 12 $(a \cdot c = 3 \cdot 4)$ whose sum is -11.

Factors of 12	Sum
−1, −12	−13
−2, −6	−8
−3, −4	−7

Because no integer factors of 12 have a sum of -11, $3x^2 - 11x + 4$ is nonfactorable over the integers.

Either method of factoring discussed in this objective will always lead to a correct factorization of trinomials of the form $ax^2 + bx + c$ that are factorable.

Example 6 Factor. A. $2x^2 - 21x + 10$ B. $10 - 17x - 6x^2$

Solution A. $2x^2 - 21x + 10$

Factors of 20	Sum
$-1, -20$	-21

▶ $a \cdot c = 2 \cdot 10 = 20$. Find two factors of 20 whose sum is -21.

$$\begin{aligned} &2x^2 - 21x + 10 \\ &= 2x^2 - 20x - x + 10 \\ &= (2x^2 - 20x) - (x - 10) \\ &= 2x(x - 10) - (x - 10) \\ &= (x - 10)(2x - 1) \end{aligned}$$

▶ Rewrite $-21x$ as $-20x - x$.
▶ Factor by grouping.

B. $10 - 17x - 6x^2$

Factors of -60	Sum
$1, -60$	-59
$-1, 60$	59
$2, -30$	-28
$-2, 30$	28
$3, -20$	-17

▶ $a \cdot c = -60$. Find two factors of -60 whose sum is -17.

$$\begin{aligned} &10 - 17x - 6x^2 \\ &= 10 - 20x + 3x - 6x^2 \\ &= (10 - 20x) + (3x - 6x^2) \\ &= 10(1 - 2x) + 3x(1 - 2x) \\ &= (1 - 2x)(10 + 3x) \end{aligned}$$

▶ Rewrite $-17x$ as $-20x + 3x$.
▶ Factor by grouping.

Problem 6 Factor. A. $6x^2 + 7x - 20$ B. $2 - x - 6x^2$

Solution See page S15. A. $(3x - 4)(2x + 5)$ B. $(1 - 2x)(2 + 3x)$

A polynomial is factored completely when it is written as a product of factors that are nonfactorable over the integers.

LOOK CLOSELY

The first step in *any* factoring problem is to determine whether the terms of the polynomial have a **common factor.** If they do, factor it out first.

⇒ Factor: $4x^3 + 12x^2 - 160x$

Factor out the GCF of the terms. The GCF of $4x^3$, $12x^2$, and $160x$ is $4x$.

$$\begin{aligned} &4x^3 + 12x^2 - 160x \\ &= 4x(x^2 + 3x - 40) \end{aligned}$$

Factor the trinomial.

$$= 4x(x + 8)(x - 5)$$

Check: $4x(x + 8)(x - 5) = (4x^2 + 32x)(x - 5) = 4x^3 + 12x^2 - 160x$

Example 7 Factor. A. $30y + 2xy - 4x^2y$ B. $12x^3y^2 + 14x^2y - 6x$

Solution A. $30y + 2xy - 4x^2y$

$$\begin{aligned} &= 2y(15 + x - 2x^2) \\ &= 2y(5 + 2x)(3 - x) \end{aligned}$$

▶ The GCF of $30y$, $2xy$, and $4x^2y$ is $2y$.
▶ Factor out the GCF.
▶ Factor the trinomial.

B. $12x^3y^2 + 14x^2y - 6x$
$= 2x(6x^2y^2 + 7xy - 3)$
$= 2x(3xy - 1)(2xy + 3)$

▶ The GCF of $12x^3y^2$, $14x^2y$, and $6x$ is $2x$.

Problem 7 Factor. A. $3a^3b^3 + 3a^2b^2 - 60ab$ B. $40a - 10a^2 - 15a^3$

Solution See page S15. A. $3ab(ab + 5)(ab - 4)$ B. $5a(2 + a)(4 - 3a)$

CONCEPT REVIEW 5.5

Determine whether the following statements are always true, sometimes true, or never true.

1. Factoring reverses the effects of multiplication. Always true
2. A trinomial will factor into the product of two binomials. Sometimes true
3. A trinomial of the form $ax^2 + bx + c$ can be factored by using trial factors. Sometimes true
4. A polynomial with four terms can be factored by grouping. Sometimes true

EXERCISES 5.5

1 1. What is the greatest common factor (GCF) of two monomials? Make up two monomials and give their GCF.

2. What does it mean to factor a polynomial?

Factor.

3. $6a^2 - 15a$ $3a(2a - 5)$
4. $32b^2 + 12b$ $4b(8b + 3)$
5. $4x^3 - 3x^2$ $x^2(4x - 3)$
6. $12a^5b^2 + 16a^4b$ $4a^4b(3ab + 4)$
7. $3a^2 - 10b^3$ Nonfactorable
8. $9x^2 + 14y^4$ Nonfactorable
9. $x^5 - x^3 - x$ $x(x^4 - x^2 - 1)$
10. $y^4 - 3y^2 - 2y$ $y(y^3 - 3y - 2)$
11. $16x^2 - 12x + 24$ $4(4x^2 - 3x + 6)$
12. $2x^5 + 3x^4 - 4x^2$ $x^2(2x^3 + 3x^2 - 4)$
13. $5b^2 - 10b^3 + 25b^4$ $5b^2(1 - 2b + 5b^2)$
14. $x^2y^4 - x^2y - 4x^2$ $x^2(y^4 - y - 4)$
15. $x^{2n} - x^n$ $x^n(x^n - 1)$
16. $a^{5n} + a^{2n}$ $a^{2n}(a^{3n} + 1)$
17. $x^{3n} - x^{2n}$ $x^{2n}(x^n - 1)$
18. $y^{4n} + y^{2n}$ $y^{2n}(y^{2n} + 1)$
19. $a^{2n+2} + a^2$ $a^2(a^{2n} + 1)$
20. $b^{n+5} - b^5$ $b^5(b^n - 1)$

21. $12x^2y^2 - 18x^3y + 24x^2y$ $6x^2y(2y - 3x + 4)$

22. $14a^4b^4 - 42a^3b^3 + 28a^3b^2$ $14a^3b^2(ab^2 - 3b + 2)$

23. $-16a^2b^4 - 4a^2b^2 + 24a^3b^2$ $4a^2b^2(-4b^2 - 1 + 6a)$

24. $10x^2y + 20x^2y^2 + 30x^2y^3$ $10x^2y(1 + 2y + 3y^2)$

25. $y^{2n+2} + y^{n+2} - y^2$ $y^2(y^{2n} + y^n - 1)$

26. $a^{2n+2} + a^{2n+1} + a^n$ $a^n(a^{n+2} + a^{n+1} + 1)$

2 Factor.

27. $x(a + 2) - 2(a + 2)$
$(a + 2)(x - 2)$

28. $3(x + y) + a(x + y)$
$(x + y)(3 + a)$

29. $a(x - 2) - b(2 - x)$
$(x - 2)(a + b)$

30. $3(a - 7) - b(7 - a)$
$(a - 7)(3 + b)$

31. $x^2 + 3x + 2x + 6$
$(x + 3)(x + 2)$

32. $x^2 - 5x + 4x - 20$
$(x - 5)(x + 4)$

33. $xy + 4y - 2x - 8$
$(x + 4)(y - 2)$

34. $ab + 7b - 3a - 21$
$(a + 7)(b - 3)$

35. $ax + bx - ay - by$
$(a + b)(x - y)$

36. $2ax - 3ay - 2bx + 3by$
$(2x - 3y)(a - b)$

37. $x^2y - 3x^2 - 2y + 6$
$(y - 3)(x^2 - 2)$

38. $a^2b + 3a^2 + 2b + 6$
$(b + 3)(a^2 + 2)$

39. $6 + 2y + 3x^2 + x^2y$
$(3 + y)(2 + x^2)$

40. $15 + 3b - 5a^2 - a^2b$
$(5 + b)(3 - a^2)$

41. $2ax^2 + bx^2 - 4ay - 2by$
$(2a + b)(x^2 - 2y)$

42. $4a^2x + 2a^2y - 6bx - 3by$
$(2x + y)(2a^2 - 3b)$

43. $x^ny - 5x^n + y - 5$
$(y - 5)(x^n + 1)$

44. $a^nx^n + 2a^n + x^n + 2$
$(x^n + 2)(a^n + 1)$

45. $x^3 + x^2 + 2x + 2$ $(x + 1)(x^2 + 2)$

46. $y^3 - y^2 + 3y - 3$ $(y - 1)(y^2 + 3)$

47. $2x^3 - x^2 + 4x - 2$ $(2x - 1)(x^2 + 2)$

48. $2y^3 - y^2 + 6y - 3$ $(2y - 1)(y^2 + 3)$

49. What is a quadratic trinomial and what does it mean to factor a quadratic trinomial?

50. What does the phrase "nonfactorable over the integers" mean?

3 Factor.

51. $x^2 - 8x + 15$
$(x - 5)(x - 3)$

52. $x^2 + 12x + 20$
$(x + 10)(x + 2)$

53. $a^2 + 12a + 11$
$(a + 11)(a + 1)$

54. $a^2 + a - 72$
$(a + 9)(a - 8)$

55. $b^2 + 2b - 35$
$(b + 7)(b - 5)$

56. $a^2 + 7a + 6$
$(a + 6)(a + 1)$

57. $y^2 - 16y + 39$
$(y - 3)(y - 13)$

58. $y^2 - 18y + 72$
$(y - 6)(y - 12)$

59. $b^2 + 4b - 32$
$(b + 8)(b - 4)$

60. $x^2 + x - 132$
$(x + 12)(x - 11)$

61. $a^2 - 15a + 56$
$(a - 7)(a - 8)$

62. $x^2 + 15x + 50$
$(x + 10)(x + 5)$

63. $y^2 + 13y + 12$
$(y + 12)(y + 1)$

64. $b^2 - 6b - 16$
$(b - 8)(b + 2)$

65. $x^2 + 4x - 5$
$(x + 5)(x - 1)$

66. $a^2 - 3ab + 2b^2$
$(a - 2b)(a - b)$

67. $a^2 + 11ab + 30b^2$
$(a + 6b)(a + 5b)$

68. $a^2 + 8ab - 33b^2$
$(a + 11b)(a - 3b)$

69. $x^2 - 14xy + 24y^2$
$(x - 12y)(x - 2y)$

70. $x^2 + 5xy + 6y^2$
$(x + 2y)(x + 3y)$

71. $y^2 + 2xy - 63x^2$
$(y + 9x)(y - 7x)$

72. $y^2 - 13y + 12$
$(y - 1)(y - 12)$

73. $x^2 - 35x - 36$
$(x - 36)(x + 1)$

74. $x^2 + 7x - 18$
$(x - 2)(x + 9)$

75. $a^2 + 13a + 36$
$(a + 9)(a + 4)$

76. $x^2 - 5x + 7$
Nonfactorable

77. $x^2 - 7x - 12$
Nonfactorable

4 Factor.

78. $2x^2 + 7x + 3$
$(2x + 1)(x + 3)$

79. $2x^2 - 11x - 40$
$(2x + 5)(x - 8)$

80. $6y^2 + 5y - 6$
$(2y + 3)(3y - 2)$

81. $4y^2 - 15y + 9$
$(4y - 3)(y - 3)$

82. $6b^2 - b - 35$
$(2b - 5)(3b + 7)$

83. $2a^2 + 13a + 6$
$(2a + 1)(a + 6)$

84. $3y^2 - 22y + 39$
$(3y - 13)(y - 3)$

85. $12y^2 - 13y - 72$
Nonfactorable

86. $6a^2 - 26a + 15$
Nonfactorable

87. $5x^2 + 26x + 5$
$(5x + 1)(x + 5)$

88. $4a^2 - a - 5$
$(4a - 5)(a + 1)$

89. $11x^2 - 122x + 11$
$(11x - 1)(x - 11)$

90. $11y^2 - 47y + 12$
$(11y - 3)(y - 4)$

91. $12x^2 - 17x + 5$
$(12x - 5)(x - 1)$

92. $12x^2 - 40x + 25$
$(6x - 5)(2x - 5)$

93. $8y^2 - 18y + 9$
$(4y - 3)(2y - 3)$

94. $4x^2 + 9x + 10$
Nonfactorable

95. $6a^2 - 5a - 2$
Nonfactorable

96. $10x^2 - 29x + 10$
$(5x - 2)(2x - 5)$

97. $2x^2 + 5x + 12$
Nonfactorable

98. $4x^2 - 6x + 1$
Nonfactorable

99. $6x^2 + 5xy - 21y^2$
$(2x - 3y)(3x + 7y)$

100. $6x^2 + 41xy - 7y^2$
$(6x - y)(x + 7y)$

101. $4a^2 + 43ab + 63b^2$
$(4a + 7b)(a + 9b)$

102. $7a^2 + 46ab - 21b^2$
$(7a - 3b)(a + 7b)$

103. $10x^2 - 23xy + 12y^2$
$(5x - 4y)(2x - 3y)$

104. $18x^2 + 27xy + 10y^2$
$(6x + 5y)(3x + 2y)$

105. $24 + 13x - 2x^2$
$(8 - x)(3 + 2x)$

106. $6 - 7x - 5x^2$
$(2 + x)(3 - 5x)$

107. $8 - 13x + 6x^2$
Nonfactorable

108. $30 + 17a - 20a^2$
Nonfactorable

109. $15 - 14a - 8a^2$
$(3 - 4a)(5 + 2a)$

110. $35 - 6b - 8b^2$
$(7 - 4b)(5 + 2b)$

111. $5y^4 - 29y^3 + 20y^2$
$y^2(5y - 4)(y - 5)$

112. $30a^2 + 85ab + 60b^2$
$5(2a + 3b)(3a + 4b)$

113. $4x^3 + 10x^2y - 24xy^2$
$2x(2x - 3y)(x + 4y)$

114. $8a^4 + 37a^3b - 15a^2b^2$
$a^2(8a - 3b)(a + 5b)$

115. $100 - 5x - 5x^2$
$5(5 + x)(4 - x)$

116. $50x^2 + 25x^3 - 12x^4$
$x^2(5 + 4x)(10 - 3x)$

117. $320x - 8x^2 - 4x^3$
$4x(10 + x)(8 - x)$

118. $96y - 16xy - 2x^2y$
$2y(12 + x)(4 - x)$

119. $20x^2 - 38x^3 - 30x^4$
$2x^2(5 + 3x)(2 - 5x)$

120. $4x^2y^2 - 32xy + 60$
$4(xy - 3)(xy - 5)$

121. $a^4b^4 - 3a^3b^3 - 10a^2b^2$
$a^2b^2(ab - 5)(ab + 2)$

122. $2a^2b^4 + 9ab^3 - 18b^2$
$b^2(2ab - 3)(ab + 6)$

123. $90a^2b^2 + 45ab + 10$
$5(18a^2b^2 + 9ab + 2)$

124. $3x^3y^2 + 12x^2y - 96x$
$3x(xy + 8)(xy - 4)$

125. $4x^4 - 45x^2 + 80$
Nonfactorable

126. $x^4 + 2x^2 + 15$
Nonfactorable

127. $2a^5 + 14a^3 + 20a$
$2a(a^2 + 5)(a^2 + 2)$

128. $3b^6 - 9b^4 - 30b^2$
$3b^2(b^2 - 5)(b^2 + 2)$

129. $3x^4y^2 - 39x^2y^2 + 120y^2$
$3y^2(x^2 - 5)(x^2 - 8)$

130. $2x^4y - 7x^3y - 30x^2y$
$x^2y(2x + 5)(x - 6)$

131. $45a^2b^2 + 6ab^2 - 72b^2$
$3b^2(3a + 4)(5a - 6)$

132. $16x^2y^3 + 36x^2y^2 + 20x^2y$
$4x^2y(4y + 5)(y + 1)$

133. $36x^3y + 24x^2y^2 - 45xy^3$
$3xy(6x - 5y)(2x + 3y)$

134. $12a^3b - 70a^2b^2 - 12ab^3$
$2ab(6a + b)(a - 6b)$

135. $48a^2b^2 - 36ab^3 - 54b^4$
$6b^2(2a - 3b)(4a + 3b)$

136. $x^{3n} + 10x^{2n} + 16x^n$
$x^n(x^n + 8)(x^n + 2)$

137. $10x^{2n} + 25x^n - 60$
$5(2x^n - 3)(x^n + 4)$

APPLYING CONCEPTS 5.5

Factor.

138. $2a^3b - ab^3 - a^2b^2$ $ab(2a + b)(a - b)$

139. $3x^3y - xy^3 - 2x^2y^2$ $xy(3x + y)(x - y)$

140. $2y^3 + 2y^5 - 24y$ $2y(y^2 + 4)(y^2 - 3)$

141. $9b^3 + 3b^5 - 30b$ $3b(b^2 + 5)(b^2 - 2)$

Find all integers k such that the trinomial can be factored over the integers.

[C] **142.** $x^2 + kx + 8$ $6, -6, 9, -9$ [C] **143.** $x^2 + kx - 6$ $5, -5, 1, -1$ [C] **144.** $2x^2 - kx + 3$ $7, -7, 5, -5$

[C] **145.** $2x^2 - kx - 5$ $3, -3, 9, -9$ [C] **146.** $3x^2 + kx + 5$ $16, -16, 8, -8$ [C] **147.** $2x^2 + kx - 3$ $1, -1, 5, -5$

148. *Geometry* Write the area of the shaded region in factored form.

a.

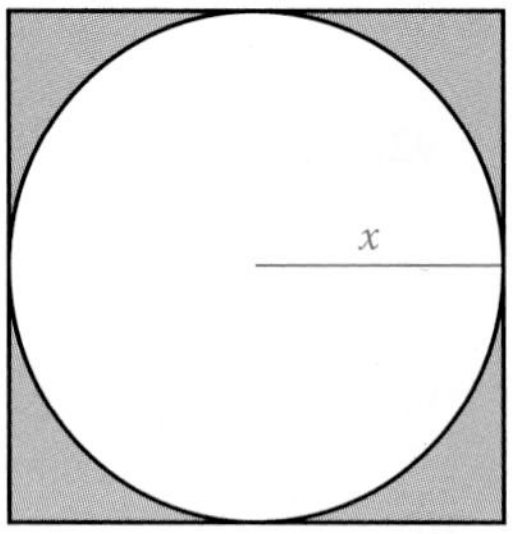

$x^2(4 - \pi)$

b.

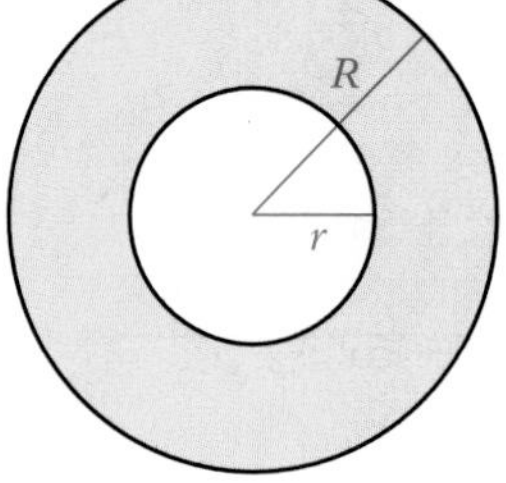

$\pi(R^2 - r^2)$

c.

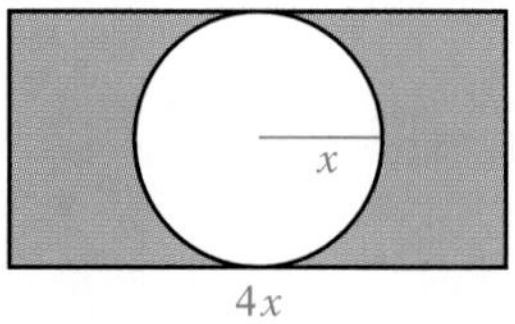

$x^2(8 - \pi)$

SECTION 5.6 Special Factoring

1 Factor the difference of two perfect squares and factor perfect-square trinomials

The product of a term and itself is called a **perfect square.** The exponents on variables of perfect squares are always even numbers.

Term		Perfect Square
5	$5 \cdot 5 =$	25
x	$x \cdot x =$	x^2
$3y^4$	$3y^4 \cdot 3y^4 =$	$9y^8$
x^n	$x^n \cdot x^n =$	x^{2n}

The **square root of a perfect square** is one of the two equal factors of the perfect square. $\sqrt{\ }$ is the symbol for square root. To find the exponent of the square root of a variable term, divide the exponent by 2. For the examples at the right, assume that x and y represent positive numbers.

$\sqrt{25} = 5$
$\sqrt{x^2} = x$
$\sqrt{9y^8} = 3y^4$
$\sqrt{x^{2n}} = x^n$

The factors of the difference of two perfect squares are the sum and difference of the square roots of the perfect squares.

LOOK CLOSELY

$(a + b)$ is the sum of the two terms a and b. $(a - b)$ is the difference of the two terms a and b. a and b are the square roots of a^2 and b^2.

Factors of the Difference of Two Perfect Squares

$$a^2 - b^2 = (a + b)(a - b)$$

The *sum* of two perfect squares, $a^2 + b^2$, is nonfactorable over the integers.

Example 1 Factor: $25x^2 - 1$

Solution $25x^2 - 1 = (5x)^2 - 1^2$ ▸ Write the binomial as the difference of two perfect squares.

$= (5x + 1)(5x - 1)$ ▸ The factors are the sum and difference of the square roots of the perfect squares.

Problem 1 Factor: $x^2 - 36y^4$

Solution See page S15. $(x + 6y^2)(x - 6y^2)$

A perfect-square trinomial is the square of a binomial.

Factors of a Perfect-Square Trinomial

$$a^2 + 2ab + b^2 = (a + b)^2$$
$$a^2 - 2ab + b^2 = (a - b)^2$$

In factoring a perfect-square trinomial, remember that the terms of the binomial are the square roots of the perfect squares of the trinomial. The sign in the binomial is the sign of the middle term of the trinomial.

➡ Factor: $4x^2 - 12x + 9$

Because $4x^2$ is a perfect square and 9 is a perfect square, try factoring $4x^2 - 12x + 9$ as the square of a binomial.

$$4x^2 - 12x + 9 \stackrel{?}{=} (2x - 3)^2$$

Check: $(2x - 3)^2 = (2x - 3)(2x - 3) = 4x^2 - 6x - 6x + 9 = 4x^2 - 12x + 9$

The factorization checks. Thus $4x^2 - 12x + 9 = (2x - 3)^2$.

It is important to check a proposed factorization as we did above. The next example illustrates the importance of this check.

INSTRUCTOR NOTE

Another example of what appears to be a perfect-square trinomial, but does not factor, is $4x^2 - 17x + 25$.

➡ Factor: $x^2 + 13x + 36$

Because x^2 is a perfect square and 36 is a perfect square, try factoring $x^2 + 13x + 36$ as the square of a binomial.

$$x^2 + 13x + 36 \stackrel{?}{=} (x + 6)^2$$

Check: $(x + 6)^2 = (x + 6)(x + 6) = x^2 + 6x + 6x + 36 = x^2 + 12x + 36$

In this case, the proposed factorization of $x^2 + 13x + 36$ does *not* check. Try another factorization. The numbers 4 and 9 are factors of 36 whose sum is 13.

$$x^2 + 13x + 36 = (x + 4)(x + 9)$$

Example 2 Factor: $4x^2 - 20x + 25$

Solution $4x^2 - 20x + 25 = (2x - 5)^2$

Problem 2 Factor: $9x^2 + 12x + 4$

Solution See page S15. $(3x + 2)^2$

2 Factor the sum or the difference of two cubes

The product of the same three factors is called a **perfect cube.** The exponents on variables of perfect cubes are always divisible by 3.

Term		Perfect Cube
2	$2 \cdot 2 \cdot 2 =$	8
$3y$	$3y \cdot 3y \cdot 3y =$	$27y^3$
x^2	$x^2 \cdot x^2 \cdot x^2 =$	x^6
x^n	$x^n \cdot x^n \cdot x^n =$	x^{3n}

The **cube root** of a perfect cube is one of the three equal factors of the perfect cube. $\sqrt[3]{}$ is the symbol for cube root. To find the exponent of the cube root of a perfect-cube variable term, divide the exponent by 3.

$$\sqrt[3]{8} = 2$$
$$\sqrt[3]{27y^3} = 3y$$
$$\sqrt[3]{x^6} = x^2$$
$$\sqrt[3]{x^{3n}} = x^n$$

Factoring the Sum or Difference of Two Cubes

$$a^3 + b^3 = (a + b)(a^2 - ab + b^2)$$
$$a^3 - b^3 = (a - b)(a^2 + ab + b^2)$$

To factor $8x^3 - 27$:

Write the binomial as the difference of two perfect cubes.

$$8x^3 - 27 = (2x)^3 - 3^3$$

The terms of the binomial factor are the cube roots of the perfect cubes. The sign of the binomial factor is the same sign as in the given binomial. The trinomial factor is obtained from the binomial factor.

$$= (2x - 3)(4x^2 + 6x + 9)$$

Square of the first term ($4x^2$)

Opposite of the product of the two terms ($6x$)

Square of the last term (9)

Example 3 Factor. A. $x^3y^3 - 1$ B. $(x + y)^3 - x^3$

Solution A. $x^3y^3 - 1$

$= (xy)^3 - 1^3$ ▶ Write the binomial as the difference of two perfect cubes.

$= (xy - 1)(x^2y^2 + xy + 1)$

B. $(x + y)^3 - x^3$
$= [(x + y) - x][(x + y)^2 + x(x + y) + x^2]$ ▸ This is a difference of two perfect cubes.
$= y(x^2 + 2xy + y^2 + x^2 + xy + x^2)$ ▸ Simplify.
$= y(3x^2 + 3xy + y^2)$

Problem 3 Factor. A. $8x^3 + y^3z^3$ B. $(x - y)^3 + (x + y)^3$

Solution See page S15. A. $(2x + yz)(4x^2 - 2xyz + y^2z^2)$
B. $2x(x^2 + 3y^2)$

3 Factor trinomials that are quadratic in form

HM[3]

Certain trinomials can be expressed as quadratic trinomials by making suitable variable substitutions. A trinomial is quadratic in form if it can be written as

$$au^2 + bu + c$$

If we let $x^2 = u$, the trinomial $x^4 + 5x^2 + 6$ can be written as shown at the right. The trinomial is quadratic in form.

$$\begin{aligned} &x^4 + 5x^2 + 6 \\ &= (x^2)^2 + 5(x^2) + 6 \\ &= u^2 + 5u + 6 \end{aligned}$$

If we let $xy = u$, the trinomial $2x^2y^2 + 3xy - 9$ can be written as shown at the right. The trinomial is quadratic in form.

$$\begin{aligned} &2x^2y^2 + 3xy - 9 \\ &= 2(xy)^2 + 3(xy) - 9 \\ &= 2u^2 + 3u - 9 \end{aligned}$$

When a trinomial that is quadratic in form is factored, the variable part of the first term in each binomial factor will be u. For example, because $x^4 + 5x^2 + 6$ is quadratic in form when $x^2 = u$, the first term in each binomial factor will be x^2.

$$\begin{aligned} x^4 + 5x^2 + 6 &= (x^2)^2 + 5(x^2) + 6 \\ &= (x^2 + 2)(x^2 + 3) \end{aligned}$$

The trinomial $x^2y^2 - 2xy - 15$ is quadratic in form when $xy = u$. The first term in each binomial factor will be xy.

$$\begin{aligned} x^2y^2 - 2xy - 15 &= (xy)^2 - 2(xy) - 15 \\ &= (xy + 3)(xy - 5) \end{aligned}$$

Example 4 Factor. A. $6x^2y^2 - xy - 12$ B. $2x^4 + 5x^2 - 12$

Solution A. $6x^2y^2 - xy - 12$
$= (3xy + 4)(2xy - 3)$ ▸ The trinomial is quadratic in form when $xy = u$.

B. $2x^4 + 5x^2 - 12$
$= (x^2 + 4)(2x^2 - 3)$ ▸ The trinomial is quadratic in form when $x^2 = u$.

Problem 4 Factor. A. $6x^2y^2 - 19xy + 10$ B. $3x^4 + 4x^2 - 4$

Solution See page S15. A. $(3xy - 2)(2xy - 5)$ B. $(x^2 + 2)(3x^2 - 2)$

4 Factor completely

20 HM[3]

When factoring a polynomial completely, ask yourself the following questions about the polynomial:

1. Is there a common factor? If so, factor out the GCF.
2. If the polynomial is a binomial, is it the difference of two perfect squares, the sum of two cubes, or the difference of two cubes? If so, factor.
3. If the polynomial is a trinomial, is it a perfect-square trinomial or the product of two binomials? If so, factor.
4. If the polynomial has four terms, can it be factored by grouping?
5. Is each factor nonfactorable over the integers? If not, factor.

Example 5 Factor. A. $x^2y + 2x^2 - y - 2$ B. $x^{4n} - y^{4n}$

Solution A. $x^2y + 2x^2 - y - 2$

$= (x^2y + 2x^2) - (y + 2)$ ▶ Factor by grouping.

$= x^2(y + 2) - (y + 2)$

$= (y + 2)(x^2 - 1)$

$= (y + 2)(x + 1)(x - 1)$ ▶ Factor the difference of two perfect squares.

B. $x^{4n} - y^{4n}$

$= (x^{2n})^2 - (y^{2n})^2$ ▶ Factor the difference of two perfect squares.

$= (x^{2n} + y^{2n})(x^{2n} - y^{2n})$

$= (x^{2n} + y^{2n})[(x^n)^2 - (y^n)^2]$ ▶ Factor the difference of two perfect squares.

$= (x^{2n} + y^{2n})(x^n + y^n)(x^n - y^n)$

Problem 5 Factor. A. $4x - 4y - x^3 + x^2y$ B. $x^{4n} - x^{2n}y^{2n}$

Solution See page S15. A. $(x - y)(2 + x)(2 - x)$
B. $x^{2n}(x^n + y^n)(x^n - y^n)$

CONCEPT REVIEW 5.6

Determine whether the following statements are always true, sometimes true, or never true.

1. $a^2b^2c^2$ is an example of a perfect-square trinomial. Never true

2. The factors of $a^3 + b^3$ are $(a + b)(a + b)(a + b)$. Never true

3. The difference of two perfect cubes is not factorable. Never true

4. Factoring out the GCF is the first thing to try in factoring polynomials. Always true

5. The polynomial $b^3 + 1$ is not factorable. Never true

EXERCISES 5.6

1 1. Explain how to factor the difference of two perfect squares.

2. What is a perfect-square trinomial?

Factor.

3. $x^2 - 16$ $(x+4)(x-4)$
4. $y^2 - 49$ $(y+7)(y-7)$
5. $4x^2 - 1$ $(2x+1)(2x-1)$
6. $81x^2 - 4$ $(9x+2)(9x-2)$
7. $b^2 - 2b + 1$ $(b-1)^2$
8. $a^2 + 14a + 49$ $(a+7)^2$
9. $16x^2 - 40x + 25$ $(4x-5)^2$
10. $49x^2 + 28x + 4$ $(7x+2)^2$
11. $x^2y^2 - 100$ $(xy+10)(xy-10)$
12. $a^2b^2 - 25$ $(ab+5)(ab-5)$
13. $x^2 + 4$ Nonfactorable
14. $a^2 + 16$ Nonfactorable
15. $x^2 + 6xy + 9y^2$ $(x+3y)^2$
16. $4x^2y^2 + 12xy + 9$ $(2xy+3)^2$
17. $4x^2 - y^2$ $(2x+y)(2x-y)$
18. $49a^2 - 16b^4$ $(7a+4b^2)(7a-4b^2)$
19. $a^{2n} - 1$ $(a^n+1)(a^n-1)$
20. $b^{2n} - 16$ $(b^n+4)(b^n-4)$
21. $a^2 + 4a + 4$ $(a+2)^2$
22. $b^2 - 18b + 81$ $(b-9)^2$
23. $x^2 - 12x + 36$ $(x-6)^2$
24. $y^2 - 6y + 9$ $(y-3)^2$
25. $16x^2 - 121$ $(4x+11)(4x-11)$
26. $49y^2 - 36$ $(7y+6)(7y-6)$
27. $1 - 9a^2$ $(1+3a)(1-3a)$
28. $16 - 81y^2$ $(4+9y)(4-9y)$
29. $4a^2 + 4a - 1$ Nonfactorable
30. $9x^2 + 12x - 4$ Nonfactorable
31. $b^2 + 7b + 14$ Nonfactorable
32. $y^2 - 5y + 25$ Nonfactorable
33. $25 - a^2b^2$ $(5+ab)(5-ab)$
34. $64 - x^2y^2$ $(8+xy)(8-xy)$
35. $25a^2 - 40ab + 16b^2$ $(5a-4b)^2$
36. $4a^2 - 36ab + 81b^2$ $(2a-9b)^2$
37. $x^{2n} + 6x^n + 9$ $(x^n+3)^2$
38. $y^{2n} - 16y^n + 64$ $(y^n-8)^2$

2 Factor.

39. $x^3 - 27$ $(x-3)(x^2+3x+9)$
40. $y^3 + 125$ $(y+5)(y^2-5y+25)$
41. $8x^3 - 1$ $(2x-1)(4x^2+2x+1)$
42. $64a^3 + 27$ $(4a+3)(16a^2-12a+9)$
43. $x^3 - y^3$ $(x-y)(x^2+xy+y^2)$
44. $x^3 - 8y^3$ $(x-2y)(x^2+2xy+4y^2)$
45. $m^3 + n^3$ $(m+n)(m^2-mn+n^2)$
46. $27a^3 + b^3$ $(3a+b)(9a^2-3ab+b^2)$
47. $64x^3 + 1$ $(4x+1)(16x^2-4x+1)$
48. $1 - 125b^3$ $(1-5b)(1+5b+25b^2)$
49. $27x^3 - 8y^3$ $(3x-2y)(9x^2+6xy+4y^2)$
50. $64x^3 + 27y^3$ $(4x+3y)(16x^2-12xy+9y^2)$
51. $x^3y^3 + 64$ $(xy+4)(x^2y^2-4xy+16)$
52. $8x^3y^3 + 27$ $(2xy+3)(4x^2y^2-6xy+9)$
53. $16x^3 - y^3$ Nonfactorable
54. $27x^3 - 8y^3$ $(3x-2y)(9x^2+6xy+4y^2)$
55. $8x^3 - 9y^3$ Nonfactorable
56. $27a^3 - 16$ Nonfactorable
57. $(a-b)^3 - b^3$ $(a-2b)(a^2-ab+b^2)$
58. $a^3 + (a+b)^3$ $(2a+b)(a^2+ab+b^2)$
59. $x^{6n} + y^{3n}$ $(x^{2n}+y^n)(x^{4n}-x^{2n}y^n+y^{2n})$
60. $x^{3n} + y^{3n}$ $(x^n+y^n)(x^{2n}-x^ny^n+y^{2n})$

61. What does it mean for a polynomial to be quadratic in form?

62. Do all polynomials that are quadratic in form factor? If not, give an example of one that does not factor.

3 Factor.

63. $x^2y^2 - 8xy + 15$
$(xy - 3)(xy - 5)$

64. $x^2y^2 - 8xy - 33$
$(xy - 11)(xy + 3)$

65. $x^2y^2 - 17xy + 60$
$(xy - 5)(xy - 12)$

66. $a^2b^2 + 10ab + 24$
$(ab + 4)(ab + 6)$

67. $x^4 - 9x^2 + 18$
$(x^2 - 6)(x^2 - 3)$

68. $y^4 - 6y^2 - 16$
$(y^2 - 8)(y^2 + 2)$

69. $b^4 - 13b^2 - 90$
$(b^2 - 18)(b^2 + 5)$

70. $a^4 + 14a^2 + 45$
$(a^2 + 9)(a^2 + 5)$

71. $x^4y^4 - 8x^2y^2 + 12$
$(x^2y^2 - 6)(x^2y^2 - 2)$

72. $a^4b^4 + 11a^2b^2 - 26$
$(a^2b^2 + 13)(a^2b^2 - 2)$

73. $x^{2n} + 3x^n + 2$
$(x^n + 2)(x^n + 1)$

74. $a^{2n} - a^n - 12$
$(a^n - 4)(a^n + 3)$

75. $3x^2y^2 - 14xy + 15$
$(3xy - 5)(xy - 3)$

76. $5x^2y^2 - 59xy + 44$
$(5xy - 4)(xy - 11)$

77. $6a^2b^2 - 23ab + 21$
$(2ab - 3)(3ab - 7)$

78. $10a^2b^2 + 3ab - 7$
$(ab + 1)(10ab - 7)$

79. $2x^4 - 13x^2 - 15$
$(2x^2 - 15)(x^2 + 1)$

80. $3x^4 + 20x^2 + 32$
$(3x^2 + 8)(x^2 + 4)$

81. $2x^{2n} - 7x^n + 3$
$(2x^n - 1)(x^n - 3)$

82. $4x^{2n} + 8x^n - 5$
$(2x^n - 1)(2x^n + 5)$

83. $6a^{2n} + 19a^n + 10$
$(2a^n + 5)(3a^n + 2)$

4 Factor.

84. $5x^2 + 10x + 5$
$5(x + 1)^2$

85. $12x^2 - 36x + 27$
$3(2x - 3)^2$

86. $3x^4 - 81x$
$3x(x - 3)(x^2 + 3x + 9)$

87. $27a^4 - a$
$a(3a - 1)(9a^2 + 3a + 1)$

88. $7x^2 - 28$
$7(x + 2)(x - 2)$

89. $20x^2 - 5$
$5(2x + 1)(2x - 1)$

90. $y^4 - 10y^3 + 21y^2$
$y^2(y - 7)(y - 3)$

91. $y^5 + 6y^4 - 55y^3$
$y^3(y + 11)(y - 5)$

92. $x^4 - 16$
$(x^2 + 4)(x + 2)(x - 2)$

93. $16x^4 - 81$
$(4x^2 + 9)(2x + 3)(2x - 3)$

94. $8x^5 - 98x^3$
$2x^3(2x + 7)(2x - 7)$

95. $16a - 2a^4$
$2a(2 - a)(4 + 2a + a^2)$

96. $x^3y^3 - x^3$
$x^3(y - 1)(y^2 + y + 1)$

97. $x^3 + 2x^2 - x - 2$
$(x + 2)(x + 1)(x - 1)$

98. $2x^3 - 3x^2 - 8x + 12$
$(2x - 3)(x + 2)(x - 2)$

99. $2x^3 + 4x^2 - 3x - 6$
$(x + 2)(2x^2 - 3)$

100. $3x^3 - 3x^2 + 4x - 4$
$(x - 1)(3x^2 + 4)$

101. $x^3 + x^2 - 16x - 16$
$(x + 1)(x + 4)(x - 4)$

102. $4x^3 + 8x^2 - 9x - 18$
$(x + 2)(2x + 3)(2x - 3)$

103. $a^3b^6 - b^3$
$b^3(ab - 1)(a^2b^2 + ab + 1)$

104. $x^6y^6 - x^3y^3$
$x^3y^3(xy - 1)(x^2y^2 + xy + 1)$

105. $x^4 - 2x^3 - 35x^2$
$x^2(x - 7)(x + 5)$

106. $x^4 + 15x^3 - 56x^2$
$x^2(x^2 + 15x - 56)$

107. $4x^2 + 4x - 1$
Nonfactorable

108. $8x^4 - 40x^3 + 50x^2$
$2x^2(2x - 5)^2$

109. $6x^5 + 74x^4 + 24x^3$
$2x^3(3x + 1)(x + 12)$

110. $x^4 - y^4$
$(x^2 + y^2)(x + y)(x - y)$

111. $16a^4 - b^4$
$(4a^2 + b^2)(2a + b)(2a - b)$

112. $x^6 + y^6$
$(x^2 + y^2)(x^4 - x^2y^2 + y^4)$

113. $x^4 - 5x^2 - 4$
Nonfactorable

114. $a^4 - 25a^2 - 144$
Nonfactorable

115. $3b^5 - 24b^2$
$3b^2(b - 2)(b^2 + 2b + 4)$

116. $16a^4 - 2a$
$2a(2a - 1)(4a^2 + 2a + 1)$

117. $x^4y^2 - 5x^3y^3 + 6x^2y^4$
$x^2y^2(x - 3y)(x - 2y)$

118. $a^4b^2 - 8a^3b^3 - 48a^2b^4$
$a^2b^2(a + 4b)(a - 12b)$

119. $16x^3y + 4x^2y^2 - 42xy^3$
$2xy(4x + 7y)(2x - 3y)$

120. $24a^2b^2 - 14ab^3 - 90b^4$
$2b^2(4a - 9b)(3a + 5b)$

121. $x^3 - 2x^2 - x + 2$
$(x - 2)(x + 1)(x - 1)$

122. $x^3 - 2x^2 - 4x + 8$
$(x - 2)^2(x + 2)$

123. $8xb - 8x - 4b + 4$ $4(b - 1)(2x - 1)$

124. $4xy + 8x + 4y + 8$ $4(x + 1)(y + 2)$

125. $4x^2y^2 - 4x^2 - 9y^2 + 9$ $(y + 1)(y - 1)(2x + 3)(2x - 3)$

126. $4x^4 - x^2 - 4x^2y^2 + y^2$ $(2x + 1)(2x - 1)(x + y)(x - y)$

127. $x^5 - 4x^3 - 8x^2 + 32$ $(x + 2)(x - 2)^2(x^2 + 2x + 4)$

128. $x^6y^3 + x^3 - x^3y^3 - 1$ $(xy + 1)(x^2y^2 - xy + 1)(x - 1)(x^2 + x + 1)$

129. $a^{2n+2} - 6a^{n+2} + 9a^2$ $a^2(a^n - 3)^2$

130. $x^{2n+1} + 2x^{n+1} + x$ $x(x^n + 1)^2$

131. $2x^{n+2} - 7x^{n+1} + 3x^n$ $x^n(2x - 1)(x - 3)$

132. $3b^{n+2} + 4b^{n+1} - 4b^n$ $b^n(3b - 2)(b + 2)$

APPLYING CONCEPTS 5.6

Find all integers k such that the trinomial is a perfect-square trinomial.

133. $4x^2 - kx + 25$ 20, −20

134. $9x^2 - kx + 1$ 6, −6

135. $16x^2 + kxy + y^2$ 8, −8

136. $49x^2 + kxy + 64y^2$ 112, −112

Factor.

137. $ax^3 + b - bx^3 - a$ $(x - 1)(x^2 + x + 1)(a - b)$

138. $xy^2 - 2b - x + 2by^2$ $(y + 1)(y - 1)(x + 2b)$

139. $y^{8n} - 2y^{4n} + 1$ $(y^{2n} + 1)^2(y^n - 1)^2(y^n + 1)^2$

140. $x^{6n} - 1$ $(x^n - 1)(x^{2n} + x^n + 1)(x^n + 1)(x^{2n} - x^n + 1)$

Solve.

[C] **141.** *Number Problem* The product of two numbers is 63. One of the two numbers is a perfect square. The other is a prime number. Find the sum of the two numbers. 16

[C] **142.** *Number Problem* What is the smallest whole number by which 250 can be multiplied so that the product will be a perfect square? 10

[C] **143.** *Number Problem* Palindromic numbers are natural numbers that remain unchanged when their digits are written in reverse order. Find all perfect squares less than 500 that are palindromic numbers. 1, 4, 9, 121, 484

[C] **144.** *Food Sciences* A large circular cookie is cut from a square piece of dough. The diameter of the cookie is x cm. The piece of dough is x cm on a side and is 1 cm deep. In terms of x, how many cubic centimeters of dough are left over? Use 3.14 for π. $0.215x^2$ cm^3

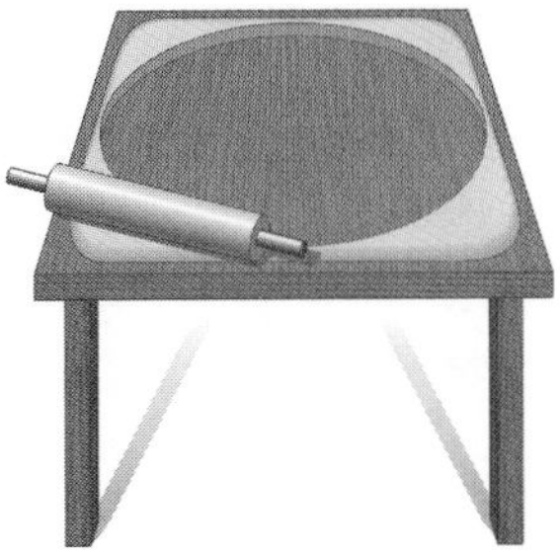

145. Factor $x^4 + 64$. (*Suggestion:* Add and subtract $16x^2$ so that the expression becomes $(x^4 + 16x^2 + 64) - 16x^2$. Now factor the difference of two squares.) $(x^2 - 4x + 8)(x^2 + 4x + 8)$

146. Using the strategy of Exercise 145, factor $x^4 + 4$. (*Suggestion:* Add and subtract $4x^2$.) $(x^2 - 2x + 2)(x^2 + 2x + 2)$

147. Can a third-degree polynomial have factors $(x - 1)$, $(x + 1)$, $(x - 3)$, and $(x + 4)$? Why or why not?

148. Given that $(x - 3)$ and $(x + 4)$ are factors of $x^3 + 6x^2 - 7x - 60$, explain how you can find a third *first-degree* factor of $x^3 + 6x^2 - 7x - 60$. Then find the factor.

SECTION 5.7

Solving Equations by Factoring

1 Solve equations by factoring

21 HM³

Consider the equation $ab = 0$. If a is not zero, then b must be zero. Conversely, if b is not zero, then a must be zero. This is summarized in the Principle of Zero Products.

Principle of Zero Products

If the product of two factors is zero, then at least one of the factors must be zero.

If $ab = 0$, then $a = 0$ or $b = 0$.

INSTRUCTOR NOTE

As an application of factoring, quadratic equations are solved by using the Principle of Zero Products. A complete discussion of quadratic equations occurs later in the text.

The Principle of Zero Products is used to solve equations.

Solve: $(x - 4)(x + 2) = 0$

By the Principle of Zero Products, if $(x - 4)(x + 2) = 0$, then $x - 4 = 0$ or $x + 2 = 0$.

$$x - 4 = 0 \qquad x + 2 = 0$$
$$x = 4 \qquad x = -2$$

Check:

$$\begin{array}{c|c} (x - 4)(x + 2) & 0 \\ \hline (4 - 4)(4 + 2) & 0 \\ 0 \cdot 6 & 0 \\ 0 = 0 & \end{array} \qquad \begin{array}{c|c} (x - 4)(x + 2) & 0 \\ \hline (-2 - 4)(-2 + 2) & 0 \\ -6 \cdot 0 & 0 \\ 0 = 0 & \end{array}$$

-2 and 4 check as solutions. The solutions are -2 and 4.

An equation of the form $ax^2 + bx + c = 0$, $a \neq 0$, is a **quadratic equation.** A quadratic equation is in standard form when the polynomial is written in descending order and is equal to zero.

Some quadratic equations can be solved by factoring and then using the Principle of Zero Products.

Solve: $2x^2 - x = 1$

$$2x^2 - x = 1$$

Write the equation in standard form. $\quad 2x^2 - x - 1 = 0$

Factor the trinomial. $\quad (2x + 1)(x - 1) = 0$

Use the Principle of Zero Products. $\quad 2x + 1 = 0 \qquad x - 1 = 0$

Solve each equation for x. $\quad 2x = -1 \qquad x = 1$

$$x = -\frac{1}{2}$$

The solutions are $-\frac{1}{2}$ and 1.

The Principle of Zero Products can be extended to more than two factors. For example, if $abc = 0$, then $a = 0$, $b = 0$, or $c = 0$.

Solve: $x^3 - x^2 - 4x + 4 = 0$

$$x^3 - x^2 - 4x + 4 = 0$$

Factor by grouping.

$$(x^3 - x^2) - (4x - 4) = 0$$
$$x^2(x - 1) - 4(x - 1) = 0$$
$$(x - 1)(x^2 - 4) = 0$$
$$(x - 1)(x + 2)(x - 2) = 0$$

Use the Principle of Zero Products. $x - 1 = 0 \quad x + 2 = 0 \quad x - 2 = 0$

Solve each equation for x. $x = 1 \quad x = -2 \quad x = 2$

The solutions are -2, 1, and 2.

Example 1 Solve.

A. $3x^2 + 5x = 2$

B. $(x + 4)(x - 3) = 8$

C. $x^3 - x^2 - 25x + 25 = 0$

Solution A.

$3x^2 + 5x = 2$

$3x^2 + 5x - 2 = 0$ ▶ Write the equation in standard form.

$(3x - 1)(x + 2) = 0$ ▶ Factor the trinomial.

$3x - 1 = 0 \quad x + 2 = 0$ ▶ Let each factor equal zero (the Principle of Zero Products).

$3x = 1 \quad x = -2$ ▶ Solve each equation for x.

$x = \frac{1}{3}$

The solutions are $\frac{1}{3}$ and -2. ▶ Write the solutions.

B. $(x + 4)(x - 3) = 8$

$x^2 + x - 12 = 8$ ▶ Write the equation in standard form by first multiplying the binomials.

$x^2 + x - 20 = 0$ ▶ Subtract 8 from each side of the equation.

$(x + 5)(x - 4) = 0$ ▶ Factor the trinomial.

$x + 5 = 0 \quad x - 4 = 0$ ▶ Let each factor equal zero.

$x = -5 \quad x = 4$ ▶ Solve each equation for x.

The solutions are -5 and 4. ▶ Write the solutions.

C. $x^3 - x^2 - 25x + 25 = 0$

$(x^3 - x^2) - (25x - 25) = 0$ ▶ Factor by grouping.

$x^2(x - 1) - 25(x - 1) = 0$

$(x^2 - 25)(x - 1) = 0$

$(x + 5)(x - 5)(x - 1) = 0$

$x + 5 = 0 \quad x - 5 = 0 \quad x - 1 = 0$ ▶ Let each factor equal zero.

$x = -5 \quad x = 5 \quad x = 1$ ▶ Solve each equation for x.

The solutions are -5, 1, and 5. ▶ Write the solutions.

Problem 1   Solve.

A. $4x^2 + 11x = 3$ B. $(x - 2)(x + 5) = 8$

C. $x^3 + 4x^2 - 9x - 36 = 0$

Solution   See page S15.   A. $\frac{1}{4}$ and -3 B. -6 and 3 C. $-4, -3,$ and 3

The Principle of Zero Products is used to find elements in the domain of a quadratic function that correspond to a given element in the range.

Example 2   Given that -1 is in the range of the function defined by $f(x) = x^2 - 3x - 5$, find two values of c for which $f(c) = -1$.

Solution
$$f(c) = -1$$
$$c^2 - 3c - 5 = -1 \qquad \blacktriangleright f(c) = c^2 - 3c - 5$$
$$c^2 - 3c - 4 = 0 \qquad \blacktriangleright \text{Solve for } c.$$
$$(c - 4)(c + 1) = 0$$

$$c - 4 = 0 \qquad c + 1 = 0$$
$$c = 4 \qquad\quad c = -1$$

The values of c are -1 and 4.

Problem 2   Given that 4 is in the range of the function defined by $s(t) = t^2 - t - 2$, find two values of c for which $s(c) = 4$.

Solution   See page S16. $-2, 3$

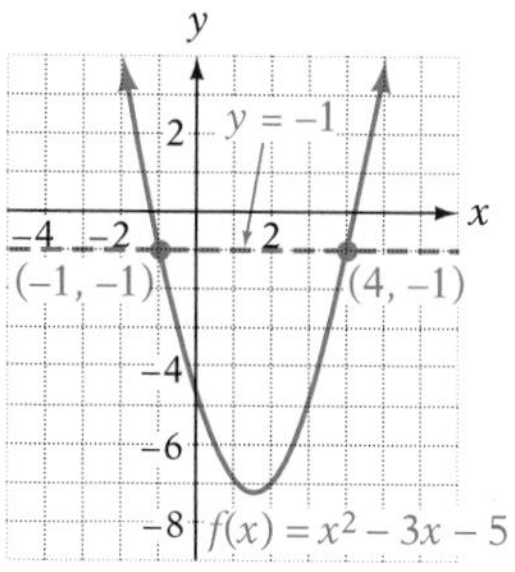

In Example 2, there are two values in the domain that can be paired with the range element -1. The two values are -1 and 4. Two ordered pairs that belong to the function are $(-1, -1)$ and $(4, -1)$. *Remember:* A function can have different first elements paired with the same second element. A function cannot have the same first element paired with different second elements.

The graph of $f(x) = x^2 - 3x - 5$ is shown at the left.

2   Application problems

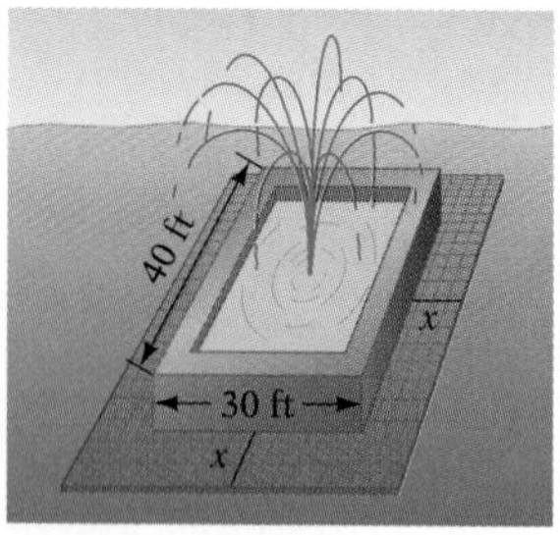

Example 3   An architect wants to design a fountain to be placed on the front lawn of a new art museum. The base of the fountain is to be 30 ft by 40 ft with a uniform brick walkway around the fountain. The total area of the fountain and the walkway is 2576 ft². How wide is the walkway?

Strategy   Let x represent the width of the brick walkway. Then the width of the fountain and walkway is $30 + 2x$, and the length of the fountain and walkway is $40 + 2x$. Use the formula $A = LW$, where $A = 2576$, to write an equation.

Solution $A = LW$

$2576 = (40 + 2x)(30 + 2x)$ ▶ Replace A, L, and W by their values.

$$\begin{aligned} 2576 &= 1200 + 140x + 4x^2 \\ 0 &= 4x^2 + 140x - 1376 \\ 0 &= x^2 + 35x - 344 \\ 0 &= (x - 8)(x + 43) \end{aligned}$$

▶ Multiply.
▶ Divide each side by 4.
▶ Factor and use the Principle of Zero Products.

$x - 8 = 0 \qquad x + 43 = 0$

$x = 8 \qquad x = -43$

A walkway -43 ft wide would not make sense.

The walkway must be 8 ft wide.

Problem 3 A diagonal of a polygon is a line from one vertex of the polygon to a nonadjacent vertex. The number of diagonals, D, of a polygon with n sides is given by $D = \frac{n(n-3)}{2}$. Find the number of sides for a polygon that has 20 diagonals.

Solution See page S16. 8 sides

CONCEPT REVIEW 5.7

Determine whether the following statements are always true, sometimes true, or never true.

1. The Principle of Zero Products can be stated as follows: If $ab = 0$, then $a = 0$ or $b = 0$. Always true

2. One side of an equation must be zero in order for us to solve the equation by factoring. Always true

3. Two solutions of an equation are found when solving an equation by the Principle of Zero Products. Sometimes true

4. The Principle of Zero Products can be used to solve a second-degree equation. Sometimes true

5. If n is an integer, then $n + 1$ and $n + 3$ represent odd integers. Sometimes true

6. By the Principle of Zero Products, if $(x + 2)(x + 3) = 4$, then $x + 2 = 4$ or $x + 3 = 4$. Never true

EXERCISES 5.7

1 1. What is a quadratic equation? How does it differ from a linear equation? Give an example of each type of equation.

2. How is the Principle of Zero Products used to solve some quadratic equations?

Solve.

3. $(y + 4)(y + 6) = 0$ $-4, -6$

4. $(a - 5)(a - 2) = 0$ $5, 2$

5. $x(x - 7) = 0$ $0, 7$

6. $b(b + 8) = 0$ $0, -8$

7. $3z(2z + 5) = 0$ $0, -\frac{5}{2}$

8. $4y(3y - 2) = 0$ $0, \frac{2}{3}$

9. $(2x + 3)(x - 7) = 0$ $-\frac{3}{2}, 7$

10. $(4a - 1)(a + 9) = 0$ $\frac{1}{4}, -9$

11. $b^2 - 49 = 0$ $7, -7$

12. $4z^2 - 1 = 0$ $\frac{1}{2}, -\frac{1}{2}$

13. $9t^2 - 16 = 0$ $\frac{4}{3}, -\frac{4}{3}$

14. $x^2 + x - 6 = 0$ $2, -3$

15. $y^2 + 4y - 5 = 0$ $-5, 1$

16. $a^2 - 8a + 16 = 0$ 4

17. $2b^2 - 5b - 12 = 0$ $-\frac{3}{2}, 4$

18. $t^2 - 8t = 0$ $0, 8$

19. $x^2 - 9x = 0$ $0, 9$

20. $2y^2 - 10y = 0$ $0, 5$

21. $3a^2 - 12a = 0$ $0, 4$

22. $b^2 - 4b = 32$ $-4, 8$

23. $z^2 - 3z = 28$ $7, -4$

24. $2x^2 - 5x = 12$ $-\frac{3}{2}, 4$

25. $3t^2 + 13t = 10$ $\frac{2}{3}, -5$

26. $4y^2 - 19y = 5$ $-\frac{1}{4}, 5$

27. $5b^2 - 17b = -6$ $\frac{2}{5}, 3$

28. $6a^2 + a = 2$ $\frac{1}{2}, -\frac{2}{3}$

29. $8x^2 - 10x = 3$ $\frac{3}{2}, -\frac{1}{4}$

30. $z(z - 1) = 20$ $5, -4$

31. $y(y - 2) = 35$ $7, -5$

32. $t(t + 1) = 42$ $-7, 6$

33. $x(x - 12) = -27$ $9, 3$

34. $x(2x - 5) = 12$ $-\frac{3}{2}, 4$

35. $y(3y - 2) = 8$ $-\frac{4}{3}, 2$

36. $2b^2 - 6b = b - 3$ $\frac{1}{2}, 3$

37. $3a^2 - 4a = 20 - 15a$ $\frac{4}{3}, -5$

38. $2t^2 + 5t = 6t + 15$ $-\frac{5}{2}, 3$

39. $(y + 5)(y - 7) = -20$ $5, -3$

40. $(x + 2)(x - 6) = 20$ $8, -4$

41. $(b + 5)(b + 10) = 6$ $-11, -4$

42. $(a - 9)(a - 1) = -7$ $2, 8$

43. $(t - 3)^2 = 1$ $4, 2$

44. $(y - 4)^2 = 4$ $6, 2$

45. $(3 - x)^2 + x^2 = 5$ $2, 1$

46. $(2 - b)^2 + b^2 = 10$ $3, -1$

47. $(a - 1)^2 = 3a - 5$ $2, 3$

48. $2x^3 + x^2 - 8x - 4 = 0$ $-2, -\frac{1}{2}, 2$

49. $x^3 + 4x^2 - x - 4 = 0$ $-4, -1, 1$

50. $12x^3 - 8x^2 - 3x + 2 = 0$ $-\frac{1}{2}, \frac{1}{2}, \frac{2}{3}$

Find the values c in the domain of f for which $f(c)$ is the indicated value.

51. $f(x) = x^2 - 3x + 3; f(c) = 1$ $1, 2$

52. $f(x) = x^2 + 4x - 2; f(c) = 3$ $-5, 1$

53. $f(x) = 2x^2 - x - 5; f(c) = -4$ $-\frac{1}{2}, 1$

54. $f(x) = 6x^2 - 5x - 9; f(c) = -3$ $-\frac{2}{3}, \frac{3}{2}$

55. $f(x) = 4x^2 - 4x + 3; f(c) = 2$ $\frac{1}{2}$

56. $f(x) = x^2 - 6x + 12; f(c) = 3$ 3

57. $f(x) = x^3 + 9x^2 - x - 14; f(c) = -5$ $-9, -1, 1$

58. $f(x) = x^3 + 3x^2 - 4x - 11; f(c) = 1$ $-3, -2, 2$

2 **59.** *Number Problem* The sum of a number and its square is 90. Find the number. -10 or 9

60. *Number Problem* The sum of a number and its square is 132. Find the number. -12 or 11

61. *Number Problem* The sum of the squares of two consecutive positive integers is equal to 145. Find the two integers. 8 and 9

62. *Number Problem* The sum of the squares of two consecutive positive odd integers is equal to 290. Find the two integers. 11 and 13

63. *Number Problem* The sum of the cube of a number and the product of the number and twelve is equal to seven times the square of the number. Find the number. 0, 3, or 4

64. *Number Problem* The sum of the cube of a number and the product of the number and seven is equal to eight times the square of the number. Find the number. 0, 1, or 7

65. *Geometry* The length of a rectangle is 5 in. more than twice the width. The area is 168 in^2. Find the width and length of the rectangle. 8 in. by 21 in.

66. *Geometry* The width of a rectangle is 5 ft less than the length. The area of the rectangle is 300 ft^2. Find the length and width of the rectangle.
15 ft by 20 ft

67. *Construction* A trough is made from a rectangular piece of metal by folding up the sides as shown in the figure at the right. What should the value of x be so that the volume is 72 m^3? 2 m or 3 m

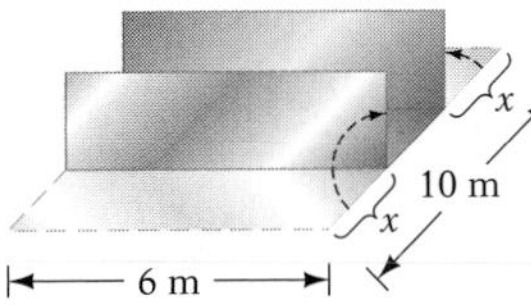

68. *Geometry* The number of possible diagonals, D, in a polygon with n sides is given by $D = \frac{n(n-3)}{2}$. Find the number of sides for a polygon with 54 diagonals. 12 sides

69. *Space Science* A small rocket is launched with an acceleration of 10 m/s^2. The velocity, v, of the rocket after it has traveled s meters is given by $v^2 = 20s$. Find the velocity of this rocket after it has traveled 500 m. 100 m/s

70. *Construction* A pole is being secured with two guy wires as shown in the figure at the right. What is the value of x so that the 20-foot guy wire will attach as shown? 3 ft

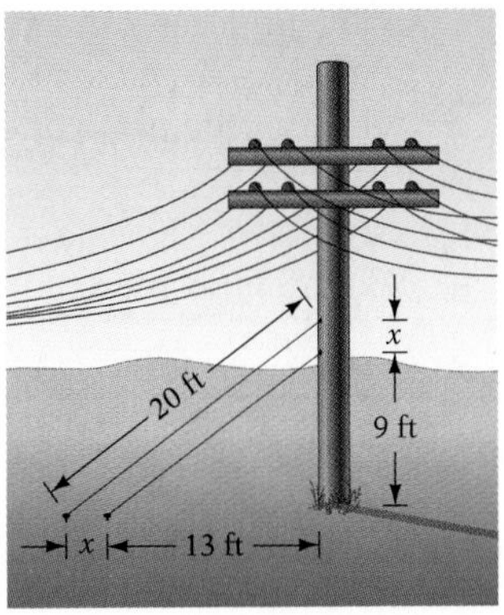

71. *Geometry* The length of the base of a triangle is three times the height. The area of the triangle is 24 cm^2. Find the base and height of the triangle.
Height: 4 cm; base: 12 cm

72. *Geometry* The height of a triangle is 4 cm more than twice the length of the base. The area of the triangle is 35 cm^2. Find the height of the triangle.
14 cm

73. *Physics* An object is thrown downward, with an initial speed of 16 ft/s, from the top of a building 480 ft high. How many seconds later will the object hit the ground? Use the equation $d = vt + 16t^2$, where d is the distance in feet, v is the initial speed in feet per second, and t is the time in seconds.
5 s

74. *Physics* A stone is thrown into an empty well with an initial speed of 8 ft/s. The well is 624 ft deep. How many seconds later will the stone hit the bottom of the well? Use the equation $d = vt + 16t^2$, where d is the distance in feet, v is the initial speed in feet per second, and t is the time in seconds. 6 s

75. *Geometry* The length of a rectangle is 6 cm, and the width is 3 cm. If both the length and the width are increased by equal amounts, the area of the rectangle is increased by 70 cm². Find the length and width of the larger rectangle. 8 cm by 11 cm

76. *Geometry* The width of a rectangle is 4 cm, and the length is 8 cm. If both the width and the length are increased by equal amounts, the area of the rectangle is increased by 64 cm². Find the length and width of the larger rectangle. 8 cm by 12 cm

APPLYING CONCEPTS 5.7

Solve for x in terms of a.

77. $x^2 + 3ax - 10a^2 = 0$ $2a$ and $-5a$

78. $x^2 + 4ax - 21a^2 = 0$ $-7a$ and $3a$

79. $x^2 - 9a^2 = 0$ $3a$ and $-3a$

80. $x^2 - 16a^2 = 0$ $4a$ and $-4a$

81. $x^2 - 5ax = 15a^2 - 3ax$ $5a$ and $-3a$

82. $x^2 + 3ax = 8ax + 24a^2$ $8a$ and $-3a$

Solve.

83. Find $3n^2 + 2n - 1$ if $n(n + 6) = 16$. 175 or 15

84. *Geometry* The perimeter of a rectangular garden is 44 m. The area of the garden is 120 m². Find the length and width of the garden.
Length: 12 m; width: 10 m

[C] **85.** *Construction* A rectangular piece of cardboard is 10 in. longer than it is wide. Squares 2 in. on a side are to be cut from each corner and then the sides folded up to make an open box with a volume of 112 in³. Find the length and width of the piece of cardboard. Length: 18 in.; width: 8 in.

[C] **86.** *Geometry* The sides of a rectangular box have areas of 16 cm², 20 cm², and 80 cm². Find the volume of the box. 160 cm³

87. Write an equation whose solutions are 3, 2, and −1.
For example, $x^3 - 4x^2 + x + 6 = 0$.

88. The following seems to show that $1 = 2$. Explain the error.

$$
\begin{aligned}
a &= b \\
a^2 &= ab && \text{▶ Multiply each side of the equation by } a. \\
a^2 - b^2 &= ab - b^2 && \text{▶ Subtract } b^2 \text{ from each side of the equation.} \\
(a - b)(a + b) &= b(a - b) && \text{▶ Factor.}
\end{aligned}
$$

$a + b = b$ ▶ Divide each side by $a - b$.
$b + b = b$ ▶ Because $a = b$, substitute b for a.
$2b = b$
$2 = 1$ ▶ Divide both sides by b.

$a - b$ is equal to zero, and division by zero is undefined.

89. *Airline Industry* The table below gives the annual profit, in billions of dollars, for the U.S. airline industry for selected years.

Years	1988	'89	'90	'92	'94	'95	'96	'97	'98
Profit (in billions of $)	1.8	0.2	−4.0	−4.4	−0.2	1.2	7.2	7.7	8.9

A quadratic function that models these data is given by the equation

$$P(x) = 0.3199x^2 - 58.4992x + 2671.3086$$

where $x = 88$ corresponds to 1988 and $P(x)$ is the profit in billions of dollars.

a. According to this model, what was the profit for the U.S. airline industry in 1997? $6.83 billion

b. What is the percent error between the value given by the model and the actual value for 1997? Round to the nearest tenth of a percent. 11.4%

$$\% \text{ error} = \left|\frac{\text{actual value} - \text{predicted value}}{\text{actual value}}\right| \times 100$$

c. What do the zeroes of this function represent in the context of this problem? Break-even point (revenue = expenditures)

d. In 1996, the actual profit for the U.S. airline industry was $7.2 billion. What is the percent error between the predicted value given by the model and the actual value? Why does the answer to this problem suggest that models must be thoroughly tested before they can be used?
The percent error is 50.2%. Some models work better for certain data points than for others.

Focus on Problem Solving

Find a Counterexample

When you are faced with an assertion, it may be that the assertion is false. For instance, consider the statement "Every prime number is an odd number." This assertion is false because the prime number 2 is an even number.

Finding an example that illustrates that an assertion is false is called finding a *counterexample*. The number 2 is a counterexample to the assertion that every prime number is an odd number.

If you are given an unfamiliar problem, one strategy to consider as a means of solving the problem is to try to find a counterexample. For each of the following problems, answer *true* if the assertion is always true. If the assertion is not true, answer *false* and find a counterexample. If terms are used that you do not understand, consult a reference to find their meaning.

1. If x is a real number, then x^2 is always positive.

2. The product of an odd integer and an even integer is an even integer.

3. If m is a positive integer, then $2m + 1$ is always a positive odd integer.
4. If $x < y$, then $x^2 < y^2$.
5. Given any three positive numbers a, b, and c, it is possible to construct a triangle whose sides have lengths a, b, and c.
6. The product of two irrational numbers is an irrational number.
7. If n is a positive integer greater than 2, then $1 \cdot 2 \cdot 3 \cdot 4 \cdot \cdots \cdot n + 1$ is a prime number.
8. Draw a polygon with more than three sides. Select two different points inside the polygon and join the points with a line segment. The line segment always lies completely inside the polygon.
9. Let A, B, and C be three points in the plane that are not collinear. Let d_1 be the distance from A to B, and let d_2 be the distance from A to C. Then the distance between B and C is less than $d_1 + d_2$.

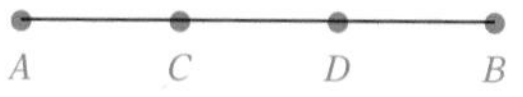

10. Consider the line segment AB shown at the left. Two points, C and D, are randomly selected on the line segment and three new segments are formed: AC, CD, and DB. The three new line segments can always be connected to form a triangle.

It may not be easy to establish that an assertion is true or to find a counterexample to the assertion. For instance, consider the assertion that every positive integer greater than 3 can be written as the sum of two primes. For example, $9 = 2 + 7$, $6 = 3 + 3$ and $8 = 3 + 5$. Is this assertion always true? (*Note:* This assertion, called Goldbach's conjecture, has never been proved, nor has a counterexample ever been found!)

Projects and Group Activities

Reverse Polish Notation

Following the Order of Operations Agreement is sometimes referred to as applying algebraic logic. It is a system used by many calculators. Algebraic logic is not the only system in use by calculators. Another system is called RPN logic. RPN stands for Reverse Polish Notation.

During the 1950s, Jan Lukasiewicz, a Polish logician, developed a parenthesis-free notational system for writing mathematical expressions. In this system, the operator (such as addition, multiplication, or division) follows the operands (the numbers to be added, multiplied, or divided). For RPN calculators, an [ENTER] key is used to store a number temporarily until another number and the operation can be entered. Here are some examples, along with the algebraic logic equivalent.

To find	RPN Logic	Algebraic Logic
1. $3 + 4$	3 [ENTER] 4 [+]	3 [+] 4 [=]
2. $5 \times 6 \times 7$	5 [ENTER] 6 [×] 7 [×]	5 [×] 6 [×] 7 [=]
3. $4 \times (7 + 3)$	4 [ENTER] 7 [ENTER] 3 [+] [×]	4 [×] [(] 7 [+] 3 [)] [=]
4. $(3 + 4) \times (5 + 2)$	3 [ENTER] 4 [+] 5 [ENTER] 2 [+] [×]	[(] 3 [+] 4 [)] [×] [(] 5 [+] 2 [)] [=]

Examples 3 and 4 above illustrate the concept of being "parenthesis-free." Note that the examples under RPN logic do not require parentheses, whereas those under algebraic logic do. If the parentheses keys are not used for Examples 3 and 4, a calculator, following algebraic logic, would use the Order of Operations Agreement. The result in Example 3 would be

$$4 \times 7 + 3 = 28 + 3 = 31 \qquad \text{instead of} \qquad 4 \times (7 + 3) = 4 \times 10 = 40$$

This system may seem quite strange, but it is actually very efficient. A glimpse of its efficiency can be seen from Example 4. For the RPN operation, nine keys were pushed, whereas twelve keys were pushed for the algebraic operation. Note also that RPN logic does not use the "equals" key. The result of an operation is displayed after the operation key is pressed.

Try the following exercises, which ask you to change from algebraic logic to RPN logic or to evaluate an RPN logic expression.

Change to RPN logic.

1. $12 \div 6$ **2.** $7 \times 5 + 6$ **3.** $(9 + 3) \div 4$

4. $(5 + 7) \div (2 + 4)$ **5.** $6 \times 7 \times 10$ **6.** $(1 + 4 \times 5) \div 7$

Evaluate each of the following by using RPN logic.

7. 18 [ENTER] 2 [÷] **8.** 6 [ENTER] 5 [ENTER] 3 [×] [+]

9. 3 [ENTER] 5 [×] 4 [+] **10.** 7 [ENTER] 4 [ENTER] 5 [×] [+] 3 [÷]

11. 228 [ENTER] 6 [ENTER] 9 [ENTER] 12 [×] [+] [÷]

12. 1 [ENTER] 1 [ENTER] 1 [ENTER] 3 [+] [÷] [+]

Pythagorean Triples

Recall that the Pythagorean Theorem states that if a and b are the legs of a right triangle and c is the length of the hypotenuse, then $a^2 + b^2 = c^2$.

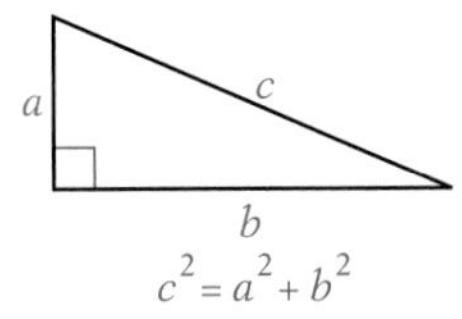

For instance, the triangle with legs 3 and 4 and hypotenuse 5 is a right triangle because $3^2 + 4^2 = 5^2$. The numbers 3, 4, and 5 are called a **Pythagorean triple** because they are natural numbers that satisfy the equation of the Pythagorean Theorem.

1. Determine whether the numbers are a Pythagorean triple.

a. 5, 7, and 9 **b.** 8, 15, and 17

c. 11, 60, and 61 **d.** 28, 45, and 53

Mathematicians have investigated Pythagorean triples and have found formulas that will generate these triples. One such set of formulas is

$$a = m^2 - n^2 \qquad b = 2mn \qquad c = m^2 + n^2, \text{ where } m > n$$

For instance, let $m = 2$ and $n = 1$. Then $a = 2^2 - 1^2 = 3$, $b = 2(2)(1) = 4$, and $c = 2^2 + 1^2 = 5$. This is the Pythagorean triple given above.

2. Find the Pythagorean triple produced by each of the following.

a. $m = 3$ and $n = 1$ **b.** $m = 5$ and $n = 2$

c. $m = 4$ and $n = 2$ **d.** $m = 6$ and $n = 1$

3. Find values of m and n that yield the Pythagorean triple 11, 60, 61.

4. Verify that $a^2 + b^2 = c^2$ when $a = m^2 - n^2$, $b = 2mn$, and $c = m^2 + n^2$.

5. The early Greek builders used a rope with 12 equally spaced knots to make right-angle corners for buildings. Explain how they used the rope.

6. Find three odd integers a, b, and c such that $a^2 + b^2 = c^2$.

Chapter Summary

Key Words

A *monomial* is a number, a variable, or a product of numbers and variables. (Objective 5.1.1)

$2, x, 3x, -4x^2, 23x^2y^3$ are monomials.

The *degree of a polynomial* is the greatest of the degrees of any of its terms. (Objective 5.2.1)

The degree of the polynomial $x^3 + 3x^2y^2 - 4xy - 3$ is 4.

A *polynomial* is a variable expression in which the terms are monomials. (Objective 5.2.1)

$x^4 - 2xy - 32x + 8$ is a polynomial. The terms are x^4, $-2xy$, $-32x$, and 8.

A polynomial of two terms is a *binomial*. (Objective 5.2.1)

$3x - y$ is a binomial.

A polynomial of three terms is a *trinomial*. (Objective 5.2.1)

$x^2 - 2x + 4$ is a trinomial.

Synthetic division is a shorter method of dividing a polynomial by a binomial of the form $x - a$. This method uses only the coefficients of the variable terms. (Objective 5.4.2)

$(3x^3 - 9x - 5) \div (x - 2)$

$$\begin{array}{r|rrrr} 2 & 3 & 0 & -9 & -5 \\ & & 6 & 12 & 6 \\ \hline & 3 & 6 & 3 & 1 \end{array}$$

$$(3x^3 - 9x - 5) \div (x - 2) = 3x^2 + 6x + 3 + \frac{1}{x-2}$$

A number written in *scientific notation* is a number written in the form $a \times 10^n$, where $1 \leq a < 10$. (Objective 5.1.3)

$0.000000023 = 2.3 \times 10^{-8}$

To *factor a polynomial* means to write the polynomial as the product of other polynomials. (Objective 5.5.1)

$x^2 + 5x + 6 = (x + 2)(x + 3)$

A *quadratic trinomial* is a polynomial of the form $ax^2 + bx + c$, where a and b are nonzero coefficients and c is a nonzero constant. (Objective 5.5.3)

$3x^2 - 2x + 8$, $a = 3$, $b = -2$, $c = 8$

To *factor a quadratic trinomial* means to express the trinomial as the product of two binomials. (Objective 5.5.3)

$3x^2 + 10x - 8 = (3x - 2)(x + 4)$

A polynomial is *nonfactorable over the integers* if it does not factor using only integers. (Objective 5.5.3)

$x^2 + x + 1$ is nonfactorable.

The product of a term and itself is a *perfect square.* (Objective 5.6.1)

$(5x)(5x) = 25x^2$; $25x^2$ is a perfect square.

The *square root of a perfect square* is one of the two equal factors of the perfect square. (Objective 5.6.1)

$\sqrt{25x^2} = 5x$

The product of three factors is called a *perfect cube.* (Objective 5.6.2)

$(2x)(2x)(2x) = 8x^3$; $8x^3$ is a perfect cube.

The *cube root* of a perfect cube is one of the three equal factors of the perfect cube. (Objective 5.6.2)

$\sqrt[3]{8x^3} = 2x$

A *quadratic equation* is an equation of the form

$$ax^2 + bx + c = 0, \text{ where } a \neq 0$$

A quadratic equation is in standard form when the polynomial is in descending order and equal to zero. (Objective 5.7.1)

$3x^2 + 3x + 8 = 0$ is a quadratic equation in standard form.

Essential Rules and Procedures

Rule for Multiplying Exponential Expressions
If m and n are integers, then $x^m \cdot x^n = x^{m+n}$. (Objective 5.1.1)

$x^2 \cdot x^5 = x^{2+5} = x^7$

Rule for Simplifying a Power of an Exponential Expression
If m and n are integers, then $(x^m)^n = x^{mn}$. (Objective 5.1.1)

$(x^4)^3 = x^{4 \cdot 3} = x^{12}$

Rule for Dividing Exponential Expressions
If m and n are integers and $x \neq 0$, then $\frac{x^m}{x^n} = x^{m-n}$. (Objective 5.1.2)

$\frac{x^{12}}{x^7} = x^{12-7} = x^5$

Rule for Simplifying Powers of Products
If m, n, and p are integers, then $(x^m \cdot y^n)^p = x^{mp}y^{np}$. (Objective 5.1.1)

$(x^2 \cdot y^5)^3 = x^{2 \cdot 3}y^{5 \cdot 3} = x^6y^{15}$

Rule for Simplifying Powers of Quotients
If m, n, and p are integers and $y \neq 0$, then $\left(\frac{x^m}{y^n}\right)^p = \frac{x^{mp}}{y^{np}}$. (Objective 5.1.2)

$\left(\frac{x^2}{y^4}\right)^5 = \frac{x^{2 \cdot 5}}{y^{4 \cdot 5}} = \frac{x^{10}}{y^{20}}$

Definition of a Negative Exponent
If $n > 0$ and $x \neq 0$, then $x^{-n} = \frac{1}{x^n}$ and $\frac{1}{x^{-n}} = x^n$. (Objective 5.1.2)

$x^{-3} = \frac{1}{x^3}$ $\qquad \frac{1}{x^{-4}} = x^4$

Definition of Zero as an Exponent
If $x \neq 0$, then $x^0 = 1$. (Objective 5.1.2)

$5^0 = 1$ $\qquad (2xy^4)^0 = 1$

Remainder Theorem
If the polynomial $P(x)$ is divided by $x - a$, the remainder is $P(a)$. (Objective 5.4.3)

$P(x) = x^3 - x^2 + x - 1$

$$\begin{array}{r|rrrr} -2 & 1 & -1 & 1 & -1 \\ & & -2 & 6 & -14 \\ \hline & 1 & -3 & 7 & -15 \end{array}$$

$P(-2) = -15$

Factors of the Difference of Two Perfect Squares
$a^2 - b^2 = (a - b)(a + b)$ (Objective 5.6.1)

$x^2 - 9 = (x - 3)(x + 3)$

Factors of a Perfect-Square Trinomial
$a^2 + 2ab + b^2 = (a + b)^2$ (Objective 5.6.1)

$4x^2 + 12x + 9 = (2x + 3)^2$

Factoring the Sum or Difference of Two Cubes
$a^3 + b^3 = (a + b)(a^2 - ab + b^2)$
$a^3 - b^3 = (a - b)(a^2 + ab + b^2)$
(Objective 5.6.2)

$x^3 + 64 = (x + 4)(x^2 - 4x + 16)$
$8b^3 - 1 = (2b - 1)(4b^2 + 2b + 1)$

The Principle of Zero Products
If $ab = 0$, then $a = 0$ or $b = 0$. (Objective 5.7.1)

$(x - 4)(x + 2) = 0$
$x - 4 = 0 \qquad x + 2 = 0$

Chapter Review Exercises

1. Add: $(3x^2 - 2x - 6) + (-x^2 - 3x + 4)$
$2x^2 - 5x - 2$

2. Subtract: $(5x^2 - 8xy + 2y^2) - (x^2 - 3y^2)$
$4x^2 - 8xy + 5y^2$

3. Multiply: $(5x^2yz^4)(2xy^3z^{-1})(7x^{-2}y^{-2}z^3)$ $70xy^2z^6$

4. Multiply: $(2x^{-1}y^2z^5)^4(-3x^3yz^{-3})$ $-\dfrac{48y^9z^{17}}{x}$

5. Simplify: $\dfrac{3x^4yz^{-1}}{-12xy^3z^2}$ $-\dfrac{x^3}{4y^2z^3}$

6. Simplify: $\dfrac{(2a^4b^{-3}c^2)^3}{(2a^3b^2c^{-1})^4}$ $\dfrac{c^{10}}{2b^{17}}$

7. Write 93,000,000 in scientific notation.
9.3×10^7

8. Write 2.54×10^{-3} in decimal notation. 0.00254

9. Simplify: $\dfrac{3 \times 10^{-3}}{15 \times 10^2}$ 2×10^{-6}

10. Given $P(x) = 2x^3 - x + 7$, evaluate $P(-2)$. -7

11. Graph: $y = x^2 + 1$

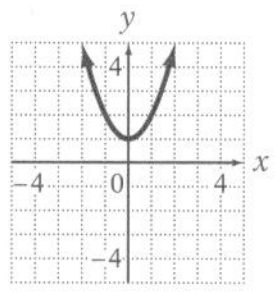

12. Identify **a.** the leading coefficient, **b.** the constant term, and **c.** the degree of the polynomial $P(x) = 3x^5 - 3x^2 + 7x + 8$. **a.** 3 **b.** 8 **c.** 5

13. Use the Remainder Theorem to evaluate $P(x) = -2x^3 + 2x^2 - 4$ when $x = -3$. 68

14. Divide: $\dfrac{15x^2 + 2x - 2}{3x - 2}$ $5x + 4 + \dfrac{6}{3x - 2}$

15. Divide: $\dfrac{12x^2 - 16x - 7}{6x + 1}$ $2x - 3 - \dfrac{4}{6x + 1}$

16. Divide: $\dfrac{4x^3 + 27x^2 + 10x + 2}{x + 6}$
$4x^2 + 3x - 8 + \dfrac{50}{x + 6}$

17. Divide: $\dfrac{x^4 - 4}{x - 4}$ $x^3 + 4x^2 + 16x + 64 + \dfrac{252}{x - 4}$

18. Multiply: $4x^2y(3x^3y^2 + 2xy - 7y^3)$
$12x^5y^3 + 8x^3y^2 - 28x^2y^4$

19. Multiply: $a^{2n+3}(a^n - 5a + 2)$
$a^{3n+3} - 5a^{2n+4} + 2a^{2n+3}$

20. Simplify: $5x^2 - 4x[x - (3x + 2) + x]$ $9x^2 + 8x$

21. Multiply: $(x^{2n} - x)(x^{n+1} - 3)$ $x^{3n+1} - 3x^{2n} - x^{n+2} + 3x$

22. Multiply: $(x + 6)(x^3 - 3x^2 - 5x + 1)$ $x^4 + 3x^3 - 23x^2 - 29x + 6$

23. Multiply: $(x - 4)(3x + 2)(2x - 3)$ $6x^3 - 29x^2 + 14x + 24$

24. Multiply: $(5a + 2b)(5a - 2b)$ $25a^2 - 4b^2$

25. Simplify: $(4x - 3y)^2$ $16x^2 - 24xy + 9y^2$

26. Factor: $18a^5b^2 - 12a^3b^3 + 30a^2b$ $6a^2b(3a^3b - 2ab^2 + 5)$

27. Factor: $5x^{n+5} + x^{n+3} + 4x^2$ $x^2(5x^{n+3} + x^{n+1} + 4)$

28. Factor: $x(y - 3) + 4(3 - y)$ $(y - 3)(x - 4)$

29. Factor: $2ax + 4bx - 3ay - 6by$ $(a + 2b)(2x - 3y)$

30. Factor: $x^2 + 12x + 35$ $(x + 5)(x + 7)$

31. Factor: $12 + x - x^2$ $(3 + x)(4 - x)$

32. Factor: $x^2 - 16x + 63$ $(x - 7)(x - 9)$

33. Factor: $6x^2 - 31x + 18$ $(3x - 2)(2x - 9)$

34. Factor: $24x^2 + 61x - 8$ $(8x - 1)(3x + 8)$

35. Factor: $x^2y^2 - 9$ $(xy + 3)(xy - 3)$

36. Factor: $4x^2 + 12xy + 9y^2$ $(2x + 3y)^2$

37. Factor: $x^{2n} - 12x^n + 36$ $(x^n - 6)^2$

38. Factor: $36 - a^{2n}$ $(6 + a^n)(6 - a^n)$

39. Factor: $64a^3 - 27b^3$ $(4a - 3b)(16a^2 + 12ab + 9b^2)$

40. Factor: $8 - y^{3n}$ $(2 - y^n)(4 + 2y^n + y^{2n})$

41. Factor: $15x^4 + x^2 - 6$ $(3x^2 + 2)(5x^2 - 3)$

42. Factor: $36x^8 - 36x^4 + 5$ $(6x^4 - 5)(6x^4 - 1)$

43. Factor: $21x^4y^4 + 23x^2y^2 + 6$ $(3x^2y^2 + 2)(7x^2y^2 + 3)$

44. Factor: $3a^6 - 15a^4 - 18a^2$ $3a^2(a^2 - 6)(a^2 + 1)$

45. Factor: $x^{4n} - 8x^{2n} + 16$ $(x^n + 2)^2(x^n - 2)^2$

46. Factor: $3a^4b - 3ab^4$ $3ab(a - b)(a^2 + ab + b^2)$

47. Solve: $x^3 - x^2 - 6x = 0$ $-2, 0, 3$

48. Solve: $6x^2 + 60 = 39x$ $\frac{5}{2}, 4$

49. Solve: $x^3 - 16x = 0$ $-4, 0, 4$

50. Solve: $y^3 + y^2 - 36y - 36 = 0$ $-1, -6, 6$

51. The most distant object visible from Earth without the aid of a telescope is the Great Galaxy of Andromeda. It takes light from the Great Galaxy of Andromeda 2.2×10^6 years to travel to Earth. Light travels about 5.9×10^{12} mph. How far from Earth is the Great Galaxy of Andromeda? 1.137048×10^{23} mi

52. Light from the sun supplies Earth with 2.4×10^{14} horsepower. Earth receives only 2.2×10^{-7} of the power generated by the sun. How much power is generated by the sun? 1.09×10^{21} horsepower

53. The length of a rectangle is $(5x + 3)$ cm. The width is $(2x - 7)$ cm. Find the area of the rectangle in terms of the variable x. $(10x^2 - 29x - 21)$ cm^2

54. The length of the side of a cube is $(3x - 1)$ ft. Find the volume of the cube in terms of the variable x. $(27x^3 - 27x^2 + 9x - 1)$ ft^3

55. Find the area of the figure shown at the right. All dimensions given are in inches. $(5x^2 + 8x - 8)$ in^2

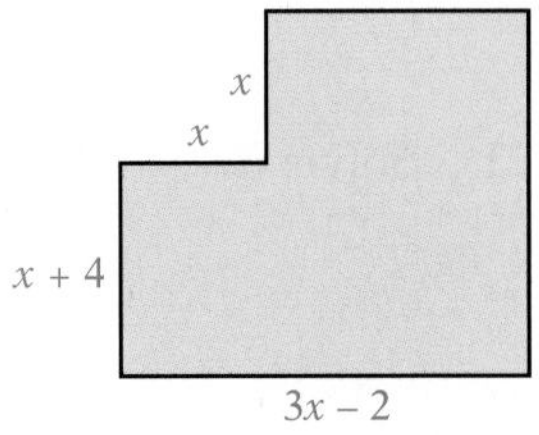

56. The sum of the squares of two consecutive even integers is 52. Find the two integers. -6 and -4, or 4 and 6

57. The sum of a number and its square is 56. Find the number. -8 or 7

58. The length of a rectangle is 2 m more than twice the width. The area of the rectangle is 60 m^2. Find the length of the rectangle. 12 m

Chapter Test

1. Subtract: $(6x^3 - 7x^2 + 6x - 7) - (4x^3 - 3x^2 + 7)$ $2x^3 - 4x^2 + 6x - 14$

2. Simplify: $(-4a^2b)^3(-ab^4)$ $64a^7b^7$

3. Simplify: $\frac{(2a^{-4}b^2)^3}{4a^{-2}b^{-1}}$ $\frac{2b^7}{a^{10}}$

4. Write the number 0.000000501 in scientific notation. 5.01×10^{-7}

5. Write the number of seconds in 1 week in scientific notation. 6.048×10^5

6. Simplify: $(2x^{-3}y)^{-4}$ $\frac{x^{12}}{16y^4}$

7. Simplify: $-5x[3 - 2(2x - 4) - 3x]$ $35x^2 - 55x$

8. Multiply: $(3a + 4b)(2a - 7b)$ $6a^2 - 13ab - 28b^2$

9. Multiply: $(3t^3 - 4t^2 + 1)(2t^2 - 5)$ $6t^5 - 8t^4 - 15t^3 + 22t^2 - 5$

10. Simplify: $(3z - 5)^2$ $9z^2 - 30z + 25$

11. Divide: $(4x^3 + x - 15) \div (2x - 3)$ $2x^2 + 3x + 5$

12. Divide: $(x^3 - 5x^2 + 5x + 5) \div (x - 3)$ $x^2 - 2x - 1 + \frac{2}{x - 3}$

13. Given $P(x) = 3x^2 - 8x + 1$, evaluate $P(2)$. -3

14. Use the Remainder Theorem to evaluate $P(x) = -x^3 + 4x - 8$ when $x = -2$. -8

15. Factor: $6a^4 - 13a^2 - 5$ $(2a^2 - 5)(3a^2 + 1)$

16. Factor: $12x^3 + 12x^2 - 45x$ $3x(2x - 3)(2x + 5)$

17. Factor: $16x^2 - 25$ $(4x - 5)(4x + 5)$

18. Factor: $16t^2 + 24t + 9$ $(4t + 3)^2$

19. Factor: $27x^3 - 8$ $(3x - 2)(9x^2 + 6x + 4)$

20. Factor: $6x^2 - 4x - 3xa + 2a$ $(3x - 2)(2x - a)$

21. Solve: $6x^2 = x + 1$ $-\frac{1}{3}, \frac{1}{2}$

22. Solve: $6x^3 + x^2 - 6x - 1 = 0$ $-1, -\frac{1}{6}, 1$

23. The length of a rectangle is $(5x + 1)$ ft. The width is $(2x - 1)$ ft. Find the area of the rectangle in terms of the variable x. $(10x^2 - 3x - 1)\text{ ft}^2$

24. A space vehicle travels 2.4×10^5 mi from Earth to the moon at an average velocity of 2×10^4 mph. How long does it take the space vehicle to reach the moon? 12 h

Cumulative Review Exercises

1. Simplify: $8 - 2[-3 - (-1)]^2 + 4$ 4

2. Evaluate $\frac{2a - b}{b - c}$ when $a = 4$, $b = -2$, and $c = 6$. $-\frac{5}{4}$

3. Identify the property that justifies the statement. $2x + (-2x) = 0$
Inverse Property of Addition

4. Simplify: $2x - 4[x - 2(3 - 2x) + 4]$ $-18x + 8$

5. Solve: $\frac{2}{3} - y = \frac{5}{6}$ $-\frac{1}{6}$

6. Solve: $8x - 3 - x = -6 + 3x - 8$ $-\frac{11}{4}$

7. Simplify: $\frac{x^3 - 3}{x - 3}$ $x^2 + 3x + 9 + \frac{24}{x - 3}$

8. Solve: $3 - |2 - 3x| = -2$ $-1, \frac{7}{3}$

9. Given $P(x) = 3x^2 - 2x + 2$, evaluate $P(-2)$.
18

10. What values of x are excluded from the domain of the function $f(x) = \frac{x + 1}{x + 2}$? -2

11. Find the range of the function given by $F(x) = 3x^2 - 4$ if the domain is $\{-2, -1, 0, 1, 2\}$.
$\{-4, -1, 8\}$

12. Find the slope of the line that contains the points $(-2, 3)$ and $(4, 2)$. $-\frac{1}{6}$

13. Find the equation of the line that contains the point $(-1, 2)$ and has slope $-\frac{3}{2}$. $y = -\frac{3}{2}x + \frac{1}{2}$

14. Find the equation of the line that contains the point $(-2, 4)$ and is perpendicular to the line $3x + 2y = 4$. $y = \frac{2}{3}x + \frac{16}{3}$

15. Solve by using Cramer's Rule:
$$\begin{aligned} 2x - 3y &= 2 \\ x + y &= -3 \end{aligned}$$
$\left(-\frac{7}{5}, -\frac{8}{5}\right)$

16. Solve by the addition method:
$$\begin{aligned} x - y + z &= 0 \\ 2x + y - 3z &= -7 \\ -x + 2y + 2z &= 5 \end{aligned}$$
$\left(-\frac{9}{7}, \frac{2}{7}, \frac{11}{7}\right)$

17. Graph $3x - 4y = 12$ by using the x- and y-intercepts.

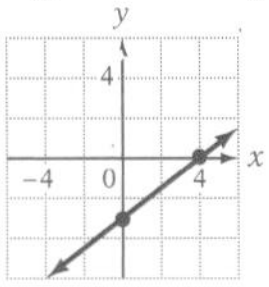

18. Graph the solution set of $-3x + 2y < 6$.

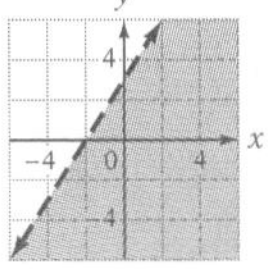

19. Solve by graphing: $x - 2y = 3$
$-2x + y = -3$ $(1, -1)$

20. Graph the solution set: $2x + y < 3$
$-6x + 3y \geq 4$

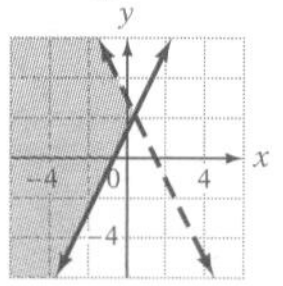

21. Simplify: $(4a^{-2}b^3)(2a^3b^{-1})^{-2}$ $\frac{b^5}{a^8}$

22. Simplify: $\frac{(5x^3y^{-3}z)^{-2}}{y^4z^{-2}}$ $\frac{y^2}{25x^6}$

23. Simplify: $3 - (3 - 3^{-1})^{-1}$ $\frac{21}{8}$

24. Multiply: $(2x + 3)(2x^2 - 3x + 1)$ $4x^3 - 7x + 3$

25. Factor: $-4x^3 + 14x^2 - 12x$ $-2x(2x - 3)(x - 2)$

26. Factor: $a(x - y) - b(y - x)$ $(x - y)(a + b)$

27. Factor: $x^4 - 16$ $(x - 2)(x + 2)(x^2 + 4)$

28. Factor: $2x^3 - 16$ $2(x - 2)(x^2 + 2x + 4)$

29. The sum of two integers is twenty-four. The difference between four times the smaller integer and nine is three less than twice the larger integer. Find the integers. 9 and 15

30. How many ounces of pure gold that costs \$360 per ounce must be mixed with 80 oz of an alloy that costs \$120 per ounce to make a mixture that costs \$200 per ounce? 40 oz

31. Two bicycles are 25 mi apart and are traveling toward each other. One cyclist is traveling at $\frac{2}{3}$ the rate of the other cyclist. They pass in 2 h. Find the rate of each cyclist. Slower cyclist, 5 mph; faster cyclist, 7.5 mph

32. If \$3000 is invested at an annual simple interest rate of 7.5%, how much additional money must be invested at an annual simple interest rate of 10% so that the total interest earned is 9% of the total investment? \$4500

33. The graph below shows the relationship between the distance traveled in miles and the time of travel in hours. Find the slope of the line between the two points on the graph. Write a sentence that states the meaning of the slope.

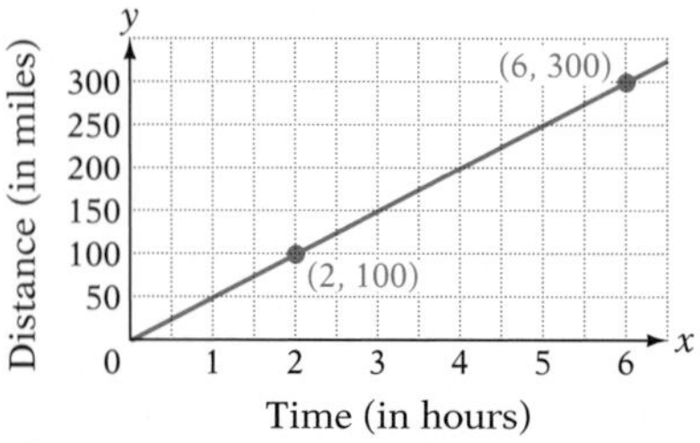

$m = 50$; a slope of 50 means the average speed was 50 mph.

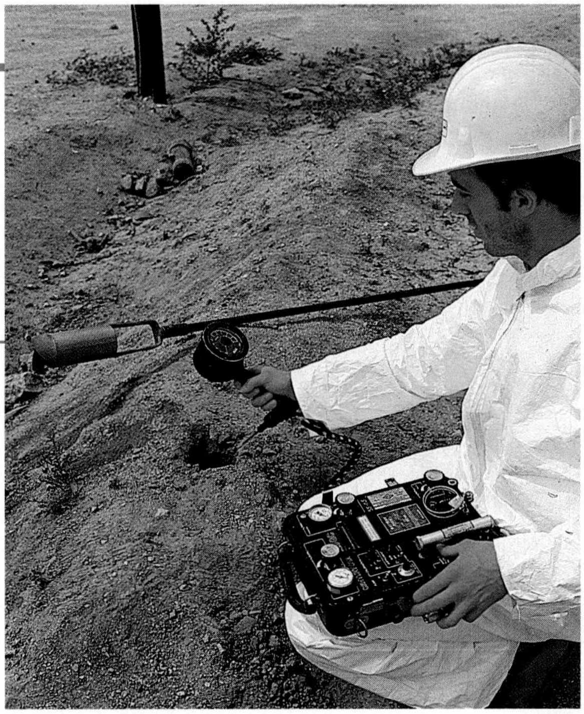

Geophysicists study the composition and physical characteristics of Earth and its magnetic, electric, and gravitational fields. A main responsibility of an exploration geophysicist *is to use techniques to locate oil and other minerals.* Hydrologists *study the waters underground and on the surface of our planet.* Geomagneticians *study Earth's magnetic field. Although these are only a few of the careers within geophysics, all physicists use equations containing rational expressions. The exercises for Section 5 provide a few examples.*

6

Rational Expressions

OBJECTIVES

6.1.1	Find the domain of a rational function
6.1.2	Simplify rational expressions
6.2.1	Multiply and divide rational expressions
6.2.2	Add and subtract rational expressions
6.3.1	Simplify complex fractions
6.4.1	Solve fractional equations
6.4.2	Work problems
6.4.3	Uniform motion problems
6.5.1	Proportions
6.5.2	Variation problems
6.6.1	Solve literal equations

The Abacus

The abacus is an ancient device used to add, subtract, multiply, and divide and to calculate square and cube roots. There are many variations of the abacus, but the one shown here has been used in China for many hundreds of years.

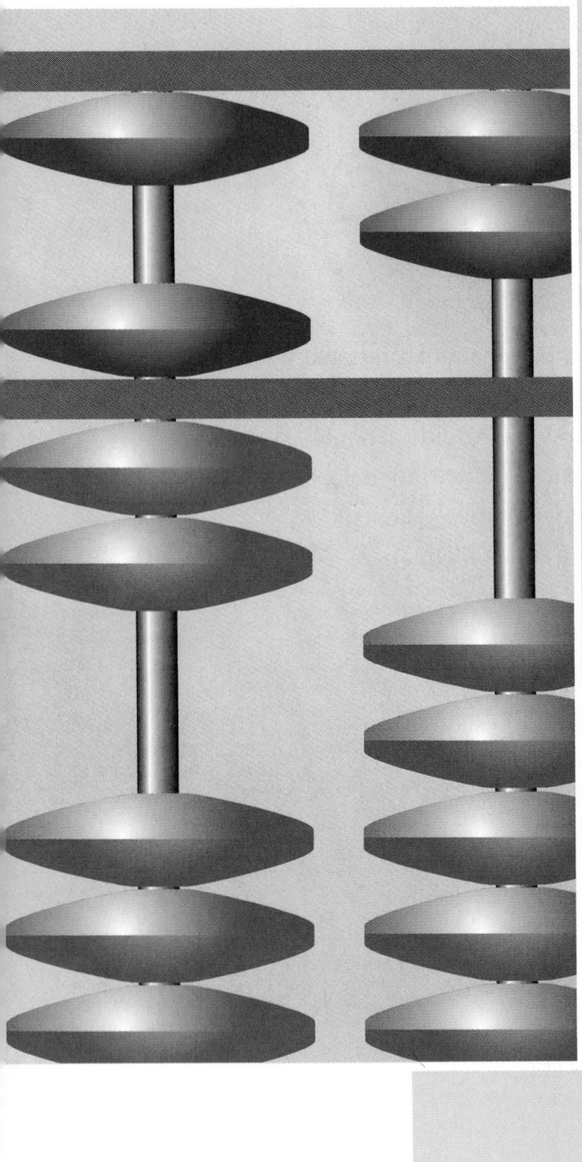

The abacus shown here consists of thirteen rows of beads. A cross-bar separates the beads. Each upper bead represents five units, and each lower bead represents one unit. The place holder of each row is shown above the abacus. The first row of beads represents numbers from 1 to 9, the second row represents numbers from 10 to 90, etc.

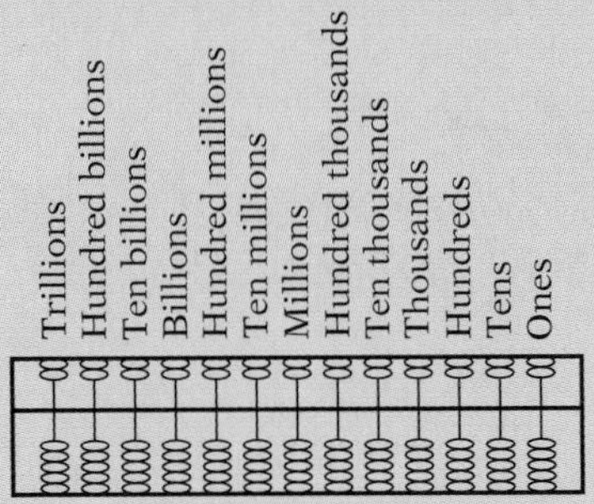

The number shown at the right is 7036.

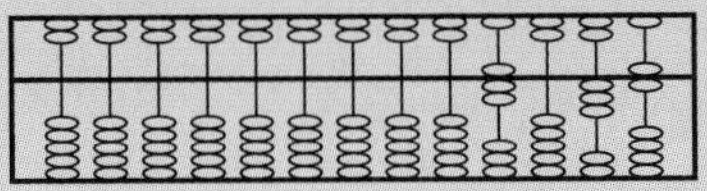

From the thousands row:
- one upper bead—5000
- two lower beads—2000

From the hundreds row:
- no beads—0

From the tens row:
- three lower beads—30

From the ones row:
- one upper bead—5
- one lower bead—1

Add: **1247 + 2516**

1247 is shown at the right.

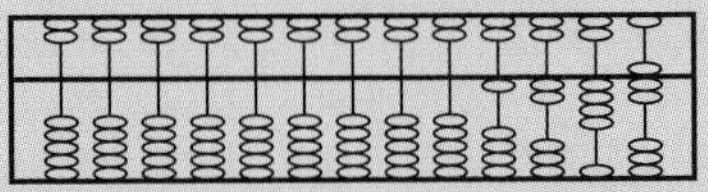

Add 6 to the ones row.

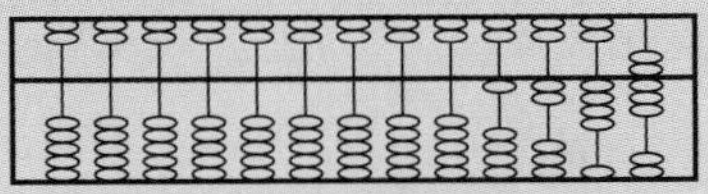

Move the 10 (the upper two beads in the ones row) to the tens row (one lower bead). This makes five beads in the lower tens row. Remove five beads from the lower row and place one bead in the upper row.

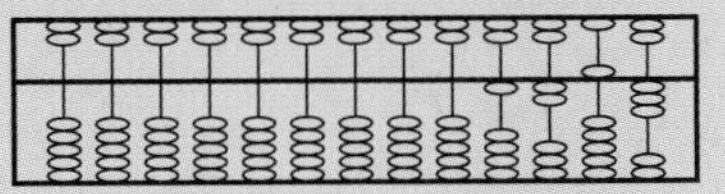

Continue adding in each row until the problem is completed. Carry if necessary.

1247 + 2516 = 3763

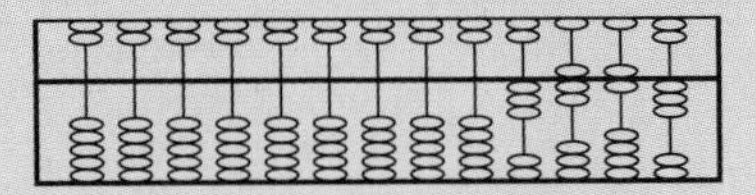

3763

SECTION 6.1

Introduction to Rational Functions

1 Find the domain of a rational function

An expression in which the numerator and denominator are polynomials is called a **rational expression.** Examples of rational expressions are shown at the right.

$$\frac{9}{z} \qquad \frac{3x+4}{2x^2+1} \qquad \frac{x^3-x+1}{x^2-3x-5}$$

INSTRUCTOR NOTE

Many students have not learned the definition of *polynomial.* It may help these students to review the definition at this time.

The expression $\frac{\sqrt{x}+3}{x}$ is not a rational expression because $\sqrt{x}+3$ is not a polynomial.

A function that is written in terms of a rational expression is a **rational function.** Each of the following equations represents a rational function.

$$f(x) = \frac{x^2+3}{2x-1} \qquad g(t) = \frac{3}{t^2-4} \qquad R(z) = \frac{z^2+3z-1}{z^2+z-12}$$

To evaluate a rational function, replace the variable by its value. Then simplify.

Example 1 Given $f(x) = \frac{3x-4}{x^2-2x+1}$, find $f(-3)$.

Solution

$$f(x) = \frac{3x-4}{x^2-2x+1}$$

$$f(-3) = \frac{3(-3)-4}{(-3)^2-2(-3)+1} \qquad \text{► Substitute } -3 \text{ for } x.$$

$$f(-3) = \frac{-9-4}{9+6+1}$$

$$f(-3) = \frac{-13}{16}$$

$$f(-3) = -\frac{13}{16}$$

Problem 1 Given $f(x) = \frac{3-5x}{x^2+5x+6}$, find $f(2)$.

Solution See page S16. $-\frac{7}{20}$

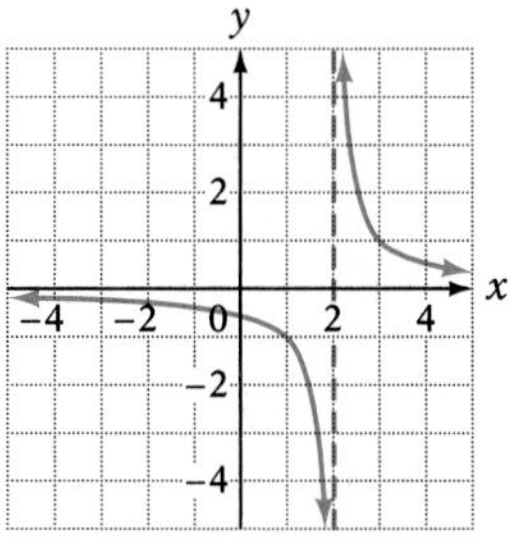

Because division by zero is not defined, the domain of a rational function must exclude those numbers for which the value of the polynomial in the denominator is zero.

The graph of $f(x) = \frac{1}{x-2}$ is shown at the left. Note that the graph never intersects the line $x = 2$ (shown as a dashed line). This value of x is excluded from the domain of $f(x) = \frac{1}{x-2}$.

⇒ Determine the domain of $g(x) = \dfrac{x^2 + 4}{3x - 6}$.

The domain of g must exclude values of x for which the denominator is zero.

Set the denominator equal to zero. $3x - 6 = 0$

Solve for x. $3x = 6$

This value must be *excluded* from the domain. $x = 2$

The domain of g is $\{x \mid x \neq 2\}$.

Example 2   Find the domain of $f(x) = \dfrac{2x - 6}{x^2 - 3x - 4}$.

Solution   The domain must exclude values of x for which $x^2 - 3x - 4 = 0$. Solve this equation for x. Because $x^2 - 3x - 4$ factors over the integers, solve the equation by using the Principle of Zero Products.

$$x^2 - 3x - 4 = 0$$
$$(x + 1)(x - 4) = 0$$

$$x + 1 = 0 \qquad x - 4 = 0$$
$$x = -1 \qquad x = 4$$

When $x = -1$ and $x = 4$, the value of the denominator is zero. Therefore, these values must be excluded from the domain of f.

The domain is $\{x \mid x \neq -1, x \neq 4\}$.

Problem 2   Find the domain of $g(x) = \dfrac{5 - x}{x^2 - 4}$.

Solution   See page S16. $\{x \mid x \neq -2, 2\}$

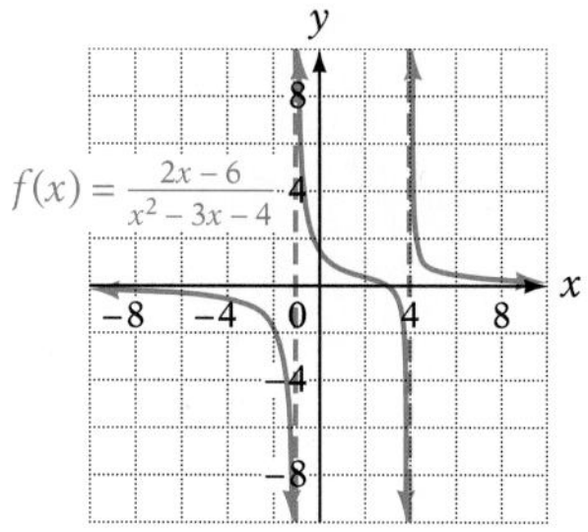

$f(x) = \dfrac{2x - 6}{x^2 - 3x - 4}$

The graph of the function in Example 2 is shown at the left. Note that the graph never intersects the lines $x = -1$ and $x = 4$ (shown as dashed lines). These are the two values of x excluded from the domain of f.

Example 3   Find the domain of $f(x) = \dfrac{3x + 2}{x^2 + 1}$.

Solution   The domain must exclude values of x for which $x^2 + 1 = 0$. It is not possible for $x^2 + 1 = 0$, because $x^2 \geq 0$, and a positive number added to a number equal to or greater than zero cannot equal zero. Therefore, there are no real numbers that must be excluded from the domain of f.

The domain is $\{x \mid x \in \text{real numbers}\}$.

Problem 3   Find the domain of $p(x) = \dfrac{6x}{x^2 + 4}$.

Solution   See page S16. $\{x \mid x \in \text{real numbers}\}$

$f(x) = \dfrac{3x + 2}{x^2 + 1}$

The graph of the function in Example 3 is shown at the left. Because the domain of f is all real numbers, there are points on the graph for all real numbers.

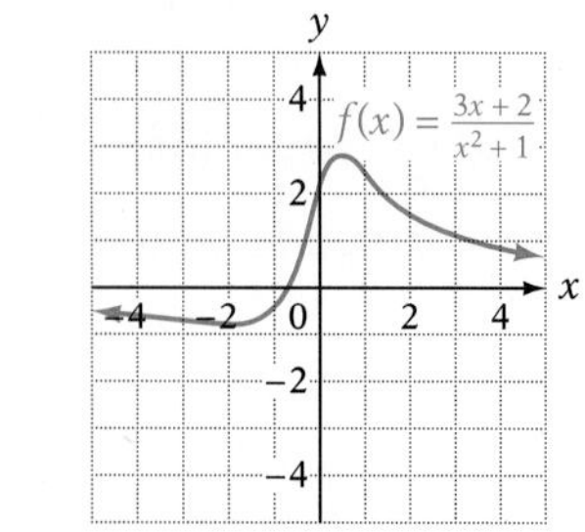

2 Simplify rational expressions

A rational expression is in **simplest form** when the numerator and denominator have no common factors other than 1. The Multiplication Property of One is used to write a rational expression in simplest form.

➡ Simplify: $\dfrac{x^2 - 25}{x^2 + 13x + 40}$

$$\dfrac{x^2 - 25}{x^2 + 13x + 40} = \dfrac{(x - 5)(x + 5)}{(x + 8)(x + 5)} = \dfrac{(x - 5)}{(x + 8)} \cdot \boxed{\dfrac{(x + 5)}{(x + 5)}} = \dfrac{x - 5}{x + 8} \cdot 1 = \dfrac{x - 5}{x + 8}, x \neq -8, -5$$

The requirement $x \neq -8, -5$ must be included because division by 0 is not allowed.

The simplification above is usually shown with slashes to indicate that a common factor has been removed:

$$\dfrac{x^2 - 25}{x^2 + 13x + 40} = \dfrac{(x - 5)\cancel{(x + 5)}^{1}}{(x + 8)\cancel{(x + 5)}_{1}} = \dfrac{x - 5}{x + 8}, x \neq -8, -5$$

We will show a simplification with slashes. We will also omit the restrictions that prevent division by zero. Nonetheless, those restrictions *always* are implied.

Example 4 Simplify.

A. $\dfrac{x^2 - 16}{x^2 + 11x + 28}$ B. $\dfrac{12 + 5x - 2x^2}{2x^2 - 3x - 20}$ C. $\dfrac{x^{2n} + x^n - 2}{x^{2n} - 1}$

Solution A. $\dfrac{x^2 - 16}{x^2 + 11x + 28} = \dfrac{(x + 4)(x - 4)}{(x + 4)(x + 7)}$ ▶ Factor the numerator and the denominator.

$$= \dfrac{\cancel{(x + 4)}^{1}(x - 4)}{\cancel{(x + 4)}_{1}(x + 7)}$$ ▶ Divide by the common factors.

$$= \dfrac{x - 4}{x + 7}$$ ▶ Write the answer in simplest form.

B. $\dfrac{12 + 5x - 2x^2}{2x^2 - 3x - 20} = \dfrac{(4 - x)(3 + 2x)}{(x - 4)(2x + 5)}$ ▶ Factor the numerator and the denominator.

$$= \dfrac{\cancel{(4 - x)}^{-1}(3 + 2x)}{\cancel{(x - 4)}_{1}(2x + 5)}$$ ▶ Divide by the common factors. Remember that $4 - x = -(x - 4)$. Therefore, $\dfrac{4 - x}{x - 4} = \dfrac{-(x - 4)}{x - 4} = \dfrac{-1}{1} = -1$.

$$= -\dfrac{2x + 3}{2x + 5}$$ ▶ Write the answer in simplest form.

C. $\dfrac{x^{2n} + x^n - 2}{x^{2n} - 1} = \dfrac{(x^n - 1)(x^n + 2)}{(x^n - 1)(x^n + 1)}$ ▶ Factor the numerator and the denominator.

$$= \dfrac{\cancel{(x^n - 1)}^{1}(x^n + 2)}{\cancel{(x^n - 1)}_{1}(x^n + 1)}$$ ▶ Divide by the common factors.

$$= \dfrac{x^n + 2}{x^n + 1}$$

Problem 4 Simplify.

A. $\dfrac{6x^4 - 24x^3}{12x^3 - 48x^2}$ B. $\dfrac{20x - 15x^2}{15x^3 - 5x^2 - 20x}$ C. $\dfrac{x^{2n} + x^n - 12}{x^{2n} - 3x^n}$

Solution See page S16. A. $\dfrac{x}{2}$ B. $-\dfrac{1}{x+1}$ C. $\dfrac{x^n + 4}{x^n}$

CONCEPT REVIEW 6.1

Determine whether the following statements are always true, sometimes true, or never true.

1. $\dfrac{x^2 + 2x + 1}{x^{\frac{1}{2}} - 2x + 4}$ is a rational expression. Never true

2. A rational expression with a variable in the denominator will have restrictions on the value of x. Sometimes true

3. When simplifying rational expressions, divide the numerator and denominator by the GCF of the numerator and denominator. Always true

4. If $x \neq y$, the rational expression $\dfrac{x - y}{y - x}$ can be reduced to -1. Always true

5. $\dfrac{a + 4}{a^2 + 6a + 8} = \dfrac{a + 4}{(a + 4)(a + 2)} = \dfrac{\cancel{a + 4}}{\cancel{(a + 4)}(a + 2)} = \dfrac{0}{a + 2} = 0$ Never true

6. For all values of a, $\dfrac{a(a + 4)}{a} = a + 4$. Sometimes true

EXERCISES 6.1

1 1. What is a rational function? Give an example of a rational function.

2. What values are excluded from the domain of a rational function?

3. Given $f(x) = \dfrac{2}{x - 3}$, find $f(4)$. 2

4. Given $f(x) = \dfrac{-7}{5 - x}$, find $f(-2)$. -1

5. Given $f(x) = \dfrac{x - 2}{x + 4}$, find $f(-2)$. -2

6. Given $f(x) = \dfrac{x - 3}{2x - 1}$, find $f(3)$. 0

7. Given $f(x) = \dfrac{1}{x^2 - 2x + 1}$, find $f(-2)$. $\dfrac{1}{9}$

8. Given $f(x) = \dfrac{-3}{x^2 - 4x + 2}$, find $f(-1)$. $-\dfrac{3}{7}$

9. Given $f(x) = \dfrac{x - 2}{2x^2 + 3x + 8}$, find $f(3)$. $\dfrac{1}{35}$

10. Given $f(x) = \dfrac{x^2}{3x^2 - 3x + 5}$, find $f(4)$ $\dfrac{16}{41}$

11. Given $f(x) = \dfrac{x^2 - 2x}{x^3 - x + 4}$, find $f(-1)$. $\dfrac{3}{4}$

12. Given $f(x) = \dfrac{8 - x^2}{x^3 - x^2 + 4}$, find $f(-3)$. $\dfrac{1}{32}$

Find the domain of the function.

13. $H(x) = \dfrac{4}{x - 3}$ $\{x \mid x \neq 3\}$

14. $G(x) = \dfrac{-2}{x + 2}$ $\{x \mid x \neq -2\}$

15. $f(x) = \dfrac{x}{x + 4}$ $\{x \mid x \neq -4\}$

16. $g(x) = \dfrac{3x}{x - 5}$ $\{x \mid x \neq 5\}$

17. $R(x) = \dfrac{5x}{3x + 9}$ $\{x \mid x \neq -3\}$

18. $p(x) = \dfrac{-2x}{6 - 2x}$ $\{x \mid x \neq 3\}$

19. $q(x) = \frac{4 - x}{(x - 4)(x + 2)}$ $\{x \mid x \neq -2, 4\}$

20. $h(x) = \frac{2x + 1}{(x + 1)(x + 5)}$ $\{x \mid x \neq -1, -5\}$

21. $V(x) = \frac{x^2}{(2x + 5)(3x - 6)}$ $\left\{x \mid x \neq -\frac{5}{2}, 2\right\}$

22. $F(x) = \frac{x^2 - 1}{(4x + 8)(3x - 1)}$ $\left\{x \mid x \neq -2, \frac{1}{3}\right\}$

23. $f(x) = \frac{x^2 + 1}{x}$ $\{x \mid x \neq 0\}$

24. $g(x) = \frac{2x^3 - x - 1}{x^2}$ $\{x \mid x \neq 0\}$

25. $k(x) = \frac{x + 1}{x^2 + 1}$ $\{x \mid x \text{ is a real number}\}$

26. $P(x) = \frac{2x + 3}{2x^2 + 3}$ $\{x \mid x \text{ is a real number}\}$

27. $f(x) = \frac{2x - 1}{x^2 + x - 6}$ $\{x \mid x \neq -3, 2\}$

28. $G(x) = \frac{3 - 4x}{x^2 + 4x - 5}$ $\{x \mid x \neq -5, 1\}$

29. $A(x) = \frac{5x + 2}{x^2 + 2x - 24}$ $\{x \mid x \neq -6, 4\}$

30. $h(x) = \frac{3x}{x^2 - 4}$ $\{x \mid x \neq -2, 2\}$

31. $f(x) = \frac{4x - 7}{3x^2 + 12}$ $\{x \mid x \text{ is a real number}\}$

32. $g(x) = \frac{x^2 + x + 1}{5x^2 + 1}$ $\{x \mid x \text{ is a real number}\}$

33. $G(x) = \frac{x^2 + 1}{6x^2 - 13x + 6}$ $\left\{x \mid x \neq \frac{2}{3}, \frac{3}{2}\right\}$

34. $A(x) = \frac{5x - 7}{x(x - 2)(x - 3)}$ $\{x \mid x \neq 0, 2, 3\}$

35. $f(x) = \frac{x^2 + 8x + 4}{2x^3 + 9x^2 - 5x}$ $\left\{x \mid x \neq 0, \frac{1}{2}, -5\right\}$

36. $H(x) = \frac{x^4 - 1}{2x^3 + 2x^2 - 24x}$ $\{x \mid x \neq 0, -4, 3\}$

2 37. When is a rational expression in simplest form?

38. Are the rational expressions $\frac{x(x - 2)}{2(x - 2)}$ and $\frac{x}{2}$ equal for all values of x? Why or why not?

Simplify.

39. $\frac{4 - 8x}{4}$ $1 - 2x$

40. $\frac{8y + 2}{2}$ $4y + 1$

41. $\frac{6x^2 - 2x}{2x}$ $3x - 1$

42. $\frac{3y - 12y^2}{3y}$ $1 - 4y$

43. $\frac{8x^2(x - 3)}{4x(x - 3)}$ $2x$

44. $\frac{16y^4(y + 8)}{12y^3(y + 8)}$ $\frac{4y}{3}$

45. $\frac{2x - 6}{3x - x^2}$ $-\frac{2}{x}$

46. $\frac{3a^2 - 6a}{12 - 6a}$ $-\frac{a}{2}$

47. $\frac{6x^3 - 15x^2}{12x^2 - 30x}$ $\frac{x}{2}$

48. $\frac{-36a^2 - 48a}{18a^3 + 24a^2}$ $-\frac{2}{a}$

49. $\frac{a^2 + 4a}{4a - 16}$ Simplest form

50. $\frac{3x - 6}{x^2 + 2x}$ Simplest form

51. $\frac{16x^3 - 8x^2 + 12x}{4x}$ $4x^2 - 2x + 3$

52. $\frac{3x^3y^3 - 12x^2y^2 + 15xy}{3xy}$ $x^2y^2 - 4xy + 5$

53. $\frac{-10a^4 - 20a^3 + 30a^2}{-10a^2}$ $a^2 + 2a - 3$

54. $\frac{-7a^5 - 14a^4 + 21a^3}{-7a^3}$ $a^2 + 2a - 3$

55. $\frac{3x^{3n} - 9x^{2n}}{12x^{2n}}$ $\frac{x^n - 3}{4}$

56. $\frac{8a^n}{4a^{2n} - 8a^n}$ $\frac{2}{a^n - 2}$

57. $\frac{x^2 - 7x + 12}{x^2 - 9x + 20}$ $\frac{x - 3}{x - 5}$

58. $\frac{x^2 - x - 20}{x^2 - 2x - 15}$ $\frac{x + 4}{x + 3}$

59. $\frac{x^2 - xy - 2y^2}{x^2 - 3xy + 2y^2}$ $\frac{x + y}{x - y}$

60. $\frac{2x^2 + 7xy - 4y^2}{4x^2 - 4xy + y^2}$ $\frac{x + 4y}{2x - y}$

61. $\frac{6 - x - x^2}{3x^2 - 10x + 8}$ $-\frac{x + 3}{3x - 4}$

62. $\frac{3x^2 + 10x - 8}{8 - 14x + 3x^2}$ $-\frac{x + 4}{4 - x}$

63. $\frac{14 - 19x - 3x^2}{3x^2 - 23x + 14}$ $-\frac{x + 7}{x - 7}$

64. $\frac{x^2 + x - 12}{x^2 - x - 12}$ Simplest form

65. $\frac{a^2 - 7a + 10}{a^2 + 9a + 14}$ Simplest form

66. $\frac{x^2 - 2x}{x^2 + 2x}$ $\frac{x - 2}{x + 2}$

67. $\frac{a^2 - b^2}{a^3 + b^3}$ $\frac{a - b}{a^2 - ab + b^2}$

68. $\frac{x^4 - y^4}{x^2 + y^2}$ $(x + y)(x - y)$

69. $\frac{x^3 + y^3}{3x^3 - 3x^2y + 3xy^2}$ $\frac{x + y}{3x}$

70. $\frac{3x^3 + 3x^2 + 3x}{9x^3 - 9}$ $\frac{x}{3(x - 1)}$

71. $\dfrac{x^3 - 4xy^2}{3x^3 - 2x^2y - 8xy^2}$ $\quad \dfrac{x + 2y}{3x + 4y}$

72. $\dfrac{4a^2 - 8ab + 4b^2}{4a^2 - 4b^2}$ $\quad \dfrac{a - b}{a + b}$

73. $\dfrac{4x^3 - 14x^2 + 12x}{24x + 4x^2 - 8x^3}$ $\quad -\dfrac{2x - 3}{2(2x + 3)}$

74. $\dfrac{6x^3 - 15x^2 - 75x}{150x + 30x^2 - 12x^3}$ $\quad -\dfrac{1}{2}$

75. $\dfrac{x^2 - 4}{a(x + 2) - b(x + 2)}$ $\quad \dfrac{x - 2}{a - b}$

76. $\dfrac{x^2(a - 2) - a + 2}{ax^2 - ax}$ $\quad \dfrac{(x + 1)(a - 2)}{ax}$

77. $\dfrac{x^4 + 3x^2 + 2}{x^4 - 1}$ $\quad \dfrac{x^2 + 2}{(x + 1)(x - 1)}$

78. $\dfrac{x^4 - 2x^2 - 3}{x^4 + 2x^2 + 1}$ $\quad \dfrac{x^2 - 3}{x^2 + 1}$

79. $\dfrac{x^2y^2 + 4xy - 21}{x^2y^2 - 10xy + 21}$ $\quad \dfrac{xy + 7}{xy - 7}$

80. $\dfrac{6x^2y^2 + 11xy + 4}{9x^2y^2 + 9xy - 4}$ $\quad \dfrac{2xy + 1}{3xy - 1}$

81. $\dfrac{a^{2n} - a^n - 2}{a^{2n} + 3a^n + 2}$ $\quad \dfrac{a^n - 2}{a^n + 2}$

82. $\dfrac{a^{2n} + a^n - 12}{a^{2n} - 2a^n - 3}$ $\quad \dfrac{a^n + 4}{a^n + 1}$

APPLYING CONCEPTS 6.1

83. Evaluate $h(x) = \dfrac{x + 2}{x - 3}$ when $x = 2.9, 2.99, 2.999$, and 2.9999. On the basis of your evaluations, complete the following sentence. As x becomes closer to 3, the values of $h(x)$ decrease.

84. Evaluate $h(x) = \dfrac{x + 2}{x - 3}$ when $x = 3.1, 3.01, 3.001$, and 3.0001. On the basis of your evaluations, complete the following sentence. As x becomes closer to 3, the values of $h(x)$ increase.

85. The relationship among the focal length (F) of a camera lens, the distance between the object and the lens (x), and the distance between the lens and the film (y) is given by $\dfrac{1}{F} = \dfrac{1}{x} + \dfrac{1}{y}$. A camera used by a professional photographer has a dial that allows the focal length to be set at a constant value. Suppose a photographer chooses a focal length of 50 mm. Substituting this value into the equation, solving for y, and using $y = f(x)$ notation yield $f(x) = \dfrac{50x}{x - 50}$.

a. Graph this equation for $50 < x \leq 50{,}000$.

b. The point whose coordinates are (2000, 51), to the nearest integer, is on the graph of the function. Give an interpretation of this ordered pair.

c. Give a reason for choosing the domain so that $x > 50$.

d. Photographers refer to depth of field as a range of distances in which an object remains in focus. Use the graph to explain why the depth of field is larger for objects that are far from the lens than for objects that are close to the lens. For $x > 1000$, $f(x)$ changes very little for large changes in x.

a.

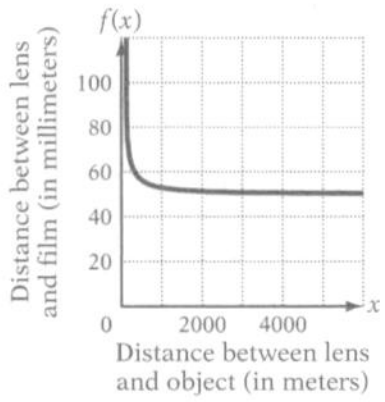

b. The ordered pair (2000, 51) means that when the distance between the object and the lens is 2000 m, the distance between the lens and the film is 51 mm.

c. For $x = 50$, the expression $\dfrac{50x}{x - 50}$ is undefined. For $0 < x < 50$, $f(x)$ is negative, and the distance cannot be negative. Therefore, the domain is $x > 50$.

86. Suppose that $F(x) = \dfrac{g(x)}{h(x)}$ and that, for some real number a, $g(a) = 0$ and $h(a) = 0$. Is $F(x)$ in simplest form? Explain your answer.

87. Why can the numerator and denominator of a rational expression be divided by their common factors? What conditions must be placed on the value of the variables when a rational expression is simplified?

SECTION 6.2

Operations on Rational Expressions

1 Multiply and divide rational expressions

22 HM³

The product of two fractions is a fraction whose numerator is the product of the numerators of the two fractions and whose denominator is the product of the denominators of the two fractions.

$$\frac{a}{b} \cdot \frac{c}{d} = \frac{ac}{bd}$$

$$\frac{5}{a+2} \cdot \frac{b-3}{3} = \frac{5(b-3)}{(a+2)3} = \frac{5b-15}{3a+6}$$

The product of two rational expressions can often be simplified by factoring the numerator and the denominator.

⇒ Simplify: $\dfrac{x^2-2x}{2x^2+x-15} \cdot \dfrac{2x^2-x-10}{x^2-4}$

$$\frac{x^2-2x}{2x^2+x-15} \cdot \frac{2x^2-x-10}{x^2-4}$$

Factor the numerator and denominator of each fraction.

$$= \frac{x(x-2)}{(x+3)(2x-5)} \cdot \frac{(x+2)(2x-5)}{(x+2)(x-2)}$$

Multiply.

$$= \frac{x(x-2)(x+2)(2x-5)}{(x+3)(2x-5)(x+2)(x-2)}$$

Divide by the common factors.

$$= \frac{x\overset{1}{\cancel{(x-2)}}\overset{1}{\cancel{(x+2)}}\overset{1}{\cancel{(2x-5)}}}{(x+3)\underset{1}{\cancel{(2x-5)}}\underset{1}{\cancel{(x+2)}}\underset{1}{\cancel{(x-2)}}}$$

Write the answer in simplest form.

$$= \frac{x}{x+3}$$

Example 1 Multiply.

A. $\dfrac{2x^2-6x}{3x-6} \cdot \dfrac{6x-12}{8x^3-12x^2}$ B. $\dfrac{6x^2+x-2}{6x^2+7x+2} \cdot \dfrac{2x^2+9x+4}{4-7x-2x^2}$

Solution A. $\dfrac{2x^2-6x}{3x-6} \cdot \dfrac{6x-12}{8x^3-12x^2} = \dfrac{2x(x-3)}{3(x-2)} \cdot \dfrac{6(x-2)}{4x^2(2x-3)}$

$$= \frac{12x(x-3)\overset{1}{\cancel{(x-2)}}}{12x^2\underset{1}{\cancel{(x-2)}}(2x-3)} = \frac{x-3}{x(2x-3)}$$

B. $\dfrac{6x^2+x-2}{6x^2+7x+2} \cdot \dfrac{2x^2+9x+4}{4-7x-2x^2} = \dfrac{(2x-1)(3x+2)}{(3x+2)(2x+1)} \cdot \dfrac{(2x+1)(x+4)}{(1-2x)(4+x)}$

$$= \frac{\overset{-1}{\cancel{(2x-1)}}\overset{1}{\cancel{(3x+2)}}\overset{1}{\cancel{(2x+1)}}\overset{1}{\cancel{(x+4)}}}{\underset{1}{\cancel{(3x+2)}}\underset{1}{\cancel{(2x+1)}}\underset{1}{\cancel{(1-2x)}}\underset{1}{\cancel{(x+4)}}} = -1$$

Problem 1 Multiply.

A. $\dfrac{12+5x-3x^2}{x^2+2x-15} \cdot \dfrac{2x^2+x-45}{3x^2+4x}$ B. $\dfrac{2x^2-13x+20}{x^2-16} \cdot \dfrac{2x^2+9x+4}{6x^2-7x-5}$

Solution See page S16–S17. A. $-\dfrac{2x-9}{x}$ B. $\dfrac{2x-5}{3x-5}$

The **reciprocal of a rational expression** is the rational expression with the numerator and denominator interchanged.

$$\text{Rational expression} \left\{ \begin{array}{cc} \dfrac{a}{b} & \dfrac{b}{a} \\[2ex] \dfrac{a^2 - 2y}{4} & \dfrac{4}{a^2 - 2y} \end{array} \right\} \text{Reciprocal}$$

To divide two rational expressions, multiply by the reciprocal of the divisor.

$$\frac{a}{b} \div \frac{c}{d} = \frac{a}{b} \cdot \frac{d}{c} = \frac{ad}{bc}$$

$$\frac{2}{a} \div \frac{5}{b} = \frac{2}{a} \cdot \frac{b}{5} = \frac{2b}{5a}$$

$$\frac{x + y}{2} \div \frac{x - y}{5} = \frac{x + y}{2} \cdot \frac{5}{x - y} = \frac{(x + y)5}{2(x - y)} = \frac{5x + 5y}{2x - 2y}$$

Example 2 Divide.

A. $\dfrac{12x^2y^2 - 24xy^2}{5z^2} \div \dfrac{4x^3y - 8x^2y}{3z^4}$ B. $\dfrac{3y^2 - 10y + 8}{3y^2 + 8y - 16} \div \dfrac{2y^2 - 7y + 6}{2y^2 + 5y - 12}$

Solution A.

$$\frac{12x^2y^2 - 24xy^2}{5z^2} \div \frac{4x^3y - 8x^2y}{3z^4} = \frac{12x^2y^2 - 24xy^2}{5z^2} \cdot \frac{3z^4}{4x^3y - 8x^2y}$$

$$= \frac{12xy^2(x - 2)}{5z^2} \cdot \frac{3z^4}{4x^2y(x - 2)}$$

$$= \frac{36xy^2z^4\cancel{(x - 2)}^{1}}{20x^2yz^2\cancel{(x - 2)}_{1}} = \frac{9yz^2}{5x}$$

B.

$$\frac{3y^2 - 10y + 8}{3y^2 + 8y - 16} \div \frac{2y^2 - 7y + 6}{2y^2 + 5y - 12} = \frac{3y^2 - 10y + 8}{3y^2 + 8y - 16} \cdot \frac{2y^2 + 5y - 12}{2y^2 - 7y + 6}$$

$$= \frac{(y - 2)(3y - 4)}{(3y - 4)(y + 4)} \cdot \frac{(y + 4)(2y - 3)}{(y - 2)(2y - 3)}$$

$$= \frac{\cancel{(y - 2)}^{1}\cancel{(3y - 4)}^{1}\cancel{(y + 4)}^{1}\cancel{(2y - 3)}^{1}}{\cancel{(3y - 4)}_{1}\cancel{(y + 4)}_{1}\cancel{(y - 2)}_{1}\cancel{(2y - 3)}_{1}} = 1$$

Problem 2 Divide.

A. $\dfrac{6x^2 - 3xy}{10ab^4} \div \dfrac{16x^2y^2 - 8xy^3}{15a^2b^2}$ B. $\dfrac{6x^2 - 7x + 2}{3x^2 + x - 2} \div \dfrac{4x^2 - 8x + 3}{5x^2 + x - 4}$

Solution See page S17. A. $\dfrac{9a}{16b^2y^2}$ B. $\dfrac{5x - 4}{2x - 3}$

2 Add and subtract rational expressions

When adding rational expressions in which the denominators are the same, add the numerators. The denominator of the sum is the common denominator. Write the answer in simplest form.

$$\frac{a}{c} + \frac{b}{c} = \frac{a + b}{c}$$

$$\frac{4x}{15} + \frac{8x}{15} = \frac{4x + 8x}{15} = \frac{12x}{15} = \frac{4x}{5}$$

$$\frac{a}{a^2 - b^2} + \frac{b}{a^2 - b^2} = \frac{a + b}{a^2 - b^2} = \frac{a + b}{(a - b)(a + b)} = \frac{\cancel{(a + b)}^{1}}{(a - b)\cancel{(a + b)}_{1}} = \frac{1}{a - b}$$

When subtracting rational expressions with the same denominators, subtract the numerators. The denominator of the difference is the common denominator. Write the answer in simplest form.

$$\frac{y}{y-3} - \frac{3}{y-3} = \frac{y-3}{y-3} = \frac{\overset{1}{\cancel{(y-3)}}}{\underset{1}{\cancel{(y-3)}}} = 1$$

$$\frac{7x-12}{2x^2+5x-12} - \frac{3x-6}{2x^2+5x-12} = \frac{(7x-12)-(3x-6)}{2x^2+5x-12} = \frac{7x-12-3x+6}{2x^2+5x-12}$$

$$= \frac{4x-6}{2x^2+5x-12}$$

$$= \frac{2(2x-3)}{(2x-3)(x+4)}$$

$$= \frac{2\overset{1}{\cancel{(2x-3)}}}{\underset{1}{\cancel{(2x-3)}}(x+4)} = \frac{2}{x+4}$$

Before two rational expressions with different denominators can be added or subtracted, both rational expressions must be expressed in terms of a common denominator. This common denominator is the LCM of the denominators of the rational expressions.

The LCM of two or more polynomials is the simplest polynomial that contains the factors of each polynomial. To find the LCM, first factor each polynomial completely. The LCM is the product of each factor the greatest number of times it occurs in any one factorization.

To find the LCM of $3x^2 + 15x$ and $6x^4 + 24x^3 - 30x^2$, factor each polynomial.

$$3x^2 + 15x = 3x(x+5)$$
$$6x^4 + 24x^3 - 30x^2 = 6x^2(x^2 + 4x - 5) = 6x^2(x-1)(x+5)$$

The LCM is the product of the LCM of the numerical coefficients and each variable factor the greatest number of times it occurs in any one factorization.

$$\text{LCM} = 6x^2(x-1)(x+5)$$

➡ Write the fractions $\dfrac{x+2}{x^2-2x}$ and $\dfrac{5x}{3x-6}$ in terms of the LCM of the denominators.

Find the LCM of the denominators.

$$x^2 - 2x = x(x-2)$$
$$3x - 6 = 3(x-2)$$

The LCM is $3x(x-2)$.

For each fraction, multiply the numerator and denominator by the factor whose product with the denominator is the LCM.

$$\frac{x+2}{x^2-2x} = \frac{x+2}{x(x-2)} \cdot \frac{3}{3} = \frac{3x+6}{3x(x-2)}$$

$$\frac{5x}{3x-6} = \frac{5x}{3(x-2)} \cdot \frac{x}{x} = \frac{5x^2}{3x(x-2)}$$

➡ Add: $\dfrac{3x}{2x-3} + \dfrac{3x+6}{2x^2+x-6}$

The LCM of the denominators is $(2x-3)(x+2)$.

$$\frac{3x}{2x-3} + \frac{3x+6}{2x^2+x-6}$$

INSTRUCTOR NOTE

One reason why adding or subtracting rational expressions is difficult for students is the number of steps. Encourage students to develop an outline similar to the following that they can use for the exercises in this objective.

- Find a common denominator.
- Express each fraction in terms of the common denominator.
- Add (or subtract) the rational expressions.
- Simplify.

Rewrite each fraction in terms of the LCM of the denominators.

$$= \frac{3x}{2x-3} \cdot \frac{x+2}{x+2} + \frac{3x+6}{(2x-3)(x+2)}$$

$$= \frac{3x^2+6x}{(2x-3)(x+2)} + \frac{3x+6}{(2x-3)(x+2)}$$

Add the fractions.

$$= \frac{(3x^2+6x)+(3x+6)}{(2x-3)(x+2)}$$

$$= \frac{3x^2+9x+6}{(2x-3)(x+2)}$$

Factor the numerator to determine whether there are common factors in the numerator and denominator.

$$= \frac{3(x^2+3x+2)}{(2x-3)(x+2)}$$

$$= \frac{3(x+2)(x+1)}{(2x-3)(x+2)}$$

$$= \frac{3\overset{1}{\cancel{(x+2)}}(x+1)}{(2x-3)\underset{1}{\cancel{(x+2)}}} = \frac{3(x+1)}{2x-3}$$

Example 3 Subtract: $\frac{x}{2x-4} - \frac{4-x}{x^2-2x}$

Solution $\frac{x}{2x-4} - \frac{4-x}{x^2-2x}$

$$= \frac{x}{2(x-2)} \cdot \frac{x}{x} - \frac{4-x}{x(x-2)} \cdot \frac{2}{2}$$

▶ Write each fraction in terms of the LCM. The LCM is $2x(x-2)$.

$$= \frac{x^2}{2x(x-2)} - \frac{8-2x}{2x(x-2)}$$

▶ Subtract the fractions.

$$= \frac{x^2-(8-2x)}{2x(x-2)}$$

$$= \frac{x^2+2x-8}{2x(x-2)}$$

$$= \frac{(x+4)(x-2)}{2x(x-2)}$$

$$= \frac{(x+4)\overset{1}{\cancel{(x-2)}}}{2x\underset{1}{\cancel{(x-2)}}} = \frac{x+4}{2x}$$

▶ Divide by the common factors.

Problem 3 Add: $\frac{a-3}{a^2-5a} + \frac{a-9}{a^2-25}$

Solution See page S17. $\frac{2a+3}{a(a+5)}$

Example 4 Simplify: $\frac{6x-23}{2x^2+x-6} + \frac{3x}{2x-3} - \frac{5}{x+2}$

Solution $\frac{6x-23}{2x^2+x-6} + \frac{3x}{2x-3} - \frac{5}{x+2}$

$$= \frac{6x-23}{(2x-3)(x+2)} + \frac{3x}{2x-3} \cdot \frac{x+2}{x+2} - \frac{5}{x+2} \cdot \frac{2x-3}{2x-3}$$

$$= \frac{6x-23}{(2x-3)(x+2)} + \frac{3x^2+6x}{(2x-3)(x+2)} - \frac{10x-15}{(2x-3)(x+2)}$$

$$= \frac{(6x-23)+(3x^2+6x)-(10x-15)}{(2x-3)(x+2)}$$

$$= \frac{6x-23+3x^2+6x-10x+15}{(2x-3)(x+2)}$$

$$= \frac{3x^2+2x-8}{(2x-3)(x+2)} = \frac{(3x-4)(x+2)}{(2x-3)(x+2)} = \frac{3x-4}{2x-3}$$

Problem 4 Simplify: $\frac{x-1}{x-2} - \frac{7-6x}{2x^2-7x+6} + \frac{4}{2x-3}$

Solution See page S17. $\frac{x+4}{x-2}$

CONCEPT REVIEW 6.2

Determine whether the following statements are always true, sometimes true, or never true.

1. To divide an expression by a nonzero rational number, multiply the expression by the reciprocal of the divisor. Always true

2. To multiply rational expressions, multiply the numerators and multiply the denominators. Always true

3. $\frac{3}{x+y} + \frac{2}{x+y} = \frac{3+2}{(x+y)+(x+y)} = \frac{5}{2x+2y}$ Never true

4. $\frac{2x}{x(x+4)} - \frac{2}{x+4} = \frac{2\not{x}}{\not{x}(x+4)} - \frac{2}{x+4} = \frac{2}{x+4} - \frac{2}{x+4} = 0$ Always true

EXERCISES 6.2

1 1. Explain how to multiply two rational expressions.

2. Explain how to divide two rational expressions.

Multiply or divide.

3. $\frac{27a^2b^5}{16xy^2} \cdot \frac{20x^2y^3}{9a^2b}$ $\frac{15b^4xy}{4}$

4. $\frac{15x^2y^4}{24ab^3} \cdot \frac{28a^2b^4}{35xy^4}$ $\frac{abx}{2}$

5. $\frac{3x-15}{4x^2-2x} \cdot \frac{20x^2-10x}{15x-75}$ 1

6. $\frac{2x^2+4x}{8x^2-40x} \cdot \frac{6x^3-30x^2}{3x^2+6x}$ $\frac{x}{2}$

7. $\frac{x^2y^3}{x^2-4x-5} \cdot \frac{2x^2-13x+15}{x^4y^3}$ $\frac{2x-3}{x^2(x+1)}$

8. $\frac{2x^2-5x+3}{x^6y^3} \cdot \frac{x^4y^4}{2x^2-x-3}$ $\frac{y(x-1)}{x^2(x+1)}$

9. $\frac{x^2-3x+2}{x^2-8x+15} \cdot \frac{x^2+x-12}{8-2x-x^2}$ $-\frac{x-1}{x-5}$

10. $\frac{x^2+x-6}{12+x-x^2} \cdot \frac{x^2+x-20}{x^2-4x+4}$ $-\frac{x+5}{x-2}$

11. $\frac{x^{n+1}+2x^n}{4x^2-6x} \cdot \frac{8x^2-12x}{x^{n+1}-x^n}$ $\frac{2(x+2)}{x-1}$

12. $\frac{x^{2n}+2x^n}{x^{n+1}+2x} \cdot \frac{x^2-3x}{x^{n+1}-3x^n}$ 1

13. $\frac{12+x-6x^2}{6x^2+29x+28} \cdot \frac{2x^2+x-21}{4x^2-9}$ $-\frac{x-3}{2x+3}$

14. $\frac{x^2+5x+4}{4+x-3x^2} \cdot \frac{3x^2+2x-8}{x^2+4x}$ $-\frac{x+2}{x}$

15. $\frac{x^{2n}-x^n-6}{x^{2n}+x^n-2} \cdot \frac{x^{2n}-5x^n-6}{x^{2n}-2x^n-3}$ $\frac{x^n-6}{x^n-1}$

16. $\frac{x^{2n}+3x^n+2}{x^{2n}-x^n-6} \cdot \frac{x^{2n}+x^n-12}{x^{2n}-1}$ $\frac{x^n+4}{x^n-1}$

17. $\frac{x^3-y^3}{2x^2+xy-3y^2} \cdot \frac{2x^2+5xy+3y^2}{x^2+xy+y^2}$ $x+y$

18. $\frac{x^4-5x^2+4}{3x^2-4x-4} \cdot \frac{3x^2-10x-8}{x^2-4}$ $\frac{(x+1)(x-1)(x-4)}{x-2}$

19. $\frac{6x^2y^4}{35a^2b^5} \div \frac{12x^3y^3}{7a^4b^5}$ $\frac{a^2y}{10x}$

20. $\frac{12a^4b^7}{13x^2y^2} \div \frac{18a^5b^6}{26xy^3}$ $\frac{4by}{3ax}$

21. $\frac{2x-6}{6x^2-15x} \div \frac{4x^2-12x}{18x^3-45x^2}$ $\frac{3}{2}$

22. $\frac{4x^2-4y^2}{6x^2y^2} \div \frac{3x^2+3xy}{2x^2y-2xy^2}$ $\frac{4(x-y)^2}{9x^2y}$

23. $\frac{2x^2-2y^2}{14x^2y^4} \div \frac{x^2+2xy+y^2}{35xy^3}$ $\frac{5(x-y)}{xy(x+y)}$

24. $\frac{8x^3+12x^2y}{4x^2-9y^2} \div \frac{16x^2y^2}{4x^2-12xy+9y^2}$ $\frac{2x-3y}{4y^2}$

25. $\frac{2x^2 - 5x - 3}{2x^2 + 7x + 3} \div \frac{2x^2 - 3x - 20}{2x^2 - x - 15}$ $\frac{(x-3)^2}{(x+3)(x-4)}$

26. $\frac{3x^2 - 10x - 8}{6x^2 + 13x + 6} \div \frac{2x^2 - 9x + 10}{4x^2 - 4x - 15}$ $\frac{x-4}{x-2}$

27. $\frac{x^2 - 8x + 15}{x^2 + 2x - 35} \div \frac{15 - 2x - x^2}{x^2 + 9x + 14}$ $-\frac{x+2}{x+5}$

28. $\frac{2x^2 + 13x + 20}{8 - 10x - 3x^2} \div \frac{6x^2 - 13x - 5}{9x^2 - 3x - 2}$ $-\frac{2x+5}{2x-5}$

29. $\frac{x^{2n} + x^n}{2x - 2} \div \frac{4x^n + 4}{x^{n+1} - x^n}$ $\frac{x^{2n}}{8}$

30. $\frac{x^{2n} - 4}{4x^n + 8} \div \frac{x^{n+1} - 2x}{4x^3 - 12x^2}$ $x(x-3)$

31. $\frac{2x^2 - 13x + 21}{2x^2 + 11x + 15} \div \frac{2x^2 + x - 28}{3x^2 + 4x - 15}$ $\frac{(x-3)(3x-5)}{(2x+5)(x+4)}$

32. $\frac{2x^2 - 13x + 15}{2x^2 - 3x - 35} \div \frac{6x^2 + x - 12}{6x^2 + 13x - 28}$ $\frac{2x-3}{2x+3}$

33. $\frac{14 + 17x - 6x^2}{3x^2 + 14x + 8} \div \frac{4x^2 - 49}{2x^2 + 15x + 28}$ -1

34. $\frac{16x^2 - 9}{6 - 5x - 4x^2} \div \frac{16x^2 + 24x + 9}{4x^2 + 11x + 6}$ -1

35. $\frac{2x^{2n} - x^n - 6}{x^{2n} - x^n - 2} \div \frac{2x^{2n} + x^n - 3}{x^{2n} - 1}$ 1

36. $\frac{x^{4n} - 1}{x^{2n} + x^n - 2} \div \frac{x^{2n} + 1}{x^{2n} + 3x^n + 2}$ $(x^n + 1)^2$

37. $\frac{6x^2 + 6x}{3x + 6x^2 + 3x^3} \div \frac{x^2 - 1}{1 - x^3}$ $-\frac{2(x^2+x+1)}{(x+1)^2}$

38. $\frac{x^3 + y^3}{2x^3 + 2x^2y} \div \frac{3x^3 - 3x^2y + 3xy^2}{6x^2 - 6y^2}$ $\frac{(x+y)(x-y)}{x^3}$

2 Add or subtract.

39. $\frac{3}{2xy} - \frac{7}{2xy} - \frac{9}{2xy}$ $-\frac{13}{2xy}$

40. $-\frac{3}{4x^2} + \frac{8}{4x^2} - \frac{3}{4x^2}$ $\frac{1}{2x^2}$

41. $\frac{x}{x^2 - 3x + 2} - \frac{2}{x^2 - 3x + 2}$ $\frac{1}{x-1}$

42. $\frac{3x}{3x^2 + x - 10} - \frac{5}{3x^2 + x - 10}$ $\frac{1}{x+2}$

43. $\frac{3}{2x^2y} - \frac{8}{5x} - \frac{9}{10xy}$ $\frac{15 - 16xy - 9x}{10x^2y}$

44. $\frac{2}{5ab} - \frac{3}{10a^2b} + \frac{4}{15ab^2}$ $\frac{12ab - 9b + 8a}{30a^2b^2}$

45. $\frac{2}{3x} - \frac{3}{2xy} + \frac{4}{5xy} - \frac{5}{6x}$ $-\frac{5y+21}{30xy}$

46. $\frac{3}{4ab} - \frac{2}{5a} + \frac{3}{10b} - \frac{5}{8ab}$ $\frac{12a - 16b + 5}{40ab}$

47. $\frac{2x - 1}{12x} - \frac{3x + 4}{9x}$ $-\frac{6x+19}{36x}$

48. $\frac{3x - 4}{6x} - \frac{2x - 5}{4x}$ $\frac{7}{12x}$

49. $\frac{3x + 2}{4x^2y} - \frac{y - 5}{6xy^2}$ $\frac{10x + 6y + 7xy}{12x^2y^2}$

50. $\frac{2y - 4}{5xy^2} + \frac{3 - 2x}{10x^2y}$ $\frac{2xy - 8x + 3y}{10x^2y^2}$

51. $\frac{2x}{x - 3} - \frac{3x}{x - 5}$ $-\frac{x^2+x}{(x-3)(x-5)}$

52. $\frac{3a}{a - 2} - \frac{5a}{a + 1}$ $-\frac{2a^2 - 13a}{(a-2)(a+1)}$

53. $\frac{3}{2a - 3} + \frac{2a}{3 - 2a}$ -1

54. $\frac{x}{2x - 5} - \frac{2}{5x - 2}$ $\frac{5x^2 - 6x + 10}{(2x-5)(5x-2)}$

55. $\frac{3}{x + 5} + \frac{2x + 7}{x^2 - 25}$ $\frac{5x - 8}{(x+5)(x-5)}$

56. $\frac{x}{4 - x} - \frac{4}{x^2 - 16}$ $-\frac{(x+2)^2}{(x+4)(x-4)}$

57. $\frac{2}{x} - 3 - \frac{10}{x - 4}$ $-\frac{3x^2 - 4x + 8}{x(x-4)}$

58. $\frac{6a}{a - 3} - 5 + \frac{3}{a}$ $\frac{a^2 + 18a - 9}{a(a-3)}$

59. $\frac{1}{2x - 3} - \frac{5}{2x} + 1$ $\frac{4x^2 - 14x + 15}{2x(2x-3)}$

60. $\frac{5}{x} - \frac{5x}{5 - 6x} + 2$ $\frac{17x^2 + 20x - 25}{x(6x-5)}$

61. $\frac{3}{x^2 - 1} + \frac{2x}{x^2 + 2x + 1}$ $\frac{2x^2 + x + 3}{(x+1)^2(x-1)}$

62. $\frac{1}{x^2 - 6x + 9} - \frac{1}{x^2 - 9}$ $\frac{6}{(x-3)^2(x+3)}$

63. $\frac{x}{x + 3} - \frac{3 - x}{x^2 - 9}$ $\frac{x+1}{x+3}$

64. $\frac{1}{x + 2} - \frac{3x}{x^2 + 4x + 4}$ $-\frac{2(x-1)}{(x+2)^2}$

65. $\frac{2x - 3}{x + 5} - \frac{x^2 - 4x - 19}{x^2 + 8x + 15}$ $\frac{x+2}{x+3}$

66. $\frac{-3x^2 + 8x + 2}{x^2 + 2x - 8} - \frac{2x - 5}{x + 4}$ $-\frac{5x^2 - 17x + 8}{(x+4)(x-2)}$

67. $\frac{x^n}{x^{2n} - 1} - \frac{2}{x^n + 1}$ $-\frac{x^n - 2}{(x^n+1)(x^n-1)}$

68. $\frac{2}{x^n - 1} + \frac{x^n}{x^{2n} - 1}$ $\frac{3x^n + 2}{(x^n+1)(x^n-1)}$

69. $\frac{2x - 2}{4x^2 - 9} - \frac{5}{3 - 2x}$ $\frac{12x + 13}{(2x+3)(2x-3)}$

70. $\frac{x^2 + 4}{4x^2 - 36} - \frac{13}{x + 3}$ $\frac{x^2 - 52x + 160}{4(x+3)(x-3)}$

71. $\frac{x - 2}{x + 1} - \frac{3 - 12x}{2x^2 - x - 3}$ $\frac{2x+3}{2x-3}$

72. $\frac{3x - 4}{4x + 1} + \frac{3x + 6}{4x^2 + 9x + 2}$ $\frac{3x-1}{4x+1}$

73. $\dfrac{x+1}{x^2+x-6} - \dfrac{x+2}{x^2+4x+3}$ $\quad \dfrac{2x+5}{(x+3)(x-2)(x+1)}$

74. $\dfrac{x+1}{x^2+x-12} - \dfrac{x-3}{x^2+7x+12}$ $\quad \dfrac{10x-6}{(x+4)(x+3)(x-3)}$

75. $\dfrac{x-1}{2x^2+11x+12} + \dfrac{2x}{2x^2-3x-9}$ $\quad \dfrac{3x^2+4x+3}{(2x+3)(x+4)(x-3)}$

76. $\dfrac{x-2}{4x^2+4x-3} + \dfrac{3-2x}{6x^2+x-2}$ $\quad -\dfrac{x^2+4x-5}{(2x-1)(2x+3)(3x+2)}$

77. $\dfrac{x}{x-3} - \dfrac{2}{x+4} - \dfrac{14}{x^2+x-12}$ $\quad \dfrac{x-2}{x-3}$

78. $\dfrac{x^2}{x^2+x-2} + \dfrac{3}{x-1} - \dfrac{4}{x+2}$ $\quad \dfrac{x^2-x+10}{(x+2)(x-1)}$

79. $\dfrac{x^2+6x}{x^2+3x-18} - \dfrac{2x-1}{x+6} + \dfrac{x-2}{3-x}$ $\quad -\dfrac{2x^2-9x-9}{(x+6)(x-3)}$

80. $\dfrac{2x^2-2x}{x^2-2x-15} - \dfrac{2}{x+3} + \dfrac{x}{5-x}$ $\quad \dfrac{x-2}{x+3}$

81. $\dfrac{4-20x}{6x^2+11x-10} - \dfrac{4}{2-3x} + \dfrac{x}{2x+5}$ $\quad \dfrac{3x^2-14x+24}{(3x-2)(2x+5)}$

82. $\dfrac{x}{4x-1} + \dfrac{2}{2x+1} + \dfrac{6}{8x^2+2x-1}$ $\quad \dfrac{x+4}{4x-1}$

83. $\dfrac{2x^2}{x^4-1} - \dfrac{1}{x^2-1} + \dfrac{1}{x^2+1}$ $\quad \dfrac{2}{x^2+1}$

84. $\dfrac{x^2-12}{x^4-16} + \dfrac{1}{x^2-4} - \dfrac{1}{x^2+4}$ $\quad \dfrac{1}{x^2+4}$

APPLYING CONCEPTS 6.2

Simplify.

85. $\dfrac{(x+1)^2}{1-2x} \cdot \dfrac{2x-1}{x+1}$ $\quad -x-1$

86. $\left(\dfrac{2m}{3}\right)^2 \div \left(\dfrac{m^2}{6} + \dfrac{m}{2}\right)$ $\quad \dfrac{8m}{3(m+3)}$

87. $\left(\dfrac{y-2}{x^2}\right)^3 \cdot \left(\dfrac{x}{2-y}\right)^2$ $\quad \dfrac{y-2}{x^4}$

88. $\dfrac{b+3}{b-1} \div \dfrac{b+3}{b-2} \cdot \dfrac{b-1}{b+4}$ $\quad \dfrac{b-2}{b+4}$

89. $\left(\dfrac{y+1}{y-1}\right)^2 - 1$ $\quad \dfrac{4y}{(y-1)^2}$

90. $\left(\dfrac{1}{3} - \dfrac{2}{a}\right) \div \left(\dfrac{3}{a} - 2 + \dfrac{a}{4}\right)$ $\quad \dfrac{4}{3(a-2)}$

91. $\dfrac{3x^2+6x}{4x^2-16} \cdot \dfrac{2x+8}{x^2+2x} \div \dfrac{3x-9}{5x-20}$ $\quad \dfrac{5(x+4)(x-4)}{2(x+2)(x-2)(x-3)}$

92. $\dfrac{5y^2-20}{3y^2-12y} \cdot \dfrac{9y^3+6y^2}{2y^2-4y} \div \dfrac{y^3+2y^2}{2y^2-8y}$ $\quad \dfrac{5(3y+2)}{y}$

93. $\dfrac{a^2+a-6}{4+11a-3a^2} \cdot \dfrac{15a^2-a-2}{4a^2+7a-2} \div \dfrac{6a^2-7a-3}{4-17a+4a^2}$ $\quad -\dfrac{(a+3)(a-2)(5a-2)}{(3a+1)(a+2)(2a-3)}$

94. $\dfrac{25x-x^3}{x^4-1} \cdot \dfrac{3-x-4x^2}{2x^2+7x-15} \div \dfrac{4x^3-23x^2+15x}{3-5x+2x^2}$ $\quad \dfrac{1}{x^2+1}$

95. $\left(\dfrac{x+1}{2x-1} - \dfrac{x-1}{2x+1}\right) \cdot \left(\dfrac{2x-1}{x} - \dfrac{2x-1}{x^2}\right)$ $\quad \dfrac{6(x-1)}{x(2x+1)}$

96. $\left(\dfrac{y-2}{3y+1} - \dfrac{y+2}{3y-1}\right) \cdot \left(\dfrac{3y+1}{y} - \dfrac{3y-1}{y^2}\right)$ $\quad -\dfrac{14(3y^2-2y+1)}{y(3y+1)(3y-1)}$

Solve.

97. Use $x = 3$ and $y = 5$ to show that $\dfrac{1}{x} + \dfrac{1}{y} \neq \dfrac{1}{x+y}$. $\quad \dfrac{8}{15} \neq \dfrac{1}{8}$

98. Use $x = 3$ and $y = 5$ to show that $\dfrac{1}{x} - \dfrac{1}{y} \neq \dfrac{1}{x-y}$. $\quad \dfrac{2}{15} \neq -\dfrac{1}{2}$

Rewrite the expression as the sum of two fractions in simplest form.

[C] 99. $\dfrac{3x+6y}{xy}$ $\quad \dfrac{3}{y} + \dfrac{6}{x}$

[C] 100. $\dfrac{5a+8b}{ab}$ $\quad \dfrac{5}{b} + \dfrac{8}{a}$

[C] 101. $\dfrac{4a^2+3ab}{a^2b^2}$ $\quad \dfrac{4}{b^2} + \dfrac{3}{ab}$

102. *Manufacturing* Manufacturers who package their product in cans would like to design the can so that the minimum amount of aluminum is needed. If a soft drink can contains 12 oz (355 cm^3), the function that relates the surface area of the can (the amount of aluminum needed) to the radius of the bottom of the can is given by the equation $f(r) = 2\pi r^2 + \dfrac{710}{r}$, where r is measured in centimeters.

a. Express the right side of this equation with a common denominator. $\quad$ **a.** $f(r) = \dfrac{2\pi r^3 + 710}{r}$

b. Graph the equation for $0 < r \leq 19$.

c. The point whose coordinates are (7, 409), to the nearest integer, is on the graph of f. Write a sentence that gives an interpretation of this ordered pair.

d. Use a graphing utility to determine the radius of the can that has a minimum surface area. Round to the nearest tenth. 3.8 cm

e. The height of the can is determined from $h = \dfrac{355}{\pi r^2}$. Use the answer to part (d) to determine the height of the can that has a minimum surface area. Round to the nearest tenth. 7.8 cm

f. Determine the minimum surface area. Round to the nearest tenth. 277.5 cm^2

b.

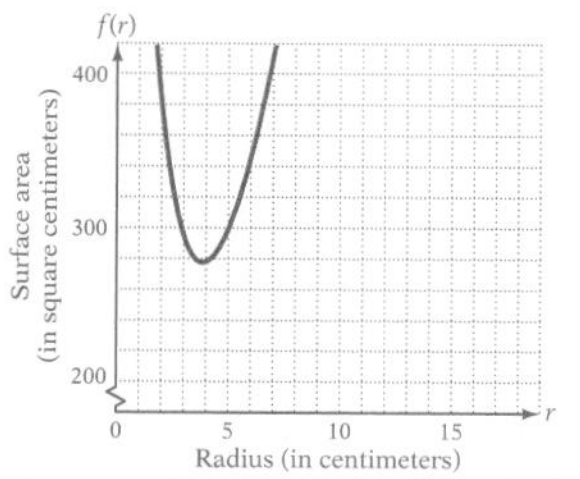

c. The ordered pair (7, 409) means that when the radius of the can is 7 cm, the surface area of the can is 409 cm^2.

103. *Manufacturing* A manufacturer wants to make square tissues and package them in a box. The manufacturer has determined that to be competitive, the box needs to hold 175 tissues, which means that the volume of the box will be 132 in^3. The amount of cardboard (surface area) that will be necessary to build this box is given by $f(x) = 2x^2 + \dfrac{528}{x}$, where x is the length of one side of the square tissue.

a. Express the right side of this equation with a common denominator.

b. Graph the equation for $0 < x \leq 10$.

c. The point whose coordinates are (4, 164) is on the graph. Write a sentence that explains the meaning of this point.

d. With a graphing utility, determine, to the nearest tenth, the height of the box that uses the minimum amount of cardboard. 5.1 in.

e. Determine the minimum amount of cardboard. Round to the nearest tenth. 155.5 in^2

a. $f(x) = \dfrac{2x^3 + 528}{x}$

b.

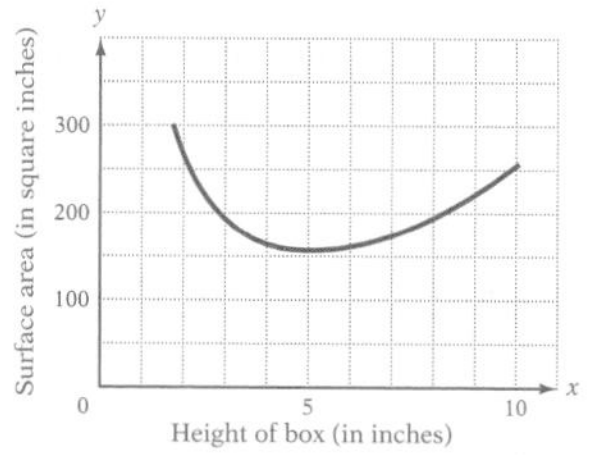

c. When the base of the box is 4 in. by 4 in., 164 in^2 of cardboard will be needed.

104. For adding or subtracting fractions, any common denominator will do. Explain the advantages and disadvantages of using the LCM of the denominators.

105. A student incorrectly tried to add the fractions $\frac{1}{5}$ and $\frac{2}{3}$ by adding the numerators and the denominators. The procedure was shown as

$$\frac{1}{5} + \frac{2}{3} = \frac{1+2}{5+3} = \frac{3}{8}$$

Write the fractions $\frac{1}{5}$, $\frac{2}{3}$, and $\frac{3}{8}$ in order from smallest to largest. Now take any two other fractions, add the numerators and the denominators, and then write the fractions in order from smallest to largest. Do you see a pattern? If so, explain it. If not, try a few more examples until you find a pattern.

SECTION 6.3 Complex Fractions

1 Simplify complex fractions

24 HM[3]

A **complex fraction** is a fraction whose numerator or denominator contains one or more fractions. Examples of complex fractions are shown below.

$$\frac{5}{2+\frac{1}{2}} \qquad \frac{5+\frac{1}{y}}{5-\frac{1}{y}} \qquad \frac{x+4+\frac{1}{x+2}}{x-2+\frac{1}{x+2}}$$

Simplify: $\dfrac{\frac{1}{x}+\frac{1}{y}}{\frac{1}{x}-\frac{1}{y}}$

Multiply the numerator and denominator of the complex fraction by the LCM of the denominators. The LCM of x and y is xy.

$$\frac{\frac{1}{x}+\frac{1}{y}}{\frac{1}{x}-\frac{1}{y}} = \frac{\frac{1}{x}+\frac{1}{y}}{\frac{1}{x}-\frac{1}{y}} \cdot \frac{xy}{xy} = \frac{\frac{1}{x}\cdot xy+\frac{1}{y}\cdot xy}{\frac{1}{x}\cdot xy-\frac{1}{y}\cdot xy} = \frac{y+x}{y-x}$$

Note that after the numerator and denominator of the complex fraction have been multiplied by the LCM of the denominators, no fraction remains in the numerator or denominator.

Example 1 Simplify. A. $\dfrac{2-\frac{11}{x}+\frac{15}{x^2}}{3-\frac{5}{x}-\frac{12}{x^2}}$ B. $\dfrac{2x-1+\frac{7}{x+4}}{3x-8+\frac{17}{x+4}}$

Solution A. $\dfrac{2-\frac{11}{x}+\frac{15}{x^2}}{3-\frac{5}{x}-\frac{12}{x^2}} = \dfrac{2-\frac{11}{x}+\frac{15}{x^2}}{3-\frac{5}{x}-\frac{12}{x^2}} \cdot \dfrac{x^2}{x^2}$ ▶ The LCM is x^2.

$$= \frac{2\cdot x^2-\frac{11}{x}\cdot x^2+\frac{15}{x^2}\cdot x^2}{3\cdot x^2-\frac{5}{x}\cdot x^2-\frac{12}{x^2}\cdot x^2}$$

$$= \frac{2x^2-11x+15}{3x^2-5x-12}$$

$$= \frac{(2x-5)(x-3)}{(3x+4)(x-3)} = \frac{2x-5}{3x+4}$$

B. $$\frac{2x-1+\frac{7}{x+4}}{3x-8+\frac{17}{x+4}} = \frac{2x-1+\frac{7}{x+4}}{3x-8+\frac{17}{x+4}} \cdot \frac{x+4}{x+4}$$ ▶ The LCM is $x + 4$.

$$= \frac{(2x-1)(x+4)+\frac{7}{x+4}(x+4)}{(3x-8)(x+4)+\frac{17}{x+4}(x+4)}$$

$$= \frac{2x^2+7x-4+7}{3x^2+4x-32+17}$$

$$= \frac{2x^2+7x+3}{3x^2+4x-15}$$

$$= \frac{(2x+1)(x+3)}{(3x-5)(x+3)} = \frac{2x+1}{3x-5}$$

Problem 1 Simplify. A. $\dfrac{3+\frac{16}{x}+\frac{16}{x^2}}{6+\frac{5}{x}-\frac{4}{x^2}}$ B. $\dfrac{2x+5+\frac{14}{x-3}}{4x+16+\frac{49}{x-3}}$

Solution See page S17–S18. A. $\dfrac{x+4}{2x-1}$ B. $\dfrac{x-1}{2x+1}$

CONCEPT REVIEW 6.3

Determine whether the following statements are always true, sometimes true, or never true.

1. The numerator of a complex fraction contains one or more fractions.
   Sometimes true

2. To simplify a complex fraction, multiply the numerator and denominator of the complex fraction by the LCM of the denominators. Always true

3. The LCM of the denominators in the complex fraction $\dfrac{1-\frac{3}{x-3}}{\frac{2}{x}+1}$ is $x(x-3)$.
   Always true

4. $\dfrac{c^{-1}}{a^{-1}+b^{-1}} = \dfrac{a+b}{c}$ Never true

EXERCISES 6.3

1 1. What is a complex fraction?

2. What is the general goal of simplifying a complex fraction?

Simplify.

3. $\dfrac{2-\frac{1}{3}}{4+\frac{11}{3}}$ $\dfrac{5}{23}$

4. $\dfrac{3+\frac{5}{2}}{8-\frac{7}{2}}$ $\dfrac{11}{9}$

5. $\dfrac{3-\frac{2}{3}}{5+\frac{5}{6}}$ $\dfrac{2}{5}$

6. $\dfrac{1+\frac{1}{x}}{1-\frac{1}{x^2}}$ $\dfrac{x}{x-1}$

7. $\dfrac{\frac{1}{y^2} - 1}{1 + \frac{1}{y}}$ $\dfrac{1-y}{y}$

8. $\dfrac{a-2}{\frac{4}{a} - a}$ $-\dfrac{a}{a+2}$

9. $\dfrac{\frac{25}{a} - a}{5 + a}$ $\dfrac{5-a}{a}$

10. $\dfrac{\frac{1}{a^2} - \frac{1}{a}}{\frac{1}{a^2} + \frac{1}{a}}$ $-\dfrac{a-1}{a+1}$

11. $\dfrac{\frac{1}{b} + \frac{1}{2}}{\frac{4}{b^2} - 1}$ $\dfrac{b}{2(2-b)}$

12. $\dfrac{2 - \frac{4}{x+2}}{5 - \frac{10}{x+2}}$ $\dfrac{2}{5}$

13. $\dfrac{4 + \frac{12}{2x-3}}{5 + \frac{15}{2x-3}}$ $\dfrac{4}{5}$

14. $\dfrac{\frac{3}{2a-3} + 2}{\frac{-6}{2a-3} - 4}$ $-\dfrac{1}{2}$

15. $\dfrac{\frac{-5}{b-5} - 3}{\frac{10}{b-5} + 6}$ $-\dfrac{1}{2}$

16. $\dfrac{\frac{x}{x+1} - \frac{1}{x}}{\frac{x}{x+1} + \frac{1}{x}}$ $\dfrac{x^2-x-1}{x^2+x+1}$

17. $\dfrac{\frac{2a}{a-1} - \frac{3}{a}}{\frac{1}{a-1} + \frac{2}{a}}$ $\dfrac{2a^2-3a+3}{3a-2}$

18. $\dfrac{\frac{3}{x}}{\frac{9}{x^2}}$ $\dfrac{x}{3}$

19. $\dfrac{\frac{x}{3x-2}}{\frac{x}{9x^2-4}}$ $3x + 2$

20. $\dfrac{\frac{a^2-b^2}{4a^2b}}{\frac{a+b}{16ab^2}}$ $\dfrac{4b(a-b)}{a}$

21. $\dfrac{1 - \frac{1}{x} - \frac{6}{x^2}}{1 - \frac{4}{x} + \frac{3}{x^2}}$ $\dfrac{x+2}{x-1}$

22. $\dfrac{1 - \frac{3}{x} - \frac{10}{x^2}}{1 + \frac{11}{x} + \frac{18}{x^2}}$ $\dfrac{x-5}{x+9}$

23. $\dfrac{\frac{15}{x^2} - \frac{2}{x} - 1}{\frac{4}{x^2} - \frac{5}{x} + 4}$ $-\dfrac{(x-3)(x+5)}{4x^2-5x+4}$

24. $\dfrac{1 - \frac{2x}{3x-4}}{x - \frac{32}{3x-4}}$ $\dfrac{1}{3x+8}$

25. $\dfrac{1 - \frac{12}{3x+10}}{x - \frac{8}{3x+10}}$ $\dfrac{1}{x+4}$

26. $\dfrac{x - 1 + \frac{2}{x-4}}{x + 3 + \frac{6}{x-4}}$ $\dfrac{x-2}{x+2}$

27. $\dfrac{x - 5 - \frac{18}{x+2}}{x + 7 + \frac{6}{x+2}}$ $\dfrac{x-7}{x+5}$

28. $\dfrac{x - 4 + \frac{9}{2x+3}}{x + 3 - \frac{5}{2x+3}}$ $\dfrac{x-3}{x+4}$

29. $\dfrac{\frac{1}{a} - \frac{3}{a-2}}{\frac{2}{a} + \frac{5}{a-2}}$ $-\dfrac{2(a+1)}{7a-4}$

30. $\dfrac{\frac{2}{b} - \frac{5}{b+3}}{\frac{3}{b} + \frac{3}{b+3}}$ $-\dfrac{b-2}{2b+3}$

31. $\dfrac{\frac{1}{y^2} - \frac{1}{xy} - \frac{2}{x^2}}{\frac{1}{y^2} - \frac{3}{xy} + \frac{2}{x^2}}$ $\dfrac{x+y}{x-y}$

32. $\dfrac{\frac{2}{b^2} - \frac{5}{ab} - \frac{3}{a^2}}{\frac{2}{b^2} + \frac{7}{ab} + \frac{3}{a^2}}$ $\dfrac{a-3b}{a+3b}$

33. $\dfrac{\frac{x-1}{x+1} - \frac{x+1}{x-1}}{\frac{x-1}{x+1} + \frac{x+1}{x-1}}$ $-\dfrac{2x}{x^2+1}$

34. $\dfrac{\frac{y}{y+2} - \frac{y}{y-2}}{\frac{y}{y+2} + \frac{y}{y-2}}$ $-\dfrac{2}{y}$

35. $4 - \dfrac{2}{2 - \frac{3}{x}}$ $\dfrac{6(x-2)}{2x-3}$

APPLYING CONCEPTS 6.3

Simplify.

36. $\dfrac{x^{-1} + y^{-1}}{x^{-1} - y^{-1}}$ $\dfrac{y+x}{y-x}$

37. $\dfrac{x^{-1}}{y^{-1}} + \dfrac{y}{x}$ $\dfrac{2y}{x}$

38. $\dfrac{x^{-1} + y}{x^{-1} - y}$ $\dfrac{1+xy}{1-xy}$

39. $\dfrac{x - \frac{1}{x}}{1 + \frac{1}{x}}$ $x - 1$

40. $1 - \dfrac{1}{1 - \frac{1}{b-2}}$ $\dfrac{1}{3-b}$

41. $2 - \dfrac{2}{2 - \frac{2}{c-1}}$ $\dfrac{c-3}{c-2}$

42. $3 - \cfrac{2}{1 - \cfrac{2}{3 - \cfrac{2}{x}}}$ $\quad \frac{3x+2}{x-2}$

43. $a + \cfrac{a}{2 + \cfrac{1}{1 - \cfrac{2}{a}}}$ $\quad \frac{2a(2a-3)}{3a-4}$

44. $a - \cfrac{1}{2 - \cfrac{2}{2 - \cfrac{2}{a}}}$ $\quad \frac{a^2-3a+1}{a-2}$

[C] **45.** $\dfrac{\frac{1}{x+h} - \frac{1}{x}}{h}$ $\quad -\frac{1}{x(x+h)}$

[C] **46.** $\dfrac{\frac{1}{(x+h)^2} - \frac{1}{x^2}}{h}$ $\quad -\frac{2x+h}{x^2(x+h)^2}$

47. Find the sum of the reciprocals of three consecutive even integers. $\frac{3n^2+12n+8}{n(n+2)(n+4)}$

48. The total resistance, R, of three resistors in parallel is given by the formula $R = \dfrac{1}{\frac{1}{R_1} + \frac{1}{R_2} + \frac{1}{R_3}}$. Find the total parallel resistance when $R_1 = 2$ ohms, $R_2 = 4$ ohms, and $R_3 = 8$ ohms. $\frac{8}{7}$ ohms

49. The interest rate on a loan to purchase a car affects the monthly payment. The function that relates the monthly payment for a 5-year loan (60-month loan) to the monthly interest rate is given by

$$P(x) = \frac{Cx}{\left[1 - \frac{1}{(x+1)^{60}}\right]}$$

where x is the monthly interest rate (as a decimal), C is the loan amount for the car, and $P(x)$ is the monthly payment.

a. Simplify the complex fraction.

b. Graph this equation for $0 < x \leq 0.019$ and $C = \$10{,}000$.

c. What is the interval of *annual* interest rates for the domain in part (b)?

d. The point whose coordinates are (0.006, 198.96), to the nearest cent, is on the graph of this equation. Write a sentence that gives an interpretation of this ordered pair.

e. Use a graphing utility to determine the monthly payment for a car with a loan amount of \$10,000 and an annual interest rate of 8%. Round to the nearest dollar. \$203

a. $P(x) = \frac{Cx(x+1)^{60}}{(x+1)^{60} - 1}$

b.

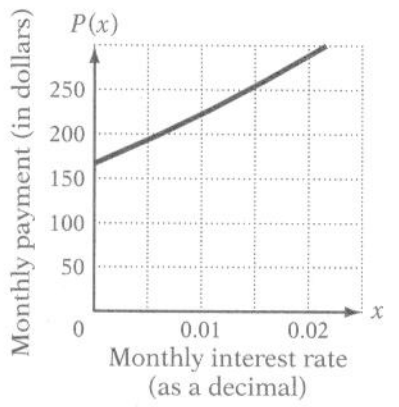

c. 0% to 22.8%

d. The ordered pair (0.006, 198.96) means that when the monthly interest rate on a car loan is 0.6%, the monthly payment on the loan is \$198.96.

50. Explain, in your own words, two methods of simplifying a complex fraction. Give examples.

51. Write a report on continued fractions. Give an example of a continued fraction that can be used to approximate $\sqrt{2}$.

52. According to the theory of relativity, the mass of a moving object is given by an equation that contains a complex fraction. The equation is $m = \dfrac{m_0}{\sqrt{1 - \frac{v^2}{c^2}}}$, where m is the mass of the moving object, m_0 is the mass of the object at rest, v is the speed of the object, and c is the speed of light. Evaluate the expression at speeds of $0.5c$, $0.75c$, $0.90c$, $0.95c$, and $0.99c$ when the mass of the object at rest is 10 g. Explain how m changes as the speed of the object becomes closer to the speed of light. Explain how this equation can be used to support the statement that an object cannot travel at the speed of light.

SECTION 6.4
Rational Equations

1 Solve fractional equations

24 HM³

To solve an equation containing fractions, **clear denominators** by multiplying each side of the equation by the LCM of the denominators. Then solve for the variable.

Solve: $\frac{3x}{x-5} = 5 - \frac{5}{x-5}$

$$\frac{3x}{x-5} = 5 - \frac{5}{x-5}$$

Multiply each side of the equation by the LCM of the denominators.

$$(x-5)\left(\frac{3x}{x-5}\right) = (x-5)\left(5 - \frac{5}{x-5}\right)$$

$$3x = (x-5)5 - (x-5)\left(\frac{5}{x-5}\right)$$

Simplify.

$$3x = 5x - 25 - 5$$
$$3x = 5x - 30$$

Solve the equation for x.

$$-2x = -30$$
$$x = 15$$

15 checks as a solution.

The solution is 15.

Occasionally, a value of the variable that appears to be a solution will make one of the denominators zero. In this case, the equation has no solution for that value of the variable.

Solve: $\frac{3x}{x-3} = 2 + \frac{9}{x-3}$

$$\frac{3x}{x-3} = 2 + \frac{9}{x-3}$$

Multiply each side of the equation by the LCM of the denominators.

$$(x-3)\left(\frac{3x}{x-3}\right) = (x-3)\left(2 + \frac{9}{x-3}\right)$$

Use the Distributive Property.

$$3x = (x-3)2 + (x-3)\left(\frac{9}{x-3}\right)$$
$$3x = 2x - 6 + 9$$
$$3x = 2x + 3$$
$$x = 3$$

Substituting 3 into the equation results in division by zero. Because division by zero is not defined, the equation has no solution.

$$\frac{3x}{x-3} = 2 + \frac{9}{x-3}$$
$$\frac{3(3)}{3-3} = 2 + \frac{9}{3-3}$$
$$\frac{9}{0} = 2 + \frac{9}{0}$$

LOOK CLOSELY

If each side of an equation is multiplied by a variable expression, it is essential that the solutions be checked. As this example shows, a proposed solution to an equation may not check when it is substituted into the original equation.

Multiplying each side of an equation by a variable expression may produce an equation with different solutions from the original equation. Thus, any time you multiply each side of an equation by a variable expression, you must check the resulting solution.

Example 1 Solve. A. $\frac{1}{4} = \frac{5}{x+5}$ B. $\frac{2x}{x-2} = \frac{1}{3x-4} + 2$

Solution A.

$$\frac{1}{4} = \frac{5}{x+5}$$

$$4(x+5)\frac{1}{4} = 4(x+5)\frac{5}{x+5}$$

▸ Multiply each side of the equation by the LCM of the denominators.

$$x + 5 = 4(5)$$
$$x + 5 = 20$$
$$x = 15$$

15 checks as a solution.

The solution is 15.

B.

$$\frac{2x}{x-2} = \frac{1}{3x-4} + 2$$

$$(x-2)(3x-4)\frac{2x}{x-2} = (x-2)(3x-4)\left(\frac{1}{3x-4} + 2\right)$$

$$(3x-4)2x = (x-2)(3x-4)\left(\frac{1}{3x-4}\right) + (x-2)(3x-4)2$$

$$6x^2 - 8x = x - 2 + 6x^2 - 20x + 16$$
$$6x^2 - 8x = 6x^2 - 19x + 14$$
$$11x = 14$$
$$x = \frac{14}{11}$$

$\frac{14}{11}$ checks as a solution.

The solution is $\frac{14}{11}$.

Problem 1 Solve. A. $\frac{5}{2x-3} = \frac{-2}{x+1}$ B. $\frac{4x+1}{2x-1} = 2 + \frac{3}{x-3}$

Solution See page S18. A. $\frac{1}{9}$ B. -2

2 Work problems

POINT OF INTEREST

The following problem was recorded in the *Jiuzhang,* a Chinese text that dates to the Han dynasty (about 200 B.C. to A.D. 200). "A reservoir has 5 channels bringing water to it. The first can fill the reservoir in $\frac{1}{3}$ day, the second in 1 day, the third in $2\frac{1}{2}$ days, the fourth in 3 days, and the fifth in 5 days. If all channels are open, how long does it take to fill the reservoir?" This problem is the earliest known work problem. The answer is $\frac{15}{74}$ day.

If a mason can build a retaining wall in 12 h, then in 1 h the mason can build $\frac{1}{12}$ of the wall. The mason's rate of work is $\frac{1}{12}$ of the wall each hour. The **rate of work** is that part of a task that is completed in one unit of time. If an apprentice can build the wall in x hours, the rate of work for the apprentice is $\frac{1}{x}$ of the wall each hour.

In solving a work problem, the goal is to determine the time it takes to complete a task. The basic equation that is used to solve work problems is

Rate of work × Time worked = Part of task completed

For example, if a pipe can fill a tank in 5 h, then in 2 h the pipe will fill $\frac{1}{5} \times 2 = \frac{2}{5}$ of the tank. In t hours, the pipe will fill $\frac{1}{5} \times t = \frac{t}{5}$ of the tank.

Solve: A mason can build a wall in 10 h. An apprentice can build a wall in 15 h. How long would it take them to build the wall if they worked together?

STRATEGY for solving a work problem

■ For each person or machine, write a numerical or variable expression for the rate of work, the time worked, and the part of the task completed. The results can be recorded in a table.

Unknown time to build the wall working together: t

	Rate of work	·	Time worked	=	Part of task completed
Mason	$\dfrac{1}{10}$	·	t	=	$\dfrac{t}{10}$
Apprentice	$\dfrac{1}{15}$	·	t	=	$\dfrac{t}{15}$

■ Determine how the parts of the task completed are related. Use the fact that the sum of the parts of the task completed must equal 1, the complete task.

The sum of the part of the task completed by the mason and the part of the task completed by the apprentice is 1.

$$\frac{t}{10} + \frac{t}{15} = 1$$
$$30\left(\frac{t}{10} + \frac{t}{15}\right) = 30(1)$$
$$3t + 2t = 30$$
$$5t = 30$$
$$t = 6$$

Working together, they would build the wall in 6 h.

Example 2   An electrician requires 12 h to wire a house. The electrician's apprentice can wire a house in 16 h. After working alone on one job for 4 h, the electrician quits, and the apprentice completes the task. How long does it take the apprentice to finish wiring the house?

Strategy   ■ Time required for the apprentice to finish wiring the house: t

	Rate	Time	Part
Electrician	$\dfrac{1}{12}$	4	$\dfrac{4}{12}$
Apprentice	$\dfrac{1}{16}$	t	$\dfrac{t}{16}$

- The sum of the part of the task completed by the electrician and the part of the task completed by the apprentice is 1.

Solution

$$\frac{4}{12} + \frac{t}{16} = 1$$
$$\frac{1}{3} + \frac{t}{16} = 1$$
$$48\left(\frac{1}{3} + \frac{t}{16}\right) = 48(1)$$
$$16 + 3t = 48$$
$$3t = 32$$
$$t = \frac{32}{3}$$

It takes the apprentice $10\frac{2}{3}$ h to finish wiring the house.

Problem 2 Two water pipes can fill a tank with water in 6 h. The larger pipe working alone can fill the tank in 9 h. How long would it take the smaller pipe, working alone, to fill the tank?

Solution See page S18. 18 h

3 Uniform motion problems

25 HM³

A car that travels constantly in a straight line at 55 mph is in uniform motion. **Uniform motion** means that the speed of an object does not change.

The basic equation used to solve uniform motion problems is

Distance = Rate × Time

An alternative form of this equation can be written by solving the equation for time. This form of the equation is used to solve the following problem.

$$\frac{\textbf{Distance}}{\textbf{Rate}} = \textbf{Time}$$

Solve: A motorist drove 150 mi on country roads before driving 50 mi on mountain roads. The rate of speed on the country roads was three times the rate on the mountain roads. The time spent traveling the 200 mi was 5 h. Find the rate of the motorist on the country roads.

STRATEGY for solving a uniform motion problem

- For each object, write a numerical or variable expression for the distance, rate, and time. The results can be recorded in a table.

The unknown rate of speed on the mountain roads: r
Rate of speed on the country roads: $3r$

	Distance	÷	Rate	=	Time
Country roads	150	÷	$3r$	=	$\dfrac{150}{3r}$
Mountain roads	50	÷	r	=	$\dfrac{50}{r}$

■ Determine how the times traveled by each object are related. For example, it may be known that the times are equal, or the total time may be known.

The total time for the trip is 5 h.

$$\frac{150}{3r} + \frac{50}{r} = 5$$

$$\frac{50}{r} + \frac{50}{r} = 5$$

$$r\left(\frac{50}{r} + \frac{50}{r}\right) = r(5)$$

$$50 + 50 = 5r$$

$$100 = 5r$$

$$20 = r$$

The rate of speed on the country roads was $3r$. Replace r with 20 and evaluate.

$$3r = 3(20) = 60$$

The rate of speed on the country roads was 60 mph.

Example 3 A marketing executive traveled 810 mi on a corporate jet in the same amount of time that it took to travel an additional 162 mi by helicopter. The rate of the jet was 360 mph greater than the rate of the helicopter. Find the rate of the jet.

Strategy ■ Rate of the helicopter: r
Rate of the jet: $r + 360$

	Distance	Rate	Time
Jet	810	$r + 360$	$\dfrac{810}{r + 360}$
Helicopter	162	r	$\dfrac{162}{r}$

■ The time traveled by jet is equal to the time traveled by helicopter.

Solution

$$\frac{810}{r+360} = \frac{162}{r}$$

$$r(r+360)\left(\frac{810}{r+360}\right) = r(r+360)\left(\frac{162}{r}\right)$$

$$810r = (r+360)162$$

$$810r = 162r + 58{,}320$$

$$648r = 58{,}320$$

$$r = 90$$

▶ The rate of the helicopter was 90 mph.

$$r + 360 = 90 + 360 = 450$$

▶ Substitute the value of r into the variable expression for the rate of the jet.

The rate of the jet was 450 mph.

Problem 3 A plane can fly at a rate of 150 mph in calm air. Traveling with the wind, the plane flew 700 mi in the same amount of time it took to fly 500 mi against the wind. Find the rate of the wind.

Solution See page S18. 25 mph

CONCEPT REVIEW 6.4

Determine whether the following statements are always true, sometimes true, or never true.

1. $\frac{3x}{x-2} + \frac{4}{x} - 3$ is a rational equation. Never true

2. Multiplying both sides of an equation by the same variable expression produces an equivalent equation. Sometimes true

3. -2 and 5 cannot be solutions of the equation $\frac{1}{x+2} - 3 = \frac{x-2}{x-5}$. Always true

4. An equation that contains a variable in a denominator can be simplified to an equivalent equation with no variables in the denominator. Sometimes true

5. If it takes 3 h to fill a tub with water, then $\frac{1}{3}$ of the tub can be filled in 1 h. Always true

EXERCISES 6.4

1 1. What is a rational equation?

2. Explain why it is necessary to check the solution of a rational equation.

Solve.

3. $\frac{x}{2} + \frac{5}{6} = \frac{x}{3}$ -5

4. $\frac{x}{5} - \frac{2}{9} = \frac{x}{15}$ $\frac{5}{3}$

5. $1 - \frac{3}{y} = 4$ -1

6. $7 + \frac{6}{y} = 5$ -3

7. $\frac{8}{2x-1} = 2$ $\frac{5}{2}$

8. $3 = \frac{18}{3x-4}$ $\frac{10}{3}$

9. $\frac{4}{x-4}=\frac{2}{x-2}$ 0

10. $\frac{x}{3}=\frac{x+1}{7}$ $\frac{3}{4}$

11. $\frac{x-2}{5}=\frac{1}{x+2}$ $-3, 3$

12. $\frac{x+4}{10}=\frac{6}{x-3}$ $-9, 8$

13. $\frac{3}{x-2}=\frac{4}{x}$ 8

14. $\frac{5}{x}=\frac{2}{x+3}$ -5

15. $\frac{3}{x-4}+2=\frac{5}{x-4}$ 5

16. $\frac{5}{y+3}-2=\frac{7}{y+3}$ -4

17. $\frac{8}{x-5}=\frac{3}{x}$ -3

18. $\frac{16}{2-x}=\frac{4}{x}$ $\frac{2}{5}$

19. $5+\frac{8}{a-2}=\frac{4a}{a-2}$ No solution

20. $\frac{-4}{a-4}=3-\frac{a}{a-4}$ No solution

21. $\frac{x}{2}+\frac{20}{x}=7$ 4, 10

22. $3x=\frac{4}{x}-\frac{13}{2}$ $-\frac{8}{3}, \frac{1}{2}$

23. $\frac{6}{x-5}=\frac{1}{x}$ -1

24. $\frac{8}{x-2}=\frac{4}{x+1}$ -4

25. $\frac{x}{x+2}=\frac{6}{x+5}$ $-3, 4$

26. $\frac{x}{x-2}=\frac{3}{x-4}$ 1, 6

27. $-\frac{5}{x+7}+1=\frac{4}{x+7}$ 2

28. $5-\frac{2}{2x-5}=\frac{3}{2x-5}$ 3

29. $\frac{2}{4y^2-9}+\frac{1}{2y-3}=\frac{3}{2y+3}$ $\frac{7}{2}$

30. $\frac{5}{x-2}-\frac{2}{x+2}=\frac{3}{x^2-4}$ $-\frac{11}{3}$

31. $\frac{5}{x^2-7x+12}=\frac{2}{x-3}+\frac{5}{x-4}$ No solution

32. $\frac{9}{x^2+7x+10}=\frac{5}{x+2}-\frac{3}{x+5}$ No solution

2 **33.** If a gardener can mow a lawn in 20 min, what portion of the lawn can the gardener mow in 1 min?

34. If one person can complete a task in 2 h and another person can complete the same task in 3 h, will it take more or less than 2 h to complete the task when both people are working? Explain your answer.

Work problems

35. A large biotech firm uses two computers to process the daily results of its research studies. One computer can process the data in 2 h; the other computer takes 3 h to do the same job. How long would it take to process the data if both computers were used? 1.2 h

36. Two college students have started their own business building computers from kits. Working alone, one student can build a computer in 20 h. When the second student helps, they can build a computer in 7.5 h. How long would it take the second student, working alone, to build the computer? 12 h

37. One solar heating panel can raise the temperature of water 1 degree in 30 min. A second solar heating panel can raise the temperature 1 degree in 45 min. How long would it take to raise the temperature of the water 1 degree with both solar panels operating? 18 min

38. One member of a gardening team can landscape a new lawn in 36 h. The other member of the team can do the job in 45 h. How long would it take to landscape a lawn if both gardeners worked together? 20 h

39. One member of a telephone crew can wire new telephone lines in 5 h. It takes 7.5 h for the other member of the crew to do the job. How long would it take to wire new telephone lines if both members of the crew worked together? 3 h

40. As the June 1998 flood waters began to recede in Texas, a young family was faced with pumping the water from their basement. One pump they were using could dispose of 9000 gal in 3 h. A second pump could dispose of the same number of gallons in 4.5 h. How many hours would it take to dispose of 9000 gal if both pumps were working? 1.8 h

41. A new machine can package transistors four times as fast as an older machine. Working together, the machines can package the transistors in 8 h. How long would it take the new machine, working alone, to package the transistors? 10 h

42. The heat wave in Texas during the summer of 1998 forced even small businesses to run their air conditioners 24 hours a day. In the office of one such business, there were two air conditioners, one older than the other. The newer one was able to cool the room by 2 degrees in 8 min. With both running, the room could be cooled by the same number of degrees in 4.8 min. How long would it take the older air conditioner, working alone, to cool the room by 2 degrees? 12 min

43. The larger of two printers being used to print the payroll for a major corporation requires 40 min to print the payroll. After both printers have been operating for 10 min, the larger printer malfunctions. The smaller printer requires 50 more minutes to complete the payroll. How long would it take the smaller printer, working alone, to print the payroll? 80 min

44. An experienced bricklayer can work twice as fast as an apprentice bricklayer. After they worked together on a job for 8 h, the experienced bricklayer quit. The apprentice required 12 more hours to finish the job. How long would it take the experienced bricklayer, working alone, to do the job? 18 h

45. A roofer requires 12 h to shingle a roof. After the roofer and an apprentice work on a roof for 3 h, the roofer moves on to another job. The apprentice requires 12 more hours to finish the job. How long would it take the apprentice, working alone, to do the job? 20 h

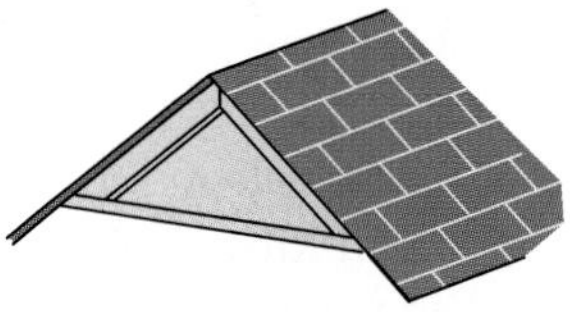

46. A welder requires 25 h to do a job. After the welder and an apprentice work on a job for 10 h, the welder quits. The apprentice finishes the job in 17 h. How long would it take the apprentice, working alone, to do the job? 45 h

47. A New York city pizza parlor hired three part-time employees to make pizzas. After a short training period of 2 days, the first employee was able to make a large pepperoni pizza in 3.5 min. The second employee took 2.5 min, and the third took 3.0 min to create the same large pizza. To the nearest minute, how long would it take to make the pizza if all three employees worked at the same time? 1 min

48. Three machines fill soda bottles. The machines can fill the daily quota of soda bottles in 12 h, 15 h, and 20 h, respectively. How long would it take to fill the daily quota of soda bottles with all three machines working? 5 h

49. With both hot and cold water running, a bathtub can be filled in 10 min. The drain will empty the tub in 15 min. A child turns both faucets on and leaves the drain open. How long will it be before the bathtub starts to overflow? 30 min

50. The inlet pipe can fill a water tank in 30 min. The outlet pipe can empty the tank in 20 min. How long would it take to empty a full tank with both pipes open? 60 min

51. An oil tank has two inlet pipes and one outlet pipe. One inlet pipe can fill the tank in 12 h, and the other inlet pipe can fill the tank in 20 h. The outlet pipe can empty the tank in 10 h. How long would it take to fill the tank with all three pipes open? 30 h

52. Water from a tank is being used for irrigation at the same time as the tank is being filled. The two inlet pipes can fill the tank in 6 h and 12 h, respectively. The outlet pipe can empty the tank in 24 h. How long would it take to fill the tank with all three pipes open? 4.8 h

3 Rate-of-wind and rate-of-current problems

53. Two skaters take off for an afternoon of rollerblading in Central Park. The first skater can cover 15 mi in the same time it takes the second skater, traveling 3 mph slower than the first skater, to cover 12 mi. Find the rate of each rollerblader. 15 mph; 12 mph

54. A commercial jet travels 1620 mi in the same amount of time it takes a corporate jet to travel 1260 mi. The rate of the commercial jet is 120 mph greater than the rate of the corporate jet. Find the rate of each jet. Commercial: 540 mph; corporate: 420 mph

55. A passenger train travels 295 mi in the same amount of time it takes a freight train to travel 225 mi. The rate of the passenger train is 14 mph greater than the rate of the freight train. Find the rate of each train. Passenger: 59 mph; freight: 45 mph

56. The rate of a bicyclist is 7 mph more than the rate of a long-distance runner. The bicyclist travels 30 mi in the same amount of time it takes the runner to travel 16 mi. Find the rate of the runner. 8 mph

57. A cyclist rode 40 mi before having a flat tire and then walking 5 mi to a service station. The cycling rate was four times the walking rate. The time spent cycling and walking was 5 h. Find the rate at which the cyclist was riding. 12 mph

58. A sales executive traveled 32 mi by car and then an additional 576 mi by plane. The rate of the plane was nine times the rate of the car. The total time of the trip was 3 h. Find the rate of the plane. 288 mph

59. A motorist drove 72 mi before running out of gas and then walking 4 mi to a gas station. The driving rate of the motorist was twelve times the walking rate. The time spent driving and walking was 2.5 h. Find the rate at which the motorist walks. 4 mph

60. An insurance representative traveled 735 mi by commercial jet and then an additional 105 mi by helicopter. The rate of the jet was four times the rate of the helicopter. The entire trip took 2.2 h. Find the rate of the jet. 525 mph

61. A business executive can travel the 480 ft between two terminals of an airport by walking on a moving sidewalk in the same time required to walk 360 ft without using the moving sidewalk. If the rate of the moving sidewalk is 2 ft/s, find the rate at which the executive can walk. 6 ft/s

62. A cyclist and a jogger start from a town at the same time and head for a destination 18 mi away. The rate of the cyclist is twice the rate of the jogger. The cyclist arrives 1.5 h before the jogger. Find the rate of the cyclist. 12 mph

63. A single-engine plane and a commercial jet leave an airport at 10 A.M. and head for an airport 960 mi away. The rate of the jet is four times the rate of the single-engine plane. The single-engine plane arrives 4 h after the jet. Find the rate of each plane. Jet: 720 mph; single-engine plane: 180 mph

64. Marlys can row a boat 3 mph faster than she can swim. She is able to row 10 mi in the same time it takes her to swim 4 mi. Find Marlys's rate swimming. 2 mph

65. A cruise ship can sail 28 mph in calm water. Sailing with the Gulf Stream, the ship can sail 170 mi in the same amount of time as it takes to sail 110 mi against the Gulf Stream. Find the rate of the Gulf Stream. 6 mph

66. A commercial jet can fly 500 mph in calm air. Traveling with the jet stream, the plane flew 2420 mi in the same amount of time as it takes to fly 1580 mi against the jet stream. Find the rate of the jet stream. 105 mph

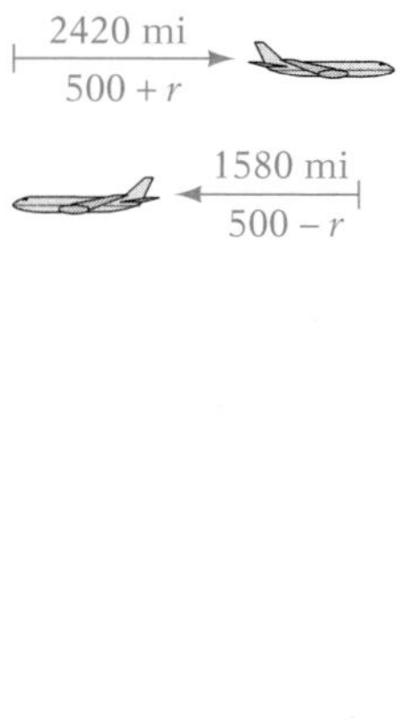

67. A tour boat used for river excursions can travel 7 mph in calm water. The amount of time it takes to travel 20 mi with the current is the same amount of time it takes to travel 8 mi against the current. Find the rate of the current. 3 mph

68. A canoe can travel 8 mph in still water. Traveling with the current of a river, the canoe can travel 15 mi in the same amount of time it takes to travel 9 mi against the current. Find the rate of the current. 2 mph

APPLYING CONCEPTS 6.4

69. The denominator of a fraction is 4 more than the numerator. If both the numerator and the denominator of the fraction are increased by 3, the new fraction is $\frac{5}{6}$. Find the original fraction. $\frac{17}{21}$

70. One pipe can fill a tank in 3 h, a second pipe can fill the tank in 4 h, and a third pipe can fill the tank in 6 h. How long would it take to fill the tank with all three pipes operating? $1\frac{1}{3}$ h

71. One printer can print a company's paychecks in 24 min, a second printer can print the checks in 16 min, and a third printer can do the job in 12 min. How long would it take to print the checks with all three printers operating?

$5\frac{1}{3}$ min

[C] 72. By increasing your speed by 10 mph, you can drive the 200-mile trip to your hometown in 40 min less time than the trip usually takes you. How fast do you usually drive? 50 mph

[C] 73. Because of weather conditions, a bus driver reduced the usual speed along a 165-mile bus route by 5 mph. The bus arrived only 15 min later than its usual arrival time. How fast does the bus usually travel? 60 mph

[C] 74. If a pump can fill a pool in A hours and a second pump can fill the pool in B hours, find a formula, in terms of A and B, for the time it takes both pumps, working together, to fill the pool. $t = \frac{AB}{A + B}$

[C] 75. If a parade is 1 mi long and is proceeding at 3 mph, how long will it take a runner, jogging at 5 mph, to run from the beginning of the parade to the end and then back to the beginning? $\frac{5}{8}$ h

76. Write a report on the Rhind Papyrus.

77. What is a unit fraction? What role did unit fractions play in early Egyptian mathematics?

SECTION 6.5 Proportions and Variation

1 Proportions

Quantities such as 3 feet, 5 liters, and 2 miles are number quantities written with units. In these examples, the units are feet, liters, and miles.

A **ratio** is the quotient of two quantities that have the same unit.

The weekly wages of a painter are \$425. The painter spends \$50 a week for food. The ratio of wages spent for food to the total weekly wages is written as shown below.

$\frac{\$50}{\$425} = \frac{50}{425} = \frac{2}{17}$ A ratio is in simplest form when the two numbers do not have a common factor. The units are not written.

A **rate** is the quotient of two quantities that have different units.

A car travels 120 mi on 3 gal of gas. The miles-to-gallon rate is written as shown below.

$\frac{120 \text{ mi}}{3 \text{ gal}} = \frac{40 \text{ mi}}{1 \text{ gal}}$ A rate is in simplest form when the two numbers do not have a common factor. The units are written as part of the rate.

A **proportion** is an equation that states that two ratios or rates are equal. For example, $\frac{90 \text{ km}}{4 \text{ L}} = \frac{45 \text{ km}}{2 \text{ L}}$ and $\frac{3}{4} = \frac{x+2}{16}$ are proportions.

Note that a proportion is a special kind of fractional equation. Many application problems can be solved by using proportions.

INSTRUCTOR NOTE

A proposition in Euclid's *Elements* gives the procedure for finding the golden ratio. A paraphrase of this proposition is "On a line segment *AB*, find a point *C* so that the ratio of *AB* to *AC* is equal to the ratio of *AC* to *CB*." Finding this ratio in numerical terms can serve as an extra-credit problem for students. The ratio is $\frac{AB}{AC} = \frac{1+\sqrt{5}}{2}$, which is the golden ratio.

The sales tax on a car that costs $4000 is $220. Find the sales tax on a car that costs $10,500.

Write a proportion using x to represent the sales tax.

$$\frac{220}{4000} = \frac{x}{10{,}500}$$

$$\frac{11}{200} = \frac{x}{10{,}500}$$

To solve the proportion, multiply each side of the equation by the denominators.

$$(200)(10{,}500)\frac{11}{200} = (200)(10{,}500)\frac{x}{10{,}500}$$

$$(10{,}500)(11) = 200x$$

$$115{,}500 = 200x$$

$$577.50 = x$$

The sales tax on the $10,500 car is $577.50.

Example 1 A stock investment of 50 shares pays a dividend of $106. At this rate, how many additional shares are needed to earn a dividend of $424?

Strategy To find the additional number of shares that are required, write and solve a proportion using x to represent the additional number of shares. Then $50 + x$ is the total number of shares of stock.

Solution

$$\frac{106}{50} = \frac{424}{50+x}$$

$$\frac{53}{25} = \frac{424}{50+x}$$

$$25(50+x)\frac{53}{25} = 25(50+x)\frac{424}{50+x}$$

$$(50+x)53 = (25)424$$

$$2650 + 53x = 10{,}600$$

$$53x = 7950$$

$$x = 150$$

An additional 150 shares of stock are required.

Problem 1 Two pounds of cashews cost $3.10. At this rate, how much would 15 lb of cashews cost?

Solution See page S18. $23.25

2 Variation problems

26 HM³

Direct variation is a special function that can be expressed as the equation $y = kx$, where k is a constant. The equation $y = kx$ is read "y varies directly as x" or "y is proportional to x." The constant k is called the **constant of variation** or the **constant of proportionality.**

The circumference (C) of a circle varies directly as the diameter (d). The direct variation equation is written $C = \pi d$. The constant of variation is π.

A nurse makes $18 per hour. The nurse's total wage (w) is directly proportional to the number of hours (h) worked. The equation of variation is $w = 18h$. The constant of proportionality is 18.

In general, a direct variation equation can be written in the form $y = kx^n$, where n is a positive number. For example, the equation $y = kx^2$ is read "y varies directly as the square of x."

The area (A) of a circle varies directly as the square of the radius (r) of the circle. The direct variation equation is $A = \pi r^2$.

Given that V varies directly as r and that $V = 20$ when $r = 4$, the constant of variation can be found by writing the basic direct variation equation, replacing V and r by the given values, and solving for the constant of variation.

$$V = kr$$
$$20 = k \cdot 4$$
$$5 = k$$

The direct variation equation can then be written by substituting the value of k into the basic direct variation equation.

$$V = 5r$$

Example 2 The amount (A) of medication prescribed for a person varies directly with the person's weight (W). For a person who weighs 50 kg, 2 ml of medication are prescribed. How many milliliters of medication are required for a person who weighs 75 kg?

Strategy To find the required amount of medication:
- Write the basic direct variation equation, replace the variables by the given values, and solve for k.
- Write the direct variation equation, replacing k by its value. Substitute 75 for W, and solve for A.

Solution

$$A = kW$$
$$2 = k \cdot 50$$
$$\frac{1}{25} = k$$

$A = \frac{1}{25}W$ ▶ This is the direct variation equation.

$A = \frac{1}{25} \cdot 75 = 3$ ▶ Replace W by 75.

The required amount of medication is 3 ml.

Problem 2 The distance (s) a body falls from rest varies directly as the square of the time (t) of the fall. An object falls 64 ft in 2 s. How far will it fall in 5 s?

Solution See page S19. 400 ft

A **joint variation** is a variation in which a variable varies directly as the product of two or more other variables. A joint variation can be expressed as the equation $z = kxy$, where k is a constant. The equation $z = kxy$ is read "z varies jointly as x and y."

The area (A) of a triangle varies jointly as the base (b) and the height (h). The joint variation equation is written $A = \frac{1}{2}bh$. The constant of variation is $\frac{1}{2}$.

An **inverse variation** is a function that can be expressed as the equation $y = \frac{k}{x}$, where k is a constant. The equation $y = \frac{k}{x}$ is read "y varies inversely as x" or "y is inversely proportional to x."

In general, an inverse variation equation can be written $y = \frac{k}{x^n}$, where n is a positive number. For example, the equation $y = \frac{k}{x^2}$ is read "y varies inversely as the square of x."

Given that P varies inversely as the square of x and that $P = 5$ when $x = 2$, the variation constant can be found by writing the basic inverse variation equation, replacing P and x by the given values, and solving for the constant of variation.

$$P = \frac{k}{x^2}$$
$$5 = \frac{k}{2^2}$$
$$5 = \frac{k}{4}$$
$$20 = k$$

The inverse variation equation can then be found by substituting the value of k into the basic inverse variation equation.

$$P = \frac{20}{x^2}$$

Example 3 A company that produces personal computers has determined that the number of computers it can sell (s) is inversely proportional to the price (P) of the computer. Two thousand computers can be sold when the price is \$2500. How many computers can be sold when the price of a computer is \$2000?

Strategy To find the number of computers:

- Write the basic inverse variation equation, replace the variables by the given values, and solve for k.
- Write the inverse variation equation, replacing k by its value. Substitute 2000 for P, and solve for s.

Solution

$$s = \frac{k}{P}$$
$$2000 = \frac{k}{2500}$$
$$5{,}000{,}000 = k$$

$$s = \frac{5{,}000{,}000}{P}$$ ▶ This is the inverse variation equation.

$$s = \frac{5{,}000{,}000}{2000} = 2500$$ ▶ Replace P by 2000.

At a price of \$2000, 2500 computers can be sold.

Problem 3 The resistance (R) to the flow of electric current in a wire of fixed length is inversely proportional to the square of the diameter (d) of the wire. If a wire of diameter 0.01 cm has a resistance of 0.5 ohm, what is the resistance in a wire that is 0.02 cm in diameter?

Solution See page S19. 0.125 ohm

A **combined variation** is a variation in which two or more types of variation occur at the same time. For example, in physics, the volume (V) of a gas varies directly as the temperature (T) and inversely as the pressure (P). This combined variation is written $V = \frac{kT}{P}$.

Example 4 The pressure (P) of a gas varies directly as the temperature (T) and inversely as the volume (V). When $T = 50°$ and $V = 275$ in^3, $P = 20$ lb/in^2. Find the pressure of a gas when $T = 60°$ and $V = 250$ in^3.

Strategy To find the pressure:

- Write the basic combined variation equation, replace the variables by the given values, and solve for k.
- Write the combined variation equation, replacing k by its value. Substitute 60 for T and 250 for V, and solve for P.

Solution

$$P = \frac{kT}{V}$$

$$20 = \frac{k(50)}{275}$$

$$110 = k$$

$$P = \frac{110T}{V}$$ ▶ This is the combined variation equation.

$$P = \frac{110(60)}{250} = 26.4$$ ▶ Replace T by 60 and V by 250.

The pressure is 26.4 lb/in^2.

Problem 4 The strength (s) of a rectangular beam varies jointly as its width (W) and as the square of its depth (d) and inversely as its length (L). If the strength of a beam 2 in. wide, 12 in. deep, and 12 ft long is 1200 lb, find the strength of a beam 4 in. wide, 8 in. deep, and 16 ft long.

Solution See page S19. 800 lb

CONCEPT REVIEW 6.5

Determine whether the following statements are always true, sometimes true, or never true.

1. If x varies inversely as y, then when x is doubled, y is doubled. Never true

2. If a varies inversely as b, then ab is a constant. Always true

3. If a varies jointly as b and c, then $a = \frac{kb}{c}$. Never true

4. If the length of a rectangle is held constant, then the area of the rectangle varies directly as the width. Always true

5. If the area of a rectangle is held constant, then the length varies directly as the width. Never true

6. The circumference of a circle varies directly as the diameter. If the diameter of a circle is doubled, then the circumference of the circle will be doubled. Always true

EXERCISES 6.5

1 1. How does a ratio differ from a rate?

2. What is a proportion?

3. *Environmental Science* In a wildlife preserve, 60 ducks are captured, tagged, and then released. Later, 200 ducks are examined, and three of the 200 ducks are found to have tags. Estimate the number of ducks in the preserve. 4000 ducks

4. *Political Science* A pre-election survey showed that 7 out of every 12 voters would vote in an election. At this rate, how many people would be expected to vote in a city of 210,000? 122,500 people

5. *Food Science* A caterer estimates that 2 gal of fruit punch will serve 30 people. How much additional punch is necessary to serve 75 people? 3 gal

6. *Taxes* In 1998, the automobile excise tax for a town in Massachusetts was based on 90% of the dealer's sticker price of a new car. The excise tax on the new Volkswagen Beetle with a list price of \$15,995 is \$359.89. At this rate, what is the excise tax for a new Porsche that has a list price of \$72,500? \$1631.26

7. *Architecture* On an architectural drawing, $\frac{1}{4}$ in. represents 1 ft. Find the dimensions of a room that measures $4\frac{1}{4}$ in. by $5\frac{1}{2}$ in. on the drawing. 17 ft by 22 ft

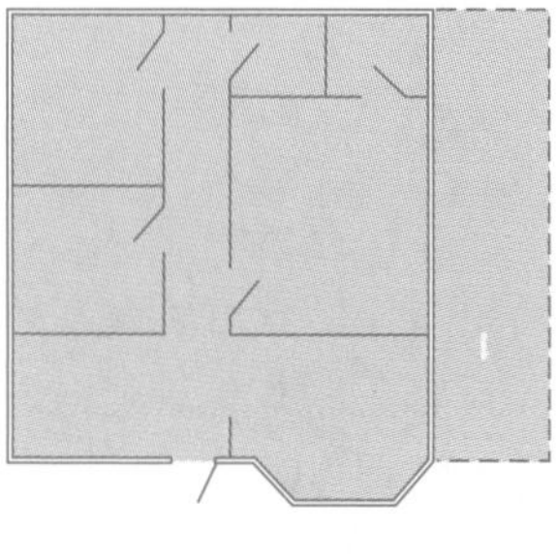

8. *Construction* A contractor estimated that 15 ft^2 of window space will be allowed for every 160 ft^2 of floor space. Using this estimate, how much window space will be allowed for 3200 ft^2 of floor space? 300 ft^2

9. *Quality Control* A quality control inspector found 5 defective diodes in a shipment of 4000 diodes. At this rate, how many diodes would be defective in a shipment of 3200 diodes? 4 diodes

10. *Construction* To tile 24 ft^2 of area, 120 ceramic tiles are needed. At this rate, how many tiles are needed to tile 300 ft^2? 1500 tiles

11. *Health Science* Three-fourths of an ounce of a medication are required for a 120-pound adult. At the same rate, how many additional ounces of medication are required for a 200-pound adult? 0.5 additional ounce

12. *Physics* An object that weighs 100 lb on Earth would weigh 90.5 lb on Venus. At this rate, what is the weight on Venus of an object that weighs 150 lb on Earth? 135.75 lb

13. *Environmental Science* Six ounces of an insecticide are mixed with 15 gal of water to make a spray for spraying an orange grove. At the same rate, how much additional insecticide is required to be mixed with 100 gal of water? 34 additional ounces

14. *Ecology* In an attempt to estimate the number of Siberian tigers in a preserve, a wildlife management team captured, tagged, and released 50 Siberian tigers. A few months later, they captured 150 Siberian tigers of which 30 had tags. Estimate the number of Siberian tigers in the region. 250 tigers

15. *Art* Leonardo da Vinci studied human proportions so that his drawings would be more accurate. He observed that the height of a person kneeling was approximately three-fourths of that person's height when standing. Using this information, find the approximate height of a person whose kneeling height is 4 ft. Round to the nearest tenth. 5.3 ft

16. *Construction* A contractor estimated that 30 ft^3 of cement is required to make a concrete floor that measures 90 ft^2. Using this estimate, find how many additional cubic feet of cement would be required to make a concrete floor that measures 120 ft^2. 10 ft^3

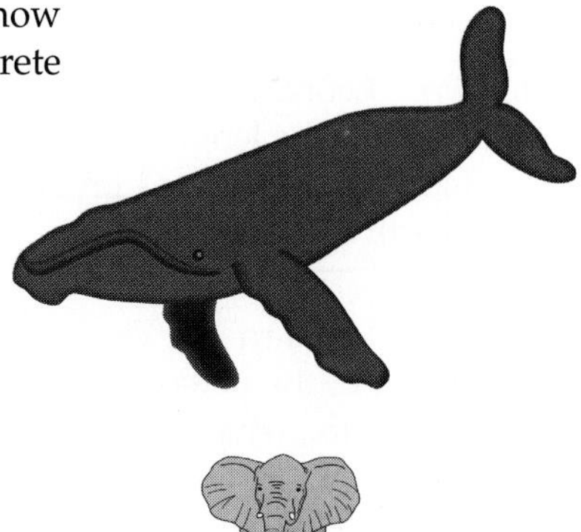

17. *Graphic Arts* The picture of a whale at the right uses a scale whereby 1 in. represents 48 ft. Estimate the actual length of the whale. 96 ft

18. *Graphic Arts* The picture of an elephant at the right uses a scale whereby 1 in. represents 12 ft. Estimate the actual height of the elephant. 9 ft

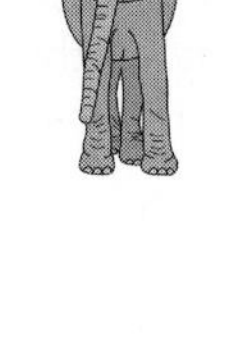

2 19. What is a direct variation?

20. What is an inverse variation?

21. *Business* The profit (P) realized by a company varies directly as the number of products it sells (s). If a company makes a profit of \$2500 on the sale of 20 products, what is the profit when the company sells 300 products? \$37,500

22. *Farming* The number of bushels of wheat (b) produced by a farm is directly proportional to the number of acres (A) planted in wheat. If a 25-acre farm yields 1125 bushels of wheat, what is the yield of a farm that has 220 acres of wheat? 9900 bushels

23. *Scuba Diving* The pressure (p) on a diver in the water varies directly as the depth (d). If the pressure is 3.6 lb/in^2 when the depth is 8 ft, what is the pressure when the depth is 30 ft? 13.5 lb/in^2

24. *Physics* The distance (d) a spring will stretch varies directly as the force (f) applied to the spring. If a force of 5 lb is required to stretch a spring 2 in., what force is required to stretch the spring 5 in.? 12.5 lb

25. *Optics* The distance (d) a person can see to the horizon from a point above the surface of the earth varies directly as the square root of the height (H). If, for a height of 500 ft, the horizon is 19 mi away, how far is the horizon from a point that is 800 ft high? Round to the nearest hundredth. 24.03 mi

26. *Physics* The period (p) of a pendulum, or the time it takes for the pendulum to make one complete swing, varies directly as the square root of the length (L) of the pendulum. If the period of a pendulum is 1.5 s when the length is 2 ft, find the period when the length is 4.5 ft. Round to the nearest hundredth. 2.25 s

27. *Physics* The distance (s) a ball will roll down an inclined plane is directly proportional to the square of the time (t). If the ball rolls 5 ft in 1 s, how far will it roll in 4 s? 80 ft

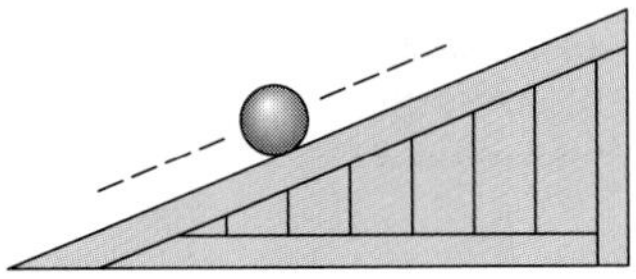

28. *Automotive Technology* The stopping distance (s) of a car varies directly as the square of its speed (v). If a car traveling 30 mph requires 60 ft to stop, find the stopping distance for a car traveling 55 mph. Round to the nearest tenth. 201.7 ft

29. *Art* Leonardo da Vinci observed that the length of a person's face varies directly as the length of the person's chin. If a person whose face length is 9 in. has a chin length of 1.5 in., what is the length of a person's face whose chin length is 1.7 in.? 10.2 in.

30. *Art* Leonardo da Vinci observed that a person's height varies directly as the width of the person's shoulders. If a person 70 in. tall has a shoulder width of 17.5 in., what is the height of a person whose shoulder width is 16 in.? 64 in.

31. *Physics* For a constant temperature, the pressure (P) of a gas varies inversely as the volume (V). If the pressure is 25 lb/in^2 when the volume is 400 ft^3, find the pressure when the volume is 150 ft^3. $66\frac{2}{3}$ lb/in^2

32. *Mechanical Engineering* The speed (v) of a gear varies inversely as the number of teeth (t). If a gear that has 48 teeth makes 20 revolutions per minute, how many revolutions per minute will a gear that has 30 teeth make? 32 revolutions

33. *Physics* The pressure (p) in a liquid varies directly as the product of the depth (d) and the density (D) of the liquid. If the pressure is 37.5 lb/in^2 when the depth is 100 in. and the density is 1.2 lb/in^2, find the pressure when the density remains the same and the depth is 60 in. 22.5 lb/in^2

34. *Electricity* The current (I) in a wire varies directly as the voltage (v) and inversely as the resistance (r). If the current is 27.5 amps when the voltage is 110 volts and the resistance is 4 ohms, find the current when the voltage is 195 volts and the resistance is 12 ohms. 16.25 amps

35. *Magnetism* The repulsive force (f) between the north poles of two magnets is inversely proportional to the square of the distance (d) between them. If the repulsive force is 18 lb when the distance is 3 in., find the repulsive force when the distance is 1.2 in. 112.5 lb

36. *Physics* The intensity (I) of a light source is inversely proportional to the square of the distance (d) from the source. If the intensity is 8 lumens at a distance of 6 m, what is the intensity when the distance is 4 m? 18 lumens

37. *Electricity* The resistance (R) of a wire varies directly as the length (L) of the wire and inversely as the square of the diameter (d). If the resistance is 9 ohms in 50 ft of wire that has a diameter of 0.05 in., find the resistance in 50 ft of a similar wire that has a diameter of 0.02 in. 56.25 ohms

38. *Music* The frequency of vibration (f) of a guitar string varies directly as the square root of the tension (T) and inversely as the length (L) of the string. If the frequency is 40 vibrations per second when the tension is 25 lb and the length of the string is 3 ft, find the frequency when the tension is 36 lb and the string is 4 ft long. 36 vibrations/s

39. *Sailing* The wind force (w) on a vertical surface varies directly as the product of the area (A) of the surface and the square of the wind velocity (v). When the wind is blowing at 30 mph, the force on an area of 10 ft^2 is 45 lb. Find the force on this area when the wind is blowing at 60 mph. 180 lb

40. *Electricity* The power (P) in an electric circuit is directly proportional to the product of the current (I) and the square of the resistance (R). If the power is 100 watts when the current is 4 amps and the resistance is 5 ohms, find the power when the current is 2 amps and the resistance is 10 ohms. 200 watts

APPLYING CONCEPTS 6.5

41. a. Graph $y = kx$ when $k = 2$.
    b. What kind of function does the graph represent?
    A linear function

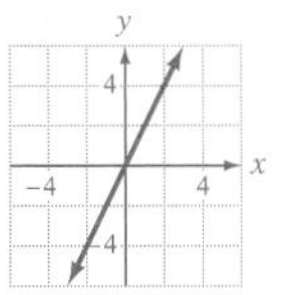

42. **a.** Graph $y = kx$ when $k = \frac{1}{2}$.

b. What kind of function does the graph represent?
A linear function

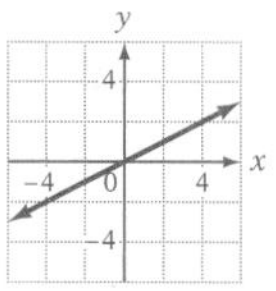

43. **a.** Graph $y = \frac{k}{x}$ when $k = 2$ and $x > 0$.

b. Is this the graph of a function?
Yes

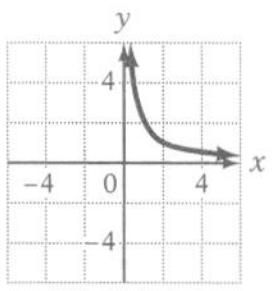

44. **a.** Graph $y = \frac{k}{x^2}$ when $k = 2$ and $x > 0$.

b. Is this the graph of a function?
Yes

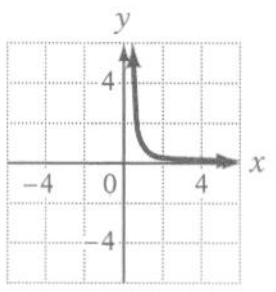

45. In the inverse variation equation $y = \frac{k}{x}$, what is the effect on x when y doubles? x is halved.

46. In the direct variation equation $y = kx$, what is the effect on y when x doubles? y doubles.

Complete using the word *directly* or *inversely*.

[C] **47.** If a varies directly as b and inversely as c, then c varies directly as b and inversely as a.

[C] **48.** If a varies inversely as b and c, then abc is constant.

[C] **49.** If the width of a rectangle is held constant, the area of the rectangle varies directly as the length.

[C] **50.** If the area of a rectangle is held constant, the length of the rectangle varies inversely as the width.

51.  One of the measures that financial analysts use to judge the strength of a company is the ratio *sales per share*. This is the ratio of the total annual sales to the number of shares of stock outstanding. The table that follows shows the sales (in millions of dollars) and the number of outstanding shares of stock (in millions) for Reynolds Metals. (*Source:* Eleven-Year Statistical Summary of Reynolds Metals.)

Year	Sales (in millions of $)	Shares of stock (in millions)
1993	5269	60.49
1994	5879	62.17
1995	7213	63.60
1996	6972	72.72
1997	6881	73.91

a. What is the unit for the sales per share ratio? Dollars per share

b. Calculate the sales per share of Reynolds Metals for the years shown in the table. 1993, \$87.11; 1994, \$94.56; 1995, \$113.41; 1996, \$95.87; 1997, \$93.10

c. A model of the sales per share of Reynolds Metals is given by the function $f(x) = 0.148x^{1.4227}$, where x is the last two digits of the year and $f(x)$ is the sales per share. What is the percent error between the value given by the model and the actual value for 1994? Use the formula

$$\% \text{ error} = \left| \frac{\text{actual value} - \text{predicted value}}{\text{actual value}} \right| \times 100$$ 0.40%

52. Write a report on financial ratios used by stock market analysts. Include the price-earnings ratio, current ratio, quick ratio, percent yield, and ratio of stockholders' equity to liabilities.

53. Explain joint variation.

SECTION 6.6

Literal equations

1 Solve literal equations

A **literal equation** is an equation that contains more than one variable. Examples of literal equations are shown at the right.

$$3x - 2y = 4$$
$$v^2 = v_0^2 + 2as$$

Formulas are used to express relationships among physical quantities. A **formula** is a literal equation that states a rule about measurement. Examples of formulas are shown below.

$$s = vt - 16t^2 \quad \text{(Physics)}$$
$$c^2 = a^2 + b^2 \quad \text{(Geometry)}$$
$$A = P(1 + r)^t \quad \text{(Business)}$$

The Addition and Multiplication Properties of Equations can be used to solve a literal equation for one of the variables. The goal is to rewrite the equation so that the variable being solved for is alone on one side of the equation, and all the other numbers and variables are on the other side.

Solve $C = \frac{5}{9}(F - 32)$ for F.

$$C = \frac{5}{9}(F - 32)$$

Use the Distributive Property to remove parentheses. $C = \frac{5}{9}F - \frac{160}{9}$

Multiply each side of the equation by the LCM of the denominators. $9C = 5F - 160$

Add 160 to each side of the equation. $9C + 160 = 5F$

Divide each side of the equation by the coefficient 5. $\frac{9C + 160}{5} = F$

Example 1 A. Solve $A = P + Prt$ for P. B. Solve $\frac{S}{S - C} = R$ for C.

Solution A.

$$A = P + Prt$$

$$A = (1 + rt)P$$ ▶ Factor P from $P + Prt$.

$$\frac{A}{1 + rt} = \frac{(1 + rt)P}{1 + rt}$$ ▶ Divide each side of the equation by $1 + rt$.

$$\frac{A}{1 + rt} = P$$

B.

$$\frac{S}{S - C} = R$$

$$(S - C)\frac{S}{S - C} = (S - C)R$$ ▶ Multiply each side of the equation by $S - C$.

$$S = SR - CR$$

$$CR + S = SR$$ ▶ Add CR to each side of the equation.

$$CR = SR - S$$ ▶ Subtract S from each side of the equation.

$$C = \frac{SR - S}{R}$$ ▶ Divide each side of the equation by R.

Problem 1 A. Solve $\frac{1}{R_1} + \frac{1}{R_2} = \frac{1}{R}$ for R. B. Solve $t = \frac{r}{r + 1}$ for r.

Solution See page S19. A. $R = \frac{R_1R_2}{R_2 + R_1}$ B. $r = \frac{t}{1 - t}$

CONCEPT REVIEW 6.6

Determine whether the following statements are always true, sometimes true, or never true.

1. A formula is a literal equation that states a rule about measurement. Always true
2. A literal equation contains two variables. Sometimes true
3. The formula $\frac{1}{R_1} + \frac{1}{R_2} = \frac{1}{R}$ is taken from physics. When we solve for R_2, we get $R_2 = R - R_1$. Never true
4. Consider the formula $P = 2L + 2W$. When we double the perimeter, the length and width will both be doubled. Sometimes true

EXERCISES 6.6

1 Solve the formula for the variable given.

1. $P = 2L + 2W$; W (Geometry) $W = \frac{P - 2L}{2}$
2. $F = \frac{9}{5}C + 32$; C (Temperature conversion) $C = \frac{5}{9}(F - 32)$
3. $S = C - rC$; C (Business) $C = \frac{S}{1 - r}$
4. $A = P + Prt$; t (Business) $t = \frac{A - P}{Pr}$

5. $PV = nRT; R$ (Chemistry) $R = \frac{PV}{nT}$

6. $A = \frac{1}{2}bh; h$ (Geometry) $h = \frac{2A}{b}$

7. $F = \frac{Gm_1m_2}{r^2}; m_2$ (Physics) $m_2 = \frac{Fr^2}{Gm_1}$

8. $\frac{P_1V_1}{T_1} = \frac{P_2V_2}{T_2}; P_2$ (Chemistry) $P_2 = \frac{P_1V_1T_2}{T_1V_2}$

9. $I = \frac{E}{R + r}; R$ (Physics) $R = \frac{E - Ir}{I}$

10. $S = V_0t - 16t^2; V_0$ (Physics) $V_0 = \frac{S + 16t^2}{t}$

11. $A = \frac{1}{2}h(b_1 + b_2); b_2$ (Geometry) $b_2 = \frac{2A - hb_1}{h}$

12. $V = \frac{1}{3}\pi r^2h; h$ (Geometry) $h = \frac{3V}{\pi r^2}$

13. $\frac{1}{R} = \frac{1}{R_1} + \frac{1}{R_2}; R_2$ (Physics) $R_2 = \frac{RR_1}{R_1 - R}$

14. $\frac{1}{f} = \frac{1}{a} + \frac{1}{b}; b$ (Physics) $b = \frac{af}{a - f}$

15. $a_n = a_1 + (n - 1)d; d$ (Mathematics) $d = \frac{a_n - a_1}{n - 1}$

16. $P = \frac{R - C}{n}; R$ (Business) $R = Pn + C$

17. $S = 2WH + 2WL + 2LH; H$ (Geometry) $H = \frac{S - 2WL}{2W + 2L}$

18. $S = 2\pi r^2 + 2\pi rH; H$ (Geometry) $H = \frac{S - 2\pi r^2}{2\pi r}$

Solve for x.

19. $ax + by + c = 0$ $x = \frac{-by - c}{a}$

20. $x = ax + b$ $x = \frac{b}{1 - a}$

21. $ax + b = cx + d$ $x = \frac{d - b}{a - c}$

22. $y - y_1 = m(x - x_1)$ $x = \frac{y - y_1 + mx_1}{m}$

23. $\frac{a}{x} = \frac{b}{c}$ $x = \frac{ac}{b}$

24. $\frac{1}{x} + \frac{1}{a} = b$ $x = \frac{a}{ab - 1}$

25. $\frac{1}{a} + \frac{1}{b} = \frac{1}{x}$ $x = \frac{ab}{a + b}$

26. $a(a - x) = b(b - x)$ $x = a + b$

APPLYING CONCEPTS 6.6

Solve for x.

27. $\frac{x - y}{y} = \frac{x + 5}{2y}$ $5 + 2y$

28. $\frac{x - 2}{y} = \frac{x + 2}{5y}$ 3

29. $\frac{x}{x + y} = \frac{2x}{4y}$ $0, y$

30. $\frac{2x}{x - 2y} = \frac{x}{2y}$ $0, 6y$

31. $\frac{x - y}{2x} = \frac{x - 3y}{5y}$ $\frac{y}{2}, 5y$

32. $\frac{x - y}{x} = \frac{2x}{9y}$ $3y, \frac{3y}{2}$

Solve for the given variable.

[C] 33. $\frac{w_1}{w_2} = \frac{f_2 - f}{f - f_1}; f$ $f = \frac{w_2f_2 + w_1f_1}{w_1 + w_2}$

[C] 34. $v = \frac{v_1 + v_2}{1 + \frac{v_1v_2}{c^2}}; v_1$ $v_1 = \frac{c^2v - c^2v_2}{c^2 - vv_2}$

Focus on Problem Solving

Implication

Sentences that are constructed using "If..., then..." occur frequently in problem-solving situations. These sentences are called **implications.** The sentence "If it rains, then I will stay home" is an implication. The phrase *it rains* is called the **antecedent** of the implication. The phrase *I will stay home* is the **consequent** of the implication.

The sentence "If $x = 4$, then $x^2 = 16$" is a true sentence. The **contrapositive** of an implication is formed by switching the antecedent and the consequent and then negating each one. The contrapositive of "If $x = 4$, then $x^2 = 16$" is "If $x^2 \neq 16$, then $x \neq 4$." This sentence is also true. It is a principle of logic that an implication and its contrapositive are either both true or both false.

The **converse** of an implication is formed by switching the antecedent and the consequent. The converse of "If $x = 4$, then $x^2 = 16$" is "If $x^2 = 16$, then $x = 4$." Note that the converse is not a true statement, because if $x^2 = 16$, then x could equal -4. The converse of an implication may or may not be true.

Those statements for which the implication and the converse are both true are very important. They can be stated in the form "x if and only if y." For instance, a number is divisible by 5 if and only if the last digit of the number is 0 or 5. The implication "If a number is divisible by 5, then it ends in 0 or 5" and the converse "If a number ends in 0 or 5, then it is divisible by 5" are both true.

For each of the following, state the contrapositive and the converse of the implication. If the converse and the implication are both true statements, write a sentence using the phrase "if and only if."

1. If a number is divisible by 8, then it is divisible by 4.
2. If $4z = 20$, then $z = 5$.
3. If p is a prime number greater than 2, then p is an odd number.
4. If x is a rational number, then $x = \frac{a}{b}$, where a and b are integers and $b \neq 0$.
5. If $\sqrt{x} = a$, then $a^2 = x$.
6. If the equation of a graph is $y = mx + b$, then the graph of the equation is a straight line.
7. If $a = 0$ or $b = 0$, then $ab = 0$.
8. If the coordinates of a point are (5, 0), then the point is on the x-axis.
9. If a quadrilateral is a square, then the quadrilateral has four sides of equal length.

Projects and Group Activities

Continued Fractions

The following complex fraction is called a **continued fraction.**

$$1 + \cfrac{1}{1 + \cfrac{1}{1 + \cfrac{1}{1 + \cfrac{1}{1 + \cdots}}}}$$

The dots indicate that the pattern continues to repeat forever.

A **convergent** for a continued fraction is an approximation of the repeated pattern. For instance,

$$c_2 = 1 + \cfrac{1}{1 + \cfrac{1}{1 + 1}} \qquad c_3 = 1 + \cfrac{1}{1 + \cfrac{1}{1 + \cfrac{1}{1 + 1}}} \qquad c_4 = 1 + \cfrac{1}{1 + \cfrac{1}{1 + \cfrac{1}{1 + \cfrac{1}{1 + 1}}}}$$

1. Calculate c_5 for the continued fraction at the bottom of the previous page.

The fraction above is related to the golden rectangle, which has been used in architectural designs as diverse as the Parthenon in Athens, built around 440 B.C., and the United Nations building. A golden rectangle is one for which

$$\frac{\text{length}}{\text{width}} = \frac{\text{length} + \text{width}}{\text{length}}$$

An example of a golden rectangle is shown below.

Here is another continued fraction, which was discovered by Leonhard Euler (1707–1793). Calculating the convergents of this continued fraction yields approximations that are closer and closer to π.

$$\pi = 3 + \cfrac{1^2}{6 + \cfrac{3^2}{6 + \cfrac{5^2}{6 + \cfrac{7^2}{6 + \cdots}}}}$$

2. Calculate $c_5 = 3 + \cfrac{1^2}{6 + \cfrac{3^2}{6 + \cfrac{5^2}{6 + \cfrac{7^2}{6 + \cfrac{9^2}{6 + 11^2}}}}}$

Graphing Rational Expressions

The domain of a function is the set of the first coordinates of all the ordered pairs of the function. When a function is given by an equation, the domain of the function is all real numbers for which the function evaluates to a real number. For rational functions, we must exclude from the domain all those values of the variable for which the denominator of the rational function is zero. The graphing calculator is a useful tool to show the graphs of functions with excluded values in the domain of the function.

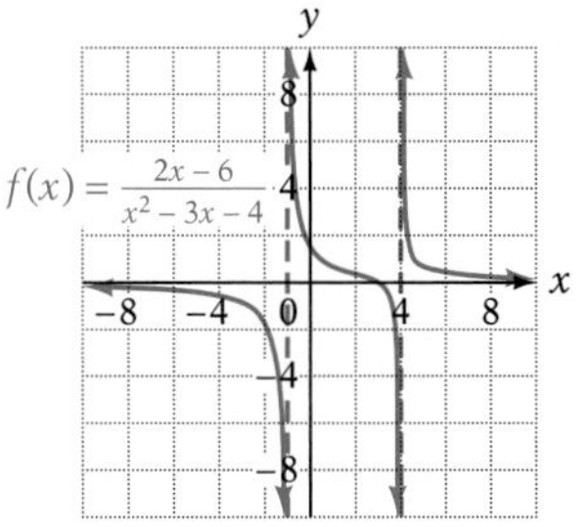

The domain is $\{x \mid x \neq -1, x \neq 4\}$.

The graph of the function $f(x) = \frac{2x - 6}{x^2 - 3x - 4}$ is shown at the left. Note that the graph never intersects the lines $x = -1$ and $x = 4$ (shown as dashed lines). These are the two values excluded from the domain of f. The graph of the function f gets closer to the dashed line $x = 4$ as x gets closer to 4. The graph of the function f also gets closer to $x = -1$ as x gets closer to -1. The lines that are "approached" by the function are called **asymptotes.**

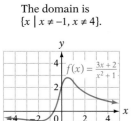

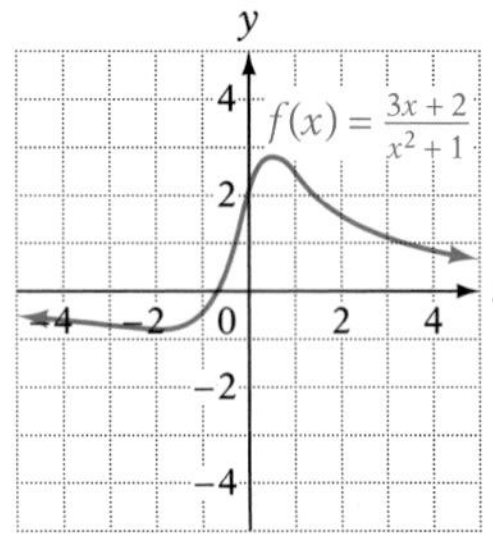

The domain is $\{x \mid x \in \text{real numbers}\}$.

The graph of the function $f(x) = \frac{3x + 2}{x^2 + 1}$ is shown at the left. The domain must exclude values of x for which $x^2 + 1 = 0$. It is not possible that $x^2 + 1 = 0$, because $x^2 \geq 0$, and a positive number added to a number equal to or greater than zero cannot equal zero. Therefore, there are no real numbers that must be excluded from the domain of f.

Exercises 13–36 in Section 6.1 ask you to find the domains of rational functions by algebraic means. Try these exercises again, this time using a graphing calculator.

Chapter Summary

Key Words

A fraction in which the numerator and denominator are polynomials is called a *rational expression.* (Objective 6.1.1)

$\frac{x^2 + 2x + 1}{x - 3}$ is a rational expression.

A rational expression is in *simplest form* when the numerator and denominator have no common factors. (Objective 6.1.2)

$$\frac{x^2 + 3x + 2}{(x + 1)^2} = \frac{(x + 1)(x + 2)}{(x + 1)(x + 1)} = \frac{x + 2}{x + 1}$$

The *reciprocal of a rational expression* is the rational expression with the numerator and denominator interchanged. (Objective 6.2.1)

The reciprocal of $\frac{a^2 - b^2}{x - y}$ is $\frac{x - y}{a^2 - b^2}$.

The *least common multiple* (LCM) *of two or more polynomials* is the simplest polynomial that contains the factors of each polynomial. (Objective 6.2.2)

$4x^2 - 12x = 4x(x - 3)$
$3x^3 - 21x^2 + 36x = 3x(x - 3)(x - 4)$
$\text{LCM} = 12x(x - 3)(x - 4)$

A *complex fraction* is a fraction whose numerator or denominator contains one or more fractions. (Objective 6.3.1)

$\dfrac{\frac{1}{x} + \frac{1}{y}}{\frac{1}{x}}$ is a complex fraction.

A *ratio* is the quotient of two quantities that have the same unit. When a ratio is in simplest form, the units are not written. (Objective 6.5.1)

$\frac{\$35}{\$100}$ written as a ratio in simplest form is $\frac{7}{20}$.

A *rate* is the quotient of two quantities that have different units. (Objective 6.5.1)

$\frac{65 \text{ mi}}{2 \text{ gal}}$ is a rate.

A *proportion* is an equation that states the equality of two ratios or rates. (Objective 6.5.1)

$\frac{75}{3} = \frac{x}{42}$ is a proportion.

A *formula* is a literal equation that states rules about measurements. (Objective 6.6.1)

$P = 2L + 2W$ is the formula for the perimeter of a rectangle.

Direct variation is a special function that can be expressed as the equation $y = kx^n$, where k is a constant called the *constant of variation* or the *constant of proportionality.* (Objective 6.5.2)

$E = mc^2$ is Einstein's formula relating energy and mass. c^2 is the constant of proportionality.

Inverse variation is a function that can be expressed as the equation $y = \frac{k}{x^n}$, where k is a constant. (Objective 6.5.2)

$I = \frac{k}{d^2}$ gives the intensity of a light source at a distance d from the source.

Joint variation is a variation in which a variable varies directly as the product of two or more variables. A joint variation can be expressed as the equation $z = kxy$, where k is a constant. (Objective 6.5.2)

$C = kAT$ is a formula for the cost of insulation, where A is the area to be insulated and T is the thickness of the insulation.

Combined variation is a variation in which two or more types of variation occur at the same time. (Objective 6.5.2)

$V = \frac{kT}{P}$ is a formula that states that the volume of a gas is directly proportional to the temperature and inversely proportional to the pressure.

Essential Rules and Procedures

To multiply fractions

$\frac{a}{b} \cdot \frac{c}{d} = \frac{ac}{bd}$ (Objective 6.2.1)

$$\frac{x^2 + x}{3} \cdot \frac{2}{5x} = \frac{(x^2 + x) \cdot 2}{3(5x)} = \frac{2x(x + 1)}{15x} = \frac{2(x + 1)}{15} = \frac{2x + 2}{15}$$

To divide fractions

$\frac{a}{b} \div \frac{c}{d} = \frac{a}{b} \cdot \frac{d}{c}$ (Objective 6.2.1)

$$\frac{3x}{x - 5} \div \frac{3}{x + 4} = \frac{3x}{x - 5} \cdot \frac{x + 4}{3} = \frac{3x(x + 4)}{3(x - 5)} = \frac{x(x + 4)}{x - 5} = \frac{x^2 + 4x}{x - 5}$$

To add fractions

$\frac{a}{c} + \frac{b}{c} = \frac{a + b}{c}$ (Objective 6.2.2)

$$\frac{2x - 7}{x^2 + 4} + \frac{x + 2}{x^2 + 4} = \frac{(2x - 7) + (x + 2)}{x^2 + 4} = \frac{3x - 5}{x^2 + 4}$$

To subtract fractions

$\frac{a}{c} - \frac{b}{c} = \frac{a - b}{c}$ (Objective 6.2.2)

$$\frac{5x - 6}{x + 2} - \frac{2x + 4}{x + 2} = \frac{(5x - 6) - (2x + 4)}{x + 2} = \frac{3x - 10}{x + 2}$$

Equation for Work Problems

Rate of work × Time worked = Part of task completed (Objective 6.4.2)

A roofer requires 24 h to shingle a roof. An apprentice can shingle the roof in 36 h. How long would it take to shingle the roof if both roofers worked together?

$$\frac{t}{24} + \frac{t}{36} = 1$$

Equation for Uniform Motion Problems

Distance = Rate × Time or $\frac{\text{Distance}}{\text{Rate}} = \text{Time}$

(Objective 6.4.3)

A motorcycle travels 195 mi in the same amount of time it takes a car to travel 159 mi. The rate of the motorcycle is 12 mph greater than the rate of the car. Find the rate of the car.

$$\frac{195}{r+12} = \frac{159}{r}$$

Chapter Review Exercises

1. Given $P(x) = \frac{x}{x-3}$, find $P(4)$. 4

2. Given $P(x) = \frac{x^2-2}{3x^2-2x+5}$, find $P(-2)$. $\frac{2}{21}$

3. Find the domain of $g(x) = \frac{2x}{x-3}$. $\{x \mid x \neq 3\}$

4. Find the domain of $f(x) = \frac{2x-7}{3x^2+3x-18}$. $\{x \mid x \neq -3, 2\}$

5. Find the domain of $F(x) = \frac{x^2-x}{3x^2+4}$. $\{x \mid x \in \text{real numbers}\}$

6. Simplify: $\frac{6a^{5n}+4a^{4n}-2a^{3n}}{2a^{3n}}$ $3a^{2n}+2a^n-1$

7. Simplify: $\frac{16-x^2}{x^3-2x^2-8x}$ $-\frac{x+4}{x(x+2)}$

8. Simplify: $\frac{x^3-27}{x^2-9}$ $\frac{x^2+3x+9}{x+3}$

9. Multiply: $\frac{a^6b^4+a^4b^6}{a^5b^4-a^4b^4} \cdot \frac{a^2-b^2}{a^4-b^4}$ $\frac{1}{a-1}$

10. Multiply: $\frac{x^3-8}{x^3+2x^2+4x} \cdot \frac{x^3+2x^2}{x^2-4}$ x

11. Multiply: $\frac{16-x^2}{6x-6} \cdot \frac{x^2+5x+6}{x^2-8x+16}$ $-\frac{(x+4)(x+3)(x+2)}{6(x-1)(x-4)}$

12. Divide: $\frac{x^{2n}-5x^n+4}{x^{2n}-2x^n-8} \div \frac{x^{2n}-4x^n+3}{x^{2n}+8x^n+12}$ $\frac{x^n+6}{x^n-3}$

13. Divide: $\frac{27x^3-8}{9x^3+6x^2+4x} \div \frac{9x^2-12x+4}{9x^2-4}$ $\frac{3x+2}{x}$

14. Divide: $\frac{3-x}{x^2+3x+9} \div \frac{x^2-9}{x^3-27}$ $-\frac{x-3}{x+3}$

15. Add: $\frac{5}{3a^2b^3} + \frac{7}{8ab^4}$ $\frac{21a+40b}{24a^2b^4}$

16. Subtract: $\frac{3x^2+2}{x^2-4} - \frac{9x-x^2}{x^2-4}$ $\frac{4x-1}{x+2}$

17. Simplify: $\frac{8}{9x^2-4} + \frac{5}{3x-2} - \frac{4}{3x+2}$ $\frac{3x+26}{(3x-2)(3x+2)}$

18. Simplify: $\frac{6x}{3x^2-7x+2} - \frac{2}{3x-1} + \frac{3x}{x-2}$ $\frac{9x^2+x+4}{(3x-1)(x-2)}$

19. Simplify: $\dfrac{x}{x-3} - 4 - \dfrac{2x-5}{x+2}$ $\quad -\dfrac{5x^2-17x-9}{(x-3)(x+2)}$

20. Simplify: $\dfrac{x-6+\dfrac{6}{x-1}}{x+3-\dfrac{12}{x-1}}$ $\quad \dfrac{x-4}{x+5}$

21. Simplify: $\dfrac{x+\dfrac{3}{x-4}}{3+\dfrac{x}{x-4}}$ $\quad \dfrac{x-1}{4}$

22. Solve: $\dfrac{5x}{2x-3} + 4 = \dfrac{3}{2x-3}$ $\quad \dfrac{15}{13}$

23. Solve: $\dfrac{x}{x-3} = \dfrac{2x+5}{x+1}$ $\quad -3, 5$

24. Solve: $\dfrac{6}{x-3} - \dfrac{1}{x+3} = \dfrac{51}{x^2-9}$ $\quad 6$

25. Solve: $\dfrac{30}{x^2+5x+4} + \dfrac{10}{x+4} = \dfrac{4}{x+1}$ No solution

26. Solve $I = \dfrac{1}{R}V$ for R. $\quad R = \dfrac{V}{I}$

27. Solve $Q = \dfrac{N-S}{N}$ for N. $\quad N = \dfrac{S}{1-Q}$

28. Solve $S = \dfrac{a}{1-r}$ for r. $\quad r = \dfrac{S-a}{S}$

29. A car uses 4 tanks of fuel to travel 1800 mi. At this rate, how many tanks of fuel would be required for a trip of 3000 mi? $6\frac{2}{3}$ tanks

30. On a certain map, 2.5 in. represents 10 mi. How many miles would be represented by 12 in.? 48 mi

31. An electrician requires 65 min to install a ceiling fan. The electrician and an apprentice, working together, take 40 min to install the fan. How long would it take the apprentice, working alone, to install the ceiling fan? 104 min

32. The inlet pipe can fill a tub in 24 min. The drain pipe can empty the tub in 15 min. How long would it take to empty a full tub with both pipes open? 40 min

33. Three students can paint a dormitory room in 8 h, 16 h, and 16 h, respectively. How long would it take to do the job if all three students worked together? 4 h

34. A canoeist can travel 10 mph in calm water. The amount of time it takes to travel 60 mi with the current is the same amount of time it takes to travel 40 mi against the current. Find the rate of the current. 2 mph

35. A bus and a cyclist leave a school at 8 A.M. and head for a stadium 90 mi away. The rate of the bus is three times the rate of the cyclist. The cyclist arrives 4 h after the bus. Find the rate of the bus. 45 mph

36. A tractor travels 10 mi in the same amount of time it takes a car to travel 15 mi. The rate of the tractor is 15 mph less than the rate of the car. Find the rate of the tractor. 30 mph

37. The pressure (p) of wind on a flat surface varies jointly as the area (A) of the surface and the square of the wind's velocity (v). If the pressure on 22 ft^2 is 10 lb when the wind's velocity is 10 mph, find the pressure on the same surface when the wind's velocity is 20 mph. 40 lb

38. The illumination (I) produced by a light varies inversely as the square of the distance (d) from the light. If the illumination produced 10 ft from a light is 12 lumens, find the illumination 2 ft from the light. 300 lumens

39. The electrical resistance (r) of a cable varies directly as its length (l) and inversely as the square of its diameter (d). If a cable 16,000 ft long and $\frac{1}{4}$ in. in diameter has a resistance of 3.2 ohms, what is the resistance of a cable that is 8000 ft long and $\frac{1}{2}$ in. in diameter? 0.4 ohm

Chapter Test

1. Simplify: $\dfrac{v^3 - 4v}{2v^2 - 5v + 2}$ $\dfrac{v(v+2)}{2v-1}$

2. Simplify: $\dfrac{2a^2 - 8a + 8}{4 + 4a - 3a^2}$ $-\dfrac{2(a-2)}{3a+2}$

3. Multiply: $\dfrac{3x^2 - 12}{5x - 15} \cdot \dfrac{2x^2 - 18}{x^2 + 5x + 6}$ $\dfrac{6(x-2)}{5}$

4. Given $P(x) = \dfrac{3 - x^2}{x^3 - 2x^2 + 4}$, find $P(-1)$. $P(-1) = 2$

5. Divide: $\dfrac{2x^2 - x - 3}{2x^2 - 5x + 3} \div \dfrac{3x^2 - x - 4}{x^2 - 1}$ $\dfrac{x+1}{3x-4}$

6. Multiply: $\dfrac{x^{2n} - x^n - 2}{x^{2n} + x^n} \cdot \dfrac{x^{2n} - x^n}{x^{2n} - 4}$ $\dfrac{x^n - 1}{x^n + 2}$

7. Simplify: $\dfrac{2}{x^2} + \dfrac{3}{y^2} - \dfrac{5}{2xy}$ $\dfrac{4y^2 + 6x^2 - 5xy}{2x^2y^2}$

8. Simplify: $\dfrac{3x}{x-2} - 3 + \dfrac{4}{x+2}$ $\dfrac{2(5x+2)}{(x-2)(x+2)}$

9. Find the domain of $f(x) = \dfrac{3x^2 - x + 1}{x^2 - 9}$. $\{x \mid x \neq -3, 3\}$

10. Subtract: $\dfrac{x+2}{x^2 + 3x - 4} - \dfrac{2x}{x^2 - 1}$ $-\dfrac{x^2 + 5x - 2}{(x+1)(x+4)(x-1)}$

11. Simplify: $\dfrac{1 - \frac{1}{x} - \frac{12}{x^2}}{1 + \frac{6}{x} + \frac{9}{x^2}}$ $\dfrac{x-4}{x+3}$

12. Simplify: $\dfrac{1 - \frac{1}{x+2}}{1 - \frac{3}{x+4}}$ $\dfrac{x+4}{x+2}$

13. Solve: $\dfrac{3}{x+1} = \dfrac{2}{x}$ 2

14. Solve: $\dfrac{4x}{2x-1} = 2 - \dfrac{1}{2x-1}$ No solution

15. Solve $ax = bx + c$ for x. $x = \frac{c}{a - b}$

16. The inlet pipe can fill a water tank in 48 min. The outlet pipe can empty the tank in 30 min. How long would it take to empty a full tank with both pipes open? 80 min

17. An interior designer uses two rolls of wallpaper for every 45 ft^2 of wall space in an office. At this rate, how many rolls of wallpaper are needed for an office that has 315 ft^2 of wall space? 14 rolls

18. One landscaper can till the soil for a lawn in 30 min, whereas it takes a second landscaper 15 min to do the same job. How long would it take to till the soil for the lawn if both landscapers worked together? 10 min

19. A cyclist travels 20 mi in the same amount of time it takes a hiker to walk 6 mi. The rate of the cyclist is 7 mph greater than the rate of the hiker. Find the rate of the cyclist. 10 mph

20. The stopping distance (s) of a car varies directly as the square of the speed (v) of the car. For a car traveling at 50 mph, the stopping distance is 170 ft. Find the stopping distance of a car that is traveling at 30 mph. 61.2 ft

Cumulative Review Exercises

1. Simplify: $8 - 4[-3 - (-2)]^2 \div 5$ $\frac{36}{5}$

2. Solve: $\frac{2x - 3}{6} - \frac{x}{9} = \frac{x - 4}{3}$ $\frac{15}{2}$

3. Solve: $5 - |x - 4| = 2$ 7, 1

4. Find the domain of $f(x) = \frac{x}{x - 3}$. $\{x \mid x \neq 3\}$

5. Given $P(x) = \frac{x - 1}{2x - 3}$, find $P(-2)$. $\frac{3}{7}$

6. Write 0.000000035 in scientific notation. 3.5×10^{-8}

7. Solve: $\frac{x}{x + 1} = 1$ No solution

8. Solve: $(9x - 1)(x - 4) = 0$ $\frac{1}{9}, 4$

9. Simplify: $\frac{(2a^{-2}b^3)}{(4a)^{-1}}$ $\frac{8b^3}{a}$

10. Solve: $x - 3(1 - 2x) \geq 1 - 4(2 - 2x)$ $\{x \mid x \leq 4\}$

11. Multiply: $(2a^2 - 3a + 1)(-2a^2)$ $-4a^4 + 6a^3 - 2a^2$

12. Factor: $2x^{2n} + 3x^n - 2$ $(2x^n - 1)(x^n + 2)$

13. Factor: $x^3y^3 - 27$ $(xy - 3)(x^2y^2 + 3xy + 9)$

14. Simplify: $\frac{x^4 + x^3y - 6x^2y^2}{x^3 - 2x^2y}$ $x + 3y$

15. Find the equation of the line that contains the point $(-2, -1)$ and is parallel to the line $3x - 2y = 6$. $y = \frac{3}{2}x + 2$

16. Simplify: $(x - x^{-1})^{-1}$ $\frac{x}{x^2 - 1}$

17. Graph $-3x + 5y = -15$ by using the x- and y-intercepts.

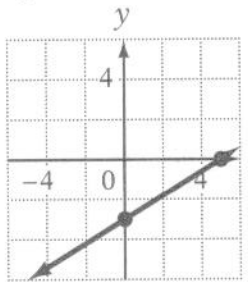

18. Graph the solution set: $x + y \leq 3$, $-2x + y > 4$

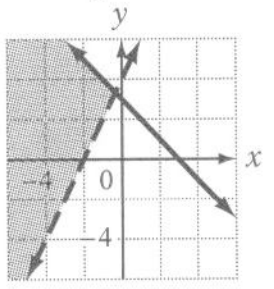

19. Divide: $\dfrac{4x^3 + 2x^2 - 10x + 1}{x - 2}$ $4x^2 + 10x + 10 + \dfrac{21}{x - 2}$

20. Divide: $\dfrac{16x^2 - 9y^2}{16x^2y - 12xy^2} \div \dfrac{4x^2 - xy - 3y^2}{12x^2y^2}$ $\dfrac{3xy}{x - y}$

21. Find the domain of $f(x) = \dfrac{x^2 + 6x + 7}{3x^2 + 5}$.
$\{x \mid x \in \text{real numbers}\}$

22. Subtract: $\dfrac{5x}{3x^2 - x - 2} - \dfrac{2x}{x^2 - 1}$ $\dfrac{x}{(3x + 2)(x + 1)}$

23. Evaluate the determinant: $\begin{vmatrix} 6 & 5 \\ 2 & -3 \end{vmatrix}$ -28

24. Simplify: $\dfrac{x - 4 + \dfrac{5}{x + 2}}{x + 2 - \dfrac{1}{x + 2}}$ $\dfrac{x - 3}{x + 3}$

25. Solve:
$$\begin{aligned} x + y + z &= 3 \\ -2x + y + 3z &= 2 \\ 2x - 4y + z &= -1 \end{aligned}$$
$(1, 1, 1)$

26. Given $f(x) = x^2 - 3x + 3$ and $f(c) = 1$, find c.
1, 2

27. Solve the proportion: $\dfrac{2}{x - 3} = \dfrac{5}{2x - 3}$ 9

28. Solve: $\dfrac{3}{x^2 - 36} = \dfrac{2}{x - 6} - \dfrac{5}{x + 6}$ 13

29. Multiply: $(a + 5)(a^3 - 3a + 4)$
$a^4 + 5a^3 - 3a^2 - 11a + 20$

30. Solve $I = \dfrac{E}{R + r}$ for r. $r = \dfrac{E - IR}{I}$

31. Are the graphs of $4x + 3y = 12$ and $3x + 4y = 16$ perpendicular? No

32. Graph: $f(x) = 1 - x + x^3$

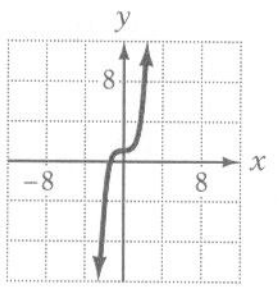

33. The sum of two integers is 15. Five times the smaller integer is five more than twice the larger integer. Find the integers. smaller integer: 5; larger integer: 10

34. How many pounds of almonds that cost \$5.40 per pound must be mixed with 50 lb of peanuts that cost \$2.60 per pound to make a mixture that costs \$4.00 per pound? 50 lb

35. A pre-election survey showed that three out of five voters would vote in an election. At this rate, how many people would be expected to vote in a city of 125,000? 75,000 people

36. A new computer can work six times as fast as an older computer. Working together, the computers can complete a job in 12 min. How long would it take the new computer, working alone, to do the job? 14 min

37. A plane can fly at a rate of 300 mph in calm air. Traveling with the wind, the plane flew 900 mi in the same amount of time it took to fly 600 mi against the wind. Find the rate of the wind. 60 mph

38. The frequency of vibration (f) in an open pipe organ varies inversely as the length (L) of the pipe. If the air in a pipe 2 m long vibrates 60 times per minute, find the frequency in a pipe that is 1.5 m long. 80 vibrations/min

7

Rational Exponents and Radicals

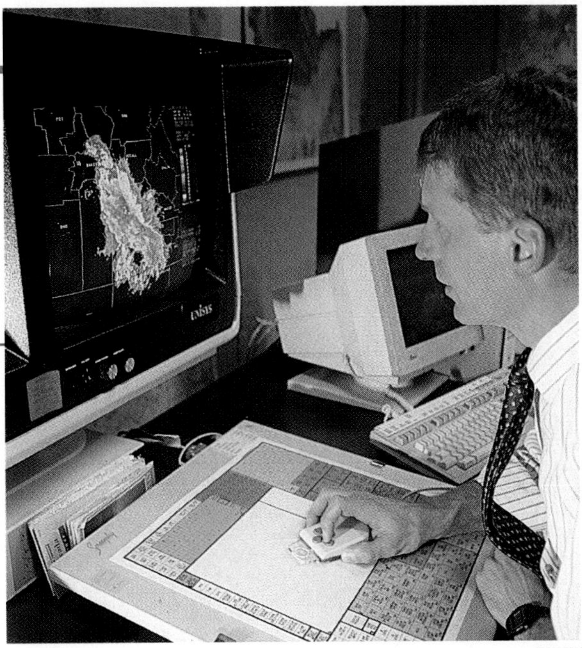

Meteorologists use data such as air pressure, temperature, humidity, and wind velocity to predict the weather. Today meteorologists are aided by computers and weather satellites. For example, weather satellites can measure the diameter of a storm. Then a meteorologist can use the formula $t = \left(\dfrac{d}{2}\right)^{\frac{3}{2}}$, where d is the diameter of a storm in miles, to approximate t, the duration of the storm in hours. The expression $t = \left(\dfrac{d}{2}\right)^{\frac{3}{2}}$ is an exponential expression with a rational exponent.

OBJECTIVES

Golden Ratio

The golden rectangle fascinated the early Greeks and appeared in much of their architecture. They considered this particular rectangle to be the most pleasing to the eye and thought that consequently, when used in the design of a building, it would make the structure pleasant to see.

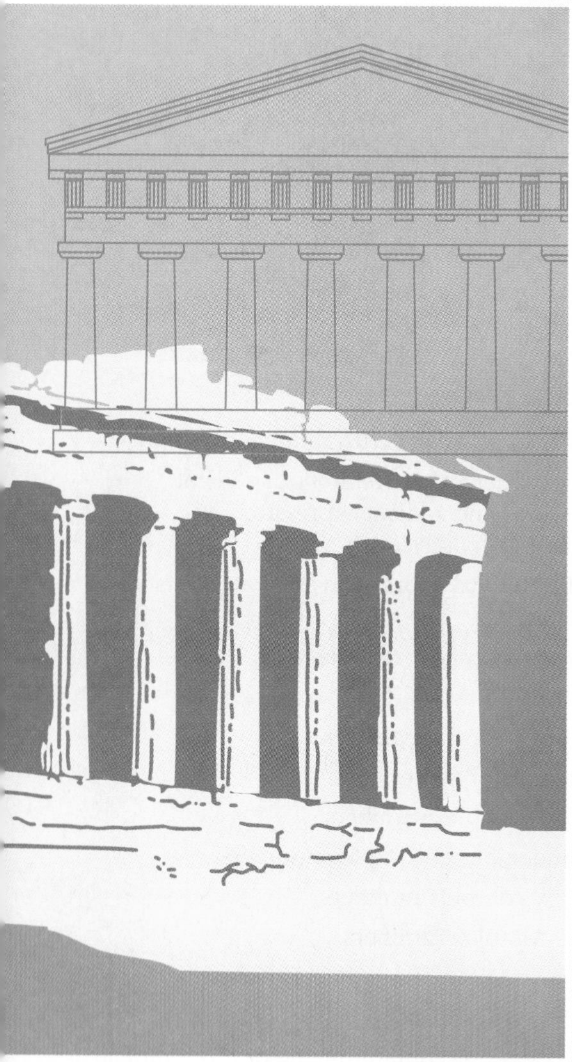

The golden rectangle is constructed from a square by drawing a line from the midpoint of the base of the square to the opposite vertex. Now extend the base of the square, starting from the midpoint, the length of the line. The resulting rectangle is called the golden rectangle.

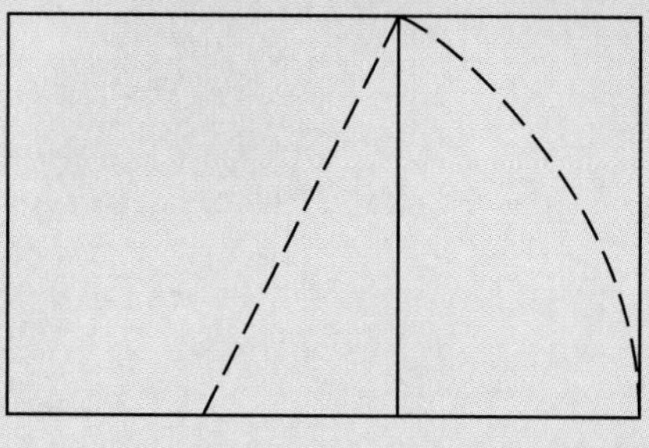

The Parthenon in Athens, Greece, is the classic example of the use of the golden rectangle in Greek architecture. A rendering of the Parthenon is shown here.

SECTION 7.1 Rational Exponents and Radical Expressions

1 Simplify expressions with rational exponents

27 HM[3]

POINT OF INTEREST

Nicolas Chuquet (ca. 1475), a French physician, wrote an algebra text in which he used a notation for expressions with fractional exponents. He wrote R^26 to mean $6^{1/2}$ and R^315 to mean $15^{1/3}$. This was an improvement over earlier notations that used words for these expressions.

INSTRUCTOR NOTE

The introduction to rational exponents excludes complex numbers. These numbers are discussed later in the chapter.

In this section, the definition of an exponent is extended beyond integers so that any rational number can be used as an exponent. The definition is expressed in such a way that the Rules of Exponents hold true for rational exponents.

Consider the expression $\left(a^{\frac{1}{n}}\right)^n$ for $a > 0$ and n a positive integer. Now simplify, assuming that the Rule for Simplifying a Power of an Exponential Expression is true.

$$\left(a^{\frac{1}{n}}\right)^n = a^{\frac{1}{n}\cdot n} = a^1 = a$$

Because $\left(a^{\frac{1}{n}}\right)^n = a$, the number $a^{\frac{1}{n}}$ is the number whose nth power is a.

$$25^{\frac{1}{2}} = 5 \text{ because } (5)^2 = 25.$$

$$8^{\frac{1}{3}} = 2 \text{ because } (2)^3 = 8.$$

In the expression $a^{\frac{1}{n}}$, if a is a negative number and n is a positive even integer, then $a^{\frac{1}{n}}$ is not a real number.

$(-4)^{\frac{1}{2}}$ is not a real number, because there is no real number whose second power is -4.

When n is a positive odd integer, a can be a positive or a negative number.

$$(-27)^{\frac{1}{3}} = -3 \text{ because } (-3)^3 = -27.$$

Using the definition of $a^{\frac{1}{n}}$ and the Rules of Exponents, it is possible to define any exponential expression that contains a rational exponent.

Definition of $a^{\frac{m}{n}}$

If m and n are positive integers and $a^{\frac{1}{n}}$ is a real number, then

$$a^{\frac{m}{n}} = \left(a^{\frac{1}{n}}\right)^m$$

The expression $a^{\frac{m}{n}}$ can also be written $a^{\frac{m}{n}} = a^{m\cdot\frac{1}{n}} = (a^m)^{\frac{1}{n}}$.

As shown above, expressions that contain rational exponents do not always represent real numbers when the base of the exponential expression is a negative number. For this reason, all variables in this chapter represent positive numbers unless otherwise stated.

Example 1   Simplify.   A. $27^{\frac{2}{3}}$ B. $32^{-\frac{2}{5}}$ C. $(-49)^{\frac{3}{2}}$

Solution   A. $27^{\frac{2}{3}} = (3^3)^{\frac{2}{3}}$ ▶ Rewrite 27 as 3^3.

$\qquad = 3^{3\left(\frac{2}{3}\right)}$ ▶ Use the Rule for Simplifying a Power of an Exponential Expression.

$\qquad = 3^2$

$\qquad = 9$ ▶ Simplify.

B. $32^{-\frac{2}{5}} = (2^5)^{-\frac{2}{5}}$ ▶ Rewrite 32 as 2^5.

$\qquad = 2^{-2}$ ▶ Use the Rule for Simplifying a Power of an Exponential Expression.

$\qquad = \frac{1}{2^2}$ ▶ Use the Rule of Negative Exponents.

$\qquad = \frac{1}{4}$ ▶ Simplify.

C. $(-49)^{\frac{3}{2}}$ ▶ The base of the exponential expression is a negative number, while the denominator of the exponent is a positive even number.

$\qquad (-49)^{\frac{3}{2}}$ is not a real number.

Problem 1   Simplify.   A. $64^{\frac{2}{3}}$ B. $16^{-\frac{3}{4}}$ C. $(-81)^{\frac{3}{4}}$

Solution   See page S19.   A. 16   B. $\frac{1}{8}$ C. Not a real number

Example 2   Simplify.   A. $b^{\frac{1}{2}} \cdot b^{\frac{2}{3}} \cdot b^{-\frac{1}{4}}$ B. $(x^6 y^4)^{\frac{3}{2}}$ C. $\left(\dfrac{8a^3 b^{-4}}{64a^{-9} b^2}\right)^{\frac{2}{3}}$

Solution   A. $b^{\frac{1}{2}} \cdot b^{\frac{2}{3}} \cdot b^{-\frac{1}{4}} = b^{\frac{1}{2}+\frac{2}{3}-\frac{1}{4}}$ ▶ Use the Rule for Multiplying Exponential Expressions.

$\qquad = b^{\frac{6}{12}+\frac{8}{12}-\frac{3}{12}}$

$\qquad = b^{\frac{11}{12}}$

B. $(x^6 y^4)^{\frac{3}{2}} = x^{6\left(\frac{3}{2}\right)} y^{4\left(\frac{3}{2}\right)}$ ▶ Use the Rule for Simplifying Powers of Products.

$\qquad = x^9 y^6$

C. $\left(\dfrac{8a^3 b^{-4}}{64a^{-9} b^2}\right)^{\frac{2}{3}} = \left(\dfrac{a^{12}}{8b^6}\right)^{\frac{2}{3}}$ ▶ Use the Rule for Dividing Exponential Expressions.

$\qquad = \left(\dfrac{a^{12}}{2^3 b^6}\right)^{\frac{2}{3}}$ ▶ Rewrite 8 as 2^3.

$\qquad = \dfrac{a^8}{2^2 b^4}$ ▶ Use the Rule for Simplifying Powers of Quotients.

$\qquad = \dfrac{a^8}{4b^4}$

Problem 2   Simplify.   A. $\dfrac{x^{\frac{1}{2}} y^{-\frac{5}{4}}}{x^{-\frac{4}{3}} y^{\frac{1}{3}}}$ B. $\left(x^{\frac{3}{4}} y^{\frac{1}{2}} z^{-\frac{2}{3}}\right)^{-\frac{4}{3}}$ C. $\left(\dfrac{16a^{-2} b^{\frac{4}{3}}}{9a^4 b^{-\frac{2}{3}}}\right)^{-\frac{1}{2}}$

Solution   See page S20.   A. $\dfrac{x^{\frac{11}{6}}}{y^{\frac{19}{12}}}$ B. $\dfrac{z^{\frac{8}{9}}}{xy^{\frac{2}{3}}}$ C. $\dfrac{3a^3}{4b}$

2 Write exponential expressions as radical expressions and radical expressions as exponential expressions

27 HM³

POINT OF INTEREST

The radical sign was introduced in 1525 in a book by Christoff Rudolff called *Coss.* He modified the symbol to indicate square roots, cube roots, and fourth roots. The idea of using an index, as we do in our modern notation, did not occur until some years later.

Recall that $a^{\frac{1}{n}}$ is the **nth root** of a. The expression $\sqrt[n]{a}$ is another symbol for the nth root of a.

Alternative notation for the nth root of a

If a is a real number and n is a positive integer, then $a^{\frac{1}{n}} = \sqrt[n]{a}$.

In the expression $\sqrt[n]{a}$, the symbol $\sqrt{\ }$ is called a **radical,** n is the **index** of the radical, and a is the **radicand.** When $n = 2$, the radical expression represents a square root and the index 2 is usually not written.

An exponential expression with a rational exponent can be written as a radical expression.

Definition of the nth root of a^m

If $a^{\frac{1}{n}}$ is a real number, then $a^{\frac{m}{n}} = a^{m \cdot \frac{1}{n}} = (a^m)^{\frac{1}{n}} = \sqrt[n]{a^m}$.

The expression $a^{\frac{m}{n}}$ can also be written $a^{\frac{m}{n}} = a^{\frac{1}{n} \cdot m} = (\sqrt[n]{a})^m$.

The exponential expression at the right has been written as a radical expression.

$$y^{\frac{2}{3}} = (y^2)^{\frac{1}{3}} = \sqrt[3]{y^2}$$

The radical expressions at the right have been written as exponential expressions.

$$\sqrt[5]{x^6} = (x^6)^{\frac{1}{5}} = x^{\frac{6}{5}}$$

$$\sqrt{17} = (17)^{\frac{1}{2}} = 17^{\frac{1}{2}}$$

➡ Write $(5x)^{\frac{2}{5}}$ as a radical expression.

The denominator of the rational exponent is the index of the radical. The numerator is the power of the radicand.

$$(5x)^{\frac{2}{5}} = \sqrt[5]{(5x)^2}$$

Simplify.

$$= \sqrt[5]{25x^2}$$

➡ Write $\sqrt[3]{x^4}$ as an exponential expression with a rational exponent.

The index of the radical is the denominator of the rational exponent. The power of the radicand is the numerator of the rational exponent.

$$\sqrt[3]{x^4} = (x^4)^{\frac{1}{3}} = x^{\frac{4}{3}}$$

Example 3   Rewrite the exponential expression as a radical expression.

A. $(3x)^{\frac{2}{3}}$ B. $-2x^{\frac{2}{3}}$

Solution   A. $(3x)^{\frac{2}{3}} = \sqrt[3]{(3x)^2}$ ▶ The denominator of the rational exponent is the index of the radical. The numerator is the power of the radicand.

$\qquad\qquad = \sqrt[3]{9x^2}$

B. $-2x^{\frac{2}{3}} = -2(x^2)^{\frac{1}{3}}$ ▶ The -2 is not raised to the power.

$\qquad\qquad = -2\sqrt[3]{x^2}$

Problem 3   Rewrite the exponential expression as a radical expression.

A. $(2x^3)^{\frac{3}{4}}$ B. $-5a^{\frac{5}{6}}$

Solution   See page S20.   A. $\sqrt[4]{8x^9}$ B. $-5\sqrt[6]{a^5}$

Example 4   Rewrite the radical expression as an exponential expression.

A. $\sqrt[7]{x^5}$ B. $\sqrt[3]{a^3 + b^3}$

Solution   A. $\sqrt[7]{x^5} = (x^5)^{\frac{1}{7}} = x^{\frac{5}{7}}$ ▶ The index of the radical is the denominator of the rational exponent. The power of the radicand is the numerator of the rational exponent.

B. $\sqrt[3]{a^3 + b^3} = (a^3 + b^3)^{\frac{1}{3}}$ ▶ Note that $(a^3 + b^3)^{\frac{1}{3}} \neq a + b$.

Problem 4   Rewrite the radical expression as an exponential expression.

A. $\sqrt[3]{3ab}$ B. $\sqrt[4]{x^4 + y^4}$

Solution   See page S20.   A. $(3ab)^{\frac{1}{3}}$ B. $(x^4 + y^4)^{\frac{1}{4}}$

---

3   ## Simplify radical expressions that are roots of perfect powers

Every positive number has two square roots, one a positive and one a negative number. For example, because $(5)^2 = 25$ and $(-5)^2 = 25$, there are two square roots of 25: 5 and -5.

The symbol $\sqrt{\ }$ is used to indicate the positive or **principal square root.** To indicate the negative square root of a number, a negative sign is placed in front of the radical.

$\sqrt{25} = 5$

$-\sqrt{25} = -5$

The square root of zero is zero.

$\sqrt{0} = 0$

The square root of a negative number is not a real number, because the square of a real number must be positive.

$\sqrt{-25}$ is not a real number.

The square root of a squared negative number is a positive number.

$\sqrt{(-5)^2} = \sqrt{25} = 5$

For any real number a, $\sqrt{a^2} = |a|$ and $-\sqrt{a^2} = -|a|$. If a is a positive real number, then $\sqrt{a^2} = a$ and $(\sqrt{a})^2 = a$.

The cube root of a positive number is positive.

$\sqrt[3]{8} = 2$, because $2^3 = 8$.

The cube root of a negative number is negative.

$\sqrt[3]{-8} = -2$, because $(-2)^3 = -8$.

For any real number a, $\sqrt[3]{a^3} = a$.

The following holds true for finding the nth root of a real number.

If n is an even integer, then $\sqrt[n]{a^n} = |a|$ and $-\sqrt[n]{a^n} = -|a|$. If n is an odd integer, then $\sqrt[n]{a^n} = a$.

For example,

$$\sqrt[6]{y^6} = |y| \qquad -\sqrt[12]{x^{12}} = -|x| \qquad \sqrt[5]{b^5} = b$$

We have agreed that all variables in this chapter represent positive numbers unless otherwise stated, so it is not necessary to use the absolute value signs.

⇒ Simplify: $\sqrt[3]{x^6y^9}$

The radicand of the radical expression $\sqrt[3]{x^6y^9}$ is a perfect cube because the exponents on the variables are divisible by 3. Write the radical expression as an exponential expression.

$$\sqrt[3]{x^6y^9} = (x^6y^9)^{\frac{1}{3}}$$

Use the Rule for Simplifying Powers of Products.

$$= x^2y^3 \qquad \blacktriangleleft$$

Example 5   Simplify.   A. $\sqrt[5]{x^{15}}$ B. $\sqrt{49x^2y^{12}}$ C. $-\sqrt[4]{16a^4b^8}$

Solution   A. $\sqrt[5]{x^{15}} = (x^{15})^{\frac{1}{5}}$

▶ The radicand is a perfect fifth power because 15 is divisible by 5. Write the radical expression as an exponential expression.

$$= x^3$$

▶ Use the Rule for Simplifying Powers of Exponential Expressions.

B. $\sqrt{49x^2y^{12}} = \sqrt{7^2x^2y^{12}}$

▶ Write the prime factorization of 49.

$$= 7xy^6$$

C. $-\sqrt[4]{16a^4b^8} = -\sqrt[4]{2^4a^4b^8}$
$$= -2ab^2$$

Problem 5   Simplify.   A. $-\sqrt[4]{x^{12}}$ B. $\sqrt{121x^{10}y^4}$ C. $\sqrt[3]{-125a^6b^9}$

Solution   See page S20.   A. $-x^3$ B. $11x^5y^2$ C. $-5a^2b^3$

CONCEPT REVIEW 7.1

Determine whether the following statements are always true, sometimes true, or never true.

1. For real numbers, $\sqrt{x^2} = x$.   Sometimes true

2. $\sqrt{(-2)^2} = -2$ Never true

3. $\sqrt[3]{(-3)^3} = -3$ Always true

4. For positive integers a and n, $\sqrt[n]{a} = a^{\frac{1}{n}}$. Always true

5. For positive integers a, m, and n, $(a^m)^{\frac{1}{n}} = a^{\frac{m}{n}}$. Always true

6. The nth root of a negative number is negative. Sometimes true

EXERCISES 7.1

1 1. What is the nth root of a number?

2. What is the meaning of the exponential expression $a^{\frac{m}{n}}$?

Simplify.

3. $8^{\frac{1}{3}}$ 2

4. $16^{\frac{1}{2}}$ 4

5. $9^{\frac{3}{2}}$ 27

6. $25^{\frac{3}{2}}$ 125

7. $27^{-\frac{2}{3}}$ $\frac{1}{9}$

8. $64^{-\frac{1}{3}}$ $\frac{1}{4}$

9. $32^{\frac{2}{5}}$ 4

10. $16^{\frac{3}{4}}$ 8

11. $(-25)^{\frac{5}{2}}$ Not a real number

12. $(-36)^{\frac{1}{4}}$ Not a real number

13. $\left(\frac{25}{49}\right)^{-\frac{3}{2}}$ $\frac{343}{125}$

14. $\left(\frac{8}{27}\right)^{-\frac{2}{3}}$ $\frac{9}{4}$

15. $x^{\frac{1}{2}}x^{\frac{1}{2}}$ x

16. $a^{\frac{1}{3}}a^{\frac{5}{3}}$ a^2

17. $y^{-\frac{1}{4}}y^{\frac{3}{4}}$ $y^{\frac{1}{2}}$

18. $x^{\frac{2}{5}} \cdot x^{-\frac{4}{5}}$ $\frac{1}{x^{\frac{2}{5}}}$

19. $x^{-\frac{2}{3}} \cdot x^{\frac{3}{4}}$ $x^{\frac{1}{12}}$

20. $x \cdot x^{-\frac{1}{2}}$ $x^{\frac{1}{2}}$

21. $a^{\frac{1}{3}} \cdot a^{\frac{3}{4}} \cdot a^{-\frac{1}{2}}$ $a^{\frac{7}{12}}$

22. $y^{-\frac{1}{6}} \cdot y^{\frac{2}{3}} \cdot y^{\frac{1}{2}}$ y

23. $\frac{a^{\frac{1}{2}}}{a^{\frac{3}{2}}}$ $\frac{1}{a}$

24. $\frac{b^{\frac{1}{3}}}{b^{\frac{4}{3}}}$ $\frac{1}{b}$

25. $\frac{y^{-\frac{3}{4}}}{y^{\frac{1}{4}}}$ $\frac{1}{y}$

26. $\frac{x^{-\frac{3}{5}}}{x^{\frac{1}{5}}}$ $\frac{1}{x^{\frac{4}{5}}}$

27. $\frac{y^{\frac{2}{3}}}{y^{-\frac{5}{6}}}$ $y^{\frac{3}{2}}$

28. $\frac{b^{\frac{3}{4}}}{b^{-\frac{3}{2}}}$ $b^{\frac{9}{4}}$

29. $(x^2)^{-\frac{1}{2}}$ $\frac{1}{x}$

30. $(a^8)^{-\frac{3}{4}}$ $\frac{1}{a^6}$

31. $\left(x^{-\frac{2}{3}}\right)^6$ $\frac{1}{x^4}$

32. $\left(y^{-\frac{5}{6}}\right)^{12}$ $\frac{1}{y^{10}}$

33. $\left(a^{-\frac{1}{2}}\right)^{-2}$ a

34. $\left(b^{-\frac{2}{3}}\right)^{-6}$ b^4

35. $\left(x^{-\frac{3}{8}}\right)^{-\frac{4}{5}}$ $x^{\frac{3}{10}}$

36. $\left(y^{-\frac{3}{2}}\right)^{-\frac{2}{9}}$ $y^{\frac{1}{3}}$

37. $\left(a^{\frac{1}{2}} \cdot a\right)^2$ a^3

38. $\left(b^{\frac{2}{3}} \cdot b^{\frac{1}{6}}\right)^6$ b^5

39. $\left(x^{-\frac{1}{2}} \cdot x^{\frac{3}{4}}\right)^{-2}$ $\frac{1}{x^{\frac{1}{2}}}$

40. $\left(a^{\frac{1}{2}} \cdot a^{-2}\right)^3$ $\frac{1}{a^{\frac{9}{2}}}$

41. $\left(y^{-\frac{1}{2}} \cdot y^{\frac{3}{2}}\right)^{\frac{2}{3}}$ $y^{\frac{2}{3}}$

42. $\left(b^{-\frac{2}{3}} \cdot b^{\frac{1}{4}}\right)^{-\frac{4}{3}}$ $b^{\frac{5}{9}}$

43. $(x^8y^2)^{\frac{1}{2}}$ x^4y

44. $(a^3b^9)^{\frac{2}{3}}$ a^2b^6

45. $(x^4y^2z^6)^{\frac{3}{2}}$ $x^6y^3z^9$

46. $(a^8b^4c^4)^{\frac{3}{4}}$ $a^6b^3c^3$

47. $(x^{-3}y^6)^{-\frac{1}{3}}$ $\frac{x}{y^2}$

48. $(a^2b^{-6})^{-\frac{1}{2}}$ $\frac{b^3}{a}$

49. $\left(x^{-2}y^{\frac{1}{3}}\right)^{-\frac{3}{4}}$ $\frac{x^{\frac{3}{2}}}{y^{\frac{1}{4}}}$

50. $\left(a^{-\frac{2}{3}}b^{\frac{2}{3}}\right)^{\frac{3}{2}}$ $\frac{b}{a}$

51. $\left(\dfrac{x^{\frac{1}{2}}}{y^{-2}}\right)^4$ x^2y^8

52. $\left(\dfrac{b^{-\frac{3}{4}}}{a^{-\frac{1}{2}}}\right)^8$ $\dfrac{a^4}{b^6}$

53. $\dfrac{x^{\frac{1}{4}} \cdot x^{-\frac{1}{2}}}{x^{\frac{2}{3}}}$ $\dfrac{1}{x^{\frac{11}{12}}}$

54. $\dfrac{b^{\frac{1}{2}} \cdot b^{-\frac{3}{4}}}{b^{\frac{1}{4}}}$ $\dfrac{1}{b^{\frac{1}{2}}}$

55. $\left(\dfrac{y^{\frac{2}{3}} \cdot y^{-\frac{5}{6}}}{y^{\frac{1}{9}}}\right)^9$ $\dfrac{1}{y^{\frac{5}{2}}}$

56. $\left(\dfrac{a^{\frac{1}{3}} \cdot a^{-\frac{2}{3}}}{a^{\frac{1}{2}}}\right)^4$ $\dfrac{1}{a^{\frac{10}{3}}}$

57. $\left(\dfrac{b^2 \cdot b^{-\frac{3}{4}}}{b^{-\frac{1}{2}}}\right)^{-\frac{1}{2}}$ $\dfrac{1}{b^{\frac{7}{8}}}$

58. $\dfrac{\left(x^{-\frac{5}{6}} \cdot x^3\right)^{-\frac{2}{3}}}{x^{\frac{4}{3}}}$ $\dfrac{1}{x^{\frac{25}{9}}}$

59. $\left(a^{\frac{2}{3}}b^2\right)^6\left(a^3b^3\right)^{\frac{1}{3}}$ a^5b^{13}

60. $\left(x^3y^{-\frac{1}{2}}\right)^{-2}\left(x^{-3}y^2\right)^{\frac{1}{6}}$ $\dfrac{y^{\frac{4}{3}}}{x^{\frac{13}{2}}}$

61. $\left(16m^{-2}n^4\right)^{-\frac{1}{2}}\left(mn^{\frac{1}{2}}\right)$ $\dfrac{m^2}{4n^{\frac{3}{2}}}$

62. $\left(27m^3n^{-6}\right)^{\frac{1}{3}}\left(m^{-\frac{1}{3}}n^{\frac{5}{6}}\right)^6$ $\dfrac{3n^3}{m}$

63. $\left(\dfrac{x^{\frac{1}{2}}y^{-\frac{3}{4}}}{y^{\frac{2}{3}}}\right)^{-6}$ $\dfrac{y^{\frac{17}{2}}}{x^3}$

64. $\left(\dfrac{x^{\frac{1}{2}}y^{-\frac{5}{4}}}{y^{-\frac{3}{4}}}\right)^{-4}$ $\dfrac{y^2}{x^2}$

65. $\left(\dfrac{2^{-6}b^{-3}}{a^{-\frac{1}{2}}}\right)^{-\frac{2}{3}}$ $\dfrac{16b^2}{a^{\frac{1}{3}}}$

66. $\left(\dfrac{49c^{\frac{5}{3}}}{a^{-\frac{1}{4}}b^{\frac{5}{6}}}\right)^{-\frac{3}{2}}$ $\dfrac{b^{\frac{5}{4}}}{343a^{\frac{3}{8}}c^{\frac{5}{2}}}$

67. $\dfrac{\left(x^{-2}y^4\right)^{\frac{1}{2}}}{\left(x^{\frac{1}{2}}\right)^4}$ $\dfrac{y^2}{x^3}$

68. $\dfrac{\left(x^{-3}\right)^{\frac{1}{3}}}{\left(x^9y^6\right)^{\frac{1}{6}}}$ $\dfrac{1}{x^{\frac{5}{2}}y}$

69. $a^{-\frac{1}{4}}\left(a^{\frac{5}{4}} - a^{\frac{9}{4}}\right)$ $a - a^2$

70. $x^{\frac{4}{3}}\left(x^{\frac{2}{3}} + x^{-\frac{1}{3}}\right)$ $x^2 + x$

71. $y^{\frac{2}{3}}\left(y^{\frac{1}{3}} + y^{-\frac{2}{3}}\right)$ $y + 1$

72. $b^{-\frac{2}{5}}\left(b^{-\frac{3}{5}} - b^{\frac{7}{5}}\right)$ $\dfrac{1}{b} - b$

73. $a^{\frac{1}{6}}\left(a^{\frac{5}{6}} - a^{-\frac{7}{6}}\right)$ $a - \dfrac{1}{a}$

74. $\left(x^{\frac{n}{3}}\right)^{3n}$ x^{n^2}

75. $\left(a^{\frac{2}{n}}\right)^{-5n}$ $\dfrac{1}{a^{10}}$

76. $x^n \cdot x^{\frac{n}{2}}$ $x^{\frac{3n}{2}}$

77. $a^{\frac{n}{2}} \cdot a^{-\frac{n}{3}}$ $a^{\frac{n}{6}}$

78. $\dfrac{y^{\frac{n}{2}}}{y^{-n}}$ $y^{\frac{3n}{2}}$

79. $\dfrac{b^{\frac{m}{3}}}{b^m}$ $\dfrac{1}{b^{\frac{2m}{3}}}$

80. $\left(x^{\frac{2}{n}}\right)^n$ x^2

81. $(x^{5n})^{2n}$ x^{10n^2}

82. $\left(x^{\frac{n}{4}}y^{\frac{n}{8}}\right)^8$ $x^{2n}y^n$

83. $\left(x^{\frac{n}{2}}y^{\frac{n}{3}}\right)^6$ $x^{3n}y^{2n}$

2 Rewrite the exponential expression as a radical expression.

84. $3^{\frac{1}{4}}$ $\sqrt[4]{3}$

85. $5^{\frac{1}{2}}$ $\sqrt{5}$

86. $a^{\frac{3}{2}}$ $\sqrt{a^3}$

87. $b^{\frac{4}{3}}$ $\sqrt[3]{b^4}$

88. $(2t)^{\frac{5}{2}}$ $\sqrt{32t^5}$

89. $(3x)^{\frac{2}{3}}$ $\sqrt[3]{9x^2}$

90. $-2x^{\frac{2}{3}}$ $-2\sqrt[3]{x^2}$

91. $-3a^{\frac{2}{5}}$ $-3\sqrt[5]{a^2}$

92. $(a^2b)^{\frac{2}{3}}$ $\sqrt[3]{a^4b^2}$

93. $(x^2y^3)^{\frac{3}{4}}$ $\sqrt[4]{x^6y^9}$

94. $(a^2b^4)^{\frac{3}{5}}$ $\sqrt[5]{a^6b^{12}}$

95. $(a^3b^7)^{\frac{3}{2}}$ $\sqrt{a^9b^{21}}$

96. $(4x + 3)^{\frac{3}{4}}$ $\sqrt[4]{(4x + 3)^3}$

97. $(3x - 2)^{\frac{1}{3}}$ $\sqrt[3]{3x - 2}$

98. $x^{-\frac{2}{3}}$ $\dfrac{1}{\sqrt[3]{x^2}}$

Rewrite the radical expression as an exponential expression.

99. $\sqrt{14}$ $14^{\frac{1}{2}}$

100. $\sqrt{7}$ $7^{\frac{1}{2}}$

101. $\sqrt[3]{x}$ $x^{\frac{1}{3}}$

102. $\sqrt[4]{y}$ $y^{\frac{1}{4}}$

103. $\sqrt[3]{x^4}$ $x^{\frac{4}{3}}$

104. $\sqrt[4]{a^3}$ $a^{\frac{3}{4}}$

105. $\sqrt[5]{b^3}$ $b^{\frac{3}{5}}$

106. $\sqrt[4]{b^5}$ $b^{\frac{5}{4}}$

107. $\sqrt[3]{2x^2}$ $(2x^2)^{\frac{1}{3}}$

108. $\sqrt[5]{4y^7}$ $(4y^7)^{\frac{1}{5}}$

109. $-\sqrt{3x^5}$ $-(3x^5)^{\frac{1}{2}}$

110. $-\sqrt[4]{4x^5}$ $-(4x^5)^{\frac{1}{4}}$

111. $3x\sqrt[3]{y^2}$ $3xy^{\frac{2}{3}}$

112. $2y\sqrt{x^3}$ $2yx^{\frac{3}{2}}$

113. $\sqrt{a^2 + 2}$ $(a^2 + 2)^{\frac{1}{2}}$

Simplify.

114. $\sqrt{x^{16}}$ x^8

115. $\sqrt{y^{14}}$ y^7

116. $-\sqrt{x^8}$ $-x^4$

117. $-\sqrt{a^6}$ $-a^3$

118. $\sqrt{x^2y^{10}}$ xy^5

119. $\sqrt{a^{14}b^6}$ a^7b^3

120. $\sqrt{25x^6}$ $5x^3$

121. $\sqrt{121y^{12}}$ $11y^6$

122. $\sqrt[3]{x^3y^9}$ xy^3

123. $\sqrt[3]{a^6b^{12}}$ a^2b^4

124. $-\sqrt[3]{x^{15}y^3}$ $-x^5y$

125. $-\sqrt[3]{a^9b^9}$ $-a^3b^3$

126. $\sqrt[3]{27a^9}$ $3a^3$

127. $\sqrt[3]{125b^{15}}$ $5b^5$

128. $\sqrt[3]{-8x^3}$ $-2x$

129. $\sqrt[3]{-a^6b^9}$ $-a^2b^3$

130. $\sqrt{16a^4b^{12}}$ $4a^2b^6$

131. $\sqrt{25x^8y^2}$ $5x^4y$

132. $\sqrt{-16x^4y^2}$ Not a real number

133. $\sqrt{-9a^6b^8}$ Not a real number

134. $\sqrt[3]{27x^9}$ $3x^3$

135. $\sqrt[3]{8a^{21}b^6}$ $2a^7b^2$

136. $\sqrt[3]{-64x^9y^{12}}$ $-4x^3y^4$

137. $\sqrt[3]{-27a^3b^{15}}$ $-3ab^5$

138. $\sqrt[4]{x^{16}}$ x^4

139. $\sqrt[4]{y^{12}}$ y^3

140. $\sqrt[4]{16x^{12}}$ $2x^3$

141. $\sqrt[4]{81a^{20}}$ $3a^5$

142. $-\sqrt[4]{x^8y^{12}}$ $-x^2y^3$

143. $-\sqrt[4]{a^{16}b^4}$ $-a^4b$

144. $\sqrt[5]{x^{20}y^{10}}$ x^4y^2

145. $\sqrt[5]{a^5b^{25}}$ ab^5

146. $\sqrt[4]{81x^4y^{20}}$ $3xy^5$

147. $\sqrt[4]{16a^8b^{20}}$ $2a^2b^5$

148. $\sqrt[5]{32a^5b^{10}}$ $2ab^2$

149. $\sqrt[5]{-32x^{15}y^{20}}$ $-2x^3y^4$

150. $\sqrt[5]{243x^{10}y^{40}}$ $3x^2y^8$

151. $\sqrt{\dfrac{16x^2}{y^{14}}}$ $\dfrac{4x}{y^7}$

152. $\sqrt{\dfrac{49a^4}{b^{24}}}$ $\dfrac{7a^2}{b^{12}}$

153. $\sqrt[3]{\dfrac{27b^3}{a^9}}$ $\dfrac{3b}{a^3}$

154. $\sqrt[3]{\dfrac{64x^{15}}{y^6}}$ $\dfrac{4x^5}{y^2}$

155. $\sqrt{(2x+3)^2}$ $2x+3$

156. $\sqrt{(4x+1)^2}$ $4x+1$

157. $\sqrt{x^2+2x+1}$ $x+1$

158. $\sqrt{x^2+4x+4}$ $x+2$

APPLYING CONCEPTS 7.1

Which of the numbers is the largest, **a**, **b**, or **c**?

159. **a.** $16^{\frac{1}{2}}$ **b.** $16^{\frac{1}{4}}$ **c.** $16^{\frac{3}{2}}$ c

160. **a.** $(-32)^{-\frac{1}{5}}$ **b.** $(-32)^{\frac{2}{5}}$ **c.** $(-32)^{\frac{1}{5}}$ b

161. **a.** $4^{\frac{1}{2}} \cdot 4^{\frac{3}{2}}$ **b.** $3^{\frac{4}{5}} \cdot 3^{\frac{6}{5}}$ **c.** $7^{\frac{1}{4}} \cdot 7^{\frac{3}{4}}$ a

162. **a.** $\dfrac{81^{\frac{3}{4}}}{81^{\frac{1}{4}}}$ **b.** $\dfrac{64^{\frac{2}{3}}}{64^{\frac{1}{3}}}$ **c.** $\dfrac{36^{\frac{5}{2}}}{36^{\frac{3}{2}}}$ c

Simplify.

163. $\sqrt[3]{\sqrt{x^6}}$ x

164. $\sqrt{\sqrt[3]{y^6}}$ y

165. $\sqrt[5]{\sqrt[3]{b^{15}}}$ b

166. $\sqrt{\sqrt{16x^{12}}}$ $2x^3$

167. $\sqrt[5]{\sqrt{a^{10}b^{20}}}$ ab^2

168. $\sqrt[3]{\sqrt{64x^{36}y^{30}}}$ $2x^6y^5$

For what value of p is the given equation true?

169. $y^p y^{\frac{2}{5}} = y$ $\frac{3}{5}$

170. $\dfrac{y^p}{y^{\frac{3}{4}}} = y^{\frac{1}{2}}$ $\frac{5}{4}$

171. $x^p x^{-\frac{1}{2}} = x^{\frac{1}{4}}$ $\frac{3}{4}$

SECTION 7.2

Operations on Radical Expressions

1   Simplify radical expressions

If a number is not a perfect power, its root can only be approximated; examples include $\sqrt{5}$ and $\sqrt[3]{3}$. These numbers are **irrational numbers.** Their decimal representations never terminate or repeat.

$$\sqrt{5} = 2.2360679\ldots \qquad \sqrt[3]{3} = 1.4422495\ldots$$

A radical expression is in simplest form when the radicand contains no factor that is a perfect power. The Product Property of Radicals is used to simplify radical expressions whose radicands are not perfect powers.

The Product Property of Radicals

If $\sqrt[n]{a}$ and $\sqrt[n]{b}$ are positive real numbers, then
$$\sqrt[n]{ab} = \sqrt[n]{a} \cdot \sqrt[n]{b} \quad \text{and} \quad \sqrt[n]{a} \cdot \sqrt[n]{b} = \sqrt[n]{ab}$$

➡ Simplify: $\sqrt{48}$

Write the prime factorization of the radicand in exponential form. $\sqrt{48} = \sqrt{2^4 \cdot 3}$

Use the Product Property of Radicals to write the expression as a product. $= \sqrt{2^4}\sqrt{3}$

Simplify. $= 2^2\sqrt{3}$
   $= 4\sqrt{3}$ ◀

➡ Simplify: $\sqrt[3]{x^7}$

Write the radicand as the product of a perfect cube and a factor that does not contain a perfect cube. $\sqrt[3]{x^7} = \sqrt[3]{x^6 \cdot x}$

Use the Product Property of Radicals to write the expression as a product. $= \sqrt[3]{x^6}\sqrt[3]{x}$

Simplify. $= x^2\sqrt[3]{x}$ ◀

Example 1 Simplify: $\sqrt[4]{32x^7}$

Solution $\sqrt[4]{32x^7} = \sqrt[4]{2^5x^7}$ ▶ Write the prime factorization of the coefficient of the radicand in exponential form.

$= \sqrt[4]{2^4x^4(2x^3)}$ ▶ Write the radicand as the product of a perfect fourth power and factors that do not contain a perfect fourth power.

$= \sqrt[4]{2^4x^4}\sqrt[4]{2x^3}$ ▶ Use the Product Property of Radicals to write the expression as a product.

$= 2x\sqrt[4]{2x^3}$ ▶ Simplify.

Problem 1 Simplify: $\sqrt[5]{128x^7}$

Solution See page S20. $2x\sqrt[5]{4x^2}$

2 Add and subtract radical expressions

INSTRUCTOR NOTE

Mention to students that adding and subtracting radicals is similar to combining like terms.

The Distributive Property is used to simplify the sum or difference of radical expressions that have the same radicand and the same index. For example,

$$3\sqrt{5} + 8\sqrt{5} = (3 + 8)\sqrt{5} = 11\sqrt{5}$$
$$2\sqrt[3]{3x} - 9\sqrt[3]{3x} = (2 - 9)\sqrt[3]{3x} = -7\sqrt[3]{3x}$$

Radical expressions that are in simplest form and have unlike radicands or different indices cannot be simplified by the Distributive Property. The following expressions cannot be simplified by the Distributive Property.

$$3\sqrt[4]{2} - 6\sqrt[4]{3}$$
$$2\sqrt[4]{4x} + 3\sqrt[3]{4x}$$

⇒ Simplify: $3\sqrt{32x^2} - 2x\sqrt{2} + \sqrt{128x^2}$

First simplify each term. Then combine like terms by using the Distributive Property.

$$\begin{aligned} 3\sqrt{32x^2} - 2x\sqrt{2} + \sqrt{128x^2} &= 3\sqrt{2^5x^2} - 2x\sqrt{2} + \sqrt{2^7x^2} \\ &= 3\sqrt{2^4x^2}\sqrt{2} - 2x\sqrt{2} + \sqrt{2^6x^2}\sqrt{2} \\ &= 3 \cdot 2^2x\sqrt{2} - 2x\sqrt{2} + 2^3x\sqrt{2} \\ &= 12x\sqrt{2} - 2x\sqrt{2} + 8x\sqrt{2} \\ &= 18x\sqrt{2} \end{aligned}$$

Example 2 Subtract.

A. $5b\sqrt[4]{32a^7b^5} - 2a\sqrt[4]{162a^3b^9}$ B. $5\sqrt[5]{2x^7y^{11}} - y\sqrt[5]{64x^7y^6}$

Solution A. $5b\sqrt[4]{32a^7b^5} - 2a\sqrt[4]{162a^3b^9}$

$$\begin{aligned} &= 5b\sqrt[4]{2^5a^7b^5} - 2a\sqrt[4]{3^4 \cdot 2a^3b^9} \\ &= 5b\sqrt[4]{2^4a^4b^4}\sqrt[4]{2a^3b} - 2a\sqrt[4]{3^4b^8}\sqrt[4]{2a^3b} \\ &= 5b \cdot 2ab\sqrt[4]{2a^3b} - 2a \cdot 3b^2\sqrt[4]{2a^3b} \\ &= 10ab^2\sqrt[4]{2a^3b} - 6ab^2\sqrt[4]{2a^3b} \\ &= 4ab^2\sqrt[4]{2a^3b} \end{aligned}$$

B. $5\sqrt[5]{2x^7y^{11}} - y\sqrt[5]{64x^7y^6} = 5\sqrt[5]{2x^7y^{11}} - y\sqrt[5]{2^6x^7y^6}$
$= 5\sqrt[5]{x^5y^{10}}\sqrt[5]{2x^2y} - y\sqrt[5]{2^5x^5y^5}\sqrt[5]{2x^2y}$
$= 5 \cdot xy^2\sqrt[5]{2x^2y} - y \cdot 2xy\sqrt[5]{2x^2y}$
$= 5xy^2\sqrt[5]{2x^2y} - 2xy^2\sqrt[5]{2x^2y}$
$= 3xy^2\sqrt[5]{2x^2y}$

Problem 2 Add or subtract.

A. $3xy\sqrt[3]{81x^5y} - \sqrt[3]{192x^8y^4}$ B. $4a\sqrt[3]{54a^7b^9} + a^2b\sqrt[3]{128a^4b^6}$

Solution See page S20. A. $5x^2y\sqrt[3]{3x^2y}$ B. $16a^3b^3\sqrt[3]{2a}$

3 Multiply radical expressions

The Product Property of Radicals is used to multiply radical expressions with the same index.

$$\sqrt{3x} \cdot \sqrt{5y} = \sqrt{3x \cdot 5y} = \sqrt{15xy}$$

Multiply: $\sqrt[3]{2a^5b}\sqrt[3]{16a^2b^2}$

Use the Product Property of Radicals to multiply the radicands. Then simplify.

$\sqrt[3]{2a^5b}\sqrt[3]{16a^2b^2} = \sqrt[3]{32a^7b^3}$
$= \sqrt[3]{2^5a^7b^3}$
$= \sqrt[3]{2^3a^6b^3}\sqrt[3]{2^2a}$
$= 2a^2b\sqrt[3]{4a}$

Multiply: $\sqrt{2x}(\sqrt{8x} - \sqrt{3})$

Use the Distributive Property to remove parentheses. Then simplify.

$\sqrt{2x}(\sqrt{8x} - \sqrt{3}) = \sqrt{16x^2} - \sqrt{6x}$
$= \sqrt{2^4x^2} - \sqrt{6x}$
$= 2^2x - \sqrt{6x}$
$= 4x - \sqrt{6x}$

Example 3 Multiply: $\sqrt{3x}(\sqrt{27x^2} - \sqrt{3x})$

Solution $\sqrt{3x}(\sqrt{27x^2} - \sqrt{3x}) = \sqrt{81x^3} - \sqrt{9x^2}$
$= \sqrt{3^4x^3} - \sqrt{3^2x^2}$
$= \sqrt{3^4x^2}\sqrt{x} - \sqrt{3^2x^2}$
$= 3^2x\sqrt{x} - 3x$
$= 9x\sqrt{x} - 3x$

Problem 3 Multiply: $\sqrt{5b}(\sqrt{3b} - \sqrt{10})$

Solution See page S20. $b\sqrt{15} - 5\sqrt{2b}$

LOOK CLOSELY

The concept of conjugate is used in a number of different instances. Make sure you understand this idea.

The conjugate of $\sqrt{3} - 4$ is $\sqrt{3} + 4$.

The conjugate of $\sqrt{3} + 4$ is $\sqrt{3} - 4$.

The conjugate of $\sqrt{5a} + \sqrt{b}$ is $\sqrt{5a} - \sqrt{b}$.

To multiply $(\sqrt[3]{x} - 1)(\sqrt[3]{x} + 7)$, use the FOIL method. Then combine like terms.

$$(\sqrt[3]{x} - 1)(\sqrt[3]{x} + 7) = \sqrt[3]{x^2} + 7\sqrt[3]{x} - \sqrt[3]{x} - 7$$
$$= \sqrt[3]{x^2} + 6\sqrt[3]{x} - 7$$

The expressions $a + b$ and $a - b$, which are the sum and difference of two terms, are called **conjugates** of each other. The product of conjugates of the form $(a + b)(a - b)$ is $a^2 - b^2$.

$$(\sqrt{x} - 3)(\sqrt{x} + 3) = (\sqrt{x})^2 - 3^2 = x - 9$$

Example 4 Multiply.

A. $(2\sqrt[3]{x} - 3)(3\sqrt[3]{x} - 4)$ B. $(\sqrt{xy} - 2)(\sqrt{xy} + 2)$

Solution A. $(2\sqrt[3]{x} - 3)(3\sqrt[3]{x} - 4)$ ▶ Use the FOIL method.

$= 6\sqrt[3]{x^2} - 8\sqrt[3]{x} - 9\sqrt[3]{x} + 12$

$= 6\sqrt[3]{x^2} - 17\sqrt[3]{x} + 12$

B. $(\sqrt{xy} - 2)(\sqrt{xy} + 2)$ ▶ Use $(a - b)(a + b) = a^2 - b^2$.

$= (\sqrt{xy})^2 - 2^2$

$= xy - 4$

Problem 4 Multiply.

A. $(2\sqrt[3]{2x} - 3)(\sqrt[3]{2x} - 5)$ B. $(2\sqrt{x} - 3)(2\sqrt{x} + 3)$

Solution See page S20. A. $2\sqrt[3]{4x^2} - 13\sqrt[3]{2x} + 15$ B. $4x - 9$

4 Divide radical expressions

29 HM³

The Quotient Property of Radicals is used to divide radical expressions with the same index.

The Quotient Property of Radicals

If $\sqrt[n]{a}$ and $\sqrt[n]{b}$ are real numbers, and $b \neq 0$, then

$$\sqrt[n]{\frac{a}{b}} = \frac{\sqrt[n]{a}}{\sqrt[n]{b}} \quad \text{and} \quad \frac{\sqrt[n]{a}}{\sqrt[n]{b}} = \sqrt[n]{\frac{a}{b}}$$

Simplify: $\sqrt[3]{\dfrac{81x^5}{y^6}}$

Use the Quotient Property of Radicals. Then simplify each radical expression.

$$\sqrt[3]{\frac{81x^5}{y^6}} = \frac{\sqrt[3]{81x^5}}{\sqrt[3]{y^6}} = \frac{\sqrt[3]{3^4x^5}}{\sqrt[3]{y^6}} = \frac{\sqrt[3]{3^3x^3}\sqrt[3]{3x^2}}{\sqrt[3]{y^6}} = \frac{3x\sqrt[3]{3x^2}}{y^2}$$

Simplify: $\dfrac{\sqrt{5a^4b^7c^2}}{\sqrt{ab^3c}}$

Use the Quotient Property of Radicals. Then simplify the radicand.

$$\frac{\sqrt{5a^4b^7c^2}}{\sqrt{ab^3c}} = \sqrt{\frac{5a^4b^7c^2}{ab^3c}} = \sqrt{5a^3b^4c} = \sqrt{a^2b^4}\sqrt{5ac} = ab^2\sqrt{5ac}$$

A radical expression is in simplest form when there is no fraction as part of the radicand and no radical remains in the denominator of the radical expression. The procedure used to remove a radical from the denominator is called **rationalizing the denominator.**

➡ Simplify: $\dfrac{2}{\sqrt{x}}$

Multiply the expression by 1 in the form $\dfrac{\sqrt{x}}{\sqrt{x}}$. $\dfrac{2}{\sqrt{x}} = \dfrac{2}{\sqrt{x}} \cdot \dfrac{\sqrt{x}}{\sqrt{x}} = \dfrac{2\sqrt{x}}{\sqrt{x^2}} = \dfrac{2\sqrt{x}}{x}$

Then simplify.

Example 5   Simplify.   A. $\dfrac{5}{\sqrt{5x}}$ B. $\dfrac{3x}{\sqrt[3]{4x}}$

Solution   A. $\dfrac{5}{\sqrt{5x}} = \dfrac{5}{\sqrt{5x}} \cdot \dfrac{\sqrt{5x}}{\sqrt{5x}}$ ▶ Multiply the expression by $\dfrac{\sqrt{5x}}{\sqrt{5x}}$.

$= \dfrac{5\sqrt{5x}}{\sqrt{5^2 x^2}}$

$= \dfrac{5\sqrt{5x}}{5x}$

$= \dfrac{\sqrt{5x}}{x}$

B. $\dfrac{3x}{\sqrt[3]{4x}} = \dfrac{3x}{\sqrt[3]{2^2 x}} \cdot \dfrac{\sqrt[3]{2x^2}}{\sqrt[3]{2x^2}}$ ▶ Multiply the expression by $\dfrac{\sqrt[3]{2x^2}}{\sqrt[3]{2x^2}}$.

$= \dfrac{3x\sqrt[3]{2x^2}}{\sqrt[3]{2^3 x^3}}$ $\sqrt[3]{4x} \cdot \sqrt[3]{2x^2} = \sqrt[3]{2^2 x} \cdot \sqrt[3]{2x^2} = \sqrt[3]{2^3 x^3}$, a perfect cube.

$= \dfrac{3x\sqrt[3]{2x^2}}{2x}$

$= \dfrac{3\sqrt[3]{2x^2}}{2}$

Problem 5   Simplify.   A. $\dfrac{y}{\sqrt{3y}}$ B. $\dfrac{3}{\sqrt[3]{3x^2}}$

Solution   See page S20.   A. $\dfrac{\sqrt{3y}}{3}$ B. $\dfrac{\sqrt[3]{9x}}{x}$

LOOK CLOSELY

Note that multiplying by $\dfrac{\sqrt[3]{4x}}{\sqrt[3]{4x}}$ will not rationalize the denominator of $\dfrac{3x}{\sqrt[3]{4x}}$.

$\dfrac{3x}{\sqrt[3]{4x}} \cdot \dfrac{\sqrt[3]{4x}}{\sqrt[3]{4x}} = \dfrac{3x\sqrt[3]{4x}}{\sqrt[3]{16x^2}}$

Because $\sqrt[3]{16x^2}$ is not a perfect cube, the denominator still contains a radical expression.

To simplify a fraction that has a square-root radical expression with two terms in the denominator, multiply the numerator and denominator by the conjugate of the denominator.

$$\dfrac{3}{5 - \sqrt{7}} = \dfrac{3}{5 - \sqrt{7}} \cdot \dfrac{5 + \sqrt{7}}{5 + \sqrt{7}} = \dfrac{15 + 3\sqrt{7}}{25 - 7} = \dfrac{3(5 + \sqrt{7})}{18} = \dfrac{5 + \sqrt{7}}{6}$$

$$\dfrac{\sqrt{x} - \sqrt{y}}{\sqrt{x} + \sqrt{y}} = \dfrac{\sqrt{x} - \sqrt{y}}{\sqrt{x} + \sqrt{y}} \cdot \dfrac{\sqrt{x} - \sqrt{y}}{\sqrt{x} - \sqrt{y}} = \dfrac{\sqrt{x^2} - \sqrt{xy} - \sqrt{xy} + \sqrt{y^2}}{(\sqrt{x})^2 - (\sqrt{y})^2} = \dfrac{x - 2\sqrt{xy} + y}{x - y}$$

LOOK CLOSELY

Here are two examples of using a conjugate to simplify a radical expression.

Example 6   Simplify.   A. $\dfrac{2 - \sqrt{5}}{3 + \sqrt{2}}$ B. $\dfrac{\sqrt{3} + \sqrt{y}}{\sqrt{3} - \sqrt{y}}$

Solution   A. $\dfrac{2 - \sqrt{5}}{3 + \sqrt{2}} = \dfrac{2 - \sqrt{5}}{3 + \sqrt{2}} \cdot \dfrac{3 - \sqrt{2}}{3 - \sqrt{2}} = \dfrac{6 - 2\sqrt{2} - 3\sqrt{5} + \sqrt{10}}{9 - 2}$

$= \dfrac{6 - 2\sqrt{2} - 3\sqrt{5} + \sqrt{10}}{7}$

B. $\dfrac{\sqrt{3}+\sqrt{y}}{\sqrt{3}-\sqrt{y}} = \dfrac{\sqrt{3}+\sqrt{y}}{\sqrt{3}-\sqrt{y}} \cdot \dfrac{\sqrt{3}+\sqrt{y}}{\sqrt{3}+\sqrt{y}} = \dfrac{\sqrt{3^2}+\sqrt{3y}+\sqrt{3y}+\sqrt{y^2}}{(\sqrt{3})^2-(\sqrt{y})^2}$

$= \dfrac{3+2\sqrt{3y}+y}{3-y}$

Problem 6 Simplify. A. $\dfrac{4+\sqrt{2}}{3-\sqrt{3}}$ B. $\dfrac{\sqrt{2}+\sqrt{x}}{\sqrt{2}-\sqrt{x}}$

Solution See page S20. A. $\dfrac{12+4\sqrt{3}+3\sqrt{2}+\sqrt{6}}{6}$ B. $\dfrac{2+2\sqrt{2x}+x}{2-x}$

CONCEPT REVIEW 7.2

Determine whether the following statements are always true, sometimes true, or never true.

1. If a is a real number, then $\sqrt{a}$ represents a real number. Sometimes true
2. If b is a real number, then $\sqrt[3]{b}$ represents a real number. Always true
3. For positive integers x and y, $\sqrt[a]{x} \cdot \sqrt[b]{y} = \sqrt[ab]{xy}$. Never true
4. The conjugate of $\sqrt{a} + \sqrt{b}$ is found by replacing b with its opposite. Never true
5. If $x < 0$, then $\sqrt{x^2} = -x$. Always true
6. For positive integers x and y, $\sqrt[a]{x} \cdot \sqrt[a]{y} = \sqrt[a]{xy}$. Always true

EXERCISES 7.2

1

1. What is an irrational number?
2. When is a radical expression in simplest form?

Simplify.

3. $\sqrt{18}$ $3\sqrt{2}$
4. $\sqrt{40}$ $2\sqrt{10}$
5. $\sqrt{98}$ $7\sqrt{2}$
6. $\sqrt{128}$ $8\sqrt{2}$
7. $\sqrt[3]{72}$ $2\sqrt[3]{9}$
8. $\sqrt[3]{54}$ $3\sqrt[3]{2}$
9. $\sqrt[3]{16}$ $2\sqrt[3]{2}$
10. $\sqrt[3]{128}$ $4\sqrt[3]{2}$
11. $\sqrt{x^4y^3z^5}$ $x^2yz^2\sqrt{yz}$
12. $\sqrt{x^3y^6z^9}$ $xy^3z^4\sqrt{xz}$
13. $\sqrt{8a^3b^8}$ $2ab^4\sqrt{2a}$
14. $\sqrt{24a^9b^6}$ $2a^4b^3\sqrt{6a}$
15. $\sqrt{45x^2y^3z^5}$ $3xyz^2\sqrt{5yz}$
16. $\sqrt{60xy^7z^{12}}$ $2y^3z^6\sqrt{15xy}$
17. $\sqrt[3]{-125x^2y^4}$ $-5y\sqrt[3]{x^2y}$
18. $\sqrt[4]{16x^9y^5}$ $2x^2y\sqrt[4]{xy}$
19. $\sqrt[3]{-216x^5y^9}$ $-6xy^3\sqrt[3]{x^2}$
20. $\sqrt[3]{a^8b^{11}c^{15}}$ $a^2b^3c^5\sqrt[3]{a^2b^2}$
21. $\sqrt[3]{a^5b^8}$ $ab^2\sqrt[3]{a^2b^2}$
22. $\sqrt[4]{64x^8y^{10}}$ $2x^2y^2\sqrt[4]{4y^2}$

2 Simplify.

23. $\sqrt{2} + \sqrt{2}$ $2\sqrt{2}$

24. $\sqrt{5} - \sqrt{5}$ 0

25. $4\sqrt[3]{7} - \sqrt[3]{7}$ $3\sqrt[3]{7}$

26. $3\sqrt[3]{11} - 8\sqrt[3]{11}$ $-5\sqrt[3]{11}$

27. $2\sqrt{x} - 8\sqrt{x}$ $-6\sqrt{x}$

28. $3\sqrt{y} + 12\sqrt{y}$ $15\sqrt{y}$

29. $\sqrt{8} - \sqrt{32}$ $-2\sqrt{2}$

30. $\sqrt{27} - \sqrt{75}$ $-2\sqrt{3}$

31. $\sqrt{128x} - \sqrt{98x}$ $\sqrt{2x}$

32. $\sqrt{48x} + \sqrt{147x}$ $11\sqrt{3x}$

33. $\sqrt{27a} - \sqrt{8a}$ $3\sqrt{3a} - 2\sqrt{2a}$

34. $\sqrt{18b} + \sqrt{75b}$ $3\sqrt{2b} + 5\sqrt{3b}$

35. $2\sqrt{2x^3} + 4x\sqrt{8x}$ $10x\sqrt{2x}$

36. $5y\sqrt{8y} + 2\sqrt{50y^3}$ $20y\sqrt{2y}$

37. $x\sqrt{75xy} - \sqrt{27x^3y}$ $2x\sqrt{3xy}$

38. $3\sqrt{8x^2y^3} - 2x\sqrt{32y^3}$ $-2xy\sqrt{2y}$

39. $2\sqrt{32x^2y^3} - xy\sqrt{98y}$ $xy\sqrt{2y}$

40. $6y\sqrt{x^3y} - 2\sqrt{x^3y^3}$ $4xy\sqrt{xy}$

41. $7b\sqrt{a^5b^3} - 2ab\sqrt{a^3b^3}$ $5a^2b^2\sqrt{ab}$

42. $2a\sqrt{27ab^5} + 3b\sqrt{3a^3b}$ $6ab^2\sqrt{3ab} + 3ab\sqrt{3ab}$

43. $\sqrt[3]{128} + \sqrt[3]{250}$ $9\sqrt[3]{2}$

44. $\sqrt[3]{16} - \sqrt[3]{54}$ $-\sqrt[3]{2}$

45. $2\sqrt[3]{3a^4} - 3a\sqrt[3]{81a}$ $-7a\sqrt[3]{3a}$

46. $2b\sqrt[3]{16b^2} + \sqrt[3]{128b^5}$ $8b\sqrt[3]{2b^2}$

47. $3\sqrt[3]{x^5y^7} - 8xy\sqrt[3]{x^2y^4}$ $-5xy^2\sqrt[3]{x^2y}$

48. $3\sqrt[4]{32a^5} - a\sqrt[4]{162a}$ $3a\sqrt[4]{2a}$

49. $2a\sqrt[4]{16ab^5} + 3b\sqrt[4]{256a^5b}$ $16ab\sqrt[4]{ab}$

50. $2\sqrt{50} - 3\sqrt{125} + \sqrt{98}$ $17\sqrt{2} - 15\sqrt{5}$

51. $3\sqrt{108} - 2\sqrt{18} - 3\sqrt{48}$ $6\sqrt{3} - 6\sqrt{2}$

52. $\sqrt{9b^3} - \sqrt{25b^3} + \sqrt{49b^3}$ $5b\sqrt{b}$

53. $\sqrt{4x^7y^5} + 9x^2\sqrt{x^3y^5} - 5xy\sqrt{x^5y^3}$ $6x^3y^2\sqrt{xy}$

54. $2x\sqrt{8xy^2} - 3y\sqrt{32x^3} + \sqrt{8x^3y^2}$ $-6xy\sqrt{2x}$

55. $5a\sqrt{3a^3b} + 2a^2\sqrt{27ab} - 4\sqrt{75a^5b}$ $-9a^2\sqrt{3ab}$

56. $\sqrt[3]{54xy^3} - 5\sqrt[3]{2xy^3} + \sqrt[3]{128xy^3}$ $2y\sqrt[3]{2x}$

3 Multiply.

57. $\sqrt{8}\sqrt{32}$ 16

58. $\sqrt{14}\sqrt{35}$ $7\sqrt{10}$

59. $\sqrt[3]{4}\sqrt[3]{8}$ $2\sqrt[3]{4}$

60. $\sqrt[3]{6}\sqrt[3]{36}$ 6

61. $\sqrt{x^2y^5}\sqrt{xy}$ $xy^3\sqrt{x}$

62. $\sqrt{a^3b}\sqrt{ab^4}$ $a^2b^2\sqrt{b}$

63. $\sqrt{2x^2y}\sqrt{32xy}$ $8xy\sqrt{x}$

64. $\sqrt{5x^3y}\sqrt{10x^3y^4}$ $5x^3y^2\sqrt{2y}$

65. $\sqrt[3]{x^2y}\sqrt[3]{16x^4y^2}$ $2x^2y\sqrt[3]{2}$

66. $\sqrt[3]{4a^2b^3}\sqrt[3]{8ab^5}$ $2ab^2\sqrt[3]{4b^2}$

67. $\sqrt[4]{12ab^3}\sqrt[4]{4a^5b^2}$ $2ab\sqrt[4]{3a^2b}$

68. $\sqrt[4]{36a^2b^4}\sqrt[4]{12a^5b^3}$ $2ab\sqrt[4]{27a^3b^3}$

69. $\sqrt{3}(\sqrt{27} - \sqrt{3})$ 6

70. $\sqrt{10}(\sqrt{10} - \sqrt{5})$ $10 - 5\sqrt{2}$

71. $\sqrt{x}(\sqrt{x} - \sqrt{2})$ $x - \sqrt{2x}$

72. $\sqrt{y}(\sqrt{y} - \sqrt{5})$ $y - \sqrt{5y}$

73. $\sqrt{2x}(\sqrt{8x} - \sqrt{32})$ $4x - 8\sqrt{x}$

74. $\sqrt{3a}(\sqrt{27a^2} - \sqrt{a})$ $9a\sqrt{a} - a\sqrt{3}$

75. $(\sqrt{x} - 3)^2$ $x - 6\sqrt{x} + 9$

76. $(\sqrt{2x} + 4)^2$ $2x + 8\sqrt{2x} + 16$

77. $(4\sqrt{5} + 2)^2$ $84 + 16\sqrt{5}$

78. $2\sqrt{3x^2} \cdot 3\sqrt{12xy^3} \cdot \sqrt{6x^3y}$ $36x^3y^2\sqrt{6}$

79. $2\sqrt{14xy} \cdot 4\sqrt{7x^2y} \cdot 3\sqrt{8xy^2}$ $672x^2y^2$

80. $\sqrt[3]{8ab}\,\sqrt[3]{4a^2b^3}\,\sqrt[3]{9ab^4}$ $2ab^2\sqrt[3]{36ab^2}$

81. $\sqrt[3]{2a^2b}\,\sqrt[3]{4a^3b^2}\,\sqrt[3]{8a^5b^6}$ $4a^3b^3\sqrt[3]{a}$

82. $(\sqrt{2} - 3)(\sqrt{2} + 4)$ $-10 + \sqrt{2}$

83. $(\sqrt{5} - 5)(2\sqrt{5} + 2)$ $-8\sqrt{5}$

84. $(\sqrt{y} - 2)(\sqrt{y} + 2)$ $y - 4$

85. $(\sqrt{x} - y)(\sqrt{x} + y)$ $x - y^2$

86. $(\sqrt{2x} - 3\sqrt{y})(\sqrt{2x} + 3\sqrt{y})$ $2x - 9y$

87. $(2\sqrt{3x} - \sqrt{y})(2\sqrt{3x} + \sqrt{y})$ $12x - y$

88. $(\sqrt{a} - 2)(\sqrt{a} - 3)$ $a - 5\sqrt{a} + 6$

89. $(\sqrt{x} + 4)(\sqrt{x} - 7)$ $x - 3\sqrt{x} - 28$

90. $(\sqrt[3]{a} + 2)(\sqrt[3]{a} + 3)$ $\sqrt[3]{a^2} + 5\sqrt[3]{a} + 6$

91. $(\sqrt[3]{x} - 4)(\sqrt[3]{x} + 5)$ $\sqrt[3]{x^2} + \sqrt[3]{x} - 20$

92. $(2\sqrt{x} - \sqrt{y})(3\sqrt{x} + \sqrt{y})$ $6x - \sqrt{xy} - y$

4 Simplify.

93. $\dfrac{\sqrt{32x^2}}{\sqrt{2x}}$ $4\sqrt{x}$

94. $\dfrac{\sqrt{60y^4}}{\sqrt{12y}}$ $y\sqrt{5y}$

95. $\dfrac{\sqrt{42a^3b^5}}{\sqrt{14a^2b}}$ $b^2\sqrt{3a}$

96. $\dfrac{\sqrt{65ab^4}}{\sqrt{5ab}}$ $b\sqrt{13b}$

97. $\dfrac{1}{\sqrt{5}}$ $\dfrac{\sqrt{5}}{5}$

98. $\dfrac{1}{\sqrt{2}}$ $\dfrac{\sqrt{2}}{2}$

99. $\dfrac{1}{\sqrt{2x}}$ $\dfrac{\sqrt{2x}}{2x}$

100. $\dfrac{2}{\sqrt{3y}}$ $\dfrac{2\sqrt{3y}}{3y}$

101. $\dfrac{5}{\sqrt{5x}}$ $\dfrac{\sqrt{5x}}{x}$

102. $\dfrac{9}{\sqrt{3a}}$ $\dfrac{3\sqrt{3a}}{a}$

103. $\sqrt{\dfrac{x}{5}}$ $\dfrac{\sqrt{5x}}{5}$

104. $\sqrt{\dfrac{y}{2}}$ $\dfrac{\sqrt{2y}}{2}$

105. $\dfrac{3}{\sqrt[3]{2}}$ $\dfrac{3\sqrt[3]{4}}{2}$

106. $\dfrac{5}{\sqrt[3]{9}}$ $\dfrac{5\sqrt[3]{3}}{3}$

107. $\dfrac{3}{\sqrt[3]{4x^2}}$ $\dfrac{3\sqrt[3]{2x}}{2x}$

108. $\dfrac{5}{\sqrt[3]{3y}}$ $\dfrac{5\sqrt[3]{9y^2}}{3y}$

109. $\dfrac{\sqrt{40x^3y^2}}{\sqrt{80x^2y^3}}$ $\dfrac{\sqrt{2xy}}{2y}$

110. $\dfrac{\sqrt{15a^2b^5}}{\sqrt{30a^5b^3}}$ $\dfrac{b\sqrt{2a}}{2a^2}$

111. $\dfrac{\sqrt{24a^2b}}{\sqrt{18ab^4}}$ $\dfrac{2\sqrt{3ab}}{3b^2}$

112. $\dfrac{\sqrt{12x^3y}}{\sqrt{20x^4y}}$ $\dfrac{\sqrt{15x}}{5x}$

113. $\dfrac{2}{\sqrt{5} + 2}$ $2\sqrt{5} - 4$

114. $\dfrac{5}{2 - \sqrt{7}}$ $-\dfrac{10 + 5\sqrt{7}}{3}$

115. $\dfrac{3}{\sqrt{y} - 2}$ $\dfrac{3\sqrt{y} + 6}{y - 4}$

116. $\dfrac{-7}{\sqrt{x} - 3}$ $-\dfrac{7\sqrt{x} + 21}{x - 9}$

117. $\dfrac{\sqrt{2} - \sqrt{3}}{\sqrt{2} + \sqrt{3}}$ $-5 + 2\sqrt{6}$

118. $\dfrac{\sqrt{3} + \sqrt{4}}{\sqrt{2} + \sqrt{3}}$ $-\sqrt{6} + 3 - 2\sqrt{2} + 2\sqrt{3}$

119. $\dfrac{4 - \sqrt{2}}{2 - \sqrt{3}}$ $8 + 4\sqrt{3} - 2\sqrt{2} - \sqrt{6}$

120. $\dfrac{3 - \sqrt{x}}{3 + \sqrt{x}}$ $\dfrac{9 - 6\sqrt{x} + x}{9 - x}$

121. $\dfrac{\sqrt{3}-\sqrt{5}}{\sqrt{2}+\sqrt{5}}$ $\dfrac{\sqrt{15}+\sqrt{10}-\sqrt{6}-5}{3}$

122. $\dfrac{\sqrt{2}+\sqrt{3}}{\sqrt{3}-\sqrt{2}}$ $5+2\sqrt{6}$

123. $\dfrac{3}{\sqrt[4]{8x^3}}$ $\dfrac{3\sqrt[4]{2x}}{2x}$

124. $\dfrac{-3}{\sqrt[4]{27y^2}}$ $-\dfrac{\sqrt[4]{3y^2}}{y}$

125. $\dfrac{4}{\sqrt[5]{16a^2}}$ $\dfrac{2\sqrt[5]{2a^3}}{a}$

126. $\dfrac{a}{\sqrt[5]{81a^4}}$ $\dfrac{\sqrt[5]{3a}}{3}$

127. $\dfrac{2x}{\sqrt[5]{64x^3}}$ $\dfrac{\sqrt[5]{16x^2}}{2}$

128. $\dfrac{3y}{\sqrt[4]{32y^2}}$ $\dfrac{3\sqrt[4]{8y^2}}{4}$

129. $\dfrac{\sqrt{a}+a\sqrt{b}}{\sqrt{a}-a\sqrt{b}}$ $\dfrac{1+2\sqrt{ab}+ab}{1-ab}$

130. $\dfrac{\sqrt{3}-3\sqrt{y}}{\sqrt{3}+3\sqrt{y}}$ $\dfrac{1-2\sqrt{3y}+3y}{1-3y}$

131. $\dfrac{3\sqrt{xy}+2\sqrt{xy}}{\sqrt{x}-\sqrt{y}}$ $\dfrac{5x\sqrt{y}+5y\sqrt{x}}{x-y}$

132. $\dfrac{2\sqrt{x}+3\sqrt{y}}{\sqrt{x}-4\sqrt{y}}$ $\dfrac{2x+11\sqrt{xy}+12y}{x-16y}$

APPLYING CONCEPTS 7.2

Simplify.

133. $(\sqrt{8}-\sqrt{2})^3$ $2\sqrt{2}$

134. $(\sqrt{27}-\sqrt{3})^3$ $24\sqrt{3}$

135. $(\sqrt{2}-3)^3$ $29\sqrt{2}-45$

136. $(\sqrt{5}+2)^3$ $38+17\sqrt{5}$

137. $\dfrac{3}{\sqrt{y+1}+1}$ $\dfrac{3\sqrt{y+1}-3}{y}$

138. $\dfrac{2}{\sqrt{x+4}+2}$ $\dfrac{2\sqrt{x+4}-4}{x}$

Rewrite as an expression with a single radical.

139. $\dfrac{\sqrt[3]{(x+y)^2}}{\sqrt{x+y}}$ $\sqrt[6]{x+y}$

140. $\dfrac{\sqrt[4]{(a+b)^3}}{\sqrt{a+b}}$ $\sqrt[4]{a+b}$

141. $\sqrt[4]{2y}\sqrt{x+3}$ $\sqrt[4]{2y(x+3)^2}$

142. $\sqrt[4]{2x}\sqrt{y-2}$ $\sqrt[4]{2x(y-2)^2}$

143. $\sqrt{a}\sqrt[3]{a+3}$ $\sqrt[6]{a^3(a+3)^2}$

144. $\sqrt{b}\sqrt[3]{b-1}$ $\sqrt[6]{b^3(b-1)^2}$

Write in exponential form and then simplify.

145. $\sqrt{16^{\frac{1}{2}}}$ $16^{\frac{1}{4}}, 2$

146. $\sqrt[3]{4^{\frac{3}{2}}}$ $4^{\frac{1}{2}}, 2$

147. $\sqrt[4]{32^{-\frac{4}{5}}}$ $32^{-\frac{1}{5}}, \frac{1}{2}$

148. $\sqrt{243^{-\frac{4}{5}}}$ $243^{-\frac{2}{5}}, \frac{1}{9}$

149. *National Debt* The approximate national debt for selected years is given in the following table (*Source:* www.publicdebt.treas.gov/opd/opdpenny.htm).

Year	1989	1991	1993	1995	1997
Debt (in trillions of dollars)	2.86	3.67	4.41	4.97	5.41

A function that models this data is $D(x) = 2.79x^{0.3}$ where $x = 1$ corresponds to the year 1989 and $D(x)$ is the national debt in trillions of dollars.

a. Write this function in radical form. $D(x) = 2.79\sqrt[10]{x^3}$

b. Find the value of the national debt predicted by the model for the year 1993. \$4.52 trillion ($x = 5$)

c. Assuming that the model is accurate past 1997, what will the national debt be in the year 2004? \$6.41 trillion ($x = 16$)

150. // By what factor must you multiply a number in order to double its square root? to triple its square root? to double its cube root? to triple its cube root? Explain.

151. // Explain in your own words what it means to rationalize the denominator of a radical expression and how to do so.

S E C T I O N **7.3**

Complex Numbers

■ 1   Simplify complex numbers

INSTRUCTOR NOTE

Complex numbers were not accepted by mathematicians until late in the 19th century. Part of the difficulty was that there were no physical instances of these numbers. The "Projects and Group Activities" section at the end of this chapter presents a geometric interpretation of complex numbers.

The radical expression $\sqrt{-4}$ is not a real number, because there is no real number whose square is -4. However, the solution of an algebraic equation is sometimes the square root of a negative number.

For example, the equation $x^2 + 1 = 0$ does not have a real number solution, because there is no real number whose square is a negative number.

$$x^2 + 1 = 0$$
$$x^2 = -1$$

During the late 17th century, a new number, called an **imaginary number,** was defined so that a negative number would have a square root. The letter i was chosen to represent the number whose square is -1.

$$i^2 = -1$$

An imaginary number is defined in terms of i.

Principal Square Root of a Negative Number

If a is a positive real number, then the principal square root of negative a is the imaginary number $i\sqrt{a}$.

$$\sqrt{-a} = i\sqrt{a}$$

POINT OF INTEREST

The first written occurrence of an imaginary number was in a book published in 1545 by Hieronimo Cardano, where he wrote (in our modern notation) $5 + \sqrt{-15}$. He went on to say that the number "is as refined as it is useless." It was not until the 20th century that applications of complex numbers were found.

Here are some examples.

$$\sqrt{-16} = i\sqrt{16} = 4i$$
$$\sqrt{-12} = i\sqrt{12} = 2i\sqrt{3}$$
$$\sqrt{-21} = i\sqrt{21}$$
$$\sqrt{-1} = i\sqrt{1} = i$$

It is customary to write i in front of a radical to avoid confusing $\sqrt{a}\,i$ with $\sqrt{ai}$.

The real numbers and imaginary numbers make up the complex numbers.

Definition of a Complex Number

A **complex number** is a number of the form $a + bi$, where a and b are real numbers and $i = \sqrt{-1}$. The number a is the **real part** of $a + bi$, and b is the **imaginary part**.

Examples of complex numbers are shown at the right.

$$\overset{\text{Real part}}{\underset{\downarrow}{}} \overset{\text{Imaginary part}}{\underset{\downarrow}{}}$$
$$a + bi$$
$$3 + 2i$$
$$8 - 10i$$

Complex numbers $a + bi$
—— Real numbers
$a + 0i$
—— Imaginary numbers
$0 + bi$

A **real number** is a complex number in which $b = 0$.

An **imaginary number** is a complex number in which $a = 0$.

Example 1   Simplify: $\sqrt{-80}$

Solution $\sqrt{-80} = i\sqrt{80} = i\sqrt{2^4 \cdot 5} = 4i\sqrt{5}$

Problem 1   Simplify: $\sqrt{-45}$

Solution   See page S20. $3i\sqrt{5}$

Simplify: $\sqrt{20} - \sqrt{-50}$

Write the complex number in the form $a + bi$.
$$\sqrt{20} - \sqrt{-50} = \sqrt{20} - i\sqrt{50}$$

Use the Product Property of Radicals to simplify each radical.
$$= \sqrt{2^2 \cdot 5} - i\sqrt{5^2 \cdot 2}$$
$$= 2\sqrt{5} - 5i\sqrt{2}$$

Example 2   Simplify: $\sqrt{25} + \sqrt{-40}$

Solution $\sqrt{25} + \sqrt{-40} = \sqrt{25} + i\sqrt{40} = \sqrt{5^2} + i\sqrt{2^2 \cdot 2 \cdot 5}$
$$= 5 + 2i\sqrt{10}$$

Problem 2   Simplify: $\sqrt{98} - \sqrt{-60}$

Solution   See page S20. $7\sqrt{2} - 2i\sqrt{15}$

2 Add and subtract complex numbers

HM³

Addition and Subtraction of Complex Numbers

To add two complex numbers, add the real parts and add the imaginary parts. To subtract two complex numbers, subtract the real parts and subtract the imaginary parts.

$$(a + bi) + (c + di) = (a + c) + (b + d)i$$

$$(a + bi) - (c + di) = (a - c) + (b - d)i$$

Example 3 Add: $(3 + 2i) + (6 - 5i)$

Solution $(3 + 2i) + (6 - 5i)$
$= (3 + 6) + (2 - 5)i$ ▶ Add the real parts and add the imaginary parts.
$= 9 - 3i$

Problem 3 Subtract: $(-4 + 2i) - (6 - 8i)$

Solution See page S20. $-10 + 10i$

➡ Add: $(3 + \sqrt{-12}) + (7 - \sqrt{-27})$

Write each complex number in the form $a + bi$.

$(3 + \sqrt{-12}) + (7 - \sqrt{-27})$
$= (3 + i\sqrt{12}) + (7 - i\sqrt{27})$

Use the Product Property of Radicals to simplify each radical.

$= (3 + i\sqrt{2^2 \cdot 3}) + (7 - i\sqrt{3^2 \cdot 3})$
$= (3 + 2i\sqrt{3}) + (7 - 3i\sqrt{3})$

Add the complex numbers.

$= 10 - i\sqrt{3}$

Example 4 Subtract: $(9 - \sqrt{-8}) - (5 + \sqrt{-32})$

Solution $(9 - \sqrt{-8}) - (5 + \sqrt{-32})$
$= (9 - i\sqrt{8}) - (5 + i\sqrt{32})$ ▶ Write each complex number in the form $a + bi$.
$= (9 - i\sqrt{2^2 \cdot 2}) - (5 + i\sqrt{2^4 \cdot 2})$ ▶ Simplify each radical.
$= (9 - 2i\sqrt{2}) - (5 + 4i\sqrt{2})$
$= 4 - 6i\sqrt{2}$

Problem 4 Subtract: $(16 - \sqrt{-45}) - (3 + \sqrt{-20})$

Solution See page S20. $13 - 5i\sqrt{5}$

3 Multiply complex numbers

HM³

When multiplying complex numbers, the term i^2 is frequently a part of the product. Recall that $i^2 = -1$.

➡ Multiply: $2i \cdot 3i$

Multiply the imaginary numbers. $2i \cdot 3i = 6i^2$

Replace i^2 by -1. Then simplify. $= 6(-1) = -6$ ◀

When multiplying square roots of negative numbers, first rewrite the radical expressions using i.

➡ Multiply: $\sqrt{-6} \cdot \sqrt{-24}$

Write each radical as the product of a real number and i. $\sqrt{-6} \cdot \sqrt{-24} = i\sqrt{6} \cdot i\sqrt{24}$

Multiply the imaginary numbers. $= i^2\sqrt{144}$

Replace i^2 by -1. $= -\sqrt{144}$

Simplify. $= -12$ ◀

Note from this example that it would have been incorrect to multiply the radicands of the two radical expressions. To illustrate,

$$\sqrt{-6} \cdot \sqrt{-24} = \sqrt{(-6)(-24)} = \sqrt{144} = 12, \, not \, -12$$

➡ Multiply: $4i(3 - 2i)$

Use the Distributive Property to remove parentheses. $4i(3 - 2i) = 12i - 8i^2$

Replace i^2 by -1. $= 12i - 8(-1)$

Write the answer in the form $a + bi$. $= 8 + 12i$ ◀

Example 5   Multiply: $\sqrt{-8}(\sqrt{6} - \sqrt{-2})$

Solution $\sqrt{-8}(\sqrt{6} - \sqrt{-2})$

$= i\sqrt{8}(\sqrt{6} - i\sqrt{2})$ ▶ Write each complex number in the form $a + bi$.

$= i\sqrt{48} - i^2\sqrt{16}$ ▶ Use the Distributive Property.

$= i\sqrt{2^4 \cdot 3} - (-1)\sqrt{2^4}$ ▶ Simplify each radical. Replace i^2 by -1.

$= 4i\sqrt{3} + 4$

$= 4 + 4i\sqrt{3}$ ▶ Write the answer in the form $a + bi$.

Problem 5   Multiply: $\sqrt{-3}(\sqrt{27} - \sqrt{-6})$

Solution   See page S20. $3\sqrt{2} + 9i$

The product of two complex numbers can be found by using the FOIL Method. For example,

$$(2 + 4i)(3 - 5i) = 6 - 10i + 12i - 20i^2$$
$$= 6 + 2i - 20i^2$$
$$= 6 + 2i - 20(-1)$$
$$= 26 + 2i$$

The conjugate of $a + bi$ is $a - bi$.

The product of conjugates of the form $(a + bi)(a - bi)$ is $a^2 + b^2$.

$$(a + bi)(a - bi) = a^2 - b^2i^2 = a^2 - b^2(-1) = a^2 + b^2$$

For example, $(2 + 3i)(2 - 3i) = 2^2 + 3^2 = 4 + 9 = 13$.

Note that the product of a complex number and its conjugate is a real number.

Example 6 Multiply.

A. $(3 - 4i)(2 + 5i)$ B. $\left(\frac{9}{10} + \frac{3}{10}i\right)\left(1 - \frac{1}{3}i\right)$

C. $(4 + 5i)(4 - 5i)$

Solution A. $(3 - 4i)(2 + 5i)$

$= 6 + 15i - 8i - 20i^2$ ▶ Use the FOIL method.

$= 6 + 7i - 20i^2$ ▶ Combine like terms.

$= 6 + 7i - 20(-1)$ ▶ Replace i^2 by -1.

$= 26 + 7i$ ▶ Write the answer in the form $a + bi$.

B. $\left(\frac{9}{10} + \frac{3}{10}i\right)\left(1 - \frac{1}{3}i\right)$

$= \frac{9}{10} - \frac{3}{10}i + \frac{3}{10}i - \frac{1}{10}i^2$ ▶ Use the FOIL method.

$= \frac{9}{10} - \frac{1}{10}i^2$ ▶ Combine like terms.

$= \frac{9}{10} - \frac{1}{10}(-1)$ ▶ Replace i^2 by -1.

$= \frac{9}{10} + \frac{1}{10} = 1$ ▶ Simplify.

C. $(4 + 5i)(4 - 5i) = 4^2 + 5^2$ ▶ The product of conjugates of the form $(a + bi)(a - bi)$ is $a^2 + b^2$.

$= 16 + 25$

$= 41$

Problem 6 Multiply.

A. $(4 - 3i)(2 - i)$ B. $(3 - i)\left(\frac{3}{10} + \frac{1}{10}i\right)$

C. $(3 + 6i)(3 - 6i)$

Solution See page S20–S21. A. $5 - 10i$ B. 1 C. 45

4 Divide complex numbers

A fraction containing one or more complex numbers is in simplest form when no imaginary number remains in the denominator.

Simplify: $\frac{4-5i}{2i}$

Multiply the expression by 1 in the form $\frac{i}{i}$. $\frac{4-5i}{2i} = \frac{4-5i}{2i} \cdot \frac{i}{i}$

$= \frac{4i-5i^2}{2i^2}$

Replace i^2 by -1. $= \frac{4i-5(-1)}{2(-1)}$

Simplify. $= \frac{5+4i}{-2}$

Write the answer in the form $a + bi$. $= -\frac{5}{2} - 2i$

Example 7 Simplify: $\frac{5+4i}{3i}$

Solution $\frac{5+4i}{3i} = \frac{5+4i}{3i} \cdot \frac{i}{i} = \frac{5i+4i^2}{3i^2} = \frac{5i+4(-1)}{3(-1)} = \frac{-4+5i}{-3} = \frac{4}{3} - \frac{5}{3}i$

Problem 7 Simplify: $\frac{2-3i}{4i}$

Solution See page S21. $-\frac{3}{4} - \frac{1}{2}i$

Simplify: $\frac{3+2i}{1+i}$

INSTRUCTOR NOTE

Students think that division of complex numbers is somehow different from division of other numbers. Show them that just as $\frac{12}{4} = 3$ because $4 \cdot 3 = 12$, $\frac{3+2i}{1+i} = \frac{5}{2} - \frac{1}{2}i$ because $(1+i)\left(\frac{5}{2} - \frac{1}{2}i\right) = 3 + 2i$.

To simplify a fraction that has a complex number in the denominator, multiply the numerator and denominator by the conjugate of the complex number. The conjugate of $1 + i$ is $1 - i$.

$\frac{3+2i}{1+i} = \frac{(3+2i)}{(1+i)} \cdot \frac{(1-i)}{(1-i)} = \frac{3-3i+2i-2i^2}{1^2+1^2}$

$= \frac{3-i-2(-1)}{2}$

$= \frac{5-i}{2} = \frac{5}{2} - \frac{1}{2}i$

Example 8 Simplify: $\frac{5-3i}{4+2i}$

Solution $\frac{5-3i}{4+2i} = \frac{(5-3i)}{(4+2i)} \cdot \frac{(4-2i)}{(4-2i)} = \frac{20-10i-12i+6i^2}{4^2+2^2} = \frac{20-22i+6(-1)}{20}$

$= \frac{14-22i}{20}$

$= \frac{7-11i}{10}$

$= \frac{7}{10} - \frac{11}{10}i$

Problem 8 Simplify: $\frac{2+5i}{3-2i}$

Solution See page S21. $-\frac{4}{13} + \frac{19}{13}i$

CONCEPT REVIEW 7.3

Determine whether the following statements are always true, sometimes true, or never true.

1. The product of two imaginary numbers is a real number. Always true

2. The sum of two complex numbers is a real number. Sometimes true

3. A complex number is a number of the form $a + bi$, where a and b are real numbers and $\sqrt{-1} = i$. Always true

4. The product of two complex numbers is a real number. Sometimes true

EXERCISES 7.3

1 1. What is an imaginary number? What is a complex number?

2. Are all real numbers also complex numbers? Are all complex numbers also real numbers?

Simplify.

3. $\sqrt{-4}$ $2i$
4. $\sqrt{-64}$ $8i$
5. $\sqrt{-98}$ $7i\sqrt{2}$
6. $\sqrt{-72}$ $6i\sqrt{2}$
7. $\sqrt{-27}$ $3i\sqrt{3}$
8. $\sqrt{-75}$ $5i\sqrt{3}$
9. $\sqrt{16} + \sqrt{-4}$ $4 + 2i$
10. $\sqrt{25} + \sqrt{-9}$ $5 + 3i$
11. $\sqrt{12} - \sqrt{-18}$ $2\sqrt{3} - 3i\sqrt{2}$
12. $\sqrt{60} - \sqrt{-48}$ $2\sqrt{15} - 4i\sqrt{3}$
13. $\sqrt{160} - \sqrt{-147}$ $4\sqrt{10} - 7i\sqrt{3}$
14. $\sqrt{96} - \sqrt{-125}$ $4\sqrt{6} - 5i\sqrt{5}$
15. $\sqrt{-4a^2}$ $2ai$
16. $\sqrt{-16b^6}$ $4b^3i$
17. $\sqrt{-49x^{12}}$ $7x^6i$
18. $\sqrt{-32x^3y^2}$ $4xyi\sqrt{2x}$
19. $\sqrt{-144a^3b^5}$ $12ab^2i\sqrt{ab}$
20. $\sqrt{-18a^{10}b^9}$ $3a^5b^4i\sqrt{2b}$
21. $\sqrt{4a} + \sqrt{-12a^2}$ $2\sqrt{a} + 2ai\sqrt{3}$
22. $\sqrt{25b} - \sqrt{-48b^2}$ $5\sqrt{b} - 4bi\sqrt{3}$
23. $\sqrt{18b^5} - \sqrt{-27b^3}$ $3b^2\sqrt{2b} - 3bi\sqrt{3b}$
24. $\sqrt{a^5b^2} - \sqrt{-a^5b^2}$ $a^2b\sqrt{a} - a^2bi\sqrt{a}$
25. $\sqrt{-50x^3y^3} + x\sqrt{25x^4y^3}$ $5x^3y\sqrt{y} + 5xyi\sqrt{2xy}$
26. $\sqrt{-121xy} + \sqrt{60x^2y^2}$ $2xy\sqrt{15} + 11i\sqrt{xy}$
27. $\sqrt{-49a^5b^2} - ab\sqrt{-25a^3}$ $2a^2bi\sqrt{a}$
28. $\sqrt{-16x^2y} - x\sqrt{-49y}$ $-3xi\sqrt{y}$
29. $\sqrt{12a^3} + \sqrt{-27b^3}$ $2a\sqrt{3a} + 3bi\sqrt{3b}$

2 Add or subtract.

30. $(2 + 4i) + (6 - 5i)$ $8 - i$
31. $(6 - 9i) + (4 + 2i)$ $10 - 7i$
32. $(-2 - 4i) - (6 - 8i)$ $-8 + 4i$
33. $(3 - 5i) + (8 - 2i)$ $11 - 7i$
34. $(8 - \sqrt{-4}) - (2 + \sqrt{-16})$ $6 - 6i$
35. $(5 - \sqrt{-25}) - (11 - \sqrt{-36})$ $-6 + i$

36. $(12 - \sqrt{-50}) + (7 - \sqrt{-8})$ $19 - 7i\sqrt{2}$

37. $(5 - \sqrt{-12}) - (9 + \sqrt{-108})$ $-4 - 8i\sqrt{3}$

38. $(\sqrt{8} + \sqrt{-18}) + (\sqrt{32} - \sqrt{-72})$ $6\sqrt{2} - 3i\sqrt{2}$

39. $(\sqrt{40} - \sqrt{-98}) - (\sqrt{90} + \sqrt{-32})$ $-\sqrt{10} - 11i\sqrt{2}$

40. $(5 - 3i) + 2i$ $5 - i$

41. $(6 - 8i) + 4i$ $6 - 4i$

42. $(7 + 2i) + (-7 - 2i)$ 0

43. $(8 - 3i) + (-8 + 3i)$ 0

44. $(9 + 4i) + 6$ $15 + 4i$

45. $(4 + 6i) + 7$ $11 + 6i$

3 Multiply.

46. $(7i)(-9i)$ 63

47. $(-6i)(-4i)$ -24

48. $\sqrt{-2}\sqrt{-8}$ -4

49. $\sqrt{-5}\sqrt{-45}$ -15

50. $\sqrt{-3}\sqrt{-6}$ $-3\sqrt{2}$

51. $\sqrt{-5}\sqrt{-10}$ $-5\sqrt{2}$

52. $2i(6 + 2i)$ $-4 + 12i$

53. $-3i(4 - 5i)$ $-15 - 12i$

54. $\sqrt{-2}(\sqrt{8} + \sqrt{-2})$ $-2 + 4i$

55. $\sqrt{-3}(\sqrt{12} - \sqrt{-6})$ $3\sqrt{2} + 6i$

56. $(5 - 2i)(3 + i)$ $17 - i$

57. $(2 - 4i)(2 - i)$ $-10i$

58. $(6 + 5i)(3 + 2i)$ $8 + 27i$

59. $(4 - 7i)(2 + 3i)$ $29 - 2i$

60. $(1 - i)\left(\frac{1}{2} + \frac{1}{2}i\right)$ 1

61. $\left(\frac{4}{5} - \frac{2}{5}i\right)\left(1 + \frac{1}{2}i\right)$ 1

62. $\left(\frac{6}{5} + \frac{3}{5}i\right)\left(\frac{2}{3} - \frac{1}{3}i\right)$ 1

63. $(2 - i)\left(\frac{2}{5} + \frac{1}{5}i\right)$ 1

64. $(4 - 3i)(4 + 3i)$ 25

65. $(8 - 5i)(8 + 5i)$ 89

66. $(3 - i)(3 + i)$ 10

67. $(7 - i)(7 + i)$ 50

4 Simplify.

68. $\frac{3}{i}$ $-3i$

69. $\frac{4}{5i}$ $-\frac{4}{5}i$

70. $\frac{2 - 3i}{-4i}$ $\frac{3}{4} + \frac{1}{2}i$

71. $\frac{16 + 5i}{-3i}$ $-\frac{5}{3} + \frac{16}{3}i$

72. $\frac{4}{5 + i}$ $\frac{10}{13} - \frac{2}{13}i$

73. $\frac{6}{5 + 2i}$ $\frac{30}{29} - \frac{12}{29}i$

74. $\frac{2}{2 - i}$ $\frac{4}{5} + \frac{2}{5}i$

75. $\frac{5}{4 - i}$ $\frac{20}{17} + \frac{5}{17}i$

76. $\frac{1 - 3i}{3 + i}$ $-i$

77. $\frac{2 + 12i}{5 + i}$ $\frac{11}{13} + \frac{29}{13}i$

78. $\frac{\sqrt{-10}}{\sqrt{8} - \sqrt{-2}}$ $-\frac{\sqrt{5}}{5} + \frac{2\sqrt{5}}{5}i$

79. $\frac{\sqrt{-2}}{\sqrt{12} - \sqrt{-8}}$ $-\frac{1}{5} + \frac{\sqrt{6}}{10}i$

80. $\frac{2 - 3i}{3 + i}$ $\frac{3}{10} - \frac{11}{10}i$

81. $\frac{3 + 5i}{1 - i}$ $-1 + 4i$

APPLYING CONCEPTS 7.3

Note the pattern when successive powers of i are simplified.

$i^1 = i$ $\qquad$ $i^5 = i \cdot i^4 = i(1) = i$

$i^2 = -1$ $\qquad$ $i^6 = i^2 \cdot i^4 = -1$

$i^3 = i^2 \cdot i = -i$ $\qquad$ $i^7 = i^3 \cdot i^4 = -i$

$i^4 = i^2 \cdot i^2 = (-1)(-1) = 1$ $\qquad$ $i^8 = i^4 \cdot i^4 = 1$

82. When the exponent on i is a multiple of 4, the power equals __1__.

Use the pattern above to simplify the power of i.

83. i^6 $\quad -1$

84. i^9 $\quad i$

85. i^{57} $\quad i$

86. i^{65} $\quad i$

87. i^{-6} $\quad -1$

88. i^{-34} $\quad -1$

89. i^{-58} $\quad -1$

90. i^{-180} $\quad 1$

The property that the product of conjugates of the form $(a + bi)(a - bi)$ is equal to $a^2 + b^2$ can be used to factor the sum of two perfect squares over the set of complex numbers. For example, $x^2 + y^2 = (x + yi)(x - yi)$. Factor over the set of complex numbers.

91. $y^2 + 1$ $\quad (y + i)(y - i)$

92. $a^2 + 4$ $\quad (a + 2i)(a - 2i)$

93. $x^2 + 25$ $\quad (x + 5i)(x - 5i)$

94. **a.** Is $3i$ a solution of $2x^2 + 18 = 0$? Yes

b. Is $-3i$ a solution of $2x^2 + 18 = 0$? Yes

95. **a.** Is $1 + 3i$ a solution of $x^2 - 2x - 10 = 0$? No

b. Is $1 - 3i$ a solution of $x^2 - 2x - 10 = 0$? No

[C] **96.** Show that $\sqrt{i} = \frac{\sqrt{2}}{2} + \frac{\sqrt{2}}{2}i$ by simplifying $\left[\frac{\sqrt{2}}{2} + \frac{\sqrt{2}}{2}i\right]^2$. Complete solution is available in the Solutions Manual.

[C] **97.** Given $\sqrt{i} = \frac{\sqrt{2}}{2} + \frac{\sqrt{2}}{2}i$, find $\sqrt{-i}$. $\quad -\frac{\sqrt{2}}{2} + \frac{\sqrt{2}}{2}i$

SECTION 7.4 Radical Functions

1 Find the domain of a radical function

HM³

A **radical function** is one that contains a variable underneath a radical or a fractional exponent. Examples of radical functions are shown at the right.

$$f(x) = 3\sqrt[4]{x^5} - 7$$

$$g(x) = 3x - 2x^{\frac{1}{2}} + 5$$

Note that these are *not* polynomial functions because polynomial functions do not contain variable radical expressions or variables raised to a fractional power.

The domain of a radical function is a set of real numbers for which the radical expression is a real number. For example, -9 is one number that would be excluded from the domain of $f(x) = \sqrt{x + 5}$ because

$$f(-9) = \sqrt{-9 + 5} = \sqrt{-4} = 2i, \text{ which is not a real number.}$$

State the domain of $f(x) = \sqrt{x + 5}$ in set-builder notation.

The value of $\sqrt{x + 5}$ is a real number when $x + 5$ is greater than or equal to zero: $x + 5 \geq 0$. Solving this inequality for x results in $x \geq -5$.

The domain of f is $\{x \mid x \geq -5\}$.

State the domain of $F(x) = \sqrt[3]{2x - 6}$.

Because the cube root of a real number is a real number, $\sqrt[3]{2x - 6}$ is a real number for all values of x. [For instance, $F(-1) = \sqrt[3]{2(-1) - 6} = \sqrt[3]{-8} = -2$.] Therefore, the domain of F is all real numbers.

These last two examples suggest the following:

If the index of a radical expression is an even number, the radicand must be greater than or equal to zero to ensure that the value of the expression will be a real number. If the index of a radical expression is an odd number, the radicand may be a positive or a negative number.

Example 1 State the domain of each function in set-builder notation.

A. $V(x) = \sqrt[4]{6 - 4x}$ B. $R(x) = \sqrt[5]{x + 4}$

Solution A. $6 - 4x \geq 0$

$-4x \geq -6$

$x \leq \frac{3}{2}$

▶ V contains an even root. Therefore, the radicand must be greater than or equal to zero.

The domain is $\left\{x \mid x \leq \frac{3}{2}\right\}$.

B. $\{x \mid x \in \text{real numbers}\}$ ▶ Because R contains an odd root, the radicand may be positive or negative.

Problem 1 State the domain of each function in interval notation.

A. $Q(x) = \sqrt[3]{6x + 12}$ B. $T(x) = (3x + 9)^{\frac{1}{2}}$

Solution See page S21. A. $(-\infty, \infty)$ B. $[-3, \infty)$

2 Graph a radical function

11 HM³

The graph of a radical function is produced in the same manner as the graph of any other function. The function is evaluated at several values in the domain of the function, and the resulting ordered pairs are graphed. Ordered pairs must be graphed until an accurate graph can be made.

LOOK CLOSELY

By first determining the domain of the function, you discover the possible values of x to use when evaluating the function. Any value of x in the domain can be used. We used "nice" values—values of x that result in a perfect-square radicand. However, any value of x can be used. For instance, if $x = 3$, then

$$f(3) = \sqrt{5} \approx 2.24$$

and the ordered pair (3, 2.24) belongs to the graph. This ordered pair is just a little more difficult to graph.

Graph: $f(x) = \sqrt{x + 2}$

Because f contains an even root, the radicand must be positive. To determine the domain of f, solve the inequality $x + 2 \geq 0$. The solution is $x \geq -2$, so the domain is $\{x \mid x \geq -2\}$. Now determine ordered pairs of the function by choosing values of x from the domain. Some possible choices are shown in the following table.

x	$f(x) = \sqrt{x + 2}$	y
-2	$f(-2) = \sqrt{-2 + 2} = \sqrt{0} = 0$	0
-1	$f(-1) = \sqrt{-1 + 2} = \sqrt{1} = 1$	1
2	$f(2) = \sqrt{2 + 2} = \sqrt{4} = 2$	2
7	$f(7) = \sqrt{7 + 2} = \sqrt{9} = 3$	3

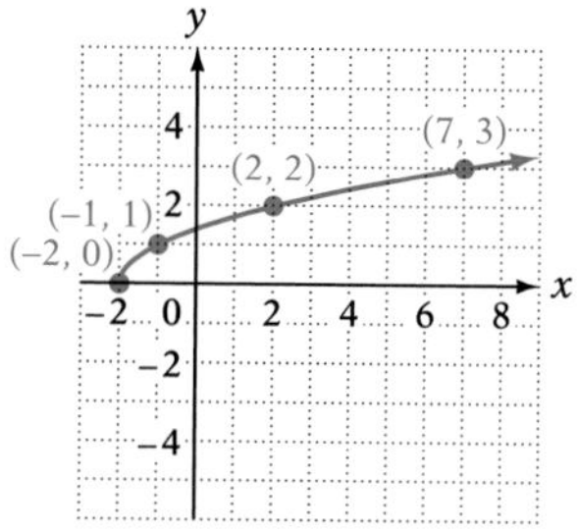

Example 2 Graph: $H(x) = \sqrt[3]{x}$

Solution Because H contains only an odd root, the domain of H is all real numbers. Choose some values of x in the domain of H, and evaluate the function for those values. Some possible choices are given in the following table.

x	$H(x) = \sqrt[3]{x}$	y
-8	$H(-8) = \sqrt[3]{-8} = -2$	-2
-1	$H(-1) = \sqrt[3]{-1} = -1$	-1
0	$H(0) = \sqrt[3]{0} = 0$	0
1	$H(1) = \sqrt[3]{1} = 1$	1
8	$H(8) = \sqrt[3]{8} = 2$	2

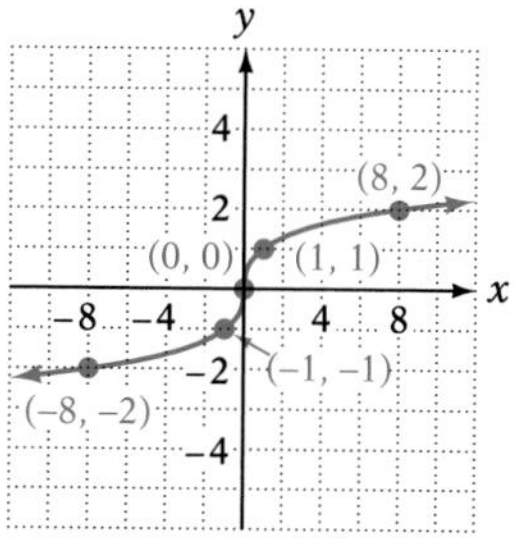

Problem 2 Graph: $F(x) = \sqrt{x - 2}$

Solution See page S21.

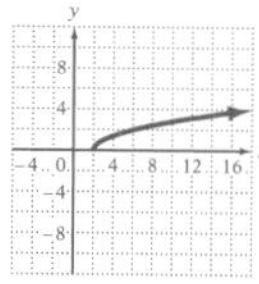

A graphing utility can be used to graph radical functions. See the Appendix for instructions on how to enter a radical function on a graphing utility.

Example 3 Graph: $Z(x) = 4 + (8 - 6x)^{\frac{1}{2}}$

Solution Because Z involves an even root, the radicand must be positive. The domain is $\left\{x \mid x \leq \frac{4}{3}\right\}$.

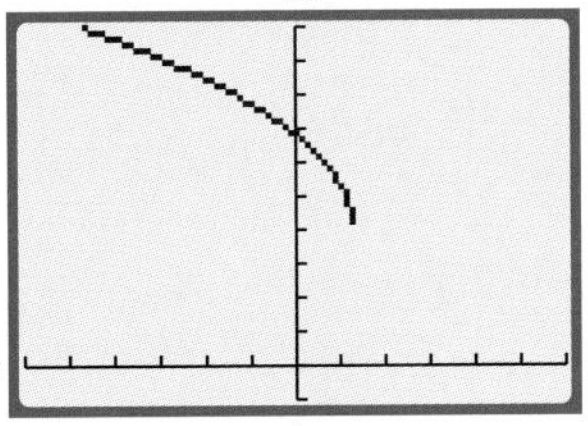

Problem 3 Graph: $y(x) = 2 - \sqrt[3]{x - 1}$

Solution See page S21.

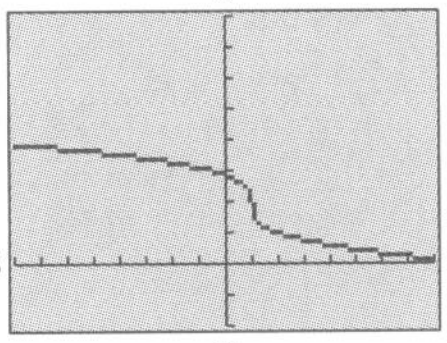

CONCEPT REVIEW 7.4

Determine whether the following statements are always true, sometimes true, or never true.

1. The number 5 is not in the domain of $f(x) = \sqrt{4 - x}$. Always true

2. The domain of $f(x) = \sqrt[3]{x - 2}$ is the set of all real numbers. Always true

3. The domain of $f(x) = \frac{3}{x^2 + 4}$ is all the real numbers except −2 and 2. Never true

4. If x is a real number, the expression $f(x) = \sqrt{-x^2}$ represents a radical function. Never true

EXERCISES 7.4

1 1. What is a radical function?

2. What is the difference between a radical function and a polynomial function?

State the domain of each function in set-builder notation.

3. $f(x) = 2x^{\frac{1}{3}}$ $\{x \mid x \in \text{real numbers}\}$

4. $r(x) = -3\sqrt[5]{2x}$ $\{x \mid x \in \text{real numbers}\}$

5. $g(x) = -2\sqrt{x + 1}$ $\{x \mid x \geq -1\}$

6. $h(x) = 3x^{\frac{1}{4}} - 2$ $\{x \mid x \geq 0\}$

7. $f(x) = 2x\sqrt{x} - 3$ $\{x \mid x \geq 0\}$

8. $y(x) = -3\sqrt[3]{1 + x}$ $\{x \mid x \in \text{real numbers}\}$

9. $C(x) = -3x^{\frac{3}{4}} + 1$ $\{x \mid x \geq 0\}$

10. $G(x) = 6x^{\frac{2}{5}} + 5$ $\{x \mid x \in \text{real numbers}\}$

11. $F(x) = 4(3x - 6)^{\frac{1}{2}}$ $\{x \mid x \geq 2\}$

State the domain of each function in interval notation.

12. $f(x) = -2(4x - 12)^{\frac{1}{2}}$
$[3, \infty)$

13. $g(x) = 2(2x - 10)^{\frac{2}{3}}$
$(-\infty, \infty)$

14. $J(x) = 4 - (3x - 3)^{\frac{2}{5}}$
$(-\infty, \infty)$

15. $V(x) = x - \sqrt{12 - 4x}$
$(-\infty, 3]$

16. $Y(x) = -6 + \sqrt{6 - x}$
$(-\infty, 6]$

17. $h(x) = 3\sqrt[4]{(x - 2)^3}$
$[2, \infty)$

18. $g(x) = \frac{2}{3}\sqrt[4]{(4 - x)^3}$
$(-\infty, 4]$

19. $f(x) = x - (4 - 6x)^{\frac{1}{2}}$
$\left(-\infty, \frac{2}{3}\right]$

20. $F(x) = (9 + 12x)^{\frac{1}{2}} - 4$
$\left[-\frac{3}{4}, \infty\right)$

2 Graph.

21. $F(x) = \sqrt{x}$

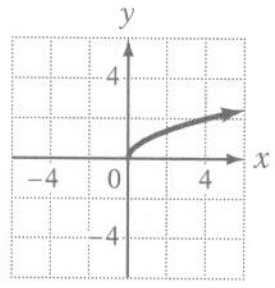

22. $G(x) = -\sqrt{x}$

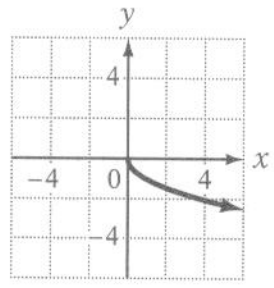

23. $h(x) = -\sqrt[3]{x}$

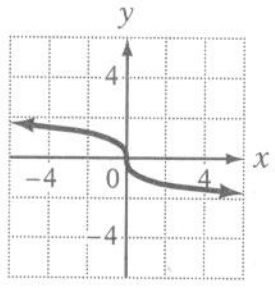

24. $K(x) = \sqrt[3]{x + 1}$

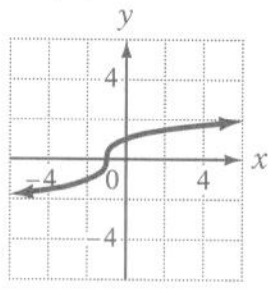

25. $f(x) = -\sqrt{x + 2}$

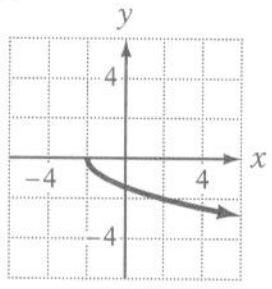

26. $g(x) = \sqrt{x - 1}$

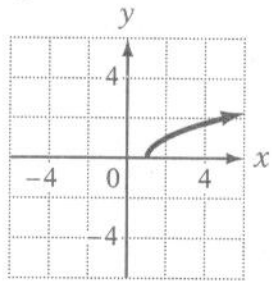

27. $S(x) = -\sqrt[4]{x}$

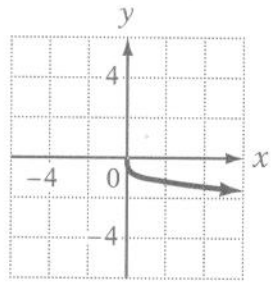

28. $C(x) = (x + 2)^{\frac{1}{4}}$

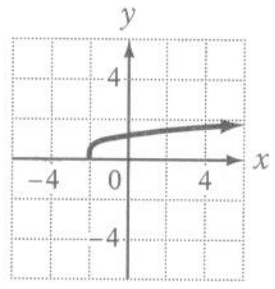

29. $F(x) = (x - 2)^{\frac{1}{2}}$

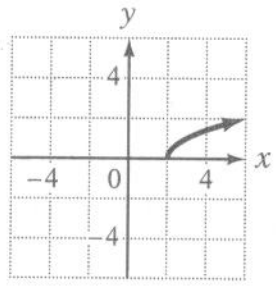

30. $f(x) = -(x - 1)^{\frac{1}{2}}$

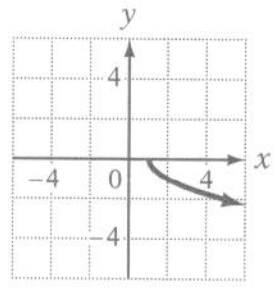

31. $Q(x) = (x - 3)^{\frac{1}{3}}$

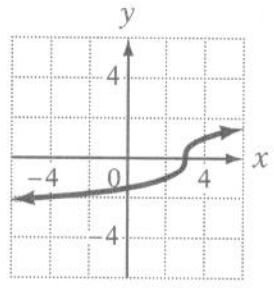

32. $H(x) = (-x)^{\frac{1}{3}}$

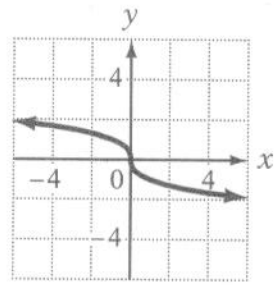

Graph.

33. $f(x) = 2x^{\frac{2}{5}} - 1$

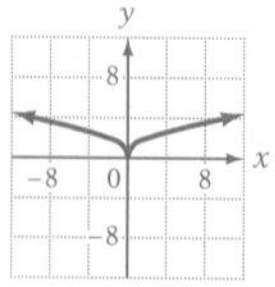

34. $h(x) = 3x^{\frac{2}{5}} + 2$

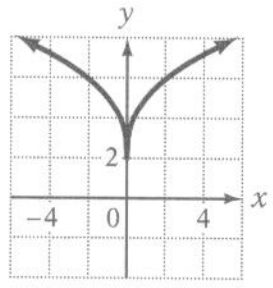

35. $g(x) = 3 - (5 - 2x)^{\frac{1}{2}}$

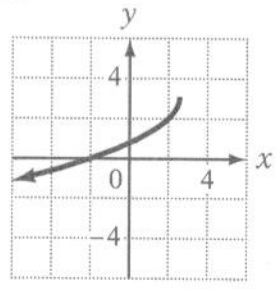

36. $y(x) = 1 + (4 - 8x)^{\frac{1}{2}}$

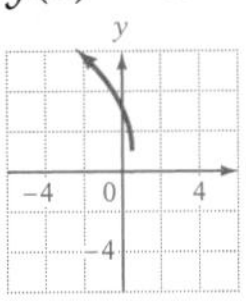

37. $V(x) = \sqrt[5]{(x - 2)^2}$

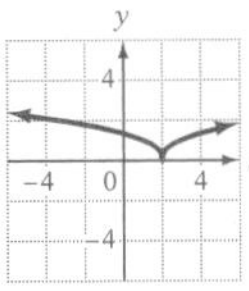

38. $A(x) = x\sqrt{3x - 9}$

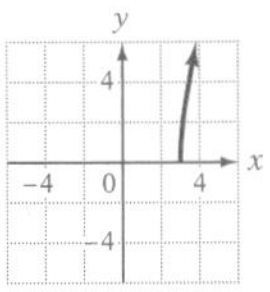

39. $F(x) = 2x - 3\sqrt[3]{x} - 1$

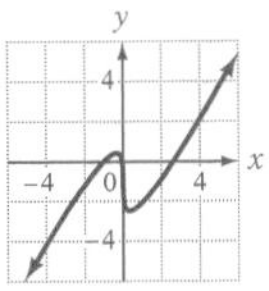

40. $f(x) = 2x - \sqrt{x} - 1$

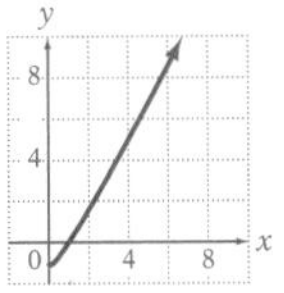

41. $f(x) = 3x\sqrt{4x + 8}$

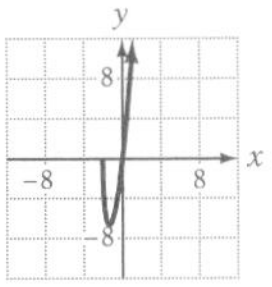

APPLYING CONCEPTS 7.4

42. *Sports* Many new major league baseball parks have a symmetric design as shown in the figure at the right. One decision that the designer must make is the shape of the outfield. One possible design uses the function

$$f(x) = k + (400 - k)\sqrt{1 - \frac{x^2}{a^2}}$$

to determine the shape of the outfield.

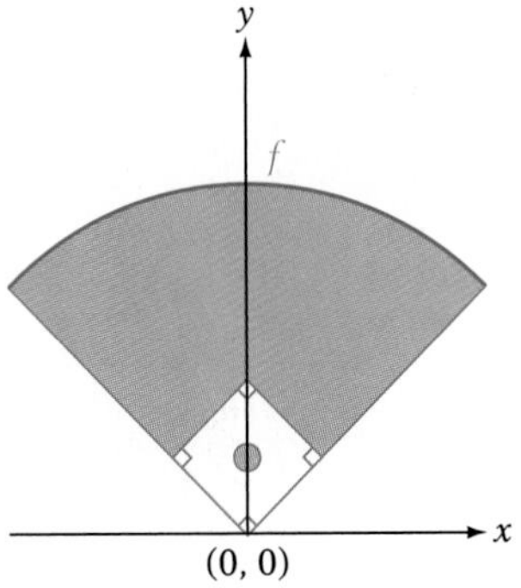

a. Graph this equation for $k = 0$, $a = 287$, and $-240 \le x \le 240$.

b. What is the maximum value of this function for the given interval? 400

c. The equation of the right-field foul line is $y = x$. Where does the foul line intersect the graph of f? That is, find the point on the graph of f for which $y = x$. Approximately (233, 233)

d. If the units on the axes are feet, what is the distance from home plate to the base of the right-field wall? 330 ft

a.

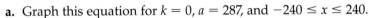

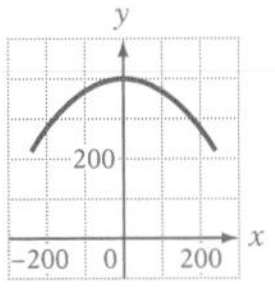

43. *Currency* According to the Bureau of Engraving and Printing, the average life span of different denominations of currency is as shown in the table at the right. The function that approximately models these data is $f(x) = 1.3x^{\frac{2}{5}}$, where x is the denomination of the bill and $f(x)$ is the average life span in years.

a. What is the domain of the function $f(x) = 1.3x^{\frac{2}{5}}$? Why is 0 not in the domain? Why are negative numbers not in the domain?

b. Use the model to approximate the life span of a \$2 bill. Round to the nearest tenth. Do you think this estimate is reasonable?

Currency	Average Life Span
\$1 bill	1.5 years
\$5 bill	2 years
\$10 bill	3 years
\$20 bill	4 years
\$50 bill	5 years
\$100 bill	9 years

44. Writing the fraction $\frac{2}{4}$ in lowest terms as $\frac{1}{2}$, it appears that $(x^2)^{\frac{1}{4}} = x^{\frac{1}{2}}$. Using $-10 \le x \le 10$, graph $f(x) = (x^2)^{\frac{1}{4}}$ and then graph $g(x) = x^{\frac{1}{2}}$. Are the graphs the same? If they are, try graphing $g(x) = x^{\frac{1}{2}}$ first and then $f(x) = (x^2)^{\frac{1}{4}}$. Explain why the graphs are not the same. In your explanation, include why $x^{\frac{2}{4}} = x^{\frac{1}{2}}$ is not always a true statement.

SECTION 7.5

Solving Equations Containing Radical Expressions

1 Solve equations containing one or more radical expressions

An equation that contains a variable expression in a radicand is a **radical equation.**

$$\sqrt[3]{2x - 5} + x = 7$$
$$\sqrt{x + 1} - \sqrt{x} = 4$$
Radical equations

The following property is used to solve a radical equation.

The Property of Raising Each Side of an Equation to a Power

If two numbers are equal, then the same powers of the numbers are equal.

If $a = b$, then $a^n = b^n$.

INSTRUCTOR NOTE

Some students may assume that the converse of this result is also true. That is, if $a^n = b^n$, then $a = b$. Use $a = -4$ and $b = 4$ to show that $(-4)^2 = 4^2$ but $-4 \neq 4$. The fact that the converse is not true is what creates the potential for extraneous solutions.

Solve: $\sqrt{x - 2} - 6 = 0$

Rewrite the equation with the radical on one side of the equation and the constant on the other side.

$$\sqrt{x - 2} - 6 = 0$$
$$\sqrt{x - 2} = 6$$

Square each side of the equation.

$$(\sqrt{x - 2})^2 = 6^2$$

Solve the resulting equation.

$$x - 2 = 36$$
$$x = 38$$

Check the solution.

Check: $\dfrac{\sqrt{x - 2} - 6 = 0}{\begin{array}{c|c} \sqrt{38 - 2} - 6 & 0 \\ \sqrt{36} - 6 & 0 \\ 6 - 6 & 0 \\ & 0 = 0 \end{array}}$

38 checks as a solution.
The solution is 38.

Example 1 Solve.   A. $\sqrt{3x - 2} - 8 = -3$ B. $\sqrt[3]{3x - 1} = -4$

Solution   A. $\sqrt{3x - 2} - 8 = -3$
$$\sqrt{3x - 2} = 5$$
▶ Rewrite the equation so that the radical is alone on one side of the equation.

$$(\sqrt{3x - 2})^2 = 5^2$$
▶ Square each side of the equation.
$$3x - 2 = 25$$
▶ Solve the resulting equation.
$$3x = 27$$
$$x = 9$$

Check: ▶ Check the solution.

$\sqrt{3x-2}-8=-3$	
$\sqrt{3\cdot 9-2}-8$	-3
$\sqrt{27-2}-8$	-3
$\sqrt{25}-8$	-3
$5-8$	-3
$-3=-3$	

The solution is 9.

B. $\sqrt[3]{3x-1}=-4$

$(\sqrt[3]{3x-1})^3=(-4)^3$ ▶ Cube each side of the equation.

$3x-1=-64$ ▶ Solve the resulting equation.

$3x=-63$

$x=-21$

Check: ▶ Check the solution.

$\sqrt[3]{3x-1}=-4$	
$\sqrt[3]{3(-21)-1}$	-4
$\sqrt[3]{-63-1}$	-4
$\sqrt[3]{-64}$	-4
$-4=-4$	

The solution is -21.

Problem 1 Solve. A. $\sqrt{4x+5}-12=-5$ B. $\sqrt[4]{x-8}=3$

Solution See page S21. A. 11 B. 89

When you raise both sides of an equation to an even power, the resulting equation may have a solution that is not a solution of the original equation. Therefore, it is necessary to check the solution of a radical equation.

Example 2 Solve. A. $x+2\sqrt{x-1}=9$ B. $\sqrt{x+7}=\sqrt{x}+1$

Solution A. $x+2\sqrt{x-1}=9$

$2\sqrt{x-1}=9-x$ ▶ Rewrite the equation with the radical on one side of the equation.

$(2\sqrt{x-1})^2=(9-x)^2$ ▶ Square each side of the equation.

$4(x-1)=81-18x+x^2$

$4x-4=81-18x+x^2$

$0=x^2-22x+85$ ▶ Write the quadratic equation in standard form.

$0=(x-5)(x-17)$ ▶ Factor.

$x-5=0$ $\quad$ $x-17=0$ ▶ Use the Principle of Zero Products.

$x=5$ $\quad$ $x=17$

LOOK CLOSELY

You must check the proposed solutions to radical equations. The proposed solutions of the equation on the right were 5 and 17. However, 17 did not check as a solution.

Check:

$$\begin{array}{r|l} x + 2\sqrt{x-1} & = 9 \\ \hline 5 + 2\sqrt{5-1} & 9 \\ 5 + 2\sqrt{4} & 9 \\ 5 + 2 \cdot 2 & 9 \\ 5 + 4 & 9 \\ 9 & = 9 \end{array} \qquad \begin{array}{r|l} x + 2\sqrt{x-1} & = 9 \\ \hline 17 + 2\sqrt{17-1} & 9 \\ 17 + 2\sqrt{16} & 9 \\ 17 + 2 \cdot 4 & 9 \\ 17 + 8 & 9 \\ 25 & \neq 9 \end{array}$$

The solution is 5.

17 does not check as a solution. It is called an **extraneous** solution of the equation.

LOOK CLOSELY

Note that

$(\sqrt{x} + 1)^2$
$= (\sqrt{x} + 1)(\sqrt{x} + 1)$
$= x + 2\sqrt{x} + 1$

B. $\sqrt{x+7} = \sqrt{x} + 1$ ▶ A radical appears on each side of the equation.

$(\sqrt{x+7})^2 = (\sqrt{x} + 1)^2$ ▶ Square each side of the equation.

$x + 7 = x + 2\sqrt{x} + 1$

$6 = 2\sqrt{x}$ ▶ Simplify the resulting equation.

$3 = \sqrt{x}$ ▶ The equation contains a radical.

$3^2 = (\sqrt{x})^2$ ▶ Square each side of the equation.

$9 = x$

Check: ▶ Check the solution.

$$\begin{array}{r|l} \sqrt{x+7} & = \sqrt{x} + 1 \\ \hline \sqrt{9+7} & \sqrt{9} + 1 \\ \sqrt{16} & 3 + 1 \\ 4 & = 4 \end{array}$$

The solution is 9.

Problem 2 Solve. A. $x + 3\sqrt{x+2} = 8$ B. $\sqrt{x+5} = 5 - \sqrt{x}$

Solution See page S21. A. 2 B. 4

2 Application problems

A right triangle contains one 90° angle. The side opposite the 90° angle is called the **hypotenuse.** The other two sides are called **legs.**

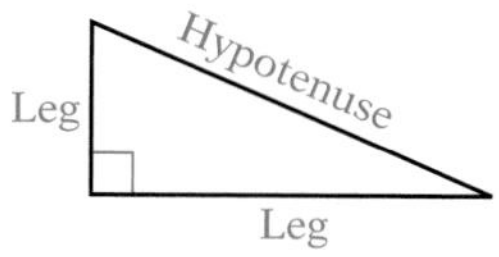

Pythagoras, a Greek mathematician, is credited with the discovery that the square of the hypotenuse of a right triangle is equal to the sum of the squares of the two legs. This is called the **Pythagorean Theorem.**

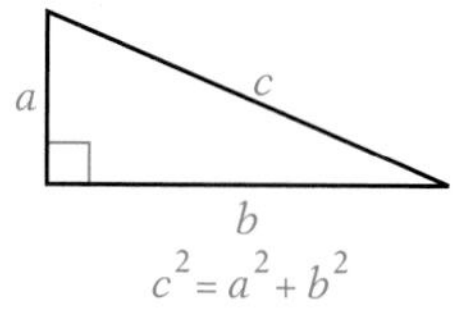

The Pythagorean Theorem

The square of the hypotenuse, c, of a right triangle is equal to the sum of the squares of the two legs, a and b.

$$c^2 = a^2 + b^2$$

Example 3 A ladder 20 ft long is leaning against a building. How high on the building will the ladder reach when the bottom of the ladder is 8 ft from the building? Round to the nearest tenth.

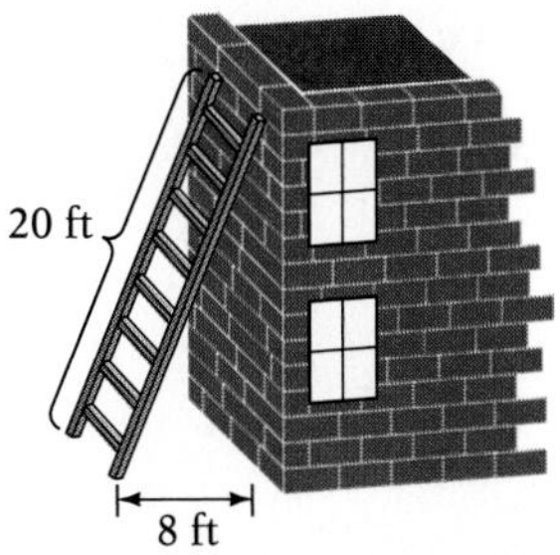

Strategy To find the distance, use the Pythagorean Theorem. The hypotenuse is the length of the ladder. One leg is the distance from the bottom of the ladder to the base of the building. The distance along the building from the ground to the top of the ladder is the unknown leg.

Solution

$$\begin{aligned} c^2 &= a^2 + b^2 \\ 20^2 &= 8^2 + b^2 \\ 400 &= 64 + b^2 \\ 336 &= b^2 \\ (336)^{\frac{1}{2}} &= (b^2)^{\frac{1}{2}} \\ \sqrt{336} &= b \\ 18.3 &\approx b \end{aligned}$$

The distance is 18.3 ft.

Problem 3 Find the diagonal of a rectangle that is 6 cm in length and 3 cm in width. Round to the nearest tenth.

Solution See page S22. 6.7 cm

CONCEPT REVIEW 7.5

Determine whether the following statements are always true, sometimes true, or never true.

1. If $a^n = b^n$, then $a = b$. Sometimes true

2. When you raise both sides of an equation to an even power, the resulting equation has a solution that is not a solution of the original equation. Sometimes true

3. If two sides of a triangle are known, the third side can be found by using the Pythagorean Theorem. Sometimes true

4. The solution of $\sqrt[3]{-x} = 8$ is -512. Always true

EXERCISES 7.5

1 1. What is the first step when solving $\sqrt{x} + 3 = 9$?

2. Why is it necessary to check the proposed solutions of a radical equation?

Solve.

3. $\sqrt{x} = 5$ 25

4. $\sqrt{y} = 2$ 4

5. $\sqrt[3]{a} = 3$ 27

6. $\sqrt[3]{y} = 5$ 125

7. $\sqrt{3x} = 12$ 48

8. $\sqrt{5x} = 10$ 20

9. $\sqrt[3]{4x} = -2$ -2

10. $\sqrt[3]{6x} = -3$ $-\frac{9}{2}$

11. $\sqrt{2x} = -4$ No solution

12. $\sqrt{5x} = -5$ No solution

13. $\sqrt{3x - 2} = 5$ 9

14. $\sqrt{5x - 4} = 9$ 17

15. $\sqrt{3 - 2x} = 7$ -23

16. $\sqrt{9 - 4x} = 4$ $-\frac{7}{4}$

17. $7 = \sqrt{1 - 3x}$ -16

18. $6 = \sqrt{8 - 7x}$ -4

19. $\sqrt[3]{4x - 1} = 2$ $\frac{9}{4}$

20. $\sqrt[3]{5x + 2} = 3$ 5

21. $\sqrt[3]{1 - 2x} = -3$ 14

22. $\sqrt[3]{3 - 2x} = -2$ $\frac{11}{2}$

23. $\sqrt[3]{9x + 1} = 4$ 7

24. $\sqrt{3x + 9} - 12 = 0$ 45

25. $\sqrt{4x - 3} - 5 = 0$ 7

26. $\sqrt{x - 2} = 4$ 18

27. $\sqrt[3]{x - 3} + 5 = 0$ -122

28. $\sqrt[3]{x - 2} = 3$ 29

29. $\sqrt[3]{2x - 6} = 4$ 35

30. $\sqrt[4]{4x + 1} = 2$ $\frac{15}{4}$

31. $\sqrt[4]{2x - 9} = 3$ 45

32. $\sqrt{2x - 3} - 2 = 1$ 6

33. $\sqrt{3x - 5} - 5 = 3$ 23

34. $\sqrt[3]{2x - 3} + 5 = 2$ -12

35. $\sqrt[3]{x - 4} + 7 = 5$ -4

36. $\sqrt{5x - 16} + 1 = 4$ 5

37. $\sqrt{3x - 5} - 2 = 3$ 10

38. $\sqrt{2x - 1} - 8 = -5$ 5

39. $\sqrt{7x + 2} - 10 = -7$ 1

40. $\sqrt[3]{4x - 3} - 2 = 3$ 32

41. $\sqrt[3]{1 - 3x} + 5 = 3$ 3

42. $1 - \sqrt{4x + 3} = -5$ $\frac{33}{4}$

43. $7 - \sqrt{3x + 1} = -1$ 21

44. $\sqrt{x + 1} = 2 - \sqrt{x}$ $\frac{9}{16}$

45. $\sqrt{2x + 4} = 3 - \sqrt{2x}$ $\frac{25}{72}$

46. $\sqrt{x^2 + 3x - 2} - x = 1$ 3

47. $\sqrt{x^2 - 4x - 1} + 3 = x$ 5

48. $\sqrt{x^2 - 3x - 1} = 3$ 5, -2

49. $\sqrt{x^2 - 2x + 1} = 3$ -2, 4

50. $\sqrt{2x + 5} - \sqrt{3x - 2} = 1$ 2

51. $\sqrt{4x + 1} - \sqrt{2x + 4} = 1$ 6

52. $\sqrt{5x - 1} - \sqrt{3x - 2} = 1$ 2, 1

53. $\sqrt{5x + 4} - \sqrt{3x + 1} = 1$ 0, 1

54. $\sqrt[4]{x^2 + 2x + 8} - 2 = 0$ -4, 2

55. $\sqrt[4]{x^2 + x - 1} - 1 = 0$ -2, 1

56. $4\sqrt{x + 1} - x = 1$ -1, 15

57. $3\sqrt{x - 2} + 2 = x$ 2, 11

58. $x + 3\sqrt{x - 2} = 12$ 6

59. $x + 2\sqrt{x + 1} = 7$ 3

2 60. *Sports* The equation $s = 16.97\sqrt[9]{n}$ can be used to predict the maximum speed, s (in feet per second), of n rowers on a scull. Find, to the nearest whole number, how many rowers are needed to travel at 20 ft/s. Does doubling the number of rowers double the maximum speed of the scull? 4 rowers; no

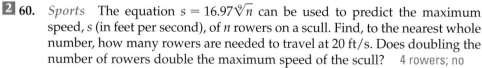

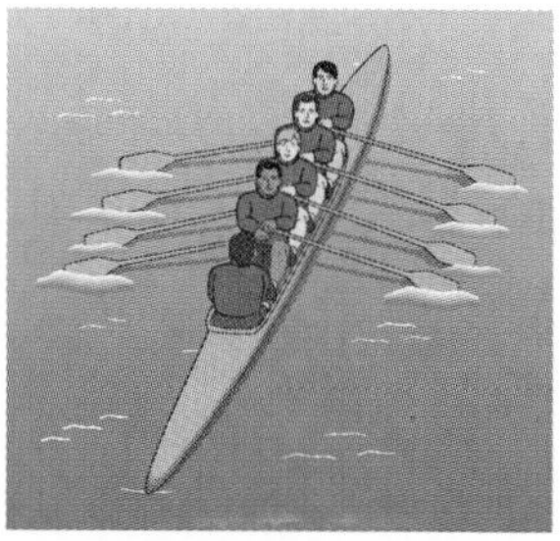

61. *Meteorology* The sustained wind velocity, v (in meters per second), in a hurricane is given by $v = 6.3\sqrt{1013 - p}$, where p is the air pressure in millibars (mb). If the velocity of the wind in a hurricane is 64 m/s, what is the air pressure? Round to the nearest tenth. What happens to wind speed in a hurricane as air pressure decreases? 909.8 mb; increases

62. *Astronomy* The time, T (in days), that it takes a planet to revolve around the sun can be approximated by the equation $T = 0.407\sqrt{d^3}$, where d is the mean distance of the planet from the sun in millions of miles. It takes Venus approximately 226 days to complete one revolution of the sun. To the nearest million miles, what is the mean distance of Venus from the sun? 68 million miles

63. *Astronomy* The time, T (in days), that it takes a moon of Saturn to revolve around Saturn can be approximated by the equation $T = 0.373\sqrt{d^3}$, where d is the mean distance of the moon from Saturn in units of 100,000 km. It takes the moon Tethys approximately 1.89 days to complete one revolution of Saturn. To the nearest 1000 km, what is the mean distance of Tethys from Saturn? 295,000 km

64. *Construction* The maximum velocity of a roller coaster depends on the vertical drop from the top of the highest hill to the bottom of that hill. The formula $v = 8\sqrt{h}$ gives the relationship between maximum velocity, v (in feet per second), and height, h (in feet). The maximum velocity of the Magnum XL-200 roller coaster in Sandusky, Ohio, is approximately 114 ft/s. How tall, to the nearest foot, is the highest hill for this roller coaster? 203 ft

65. *Health Science* The number of calories an animal uses per day (called the metabolic rate of the animal) can be approximated by $M = 126.4\sqrt[4]{W^3}$, where M is the metabolic rate and W is the weight of the animal in pounds. Find, to the nearest hundred pounds, the weight of an elephant whose metabolic rate is 60,000 calories per day. 3700 lb

66. *Construction* A 12-foot ladder is leaning against a building. How high on the building will the ladder reach when the bottom of the ladder is 4 ft from the building? Round to the nearest tenth. 11.3 ft

67. *Construction* A 26-foot ladder is leaning against a building. How far is the bottom of the ladder from the wall when the ladder reaches a height of 24 ft on the building? 10 ft

68. *Oceanography* How far would a submarine periscope have to be above the water to locate a ship 3.2 mi away? The equation for the distance in miles that the lookout can see is $d = \sqrt{1.5h}$, where h is the height in feet above the surface of the water. Round to the nearest hundredth. 6.83 ft

69. *Oceanography* How far would a submarine periscope have to be above the water to locate a ship 3.5 mi away? The equation for the distance in miles that the lookout can see is $d = \sqrt{1.5h}$, where h is the height in feet above the surface of the water. Round to the nearest hundredth. 8.17 ft

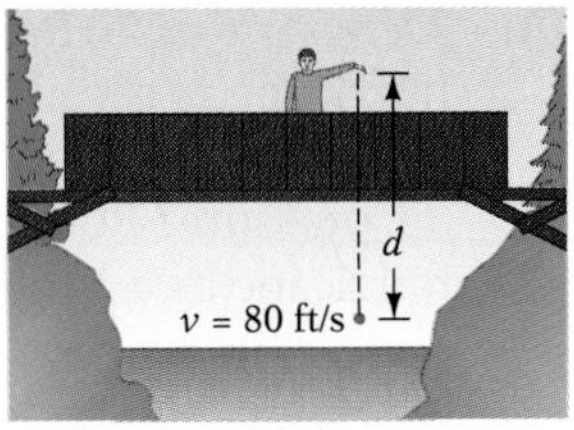

70. *Physics* An object is dropped from a bridge. Find the distance the object has fallen when the speed reaches 80 ft/s. Use the equation $v = 8\sqrt{d}$, where v is the speed of the object and d is the distance. 100 ft

71. *Physics* An object is dropped from a high building. Find the distance the object has fallen when the speed reaches 120 ft/s. Use the equation $v = 8\sqrt{d}$, where v is the speed of the object and d is the distance. 225 ft

72. *Automotive Technology* Find the distance required for a car to reach a velocity of 60 m/s when the acceleration is 10 m/s^2. Use the equation $v = \sqrt{2as}$, where v is the velocity, a is the acceleration, and s is the distance. 180 m

73. *Automotive Technology* Find the distance required for a car to reach a velocity of 48 ft/s when the acceleration is 12 ft/s^2. Use the equation $v = \sqrt{2as}$, where v is the velocity, a is the acceleration, and s is the distance. 96 ft

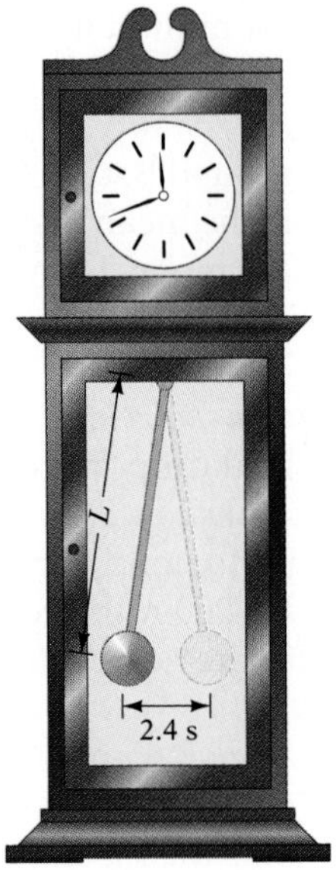

74. *Clocks* Find the length of a pendulum that makes one swing in 3 s. The equation for the time of one swing of a pendulum is $T = 2\pi\sqrt{\frac{L}{32}}$, where T is the time in seconds and L is the length in feet. Round to the nearest hundredth. 7.30 ft

75. *Clocks* Find the length of a pendulum that makes one swing in 2.4 s. The equation for the time of one swing of a pendulum is $T = 2\pi\sqrt{\frac{L}{32}}$, where T is the time in seconds and L is the length in feet. Round to the nearest hundredth. 4.67 ft

APPLYING CONCEPTS 7.5

Solve.

76. $x^{\frac{3}{4}} = 8$ 16 **77.** $x^{\frac{2}{3}} = 9$ 27 **78.** $x^{\frac{5}{4}} = 32$ 16

Solve the formula for the given variable.

79. $v = \sqrt{64d}$; d $d = \frac{v^2}{64}$

80. $a^2 + b^2 = c^2$; a $a = \sqrt{c^2 - b^2}$

81. $V = \pi r^2 h$; r $r = \frac{\sqrt{V\pi h}}{\pi h}$

82. $V = \frac{4}{3}\pi r^3$; r $r = \frac{\sqrt[3]{6V\pi^2}}{2\pi}$

Solve.

[C] **83.** $\sqrt{3x - 2} = \sqrt{2x - 3} + \sqrt{x - 1}$ 2

[C] **84.** $\sqrt{2x + 3} + \sqrt{x + 2} = \sqrt{x + 5}$ −1

[C] **85.** *Geometry* A box has a base that measures 4 in. by 6 in. The height of the box is 3 in. Find the greatest distance between two corners. Round to the nearest hundredth. 7.81 in.

[C] **86.** *Number Problem* Find three odd integers a, b, and c such that $a^2 + b^2 = c^2$. Impossible; at least one of the integers must be even.

87. *Geometry* Solve the following equations. Describe the solution by using the following terms: integer, rational number, irrational number, real number, and imaginary number. Note that more than one term may be used to describe the answer.

a. $x^2 + 3 = 7$ 2, −2; integer, rational number, real number **b.** $x^2 + 1 = 0$ $i, -i$; imaginary

c. $\frac{5}{8}x = \frac{2}{3}$ $\frac{16}{15}$; rational number, real number **d.** $x^2 + 1 = 9$ $2\sqrt{2}, -2\sqrt{2}$; irrational number, real number

e. $x^{\frac{3}{4}} = 8$ 16; integer, rational number, real number **f.** $\sqrt[3]{x} = -27$ −19,683; integer, rational number, real number

88. *Geometry* Beginning with a square with each side of length s, draw a line segment from the midpoint of the base of the square to a vertex as shown. Call this distance a. Make a rectangle whose width is that of the square and whose length is $\frac{s}{2} + a$.

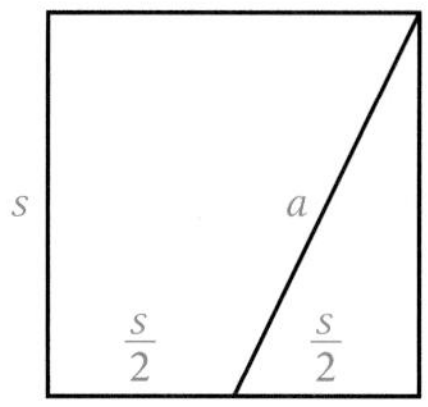

a. Find the area of the rectangle in terms of s. **a.** $A = \frac{s^2 + s^2\sqrt{5}}{2}$

b. Find the ratio of the length to the width. **b.** $\frac{1 + \sqrt{5}}{2}$

The rectangle formed in this way is called a golden rectangle. It was described at the beginning of this chapter.

89. *Geometry* Find the length of the side labeled x. $\sqrt{6}$

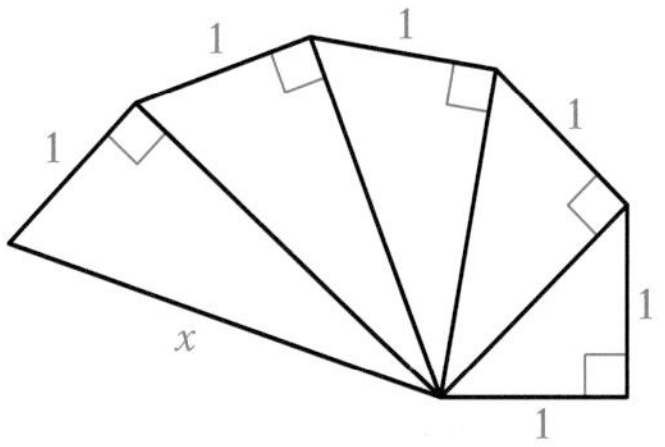

90. *Computers* As the number of computer users has grown, there have been increasing incidents of people obtaining unauthorized access to computer systems. The table below shows the number of security violations reported to the Computer Emergency Response Team for 1993 through 1997 (*Source:* www.cert.org/stats/cert_stats.html).

Year	1993	1994	1995	1996	1997
Incidents	1334	2341	2412	2573	2134

A function that models these data is given by $f(x) = 1544x^{0.3}$, where $x = 1$ corresponds to the year 1993 and $f(x)$ is the number of security incidents.

a. Write $1544x^{0.3}$ as a radical expression. $1544\sqrt[10]{x^3}$

b. Assuming that this model is valid through the year 2000, what does the model predict will be the number of security incidents in the year 2000?
2881 security incidents

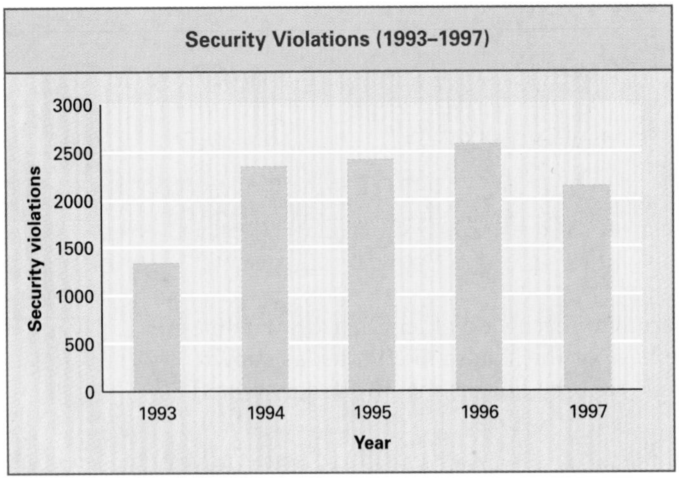

91. If a and b are nonnegative real numbers, is the equation always true, sometimes true, or never true?

$$\sqrt{a^2 + b^2} = a + b$$

Write a report that supports your answer.

92. If a and b are both positive real numbers and $a > b$, is a^b less than, equal to, or greater than b^a? Does your answer depend on a and b?

Focus on Problem Solving

Another Look at Polya's Four-Step Process

Polya's four general steps to follow when attempting to solve a problem are to understand the problem, devise a plan, carry out the plan, and review the solution. (See the Focus on Problem Solving in the chapter entitled Review of Real Numbers.) In the process of devising a plan (Step 2), it may be appropriate to write a mathematical expression or an equation. We will illustrate this with the following problem.

Number the letters of the alphabet in sequence from 1 to 26. (See the list on the next page.) Find a word for which the product of the numerical values of the letters of the word equals 1,000,000. We will agree that a "word" is any sequence of letters that contains at least one vowel but is not necessarily in the dictionary.

Understand the Problem

Consider REZB. The product of the letters is $18 \cdot 5 \cdot 26 \cdot 2 = 4680$. This "word" is a sequence of letters with at least one vowel. However, the product of the numerical values of the letters is not 1,000,000. Thus this word does not solve our problem.

A = 1
B = 2
C = 3
D = 4
E = 5
F = 6
G = 7
H = 8
I = 9
J = 10
K = 11
L = 12
M = 13
N = 14
O = 15
P = 16
Q = 17
R = 18
S = 19
T = 20
U = 21
V = 22
W = 23
X = 24
Y = 25
Z = 26

Devise a Plan

Actually, we should have known that the product of the letters in REZB could not equal 1,000,000. The letter R has a factor of 9, and the letter Z has a factor of 13. Neither of these two numbers is a factor of 1,000,000. Consequently, R and Z cannot be letters in the word we are trying to find. This observation leads to an important observation: each of the letters that make up our word must be factors of 1,000,000. To find these letters, consider the prime factorization of 1,000,000.

$$1{,}000{,}000 = 2^6 \cdot 5^6$$

Looking at the prime factorization, we note that only letters that contain only 2 or 5 as factors are possible candidates. These letters are B, D, E, H, J, P, T, and Y. One additional point: Because 1 times any number is the number, A can be part of any word we construct.

Our task is now to construct a word from these letters such that the product is 1,000,000. From the prime factorization above, we must have 2 as a factor 6 times and 5 as a factor 6 times.

Carry Out the Plan

We must construct a word with the characteristics described in our plan. Here is a possibility:

THEBEYE

Review the Solution

You should multiply the values of all of the letters and verify that the product is 1,000,000. To ensure that you have an understanding of the problem, try to find other "words" that satisfy the condition of the problem.

Projects and Group Activities

Graphing Complex Numbers

That a negative number might be the solution of an equation was not readily accepted by mathematicians until the middle of the 16th century. Hieronimo Cardano was the first to take notice of these numbers in his book *Ars Magna* (*The Great Art*), which was published in 1545. However, he called these solutions fictitious. In that same book, in one case, he also considered a solution of an equation of the form $5 + \sqrt{-15}$, a number we now call a complex number.

Complex numbers were thought to have no real or geometric interpretation and therefore to be "imaginary." One mathematician called them "ghosts of real numbers." Then, in 1806, Jean-Robert Argand published a book that included a geometric description of complex numbers. (Actually, Caspar Wessel first proposed this idea in 1797. However, his work went unnoticed.)

An Argand diagram consists of a horizontal and a vertical axis. The horizontal axis is the *real axis;* the vertical axis is the *imaginary axis.* A complex number of the form $a + bi$ is shown as the point in the plane whose coordinates are (a, b). The graphs of $2 + 4i$, $-4 - 3i$, 4, which is $4 + 0i$, and $2i$, which is $0 + 2i$, are shown at the right.

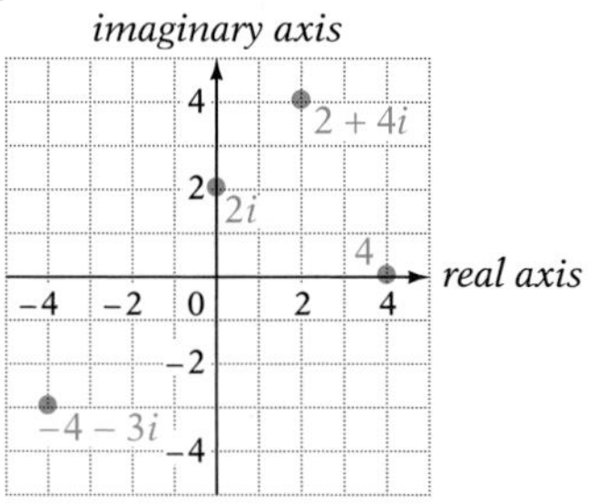

By using the Argand diagram, it is possible to give various interpretations of complex numbers and the operations on them. We will focus on only one small aspect: a geometric explanation of multiplying by i, the imaginary unit.

Before doing this, recall that real number multiplication is repeated addition. For instance, $2 \cdot 3 = 2 + 2 + 2$. Geometrically, this can be shown on a number line.

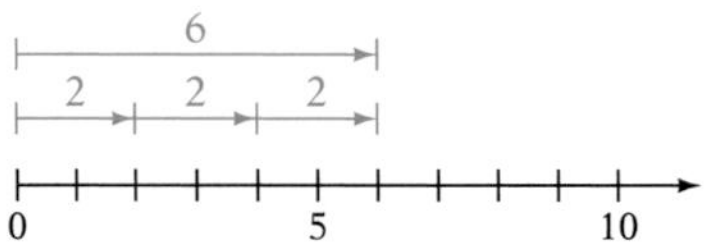

Thus, multiplying by 3 triples the size of the arrow that represents 2. In general, multiplying a real number by a number greater than 1 increases the magnitude of the number.

Consider the complex number $3 + 4i$. Multiply this number by i.

$$i(3 + 4i) = -4 + 3i$$

The two complex numbers are shown in the graph at the right.

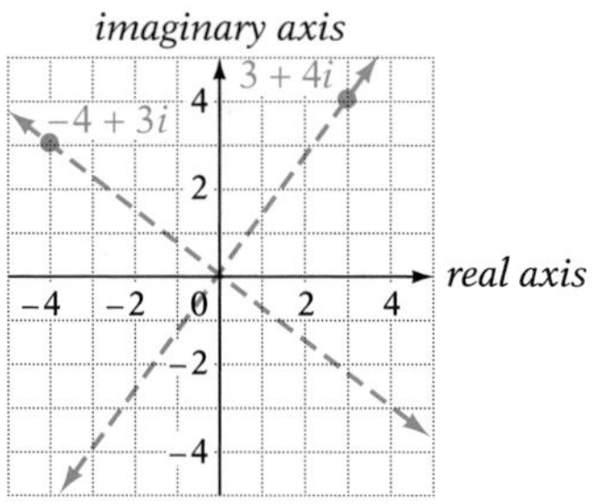

1. Find the slope of the line that passes through the origin and the point (3, 4).
2. Find the slope of the line that passes through the origin and the point (−4, 3).
3. Find the product of the slopes of the two lines. What does this product tell you?
4. Find the distance of the points (3, 4) and (−4, 3) from the origin.

A conclusion that can be drawn from the answers to the foregoing questions is that multiplying a number by i rotates that number 90°. This is shown in the figure at the right.

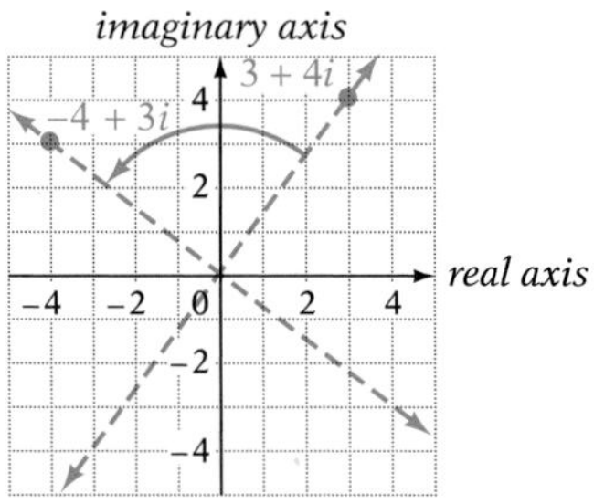

This property of complex numbers is important in the study of electricity and magnetism. Complex numbers are used extensively by electrical engineers to study electrical circuits. (Visit www.coolmath.com to find more information about how math is used in many other career choices.)

Solving Radical Equations with a Graphing Calculator

The radical equation $\sqrt{x-2}-6=0$ was solved algebraically at the beginning of this section. To solve $\sqrt{x-2}-6=0$ with a graphing calculator, use the left side of the equation and write an equation in two variables.

$$y=\sqrt{x-2}-6$$

The graph of $y=\sqrt{x-2}-6$ is shown below. The solution set of the equation $y=\sqrt{x-2}-6$ is the set of ordered pairs (x, y) whose coordinates make the equation a true statement. The x-coordinate where $y=0$ is the solution of the equation $\sqrt{x-2}-6=0$. The solution is the x-intercept of the curve given by the equation $y=\sqrt{x-2}-6$. The solution is 38.

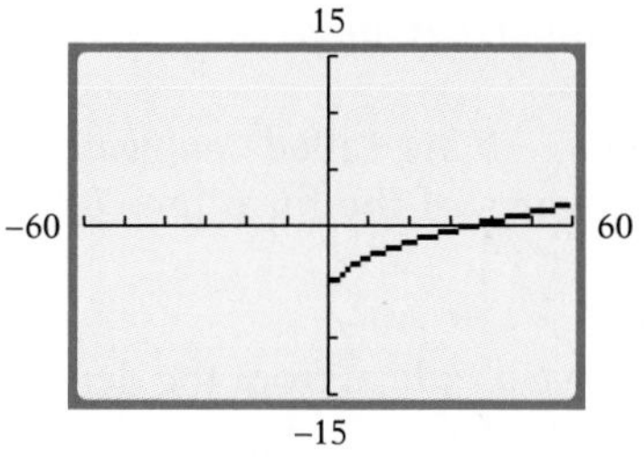

The solution of the radical equation $\sqrt[3]{x-2}=-0.8$ is 1.488. A graphing calculator with the trace feature can find any rational or irrational solution to a predetermined degree of accuracy. The graph of $y=\sqrt[3]{x-2}+0.8$ is shown below.

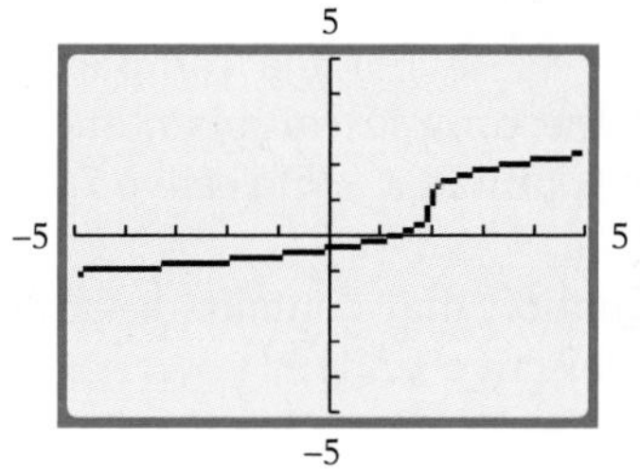

You were to solve Exercises 3–59 in this section by algebraic means. Try some of those exercises again, this time using a graphing calculator. Guidelines for using graphing calculators are found in the Appendix.

Use a graphing calculator to find the solution of each equation correct to the nearest thousandth.

1. $\sqrt{x+0.3}=1.3$

2. $\sqrt[3]{x+1.2}=-1.1$

3. $\sqrt[4]{3x-1.5}=1.4$

Chapter Summary

Key Words

The *nth root of a* is $a^{\frac{1}{n}}$. The expression $\sqrt[n]{a}$ is another symbol for the *n*th root of *a*. In the expression $\sqrt[n]{a}$, the symbol $\sqrt{}$ is called a *radical*, *n* is the *index* of the radical, and *a* is the *radicand*. (Objective 7.1.2)

$125^{\frac{1}{3}} = \sqrt[3]{125} = 5$

The index is 3 and the radicand is 125.

If $a^{\frac{1}{n}}$ is a real number, then $a^{\frac{m}{n}} = \sqrt[n]{a^m} = \left(\sqrt[n]{a}\right)^m$. (Objective 7.1.2)

$8^{\frac{2}{3}} = \sqrt[3]{8^2} = (\sqrt[3]{8})^2 = 2^2 = 4$

The symbol $\sqrt{}$ is used to indicate the positive square root, or *principal square root*, of a number. (Objective 7.1.3)

$\sqrt{16} = 4$
$-\sqrt{16} = -4$

The expressions $a + b$ and $a - b$ are called *conjugates* of each other. The product of conjugates of the form $(a + b)(a - b)$ is $a^2 - b^2$. (Objective 7.2.3)

$(x - 3)(x + 3) = x^2 - 3^2 = x^2 - 9$

The procedure used to remove a radical from the denominator of a radical expression is called *rationalizing the denominator*. (Objective 7.2.4)

$$\frac{2}{1-\sqrt{3}} = \frac{2}{1-\sqrt{3}} \cdot \frac{1+\sqrt{3}}{1+\sqrt{3}} = \frac{2(1+\sqrt{3})}{(1-\sqrt{3})(1+\sqrt{3})}$$
$$= \frac{2+2\sqrt{3}}{1-3}$$
$$= -1 - \sqrt{3}$$

A *complex number* is a number of the form $a + bi$, where *a* and *b* are real numbers and $i = \sqrt{-1}$. For the complex number $a + bi$, *a* is the *real part* of the complex number, and *b* is the *imaginary part* of the complex number. (Objective 7.3.1)

$3 + 2i$ is a complex number.

A *radical equation* is an equation that contains a variable expression in a radicand. (Objective 7.5.1)

$\sqrt{x-2} - 3 = 6$ is a radical equation.

Essential Rules and Procedures

Definition of $a^{\frac{m}{n}}$
If *m* and *n* are positive integers and $a^{\frac{1}{n}}$ is a real number, then $a^{\frac{m}{n}} = (a^{\frac{1}{n}})^m$. (Objective 7.1.1)

$8^{\frac{2}{3}} = \left(8^{\frac{1}{3}}\right)^2 = 2^2 = 4$

Definition of the *n*th root of *a*
If *a* is a real number, then $a^{\frac{1}{n}} = \sqrt[n]{a}$. (Objective 7.1.2)

$x^{\frac{1}{3}} = \sqrt[3]{x}$

Definition of the *n*th root of a^m
If $a^{\frac{1}{n}}$ is a real number, then $a^{\frac{m}{n}} = \sqrt[n]{a^m}$. (Objective 7.1.2)

$b^{\frac{3}{4}} = \sqrt[4]{b^3}$

The Product Property of Radicals
If $\sqrt[n]{a}$ and $\sqrt[n]{b}$ are positive real numbers, then $\sqrt[n]{ab} = \sqrt[n]{a}\,\sqrt[n]{b}$. (Objective 7.2.1)

$\sqrt{9 \cdot 7} = \sqrt{9}\,\sqrt{7} = 3\sqrt{7}$

The Quotient Property of Radicals
If $\sqrt[n]{a}$ and $\sqrt[n]{b}$ are positive real numbers, and $b \neq 0$, then $\sqrt[n]{\frac{a}{b}} = \frac{\sqrt[n]{a}}{\sqrt[n]{b}}$. (Objective 7.2.4)

$$\sqrt[3]{\frac{5}{27}} = \frac{\sqrt[3]{5}}{\sqrt[3]{27}} = \frac{\sqrt[3]{5}}{3}$$

Principal Square Root of a Negative Number
If a is a positive real number, then $\sqrt{-a} = i\sqrt{a}$. (Objective 7.3.1)

$$\sqrt{-8} = i\sqrt{8} = 2i\sqrt{2}$$

Addition Property of Complex Numbers
If $a + bi$ and $c + di$ are complex numbers, then $(a + bi) + (c + di) = (a + c) + (b + d)i$. (Objective 7.3.2)

$$(2 + 4i) + (3 + 6i) = (2 + 3) + (4 + 6)i = 5 + 10i$$

Subtraction Property of Complex Numbers
If $a + bi$ and $c + di$ are complex numbers, then $(a + bi) - (c + di) = (a - c) + (b - d)i$. (Objective 7.3.2)

$$(4 + 3i) - (7 + 4i) = (4 - 7) + (3 - 4)i = -3 - i$$

The Property of Raising Both Sides of an Equation to a Power
If a and b are real numbers and $a = b$, then $a^n = b^n$. (Objective 7.5.1)

If $x = 4$, then $x^2 = 16$.

The Pythagorean Theorem
$c^2 = a^2 + b^2$ (Objective 7.5.2)

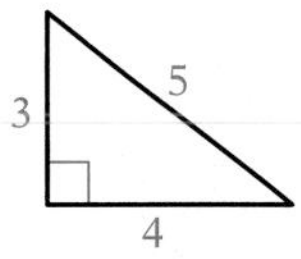

$3^2 + 4^2 = 5^2$

Chapter Review Exercises

1. Simplify: $81^{-\frac{1}{4}}$ $\frac{1}{3}$

2. Simplify: $\frac{x^{-\frac{3}{2}}}{x^{\frac{7}{2}}}$ $\frac{1}{x^5}$

3. Simplify: $(a^{16})^{-\frac{5}{8}}$ $\frac{1}{a^{10}}$

4. Simplify: $(16x^{-4}y^{12})(100x^6y^{-2})^{\frac{1}{2}}$ $\frac{160y^{11}}{x}$

5. Rewrite $3x^{\frac{3}{4}}$ as a radical expression. $3\sqrt[4]{x^3}$

6. Rewrite $7y\sqrt[3]{x^2}$ as an exponential expression. $7x^{\frac{2}{3}}y$

7. Simplify: $\sqrt[4]{81a^8b^{12}}$ $3a^2b^3$

8. Simplify: $-\sqrt{49x^6y^{16}}$ $-7x^3y^8$

9. Simplify: $\sqrt[3]{-8a^6b^{12}}$ $-2a^2b^4$

10. Simplify: $\sqrt{18a^3b^6}$ $3ab^3\sqrt{2a}$

11. Simplify: $\sqrt[5]{-64a^8b^{12}}$ $-2ab^2\sqrt[5]{2a^3b^2}$

12. Simplify: $\sqrt[4]{x^6y^8z^{10}}$ $xy^2z^2\sqrt{xz}$

13. Add: $\sqrt{54} + \sqrt{24}$ $5\sqrt{6}$

14. Subtract: $\sqrt{48x^5y} - x\sqrt{80x^3y}$ $4x^2\sqrt{3xy} - 4x^2\sqrt{5xy}$

15. Subtract: $\sqrt{50a^4b^3} - ab\sqrt{18a^2b}$ $2a^2b\sqrt{2b}$

16. Simplify: $4x\sqrt{12x^2y} + \sqrt{3x^4y} - x^2\sqrt{27y}$ $6x^2\sqrt{3y}$

17. Multiply: $\sqrt{32}\sqrt{50}$ 40

18. Multiply: $\sqrt[3]{16x^4y}\sqrt[3]{4xy^5}$ $4xy^2\sqrt[3]{x^2}$

19. Multiply: $\sqrt{3x}(3 + \sqrt{3x})$ $3x + 3\sqrt{3x}$

20. Multiply: $(5 - \sqrt{6})^2$ $31 - 10\sqrt{6}$

21. Multiply: $(\sqrt{3} + 8)(\sqrt{3} - 2)$ $-13 + 6\sqrt{3}$

22. Simplify: $\dfrac{\sqrt{125x^6}}{\sqrt{5x^3}}$ $5x\sqrt{x}$

23. Simplify: $\dfrac{8}{\sqrt{3y}}$ $\dfrac{8\sqrt{3y}}{3y}$

24. Simplify: $\dfrac{x + 2}{\sqrt{x} + \sqrt{2}}$ $\dfrac{x\sqrt{x} - x\sqrt{2} + 2\sqrt{x} - 2\sqrt{2}}{x - 2}$

25. Simplify: $\dfrac{\sqrt{x} + \sqrt{y}}{\sqrt{x} - \sqrt{y}}$ $\dfrac{x + 2\sqrt{xy} + y}{x - y}$

26. Simplify: $\sqrt{-36}$ $6i$

27. Simplify: $\sqrt{-50}$ $5i\sqrt{2}$

28. Simplify: $\sqrt{49} - \sqrt{-16}$ $7 - 4i$

29. Simplify: $\sqrt{200} + \sqrt{-12}$ $10\sqrt{2} + 2i\sqrt{3}$

30. Add: $(5 + 2i) + (4 - 3i)$ $9 - i$

31. Subtract: $(-8 + 3i) - (4 - 7i)$ $-12 + 10i$

32. Add: $(9 - \sqrt{-16}) + (5 + \sqrt{-36})$ $14 + 2i$

33. Subtract: $(\sqrt{50} + \sqrt{-72}) - (\sqrt{162} - \sqrt{-8})$ $-4\sqrt{2} + 8i\sqrt{2}$

34. Add: $(3 - 9i) + 7$ $10 - 9i$

35. Multiply: $(8i)(2i)$ -16

36. Multiply: $i(3 - 7i)$ $7 + 3i$

37. Multiply: $\sqrt{-12}\sqrt{-6}$ $-6\sqrt{2}$

38. Multiply: $(6 - 5i)(4 + 3i)$ $39 - 2i$

39. Simplify: $\dfrac{-6}{i}$ $6i$

40. Simplify: $\dfrac{5 + 2i}{3i}$ $\dfrac{2}{3} - \dfrac{5}{3}i$

41. Simplify: $\dfrac{7}{2 - i}$ $\dfrac{14}{5} + \dfrac{7}{5}i$

42. Simplify: $\dfrac{\sqrt{16}}{\sqrt{4} - \sqrt{-4}}$ $1 + i$

43. Simplify: $\dfrac{5 + 9i}{1 - i}$ $-2 + 7i$

44. Solve: $\sqrt[3]{9x} = -6$ -24

45. State the domain of $f(x) = \sqrt[4]{3x - 2}$ in set-builder notation. $\left\{x \mid x \geq \dfrac{2}{3}\right\}$

46. State the domain of $f(x) = \sqrt[3]{x - 5}$ in set-builder notation. $\{x \mid x \in \text{real numbers}\}$

47. Graph: $g(x) = 3x^{\frac{1}{3}}$

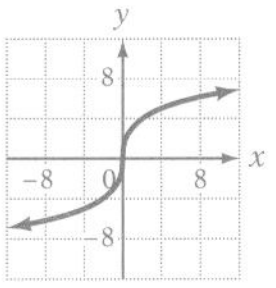

48. Graph: $f(x) = \sqrt{x - 3}$

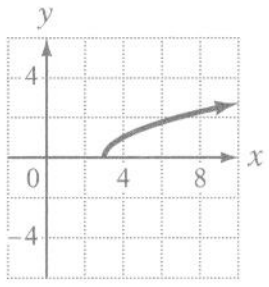

49. Solve: $\sqrt[3]{3x - 5} = 2$ $\frac{13}{3}$

50. Solve: $\sqrt{4x + 9} + 10 = 11$ -2

51. Find the width of a rectangle that has a diagonal of 13 in. and a length of 12 in.   5 in.

52. The velocity of the wind determines the amount of power generated by a windmill. A typical equation for this relationship is $v = 4.05\sqrt[3]{P}$, where v is the velocity in miles per hour and P is the power in watts. Find the amount of power generated by a wind of 20 mph. Round to the nearest whole number. 120 watts

53. Find the distance required for a car to reach a velocity of 88 ft/s when the acceleration is 16 ft/s^2. Use the equation $v = \sqrt{2as}$, where v is the velocity, a is the acceleration, and s is the distance.   242 ft

54. A 12-foot ladder is leaning against a building. How far from the building is the bottom of the ladder when the top of the ladder touches the building 10 ft above the ground? Round to the nearest hundredth.   6.63 ft

Chapter Test

1. Simplify: $\dfrac{r^{\frac{2}{3}}r^{-1}}{r^{-\frac{1}{2}}}$ $r^{\frac{1}{6}}$

2. Simplify: $\dfrac{\left(2x^{\frac{1}{3}}y^{-\frac{2}{3}}\right)^6}{(x^{-4}y^8)^{\frac{1}{4}}}$ $\dfrac{64x^3}{y^6}$

3. Simplify: $\left(\dfrac{4a^4}{b^2}\right)^{-\frac{3}{2}}$ $\dfrac{b^3}{8a^6}$

4. Rewrite $3y^{\frac{2}{5}}$ as a radical expression. $3\sqrt[5]{y^2}$

5. Rewrite $\frac{1}{2}\sqrt[4]{x^3}$ as an exponential expression. $\frac{1}{2}x^{\frac{3}{4}}$

6. State the domain of $f(x) = \sqrt{4 - x}$ in set-builder notation. $\{x \mid x \le 4\}$

7. State the domain of $f(x) = (2x - 3)^{\frac{1}{3}}$ in interval notation. $(-\infty, \infty)$

8. Simplify: $\sqrt[3]{27a^4b^3c^7}$ $3abc^2\sqrt[3]{ac}$

9. Add: $\sqrt{18a^3} + a\sqrt{50a}$ $8a\sqrt{2a}$

10. Subtract: $\sqrt[3]{54x^7y^3} - x\sqrt[3]{128x^4y^3} - x^2\sqrt[3]{2xy^3}$ $-2x^2y\sqrt[3]{2x}$

11. Multiply: $\sqrt{3x}(\sqrt{x} - \sqrt{25x})$ $-4x\sqrt{3}$

12. Multiply: $(2\sqrt{3} + 4)(3\sqrt{3} - 1)$ $14 + 10\sqrt{3}$

13. Multiply: $(\sqrt{a} - 3\sqrt{b})(2\sqrt{a} + 5\sqrt{b})$
$2a - \sqrt{ab} - 15b$

14. Multiply: $(2\sqrt{x} + \sqrt{y})^2$ $4x + 4\sqrt{xy} + y$

15. Simplify: $\dfrac{\sqrt{32x^5y}}{\sqrt{2xy^3}}$ $\dfrac{4x^2}{y}$

16. Simplify: $\dfrac{4 - 2\sqrt{5}}{2 - \sqrt{5}}$ 2

17. Simplify: $\dfrac{\sqrt{x}}{\sqrt{x} - \sqrt{y}}$ $\dfrac{x + \sqrt{xy}}{x - y}$

18. Multiply: $(\sqrt{-8})(\sqrt{-2})$ -4

19. Subtract: $(5 - 2i) - (8 - 4i)$ $-3 + 2i$

20. Multiply: $(2 + 5i)(4 - 2i)$ $18 + 16i$

21. Simplify: $\dfrac{2 + 3i}{1 - 2i}$ $-\dfrac{4}{5} + \dfrac{7}{5}i$

22. Add: $(2 + i) + (2 - i)$ 4

23. Solve: $\sqrt{x + 12} - \sqrt{x} = 2$ 4

24. Solve: $\sqrt[3]{2x - 2} + 4 = 2$ -3

25. An object is dropped from a high building. Find the distance the object has fallen when the speed reaches 192 ft/s. Use the equation $v = \sqrt{64d}$, where v is the speed of the object and d is the distance. 576 ft

Cumulative Review Exercises

1. Identify the property that justifies the statement.
$(a + 2)b = ab + 2b$ The Distributive Property

2. Simplify: $2x - 3[x - 2(x - 4) + 2x]$ $-x - 24$

3. Find $A \cap B$ given $A = \{2, 4, 6\}$ and $B = \{1, 3, 5\}$.
$\varnothing$

4. Solve: $\sqrt[3]{2x - 5} + 3 = 6$ 16

5. Solve: $5 - \frac{2}{3}x = 4$ $\frac{3}{2}$

6. Solve: $2[4 - 2(3 - 2x)] = 4(1 - x)$ $\frac{2}{3}$

7. Solve: $3x - 4 \le 8x + 1$
Write the solution set in set-builder notation.
$\{x \mid x \ge -1\}$

8. Solve: $5 < 2x - 3 < 7$
Write the solution set in set-builder notation.
$\{x \mid 4 < x < 5\}$

9. Solve: $|7 - 3x| > 1$ $\left\{x \mid x < 2 \text{ or } x > \frac{8}{3}\right\}$

10. Factor: $64a^2 - b^2$ $(8a + b)(8a - b)$

11. Factor: $x^5 + 2x^3 - 3x$ $x(x^2 + 3)(x + 1)(x - 1)$

12. Solve: $3x^2 + 13x - 10 = 0$ $\frac{2}{3}, -5$

13. Graph: $g(x) = \sqrt{1 - x}$

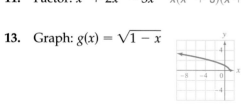

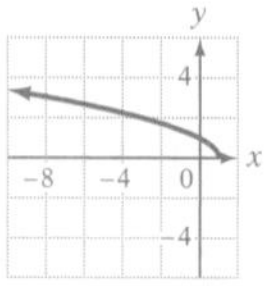

14. Is the line $x - 2y = 4$ perpendicular to the line $2x + y = 4$? Yes

15. Simplify: $(3^{-1}x^3y^{-5})(3^{-1}y^{-2})^{-2}$ $\frac{3x^3}{y}$

16. Simplify: $\left(\frac{x^{-\frac{1}{2}}y^{\frac{3}{4}}}{y^{-\frac{5}{4}}}\right)^4$ $\frac{y^8}{x^2}$

17. Subtract: $\sqrt{20x^3} - x\sqrt{45x}$ $-x\sqrt{5x}$

18. Multiply: $(\sqrt{5} - 3)(\sqrt{5} - 2)$ $11 - 5\sqrt{5}$

19. Simplify: $\frac{\sqrt[3]{4x^5y^4}}{\sqrt[3]{8x^2y^5}}$ $\frac{x\sqrt[3]{4y^2}}{2y}$

20. Simplify: $\frac{3i}{2 - i}$ $-\frac{3}{5} + \frac{6}{5}i$

21. Graph: $\{x \mid x > -1\} \cap \{x \mid x \le 3\}$

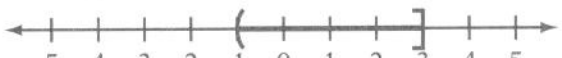

22. State the domain of $g(x) = \sqrt[5]{2x + 5}$ in set-builder notation. $\{x \mid x \in \text{real numbers}\}$

23. Given $f(x) = 3x^2 - 2x + 1$, evaluate $f(-3)$. 34

24. Find an equation of the line that passes through the points (2, 3) and (−1, 2). $y = \frac{1}{3}x + \frac{7}{3}$

25. Evaluate the determinant: $\begin{vmatrix} 1 & 2 & -3 \\ 0 & -1 & 2 \\ 3 & 1 & -2 \end{vmatrix}$ 3

26. Solve by using Cramer's Rule:

$$2x - y = 4$$
$$-2x + 3y = 5$$

$\left(\frac{17}{4}, \frac{9}{2}\right)$

27. Find the slope and y-intercept of the equation $3x - 2y = -6$ and then graph the line.

$m = \frac{3}{2}$

$b = 3$

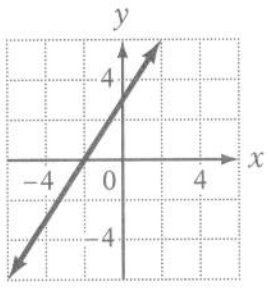

28. Graph the solution set of $3x + 2y \le 4$.

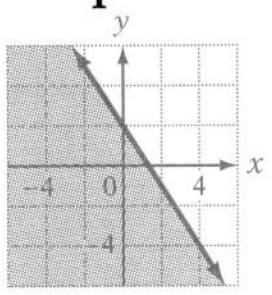

29. A collection of thirty stamps consists of 13¢ stamps and 18¢ stamps. The total value of the stamps is \$4.85. Find the number of 18¢ stamps. 19 stamps

30. An investment of \$2500 is made at an annual simple interest rate of 7.2%. How much additional money must be invested at an annual simple interest rate of 8.4% so that the total interest earned is \$516? \$4000

31. The width of a rectangle is 6 ft less than the length. The area of the rectangle is 72 ft^2. Find the length and width of the rectangle. Length: 12 ft; width: 6 ft

32. A sales executive traveled 25 mi by car and then an additional 625 mi by plane. The rate by plane was five times the rate by car. The total time of the trip was 3 h. Find the rate of the plane. 250 mph

33. How long does it take light to travel to Earth from the moon when the moon is 232,500 mi from Earth? Light travels 1.86×10^5 mi/s. 1.25 s

34. How far would a submarine periscope have to be above the water to locate a ship 7 mi away? The equation for the distance in miles that the lookout can see is $d = \sqrt{1.5h}$, where h is the height in feet above the surface of the water. Round to the nearest tenth. 32.7 ft

35. The graph shows the amount invested and the annual interest income from an investment. Find the slope of the line between the two points shown on the graph. Then write a sentence that states the meaning of the slope.

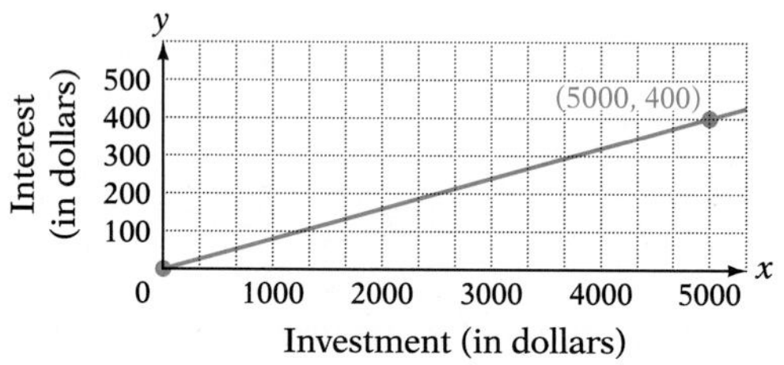

$m = 0.08$; a slope of 0.08 means that the annual interest income is 8% of the investment.

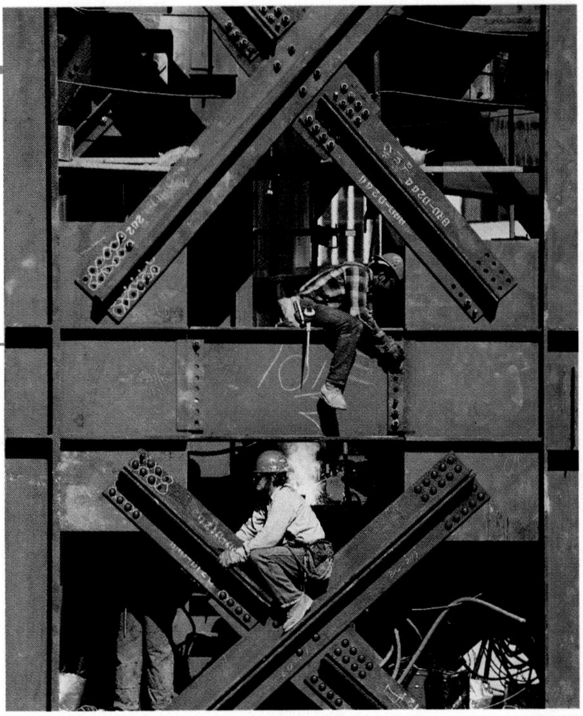

Building inspectors assess the structural integrity of buildings. Inspectors may specialize in structural steel or reinforced concrete structures. Before construction begins, plan examiners determine whether the plans for the building or other structure comply with building code regulations and are suited to the engineering and environmental demands of the building site. Some of the equations that determine the loads various beams can safely carry are quadratic equations, the subject of this chapter.

8

Quadratic Equations and Inequalities

OBJECTIVES

- 8.1.1 Solve quadratic equations by factoring
- 8.1.2 Write a quadratic equation given its solutions
- 8.1.3 Solve quadratic equations by taking square roots
- 8.2.1 Solve quadratic equations by completing the square
- 8.2.2 Solve quadratic equations by using the quadratic formula
- 8.3.1 Equations that are quadratic in form
- 8.3.2 Radical equations
- 8.3.3 Fractional equations
- 8.4.1 Application problems
- 8.5.1 Solve inequalities by factoring
- 8.5.2 Solve rational inequalities
- 8.6.1 Graph a quadratic function
- 8.6.2 Find the x-intercepts of a parabola
- 8.7.1 Minimum and maximum problems
- 8.7.2 Applications of minimum and maximum

Complex Numbers

Negative numbers were not universally accepted in the mathematical community until well into the 14th century. It is no wonder, then, that *imaginary numbers* took an even longer time to gain acceptance.

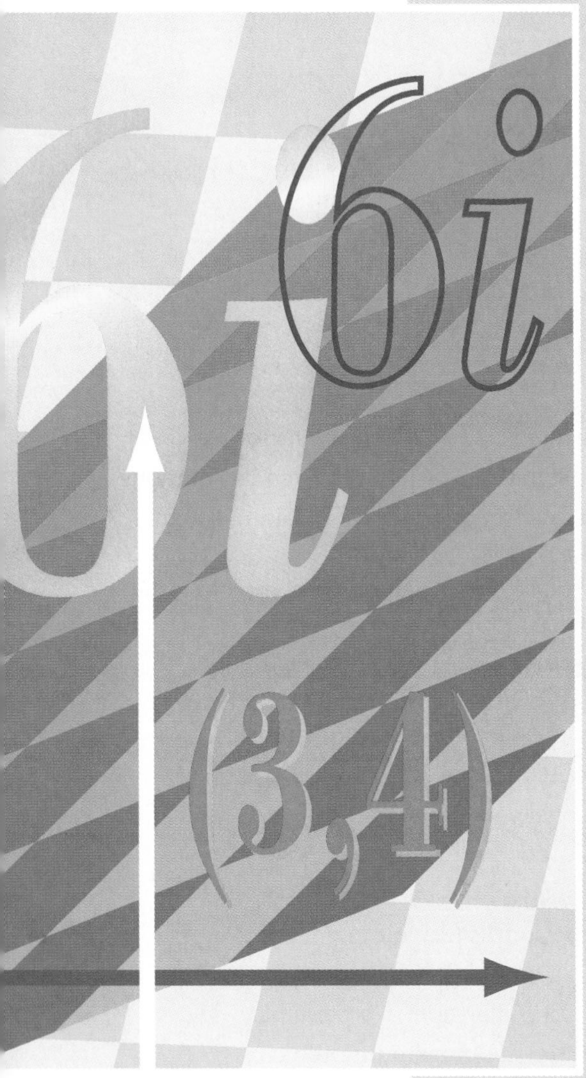

In the mid-16th century, mathematicians were beginning to integrate imaginary numbers into their writings. One notation for $3i$ was R (0 m 3). Literally this was interpreted as $\sqrt{0-3}$.

By the mid-18th century, the symbol i was introduced. Still later it was shown that complex numbers could be thought of as points in the plane. The complex number $3 + 4i$ was associated with the point $(3, 4)$.

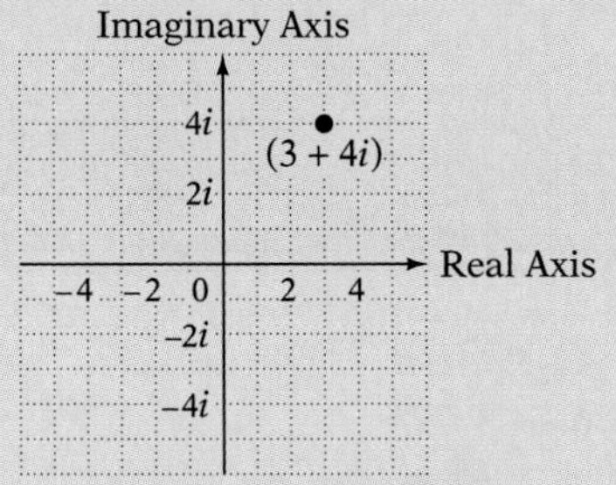

By the end of the 19th century, complex numbers were fully integrated into mathematics. This was due in large part to some eminent mathematicians who used complex numbers to prove theorems that had previously eluded proof.

Solving Quadratic Equations by Factoring or by Taking Square Roots

1 Solve quadratic equations by factoring

A **quadratic equation** is an equation of the form $ax^2 + bx + c = 0$, where a and b are coefficients, c is a constant, and $a \neq 0$.

$$3x^2 - x + 2 = 0, \quad a = 3, \quad b = -1, \quad c = 2$$
$$-x^2 + 4 = 0, \quad a = -1, \quad b = 0, \quad c = 4$$
$$6x^2 - 5x = 0, \quad a = 6, \quad b = -5, \quad c = 0$$

A quadratic equation is in **standard form** when the polynomial is in descending order and equal to zero.

Because the degree of the polynomial $ax^2 + bx + c$ is 2, a quadratic equation is also called a **second-degree equation.**

In the chapter entitled "Polynomials and Exponents," the Principle of Zero Products was used to solve some quadratic equations. That procedure is reviewed here.

The Principle of Zero Products

If the product of two factors is zero, then at least one of the factors must be zero.

If $ab = 0$, then $a = 0$ or $b = 0$.

➡ Solve by factoring: $x^2 - 6x = -9$

$$x^2 - 6x = -9$$

Write the equation in standard form. $x^2 - 6x + 9 = 0$

$$(x - 3)(x - 3) = 0$$

Use the Principle of Zero Products. $x - 3 = 0 \qquad x - 3 = 0$

Solve each equation. $x = 3 \qquad x = 3$

3 checks as a solution.

Write the solutions.   The solution is 3.

When a quadratic equation has two solutions that are the same number, the solution is called a **double root** of the equation. The solution 3 is a double root of the equation $x^2 - 6x = -9$.

Example 1 Solve for x by factoring: $x^2 - 4ax - 5a^2 = 0$

Solution $x^2 - 4ax - 5a^2 = 0$ ▶ This is a literal equation. Solve for x in terms of a.

$(x + a)(x - 5a) = 0$ ▶ Factor.

$x + a = 0 \qquad x - 5a = 0$

$x = -a \qquad x = 5a$

The solutions are $-a$ and $5a$.

Problem 1 Solve for x by factoring: $x^2 - 3ax - 4a^2 = 0$

Solution See page S22. $-a, 4a$

2 Write a quadratic equation given its solutions

31 HM³

INSTRUCTOR NOTE

Here is an extension of these ideas.

$(x - r_1)(x - r_2)$
$= x^2 - (r_1 + r_2)x + r_1r_2$

Thus the coefficient of x is the opposite of the sum of the roots, and the constant term is the product of the roots. This can be used as a quick way of checking the solution of a quadratic equation in the next section. For instance, $1 + \sqrt{2}$ and $1 - \sqrt{2}$ are solutions of the equation $x^2 - 2x - 1 = 0$ because $-[(1 + \sqrt{2}) + (1 - \sqrt{2})] = -2$, the coefficient of x, and $(1 + \sqrt{2})(1 - \sqrt{2}) = -1$, the constant term.

As shown below, the solutions of the equation $(x - r_1)(x - r_2) = 0$ are r_1 and r_2.

$(x - r_1)(x - r_2) = 0$

$x - r_1 = 0 \qquad x - r_2 = 0$

$x = r_1 \qquad x = r_2$

Check:

$$\begin{array}{c|c} (x - r_1)(x - r_2) = 0 & \\ \hline (r_1 - r_1)(r_1 - r_2) & 0 \\ 0 \cdot (r_1 - r_2) & \\ 0 = 0 & \end{array} \qquad \begin{array}{c|c} (x - r_1)(x - r_2) = 0 & \\ \hline (r_2 - r_1)(r_2 - r_2) & 0 \\ (r_2 - r_1) \cdot 0 & \\ 0 = 0 & \end{array}$$

Using the equation $(x - r_1)(x - r_2) = 0$ and the fact that r_1 and r_2 are solutions of this equation, it is possible to write a quadratic equation given its solutions.

⇨ Write a quadratic equation that has solutions 4 and −5.

$(x - r_1)(x - r_2) = 0$

Replace r_1 by 4 and r_2 by −5. $(x - 4)[x - (-5)] = 0$

Simplify. $(x - 4)(x + 5) = 0$

Multiply. $x^2 + x - 20 = 0$

Example 2 Write a quadratic equation that has integer coefficients and has solutions $\frac{2}{3}$ and $\frac{1}{2}$.

Solution $(x - r_1)(x - r_2) = 0$

$\left(x - \frac{2}{3}\right)\left(x - \frac{1}{2}\right) = 0$ ▶ Replace r_1 by $\frac{2}{3}$ and r_2 by $\frac{1}{2}$.

$x^2 - \frac{7}{6}x + \frac{1}{3} = 0$ ▶ Multiply.

$6\left(x^2 - \frac{7}{6}x + \frac{1}{3}\right) = 6 \cdot 0$ ▶ Multiply each side of the equation by the LCM of the denominators.

$6x^2 - 7x + 2 = 0$

Problem 2 Write a quadratic equation that has integer coefficients and has solutions $-\frac{2}{3}$ and $\frac{1}{6}$.

Solution See page S22. $18x^2 + 9x - 2 = 0$

31 HM³

3 Solve quadratic equations by taking square roots

The solution of the quadratic equation $x^2 = 16$ is shown at the right.

$$x^2 = 16$$
$$x^2 - 16 = 0$$
$$(x + 4)(x - 4) = 0$$

Note that the solution is the positive or the negative square root of 16, 4 or -4.

$$x + 4 = 0 \qquad x - 4 = 0$$
$$x = -4 \qquad x = 4$$

INSTRUCTOR NOTE

Another way to demonstrate that $x^2 = a$ implies that $x = \pm\sqrt{a}$ is to rely on the fact that $\sqrt{x^2} = |x|$. Thus $x^2 = a$ implies that $|x| = \sqrt{a}$, or that $x = \pm\sqrt{a}$.

The solution can also be found by taking the square root of each side of the equation and writing the positive and the negative square roots of the number. The notation $x = \pm 4$ means $x = 4$ or $x = -4$.

$$x^2 = 16$$
$$\sqrt{x^2} = \sqrt{16}$$
$$x = \pm\sqrt{16}$$
$$x = \pm 4$$

The solutions are 4 and -4.

Solve by taking square roots: $3x^2 = 54$

$$3x^2 = 54$$

Solve for x^2.

$$x^2 = 18$$

Take the square root of each side of the equation.

$$\sqrt{x^2} = \sqrt{18}$$
$$x = \pm\sqrt{18}$$

Simplify.

$$x = \pm 3\sqrt{2}$$

$3\sqrt{2}$ and $-3\sqrt{2}$ check as solutions.

Write the solutions.

The solutions are $3\sqrt{2}$ and $-3\sqrt{2}$.

Solving a quadratic equation by taking the square root of each side of the equation can lead to solutions that are complex numbers.

Solve by taking square roots: $2x^2 + 18 = 0$

Solve for x^2.

$$2x^2 + 18 = 0$$
$$2x^2 = -18$$
$$x^2 = -9$$

Take the square root of each side of the equation.

$$\sqrt{x^2} = \sqrt{-9}$$
$$x = \pm\sqrt{-9}$$

Simplify.

$$x = \pm 3i$$

$3i$ and $-3i$ check as solutions.

Write the solutions.

The solutions are $3i$ and $-3i$.

An equation containing the square of a binomial can be solved by taking square roots.

Example 3 Solve by taking square roots: $3(x-2)^2 + 12 = 0$

Solution

$$3(x-2)^2 + 12 = 0$$
$$3(x-2)^2 = -12 \quad \text{▶ Solve for } (x-2)^2.$$
$$(x-2)^2 = -4$$
$$\sqrt{(x-2)^2} = \sqrt{-4} \quad \text{▶ Take the square root of each side of the equation. Then simplify.}$$
$$x-2 = \pm\sqrt{-4}$$
$$x-2 = \pm 2i$$
$$x-2 = 2i \qquad x-2 = -2i \quad \text{▶ Solve for } x.$$
$$x = 2+2i \qquad x = 2-2i$$

The solutions are $2+2i$ and $2-2i$.

Problem 3 Solve by taking square roots: $2(x+1)^2 + 24 = 0$

Solution See page S22. $-1 \pm 2i\sqrt{3}$

CONCEPT REVIEW 8.1

Determine whether the following statements are always true, sometimes true, or never true.

1. A quadratic equation has two real roots. Sometimes true
2. A quadratic equation is also called a second-degree equation. Always true
3. If $(x-3)(x+4) = 8$, then $x-3 = 8$ or $x+4 = 8$. Never true
4. A quadratic equation has one root. Sometimes true

EXERCISES 8.1

1 **1.** What is the Principle of Zero Products?

2. How can the Principle of Zero Products be used to solve a quadratic equation?

Solve by factoring.

3. $x^2 - 4x = 0$ 0, 4

4. $y^2 + 6y = 0$ −6, 0

5. $t^2 - 25 = 0$ ±5

6. $p^2 - 81 = 0$ ±9

7. $s^2 - s - 6 = 0$ −2, 3

8. $v^2 + 4v - 5 = 0$ −5, 1

9. $y^2 - 6y + 9 = 0$ 3

10. $x^2 + 10x + 25 = 0$ −5

11. $9z^2 - 18z = 0$ 0, 2

12. $4y^2 + 20y = 0$ 0, −5

13. $r^2 - 3r = 10$ −2, 5

14. $p^2 + 5p = 6$ 1, −6

15. $v^2 + 10 = 7v$ 2, 5

16. $t^2 - 16 = 15t$ −1, 16

17. $2x^2 - 9x - 18 = 0$ $6, -\frac{3}{2}$

18. $3y^2 - 4y - 4 = 0$ $-\frac{2}{3}, 2$

19. $4z^2 - 9z + 2 = 0$ $\frac{1}{4}, 2$

20. $2s^2 - 9s + 9 = 0$ $\frac{3}{2}, 3$

21. $3w^2 + 11w = 4$ $-4, \frac{1}{3}$

22. $2r^2 + r = 6$ $\frac{3}{2}, -2$

23. $6x^2 = 23x + 18$ $-\frac{2}{3}, \frac{9}{2}$

24. $6x^2 = 7x - 2$ $\frac{1}{2}, \frac{2}{3}$

25. $4 - 15u - 4u^2 = 0$ $-4, \frac{1}{4}$

26. $3 - 2y - 8y^2 = 0$ $-\frac{3}{4}, \frac{1}{2}$

27. $x + 18 = x(x - 6)$ $-2, 9$

28. $t + 24 = t(t + 6)$ $-8, 3$

29. $4s(s + 3) = s - 6$ $-2, -\frac{3}{4}$

30. $3v(v - 2) = 11v + 6$ $-\frac{1}{3}, 6$

31. $u^2 - 2u + 4 = (2u - 3)(u + 2)$ $2, -5$

32. $(3v - 2)(2v + 1) = 3v^2 - 11v - 10$ $-\frac{4}{3}, -2$

33. $(3x - 4)(x + 4) = x^2 - 3x - 28$ $-\frac{3}{2}, -4$

Solve for x by factoring.

34. $x^2 + 14ax + 48a^2 = 0$ $-6a, -8a$

35. $x^2 - 9bx + 14b^2 = 0$ $2b, 7b$

36. $x^2 + 9xy - 36y^2 = 0$ $3y, -12y$

37. $x^2 - 6cx - 7c^2 = 0$ $-c, 7c$

38. $x^2 - ax - 20a^2 = 0$ $-4a, 5a$

39. $2x^2 + 3bx + b^2 = 0$ $-\frac{b}{2}, -b$

40. $3x^2 - 4cx + c^2 = 0$ $\frac{c}{3}, c$

41. $3x^2 - 14ax + 8a^2 = 0$ $\frac{2a}{3}, 4a$

42. $3x^2 - 11xy + 6y^2 = 0$ $\frac{2y}{3}, 3y$

2 Write a quadratic equation that has integer coefficients and has as solutions the given pair of numbers.

43. 2 and 5 $x^2 - 7x + 10 = 0$

44. 3 and 1 $x^2 - 4x + 3 = 0$

45. -2 and -4 $x^2 + 6x + 8 = 0$

46. -1 and -3 $x^2 + 4x + 3 = 0$

47. 6 and -1 $x^2 - 5x - 6 = 0$

48. -2 and 5 $x^2 - 3x - 10 = 0$

49. 3 and -3 $x^2 - 9 = 0$

50. 5 and -5 $x^2 - 25 = 0$

51. 4 and 4 $x^2 - 8x + 16 = 0$

52. 2 and 2 $x^2 - 4x + 4 = 0$

53. 0 and 5 $x^2 - 5x = 0$

54. 0 and -2 $x^2 + 2x = 0$

55. 0 and 3 $x^2 - 3x = 0$

56. 0 and -1 $x^2 + x = 0$

57. 3 and $\frac{1}{2}$ $2x^2 - 7x + 3 = 0$

58. 2 and $\frac{2}{3}$ $3x^2 - 8x + 4 = 0$

59. $-\frac{3}{4}$ and 2 $4x^2 - 5x - 6 = 0$

60. $-\frac{1}{2}$ and 5 $2x^2 - 9x - 5 = 0$

61. $-\frac{5}{3}$ and -2 $3x^2 + 11x + 10 = 0$

62. $-\frac{3}{2}$ and -1 $2x^2 + 5x + 3 = 0$

63. $-\frac{2}{3}$ and $\frac{2}{3}$ $9x^2 - 4 = 0$

64. $-\frac{1}{2}$ and $\frac{1}{2}$ $4x^2 - 1 = 0$

65. $\frac{1}{2}$ and $\frac{1}{3}$ $6x^2 - 5x + 1 = 0$

66. $\frac{3}{4}$ and $\frac{2}{3}$ $12x^2 - 17x + 6 = 0$

67. $\frac{6}{5}$ and $-\frac{1}{2}$ $10x^2 - 7x - 6 = 0$

68. $\frac{3}{4}$ and $-\frac{3}{2}$ $8x^2 + 6x - 9 = 0$

69. $-\frac{1}{4}$ and $-\frac{1}{2}$ $8x^2 + 6x + 1 = 0$

70. $-\frac{5}{6}$ and $-\frac{2}{3}$ $18x^2 + 27x + 10 = 0$

71. $\frac{3}{5}$ and $-\frac{1}{10}$ $50x^2 - 25x - 3 = 0$

72. $\frac{7}{2}$ and $-\frac{1}{4}$ $8x^2 - 26x - 7 = 0$

3 Solve by taking square roots.

73. $y^2 = 49$ ± 7 **74.** $x^2 = 64$ ± 8 **75.** $z^2 = -4$ $\pm 2i$ **76.** $v^2 = -16$ $\pm 4i$

77. $s^2 - 4 = 0$ ± 2 **78.** $r^2 - 36 = 0$ ± 6 **79.** $4x^2 - 81 = 0$ $\pm\frac{9}{2}$ **80.** $9x^2 - 16 = 0$ $\pm\frac{4}{3}$

81. $y^2 + 49 = 0$ $\pm 7i$ **82.** $z^2 + 16 = 0$ $\pm 4i$ **83.** $v^2 - 48 = 0$ $\pm 4\sqrt{3}$ **84.** $s^2 - 32 = 0$ $\pm 4\sqrt{2}$

85. $r^2 - 75 = 0$ $\pm 5\sqrt{3}$ **86.** $u^2 - 54 = 0$ $\pm 3\sqrt{6}$ **87.** $z^2 + 18 = 0$ $\pm 3i\sqrt{2}$

88. $t^2 + 27 = 0$ $\pm 3i\sqrt{3}$ **89.** $(x - 1)^2 = 36$ $7, -5$ **90.** $(x + 2)^2 = 25$ $-7, 3$

91. $3(y + 3)^2 = 27$ $0, -6$ **92.** $4(s - 2)^2 = 36$ $5, -1$ **93.** $5(z + 2)^2 = 125$ $-7, 3$

94. $(x - 2)^2 = -4$ $2 \pm 2i$ **95.** $(x + 5)^2 = -25$ $-5 \pm 5i$ **96.** $(x - 8)^2 = -64$ $8 \pm 8i$

97. $3(x - 4)^2 = -12$ $4 \pm 2i$ **98.** $5(x + 2)^2 = -125$ $-2 \pm 5i$ **99.** $3(x - 9)^2 = -27$ $9 \pm 3i$

100. $2(y - 3)^2 = 18$ $0, 6$ **101.** $\left(v - \frac{1}{2}\right)^2 = \frac{1}{4}$ $0, 1$ **102.** $\left(r + \frac{2}{3}\right)^2 = \frac{1}{9}$ $-1, -\frac{1}{3}$

103. $\left(x - \frac{2}{5}\right)^2 = \frac{9}{25}$ $-\frac{1}{5}, 1$ **104.** $\left(y + \frac{1}{3}\right)^2 = \frac{4}{9}$ $-1, \frac{1}{3}$ **105.** $\left(a + \frac{3}{4}\right)^2 = \frac{9}{16}$ $-\frac{3}{2}, 0$

106. $4\left(x - \frac{1}{2}\right)^2 = 1$ $0, 1$ **107.** $3\left(x - \frac{5}{3}\right)^2 = \frac{4}{3}$ $\frac{7}{3}, 1$ **108.** $2\left(x + \frac{3}{5}\right)^2 = \frac{8}{25}$ $-1, -\frac{1}{5}$

109. $(x + 5)^2 - 6 = 0$ $-5 \pm \sqrt{6}$ **110.** $(t - 1)^2 - 15 = 0$ $1 \pm \sqrt{15}$ **111.** $(s - 2)^2 - 24 = 0$ $2 \pm 2\sqrt{6}$

112. $(y + 3)^2 - 18 = 0$ $-3 \pm 3\sqrt{2}$ **113.** $(z + 1)^2 + 12 = 0$ $-1 \pm 2i\sqrt{3}$ **114.** $(r - 2)^2 + 28 = 0$ $2 \pm 2i\sqrt{7}$

115. $(v - 3)^2 + 45 = 0$ $3 \pm 3i\sqrt{5}$ **116.** $(x + 5)^2 + 32 = 0$ $-5 \pm 4i\sqrt{2}$ **117.** $\left(u + \frac{2}{3}\right)^2 - 18 = 0$ $\frac{-2 \pm 9\sqrt{2}}{3}$

118. $\left(z - \frac{1}{2}\right)^2 - 20 = 0$ $\frac{1 \pm 4\sqrt{5}}{2}$ **119.** $\left(x + \frac{1}{2}\right)^2 + 40 = 0$ $-\frac{1}{2} \pm 2i\sqrt{10}$ **120.** $\left(r - \frac{3}{2}\right)^2 + 48 = 0$ $\frac{3}{2} \pm 4i\sqrt{3}$

APPLYING CONCEPTS 8.1

Write a quadratic equation that has as solutions the given pair of numbers.

121. $\sqrt{2}$ and $-\sqrt{2}$ $x^2 - 2 = 0$ **122.** $\sqrt{5}$ and $-\sqrt{5}$ $x^2 - 5 = 0$ **123.** i and $-i$ $x^2 + 1 = 0$

124. $2i$ and $-2i$ $x^2 + 4 = 0$ **125.** $2\sqrt{2}$ and $-2\sqrt{2}$ $x^2 - 8 = 0$ **126.** $3\sqrt{2}$ and $-3\sqrt{2}$ $x^2 - 18 = 0$

127. $2\sqrt{3}$ and $-2\sqrt{3}$ $x^2 - 12 = 0$ **128.** $i\sqrt{2}$ and $-i\sqrt{2}$ $x^2 + 2 = 0$ **129.** $2i\sqrt{3}$ and $-2i\sqrt{3}$ $x^2 + 12 = 0$

Solve for x.

130. $2a^2x^2 = 32b^2$ $\pm\frac{4b}{a}$

131. $5y^2x^2 = 125z^2$ $\pm\frac{5z}{y}$

132. $(x + a)^2 - 4 = 0$ $-a \pm 2$

133. $2(x - y)^2 - 8 = 0$ $y \pm 2$

134. $(2x - 1)^2 = (2x + 3)^2$ $-\frac{1}{2}$

135. $(x - 4)^2 = (x + 2)^2$ 1

Solve.

[C] **136.** Show that the solutions of the equation $ax^2 + bx = 0$ are 0 and $-\frac{b}{a}$. Complete solution is available in the Solutions Manual.

[C] **137.** Show that the solutions of the equation $ax^2 + c = 0$, $a > 0$, $c > 0$, are $\frac{\sqrt{ca}}{a}i$ and $-\frac{\sqrt{ca}}{a}i$. Complete solution is available in the Solutions Manual.

138. *Chemistry* A chemical reaction between carbon monoxide and water vapor is used to increase the ratio of hydrogen gas in certain gas mixtures. In the process, carbon dioxide is also formed. For a certain reaction, the concentration of carbon dioxide, x, in moles per liter, is given by $0.58 = \frac{x^2}{(0.02 - x)^2}$. Solve this equation for x. Round to the nearest ten-thousandth. (*Source: General Chemistry*, Ebbing/Dale, Houghton Mifflin) 0.0086 mole/liter

139. Explain why the restriction that $a \neq 0$ is given in the definition of a quadratic equation.

140. If $x^2 = 16$, then $x = \pm 4$. Explain why the $\pm$ sign is necessary.

141. Discuss deficient numbers, abundant numbers, and amicable numbers.

142. Write a paper on the history of solving quadratic equations. Include a discussion of methods used by the Babylonians, the Egyptians, the Greeks, and the Hindus.

143. Let r_1 and r_2 be two solutions of $ax^2 + bx + c = 0$, where $a \neq 0$. Show that $r_1 + r_2 = -\frac{b}{a}$ and $r_1r_2 = \frac{c}{a}$. Explain how these relationships can be used to check the solutions of a quadratic equation.

144. Show that if the complex number $z_1 + z_2i$ is a solution of the equation $ax^2 + bx + c = 0$, where a, b, and c are real numbers and $a \neq 0$. Then $z_1 - z_2i$, the complex conjugate, is also a solution of the equation.

SECTION 8.2

Solving Quadratic Equations by Completing the Square and by Using the Quadratic Formula

1 Solve quadratic equations by completing the square

32 HM³

Recall that a perfect-square trinomial is the square of a binomial. Some examples of perfect-square trinomials follow.

INSTRUCTOR NOTE

Completing the square is very difficult for students. It may be necessary to give quite a few examples before students can do this effectively. The following Point of Interest may help some students.

Perfect-square trinomial		Square of a binomial
$x^2 + 8x + 16$	=	$(x + 4)^2$
$x^2 - 10x + 25$	=	$(x - 5)^2$
$x^2 + 2ax + a^2$	=	$(x + a)^2$

For each perfect-square trinomial, the square of $\frac{1}{2}$ the coefficient of x equals the constant term.

$$\left(\frac{1}{2}\text{ coefficient of } x\right)^2 = \text{Constant term}$$

$$x^2 + 8x + 16, \quad \left(\frac{1}{2}\cdot 8\right)^2 = 16$$

$$x^2 - 10x + 25, \quad \left[\frac{1}{2}(-10)\right]^2 = 25$$

$$x^2 + 2ax + a^2, \quad \left(\frac{1}{2}\cdot 2a\right)^2 = a^2$$

To complete the square on $x^2 + bx$, add $\left(\frac{1}{2}b\right)^2$ to $x^2 + bx$.

POINT OF INTEREST

Early attempts to solve quadratic equations were primarily geometric. The Persian mathematician al-Khwarizmi (c. A.D. 800) essentially completed a square of $x^2 + 12x$ as shown below.

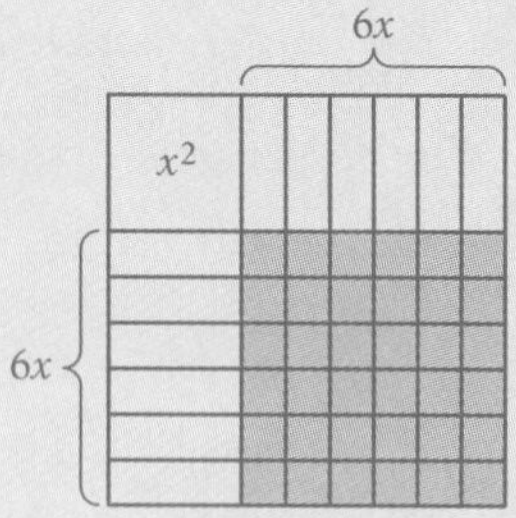

⇨ Complete the square on $x^2 - 12x$. Write the resulting perfect-square trinomial as the square of a binomial.

Find the constant term. $\left[\frac{1}{2}(-12)\right]^2 = (-6)^2 = 36$

Complete the square on $x^2 - 12x$ by adding the constant term. $x^2 - 12x + 36$

Write the resulting perfect-square trinomial as the square of a binomial. $x^2 - 12x + 36 = (x - 6)^2$

Complete the square on $z^2 + 3z$. Write the resulting perfect-square trinomial as the square of a binomial.

Find the constant term. $\left(\frac{1}{2} \cdot 3\right)^2 = \left(\frac{3}{2}\right)^2 = \frac{9}{4}$

Complete the square on $z^2 + 3z$ by adding the constant term. $z^2 + 3z + \frac{9}{4}$

Write the resulting perfect-square trinomial as the square of a binomial. $z^2 + 3z + \frac{9}{4} = \left(z + \frac{3}{2}\right)^2$

Though not all quadratic equations can be solved by factoring, any quadratic equation can be solved by completing the square. Add to each side of the equation the term that completes the square. Rewrite the equation in the form $(x + a)^2 = b$. Then take the square root of each side of the equation.

Solve by completing the square: $x^2 - 4x - 14 = 0$

$$x^2 - 4x - 14 = 0$$

Add 14 to each side of the equation.

$$x^2 - 4x = 14$$

Add the constant term that completes the square on $x^2 - 4x$ to each side of the equation.

$\left[\frac{1}{2}(-4)\right]^2 = 4$

$$x^2 - 4x + 4 = 14 + 4$$

Factor the perfect-square trinomial.

$$(x - 2)^2 = 18$$

Take the square root of each side of the equation.

$$\sqrt{(x - 2)^2} = \sqrt{18}$$
$$x - 2 = \pm\sqrt{18}$$

Simplify.

$$x - 2 = \pm 3\sqrt{2}$$

Solve for x.

$$x - 2 = 3\sqrt{2} \qquad x - 2 = -3\sqrt{2}$$
$$x = 2 + 3\sqrt{2} \qquad x = 2 - 3\sqrt{2}$$

INSTRUCTOR NOTE

An Instructor Note in the previous section suggested an alternative method of checking the solutions of a quadratic equation. You may want to use it to check the solutions here.

Check:

$x^2 - 4x - 14 = 0$	
$(2 + 3\sqrt{2})^2 - 4(2 + 3\sqrt{2}) - 14$	0
$4 + 12\sqrt{2} + 18 - 8 - 12\sqrt{2} - 14$	0
$0 = 0$	

$x^2 - 4x - 14 = 0$	
$(2 - 3\sqrt{2})^2 - 4(2 - 3\sqrt{2}) - 14$	0
$4 - 12\sqrt{2} + 18 - 8 + 12\sqrt{2} - 14$	0
$0 = 0$	

Write the solutions. The solutions are $2 + 3\sqrt{2}$ and $2 - 3\sqrt{2}$.

INSTRUCTOR NOTE

To say that students will find solving quadratic equations by completing the square a daunting task is an understatement. The positive note is that the procedure is the same each time.

- Isolate $ax^2 + bx$.
- Multiply by $\frac{1}{a}$.
- Complete the square.
- Factor.
- Take the square root.
- Solve for x.

When a, the coefficient of the x^2 term, is not 1, divide each side of the equation by a before completing the square.

Solve by completing the square: $2x^2 - x = 2$

$$2x^2 - x = 2$$

Divide each side of the equation by the coefficient of x^2.

$$\frac{2x^2 - x}{2} = \frac{2}{2}$$

The coefficient of the x^2 term is now 1.

$$x^2 - \frac{1}{2}x = 1$$

Add the term that completes the square on $x^2 - \frac{1}{2}x$ to each side of the equation. Factor the perfect-square trinomial.

$$x^2 - \frac{1}{2}x + \frac{1}{16} = 1 + \frac{1}{16}$$

$$\left(x - \frac{1}{4}\right)^2 = \frac{17}{16}$$

Take the square root of each side of the equation.

$$\sqrt{\left(x - \frac{1}{4}\right)^2} = \sqrt{\frac{17}{16}}$$

$$x - \frac{1}{4} = \pm\sqrt{\frac{17}{16}}$$

Simplify.

$$x - \frac{1}{4} = \pm\frac{\sqrt{17}}{4}$$

Solve for x.

$$x - \frac{1}{4} = \frac{\sqrt{17}}{4} \qquad x - \frac{1}{4} = -\frac{\sqrt{17}}{4}$$

$$x = \frac{1}{4} + \frac{\sqrt{17}}{4} \qquad x = \frac{1}{4} - \frac{\sqrt{17}}{4}$$

$\frac{1 + \sqrt{17}}{4}$ and $\frac{1 - \sqrt{17}}{4}$ check as solutions.

Write the solutions.

The solutions are $\frac{1 + \sqrt{17}}{4}$ and $\frac{1 - \sqrt{17}}{4}$.

POINT OF INTEREST

Mathematicians have studied quadratic equations for centuries. Many of the initial equations were a result of trying to solve a geometry problem. One of the most famous, which dates from around 500 B.C., is "squaring the circle." The question was "Is it possible to construct a square whose area is that of a given circle?" For these early mathematicians, to *construct* meant to draw with only a straightedge and a compass. It was approximately 2300 years later that mathematicians were able to prove that such a construction is impossible.

Example 1 Solve by completing the square.

A. $4x^2 - 8x + 1 = 0$ B. $x^2 + 4x + 5 = 0$

Solution A. $4x^2 - 8x + 1 = 0$

$$4x^2 - 8x = -1$$

► Subtract 1 from each side of the equation.

$$\frac{4x^2 - 8x}{4} = \frac{-1}{4}$$

► The coefficient of the x^2 term must be 1. Divide each side of the equation by 4.

$$x^2 - 2x = -\frac{1}{4}$$

$$x^2 - 2x + 1 = -\frac{1}{4} + 1$$

► Complete the square. $\left[\frac{1}{2}(-2)\right]^2 = 1$

$$(x - 1)^2 = \frac{3}{4}$$

► Factor the perfect-square trinomial.

$$\sqrt{(x - 1)^2} = \sqrt{\frac{3}{4}}$$

► Take the square root of each side of the equation.

$$x - 1 = \pm\sqrt{\frac{3}{4}}$$

$$x - 1 = \pm\frac{\sqrt{3}}{2}$$ ▶ Simplify.

$$x - 1 = \frac{\sqrt{3}}{2} \qquad x - 1 = -\frac{\sqrt{3}}{2}$$ ▶ Solve for x.

$$x = 1 + \frac{\sqrt{3}}{2} \qquad x = 1 - \frac{\sqrt{3}}{2}$$

$$x = \frac{2 + \sqrt{3}}{2} \qquad x = \frac{2 - \sqrt{3}}{2}$$

The solutions are $\frac{2 + \sqrt{3}}{2}$ and $\frac{2 - \sqrt{3}}{2}$.

B. $x^2 + 4x + 5 = 0$

$$x^2 + 4x = -5$$ ▶ Subtract 5 from each side of the equation.

$$x^2 + 4x + 4 = -5 + 4$$ ▶ Complete the square.

$$(x + 2)^2 = -1$$ ▶ Factor the perfect-square trinomial.

$$\sqrt{(x + 2)^2} = \sqrt{-1}$$ ▶ Take the square root of each side of the equation.

$$x + 2 = \pm\sqrt{-1}$$

$$x + 2 = \pm i$$ ▶ Simplify.

$$x + 2 = i \qquad x + 2 = -i$$ ▶ Solve for x.

$$x = -2 + i \qquad x = -2 - i$$

The solutions are $-2 + i$ and $-2 - i$.

Problem 1 Solve by completing the square.

A. $4x^2 - 4x - 1 = 0$ B. $2x^2 + x - 5 = 0$

Solution See page S22–S23. A. $\frac{1 \pm \sqrt{2}}{2}$ B. $\frac{-1 \pm \sqrt{41}}{4}$

2 Solve quadratic equations by using the quadratic formula

32 HM³

A general formula known as the **quadratic formula** can be derived by applying the method of completing the square to the standard form of a quadratic equation. This formula can be used to solve any quadratic equation.

The solution of the equation $ax^2 + bx + c = 0$ by completing the square is shown below.

$$ax^2 + bx + c = 0$$

Subtract the constant term from each side of the equation.

$$ax^2 + bx + c - c = 0 - c$$

$$ax^2 + bx = -c$$

Divide each side of the equation by a, the coefficient of x^2.

$$\frac{ax^2 + bx}{a} = \frac{-c}{a}$$

$$x^2 + \frac{b}{a}x = -\frac{c}{a}$$

Complete the square by adding $\left(\frac{1}{2}\cdot\frac{b}{a}\right)^2$ to each side of the equation.

$$x^2 + \frac{b}{a}x + \left(\frac{1}{2}\cdot\frac{b}{a}\right)^2 = \left(\frac{1}{2}\cdot\frac{b}{a}\right)^2 - \frac{c}{a}$$

$$x^2 + \frac{b}{a}x + \frac{b^2}{4a^2} = \frac{b^2}{4a^2} - \frac{c}{a}$$

Simplify the right side of the equation.

$$x^2 + \frac{b}{a}x + \frac{b^2}{4a^2} = \frac{b^2}{4a^2} - \left(\frac{c}{a}\cdot\frac{4a}{4a}\right)$$

$$x^2 + \frac{b}{a}x + \frac{b^2}{4a^2} = \frac{b^2}{4a^2} - \frac{4ac}{4a^2}$$

$$x^2 + \frac{b}{a}x + \frac{b^2}{4a^2} = \frac{b^2 - 4ac}{4a^2}$$

Factor the perfect-square trinomial on the left side of the equation.

$$\left(x + \frac{b}{2a}\right)^2 = \frac{b^2 - 4ac}{4a^2}$$

Take the square root of each side of the equation.

$$\sqrt{\left(x + \frac{b}{2a}\right)^2} = \sqrt{\frac{b^2 - 4ac}{4a^2}}$$

$$x + \frac{b}{2a} = \pm\sqrt{\frac{b^2 - 4ac}{4a^2}}$$

$$x + \frac{b}{2a} = \pm\frac{\sqrt{b^2 - 4ac}}{2a}$$

Solve for x.

$$x + \frac{b}{2a} = \frac{\sqrt{b^2 - 4ac}}{2a} \qquad x + \frac{b}{2a} = -\frac{\sqrt{b^2 - 4ac}}{2a}$$

$$x = -\frac{b}{2a} + \frac{\sqrt{b^2 - 4ac}}{2a} \qquad x = -\frac{b}{2a} - \frac{\sqrt{b^2 - 4ac}}{2a}$$

$$= \frac{-b + \sqrt{b^2 - 4ac}}{2a} \qquad = \frac{-b - \sqrt{b^2 - 4ac}}{2a}$$

POINT OF INTEREST

Although mathematicians have studied quadratic equations since around 500 B.C., it was not until the 18th century that the formula was written as it is today. Of further note, the word *quadratic* has the same Latin root as does the word *square*.

The Quadratic Formula

The solutions of $ax^2 + bx + c = 0, a \neq 0$, are

$$\frac{-b + \sqrt{b^2 - 4ac}}{2a} \quad \text{and} \quad \frac{-b - \sqrt{b^2 - 4ac}}{2a}$$

The quadratic formula is frequently written in the form

$$x = \frac{-b \pm \sqrt{b^2 - 4ac}}{2a}$$

INSTRUCTOR NOTE

One of the difficulties students have with the quadratic formula is correct substitution. Have them first make sure that the equation is in standard form. Then circling *a*, *b*, and *c* may help. They must remember that the sign that precedes a number is the sign of the number. For the equation at the right, we have

$$\overset{a}{(4)}x^2\overset{b}{(-8)}x\overset{c}{(+13)} = 0$$

Solve by using the quadratic formula: $4x^2 = 8x - 13$

$$4x^2 = 8x - 13$$

Write the equation in standard form.

$$4x^2 - 8x + 13 = 0$$

$$a = 4, b = -8, c = 13$$

Replace a, b, and c in the quadratic formula by their values.

$$x = \frac{-b \pm \sqrt{b^2 - 4ac}}{2a}$$

$$= \frac{-(-8) \pm \sqrt{(-8)^2 - 4\cdot 4\cdot 13}}{2\cdot 4}$$

Simplify.

$$= \frac{8 \pm \sqrt{64 - 208}}{8}$$

$$= \frac{8 \pm \sqrt{-144}}{8}$$

Write the answer in the form $a + bi$.

$$= \frac{8 \pm 12i}{8} = \frac{2 \pm 3i}{2} = 1 \pm \frac{3}{2}i$$

Check:

$$4x^2 = 8x - 13$$

$4\left(1 + \frac{3}{2}i\right)^2$	$8\left(1 + \frac{3}{2}i\right) - 13$
$4\left(1 + 3i - \frac{9}{4}\right)$	$8 + 12i - 13$
$4\left(-\frac{5}{4} + 3i\right)$	$-5 + 12i$

$$-5 + 12i = -5 + 12i$$

$$4x^2 = 8x - 13$$

$4\left(1 - \frac{3}{2}i\right)^2$	$8\left(1 - \frac{3}{2}i\right) - 13$
$4\left(1 - 3i - \frac{9}{4}\right)$	$8 - 12i - 13$
$4\left(-\frac{5}{4} - 3i\right)$	$-5 - 12i$

$$-5 - 12i = -5 - 12i$$

The solutions are $1 + \frac{3}{2}i$ and $1 - \frac{3}{2}i$.

Example 2 Solve by using the quadratic formula.

A. $4x^2 + 12x + 9 = 0$ B. $2x^2 - x + 5 = 0$

Solution A. $4x^2 + 12x + 9 = 0$

▶ $a = 4, b = 12, c = 9$

$$x = \frac{-b \pm \sqrt{b^2 - 4ac}}{2a}$$

▶ Replace a, b, and c in the quadratic formula by their values. Then simplify.

$$= \frac{-12 \pm \sqrt{12^2 - 4 \cdot 4 \cdot 9}}{2 \cdot 4}$$

$$= \frac{-12 \pm \sqrt{144 - 144}}{8}$$

$$= \frac{-12 \pm \sqrt{0}}{8} = \frac{-12}{8} = -\frac{3}{2}$$

▶ The equation has a double root.

The solution is $-\frac{3}{2}$.

B. $2x^2 - x + 5 = 0$

▶ $a = 2, b = -1, c = 5$

$$x = \frac{-b \pm \sqrt{b^2 - 4ac}}{2a}$$

▶ Replace a, b, and c in the quadratic formula by their values. Then simplify.

$$= \frac{-(-1) \pm \sqrt{(-1)^2 - 4 \cdot 2 \cdot 5}}{2 \cdot 2}$$

$$= \frac{1 \pm \sqrt{1 - 40}}{4}$$

$$= \frac{1 \pm \sqrt{-39}}{4} = \frac{1 \pm i\sqrt{39}}{4}$$

The solutions are $\frac{1}{4} + \frac{\sqrt{39}}{4}i$ and $\frac{1}{4} - \frac{\sqrt{39}}{4}i$.

Problem 2 Solve by using the quadratic formula.

A. $x^2 + 6x - 9 = 0$ B. $4x^2 = 4x - 1$

Solution See page S23. A. $-3 \pm 3\sqrt{2}$ B. $\frac{1}{2}$

In Example 2A, the solution of the equation is a double root, and in Example 2B, the solutions are complex numbers.

In the quadratic formula, the quantity $b^2 - 4ac$ is called the **discriminant.** When a, b, and c are real numbers, the discriminant determines whether a quadratic equation will have a double root, two real number solutions that are not equal, or two complex number solutions.

The Effect of the Discriminant on the Solutions of a Quadratic Equation

1. If $b^2 - 4ac = 0$, the equation has one real number solution, a double root.
2. If $b^2 - 4ac > 0$, the equation has two real number solutions that are not equal.
3. If $b^2 - 4ac < 0$, the equation has two complex number solutions.

The equation $x^2 - 4x - 5 = 0$ has two real number solutions because the discriminant is greater than zero.

$a = 1, b = -4, c = -5$

$$\begin{aligned} b^2 - 4ac &= (-4)^2 - 4(1)(-5) \\ &= 16 + 20 \\ &= 36 \end{aligned}$$

$36 > 0$

Example 3 Use the discriminant to determine whether $4x^2 - 2x + 5 = 0$ has one real number solution, two real number solutions, or two complex number solutions.

Solution

$$\begin{aligned} b^2 - 4ac &= (-2)^2 - 4(4)(5) \\ &= 4 - 80 \\ &= -76 \end{aligned}$$

▶ $a = 4, b = -2, c = 5$

$-76 < 0$ ▶ The discriminant is less than 0.

The equation has two complex number solutions.

Problem 3 Use the discriminant to determine whether $3x^2 - x - 1 = 0$ has one real number solution, two real number solutions, or two complex number solutions.

Solution See page S23. Two real number solutions

CONCEPT REVIEW 8.2

Determine whether the following statements are always true, sometimes true, or never true.

1. A quadratic equation can be solved by completing the square. Always true

2. The expression $b^2 - 4ac$ is called the discriminant. Always true

3. If $b^2 > 4ac$, the quadratic equation has two complex number solutions. Never true

4. The quadratic formula can be used to solve quadratic equations that have irrational roots. Always true

5. $x^2 + 5x + 25$ is a perfect-square trinomial. Never true

6. To complete the square on $3x^2 + 6x$, add 9. Never true

EXERCISES 8.2

1

1. What is the next step when using completing the square to solve $x^2 + 6x = 4$?

2. Are there some quadratic equations that cannot be solved by completing the square? Why or why not?

Solve by completing the square.

3. $x^2 - 4x - 5 = 0$ $5, -1$
4. $y^2 + 6y + 5 = 0$ $-5, -1$
5. $v^2 + 8v - 9 = 0$ $-9, 1$
6. $w^2 - 2w - 24 = 0$ $6, -4$
7. $z^2 - 6z + 9 = 0$ 3
8. $u^2 + 10u + 25 = 0$ -5
9. $r^2 + 4r - 7 = 0$ $-2 \pm \sqrt{11}$
10. $s^2 + 6s - 1 = 0$ $-3 \pm \sqrt{10}$
11. $x^2 - 6x + 7 = 0$ $3 \pm \sqrt{2}$
12. $y^2 + 8y + 13 = 0$ $-4 \pm \sqrt{3}$
13. $z^2 - 2z + 2 = 0$ $1 \pm i$
14. $t^2 - 4t + 8 = 0$ $2 \pm 2i$

First try to solve by factoring. If you are unable to solve the equation by factoring, solve the equation by completing the square.

15. $t^2 - t - 1 = 0$ $\frac{1 \pm \sqrt{5}}{2}$
16. $u^2 - u - 7 = 0$ $\frac{1 \pm \sqrt{29}}{2}$
17. $y^2 - 6y = 4$ $3 \pm \sqrt{13}$
18. $w^2 + 4w = 2$ $-2 \pm \sqrt{6}$
19. $x^2 = 8x - 15$ $3, 5$
20. $z^2 = 4z - 3$ $3, 1$
21. $v^2 = 4v - 13$ $2 \pm 3i$
22. $x^2 = 2x - 17$ $1 \pm 4i$
23. $p^2 + 6p = -13$ $-3 \pm 2i$
24. $x^2 + 4x = -20$ $-2 \pm 4i$
25. $y^2 - 2y = 17$ $1 \pm 3\sqrt{2}$
26. $x^2 + 10x = 7$ $-5 \pm 4\sqrt{2}$
27. $z^2 = z + 4$ $\frac{1 \pm \sqrt{17}}{2}$
28. $r^2 = 3r - 1$ $\frac{3 \pm \sqrt{5}}{2}$
29. $x^2 + 13 = 2x$ $1 \pm 2i\sqrt{3}$

30. $x^2 + 27 = 6x$ $3 \pm 3i\sqrt{2}$

31. $2y^2 + 3y + 1 = 0$ $-\dfrac{1}{2}, -1$

32. $2t^2 + 5t - 3 = 0$ $\dfrac{1}{2}, -3$

33. $4r^2 - 8r = -3$ $\dfrac{1}{2}, \dfrac{3}{2}$

34. $4u^2 - 20u = -9$ $\dfrac{1}{2}, \dfrac{9}{2}$

35. $6y^2 - 5y = 4$ $-\dfrac{1}{2}, \dfrac{4}{3}$

36. $6v^2 - 7v = 3$ $-\dfrac{1}{3}, \dfrac{3}{2}$

37. $4x^2 - 4x + 5 = 0$ $\dfrac{1}{2} \pm i$

38. $4t^2 - 4t + 17 = 0$ $\dfrac{1}{2} \pm 2i$

39. $9x^2 - 6x + 2 = 0$ $\dfrac{1}{3} \pm \dfrac{1}{3}i$

40. $9y^2 - 12y + 13 = 0$ $\dfrac{2}{3} \pm i$

41. $2s^2 = 4s + 5$ $\dfrac{2 \pm \sqrt{14}}{2}$

42. $3u^2 = 6u + 1$ $\dfrac{3 \pm 2\sqrt{3}}{3}$

43. $2r^2 = 3 - r$ $-\dfrac{3}{2}, 1$

44. $2x^2 = 12 - 5x$ $\dfrac{3}{2}, -4$

45. $y - 2 = (y - 3)(y + 2)$ $1 \pm \sqrt{5}$

46. $8s - 11 = (s - 4)(s - 2)$ $7 \pm \sqrt{30}$

47. $6t - 2 = (2t - 3)(t - 1)$ $\dfrac{1}{2}, 5$

48. $2z + 9 = (2z + 3)(z + 2)$ $\dfrac{1}{2}, -3$

49. $(x - 4)(x + 1) = x - 3$ $2 \pm \sqrt{5}$

50. $(y - 3)^2 = 2y + 10$ $4 \pm \sqrt{17}$

Solve by completing the square. Approximate the solutions to the nearest thousandth.

51. $z^2 + 2z = 4$ $-3.236, 1.236$

52. $t^2 - 4t = 7$ $5.317, -1.317$

53. $2x^2 = 4x - 1$ $1.707, 0.293$

54. $3y^2 = 5y - 1$ $1.434, 0.232$

55. $4z^2 + 2z - 1 = 0$ $0.309, -0.809$

56. $4w^2 - 8w = 3$ $2.323, -0.323$

2 **57.** ✍ Suppose you must solve the quadratic equation $x^2 = 3x + 5$. Does it matter whether you rewrite the equation as $x^2 - 3x - 5 = 0$ or as $0 = -x^2 + 3x + 5$ before you begin?

58. ✍ Which method of solving a quadratic equation do you prefer: completing the square or using the quadratic formula? Why?

Solve by using the quadratic formula.

59. $x^2 - 3x - 10 = 0$ $5, -2$

60. $z^2 - 4z - 8 = 0$ $2 \pm 2\sqrt{3}$

61. $y^2 + 5y - 36 = 0$ $4, -9$

62. $z^2 - 3z - 40 = 0$ $8, -5$

63. $w^2 = 8w + 72$ $4 \pm 2\sqrt{22}$

64. $t^2 = 2t + 35$ $7, -5$

65. $v^2 = 24 - 5v$ $3, -8$

66. $x^2 = 18 - 7x$ $2, -9$

67. $2y^2 + 5y - 3 = 0$ $\dfrac{1}{2}, -3$

68. $4p^2 - 7p + 3 = 0$ $\dfrac{3}{4}, 1$

69. $8s^2 = 10s + 3$ $-\dfrac{1}{4}, \dfrac{3}{2}$

70. $12t^2 = 5t + 2$ $\dfrac{2}{3}, -\dfrac{1}{4}$

First try to solve by factoring. If you are unable to solve the equation by factoring, solve the equation by using the quadratic formula.

71. $v^2 - 2v - 7 = 0$ $1 \pm 2\sqrt{2}$

72. $t^2 - 2t - 11 = 0$ $1 \pm 2\sqrt{3}$

73. $y^2 - 8y - 20 = 0$ $10, -2$

74. $x^2 = 14x - 24$ $2, 12$

75. $v^2 = 12v - 24$ $6 \pm 2\sqrt{3}$

76. $2z^2 - 2z - 1 = 0$ $\dfrac{1 \pm \sqrt{3}}{2}$

77. $4x^2 - 4x - 7 = 0$ $\dfrac{1 \pm 2\sqrt{2}}{2}$

78. $2p^2 - 8p + 5 = 0$ $\dfrac{4 \pm \sqrt{6}}{2}$

79. $2s^2 - 3s + 1 = 0$ $\dfrac{1}{2}, 1$

80. $4w^2 - 4w - 1 = 0$ $\frac{1 \pm \sqrt{2}}{2}$

81. $3x^2 + 10x + 6 = 0$ $\frac{-5 \pm \sqrt{7}}{3}$

82. $3v^2 = 6v - 2$ $\frac{3 \pm \sqrt{3}}{3}$

83. $6w^2 = 19w - 10$ $\frac{2}{3}, \frac{5}{2}$

84. $z^2 + 2z + 2 = 0$ $-1 \pm i$

85. $p^2 - 4p + 5 = 0$ $2 \pm i$

86. $y^2 - 2y + 5 = 0$ $1 \pm 2i$

87. $x^2 + 6x + 13 = 0$ $-3 \pm 2i$

88. $s^2 - 4s + 13 = 0$ $2 \pm 3i$

89. $t^2 - 6t + 10 = 0$ $3 \pm i$

90. $2w^2 - 2w + 5 = 0$ $\frac{1}{2} \pm \frac{3}{2}i$

91. $4v^2 + 8v + 3 = 0$ $-\frac{3}{2}, -\frac{1}{2}$

92. $2x^2 + 6x + 5 = 0$ $-\frac{3}{2} \pm \frac{1}{2}i$

93. $2y^2 + 2y + 13 = 0$ $-\frac{1}{2} \pm \frac{5}{2}i$

94. $4t^2 - 6t + 9 = 0$ $\frac{3}{4} \pm \frac{3\sqrt{3}}{4}i$

95. $3v^2 + 6v + 1 = 0$ $\frac{-3 \pm \sqrt{6}}{3}$

96. $2r^2 = 4r - 11$ $1 \pm \frac{3\sqrt{2}}{2}i$

97. $3y^2 = 6y - 5$ $1 \pm \frac{\sqrt{6}}{3}i$

98. $2x(x - 2) = x + 12$ $-\frac{3}{2}, 4$

99. $10y(y + 4) = 15y - 15$ $-\frac{3}{2}, -1$

100. $(3s - 2)(s + 1) = 2$ $-\frac{4}{3}, 1$

101. $(2t + 1)(t - 3) = 9$ $-\frac{3}{2}, 4$

Use the discriminant to determine whether the quadratic equation has one real number solution, two real number solutions, or two complex number solutions.

102. $2z^2 - z + 5 = 0$
Two complex

103. $3y^2 + y + 1 = 0$
Two complex

104. $9x^2 - 12x + 4 = 0$
One real

105. $4x^2 + 20x + 25 = 0$
One real

106. $2v^2 - 3v - 1 = 0$
Two real

107. $3w^2 + 3w - 2 = 0$
Two real

108. $2p^2 + 5p + 1 = 0$
Two real

109. $2t^2 + 9t + 3 = 0$
Two real

110. $5z^2 + 2 = 0$
Two complex

Solve by using the quadratic formula. Approximate the solutions to the nearest thousandth.

111. $x^2 + 6x - 6 = 0$ 0.873, −6.873

112. $p^2 - 8p + 3 = 0$ 7.606, 0.394

113. $r^2 - 2r - 4 = 0$ 3.236, −1.236

114. $w^2 + 4w - 1 = 0$ 0.236, −4.236

115. $3t^2 = 7t + 1$ 2.468, −0.135

116. $2y^2 = y + 5$ 1.851, −1.351

APPLYING CONCEPTS 8.2

Solve.

117. $\sqrt{2}y^2 + 3y - 2\sqrt{2} = 0$ $\frac{\sqrt{2}}{2}, -2\sqrt{2}$

118. $\sqrt{3}z^2 + 10z - 3\sqrt{3} = 0$ $\frac{-5\sqrt{3} \pm \sqrt{102}}{3}$

119. $\sqrt{2}x^2 + 5x - 3\sqrt{2} = 0$ $-3\sqrt{2}, \frac{\sqrt{2}}{2}$

120. $\sqrt{3}w^2 + w - 2\sqrt{3} = 0$ $-\sqrt{3}, \frac{2\sqrt{3}}{3}$

121. $t^2 - t\sqrt{3} + 1 = 0$ $\frac{\sqrt{3}}{2} \pm \frac{1}{2}i$

122. $y^2 + y\sqrt{7} + 2 = 0$ $-\frac{\sqrt{7}}{2} \pm \frac{1}{2}i$

Solve for x.

123. $x^2 - ax - 2a^2 = 0$ $2a, -a$

124. $x^2 - ax - 6a^2 = 0$ $3a, -2a$

125. $2x^2 + 3ax - 2a^2 = 0$ $\frac{a}{2}, -2a$

126. $2x^2 - 7ax + 3a^2 = 0$ $\frac{a}{2}, 3a$

[C] **127.** $x^2 - 2x - y = 0$ $1 \pm \sqrt{y+1}$

[C] **128.** $x^2 - 4xy - 4 = 0$ $2y \pm 2\sqrt{y^2+1}$

For what values of p does the quadratic equation have two real number solutions that are not equal? Write the answer in set-builder notation.

[C] **129.** $x^2 - 6x + p = 0$ $\{p \mid p < 9, p \in \text{real numbers}\}$

[C] **130.** $x^2 + 10x + p = 0$ $\{p \mid p < 25, p \in \text{real numbers}\}$

For what values of p does the quadratic equation have two complex number solutions? Write the answer in set-builder notation.

[C] **131.** $x^2 - 2x + p = 0$ $\{p \mid p > 1, p \in \text{real numbers}\}$

[C] **132.** $x^2 + 4x + p = 0$ $\{p \mid p > 4, p \in \text{real numbers}\}$

Solve.

[C] **133.** Show that the equation $x^2 + bx - 1 = 0$ always has real number solutions regardless of the value of b. $b^2 + 4 > 0$ for any real number b.

[C] **134.** Show that the equation $2x^2 + bx - 2 = 0$ always has real number solutions regardless of the value of b. $b^2 + 16 > 0$ for any real number b.

135. *Sports* The height, h (in feet), of a baseball above the ground t seconds after it is hit can be approximated by the equation $h = -16t^2 + 70t + 4$. Using this equation, determine when the ball will hit the ground. Round to the nearest hundredth. (*Hint:* The ball hits the ground when $h = 0$.) 4.43 s

136. *Sports* After a baseball is hit, there are two equations that can be considered. One gives the height, h (in feet), the ball is above the ground t seconds after it is hit. The second is the distance, s (in feet), the ball is from home plate t seconds after it is hit. A model of this situation is given by $h = -16t^2 + 70t + 4$ and $s = 44.5t$. Using this model, determine whether the ball will clear a fence 325 ft from home plate. Round to the nearest tenth. No. The ball will have gone only 197.2 ft when it hits the ground.

137. *Conservation* The National Forest Management Act of 1976 specifies that harvesting timber in national forests must be accomplished in conjunction with environmental considerations. One such consideration is providing a habitat for the spotted owl. One model of the survival of the spotted owl requires the solution of the equation $x^2 - s_a x - s_j s_s f = 0$ for x. Different values of s_a, s_j, s_s, and f are given in the table at the right. The values are particularly important because they relate to the survival of the owl. If $x > 1$, then the model predicts a growth in the population; if $x = 1$, the population remains steady; if $x < 1$, the population decreases. The important solution of the equation is the larger of the two roots of the equation.

a. Determine the larger root of this equation for values provided by the U.S. Forest Service. Round to the nearest hundredth. Does it predict that the population will increase, remain steady, or decrease? 1.05; increase

b. Determine the larger root of this equation for the values provided by R. Lande in *Oecologia* (Vol. 75, 1988). Round to the nearest hundredth. Does it predict that the population will increase, remain steady, or decrease? 0.96; decrease

	U.S. Forest Service	Lande
s_j	0.34	0.11
s_s	0.97	0.71
s_a	0.97	0.94
f	0.24	0.24

Source: Charles Biles, and Barry Noon. The Spotted Owl. *The Journal of Undergraduate Mathematics and Its Application* Vol. 11, No. 2, 1990.

138. *Social Security* There are many models of the solvency of the current Social Security System. Based on data from a 1995 Social Securities trustees' report, one possible model of the amount of taxes received is given by the function $A(x) = 6.145x^2 - 73.954x + 422.744$. A model for the amount of benefits paid out is given by $B(x) = 10.287x^2 - 156.632x + 486.345$. For each of these models, x is the number of years after 1994 and the amounts are in billions of dollars. According to these models, during what year after 1996 will the amount paid in benefits first exceed the amount received in taxes?
2013

139. Explain why the discriminant determines whether a quadratic equation has one real number solution, two real number solutions, or two complex number solutions.

140. Here is an outline for an alternative method for deriving the quadratic formula. Fill in the details of this derivation. Begin with the equation $ax^2 + bx + c = 0$. Subtract c from each side of the equation. Multiply each side of the equation by $4a$. Now add b^2 to each side. Supply the remaining steps until you have reached the quadratic formula.

141. Explain how to complete the square of $x^2 + bx$.

142. Write out the steps for solving a quadratic equation by completing the square.

SECTION 8.3

Equations That Are Reducible to Quadratic Equations

1 Equations that are quadratic in form

Certain equations that are not quadratic equations can be expressed in quadratic form by making suitable substitutions. An equation is **quadratic in form** if it can be written as $au^2 + bu + c = 0$.

To see that the equation at the right is quadratic in form, let $x^2 = u$. Replace x^2 by u. The equation is quadratic in form.

$$x^4 - 4x^2 - 5 = 0$$
$$(x^2)^2 - 4(x^2) - 5 = 0$$
$$u^2 - 4u - 5 = 0$$

To see that the equation at the right is quadratic in form, let $y^{\frac{1}{2}} = u$. Replace $y^{\frac{1}{2}}$ by u. The equation is quadratic in form.

$$y - y^{\frac{1}{2}} - 6 = 0$$
$$(y^{\frac{1}{2}})^2 - (y^{\frac{1}{2}}) - 6 = 0$$
$$u^2 - u - 6 = 0$$

The key to recognizing equations that are quadratic in form is that when the equation is written in standard form, the exponent on one variable term is $\frac{1}{2}$ the exponent on the other variable term.

⇒ Solve: $z + 7z^{\frac{1}{2}} - 18 = 0$

The equation $z + 7z^{\frac{1}{2}} - 18 = 0$ is quadratic in form.

$$z + 7z^{\frac{1}{2}} - 18 = 0$$
$$(z^{\frac{1}{2}})^2 + 7(z^{\frac{1}{2}}) - 18 = 0$$

To solve this equation, let $z^{\frac{1}{2}} = u$.

$$u^2 + 7u - 18 = 0$$

Solve for u by factoring.

$$(u - 2)(u + 9) = 0$$

$$
\begin{array}{ll}
u - 2 = 0 & u + 9 = 0 \\
u = 2 & u = -9 \\
\end{array}
$$

Replace u by $z^{\frac{1}{2}}$.

$$
\begin{array}{ll}
z^{\frac{1}{2}} = 2 & z^{\frac{1}{2}} = -9 \\
\end{array}
$$

Solve for z by squaring each side of the equation.

$$
\begin{array}{ll}
(z^{\frac{1}{2}})^2 = 2^2 & (z^{\frac{1}{2}})^2 = (-9)^2 \\
z = 4 & z = 81 \\
\end{array}
$$

Check the solution. When each side of an equation has been squared, the resulting equation may have a solution that is not a solution of the original equation.

Check: $\dfrac{z + 7z^{\frac{1}{2}} - 18 = 0}{\begin{array}{c|c} 4 + 7(4)^{\frac{1}{2}} - 18 & 0 \\ 4 + 7 \cdot 2 - 18 & \\ 4 + 14 - 18 & \\ & 0 = 0 \end{array}}$

$\dfrac{z + 7z^{\frac{1}{2}} - 18 = 0}{\begin{array}{c|c} 81 + 7(81)^{\frac{1}{2}} - 18 & 0 \\ 81 + 7 \cdot 9 - 18 & \\ 81 + 63 - 18 & \\ & 126 \neq 0 \end{array}}$

4 checks as a solution, but 81 does not.

Write the solution.   The solution is 4.

Example 1   Solve.   A. $x^4 + x^2 - 12 = 0$ B. $x^{\frac{2}{3}} - 2x^{\frac{1}{3}} - 3 = 0$

Solution   A.
$$
\begin{aligned}
x^4 + x^2 - 12 &= 0 \\
(x^2)^2 + (x^2) - 12 &= 0 \\
u^2 + u - 12 &= 0 \\
(u - 3)(u + 4) &= 0
\end{aligned}
$$

▶ The equation is quadratic in form.
▶ Let $x^2 = u$.
▶ Solve for u by factoring.

$$
\begin{array}{ll}
u - 3 = 0 & u + 4 = 0 \\
u = 3 & u = -4 \\
\end{array}
$$

$$
\begin{array}{ll}
x^2 = 3 & x^2 = -4 \\
\sqrt{x^2} = \sqrt{3} & \sqrt{x^2} = \sqrt{-4} \\
x = \pm\sqrt{3} & x = \pm 2i \\
\end{array}
$$

▶ Replace u by x^2.
▶ Solve for x by taking square roots.

The solutions are $\sqrt{3}, -\sqrt{3}, 2i,$ and $-2i$.

B.
$$
\begin{aligned}
x^{\frac{2}{3}} - 2x^{\frac{1}{3}} - 3 &= 0 \\
(x^{\frac{1}{3}})^2 - 2(x^{\frac{1}{3}}) - 3 &= 0 \\
u^2 - 2u - 3 &= 0 \\
(u - 3)(u + 1) &= 0
\end{aligned}
$$

▶ The equation is quadratic in form.
▶ Let $x^{\frac{1}{3}} = u$.
▶ Solve for u by factoring.

$$
\begin{array}{ll}
u - 3 = 0 & u + 1 = 0 \\
u = 3 & u = -1 \\
\end{array}
$$

$x^{\frac{1}{3}} = 3$ $\quad$ $x^{\frac{1}{3}} = -1$ $\quad$ ▶ Replace u by $x^{\frac{1}{3}}$.

$(x^{\frac{1}{3}})^3 = 3^3$ $\quad$ $(x^{\frac{1}{3}})^3 = (-1)^3$ $\quad$ ▶ Solve for x by cubing both sides of the equation.

$x = 27$ $\quad$ $x = -1$

The solutions are 27 and −1.

Problem 1 Solve. A. $x - 5x^{\frac{1}{2}} + 6 = 0$ B. $4x^4 + 35x^2 - 9 = 0$

Solution See page S23. A. 4, 9 B. $\pm\frac{1}{2}, \pm 3i$

2 Radical equations

33 HM³

Certain equations containing a radical can be solved by first solving the equation for the radical expression and then squaring each side of the equation.

Remember that when each side of an equation has been squared, the resulting equation may have a solution that is not a solution of the original equation. Therefore, the solutions of a radical equation must be checked.

➡ Solve: $\sqrt{x+2} + 4 = x$

$$\sqrt{x+2} + 4 = x$$

Solve for the radical expression. $\quad \sqrt{x+2} = x - 4$

Square each side of the equation. $\quad (\sqrt{x+2})^2 = (x-4)^2$

Simplify. $\quad x + 2 = x^2 - 8x + 16$

Write the equation in standard form. $\quad 0 = x^2 - 9x + 14$

Solve for x by factoring. $\quad 0 = (x-7)(x-2)$

$x - 7 = 0 \qquad x - 2 = 0$

$x = 7 \qquad x = 2$

Check the solution.

Check: $\sqrt{x+2} + 4 = x$

$\sqrt{7+2} + 4 \mid 7$

$\sqrt{9} + 4$

$3 + 4$

$7 = 7$

$\sqrt{x+2} + 4 = x$

$\sqrt{2+2} + 4 \mid 2$

$\sqrt{4} + 4$

$2 + 4$

$6 \neq 2$

7 checks as a solution, but 2 does not.

Write the solution. $\quad$ The solution is 7.

Example 2 Solve: $\sqrt{7y-3} + 3 = 2y$

Solution

$\sqrt{7y-3} + 3 = 2y$

$\sqrt{7y-3} = 2y - 3$ $\quad$ ▶ Solve for the radical expression.

$(\sqrt{7y-3})^2 = (2y-3)^2$ $\quad$ ▶ Square each side of the equation.

$7y - 3 = 4y^2 - 12y + 9$

$0 = 4y^2 - 19y + 12$ $\quad$ ▶ Write the equation in standard form.

$$0 = (4y - 3)(y - 4)$$ ▶ Solve for y by factoring.

$$4y - 3 = 0 \qquad y - 4 = 0$$
$$4y = 3 \qquad y = 4$$ ▶ 4 checks as a solution.
$$y = \frac{3}{4}$$ $\frac{3}{4}$ does not check as a solution.

The solution is 4.

Problem 2 Solve: $\sqrt{2x + 1} + x = 7$

Solution See page S23. 4

If an equation contains more than one radical, the procedure of solving for the radical expression and squaring each side of the equation may have to be repeated.

Example 3 Solve: $\sqrt{2y + 1} - \sqrt{y} = 1$

Solution

$$\sqrt{2y + 1} - \sqrt{y} = 1$$
$$\sqrt{2y + 1} = \sqrt{y} + 1$$ ▶ Solve for one of the radical expressions.
$$(\sqrt{2y + 1})^2 = (\sqrt{y} + 1)^2$$ ▶ Square each side of the equation.
$$2y + 1 = y + 2\sqrt{y} + 1$$
$$y = 2\sqrt{y}$$ ▶ Solve for the radical expression.
$$y^2 = (2\sqrt{y})^2$$ ▶ Square each side of the equation.
$$y^2 = 4y$$
$$y^2 - 4y = 0$$
$$y(y - 4) = 0$$
$$y = 0 \qquad y - 4 = 0$$
$$y = 4$$ ▶ 0 and 4 check as solutions.

The solutions are 0 and 4.

Problem 3 Solve: $\sqrt{2x - 1} + \sqrt{x} = 2$

Solution See page S24. 1

3 Fractional equations

33 HM³

After each side of a fractional equation has been multiplied by the LCM of the denominators, the resulting equation is sometimes a quadratic equation. The solutions to the resulting equation must be checked, because multiplying each side of an equation by a variable expression may produce an equation that has a solution that is not a solution of the original equation.

Solve: $\frac{1}{r} + \frac{1}{r+1} = \frac{3}{2}$

$$\frac{1}{r} + \frac{1}{r+1} = \frac{3}{2}$$

Multiply each side of the equation by the LCM of the denominators.

$$2r(r+1)\left(\frac{1}{r} + \frac{1}{r+1}\right) = 2r(r+1) \cdot \frac{3}{2}$$
$$2(r+1) + 2r = r(r+1) \cdot 3$$
$$2r + 2 + 2r = 3r^2 + 3r$$
$$4r + 2 = 3r^2 + 3r$$

Write the equation in standard form.

$$0 = 3r^2 - r - 2$$

Solve for r by factoring.

$$0 = (3r+2)(r-1)$$

$$3r + 2 = 0 \qquad r - 1 = 0$$
$$3r = -2 \qquad r = 1$$
$$r = -\frac{2}{3}$$

$-\frac{2}{3}$ and 1 check as solutions.

Write the solutions.

The solutions are $-\frac{2}{3}$ and 1.

Example 4 Solve: $\frac{18}{2a-1} + 3a = 17$

Solution

$$\frac{18}{2a-1} + 3a = 17$$

▶ The LCM of the denominators is $2a - 1$.

$$(2a-1)\left(\frac{18}{2a-1} + 3a\right) = (2a-1)17$$
$$(2a-1)\frac{18}{2a-1} + (2a-1)(3a) = (2a-1)17$$
$$18 + 6a^2 - 3a = 34a - 17$$
$$6a^2 - 37a + 35 = 0$$

▶ Write the equation in standard form.

$$(6a-7)(a-5) = 0$$

▶ Solve for x by factoring.

$$6a - 7 = 0 \qquad a - 5 = 0$$
$$6a = 7 \qquad a = 5$$
$$a = \frac{7}{6}$$

The solutions are $\frac{7}{6}$ and 5.

Problem 4 Solve: $3y + \frac{25}{3y-2} = -8$

Solution See page S24. -1

CONCEPT REVIEW 8.3

Determine whether the following statements are always true, sometimes true, or never true.

1. An equation that is quadratic in form can be solved by the quadratic formula.
Always true
2. Squaring both sides on a radical equation produces an extraneous root.
Sometimes true
3. $x + 3\sqrt{x} - 8 = 0$ is quadratic in form. Always true
4. $\sqrt[4]{x} + 2\sqrt{3x} - 8 = 0$ is quadratic in form. Never true

EXERCISES 8.3

1 1. What does it mean for an equation to be quadratic in form?

2. Explain how to show that $x^4 - 2x^2 - 3 = 0$ is quadratic in form.

Solve.

3. $x^4 - 13x^2 + 36 = 0$ $\pm 3, \pm 2$
4. $y^4 - 5y^2 + 4 = 0$ $\pm 2, \pm 1$
5. $z^4 - 6z^2 + 8 = 0$ $\pm 2, \pm\sqrt{2}$
6. $t^4 - 12t^2 + 27 = 0$ $\pm 3, \pm\sqrt{3}$
7. $p - 3p^{\frac{1}{2}} + 2 = 0$ 1, 4
8. $v - 7v^{\frac{1}{2}} + 12 = 0$ 9, 16
9. $x - x^{\frac{1}{2}} - 12 = 0$ 16
10. $w - 2w^{\frac{1}{2}} - 15 = 0$ 25
11. $z^4 + 3z^2 - 4 = 0$ $\pm 1, \pm 2i$
12. $y^4 + 5y^2 - 36 = 0$ $\pm 2, \pm 3i$
13. $x^4 + 12x^2 - 64 = 0$ $\pm 2, \pm 4i$
14. $x^4 - 81 = 0$ $\pm 3, \pm 3i$
15. $p + 2p^{\frac{1}{2}} - 24 = 0$ 16
16. $v + 3v^{\frac{1}{2}} - 4 = 0$ 1
17. $y^{\frac{2}{3}} - 9y^{\frac{1}{3}} + 8 = 0$ 512, 1
18. $z^{\frac{2}{3}} - z^{\frac{1}{3}} - 6 = 0$ 27, −8
19. $x^6 - 9x^3 + 8 = 0$
$2, 1, -1 \pm i\sqrt{3}, -\frac{1}{2} \pm \frac{\sqrt{3}}{2}i$
20. $y^6 + 9y^3 + 8 = 0$
$-2, -1, 1 \pm i\sqrt{3}, \frac{1}{2} \pm \frac{\sqrt{3}}{2}i$
21. $z^8 - 17z^4 + 16 = 0$
$\pm 1, \pm 2, \pm i, \pm 2i$
22. $v^4 - 15v^2 - 16 = 0$ $\pm 4, \pm i$
23. $p^{\frac{2}{3}} + 2p^{\frac{1}{3}} - 8 = 0$ −64, 8
24. $w^{\frac{2}{3}} + 3w^{\frac{1}{3}} - 10 = 0$ −125, 8
25. $2x - 3x^{\frac{1}{2}} + 1 = 0$ $\frac{1}{4}, 1$
26. $3y - 5y^{\frac{1}{2}} - 2 = 0$ 4

2 Solve.

27. $\sqrt{x + 1} + x = 5$ 3
28. $\sqrt{x - 4} + x = 6$ 5
29. $x = \sqrt{x} + 6$ 9

30. $\sqrt{2y-1}=y-2$ 5

31. $\sqrt{3w+3}=w+1$ 2, −1

32. $\sqrt{2s+1}=s-1$ 4

33. $\sqrt{4y+1}-y=1$ 0, 2

34. $\sqrt{3s+4}+2s=12$ 4

35. $\sqrt{10x+5}-2x=1$ $-\frac{1}{2}, 2$

36. $\sqrt{t+8}=2t+1$ 1

37. $\sqrt{p+11}=1-p$ −2

38. $x-7=\sqrt{x-5}$ 9

39. $\sqrt{x-1}-\sqrt{x}=-1$ 1

40. $\sqrt{y}+1=\sqrt{y+5}$ 4

41. $\sqrt{2x-1}=1-\sqrt{x-1}$ 1

42. $\sqrt{x+6}+\sqrt{x+2}=2$ −2

43. $\sqrt{t+3}+\sqrt{2t+7}=1$ −3

44. $\sqrt{5-2x}=\sqrt{2-x}+1$ ±2

3 Solve.

45. $x=\frac{10}{x-9}$ 10, −1

46. $z=\frac{5}{z-4}$ 5, −1

47. $\frac{t}{t+1}=\frac{-2}{t-1}$ $-\frac{1}{2}\pm\frac{\sqrt{7}}{2}i$

48. $\frac{2v}{v-1}=\frac{5}{v+2}$ $\frac{1}{4}\pm\frac{\sqrt{39}}{4}i$

49. $\frac{y-1}{y+2}+y=1$ 1, −3

50. $\frac{2p-1}{p-2}+p=8$ 3, 5

51. $\frac{3r+2}{r+2}-2r=1$ 0, −1

52. $\frac{2v+3}{v+4}+3v=4$ $-\frac{13}{3}, 1$

53. $\frac{2}{2x+1}+\frac{1}{x}=3$ $\frac{1}{2}, -\frac{1}{3}$

54. $\frac{3}{s}-\frac{2}{2s-1}=1$ $\frac{3}{2}, 1$

55. $\frac{16}{z-2}+\frac{16}{z+2}=6$ $6, -\frac{2}{3}$

56. $\frac{2}{y+1}+\frac{1}{y-1}=1$ 0, 3

57. $\frac{t}{t-2}+\frac{2}{t-1}=4$ $\frac{4}{3}, 3$

58. $\frac{4t+1}{t+4}+\frac{3t-1}{t+1}=2$ $1, -\frac{11}{5}$

59. $\frac{5}{2p-1}+\frac{4}{p+1}=2$ $-\frac{1}{4}, 3$

60. $\frac{3w}{2w+3}+\frac{2}{w+2}=1$ 0, −3

61. $\frac{2v}{v+2}+\frac{3}{v+4}=1$ $\frac{-5\pm\sqrt{33}}{2}$

62. $\frac{x+3}{x+1}-\frac{x-2}{x+3}=5$ $\frac{-13\pm\sqrt{89}}{10}$

APPLYING CONCEPTS 8.3

Solve.

63. $\frac{x^2}{4}+\frac{x}{2}=6$ 4, −6

64. $3\left(\frac{x+1}{2}\right)^2=54$ $-1\pm6\sqrt{2}$

65. $\frac{x+2}{3}+\frac{2}{x-2}=3$ 5, 4

66. $\frac{x^4}{4}+1=\frac{5x^2}{4}$ ±1, ±2

67. $\frac{x^4}{3}-\frac{8x^2}{3}=3$ ±3, ±i

68. $\frac{x^2}{4}+\frac{x}{2}+\frac{1}{8}=0$ $\frac{-2\pm\sqrt{2}}{2}$

69. $\frac{x^4}{8}+\frac{x^2}{4}=3$ $\pm2, \pm i\sqrt{6}$

70. $\sqrt{x^4-2}=x$ $\sqrt{2}, i$

71. $\sqrt{x^4+4}=2x$ $\sqrt{2}$

[C] **72.** $(\sqrt{x}-2)^2-5\sqrt{x}+14=0$ (*Hint:* Let $u=\sqrt{x}-2$.) 9, 36

[C] **73.** $(\sqrt{x}+3)^2-4\sqrt{x}-17=0$ (*Hint:* Let $u=\sqrt{x}+3$.) 4

74. *Sports* According to the Compton's Interactive Encyclopedia, the minimum dimensions of a football used in the National Football Association games are 10.875 in. long and 20.75 in. in circumference at the center. A possible model for the cross section of a football is given by $y=\pm3.3041\sqrt{1-\frac{x^2}{29.7366}}$, where x is the distance from the center of the football and y is the radius of the football at x. See the graph at the right.

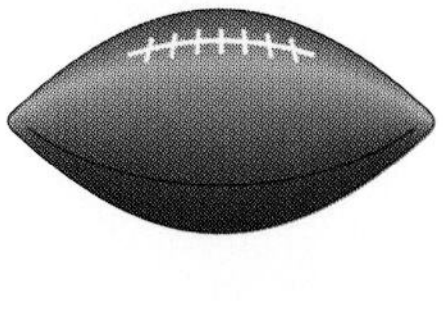

a. What is the domain of the equation? $\{x \mid -\sqrt{29.7366}\le x\le\sqrt{29.7366}\}$

b. Graph $y = 3.3041\sqrt{1 - \frac{x^2}{29.7366}}$ and $y = -3.3041\sqrt{1 - \frac{x^2}{29.7366}}$ on the same coordinate axes. Explain why the $\pm$ symbol occurs in the equation.

c. Determine the radius of the football when x is 3 in. Round to the nearest ten-thousandth. 2.7592 in.

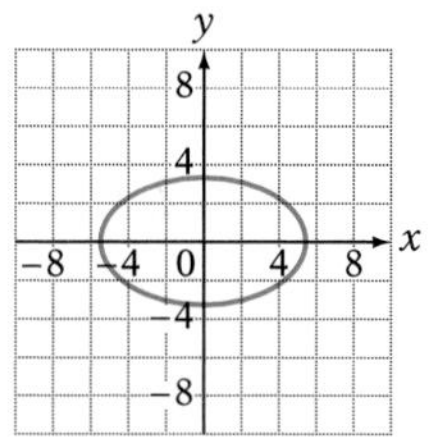

b. The $\pm$ symbol occurs in the equation so that the graph pictures the entire shape of the football.

SECTION 8.4 Applications of Quadratic Equations

1 Application problems

The application problems in this section are similar to those problems that were solved earlier in the text. Each of the strategies for the problems in this section will result in a quadratic equation.

Solve: A small pipe takes 16 min longer to empty a tank than does a larger pipe. Working together, the pipes can empty the tank in 6 min. How long would it take each pipe, working alone, to empty the tank?

STRATEGY for solving an application problem

- Determine the type of problem. Is it a uniform motion problem, a geometry problem, an integer problem, or a work problem?

The problem is a work problem.

- Choose a variable to represent the unknown quantity. Write numerical or variable expressions for all the remaining quantities. These results can be recorded in a table.

The unknown time of the larger pipe: t
The unknown time of the smaller pipe: $t + 16$

	Rate of work	$\cdot$	Time worked	=	Part of task completed
Larger pipe	$\frac{1}{t}$	$\cdot$	6	=	$\frac{6}{t}$
Smaller pipe	$\frac{1}{t + 16}$	$\cdot$	6	=	$\frac{6}{t + 16}$

■ Determine how the quantities are related. If necessary, review the strategies presented in earlier chapters.

The sum of the parts of the task completed must equal 1.

$$\frac{6}{t} + \frac{6}{t+16} = 1$$

$$t(t+16)\left(\frac{6}{t} + \frac{6}{t+16}\right) = t(t+16) \cdot 1$$

$$(t+16)6 + 6t = t^2 + 16t$$

$$6t + 96 + 6t = t^2 + 16t$$

$$0 = t^2 + 4t - 96$$

$$0 = (t+12)(t-8)$$

$$t + 12 = 0 \qquad t - 8 = 0$$
$$t = -12 \qquad t = 8$$

The solution $t = -12$ is not possible because time cannot be a negative number.

The time for the smaller pipe is $t + 16$. Replace t by 8 and evaluate.

$$t + 16 = 8 + 16 = 24$$

The larger pipe requires 8 min to empty the tank.
The smaller pipe requires 24 min to empty the tank.

Example 1   In 8 h, two campers rowed 15 mi down a river and then rowed back to their campsite. The rate of the river's current was 1 mph. Find the rate at which the campers row in calm water.

Strategy   ■ This is a uniform motion problem.
■ Unknown rowing rate of the campers: r

	Distance	÷	Rate	=	Time
Down river	15	÷	$r+1$	=	$\frac{15}{r+1}$
Up river	15	÷	$r-1$	=	$\frac{15}{r-1}$

■ The total time of the trip was 8 h.

Solution
$$\frac{15}{r+1} + \frac{15}{r-1} = 8$$

$$(r+1)(r-1)\left(\frac{15}{r+1} + \frac{15}{r-1}\right) = (r+1)(r-1)8$$

$$(r-1)15 + (r+1)15 = (r^2-1)8$$

$$15r - 15 + 15r + 15 = 8r^2 - 8$$

$$30r = 8r^2 - 8$$

$$0 = 8r^2 - 30r - 8$$

$$0 = 2(4r^2 - 15r - 4)$$

$$0 = 2(4r+1)(r-4)$$

$$4r + 1 = 0 \qquad r - 4 = 0$$
$$4r = -1 \qquad r = 4$$
$$r = -\frac{1}{4}$$

▶ The solution $r = -\frac{1}{4}$ is not possible because the rate cannot be a negative number.

The rowing rate is 4 mph.

Problem 1 The length of a rectangle is 3 m more than the width. The area is 54 m^2. Find the length of the rectangle.

Solution See page S24. 9 m

CONCEPT REVIEW 8.4

Determine whether the following statements are always true, sometimes true, or never true.

1. Let x be an integer. Then x^2 and $(x + 2)^2$ represent the squares of two consecutive odd integers. Sometimes true

2. Let x be an integer. Then x^2, $(x + 1)^2$, and $(x + 2)^2$ represent the squares of three consecutive integers. Always true

3. If t is the time to complete an amount of work, then $\frac{1}{t}$ is the rate of doing work. Always true
4. Let r be the rate of rowing a boat in calm water. If the river has a current of 2 mph, then the rate of rowing down the river is $r + 2$, and the rate of rowing up the river is $r - 2$. Always true

EXERCISES 8.4

1 Solve.

1. *Farming* A rectangular corral is constructed in a pasture. The length of the rectangle is 8 ft more than twice the width. The area of the rectangle is to be 640 ft^2. What are the length and width of the rectangle?
length: 40 ft; width: 16 ft

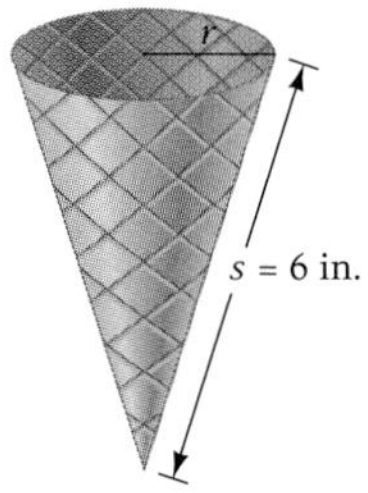

2. *Manufacturing* The surface area of the ice cream cone shown at the right is given by $A = \pi r^2 + \pi rs$, where r is the radius of the circular top of the ice cream cone and s is the slant height of the cone. If the area of the cone is 11.25π in^2 and the slant height of the cone is 6 in., find the radius of the cone.
1.5 in.

3. *Geometry* The length of a rectangle is 2 ft less than three times the width of the rectangle. The area of the rectangle is 65 ft^2. Find the length and width of the rectangle. Length: 13 ft; width: 5 ft

4. *Geometry* The length of a rectangle is 2 cm less than twice the width. The area of the rectangle is 180 cm^2. Find the length and width of the rectangle. Length: 18 cm; width: 10 cm

5. *Transportation* An 18-wheeler left a grain depot with a load of wheat and traveled 550 mi to deliver the wheat before returning to the depot. Because of the lighter load on the return trip, the average speed of the truck returning was 5 mph faster than its average speed going. Find the rate returning if the entire trip, not counting unloading time or rest stops, was 21 h. 55 mph

6. *Racing* The *Tour de France* is a bicycle race that begins in Dublin, Ireland, but is primarily a race through France. On each day, racers cover a certain distance depending on the steepness of the terrain. Suppose that on one particular day, the racers must complete 210 mi. One cyclist, traveling 10 mph faster than a second cyclist, covers this distance in 2.4 h less time than the second cyclist. Find the rate of the first cyclist. 35 mph

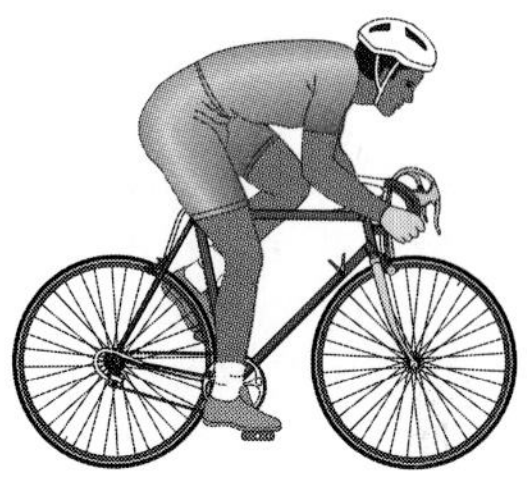

7. *Physics* The height of a projectile fired upward is given by the formula $s = v_0t - 16t^2$, where s is the height, v_0 is the initial velocity, and t is the time. Find the time for a projectile to return to Earth if it has an initial velocity of 200 ft/s. 12.5 s

8. *Physics* The height of a projectile fired upward is given by the formula $s = v_0t - 16t^2$, where s is the height, v_0 is the initial velocity, and t is the time. Find the times at which a projectile with an initial velocity of 128 ft/s will be 64 ft above the ground. Round to the nearest hundredth of a second. 7.46 s and 0.54 s

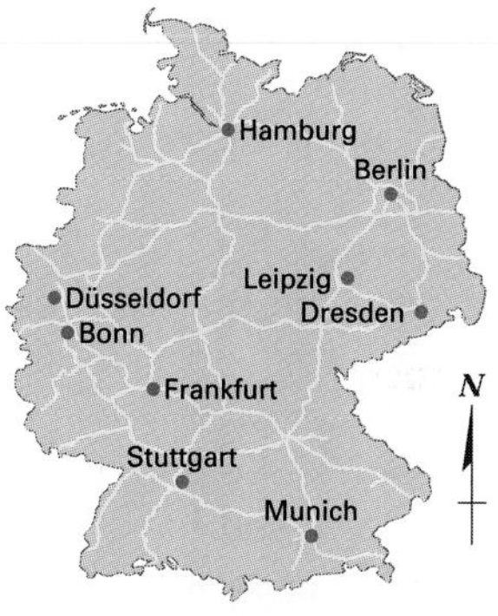

German Autobahn System

9. *Automotive Technology* In Germany, there are no speed limits on some portions of the autobahn (highway). Other portions have a speed limit of 180 km/h (approximately 112 mph). The distance, d (in meters), required to stop a car traveling at v kilometers per hour is $d = 0.019v^2 + 0.69v$. Approximate, to the nearest tenth, the maximum speed a driver can be going and still be able to stop within 150 m. 72.5 km/h

10. *Sports* A penalty kick in soccer is made from a penalty mark that is 36 ft from a goal that is 8 ft high. A possible equation for the flight of a penalty kick is $h = -0.002x^2 + 0.35x$, where h is the height (in feet) of the ball x feet from the penalty mark. Assuming that the flight of the kick is toward the goal and that it is not touched by the goalie, will the ball land in the net? No

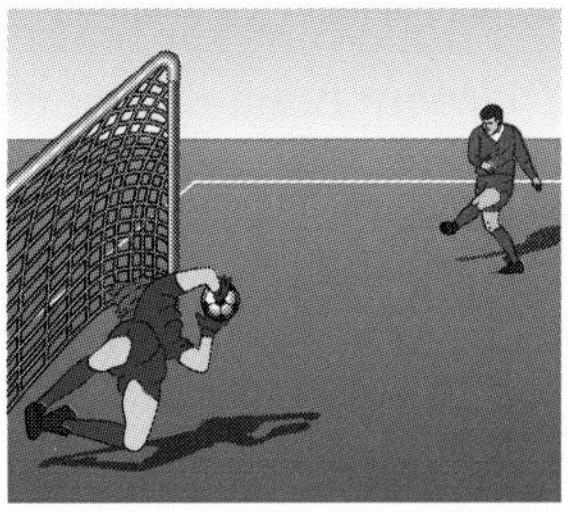

11. *Aeronautics* A model rocket is launched with an initial velocity of 200 ft/s. The height, h, of the rocket t seconds after the launch is given by $h = -16t^2 + 200t$. How many seconds after launch will the rocket be 300 ft above the ground? Round to the nearest hundredth of a second. 1.74 s and 10.76 s

12. *Sports* A diver jumps from a platform that is 10 m high. The height of the diver t seconds after the beginning of the dive is given by $h = -4.9t^2 + 3.2t + 10.5$. To the nearest hundredth of a second, how long after the beginning of the dive will the diver enter the water? 1.83 s

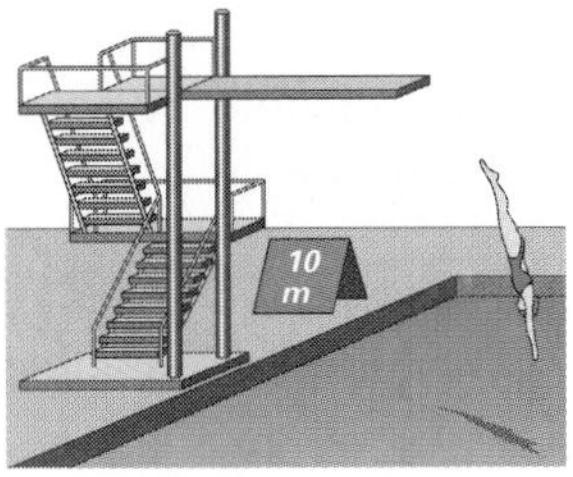

13. *Petroleum Engineering* A small pipe can fill an oil tank in 6 min more time than it takes a larger pipe to fill the same tank. Working together, both pipes can fill the tank in 4 min. How long would it take each pipe, working alone, to fill the tank? Smaller pipe: 12 min; larger pipe: 6 min

14. *Metallurgy* A small heating unit takes 8 h longer to melt a piece of iron than does a larger unit. Working together, the heating units can melt the iron in 3 h. How long would it take each heating unit, working alone, to melt the iron? Smaller unit: 12 h; larger unit: 4 h

15. *Recreation* A cruise ship made a trip of 100 mi in 8 h. The ship traveled the first 40 mi at a constant rate before increasing its speed by 5 mph. Another 60 mi was traveled at the increased speed. Find the rate of the cruise ship for the first 40 mi. 10 mph

16. *Sports* A cyclist traveled 60 mi at a constant rate before reducing the speed by 2 mph. Another 40 mi was traveled at the reduced speed. The total time for the 100-mile trip was 9 h. Find the rate during the first 60 mi. 12 mph

17. *Aeronautics* The rate of a single-engine plane in calm air is 100 mph. Flying with the wind, the plane can fly 240 mi in 1 h less time than is required to make the return trip of 240 mi. Find the rate of the wind. 20 mph

18. *Transportation* A car travels 120 mi. A second car, traveling 10 mph faster than the first car, makes the same trip in 1 h less time. Find the speed of each car. 1st car: 30 mph; 2nd car: 40 mph

19. *Sports* To prepare for an upcoming race, a sculling crew rowed 16 mi down a river and back in 6 h. If the rate of the river's current is 2 mph, find the sculling crew's rate of rowing in calm water. 6 mph

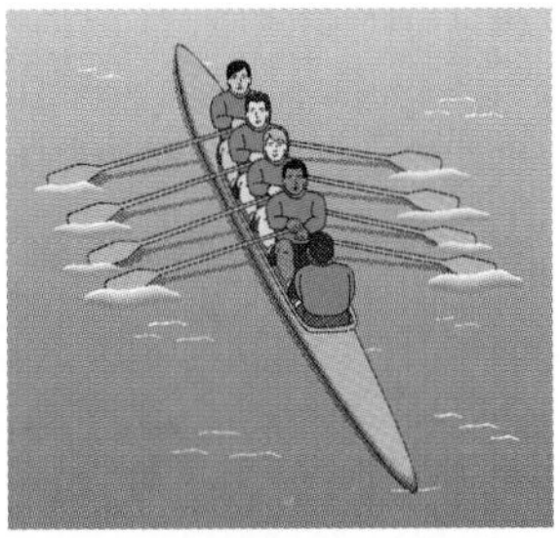

20. *Sports* A fishing boat traveled 30 mi down a river and then returned. The total time for the round trip was 4 h, and the rate of the river's current was 4 mph. Find the rate of the boat in still water. 16 mph

APPLYING CONCEPTS 8.4

Solve.

21. *Number Problem* The sum of a number and twice its reciprocal is $\frac{33}{4}$. Find the number. $\frac{1}{4}$ or 8

22. *Number Problem* The numerator of a fraction is 3 less than the denominator. The sum of the fraction and four times its reciprocal is $\frac{17}{2}$. Find the fraction. $\frac{3}{6}$

23. *Integer Problem* Find two consecutive integers whose cubes differ by 127.
6 and 7 or −7 and −6

24. *Integer Problem* Find two consecutive even integers whose cubes differ by 488. 8 and 10 or −10 and −8

[C] **25.** *Geometry* An open box is formed from a rectangular piece of cardboard whose length is 8 cm more than its width by cutting squares whose sides are 2 cm in length from each corner and then folding up the sides. Find the dimensions of the box if its volume is 256 cm^3. 2 cm × 8 cm × 16 cm

26. *Construction* The height of an arch is given by the equation

$$h(x) = -\frac{3}{64}x^2 + 27, \ -24 \le x \le 24$$

where $|x|$ is the distance in feet from the center of the arch.

a. What is the maximum height of the arch? 27 ft

b. What is the height of the arch 8 ft to the right of center? 24 ft

c. How far from the center is the arch 8 ft tall? 20.13 ft

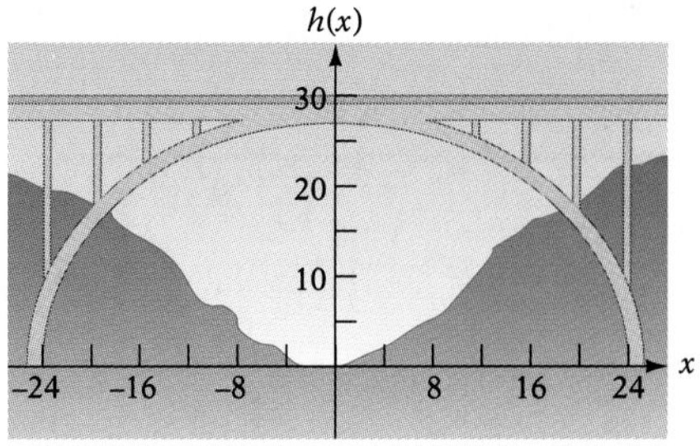

27. *Business* Today Microsoft Corporation produces operating systems (OS) that run the majority of PCs around the world. The accompanying bar chart shows Microsoft OS product shipments worldwide growing substantially as we approach the millennium. (*Source:* http://gartnerweb.com/dq/static/about/press/pr-b9829.html dq = dataquest Inc.) A quadratic model of this growth is given by the equation

$$G(x) = \frac{1}{2}x^2 + 15.5x + 78$$

where $G(x)$ equals millions of units shipped and $x = 1$ corresponds to the year 1997. According to this model, when will Microsoft OS first ship more than 150 million units in a year?

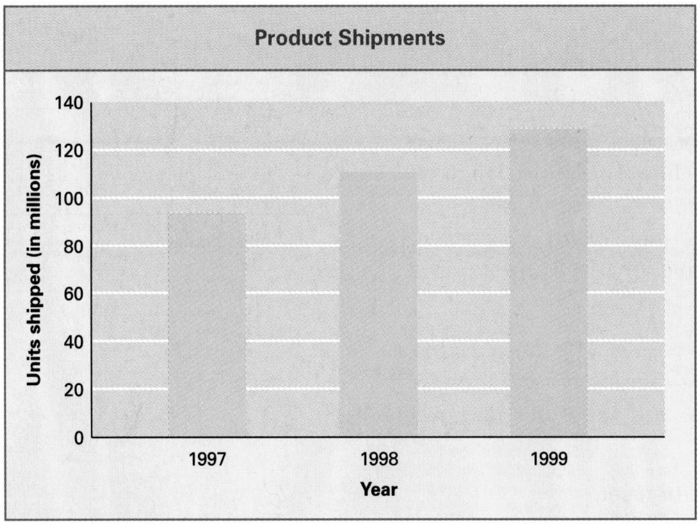

2001 (168 million units)

28. *Geometry* The volumes of two spheres differ by 372π cm^3. The radius of the larger sphere is 3 cm more than the radius of the smaller sphere. Find the radius of the larger sphere. (*Hint:* The formula for the volume of a sphere is $V = \frac{4}{3}\pi r^3$.) 7 cm

[C] **29.** *Geometry* A perfectly spherical scoop of mint chocolate chip ice cream is placed in a cone as shown at the right. How far is the bottom of the scoop of ice cream from the bottom of the cone? (*Hint:* A line segment from the center of the scoop of ice cream to the point at which the ice cream touches the cone is perpendicular to the edge of the cone.) 2.3 in.

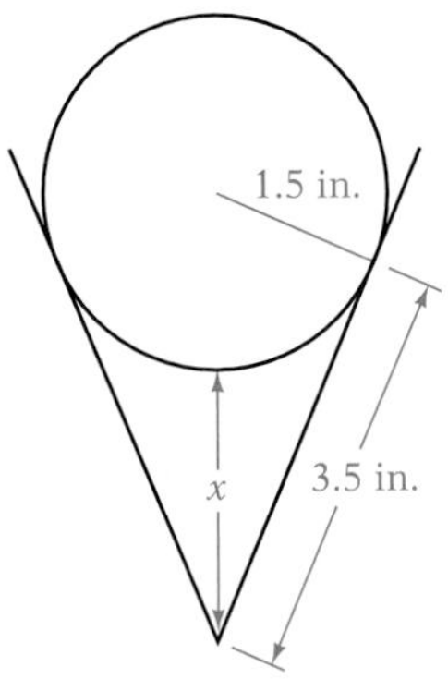

SECTION **8.5**

Nonlinear Inequalities

1 Solve inequalities by factoring

A **quadratic inequality in one variable** is one that can be written in the form $ax^2 + bx + c < 0$ or $ax^2 + bx + c > 0$, where $a \neq 0$. The symbols $\leq$ and $\geq$ can also be used.

Quadratic inequalities can be solved by algebraic means. However, it is often easier to use a graphical method to solve these inequalities. The graphical method is used in the example that follows.

Solve and graph the solution set of $x^2 - x - 6 < 0$.

Factor the trinomial.

$$x^2 - x - 6 < 0$$
$$(x - 3)(x + 2) < 0$$

On a number line, draw vertical lines at the numbers that make each factor equal to zero.

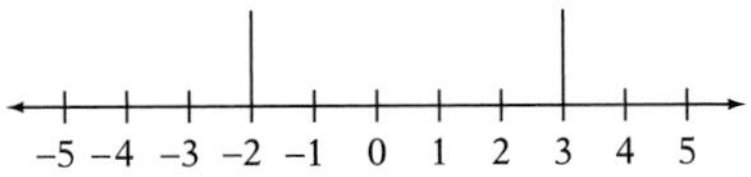

$$x - 3 = 0 \qquad x + 2 = 0$$
$$x = 3 \qquad x = -2$$

For each factor, place plus signs above the number line for those regions where the factor is positive and negative signs where the factor is negative. $x - 3$ is positive for $x > 3$ and $x + 2$ is positive for $x > -2$.

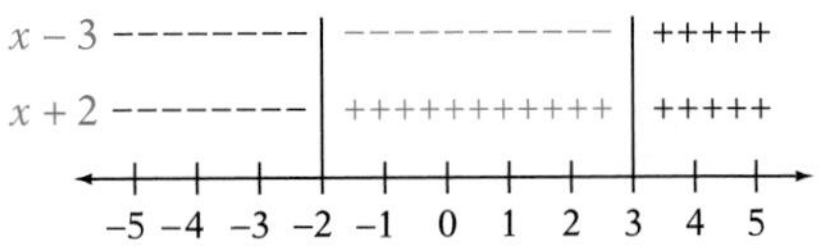

Because $x^2 - x - 6 < 0$, the solution set will be the regions where one factor is positive and the other factor is negative.

Write the solution set.

$$\{x \mid -2 < x < 3\}$$

The graph of the solution set of the inequality $x^2 - x - 6 < 0$ is shown at the right.

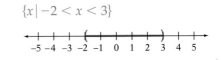

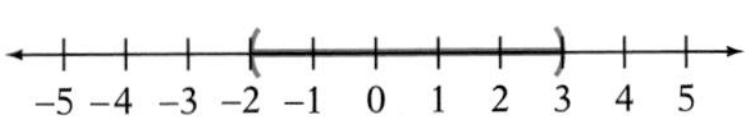

This method of solving quadratic inequalities can be used on any polynomial that can be factored into linear factors.

Solve and graph the solution set of $x^3 - 4x^2 - 4x + 16 > 0$.

Factor the polynomial by grouping.

$$\begin{aligned} x^3 - 4x^2 - 4x + 16 &> 0 \\ x^2(x - 4) - 4(x - 4) &> 0 \\ (x^2 - 4)(x - 4) &> 0 \\ (x - 2)(x + 2)(x - 4) &> 0 \end{aligned}$$

On a number line, identify for each factor the regions where the factor is positive and where the factor is negative.

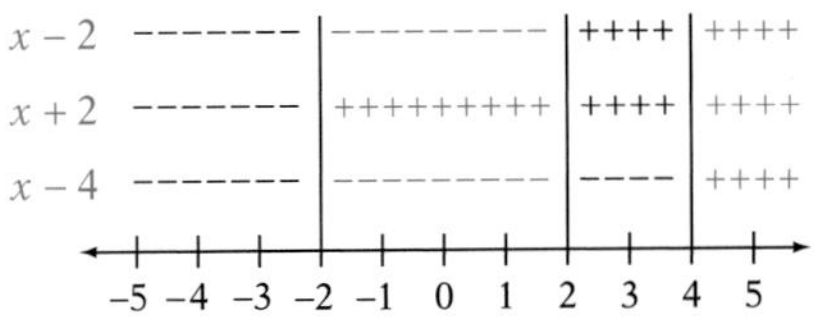

There are two regions where the product of the three factors is positive.

Write the solution set.

$\{x \mid -2 < x < 2 \text{ or } x > 4\}$

The graph of the solution set of the inequality $x^3 - 4x^2 - 4x + 16 > 0$ is shown at the right.

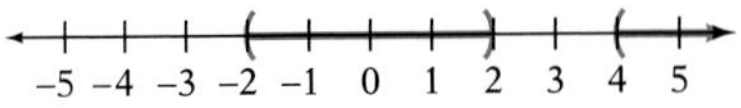

Example 1 Solve and graph the solution set of $2x^2 - x - 3 \geq 0$.

Solution

$$\begin{aligned} 2x^2 - x - 3 &\geq 0 \\ (2x - 3)(x + 1) &\geq 0 \end{aligned}$$

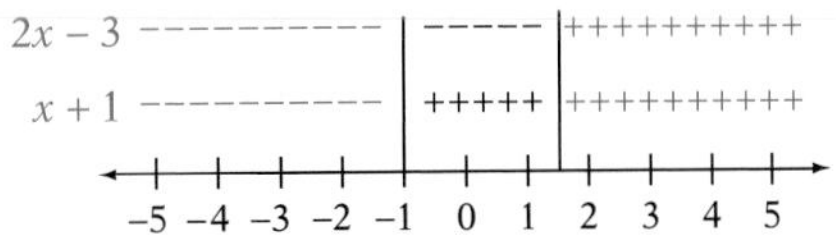

$\left\{x \mid x \leq -1 \text{ or } x \geq \frac{3}{2}\right\}$

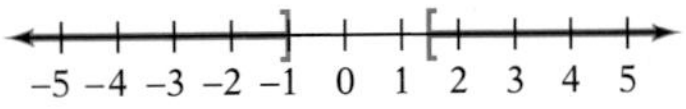

Problem 1 Solve and graph the solution set of $2x^2 - x - 10 \leq 0$.

Solution See page S24. $\left\{x \mid -2 \leq x \leq \frac{5}{2}\right\}$

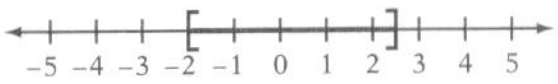

2 Solve rational inequalities

The graphical method used in the last objective can be used to solve rational inequalities.

Solve: $\dfrac{2x - 5}{x - 4} \leq 1$

$$\frac{2x - 5}{x - 4} \leq 1$$

Rewrite the inequality so that 0 appears on the right side of the inequality.

$$\frac{2x - 5}{x - 4} - 1 \leq 0$$

Then simplify.

$$\frac{2x - 5}{x - 4} - \frac{x - 4}{x - 4} \leq 0$$

$$\frac{x - 1}{x - 4} \leq 0$$

On a number line, identify for each factor of the numerator and each factor of the denominator the regions where the factor is positive and where the factor is negative.

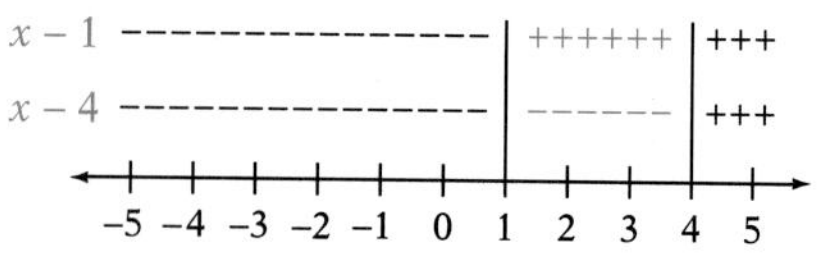

The region where the quotient of the two factors is negative is between 1 and 4.

Write the solution set. $\{x \mid 1 \le x < 4\}$

Note that 1 is part of the solution set, but 4 is not part of the solution set because the denominator of the rational expression is zero when $x = 4$.

Example 2 Solve and graph the solution set of $\frac{x+4}{x-3} \ge 0$.

Solution $\frac{x+4}{x-3} \ge 0$

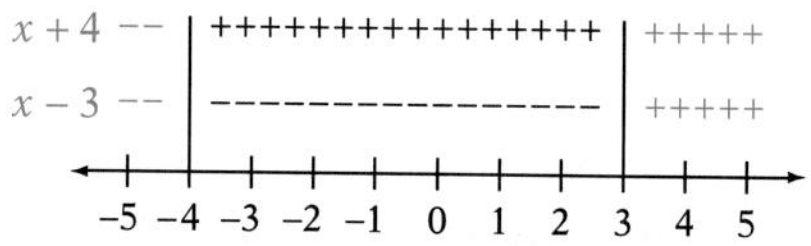

$\{x \mid x > 3 \text{ or } x \le -4\}$

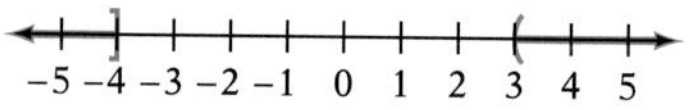

Problem 2 Solve and graph the solution set of $\frac{x}{x-2} \le 0$.

Solution See page S24. $\{x \mid 0 \le x < 2\}$

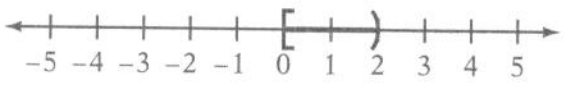

CONCEPT REVIEW 8.5

Determine whether the following statements are always true, sometimes true, or never true.

1. The end points of a solution set of a quadratic inequality are not included in the solution set. Sometimes true

2. The solution set of $(x-3)(x-2)(x+2) > 0$ is 3, 2, and −2. Never true

3. The solution set of a quadratic inequality is an infinite set. Sometimes true

4. The solution set of $\frac{1}{x-1} > 0$ is $\{x \mid x > 1\}$. Always true

EXERCISES 8.5

1 **1.** If $(x - 3)(x - 5) > 0$, what must be true of the values of $x - 3$ and $x - 5$?

2. For the nonlinear inequality $\frac{x - 2}{x - 3} \geq 1$, is 3 a possible element of the solution set? Why or why not?

Solve and graph the solution set.

3. $(x - 4)(x + 2) > 0$ $\{x \mid x < -2 \text{ or } x > 4\}$

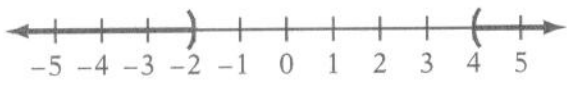

4. $(x + 1)(x - 3) > 0$ $\{x \mid x < -1 \text{ or } x > 3\}$

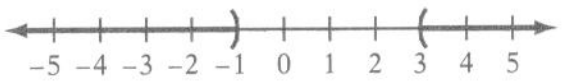

5. $x^2 - 3x + 2 \geq 0$ $\{x \mid x \leq 1 \text{ or } x \geq 2\}$

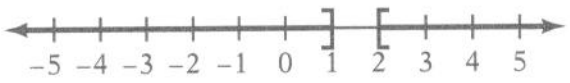

6. $x^2 + 5x + 6 > 0$ $\{x \mid x < -3 \text{ or } x > -2\}$

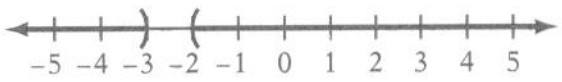

7. $x^2 - x - 12 < 0$ $\{x \mid -3 < x < 4\}$

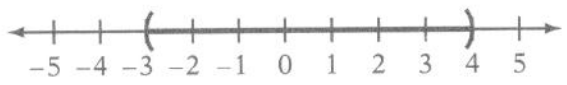

8. $x^2 + x - 20 < 0$ $\{x \mid -5 < x < 4\}$

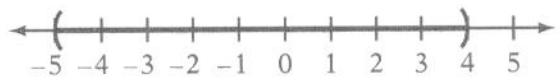

9. $(x - 1)(x + 2)(x - 3) < 0$
$\{x \mid x < -2 \text{ or } 1 < x < 3\}$

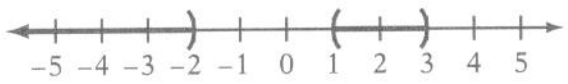

10. $(x + 4)(x - 2)(x - 1) \geq 0$
$\{x \mid -4 \leq x \leq 1 \text{ or } x \geq 2\}$

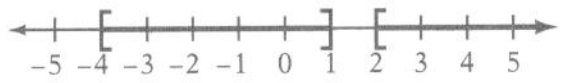

Solve.

11. $x^2 - 16 > 0$
$\{x \mid x < -4 \text{ or } x > 4\}$

12. $x^2 - 4 \geq 0$
$\{x \mid x \leq -2 \text{ or } x \geq 2\}$

13. $x^2 - 4x + 4 > 0$
$\{x \mid x < 2 \text{ or } x > 2\}$

14. $x^2 + 6x + 9 > 0$
$\{x \mid x < -3 \text{ or } x > -3\}$

15. $x^2 - 9x \leq 36$
$\{x \mid -3 \leq x \leq 12\}$

16. $x^2 + 4x > 21$
$\{x \mid x < -7 \text{ or } x > 3\}$

17. $2x^2 - 5x + 2 \geq 0$
$\left\{x \mid x \leq \frac{1}{2} \text{ or } x \geq 2\right\}$

18. $4x^2 - 9x + 2 < 0$
$\left\{x \mid \frac{1}{4} < x < 2\right\}$

19. $4x^2 - 8x + 3 < 0$
$\left\{x \mid \frac{1}{2} < x < \frac{3}{2}\right\}$

20. $2x^2 + 11x + 12 \geq 0$
$\left\{x \mid x \leq -4 \text{ or } x \geq -\frac{3}{2}\right\}$

21. $(x - 6)(x + 3)(x - 2) \leq 0$
$\{x \mid x \leq -3 \text{ or } 2 \leq x \leq 6\}$

22. $(x + 5)(x - 2)(x - 3) > 0$
$\{x \mid -5 < x < 2 \text{ or } x > 3\}$

23. $(2x - 1)(x - 4)(2x + 3) > 0$
$\left\{x \mid -\frac{3}{2} < x < \frac{1}{2} \text{ or } x > 4\right\}$

24. $(x - 2)(3x - 1)(x + 2) \leq 0$
$\left\{x \mid x \leq -2 \text{ or } \frac{1}{3} \leq x \leq 2\right\}$

25. $x^3 + 3x^2 - x - 3 \leq 0$
$\{x \mid x \leq -3 \text{ or } -1 \leq x \leq 1\}$

26. $x^3 + x^2 - 9x - 9 < 0$
$\{x \mid x < -3 \text{ or } -1 < x < 3\}$

27. $x^3 - x^2 - 4x + 4 \geq 0$
$\{x \mid -2 \leq x \leq 1 \text{ or } x \geq 2\}$

28. $2x^3 + 3x^2 - 8x - 12 \geq 0$
$\left\{x \mid -2 \leq x \leq -\frac{3}{2} \text{ or } x \geq 2\right\}$

2 Solve and graph the solution set.

29. $\frac{x - 4}{x + 2} > 0$
$\{x \mid x < -2 \text{ or } x > 4\}$

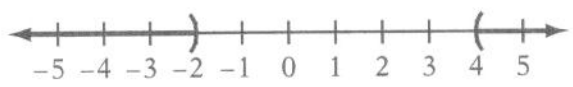

30. $\frac{x + 2}{x - 3} > 0$
$\{x \mid x < -2 \text{ or } x > 3\}$

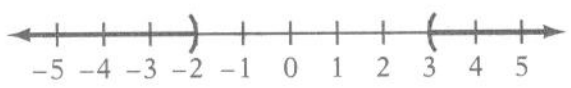

31. $\frac{x - 3}{x + 1} \leq 0$
$\{x \mid -1 < x \leq 3\}$

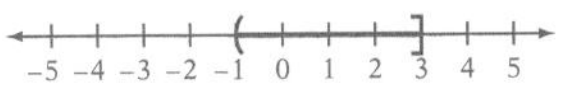

32. $\dfrac{x-1}{x} > 0$

$\{x \mid x < 0 \text{ or } x > 1\}$

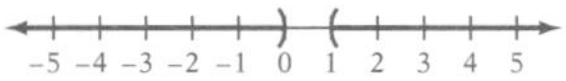

33. $\dfrac{(x-1)(x+2)}{x-3} \le 0$

$\{x \mid x \le -2 \text{ or } 1 \le x < 3\}$

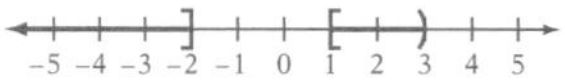

34. $\dfrac{(x+3)(x-1)}{x-2} \ge 0$

$\{x \mid -3 \le x \le 1 \text{ or } x > 2\}$

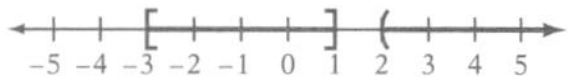

Solve.

35. $\dfrac{3x}{x-2} > 1$ $\{x \mid x < -1 \text{ or } x > 2\}$

36. $\dfrac{2x}{x+1} < 1$ $\{x \mid -1 < x < 1\}$

37. $\dfrac{2}{x+1} \ge 2$ $\{x \mid -1 < x \le 0\}$

38. $\dfrac{3}{x-1} < 2$ $\left\{x \mid x < 1 \text{ or } x > \dfrac{5}{2}\right\}$

39. $\dfrac{x}{(x-1)(x+2)} \ge 0$ $\{x \mid -2 < x \le 0 \text{ or } x > 1\}$

40. $\dfrac{x-2}{(x+1)(x-1)} \le 0$ $\{x \mid x < -1 \text{ or } 1 < x \le 2\}$

41. $\dfrac{1}{x} < 2$ $\left\{x \mid x < 0 \text{ or } x > \dfrac{1}{2}\right\}$

42. $\dfrac{x}{2x-1} \ge 1$ $\left\{x \mid \dfrac{1}{2} < x \le 1\right\}$

APPLYING CONCEPTS 8.5

Graph the solution set.

43. $(x+2)(x-3)(x+1)(x+4) > 0$

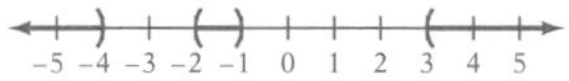

44. $(x-1)(x+3)(x-2)(x-4) \ge 0$

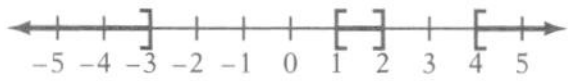

45. $(x^2 + 2x - 8)(x^2 - 2x - 3) < 0$

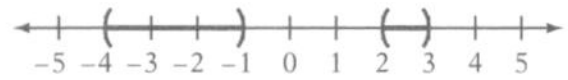

46. $(x^2 + 2x - 3)(x^2 + 3x + 2) \ge 0$

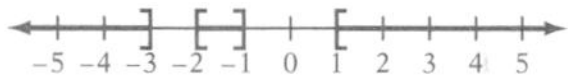

47. $(x^2 + 1)(x^2 - 3x + 2) > 0$

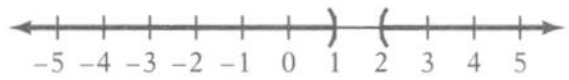

48. $(x^2 - 9)(x^2 + 5x + 6) \le 0$

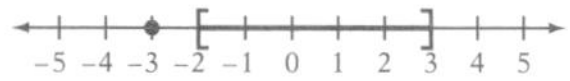

49. $\dfrac{x^2(3-x)(2x+1)}{(x+4)(x+2)} \ge 0$

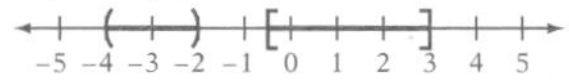

[C] 50. $x < x^2$

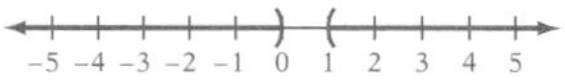

[C] 51. $x^3 > x$

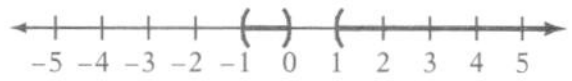

[C] 52. $\dfrac{1}{x} + x > 2$

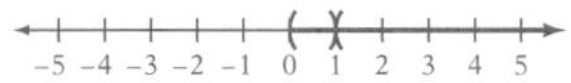

[C] 53. $3x - \dfrac{1}{x} \le 2$

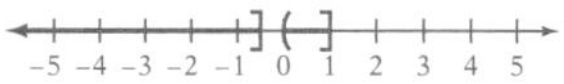

[C] 54. $x^2 - x < \dfrac{1-x}{x}$

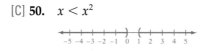

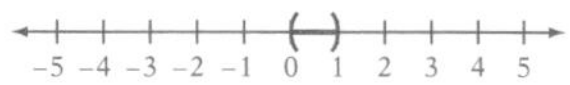

S E C T I O N **8.6**

Properties of Quadratic Functions

1 Graph a quadratic function

Recall that a linear function is one that can be expressed by the equation $f(x) = mx + b$. The graph of a linear function has certain characteristics. It is a straight line with slope m and y-intercept $(0, b)$. A **quadratic function** is one that can be expressed by the equation $f(x) = ax^2 + bx + c$, $a \ne 0$. The graph of this function, which is called a **parabola**, also has certain characteristics. The graph of a quadratic function can be drawn by finding ordered pairs that belong to the function.

⯈ Graph: $f(x) = x^2 - 2x - 3$

By evaluating the function for various values of x, find enough ordered pairs to determine the shape of the graph.

x	$f(x) = x^2 - 2x - 3$	$f(x)$	(x, y)
-2	$f(-2) = (-2)^2 - 2(-2) - 3$	5	$(-2, 5)$
-1	$f(-1) = (-1)^2 - 2(-1) - 3$	0	$(-1, 0)$
0	$f(0) = (0)^2 - 2(0) - 3$	-3	$(0, -3)$
1	$f(1) = (1)^2 - 2(1) - 3$	-4	$(1, -4)$
2	$f(2) = (2)^2 - 2(2) - 3$	-3	$(2, -3)$
3	$f(3) = (3)^2 - 2(3) - 3$	0	$(3, 0)$
4	$f(4) = (4)^2 - 2(4) - 3$	5	$(4, 5)$

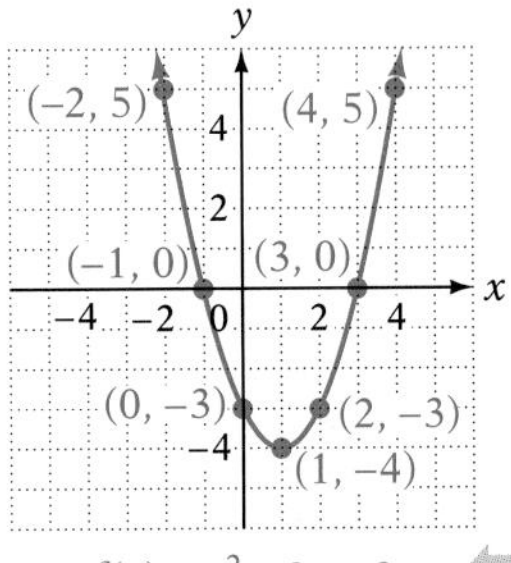

$f(x) = x^2 - 2x - 3$

Because the value of $f(x) = x^2 - 2x - 3$ is a real number for all values of x, the domain of f is all real numbers. From the graph, it appears that no value of y is less than -4. Thus the range is $\{y \mid y \geq -4\}$. The range can also be determined algebraically, as shown below, by completing the square.

$$f(x) = x^2 - 2x - 3$$

Group the variable terms. $f(x) = (x^2 - 2x) - 3$

Complete the square of $x^2 - 2x$. Add and subtract $\left[\frac{1}{2}(-2)\right]^2 = 1$ to $x^2 - 2x$. $f(x) = (x^2 - 2x + 1) - 1 - 3$

Factor and combine like terms. $f(x) = (x - 1)^2 - 4$

The square of a number is always positive. $(x - 1)^2 \geq 0$

Subtract 4 from each side of the inequality. $(x - 1)^2 - 4 \geq -4$

Replace $(x - 1)^2 - 4$ with $f(x)$. $f(x) \geq -4$

$$y \geq -4$$

From the last inequality, the range is $\{y \mid y \geq -4\}$.

In general, the graph of $f(x) = ax^2 + bx + c$, $a \neq 0$, resembles a "cup" shape as shown below. The parabola opens up when $a > 0$ and opens down when $a < 0$. The vertex of the parabola is the point with the smallest y-coordinate when $a > 0$ and is the point with the largest y-coordinate when $a < 0$.

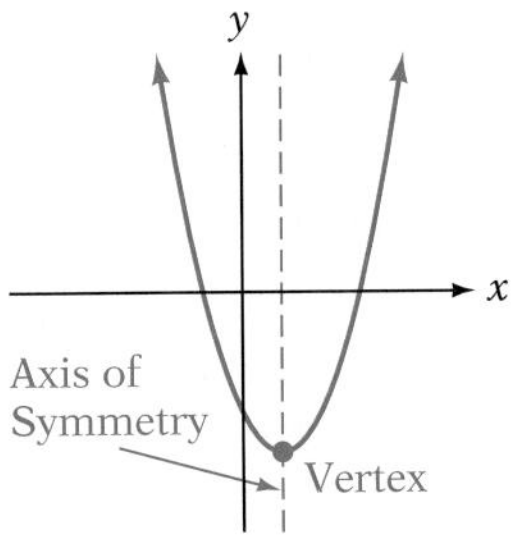

$a > 0$

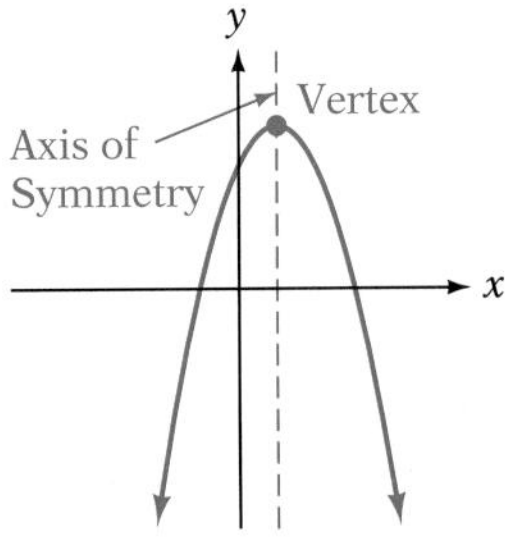

$a < 0$

The **axis of symmetry** is a line that passes through the vertex of the parabola and is parallel to the y-axis. To understand the axis of symmetry, think of folding the graph along that line. The two portions of the graph will match up.

The vertex and axis of symmetry of a parabola can be found by completing the square.

INSTRUCTOR NOTE

The argument to determine the vertex is difficult for students. One approach is to have students evaluate $x^2 + 3$ and $x^2 - 1$ for various values of x and discover that the least value of the expression occurs when $x = 0$. Then repeat this process for $(x + 2)^2 + 3$ and $(x + 2)^2 - 1$.

Find the vertex and the axis of symmetry of the graph of $F(x) = x^2 + 4x + 3$.

To find the coordinates of the vertex, complete the square. $F(x) = x^2 + 4x + 3$

Group the variable terms. $F(x) = (x^2 + 4x) + 3$

Complete the square of $x^2 + 4x$. Add and subtract $\left[\frac{1}{2}(4)\right]^2 = 4$ to and from $x^2 + 4x$. $F(x) = (x^2 + 4x + 4) - 4 + 3$

Factor and combine like terms. $F(x) = (x + 2)^2 - 1$

Because a, the coefficient of x^2, is positive ($a = 1$), the parabola opens up and the vertex is the point with the least y-coordinate. Because $(x + 2)^2 \geq 0$ for all values of x, the least y-coordinate occurs when $(x + 2)^2 = 0$. When $x = -2$, this expression equals zero.

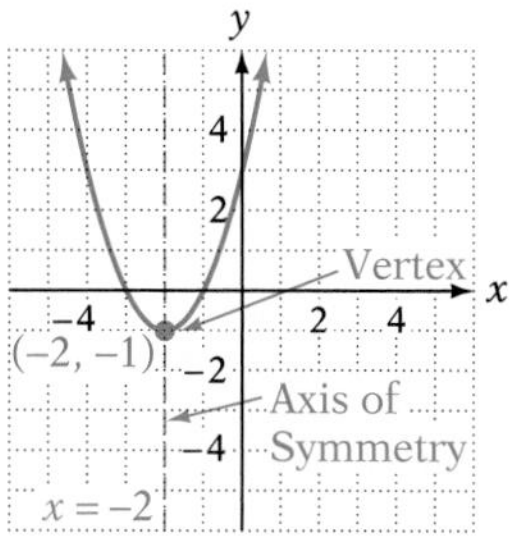

The x-coordinate of the vertex is -2.

To find the y-coordinate of the vertex, evaluate the function at $x = -2$.

$$F(x) = (x + 2)^2 - 1$$
$$F(-2) = (-2 + 2)^2 - 1 = -1$$

The y-coordinate of the vertex is -1.

From these results, the coordinates of the vertex are $(-2, -1)$. The axis of symmetry is the line whose equation is $x = -2$.

By following the process illustrated in the last example and completing the square of $f(x) = ax^2 + bx + c$, we can find a formula for the coordinates of the vertex of a parabola.

Vertex of a Parabola

Let $f(x) = ax^2 + bx + c$ be the equation of a parabola. The coordinates of the vertex are $\left(-\frac{b}{2a}, f\left(-\frac{b}{2a}\right)\right)$. The equation of the axis of symmetry is $x = -\frac{b}{2a}$.

The coordinates of the vertex and the axis of symmetry can be used to graph a parabola.

Graph $y = x^2 + 2x - 3$ using the vertex and the axis of symmetry.

Find the x-coordinate of the vertex. $a = 1$ and $b = 2$.

$$x = -\frac{b}{2a} = -\frac{2}{2(1)} = -1$$

Find the y-coordinate of the vertex by replacing x with -1 and solving for y.

$$y = x^2 + 2x - 3$$
$$y = (-1)^2 + 2(-1) - 3$$
$$y = 1 - 2 - 3$$
$$y = -4$$

The vertex is $(-1, -4)$.

The axis of symmetry is the line $x = -1$.

Find some ordered-pair solutions of the equation and record these in a table. Because the graph is symmetric to the line $x = -1$, choose values of x greater than -1.

x	y
0	-3
1	0
2	5

Graph the ordered-pair solutions on a rectangular coordinate system. Use symmetry to locate points of the graph on the other side of the axis of symmetry. Remember that corresponding points on the graph are the same distance from the axis of symmetry.

Draw a parabola through the points.

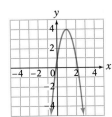

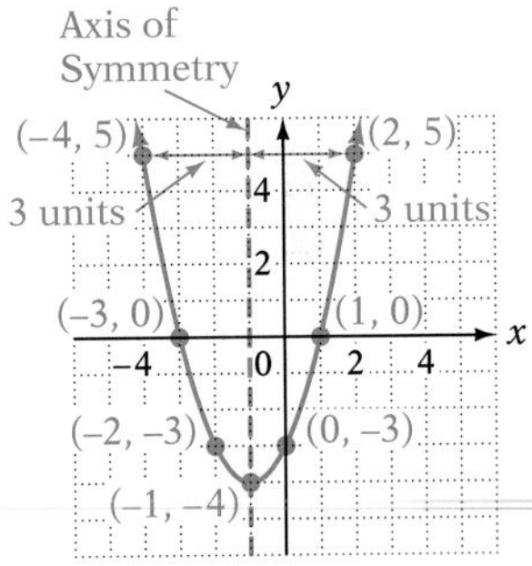

Using a graphing utility, enter the equation $y = x^2 + 2x - 3$ and verify the graph shown above. Now trace along the graph and verify that $(-1, -4)$ are the coordinates of the vertex.

Example 1 Find the vertex and the axis of symmetry of the parabola whose equation is $y = -3x^2 + 6x + 1$. Then sketch its graph.

Solution $-\frac{b}{2a} = -\frac{6}{2(-3)} = 1$

▸ Find the x-coordinate of the vertex. $a = -3$ and $b = 6$.

$$y = -3x^2 + 6x + 1$$
$$y = -3(1)^2 + 6(1) + 1$$
$$y = 4$$

▸ Find the y-coordinate of the vertex by replacing x by 1 and solving for y.

The vertex is $(1, 4)$.

The axis of symmetry is the line $x = 1$.

▸ The axis of symmetry is the line $x = -\frac{b}{2a}$.

▸ Because a is negative, the parabola opens down. Find a few ordered pairs, and use symmetry to sketch the graph.

Problem 1 Find the vertex and the axis of symmetry of the parabola whose equation is $y = x^2 - 2$. Then sketch its graph.

Solution See page S24. Vertex: $(0, -2)$
Axis of symmetry: $x = 0$

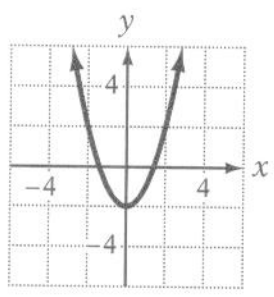

Using a graphing utility, enter the equation $y = -3x^2 + 6x + 1$ and verify the graph drawn in Example 1. Now trace along the graph and verify that the vertex is (1, 4). Follow the same procedure for Problem 1.

Example 2 Graph $f(x) = -2x^2 - 4x + 3$. State the domain and range of the function.

Solution Because a is negative ($a = -2$), the graph of f will open down. The x-coordinate of the vertex is $x = -\frac{b}{2a} = -\frac{-4}{2(-2)} = -1$. The y-coordinate of the vertex is $f(-1) = -2(-1)^2 - 4(-1) + 3 = 5$. The vertex is $(-1, 5)$.

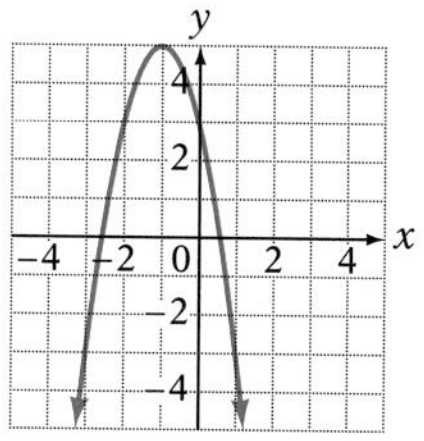

Evaluate $f(x)$ for various values of x, and use symmetry to draw the graph.

Because $f(x) = -2x^2 - 4x + 3$ is a real number for all values of x, the domain of the function is $\{x \mid x \in \text{real numbers}\}$. The vertex of the parabola is the highest point on the graph. Because the y-coordinate at that point is 5, the range is $\{y \mid y \leq 5\}$.

LOOK CLOSELY

Once the coordinates of the vertex are found, the range of the quadratic function can be determined.

Problem 2 Graph $g(x) = x^2 + 4x - 2$. State the domain and range of the function.

Solution See page S24.

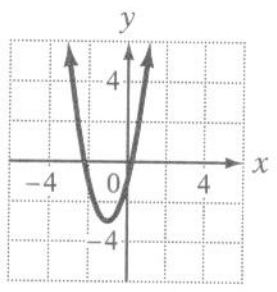

D: $\{x \mid x \in \text{real numbers}\}$
R: $\{y \mid y \geq -6\}$

2 Find the x-intercepts of a parabola

INSTRUCTOR NOTE

Intercepts were discussed in the context of linear functions. Making a connection to those ideas may help students realize that this is not a new concept.

Recall that a point at which a graph crosses the x- or y-axis is called an *intercept* of the graph. The x-intercepts of the graph of an equation occur when $y = 0$; the y-intercepts occur when $x = 0$.

The graph of $y = x^2 + 3x - 4$ is shown at the right. The points whose coordinates are $(-4, 0)$ and $(1, 0)$ are x-intercepts of the graph.

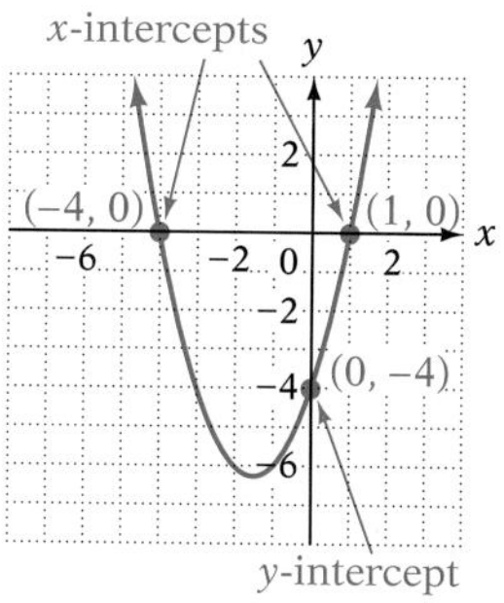

➡ Find the x-intercepts for the parabola whose equation is $y = 4x^2 - 4x + 1$.

$$y = 4x^2 - 4x + 1$$

To find x-intercepts, let $y = 0$.

$$0 = 4x^2 - 4x + 1$$

Solve for x by factoring and using the Principle of Zero Products.

$$0 = (2x - 1)(2x - 1)$$

$$2x - 1 = 0 \qquad 2x - 1 = 0$$
$$2x = 1 \qquad 2x = 1$$
$$x = \frac{1}{2} \qquad x = \frac{1}{2}$$

The x-intercept is $\left(\frac{1}{2}, 0\right)$.

In this last example, the parabola has only one x-intercept. In this case, the parabola is said to be *tangent* to the x-axis at $x = \frac{1}{2}$.

➡ Find the x-intercepts of $y = 2x^2 - x - 6$.

$$y = 2x^2 - x - 6$$

To find the x-intercepts, let $y = 0$.

$$0 = 2x^2 - x - 6$$

Solve for x by factoring and using the Principle of Zero Products.

$$0 = (2x + 3)(x - 2)$$

$$2x + 3 = 0 \qquad x - 2 = 0$$
$$x = -\frac{3}{2} \qquad x = 2$$

The x-intercepts are $\left(-\frac{3}{2}, 0\right)$ and $(2, 0)$.

LOOK CLOSELY

A real zero of a function is the x-coordinate of the x-intercept of the graph of the function. Because the x-intercepts of the graph of $f(x) = 2x^2 - x - 6$ are $\left(-\frac{3}{2}, 0\right)$ and $(2, 0)$, the zeros are $-\frac{3}{2}$ and 2.

If the equation above, $y = 2x^2 - x - 6$, were written in functional notation as $f(x) = 2x^2 - x - 6$, then to find the x-intercepts you would let $f(x) = 0$ and solve for x. A value of x for which $f(x) = 0$ is a *zero* of the function. Thus, $-\frac{3}{2}$ and 2 are *zeros* of $f(x) = 2x^2 - x - 6$.

➡ Find the zeros of $f(x) = x^2 - 2x - 1$.

$$f(x) = x^2 - 2x - 1$$

To find the zeros, let $f(x) = 0$ and solve for x.

$$0 = x^2 - 2x - 1$$

Because $x^2 - 2x - 1$ does not easily factor, use the quadratic formula to solve for x.

$a = 1$, $b = -2$, and $c = -1$.

$$x = \frac{-b \pm \sqrt{b^2 - 4ac}}{2a}$$
$$= \frac{-(-2) \pm \sqrt{(-2)^2 - 4(1)(-1)}}{2(1)}$$
$$= \frac{2 \pm \sqrt{4 + 4}}{2}$$
$$= \frac{2 \pm \sqrt{8}}{2}$$
$$= \frac{2 \pm 2\sqrt{2}}{2}$$
$$= 1 \pm \sqrt{2}$$

The zeros of the function are $1 - \sqrt{2}$ and $1 + \sqrt{2}$.

The graph of $f(x) = x^2 - 2x - 1$ is shown at the right. Note that the zeros are the x-intercepts of the graph of f.

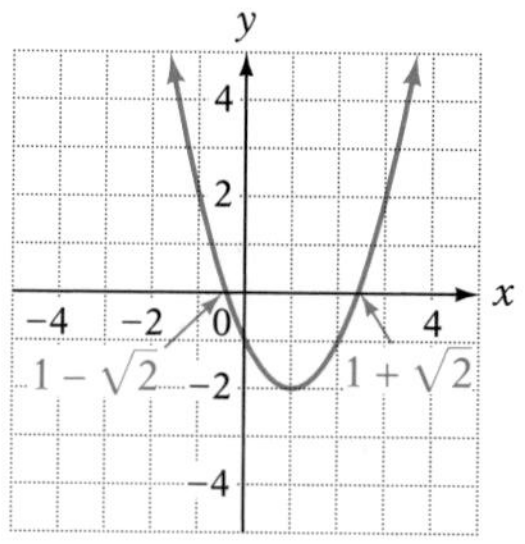

INSTRUCTOR NOTE

This is difficult for students. Giving additional simple examples may help.

Find the x-intercept of $y = 2x - 6$.

Find the zero of $f(x) = 2x - 6$.

Solve $2x - 6 = 0$.

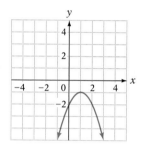

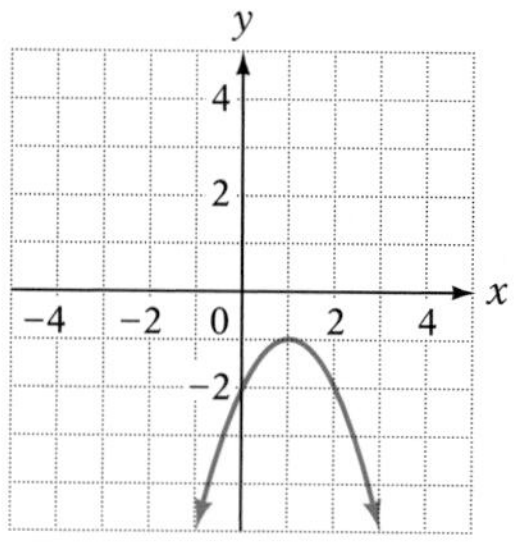

The preceding examples suggest that there is a relationship among the x-intercepts of the graph of a function, the zeros of a function, and the solutions of an equation. In fact, those three concepts are different ways of discussing the same number. The choice depends on the focus of the discussion. If we are discussing graphing, then intercept is our focus; if we are discussing functions, then the zero of the function is our focus; and if we are discussing equations, then the solution of the equation is our focus.

The graph of a parabola may not have x-intercepts. The graph of $y = -x^2 + 2x - 2$ is shown at the left. Note that the graph does not pass through the x-axis and thus there are no x-intercepts. This means that there are no real number zeros of $f(x) = -x^2 + 2x - 2$ and no real number solutions of $-x^2 + 2x - 2 = 0$.

Using the quadratic formula, we find that the solutions of the equation $-x^2 + 2x - 2 = 0$ are $1 - i$ and $1 + i$. Thus the zeros of $f(x) = -x^2 + 2x - 2$ are the complex numbers $1 - i$ and $1 + i$.

Example 3 Find the x-intercepts of the parabola given by each equation.

A. $y = x^2 + 2x - 2$ B. $y = 4x^2 + 4x + 1$

Solution A. $y = x^2 + 2x - 2$

$0 = x^2 + 2x - 2$ ▶ Let $y = 0$.

$$x = \frac{-b \pm \sqrt{b^2 - 4ac}}{2a}$$

▶ The equation is nonfactorable over the integers. Use the quadratic formula to solve for x.

$$= \frac{-(2) \pm \sqrt{(2)^2 - 4(1)(-2)}}{2 \cdot 1}$$

$$= \frac{-2 \pm \sqrt{4 + 8}}{2}$$

$$= \frac{-2 \pm \sqrt{12}}{2}$$

$$= \frac{-2 \pm 2\sqrt{3}}{2} = -1 \pm \sqrt{3}$$

The x-intercepts are $(-1 + \sqrt{3}, 0)$ and $(-1 - \sqrt{3}, 0)$.

B. $y = 4x^2 + 4x + 1$

$0 = 4x^2 + 4x + 1$ ▶ Let $y = 0$.

$0 = (2x + 1)(2x + 1)$ ▶ Solve for x by factoring.

$$2x + 1 = 0 \qquad 2x + 1 = 0$$
$$2x = -1 \qquad 2x = -1$$
$$x = -\frac{1}{2} \qquad x = -\frac{1}{2}$$

▶ The equation has a double root.

The x-intercept is $\left(-\frac{1}{2}, 0\right)$.

Problem 3 Find the x-intercepts of the parabola given by each equation.

A. $y = 2x^2 - 5x + 2$ B. $y = x^2 + 4x + 4$

Solution See page S25. A. $\left(\frac{1}{2}, 0\right)$, $(2, 0)$ B. $(-2, 0)$

The zeros of a quadratic function can always be determined exactly by solving a quadratic equation. However, it is possible to approximate the real zeros by graphing the function and determining the x-intercepts.

Example 4 Graph $f(x) = 2x^2 - x - 5$. Estimate the real zeros of the function to the nearest tenth.

Solution Graph the function. Then use the features of a graphing utility to estimate the zeros. The approximate values of the zeros are -1.4 and 1.9.

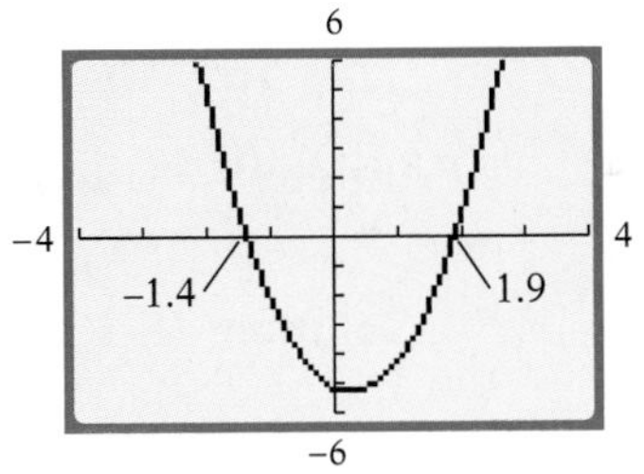

Problem 4 Graph $f(x) = -x^2 - 4x + 6$. Estimate the real zeros of the function to the nearest tenth.

Solution See page S25. -5.2, 1.2

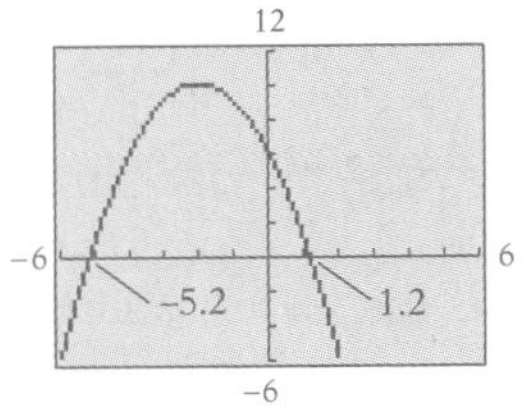

INSTRUCTOR NOTE

The discriminant was discussed in conjunction with the solutions of quadratic equations. It may be necessary to review that material at this time.

Recall that the discriminant of $ax^2 + bx + c$ is the expression $b^2 - 4ac$ and that this expression can be used to determine whether $ax^2 + bx + c = 0$ has zero, one, or two real number solutions. Because there is a connection between the solutions of $ax^2 + bx + c = 0$ and the x-intercepts of the graph of $y = ax^2 + bx + c$, the discriminant can be used to determine the number of x-intercepts of a parabola.

The Effect of the Discriminant on the Number of x-Intercepts of a Parabola

1. If $b^2 - 4ac = 0$, the parabola has one x-intercept.
2. If $b^2 - 4ac > 0$, the parabola has two x-intercepts.
3. If $b^2 - 4ac < 0$, the parabola has no x-intercepts.

Use the discriminant to determine the number of x-intercepts of the parabola whose equation is $y = 2x^2 - x + 2$.

$a = 2, b = -1, c = 2$

Evaluate the discriminant.

$$\begin{aligned} b^2 - 4ac &= (-1)^2 - 4(2)(2) \\ &= 1 - 16 \\ &= -15 \end{aligned}$$

$-15 < 0$

The discriminant is less than zero, so the parabola has no x-intercepts.

Example 5 Use the discriminant to determine the number of x-intercepts of the parabola whose equation is $y = x^2 - 6x + 9$.

Solution

$$\begin{aligned} b^2 - 4ac &= (-6)^2 - 4(1)(9) && \blacktriangleright a = 1, b = -6, c = 9 \\ &= 36 - 36 \\ &= 0 && \blacktriangleright \text{The discriminant is equal to zero.} \end{aligned}$$

The parabola has one x-intercept.

Problem 5 Use the discriminant to determine the number of x-intercepts of the parabola whose equation is $y = x^2 - x - 6$.

Solution See page S25. Two

CONCEPT REVIEW 8.6

Determine whether the following statements are always true, sometimes true, or never true.

1. The axis of symmetry of a parabola goes through the origin. Sometimes true
2. The vertex is the lowest point on the graph of a quadratic function. Sometimes true
3. A parabola has two x-intercepts. Sometimes true
4. If the discriminant is zero, the graph of a quadratic function has one x-intercept. Always true

5. The graph of a quadratic function has one y-intercept. Always true

6. The domain of a quadratic function is the real numbers. Always true

EXERCISES 8.6

1 1. What is a quadratic function?

2. What are the vertex and axis of symmetry of the graph of a parabola?

Find the vertex and axis of symmetry of the parabola given by each equation. Then sketch its graph.

3. $y = x^2$
Vertex: (0, 0)
Axis of symmetry: $x = 0$

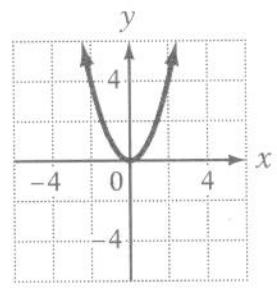

4. $y = -x^2$
Vertex: (0, 0)
Axis of symmetry: $x = 0$

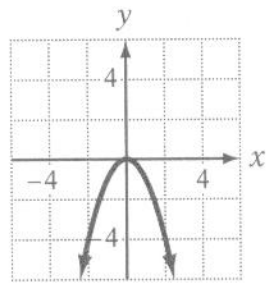

5. $y = x^2 - 2$
Vertex: (0, −2)
Axis of symmetry: $x = 0$

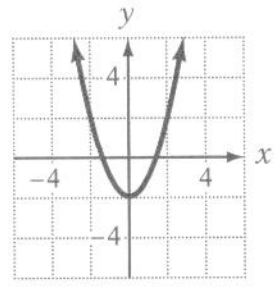

6. $y = x^2 + 2$
Vertex: (0, 2)
Axis of symmetry: $x = 0$

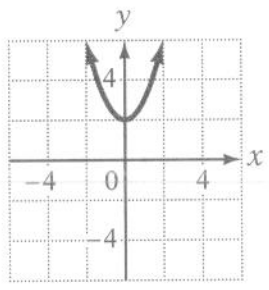

7. $y = -x^2 + 3$
Vertex: (0, 3)
Axis of symmetry: $x = 0$

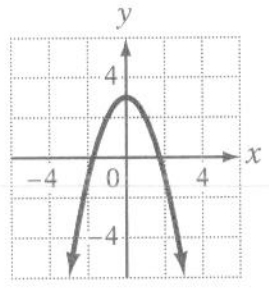

8. $y = -x^2 - 1$
Vertex: (0, −1)
Axis of symmetry: $x = 0$

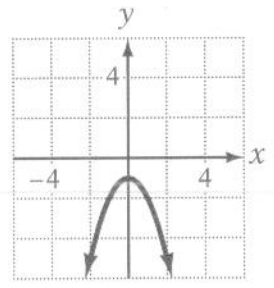

9. $y = \frac{1}{2}x^2$
Vertex: (0, 0)
Axis of symmetry: $x = 0$

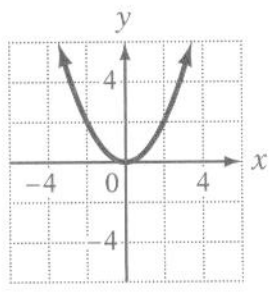

10. $y = 2x^2$
Vertex: (0, 0)
Axis of symmetry: $x = 0$

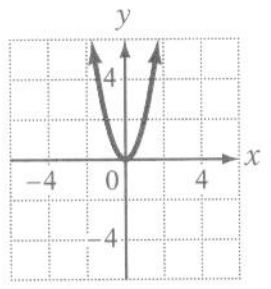

11. $y = 2x^2 - 1$
Vertex: (0, −1)
Axis of symmetry: $x = 0$

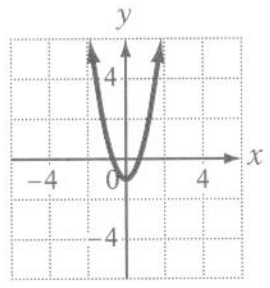

12. $y = -\frac{1}{2}x^2 + 2$
Vertex: (0, 2)
Axis of symmetry: $x = 0$

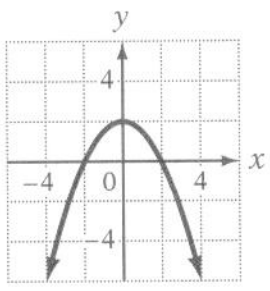

13. $y = x^2 - 2x$
Vertex: (1, −1)
Axis of symmetry: $x = 1$

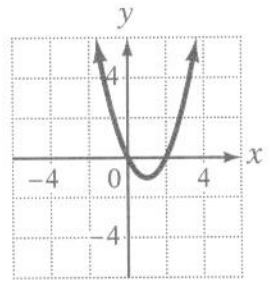

14. $y = x^2 + 2x$
Vertex: (−1, −1)
Axis of symmetry: $x = -1$

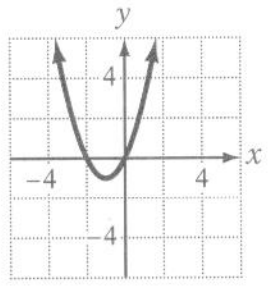

15. $y = -2x^2 + 4x$
Vertex: (1, 2)
Axis of symmetry: $x = 1$

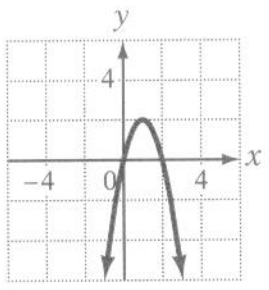

16. $y = \frac{1}{2}x^2 - x$
Vertex: $\left(1, -\frac{1}{2}\right)$
Axis of symmetry: $x = 1$

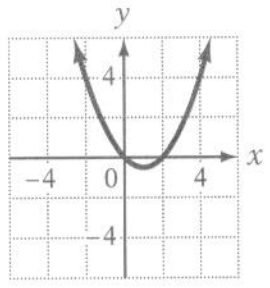

17. $y = x^2 - x - 2$
Vertex: $\left(\frac{1}{2}, -\frac{9}{4}\right)$
Axis of symmetry: $x = \frac{1}{2}$

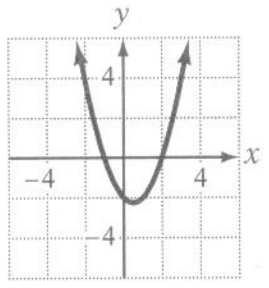

18. $y = x^2 - 3x + 2$
Vertex: $\left(\frac{3}{2}, -\frac{1}{4}\right)$
Axis of symmetry: $x = \frac{3}{2}$

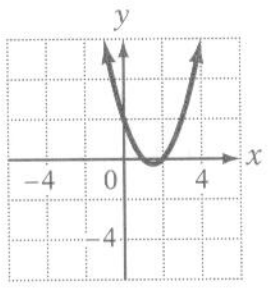

19. $y = 2x^2 - x - 5$
Vertex: $\left(\frac{1}{4}, -\frac{41}{8}\right)$
Axis of symmetry: $x = \frac{1}{4}$

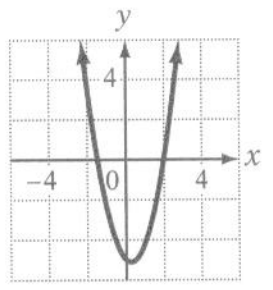

20. $y = 2x^2 - x - 3$
Vertex: $\left(\frac{1}{4}, -\frac{25}{8}\right)$
Axis of symmetry: $x = \frac{1}{4}$

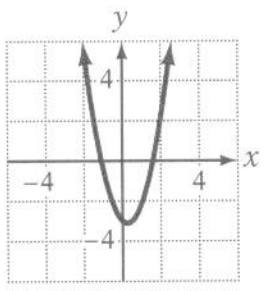

Graph the function. State the domain and range of the function.

21. $f(x) = 2x^2 - 4x - 5$

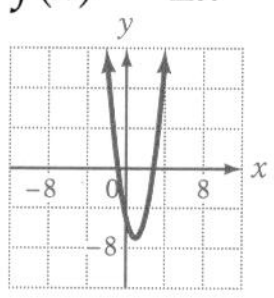

D: $\{x \mid x \in \text{real numbers}\}$
R: $\{y \mid y \geq -7\}$

22. $f(x) = 2x^2 + 8x + 3$

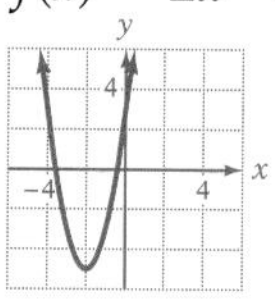

D: $\{x \mid x \in \text{real numbers}\}$
R: $\{y \mid y \geq -5\}$

23. $f(x) = -2x^2 - 3x + 2$

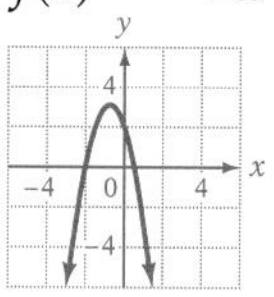

D: $\{x \mid x \in \text{real numbers}\}$
R: $\left\{y \mid y \leq \frac{25}{8}\right\}$

24. $f(x) = 2x^2 - 7x + 3$

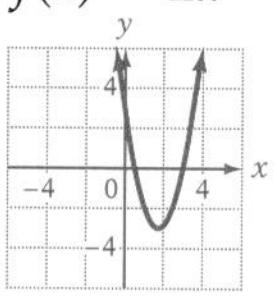

D: $\{x \mid x \in \text{real numbers}\}$
R: $\left\{y \mid y \geq -\frac{25}{8}\right\}$

25. $f(x) = x^2 - 4x + 4$

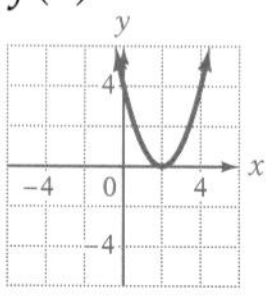

D: $\{x \mid x \in \text{real numbers}\}$
R: $\{y \mid y \geq 0\}$

26. $f(x) = -x^2 + 6x - 9$

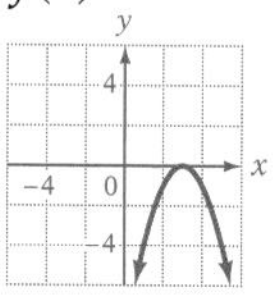

D: $\{x \mid x \in \text{real numbers}\}$
R: $\{y \mid y \leq 0\}$

27. $f(x) = x^2 + 4x - 3$

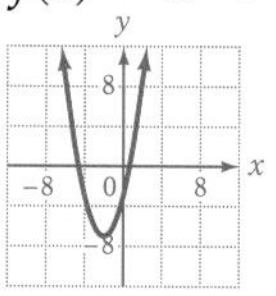

D: $\{x \mid x \in \text{real numbers}\}$
R: $\{y \mid y \geq -7\}$

28. $f(x) = x^2 - 2x - 2$

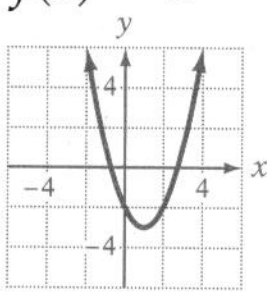

D: $\{x \mid x \in \text{real numbers}\}$
R: $\{y \mid y \geq -3\}$

29. $f(x) = -x^2 - 4x - 5$

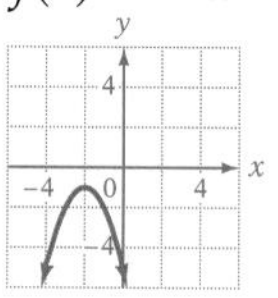

D: $\{x \mid x \in \text{real numbers}\}$
R: $\{y \mid y \leq -1\}$

30. $f(x) = -x^2 + 4x + 1$

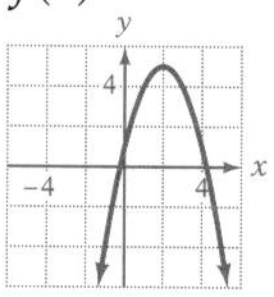

D: $\{x \mid x \in \text{real numbers}\}$
R: $\{y \mid y \leq 5\}$

2 **31.** How can you find the x-intercepts for the graph of a quadratic function?

32. How can the discriminant be used to determine the number of x-intercepts for the graph of a quadratic function?

Find the x-intercepts of the parabola given by each equation.

33. $y = x^2 - 4$ $(2, 0), (-2, 0)$

34. $y = x^2 - 9$ $(3, 0), (-3, 0)$

35. $y = 2x^2 - 4x$ $(0, 0), (2, 0)$

36. $y = 3x^2 + 6x$ $(0, 0), (-2, 0)$

37. $y = x^2 - x - 2$ $(2, 0), (-1, 0)$

38. $y = x^2 - 2x - 8$ $(4, 0), (-2, 0)$

39. $y = 2x^2 - 5x - 3$ $(3, 0), \left(-\frac{1}{2}, 0\right)$

40. $y = 4x^2 + 11x + 6$ $\left(-\frac{3}{4}, 0\right), (-2, 0)$

41. $y = 3x^2 - 19x - 14$ $\left(-\frac{2}{3}, 0\right), (7, 0)$

42. $y = 6x^2 + 7x + 2$ $\left(-\frac{1}{2}, 0\right), \left(-\frac{2}{3}, 0\right)$

43. $y = 3x^2 - 19x + 20$ $(5, 0), \left(\frac{4}{3}, 0\right)$

44. $y = 3x^2 + 19x + 28$ $(-4, 0), \left(-\frac{7}{3}, 0\right)$

45. $y = 9x^2 - 12x + 4$ $\left(\frac{2}{3}, 0\right)$

46. $y = x^2 - 2$ $(\sqrt{2}, 0), (-\sqrt{2}, 0)$

47. $y = 9x^2 - 2$ $\left(\frac{\sqrt{2}}{3}, 0\right), \left(-\frac{\sqrt{2}}{3}, 0\right)$

48. $y = 2x^2 - x - 1$ $\left(-\frac{1}{2}, 0\right), (1, 0)$

49. $y = 2x^2 - 5x - 3$ $\left(-\frac{1}{2}, 0\right), (3, 0)$

50. $y = x^2 + 2x - 1$ $(-1 + \sqrt{2}, 0), (-1 - \sqrt{2}, 0)$

51. $y = x^2 + 4x - 3$ $(-2 + \sqrt{7}, 0), (-2 - \sqrt{7}, 0)$

52. $y = x^2 + 6x + 10$ No x-intercepts

53. $y = -x^2 - 4x - 5$ No x-intercepts

54. $y = x^2 - 2x - 2$ $(1 + \sqrt{3}, 0), (1 - \sqrt{3}, 0)$

55. $y = -x^2 - 2x + 1$ $(-1 + \sqrt{2}, 0), (-1) - \sqrt{2}, 0)$

56. $y = -x^2 + 4x + 1$ $(2 + \sqrt{5}, 0), (2 - \sqrt{5}, 0)$

Graph the function. Estimate the real zeros of the function to the nearest tenth.

57. $f(x) = x^2 + 3x - 1$ −3.3, 0.3

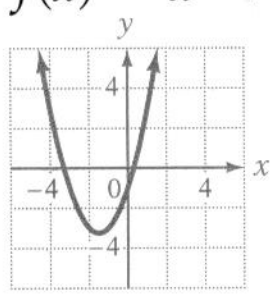

58. $f(x) = x^2 - 2x - 4$ −1.2, 3.2

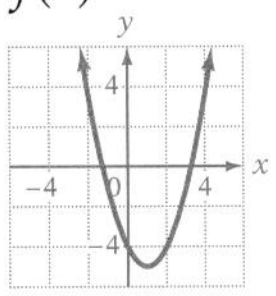

59. $f(x) = 2x^2 - 3x - 7$ −1.3, 2.8

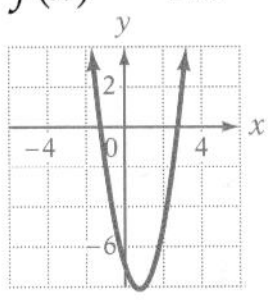

60. $f(x) = -2x^2 - x + 2$ −1.3, 0.8

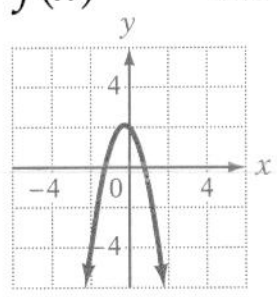

61. $f(x) = x^2 + 6x + 12$ No real zeros

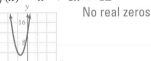

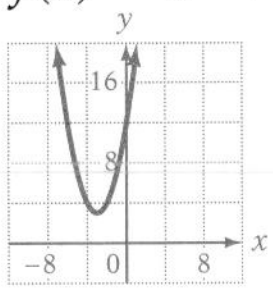

62. $f(x) = x^2 - 3x + 9$ No real zeros

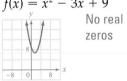

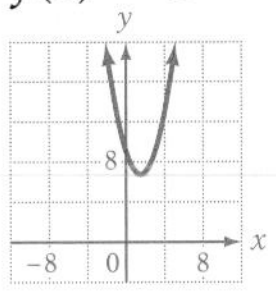

Use the discriminant to determine the number of x-intercepts of the graph of the parabola.

63. $y = 2x^2 + x + 1$ No x-intercepts

64. $y = 2x^2 + 2x - 1$ Two

65. $y = -x^2 - x + 3$ Two

66. $y = -2x^2 + x + 1$ Two

67. $y = x^2 - 8x + 16$ One

68. $y = x^2 - 10x + 25$ One

69. $y = -3x^2 - x - 2$ No x-intercepts

70. $y = -2x^2 + x - 1$ No x-intercepts

71. $y = 4x^2 - x - 2$ Two

72. $y = 2x^2 + x + 4$ No x-intercepts

73. $y = -2x^2 - x - 5$ No x-intercepts

74. $y = -3x^2 + 4x - 5$ No x-intercepts

75. $y = x^2 + 8x + 16$ One

76. $y = x^2 - 12x + 36$ One

77. $y = x^2 + x - 3$ Two

APPLYING CONCEPTS 8.6

Find the value of k such that the graph of the equation contains the given point.

78. $y = x^2 - 3x + k; (2, 5)$ 7

79. $y = x^2 + 2x + k; (-3, 1)$ −2

80. $y = 2x^2 + kx - 3; (4, -3)$ −8

81. $y = 3x^2 + kx - 6; (-2, 4)$ 1

Solve.

82. Complete the graph at the right in such a way that the entire graph is symmetric about the y-axis.

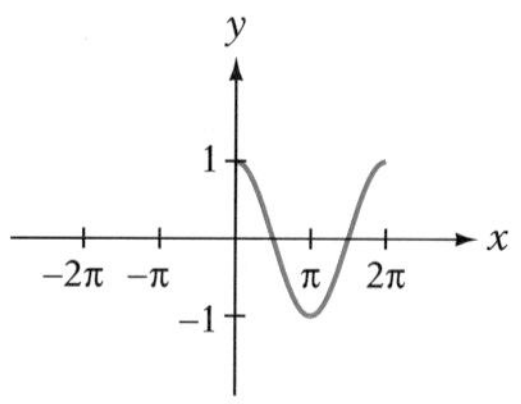

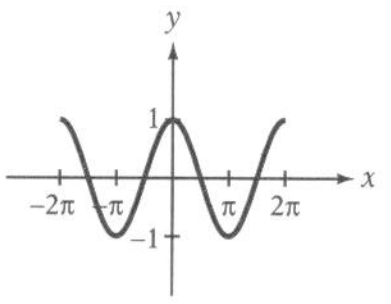

83. The point (x_1, y_1) lies in quadrant I and is a solution of the equation $y = 3x^2 - 2x - 1$. Given $y_1 = 5$, find x_1. $\frac{1 + \sqrt{19}}{3}$

84. The point (x_1, y_1) lies in quadrant II and is a solution of the equation $y = 2x^2 + 5x - 3$. Given $y_1 = 9$, find x_1. -4

85. What effect does increasing the coefficient of x^2 have on the graph of $y = ax^2 + bx + c, a > 0$? The graph becomes thinner.

86. What effect does decreasing the coefficient of x^2 have on the graph of $y = ax^2 + bx + c, a > 0$? The graph becomes wider.

87. What effect does increasing the constant term have on the graph of $y = ax^2 + bx + c$? The graph is higher on the rectangular coordinate system.

88. What effect does decreasing the constant term have on the graph of $y = ax^2 + bx + c$? The graph is lower on the rectangular coordinate system.

An equation of the form $y = ax^2 + bx + c$ can be written in the form $y = a(x - h)^2 + k$, where (h, k) are the coordinates of the vertex of the parabola. Use the process of completing the square to rewrite the equation in the form $y = a(x - h)^2 + k$. Find the vertex. (*Hint:* Review the example at the bottom of page 468.)

89. $y = x^2 - 4x + 7$
$y = (x - 2)^2 + 3$; vertex: $(2, 3)$

90. $y = x^2 - 2x - 2$
$y = (x - 1)^2 - 3$; vertex: $(1, -3)$

91. $y = x^2 + x + 2$
$y = \left(x + \frac{1}{2}\right)^2 + \frac{7}{4}$; vertex: $\left(-\frac{1}{2}, \frac{7}{4}\right)$

92. $y = x^2 - x - 3$
$y = \left(x - \frac{1}{2}\right)^2 - \frac{13}{4}$; vertex: $\left(\frac{1}{2}, -\frac{13}{4}\right)$

Using $y = a(x - h)^2 + k$ as the equation of a parabola with the vertex at (h, k), find the equation of the parabola satisfying the given information. Write the final equation in the form $y = ax^2 + bx + c$.

[C] **93.** Vertex $(1, 2)$; the graph passes through $P(2, 5)$ $y = 3x^2 - 6x + 5$

[C] **94.** Vertex $(0, -3)$; the graph passes through $P(3, -2)$ $y = \frac{1}{9}x^2 - 3$

95. The shape formed by rotating a parabola about its axis of symmetry is called a paraboloid. It is a common shape for mirrors of reflecting telescopes. Write an essay on reflecting telescopes and how paraboloids are used to focus light.

96. Graph $y = \sqrt{x + 2}$ and $y^2 = x + 2$. Discuss the similarities and differences between the two graphs.

97. In your own words, explain how x-intercepts of a graph and real zeros of a function are related.

SECTION 8.7 Applications of Quadratic Functions

1 Minimum and maximum problems

INSTRUCTOR NOTE

One difficulty students have with this concept is in making the distinction between the maximum (or minimum) *value of the function* and the *value of x* that produces the maximum (or minimum).

The graph of $f(x) = x^2 - 2x + 3$ is shown at the right. Because a is positive, the parabola opens up. The vertex of the parabola is the lowest point on the parabola. It is the point that has the minimum y-coordinate. Therefore, the value of the function at this point is a **minimum.**

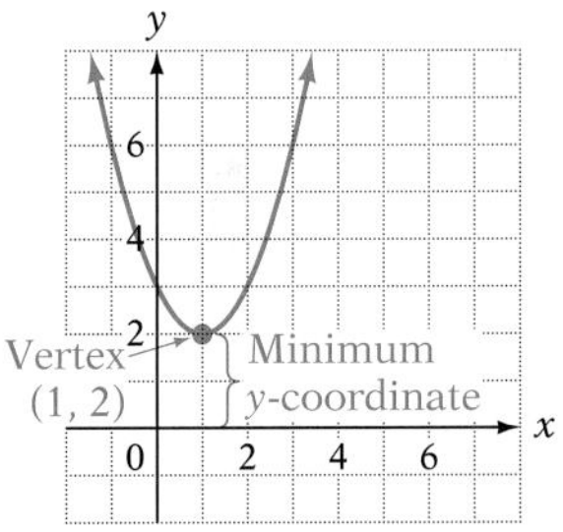

POINT OF INTEREST

Calculus is a branch of mathematics that demonstrates, among other things, how to find the maximum or minimum of functions other than quadratic functions. These are very important problems in applied mathematics. For instance, an automotive engineer wants to design a car whose shape will *minimize* the effect of air flow. The same engineer tries to *maximize* the efficiency of a car's engine. Similarly, an economist may try to determine what business practices will *minimize* cost and *maximize* profit.

The graph of $f(x) = -x^2 + 2x + 1$ is shown at the right. Because a is negative, the parabola opens down. The vertex of the parabola is the highest point on the parabola. It is the point that has the maximum y-coordinate. Therefore, the value of the function at this point is a **maximum.**

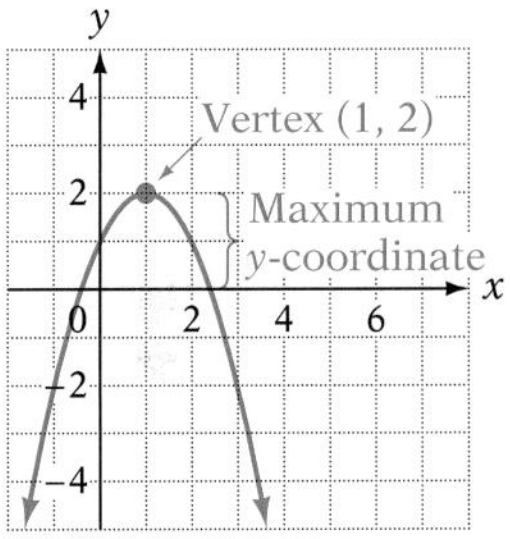

To find the minimum or maximum value of a quadratic function, first find the x-coordinate of the vertex. Then evaluate the function at that value.

Example 1 Find the maximum value of $f(x) = -2x^2 + 4x + 3$.

Solution $x = -\frac{b}{2a} = -\frac{4}{2(-2)} = 1$ ▸ Find the x-coordinate of the vertex. $a = -2, b = 4$.

$f(x) = -2x^2 + 4x + 3$ ▸ Evaluate the function at $x = 1$.
$f(1) = -2(1)^2 + 4(1) + 3$
$f(1) = 5$

The maximum value of the function is 5.

Problem 1 Find the minimum value of $f(x) = 2x^2 - 3x + 1$.

Solution See page S25. $-\frac{1}{8}$

2   Applications of minimum and maximum

Example 2   A mining company has determined that the cost in dollars (c) per ton of mining a mineral is given by

$$c(x) = 0.2x^2 - 2x + 12$$

where x is the number of tons of the mineral that is mined. Find the number of tons of the mineral that should be mined to minimize the cost. What is the minimum cost?

Strategy   ■ To find the number of tons that will minimize the cost, find the x-coordinate of the vertex.
   ■ To find the minimum cost, evaluate the function at the x-coordinate of the vertex.

Solution $x = -\dfrac{b}{2a} = -\dfrac{-2}{2(0.2)} = 5$

To minimize the cost, 5 tons should be mined.

$c(x) = 0.2x^2 - 2x + 12$
$c(5) = 0.2(5)^2 - 2(5) + 12 = 5 - 10 + 12 = 7$

The minimum cost per ton is $7.

Problem 2   The height in feet (s) of a ball thrown straight up is given by $s(t) = -16t^2 + 64t$, where t is the time in seconds. Find the time it takes the ball to reach its maximum height. What is the maximum height?

Solution   See page S25.   2 s; 64 ft

Example 3   Find two numbers whose difference is 10 and whose product is a minimum.

Strategy   Let x represent one number. Because the difference between the two numbers is 10, $x + 10$ represents the other number. Then their product is represented by $x(x + 10) = x^2 + 10x$.

   ■ To find one of the two numbers, find the x-coordinate of the vertex of $f(x) = x^2 + 10x$.
   ■ To find the other number, replace x in $x + 10$ by the x-coordinate of the vertex and evaluate.

Solution $x = -\dfrac{b}{2a} = -\dfrac{10}{2(1)} = -5$
   $x + 10 = -5 + 10 = 5$

The numbers are -5 and 5.

Problem 3   A mason is forming a rectangular floor for a storage shed. The perimeter of the rectangle is 44 ft. What dimensions would give the floor a maximum area?

Solution   See page S25.   11 ft by 11 ft

CONCEPT REVIEW 8.7

Determine whether the following statements are always true, sometimes true, or never true.

1. A quadratic function has a minimum value. Sometimes true

2. The y-coordinate of the vertex of a quadratic function is the minimum value of the function. Sometimes true

3. The maximum or minimum of a quadratic function is the value of the function evaluated at the x-coordinate of the vertex. Always true

4. The x-value of the vertex of $y = ax^2 + bx + c, a \neq 0$, is given by $-\frac{b}{2a}$. Always true

5. The maximum value of a quadratic function is a positive number. Sometimes true

6. The function $f(x) = -x^2 + 3x - 2$ will have a maximum value. Always true

EXERCISES 8.7

1 1. What is the minimum value or the maximum value of a quadratic function?

2. How can you find the minimum or maximum value of a quadratic function?

Find the minimum or maximum value of each quadratic function.

3. $f(x) = x^2 - 2x + 3$ Minimum: 2

4. $f(x) = x^2 + 3x - 4$ Minimum: $-\frac{25}{4}$

5. $f(x) = -2x^2 + 4x - 3$ Maximum: -1

6. $f(x) = -2x^2 - 3x + 4$ Maximum: $\frac{41}{8}$

7. $f(x) = 2x^2 + 4x$ Minimum: -2

8. $f(x) = -2x^2 - 3x$ Maximum: $\frac{9}{8}$

9. $f(x) = -2x^2 + 4x - 5$ Maximum: -3

10. $f(x) = -3x^2 + x - 6$ Maximum: $-\frac{71}{12}$

11. $f(x) = 2x^2 + 3x - 8$ Minimum: $-\frac{73}{8}$

12. $f(x) = -x^2 - x + 2$ Maximum: $\frac{9}{4}$

13. $f(x) = 3x^2 + 3x - 2$ Minimum: $-\frac{11}{4}$

14. $f(x) = x^2 - 5x + 3$ Minimum: $-\frac{13}{4}$

15. $f(x) = -3x^2 + 4x - 2$ Maximum: $-\frac{2}{3}$

16. $f(x) = -2x^2 - 5x + 1$ Maximum: $\frac{33}{8}$

17. $f(x) = 3x^2 + 5x + 2$ Minimum: $-\frac{1}{12}$

2 Solve.

18. *Physics* The height in feet (s) of a rock thrown upward at an initial speed of 64 ft/s from a cliff 50 ft above an ocean beach is given by the function $s(t) = -16t^2 + 64t + 50$, where t is the time in seconds. Find the maximum height above the beach that the rock will attain. 114 ft

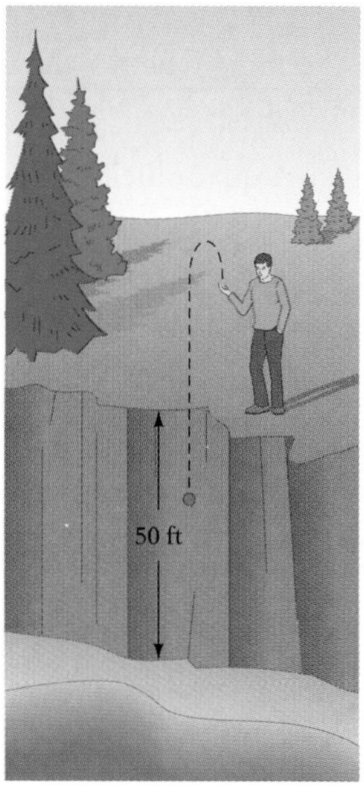

19. *Physics* The height in feet (s) of a ball thrown upward at an initial speed of 80 ft/s from a platform 50 ft high is given by the function $s(t) = -16t^2 + 80t + 50$, where t is the time in seconds. Find the maximum height above the ground that the ball will attain. 150 ft

20. *Business* A manufacturer of microwave ovens believes that the revenue in dollars (R) the company receives is related to the price in dollars (P) of an oven by the function $R(P) = 125P - \frac{1}{4}P^2$. What price will give the maximum revenue? $250

21. *Business* A manufacturer of camera lenses estimated that the average monthly cost (C) of a lens is given by the function $C(x) = 0.1x^2 - 20x + 2000$, where x is the number of lenses produced each month. Find the number of lenses the company should produce in order to minimize the average cost. 100 lenses

22. *Chemistry* A pool is treated with a chemical to reduce the amount of algae. The amount of algae in the pool t days after the treatment can be approximated by the function $A(t) = 40t^2 - 400t + 500$. How many days after treatment will the pool have the least amount of algae? 5 days

23. *Structural Engineering* The suspension cable that supports a small footbridge hangs in the shape of a parabola. The height in feet (h) of the cable above the bridge is given by the function $h(x) = 0.25x^2 - 0.8x + 25$, where x is the distance in feet from one end of the bridge. What is the minimum height of the cable above the bridge? 24.36 ft

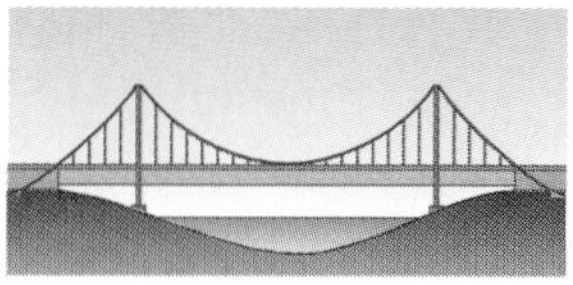

24. *Medicine* The net annual income of a family physician can be modeled by $I(x) = -290(x - 48)^2 + 148{,}000$, where x is the age of the physician and $27 \le x \le 70$. Find the age at which the physician's income will be a maximum. What is the physician's maximum income? 48, $148,000

25. *Physics* Karen is throwing an orange to her brother Saul, who is standing on the balcony of their home. The height, h (in feet), of the orange above the ground t seconds after it is thrown is given by $h(t) = -16t^2 + 32t + 4$. If Saul's outstretched arms are 18 ft above the ground, will the orange ever be high enough so that he can catch it? Yes

26. *Sports* Some football fields are built in a parabolic mound shape so that water will drain off the field. A model for the shape of the field is given by $h(x) = -0.00023475x^2 + 0.0375x$, where h is the height of the field in feet at a distance of x feet from the sideline. What is the maximum height? Round to the nearest tenth. 1.5 ft

27. *Art* The Buningham Fountain in Chicago shoots water from a nozzle at the base of the fountain. The height, h (in feet), of the water above the ground t seconds after it leaves the nozzle is given by $h(t) = -16t^2 + 90t + 15$. What is the maximum height of the water spout to the nearest tenth of a foot?
141.6 ft

28. *Civil Engineering* On wet concrete, the stopping distance, s (in feet), of a car traveling v miles per hour is given by $s(v) = 0.055v^2 + 1.1v$. At what speed could a car be traveling and still stop at a stop sign 44 ft away? 20 mph

29. *Mechanical Engineering* The fuel efficiency, E (in miles per gallon), of an average car is given by $E(v) = -0.018v^2 + 1.476v + 3.4$, where v is the speed of the car in miles per hour. What speed will yield the maximum fuel efficiency? What is the maximum fuel efficiency? 41 mph, 33.658 mi/gal

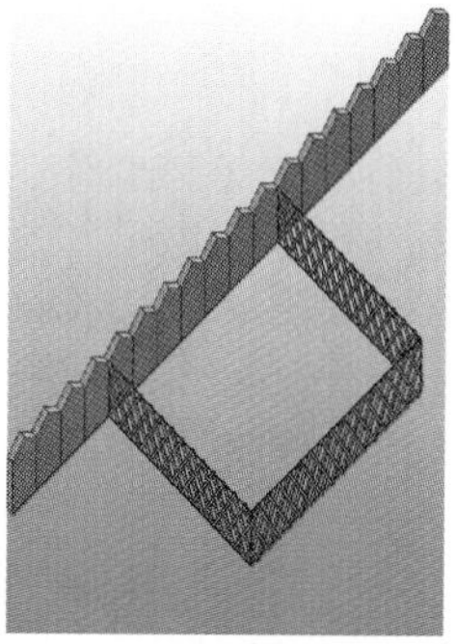

30. *Farming* A rancher has 200 ft of fencing to build a rectangular corral along the side of an existing fence. Determine the dimensions of the corral that will maximize the enclosed area. Length: 100 ft; width: 50 ft

APPLYING CONCEPTS 8.7

Use a graphing utility to find the minimum or maximum value of the function. Round to the nearest tenth. See the Graphing Calculator Appendix for assistance.

31. $f(x) = x^4 - 2x^2 + 4$ 3.0

32. $f(x) = x^4 + 2x^3 + 1$ −0.7

33. $f(x) = -x^6 + x^4 - x^3 + x$ 0.5

34. $f(x) = -x^8 + x^6 - x^4 + 5x^2 + 7$ 11.2

Solve.

35. *Traffic Control* Traffic engineers try to determine the effect a traffic light has at an intersection. By gathering data about the intersection, engineers can determine the approximate number of cars that enter the intersection in the horizontal direction and those that enter in the vertical direction. The engineers would also collect information on the time it takes a stopped car to regain the normal posted speed limit. One model of this situation is $T = \left(\frac{H+V}{2}\right)R^2 + (0.08H - 1.08V)R + 0.58V$, where H is the number of cars arriving at the intersection from the horizontal direction, V is the number of cars arriving at the intersection from the vertical direction, and R is the percent of time the light is red in the horizontal direction. T is the total delay time for all cars and is measured as the number of times the traffic light changes from red to green and back to red.

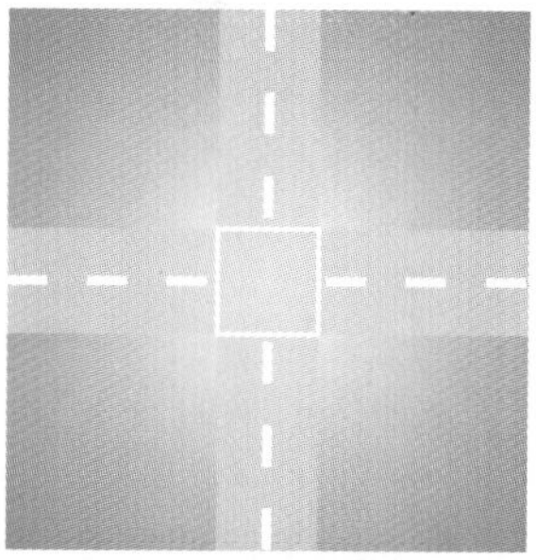

a. Graph this equation for $H = 100$, $V = 150$, and $0 \le R \le 1$.

b. Write a sentence that explains why the graph is drawn only for $0 \le R \le 1$.

c. What percent of the time should the traffic light remain red in the horizontal direction to minimize T? Round to the nearest whole percent.
62%

a.

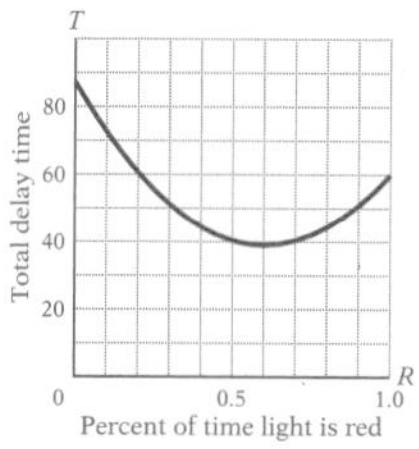

b. The percent cannot be less than 0% or more than 100%.

36. Use a graphing utility to graph $f(x) = |x|$, $g(x) = |x - 2|$, and $h(x) = |x + 2|$ on the same set of axes; describe the effect of the number inside the absolute value symbol. Graph $f(x) = |x|$, $g(x) = |x| - 2$, and $h(x) = |x| + 2$ on the same set of axes; describe the effect of the number added to the absolute value expression. Graph $f(x) = |x|$, $g(x) = 2|x|$, and $h(x) = 4|x|$ on the same set of axes; describe the effect of the number multiplying the absolute value expression. In each case, discuss the minimum values of the functions graphed.

37. Write a paper on the history of higher-degree equations. Include discoveries made by Del Farro, Ferrari, and Ruffini.

38. On the basis of Exercises 31–34, make a conjecture about the relationship between the sign of the leading coefficient of a polynomial function of even degree and whether that function will have a minimum or a maximum value.

Focus on Problem Solving

Inductive and Deductive Reasoning

Consider the following sums of odd positive integers.

$1 + 3 = 4 = 2^2$	Sum of the first two odd numbers is 2^2.
$1 + 3 + 5 = 9 = 3^2$	Sum of the first three odd numbers is 3^2.
$1 + 3 + 5 + 7 = 16 = 4^2$	Sum of the first four odd numbers is 4^2.
$1 + 3 + 5 + 7 + 9 = 25 = 5^2$	Sum of the first five odd numbers is 5^2.

1. Make a conjecture about the value of the sum

$$1 + 3 + 5 + 7 + 9 + 11 + 13 + 15$$

without adding the numbers.

If the pattern continues, a possible conjecture is that the sum of the first eight odd positive integers is $8^2 = 64$. By adding the numbers, you can verify that this is correct. Inferring that the pattern established by a few cases is valid for all cases is an example of **inductive reasoning.**

2. Use inductive reasoning to find the next figure in the sequence below.

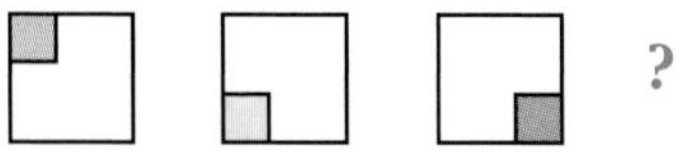

The fact that a pattern appears to be true does not *prove* that it is true for all cases. For example, consider the polynomial $n^2 - n + 41$. If we begin substituting positive integer values for n and evaluating the expression, we produce the following table.

n	$n^2 - n + 41$	
1	$1^2 - 1 + 41 = 41$	41 is a prime number.
2	$2^2 - 2 + 41 = 43$	43 is a prime number.
3	$3^2 - 3 + 41 = 47$	47 is a prime number.
4	$4^2 - 4 + 41 = 53$	53 is a prime number.

Even if we try a number like 30, we have $30^2 - 30 + 41 = 911$ and 911 is a prime number. Thus it appears that the value of this polynomial is a prime number for any value of n. However, the conjecture is not true. For instance, if $n = 41$, we have $41^2 - 41 + 41 = 41^2 = 1681$, which is not a prime number because it is divisible by 41. This illustrates that inductive reasoning may lead to incorrect conclusions and that an inductive proof must be available to prove conjectures. There is such a proof, called *mathematical induction*, that you may study in a future math course.

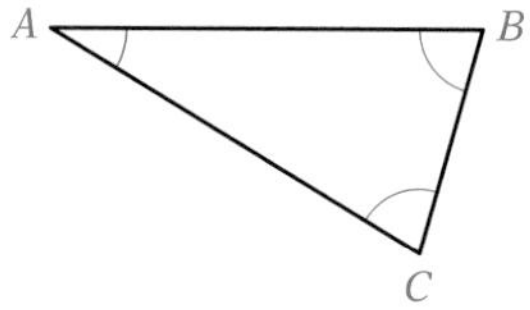

$\angle A + \angle B + \angle C = 180°$

Now consider the true statement that the sum of the measures of the interior angles of a triangle is 180°. The figure on the left is a triangle; therefore the sum of the measures of the interior angles must be 180°. This is an example of *deductive reasoning*. **Deductive reasoning** uses a rule or statement of fact to reach a conclusion.

3. Use deductive reasoning to complete the following sentence. All even numbers are divisible by 2. Because 41,386 is an even number,...

For Exercises 4 and 5, determine whether inductive or deductive reasoning is being used.

4. The tenth number in the list 1, 4, 9, 16, 25,... is 100.

5. All quadrilaterals have four sides. A square is a quadrilateral. Therefore, a square has four sides.

6. Explain the difference between inductive and deductive reasoning.

Projects and Group Activities

Completing the Square

Essentially all of the investigations into mathematics before the Renaissance were geometric. The solutions of quadratic equations were calculated from a construction of a certain area. Proofs of theorems, even theorems about numbers, were based entirely on geometry. In this project, we will examine the solution of a quadratic equation.

Solve: $x^2 + 6x = 7$

Begin with a line of unknown length, x, and one of length 6, the coefficient of x. Using these lines, construct a rectangle as shown in Figure 1.

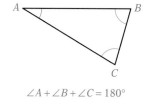

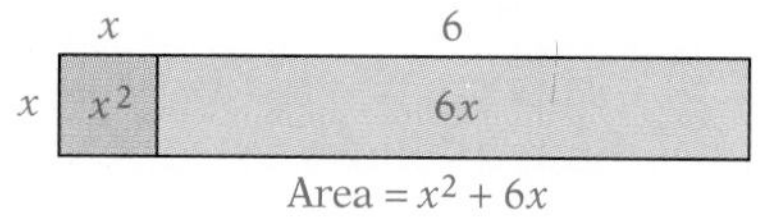

Figure 1

Now draw another area that has exactly the same area as Figure 1 by cutting one-half of the rectangle off and placing it on the bottom of the square. See Figure 2.

The unshaded area in Figure 2 has exactly the same area as Figure 1. However, when the shaded area is added to Figure 2 to make a square, the area of the square is 9 square units larger than that of Figure 1. In equation form,

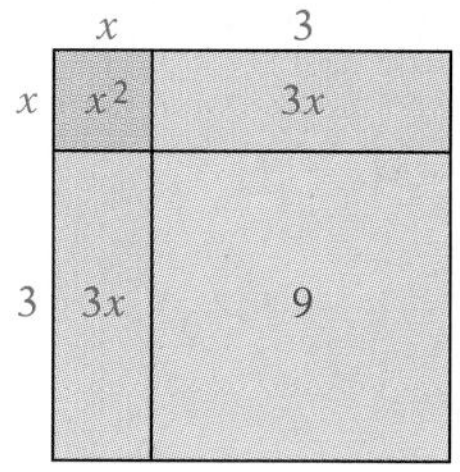

Area = $(x + 3)^2$

Figure 2

(Area of Figure 1) + 9 = area of Figure 2

or

$$x^2 + 6x + 9 = (x + 3)^2$$

From the original equation, $x^2 + 6x = 7$. Thus,

$$\begin{aligned} x^2 + 6x + 9 &= (x + 3)^2 \\ 7 + 9 &= (x + 3)^2 && \blacktriangleright\ x^2 + 6x = 7 \\ 16 &= (x + 3)^2 \\ 4 &= x + 3 && \blacktriangleright\ \text{See note below.} \\ 1 &= x \end{aligned}$$

Note: Although early mathematicians knew that a quadratic equation may have two solutions, both solutions were allowed only if they were positive. After all, a geometric construction could not have a negative length. Therefore, the solution of this equation was 1; the solution −7 would have been dismissed as *fictitious*, the actual word that was frequently used through the 15th century for negative-number solutions of an equation.

Try to solve the quadratic equation $x^2 + 4x = 12$ by geometrically completing the square.

Using a Graphing Calculator to Solve a Quadratic Equation

Recall that an x-intercept of the graph of an equation is a point at which the graph crosses the x-axis. For the graph in Figure 3, the x-intercepts are $(-2, 0)$ and $(3, 0)$.

Recall also that to find the x-intercept of a graph, set $y = 0$ and then solve for x. For the equation in Figure 3, if we set $y = 0$, the resulting equation is $0 = x^2 - x - 6$, which is a quadratic equation. Solving this equation by factoring, we have

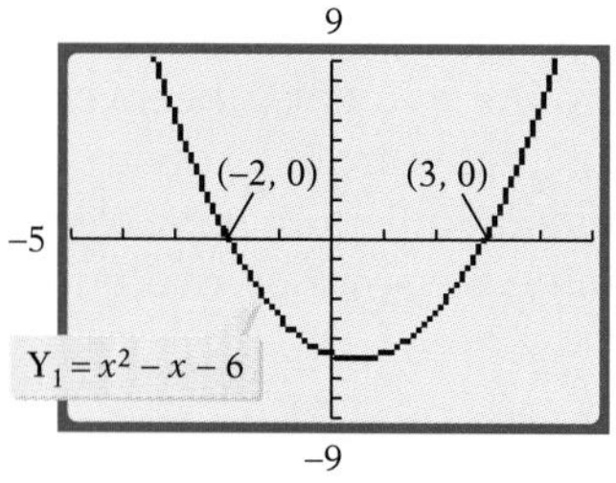

Figure 3

$$0 = x^2 - x - 6$$
$$0 = (x + 2)(x - 3)$$

$$\begin{aligned} x + 2 &= 0 & \qquad x - 3 &= 0 \\ x &= -2 & x &= 3 \end{aligned}$$

Thus the solutions of the equation are the x-coordinates of the x-intercepts of the graph.

This connection between the solutions of an equation and the x-intercepts of a graph allows us to find graphically the real number solutions of an equation.

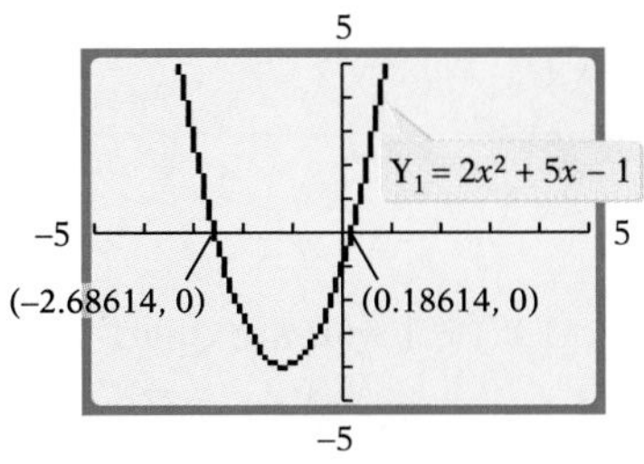

Figure 4

Using a graphing calculator, graph the equation and estimate the x-intercepts. For instance, to approximate graphically the solutions of $2x^2 + 5x - 1 = 0$, graph $y = 2x^2 + 5x - 1$ and approximate the x-coordinates of the x-intercepts. You can use the following keystrokes to graph the equation and approximate the solutions of the equation. The result is the graph in Figure 4.

TI-83 [Y=] [CLEAR] 2 [X,T,θ,n] [∧] 2 [+] 5 [X,T,θ,n] [−] 1 [2nd] QUIT [ZOOM] 5 [ENTER] [2nd] CALC 2

Use the arrow keys to move to the left of the leftmost x-intercept. Then press ENTER. Now use the arrow keys to move to a point just to the right of the leftmost x-intercept. Then press ENTER. Press ENTER again. The x-coordinate at the bottom of the screen is one approximate solution of the equation. To find the other solution, proceed in the same manner for the next x-intercept.

SHARP EL-9360 [Y=] [CL] 2 [X/θ/T/n] [a^b] 2 [▶] [+] 5 [X/θ/T/n] [−] 1 [ZOOM] 5.

This will produce the graph. Now press [2nd] CALC 5 [ENTER]. After you press 5, the x-intercept will show on the bottom of the screen. Now press [2nd] CALC 5 [ENTER] to find the other x-intercept.

CASIO CFX-9850 [MENU] 5 [F2] [F1] 2 [X/θ/T] [x²] [+] 5 [X/θ/T] [−] 1 [EXE] [F6] [SHIFT] G-Solv [F1]. Now press [▶] to find the other x-intercept.

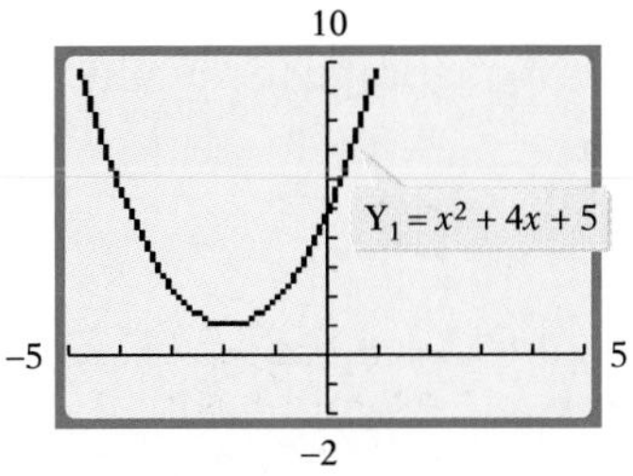

Figure 5

Attempting to find the solutions of an equation graphically will not necessarily yield all the solutions. Because the x-coordinates of the x-intercepts of a graph are *real* numbers, only real number solutions can be found. For instance, consider the equation $x^2 + 4x + 5 = 0$. The graph of $y = x^2 + 4x + 5$ is shown in Figure 5. Note that the graph has no x-intercepts and that consequently it has no real number solutions. However, $x^2 + 4x + 5 = 0$ does have complex number solutions that can be obtained by using the quadratic formula. They are $-2 + i$ and $-2 - i$.

Solving Nonlinear Inequalities Using a Graphing Calculator

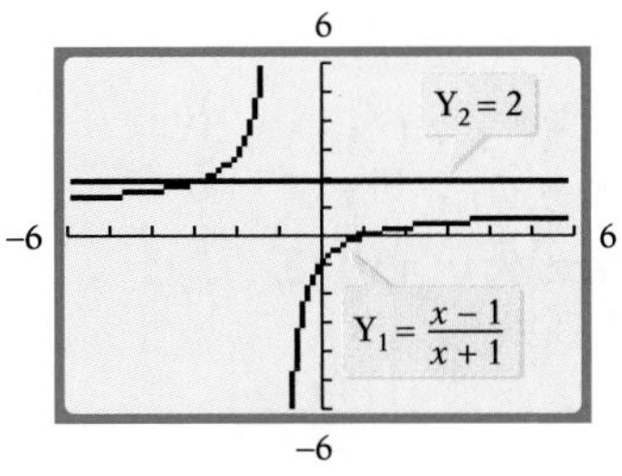

Figure 6

Inequalities such as those in Section 5 can also be solved by using a graphing calculator. For example, to solve $\frac{x-1}{x+1} < 2$, graph $y = \frac{x-1}{x+1}$ (the left side of the inequality) and $y = 2$ (the right side of the inequality). Because $\frac{x-1}{x+1}$ is required to be less than 2, the solution set is those values of x for which the graph of $y = \frac{x-1}{x+1}$ is below the graph of $y = 2$. (See Figure 6.) By moving the cursor to the point of intersection, you can determine the approximate endpoints of the intervals for the solution set. (See Projects and Group Activities in the chapter entitled "Systems of Linear Equations and Inequalities" for information on graphing two equations on the same screen.) From the graph, the solution set is $\{x \mid x < -3 \text{ or } x > -1\}$.

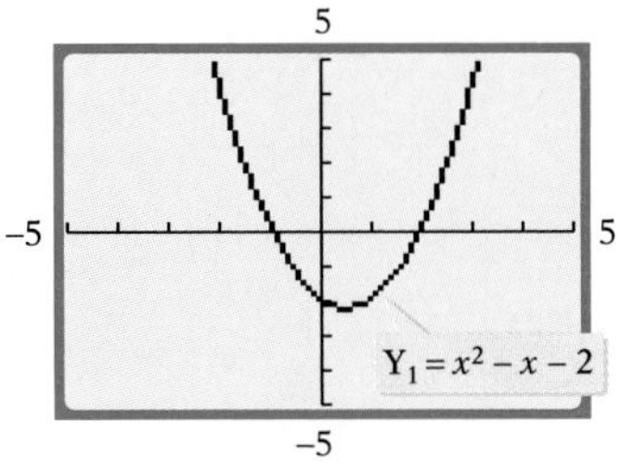

Figure 7

The solution set of a quadratic inequality such as $x^2 - x - 2 < 0$ can be found in a similar manner. Graph $y = x^2 - x - 2$ and determine those values of x for which the graph is below the x-axis ($y = 0$) by finding the x-intercepts. From Figure 7, the solution set is $\{x \mid -1 < x < 2\}$.

Chapter Summary

Key Words

A *quadratic equation* is an equation of the form $ax^2 + bx + c = 0$, where $a \neq 0$. A quadratic equation is also called a *second-degree equation*. (Objective 8.1.1)

$3x^2 + 4x - 7 = 0$ and $x^2 - 1 = 0$ are quadratic equations.

A quadratic equation is in *standard form* when the polynomial is in descending order and equal to zero. (Objective 8.1.1)

$x^2 - 5x + 6 = 0$ is a quadratic equation in standard form.

When a quadratic equation has two solutions that are the same number, the solution is called a *double root* of the equation. (Objective 8.1.1)

$$x^2 - 4x + 4 = 0$$
$$(x - 2)(x - 2) = 0$$

$$x - 2 = 0 \qquad x - 2 = 0$$
$$x = 2 \qquad x = 2$$

2 is a double root.

For an equation of the form $ax^2 + bx + c = 0$, the quantity $b^2 - 4ac$ is called the *discriminant*. (Objective 8.2.2)

$2x^2 - 3x + 2 = 0 \quad a = 2, b = -3, c = 2$
$b^2 - 4ac = (-3)^2 - 4(2)(2) = 9 - 16 = -7$

A *quadratic inequality in one variable* is one that can be written in the form $ax^2 + bx + c > 0$ or $ax^2 + bx + c < 0$, where $a \neq 0$. The symbols $\leq$ and $\geq$ can also be used. (Objective 8.5.1)

$3x^2 + 5x - 8 \leq 0$ is a quadratic inequality.

The *x-coordinate of the vertex of a parabola* is given by $-\frac{b}{2a}$. (Objective 8.6.1)

$y = 2x^2 - 8x + 5 \quad a = 2, b = -8$

The x-coordinate of the vertex is $-\frac{-8}{2(2)} = 2$.

The *x-intercepts* of the graph of a parabola occur when $y = 0$. (Objective 8.6.2)

$$x^2 - 5x + 6 = 0$$
$$(x - 3)(x - 2) = 0$$

$$x - 3 = 0 \qquad x - 2 = 0$$
$$x = 3 \qquad x = 2$$

The x-intercepts are (3, 0) and (2, 0).

The graph of a *quadratic function* $f(x) = ax^2 + bx + c$ is a *parabola* that opens up when $a > 0$ and down when $a < 0$. The *axis of symmetry* is a line that passes through the vertex of the parabola. (Objective 8.6.1)

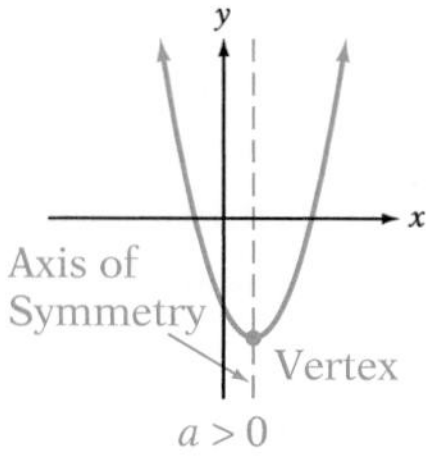

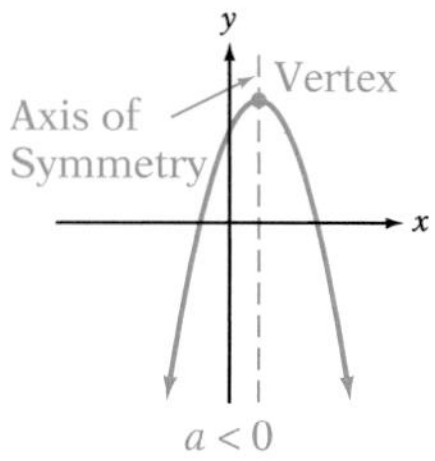

The graph of $f(x) = ax^2 + bx + x$ has a *minimum* value if $a > 0$ and a *maximum* value if $a < 0$. (Objective 8.7.1)

Essential Rules and Procedures

The Principle of Zero Products
If $ab = 0$, then $a = 0$ or $b = 0$. (Objective 8.1.1)

$(x - 3)(x + 4) = 0$

$x - 3 = 0 \qquad x + 4 = 0$
$x = 3 \qquad x = -4$

A quadratic equation can be solved by **taking square roots of each side of the equation.** (Objective 8.1.3)

$$(x + 2)^2 - 9 = 0$$
$$(x + 2)^2 = 9$$
$$\sqrt{(x + 2)^2} = \sqrt{9}$$
$$x + 2 = \pm\sqrt{9}$$
$$x + 2 = \pm 3$$

$x + 2 = 3 \qquad x + 2 = -3$
$x = 1 \qquad x = -5$

A quadratic equation can be solved by **completing the square.** (Objective 8.2.1)

$$x^2 + 4x - 1 = 0$$
$$x^2 + 4x = 1$$
$$x^2 + 4x + 4 = 1 + 4$$
$$(x + 2)^2 = 5$$
$$\sqrt{(x + 2)^2} = \sqrt{5}$$
$$x + 2 = \pm\sqrt{5}$$

$x + 2 = \sqrt{5} \qquad x + 2 = -\sqrt{5}$
$x = -2 + \sqrt{5} \qquad x = -2 - \sqrt{5}$

The Quadratic Formula
$x = \dfrac{-b \pm \sqrt{b^2 - 4ac}}{2a}$ (Objective 8.2.2)

$2x^2 - 3x + 4 = 0 \quad a = 2, b = -3, c = 4$

$$x = \frac{-(-3) \pm \sqrt{(-3)^2 - 4(2)(4)}}{2(2)}$$
$$= \frac{3 \pm \sqrt{9 - 32}}{4}$$
$$= \frac{3 \pm i\sqrt{23}}{4} = \frac{3}{4} \pm \frac{\sqrt{23}}{4}i$$

Vertex of a parabola
The coordinates of the vertex of a parabola are $\left(-\frac{b}{2a}, f\left(-\frac{b}{2a}\right)\right)$. (Objective 8.6.1)

$f(x) = x^2 - 2x - 4 \quad a = 1, b = -2$

$$-\frac{b}{2a} = -\frac{-2}{2(1)} = 1$$

$f(1) = 1^2 - 2(1) - 4 = -5$

The coordinates of the vertex are $(1, -5)$.

The Discriminant (Objective 8.6.2)
If $b^2 - 4ac = 0$, the equation has a double root.

$x^2 + 8x + 16 = 0$ has a double root because

$$b^2 - 4ac = 8^2 - 4(1)(16) = 0$$

If $b^2 - 4ac > 0$, the equation has two real number solutions that are not equal.

$2x^2 + 3x - 5 = 0$ has two unequal real number solutions because

$$b^2 - 4ac = 3^2 - 4(2)(-5) = 49$$

If $b^2 - 4ac < 0$, the equation has two complex number solutions.

$3x^2 + 2x + 4 = 0$ has two complex number solutions because

$$b^2 - 4ac = 2^2 - 4(3)(4) = 4 - 48 = -44$$

Chapter Review Exercises

1. Solve: $2x^2 - 3x = 0$ $0, \frac{3}{2}$

2. Solve for x: $6x^2 + 9xc = 6c^2$ $-2c, \frac{c}{2}$

3. Solve: $x^2 = 48$ $\pm 4\sqrt{3}$

4. Solve: $\left(x + \frac{1}{2}\right)^2 + 4 = 0$ $-\frac{1}{2} \pm 2i$

5. Find the minimum value of the function $f(x) = x^2 - 7x + 8$. $-\frac{17}{4}$

6. Find the maximum value of the function $f(x) = -2x^2 + 4x + 1$. 3

7. Write a quadratic equation that has integer coefficients and has solutions $\frac{1}{3}$ and -3. $3x^2 + 8x - 3 = 0$

8. Solve: $2x^2 + 9x = 5$ $-5, \frac{1}{2}$

9. Solve: $2(x + 1)^2 - 36 = 0$ $-1 \pm 3\sqrt{2}$

10. Solve: $x^2 + 6x + 10 = 0$ $-3 \pm i$

11. Solve: $\frac{2}{x - 4} + 3 = \frac{x}{2x - 3}$ $2, 3$

12. Solve: $x^4 - 6x^2 + 8 = 0$ $\pm 2, \pm\sqrt{2}$

13. Solve: $\sqrt{2x - 1} + \sqrt{2x} = 3$ $\frac{25}{18}$

14. Solve: $2x^{\frac{2}{3}} + 3x^{\frac{1}{3}} - 2 = 0$ $-8, \frac{1}{8}$

15. Solve: $\sqrt{3x - 2} + 4 = 3x$ 2

16. Solve: $x^2 - 6x - 2 = 0$ $3 \pm \sqrt{11}$

17. Solve: $\frac{2x}{x - 4} + \frac{6}{x + 1} = 11$ $5, -\frac{4}{9}$

18. Solve: $2x^2 - 2x = 1$ $\frac{1 \pm \sqrt{3}}{2}$

19. Solve: $2x = 4 - 3\sqrt{x - 1}$ $\frac{5}{4}$

20. Solve: $3x = \frac{9}{x - 2}$ $3, -1$

21. Solve: $\frac{3x + 7}{x + 2} + x = 3$ -1

22. Solve: $\frac{x - 2}{2x + 3} - \frac{x - 4}{x} = 2$ $\frac{-3 \pm \sqrt{249}}{10}$

23. Solve: $1 - \frac{x + 4}{2 - x} = \frac{x - 3}{x + 2}$ $\frac{-11 \pm \sqrt{129}}{2}$

24. Find the axis of symmetry of the parabola whose equation is $y = -x^2 + 6x - 5$. $x = 3$

25. Find the vertex of the parabola whose equation is $y = -x^2 + 3x - 2$. $\left(\frac{3}{2}, \frac{1}{4}\right)$

26. Use the discriminant to determine the number of x-intercepts of the parabola whose equation is $y = -2x^2 + 2x - 3$. No x-intercepts

27. Use the discriminant to determine the number of x-intercepts of the parabola whose equation is $y = 3x^2 - 2x - 4$. Two

28. Find the x-intercepts of the parabola whose equation is $y = 4x^2 + 12x + 4$. $\left(\frac{-3 - \sqrt{5}}{2}, 0\right), \left(\frac{-3 + \sqrt{5}}{2}, 0\right)$

29. Find the x-intercepts of the parabola whose equation is $y = -2x^2 - 3x + 2$. $(-2, 0), \left(\frac{1}{2}, 0\right)$

30. Find the zeros of $f(x) = 3x^2 + 2x + 2$. $-\frac{1}{3} \pm \frac{\sqrt{5}}{3}i$

31. Solve: $(x + 3)(2x - 5) < 0$
$\left\{x \mid -3 < x < \frac{5}{2}\right\}$

32. Solve: $(x - 2)(x + 4)(2x + 3) \le 0$
$\left\{x \mid x \le -4 \text{ or } -\frac{3}{2} \le x \le 2\right\}$

33. Solve and graph the solution set of $\frac{x - 2}{2x - 3} \ge 0$.
$\left\{x \mid x < \frac{3}{2} \text{ or } x \ge 2\right\}$

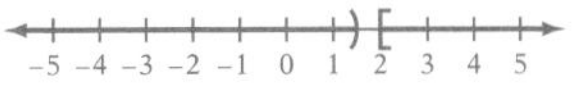

34. Solve and graph the solution set of
$\frac{(2x - 1)(x + 3)}{x - 4} \le 0$.
$\left\{x \mid x \le -3 \text{ or } \frac{1}{2} \le x < 4\right\}$

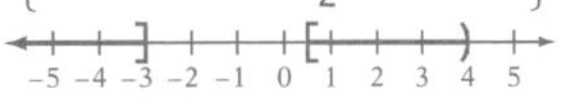

35. Graph: $f(x) = x^2 + 2x - 4$. State the domain and range.
D: $\{x \mid x \in \text{real numbers}\}$
R: $\{y \mid y \ge -5\}$

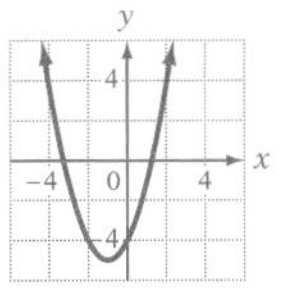

36. Find the vertex and axis of symmetry of the parabola whose equation is $y = x^2 - 2x + 3$. Then graph the equation.
Vertex: (1, 2)
Axis of symmetry: $x = 1$

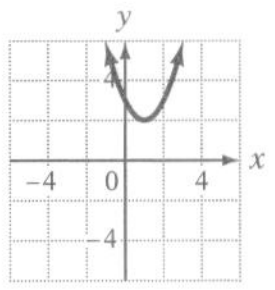

37. The length of a rectangle is 2 cm more than twice the width. The area of the rectangle is 60 cm^2. Find the length and width of the rectangle.
Length: 12 cm; width: 5 cm

38. The sum of the squares of three consecutive even integers is fifty-six. Find the three integers. 2, 4, and 6 or −6, −4, and −2

39. An older computer requires 12 min longer to print the payroll than does a newer computer. Together the computers can print the payroll in 8 min. Find the time for the newer computer, working alone, to complete the payroll.
12 min

40. A car travels 200 mi. A second car, traveling 10 mph faster than the first car, makes the same trip in 1 h less time. Find the speed of each car.
First car: 40 mph; second car: 50 mph

Chapter Test

1. Solve: $2x^2 + x = 6$ $\quad \frac{3}{2}, -2$

2. Solve: $12x^2 + 7x - 12 = 0$ $\quad \frac{3}{4}, -\frac{4}{3}$

3. Find the maximum value of the function $f(x) = -x^2 + 8x - 7$. 9

4. Write a quadratic equation that has integer coefficients and has solutions $-\frac{1}{3}$ and 3.
$3x^2 - 8x - 3 = 0$

5. Solve: $2(x + 3)^2 - 36 = 0$ $\quad -3 \pm 3\sqrt{2}$

6. Solve: $x^2 + 4x - 1 = 0$ $\quad -2 \pm \sqrt{5}$

7. Find the zeros of $g(x) = x^2 + 3x - 8$. $\quad \frac{-3 \pm \sqrt{41}}{2}$

8. Solve: $3x^2 - x + 8 = 0$ $\quad \frac{1}{6} \pm \frac{\sqrt{95}}{6}i$

9. Solve: $\frac{2x}{x-1} + \frac{3}{x+2} = 1$ $-3 \pm \sqrt{10}$

10. Solve: $2x + 7x^{\frac{1}{2}} - 4 = 0$ $\frac{1}{4}$

11. Solve: $x^4 - 11x^2 + 18 = 0$ $\pm 3, \pm\sqrt{2}$

12. Solve: $\sqrt{2x+1} + 5 = 2x$ 4

13. Solve: $\sqrt{x} - 2 = \sqrt{x} - 2$ No solution

14. Use the discriminant to determine the number of x-intercepts of the parabola $y = 3x^2 + 2x - 4$.
Two

15. Find the x-intercepts of the parabola whose equation is $y = 2x^2 + 5x - 12$.
$(-4, 0), \left(\frac{3}{2}, 0\right)$

16. Find the axis of symmetry of the parabola whose equation is $y = 2x^2 + 6x + 3$.
$x = -\frac{3}{2}$

17. Graph $y = \frac{1}{2}x^2 + x - 4$. State the domain and range.
D: $\{x \mid x \in \text{real numbers}\}$
R: $\{y \mid y \geq -4.5\}$

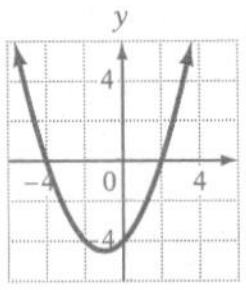

18. Solve and graph the solution set of $\frac{2x-3}{x+4} \leq 0$.
$\left\{x \mid -4 < x \leq \frac{3}{2}\right\}$

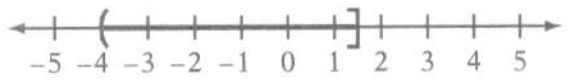

19. The base of a triangle is 3 ft more than three times the height. The area of the triangle is 30 ft². Find the base and height of the triangle.
Base: 15 ft; height: 4 ft

20. The rate of a river's current is 2 mph. A canoe was rowed 6 mi down the river and back in 4 h. Find the rowing rate in calm water. 4 mph

Cumulative Review Exercises

1. Evaluate $2a^2 - b^2 \div c^2$ when $a = 3$, $b = -4$, and $c = -2$. 14

2. Solve: $\frac{2x-3}{4} - \frac{x+4}{6} = \frac{3x-2}{8}$ -28

3. Find the slope of the line that contains the points $(3, -4)$ and $(-1, 2)$. $-\frac{3}{2}$

4. Find the equation of the line that contains the point $(1, 2)$ and is parallel to the line $x - y = 1$.
$y = x + 1$

5. Factor: $-3x^3y + 6x^2y^2 - 9xy^3$
$-3xy(x^2 - 2xy + 3y^2)$

6. Factor: $6x^2 - 7x - 20$ $(2x - 5)(3x + 4)$

7. Factor: $a^nx + a^ny - 2x - 2y$ $(x + y)(a^n - 2)$

8. Divide: $(3x^3 - 13x^2 + 10) \div (3x - 4)$
$x^2 - 3x - 4 - \frac{6}{3x-4}$

9. Multiply: $\dfrac{x^2+2x+1}{8x^2+8x} \cdot \dfrac{4x^3-4x^2}{x^2-1}$ $\dfrac{x}{2}$

10. Find the distance between the points $(-2, 3)$ and $(2, 5)$. $2\sqrt{5}$

11. Solve $S = \dfrac{n}{2}(a+b)$ for b. $b = \dfrac{2S - an}{n}$

12. Multiply: $-2i(7-4i)$ $-8-14i$

13. Simplify: $a^{-\frac{1}{2}}(a^{\frac{1}{2}} - a^{\frac{3}{2}})$ $1-a$

14. Simplify: $\dfrac{\sqrt[3]{8x^4y^5}}{\sqrt[3]{16xy^6}}$ $\dfrac{x\sqrt[3]{4y^2}}{2y}$

15. Solve: $\dfrac{x}{x+2} - \dfrac{4x}{x+3} = 1$ $-\dfrac{3}{2}, -1$

16. Solve: $\dfrac{x}{2x+3} - \dfrac{3}{4x^2-9} = \dfrac{x}{2x-3}$ $-\dfrac{1}{2}$

17. Solve: $x^4 - 6x^2 + 8 = 0$ $\pm 2, \pm\sqrt{2}$

18. Solve: $\sqrt{3x+1} - 1 = x$ $0, 1$

19. Solve: $|3x-2| < 8$ $\left\{x \mid -2 < x < \dfrac{10}{3}\right\}$

20. Find the x- and y-intercepts of the graph of $6x - 5y = 15$. $\left(\dfrac{5}{2}, 0\right), (0, -3)$

21. Graph the solution set: $x + y \le 3$
$2x - y < 4$

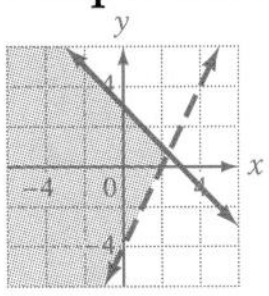

22. Solve the system of equations:
$$x + y + z = 2$$
$$-x + 2y - 3z = -9$$
$$x - 2y - 2z = -1$$
$(1, -1, 2)$

23. Given $f(x) = \dfrac{2x-3}{x^2-1}$ find $f(-2)$. $f(-2) = -\dfrac{7}{3}$

24. Find the domain of the function $f(x) = \dfrac{x-2}{x^2-2x-15}$. $\{x \mid x \ne 5, -3\}$

25. Solve and graph the solution set of $x^3 + x^2 - 6x < 0$. $\{x \mid x < -3 \text{ or } 0 < x < 2\}$

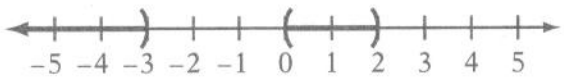

26. Solve and graph the solution set of $\dfrac{(x-1)(x-5)}{x+3} \ge 0$. $\{x \mid -3 < x \le 1 \text{ or } x \ge 5\}$

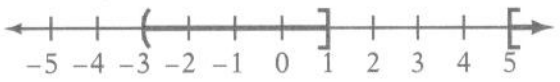

27. A piston rod for an automobile is $9\frac{3}{8}$ in. with a tolerance of $\frac{1}{64}$ in. Find the lower and upper limits of the length of the piston rod.
Lower limit: $9\frac{23}{64}$ in.; upper limit: $9\frac{25}{64}$ in.

28. The base of a triangle is $(x+8)$ ft. The height is $(2x-4)$ ft. Find the area of the triangle in terms of the variable x. $(x^2 + 6x - 16)$ ft^2

29. Use the discriminant to determine whether $2x^2 + 4x + 3 = 0$ has one real number solution, two real number solutions, or two complex number solutions.
Two complex number solutions

30. The graph shows the relationship between the cost of a building and the depreciation allowed for income tax purposes. Find the slope of the line between the two points shown on the graph. Write a sentence that states the meaning of the slope.

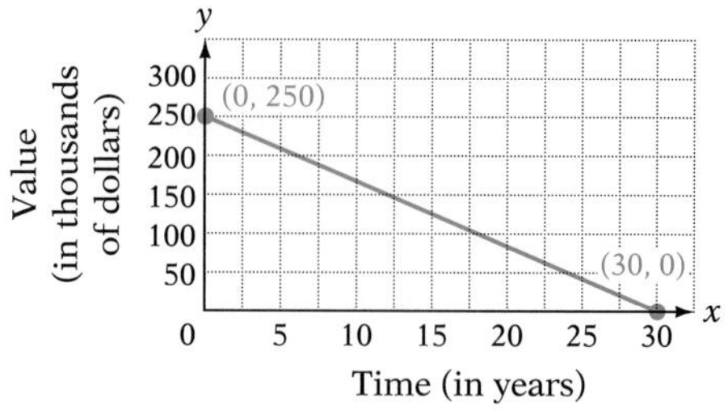

$m = -\frac{25,000}{3}$; a slope of $-\frac{25,000}{3}$ means that the value of the building decreases $\frac{\$25,000}{3}$ each year.

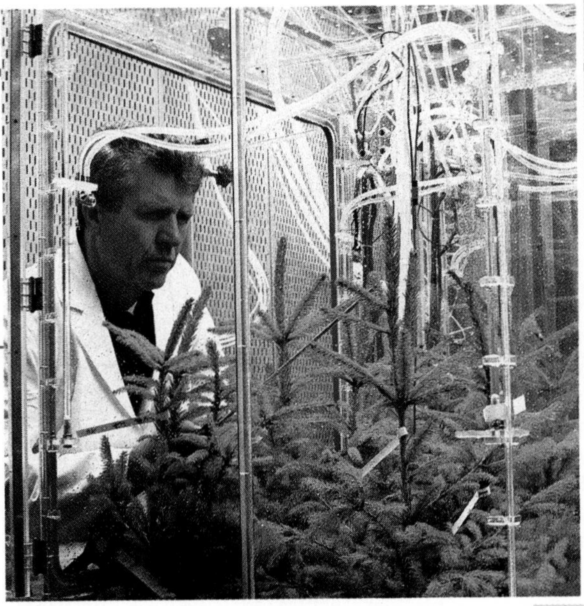

Forestry technicians compile data on the characteristics of forest land tracts, such as species and population of trees, disease and insect damage, tree seedling mortality, and conditions that may cause fire danger. They also train and lead conservation workers in seasonal activities such as planting tree seedlings, putting out forest fires, and maintaining recreational facilities. These technicians may use functions to model, for instance, the progress of a tree disease as it spreads through a tract of forest.

9

Functions and Relations

OBJECTIVES

9.1.1 Graph functions
9.2.1 Graph by using translations
9.3.1 Perform operations on functions
9.3.2 Find the composition of two functions
9.4.1 Determine whether a function is one-to-one
9.4.2 Find the inverse of a function

History of Equals Signs

A portion of a page of the first book that used an equals sign, =, is shown below. This book was written in 1557 by Robert Recorde and was titled *The Whetstone of Witte.*

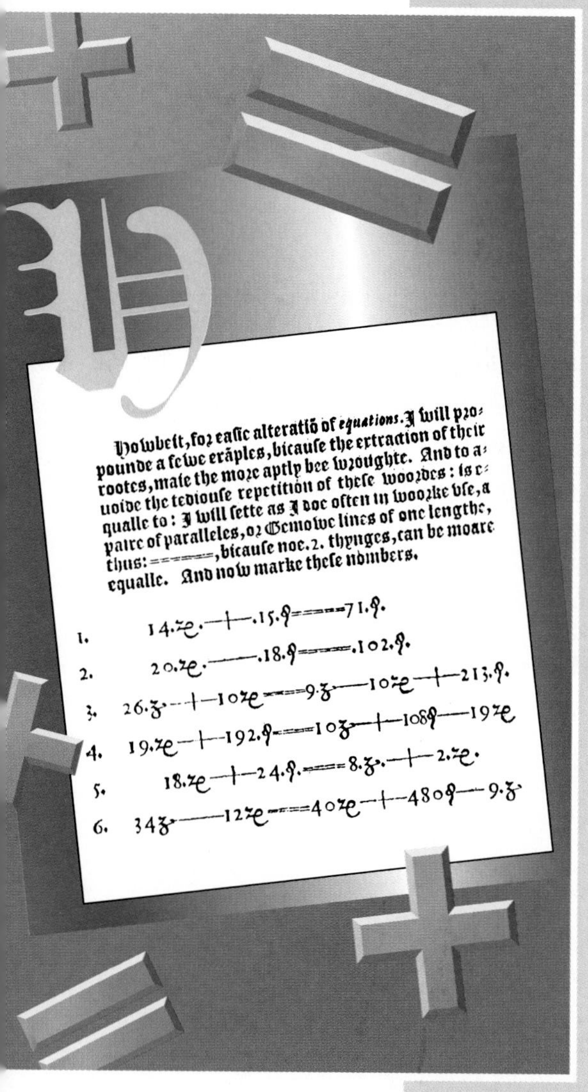

Notice in the illustration the words "bicause noe 2 thynges can be moare equalle." Recorde decided that two things could not be more equal than two parallel lines of the same length. Therefore, it made sense to use this symbol to show equality.

This page also illustrates the use of the plus sign, +, and the minus sign, −. These symbols had been widely used for only about 100 years when *The Whetstone of Witte* was written.

SECTION 9.1
Graphs of Functions

1 Graph functions

The graphs of the polynomial functions $L(x) = mx + b$ (a straight line) and $Q(x) = ax^2 + bx + c, a \neq 0$ (a parabola) have been discussed earlier. The graphs of other functions can also be drawn by finding ordered pairs that belong to the function, plotting the points that correspond to the ordered pairs, and then drawing a curve through the points.

LOOK CLOSELY

The units along the x-axis are different from those along the y-axis. If the units along the x-axis were the same as those along the y-axis, the graph would appear narrower than the one shown.

Graph: $F(x) = x^3$

Select several values of x and evaluate the function.

x	$F(x) = x^3$	$F(x)$	(x, y)
-2	$(-2)^3$	-8	$(-2, -8)$
-1	$(-1)^3$	-1	$(-1, -1)$
0	0^3	0	$(0, 0)$
1	1^3	1	$(1, 1)$
2	2^3	8	$(2, 8)$

Plot the ordered pairs and draw a graph through the points.

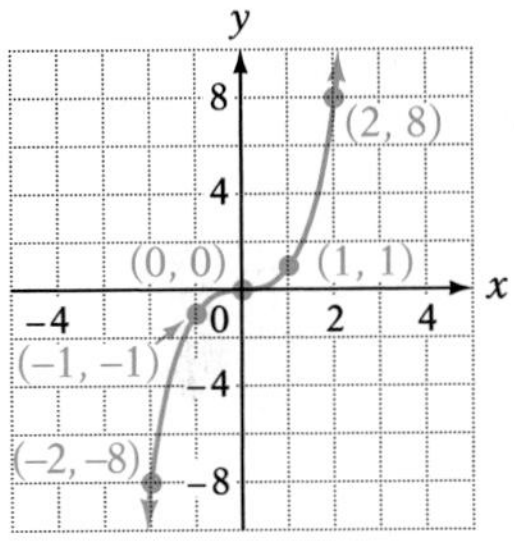

INSTRUCTOR NOTE

The general idea here is that all functions have characteristic graphs and that knowing those characteristics can be useful in checking graphs. For instance, if a student uses a graphing utility to graph a cubic polynomial and the graph below appears, there is some type of error.

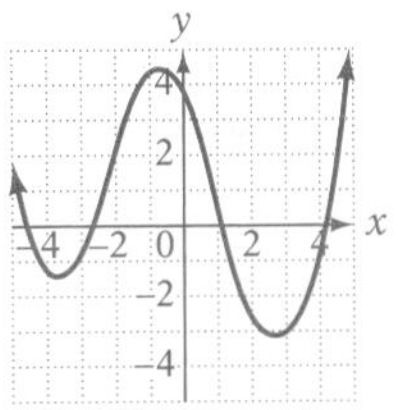

Graph: $g(x) = x^3 - 4x + 5$

Select several values of x and evaluate the function.

x	$g(x) = x^3 - 4x + 5$	$g(x)$	(x, y)
-3	$(-3)^3 - 4(-3) + 5$	-10	$(-3, -10)$
-2	$(-2)^3 - 4(-2) + 5$	5	$(-2, 5)$
-1	$(-1)^3 - 4(-1) + 5$	8	$(-1, 8)$
0	$(0)^3 - 4(0) + 5$	5	$(0, 5)$
1	$(1)^3 - 4(1) + 5$	2	$(1, 2)$
2	$(2)^3 - 4(2) + 5$	5	$(2, 5)$

Plot the ordered pairs and draw a graph through the points.

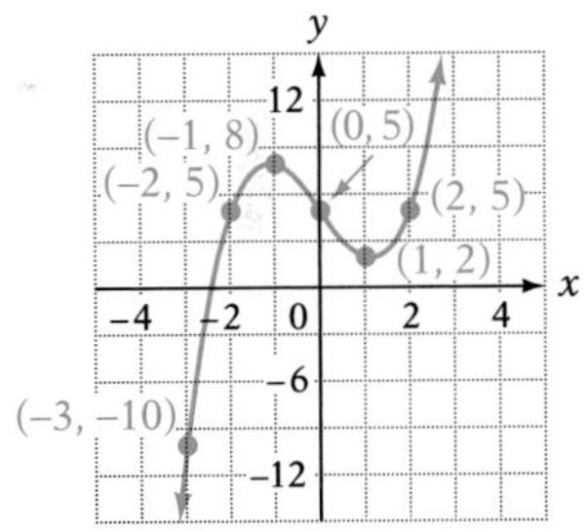

Note from the graphs of the two different cubic functions that the shapes of the graphs can be quite different. The following graphs of typical cubic polynomial functions show the general shape of the graph of a cubic polynomial.

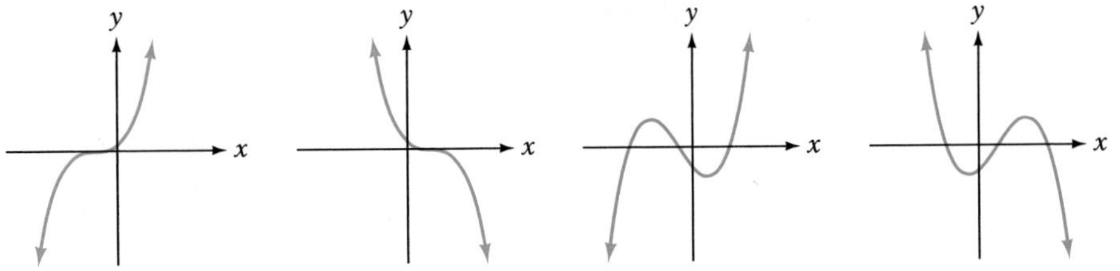

Graph: $f(x) = |x + 2|$

This is an absolute value function.

x	$f(x) = \|x + 2\|$	$f(x)$	(x, y)
-3	$\|-3 + 2\|$	1	$(-3, 1)$
-2	$\|-2 + 2\|$	0	$(-2, 0)$
-1	$\|-1 + 2\|$	1	$(-1, 1)$
0	$\|0 + 2\|$	2	$(0, 2)$
1	$\|1 + 2\|$	3	$(1, 3)$
2	$\|2 + 2\|$	4	$(2, 4)$

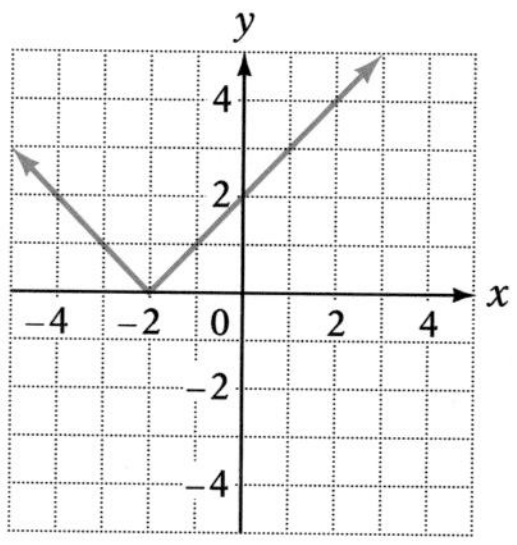

In general, the graph of the absolute value of a linear polynomial is V-shaped.

Graph: $R(x) = \sqrt{2x - 4}$

This is a radical function. Because the square root of a negative number is not a real number, the domain of this function requires that $2x - 4 \geq 0$. Solve this inequality for x.

$$2x - 4 \geq 0$$
$$2x \geq 4$$
$$x \geq 2$$

The domain is $\{x | x \geq 2\}$. This means that only values of x that are greater than or equal to 2 can be chosen as values at which to evaluate the function. In this case, some of the y-coordinates must be approximated.

x	$f(x) = \sqrt{2x - 4}$	$f(x)$	(x, y)
2	$\sqrt{2(2) - 4}$	0	$(2, 0)$
3	$\sqrt{2(3) - 4}$	1.41	$(3, 1.41)$
4	$\sqrt{2(4) - 4}$	2	$(4, 2)$
5	$\sqrt{2(5) - 4}$	2.45	$(5, 2.45)$
6	$\sqrt{2(6) - 4}$	2.83	$(6, 2.83)$

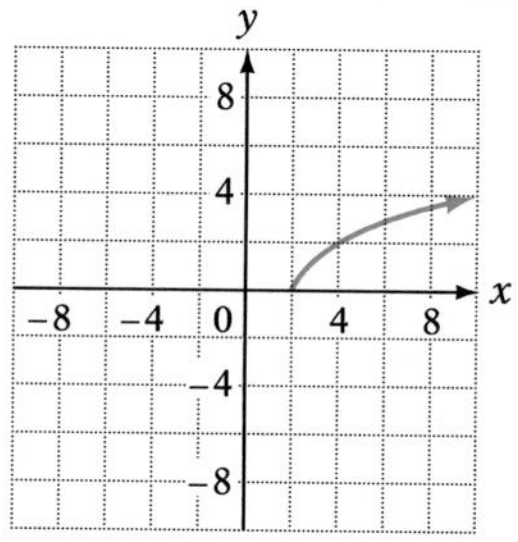

Recall that a function is a special type of relation, one for which no two ordered pairs have the same first coordinate. Graphically, this means that the graph of a function cannot pass through two points that have the same x-coordinate and different y-coordinates. For instance, the graph at the right is not the graph of a function because there are ordered pairs with the same x-coordinate and different y-coordinates.

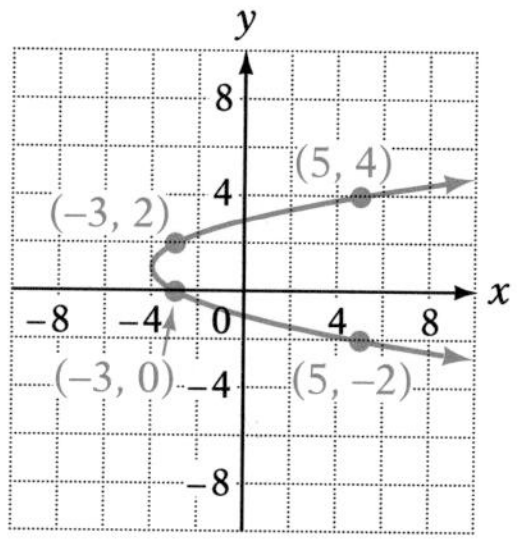

The last graph illustrates a general statement that can be made about whether a graph defines a function. It is called the *vertical-line test.*

Vertical-Line Test

A graph defines a function if any vertical line intersects the graph at no more than one point.

For example, the graph of a nonvertical straight line is the graph of a function. Any vertical line intersects the graph no more than once. The graph of a circle, however, is not the graph of a function. There are vertical lines that intersect the graph at more than one point.

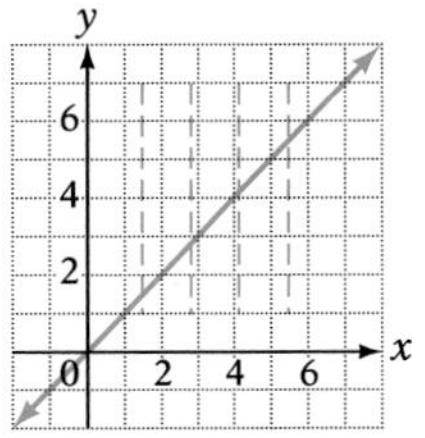

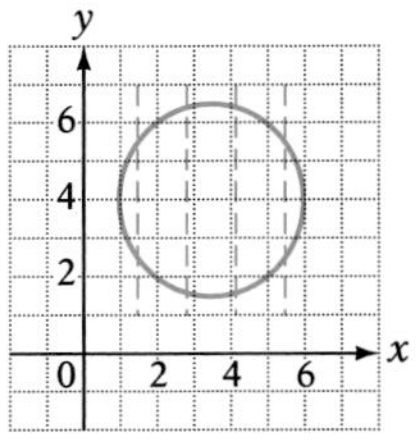

There may be practical situations in which a graph is not the graph of a function. The problem below is an example.

One of the causes of smog is an inversion layer where temperatures at higher altitudes are warmer than those at lower altitudes. The graph at the right shows the altitudes at which various temperatures were recorded. As shown by the dashed lines in the graph, there are two altitudes at which the temperature was 25°C. This means that there are two ordered pairs (shown in the graph) with the same first coordinate but different second coordinates. The graph does not define a function.

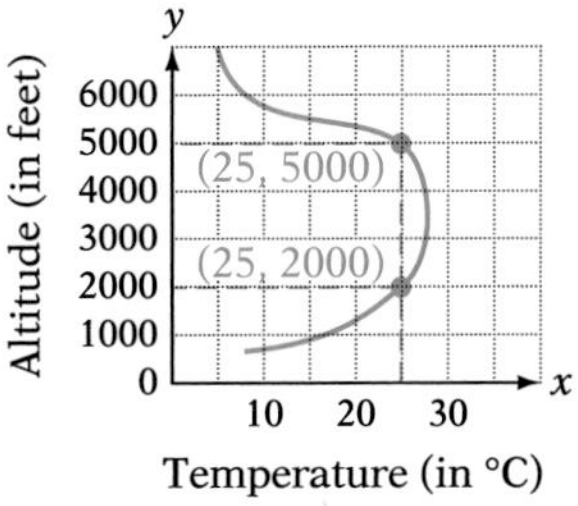

When a graph does define a function, the domain and range can be estimated from the graph.

LOOK CLOSELY

In this first example, the domain and range are given in set-builder notation. For the second example, the domain and range are given in interval notation. Either method may be used.

Determine the domain and range of the function given by the graph at the right. Write the answer in set-builder notation.

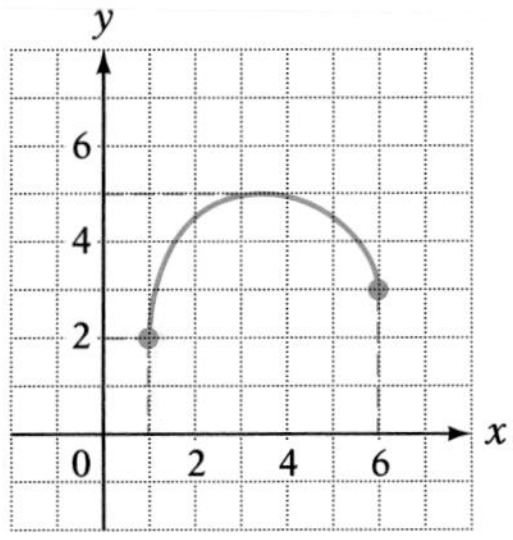

The solid dots on the graph indicate its beginning and ending points.

The domain is the set of x-coordinates. The domain is $\{x \mid 1 \le x \le 6\}$.

The range is the set of y-coordinates. The range is $\{y \mid 2 \le y \le 5\}$.

Determine the domain and range of the function given by the graph at the right. Write the answer in interval notation.

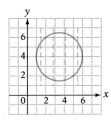

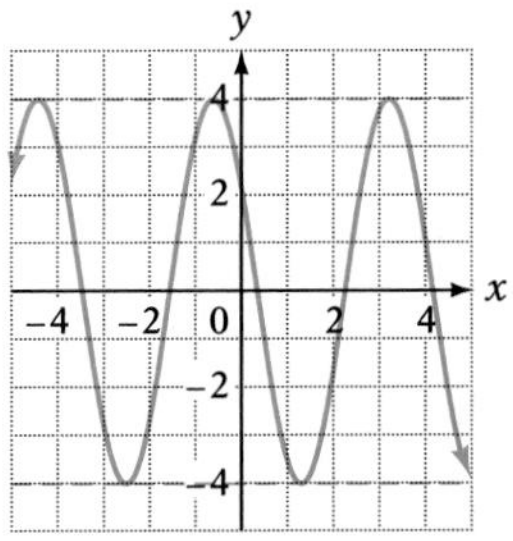

The arrows on the graph indicate that the graph continues in the same manner.

The domain is the set of x-coordinates. The domain is $(-\infty, \infty)$.

The range is the set of y-coordinates. The range is $[-4, 4]$.

Example 1 Use the vertical-line test to determine whether the graph shown at the right is the graph of a function.

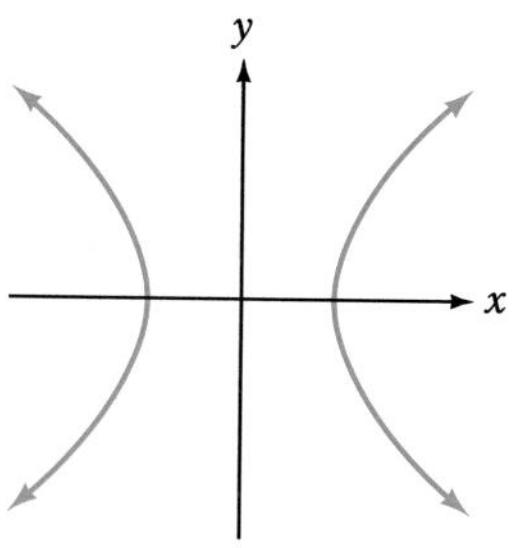

Solution

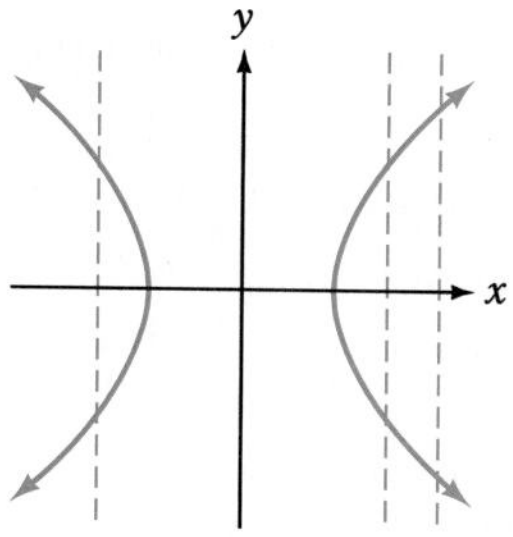

▶ A vertical line can intersect the graph at more than one point. Therefore, the relation includes ordered pairs with the same first component and different second components.

The graph is not the graph of a function.

Problem 1 Use the vertical-line test to determine whether the graph shown at the right is the graph of a function.

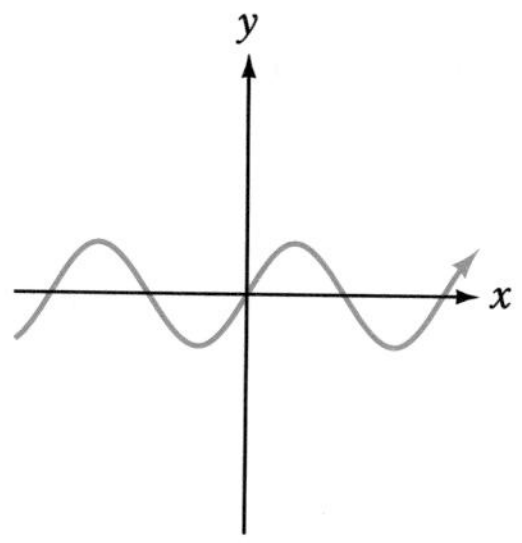

Solution See page S26. Yes

Example 2 Graph. Use set-builder notation to state the domain and range of the function defined by each equation.

A. $f(x) = |x| + 2$ B. $f(x) = -x^3 - 2x^2 + 2$

C. $f(x) = \sqrt{2 - x}$

Solution A.

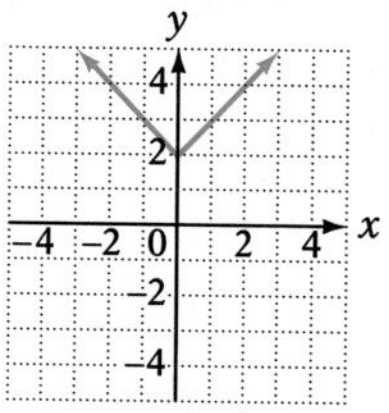

▶ This is an absolute value function. The graph is V-shaped.

The domain is $\{x \mid x \in \text{real numbers}\}$.
The range is $\{y \mid y \geq 2\}$.

B.

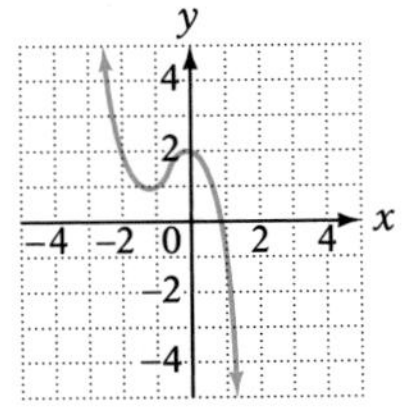

▶ This is a cubic function. Some ordered pairs are $(-2, 2)$, $(-1, 1)$, $(0, 2)$, and $(1, -1)$.

The domain is $\{x \mid x \in \text{real numbers}\}$.
The range is $\{y \mid y \in \text{real numbers}\}$.

C.

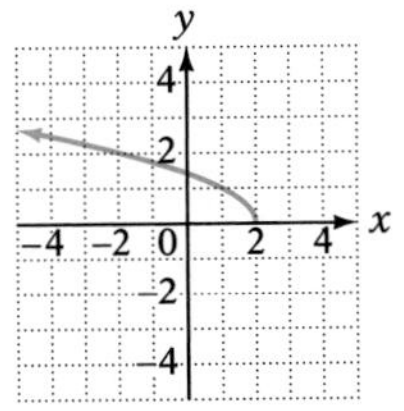

▶ This is a radical function. The numbers under the radicand must be positive.

$$2 - x \geq 0$$
$$-x \geq -2$$
$$x \leq 2$$

The domain is $\{x \mid x \leq 2\}$.
The range is $\{y \mid y \geq 0\}$.

Problem 2 Graph. Use set-builder notation to state the domain and range of the function defined by each equation.

A. $f(x) = |x + 2|$ B. $f(x) = x^3 + 2x + 1$

C. $f(x) = -\sqrt{x - 1}$

Solution See page S26.

A.

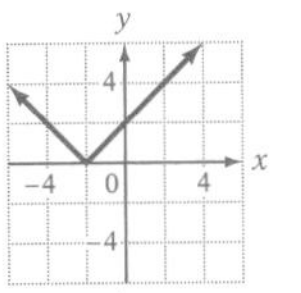

D: $\{x \mid x \in \text{real numbers}\}$
R: $\{y \mid y \geq 0\}$

B.

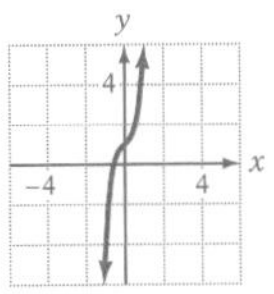

D: $\{x \mid x \in \text{real numbers}\}$
R: $\{y \mid y \in \text{real numbers}\}$

C.

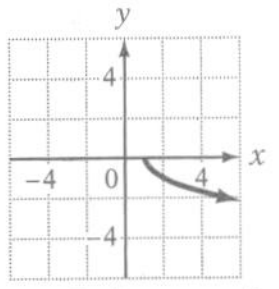

D: $\{x \mid x \geq 1\}$
R: $\{y \mid y \leq 0\}$

CONCEPT REVIEW 9.1

Determine whether the following statements are always true, sometimes true, or never true.

1. The value of an absolute value function is positive. Sometimes true

2. A graph defines a function if any vertical line intersects the graph at no more than one point. Always true

3. The range of a polynomial function is the real numbers. Sometimes true

4. A relation is a function. Sometimes true

5. The symbol $f(x)$ represents a value in the range of the function. Always true

6. The domain of a cubic polynomial is the real numbers. Always true

EXERCISES 9.1

1 **1.** What is the vertical-line test?

2. When we describe the domain or range of a function in interval notation, what is the difference between using a bracket and using a parenthesis?

Use the vertical-line test in order to determine whether the graph is the graph of a function.

3.

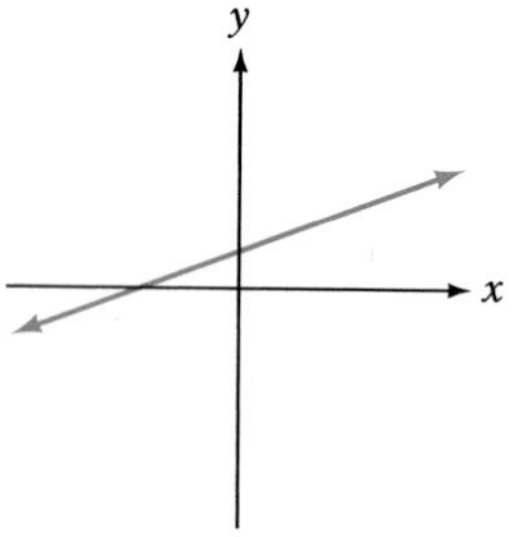

Yes

4.

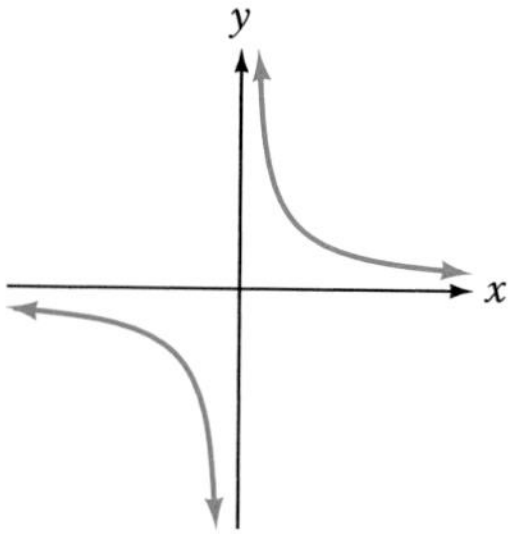

Yes

5.

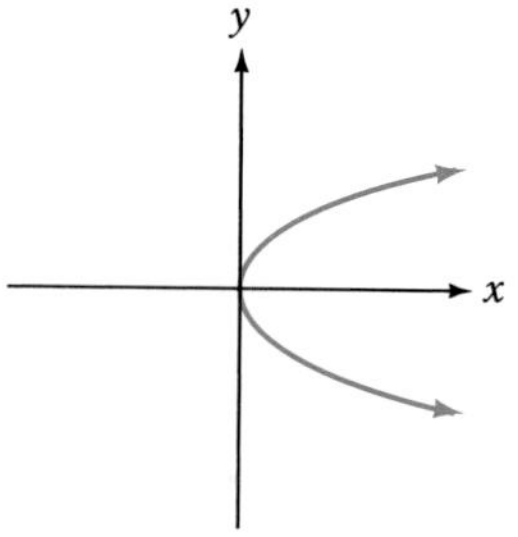

No

6.

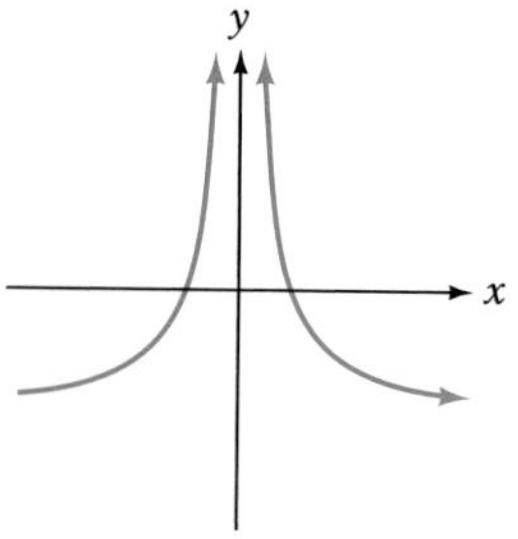

Yes

7.

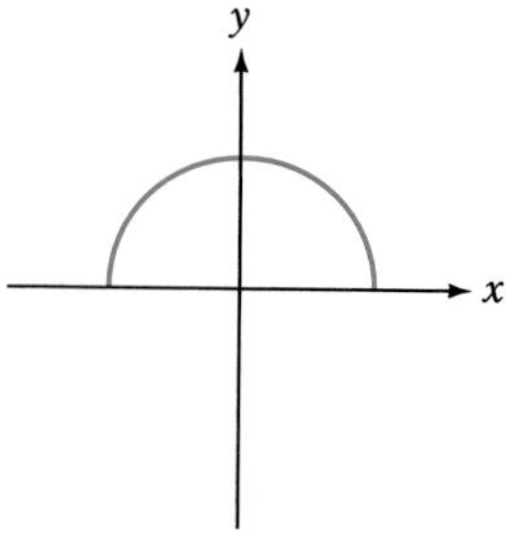

Yes

8.

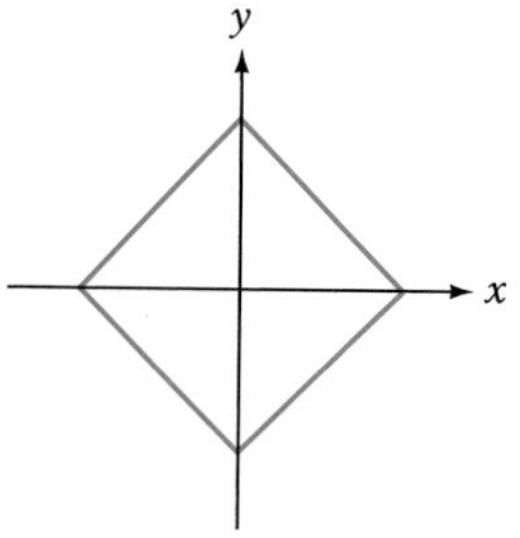

No

Graph the function defined by each equation. Use set-builder notation to state the domain and range of the function.

9. $f(x) = 2x - 1$

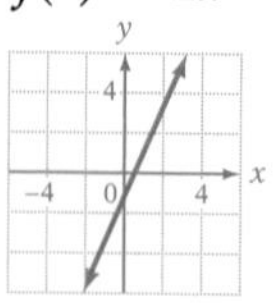

D: $\{x | x \in \text{real numbers}\}$
R: $\{y | y \in \text{real numbers}\}$

10. $f(x) = 4$

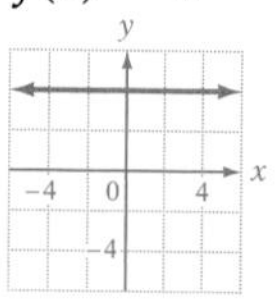

D: $\{x | x \in \text{real numbers}\}$
R: $\{y | y = 4\}$

11. $f(x) = 2x^2 - 1$

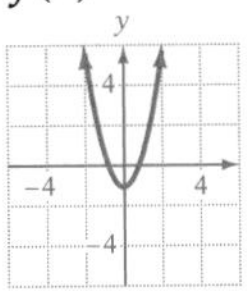

D: $\{x \mid x \in \text{real numbers}\}$
R: $\{y \mid y \geq -1\}$

12. $f(x) = 2x^2 - 3$

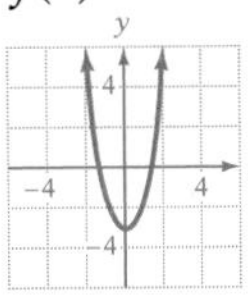

D: $\{x \mid x \in \text{real numbers}\}$
R: $\{y \mid y \geq -3\}$

13. $f(x) = |x - 1|$

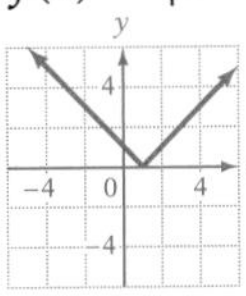

D: $\{x \mid x \in \text{real numbers}\}$
R: $\{y \mid y \geq 0\}$

14. $f(x) = |x - 3|$

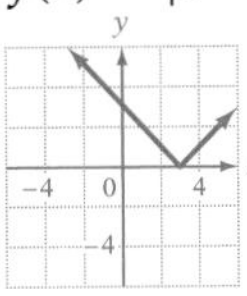

D: $\{x \mid x \in \text{real numbers}\}$
R: $\{y \mid y \geq 0\}$

15. $f(x) = x^3 - 1$

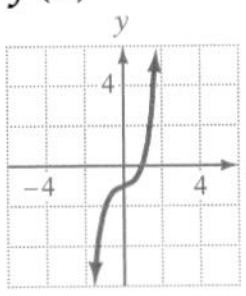

D: $\{x \mid x \in \text{real numbers}\}$
R: $\{y \mid y \in \text{real numbers}\}$

16. $f(x) = -x^3 + 1$

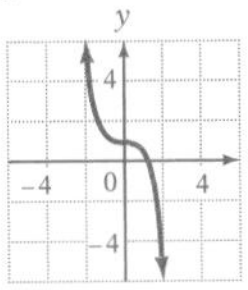

D: $\{x \mid x \in \text{real numbers}\}$
R: $\{y \mid y \in \text{real numbers}\}$

17. $f(x) = \sqrt{1 + x}$

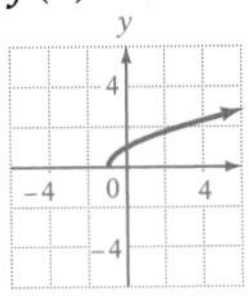

D: $\{x \mid x \geq -1\}$
R: $\{y \mid y \geq 0\}$

18. $f(x) = \sqrt{4 - x}$

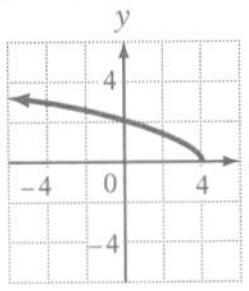

D: $\{x \mid x \leq 4\}$
R: $\{y \mid y \geq 0\}$

19. $f(x) = \frac{1}{2}x^2$

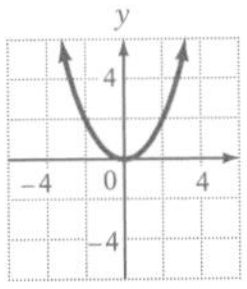

D: $\{x \mid x \in \text{real numbers}\}$
R: $\{y \mid y \geq 0\}$

20. $f(x) = \frac{1}{3}x^2$

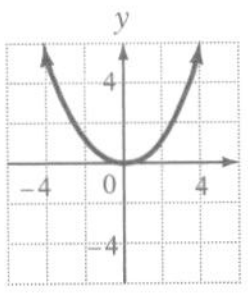

D: $\{x \mid x \in \text{real numbers}\}$
R: $\{y \mid y \geq 0\}$

21. $f(x) = |x| + 1$

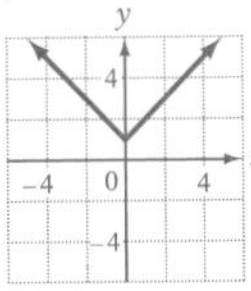

D: $\{x \mid x \in \text{real numbers}\}$
R: $\{y \mid y \geq 1\}$

22. $f(x) = |x| - 1$

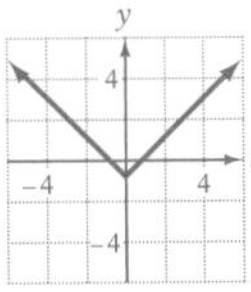

D: $\{x \mid x \in \text{real numbers}\}$
R: $\{y \mid y \geq -1\}$

23. $f(x) = x^3 + 2x^2$

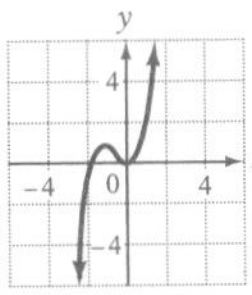

D: $\{x \mid x \in \text{real numbers}\}$
R: $\{y \mid y \in \text{real numbers}\}$

24. $f(x) = x^3 - 3x^2$

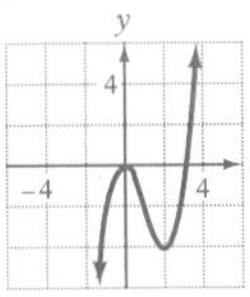

D: $\{x \mid x \in \text{real numbers}\}$
R: $\{y \mid y \in \text{real numbers}\}$

25. $f(x) = -\sqrt{x + 2}$

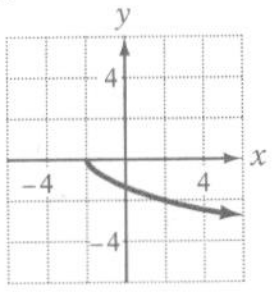

D: $\{x \mid x \geq -2\}$
R: $\{y \mid y \leq 0\}$

26. $f(x) = -\sqrt{x - 3}$

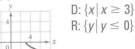

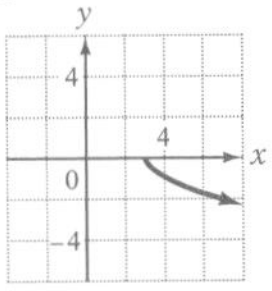

D: $\{x \mid x \geq 3\}$
R: $\{y \mid y \leq 0\}$

27. $f(x) = 2x^2 - 4x - 3$

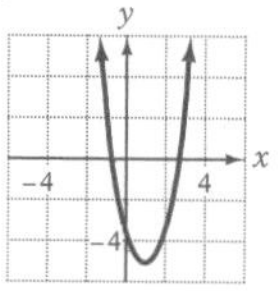

D: $\{x \mid x \in \text{real numbers}\}$
R: $\{y \mid y \geq -5\}$

28. $f(x) = -1$

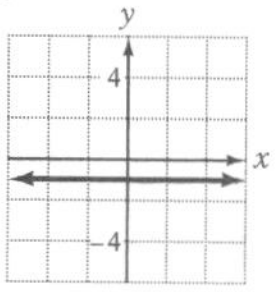

D: $\{x \mid x \in \text{real numbers}\}$
R: $\{y \mid y = -1\}$

29. $f(x) = 2|x| - 1$

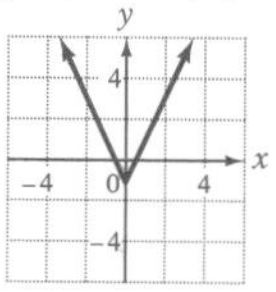

D: $\{x \mid x \in \text{real numbers}\}$
R: $\{y \mid y \geq -1\}$

30. $f(x) = 2|x| + 2$

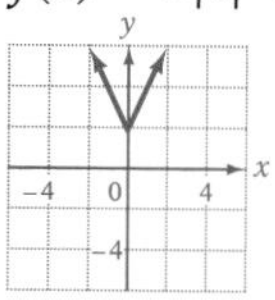

D: $\{x \mid x \in \text{real numbers}\}$
R: $\{y \mid y \geq 2\}$

31. $f(x) = 2x^3 + 3x$

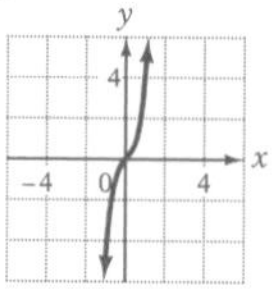

D: $\{x \mid x \in \text{real numbers}\}$
R: $\{y \mid y \in \text{real numbers}\}$

32. $f(x) = -2x^3 + 4x$

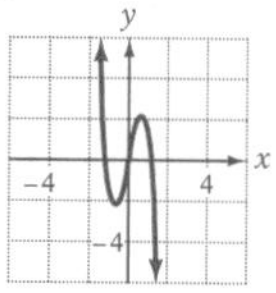

D: $\{x \mid x \in \text{real numbers}\}$
R: $\{y \mid y \in \text{real numbers}\}$

33. $f(x) = -\frac{1}{2}x - 2$

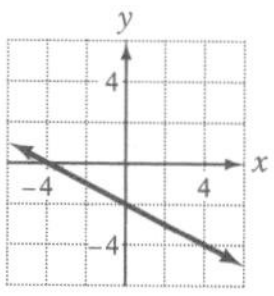

D: $\{x \mid x \in \text{real numbers}\}$
R: $\{y \mid y \in \text{real numbers}\}$

34. $f(x) = -3$

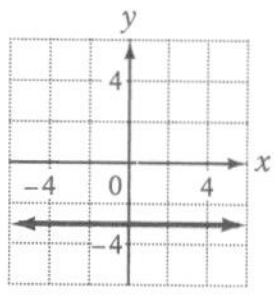

D: $\{x \mid x \in \text{real numbers}\}$
R: $\{y \mid y = -3\}$

35. $f(x) = x^2 + 2x - 4$

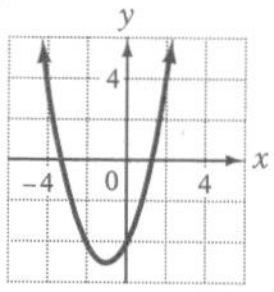

D: $\{x \mid x \in \text{real numbers}\}$
R: $\{y \mid y \geq -5\}$

36. $f(x) = -x^2 + 2x - 1$

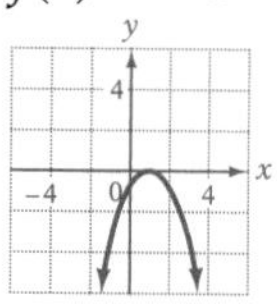

D: $\{x \mid x \in \text{real numbers}\}$
R: $\{y \mid y \leq 0\}$

37. $f(x) = -|x|$

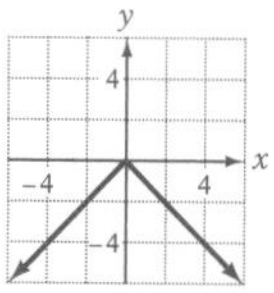

D: $\{x \mid x \in \text{real numbers}\}$
R: $\{y \mid y \leq 0\}$

38. $f(x) = -|x + 2|$

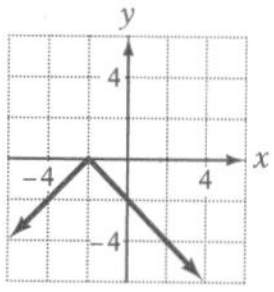

D: $\{x \mid x \in \text{real numbers}\}$
R: $\{y \mid y \leq 0\}$

39. $f(x) = x^3 - x^2 + x - 1$

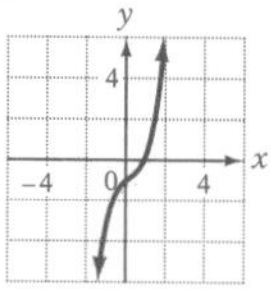

D: $\{x \mid x \in \text{real numbers}\}$
R: $\{y \mid y \in \text{real numbers}\}$

40. $f(x) = x^3 + x^2 - x + 1$

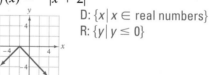

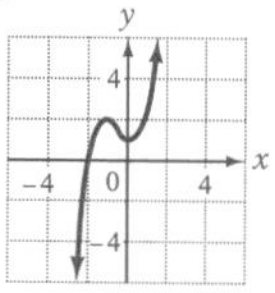

D: $\{x \mid x \in \text{real numbers}\}$
R: $\{y \mid y \in \text{real numbers}\}$

APPLYING CONCEPTS 9.1

Which of the following relations are not functions?

41. **a.** $f(x) = x$ **b.** $f(x) = \left|\frac{x}{2}\right|$ **c.** $\{(3, 1), (1, 3), (3, 0), (0, 3)\}$ c

42. **a.** $f(x) = -x$ **b.** $f(x) = \frac{2}{\sqrt{x}}$ **c.** $\{(1, 4), (4, 1), (1, -4), (-4, 1)\}$ c

43. If $f(x) = \sqrt{x-2}$ and $f(a) = 4$, find a. $a = 18$

44. Let $f(a, b) =$ the sum of a and b.
Let $g(a, b) =$ the product of a and b.
Find $f(2, 5) + g(2, 5)$. 17

45. Let $f(a, b) =$ the greatest common divisor of a and b.
Let $g(a, b) =$ the least common multiple of a and b.
Find $f(14, 35) + g(14, 35)$. 77

46. Given $f(x) = (x+2)(x-2)$, for what values of x is $f(x)$ negative? Write your answer in set-builder notation. $\{x \mid -2 < x < 2\}$

47. Given $f(x) = -|x+3|$, for what value of x is $f(x)$ greatest? $x = -3$

48. Given $f(x) = |2x-2|$, for what value of x is $f(x)$ smallest? $x = 1$

[C] **49.** **a.** For what real value of x does $x^3 = 0$? 0
b. What is the x-intercept of the graph of $f(x) = x^3$? (0, 0)

[C] **50.** **a.** For what real value of x does $x^3 - 1 = 0$? 1
b. What is the x-intercept of the graph of $f(x) = x^3 - 1$? (1, 0)

A *turning point* in the graph of a function is a point where the graph changes direction. For instance, the graph of $f(x) = x^3 - 4x + 5$ shown at the right has two turning points. In general, the number of turning points that the graph of a polynomial has depends on the degree of the polynomial. However, not all graphs of cubic functions have two turning points. For instance, the graph of $f(x) = x^3$ has no turning points.

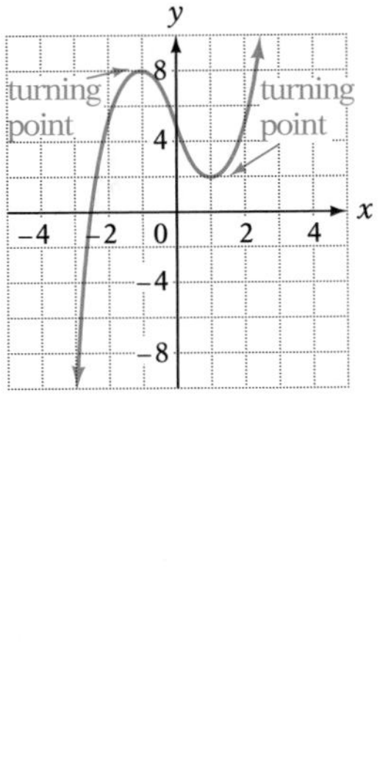

By using a graphing utility, graph the following and answer the questions. You may need to set Ymin and Ymax to fairly large numbers to accommodate these graphs.

51. Graph $f(x) = x^4 - 5$. How many turning points does the graph have? 1

52. Graph $f(x) = x^4 - 3x^3 - 7x^2 + x + 6$. How many turning points does the graph have? 3

53. Make up six more fourth-degree polynomial functions and graph each one. Record the number of turning points for each graph. Make a conjecture about the number of turning points the graph of a fourth-degree polynomial will have. Graph a few more fourth-degree polynomials and determine whether your conjecture is correct for those graphs. If not, modify your conjecture and try a few more graphs.

SECTION 9.2 Translations of Graphs

1 Graph by using translations

The graphs of $f(x) = |x|$ and $g(x) = |x| + 2$ are shown in Figure 1. Note that for a given x-coordinate, the y-coordinate on the graph of g is 2 units higher than that on the graph of f. The graph of g is said to be a **vertical translation,** or **vertical shift,** of the graph of f.

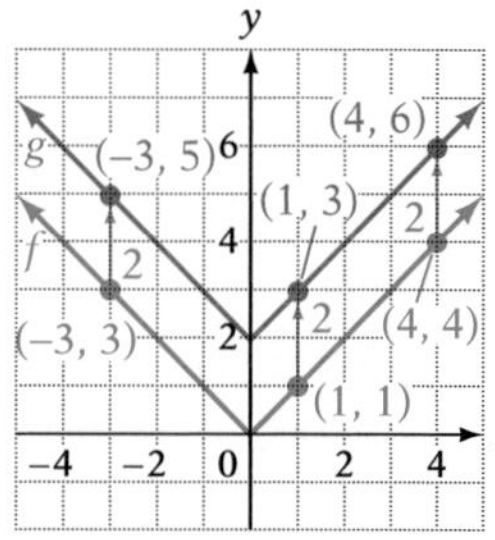

Figure 1

LOOK CLOSELY

Remember that $g(x)$ and $f(x)$ are y-coordinates of an ordered pair. The fact that $g(x)$ is 2 units more than $f(x)$ means that, for a given value of x, the y values for g are 2 units higher than the y values for f.

Note that because $f(x) = |x|$,

$$g(x) = |x| + 2$$
$$= f(x) + 2$$

Thus $g(x)$ is 2 units more than $f(x)$.

Now consider the graphs of the functions $f(x) = |x|$ and $h(x) = |x| - 3$, shown in Figure 2. Note that for a given x-coordinate, the y-coordinate on the graph of h is 3 units lower than that on the graph of f. The graph of h is a vertical translation of the graph of f.

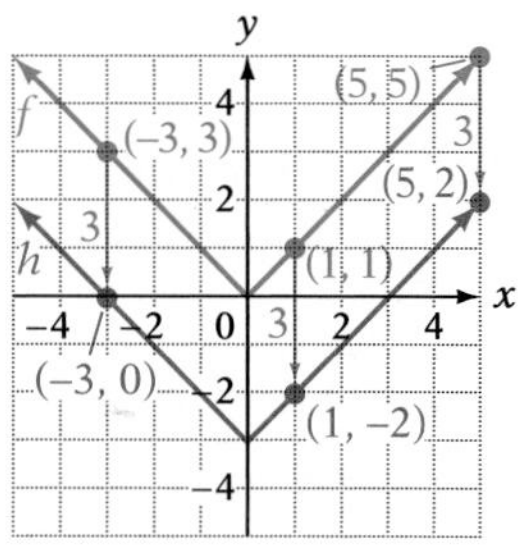

Figure 2

In Figure 1, the graph of $g(x) = |x| + 2$ is the graph of f shifted *up* 2 units, whereas in Figure 2, the graph of $h(x) = |x| - 3$ is the graph of f shifted *down* 3 units.

Vertical Translation

If f is a function and c is a positive constant, then

$y = f(x) + c$ is the graph of $y = f(x)$ shifted up c units.

$y = f(x) - c$ is the graph of $y = f(x)$ shifted down c units.

Example 1 Given the graph of the function $f(x) = \sqrt[3]{x}$ shown at the right in blue, graph $g(x) = \sqrt[3]{x} + 3$ by using a vertical translation.

Solution The graph of $g(x) = \sqrt[3]{x} + 3$ is the graph of $f(x) = \sqrt[3]{x}$ shifted 3 units up. This graph is shown in red at the right.

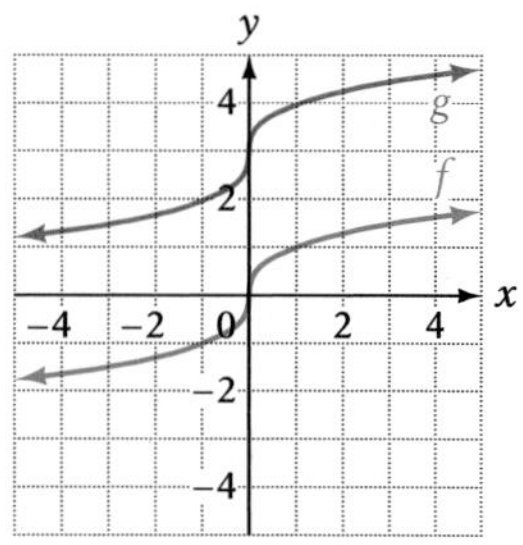

Problem 1   Given the graph of the function $f(x) = x^2 - 2x$ shown at the right, graph $g(x) = x^2 - 2x - 3$ by using a vertical translation.

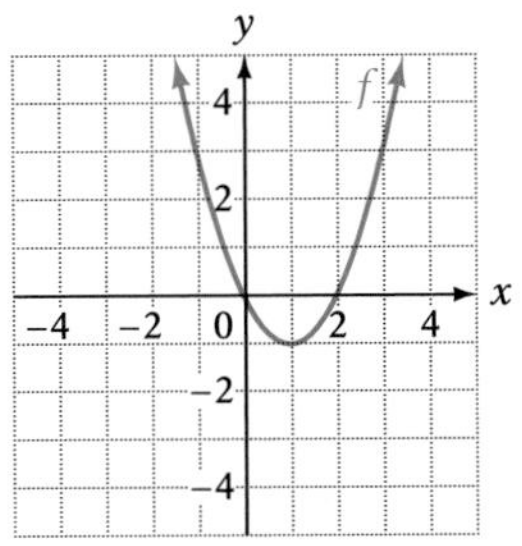

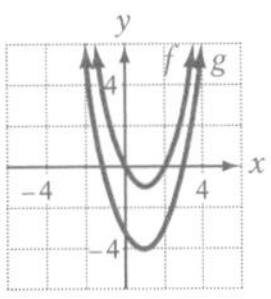

Solution   See page S26.

The graphs of $f(x) = |x|$ and $g(x) = |x - 2|$ are shown in Figure 3. Note that the graph of g is the graph of f shifted 2 units to the right. The graph of g is a **horizontal translation,** or **horizontal shift,** of the graph of f. In this situation, each x-coordinate is moved to the right 2 units, but the y-coordinates are unchanged.

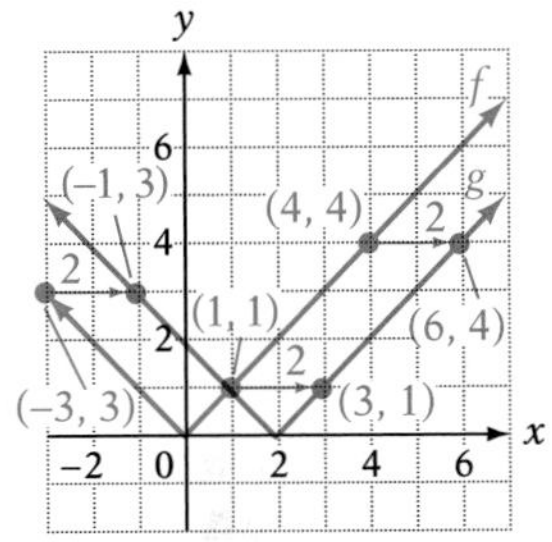

Figure 3

The graphs of $f(x) = |x|$ and $h(x) = |x + 3|$ are shown in Figure 4. In this case, the graph of f is translated 3 units to the left.

For horizontal translations, note that when $f(x) = |x|$,

$$f(x - 2) = |x - 2| = g(x)$$

is a translation to the right and

$$f(x + 3) = |x + 3| = h(x)$$

is a translation to the left.

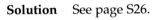

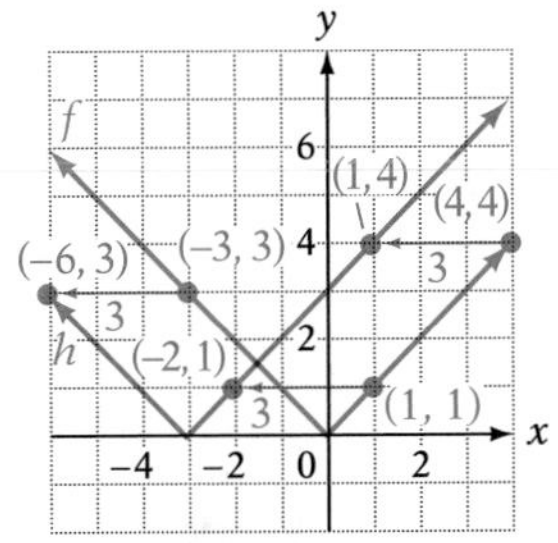

Figure 4

One way to determine the direction of the horizontal shift is to ask, "For what value of x is $y = 0$?" For instance, the horizontal shift of g in Figure 3 is 2 units to the right because $y = 0$ when $x = 2$.

$$g(x) = |x - 2|$$
$$g(2) = |2 - 2| = 0$$

Horizontal Translation

If f is a function and c is a positive constant, then

$y = f(x - c)$ is the graph of $y = f(x)$ shifted horizontally right c units.

$y = f(x + c)$ is the graph of $y = f(x)$ shifted horizontally left c units.

LOOK CLOSELY

Recall that evaluation of a function is accomplished by replacing a variable by another quantity.

$$f(x) = |x|$$
$$f(-3) = |-3|$$
$$f(x - 2) = |x - 2| = g(x)$$
$$f(x + 3) = |x + 3| = h(x)$$

Example 2 Given the graph of the function $F(x) = x^3$ shown at the right in blue, graph $G(x) = (x + 2)^3$ by using a horizontal translation.

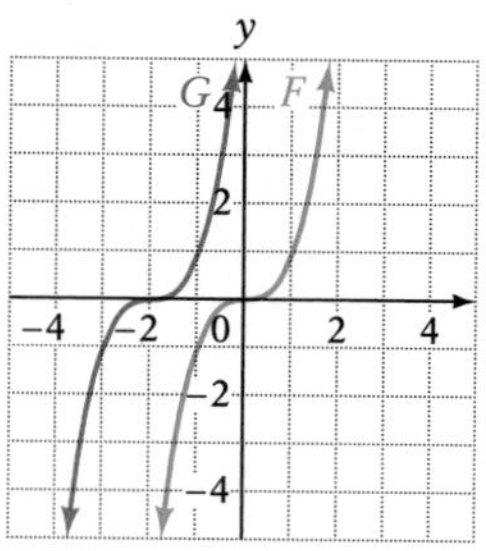

Solution The graph of G (shown in red) is the graph of F shifted horizontally 2 units to the left. Note that if

$$(x + 2)^3 = 0, \text{ then } x = -2.$$

The result, -2, is one confirmation that the graph should be shifted 2 units to the left.

Problem 2 Given the graph of the function $f(x) = \sqrt{x}$ shown at the right, graph $g(x) = \sqrt{x - 3}$ by using a horizontal translation.

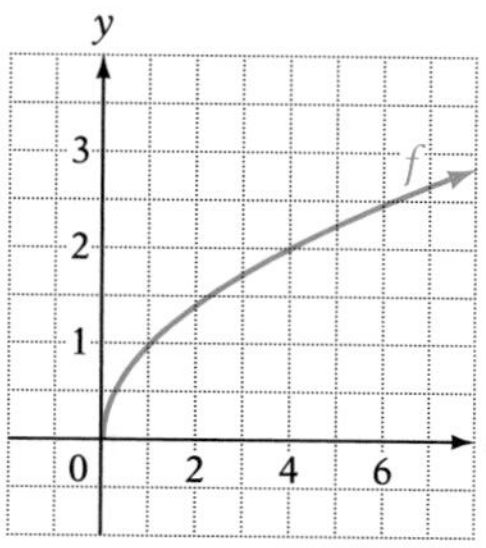

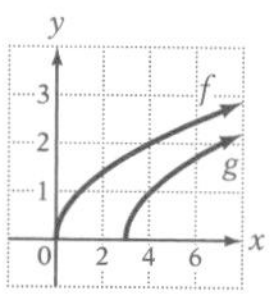

Solution See page S26.

It is possible for a graph to involve both a horizontal and a vertical translation.

Example 3 Given the graph of the function $f(x) = |x|$ shown at the right in blue, graph $A(x) = |x + 1| - 3$ by using both a horizontal and a vertical translation.

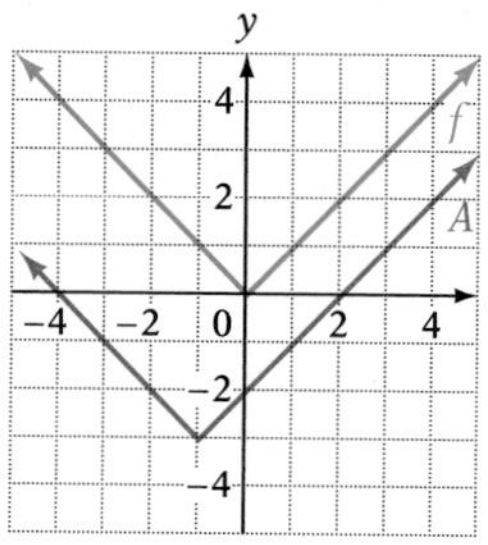

Solution The graph of A includes a horizontal shift of 1 unit to the left ($|x + 1| = 0$ means that $x = -1$) and a vertical shift of 3 units down. The graph of A is shown in red at the right.

Problem 3 Given the graph of the function $g(x) = x^3$ shown at the right, graph $V(x) = (x - 2)^3 + 1$ by using both a horizontal and a vertical translation.

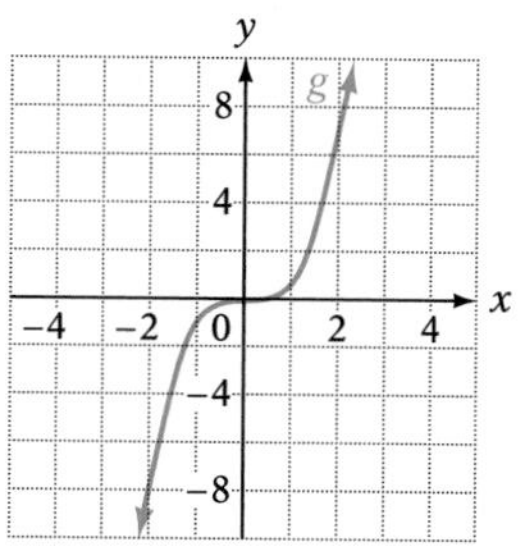

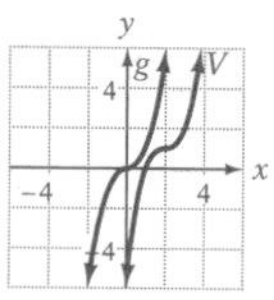

Solution See page S26.

CONCEPT REVIEW 9.2

Determine whether the following statements are always true, sometimes true, or never true.

1. If c is a positive constant, the graph of $y = f(x) + c$ is the graph of $y = f(x)$ shifted upward c units. Always true

2. If c is a positive constant, the graph of $y = f(x + c)$ is the graph of $y = f(x)$ translated c units to the right. Never true

3. The graph of $y = f(x - 2)$ is the graph of $y = f(x)$ translated 2 units to the right. Always true

4. The graph of $y = |x| + 2$ is the graph of $y = |x|$ translated 2 units to the left. Never true

5. If c_1 and c_2 are positive constants, the graph of $y = f(x + c_1) + c_2$ is the graph of $y = f(x)$ translated c_1 units to the right and shifted up c_2 units. Never true

6. The graph of $y = (x - 2)^3 + 3$ is the graph of $y = x^3$ translated 2 units to the right and shifted 3 units up. Always true

EXERCISES 9.2

1 1. Explain the meaning of a vertical or horizontal translation of a graph.

2. How does the graph of $y = f(x - 2) + 1$ differ from the graph of $y = f(x)$?

In Exercises 3–28, use translations to draw the graphs. Many of these exercises refer to Figures 1–3 below.

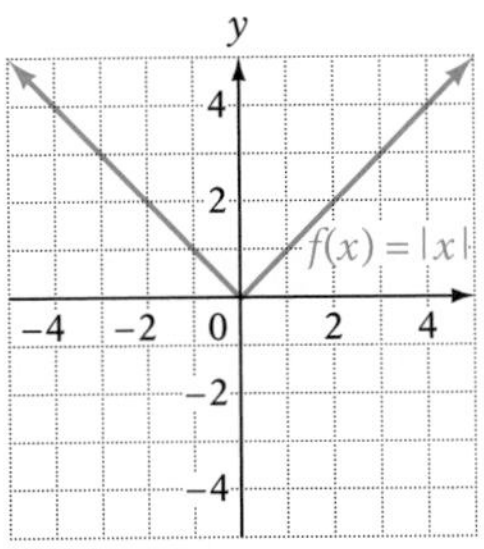

Figure 1

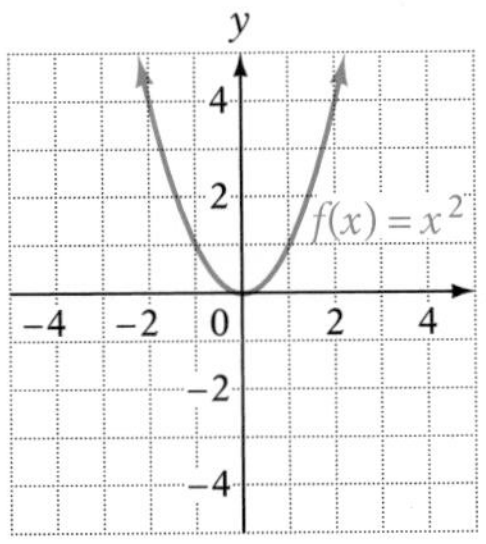

Figure 2

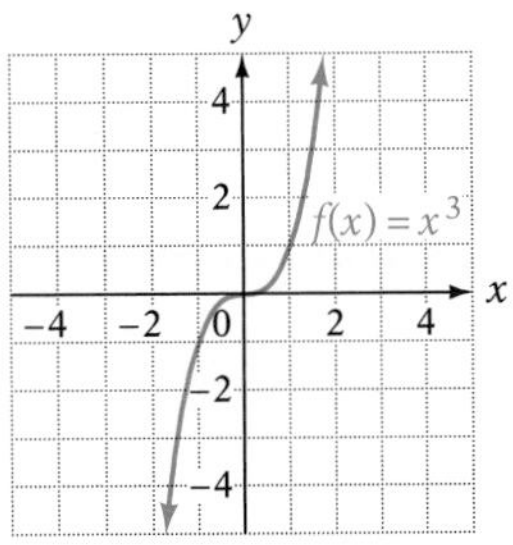

Figure 3

3. Given the graph of $f(x) = |x|$ shown in Figure 1 above, graph $g(x) = |x| + 4$.

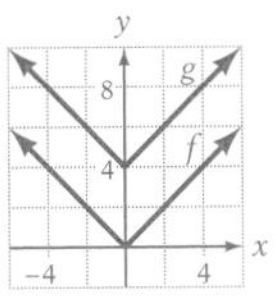

4. Given the graph of $f(x) = |x|$ shown in Figure 1 above, graph $g(x) = |x| - 3$.

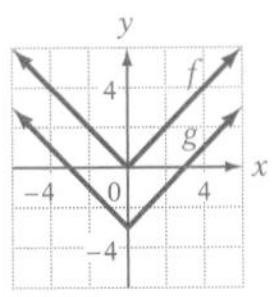

5. Given the graph of $f(x) = x^2$ shown in Figure 2 above, graph $g(x) = x^2 - 2$.

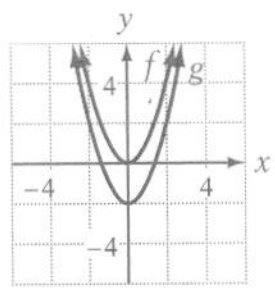

6. Given the graph of $f(x) = x^2$ shown in Figure 2 above, graph $g(x) = x^2 + 1$.

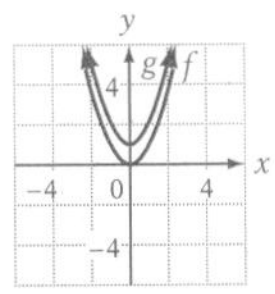

7. Given the graph of $f(x) = x^3$ shown in Figure 3 on page 509, graph $g(x) = x^3 + 1$.

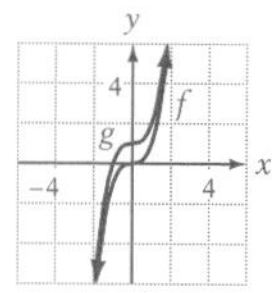

8. Given the graph of $f(x) = x^3$ shown in Figure 3 on page 509, graph $g(x) = x^3 - 2$.

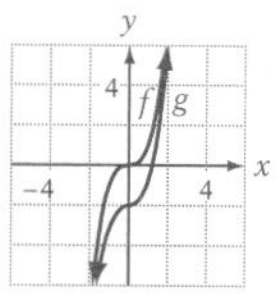

9. Given the graph of $f(x) = x^2$ shown in Figure 2 on page 509, graph $g(x) = (x - 3)^2$.

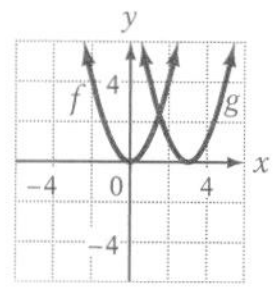

10. Given the graph of $f(x) = x^2$ shown in Figure 2 on page 509, graph $g(x) = (x + 2)^2$.

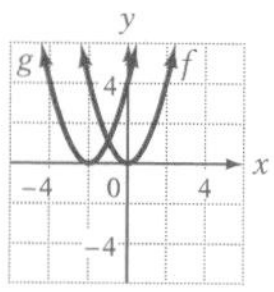

11. Given the graph of $f(x) = |x|$ shown in Figure 1 on page 509, graph $g(x) = |x - 1|$.

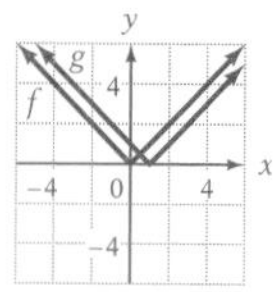

12. Given the graph of $f(x) = |x|$ shown in Figure 1 on page 509, graph $g(x) = |x + 4|$.

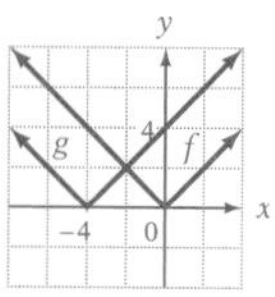

13. Given the graph of $f(x) = x^3$ shown in Figure 3 on page 509, graph $g(x) = (x - 4)^3$.

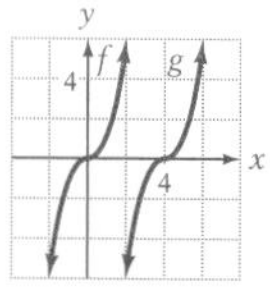

14. Given the graph of $f(x) = x^3$ shown in Figure 3 on page 509, graph $g(x) = (x + 3)^3$.

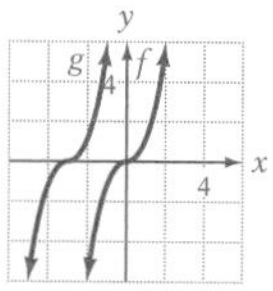

15. Given the graph of $f(x) = -x^2$ shown below, graph $g(x) = -x^2 + 2$.

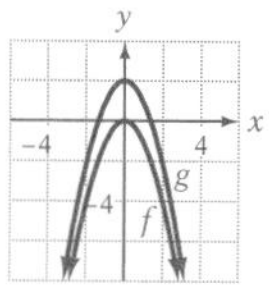

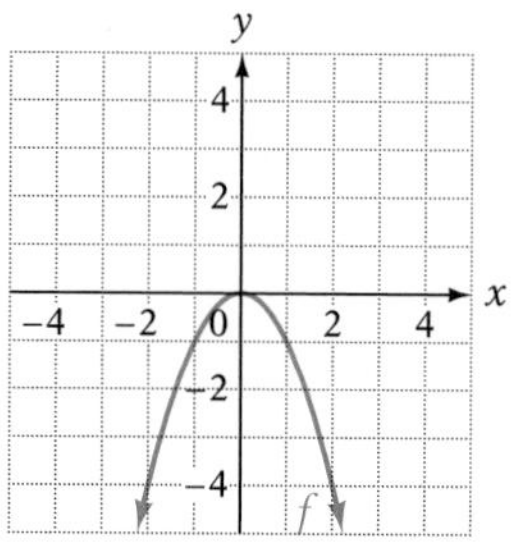

16. Given the graph of $f(x) = -|x|$ shown below, graph $g(x) = -|x| - 2$.

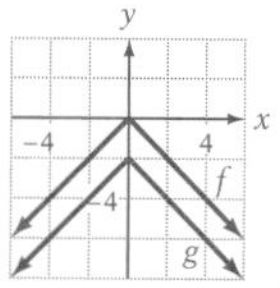

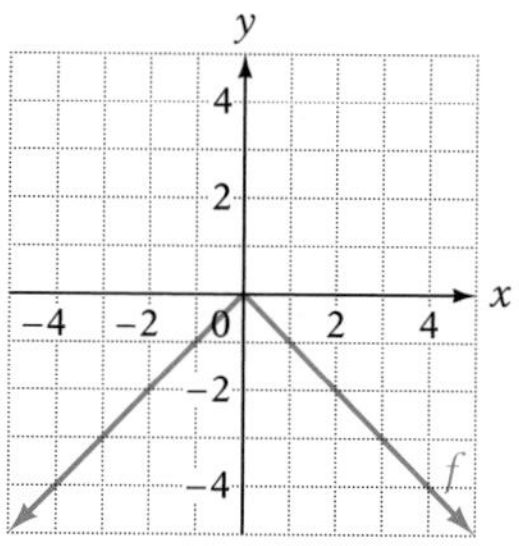

17. Given the graph of $f(x) = x^3$ shown in Figure 3 on page 509, graph $g(x) = (x + 1)^3 - 2$.

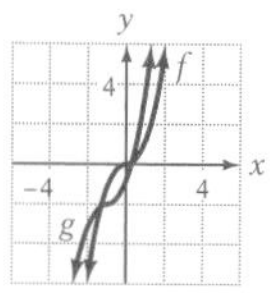

18. Given the graph of $f(x) = x^3$ shown in Figure 3 on page 509, graph $g(x) = (x - 2)^3 + 1$.

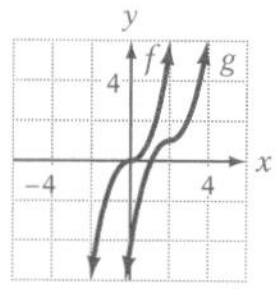

19. Given the graph of $f(x) = x^2$ shown in Figure 2 on page 509, graph $g(x) = (x - 3)^2 - 4$.

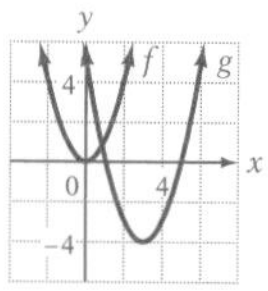

20. Given the graph of $f(x) = x^2$ shown in Figure 2 on page 509, graph $g(x) = (x + 2)^2 + 4$.

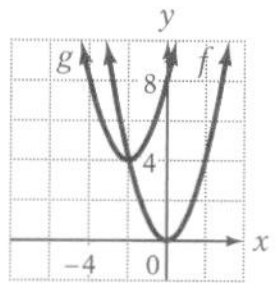

21. Given the graph of $f(x) = \sqrt{x}$ shown below, graph $g(x) = \sqrt{x - 1} + 2$.

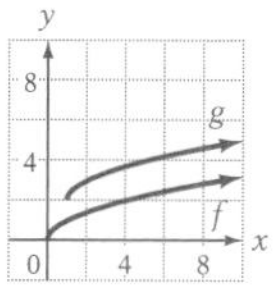

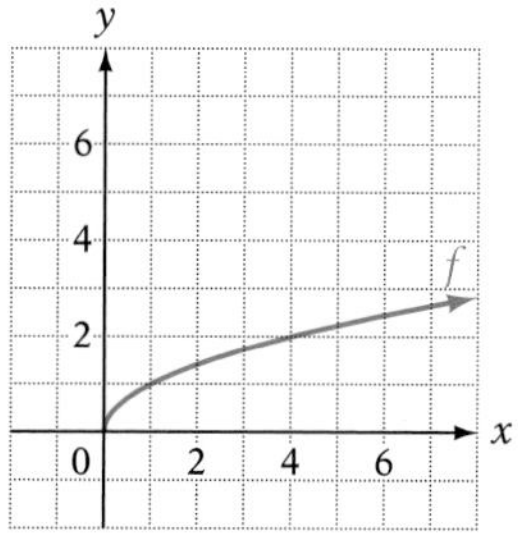

22. Given the graph of $f(x) = -\sqrt{x}$ shown below, graph $g(x) = -\sqrt{x + 1} - 3$.

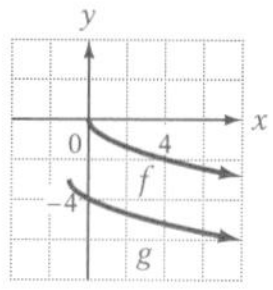

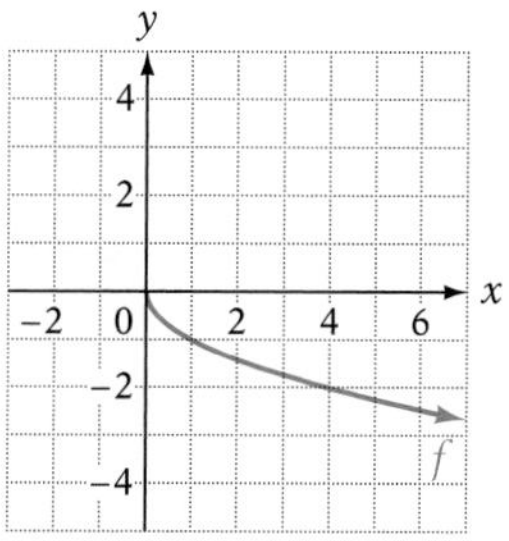

23. Given the graph of $f(x) = x^2 - 2x$ shown below, graph $g(x) = x^2 - 2x - 1$.

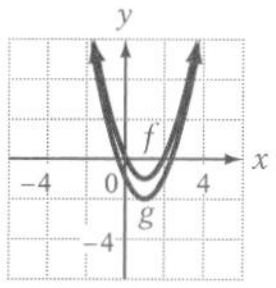

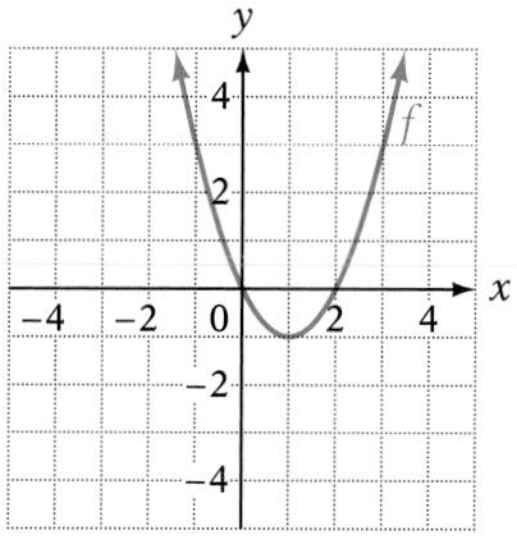

24. Given the graph of $f(x) = x^3 - 4x$ shown below, graph $g(x) = x^3 - 4x + 2$.

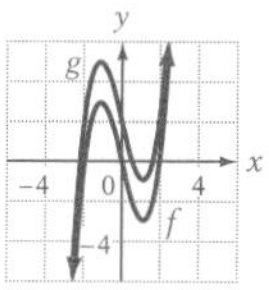

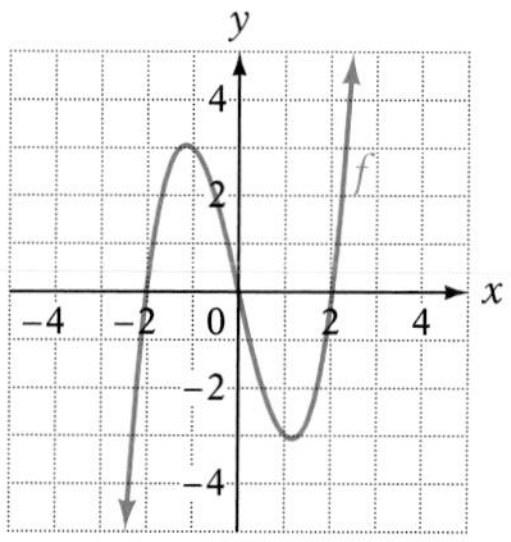

25. Given the graph of $f(x) = -x^2$ shown in Exercise 15, graph $g(x) = -(x + 1)^2 - 2$.

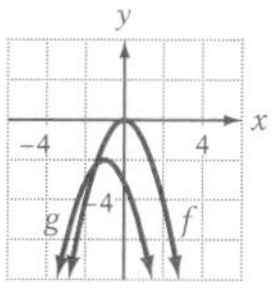

26. Given the graph of $f(x) = -x^2$ shown in Exercise 15, graph $g(x) = -(x - 1)^2 + 1$.

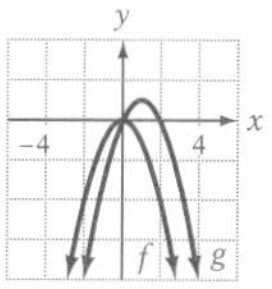

27. Given the graph of $f(x) = -\sqrt{x}$ shown in Exercise 22, graph $g(x) = -\sqrt{x - 2} + 1$.

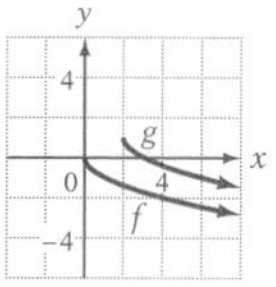

28. Given the graph of $f(x) = \sqrt{x}$ shown in Exercise 21, graph $g(x) = \sqrt{x + 2} - 3$.

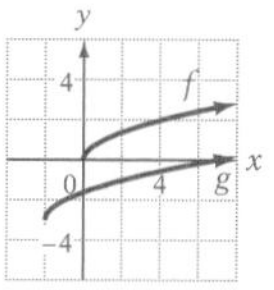

APPLYING CONCEPTS 9.2

In Exercises 29–36, use translations to draw the graph. These exercises refer to Figures 4–6 below. No equations are given for the graphs.

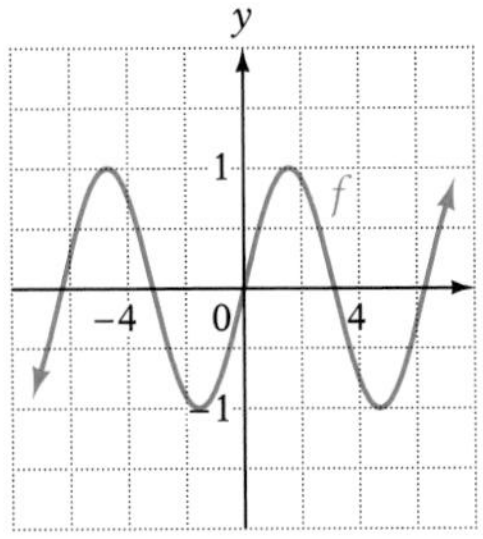

Figure 4

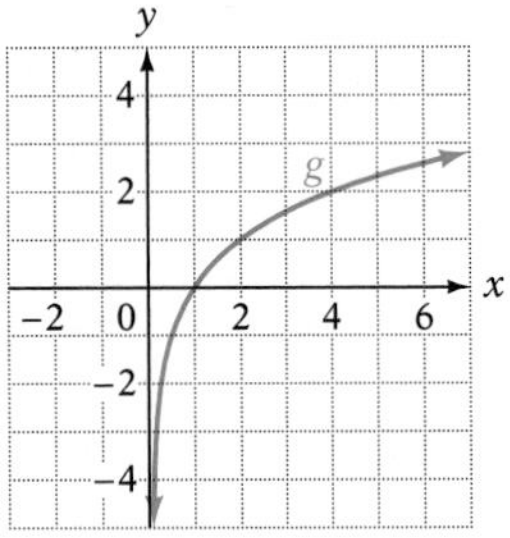

Figure 5

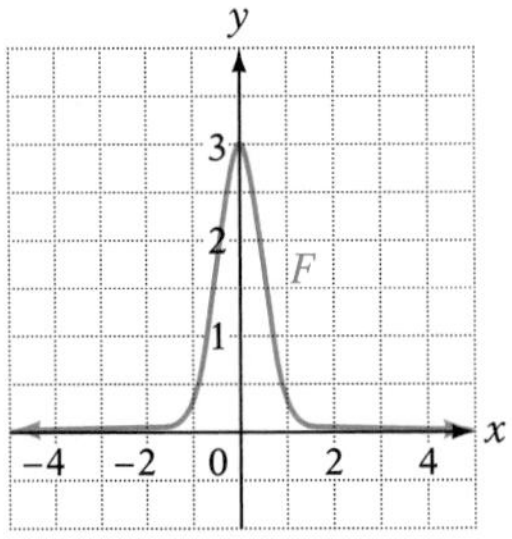

Figure 6

29. Given the graph of $y = f(x)$ shown in Figure 4 above, graph $y = f(x) + 1$.

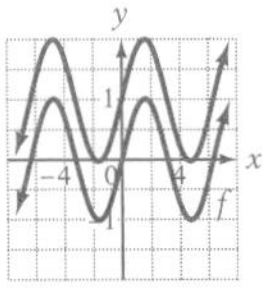

30. Given the graph of $y = f(x)$ shown in Figure 4 above, graph $y = f(x) - 2$.

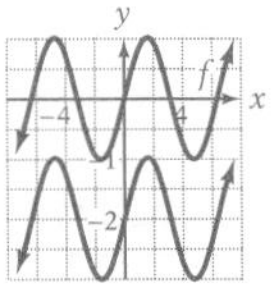

31. Given the graph of $y = g(x)$ shown in Figure 5 above, graph $y = g(x) - 3$.

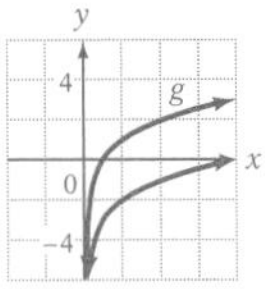

32. Given the graph of $y = g(x)$ shown in Figure 5 above, graph $y = g(x) + 1$.

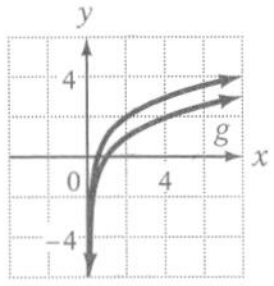

33. Given the graph of $y = f(x)$ shown in Figure 4 above, graph $y = f(x - 2)$.

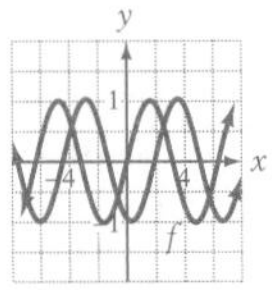

34. Given the graph of $y = f(x)$ shown in Figure 4 above, graph $y = f(x + 1)$.

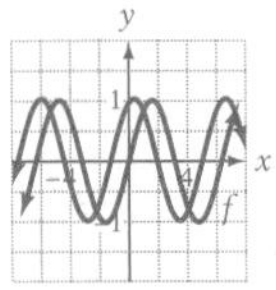

35. Given the graph of $y = F(x)$ shown in Figure 6 above, graph $y = F(x + 3)$.

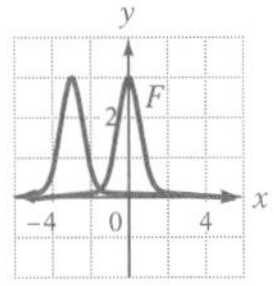

36. Given the graph of $y = F(x)$ shown in Figure 6 above, graph $y = F(x - 1)$.

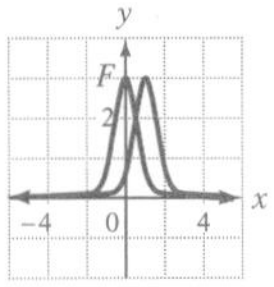

37. Given the triangle shown below, draw the triangle for which every value of x is translated 3 units to the right and every value of y is translated 2 units up.

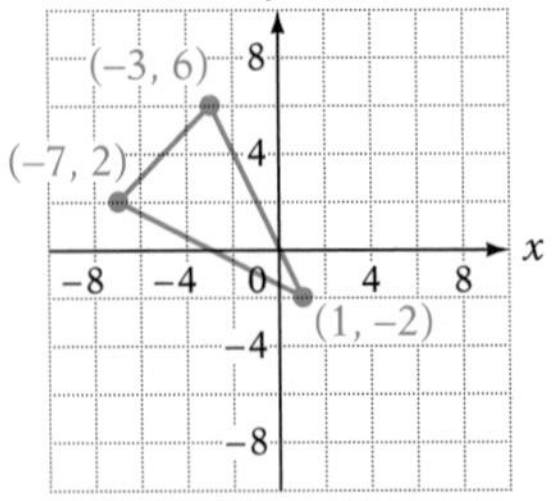

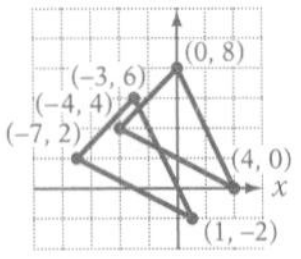

38. Given the rectangle shown below, draw the rectangle for which every value of x is translated 2 units to the right and every value of y is translated 1 unit up.

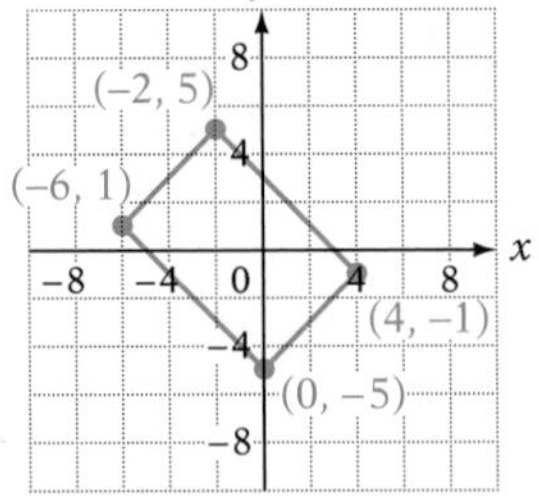

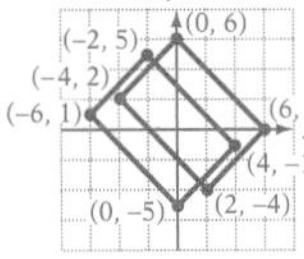

SECTION 9.3 Algebra of Functions

1 Perform operations on functions

36 HM³

The operations of addition, subtraction, multiplication, and division of functions are defined as follows.

Operations on Functions

If f and g are functions and x is an element of the domain of each function, then

$$(f+g)(x) = f(x) + g(x)$$
$$(f-g)(x) = f(x) - g(x)$$
$$(f \cdot g)(x) = f(x) \cdot g(x)$$
$$\left(\frac{f}{g}\right)(x) = \frac{f(x)}{g(x)},\ g(x) \neq 0$$

INSTRUCTOR NOTE

Definitions are usually difficult for students. After you have given some examples of operations on functions, you might point out the requirement in the definition that x must be an element of the domain of each function. For instance, $(f+g)(1)$ is not a real number if $f(x) = x^2$ and $g(x) = \sqrt{x-2}$.

Given $f(x) = x^2 + 1$ and $g(x) = 3x - 2$, find $(f+g)(3)$ and $(f \cdot g)(-1)$.

$$\begin{aligned}(f+g)(3) &= f(3) + g(3)\\ &= [(3)^2 + 1] + [3(3) - 2]\\ &= 10 + 7 = 17\end{aligned}$$

$$\begin{aligned}(f \cdot g)(-1) &= f(-1) \cdot g(-1)\\ &= [(-1)^2 + 1] \cdot [3(-1) - 2]\\ &= 2 \cdot (-5) = -10\end{aligned}$$

Consider the functions f and g above. Find $S(x)$, the sum of the two functions.

Use the definition of addition. $\quad S(x) = (f+g)(x) = f(x) + g(x)$

Substitute $x^2 + 1$ for $f(x)$ and $3x - 2$ for $g(x)$. $\quad = (x^2 + 1) + (3x - 2)$

$S(x) = x^2 + 3x - 1$

Now evaluate $S(3)$.

$$\begin{aligned}S(3) &= (3)^2 + 3(3) - 1\\ &= 9 + 9 - 1\\ &= 17 = (f+g)(3)\end{aligned}$$

Note that $S(3) = 17$ and $(f+g)(3) = 17$. This shows that adding $f(x) + g(x)$ and then evaluating is the same as evaluating $f(x)$ and $g(x)$ and then adding. The same is true for the other operations on functions. For instance, let $P(x)$ be the product of the functions f and g. Then

$$\begin{aligned}P(x) &= (f \cdot g)(x)\\ &= f(x) \cdot g(x)\\ &= (x^2 + 1)(3x - 2)\\ &= 3x^3 - 2x^2 + 3x - 2\end{aligned}$$

$$\begin{aligned}P(-1) &= 3(-1)^3 - 2(-1)^2 + 3(-1) - 2\\ &= -3 - 2 - 3 - 2\\ &= -10 = (f \cdot g)(-1)\end{aligned}$$

Given $f(x) = 2x^2 - 5x + 3$ and $g(x) = x^2 - 1$, find $\left(\frac{f}{g}\right)(1)$.

$$\left(\frac{f}{g}\right)(1) = \frac{f(1)}{g(1)}$$
$$= \frac{2(1)^2 - 5(1) + 3}{(1)^2 - 1}$$
$$= \frac{0}{0} \leftarrow \text{This is not a real number.}$$

Because $\frac{0}{0}$ is not defined, the expression $\left(\frac{f}{g}\right)(1)$ cannot be evaluated.

Example 1 Given $f(x) = x^2 - x + 1$ and $g(x) = x^3 - 4$, find $(f - g)(3)$.

Solution
$$(f - g)(3) = f(3) - g(3)$$
$$= (3^2 - 3 + 1) - (3^3 - 4)$$
$$= 7 - 23$$
$$= -16$$

$(f - g)(3) = -16$

Problem 1 Given $f(x) = x^2 + 2x$ and $g(x) = 5x - 2$, find $(f + g)(-2)$.

Solution See page S26. -12

Example 2 Given $f(x) = x^2 + 2$ and $g(x) = 2x + 3$, find $(f \cdot g)(-2)$.

Solution
$$(f \cdot g)(-2) = f(-2) \cdot g(-2)$$
$$= [(-2)^2 + 2] \cdot [2(-2) + 3]$$
$$= 6 \cdot [-1]$$
$$= -6$$

$(f \cdot g)(-2) = -6$

Problem 2 Given $f(x) = 4 - x^2$ and $g(x) = 3x - 4$, find $(f \cdot g)(3)$.

Solution See page S26. -25

Example 3 Given $f(x) = x^2 + 4x + 4$ and $g(x) = x^3 - 2$, find $\left(\frac{f}{g}\right)(3)$.

Solution
$$\left(\frac{f}{g}\right)(3) = \frac{f(3)}{g(3)}$$
$$= \frac{3^2 + 4(3) + 4}{3^3 - 2}$$
$$= \frac{25}{25}$$
$$= 1$$

$\left(\frac{f}{g}\right)(3) = 1$

Problem 3 Given $f(x) = x^2 - 4$ and $g(x) = x^2 + 2x + 1$, find $\left(\frac{f}{g}\right)(4)$.

Solution See page S26. $\frac{12}{25}$

2 Find the composition of two functions

INSTRUCTOR NOTE
This is very difficult for many students. Using open brackets may help these students.

$$f[\;\rule{0.6cm}{0.15mm}\;] = 2[\;\rule{0.6cm}{0.15mm}\;] + 7$$
$$\downarrow \qquad\qquad \downarrow$$
$$f[\;g(-2)\;] = 2[\;g(-2)\;] + 7$$
$$f[\,g(-2)\,] = 2[5] + 7 = 17$$

A function can be evaluated at the value of another function. Consider

$$f(x) = 2x + 7 \qquad \text{and} \qquad g(x) = x^2 + 1$$

The expression $f[g(-2)]$ means to evaluate the function f at $g(-2)$.

$$g(-2) = (-2)^2 + 1 = 4 + 1 = 5$$

$$f[g(-2)] = f(5) = 2(5) + 7 = 10 + 7 = 17$$

Definition of the Composition of Two Functions

> Let f and g be two functions such that $g(x)$ is in the domain of f for all x in the domain of g. Then the **composition** of the two functions, denoted by $f \circ g$, is the function whose value at x is given by $(f \circ g)(x) = f[g(x)]$.

The function defined by $f[g(x)]$ is called the **composite** of f and g.

The requirement in the definition of the composition of two functions that $g(x)$ be in the domain of f for all x in the domain of g is important. For instance, let

$$f(x) = \frac{1}{x - 1} \qquad \text{and} \qquad g(x) = 3x - 5$$

When $x = 2$,

$$g(2) = 3(2) - 5 = 1$$

$$f[g(2)] = f(1) = \frac{1}{1 - 1} = \frac{1}{0} \quad \leftarrow \text{This is not a real number.}$$

In this case, $g(2)$ is not in the domain of f. Thus the composition is not defined at 2.

➡ Given $f(x) = x^3 - x + 1$ and $g(x) = 2x^2 - 10$, evaluate $(g \circ f)(2)$.

$$f(2) = (2)^3 - (2) + 1$$
$$= 7$$

$$(g \circ f)(2) = g[f(2)]$$
$$= g(7)$$
$$= 2(7)^2 - 10 = 88$$

$$(g \circ f)(2) = 88$$

➡ Given $f(x) = 3x - 2$ and $g(x) = x^2 - 2x$, find $(f \circ g)(x)$.

$$(f \circ g)(x) = f[g(x)] = 3(x^2 - 2x) - 2$$
$$= 3x^2 - 6x - 2$$

In general, the composition of functions is not a commutative operation. That is, $(f \circ g)(x) \neq (g \circ f)(x)$. To show this, let $f(x) = x + 3$ and $g(x) = x^2 + 1$. Then

$$(f \circ g)(x) = f[g(x)] \qquad\qquad (g \circ f)(x) = g[f(x)]$$
$$= (x^2 + 1) + 3 \qquad\qquad\quad = (x + 3)^2 + 1$$
$$= x^2 + 4 \qquad\qquad\qquad\quad = x^2 + 6x + 10$$

Thus $(f \circ g)(x) \neq (g \circ f)(x)$.

Example 4 Given $f(x) = x^2 - 1$ and $g(x) = 3x + 4$, evaluate each composite function.

A. $f[g(0)]$ B. $g[f(x)]$

Solution A. $g(x) = 3x + 4$

$g(0) = 3(0) + 4 = 4$ ▶ To evaluate $f[g(0)]$, first evaluate $g(0)$.

$f(x) = x^2 - 1$

$f[g(0)] = f(4) = 4^2 - 1 = 15$ ▶ Substitute the value of $g(0)$ for x in $f(x)$; $g(0) = 4$.

$f[g(0)] = 15$

B. $g[f(x)] = g(x^2 - 1)$ ▶ $f(x) = x^2 - 1$

$= 3(x^2 - 1) + 4$ ▶ Substitute $x^2 - 1$ for x in the function $g(x)$; $g(x) = 3x + 4$.

$= 3x^2 - 3 + 4$

$= 3x^2 + 1$

Problem 4 Given $g(x) = 3x - 2$ and $h(x) = x^2 + 1$, evaluate each composite function.

A. $g[h(0)]$ B. $h[g(x)]$

Solution See page S26. A. $g[h(0)] = 1$ B. $h[g(x)] = 9x^2 - 12x + 5$

CONCEPT REVIEW 9.3

Determine whether the following statements are always true, sometimes true, or never true.

1. $(f \circ g)(x) = f[g(x)]$ Always true

2. If $f(x) = 2x$ and $g(x) = \frac{1}{2}x$, then $(f \circ g)(x) = (g \circ f)(x)$. Always true

3. If $f(x) = 2x + 1$ and $g(x) = x + 4$, then $(f \circ g)(x) = (2x + 1)(x + 4)$. Never true

4. If $f(x) = x^2 - 4$ and $g(x) = x^3$, then $(f \circ g)(x) = (x^2 - 4)^3$. Never true

EXERCISES 9.3

1 Given $f(x) = 2x^2 - 3$ and $g(x) = -2x + 4$, find:

1. $(f - g)(2)$ 5
2. $(f - g)(3)$ 17
3. $(f + g)(0)$ 1
4. $(f + g)(1)$ 1
5. $(f \cdot g)(2)$ 0
6. $(f \cdot g)(-1)$ −6
7. $\left(\frac{f}{g}\right)(4)$ $-\frac{29}{4}$
8. $\left(\frac{f}{g}\right)(-1)$ $-\frac{1}{6}$
9. $\left(\frac{g}{f}\right)(-3)$ $\frac{2}{3}$

Given $f(x) = 2x^2 + 3x - 1$ and $g(x) = 2x - 4$, find:

10. $(f + g)(-3)$ -2
11. $(f + g)(1)$ 2
12. $(f + g)(-2)$ -7
13. $(f - g)(4)$ 39
14. $(f \cdot g)(-2)$ -8
15. $(f \cdot g)(1)$ -8
16. $\left(\frac{f}{g}\right)(2)$ Undefined
17. $\left(\frac{f}{g}\right)(-3)$ $-\frac{4}{5}$
18. $(f \cdot g)\left(\frac{1}{2}\right)$ -3

Given $f(x) = x^2 + 3x - 5$ and $g(x) = x^3 - 2x + 3$, find:

19. $(f - g)(2)$ -2
20. $(f \cdot g)(-3)$ 90
21. $\left(\frac{f}{g}\right)(-2)$ 7

2 22. Explain the meaning of the notation $f[g(2)]$.

23. What is the meaning of the notation $(f \circ g)(x)$?

Given $f(x) = 2x - 3$ and $g(x) = 4x - 1$, evaluate the composite function.

24. $f[g(0)]$ -5
25. $g[f(0)]$ -13
26. $f[g(2)]$ 11
27. $g[f(-2)]$ -29
28. $f[g(x)]$ $8x - 5$
29. $g[f(x)]$ $8x - 13$

Given $g(x) = x^2 + 3$ and $h(x) = x - 2$, evaluate the composite function.

30. $g[h(0)]$ 7
31. $h[g(0)]$ 1
32. $g[h(4)]$ 7
33. $h[g(-2)]$ 5
34. $g[h(x)]$ $x^2 - 4x + 7$
35. $h[g(x)]$ $x^2 + 1$

Given $f(x) = x^2 + x + 1$ and $h(x) = 3x + 2$, evaluate the composite function.

36. $f[h(0)]$ 7
37. $h[f(0)]$ 5
38. $f[h(-1)]$ 1
39. $h[f(-2)]$ 11
40. $f[h(x)]$ $9x^2 + 15x + 7$
41. $h[f(x)]$ $3x^2 + 3x + 5$

Given $f(x) = x - 2$ and $g(x) = x^3$, evaluate the composite function.

42. $f[g(2)]$ 6
43. $f[g(-1)]$ -3
44. $g[f(2)]$ 0
45. $g[f(-1)]$ -27
46. $f[g(x)]$ $x^3 - 2$
47. $g[f(x)]$ $x^3 - 6x^2 + 12x - 8$

APPLYING CONCEPTS 9.3

Given $g(x) = x^2 - 1$, find:

48. $g(2 + h)$ $3 + 4h + h^2$
49. $g(3 + h) - g(3)$ $6h + h^2$
50. $g(-1 + h) - g(-1)$ $-2h + h^2$
51. $\frac{g(1 + h) - g(1)}{h}$ $2 + h$
52. $\frac{g(-2 + h) - g(-2)}{h}$ $-4 + h$
53. $\frac{g(a + h) - g(a)}{h}$ $h + 2a$

Given $f(x) = 2x$, $g(x) = 3x - 1$, and $h(x) = x - 2$, find:

54. $f(g[h(2)])$ -2
55. $g(h[f(1)])$ -1
56. $h(g[f(-1)])$ -9
57. $f(h[g(0)])$ -6
58. $f(g[h(x)])$ $6x - 14$
59. $g(f[h(x)])$ $6x - 13$

The graphs of the functions f and g are shown at the right. Use these graphs to determine the values of the following composite functions.

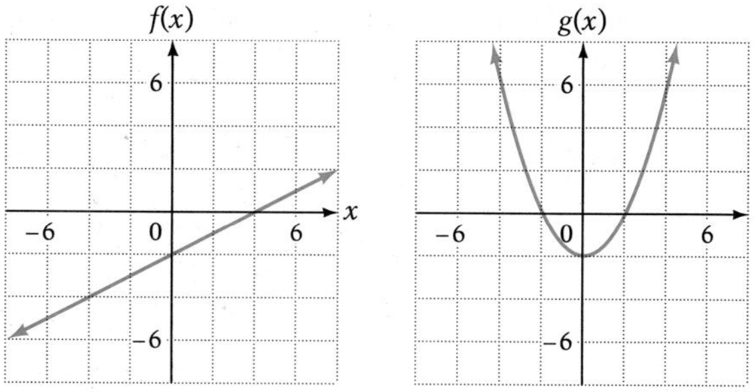

[C] **60.** $f[g(0)]$ −3

[C] **61.** $f[g(2)]$ −2

[C] **62.** $f[g(4)]$ 1

[C] **63.** $f[g(-2)]$ −2

[C] **64.** $f[g(-4)]$ 1

[C] **65.** $g[f(0)]$ 0

[C] **66.** $g[f(4)]$ −2

[C] **67.** $g[f(-4)]$ 6

SECTION 9.4

One-to-One and Inverse Functions

1 Determine whether a function is one-to-one

36 HM³

Recall that a function is a set of ordered pairs in which no two ordered pairs that have the same first component have different second components. This means that given any x there is only one y that can be paired with that x. A **one-to-one function** satisfies the additional condition that given any y, there is only one x that can be paired with the given y. One-to-one functions are commonly expressed by writing 1–1.

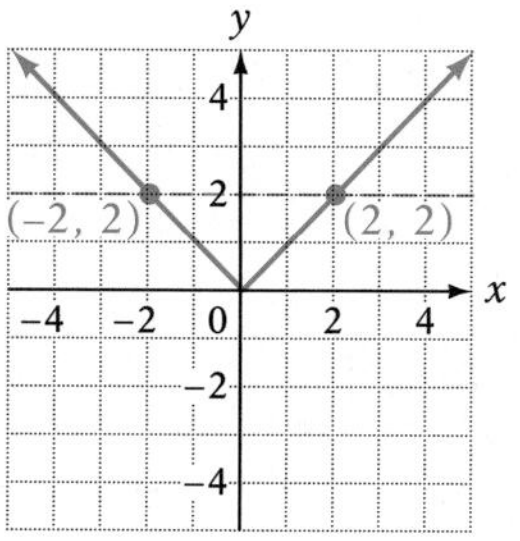

The function given by the equation $y = |x|$ is not a 1–1 function since, given $y = 2$, there are two possible values of x, 2 and −2, which can be paired with the given y-value. The graph at the left illustrates that a horizontal line intersects the graph more than once.

Just as the vertical-line test can be used to determine if a graph represents a function, a **horizontal-line test** can be used to determine if the graph of a function represents a 1–1 function.

Horizontal-Line Test

A graph of a function is the graph of a 1–1 function if any horizontal line intersects the graph at no more than one point.

The graph of a quadratic function is shown at the right. Note that a horizontal line can intersect the graph at more than one point. Therefore, this graph is not the graph of a 1–1 function. In general, $f(x) = ax^2 + bx + c$, $a \neq 0$, is not a 1–1 function.

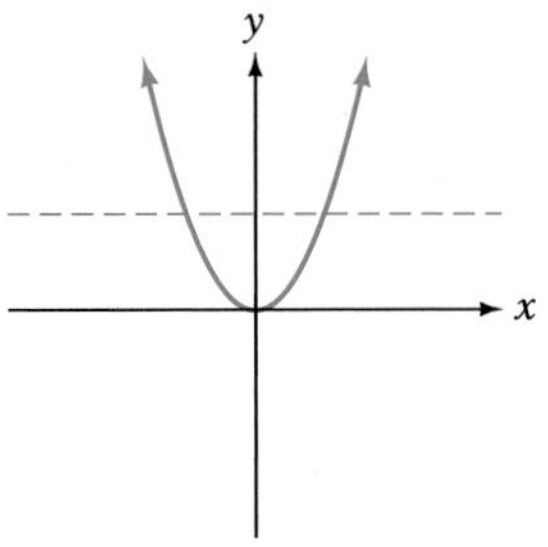

Since any vertical line will intersect the graph at the right at no more than one point, the graph is the graph of a function. Since any horizontal line will intersect the graph at no more than one point, the graph is the graph of a 1–1 function.

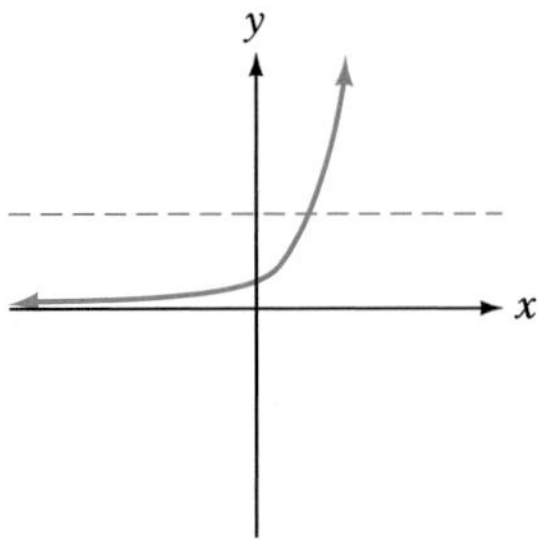

Example 1 Determine whether or not the graph represents the graph of a 1–1 function.

A.

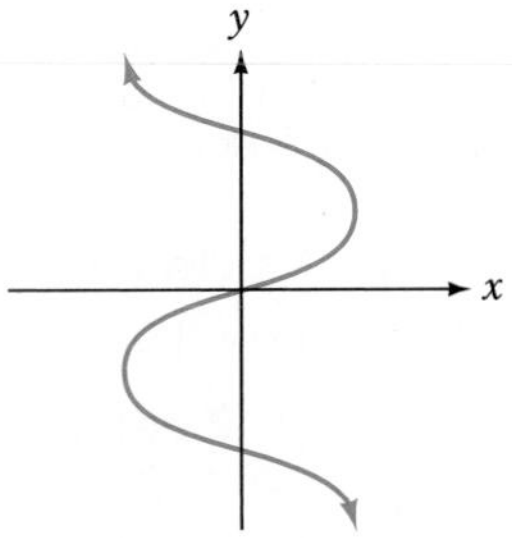

B.

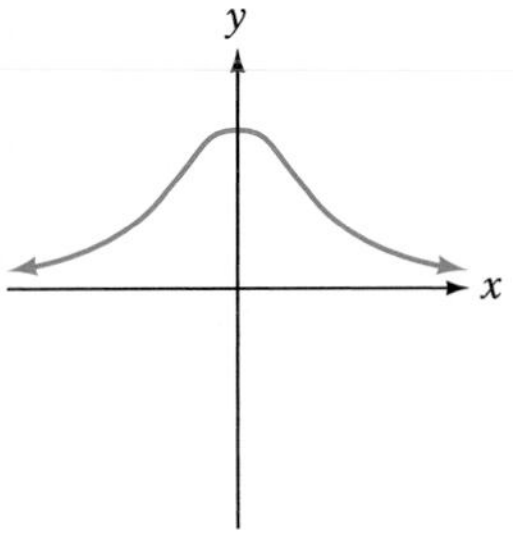

Solution A.

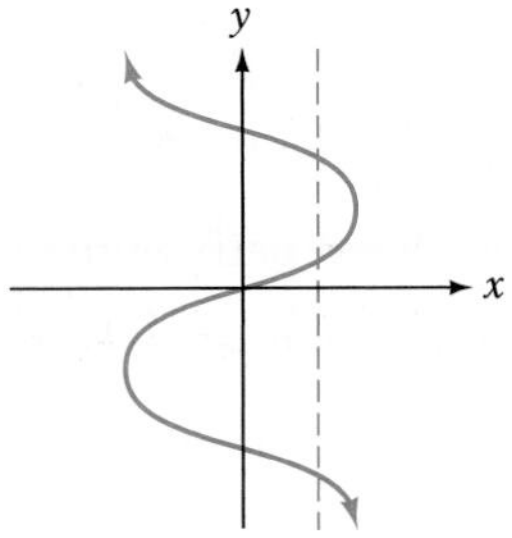

▸ A vertical line can intersect the graph at more than one point. The graph does not represent a function.

This is not the graph of a 1–1 function.

B.

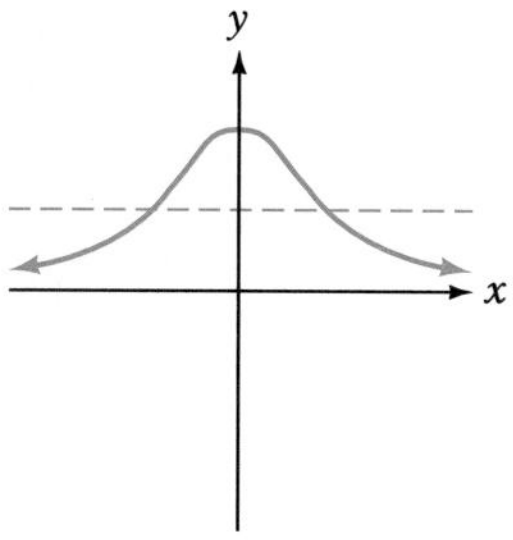

▶ A horizontal line can intersect the curve at more than one point.

This is not the graph of a 1–1 function.

Problem 1   Determine whether or not the graph represents the graph of a 1–1 function.

A.

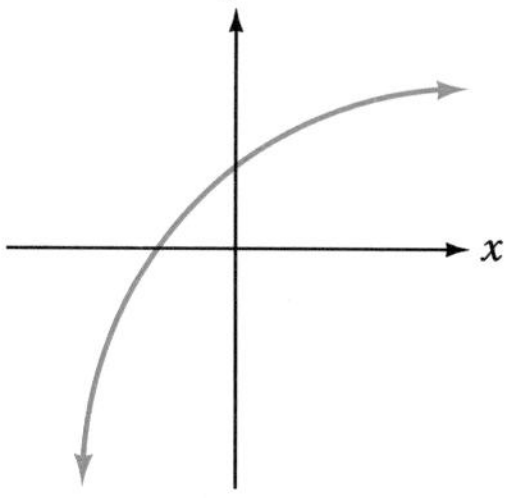

B.

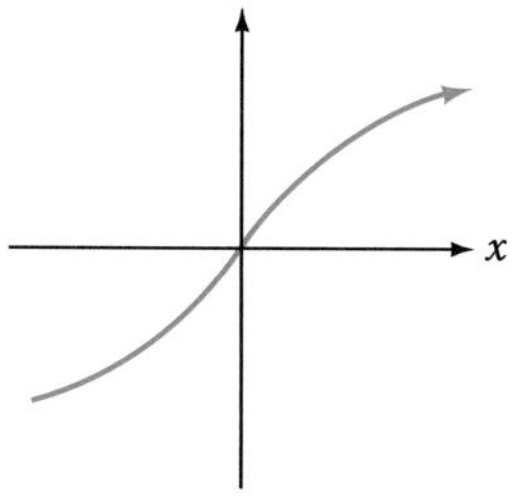

Solution   See page S27.   A. Yes   B. Yes

2   Find the inverse of a function

The **inverse of a function** is the set of ordered pairs formed by reversing the coordinates of each ordered pair of the function.

For example, the set of ordered pairs of the function defined by $f(x) = 2x$ with domain $\{-2, -1, 0, 1, 2\}$ is $\{(-2, -4), (-1, -2), (0, 0), (1, 2), (2, 4)\}$. The set of ordered pairs of the inverse function is $\{(-4, -2), (-2, -1), (0, 0), (2, 1), (4, 2)\}$.

From the ordered pairs of f, we have

$$\text{Domain} = \{-2, -1, 0, 1, 2\} \quad \text{and} \quad \text{Range} = \{-4, -2, 0, 2, 4\}$$

From the ordered pairs of the inverse function, we have

$$\text{Domain} = \{-4, -2, 0, 2, 4\} \quad \text{and} \quad \text{Range} = \{-2, -1, 0, 1, 2\}$$

Note that the domain of the inverse function is the range of the function, and the range of the inverse function is the domain of the function.

Now consider the function defined by $g(x) = x^2$ with domain $\{-2, -1, 0, 1, 2\}$. The set of ordered pairs of this function is $\{(-2, 4), (-1, 1), (0, 0), (1, 1), (2, 4)\}$. Reversing the ordered pairs gives $\{(4, -2), (1, -1), (0, 0), (1, 1), (4, 2)\}$. These ordered pairs do not satisfy the condition of a function, because there are ordered pairs with the same first coordinate and different second coordinates. This example illustrates that not all functions have an inverse function.

The graphs of $f(x) = 2x$ and $g(x) = x^2$ with the set of real numbers as the domain are shown below.

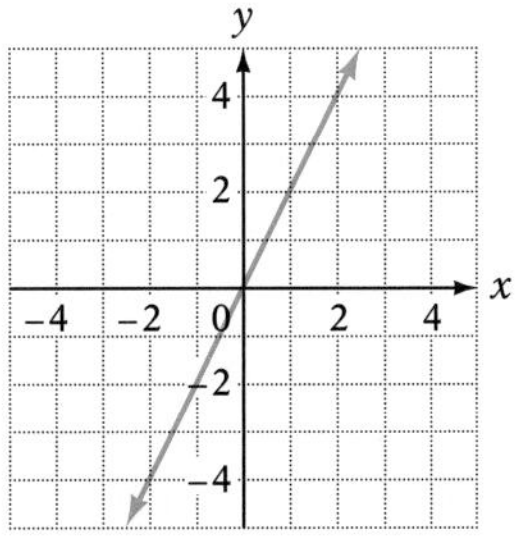

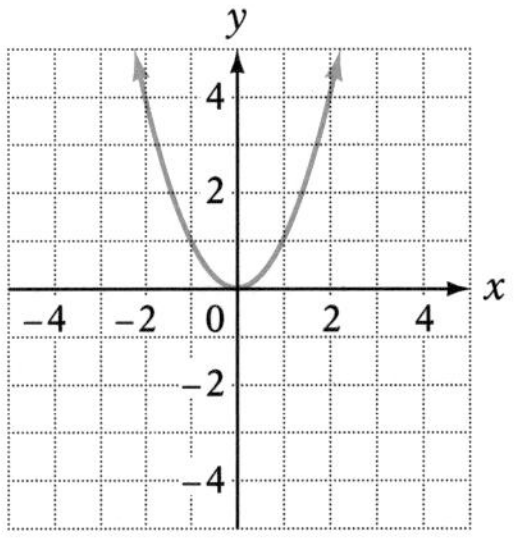

By the horizontal-line test, f is a 1–1 function but g is not.

Condition for an Inverse Function

A function f has an inverse function if and only if f is a 1–1 function.

The symbol f^{-1} is used to denote the inverse of the function f. The symbol $f^{-1}(x)$ is read "f inverse of x."

$f^{-1}(x)$ is *not* the reciprocal of $f(x)$ but is the notation for the inverse of a 1–1 function.

To find the inverse of a function, interchange x and y. Then solve for y.

⇨ Find the inverse of the function defined by $f(x) = 3x + 6$.

$$f(x) = 3x + 6$$

Replace $f(x)$ by y.	$y = 3x + 6$
Interchange x and y.	$x = 3y + 6$
Solve for y.	$x - 6 = 3y$
	$\frac{1}{3}x - 2 = y$
Replace y by $f^{-1}(x)$.	$f^{-1}(x) = \frac{1}{3}x - 2$

The inverse of the function is given by $f^{-1}(x) = \frac{1}{3}x - 2$.

The fact that the ordered pairs of the inverse of a function are the reverse of those of the function has a graphical interpretation. The function graphed at the left on the next page includes the points $(-2, 0)$, $(-1, 2)$, $(1, 4)$, and $(5, 6)$. In the graph in the middle, the points with these coordinates reversed are plotted. The inverse function is graphed by drawing a smooth curve through those points, as shown in the figure on the right.

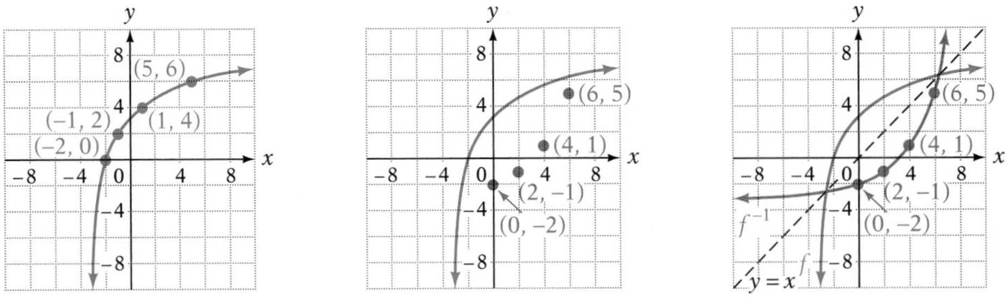

Note the dashed graph of $y = x$ is shown in the figure on the right above. If two functions are inverses of each other, their graphs are mirror images with respect to the graph of the line $y = x$.

The composition of a function and its inverse have a special property.

Composition of Inverse Functions Property

$$f^{-1}[f(x)] = x \quad \text{and} \quad f[f^{-1}(x)] = x$$

This property can be used to determine whether two functions are inverses of each other.

➡ Are $f(x) = 2x - 4$ and $g(x) = \frac{1}{2}x + 2$ inverses of each other?

To determine whether the functions are inverses, use the Composition of Inverse Functions Property.

$$f[g(x)] = 2\left(\frac{1}{2}x + 2\right) - 4 \qquad g[f(x)] = \frac{1}{2}(2x - 4) + 2$$
$$= x + 4 - 4 \qquad\qquad\qquad = x - 2 + 2$$
$$= x \qquad\qquad\qquad\qquad = x$$

Because $f[g(x)] = x$ and $g[f(x)] = x$, the functions are inverses of each other. ◀

Example 2   Find the inverse of the function defined by the equation $f(x) = 2x - 4$.

Solution $f(x) = 2x - 4$

$\quad\quad\; y = 2x - 4$ ▸ Think of the function as the equation $y = 2x - 4$.

$\quad\quad\; x = 2y - 4$ ▸ Interchange x and y.

$\quad\; 2y = x + 4$ ▸ Solve for y.

$\quad\quad\; y = \dfrac{1}{2}x + 2$

$\quad f^{-1}(x) = \dfrac{1}{2}x + 2$

Problem 2   Find the inverse of the function defined by the equation $f(x) = 4x + 2$.

Solution   See page S27. $f^{-1}(x) = \dfrac{1}{4}x - \dfrac{1}{2}$

Example 3 Are the functions defined by the equations $f(x) = -2x + 3$ and $g(x) = -\frac{1}{2}x + \frac{3}{2}$ inverses of each other?

Solution

$$f[g(x)] = f\left(-\frac{1}{2}x + \frac{3}{2}\right)$$
$$= -2\left(-\frac{1}{2}x + \frac{3}{2}\right) + 3$$
$$= x - 3 + 3$$
$$= x$$

▸ Use the property that for inverses, $f[f^{-1}(x)] = f^{-1}[f(x)] = x$.

▸ $f[g(x)] = x$

$$g[f(x)] = g(-2x + 3)$$
$$= -\frac{1}{2}(-2x + 3) + \frac{3}{2}$$
$$= x - \frac{3}{2} + \frac{3}{2}$$
$$= x$$

▸ $g[f(x)] = x$

The functions are inverses of each other.

Problem 3 Are the functions defined by the equations $h(x) = 4x + 2$ and $g(x) = \frac{1}{4}x - \frac{1}{2}$ inverses of each other?

Solution See page S27. Yes

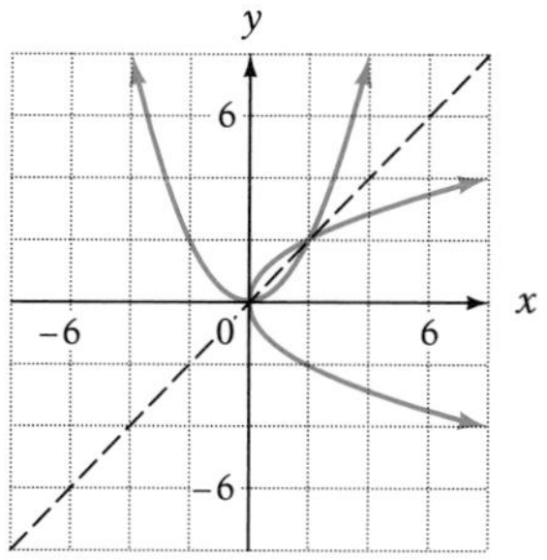

The function given by the equation $f(x) = \frac{1}{2}x^2$ does not have an inverse that is a function. Two of the ordered-pair solutions of this function are $(4, 8)$ and $(-4, 8)$.

The graph of $f(x) = \frac{1}{2}x^2$ is shown at the left. This graph does not pass the horizontal-line test for the graph of a 1–1 function. The mirror image of the graph with respect to the line $y = x$ is also shown. This graph does not pass the vertical-line test for the graph of a function.

A quadratic function with domain the real numbers does not have an inverse function.

CONCEPT REVIEW 9.4

Determine whether the following statements are always true, sometimes true, or never true.

1. The graph of a function represents the graph of a 1–1 function if any horizontal line intersects the graph at no more than one point. Always true

2. A function has an inverse if and only if it is a 1–1 function. Always true

3. The inverse of the function $\{(2, 3), (4, 5), (6, 3)\}$ is the function $\{(3, 2), (5, 4), (3, 6)\}$. Never true

4. The inverse of a function is a relation. Always true

5. The inverse of a function is a function. Sometimes true

6. The graph of $y = x^2$ is the graph of a 1–1 function. Never true

EXERCISES 9.4

1 **1.** What is a 1–1 function?

2. What is the horizontal-line test?

Determine whether the graph represents the graph of a 1–1 function.

3.
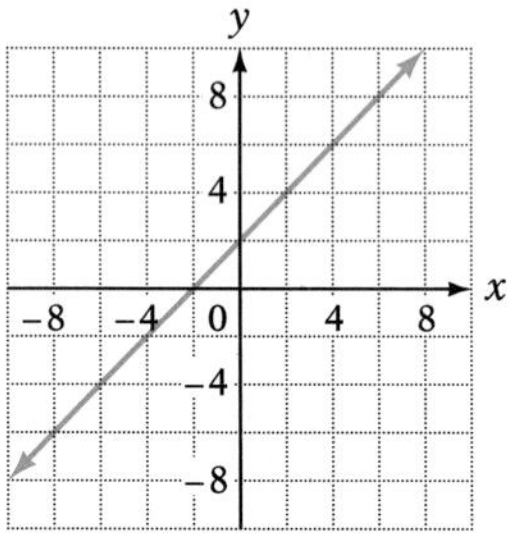
Yes

4.
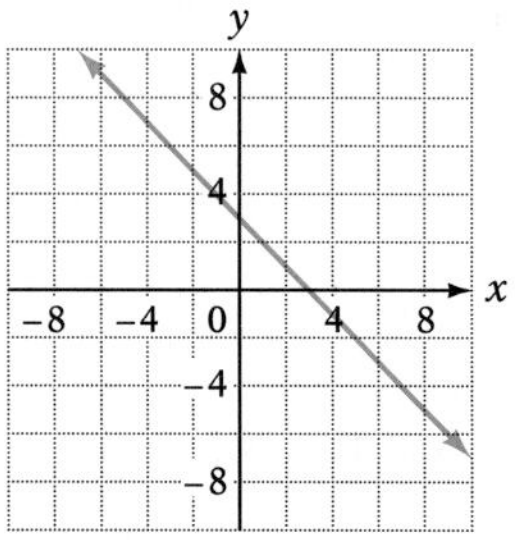
Yes

5.
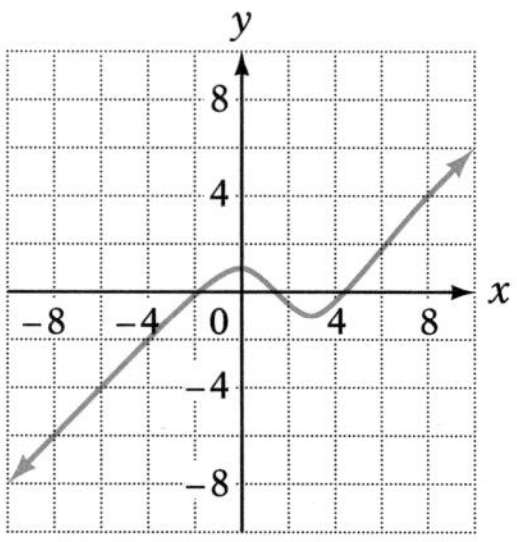
No

6.
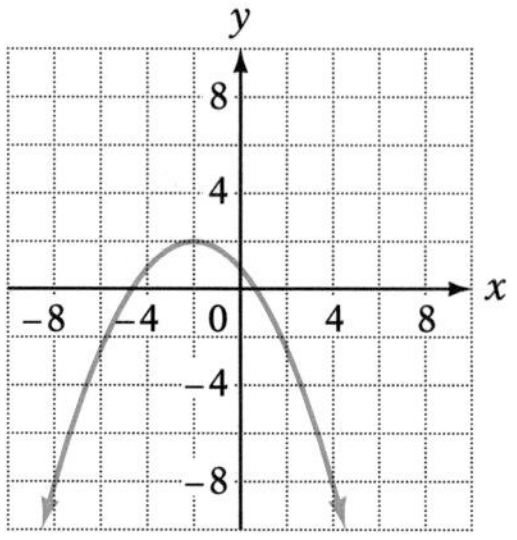
No

7.
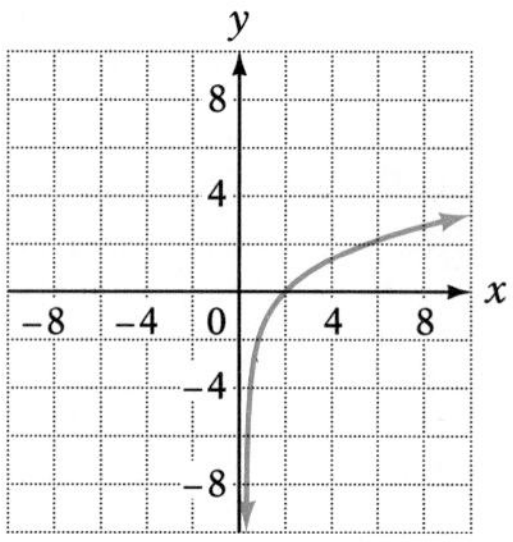
Yes

8.
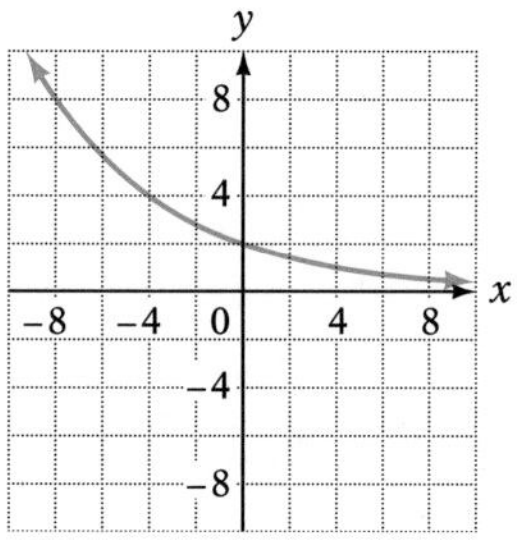
Yes

9.
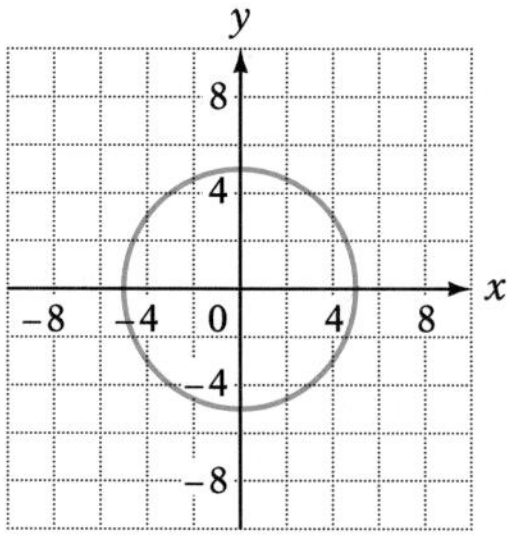
No

10.
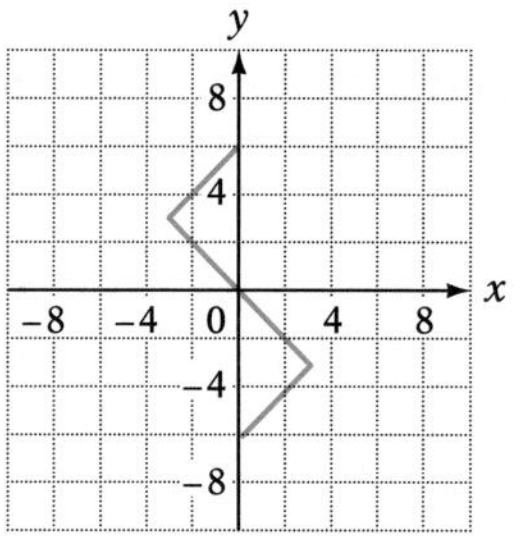
No

11.
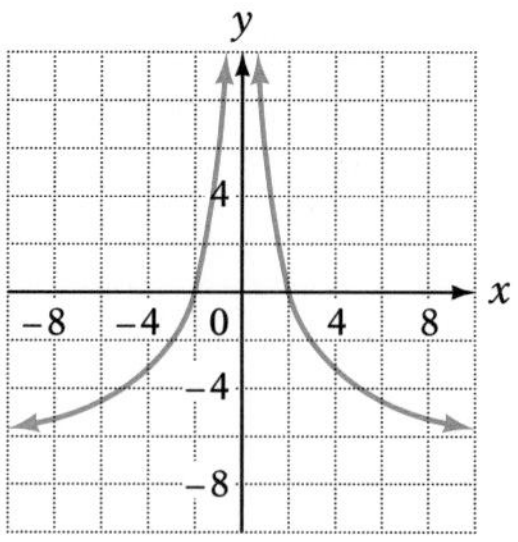
No

12.
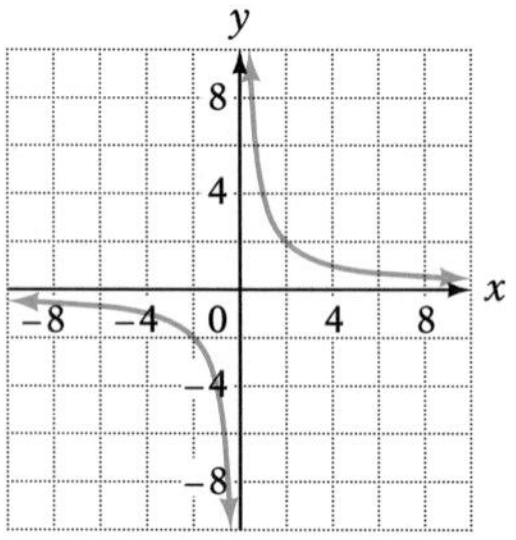
Yes

13.
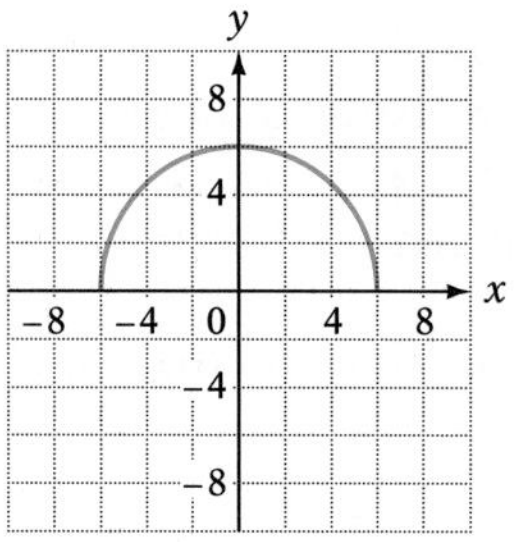
No

14.
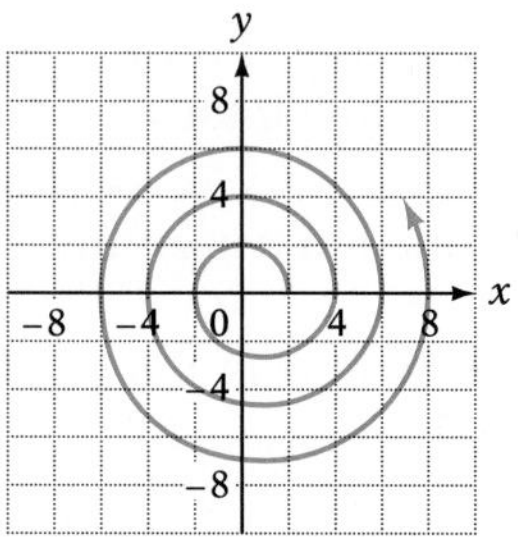
No

2 15. How are the ordered pairs of the inverse of a function related to the function?

16. If f and f^{-1} are inverse functions of one another, is it possible to determine the value of $f[f^{-1}(4)]$? If so, what is its value?

Find the inverse of the function. If the function does not have an inverse, write "No inverse."

17. $\{(1, 0), (2, 3), (3, 8), (4, 15)\}$ $\{(0, 1), (3, 2), (8, 3), (15, 4)\}$

18. $\{(1, 0), (2, 1), (-1, 0), (-2, 1)\}$ No inverse

19. $\{(3, 5), (-3, -5), (2, 5), (-2, -5)\}$ No inverse

20. $\{(-5, -5), (-3, -1), (-1, 3), (1, 7)\}$ $\{(-5, -5), (-1, -3), (3, -1), (7, 1)\}$

21. $f(x) = 4x - 8$ $f^{-1}(x) = \frac{1}{4}x + 2$

22. $f(x) = 3x + 6$ $f^{-1}(x) = \frac{1}{3}x - 2$

23. $f(x) = x^2 - 1$ No inverse

24. $f(x) = 2x + 4$ $f^{-1}(x) = \frac{1}{2}x - 2$

25. $f(x) = x - 5$ $f^{-1}(x) = x + 5$

26. $f(x) = \frac{1}{2}x - 1$ $f^{-1}(x) = 2x + 2$

27. $f(x) = \frac{1}{3}x + 2$ $f^{-1}(x) = 3x - 6$

28. $f(x) = -2x + 2$ $f^{-1}(x) = -\frac{1}{2}x + 1$

29. $f(x) = -3x - 9$ $f^{-1}(x) = -\frac{1}{3}x - 3$

30. $f(x) = 2x^2 + 2$ No inverse

31. $f(x) = \frac{2}{3}x + 4$ $f^{-1}(x) = \frac{3}{2}x - 6$

32. $f(x) = \frac{3}{4}x - 4$ $f^{-1}(x) = \frac{4}{3}x + \frac{16}{3}$

33. $f(x) = -\frac{1}{3}x + 1$ $f^{-1}(x) = -3x + 3$

34. $f(x) = -\frac{1}{2}x + 2$ $f^{-1}(x) = -2x + 4$

35. $f(x) = 2x - 5$ $f^{-1}(x) = \frac{1}{2}x + \frac{5}{2}$

36. $f(x) = 3x + 4$ $f^{-1}(x) = \frac{1}{3}x - \frac{4}{3}$

37. $f(x) = x^2 + 3$ No inverse

38. $f(x) = 5x - 2$ $f^{-1}(x) = \frac{1}{5}x + \frac{2}{5}$

Are the functions inverses of each other?

39. $f(x) = 4x; g(x) = \frac{x}{4}$ Yes

40. $g(x) = x + 5; h(x) = x - 5$ Yes

41. $f(x) = 3x, h(x) = \frac{1}{3x}$ No

42. $h(x) = x + 2; g(x) = 2 - x$ No

43. $g(x) = 3x + 2; f(x) = \frac{1}{3}x - \frac{2}{3}$ Yes

44. $h(x) = 4x - 1; f(x) = \frac{1}{4}x + \frac{1}{4}$ Yes

45. $f(x) = \frac{1}{2}x - \frac{3}{2}; g(x) = 2x + 3$ Yes

46. $g(x) = -\frac{1}{2}x - \frac{1}{2}; h(x) = -2x + 1$ No

Complete.

47. The domain of the inverse function f^{-1} is the ___range___ of f.

48. The range of the inverse function f^{-1} is the ___domain___ of f.

49. For any function f and its inverse f^{-1}, $f[f^{-1}(3)] =$ ___3___.

50. For any function f and its inverse f^{-1}, $f^{-1}[f(-4)] =$ ___−4___.

APPLYING CONCEPTS 9.4

If f is a 1–1 function and $f(0) = 5$, $f(1) = 7$, and $f(2) = 9$, find:

51. $f^{-1}(5)$ 0

52. $f^{-1}(7)$ 1

53. $f^{-1}(9)$ 2

Given $f(x) = 3x - 5$, find:

54. $f^{-1}(0)$ $\frac{5}{3}$

55. $f^{-1}(2)$ $\frac{7}{3}$

56. $f^{-1}(4)$ 3

Given the graph of the 1–1 function, draw the graph of the inverse of the function by using the technique shown in this section.

57.

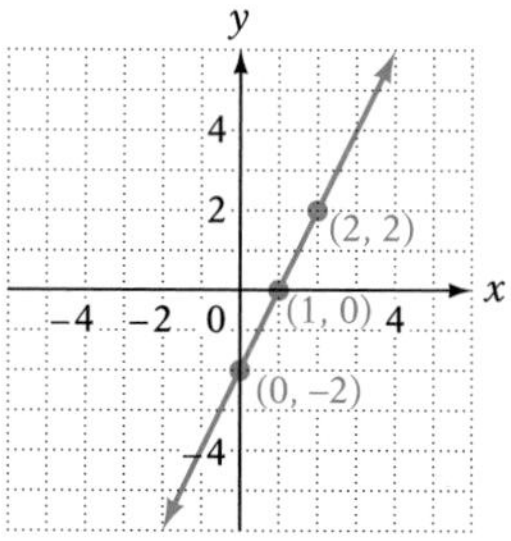

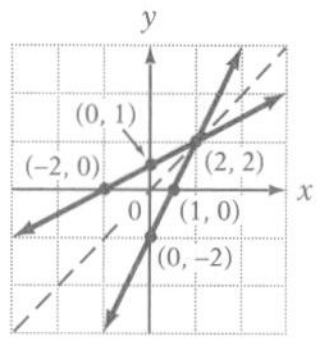

58.

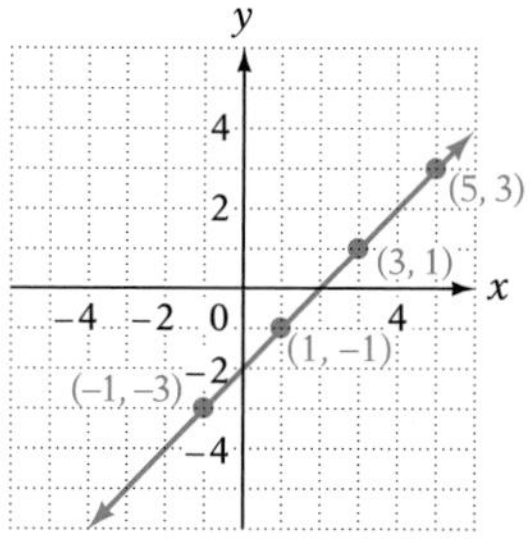

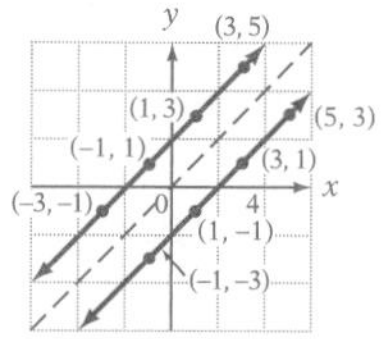

59.

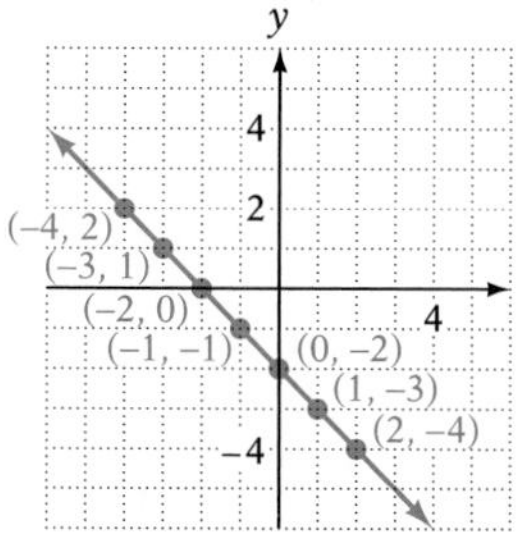

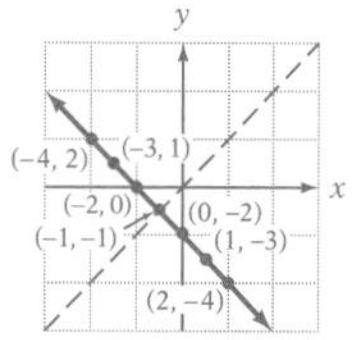

The function and its inverse have the same graph.

60.

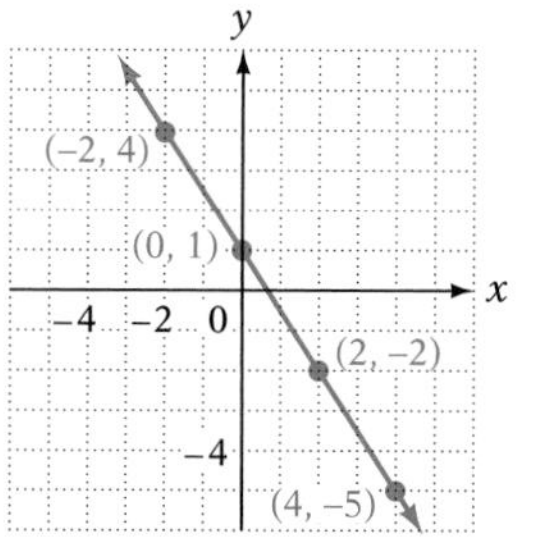

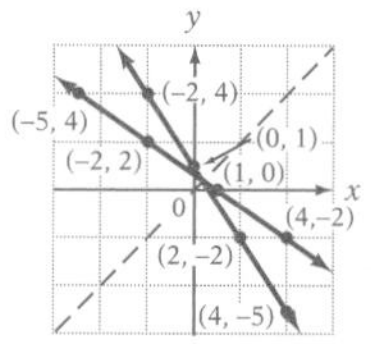

61.

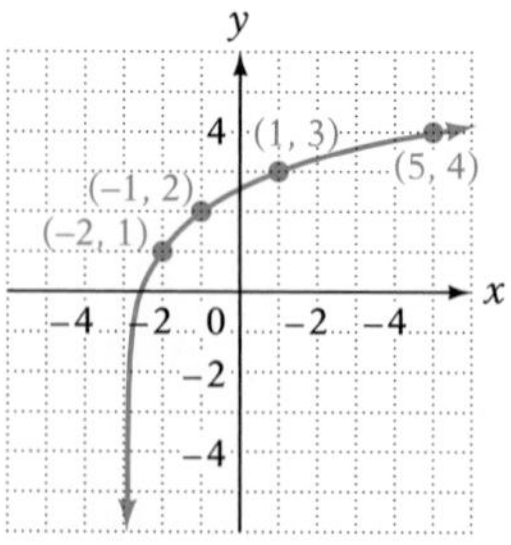

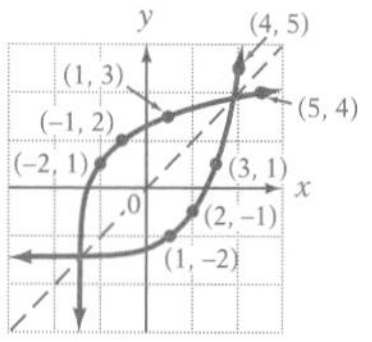

62.

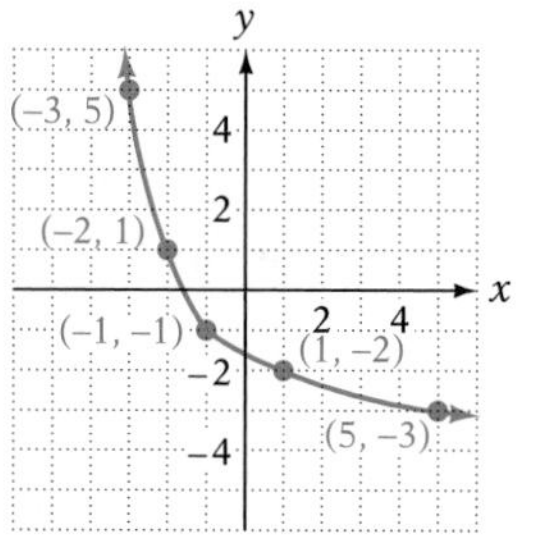

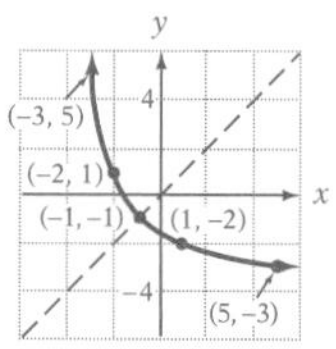

The function and its inverse have the same graph.

Each of the following tables defines a function. Is the inverse of the function a function? Explain your answer.

63. Grading Scale Table No

Score	Grade
90–100	A
80–89	B
70–79	C
60–69	D
0–59	F

64. First-Class Postage No

Weight (in oz)	Cost
$0 < w \leq 1$	\$.33
$1 < w \leq 2$	\$.55
$2 < w \leq 3$	\$.77
$3 < w \leq 4$	\$.99

65. Is the inverse of a constant function a function? Explain your answer.

66. The graphs of all functions given by $f(x) = mx + b$, $m \neq 0$, are straight lines. Are all of these functions 1–1 functions? If so, explain why. If not, given an example of a linear function that is not 1–1.

Focus on Problem Solving

Proof in Mathematics

Mathematics is based on the concept of proof, which in turn is based on principles of logic. It was Aristotle (384 B.C.–322 B.C.) who took the logical principles that had developed over time and set them in a structured form. From this structured form, several methods of proof were developed.

A **deductive proof** is one that proceeds in a carefully ordered sequence of steps. It is the type of proof you may have seen in a geometry class. Deductive proofs have a model as shown at the left.

implies implies
A true → B true → C true

implies
A true → C true

Here is an example of a deductive proof. It proves that the product of a negative and a positive number is a negative number. The proof is based on the fact that every real number has exactly one additive inverse. For instance, the additive inverse of 4 is -4 and the additive inverse of $-\frac{2}{3}$ is $\frac{2}{3}$.

Theorem: If a and b are positive real numbers, then $(-a)b = -(ab)$.

Proof: $(-a)b + ab = (-a + a)b$ ▸ Distributive Property
$= 0 \cdot b$ ▸ Inverse Property of Addition
$= 0$ ▸ Multiplication Property of Zero

Since $(-a)b + ab = 0$, $(-a)b$ is the additive inverse of ab. But $-(ab)$ is also the additive inverse of ab and because every number has exactly one additive inverse, $(-a)b = -(ab)$.

Here is another example of a deductive proof. It is a proof of part of the Pythagorean Theorem.

Theorem: If a and b are the lengths of the legs of a right triangle and c is the length of the hypotenuse, then $a^2 + b^2 = c^2$.

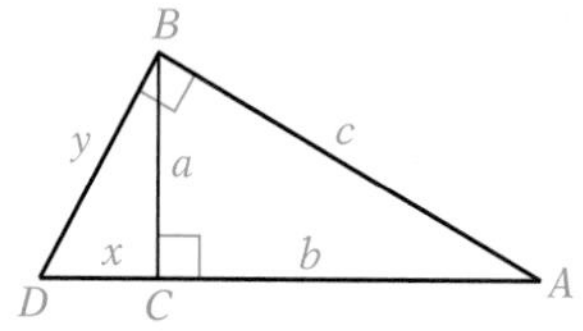

Proof: Let $\triangle DBA$ be a right triangle and let BC be the altitude to the hypotenuse of $\triangle DBA$, as shown at the left. Then $\triangle BCD$ and $\triangle BCA$ are right triangles.

1. $\triangle DBA$ is a right triangle with height BC.	Given.
2. $\triangle DBC$ is similar to $\triangle BCA$ and $\triangle DBA$ is similar to $\triangle BCA$.	Theorem from geometry: the altitude to the hypotenuse of a right triangle forms triangles that are similar to each other.
3. $\frac{a}{b} = \frac{x}{a}$	The ratios of corresponding sides of similar triangles are equal.
4. $a^2 = bx$	Multiply each side by ab.
5. $\frac{x + b}{c} = \frac{c}{b}$	The ratios of corresponding sides of similar triangles are equal.
6. $bx + b^2 = c^2$	Multiply each side by cb.
7. $a^2 + b^2 = c^2$	Substitution Property from (**4**).

Because the first statement of the proof is true and each subsequent statement of the proof is true, the first statement means the last statement is true. Therefore we have proved: If a and b are the lengths of the legs of a right triangle and c is the length of the hypotenuse, then $a^2 + b^2 = c^2$.

Projects and Group Activities

Derivative Investments

During one year, Orange County, California, lost approximately \$2 billion in its investment portfolio as a result of making investments in *derivatives*. There are many different types of derivative investments. The one discussed in this project deals with interest rates, which we will assume are simple interest rates.

We begin by discussing the traditional (nonderivative) method of investing. Suppose you invest \$5000 in an account that earns a 6% annual simple interest rate. The interest you will earn the first year is

$$I = Prt$$
$$I = (5000)(0.06)(1)$$
$$I = 300$$

If the annual interest rate increases the next year to 8%, then the interest earned for the second year is

$$I = Prt$$
$$I = (5000)(0.08)(1)$$
$$I = 400$$

Thus higher interest rates yield a higher amount of earned interest. Over the 2-year period, you have earned \$300 + \$400 = \$700.

For a certain derivative investment, the interest rate for the investment is calculated by multiplying the interest rate on a certain day—say, January 1, 1999—by 2 and then subtracting the current interest rate. Suppose that you invest $5000 in this investment on January 1, 1999 and the interest rate that day is 6%. Because the current interest rate is also 6%, the interest rate you will receive is

Interest rate on January 1, 1999 → 2(6%); Current interest rate → 6%

$$2(6\%) - 6\% = 12\% - 6\% = 6\%$$

This is the same interest rate that you would receive if the traditional method were used. The interest you will earn the first year is

$$I = Prt$$
$$I = (5000)(0.06)(1)$$
$$I = 300$$

This is the same amount of interest that you would earn with the traditional method.

Now, however, suppose that next year the current interest rate increases to 8%. The interest rate you receive is

Interest rate on January 1, 1999 → 2(6%); Current interest rate → 8%

$$2(6\%) - 8\% = 12\% - 8\% = 4\%$$

The interest you will receive for the second year of the derivative investment is

$$I = Prt$$
$$I = (5000)(0.04)(1)$$
$$I = 200$$

Over the 2-year period, you have earned $300 + $200 = $500. This is $200 less than the amount you would have earned with a traditional investment.

From what we have discussed so far, it appears that the derivative investment pays less interest over the 2-year period. Consequently, it appears that it is not as good an investment as the traditional investment.

But now suppose that the current interest rate during the second year decreases to 4%. The interest you earn the second year of the traditional investment is

$$I = Prt$$
$$I = (5000)(0.04)(1)$$
$$I = 200$$

Over the 2-year period, the traditional investment earned $300 + $200 = $500.

The interest rate for the derivative investment is

Interest rate on January 1, 1999 → 2(6%); Current interest rate → 4%

$$2(6\%) - 4\% = 12\% - 4\% = 8\%$$

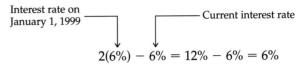

The interest earned the second year for the derivative investment is

$$I = Prt$$
$$I = (5000)(0.08)(1)$$
$$I = 400$$

Over the 2-year period, the derivative investment earned \$300 + \$400 = \$700. This is \$200 more than the traditional investment.

Note that the value of the traditional investment increases as interest rates increase and decreases as interest rates decrease. The derivative investment acts exactly opposite. That is, the value of the derivative investment increases as interest rates decrease, and its value decreases as interest rates increase. One of the problems that faced Orange County was that interest rates increased, so the value of some of its derivative investments decreased.

Suppose that the interest rate on January 1, 1999 is 5% and that on that day you have the option of purchasing a \$10,000 traditional investment or a \$10,000 derivative investment.

1. If current interest rates increase 1% each year for the next 6 years, what will be the value of the traditional investment after 6 years?
2. If current interest rates increase 1% each year for the next 6 years, what will be the value of the derivative investment after 6 years? (If interest rates are negative, use 0% as the interest rate. This is the method that is actually used for these investments.)
3. If current interest rates decrease 1% each year for the next 6 years, what will be the value of the traditional investment after 6 years? (If interest rates are negative, use 0% as the interest rate.)
4. If current interest rates decrease 1% each year for the next 6 years, what will be the value of the derivative investment after 6 years? (If interest rates are negative, use 0% as the interest rate.)
5. Visit the National Financial Services Network web site (www.nfsn.com) and find the history of the prime rate for the first day of each month over the past decade (click on Interest Rates, scroll down and click on Prime Rate). Discuss the pros and cons of derivative investments during this time period.

Properties of Polynomials

TAKE NOTE

Suggestions for graphing equations can be found in the Appendix: Guidelines for Using Graphing Calculators. For this particular activity, it may be necessary to adjust the viewing window so that you can get a reasonably accurate graph of the equation. Try Ymax = 100, Ymin = −100, and Yscl = 10.

Recall that an x-coordinate of the x-intercept of the graph of $y = f(x)$ is a real-number solution of the equation $f(x) = 0$. For instance, the graph of $P(x) = x^5 - 4x^4 - 16x^3 + 46x^2 + 63x - 90$ is shown at the right. The x-intercepts of the graph at the right are $(-3, 0)$, $(-2, 0)$, $(1, 0)$, $(3, 0)$, and $(5, 0)$, and the roots of the equation $x^5 - 4x^4 - 16x^3 + 46x^2 + 63x - 90 = 0$ are -3, -2, 1, 3, and 5.

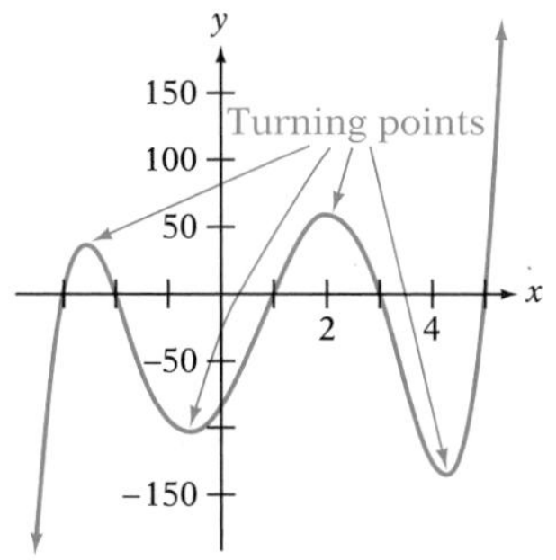

1. Graph $P(x) = x^5 - 6x^4 + 24x^2 - x - 30$ and determine the x-intercepts. Use Xmin = −4, Xmax = 6, Xscl = 1, Ymin = −300, Ymax = 300, Yscl = 100.

2. What are the real-number solutions of $x^5 - 6x^4 + 24x^2 - x - 30 = 0$?
3. Graph $P(x) = x^5 - 2x^4 + 5x^3 - 10x^2 + 4x - 8$ and determine the x-intercepts. Use Xmin $= -4$, Xmax $= 4$, Xscl $= 1$, Ymin $= -100$, Ymax $= 100$, Yscl $= 50$.
4. What are the real-number solutions of $x^5 - 2x^4 + 5x^3 - 10x^2 + 4x - 8 = 0$?
5. On the basis of the graph shown above and exercises 2 and 4, make a conjecture as to how many real-number solutions a fifth-degree equation may have.
6. Make up a few more fifth-degree equations and produce their graphs. Do the number of real-number solutions of these equations satisfy your conjecture?

The Greatest-Integer Function

The greatest-integer function is defined by the equation $f(x) = \lfloor x \rfloor$, where the symbol $\lfloor x \rfloor$ means the greatest integer less than or equal to x. Here are some examples.

$\lfloor 4.2 \rfloor = 4$ $\quad$ $\lfloor 5.9999 \rfloor = 5$ $\quad$ $\lfloor \pi \rfloor = 3$ $\quad$ $\lfloor -2.3 \rfloor = -3$ $\quad$ $\lfloor 2 \rfloor = 2$

In each of these examples, the value of the function is an integer that is less than or equal to the number in the symbol $\lfloor\ \rfloor$.

The graph of the greatest-integer function is shown at the right. A closed dot indicates that the point is part of the graph; an open dot means that the point is not part of the graph.

The greatest-integer function is sometimes called a step function because its graph resembles steps.

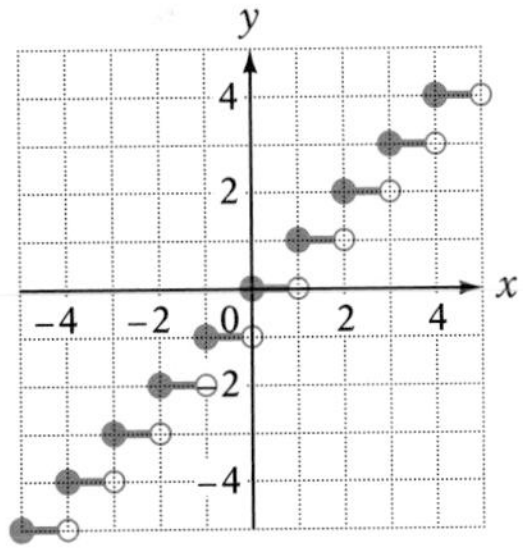

Graphing calculators have the greatest-integer function programmed into the calculator. It is usually symbolized by Int and must be accessed through one of the menus. For the TI-83, press MATH ▶ 4; for the Sharp 9600, press MATH B 5; and for the Casio CFX-9850, MENU 1 OPTN F6 F4 F2.

1. Using a viewing window of $[-4.7, 4.7]$ (use $[-6.3, 6.3]$ for the Casio CFX-9850 and the Sharp 9600) by $[-5, 5]$, graph the greatest-integer function. You may notice that the graph is drawn with jagged lines connecting the steps. These lines are *not* part of the graph. They are a result of the method by which calculators draw a graph. You can get a better idea of the graph by graphing it in *dot* mode. Use the keystrokes that follow.

To change to dot mode, use these keystrokes.

TI-83	*Sharp EL-9600*	*Casio CFX-9850*
Press MODE, use the arrow keys to select Dot, then press ENTER.	2nd F FORMAT E 2 2nd F QUIT	MENU 5 SHIFT SET UP F2

To change back to connected mode, use these keystrokes.

TI-83	*Sharp EL-9600*	*Casio CFX-9850*
Press [MODE], use the arrow keys to select Connected, then press [ENTER].	[2nd F] FORMAT E 1 [2nd F] QUIT	[MENU] 5 [SHIFT] SET UP [F1]

The greatest-integer function can be combined with other functions to produce interesting graphs. The function defined by $f(x) = |x| - \lfloor x \rfloor$ is sometimes called a *ramp* function because its graph appears as a series of ramps. The ramp function plays an important role in the study of electronics.

2. Change your calculator to *dot* mode. Using a viewing window of $[-0.1, 4.6]$ (use $[-0.1, 6.2]$ for the Casio CFX-9850 and the Sharp 9600) by $[-0.1, 1.5]$, graph the ramp function. Here are some suggested keystrokes.

TI-83	*Sharp EL-9600*	*Casio CFX-9850*
[Y=] [CLEAR] [MATH] [▶] 1 [X,T,θ,n] [)] − [MATH] [▶] 5 [X,T,θ,n] [)] GRAPH	[Y=] [CL] [MATH] B 1 [X/θ/T/n] [▶] [−] [MATH] B 5 [X/θ/T/n] [ENTER]	5 [F2] [F1] [OPTN] [F5] [F1] [X/θ/T] [−] [OPTN] [F5] [F2] [X/θ/T] [EXE] [F6]

Chapter Summary

Key Words

The *inverse of a function* is the set of ordered pairs formed by reversing the coordinates of each ordered pair of the function. (Objective 9.4.2)

Function: {(2, 4), (3, 5), (6, 7), (8, 9)}
Inverse: {(4, 2), (5, 3), (7, 6), (9, 8)}

Essential Rules and Procedures

Vertical-Line Test
A graph defines a function if any vertical line intersects the graph at no more than one point. (Objective 9.1.1)

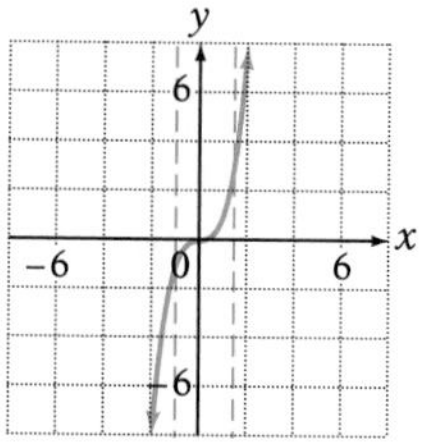

Function

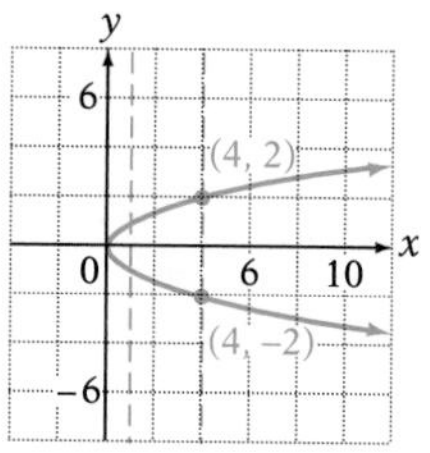

Not a function

Vertical Translation (Objective 9.2.1)
The graph of $y = f(x) + c, c > 0$, is the graph of $y = f(x)$ shifted up c units.

The graph of $y = f(x) - c, c > 0$, is the graph of $y = f(x)$ shifted down c units.

The graph $g(x) = x^2 + 1$ is the graph of $f(x) = x^2$ shifted up 1 unit.

The graph of $G(x) = x^3 - 3$ is the graph of $F(x) = x^3$ shifted down 3 units.

Horizontal Translation (Objective 9.2.1)
The graph of $y = f(x - c)$, $c > 0$, is the graph of $y = f(x)$ shifted horizontally c units to the right.

The graph of $y = f(x + c)$, $c > 0$, is the graph of $y = f(x)$ shifted horizontally c units to the left.

The graph $g(x) = (x - 2)^2$ is the graph of $f(x) = x^2$ shifted 2 units to the right.

The graph of $G(x) = (x + 4)^3$ is the graph of $F(x) = x^3$ shifted 4 units to the left.

Operations on Functions (Objective 9.3.1)
If f and g are functions and x is an element of the domain of each function, then

$(f + g)(x) = f(x) + g(x)$

$(f - g)(x) = f(x) - g(x)$

$(f \cdot g)(x) = f(x) \cdot g(x)$

$\left(\frac{f}{g}\right)(x) = \frac{f(x)}{g(x)}, \; g(x) \neq 0$

Given $f(x) = x + 2$ and $g(x) = 2x$, then

$$\begin{aligned}(f + g)(4) &= f(4) + g(4)\\ &= (4 + 2) + 2(4) = 6 + 8 = 14\end{aligned}$$

$$\begin{aligned}(f - g)(4) &= f(4) - g(4)\\ &= (4 + 2) - 2(4) = 6 - 8 = -2\end{aligned}$$

$$\begin{aligned}(f \cdot g)(4) &= f(4) \cdot g(4)\\ &= (4 + 2) \cdot (2)(4) = 6 \cdot 8 = 48\end{aligned}$$

$$\left(\frac{f}{g}\right)(4) = \frac{f(4)}{g(4)} = \frac{4 + 2}{2(4)} = \frac{6}{8} = \frac{3}{4}$$

Composition of Two Functions
$(f \circ g)(x) = f[g(x)]$ (Objective 9.3.2)

Given $f(x) = x - 4$ and $g(x) = 4x$, then

$$\begin{aligned}(f \circ g)(2) &= f[g(2)]\\ &= f(8) \qquad \text{because } g(2) = 8\\ &= 8 - 4 = 4\end{aligned}$$

Horizontal-Line Test
A function f is a 1–1 function if any horizontal line intersects the graph at no more than one point. (Objective 9.4.1)

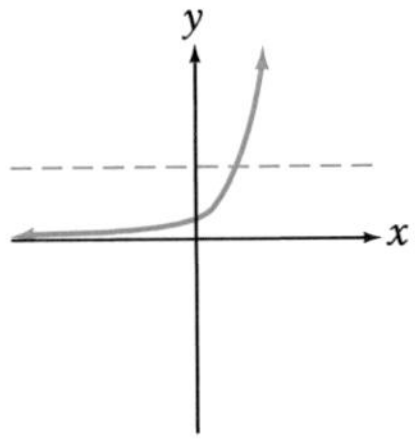

A 1–1 function

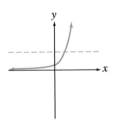

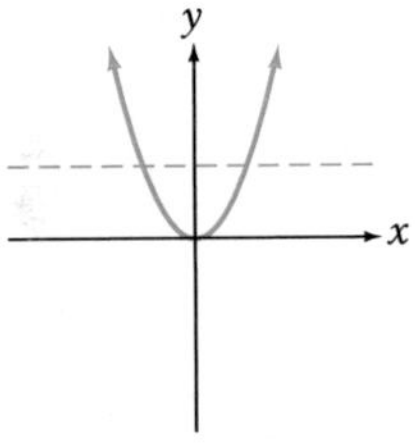

Not a 1–1 function

Condition for an Inverse Function
A function f has an inverse if and only if f is a 1–1 function. (Objective 9.4.2)

The function $f(x) = x^2$ does not have an inverse function. When $y = 4$, $x = 2$ or -2; therefore the function $f(x)$ is not a 1–1 function.

Composition of Inverse Functions Property
$f^{-1}[f(x)] = x$ and $f[f^{-1}(x)] = x$ (Objective 9.4.2)

$f(x) = 2x - 3 \qquad f^{-1}(x) = \frac{1}{2}x + \frac{3}{2}$

$$f^{-1}[f(x)] = \frac{1}{2}(2x - 3) + \frac{3}{2} = x$$

$$f[f^{-1}(x)] = 2\left(\frac{1}{2}x + \frac{3}{2}\right) - 3 = x$$

Chapter Review Exercises

1. Determine whether the graph is the graph of a function.

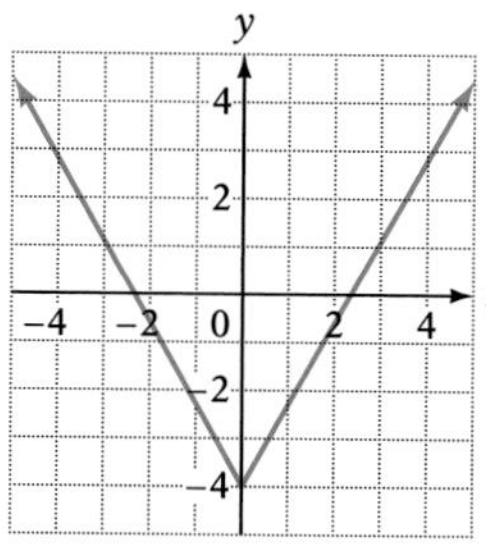

Yes

2. Graph $f(x) = |x| - 3$. Use set-builder notation to state the domain and range.

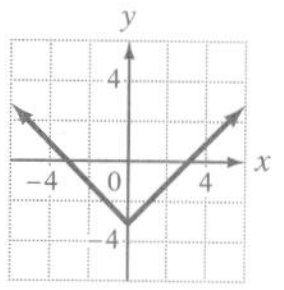

D: $\{x \mid x \in \text{real numbers}\}$
R: $\{y \mid y \geq -3\}$

3. Graph $f(x) = 3x^3 - 2$. Use set-builder notation to state the domain and range.

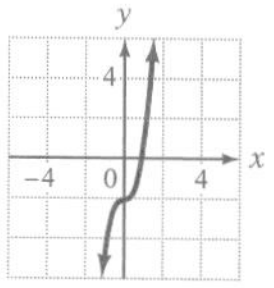

D: $\{x \mid x \in \text{real numbers}\}$
R: $\{y \mid y \in \text{real numbers}\}$

4. Graph $f(x) = \sqrt{x + 4}$. Use interval notation to state the domain and range.

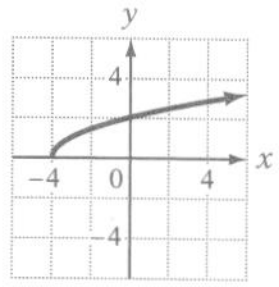

D: $[-4, \infty]$
R: $[0, \infty]$

5. Use the graph of $f(x) = x^2$ shown below to graph $g(x) = (x - 1)^2$.

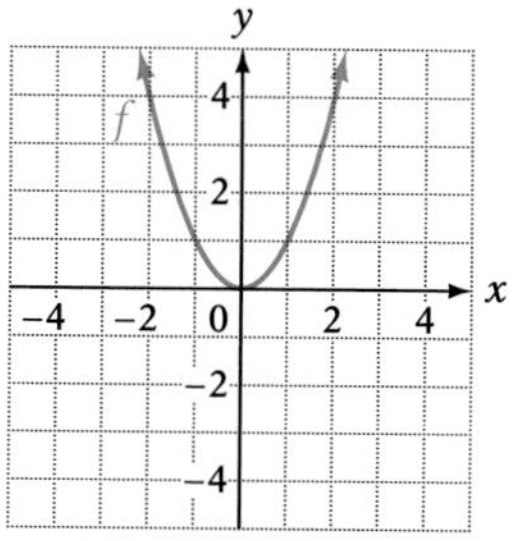

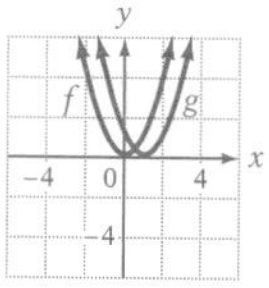

6. Use the graph of $f(x) = x^3$ shown below to graph $g(x) = x^3 + 3$.

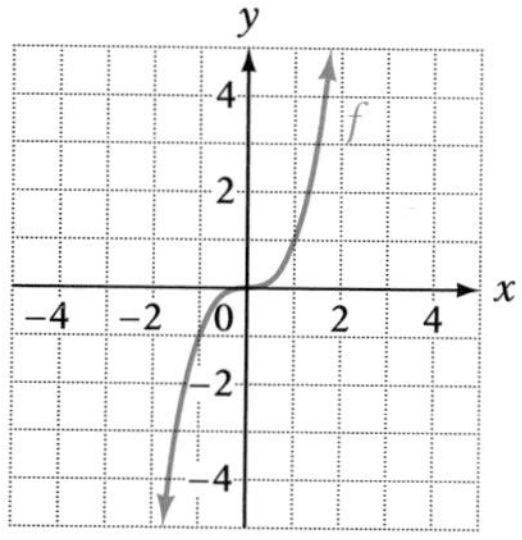

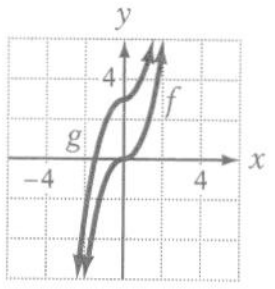

7. Use the graph of $f(x) = |x|$ shown below to graph $g(x) = |x - 1| + 2$.

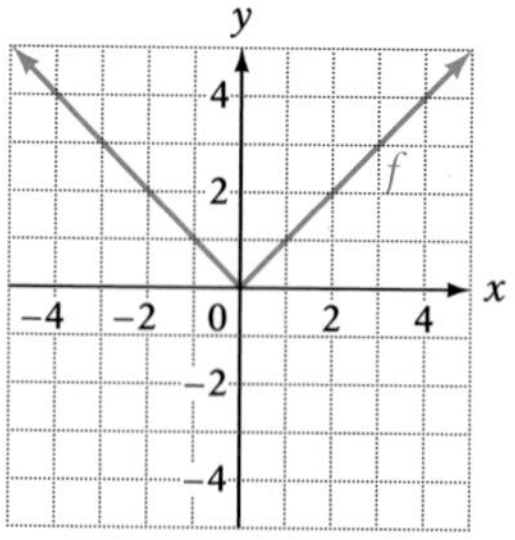

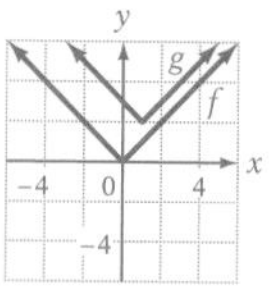

8. Use the graph of $f(x) = \sqrt{x}$ shown below to graph $g(x) = \sqrt{x + 1} - 2$.

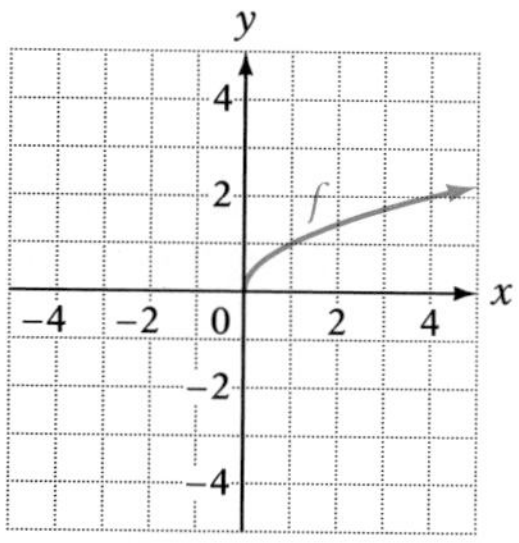

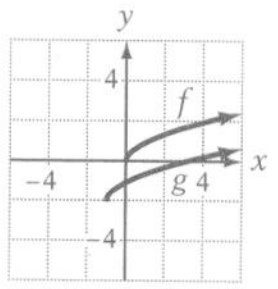

In Exercises 9–12, given $f(x) = x^2 + 2x - 3$ and $g(x) = x^2 - 2$, find:

9. $(f + g)(2)$ 7

10. $(f - g)(-4)$ −9

11. $(f \cdot g)(-4)$ 70

12. $\left(\frac{f}{g}\right)(3)$ $\frac{12}{7}$

13. Given $f(x) = 3x^2 - 4$ and $g(x) = 2x + 1$, find $f[g(x)]$. $12x^2 + 12x - 1$

14. Given $f(x) = x^2 + 4$ and $g(x) = 4x - 1$, find $f[g(0)]$. 5

15. Given $f(x) = 6x + 8$ and $g(x) = 4x + 2$, find $g[f(-1)]$. 10

16. Given $f(x) = 2x^2 + x - 5$ and $g(x) = 3x - 1$, find $g[f(x)]$. $6x^2 + 3x - 16$

17. Is the set of ordered pairs a function?

$\{(-3, 0), (-2, 0), (-1, 1), (0, 1)\}$ Yes

18. Find the inverse of the function $\{(-2, 1), (2, 3), (5, -4), (7, 9)\}$.
$\{(1, -2), (3, 2), (-4, 5), (9, 7)\}$

19. Determine whether the graph is the graph of a 1–1 function.

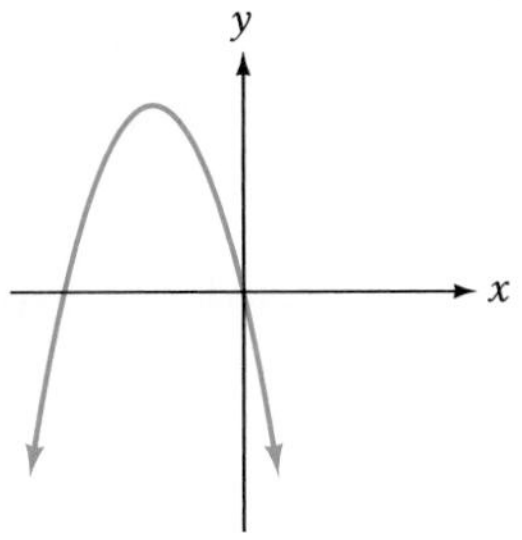

No

20. Determine whether the graph is the graph of a 1–1 function.

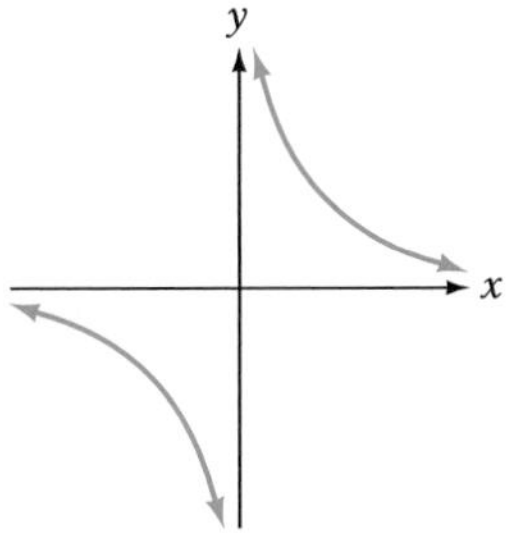

Yes

21. Find the inverse of the function $f(x) = \frac{1}{2}x + 8$.
$f^{-1}(x) = 2x - 16$

22. Find the inverse of the function $f(x) = -6x + 4$.
$f^{-1}(x) = -\frac{1}{6}x + \frac{2}{3}$

23. Find the inverse of the function $f(x) = \frac{2}{3}x - 12$.
$f^{-1}(x) = \frac{3}{2}x + 18$

24. Are the functions $f(x) = -\frac{1}{4}x + \frac{5}{4}$ and $g(x) = -4x + 5$ inverses of each other? Yes

Chapter Test

1. Determine whether the graph represents the graph of a function.

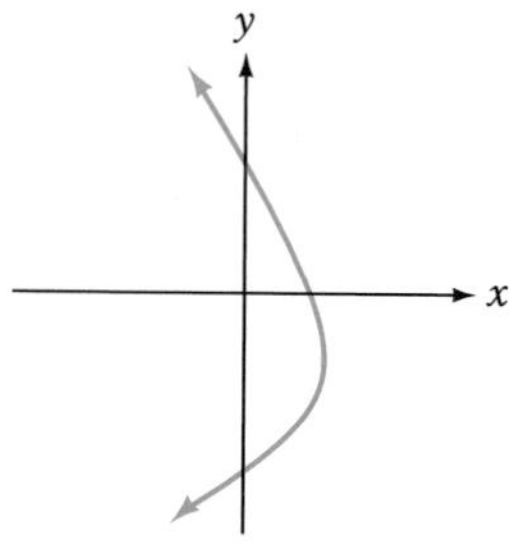

No

2. Given $f(x) = x^2 + 2x - 3$ and $g(x) = x^3 - 1$, find $(f - g)(2)$. -2

3. Given $f(x) = x^3 + 1$ and $g(x) = 2x - 3$ find $(f \cdot g)(-3)$. 234

4. Given $f(x) = 4x - 5$ and $g(x) = x^2 + 3x + 4$, find $\left(\frac{f}{g}\right)(-2)$. $-\frac{13}{2}$

5. Given $f(x) = x^2 + 4$ and $g(x) = 2x^2 + 2x + 1$, find $(f - g)(-4)$. -5

6. Given $f(x) = 4x + 2$ and $g(x) = \frac{x}{x+1}$, find $f[g(3)]$. 5

7. Use the graph of $f(x) = -x^3$ shown below to graph $g(x) = -x^3 - 2$.

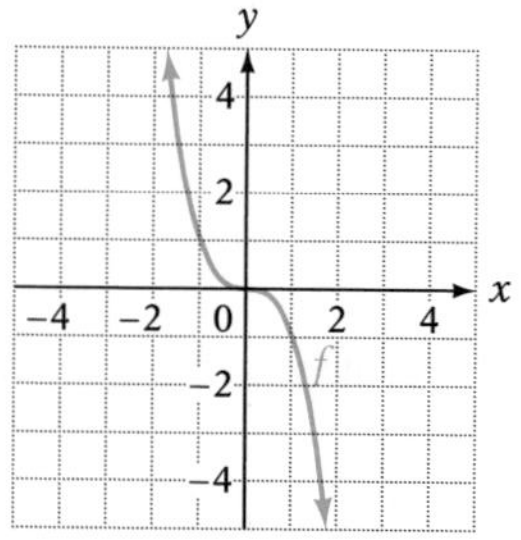

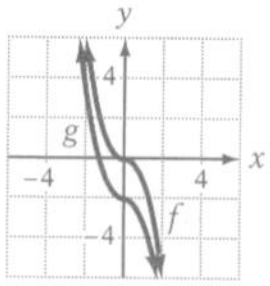

8. Use the graph of $f(x) = -|x|$ shown below to graph $g(x) = -|x| + 3$.

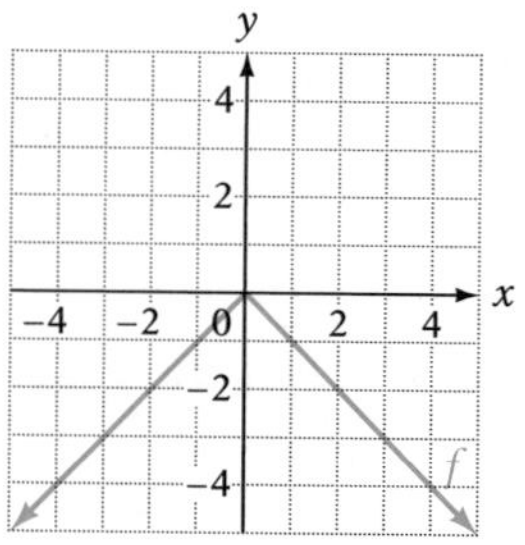

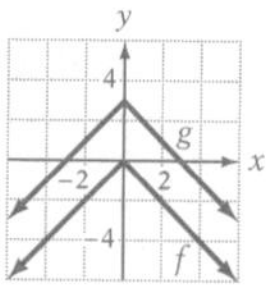

9. Graph $f(x) = -\sqrt{3 - x}$. Use interval notation to state the domain and range.

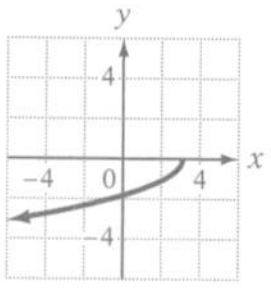

D: $(-\infty, 3]$
R: $(-\infty, 0]$

10. Graph $f(x) = \left|\frac{1}{2}x\right| - 2$. Use set-builder notation to state the domain and range.

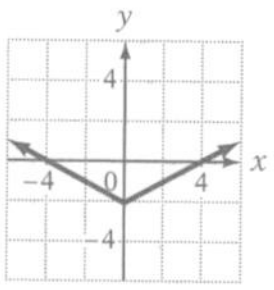

D: $\{x \mid x \in \text{real numbers}\}$
R: $\{y \mid y \geq -2\}$

11. Use the graph of $f(x) = |x|$ shown below to graph $g(x) = |x + 2| - 3$.

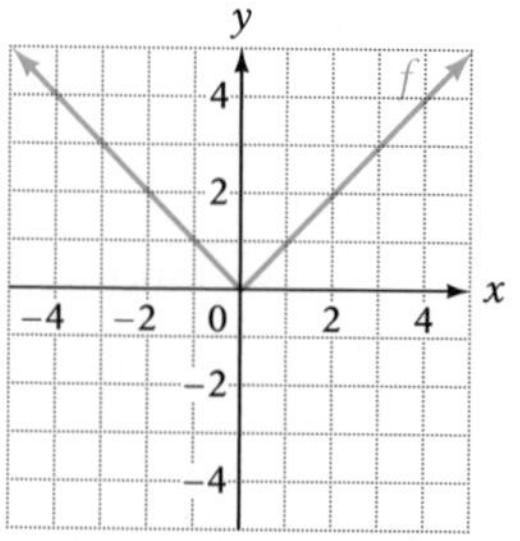

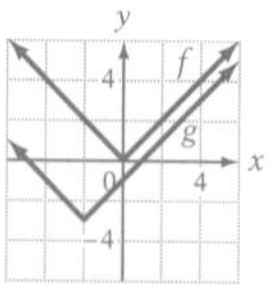

12. Use the graph of $f(x) = -x^2$ shown below to graph $g(x) = -(x - 2)^2 + 2$.

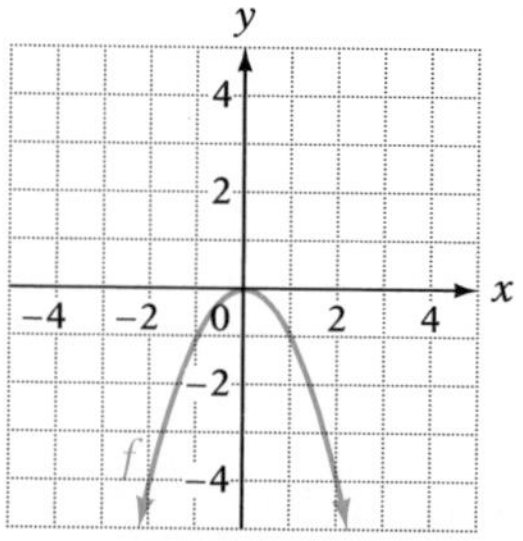

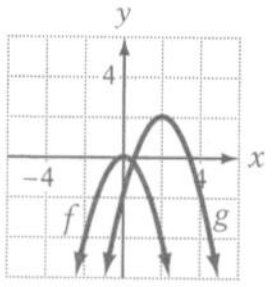

13. Find the inverse of the function $f(x) = \frac{1}{4}x - 4$.
$f^{-1}(x) = 4x + 16$

14. Find the inverse of the function $\{(2, 6), (3, 5), (4, 4), (5, 3)\}$.
$\{(6, 2), (5, 3), (4, 4), (3, 5)\}$

15. Are the functions $f(x) = \frac{1}{2}x + 2$ and $g(x) = 2x - 4$ inverses of each other? Yes

16. Given $f(x) = 2x^2 - 7$ and $g(x) = x - 1$, find $f[g(x)]$. $2x^2 - 4x - 5$

17. Find the inverse of the function $f(x) = \frac{1}{2}x - 3$.
$f^{-1}(x) = 2x + 6$

18. Are the functions $f(x) = \frac{2}{3}x + 3$ and $g(x) = \frac{3}{2}x - 3$ inverses of each other? No

19. Graph $f(x) = x^3 - 3x + 2$. Use set-builder notation to state the domain and range.

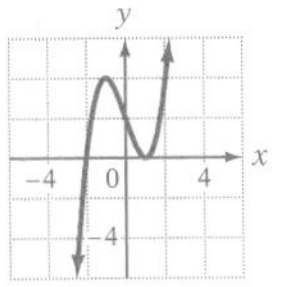

D: $\{x \mid x \in \text{real numbers}\}$
R: $\{y \mid y \in \text{real numbers}\}$

20. Determine whether the graph is the graph of a 1–1 function.

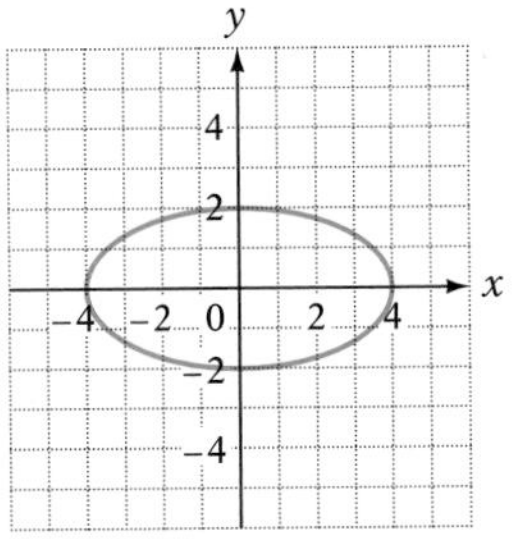

No

Cumulative Review Exercises

1. Evaluate $-3a + \left|\frac{3b - ab}{3b - c}\right|$ when $a = 2$, $b = 2$, and $c = -2$. $-\frac{23}{4}$

2. Graph: $\{x \mid x < -3\} \cap \{x \mid x > -4\}$

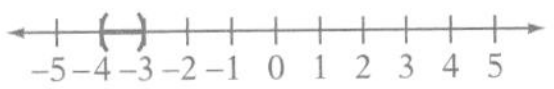

3. Solve: $\frac{3x - 1}{6} - \frac{5 - x}{4} = \frac{5}{6}$ 3

4. Solve: $4x - 2 < -10$ or $3x - 1 > 8$
$\{x \mid x < -2 \text{ or } x > 3\}$

5. Graph $f(x) = \frac{1}{4}x^2$. Find the vertex and axis of symmetry.

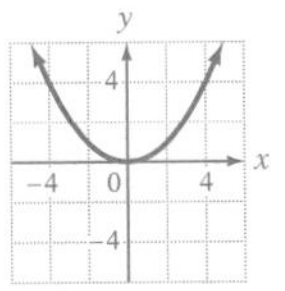

Vertex: $(0, 0)$
Axis of symmetry: $x = 0$

6. Graph the solution set of $3x - 4y \geq 8$.

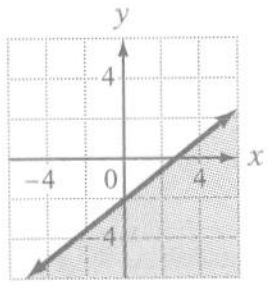

7. Solve: $|8 - 2x| \geq 0$ $\{x \mid x \in \text{real numbers}\}$

8. Simplify: $\left(\frac{3a^3b}{2a}\right)^2\left(\frac{a^2}{-3b^2}\right)^3$ $-\frac{a^{10}}{12b^4}$

9. Multiply: $(x - 4)(2x^2 + 4x - 1)$
$2x^3 - 4x^2 - 17x + 4$

10. Factor: $a^4 - 2a^2 - 8$ $(a + 2)(a - 2)(a^2 + 2)$

11. Factor: $x^3y + x^2y^2 - 6xy^3$ $xy(x - 2y)(x + 3y)$

12. Solve: $(b + 2)(b - 5) = 2b + 14$ $-3, 8$

13. Solve: $x^2 - 2x > 15$ $\{x \mid x < -3 \text{ or } x > 5\}$

14. Subtract: $\frac{x^2 + 4x - 5}{2x^2 - 3x + 1} - \frac{x}{2x - 1}$ $\frac{5}{2x - 1}$

15. Solve: $\frac{5}{x^2 + 7x + 12} = \frac{9}{x + 4} - \frac{2}{x + 3}$ -2

16. Simplify: $\frac{4 - 6i}{2i}$ $-3 - 2i$

17. Find the equation of the line that contains the points $(-3, 4)$ and $(2, -6)$. $y = -2x - 2$

18. Find the equation of the line that contains the point $(-3, 1)$ and is perpendicular to the line $2x - 3y = 6$. $y = -\frac{3}{2}x - \frac{7}{2}$

19. Solve: $3x^2 = 3x - 1$ $\frac{1}{2} + \frac{\sqrt{3}}{6}i$ and $\frac{1}{2} - \frac{\sqrt{3}}{6}i$

20. Solve: $\sqrt{8x + 1} = 2x - 1$ 3

21. Find the minimum value of the function $f(x) = 2x^2 - 3$. -3

22. Find the range of $f(x) = |3x - 4|$ if the domain is $\{0, 1, 2, 3\}$. $\{1, 2, 4, 5\}$

23. Is this set of ordered pairs a function? $\{(-3, 0), (-2, 0), (-1, 1), (0, 1)\}$ Yes

24. Solve: $\sqrt[3]{5x - 2} = 2$ 2

25. Given $g(x) = 3x - 5$ and $h(x) = \frac{1}{2}x + 4$, find $g[h(2)]$. $g[h(2)] = 10$

26. Find the inverse of the function given by $f(x) = -3x + 9$. $f^{-1}(x) = -\frac{1}{3}x + 3$

27. Find the cost per pound of a tea mixture made from 30 lb of tea costing \$4.50 per pound and 45 lb of tea costing \$3.60 per pound. \$3.96

28. How many pounds of an 80% copper alloy must be mixed with 50 lb of a 20% copper alloy to make an alloy that is 40% copper? 25 lb

29. Six ounces of an insecticide are mixed with 16 gal of water to make a spray for spraying an orange grove. How much additional insecticide is required if it is to be mixed with 28 gal of water? 4.5 oz

30. A large pipe can fill a tank in 8 min less time than it takes a smaller pipe to fill the same tank. Working together, both pipes can fill the tank in 3 min. How long would it take the larger pipe, working alone, to fill the tank? 4 min

31. The distance (d) that a spring stretches varies directly as the force (f) used to stretch the spring. If a force of 50 lb can stretch the spring 30 in., how far can a force of 40 lb stretch the spring? 24 in.

32. The frequency of vibration (f) in an open pipe organ varies inversely as the length (L) of the pipe. If the air in a pipe 2 m long vibrates 60 times per minute, find the frequency in a pipe that is 1.5 m long.
80 vibrations per minute

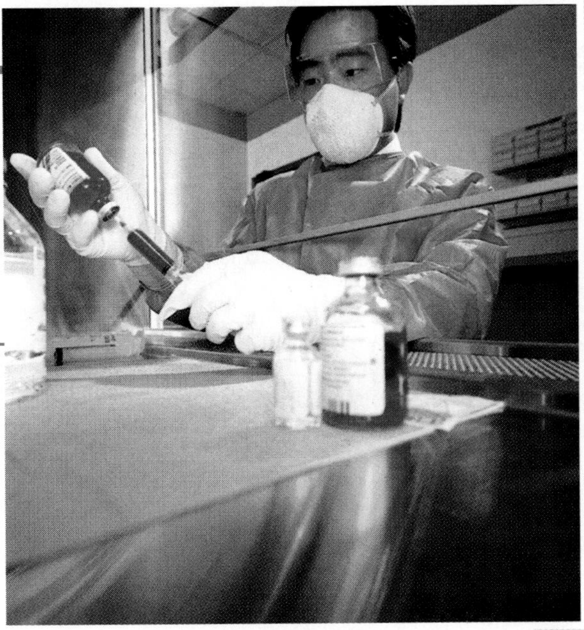

Nuclear medicine technologists operate cameras that detect and map the presence of a radioactive drug in a patient's body. They prepare a dosage of the radiopharmaceutical and administer it to the patient. In the preparation of radiopharmaceuticals, exponential equations are used to model the amount of radiation in the radiopharmaceutical.

10

Exponential and Logarithmic Functions

OBJECTIVES

10.1.1	Evaluate exponential functions
10.1.2	Graph exponential functions
10.2.1	Write equivalent exponential and logarithmic equations
10.2.2	The Properties of Logarithms
10.3.1	Graph logarithmic functions
10.4.1	Solve exponential equations
10.4.2	Solve logarithmic equations
10.5.1	Application problems

Napier Rods

The labor that is involved in calculating the products of large numbers has led many people to devise ways to short-cut the procedure. One such way was first described in the early 1600s by John Napier and is based on Napier Rods.

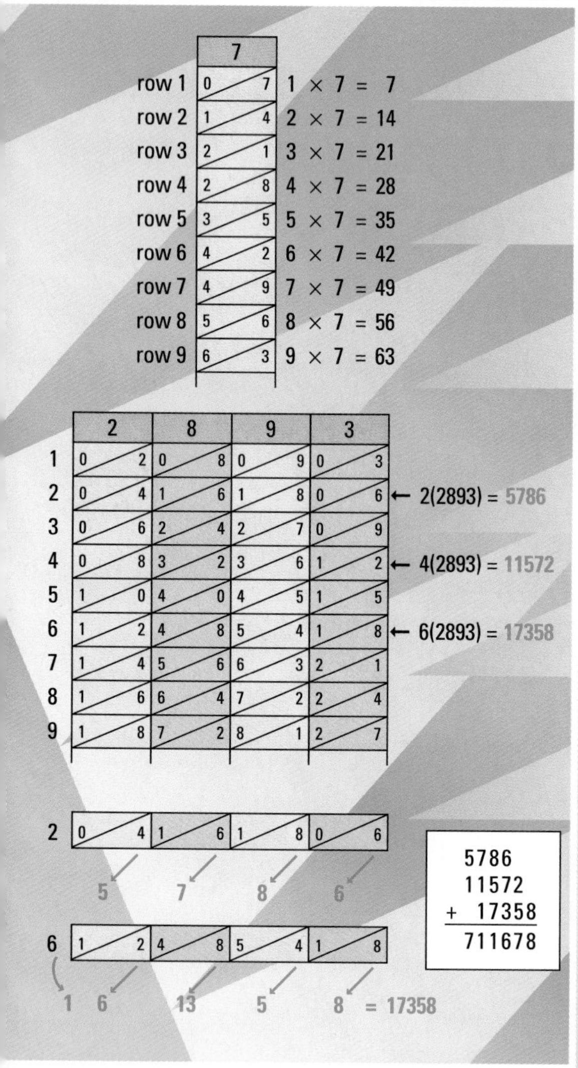

A Napier Rod is made by placing a number and the first 9 multiples of that number on a rectangular piece of paper (Napier's Rod). This is done for the first 9 positive integers. A rod for 7 is shown at the left.

To illustrate how Napier Rods were used to multiply, let's look at an example.

Multiply: 2893
         × 246

Place rods for 2, 8, 9, and 3 next to one another. The products of 2893 with 2, 4, and 6 are found by using the numbers along the 2nd, 4th, and 6th rows.

The digits of each product are found by adding the numbers along the diagonals of the corresponding row of Napier Rods, carrying to the next diagonal when necessary. The products for 2 and 6 are shown at the left.

In the product for 6, note that because the sum of the 3rd diagonal is 13, carrying to the next diagonal is necessary.

The final product is then found by addition.

John Napier is also credited with the invention of logarithms, which were used to ease the drudgery of lengthy calculations before the invention of the electronic calculator.

Rod for 7

	7	
row 1	0 / 7	1 × 7 = 7
row 2	1 / 4	2 × 7 = 14
row 3	2 / 1	3 × 7 = 21
row 4	2 / 8	4 × 7 = 28
row 5	3 / 5	5 × 7 = 35
row 6	4 / 2	6 × 7 = 42
row 7	4 / 9	7 × 7 = 49
row 8	5 / 6	8 × 7 = 56
row 9	6 / 3	9 × 7 = 63

Rods for 2, 8, 9, 3

	2	8	9	3	
1	0 / 2	0 / 8	0 / 9	0 / 3	
2	0 / 4	1 / 6	1 / 8	0 / 6	← 2(2893) = 5786
3	0 / 6	2 / 4	2 / 7	0 / 9	
4	0 / 8	3 / 2	3 / 6	1 / 2	← 4(2893) = 11572
5	1 / 0	4 / 0	4 / 5	1 / 5	
6	1 / 2	4 / 8	5 / 4	1 / 8	← 6(2893) = 17358
7	1 / 4	5 / 6	6 / 3	2 / 1	
8	1 / 6	6 / 4	7 / 2	2 / 4	
9	1 / 8	7 / 2	8 / 1	2 / 7	

Row 2: 2 | 0 / 4 | 1 / 6 | 1 / 8 | 0 / 6

5 7 8 6

Row 6: 6 | 1 / 2 | 4 / 8 | 5 / 4 | 1 / 8

1 6 13 5 8 = 17358

```
   5786
  11572
+ 17358
```

SECTION 10.1
Exponential Functions

1 Evaluate exponential functions

The growth of a \$500 savings account that earns 5% annual interest compounded daily is shown at the right. In 14 years, the savings account contains approximately \$1000, twice the initial amount. The growth of this savings account is an example of an exponential function.

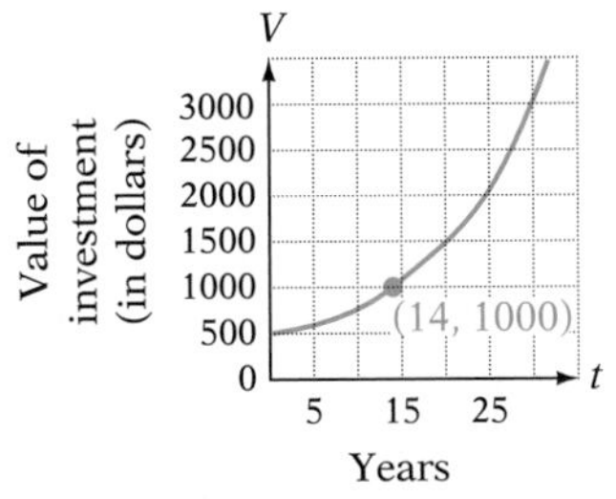

The pressure of the atmosphere at a certain height is shown in the graph at the right. This is another example of an exponential function. From the graph, we read that the air pressure is approximately 6.5 lb/in^2 at an altitude of 20,000 ft.

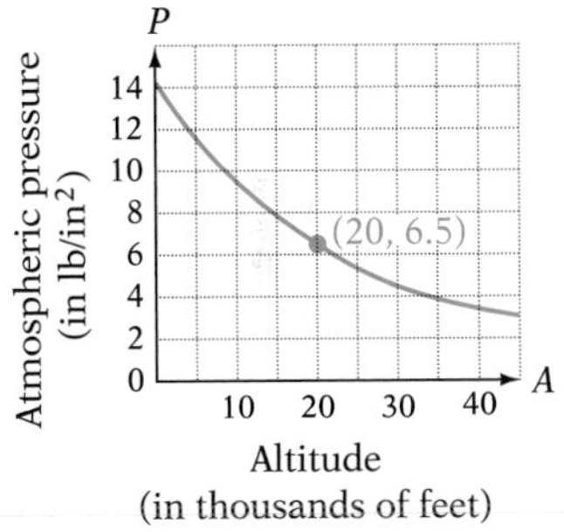

INSTRUCTOR NOTE

It is important for students to distinguish between $F(x) = 2^x$ and $P(x) = x^2$. The first is an exponential function; the second is a polynomial function. Exponential functions are characterized by a constant base and a variable exponent. Polynomial functions have a variable base and a constant exponent.

Definition of an Exponential Function

> The **exponential function** with base b is defined by
>
> $$f(x) = b^x$$
>
> where $b > 0$, $b \neq 1$, and x is any real number.

In the definition of an exponential function, b, the base, is required to be positive. If the base were a negative number, the value of the function would be a complex number for some values of x. For instance, the value of $f(x) = (-4)^x$ when $x = \frac{1}{2}$ is $f\left(\frac{1}{2}\right) = (-4)^{\frac{1}{2}} = \sqrt{-4} = 2i$. To avoid complex number values of a function, the base of the exponential function is a positive number.

Evaluate $f(x) = 2^x$ at $x = 3$ and $x = -2$.

Substitute 3 for x and simplify. $f(3) = 2^3 = 8$

Substitute -2 for x and simplify. $f(-2) = 2^{-2} = \frac{1}{2^2} = \frac{1}{4}$

To evaluate an exponential expression for an irrational number such as $\sqrt{2}$, we obtain an approximation to the value of the function by approximating the irrational number. For instance, the value of $f(x) = 4^x$ when $x = \sqrt{2}$ can be approximated by using an approximation of $\sqrt{2}$.

$$f(\sqrt{2}) = 4^{\sqrt{2}} \approx 4^{1.4142} \approx 7.1029$$

Because $f(x) = b^x$ ($b > 0$, $b \neq 1$) can be evaluated at both rational and irrational numbers, the domain of f is all real numbers. And because $b^x > 0$ for all values of x, the range of f is the positive real numbers.

Example 1 Evaluate $f(x) = \left(\frac{1}{2}\right)^x$ at $x = 2$ and $x = -3$.

Solution $f(x) = \left(\frac{1}{2}\right)^x$

$f(2) = \left(\frac{1}{2}\right)^2 = \frac{1}{4}$ $\quad f(-3) = \left(\frac{1}{2}\right)^{-3} = 2^3 = 8$

Problem 1 Evaluate $f(x) = \left(\frac{2}{3}\right)^x$ at $x = 3$ and $x = -2$.

Solution See page S27. $\frac{8}{27}; \frac{9}{4}$

Example 2 Evaluate $f(x) = 2^{3x-1}$ at $x = 1$ and $x = -1$.

Solution $f(x) = 2^{3x-1}$

$f(1) = 2^{3(1)-1} = 2^2 = 4$ $\quad f(-1) = 2^{3(-1)-1} = 2^{-4} = \frac{1}{2^4} = \frac{1}{16}$

Problem 2 Evaluate $f(x) = 2^{2x+1}$ at $x = 0$ and $x = -2$.

Solution See page S27. $2; \frac{1}{8}$

Example 3 Evaluate $f(x) = (\sqrt{5})^x$ at $x = 4$, $x = -2.1$, and $x = \pi$. Round to the nearest ten-thousandth.

Solution $f(x) = (\sqrt{5})^x$

$f(4) = (\sqrt{5})^4 = 25$ $\quad f(-2.1) = (\sqrt{5})^{-2.1} \approx 0.1845$ $\quad f(\pi) = (\sqrt{5})^{\pi} \approx 12.5297$

Problem 3 Evaluate $f(x) = \pi^x$ at $x = 3$, $x = -2$, and $x = \pi$. Round to the nearest ten-thousandth.

Solution See page S27. 31.0063; 0.1013; 36.4622

POINT OF INTEREST

The natural exponential function is an extremely important function. It is used extensively in applied problems in virtually all disciplines from archeology to zoology. Leonard Euler (1707–1783) was the first to use the letter *e* as the base of the natural exponential function.

A frequently used base in applications of exponential functions is an irrational number designated by e. The number e is approximately 2.71828183. It is an irrational number, so it has a nonterminating, nonrepeating decimal representation.

Natural Exponential Function

The function defined by $f(x) = e^x$ is called the **natural exponential function.**

The $\boxed{e^x}$ key on a calculator can be used to evaluate the natural exponential function.

INSTRUCTOR NOTE

If time permits, show students some of the remarkable relationships that exist among i, e, and π. For instance, $e^{\pi i} = -1$ and $i^{-i} = e^{\frac{\pi}{2}}$.

Example 4 Evaluate $f(x) = e^x$ at $x = 2$, $x = -3$, and $x = \pi$. Round to the nearest ten-thousandth.

Solution $f(x) = e^x$

$f(2) = e^2 \approx 7.3891$ $\quad f(-3) = e^{-3} \approx 0.0498$ $\quad f(\pi) = e^{\pi} \approx 23.1407$

Problem 4 Evaluate $f(x) = e^x$ at $x = 1.2$, $x = -2.5$, and $x = e$. Round to the nearest ten-thousandth.

Solution See page S27. 3.3201; 0.0821; 15.1543

2 Graph exponential functions

37 HM³

Some of the properties of an exponential function can be seen by considering its graph.

INSTRUCTOR NOTE

Graph $g(x) = x^2$ so that students can see the difference between this graph and that of $f(x) = 2^x$.

To graph $f(x) = 2^x$, think of the function as the equation $y = 2^x$.

Choose values of x, and find the corresponding values of y. The results can be recorded in a table.

Graph the ordered pairs on a rectangular coordinate system.

Connect the points with a smooth curve.

x	$f(x) = y$
-2	$2^{-2} = \frac{1}{4}$
-1	$2^{-1} = \frac{1}{2}$
0	$2^0 = 1$
1	$2^1 = 2$
2	$2^2 = 4$
3	$2^3 = 8$

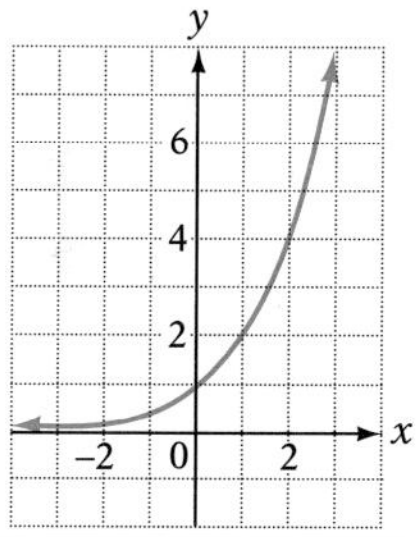

Note that a vertical line would intersect the graph at only one point. Therefore, by the vertical-line test, $f(x) = 2^x$ is the graph of a function. Also note that a horizontal line would intersect the graph at only one point. Therefore, $f(x) = 2^x$ is the graph of a 1–1 function.

INSTRUCTOR NOTE

After you discuss the graph of the exponential function, you might give students an extra-credit problem of graphing $f(x) = x^x$. A discussion of the domain of this function may lead to students gaining a better understanding of that topic.

Graph: $f(x) = \left(\frac{1}{2}\right)^x$

Think of the function as the equation $y = \left(\frac{1}{2}\right)^x$.

Choose values of x, and find the corresponding values of y.

Graph the ordered pairs on a rectangular coordinate system.

Connect the points with a smooth curve.

x	$f(x) = y$
-3	$\left(\frac{1}{2}\right)^{-3} = 8$
-2	$\left(\frac{1}{2}\right)^{-2} = 4$
-1	$\left(\frac{1}{2}\right)^{-1} = 2$
0	$\left(\frac{1}{2}\right)^{0} = 1$
1	$\left(\frac{1}{2}\right)^{1} = \frac{1}{2}$
2	$\left(\frac{1}{2}\right)^{2} = \frac{1}{4}$

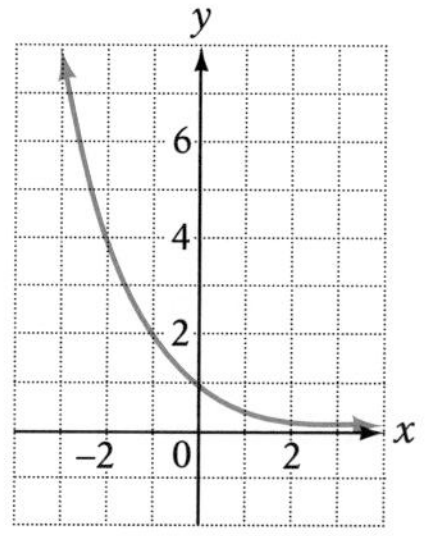

Applying the vertical- and horizontal-line tests reveals that $f(x) = \left(\frac{1}{2}\right)^x$ is also the graph of a 1–1 function.

The graph of $f(x) = 2^{-x}$ is shown at the right.

Note that because $2^{-x} = (2^{-1})^x = \left(\frac{1}{2}\right)^x$, the graphs of $f(x) = 2^{-x}$ and $f(x) = \left(\frac{1}{2}\right)^x$ are the same.

x	y
-3	8
-2	4
-1	2
0	1
1	$\frac{1}{2}$
2	$\frac{1}{4}$

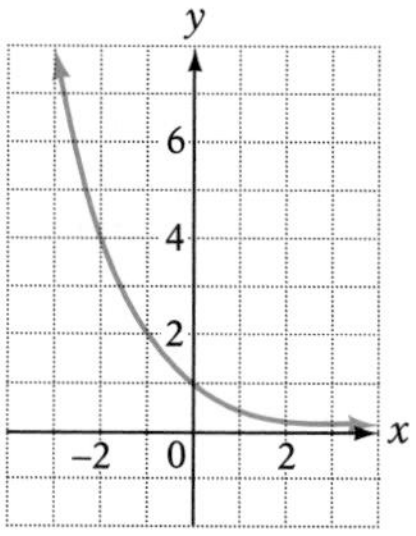

Example 5 Graph. A. $f(x) = 3^{\frac{1}{2}x-1}$ B. $f(x) = 2^x - 1$

Solution A.

x	y
-2	$\frac{1}{9}$
0	$\frac{1}{3}$
2	1
4	3

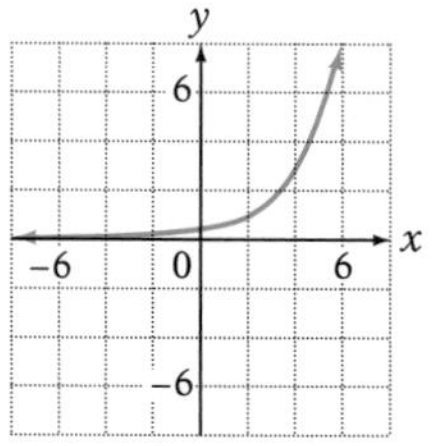

B.

x	y
-2	$-\frac{3}{4}$
-1	$-\frac{1}{2}$
0	0
1	1
2	3
3	7

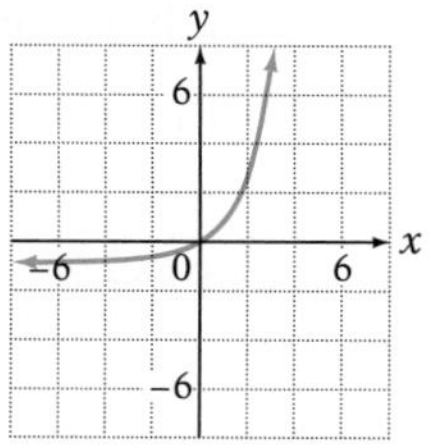

Problem 5 Graph. A. $f(x) = 2^{-\frac{1}{2}x}$ B. $f(x) = 2^x + 1$

Solution See page S27. A.

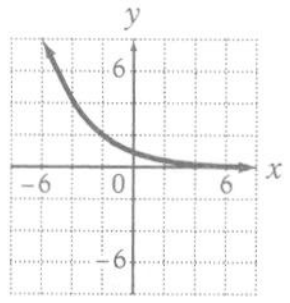

B.

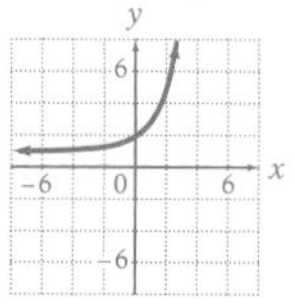

Example 6 Graph. A. $f(x) = 3^{x+1}$ B. $f(x) = e^x - 1$

Solution A.

x	y
-3	$\frac{1}{9}$
-2	$\frac{1}{3}$
-1	1
0	3
1	9

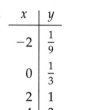

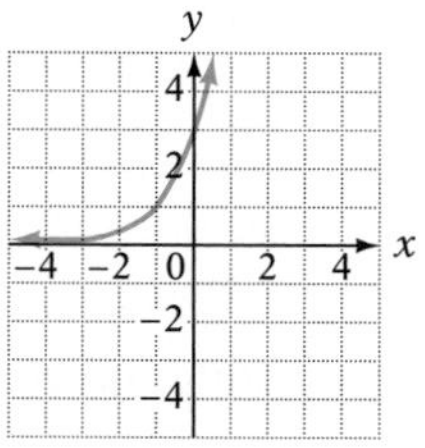

B.

x	y
-2	-0.86
-1	-0.63
0	0
1	1.72
2	6.39

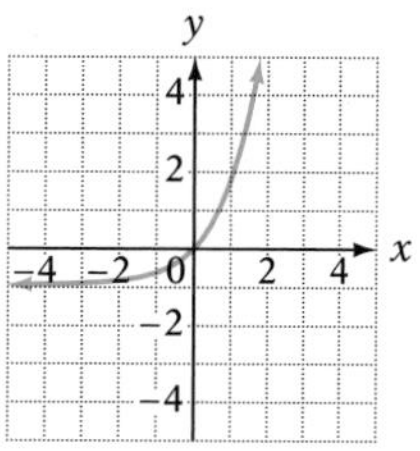

Problem 6 Graph. A. $f(x) = 2^{-x} + 2$ B. $f(x) = e^{-2x} - 4$

Solution See page S27. A.

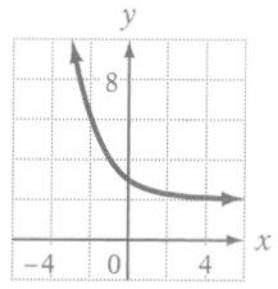

B.

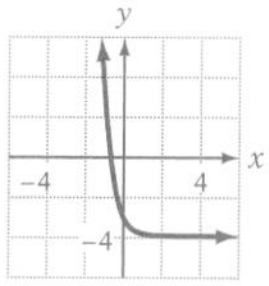

Example 7 Graph $f(x) = -\frac{1}{3}e^{2x} + 2$ and approximate the zero of f to the nearest tenth.

Solution

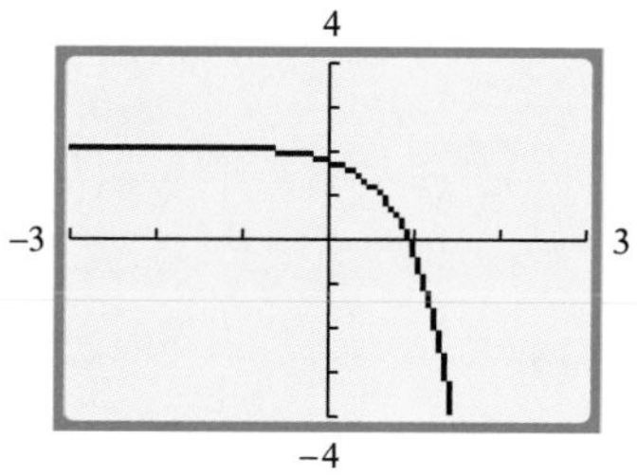

▶ Recall that a zero of f is a value of x for which $f(x) = 0$. Use the features of a graphing utility to determine the x-intercept of the graph, which is the zero of f.

The zero of f is 0.9 to the nearest tenth.

Problem 7 Graph $f(x) = 2\left(\frac{3}{4}\right)^x - 3$ and approximate, to the nearest tenth, the value of x for which $f(x) = 1$.

Solution See page S27.

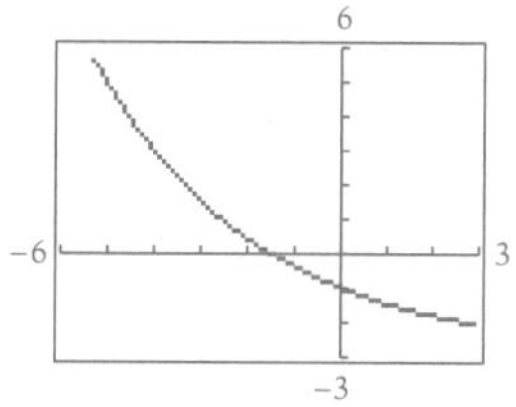

-2.4

CONCEPT REVIEW 10.1

Determine whether the following statements are always true, sometimes true, or never true.

1. The domain of an exponential function $f(x) = b^x$, $b > 0$, $b \neq 1$, is the set of positive numbers. Never true

2. An exponential function $f(x) = b^x$, $b > 0$, $b \neq 1$, is a 1–1 function. Always true

3. The graph of the exponential function $f(x) = b^x, b > 0, b \neq 1$, passes through the point $(0, 0)$.   Never true

4. For the function $f(x) = b^x, b > 0, b \neq 1$, the base b is a positive integer. Sometimes true

5. An exponential function $f(x) = b^x, b > 0, b \neq 1$, has two x-intercepts. Never true

---

EXERCISES 10.1

1. How does an exponential function differ from a polynomial function?

2. Why are the conditions $b > 0, b \neq 1$ given for $f(x) = b^x$?

3. Given $f(x) = 3^x$, evaluate:
   a. $f(2)$ **b.** $f(0)$ **c.** $f(-2)$
     9   1 $\frac{1}{9}$

4. Given $H(x) = 2^x$, evaluate:
   a. $H(-3)$ **b.** $H(0)$ **c.** $H(2)$
     $\frac{1}{8}$ 1   4

5. Given $g(x) = 2^{x+1}$, evaluate:
   a. $g(3)$ **b.** $g(1)$ **c.** $g(-3)$
     16   4 $\frac{1}{4}$

6. Given $F(x) = 3^{x-2}$, evaluate:
   a. $F(-4)$ **b.** $F(-1)$ **c.** $F(0)$
     $\frac{1}{729}$ $\frac{1}{27}$ $\frac{1}{9}$

7. Given $P(x) = \left(\frac{1}{2}\right)^{2x}$, evaluate:
   a. $P(0)$ **b.** $P\left(\frac{3}{2}\right)$ **c.** $P(-2)$
     1         $\frac{1}{8}$ 16

8. Given $R(t) = \left(\frac{1}{3}\right)^{3t}$, evaluate:
   a. $R\left(-\frac{1}{3}\right)$ **b.** $R(1)$ **c.** $R(-2)$
     3         $\frac{1}{27}$ 729

9. Given $G(x) = e^{\frac{x}{2}}$, evaluate the following. Round to the nearest ten-thousandth.
   a. $G(4)$ **b.** $G(-2)$ **c.** $G\left(\frac{1}{2}\right)$
     7.3891   0.3679   1.2840

10. Given $f(x) = e^{2x}$, evaluate the following. Round to the nearest ten-thousandth.
   a. $f(-2)$ **b.** $f\left(-\frac{2}{3}\right)$ **c.** $f(2)$
     0.0183   0.2636   54.5982

11. Given $H(r) = e^{-r+3}$, evaluate the following. Round to the nearest ten-thousandth.
   a. $H(-1)$ **b.** $H(3)$ **c.** $H(5)$
     54.5982   1   0.1353

12. Given $P(t) = e^{-\frac{1}{2}t}$, evaluate the following. Round to the nearest ten-thousandth.
   a. $P(-3)$ **b.** $P(4)$ **c.** $P\left(\frac{1}{2}\right)$
     4.4817   0.1353   0.7788

13. Given $F(x) = 2^{x^2}$, evaluate:
   a. $F(2)$ **b.** $F(-2)$ **c.** $F\left(\frac{3}{4}\right)$
     16   16   1.4768

14. Given $Q(x) = 2^{-x^2}$, evaluate:
   a. $Q(3)$ **b.** $Q(-1)$ **c.** $Q(-2)$
     $\frac{1}{512}$ $\frac{1}{2}$ $\frac{1}{16}$

15. Given $f(x) = e^{-\frac{x^2}{2}}$, evaluate the following. Round to the nearest ten-thousandth.
   a. $f(-2)$ **b.** $f(2)$ **c.** $f(-3)$
     0.1353   0.1353   0.0111

16. Given $f(x) = e^{-2x} + 1$, evaluate the following. Round to the nearest ten-thousandth.
   a. $f(-1)$ **b.** $f(3)$ **c.** $f(-2)$
     8.3891   1.0025   55.5982

2 Graph.

17. $f(x) = 3^x$

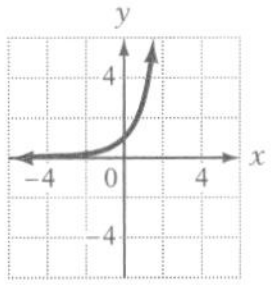

18. $f(x) = 3^{-x}$

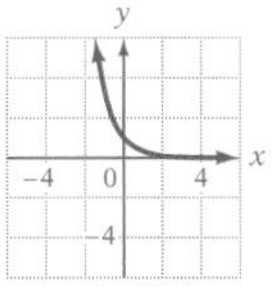

19. $f(x) = 2^{x+1}$

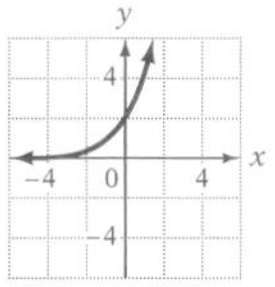

20. $f(x) = 2^{x-1}$

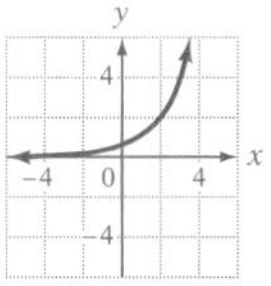

21. $f(x) = \left(\frac{1}{3}\right)^x$

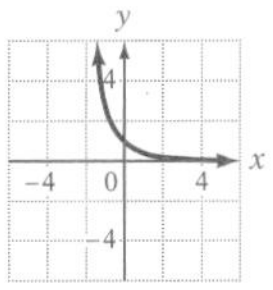

22. $f(x) = \left(\frac{2}{3}\right)^x$

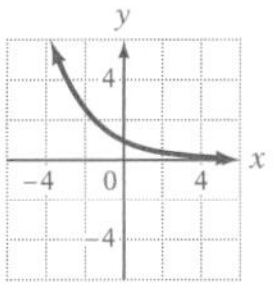

23. $f(x) = 2^{-x} + 1$

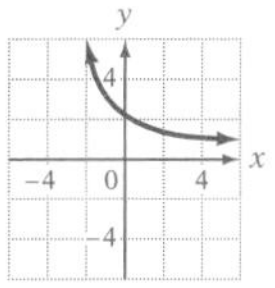

24. $f(x) = 2^x - 3$

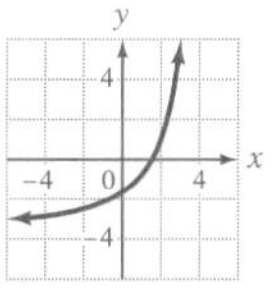

25. $f(x) = \left(\frac{1}{3}\right)^{-x}$

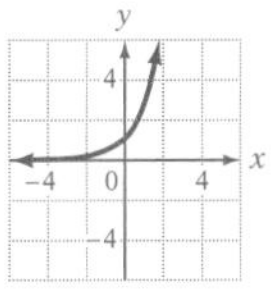

26. $f(x) = \left(\frac{3}{2}\right)^{-x}$

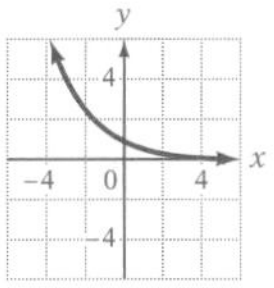

27. $f(x) = \left(\frac{1}{2}\right)^{-x} + 2$

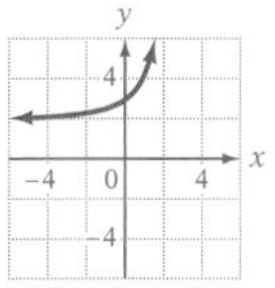

28. $f(x) = \left(\frac{1}{2}\right)^x - 1$

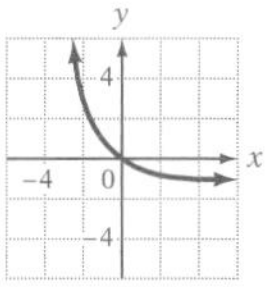

Graph.

29. $f(x) = x(2^x)$

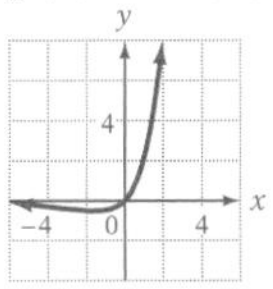

30. $f(x) = x + 2^x$

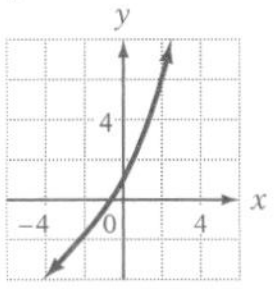

31. $f(x) = e^x - x$

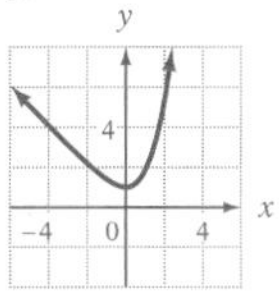

32. $f(x) = 2xe^x$

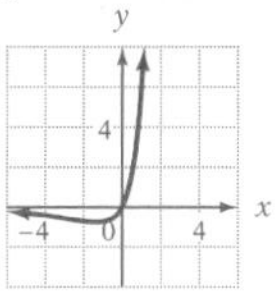

Solve.

33. Graph $f(x) = 2^x - 3$ and approximate the zero of f to the nearest tenth.

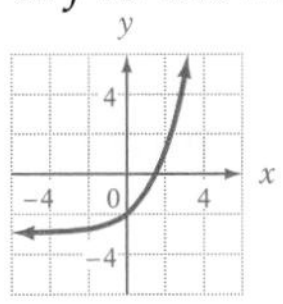

1.6

34. Graph $f(x) = 5 - 3^x$ and approximate the zero of f to the nearest tenth.

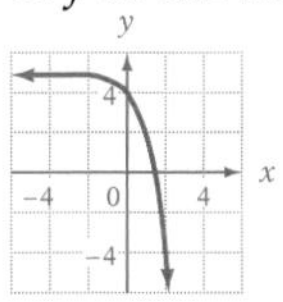

1.5

35. Graph $f(x) = e^x$ and approximate, to the nearest tenth, the value of x for which $f(x) = 3$.

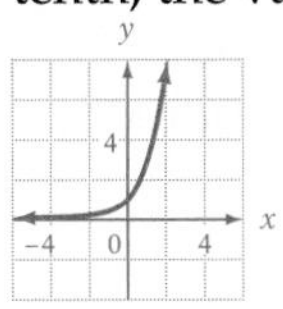

1.1

36. Graph $f(x) = e^{-2x-3}$ and approximate, to the nearest tenth, the value of x for which $f(x) = 2$.

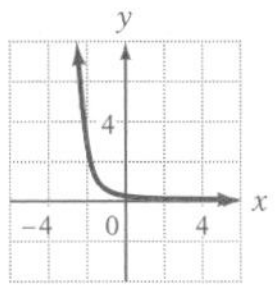

−1.8

37. *Investments* The exponential function given by $F(n) = 500(1.00021918)^{365n}$ gives the value in n years of a \$500 investment in a certificate of deposit that earns 8% annual interest compounded daily. Graph F and determine in how many years the investment will be worth \$1000. 9 years

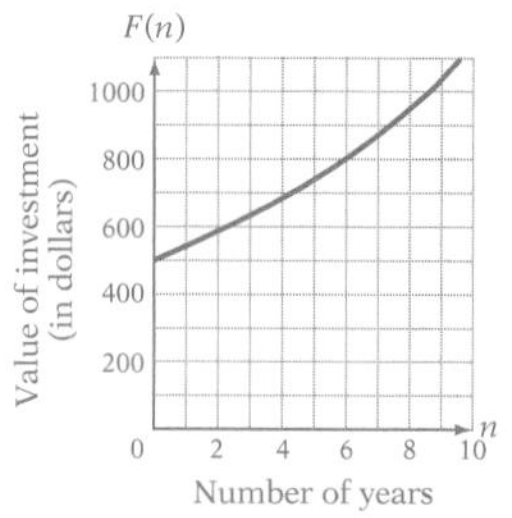

38. *Population* Assuming that the current population of Earth is 5.6 billion people and that Earth's population is growing at an annual rate of 1.5%, the exponential equation $P(t) = 5.6(1.015)^t$ gives the size, in billions of people, of the population t years from now. Graph P and determine the number of years before Earth's population reaches 7 billion people. 15 years

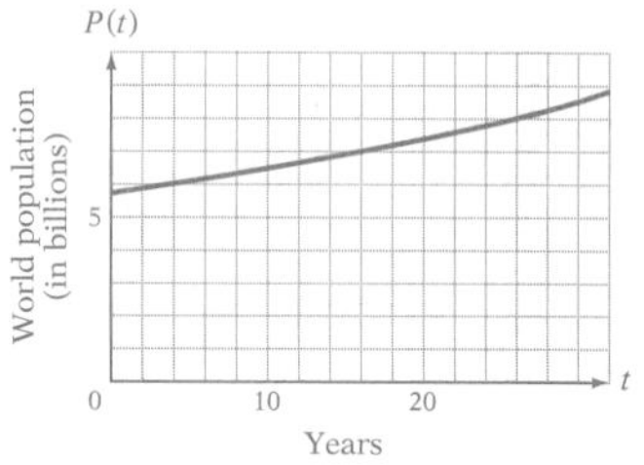

39. *Oceanography* The percent of light that reaches m meters below the surface of the ocean is given by the equation $P(m) = 100e^{-1.38m}$. Graph P and determine the depth to which 50% of the light will reach. 0.5 m

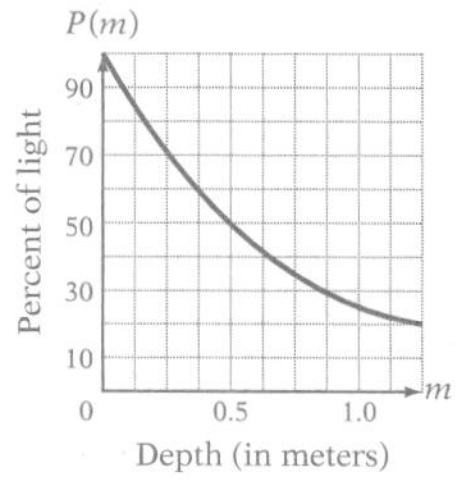

40. *Chemistry* The number of grams of radioactive cesium that remain after t years from an original sample of 30 g is given by $N(t) = 30(2^{-0.0322t})$. Graph N and determine in how many years there will be 20 g of cesium remaining. 18 years

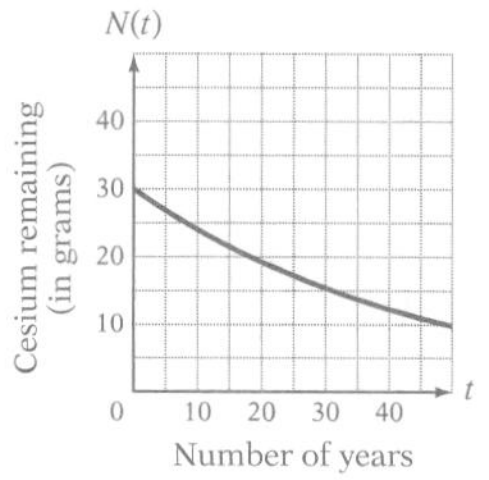

APPLYING CONCEPTS 10.1

Solve.

41. Graph $f(x) = x^2e^x$ and determine the minimum value of f to the nearest hundredth. 0

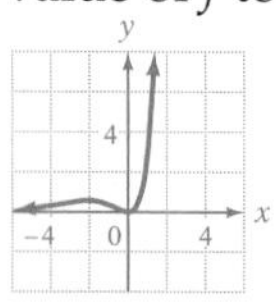

42. Graph $f(x) = 2x^2(3^x)$ and determine the minimum value of f to the nearest hundredth. 0

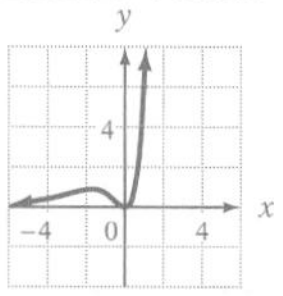

43. Graph $f(x) = x^2 + 3^x$ and determine the minimum value of f to the nearest hundredth. 0.80

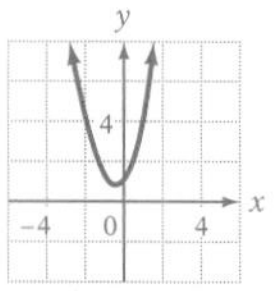

44. Graph $f(x) = 2^x - x$ and determine the minimum value of f to the nearest hundredth. 0.91

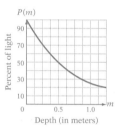

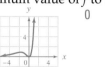

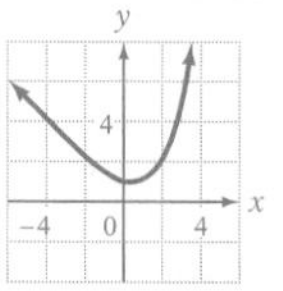

45. Evaluate $\left(1 + \frac{1}{n}\right)^n$ for $n = 100$, 1000, 10,000, and 100,000 and compare the results with the value of e, the base of the natural exponential function. On the basis of your evaluation, complete the following sentence:
As n increases, $\left(1 + \frac{1}{n}\right)^n$ becomes closer to $\underline{\quad e \quad}$.

46. *Population* According to population studies, the population of India can be approximated by the equation $P(t) = 0.984(1.02)^t$, where $t = 0$ corresponds to 1998 and $P(t)$ is the population, in billions, of India in t years.

a. Graph this equation. *Suggestion:* Use Xmin = 0, Xmax = 20, Xscl = 1.0, Ymin = 0, Ymax = 1.5, and Yscl = 0.1.

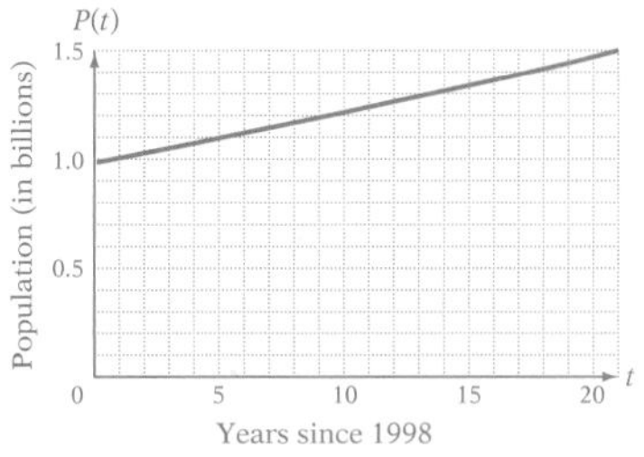

b. The point whose approximate coordinates are (5, 1.086) is on the graph. Write a sentence that explains the meaning of these coordinates.
The point (5, 1.086) means that in the year 2003, the population of India will be approximately 1.086 billion.

47. *Physics* If air resistance is ignored, the speed v, in feet per second, of an object t seconds after it has been dropped is given by the equation $v = 32t$. This is true regardless of the mass of the object. However, if air resistance is considered, then the speed depends on the mass (and on other things). For a certain mass, the speed t seconds after it has been dropped is given by $v = 64(1 - e^{-\frac{t}{2}})$.

a. Graph this equation. *Suggestion:* Use Xmin = −0.5, Xmax = 10, Xscl = 1, Ymin = −0.5, Ymax = 70, and Yscl = 10.

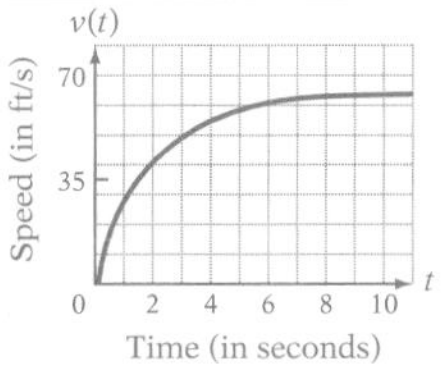

b. The point whose approximate coordinates are (4, 55.3) is on the graph. Write a sentence that explains the meaning of these coordinates.
The point (4, 55.3) means that after 4 s, the object is moving at a speed of 55.3 ft/s.

48. *Investments* An annuity is a fixed amount of money that is either paid or received over equal intervals of time. A retirement plan in which a certain amount is deposited each month is an example of an annuity; equal deposits are made over equal intervals of time (monthly). The equation that relates the amount of money available for retirement to the monthly deposit is $V = P\left[\frac{(1 + i)^x - 1}{i}\right]$, where i is the interest rate per month, x is the number of months deposits are made, P is the payment, and V is the value (called the *future value*) of the retirement fund after x payments.

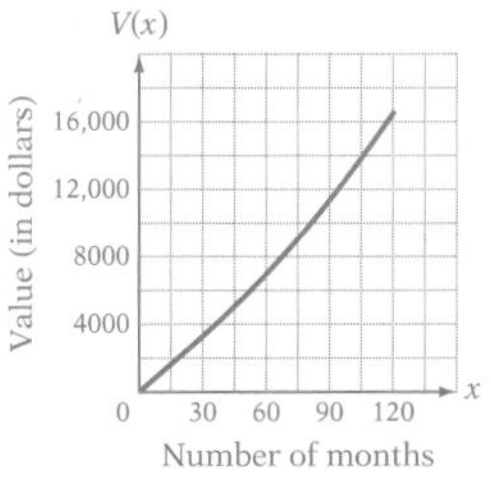

a. Suppose \$100 is deposited each month into an account that earns interest at the rate of 0.5% per month (6% per year). Graph the equation for $0 \le x \le 120$.

b. The point whose coordinates are (60, 6977), to the nearest integer, is on the graph of the equation. Write a sentence that gives an interpretation of this ordered pair. The ordered pair (60, 6977) means that after 60 months, the account will be worth \$6977.

c. For how many years must the investor make deposits in order to have a retirement account worth \$20,000? 12 years

49. Graph $f(x) = e^x$, $g(x) = e^{x-2}$, and $h(x) = e^{x+2}$ on the same rectangular coordinate system. Explain how the graphs are similar and how they are different.

50. Graph $f(x) = 2^x$, $g(x) = 2^x - 2$, and $h(x) = 2^x + 2$ on the same rectangular coordinate system. Explain how the graphs are similar and how they are different.

51. Suppose that $f(x)$ is positive and that for all real numbers x and y, $f(x + y) = f(x) \cdot f(y)$. Show that $f(0) = 1$ and that $f(2x) = [f(x)]^2$. Give an example of a function that has the properties of f.

SECTION 10.2 Introduction to Logarithms

1 Write equivalent exponential and logarithmic equations

38 HM³

Because the exponential function is a 1–1 function, it has an inverse function that is called a *logarithm.* A logarithm is used to answer a question similar to the following one: "If $16 = 2^y$, what is the value of y?" Because $16 = 2^4$, the logarithm, base 2, of 16 is 4. This is written $\log_2 16 = 4$. Note that a logarithm is an exponent that solves a certain equation.

Definition of Logarithm

For $b > 0, b \neq 1, y = \log_b x$ is equivalent to $x = b^y$.

INSTRUCTOR NOTE

Some students see the definition of logarithm as artificial. The following may help.

Defining a logarithm as the inverse of the exponential function is similar to defining square roots as the inverse of square.

"If $49 = x^2$, what is x?" The answer is the square root of 49.

Now consider "If $8 = 2^x$, what is x?" The answer is the logarithm, base 2, of 8.

This analogy can be extended to explain the need for tables or calculators.

"If $19 = x^2$, what is x?"

"If $21 = 10^x$, what is x?"

Read $\log_b x$ as "the logarithm of x, base b" or "log base b of x."

The table below shows equivalent statements written in both exponential and logarithmic form.

Exponential Form		Logarithmic Form
$2^4 = 16$	⟷	$\log_2 16 = 4$
$\left(\frac{2}{3}\right)^2 = \frac{4}{9}$	⟷	$\log_{\frac{2}{3}}\left(\frac{4}{9}\right) = 2$
$10^{-1} = 0.1$	⟷	$\log_{10}(0.1) = -1$

Example 1 Write $4^5 = 1024$ in logarithmic form.

Solution $4^5 = 1024$ is equivalent to $\log_4 1024 = 5$.

Problem 1 Write $3^{-4} = \frac{1}{81}$ in logarithmic form.

Solution See page S27. $\log_3 \frac{1}{81} = -4$

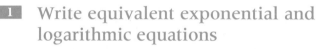

Example 2 Write $\log_7 343 = 3$ in exponential form.

Solution $\log_7 343 = 3$ is equivalent to $7^3 = 323$.

Problem 2 Write $\log_{10} 0.0001 = -4$ in exponential form.

Solution See page S27. $10^{-4} = 0.0001$

Recalling the equations $y = \log_b x$ and $x = b^y$ from the definition of a logarithm, note that because $b^y > 0$ for all values of y, x is always a positive number. Therefore, in the equation $y = \log_b x$, x is a positive number. The logarithm of a negative number is not a real number.

The 1–1 property of exponential functions can be used to evaluate some logarithms.

Equality of Exponents Property

For $b > 0, b \neq 1$, if $b^u = b^v$, then $u = v$.

Evaluate: $\log_2 8$

Write an equation.	$\log_2 8 = x$
Write the equation in its equivalent exponential form.	$8 = 2^x$
Write 8 in exponential form using 2 as the base.	$2^3 = 2^x$
Use the Equality of Exponents Property.	$3 = x$
	$\log_2 8 = 3$

Example 3 Evaluate: $\log_3\left(\frac{1}{9}\right)$

Solution

$\log_3\left(\frac{1}{9}\right) = x$ ▶ Write an equation.

$\frac{1}{9} = 3^x$ ▶ Write the equation in its equivalent exponential form.

$3^{-2} = 3^x$ ▶ Write $\frac{1}{9}$ in exponential form using 3 as the base.

$-2 = x$ ▶ Solve for x using the Equality of Exponents Property.

$\log_3\left(\frac{1}{9}\right) = -2$

Problem 3 Evaluate: $\log_4 64$

Solution See page S27. 3

Solve $\log_4 x = -2$ for x.

Write the equation in its equivalent exponential form.

$$\log_4 x = -2$$
$$4^{-2} = x$$

Solve for x.

$$\frac{1}{16} = x$$

The solution is $\frac{1}{16}$.

INSTRUCTOR NOTE

The concept of antilogarithm is nothing other than a restatement of the equivalence of a logarithmic expression and an exponential expression. This relationship can be illustrated by using a calculator. To solve the equation $\log_{10} x = \frac{1}{2}$, the student uses the [10^x] key on a calculator. To solve $10^x = 3$, the student uses the [Log x] key. The same is true for the natural exponential function and natural logarithm. Some students will not have a [10^x] key on their calculator but will have an [INV] (inverse) key instead. This affords a good opportunity to reinforce the fact that logarithmic and exponential functions are inverses of each other.

Example 4 Solve $\log_6 x = 2$ for x.

Solution $\log_6 x = 2$

$6^2 = x$ ▸ Write $\log_6 x = 2$ in its equivalent exponential form.

$36 = x$

The solution is 36.

Problem 4 Solve $\log_2 x = -4$ for x.

Solution See page S28. $\frac{1}{16}$

In Example 4, 36 is called the **antilogarithm** base 6 of 2. In general, if $\log_b M = N$, then M is the antilogarithm base b of N. The antilogarithm of a number can be determined by rewriting the logarithmic equation in exponential form. For instance, if $\log_5 x = 3$, then x, which is the antilogarithm base 5 of 3, is $x = 5^3 = 125$.

Definition of Antilogarithm

If $\log_b M = N$, the **antilogarithm,** base b, of N is M. In exponential form,

$$M = b^N$$

Logarithms with base 10 are called **common logarithms.** Usually the base, 10, is omitted when the common logarithm of a number is written. Therefore, $\log_{10} x$ is written $\log x$. To find the common logarithm of most numbers, a calculator or table is necessary. Because the logarithms of most numbers are irrational numbers, the value in the display of a calculator is an approximation of the logarithm of the number.

Using a calculator,

mantissa

$$\log 384 \approx 2.584331224$$

characteristic

The decimal part of a *common logarithm* is called the **mantissa;** the integer part is called the **characteristic.**

When e (the base of the natural exponential function) is used as a base of a logarithm, the logarithm is referred to as the **natural logarithm** and is abbreviated $\ln x$. This is read "el en x." Using a calculator, we find that

$$\ln 23 \approx 3.135494216$$

The integer and decimal parts of a natural logarithm do not have names associated with them as they do in common logarithms.

Example 5 Solve $\ln x = -1$ for x. Round to the nearest ten-thousandth.

Solution

$\ln x = -1$ ▸ $\ln x$ is the abbreviation for $\log_e x$.

$e^{-1} = x$ ▸ Write the equation in its equivalent exponential form.

$0.3679 \approx x$

The solution is 0.3679.

Problem 5 Solve $\log x = 1.5$ for x. Round to the nearest ten-thousandth.

Solution See page S28. 31.6228

2 The Properties of Logarithms

38 HM³

Because a logarithm is an exponent, the Properties of Logarithms are similar to the Properties of Exponents.

The table at the right shows some powers of 2 and the equivalent logarithmic form.

$2^0 = 1$	$\log_2 1 = 0$
$2^1 = 2$	$\log_2 2 = 1$
$2^2 = 4$	$\log_2 4 = 2$
$2^3 = 8$	$\log_2 8 = 3$
$2^4 = 16$	$\log_2 16 = 4$
$2^5 = 32$	$\log_2 32 = 5$

The table can be used to show that $\log_2 4 + \log_2 8$ equals $\log_2 32$.

$$\log_2 4 + \log_2 8 = 2 + 3 = 5$$

$$\log_2 32 = 5$$

$$\log_2 4 + \log_2 8 = \log_2 32$$

Note that $\log_2 32 = \log_2(4 \times 8) = \log_2 4 + \log_2 8$.

The property of logarithms that states that the logarithm of the product of two numbers equals the sum of the logarithms of the two numbers is similar to the property of exponents that states that to multiply two exponential expressions with the same base, we add the exponents.

LOOK CLOSELY

Pay close attention to this theorem. Note, for instance, that this theorem states that

$$\log_3(4p) = \log_3 4 + \log_3 p$$

It also states that

$$\log_5 9 + \log_5 z = \log_5(9z)$$

It does *not* state any relationship that involves $\log_b(x + y)$. **This expression cannot be simplified.**

The Product Property of Logarithms

For any positive real numbers x, y, and b, $b \neq 1$,

$$\log_b(xy) = \log_b x + \log_b y$$

Proof:

Let $\log_b x = m$ and $\log_b y = n$.

Write each equation in its equivalent exponential form. $\quad x = b^m \qquad y = b^n$

Use substitution and the Properties of Exponents.

$$xy = b^m b^n$$
$$xy = b^{m+n}$$

Write the equation in its equivalent logarithmic form. $\quad \log_b xy = m + n$

Substitute $\log_b x$ for m and $\log_b y$ for n. $\quad \log_b xy = \log_b x + \log_b y$

The Logarithm Property of Products is used to rewrite logarithmic expressions.

The $\log_b 6z$ is written in **expanded form** as $\log_b 6 + \log_b z$.

$$\log_b 6z = \log_b 6 + \log_b z$$

The $\log_b 12 + \log_b r$ is written as a single logarithm as $\log_b 12r$.

$$\log_b 12 + \log_b r = \log_b 12r$$

The Logarithm Property of Products can be extended to include the logarithm of the product of more than two factors. For example,

$$\log_b xyz = \log_b(xy)z = \log_b xy + \log_b z = \log_b x + \log_b y + \log_b z$$

To write $\log_b 5st$ in expanded form, use the Logarithm Property of Products.

$$\log_b 5st = \log_b 5 + \log_b s + \log_b t$$

A second property of logarithms involves the logarithm of the quotient of two numbers. This property of logarithms is also based on the fact that a logarithm is an exponent and that to divide two exponential expressions with the same base, we subtract the exponents.

The Quotient Property of Logarithms

> For any positive real numbers x, y, and b, $b \neq 1$,
>
> $$\log_b \frac{x}{y} = \log_b x - \log_b y$$

Proof:
Let $\log_b x = m$ and $\log_b y = n$.

Write each equation in its equivalent exponential form.

$$x = b^m \qquad y = b^n$$

Use substitution and the Properties of Exponents.

$$\frac{x}{y} = \frac{b^m}{b^n}$$

$$\frac{x}{y} = b^{m-n}$$

Write the equation in its equivalent logarithmic form.

$$\log_b \frac{x}{y} = m - n$$

Substitute $\log_b x$ for m and $\log_b y$ for n.

$$\log_b \frac{x}{y} = \log_b x - \log_b y$$

The Logarithm Property of Quotients is used to rewrite logarithmic expressions.

The $\log_b \frac{p}{8}$ is written in expanded form as $\log_b p - \log_b 8$.

$$\log_b \frac{p}{8} = \log_b p - \log_b 8$$

The $\log_b y - \log_b v$ is written as a single logarithm as $\log_b \frac{y}{v}$.

$$\log_b y - \log_b v = \log_b \frac{y}{v}$$

POINT OF INTEREST

Logarithms were developed independently by Jobst Burgi (1552–1632) and John Napier (1550–1617) as a means of simplifying the calculations of astronomers. The idea was to devise a method by which two numbers could be multiplied by performing additions. Napier is usually given credit for logarithms because he published his result first.

In Napier's original work, the logarithm of 10,000,000 was 0. After this work was published, Napier, in discussions with Henry Briggs (1561–1631), decided that tables of logarithms would be easier to use if the logarithm of 1 was 0. Napier died before new tables could be determined, and Briggs took on the task. His table consisted of logarithms accurate to 30 decimal places, all accomplished without a calculator!

The logarithms Briggs calculated are the common logarithms mentioned earlier.

A third property of logarithms, useful especially in the computation of a power of a number, is based on the fact that a logarithm is an exponent and that the power of an exponential expression is found by multiplying the exponents.

The table of the powers of 2 shown earlier can be used to show that $\log_2 2^3$ equals $3 \log_2 2$.

$$\log_2 2^3 = \log_2 8 = 3$$
$$3 \log_2 2 = 3 \cdot 1 = 3$$
$$\log_2 2^3 = 3 \log_2 2$$

The Power Property of Logarithms

For any positive real numbers x and b, $b \neq 1$, and for any real number r, $\log_b x^r = r \log_b x$.

Proof:
Let $\log_b x = m$.

Write the equation in its equivalent exponential form.	$x = b^m$
Raise each side to the r power.	$x^r = (b^m)^r$ $x^r = b^{mr}$
Write the equation in its equivalent logarithmic form.	$\log_b x^r = mr$
Substitute $\log_b x$ for m.	$\log_b x^r = r \log_b x$

The Logarithm Property of Powers is used to rewrite logarithmic expressions.

The $\log_b x^3$ is written in terms of $\log_b x$ as $3 \log_b x$.

$$\log_b x^3 = 3 \log_b x$$

$\frac{2}{3} \log_4 z$ is written with a coefficient of 1 as $\log_4 z^{\frac{2}{3}}$.

$$\frac{2}{3} \log_4 z = \log_4 z^{\frac{2}{3}}$$

The Properties of Logarithms can be used in combination to simplify expressions that contain logarithms.

INSTRUCTOR NOTE

Exercises such as these give students practice in applying the Properties of Logarithms. Facility with these operations will help students when they solve exponential and logarithmic equations later in the text.

Example 6 Write the logarithm in expanded form.

A. $\log_b \frac{xy}{z}$ B. $\ln \frac{x^2}{y^3}$ C. $\log_8 \sqrt{x^3 y}$

Solution A. $\log_b \frac{xy}{z} = \log_b(xy) - \log_b z$ ► Use the Quotient Property of Logarithms.

$= \log_b x + \log_b y - \log_b z$ ► Use the Product Property of Logarithms.

B. $\ln \frac{x^2}{y^3} = \ln x^2 - \ln y^3$ ► Use the Quotient Property of Logarithms.

$= 2 \ln x - 3 \ln y$ ► Use the Power Property of Logarithms.

C. $\log_8\sqrt{x^3y} = \log_8(x^3y)^{\frac{1}{2}}$ ▶ Write the radical expression as an exponential expression.

$= \frac{1}{2}\log_8 x^3y$ ▶ Use the Power Property of Logarithms.

$= \frac{1}{2}(\log_8 x^3 + \log_8 y)$ ▶ Use the Product Property of Logarithms.

$= \frac{1}{2}(3\log_8 x + \log_8 y)$ ▶ Use the Power Property of Logarithms.

$= \frac{3}{2}\log_8 x + \frac{1}{2}\log_8 y$ ▶ Use the Distributive Property.

Problem 6 Write the logarithm in expanded form.

A. $\log_b \frac{x^2}{y}$ B. $\ln y^{\frac{1}{3}}z^3$ C. $\log_8\sqrt[3]{xy^2}$

Solution See page S28.

A. $2\log_b x - \log_b y$ B. $\frac{1}{3}\ln y + 3\ln z$ C. $\frac{1}{3}\log_8 x + \frac{2}{3}\log_8 y$

Example 7 Express as a single logarithm with a coefficient of 1.

A. $3\log_5 x + \log_5 y - 2\log_5 z$

B. $\frac{1}{2}(\log_3 x - 3\log_3 y + \log_3 z)$

C. $\frac{1}{3}(2\ln x - 4\ln y)$

Solution A. $3\log_5 x + \log_5 y - 2\log_5 z$

$= \log_5 x^3 + \log_5 y - \log_5 z^2$ ▶ Use the Power Property of Logarithms.

$= \log_5 x^3y - \log_5 z^2$ ▶ Use the Product Property of Logarithms.

$= \log_5 \frac{x^3y}{z^2}$ ▶ Use the Quotient Property of Logarithms.

B. $\frac{1}{2}(\log_3 x - 3\log_3 y + \log_3 z)$

$= \frac{1}{2}(\log_3 x - \log_3 y^3 + \log_3 z)$ ▶ Use the Power Property of Logarithms.

$= \frac{1}{2}\left(\log_3 \frac{x}{y^3} + \log_3 z\right)$ ▶ Use the Quotient Property of Logarithms.

$= \frac{1}{2}\left(\log_3 \frac{xz}{y^3}\right)$ ▶ Use the Product Property of Logarithms.

$= \log_3\left(\frac{xz}{y^3}\right)^{\frac{1}{2}} = \log_3\sqrt{\frac{xz}{y^3}}$ ▶ Use the Power Property of Logarithms. Write the exponential expression as a radical expression.

C. $\frac{1}{3}(2 \ln x - 4 \ln y)$

$= \frac{1}{3}(\ln x^2 - \ln y^4)$ ▶ Use the Power Property of Logarithms.

$= \frac{1}{3}\left(\ln \frac{x^2}{y^4}\right)$ ▶ Use the Quotient Property of Logarithms.

$= \ln\left(\frac{x^2}{y^4}\right)^{\frac{1}{3}} = \ln \sqrt[3]{\frac{x^2}{y^4}}$ ▶ Use the Power Property of Logarithms. Write the exponential expression as a radical expression.

Problem 7 Express as a single logarithm with a coefficient of 1.

A. $2 \log_b x - 3 \log_b y - \log_b z$

B. $\frac{1}{3}(\log_4 x - 2 \log_4 y + \log_4 z)$

C. $\frac{1}{2}(2 \ln x - 5 \ln y)$

Solution See page S28. A. $\log_b \frac{x^2}{y^3 z}$ B. $\log_4 \sqrt[3]{\frac{xz}{y^2}}$ C. $\ln \sqrt{\frac{x^2}{y^5}}$

There are three other properties of logarithms that are useful in simplifying logarithmic expressions.

Properties of Logarithms

The Logarithmic Property of One
For any positive real number b, $b \neq 1$, $\log_b 1 = 0$.

The Inverse Property of Logarithms
For any positive real numbers x and b, $b \neq 1$, $\log_b b^x = x$.

The 1–1 Property of Logarithms
For any positive real numbers x, y, and b, $b \neq 1$, if $\log_b x = \log_b y$, then $x = y$.

Example 8 Simplify. A. $8 \log_4 4$ B. $\log_8 1$

Solution A. $8 \log_4 4 = \log_4 4^8$ ▶ Use the Power Property of Logarithms.

$= 8$ ▶ Use the Inverse Property of Logarithms.

B. $\log_8 1 = 0$ ▶ Use the Logarithmic Property of One.

Problem 8 Simplify. A. $\log_{16} 1$ B. $12 \log_3 3$

Solution See page S28. A. 0 B. 12

Although only common logarithms and natural logarithms are programmed into a calculator, the logarithms for other positive bases can be found.

Evaluate $\log_5 22$. Round to the nearest ten-thousandth.

Write an equation.	$\log_5 22 = x$
Write the equation in its equivalent exponential form.	$5^x = 22$
Apply the common logarithm to each side of the equation.	$\log 5^x = \log 22$
Use the Power Property of Logarithms.	$x \log 5 = \log 22$
This is an exact answer.	$x = \dfrac{\log 22}{\log 5}$
This is an approximate answer.	$x \approx 1.9206$

$\log_5 22 \approx 1.9206$

INSTRUCTOR NOTE

Because graphing calculators have only preprogrammed common and natural logarithms, the change-of-base formula is used to graph logarithms with other bases. For instance, to graph $f(x) = \log_2(x - 3)$, the student must enter

$$\frac{1}{\log 2}\log(x - 3)$$

or an equivalent expression using natural logarithms.

In the third step above, the natural logarithm, instead of the common logarithm, could have been applied to each side of the equation. The same result would have been obtained.

Using a procedure similar to the one used to evaluate $\log_5 22$, a formula for changing bases can be derived.

Change-of-Base Formula

$$\log_a N = \frac{\log_b N}{\log_b a}$$

Example 9 Evaluate $\log_7 32$. Round to the nearest ten-thousandth.

Solution $\log_7 32 = \dfrac{\ln 32}{\ln 7} \approx 1.7810$ ▶ Use the Change-of-Base Formula. $N = 32, a = 7, b = e$

Problem 9 Evaluate $\log_4 2.4$. Round to the nearest ten-thousandth.

Solution See page S28. 0.6315

Example 10 Rewrite $f(x) = -3\log_7(2x - 5)$ in terms of natural logarithms.

Solution

$$\begin{aligned} f(x) &= -3\log_7(2x - 5) \\ &= -3\frac{\ln(2x - 5)}{\ln 7} \\ &= -\frac{3}{\ln 7}\ln(2x - 5) \end{aligned}$$

▶ Use the Change-of-Base Formula to rewrite $\log_7(2x - 5)$ as $\dfrac{\ln(2x - 5)}{\ln 7}$.

Problem 10 Rewrite $f(x) = 4\log_8(3x + 4)$ in terms of common logarithms.

Solution See page S28. $\dfrac{4}{\log 8}\log(3x + 4)$

In Example 10, it is important to understand that $-\dfrac{3}{\ln 7}\ln(2x - 5)$ and $-3\log_7(2x - 5)$ are *exactly* equal. If common logarithms had been used, the result would have been $f(x) = -\dfrac{3}{\log 7}\log(2x - 5)$. The expressions $-\dfrac{3}{\log 7}\log(2x - 5)$

and $-3 \log_7(2x - 5)$ are also *exactly* equal. If you are working in a base other than base 10 or base e, the Change-of-Base Formula will enable you to calculate the value of the logarithm in that base just as if that base were programmed into the calculator.

CONCEPT REVIEW 10.2

Determine whether the following statements are always true, sometimes true, or never true.

1. For $b > 0, b \neq 1, y = \log_b x$ is equivalent to $b^y = x$. Always true
2. $\log_b \frac{x}{y} = \frac{\log_b x}{\log_b y}$ Never true
3. $\frac{\log x}{\log y} = \frac{x}{y}$ Never true
4. $\log(x + y) = \log x + \log y$ Never true
5. $\log_b \sqrt{x} = \frac{1}{2} \log_b x$ Always true
6. $\log(x^{-1}) = \frac{1}{\log x}$ Never true
7. The inverse of an exponential function is a logarithmic function. Always true
8. If x and y are positive real numbers, $x < y$, and $b > 0$, then $\log_b x < \log_b y$. Always true

EXERCISES 10.2

1 1. If $y = \log_b x$, express the antilogarithm of y as an equation.

2. If $b > 0, b \neq 1$, and $b^x = b^y$, what can be said about x and y? If $b = 1$, what can be said about x and y?

Write the exponential equation in logarithmic form.

3. $5^2 = 25$ $\log_5 25 = 2$
4. $10^3 = 1000$ $\log_{10} 1000 = 3$
5. $4^{-2} = \frac{1}{16}$ $\log_4 \frac{1}{16} = -2$
6. $3^{-3} = \frac{1}{27}$ $\log_3 \frac{1}{27} = -3$
7. $10^y = x$ $\log_{10} x = y$
8. $e^y = x$ $\log_e x = y$
9. $a^x = w$ $\log_a w = x$
10. $b^y = c$ $\log_b c = y$

Write the logarithmic equation in exponential form.

11. $\log_3 9 = 2$ $3^2 = 9$
12. $\log_2 32 = 5$ $2^5 = 32$
13. $\log 0.01 = -2$ $10^{-2} = 0.01$
14. $\log_5 \frac{1}{5} = -1$ $5^{-1} = \frac{1}{5}$
15. $\ln x = y$ $e^y = x$
16. $\log x = y$ $10^y = x$
17. $\log_b u = v$ $b^v = u$
18. $\log_c x = y$ $c^y = x$

Evaluate.

19. $\log_3 81$ 4
20. $\log_7 49$ 2
21. $\log_2 128$ 7
22. $\log_5 125$ 3
23. $\log 100$ 2
24. $\log 0.001$ -3
25. $\ln e^3$ 3
26. $\ln e^2$ 2
27. $\log_8 1$ 0
28. $\log_3 243$ 5
29. $\log_5 625$ 4
30. $\log_2 64$ 6

Solve for x.

31. $\log_3 x = 2$ 9
32. $\log_5 x = 1$ 5
33. $\log_4 x = 3$ 64
34. $\log_2 x = 6$ 64
35. $\log_7 x = -1$ $\frac{1}{7}$
36. $\log_8 x = -2$ $\frac{1}{64}$
37. $\log_6 x = 0$ 1
38. $\log_4 x = 0$ 1

Solve for x. Round to the nearest hundredth.

39. $\log x = 2.5$ 316.23
40. $\log x = 3.2$ 1584.89
41. $\log x = -1.75$ 0.02
42. $\log x = -2.1$ 0.01
43. $\ln x = 2$ 7.39
44. $\ln x = 4$ 54.60
45. $\ln x = -\frac{1}{2}$ 0.61
46. $\ln x = -1.7$ 0.18

2 **47.** What is the Product Property of Logarithms?

48. What is the Quotient Property of Logarithms?

Write the logarithm in expanded form.

49. $\log_8(xz)$ $\log_8 x + \log_8 z$
50. $\log_7(4y)$ $\log_7 4 + \log_7 y$
51. $\log_3 x^5$ $5\log_3 x$
52. $\log_2 y^7$ $7\log_2 y$
53. $\ln\left(\frac{r}{s}\right)$ $\ln r - \ln s$
54. $\ln\left(\frac{z}{4}\right)$ $\ln z - \ln 4$
55. $\log_3(x^2y^6)$ $2\log_3 x + 6\log_3 y$
56. $\log_4(t^4u^2)$ $4\log_4 t + 2\log_4 u$
57. $\log_7\left(\frac{u^3}{v^4}\right)$ $3\log_7 u - 4\log_7 v$
58. $\log\left(\frac{s^5}{t^2}\right)$ $5\log s - 2\log t$
59. $\log_2(rs)^2$ $2\log_2 r + 2\log_2 s$
60. $\log_3(x^2y)^3$ $6\log_3 x + 3\log_3 y$
61. $\log_9 x^2yz$ $2\log_9 x + \log_9 y + \log_9 z$
62. $\log_6 xy^2z^3$ $\log_6 x + 2\log_6 y + 3\log_6 z$
63. $\ln\left(\frac{xy^2}{z^4}\right)$ $\ln x + 2\ln y - 4\ln z$
64. $\ln\left(\frac{r^2s}{t^3}\right)$ $2\ln r + \ln s - 3\ln t$
65. $\log_8\left(\frac{x^2}{yz^2}\right)$ $2\log_8 x - \log_8 y - 2\log_8 z$
66. $\log_9\left(\frac{x}{y^2z^3}\right)$ $\log_9 x - 2\log_9 y - 3\log_9 z$
67. $\log_7\sqrt{xy}$ $\frac{1}{2}\log_7 x + \frac{1}{2}\log_7 y$
68. $\log_8\sqrt[3]{xz}$ $\frac{1}{3}\log_8 x + \frac{1}{3}\log_8 z$
69. $\log_2\sqrt{\frac{x}{y}}$ $\frac{1}{2}\log_2 x - \frac{1}{2}\log_2 y$
70. $\log_3\sqrt[3]{\frac{r}{s}}$ $\frac{1}{3}\log_3 r - \frac{1}{3}\log_3 s$
71. $\ln\sqrt{x^3y}$ $\frac{3}{2}\ln x + \frac{1}{2}\ln y$
72. $\ln\sqrt{x^5y^3}$ $\frac{5}{2}\ln x + \frac{3}{2}\ln y$
73. $\log_7\sqrt{\frac{x^3}{y}}$ $\frac{3}{2}\log_7 x - \frac{1}{2}\log_7 y$
74. $\log_b\sqrt[3]{\frac{r^2}{t}}$ $\frac{2}{3}\log_b r - \frac{1}{3}\log_b t$

Express as a single logarithm with a coefficient of 1.

75. $\log_3 x^3 - \log_3 y$ $\log_3 \frac{x^3}{y}$

76. $\log_7 t + \log_7 v^2$ $\log_7 tv^2$

77. $\log_8 x^4 + \log_8 y^2$ $\log_8 x^4y^2$

78. $\log_2 r^2 + \log_2 s^3$ $\log_2 r^2s^3$

79. $3 \ln x$ $\ln x^3$

80. $4 \ln y$ $\ln y^4$

81. $3 \log_5 x + 4 \log_5 y$ $\log_5 x^3y^4$

82. $2 \log_6 x + 5 \log_6 y$ $\log_6 x^2y^5$

83. $-2 \log_4 x$ $\log_4 \frac{1}{x^2}$

84. $-3 \log_2 y$ $\log_2 \frac{1}{y^3}$

85. $2 \log_3 x - \log_3 y + 2 \log_3 z$ $\log_3 \frac{x^2z^2}{y}$

86. $4 \log_5 r - 3 \log_5 s + \log_5 t$ $\log_5 \frac{r^4t}{s^3}$

87. $\log_b x - (2 \log_b y + \log_b z)$ $\log_b \frac{x}{y^2z}$

88. $2 \log_2 x - (3 \log_2 y + \log_2 z)$ $\log_2 \frac{x^2}{y^3z}$

89. $2(\ln x + \ln y)$ $\ln x^2y^2$

90. $3(\ln r + \ln t)$ $\ln r^3t^3$

91. $\frac{1}{2}(\log_6 x - \log_6 y)$ $\log_6 \sqrt{\frac{x}{y}}$

92. $\frac{1}{3}(\log_8 x - \log_8 y)$ $\log_8 \sqrt[3]{\frac{x}{y}}$

93. $2(\log_4 s - 2 \log_4 t + \log_4 r)$ $\log_4 \frac{s^2r^2}{t^4}$

94. $3(\log_9 x + 2 \log_9 y - 2 \log_9 z)$ $\log_9 \frac{x^3y^6}{z^6}$

95. $\log_5 x - 2(\log_5 y + \log_5 z)$ $\log_5 \frac{x}{y^2z^2}$

96. $\log_4 t - 3(\log_4 u + \log_4 v)$ $\log_4 \frac{t}{u^3v^3}$

97. $3 \ln t - 2(\ln r - \ln v)$ $\ln \frac{t^3v^2}{r^2}$

98. $2 \ln x - 3(\ln y - \ln z)$ $\ln \frac{x^2z^3}{y^3}$

99. $\frac{1}{2}(3 \log_4 x - 2 \log_4 y + \log_4 z)$ $\log_4 \sqrt{\frac{x^3z}{y^2}}$

100. $\frac{1}{3}(4 \log_5 t - 3 \log_5 u - 3 \log_5 v)$ $\log_5 \sqrt[3]{\frac{t^4}{u^3v^3}}$

Evaluate. Round to the nearest ten-thousandth.

101. $\ln 4$ 1.3863

102. $\ln 6$ 1.7918

103. $\ln\left(\frac{17}{6}\right)$ 1.0415

104. $\ln\left(\frac{13}{17}\right)$ −0.2683

105. $\log_8 6$ 0.8617

106. $\log_4 8$ 1.5000

107. $\log_5 30$ 2.1133

108. $\log_6 28$ 1.8597

109. $\log_3(0.5)$ −0.6309

110. $\log_5(0.6)$ −0.3174

111. $\log_7(1.7)$ 0.2727

112. $\log_6(3.2)$ 0.6492

113. $\log_5 15$ 1.6826

114. $\log_3 25$ 2.9299

115. $\log_{12} 120$ 1.9266

116. $\log_9 90$ 2.0480

Rewrite each function in terms of common logarithms.

117. $f(x) = \log_3(3x - 2)$ $\frac{\log(3x - 2)}{\log 3}$

118. $f(x) = \log_5(x^2 + 4)$ $\frac{\log(x^2 + 4)}{\log 5}$

119. $f(x) = \log_8(4 - 9x)$ $\frac{\log(4 - 9x)}{\log 8}$

120. $f(x) = \log_7(3x^2)$ $\frac{\log 3x^2}{\log 7}$

121. $f(x) = 5 \log_9(6x + 7)$ $\frac{5}{\log 9} \log(6x + 7)$

122. $f(x) = 3 \log_2(2x^2 - x)$ $\frac{3}{\log 2}(\log 2x^2 - x)$

Rewrite each function in terms of natural logarithms.

123. $f(x) = \log_2(x + 5)$ $\frac{\ln(x + 5)}{\ln 2}$

124. $f(x) = \log_4(3x + 4)$ $\frac{\ln(3x + 4)}{\ln 4}$

125. $f(x) = \log_3(x^2 + 9)$ $\frac{\ln(x^2 + 9)}{\ln 3}$

126. $f(x) = \log_7(9 - x^2)$ $\frac{\ln(9 - x^2)}{\ln 7}$

127. $f(x) = 7 \log_8(10x - 7)$ $\frac{7}{\ln 8} \ln(10x - 7)$

128. $f(x) = 7 \log_3(2x^2 - x)$ $\frac{7}{\ln 3} \ln(2x^2 - x)$

APPLYING CONCEPTS 10.2

Use the Properties of Logarithms to solve for x.

129. $\log_8 x = 3\log_8 2$ 8

130. $\log_5 x = 2\log_5 3$ 9

131. $\log_4 x = \log_4 2 + \log_4 3$ 6

132. $\log_3 x = \log_3 4 + \log_3 7$ 28

133. $\log_6 x = 3\log_6 2 - \log_6 4$ 2

134. $\log_9 x = 5\log_9 2 - \log_9 8$ 4

135. $\log x = \frac{1}{3}\log 27$ 3

136. $\log_2 x = \frac{3}{2}\log_2 4$ 8

[C] **137.** Using the Properties of Logarithms, show that $\log_a a^x = x, a > 0$.
Complete solution is available in the Solutions Manual.

[C] **138.** Using the Properties of Logarithms, show that $a^{\log_a x} = x, a > 0, x > 0$.
Complete solution is available in the Solutions Manual.

139. Complete each statement using the equation $\log_a b = c$.

a. $a^c =$ ___b___ **b.** $\text{antilog}_a(\log_a b) =$ ___b___

140. Given $f(x) = 3\log_6(2x - 1)$, determine $f(7)$ to the nearest hundredth. 4.29

141. Given $S(t) = 8\log_5(6t + 2)$, determine $S(2)$ to the nearest hundredth. 13.12

142. Given $P(v) = -3\log_6(4 - 2v)$, determine $P(-4)$ to the nearest hundredth.
−4.16

143. Given $G(x) = -5\log_7(2x + 19)$, determine $G(-3)$ to the nearest hundredth.
−6.59

144. Solve for x: $\log_2(\log_2 x) = 3$ 256

145. Solve for x: $\ln(\ln x) = 1$ 15.1543

146. Show that $\log_b a = \dfrac{1}{\log_a b}$. Complete solution is available in the Solutions Manual.

147. Show that $\log\left(\dfrac{x - \sqrt{x^2 - a^2}}{a^2}\right) = -\log(x + \sqrt{x^2 - a^2})$ where $x > a > 0$.

Complete solution is available in the Solutions Manual.

148. *Biology* To discuss the variety of species that live in a certain environment, a biologist needs a precise definition of *diversity*. Let $p_1, p_2, \ldots, p_n$ be the proportions of n species that live in an environment. The biological diversity, D, of this system is

$$D = -(p_1 \log_2 p_1 + p_2 \log_2 p_2 + \cdots + p_n \log_2 p_n)$$

The larger the value of D, the greater the diversity of the system. Suppose an ecosystem has exactly five different varieties of grass: rye (R), bermuda (B), blue (L), fescue (F), and St. Augustine (A).

a. Calculate the diversity of this ecosystem if the proportions are as in Table 1. 2.3219281

b. Because bermuda and St. Augustine are virulent grasses, after a time the proportions are as in Table 2. Does this system have more or less diversity than the one given in Table 1? Less

Table 1

R	B	L	F	A
$\frac{1}{5}$	$\frac{1}{5}$	$\frac{1}{5}$	$\frac{1}{5}$	$\frac{1}{5}$

Table 2

R	B	L	F	A
$\frac{1}{8}$	$\frac{3}{8}$	$\frac{1}{16}$	$\frac{1}{8}$	$\frac{5}{16}$

c. After an even longer time period, the bermuda and St. Augustine completely overrun the environment, and the proportions are as in Table 3. Calculate the diversity of this system. (*Note:* For purposes of the diversity definition, $0 \log_2 0 = 0$.) Does it have more or less diversity than the system given in Table 2? Less

Table 3

R	B	L	F	A
0	$\frac{1}{4}$	0	0	$\frac{3}{4}$

d. Finally, the St. Augustine overruns the bermuda, and the proportions are as in Table 4. Calculate the diversity of this system. Write a sentence that describes your answer.
0; this system has only one species, so there is no diversity in the system.

Table 4

R	B	L	F	A
0	0	0	0	1

149. Explain how common logarithms can be used to determine the number of digits in the expansion of $9^{(9^9)}$.

SECTION 10.3 Graphs of Logarithmic Functions

1 Graph logarithmic functions

The graph of a logarithmic function can be drawn by using the relationship between the exponential and logarithmic functions.

To graph $f(x) = \log_2 x$, think of the function as the equation $y = \log_2 x$.

$$f(x) = \log_2 x$$
$$y = \log_2 x$$

Write the equivalent exponential equation.

$$x = 2^y$$

Because the equation is solved for x in terms of y, it is easier to choose values of y and find the corresponding values of x. The results can be recorded in a table.

x	y
$\frac{1}{4}$	-2
$\frac{1}{2}$	-1
1	0
2	1
4	2

Graph the ordered pairs on a rectangular coordinate system.

Connect the points with a smooth curve.

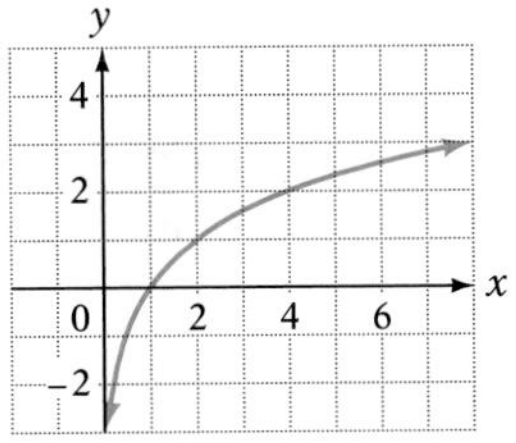

Applying the vertical- and horizontal-line tests reveals that $f(x) = \log_2 x$ is the graph of a 1–1 function.

> **POINT OF INTEREST**
>
> Although logarithms were originally developed to assist with computations, logarithmic functions have a much broader use today. These functions occur in geology, acoustics, chemistry, and economics, for example.

Recall that the graph of the inverse of a function f is the mirror image of f with respect to the line $y = x$. The graph of $f(x) = 2^x$ was shown earlier. Because $g(x) = \log_2 x$ is the inverse of $f(x) = 2^x$, the graphs of these functions are mirror images of each other with respect to the line $y = x$.

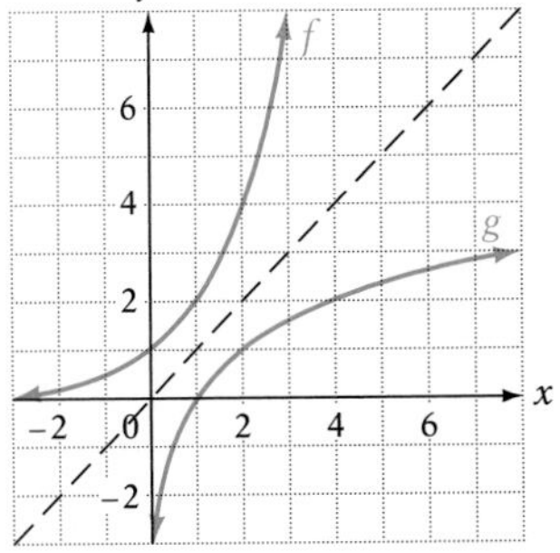

Graph: $f(x) = \log_2 x + 1$

Think of the function as the equation $y = \log_2 x + 1$.

$$f(x) = \log_2 x + 1$$
$$y = \log_2 x + 1$$

Solve the equation for $\log_2 x$.

$$y - 1 = \log_2 x$$

Write the equivalent exponential equation.

$$2^{y-1} = x$$

Choose values of y, and find the corresponding values of x.

Graph the ordered pairs on a rectangular coordinate system.

Connect the points with a smooth curve.

x	y
$\frac{1}{4}$	−1
$\frac{1}{2}$	0
1	1
2	2
4	3

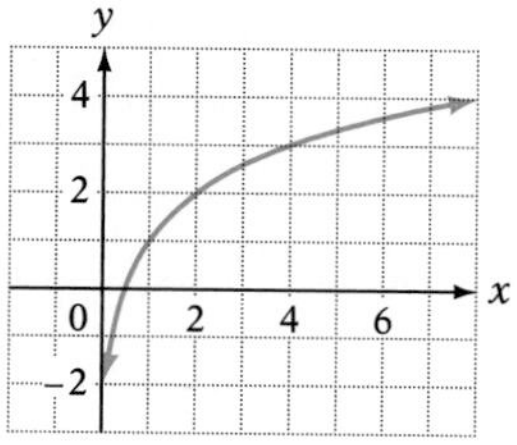

Example 1 Graph. A. $f(x) = \log_3 x$ B. $f(x) = 2\log_3 x$

Solution A. $f(x) = \log_3 x$

$y = \log_3 x$ ▶ Substitute y for $f(x)$.

$x = 3^y$ ▶ Write the equivalent exponential equation.

▶ Choose values of y, and find the corresponding values of x. Graph the ordered pairs on a rectangular coordinate system. Connect the points with a smooth curve.

x	y
$\frac{1}{9}$	−2
$\frac{1}{3}$	−1
1	0
3	1

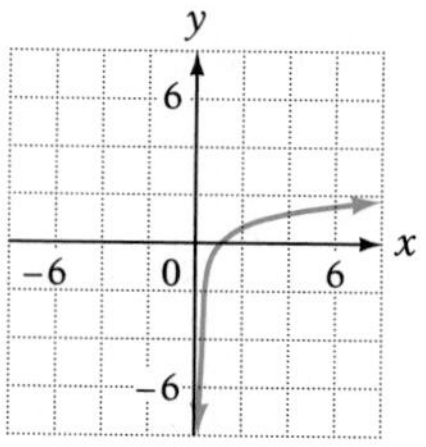

B. $f(x) = 2\log_3 x$

$y = 2\log_3 x$ ▶ Substitute y for $f(x)$.

$\frac{y}{2} = \log_3 x$ ▶ Solve the equation for $\log_3 x$.

$x = 3^{\frac{y}{2}}$ ▶ Write the equivalent exponential equation.

▶ Choose values of y, and find the corresponding values of x. Graph the ordered pairs on a rectangular coordinate system. Connect the points with a smooth curve.

x	y
$\frac{1}{9}$	−4
$\frac{1}{3}$	−2
1	0
3	2

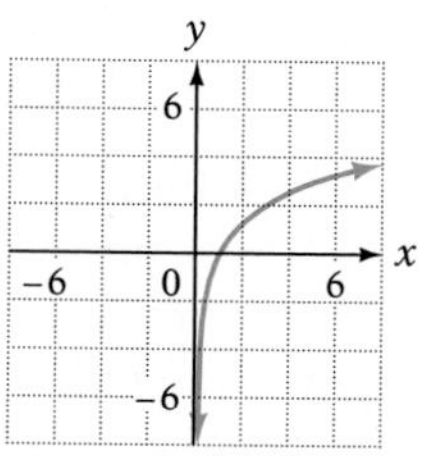

Problem 1 Graph. A. $f(x) = \log_2(x - 1)$ B. $f(x) = \log_3 2x$

Solution See page S28. A.

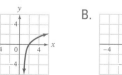

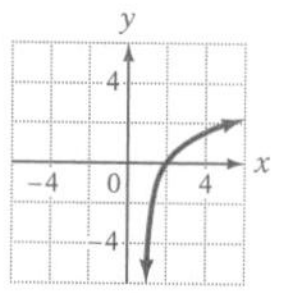

B.

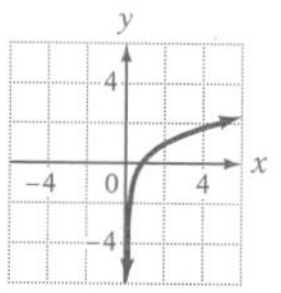

Example 2 Graph: $f(x) = 2 \ln x + 3$

Solution The graph is shown below. To verify the accuracy of the graph, evaluate $f(x)$ for a few values of x, and compare the results to values found by using features of a graphing utility to trace along the graph. For example, $f(1) = 2 \ln(1) + 3 = 3$, so $(1, 3)$ is a point on the graph. When $x = 3$, $f(3) = 2 \ln(3) + 3 \approx 5.2$, so $(3, 5.2)$ is a point on the graph.

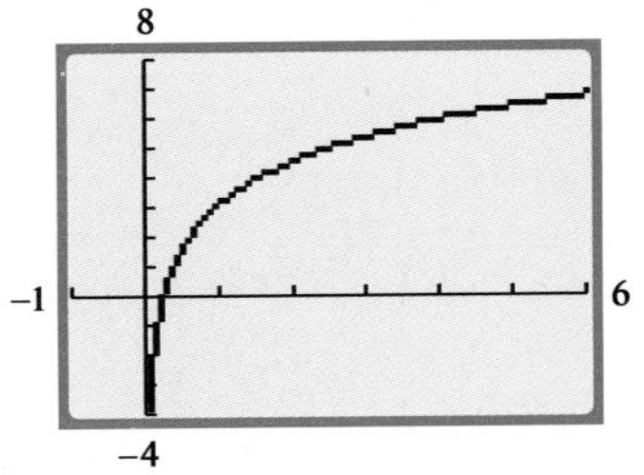

Problem 2 Graph: $f(x) = 10 \log(x - 2)$

Solution See page S29.

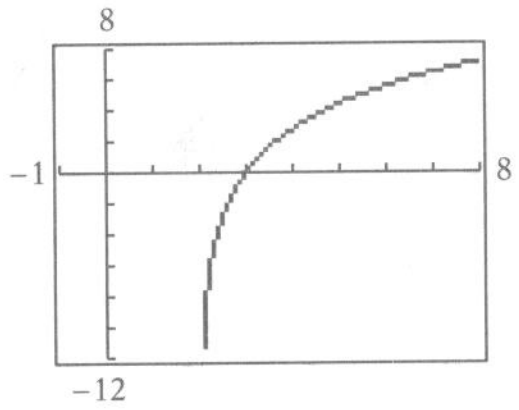

The graphs of logarithmic functions to other than base e or base 10 can be drawn with a graphing utility by first using the Change-of-Base Formula $\log_a N = \frac{\log_b N}{\log_b a}$ to rewrite the logarithmic function in terms of base e or base 10.

Graph: $f(x) = \log_3 x$

Use the Change-of-Base Formula to rewrite $\log_3 x$ in terms of $\log x$ or $\ln x$. The natural logarithm function $\ln x$ is used here.

$$\log_3 x = \frac{\ln x}{\ln 3}$$

To graph $f(x) = \log_3 x$ using a graphing utility, use the equivalent form $f(x) = \frac{\ln x}{\ln 3}$. The graph is shown at the right.

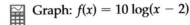

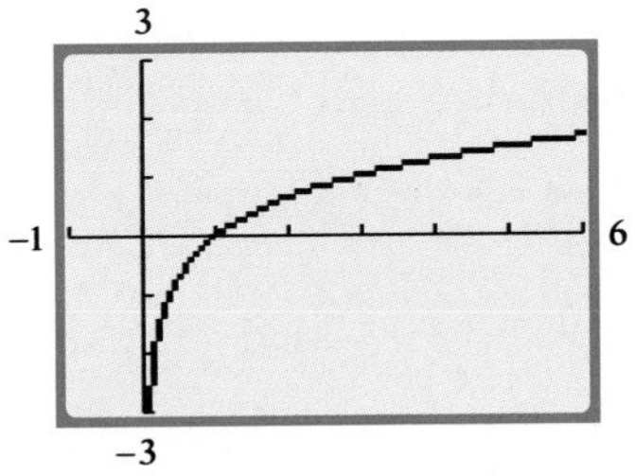

The graph of $f(x) = \log_3 x$ could have been drawn by rewriting $\log_3 x$ in terms of $\log x$, $\log_3 x = \frac{\log x}{\log 3}$. The graph of $f(x) = \log_3 x$ is identical to the graph of $f(x) = \frac{\log x}{\log 3}$.

The examples that follow were graphed by rewriting the logarithmic function in terms of the natural logarithmic function. The common logarithmic function could also have been used.

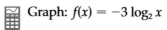

Example 3 Graph: $f(x) = -3 \log_2 x$

Solution $f(x) = -3 \log_2 x$

$$= -3\frac{\ln x}{\ln 2} = -\frac{3}{\ln 2} \ln x$$

▶ Rewrite $\log_2 x$ in terms of $\ln x$. $\log_2 x = \frac{\ln x}{\ln 2}$

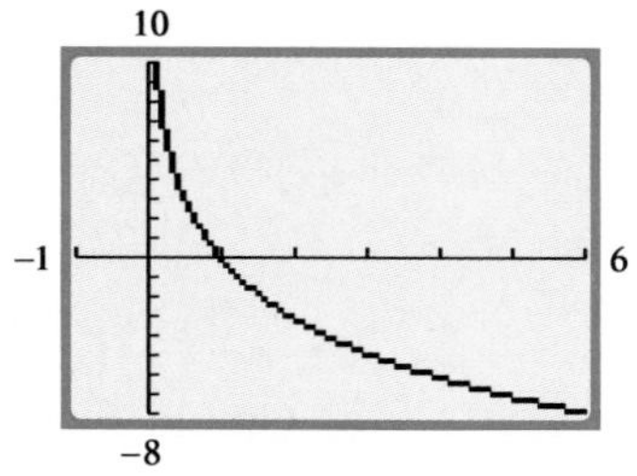

▶ The graph of $f(x) = -3 \log_2 x$ is the same as the graph of $f(x) = -\frac{3}{\ln 2} \ln x$.

Problem 3 Graph: $f(x) = 2 \log_4 x$

Solution See page S29.

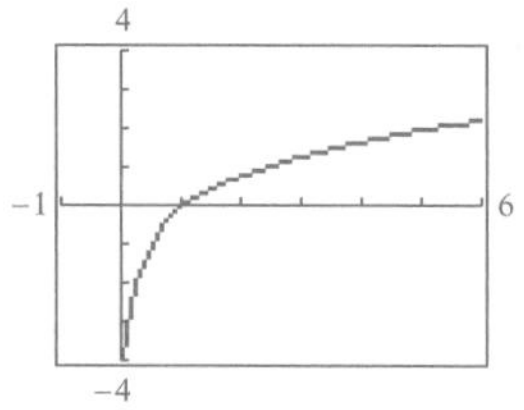

Example 4 Graph $f(x) = 3 \log_4(x + 3)$ and estimate, to the nearest tenth, the value of x for which $f(x) = 4$.

Solution $f(x) = 3 \log_4(x + 3)$

$$= \left(\frac{3}{\ln 4}\right) \ln(x + 3)$$

Using the features of a graphing utility, $f(x) = 4$ when $x \approx 3.3$.

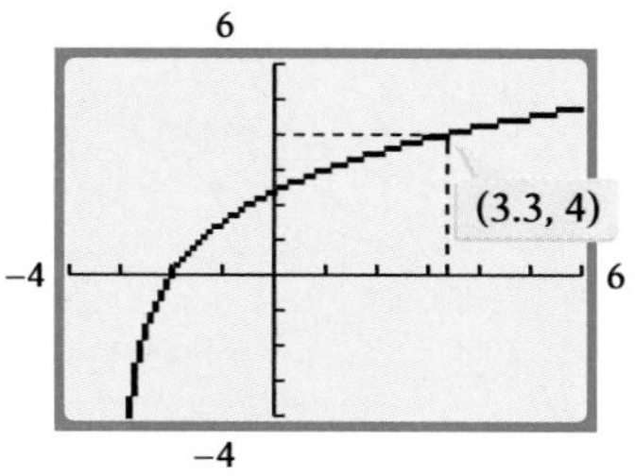

Problem 4 Graph $f(x) = -2 \log_5(3x - 4)$ and estimate, to the nearest tenth, the value of x for which $f(x) = 1$.

Solution See page S29.

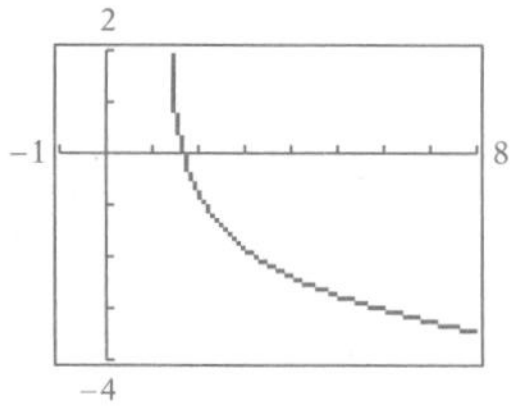

1.5

In Example 4, an algebraic solution can be determined by using the relationship between the exponential function and the logarithmic function.

	$f(x) = 3\log_4(x + 3)$
Replace $f(x)$ by 4.	$4 = 3\log_4(x + 3)$
Solve for $\log_4(x + 3)$.	$\frac{4}{3} = \log_4(x + 3)$
Rewrite the logarithmic equation in exponential form.	$4^{\frac{4}{3}} = x + 3$
Solve for x.	$x = 4^{\frac{4}{3}} - 3 \approx 3.3$

The algebraic solution confirms the graphical solution.

CONCEPT REVIEW 10.3

Determine whether the following statements are always true, sometimes true, or never true.

1. The domain of $f(x) = \log_b x$, $b > 0$, $b \neq 1$, is the real numbers. Never true
2. The range of $f(x) = \log_b x$, $b > 0$, $b \neq 1$, is the real numbers. Always true
3. The graphs of $x = 2^y$ and $y = \log_2 x$ are identical graphs. Always true
4. The range of $f(x) = \ln x$ is the real numbers greater than zero. Never true

EXERCISES 10.3

1 1. Name two characteristics of the graph of $y = \log_b x$, where $b > 1$.

2. What is the relationship between the graph of $y = 3^x$ and $y = \log_3 x$?

Graph.

3. $f(x) = \log_4 x$

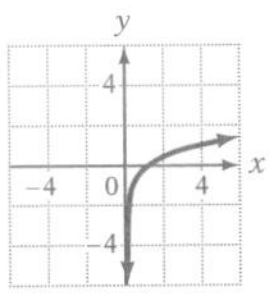

4. $f(x) = \log_2(x + 1)$

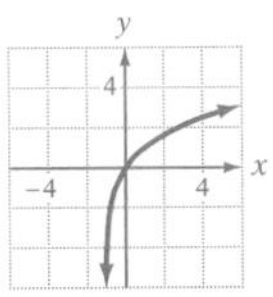

5. $f(x) = \log_3(2x - 1)$

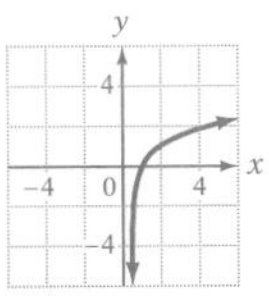

6. $f(x) = \log_2\left(\frac{1}{2}x\right)$

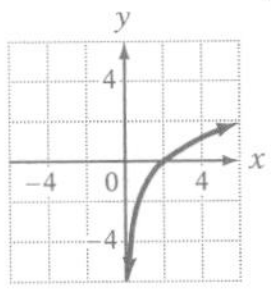

7. $f(x) = 3\log_2 x$

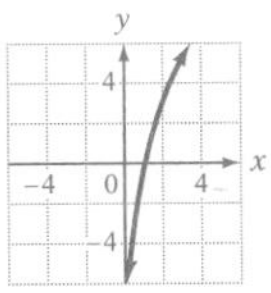

8. $f(x) = \frac{1}{2}\log_2 x$

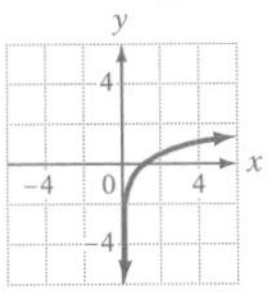

9. $f(x) = -\log_2 x$

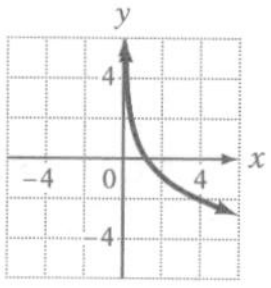

10. $f(x) = -\log_3 x$

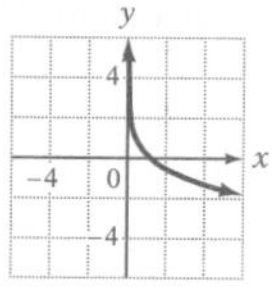

11. $f(x) = \log_2(x - 1)$

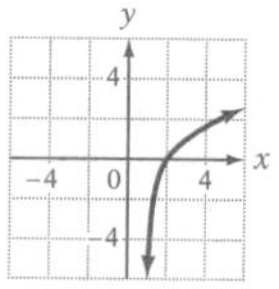

12. $f(x) = \log_3(2 - x)$

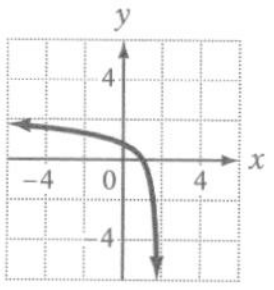

13. $f(x) = -\log_2(x - 1)$

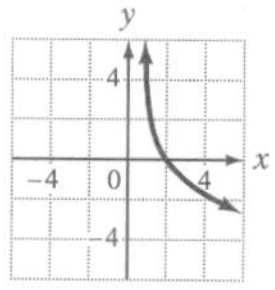

14. $f(x) = -\log_2(1 - x)$

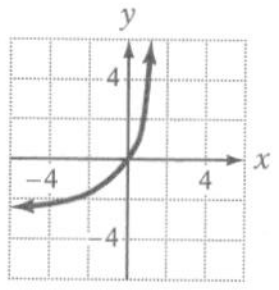

Use a graphing utility to graph the following.

15. $f(x) = \log_2 x - 3$

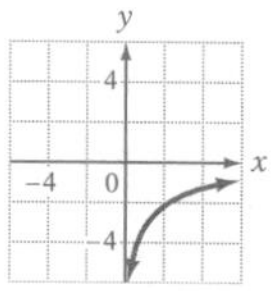

16. $f(x) = \log_3 x + 2$

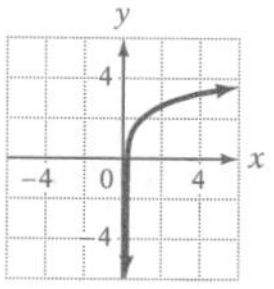

17. $f(x) = -\log_2 x + 2$

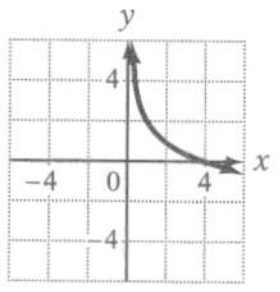

18. $f(x) = -\frac{1}{2}\log_2 x - 1$

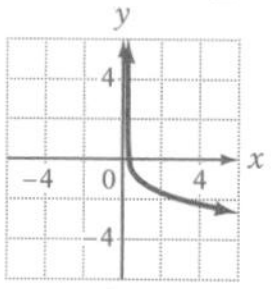

19. $f(x) = x - \log_2(1 - x)$

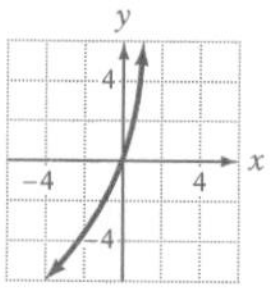

20. $f(x) = x + \log_3(2 - x)$

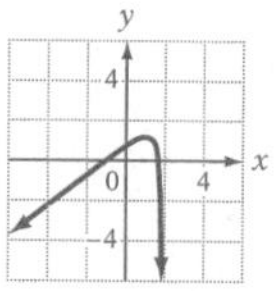

21. $f(x) = \frac{x}{2} - 2\log_2(x + 1)$

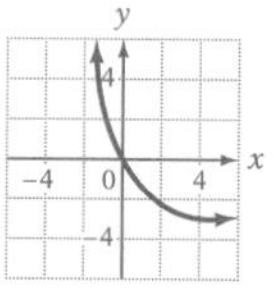

22. $f(x) = \frac{x}{3} - 3\log_2(x + 3)$

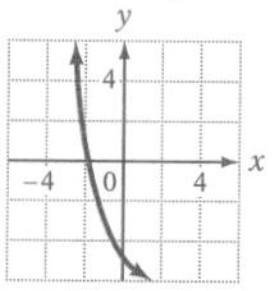

23. $f(x) = x^2 - 10\ln(x - 1)$

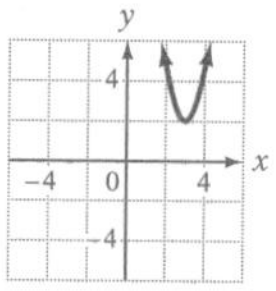

Graph the functions on the same rectangular coordinate system.

24. $f(x) = 3^x$; $g(x) = \log_3 x$

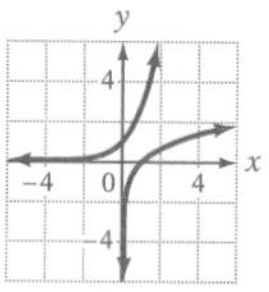

25. $f(x) = 4^x$; $g(x) = \log_4 x$

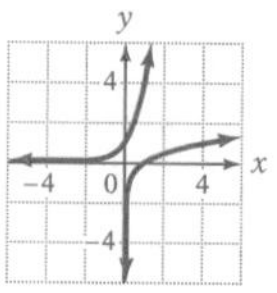

26. $f(x) = 10^x$; $g(x) = \log_{10} x$

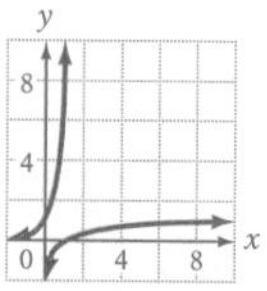

27. $f(x) = \left(\frac{1}{2}\right)^x$; $g(x) = \log_{\frac{1}{2}} x$

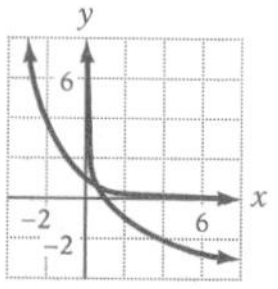

APPLYING CONCEPTS 10.3

For Exercises 28–33, determine the domain. Recall that the logarithm of a negative number is not defined.

28. $f(x) = \log_3(x - 4)$ $\{x \mid x > 4\}$

29. $f(x) = \log_2(x + 2)$ $\{x \mid x > -2\}$

30. $f(x) = \ln(x^2 - 4)$ $\{x \mid x < -2 \text{ or } x > 2\}$

31. $f(x) = \ln(x^2 + 4)$ $\{x \mid x \in \text{real numbers}\}$

32. $f(x) = \log_2 x + \log_2(x - 1)$ $\{x \mid x > 1\}$

33. $f(x) = \log_4\left(\frac{x}{x+2}\right)$ $\{x \mid x < -2 \text{ or } x > 0\}$

34. Let $f(x) = e^{2x} - 1$. Find $f^{-1}(x)$. $\frac{\ln(x+1)}{2}$

35. Let $f(x) = e^{-x+2}$. Find $f^{-1}(x)$. $-\ln x + 2$

36. Let $f(x) = \ln(2x + 3)$. Find $f^{-1}(x)$. $\frac{e^x - 3}{2}$

37. Let $f(x) = \ln(2x) + 3$. Find $f^{-1}(x)$. $\frac{e^{x-3}}{2}$

38. Graph $f(x) = \frac{\ln x}{x}$ and determine, to the nearest tenth, the maximum value of f. 0.4

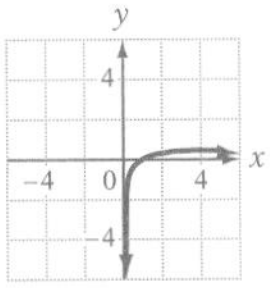

39. Graph $f(x) = x^2 - \ln x$ and determine, to the nearest tenth, the minimum value of f. 0.8

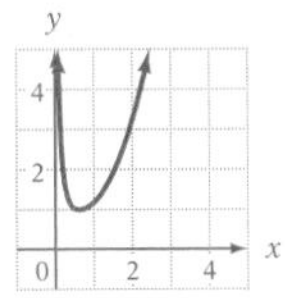

40. *Astronomy* Astronomers use the *distance modulus* of a star as a method of determining the star's distance from Earth. The formula is $M = 5 \log s - 5$, where M is the distance modulus and s is the star's distance from Earth in parsecs. (One parsec $\approx 2.1 \times 10^{13}$ miles.)

a. Graph the equation.

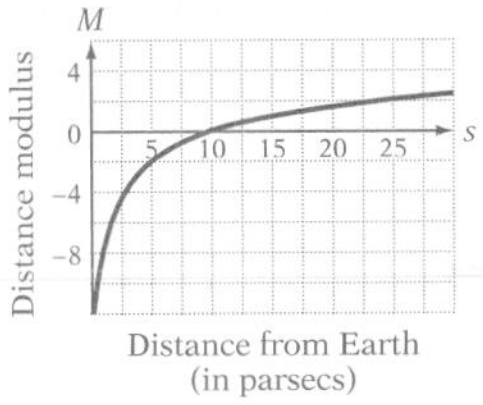

b. The point whose approximate coordinates are (25.1, 2) is on the graph. Write a sentence that describes the meaning of this ordered pair.
The point (25.1, 2) means that a star that is 25.1 parsecs from Earth has a distance modulus of 2.

41. *Employment* The proficiency of a typist decreases (without practice) over time. An equation that approximates this decrease is given by $S = 60 - 7\ln(t + 1)$, where S is the typing speed in words per minute and t is the number of months without typing.

a. Graph the equation.

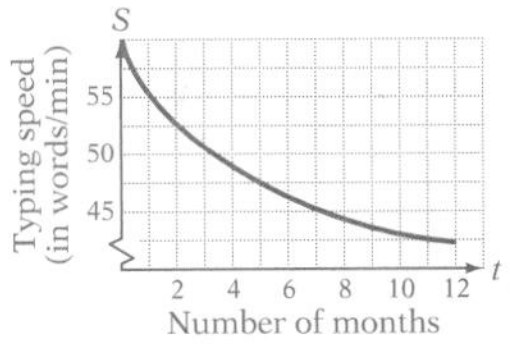

b. The point whose approximate coordinates are (4, 49) is on the graph. Write a sentence that describes the meaning of this ordered pair.
The point (4, 49) means that after 4 months, the typist's proficiency has dropped to 49 words per minute.

42. *Environmental Science* According to the U.S. Environmental Protection Agency, the amount of garbage generated per person has been increasing over the last few decades. The table below shows the per capita garbage, in pounds per day, generated in the United States.

Year	Pounds per Day
1960	2.66
1970	3.27
1980	3.61
1990	4.00

a. Draw a scatter diagram for the data in the table.

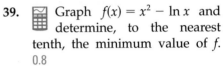

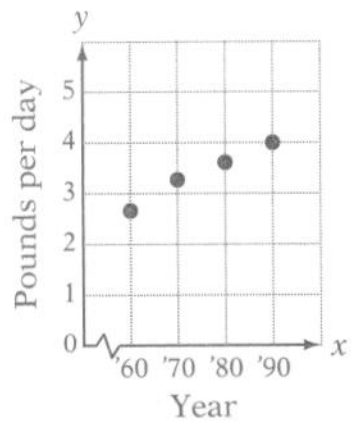

b. Would the equation that best fits the points graphed be the equation of a linear function, an exponential function, or a logarithmic function? A linear function

43. *Interest Rates* General interest rate theory suggests that short-term interest rates (less than 2 years) are lower than long-term interest rates (more than 10 years) because short-term securities are less risky than long-term ones. In periods of high inflation, however, the situation is reversed and economists discuss *inverted-yield* curves. During the early 1980s, inflation was very high in the United States. The rates for short- and long-term U.S. Treasury securities during 1980 are shown in the table at the right. An equation that models these data is given by $y = 14.33759 - 0.62561 \ln x$, where x is the term of the security in years and y is the interest rate as a percent.

Term (in years)	Interest Rate
0.5	15.0%
1	14.0%
5	13.5%
10	12.8%
20	12.5%

a. Graph the equation.

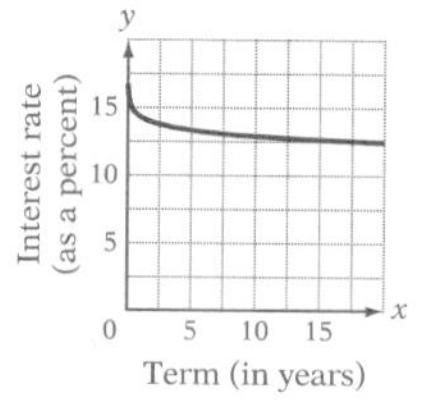

b. According to this model, what is the term, to the nearest tenth of a year, of a security that has a yield of 13%? 8.5 years

c. Determine the interest rate, to the nearest tenth of a percent, that this model predicts for a security that has a 30-year maturity. 12.2%

44. Using the Power Property of Logarithms, $\ln x^2 = 2 \ln x$. Graph the equations $f(x) = \ln x^2$ and $g(x) = 2 \ln x$ on the same rectangular coordinate system. Are the graphs the same? Why or why not?

45. Because $f(x) = e^x$ and $g(x) = \ln x$ are inverse functions of each other, $f[g(x)] = x$ and $g[f(x)] = x$. Graph $f[g(x)] = e^{\ln x}$ and $g[f(x)] = \ln e^x$. Explain why the graphs are different even though $f[g(x)] = g[f(x)]$.

SECTION 10.4 Exponential and Logarithmic Equations

1 Solve exponential equations

37 HM³

An **exponential equation** is one in which a variable occurs in the exponent. The examples at the right are exponential equations.

$6^{2x+1} = 6^{3x-2}$
$4^x = 3$
$2^{x+1} = 7$

An exponential equation in which both sides of the equation can be expressed in terms of the same base can be solved by using the Equality of Exponents Property.

Recall the 1–1 property of an exponential function.

$$\text{If } b^x = b^y, \text{ then } x = y.$$

INSTRUCTOR NOTE

There are two methods presented here for solving an exponential equation. The first method is used when both sides of the equation can be expressed in terms of the same base. The second method involves logarithms. The logarithm method can always be used but is more difficult. You can show this by solving $9^{x+1} = 27^{x-1}$ by using logarithms. (The analogy to quadratic equations may help. It is generally easier to solve a quadratic equation by factoring than it is to use the quadratic formula.)

Example 1 Solve and check: $9^{x+1} = 27^{x-1}$

Solution

$9^{x+1} = 27^{x-1}$

$(3^2)^{x+1} = (3^3)^{x-1}$ ▶ Rewrite each side of the equation using the same base.

$3^{2x+2} = 3^{3x-3}$

$2x + 2 = 3x - 3$ ▶ Use the Equality of Exponents Property to equate the exponents.

$2 = x - 3$ ▶ Solve the resulting equation.

$5 = x$

Check: $9^{x+1} = 27^{x-1}$

9^{5+1}	27^{5-1}
9^6	27^4
$(3^2)^6$	$(3^3)^4$
3^{12} =	3^{12}

The solution is 5.

Problem 1 Solve and check: $10^{3x+5} = 10^{x-3}$

Solution See page S29. −4

When both sides of an exponential equation cannot easily be expressed in terms of the same base, logarithms are used to solve the exponential equation.

Example 2 Solve for x. Round to the nearest ten-thousandth.

A. $4^x = 7$ B. $3^{2x} = 4$

Solution A.

$4^x = 7$

$\log 4^x = \log 7$ ▶ Take the common logarithm of each side of the equation.

$x \log 4 = \log 7$ ▶ Rewrite using the Properties of Logarithms.

$x = \frac{\log 7}{\log 4}$ ▶ Solve for x.

$x \approx 1.4037$

The solution is 1.4037.

B.

$3^{2x} = 4$

$\log 3^{2x} = \log 4$ ▶ Take the common logarithm of each side of the equation.

$2x \log 3 = \log 4$ ▶ Rewrite using the Properties of Logarithms.

$2x = \frac{\log 4}{\log 3}$ ▶ Solve for x.

$x = \frac{\log 4}{2 \log 3}$

$x \approx 0.6309$

The solution is 0.6309.

LOOK CLOSELY

When evaluating $\frac{\log 7}{\log 4}$, first find the logarithm of each number. Then divide the logarithms.

$$\frac{\log 7}{\log 4} \approx \frac{0.845098}{0.602060} \approx 1.4037$$

Problem 2 Solve for x. Round to the nearest ten-thousandth.

A. $4^{3x} = 25$ B. $(1.06)^x = 1.5$

Solution See page S29. A. 0.7740 B. 6.9585

The equations in Example 2 can be solved by graphing. For Example 2A, by subtracting 7 from each side of the equation $4^x = 7$, the equation can be written as $4^x - 7 = 0$. The graph of $f(x) = 4^x - 7$ is shown below. The values of x for which $f(x) = 0$ are the solutions of the equation $4^x = 7$. These values of x are the zeros of the function. By using the features of a graphing utility, it is possible to determine a very accurate solution. The solution to the nearest tenth is shown in the graph.

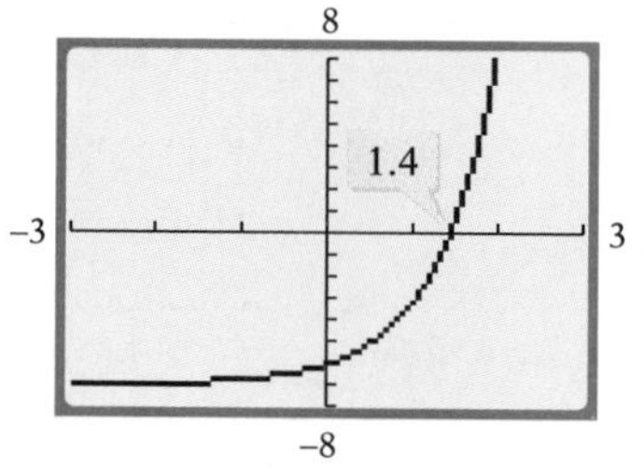

Example 3 Solve $e^x = 2x + 1$ for x. Round to the nearest hundredth.

Solution Rewrite the equation by subtracting $2x + 1$ from each side and writing the equation as $e^x - 2x - 1 = 0$. The zeros of $f(x) = e^x - 2x - 1$ are the solutions of $e^x = 2x + 1$. Graph f and use the features of a graphing utility to estimate the solutions to the nearest hundredth.

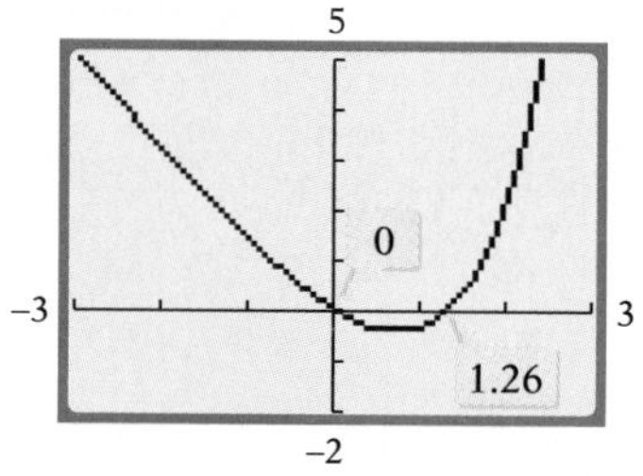

The solutions are 0 and 1.26.

Problem 3 Solve $e^x = x$ for x. Round to the nearest hundredth.

Solution See page S29. No real number solution

2 Solve logarithmic equations

A logarithmic equation can be solved by using the Properties of Logarithms.

Solve: $\log_9 x + \log_9(x - 8) = 1$

Use the Product Property of Logarithms to rewrite the left side of the equation.

$$\log_9 x + \log_9(x - 8) = 1$$
$$\log_9 x(x - 8) = 1$$

INSTRUCTOR NOTE

Solving logarithmic equations normally requires using some of the Properties of Logarithms. These properties are valid only if the arguments are positive numbers. Thus it is necessary to check the solutions of equations to ensure that this condition is met.

Write the equation in exponential form. $9^1 = x(x - 8)$

Simplify and solve for x.

$$9 = x^2 - 8x$$
$$0 = x^2 - 8x - 9$$
$$0 = (x - 9)(x + 1)$$

$$x - 9 = 0 \qquad x + 1 = 0$$
$$x = 9 \qquad x = -1$$

When x is replaced by 9 in the original equation, 9 checks as a solution. When x is replaced by -1, the original equation contains the expression $\log_9(-1)$. Because the logarithm of a negative number is not a real number, -1 does not check as a solution. Therefore, the solution of the equation is 9.

Some logarithmic equations can be solved by using the 1–1 Property of Logarithms. The use of this property is illustrated in Example 4B.

Example 4 Solve for x.

A. $\log_3(2x - 1) = 2$

B. $\log_2 x - \log_2(x - 1) = \log_2 2$

Solution A. $\log_3(2x - 1) = 2$

$$3^2 = 2x - 1$$ ▶ Rewrite in exponential form.

$$9 = 2x - 1$$ ▶ Solve for x.

$$10 = 2x$$
$$5 = x$$

The solution is 5.

B. $\log_2 x - \log_2(x - 1) = \log_2 2$

$$\log_2\left(\frac{x}{x-1}\right) = \log_2 2$$ ▶ Use the Quotient Property of Logarithms.

$$\frac{x}{x-1} = 2$$ ▶ Use the 1–1 Property of Logarithms.

$$(x - 1)\left(\frac{x}{x-1}\right) = (x - 1)2$$ ▶ Solve for x.

$$x = 2x - 2$$
$$-x = -2$$
$$x = 2$$

The solution is 2.

Problem 4 Solve for x.

A. $\log_4(x^2 - 3x) = 1$

B. $\log_3 x + \log_3(x + 3) = \log_3 4$

Solution See page S29. A. $-1, 4$ B. 1

Some logarithmic equations cannot be solved algebraically. In these cases, a graphical approach may be appropriate.

Example 5 Solve $\ln(2x + 4) = x^2$ for x. Round to the nearest hundredth.

Solution Rewrite the equation by subtracting x^2 from each side and writing the equation as $\ln(2x + 4) - x^2 = 0$. The zeros of the function defined by $f(x) = \ln(2x + 4) - x^2$ are the solutions of the equation. Graph f and then use the features of a graphing utility to estimate the solutions to the nearest hundredth.

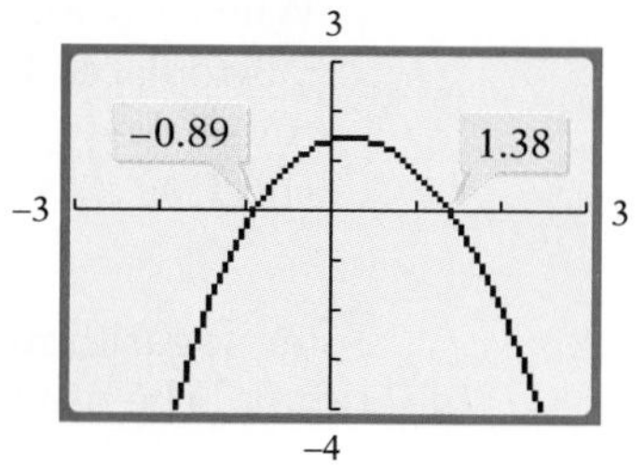

The solutions are -0.89 and 1.38.

Problem 5 Solve $\log(3x - 2) = -2x$ for x. Round to the nearest hundredth.

Solution See page S30. 0.68

Recall the equation $\log_9 x + \log_9(x - 8) = 1$ from the beginning of this objective. The algebraic solution showed that although -1 and 9 may be solutions, only 9 satisfies the equation. The extraneous solution was introduced at the second step. The Product Property of Logarithms, $\log_b(xy) = \log_b x + \log_b y$, applies only when both x and y are positive numbers. This occurs when $x > 8$. Therefore, a solution to this equation must be greater than 8. The graphs of $f(x) = \log_9 x + \log_9(x - 8) - 1$ and $g(x) = \log_9 x(x - 8) - 1$ are shown below.

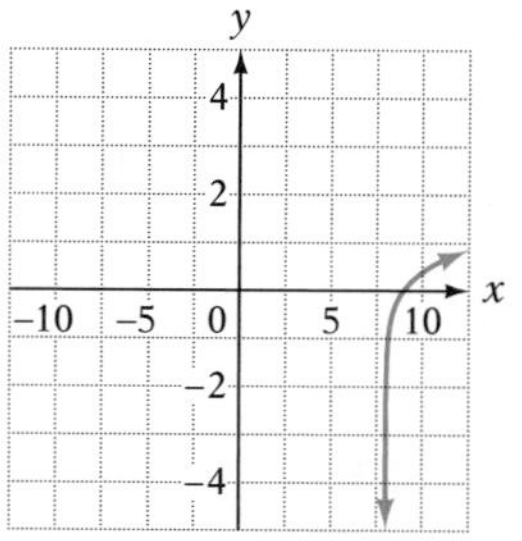

$f(x) = \log_9 x + \log_9(x - 8) - 1$

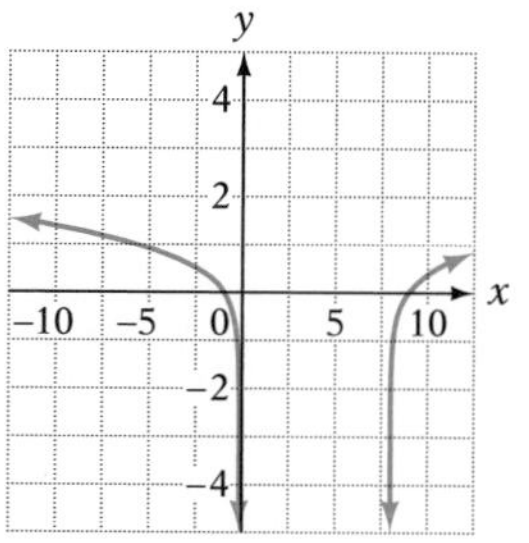

$g(x) = \log_9 x(x - 8) - 1$

Note that the only zero of f is 9, whereas the zeros of g are -1 and 9.

CONCEPT REVIEW 10.4

Determine whether the following are always true, sometimes true, or never true.

1. Suppose $b > 0, b \neq 1$. If $b^x = b^y$, then $x = y$. Always true

2. $\log x + \log(x + 2) = \log(2x + 2)$ Never true

3. $\log_2(-4) = -2$ Never true

4. $\log(2x - 2) - \log x = 4$ is equivalent to $\log \frac{2x - 2}{x} = 4$. Always true

5. $\log(2x) + \log 4 = 6$ is equivalent to $\log(8x) = 6$. Always true

6. $\log_{10} 10^4 = 4$ Always true

EXERCISES 10.4

1 1. What is an exponential equation?

2. Explain how to solve the equation $7^{x+1} = 7^5$.

Solve for x. Round to the nearest ten-thousandth.

3. $5^{4x-1} = 5^{x+2}$ 1

4. $7^{4x-3} = 7^{2x+1}$ 2

5. $8^{x-4} = 8^{5x+8}$ −3

6. $10^{4x-5} = 10^{x+4}$ 3

7. $5^x = 6$ 1.1133

8. $7^x = 10$ 1.1833

9. $12^x = 6$ 0.7211

10. $10^x = 5$ 0.6990

11. $\left(\frac{1}{2}\right)^x = 3$ −1.5850

12. $\left(\frac{1}{3}\right)^x = 2$ −0.6309

13. $(1.5)^x = 2$ 1.7095

14. $(2.7)^x = 3$ 1.1061

15. $10^x = 21$ 1.3222

16. $10^x = 37$ 1.5682

17. $2^{-x} = 7$ −2.8074

18. $3^{-x} = 14$ −2.4022

19. $2^{x-1} = 6$ 3.5850

20. $4^{x+1} = 9$ 0.5850

21. $3^{2x-1} = 4$ 1.1309

22. $4^{-x+2} = 12$ 0.2075

23. $9^x = 3^{x+1}$ 1

24. $2^{x-1} = 4^x$ −1

25. $8^{x+2} = 16^x$ 6

26. $9^{3x} = 81^{x-4}$ −8

27. $5^{x^2} = 21$ 1.3754, −1.3754

28. $3^{x^2} = 40$ 1.8324, −1.8324

29. $2^{4x-2} = 20$ 1.5805

30. $4^{3x+8} = 12$ −2.0692

31. $3^{-x+2} = 18$ −0.6309

32. $5^{-x+1} = 15$ −0.6826

33. $4^{2x} = 100$ 1.6610

34. $3^{3x} = 1000$ 2.0959

35. $2.5^{-x} = 4$ −1.5129

36. $3.25^{x+1} = 4.2$ 0.2176

37. $0.25^x = 0.125$ 1.5

38. $0.1^{5x} = 10^{-2}$ 0.4

Solve for x by graphing. Round to the nearest hundredth.

39. $3^x = 2$ 0.63

40. $5^x = 9$ 1.37

41. $2^x = 2x + 4$ −1.86, 3.44

42. $3^x = -x - 1$ −1.25

43. $e^x = -2x - 2$ −1.16

44. $e^x = 3x + 4$ −1.24, 2.42

2 **45.** What is a logarithmic equation?

46. Explain how to solve the equation $2 = \log_3 x$.

Solve for x.

47. $\log_3(x + 1) = 2$ 8

48. $\log_5(x - 1) = 1$ 6

49. $\log_2(2x - 3) = 3$ $\frac{11}{2}$

50. $\log_4(3x + 1) = 2$ 5

51. $\log_2(x^2 + 2x) = 3$ 2, −4

52. $\log_3(x^2 + 6x) = 3$ −9, 3

53. $\log_5\left(\frac{2x}{x - 1}\right) = 1$ $\frac{5}{3}$

54. $\log_6 \frac{3x}{x + 1} = 1$ −2

55. $\log_7 x = \log_7(1 - x)$ $\frac{1}{2}$

56. $\frac{3}{4}\log x = 3$ 10,000

57. $\frac{2}{3}\log x = 6$ 1,000,000,000

58. $\log(x - 2) - \log x = 3$ No solution

59. $\log_2(x - 3) + \log_2(x + 4) = 3$ 4

60. $\log x - 2 = \log(x - 4)$ $\frac{400}{99}$

61. $\log_3 x + \log_3(x - 1) = \log_3 6$ 3

62. $\log_4 x + \log_4(x - 2) = \log_4 15$ 5

63. $\log_2(8x) - \log_2(x^2 - 1) = \log_2 3$ 3

64. $\log_5(3x) - \log_5(x^2 - 1) = \log_5 2$ 2

65. $\log_9 x + \log_9(2x - 3) = \log_9 2$ 2

66. $\log_6 x + \log_6(3x - 5) = \log_6 2$ 2

67. $\log_8(6x) = \log_8 2 + \log_8(x - 4)$ No solution

68. $\log_7(5x) = \log_7 3 + \log_7(2x + 1)$ No solution

69. $\log_9(7x) = \log_9 2 + \log_9(x^2 - 2)$ 4

70. $\log_3 x = \log_3 2 + \log_3(x^2 - 3)$ 2

71. $\log(x^2 + 3) - \log(x + 1) = \log 5$ $\frac{5 \pm \sqrt{33}}{2}$

72. $\log(x + 3) + \log(2x - 4) = \log 3$ $\frac{-1 + \sqrt{31}}{2}$

Solve for x by graphing. Round to the nearest hundredth.

73. $\log x = -x + 2$ 1.76

74. $\log x = -2x$ 0.28

75. $\log(2x - 1) = -x + 3$ 2.42

76. $\log(x + 4) = -2x + 1$ 0.19

77. $\ln(x + 2) = x^2 - 3$ −1.51, 2.10

78. $\ln x = -x^2 + 1$ 1.00

APPLYING CONCEPTS 10.4

Solve for x. Round to the nearest ten-thousandth.

79. $8^{\frac{x}{2}} = 6$ 1.7233

80. $4^{\frac{x}{3}} = 2$ 1.5

81. $5^{\frac{3x}{2}} = 7$ 0.8060

82. $9^{\frac{2x}{3}} = 8$ 1.4196

83. $1.2^{\frac{x}{2}-1} = 1.4$ 5.6910

84. $5.6^{\frac{x}{3}+1} = 7.8$ 0.5770

Solve the system of equations.

[C] **85.** $2^x = 8^y$
$x + y = 4$ (3, 1)

[C] **86.** $3^{2y} = 27^x$
$y - x = 2$ (4, 6)

[C] **87.** $\log(x + y) = 3$
$x = y + 4$ (502, 498)

[C] **88.** $\log(x + y) = 3$
$x - y = 20$ (510, 490)

[C] **89.** $8^{3x} = 4^{2y}$
$x - y = 5$ $(-4, -9)$

[C] **90.** $9^{3x} = 81^{3y}$
$x + y = 3$ (2, 1)

91. *Physics* A model for the distance s (in feet) an object that is experiencing air resistance will fall in t seconds is given by $s = 312.5 \ln\left(\frac{e^{0.32t} + e^{-0.32t}}{2}\right)$.

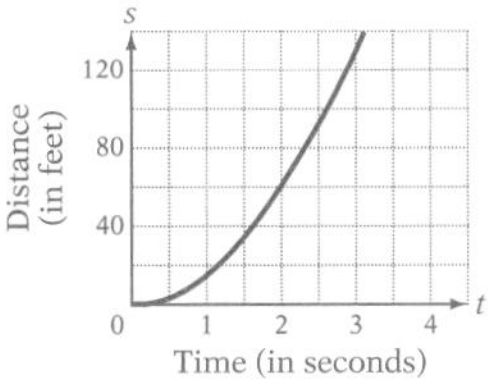

a. Graph this equation. *Suggestion:* Use Xmin $= -0.5$, Xmax $= 5$, Xscl $= 0.5$, Ymin $= -0.5$, Ymax $= 150$, and Yscl $= 10$.

b. Determine, to the nearest hundredth of a second, the time it takes the object to travel 100 ft. It will take the object approximately 2.64 s.

92. *Demography* The U.S. Census Bureau provides information on the various segments of the population in the United States. The following table gives the number of people, in millions, aged 80 and older at the beginning of each decade from 1900 to 1990.

Year	1900	1910	1920	1930	1940	1950	1960	1970	1980	1990
80-year-olds (in millions)	0.3	0.3	0.4	0.5	0.8	1.1	1.6	2.3	2.9	3.9

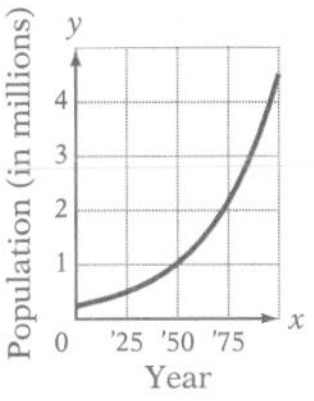

a. The equation that approximately models the data in the table is given by $y = 0.235338(1.03)^x$, where x is the last two digits of the year and y is the population, in millions, of people aged 80 and over. Use a graphing utility to graph this equation.

b. According to the model, what is the predicted population of this age group in 1990? 3.4 million

c. According to the model, what is the predicted population of this age group in the year 2000? Round to the nearest tenth of a million. (*Hint:* You will need to determine what the x-value is when the year is 2000.) 4.5 million

d. In what year does this model predict that the population of this age group will be 5 million? Round to the nearest year. 2003

93. The following "proof" appears to show that $0.04 < 0.008$. Explain the error.

$$2 < 3$$
$$2 \log 0.2 < 3 \log 0.2$$
$$\log(0.2)^2 < \log(0.2)^3$$
$$(0.2)^2 < (0.2)^3$$
$$0.04 < 0.008$$

94. Logarithms were originally invented to simplify calculations. The abacus is another invention that was invented (and is still used) to assist with computations. Write a report on the abacus.

SECTION 10.5 Applications of Exponential and Logarithmic Functions

1 Application problems

37 38 HM³

A biologist places one single-celled bacterium in a culture, and each hour that particular species of bacterium divides into two bacteria. After one hour there will be two bacteria. After two hours, each of the two bacteria will divide and there will be four bacteria. After three hours, each of the four bacteria will divide and there will be eight bacteria.

Time, t	Number of Bacteria, N
0	1
1	2
2	4
3	8
4	16

The table at the left shows the number of bacteria in the culture after various intervals of time, t, in hours. Values in this table could also be found by using the exponential equation $N = 2^t$.

The equation $N = 2^t$ is an example of an **exponential growth equation.** In general, any equation that can be written in the form $A = A_0 b^{kt}$, where A is the size at time t, A_0 is the initial size, $b > 1$, and k is a positive real number, is an exponential growth equation. These equations are important not only in population growth studies but also in physics, chemistry, psychology, and economics.

Recall that interest is the amount of money one pays (or receives) when borrowing (or investing) money. **Compound interest** is interest that is computed not only on the original principal but also on the interest already earned. The compound interest formula is an exponential equation.

INSTRUCTOR NOTE

Another exponential function from finance that will be of interest to students is the one that enables them to calculate the amount of an amortized loan payment, such as on a car loan.

$$P = B\left[\frac{\frac{i}{12}}{1 - \left[1 + \left(\frac{i}{12}\right)\right]^{-n}}\right]$$

B is the amount borrowed, *i* is the annual interest rate as a decimal, and *n* is the number of months to repay the loan.

The **compound interest formula** is $P = A(1 + i)^n$, where A is the original value of an investment, i is the interest rate per compounding period, n is the total number of compounding periods, and P is the value of the investment after n periods.

An investment broker deposits $1000 into an account that earns 12% annual interest compounded quarterly. What is the value of the investment after 2 years? Round to the nearest dollar.

Find i, the interest rate per quarter. The quarterly rate is the annual rate divided by 4, the number of quarters in 1 year. $\quad i = \frac{12\%}{4} = \frac{0.12}{4} = 0.03$

Find n, the number of compounding periods. The investment is compounded quarterly, 4 times a year, for 2 years. $\quad n = 4 \cdot 2 = 8$

Use the compound interest formula. $\quad P = A(1 + i)^n$

Replace A, i, and n by their values. $\quad P = 1000(1 + 0.03)^8$

Solve for P. $\quad P \approx 1267$

The value of the investment after 2 years is $1267.

Time, t	Amount, A
0	10
5	5
10	2.5
15	1.25
20	0.625

Exponential decay is another important example of an exponential equation. One of the most common illustrations of exponential decay is the decay of a radioactive substance.

A radioactive isotope of cobalt has a half-life of approximately 5 years. This means that one-half of any given amount of the cobalt isotope will disintegrate in 5 years.

The table at the left indicates the amount of the initial 10 mg of a cobalt isotope that remains after various intervals of time, t, in years. Values in this table could also be found by using the exponential equation $A = 10\left(\frac{1}{2}\right)^{\frac{t}{5}}$.

The equation $A = 10\left(\frac{1}{2}\right)^{\frac{t}{5}}$ is an example of an **exponential decay equation.**

Compare this equation to the exponential growth equation, and note that for exponential growth, the base of the exponential equation is greater than 1, whereas for exponential decay, the base is between 0 and 1.

A method by which an archeologist can measure the age of a bone is called **carbon dating.** Carbon dating is based on a radioactive isotope of carbon called carbon-14, which has a half-life of approximately 5570 years. The exponential decay equation is given by $A = A_0\left(\frac{1}{2}\right)^{\frac{t}{5570}}$, where A_0 is the original amount of carbon-14 present in the bone, t is the age of the bone, and A is the amount present after t years.

A bone that originally contained 100 mg of carbon-14 now has 70 mg of carbon-14. What is the approximate age of the bone? Round to the nearest year.

Use the exponential decay equation.

$$A = A_0\left(\frac{1}{2}\right)^{\frac{t}{5570}}$$

Replace A_0 and A by their given values, and solve for t.

$$70 = 100\left(\frac{1}{2}\right)^{\frac{t}{5570}}$$

$$70 = 100(0.5)^{\frac{t}{5570}}$$

Divide each side of the equation by 100.

$$0.7 = (0.5)^{\frac{t}{5570}}$$

Take the common logarithm of each side of the equation. Then simplify.

$$\log 0.7 = \log(0.5)^{\frac{t}{5570}}$$

$$\log 0.7 = \frac{t}{5570}\log 0.5$$

$$\frac{5570 \log 0.7}{\log 0.5} = t$$

$$2866 \approx t$$

The age of the bone is approximately 2866 years.

Example 1 The number of words per minute a student can type will increase with practice and can be approximated by the equation $N = 100[1 - (0.9)^t]$, where N is the number of words typed per minute after t days of instruction. In how many days will the student be able to type 60 words per minute?

Strategy To find the number of days, replace N by 60 and solve for t.

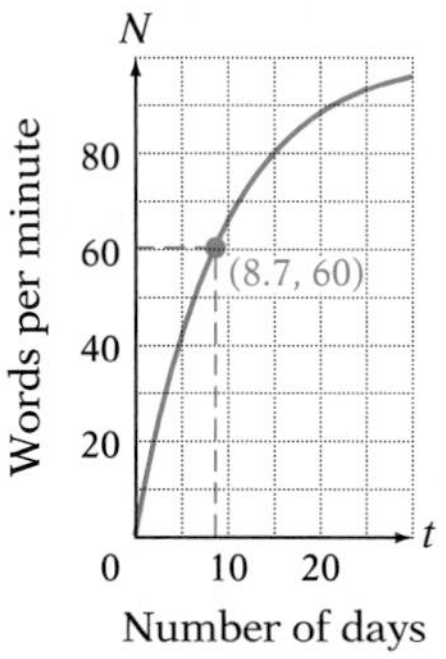

Solution

$$N = 100[1 - (0.9)^t]$$
$$60 = 100[1 - (0.9)^t]$$
$$0.6 = 1 - (0.9)^t$$

▶ Divide each side of the equation by 100.

$$-0.4 = -(0.9)^t$$

▶ Subtract 1 from each side of the equation.

$$0.4 = (0.9)^t$$
$$\log 0.4 = \log(0.9)^t$$
$$\log 0.4 = t \log 0.9$$
$$t = \frac{\log 0.4}{\log 0.9} \approx 8.6967184$$

After approximately 9 days the student will type 60 words per minute.

The graph at the left is the graph of $N = 100[1 - (0.9)^t]$. Note that t is approximately 8.7 when $N = 60$.

Problem 1 In 1962, the cost of a first-class stamp was \$.04. In 1999, the cost was \$.33. The increase in cost can be modeled by the equation $C = 0.04e^{0.057t}$, where C is the cost and t is the number of years after 1962. According to this model, in what year did a first-class stamp cost \$.22?

Solution See page S30. 1992

The first applications of logarithms (and the main reason why they were developed) were to reduce computational drudgery. Today, with the widespread use of calculators and computers, the computational uses of logarithms have diminished. However, a number of other applications of logarithms have emerged.

A chemist measures the acidity or alkalinity of a solution by the formula **pH = $-\log(H^+)$,** where H^+ is the concentration of hydrogen ions in the solution. A neutral solution such as distilled water has a pH of 7, acids have a pH less than 7, and alkaline solutions (also called basic solutions) have a pH greater than 7.

Find the pH of vinegar for which $H^+ = 1.26 \times 10^{-3}$. Round to the nearest tenth.

Use the pH equation.

$H^+ = 1.26 \times 10^{-3}$.

$$\begin{aligned} pH &= -\log(H^+) \\ &= -\log(1.26 \times 10^{-3}) \\ &= -(\log 1.26 + \log 10^{-3}) \\ &\approx -[0.1004 + (-3)] = 2.8996 \end{aligned}$$

The pH of vinegar is approximately 2.9.

The **Richter scale** measures the magnitude, M, of an earthquake in terms of the intensity, I, of its shock waves. This can be expressed as the logarithmic equation $M = \log \frac{I}{I_0}$, where I_0 is a constant.

How many times stronger is an earthquake that has magnitude 4 on the Richter scale than one that has magnitude 2 on the scale?

Let I_1 represent the intensity of the earthquake that has magnitude 4, and let I_2 represent the intensity of the earthquake that has magnitude 2. The ratio of I_1 to I_2, $\frac{I_1}{I_2}$, measures how much stronger I_1 is than I_2.

$$4 = \log \frac{I_1}{I_0}$$
$$2 = \log \frac{I_2}{I_0}$$

Use the Richter equation to write a system of equations, one equation for magnitude 4 and one for magnitude 2. Then rewrite the system using the Properties of Logarithms.

$$4 = \log I_1 - \log I_0$$
$$2 = \log I_2 - \log I_0$$

Use the addition method to eliminate $\log I_0$.

$$2 = \log I_1 - \log I_2$$

Rewrite the equations using the Properties of Logarithms.

$$2 = \log \frac{I_1}{I_2}$$

Solve for the ratio using the relationship between logarithms and exponents.

$$\frac{I_1}{I_2} = 10^2 = 100$$

An earthquake that has magnitude 4 on the Richter scale is 100 times stronger than an earthquake that has magnitude 2.

The graph below is a *seismogram*, which is used to measure the magnitude of an earthquake. The magnitude, M, is determined by the amplitude, A, of the wave and the difference in time, t, between the occurrence of two types of waves, called *s-waves* and *p-waves*. The equation is given by $M = \log A + 3 \log 8t - 2.92$.

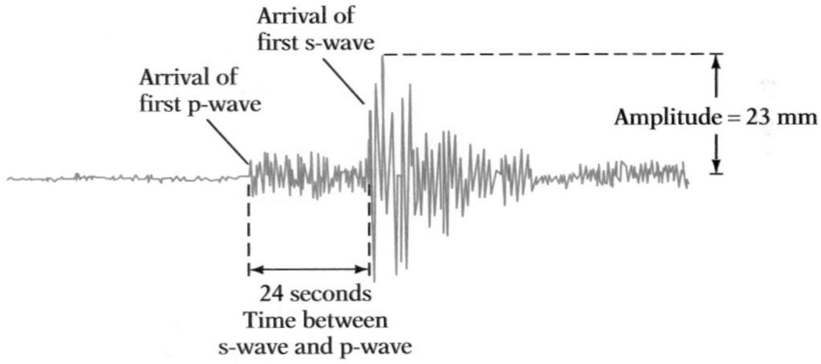
Arrival of first s-wave
Arrival of first p-wave
Amplitude = 23 mm
24 seconds
Time between s-wave and p-wave

Determine the magnitude of the earthquake for the seismogram given in the figure. Round to the nearest tenth.

$$M = \log A + 3 \log 8t - 2.92$$

Replace A by 23 and t by 24.

$$= \log 23 + 3 \log[8(24)] - 2.92$$

Simplify.

$$\approx 1.36173 + 6.84990 - 2.92$$
$$\approx 5.3$$

The earthquake has a magnitude of 5.3 on the Richter scale.

The percent of light that will pass through a substance is given by the equation $\mathbf{\log P = -kd}$, where P is the percent of light passing through the substance, k is a constant that depends on the substance, and d is the thickness of the substance in centimeters.

Find the percent of light that will pass through translucent glass for which $k = 0.4$ and d is 0.5 cm.

$\log P = -kd$

Replace k and d in the equation by their given values, and solve for P.

$\log P = -(0.4)(0.5)$

$\log P = -0.2$

Use the relationship between the logarithmic and exponential functions.

$P = 10^{-0.2}$

$P \approx 0.6310$

Approximately 63.1% of the light will pass through the glass.

Example 2 Astronomers use the *distance modulus* of a star as a method of determining how far the star is from Earth. The formula is $M = 5 \log r - 5$, where M is the distance modulus and r is the distance the star is from Earth in parsecs. (One parsec is approximately 3.3 light-years, or 1.9×10^{13} mi.) How many parsecs from Earth is a star that has a distance modulus of 4?

Strategy To find the number of parsecs, replace M by 4 and solve for r.

Solution

$$M = 5 \log r - 5$$
$$4 = 5 \log r - 5$$
$$9 = 5 \log r$$
$$\frac{9}{5} = \log r$$
$$r = 10^{\frac{9}{5}} \approx 63.095734$$

The star is approximately 63 parsecs from Earth.

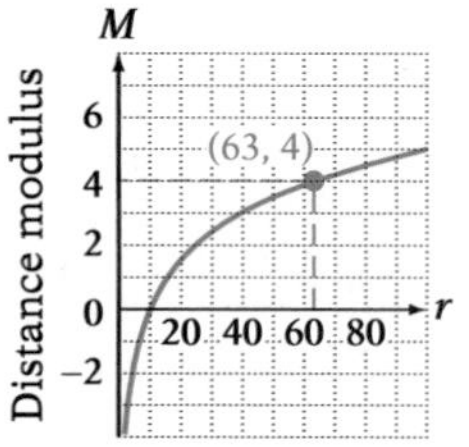

Distance from Earth (in parsecs)

The graph at the left is the graph of $M = 5 \log r - 5$. Note that r is approximately 63 when $M = 4$.

Problem 2 The *expiration time* T of a natural resource is the time remaining before it is completely consumed. A model for the expiration time of the world's oil supply is given by $T = 14.29 \ln(0.00411r + 1)$, where r is the estimated number of billions of barrels of oil remaining in the world's oil supply. According to this model, how many billion barrels of oil are needed to last 25 years?

Solution See page S30. 1156 billion barrels

CONCEPT REVIEW 10.5

Determine whether the following statements are always true, sometimes true, or never true.

1. In the exponential growth equation $A = A_0 b^{kt}$, A_0 is the initial size of the growth material. Always true

2. An investor deposits \$1000 into an account that earns 8% compounded quarterly. The investor receives interest of \$20 every quarter. Never true

3. The equation $A = A_0e^{kt}$ is an exponential growth equation. An equivalent equation to this exponential growth equation is $t = \frac{1}{k}\ln\frac{A}{A_0}$. Always true

4. The hydrogen ion concentration of a solution is given by the equation $\text{pH} = -\log(\text{H}^+)$. Solving this equation for H^+ gives the equivalent equation $\text{H}^+ = 10^{-\text{pH}}$. Always true

EXERCISES 10.5

1 1. What is meant by the phrase *exponential decay*?

2. Explain how compound interest differs from simple interest. Is compound interest an example of exponential decay or exponential growth?

For Exercises 3 through 6, use the compound interest formula $P = A(1 + i)^n$, where A is the original value of an investment, i is the interest rate per compounding period, n is the total number of compounding periods, and P is the value of the investment after n periods. Round to the nearest cent.

3. *Finance* A computer network specialist deposits \$2500 into a retirement account that earns 7.5% annual interest compounded daily. What is the value of the investment after 20 years? \$11,202.50

4. *Finance* A \$10,000 certificate of deposit (CD) earns 5% annual interest compounded daily. What is the value of the CD after 5 years? \$12,840.03

5. *Finance* The plant manager of a printing company estimates that in 3 years the company will need to purchase a new printing machine at a cost of \$575,000. How much money must be deposited in an account that earns 6% annual interest compounded monthly so that the value of the account in 3 years will be \$575,000? \$480,495.83

6. *Finance* The comptroller of a company has determined that it will be necessary to purchase a new computer in 3 years. The estimated cost of the computer is \$10,000. How much money must be deposited in an account that earns 9% annual interest compounded quarterly so that the value of the account in 3 years will be \$10,000? \$7657

For Exercises 7 through 10, use the exponential decay equation $A = A_0\left(\frac{1}{2}\right)^{t/k}$, where A is the amount of a radioactive material present after time t, k is the half-life, and A_0 is the original amount of radioactive substance. Round to the nearest tenth.

7. *Biology* An isotope of technetium is used to prepare images of internal body organs. This isotope has a half-life of approximately 6 h. If a patient is injected with 30 mg of this isotope, how long (in hours) will it take for the technetium level to reach 20 mg? 3.5 h

8. *Biology* Iodine-131 is an isotope of iodine that is used to study the functioning of the thyroid gland. This isotope has a half-life of approximately 8 days. If a patient is given an injection that contains 8 micrograms of iodine-131, how long (in days) will it take for the iodine level to reach 5 micrograms?
5.4 days

9. *Physics* A sample of promethium-147 (used in some luminous paints) contains 25 mg. One year later, the sample contains 18.95 mg. What is the half-life of promethium-147, in years? 2.5 years

10. *Physics* Francium-223 is a very rare radioactive isotope discovered in 1939 by Marguerite Percy. A 3-microgram sample of francium-223 decays to 2.54 micrograms in 5 min. What is the half-life of francium-223, in minutes?
20.8 min

For Exercises 11 and 12, consider that the percent of correct welds that a student can make will increase with practice and can be approximated by the equation $P = 100[1 - (0.75)^t]$, where P is the percent of correct welds and t is the number of weeks of practice.

11. *Education* How many weeks of practice are necessary before a student will make 80% of the welds correctly? Round to the nearest whole number.
6 weeks

12. *Education* Find the percent of correct welds that a student will make after 4 weeks of practice. Round to the nearest percent. 68%

For Exercises 13 and 14, use the equation $\text{pH} = -\log(\text{H}^+)$, where H^+ is the hydrogen ion concentration of a solution. Round to the nearest hundredth.

13. *Chemistry* Find the pH of the digestive solution of the stomach, for which the hydrogen ion concentration is 0.045. 1.35

14. *Chemistry* Find the pH of a morphine solution used to relieve pain, for which the hydrogen ion concentration is 3.2×10^{-10}. 9.49

The percent of light that will pass through a material is given by the equation $\log P = -kd$, where P is the percent of light passing through the material, k is a constant that depends on the material, and d is the thickness of the material in centimeters.

15. *Optics* The constant k for a piece of translucent glass that is 0.5 cm thick is 0.2. Find the percent of light that will pass through the glass. Round to the nearest percent. 79%

16. *Optics* The constant k for a piece of tinted glass is 0.5. How thick is a piece of this glass that allows 60% of the light incident to the glass to pass through it? Round to the nearest hundredth. 0.44 cm

The intensity I of an x-ray after it has passed through a material that is x centimeters thick is given by $I = I_0 e^{-kx}$, where I_0 is the initial intensity and k is a number that depends on the material. Use this equation for Exercises 17 and 18.

17. *Physics* The constant k for copper is 3.2. Find the thickness of copper that is needed so that the intensity of the x-ray after passing through the copper is 25% of the original intensity. Round to the nearest tenth. 0.4 cm

18. *Medicine* Radiologists (physicians who specialize in the use of radioactive substances in diagnosis and treatment of disease) wear lead shields when giving a patient an x-ray. The constant k for lead is 43. Explain, using the given equation, why a piece of lead the same thickness as a piece of copper ($k = 3.2$) makes a better shield than the piece of copper.
Answers will vary.

For Exercises 19 and 20, use the equation $M = \log A + 3 \log 8t - 2.92$ given in this section.

19. *Geology* Find the magnitude of an earthquake that has a seismogram with an amplitude of 30 mm and for which t is 21 s. 5.2

20. *Geology* Find the magnitude of an earthquake that has a seismogram with an amplitude of 28 and for which t is 28 s. 5.6

Paleobiologists can estimate the weight of a dinosaur by examining characteristics of animals that exist today. Measuring the circumference of the femur of various animals, they have arrived at an equation with which to estimate the weight, W (in kilograms), of an animal. It is $W = 0.00031C^{2.9}$, where C is the circumference of the femur in millimeters. Use this equation for Exercises 21 and 22.

21. *Biology* Estimate the weight of a Tyrannosaurus for which the circumference of the femur is 534 mm. Round to the nearest kilogram. 25,190 kg

22. *Biology* A polar bear weighs approximately 450 kg. Estimate the circumference of the femur of a polar bear. Round to the nearest millimeter.
133 mm

For Exercises 23 and 24, use the Richter equation $M = \log \frac{I}{I_0}$, where M is the magnitude of an earthquake, I is the intensity of its shock waves, and I_0 is a constant. Round to the nearest tenth.

23. *Geology* On March 2, 1933, the largest earthquake ever recorded struck Japan. The earthquake measured 8.9 on the Richter scale. In October 1989, an earthquake of magnitude 7.1 on the Richter scale struck San Francisco, California. How many times stronger was the earthquake in Japan than the San Francisco earthquake? Round to the nearest tenth. 63.1 times

24. *Geology* An earthquake that occurred in China in 1978 measured 8.2 on the Richter scale. In 1988, an earthquake in Armenia measured 6.9 on the Richter scale. How many times stronger was the earthquake in China? Round to the nearest tenth. 20.0 times

The number of decibels, D, of a sound can be indicated by the equation $D = 10(\log I + 16)$, where I is the power of a sound measured in watts. Round to the nearest whole number.

25. *Acoustics* Find the number of decibels of normal conversation. The power of the sound of normal conversation is approximately 3.2×10^{-10} watts.
65 decibels
26. *Acoustics* The loudest sound made by any animal is made by the blue whale and can be heard over 500 mi away. The power of the sound is 630 watts. Find the number of decibels of the sound emitted by the blue whale. 188 decibels

Astronomers use the distance modulus formula $M = 5 \log r - 5$, where M is the distance modulus and r is the distance of a star from Earth in parsecs. (One parsec is approximately 1.92×10^{13} miles or approximately 20 trillion miles.) Use this information in Exercises 27 through 30. Round to the nearest tenth.

27. *Astronomy* The distance modulus of the star Betelgeuse is 5.89. How many parsecs from Earth is this star? 150.7 parsecs

28. *Astronomy* The distance modulus of Alpha Centauri is -1.11. How many parsecs from Earth is this star? 6.0 parsecs

29. *Astronomy* The distance modulus of the star Antares is 5.4, which is twice that of the star Pollux. Is Antares twice as far from Earth as Pollux is? If not, how many times farther is Antares from Earth than Pollux is?
No; 3.5 times farther
30. *Astronomy* If one star has a distance modulus of 2 and a second star has a distance modulus of 6, is the second star three times as far from Earth as the first star? No; 6.3 times farther

One model for the time it will take for the world's oil supply to be depleted is given by the equation $T = 14.29 \ln(0.00411r + 1)$, where r is the estimated world oil reserves in billions of barrels and T is the time before that amount of oil is depleted. Use this model for Exercises 31 and 32. Round to the nearest tenth.

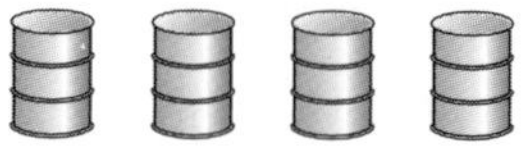

31. *Ecology* How many barrels of oil are necessary to last 20 years?
742.9 billion barrels
32. *Ecology* How many barrels of oil are necessary to last 50 years?
7805.5 billion barrels

APPLYING CONCEPTS 10.5

33. *Biology* At 9 A.M., a culture of bacteria had a population of 1.5×10^6. At noon, the population was 3.0×10^6. If the population is growing exponentially, at what time will the population be 9×10^6? Round to the nearest hour.
5 P.M.
34. *Inflation* If the average annual rate of inflation is 5%, in how many years will prices double? Round to the nearest whole number. 14 years

35. *Investments* An investment of \$1000 earns \$177.23 in interest in 2 years. If the interest is compounded annually, find the annual interest rate. Round to the nearest tenth of a percent. 8.5%

36. *Investments* Some banks now use continuous compounding of an amount invested. In this case, the equation that relates the value of an initial investment of A dollars in t years at an annual interest rate of r is given by $P = Ae^{rt}$. Using this equation, find the value in 5 years of an investment of \$2500 into an account that earns 5% annual interest. \$3210.06

37. Solve the equation $T = 14.29 \ln(0.00411r + 1)$ for r. $r = \frac{1}{0.00411}(e^{\frac{T}{14.29}} - 1)$

38. Exponential equations of the form $y = Ab^{kt}$ are frequently rewritten in the form $y = Ae^{mt}$, where the base e is used rather than the base b. Rewrite $y = 10(2^{0.14t})$ in the form $y = Ae^{mt}$. $y = 10e^{0.09704t}$

39. Rewrite $y = A2^{kt}$ in the form $y = Ae^{mt}$. (See Exercise 38.) $y = Ae^{(k \ln 2)t}$

40. *Investments* The value of an investment in an account that earns an annual interest rate of 10% compounded daily grows according to the equation $A = A_0\left(1 + \frac{0.10}{365}\right)^{365t}$. Find the time for the investment to double in value. Round to the nearest year. 7 years

When a rock is tossed into the air, the mass of the rock remains constant and a reasonable model for the height of the rock can be given by a quadratic function. However, when a rocket is launched straight up from Earth's surface, the rocket is burning fuel, so the mass of the rocket is always changing. The height of the rocket above Earth can be approximated by the equation

$$y(t) = At - 16t^2 + \frac{A}{k}(M + m - kt)\ln\left(1 - \frac{k}{M + m}t\right)$$

where M is the mass of the rocket, m is the mass of the fuel, A is the rate at which fuel is ejected from the engines, k is the rate at which fuel is burned, t is the time in seconds, and $y(t)$ is the height in feet after t seconds.

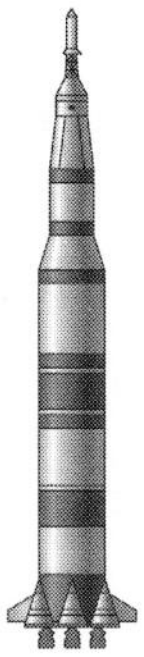

41. *Rocketry* During the development of the V-2 rocket program in the United States, approximate values for a V-2 rocket were $M = 8000$ lb, $m = 16{,}000$ lb, $A = 8000$ ft/s, and $k = 250$ lb/s.

 a. Use a graphing utility to estimate, to the nearest second, the time required for the rocket to reach a height of 1 mi (5280 ft). 14 s

 b. Use $v(t) = -32t + A\ln\left(\frac{M + m}{M + m - kt}\right)$, and the answer to part (a), to determine the velocity of the rocket. Round to the nearest whole number. 813 ft/s

 c. Determine the domain of the velocity function. $[0, 96)$

42. *Mortgages* When you purchase a car or home and make monthly payments on the loan, you are amortizing the loan. Part of each monthly payment is interest on the loan, and the remaining part of the payment is a repayment of the loan amount. The amount remaining to be repaid on the loan after x months is given by $y = A(1 + i)^x + B$, where y is the amount of the loan to be repaid. In this equation, $A = \frac{Pi - M}{i}$ and $B = \frac{M}{i}$, where P is the original loan amount, i is the monthly interest rate $\left(\frac{\text{annual interest rate}}{12}\right)$, and M is the monthly payment. For a 30-year home mortgage of \$100,000 with an annual interest rate of 8%, $i = 0.00667$ and $M = 733.76$.

a. How many months are required to reduce the loan amount to \$90,000? Round to the nearest month. 104 months

b. How many months are required to reduce the loan amount to one-half the original amount? Round to the nearest month. 269 months

c. The total amount of interest that is paid after x months is given by $I = Mx + A(1 + i)^x + B - P$. Determine the month in which the total interest paid exceeds \$100,000. Round to the nearest month. Month 163

43. One scientific study suggested that the *carrying capacity* of Earth is around 10 billion people. What is meant by "carrying capacity"? Find the current world population and project when Earth's population would reach 10 billion, assuming population growth rates of 1%, 2%, 3%, 4%, and 5%. Find the current rate of world population growth and use that number to determine when, according to your model, the population will reach 10 billion.

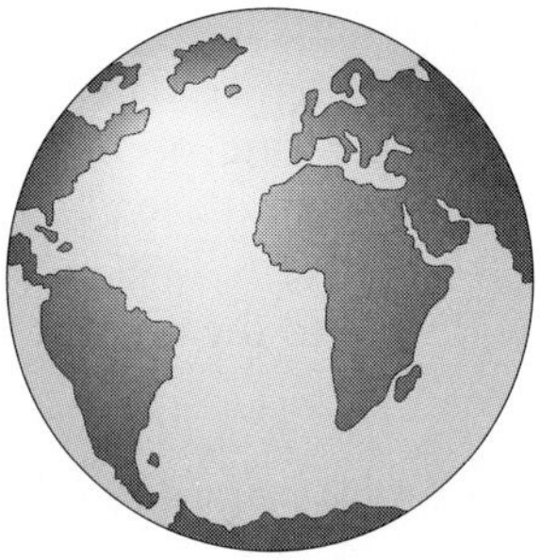

44. Radioactive carbon dating can be used only in certain situations. Other methods of radioactive dating, such as the rubidium-strontium method, are appropriate in other situations. Write a report on radioactive dating. Include in your report when each form of dating is appropriate.

Focus on Problem Solving

Proof by Contradiction

The four-step plan for solving problems that we have used before is restated here.

1. Understand the problem.
2. Devise a plan.
3. Carry out the plan.
4. Review the solution.

One of the techniques that can be used in the second step is a method called *proof by contradiction.* In this method, you assume that the conditions of the problem you are trying to solve can be met and then show that your assumption leads to a condition you already know is not true.

To illustrate this method, suppose we try to prove that $\sqrt{2}$ is a rational number. We begin by recalling that a rational number is one that can be written as the quotient of integers. Therefore, let a and b be two integers with *no common factors.* If $\sqrt{2}$ is a rational number, then

$$\sqrt{2} = \frac{a}{b}$$

$$(\sqrt{2})^2 = \left(\frac{a}{b}\right)^2 \quad \blacktriangleright \text{ Square each side.}$$

$$2 = \frac{a^2}{b^2}$$

$$2b^2 = a^2 \quad \blacktriangleright \text{ Multiply both sides by } b^2.$$

The last equation states that a^2 is an even number. Because a^2 is an even number, then a is an even number. Now divide each side of the last equation by 2.

$$2b^2 = a^2 = a \cdot a$$

$$b^2 = a \cdot x \quad \blacktriangleright\ x = \frac{a}{2}$$

Because a is an even number, $a \cdot x$ is an even number. Because $a \cdot x$ is an even number, b^2 is an even number, and this in turn means that b is an even number. This result, however, contradicts the assumption that a and b are two integers with *no common factors.* Because this assumption is now known not to be true, we conclude that our original assumption, that $\sqrt{2}$ is a rational number, is false. This proves that $\sqrt{2}$ is an irrational number.

Try a proof by contradiction for the following problem: "Is it possible to write numbers using each of the digits 0, 1, 2, 3, 4, 5, 6, 7, 8, and 9 exactly once so that the sum of the numbers is exactly 100?"[1] Some suggestions for the proof are outlined below.

First note that the sum of the ten digits is 45. This means that some of the digits must be used as tens digits. Let x be the sum of those digits.

1. What is the sum of the remaining units digits?
2. Express "the sum of the units digits and the tens digits equals 100" as an equation. (*Note:* Writing an equation is one of the possible problem-solving strategies.)
3. Solve the equation for x.
4. Explain why this means that it is impossible to satisfy both conditions of the problem.

Projects and Group Activities

Fractals*

Fractals have a wide variety of applications. They have been used to create special effects for movies such as *Star Trek II: The Wrath of Khan* and to explain the

[1]G. Polya, *How To Solve It: A New Aspect of Mathematical Method,* copyright © 1957 by G. Polya. Reprinted by permission of Princeton University Press.

*Adapted with permission from "Student Math Notes," by Tami Martin, *News Bulletin,* November 1991.

behavior of some biological and economic systems. One aspect of fractals that has fascinated mathematicians is that they apparently have *fractional dimension.*

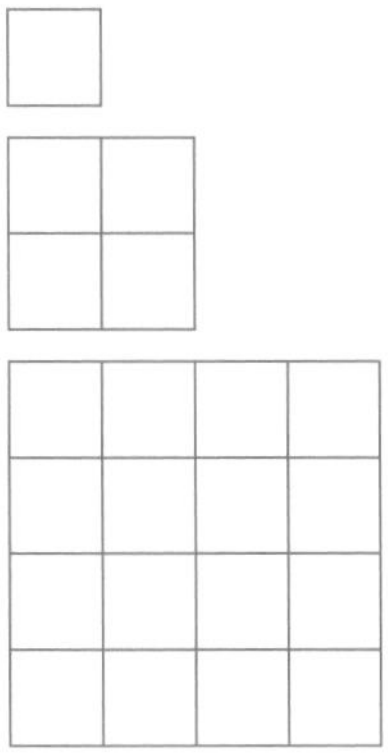

To understand the idea of fractional dimension, one must first understand the terms "scale factor" and "size." Consider a unit square (a square of length 1). By joining four of these squares, we can create another square, the length of which is 2, and the size of which is 4. (Here, size = number of square units.) Four of these larger squares can in turn be put together to make a third square of length 4 and size 16. This process of grouping together four squares can in theory be done an infinite number of times; yet at each step, the following quantities will be the same:

$$\text{Scale factor} = \frac{\text{new length}}{\text{old length}} \qquad \text{Size ratio} = \frac{\text{new size}}{\text{old size}}$$

Consider the unit square as Step 1, the 4-unit squares as Step 2, etc.

1. Calculate the scale factor going from **a.** Step 1 to Step 2, **b.** Step 2 to Step 3, and **c.** Step 3 to Step 4.

2. Calculate the size ratio going from **a.** Step 1 to Step 2, **b.** Step 2 to Step 3, and **c.** Step 3 to Step 4.

3. What is **a.** the scale factor and **b.** the size ratio going from Step n to Step $n + 1$?

Mathematicians have defined dimension using the formula $d = \dfrac{\log(\text{size ratio})}{\log(\text{scale factor})}$.

For the squares discussed above, $d = \dfrac{\log(\text{size ratio})}{\log(\text{scale factor})} = \dfrac{\log 4}{\log 2} = 2.$

So by this definition of dimension, squares are two-dimensional figures.

Now consider a unit cube (Step 1). Group eight unit cubes to form a cube that is 2 units on each side (Step 2). Group eight of the cubes from Step 2 to form a cube that is 4 units on each side (Step 3).

4. Calculate **a.** the scale factor and **b.** the size ratio for this process.

5. Show that the cubes are three-dimensional figures.

In each of the above examples, if the process is continued indefinitely, we still have a square or a cube. Consider a process that is more difficult to envision. Let Step 1 be an equilateral triangle whose base has length 1 unit, and let Step 2 be a grouping of three of these equilateral triangles, such that the space between them is another equilateral triangle with a base of length 1 unit. Three shapes from Step 2 are arranged with an equilateral triangle in their center, and so on. It is hard to imagine the result if this is done an infinite number of times, but mathematicians have shown that the result is a single figure of fractional dimension. (Similar processes have been used to create fascinating artistic patterns and to explain scientific phenomena.)

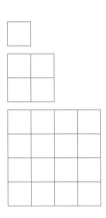

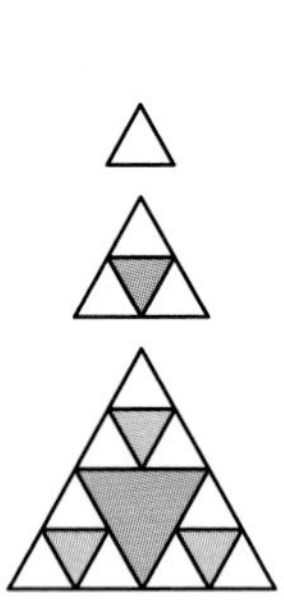

6. Show that for this process **a.** the scale factor is 2 and **b.** the size ratio is 3.

7. Calculate the dimension of the fractal. (Note that it is a *fractional* dimension!)

Chapter Summary

Key Words

A function of the form $f(x) = b^x$, where b is a positive real number not equal to 1, is an *exponential function.* The number b is the *base* of the exponential function. (Objective 10.1.1)

$f(x) = 3^x$; 3 is the base of the function.

The function $f(x) = e^x$ is called the *natural exponential function.* (Objective 10.1.1)

$f(x) = e^x$; e is an irrational number approximately equal to 2.71828183.

For $b > 0$, $b \neq 1$, $y = \log_b x$ is equivalent to $x = b^y$. Read $\log_b x$ as "the *logarithm* of x, base b." (Objective 10.2.1)

$\log_2 8 = 3$ is equivalent to $8 = 2^3$.

Common logarithms are logarithms to the base 10. (Objective 10.2.1)

$\log_{10} 100 = 2$ is usually written as $\log 100 = 2$.

Natural logarithms are logarithms to the base e. (Objective 10.2.1)

$\log_e 100 \approx 4.61$ is usually written as $\ln 100 \approx 4.61$.

An *exponential equation* is an equation in which a variable occurs in the exponent. (Objective 10.4.1)

$2^x = 12$ is an exponential equation.

Essential Rules and Procedures

Equality of Exponents Property
For $b > 0$, $b \neq 1$, if $b^u = b^v$, then $u = v$. (Objective 10.2.1)

If $b^x = b^5$, then $x = 5$.

The Product Property of Logarithms
For any positive real numbers x, y, and b, $b \neq 1$, $\log_b(xy) = \log_b x + \log_b y$. (Objective 10.2.2)

$\log_b(3x) = \log_b 3 + \log_b x$

The Quotient Property of Logarithms
For any positive real numbers x, y, and b, $b \neq 1$, $\log_b \frac{x}{y} = \log_b x - \log_b y$. (Objective 10.2.2)

$\log_b \frac{x}{20} = \log_b x - \log_b 20$

The Power Property of Logarithms
For any positive real numbers x and b, $b \neq 1$, and for any real number r, $\log_b x^r = r \log_b x$. (Objective 10.2.2)

$\log_b x^5 = 5 \log_b x$

The Inverse Property of Logarithms
For any positive real numbers x and b, $b \neq 1$, $\log_b b^x = x$. (Objective 10.2.2)

$\log_3 3^4 = 4$

The Logarithmic Property of One
For any positive real numbers b, $b \neq 1$, $\log_b 1 = 0$. (Objective 10.2.2)

$\log_6 1 = 0$

The 1–1 Property of Logarithms
For any positive real numbers x, y, and b, $b \neq 1$, if $\log_b x = \log_b y$, then $x = y$. (Objective 10.2.2)

If $\log_5(x - 2) = \log_5 3$, then $x - 2 = 3$.

Change-of-Base Formula
$\log_a N = \frac{\log_b N}{\log_b a}$ (Objective 10.2.2)

$\log_3 12 = \frac{\log 12}{\log 3}$ $\log_6 16 = \frac{\ln 16}{\ln 6}$

Chapter Review Exercises

1. Evaluate: $\log_4 16$ 2

2. Write $\frac{1}{2}(\log_3 x - \log_3 y)$ as a single logarithm with a coefficient of 1. $\log_3 \sqrt{\frac{x}{y}}$

3. Evaluate the function $f(x) = e^{x-2}$ at $x = 2$. 1

4. Solve for x: $8^x = 2^{x-6}$ -3

5. Evaluate $f(x) = \left(\frac{2}{3}\right)^x$ at $x = 0$. 1

6. Solve for x: $\log_3 x = -2$ $\frac{1}{9}$

7. Write $2^5 = 32$ in logarithmic form. $\log_2 32 = 5$

8. Solve for x: $\log x + \log(x - 4) = \log 12$ 6

9. Write $\log_6 \sqrt{xy^3}$ in expanded form.
$\frac{1}{2}\log_6 x + \frac{3}{2}\log_6 y$

10. Solve for x: $4^{5x-2} = 4^{3x+2}$ 2

11. Solve for x: $3^{7x+1} = 3^{4x-5}$ -2

12. Evaluate $f(x) = 3^{x+1}$ at $x = -2$. $\frac{1}{3}$

13. Evaluate: $\log_2 16$ 4

14. Solve for x: $\log_6 2x = \log_6 2 + \log_6(3x - 4)$ 2

15. Evaluate $\log_2 5$. Round to the nearest ten-thousandth. 2.3219

16. Evaluate $\log_6 22$. Round to the nearest ten-thousandth. 1.7251

17. Solve for x: $4^x = 8^{x-1}$ 3

18. Evaluate $f(x) = \left(\frac{1}{4}\right)^x$ at $x = -1$. 4

19. Write $\log_5 \sqrt{\frac{x}{y}}$ in expanded form.
$\frac{1}{2}\log_5 x - \frac{1}{2}\log_5 y$

20. Solve for x: $\log_5 \frac{7x + 2}{3x} = 1$ $\frac{1}{4}$

21. Solve for x: $\log_5 x = 3$ 125

22. Solve for x: $\log x + \log(2x + 3) = \log 2$ $\frac{1}{2}$

23. Write $3 \log_b x - 5 \log_b y$ as a single logarithm with a coefficient of 1. $\log_b \frac{x^3}{y^5}$

24. Evaluate $f(x) = 2^{-x-1}$ at $x = -3$. 4

25. Evaluate $\log_3 19$. Round to the nearest ten-thousandth. 2.6801

26. Solve $3^{x+2} = 5$ for x. Round to the nearest ten-thousandth. -0.5350

27. Graph: $f(x) = 2^x - 3$

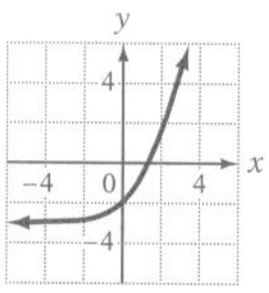

28. Graph: $f(x) = \left(\frac{1}{2}\right)^x + 1$

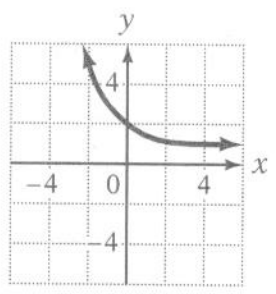

29. Graph: $f(x) = \log_2(2x)$

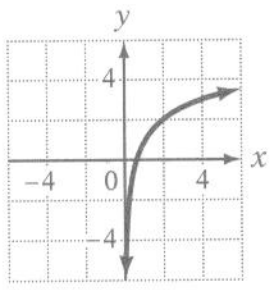

30. Graph: $f(x) = \log_2 x - 1$

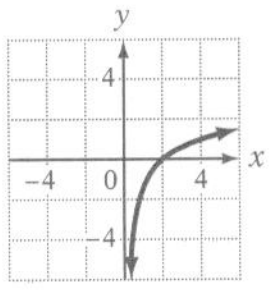

31. Graph $f(x) = \dfrac{\log x}{x}$ and approximate the zeros of f to the nearest tenth. 1.0

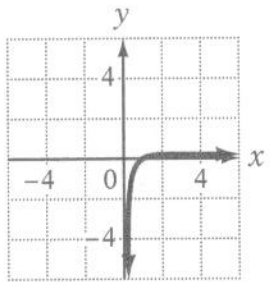

32. Use the exponential decay equation $A = A_0\left(\frac{1}{2}\right)^{\frac{t}{k}}$, where A is the amount of a radioactive material present after time t, k is the half-life, and A_0 is the original amount of radioactive material, to find the half-life of a material that decays from 10 mg to 9 mg in 5 h. Round to the nearest whole number. 33 h

33. The percent of light that will pass through a translucent material is given by the equation $\log P = -0.5d$, where P is the percent of light that passes through the material, and d is the thickness of the material in centimeters. How thick must this translucent material be so that only 50% of the light that is incident to the material will pass through it? Round to the nearest thousandth. 0.602 cm

Chapter Test

1. Evaluate $f(x) = \left(\frac{3}{4}\right)^x$ at $x = 0$. 1

2. Evaluate $f(x) = 4^{x-1}$ at $x = -2$. $\frac{1}{64}$

3. Evaluate: $\log_4 64$ 3

4. Solve for x: $\log_4 x = -2$ $\frac{1}{16}$

5. Write $\log_6 \sqrt[3]{x^2y^5}$ in expanded form.
$\frac{2}{3}\log_6 x + \frac{5}{3}\log_6 y$

6. Write $\frac{1}{2}(\log_5 x - \log_5 y)$ as a single logarithm with a coefficient of 1. $\log_5 \sqrt{\frac{x}{y}}$

7. Solve for x: $\log_6 x + \log_6(x - 1) = 1$ 3

8. Evaluate $f(x) = 3^{x+1}$ at $x = -2$. $\frac{1}{3}$

9. Solve for x: $3^x = 17$ 2.5789

10. Solve for x: $\log_2 x + 3 = \log_2(x^2 - 20)$ 10

11. Solve for x: $5^{6x-2} = 5^{3x+7}$ 3

12. Solve for x: $4^x = 2^{3x+4}$ -4

13. Solve for x: $\log(2x + 1) + \log x = \log 6$ $\frac{3}{2}$

14. Graph: $f(x) = 2^x - 1$

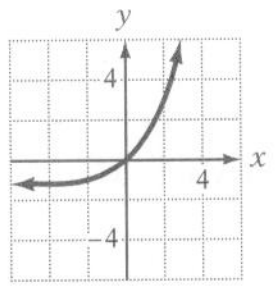

15. Graph: $f(x) = 2^x + 2$

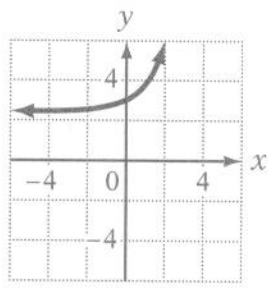

16. Graph: $f(x) = \log_2(3x)$

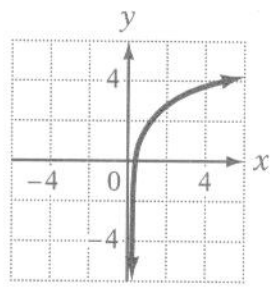

17. Graph: $f(x) = \log_3(x + 1)$

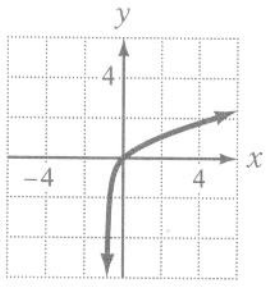

18. Graph $f(x) = 3 - 2^x$ and approximate the zeros of f to the nearest tenth.

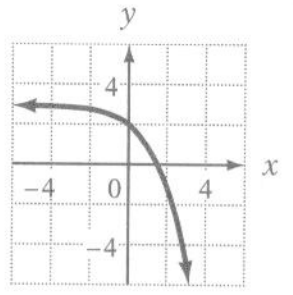

1.6

19. Find the value of \$10,000 invested for 6 years at 7.5% compounded monthly. Use the compound interest formula $P = A(1 + i)^n$, where A is the original value of the investment, i is the interest rate per compounding period, and n is the number of compounding periods. Round to the nearest dollar. \$15,661

20. Use the exponential decay equation $A = A_0\left(\frac{1}{2}\right)^{\frac{t}{k}}$ where A is the amount of a radioactive material present after time t, k is the half-life, and A_0 is the original amount of radioactive material, to find the half-life of a material that decays from 40 mg to 30 mg in 10 h. Round to the nearest whole number.
24 h

Cumulative Review Exercises

1. Solve: $4 - 2[x - 3(2 - 3x) - 4x] = 2x$ $\frac{8}{7}$

2. Solve $S = 2WH + 2WL + 2LH$ for L.
$L = \frac{S - 2WH}{2W + 2H}$

3. Solve: $|2x - 5| \leq 3$ $\{x | 1 \leq x \leq 4\}$

4. Factor: $4x^{2n} + 7x^n + 3$ $(4x^n + 3)(x^n + 1)$

5. Solve: $x^2 + 4x - 5 \leq 0$ $\{x | -5 \leq x \leq 1\}$

6. Simplify: $\dfrac{1 - \frac{5}{x} + \frac{6}{x^2}}{1 + \frac{1}{x} - \frac{6}{x^2}}$ $\frac{x - 3}{x + 3}$

7. Simplify: $\dfrac{\sqrt{xy}}{\sqrt{x} - \sqrt{y}}$ $\frac{x\sqrt{y} + y\sqrt{x}}{x - y}$

8. Simplify: $y\sqrt{18x^5y^4} - x\sqrt{98x^3y^6}$ $-4x^2y^3\sqrt{2x}$

9. Simplify: $\dfrac{i}{2 - i}$ $-\frac{1}{5} + \frac{2}{5}i$

10. Find the equation of the line that contains the point $(2, -2)$ and is parallel to the line $2x - y = 5$. $y = 2x - 6$

11. Write a quadratic equation that has integer coefficients and has as solutions $\frac{1}{3}$ and -3.
$3x^2 + 8x - 3 = 0$

12. Solve: $x^2 - 4x - 6 = 0$ $2 \pm \sqrt{10}$

13. Find the range of $f(x) = x^2 - 3x - 4$ if the domain is $\{-1, 0, 1, 2, 3\}$. $\{-6, -4, 0\}$

14. Given $f(x) = x^2 + 2x + 1$ and $g(x) = 2x - 3$, find $f[g(0)]$. 4

15. Solve by the addition method: $(0, -1, 2)$

$$3x - y + z = 3$$
$$x + y + 4z = 7$$
$$3x - 2y + 3z = 8$$

16. Solve: $y = -2x - 3$, $y = 2x - 1$ $\left(-\frac{1}{2}, -2\right)$

17. Evaluate $f(x) = 3^{-x+1}$ at $x = -4$. 243

18. Solve for x: $\log_4 x = 3$ 64

19. Solve for x: $2^{3x+2} = 4^{x+5}$ 8

20. Solve for x: $\log x + \log(3x + 2) = \log 5$ 1

21. Graph: $\{x \mid x < 0\} \cap \{x \mid x > -4\}$

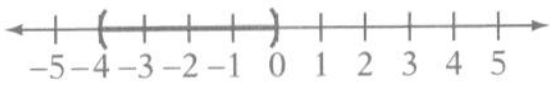

22. Graph the solution set of $\frac{x+2}{x-1} \geq 0$.

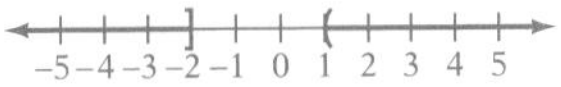

23. Graph: $y = -x^2 - 2x + 3$

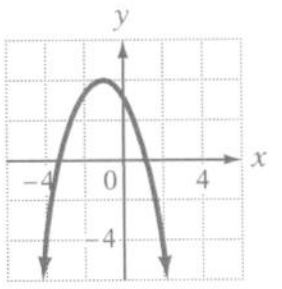

24. Graph: $f(x) = |x| - 2$

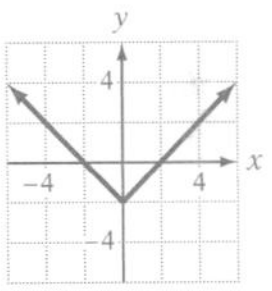

25. Graph: $f(x) = \left(\frac{1}{2}\right)^x - 1$

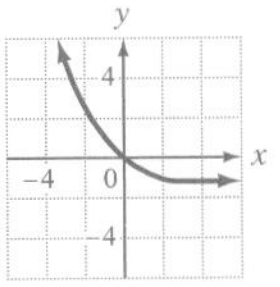

26. Graph: $f(x) = \log_2 x + 1$

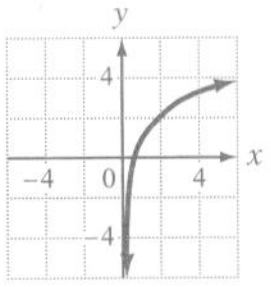

27. Graph $f(x) = -3 \ln x + x$ and determine, to the nearest tenth, the minimum value of f. −0.3

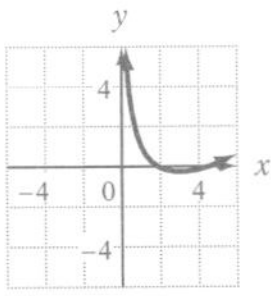

28. An alloy containing 25% tin is mixed with an alloy containing 50% tin. How much of each were used to make 2000 lb of an alloy containing 40% tin?
25% alloy: 800 lb; 50% alloy: 1200 lb

29. An account executive earns \$500 per month plus 8% commission on the amount of sales. The executive's goal is to earn a minimum of \$3000 per month. What amount of sales will enable the executive to earn \$3000 or more per month? \$31,250 or more

30. A new printer can print checks three times faster than an older printer. The old printer can print the checks in 30 min. How long would it take to print the checks with both printers operating? 7.5 min

31. For a constant temperature, the pressure (P) of a gas varies inversely as the volume (V). If the pressure is 50 lb/in^2 when the volume is 250 ft^3, find the pressure when the volume is 25 ft^3. 500 lb/in^2

32. A contractor buys 45 yd of nylon carpet and 30 yd of wool carpet for $1170. A second purchase, at the same prices, includes 25 yd of nylon carpet and 80 yd of wool carpet for $1410. Find the cost per yard of the wool carpet. $12

33. An investor deposits $10,000 into an account that earns 9% interest compounded monthly. What is the value of the investment after 5 years? Use the compound interest formula $P = A(1 + i)^n$, where A is the original value of an investment, i is the interest rate per compounding period, n is the total number of compounding periods, and P is the value of the investment after n periods. Round to the nearest dollar. $15,657

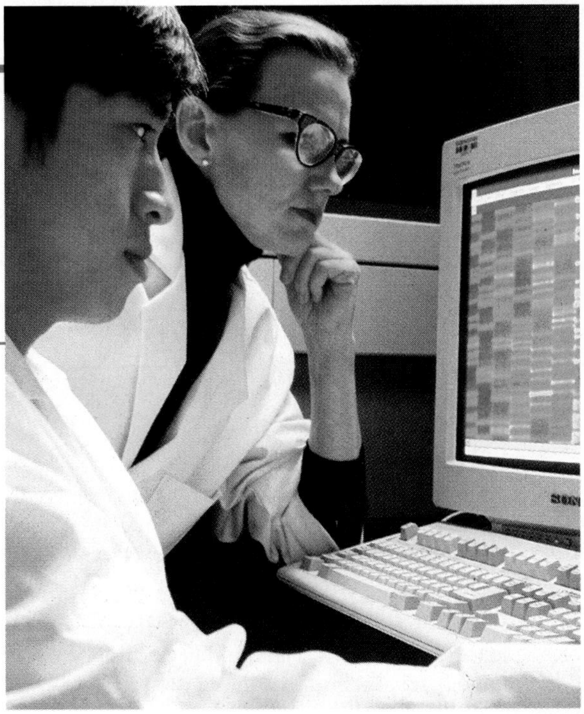

As you will learn from studying this chapter, sequences are patterns of numbers. Discovering patterns is very important in the scientific community. It is by looking for patterns that geneticists have identified patterns in DNA, thereby enabling them to decipher the genetic codes carried by the DNA.

11

Sequences and Series

OBJECTIVES

Tower of Hanoi

The Tower of Hanoi is a puzzle that has the following form.

Three pegs are placed in a board. A number of disks, graded in size, are stacked on one of the pegs with the largest disk on the bottom and the succeeding smaller disks placed on top.

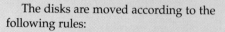

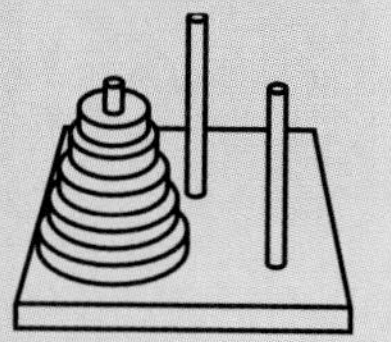

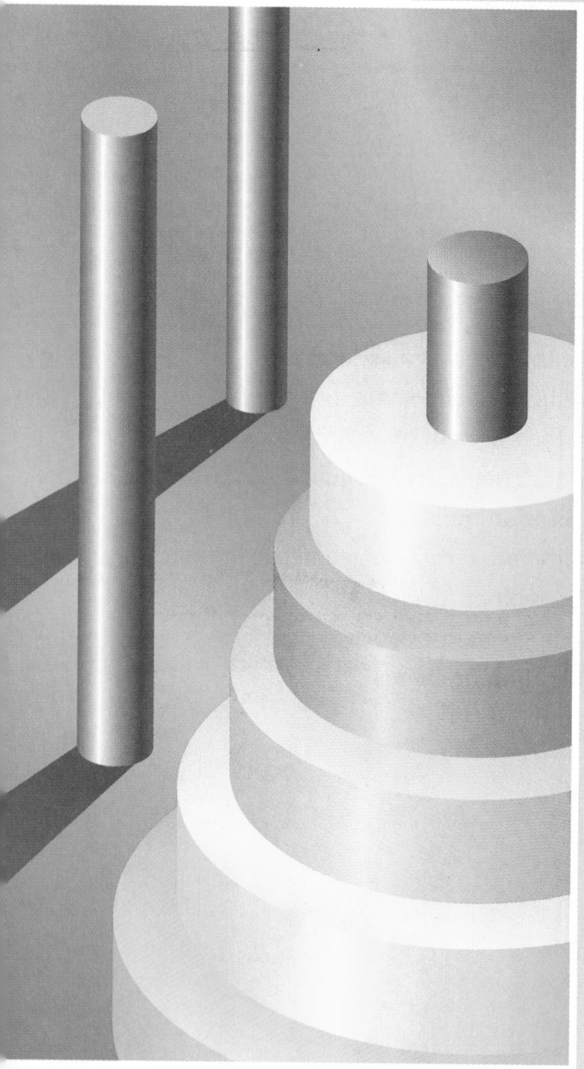

The disks are moved according to the following rules:

1. Only one disk at a time may be moved.
2. A larger disk may not be placed over a smaller disk.

The object of the puzzle is to transfer all the disks from one peg to one of the other two pegs. If initially there is only one disk, then only one move is required. With two disks initially, three moves are required, and with three disks, seven moves are required.

You can try this puzzle using playing cards. Select a number of playing cards, in sequence, from a deck. The number of cards corresponds to the number of disks, and the number on the card corresponds to the size of the disk. For example, an "ace" would correspond to the smallest disk, a "two" would correspond to the next largest disk, and so on for five cards. Now place three coins on a table to be used as the pegs. Pile the cards on one of the coins (in order) and try to move the pile to a second coin. Remember that a numerically larger card cannot be placed on a numerically smaller card. Try this for five cards. The minimum number of moves is thirty-one.

The chart below shows the minimum number of moves required for an initial number of disks. The increase in the number of moves required for each increase in the number of initial disks is also given.

1		2		3		4		5		6		7		8
1		3		7		15		31		63		127		255
	2		4		8		16		32		64		128	

For this last list of numbers, each succeeding number can be found by multiplying the preceding number by a constant (in this case 2). Such a list of numbers is called a *geometric sequence.*

The formula for the minimum number of moves is given by $M = 2^n - 1$, where M is the number of moves and n is the number of disks. This equation is an exponential equation.

Here's a hint for solving the Tower of Hanoi puzzle with five disks. First solve the puzzle for three disks. Now solve the puzzle for four disks by first moving the top three disks to one peg. (You just did this when you solved the puzzle for three disks.) Now move the fourth disk to a peg, and then again use your solution for three disks to move the disks back to the peg with the fourth disk. Now use this solution for four disks to solve the five-disk problem.

Introduction to Sequences and Series

1 Write the terms of a sequence

An investor deposits $100 in an account that earns 10% interest compounded annually. The amount of interest earned each year can be determined by using the compound interest formula. The amount of interest earned in each of the first four years of the investment is shown below.

Year	1	2	3	4
Interest Earned	$10	$11	$12.10	$13.31

The list of numbers 10, 11, 12.10, 13.31 is called a sequence. A **sequence** is an ordered list of numbers. The list 10, 11, 12.10, 13.31 is ordered because the position of a number in this list indicates the year in which the amount of interest was earned. Each of the numbers of a sequence is called a **term** of the sequence.

Examples of other sequences are shown at the right. These sequences are separated into two groups. A **finite sequence** contains a finite number of terms. An **infinite sequence** contains an infinite number of terms.

$$1, 1, 2, 3, 5, 8$$
$$1, 2, 3, 4, 5, 6, 7, 8$$ Finite sequences
$$1, -1, 1, -1$$

$$1, 3, 5, 7, \ldots$$
$$1, \frac{1}{2}, \frac{1}{4}, \frac{1}{8}, \ldots$$ Infinite sequences
$$1, 1, 2, 3, 5, 8, \ldots$$

For the sequence at the right, the first term is 2, the second term is 4, the third term is 6, and the fourth term is 8.

$$2, 4, 6, 8, \ldots$$

A general sequence is shown at the right. The first term is a_1, the second term is a_2, the third term is a_3, and the nth term, also called the **general term** of the sequence, is a_n. Note that each term of the sequence is paired with a natural number.

$$a_1, a_2, a_3, \ldots, a_n, \ldots$$

Frequently, a sequence has a definite pattern that can be expressed by a formula.

Each term of the sequence shown at the right is paired with a natural number by the formula $a_n = 3n$. The first term, a_1, is 3. The second term, a_2, is 6. The third term, a_3, is 9. The nth term, a_n, is $3n$.

$$a_n = 3n$$

$$a_1, \quad a_2, \quad a_3, \ldots, \quad a_n, \ldots$$
$$3(1), 3(2), 3(3), \ldots, 3(n), \ldots$$
$$3, \quad 6, \quad 9, \ldots, \quad 3n, \ldots$$

INSTRUCTOR NOTE

Although subscripting a variable has been shown before, the idea is confusing to students. To help these students, you might try writing the nth term of a sequence with open parentheses as shown below.

$$a_n = 2n$$
$$a_{(\ )} = 2(\quad)$$

$$a_n = n^2$$
$$a_{(\ )} = (\quad)^2$$

Substitute the term number into the parentheses and then simplify.

Example 1 Write the first three terms of the sequence whose nth term is given by the formula $a_n = 2n - 1$.

Solution
$a_n = 2n - 1$
$a_1 = 2(1) - 1 = 1$ ▶ Replace n by 1.
$a_2 = 2(2) - 1 = 3$ ▶ Replace n by 2.
$a_3 = 2(3) - 1 = 5$ ▶ Replace n by 3.

The first term is 1, the second term is 3, and the third term is 5.

Problem 1 Write the first four terms of the sequence whose nth term is given by the formula $a_n = n(n + 1)$.

Solution See page S30. 2, 6, 12, 20

Example 2 Find the eighth and tenth terms of the sequence whose nth term is given by the formula $a_n = \frac{n}{n + 1}$.

Solution
$a_n = \frac{n}{n + 1}$

$a_8 = \frac{8}{8 + 1} = \frac{8}{9}$ ▶ Replace n by 8.

$a_{10} = \frac{10}{10 + 1} = \frac{10}{11}$ ▶ Replace n by 10.

The eighth term is $\frac{8}{9}$, and the tenth term is $\frac{10}{11}$.

Problem 2 Find the sixth and ninth terms of the sequence whose nth term is given by the formula $a_n = \frac{1}{n(n + 2)}$.

Solution See page S30. $a_6 = \frac{1}{48}$; $a_9 = \frac{1}{99}$

2 Find the sum of a series

HM³

At the beginning of this section, the sequence 10, 11, 12.10, 13.31 was shown to represent the amount of interest earned in each of 4 years of an investment.

10, 11, 12.10, 13.31

The sum of the terms of this sequence represents the total interest earned by the investment over the 4-year period.

$10 + 11 + 12.10 + 13.31 = 46.41$
The total interest earned over the 4-year period is \$46.41.

The indicated sum of the terms of a sequence is called a **series.** Given the sequence 10, 11, 12.10, 13.31, the series $10 + 11 + 12.10 + 13.31$ can be written.

S_n is used to indicate the sum of the first n terms of a sequence.

For the preceding example, the sums of the series S_1, S_2, S_3, and S_4 represent the total interest earned for 1, 2, 3, and 4 years, respectively.

$S_1 = 10 = 10$
$S_2 = 10 + 11 = 21$
$S_3 = 10 + 11 + 12.10 = 33.10$
$S_4 = 10 + 11 + 12.10 + 13.31 = 46.41$

For the general sequence $a_1, a_2, a_3, \ldots, a_n$, the series S_1, S_2, S_3, and S_n are shown at the right.

$$S_1 = a_1$$
$$S_2 = a_1 + a_2$$
$$S_3 = a_1 + a_2 + a_3$$
$$S_n = a_1 + a_2 + a_3 + \cdots + a_n$$

INSTRUCTOR NOTE

Whether a student takes statistics, finite math, or any other math course after completing this one, summation notation will be a part of that course. This notation is not easy for students, so many examples may be necessary. One important aspect of this notation is that the index is *always* incremented by 1.

It is convenient to represent a series in a compact form called **summation notation,** or **sigma notation.** The Greek letter sigma, Σ, is used to indicate a sum.

The first four terms of the sequence whose nth term is given by the formula $a_n = 2n$ are 2, 4, 6, 8. The corresponding series is shown at the right written in summation notation and is read "the summation from 1 to 4 of $2n$." The letter n is called the **index** of the summation.

$$\sum_{n=1}^{4} 2n$$

To write the terms of the series, replace n by the consecutive integers from 1 to 4.

$$\sum_{n=1}^{4} 2n = 2(1) + 2(2) + 2(3) + 2(4)$$

The series is $2 + 4 + 6 + 8$. $\qquad = 2 + 4 + 6 + 8$

The sum of the series is 20. $\qquad = 20$

LOOK CLOSELY

The placement of the parentheses in this example is important. For instance,

$$\sum_{i=1}^{3} 2i - 1$$
$$= 2(1) + 2(2) + 2(3) - 1$$
$$= 2 + 4 + 6 - 1 = 11$$

This is *not* the same result as for $\sum_{i=1}^{3} (2i - 1)$.

Example 3 Find the sum of the series. A. $\sum_{i=1}^{3} (2i - 1)$ B. $\sum_{n=3}^{6} \frac{1}{2}n$

Solution A. $\sum_{i=1}^{3} (2i - 1)$

$= [2(1) - 1] + [2(2) - 1] + [2(3) - 1]$ ▶ Replace i by 1, 2, and 3.

$= 1 + 3 + 5$ ▶ Write the series.

$= 9$ ▶ Find the sum of the series.

B. $\sum_{n=3}^{6} \frac{1}{2}n$

$= \frac{1}{2}(3) + \frac{1}{2}(4) + \frac{1}{2}(5) + \frac{1}{2}(6)$ ▶ Replace n by 3, 4, 5, and 6.

$= \frac{3}{2} + 2 + \frac{5}{2} + 3$ ▶ Write the series.

$= 9$ ▶ Find the sum of the series.

Problem 3 Find the sum of the series. A. $\sum_{n=1}^{4} (7 - n)$ B. $\sum_{i=3}^{6} (i^2 - 2)$

Solution See page S30. A. 18 B. 78

Example 4 Write $\sum_{i=1}^{5} x^i$ in expanded form.

Solution $\sum_{i=1}^{5} x^i = x + x^2 + x^3 + x^4 + x^5$ ▶ This is a variable series. Replace i by 1, 2, 3, 4, and 5.

Problem 4 Write $\sum_{n=1}^{5} nx$ in expanded form.

Solution See page S30. $x + 2x + 3x + 4x + 5x$

CONCEPT REVIEW 11.1

State whether the following statements are always true, sometimes true, or never true.

1. A sequence is an ordered list of numbers. Always true
2. The indicated sum of the terms of a sequence is called a series. Always true
3. The first term of the sequence $a_n = \frac{n}{n+3}$ is $\frac{1}{4}$. Always true
4. The sum of the series $\sum_{i=1}^{3} 2i + 2$ is 18. Never true

EXERCISES 11.1

1 1. What is a sequence?

2. What is the difference between a finite sequence and an infinite sequence?

Write the first four terms of the sequence whose nth term is given by the formula.

3. $a_n = n + 1$ 2, 3, 4, 5
4. $a_n = n - 1$ 0, 1, 2, 3
5. $a_n = 2n + 1$ 3, 5, 7, 9
6. $a_n = 3n - 1$ 2, 5, 8, 11
7. $a_n = 2 - 2n$ 0, −2, −4, −6
8. $a_n = 1 - 2n$ −1, −3, −5, −7
9. $a_n = 2^n$ 2, 4, 8, 16
10. $a_n = 3^n$ 3, 9, 27, 81
11. $a_n = n^2 + 1$ 2, 5, 10, 17
12. $a_n = n^2 - 1$ 0, 3, 8, 15
13. $a_n = \frac{n}{n^2+1}$ $\frac{1}{2}, \frac{2}{5}, \frac{3}{10}, \frac{4}{17}$
14. $a_n = \frac{n^2-1}{n}$ $0, \frac{3}{2}, \frac{8}{3}, \frac{15}{4}$
15. $a_n = n - \frac{1}{n}$ $0, \frac{3}{2}, \frac{8}{3}, \frac{15}{4}$
16. $a_n = n^2 - \frac{1}{n}$ $0, \frac{7}{2}, \frac{26}{3}, \frac{63}{4}$
17. $a_n = (-1)^{n+1}n$ 1, −2, 3, −4
18. $a_n = \frac{(-1)^{n+1}}{n+1}$ $\frac{1}{2}, -\frac{1}{3}, \frac{1}{4}, -\frac{1}{5}$
19. $a_n = \frac{(-1)^{n+1}}{n^2+1}$ $\frac{1}{2}, -\frac{1}{5}, \frac{1}{10}, -\frac{1}{17}$
20. $a_n = (-1)^n(n^2 + 2n + 1)$ −4, 9, −16, 25
21. $a_n = (-1)^n 2^n$ −2, 4, −8, 16
22. $a_n = \frac{1}{3}n^3 + 1$ $\frac{4}{3}, \frac{11}{3}, 10, \frac{67}{3}$
23. $a_n = 2\left(\frac{1}{3}\right)^{n+1}$ $\frac{2}{9}, \frac{2}{27}, \frac{2}{81}, \frac{2}{243}$

Find the indicated term of the sequence whose nth term is given by the formula.

24. $a_n = 3n + 4; a_{12}$ 40

25. $a_n = 2n - 5; a_{10}$ 15

26. $a_n = n(n - 1); a_{11}$ 110

27. $a_n = \frac{n}{n+1}; a_{12}$ $\frac{12}{13}$

28. $a_n = (-1)^{n-1}n^2; a_{15}$ 225

29. $a_n = (-1)^{n-1}(n - 1); a_{25}$ 24

30. $a_n = \left(\frac{1}{2}\right)^n; a_8$ $\frac{1}{256}$

31. $a_n = \left(\frac{2}{3}\right)^n; a_5$ $\frac{32}{243}$

32. $a_n = (n + 2)(n + 3); a_{17}$ 380

33. $a_n = (n + 4)(n + 1); a_7$ 88

34. $a_n = \frac{(-1)^{2n-1}}{n^2}; a_6$ $-\frac{1}{36}$

35. $a_n = \frac{(-1)^{2n}}{n+4}; a_{16}$ $\frac{1}{20}$

36. $a_n = \frac{3}{2}n^2 - 2; a_8$ 94

37. $a_n = \frac{1}{3}n + n^2; a_6$ 38

2 **38.** What is a series?

39. What is summation notation or sigma notation?

Find the sum of the series.

40. $\sum_{n=1}^{5}(2n + 3)$ 45

41. $\sum_{i=1}^{7}(i + 2)$ 42

42. $\sum_{i=1}^{4}2i$ 20

43. $\sum_{n=1}^{7}n$ 28

44. $\sum_{i=1}^{6}i^2$ 91

45. $\sum_{i=1}^{5}(i^2 + 1)$ 60

46. $\sum_{n=1}^{6}(-1)^n$ 0

47. $\sum_{n=1}^{4}\frac{1}{2n}$ $\frac{25}{24}$

48. $\sum_{i=3}^{6}i^3$ 432

49. $\sum_{n=2}^{4}2^n$ 28

50. $\sum_{n=3}^{7}\frac{n}{n-1}$ $\frac{129}{20}$

51. $\sum_{i=3}^{6}\frac{i+1}{i}$ $\frac{99}{20}$

52. $\sum_{i=1}^{4}\frac{1}{2^i}$ $\frac{15}{16}$

53. $\sum_{i=1}^{5}\frac{1}{2i}$ $\frac{137}{120}$

54. $\sum_{n=1}^{4}(-1)^{n-1}n^2$ -10

55. $\sum_{i=1}^{4}(-1)^{i-1}(i + 1)$ -2

56. $\sum_{n=3}^{5}\frac{(-1)^{n-1}}{n-2}$ $\frac{5}{6}$

57. $\sum_{n=4}^{7}\frac{(-1)^{n-1}}{n-3}$ $-\frac{7}{12}$

Write the series in expanded form.

58. $\sum_{n=1}^{5}2x^n$ $2x + 2x^2 + 2x^3 + 2x^4 + 2x^5$

59. $\sum_{n=1}^{4}\frac{2n}{x}$ $\frac{2}{x} + \frac{4}{x} + \frac{6}{x} + \frac{8}{x}$

60. $\sum_{i=1}^{5}\frac{x^i}{i}$ $x + \frac{x^2}{2} + \frac{x^3}{3} + \frac{x^4}{4} + \frac{x^5}{5}$

61. $\sum_{i=1}^{4}\frac{x^i}{i+1}$ $\frac{x}{2} + \frac{x^2}{3} + \frac{x^3}{4} + \frac{x^4}{5}$

62. $\sum_{i=3}^{5}\frac{x^i}{2i}$ $\frac{x^3}{6} + \frac{x^4}{8} + \frac{x^5}{10}$

63. $\sum_{i=2}^{4}\frac{x^i}{2i-1}$ $\frac{x^2}{3} + \frac{x^3}{5} + \frac{x^4}{7}$

64. $\sum_{n=1}^{5}x^{2n}$ $x^2 + x^4 + x^6 + x^8 + x^{10}$

65. $\sum_{n=1}^{4}x^{2n-1}$ $x + x^3 + x^5 + x^7$

APPLYING CONCEPTS 11.1

Write a formula for the nth term of the sequence.

66. The sequence of the natural numbers $a_n = n$

67. The sequence of the odd natural numbers $a_n = 2n - 1$

68. The sequence of the negative even integers $a_n = -2n$

69. The sequence of the negative odd integers $a_n = -2n + 1$

70. The sequence of the positive multiples of 7 $a_n = 7n$

71. The sequence of the positive integers that are divisible by 4 $a_n = 4n$

Find the sum of the series. Write your answer as a single logarithm.

72. $\sum_{n=1}^{5} \log n$ log 120

73. $\sum_{i=1}^{4} \log 2i$ log 384

Solve.

[C] **74.** *Number Problem* The first 22 numbers in the sequence 4, 44, 444, 4444, ... are added together. What digit is the thousands place of the sum? 4

[C] **75.** *Number Problem* The first 31 numbers in the sequence 6, 66, 666, 6666, ... are added together. What digit is in the hundreds place of the sum? 3

A recursive sequence is one for which each term of the sequence is defined by using preceding terms. Find the first three terms of each recursively defined sequence.

[C] **76.** $a_1 = 1, a_n = na_{n-1}, n \geq 2$ 2, 6, 24

[C] **77.** $a_1 = 1, a_2 = 1, a_n = a_{n-1} + a_{n-2}, n \geq 3$ 2, 3, 5

[C] **78.** In the first box below, $\frac{1}{2}$ of the box is shaded.
In successive boxes the sums $\frac{1}{2} + \frac{1}{4}$ and $\frac{1}{2} + \frac{1}{4} + \frac{1}{8}$ are shown.
Can you identify the sum $\frac{1}{2} + \frac{1}{4} + \frac{1}{8} + \frac{1}{16} + \cdots$? 1

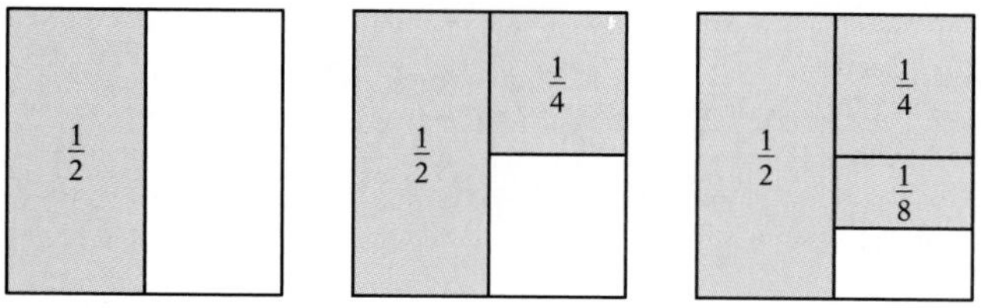

79. Rewrite $1 + \frac{1}{2} + \frac{1}{3} + \cdots + \frac{1}{n}$ using sigma notation. $\sum_{i=1}^{n} \frac{1}{i}$

80. *Medicine* A model used by epidemiologists (people who study epidemics) to study the spread of a virus suggests that the number of people in a population newly infected on a given day is proportional to the number not yet exposed on the previous day. This can be described by a recursive sequence (see Exercises 76 and 77) defined by $a_n - a_{n-1} = k(P - a_{n-1})$, where P is the number of people in the original population, a_n is the number of people exposed to the virus n days after it begins, a_{n-1} is the number of people exposed on the previous day, and k is a constant that depends on the contagiousness of the disease and is determined from experimental evidence.

a. Suppose that a population of 5000 people is exposed to a virus and 150 people become ill ($a_0 = 150$). The next day, 344 people are ill ($a_1 = 344$). Determine the value of k. 0.04

b. Substitute the values of k and P into the recursion equation, and solve for a_n. $a_n = 200 + 0.96a_{n-1}$

c. How many people are infected after 4 days? 709 people

81. Explain the difference between a sequence and a series.

82. Write a paper on Joseph Fourier.

83. Write a paragraph about the history of the sigma notation used for summation.

SECTION 11.2 Arithmetic Sequences and Series

1 Find the *n*th term of an arithmetic sequence

A company's expenses for training a new employee are quite high. To encourage employees to continue their employment with the company, a company that has a 6-month training program offers a starting salary of \$800 a month and then a \$100-per-month pay increase each month during the training period.

The sequence below shows the employee's monthly salaries during the training period. Each term of the sequence is found by adding \$100 to the previous term.

Month	1	2	3	4	5	6
Salary	800	900	1000	1100	1200	1300

> **LOOK CLOSELY**
>
> Arithmetic sequences are a special type of sequence, one for which the difference between *any* two successive terms is the same constant. For instance, 5, 10, 15, 20, 25, ... is an arithmetic sequence. The difference between any two successive terms is 5. The sequence 1, 4, 9, 16, ... is not an arithmetic sequence because $4 - 1 \neq 9 - 4$.

The sequence 800, 900, 1000, 1100, 1200, 1300 is called an arithmetic sequence. An **arithmetic sequence,** or **arithmetic progression,** is one in which the difference between any two consecutive terms is constant. The difference between consecutive terms is called the **common difference** of the sequence.

Each sequence shown below is an arithmetic sequence. To find the common difference of an arithmetic sequence, subtract the first term from the second term.

$2, 7, 12, 17, 22, \ldots$ Common difference: 5

$3, 1, -1, -3, -5, \ldots$ Common difference: -2

$1, \frac{3}{2}, 2, \frac{5}{2}, 3, \frac{7}{2}$ Common difference: $\frac{1}{2}$

Consider an arithmetic sequence in which the first term is a_1 and the common difference is d. Adding the common difference to each successive term of the arithmetic sequence yields a formula for the nth term.

The first term is a_1. $\quad a_1 = a_1$

To find the second term, add the common difference d to the first term. $\quad a_2 = a_1 + d$

To find the third term, add the common difference d to the second term.

$$a_3 = a_2 + d = (a_1 + d) + d$$
$$a_3 = a_1 + 2d$$

To find the fourth term, add the common difference d to the third term.

$$a_4 = a_3 + d = (a_1 + 2d) + d$$
$$a_4 = a_1 + 3d$$

Note the relationship between the term number and the number that multiplies d. The multiplier of d is 1 less than the term number.

$$a_n = a_1 + (n - 1)d$$

INSTRUCTOR NOTE

After introducing the formula $a_n = a_1 + (n - 1)d$ for the nth term of an arithmetic sequence, have students use the formula to verify the terms of a sequence. For instance, use $4, 7, 10, 13, \ldots$.

$a_1 = 4 + (1 - 1)(3) = 4$
$a_2 = 4 + (2 - 1)(3) = 7$
$a_3 = 4 + (3 - 1)(3) = 10$

INSTRUCTOR NOTE

Point out that when $d > 0$, the terms of the sequence increase and that when $d < 0$, the terms of the sequence decrease.

The Formula for the nth Term of an Arithmetic Sequence

> The nth term of an arithmetic sequence with a common difference of d is given by $a_n = a_1 + (n - 1)d$.

Example 1   Find the 27th term of the arithmetic sequence $-4, -1, 2, 5, 8, \ldots$.

Solution $d = a_2 - a_1 = -1 - (-4) = 3$ ▸ Find the common difference.

$a_n = a_1 + (n - 1)d$ ▸ Use the Formula for the nth Term
$a_{27} = -4 + (27 - 1)3$ of an Arithmetic Sequence to find
$\quad = -4 + (26)3$ the 27th term. $n = 27$, $a_1 = -4$,
$\quad = -4 + 78$ $d = 3$
$\quad = 74$

Problem 1   Find the 15th term of the arithmetic sequence $9, 3, -3, -9, \ldots$.

Solution   See page S31. -75

Example 2   Find the formula for the nth term of the arithmetic sequence $-5, -1, 3, 7, \ldots$.

Solution $d = a_2 - a_1 = -1 - (-5) = 4$ ▸ Find the common difference.

$a_n = a_1 + (n - 1)d$ ▸ Use the Formula for the nth Term
$a_n = -5 + (n - 1)4$ of an Arithmetic Sequence.
$a_n = -5 + 4n - 4$ $a_1 = -5$, $d = 4$
$a_n = 4n - 9$

Problem 2   Find the formula for the nth term of the arithmetic sequence $-3, 1, 5, 9, \ldots$.

Solution   See page S31. $a_n = 4n - 7$

Example 3   Find the number of terms in the finite arithmetic sequence $7, 10, 13, \ldots, 55$.

Solution $d = a_2 - a_1 = 10 - 7 = 3$ ▸ Find the common difference.

$a_n = a_1 + (n - 1)d$ ▸ Use the Formula for the nth Term of
$55 = 7 + (n - 1)3$ an Arithmetic Sequence.
$55 = 7 + 3n - 3$ $a_n = 55$, $a_1 = 7$, $d = 3$
$55 = 3n + 4$ ▸ Solve for n.
$51 = 3n$
$17 = n$

There are 17 terms in the sequence.

Problem 3   Find the number of terms in the finite arithmetic sequence $7, 9, 11, \ldots, 59$.

Solution   See page S31. 27

2 Find the sum of an arithmetic series

The indicated sum of the terms of an arithmetic sequence is called an **arithmetic series.** The sum of an arithmetic series can be found by using a formula.

POINT OF INTEREST

This formula was proved in *Aryabhatiya*, which was written by Aryabhata around 499. The book is the earliest known Indian mathematical work by an identifiable author. Although the proof of the formula appears in that text, the formula was known before Aryabhata's time.

The Formula for the Sum of n Terms of an Arithmetic Series

Let a_1 be the first term of a finite arithmetic sequence, let n be the number of terms, and let a_n be the last term of the sequence. Then the sum of the series S_n is given by $S_n = \frac{n}{2}(a_1 + a_n)$.

Each term of the arithmetic sequence shown at the right was found by adding 3 to the previous term. 2, 5, 8, . . . , 17, 20

Each term of the reverse arithmetic sequence can be found by subtracting 3 from the previous term. 20, 17, 14, . . . , 5, 2

This idea is used in the following proof of the Formula for the Sum of n Terms of an Arithmetic Series.

Let S_n represent the sum of the series.

$$S_n = a_1 + (a_1 + d) + (a_1 + 2d) + \cdots + a_n$$

Write the terms of the sum of the series in reverse order. The sum is the same.

$$S_n = a_n + (a_n - d) + (a_n - 2d) + \cdots + a_1$$

Add the two equations.

$$2S_n = (a_1 + a_n) + (a_1 + a_n) + (a_1 + a_n) + \cdots + (a_1 + a_n)$$

Simplify the right side of the equation by using the fact that there are n terms in the sequence.

$$2S_n = n(a_1 + a_n)$$

Solve for S_n.

$$S_n = \frac{n}{2}(a_1 + a_n)$$

Example 4 Find the sum of the first 10 terms of the arithmetic sequence 2, 4, 6, 8, ...

Solution $d = a_2 - a_1 = 4 - 2 = 2$ ▶ Find the common difference.

$$a_n = a_1 + (n - 1)d$$
$$a_{10} = 2 + (10 - 1)2 = 2 + (9)2 = 2 + 18 = 20$$

▶ Use the Formula for the nth Term of an Arithmetic Sequence to find the 10th term.

$$S_n = \frac{n}{2}(a_1 + a_n)$$

▶ Use the Formula for the Sum of n Terms of an Arithmetic Series. $n = 10, a_1 = 2, a_n = 20$

$$S_{10} = \frac{10}{2}(2 + 20) = 5(22) = 110$$

Problem 4 Find the sum of the first 25 terms of the arithmetic sequence $-4, -2, 0, 2, 4, \ldots$.

Solution See page S31. 500

INSTRUCTOR NOTE

The problem at the right with $\sum_{n=1}^{25}(3n+1)$ shows the efficiency of summation notation. Writing out the series would be very time-consuming.

A point worth making in connection with the same series is that the parentheses are important. As a class exercise, have the students evaluate $\sum_{n=1}^{25} 3n + 1$.

Example 5 Find the sum of the arithmetic series $\sum_{n=1}^{25}(3n+1)$.

Solution

$a_n = 3n + 1$

$a_1 = 3(1) + 1 = 4$ ▶ Find the first term.

$a_{25} = 3(25) + 1 = 76$ ▶ Find the 25th term.

$S_n = \frac{n}{2}(a_1 + a_n)$ ▶ Use the Formula for the Sum of n Terms of an Arithmetic Series. $n = 25, a_1 = 4, a_n = 76$

$S_{25} = \frac{25}{2}(4 + 76)$

$= \frac{25}{2}(80)$

$= 1000$

Problem 5 Find the sum of the arithmetic series $\sum_{n=1}^{18}(3n-2)$.

Solution See page S31. 477

3 Application problems

HM³

Example 6 The distance a ball rolls down a ramp each second is given by an arithmetic sequence. The distance in feet traveled by the ball during the nth second is given by $2n - 1$. Find the distance the ball will travel during the first 10 s.

Strategy To find the distance:

- Find the first and second terms of the sequence.
- Find the common difference of the arithmetic sequence.
- Find the tenth term of the sequence.
- Use the Formula for the Sum of n Terms of an Arithmetic Series to find the sum of the first ten terms.

Solution

$a_n = 2n - 1$

$a_1 = 2(1) - 1 = 1$

$a_2 = 2(2) - 1 = 3$

$d = a_2 - a_1 = 3 - 1 = 2$

$a_n = a_1 + (n - 1)d$

$a_{10} = 1 + (10 - 1)2 = 1 + (9)2 = 19$

$S_{10} = \frac{10}{2}(1 + 19) = 5(20) = 100$

The ball will roll 100 ft during the first 10 s.

Problem 6 A contest offers 20 prizes. The first prize is $10,000, and each successive prize is $300 less than the preceding prize. What is the value of the 20th-place prize? What is the total amount of prize money that is being awarded?

Solution See page S31.
20th-place prize: $4300; total amount being awarded: $143,000

CONCEPT REVIEW 11.2

State whether the following statements are always true, sometimes true, or never true.

1. The sum of the successive terms of an arithmetic series increases in value. Sometimes true
2. Each successive term in an arithmetic sequence is found by adding a fixed number. Always true
3. The first term of an arithmetic sequence is a positive number. Sometimes true
4. The *n*th term of an arithmetic sequence is found by using the formula $a_n = a_1 + (n - 1)d$. Always true

EXERCISES 11.2

1 **1.** How does an arithmetic sequence differ from any other type of sequence?

2. Explain how to find the common difference for an arithmetic sequence.

Find the indicated term of the arithmetic sequence.

3. $1, 11, 21, \ldots; a_{15}$ 141
4. $3, 8, 13, \ldots; a_{20}$ 98
5. $-6, -2, 2, \ldots; a_{15}$ 50
6. $-7, -2, 3, \ldots; a_{14}$ 58
7. $3, 7, 11, \ldots; a_{18}$ 71
8. $-13, -6, 1, \ldots; a_{31}$ 197
9. $-\frac{3}{4}, 0, \frac{3}{4}, \ldots; a_{11}$ $\frac{27}{4}$
10. $\frac{3}{8}, 1, \frac{13}{8}, \ldots; a_{17}$ $\frac{83}{8}$
11. $2, \frac{5}{2}, 3, \ldots; a_{31}$ 17
12. $1, \frac{5}{4}, \frac{3}{2}, \ldots; a_{17}$ 5
13. $6, 5.75, 5.50, \ldots; a_{10}$ 3.75
14. $4, 3.7, 3.4, \ldots; a_{12}$ 0.7

Find the formula for the *n*th term of the arithmetic sequence.

15. $1, 2, 3, \ldots$ $a_n = n$
16. $1, 4, 7, \ldots$ $a_n = 3n - 2$
17. $6, 2, -2, \ldots$ $a_n = -4n + 10$
18. $3, 0, -3, \ldots$ $a_n = -3n + 6$
19. $2, \frac{7}{2}, 5, \ldots$ $a_n = \frac{3n + 1}{2}$
20. $7, 4.5, 2, \ldots$ $a_n = -2.5n + 9.5$
21. $-8, -13, -18, \ldots$ $a_n = -5n - 3$
22. $17, 30, 43, \ldots$ $a_n = 13n + 4$
23. $26, 16, 6, \ldots$ $a_n = -10n + 36$

Find the number of terms in the finite arithmetic sequence.

24. $-2, 1, 4, \ldots, 73$ 26

25. $7, 11, 15, \ldots, 171$ 42

26. $-\frac{1}{2}, \frac{3}{2}, \frac{7}{2}, \ldots, \frac{71}{2}$ 19

27. $\frac{1}{3}, \frac{5}{3}, 3, \ldots, \frac{61}{3}$ 16

28. $1, 5, 9, \ldots, 81$ 21

29. $3, 8, 13, \ldots, 98$ 20

30. $2, 0, -2, \ldots, -56$ 30

31. $1, -3, -7, \ldots, -75$ 20

32. $\frac{5}{2}, 3, \frac{7}{2}, \ldots, 13$ 22

33. $\frac{7}{3}, \frac{13}{3}, \frac{19}{3}, \ldots, \frac{79}{3}$ 13

34. $1, 0.75, 0.50, \ldots, -4$ 21

35. $3.5, 2, 0.5, \ldots, -25$ 20

2 Find the sum of the indicated number of terms of the arithmetic sequence.

36. $1, 3, 5, \ldots; n = 50$ 2500

37. $2, 4, 6, \ldots; n = 25$ 650

38. $20, 18, 16, \ldots; n = 40$ -760

39. $25, 20, 15, \ldots; n = 22$ -605

40. $\frac{1}{2}, 1, \frac{3}{2}, \ldots; n = 27$ 189

41. $2, \frac{11}{4}, \frac{7}{2}, \ldots; n = 10$ $\frac{215}{4}$

Find the sum of the arithmetic series.

42. $\sum_{i=1}^{15} (3i - 1)$ 345

43. $\sum_{i=1}^{15} (3i + 4)$ 420

44. $\sum_{n=1}^{17} \left(\frac{1}{2}n + 1\right)$ $\frac{187}{2}$

45. $\sum_{n=1}^{10} (1 - 4n)$ -210

46. $\sum_{i=1}^{15} (4 - 2i)$ -180

47. $\sum_{n=1}^{10} (5 - n)$ -5

3 Solve.

48. *Physics* The distance that an object dropped from a cliff will fall is 16 ft the first second, 48 ft the next second, 80 ft the third second, and so on in an arithmetic sequence. What is the total distance the object will fall in 6 s? 576 ft

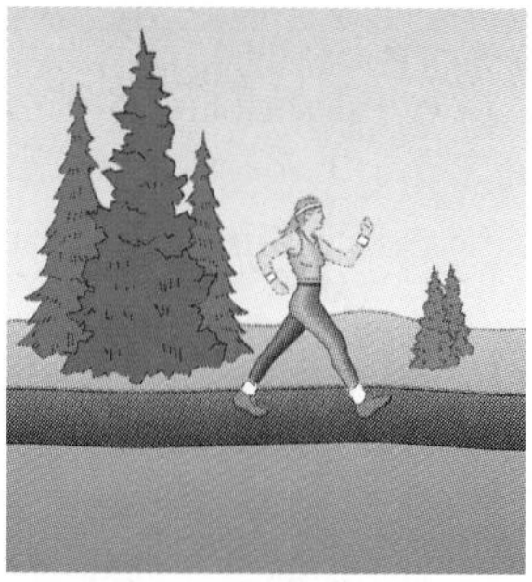

49. *Health Science* An exercise program calls for walking 12 min each day for a week. Each week thereafter, the amount of time spent walking increases by 6 min per day. In how many weeks will a person be walking 60 min each day? 9 weeks

50. *Business* A display of cans in a grocery store consists of 20 cans in the bottom row, 18 cans in the next row, and so on in an arithmetic sequence. The top row has 4 cans. Find the total number of cans in the display. 108 cans

51. *Construction* A theater in the round has 52 seats in the first row, 58 seats in the second row, 64 seats in the third row, and so on in an arithmetic sequence. Find the total number of seats in the theater if there are 20 rows of seats. 2180 seats

52. *Construction* The loge seating section in a concert hall consists of 26 rows of chairs. There are 65 seats in the first row, 71 seats in the second row, 77 seats in the third row, and so on in an arithmetic sequence. How many seats are in the loge seating section? 3640 seats

53. *Human Resources* The salary schedule for an engineering assistant is $1800 for the first month and a $150-per month salary increase for the next 9 months. Find the monthly salary during the tenth month. Find the total salary for the 10-month period. $3150, $24,750

54. *Sports* The International Amateur Athletic Federation (IAAF) specifies the design of tracks on which world records can be set. A typical design is shaped like a rectangle with semicircles on either end, as shown at the right. There are eight lanes, and each lane is 1.22 m wide. The distance around the track for each lane is measured from the center of that lane. Find a formula for the sequence of radii. Round to the nearest hundredth. $a_n = 1.22n + 35.89$

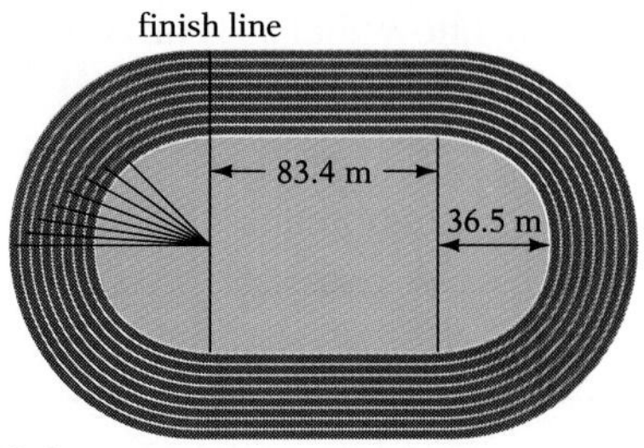

55. *Sports* Using the information in the preceding exercise, write a formula for the sequence of distances around the track for each lane. Round to the nearest hundredth and use 3.14 as an approximation for π. Remember that the distance around the circular portion of the track is measured from the center of the lane. Explain why the IAAF staggers the starting position for runners in a 400-meter race. $a_n = 7.66n + 392.19$

APPLYING CONCEPTS 11.2

Solve.

56. Find the sum of the first 50 positive integers. 1275

57. How many terms of the arithmetic sequence $-3, 2, 7, \ldots$ must be added together for the sum of the series to be 116? 8

58. Given $a_1 = -9$, $a_n = 21$, and $S_n = 36$, find d and n. $d = 6; n = 6$

59. The fourth term of an arithmetic sequence is 9, and the ninth term is 29. Find the first term. -3

[C] 60. Show that $f(n) = mn + b$, n a natural number, is an arithmetic sequence. Complete solution available in the Solutions Manual.

[C] 61. *Geometry* The sum of the interior angles of a triangle is 180°. The sum is 360° for a quadrilateral and 540° for a pentagon. Assuming that this pattern continues, find the sum of the angles of a dodecagon (12-sided figure). Find a formula for the sum of the angles of an n-sided polygon. 1800°; $180(n - 2)$

62. Write $\sum_{i=1}^{2} \log 2i$ as a single logarithm with a coefficient of 1. $\log 8$

63. Write a formula for the nth term of the sequence of natural numbers that are multiples of 3. $a_n = 3n$

64. *Business* Straight-line depreciation is used by some companies to determine the value of an asset. A model for this depreciation method is $a_n = V - dn$, where a_n is the value of the asset after n years, V is the original value of the asset, d is the annual decrease in the asset's value, and n is the number of years. Suppose that an asset has an original value of \$20,000 and that the annual decrease in value is \$3000.

   a. Substitute the values of V and d into the equation, and write an expression for a_n. $a_n = 20{,}000 - 3000n$

b. Show that a_n is an arithmetic sequence. Rewrite $a_n = 20{,}000 - 3000n$ as $a_n = 20{,}000 + n(-3000)$, which is of the form $a_n = a_1 + (n-1)d$, where $a_1 = 20{,}000$, $n - 1$ replaces n, and $d = -3000$.

65. Write an essay on the Fibonacci sequence and its relationship to natural phenomena.

SECTION 11.3 Geometric Sequences and Series

1 Find the *n*th term of a geometric sequence

HM³

INSTRUCTOR NOTE

After introducing the concept of geometric sequence, give students examples of sequences and ask them to identify each sequence as arithmetic, geometric, or neither. Three possible sequences are

$1, 4, 9, 16, \ldots, n^2, \ldots$
$2, 4, 8, 16, \ldots, 2^n, \ldots$
$2, 4, 6, 8, \ldots, 2n, \ldots$

An ore sample contains 20 mg of a radioactive material with a half-life of 1 week. The amount of the radioactive material that the sample contains at the beginning of each week can be determined by using an exponential decay equation.

The sequence below represents the amount in the sample at the beginning of each week. Each term of the sequence is found by multiplying the preceding term by $\frac{1}{2}$.

Week	1	2	3	4	5
Amount	20	10	5	2.5	1.25

The sequence 20, 10, 5, 2.5, 1.25 is called a geometric sequence. A **geometric sequence, or geometric progression,** is one in which each successive term of the sequence is the same nonzero constant multiple of the preceding term. The common multiple is called the **common ratio** of the sequence.

LOOK CLOSELY

Geometric sequences are different from arithmetic sequences. For a geometric sequence, every two successive terms have the same *ratio.* For an arithmetic sequence, every two successive terms have the same *difference.*

Each of the sequences shown below is a geometric sequence. To find the common ratio of a geometric sequence, divide the second term of the sequence by the first term.

$3, 6, 12, 24, 48, \ldots$ Common ratio: 2
$4, -12, 36, -108, 324, \ldots$ Common ratio: -3
$6, 4, \frac{8}{3}, \frac{16}{9}, \frac{32}{27}, \ldots$ Common ratio: $\frac{2}{3}$

Consider a geometric sequence in which the first term is a_1 and the common ratio is r. Multiplying each successive term of the geometric sequence by the common ratio yields a formula for the nth term.

The first term is a_1. $\quad a_1 = a_1$

To find the second term, multiply the first term by the common ratio r. $\quad a_2 = a_1 r$

To find the third term, multiply the second term by the common ratio r. $\quad a_3 = (a_2)r = (a_1 r)r$, $a_3 = a_1 r^2$

To find the fourth term, multiply the third term by the common ratio r. $\quad a_4 = (a_3)r = (a_1 r^2)r$, $a_4 = a_1 r^3$

Note the relationship between the term number and the number that is the exponent on r. The exponent on r is 1 less than the term number. $\quad a_n = a_1 r^{n-1}$

INSTRUCTOR NOTE

Some students may assume that a sequence must be either arithmetic or geometric. The sequence $a_n = \frac{1}{n}$ can be used to dispel that belief. As an extra-credit problem, have students prove that this sequence is neither arithmetic nor geometric.

The Formula for the nth Term of a Geometric Sequence

The nth term of a geometric sequence with first term a_1 and common ratio r is given by $a_n = a_1 r^{n-1}$.

Example 1   Find the 6th term of the geometric sequence 3, 6, 12, ...

Solution $r = \dfrac{a_2}{a_1} = \dfrac{6}{3} = 2$ ▶ Find the common ratio.

$a_n = a_1 r^{n-1}$ ▶ Use the Formula for the nth Term of a Geometric
$a_6 = 3(2)^{6-1}$ Sequence. $n = 6, a_1 = 3, r = 2$
$ = 3(2)^5$
$ = 3(32)$
$ = 96$

Problem 1   Find the 5th term of the geometric sequence, 5, 2, $\dfrac{4}{5}$, ...

Solution   See page S31. $\dfrac{16}{125}$

INSTRUCTOR NOTE

Point out to students that the terms of a geometric sequence for which $r > 0$ all have the same sign. If $r < 0$, the signs of the terms alternate.

Example 2   Find a_3 for the geometric sequence 8, a_2, a_3, 27, ...

Solution $a_n = a_1 r^{n-1}$
$a_4 = a_1 r^{4-1}$ ▶ Find the common ratio. $a_4 = 27, a_1 = 8, n = 4$
$27 = 8r^{4-1}$

$\dfrac{27}{8} = r^3$

$\dfrac{3}{2} = r$

$a_3 = 8\left(\dfrac{3}{2}\right)^{3-1}$ ▶ Use the Formula for the nth Term of a Geometric
    Sequence.

$ = 8\left(\dfrac{3}{2}\right)^2$

$ = 8\left(\dfrac{9}{4}\right) = 18$

Problem 2   Find a_3 for the geometric sequence 3, a_2, a_3, -192, ...

Solution   See page S31.   48

2   Finite geometric series

INSTRUCTOR NOTE

Geometric series are used extensively in the mathematics of finance. Finite geometric series are used to calculate loan balances and monthly payments for amortized loans.

The indicated sum of the terms of a geometric sequence is called a **geometric series.** The sum of a geometric series can be found by a formula.

The Formula for the Sum of n Terms of a Finite Geometric Series

Let a_1 be the first term of a finite geometric sequence, let n be the number of terms, and let r be the common ratio. Then the sum of the series S_n is given by $S_n = \dfrac{a_1(1 - r^n)}{1 - r}$.

Proof of the Formula for the Sum of n Terms of a Finite Geometric Series:

Let S_n represent the sum of n terms of the sequence. $$S_n = a_1 + a_1r + a_1r^2 + \cdots + a_1r^{n-2} + a_1r^{n-1}$$

Multiply each side of the equation by r. $$rS_n = a_1r + a_1r^2 + a_1r^3 + \cdots + a_1r^{n-1} + a_1r^n$$

Subtract the two equations. $$S_n - rS_n = a_1 - a_1r^n$$

Assuming that $r \neq 1$, solve for S_n. $$(1 - r)S_n = a_1(1 - r^n)$$ $$S_n = \frac{a_1(1 - r^n)}{1 - r}$$

Example 3 Find the sum of the geometric sequence 2, 8, 32, 128, 512.

Solution $r = \frac{a_2}{a_1} = \frac{8}{2} = 4$ ► Find the common ratio.

$S_n = \frac{a_1(1 - r^n)}{1 - r}$ ► Use the Formula for the Sum of n Terms of a Finite Geometric Series. $n = 5$, $a_1 = 2$, $r = 4$.

$$S_5 = \frac{2(1 - 4^5)}{1 - 4}$$
$$= \frac{2(1 - 1024)}{-3}$$
$$= \frac{2(-1023)}{-3}$$
$$= \frac{-2046}{-3}$$
$$= 682$$

Problem 3 Find the sum of the geometric sequence $1, -\frac{1}{3}, \frac{1}{9}, -\frac{1}{27}$.

Solution See page S32. $\frac{20}{27}$

Example 4 Find the sum of the geometric series $\sum_{n=1}^{10} (-20)(-2)^{n-1}$.

Solution

$a_n = (-20)(-2)^{n-1}$

$a_1 = (-20)(-2)^{1-1}$ ► Find the first term.

$= (-20)(-2)^0$

$= (-20)(1) = -20$

$a_2 = (-20)(-2)^{2-1}$ ► Find the second term.

$= (-20)(-2)^1$

$= (-20)(-2) = 40$

$r = \frac{a_2}{a_1} = \frac{40}{-20} = -2$ ► Find the common ratio

$$S_n = \frac{a_1(1 - r^n)}{1 - r}$$

▶ Use the Formula for the Sum of n Terms of a Finite Geometric Series. $n = 10$, $a_1 = -20$, $r = -2$

$$S_{10} = \frac{-20[1 - (-2)^{10}]}{1 - (-2)} = \frac{-20(1 - 1024)}{3} = \frac{-20(-1023)}{3} = \frac{20{,}460}{3} = 6820$$

Problem 4 Find the sum of the geometric series $\sum_{n=1}^{5} \left(\frac{1}{2}\right)^n$.

Solution See page S32. $\frac{31}{32}$

3 Infinite geometric series

HM[3]

When the absolute value of the common ratio of a geometric sequence is less than 1, $|r| < 1$, then as n becomes larger, r^n becomes closer to zero.

Examples of geometric sequences for which $|r| < 1$ are shown at the right. As the number of terms increases, the absolute value of the last term listed gets closer to zero.

$$1, \frac{1}{3}, \frac{1}{9}, \frac{1}{27}, \frac{1}{81}, \frac{1}{243}, \ldots$$

$$1, -\frac{1}{2}, \frac{1}{4}, -\frac{1}{8}, \frac{1}{16}, -\frac{1}{32}, \ldots$$

INSTRUCTOR NOTE

One application of the sum of an infinite series is again a financial model. According to Myron Gordon, the price, P, that an investor should be willing to pay for a stock that pays a dividend, D, is

$$P = \sum_{n=0}^{\infty} D\left[\frac{1 + g}{1 + k}\right]^n$$

where g is the investor's desired rate of return and k is a constant greater than g. (See Exercise 72 in this section.)

The indicated sum of the terms of an infinite geometric sequence is called an **infinite geometric series.**

An example of an infinite geometric series is shown at the right. The first term is 1. The common ratio is $\frac{1}{3}$.

$$1 + \frac{1}{3} + \frac{1}{9} + \frac{1}{27} + \frac{1}{81} + \frac{1}{243} + \cdots$$

The sum of the first 5, 7, 12, and 15 terms, along with the values of r^n, are shown at the right. Note that as n increases, the sum of the terms gets closer to 1.5, and the value of r^n gets closer to zero.

n	S_n	r^n
5	1.4938272	0.0041152
7	1.4993141	0.0004572
12	1.4999972	0.0000019
15	1.4999999	0.0000001

Using the Formula for the Sum of n Terms of a Geometric Series and the fact that r^n approaches zero when $|r| < 1$ and n increases, a formula for an infinite geometric series can be found.

The sum of the first n terms of a geometric series is shown at the right. If $|r| < 1$, then r^n can be made very close to zero by using larger and larger values of n. Therefore, the sum of the first n terms is approximately $\frac{a_1}{1 - r}$.

Approximately zero

$$S_n = \frac{a_1(1 - r^n)}{1 - r}$$

$$S_n \approx \frac{a_1(1 - 0)}{1 - r} = \frac{a_1}{1 - r}$$

The Formula for the Sum of an Infinite Geometric Series

The sum of an infinite geometric series in which $|r| < 1$ is $S = \frac{a_1}{1-r}$.

When $|r| \geq 1$, the infinite geometric series does not have a finite sum. For example, the sum of the infinite geometric series $1 + 2 + 4 + 8 + \cdots$ increases without limit.

Example 5 Find the sum of the infinite geometric sequence $1, -\frac{1}{2}, \frac{1}{4}, -\frac{1}{8}, \ldots$.

Solution

$$S = \frac{a_1}{1-r} = \frac{1}{1-\left(-\frac{1}{2}\right)} = \frac{1}{\frac{3}{2}} = \frac{2}{3}$$

▶ The common ratio is $-\frac{1}{2}$. $\left|-\frac{1}{2}\right| < 1$. Use the Formula for the Sum of an Infinite Geometric Series.

Problem 5 Find the sum of the infinite geometric sequence $3, -2, \frac{4}{3}, -\frac{8}{9}, \ldots$.

Solution See page S32. $\frac{9}{5}$

The sum of an infinite geometric series can be used to find a fraction that is equivalent to a nonterminating repeating decimal.

The repeating decimal shown at the right has been rewritten as an infinite geometric series, with first term $\frac{3}{10}$ and common ratio $\frac{1}{10}$.

$$0.\overline{3} = 0.3 + 0.03 + 0.003 + \cdots = \frac{3}{10} + \frac{3}{100} + \frac{3}{1000} + \cdots$$

Use the Formula for the Sum of an Infinite Geometric Series.

$$S = \frac{a_1}{1-r} = \frac{\frac{3}{10}}{1-\frac{1}{10}} = \frac{\frac{3}{10}}{\frac{9}{10}} = \frac{3}{9} = \frac{1}{3}$$

$\frac{1}{3}$ is equivalent to the nonterminating, repeating decimal $0.\overline{3}$.

Example 6 Find an equivalent fraction for $0.1\overline{2}$.

Solution Write the decimal as an infinite geometric series. The geometric series does not begin with the first term. The series begins with $\frac{2}{100}$. The common ratio is $\frac{1}{10}$.

$$0.1\overline{2} = 0.1 + 0.02 + 0.002 + 0.0002 + \cdots = \frac{1}{10} + \frac{2}{100} + \frac{2}{1000} + \frac{2}{10{,}000} + \cdots$$

$$S = \frac{a_1}{1 - r} = \frac{\frac{2}{100}}{1 - \frac{1}{10}} = \frac{\frac{2}{100}}{\frac{9}{10}} = \frac{2}{90}$$

▶ Use the Formula for the Sum of an Infinite Geometric Series.

$$0.1\overline{2} = \frac{1}{10} + \frac{2}{90} = \frac{11}{90}$$

▶ Add $\frac{1}{10}$ to the sum of the series.

An equivalent fraction is $\frac{11}{90}$.

INSTRUCTOR NOTE

As a class demonstration, show students that $0.\overline{9} = 1$.

Problem 6 Find an equivalent fraction for $0.\overline{36}$.

Solution See page S32. $\frac{4}{11}$

4 Application problems

HM³

Example 7 On the first swing, the length of the arc through which a pendulum swings is 16 in. The length of each successive swing is $\frac{7}{8}$ of the preceding swing. Find the length of the arc on the fifth swing. Round to the nearest tenth.

Strategy To find the length of the arc on the fifth swing, use the Formula for the nth Term of a Geometric Sequence. $n = 5$, $a_1 = 16$, $r = \frac{7}{8}$

Solution $a_n = a_1 r^{n-1}$

$$a_5 = 16\left(\frac{7}{8}\right)^{5-1} = 16\left(\frac{7}{8}\right)^4 = 16\left(\frac{2401}{4096}\right) = \frac{38{,}416}{4096} \approx 9.4$$

The length of the arc on the fifth swing is 9.4 in.

Problem 7 You start a chain letter and send it to three friends. Each of the three friends sends the letter to three other friends, and the sequence is repeated. If no one breaks the chain, how many letters will have been mailed from the first through the sixth mailings?

Solution See page S32. 1092 letters

CONCEPT REVIEW 11.3

Determine whether the following statements are always true, sometimes true, or never true.

1. Each successive term of a geometric sequence is found by multiplying the preceding term by a nonzero constant. Always true

2. The sum of an infinite geometric series is infinity. Sometimes true

3. The sum of a geometric series increases in value as more terms are added. Sometimes true

4. $1, 8, 27, \ldots, k^3, \ldots$ is a geometric sequence. Never true

5. $\sqrt{1}, \sqrt{2}, \sqrt{3}, \ldots, \sqrt{n}, \ldots$ is a geometric sequence. Never true

EXERCISES 11.3

1 1. How does a geometric sequence differ from any other type of sequence?

2. Explain how to find the common ratio for a geometric sequence.

Find the indicated term of the geometric sequence.

3. $2, 8, 32, \ldots; a_9$ 131,072

4. $4, 3, \frac{9}{4}, \ldots; a_8$ $\frac{2187}{4096}$

5. $6, -4, \frac{8}{3}, \ldots; a_7$ $\frac{128}{243}$

6. $-5, 15, -45, \ldots; a_7$ -3645

7. $1, \sqrt{2}, 2, \ldots; a_9$ 16

8. $3, 3\sqrt{3}, 9, \ldots; a_8$ $81\sqrt{3}$

Find a_2 and a_3 for the geometric sequence.

9. $9, a_2, a_3, \frac{8}{3}, \ldots$ 6, 4

10. $8, a_2, a_3, \frac{27}{8}, \ldots$ $6, \frac{9}{2}$

11. $3, a_2, a_3, -\frac{8}{9}, \ldots$ $-2, \frac{4}{3}$

12. $6, a_2, a_3, -48, \ldots$ $-12, 24$

13. $-3, a_2, a_3, 192, \ldots$ $12, -48$

14. $5, a_2, a_3, 625, \ldots$ 25, 125

2 Find the sum of the indicated number of terms of the geometric sequence.

15. $2, 6, 18, \ldots; n = 7$ 2186

16. $-4, 12, -36, \ldots; n = 7$ -2188

17. $12, 9, \frac{27}{4}, \ldots; n = 5$ $\frac{2343}{64}$

18. $3, 3\sqrt{2}, 6, \ldots; n = 12$ $189 + 189\sqrt{2}$

Find the sum of the geometric series.

19. $\sum_{i=1}^{5} (2)^i$ 62

20. $\sum_{n=1}^{6} \left(\frac{3}{2}\right)^n$ $\frac{1995}{64}$

21. $\sum_{i=1}^{5} \left(\frac{1}{3}\right)^i$ $\frac{121}{243}$

22. $\sum_{n=1}^{6} \left(\frac{2}{3}\right)^n$ $\frac{1330}{729}$

23. $\sum_{i=1}^{5} (4)^i$ 1364

24. $\sum_{n=1}^{8} (3)^n$ 9840

25. $\sum_{i=1}^{4} (7)^i$ 2800

26. $\sum_{n=1}^{5} (5)^n$ 3905

27. $\sum_{i=1}^{5} \left(\frac{3}{4}\right)^i$ $\frac{2343}{1024}$

28. $\sum_{n=1}^{3} \left(\frac{7}{4}\right)^n$ $\frac{651}{64}$

29. $\sum_{i=1}^{4} \left(\frac{5}{3}\right)^i$ $\frac{1360}{81}$

30. $\sum_{n=1}^{6} \left(\frac{1}{2}\right)^n$ $\frac{63}{64}$

3 Find the sum of the infinite geometric series.

31. $3 + 2 + \frac{4}{3} + \cdots$ 9

32. $2 - \frac{1}{4} + \frac{1}{32} + \cdots$ $\frac{16}{9}$

33. $6 - 4 + \frac{8}{3} + \cdots$ $\frac{18}{5}$

34. $\frac{1}{10} + \frac{1}{100} + \frac{1}{1000} + \cdots$ $\frac{1}{9}$

35. $\frac{7}{10} + \frac{7}{100} + \frac{7}{1000} + \cdots$ $\frac{7}{9}$

36. $\frac{5}{100} + \frac{5}{10{,}000} + \frac{5}{1{,}000{,}000} + \cdots$ $\frac{5}{99}$

Find an equivalent fraction for the repeating decimal.

37. $0.88\overline{8}$ $\frac{8}{9}$
38. $0.55\overline{5}$ $\frac{5}{9}$
39. $0.22\overline{2}$ $\frac{2}{9}$
40. $0.99\overline{9}$ 1

41. $0.4\overline{545}$ $\frac{5}{11}$
42. $0.1\overline{818}$ $\frac{2}{11}$
43. $0.166\overline{6}$ $\frac{1}{6}$
44. $0.833\overline{3}$ $\frac{5}{6}$

4 **45.** *Art* The fabric designer Jhane Barnes created a fabric pattern based on the *Sierpinski triangle.* This triangle is a *fractal,* which is a geometric pattern that is repeated at ever smaller scales to produce irregular shapes. The first four stages in the construction of a Sierpinski triangle are shown at the right. The initial triangle is an equilateral triangle with sides 1 unit long. The cut-out triangles are formed by connecting the midpoints of the sides of the unshaded triangles. This pattern is repeated indefinitely. Find a formula for the nth term of the number of unshaded triangles. $a_n = 3^{n-1}$

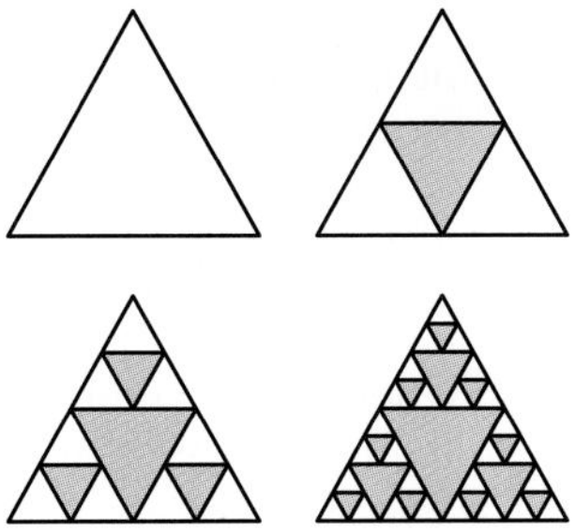

46. *Art* A *Sierpinski carpet* is similar to a Sierpinski triangle (see Exercise 45) except that all of the unshaded squares must be divided into nine congruent smaller squares with the one in the center shaded. The first three stages of the pattern are shown at the right. Find a formula for the nth term of the number of unshaded squares. $a_n = 8^{n-1}$

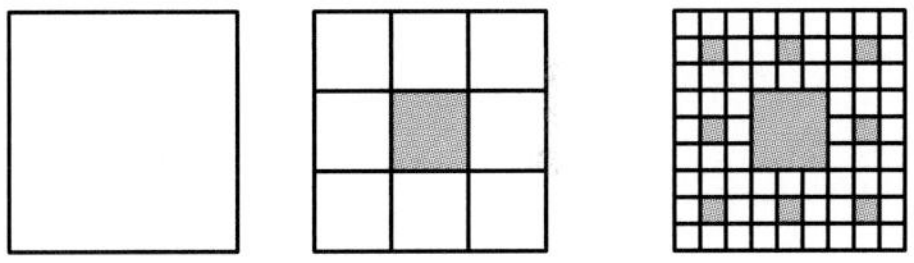

47. *Physics* A laboratory ore sample contains 500 mg of a radioactive material with a half-life of 1 day. Find the amount of radioactive material in the sample at the beginning of the seventh day. 7.8125 mg

48. *Recreation* On the first swing, the length of the arc through which a pendulum swings is 18 in. The length of each successive swing is $\frac{3}{4}$ of the preceding swing. What is the total distance the pendulum has traveled during the first five swings? Round to the nearest tenth. 54.9 in.

49. *Sports* To test the bounce of a tennis ball, the ball is dropped from a height of 8 ft. The ball bounces to 80% of its previous height with each bounce. How high does the ball bounce on the fifth bounce? Round to the nearest tenth. 2.6 ft

50. *Physics* The temperature of a hot water spa is 75°F. Each hour, the temperature is 10% higher than during the previous hour. Find the temperature of the spa after 3 h. Round to the nearest tenth. 99.8°F

51. *Real Estate* A real estate broker estimates that a piece of land will increase in value at a rate of 12% each year. If the original value of the land is $15,000, what will be its value in 15 years? $82,103.49

52. *Business* Suppose an employee receives a wage of 1¢ the first day of work, 2¢ the second day, 4¢ the third day, and so on in a geometric sequence. Find the total amount of money earned for working 30 days. $10,737,418.23

53. *Real Estate* Assume the average value of a home increases 5% per year. How much would a house costing $100,000 be worth in 30 years? $432,194.24

54. *Biology* A culture of bacteria doubles every 2 h. If there are 500 bacteria at the beginning, how many bacteria will there be after 24 h? 2,048,000 bacteria

APPLYING CONCEPTS 11.3

State whether the sequence is arithmetic (A), geometric (G), or neither (N), and write the next term in the sequence.

55. $4, -2, 1, \ldots$ G, $-\frac{1}{2}$

56. $-8, 0, 8, \ldots$ A, 16

57. $5, 6.5, 8, \ldots$ A, 9.5

58. $-7, 14, -28, \ldots$ G, 56

59. $1, 4, 9, 16, \ldots$ N, 25

60. $\sqrt{1}, \sqrt{2}, \sqrt{3}, \sqrt{4}$ N, $\sqrt{5}$

61. $x^8, x^6, x^4, \ldots$ G, x^2

62. $5a^2, 3a^2, a^2, \ldots$ A, $-a^2$

63. $\log x, 2\log x, 3\log x, \ldots$ A, $4\log x$

64. $\log x, 3\log x, 9\log x, \ldots$ G, $27\log x$

Solve.

65. The third term of a geometric sequence is 3, and the sixth term is $\frac{1}{9}$. Find the first term. 27

66. Given $a_n = 162$, $r = -3$, and $S_n = 122$ for a geometric sequence, find a_1 and n. $a_1 = 2$; $n = 5$

[C] 67. For the geometric sequence given by $a_n = 2^n$, show that the sequence $b_n = \log a_n$ is an arithmetic sequence. Common difference is log 2.

[C] 68. For the geometric sequence given by $a_n = e^n$, show that the sequence $b_n = \ln a_n$ is an arithmetic sequence. Common difference is 1.

[C] 69. For the arithmetic sequence given by $a_n = 3n - 2$, show that the sequence $b_n = 2^{a_n}$ is a geometric sequence. Common ratio is 8.

[C] 70. For $f(n) = ab^n$, n a natural number, show that $f(n)$ is a geometric sequence. Common ratio is b.

71. *Finance* A car loan is normally structured so that each month, part of the payment reduces the loan amount and the remainder of the payment pays interest on the loan. You pay interest only on the loan amount that remains to be paid (the unpaid balance). If you have a car loan of $5000 at an annual interest rate of 9%, your monthly payment for a 5-year loan is $103.79. The amount of the loan repaid, R_n, in the nth payment of the loan is a geometric

sequence given by $R_n = R_1(1.0075)^{n-1}$. For the situation described above, $R_1 = 66.29$.

a. How much of the loan is repaid in the 27th payment? $80.50

b. The total amount, T, of the loan repaid after n payments is the sum of a geometric sequence, $T = \sum_{k=1}^{n} R_1(1.0075)^{k-1}$. Find the total amount repaid after 20 payments. $1424.65

c. Determine the unpaid balance on the loan after 20 payments. $3575.35

72. *Stock Market* A number of factors influence the price of a share of stock on the stock market. One of the factors is the growth of the dividend that is paid on each share of stock. One model to predict the value of a share of stock whose dividend grows at a constant rate is the *Gordon Model,* after Myron J. Gordon. According to this model, the value of a share of stock that pays a dividend, D, and has an expected growth rate of g percent per year, is the sum of an infinite geometric series given by $\sum_{n=0}^{\infty} D\left[\frac{1+g}{1+k}\right]^n$, where k is a constant greater than g and is the investor's desired rate of return.

a. Why must k be greater than g? So that $|r| < 1$

b. According to this model, what is the value of a stock that has a dividend of $2.25, that has an expected growth rate of 5%, and for which $k = 0.10$? $49.50

73. Write an essay on the Lucas sequence. Include a description of its relationship to the Fibonacci sequence.

74. One of Zeno's paradoxes can be described as follows: A tortoise and a hare are going to race on a 200-meter course. Because the hare can run 10 times faster than the tortoise, the tortoise is given a 100-meter head start. The gun sounds to start the race. Zeno reasoned, "By the time the hare reaches the starting point of the tortoise, the tortoise will be 10 m ahead of the hare. When the hare covers those 10 m, the tortoise will be 1 m ahead. When the hare covers the 1 m, the tortoise will be 0.1 m ahead, and so on. Therefore, the hare can never catch the tortoise!" Explain what this paradox has to do with a geometric sequence.

SECTION 11.4 Binomial Expansions

1 Expand $(a + b)^n$

HM³

By carefully observing the expansion of the binomial $(a + b)^n$ shown below, it is possible to identify some interesting patterns.

$$(a + b)^1 = a + b$$
$$(a + b)^2 = a^2 + 2ab + b^2$$
$$(a + b)^3 = a^3 + 3a^2b + 3ab^2 + b^3$$
$$(a + b)^4 = a^4 + 4a^3b + 6a^2b^2 + 4ab^3 + b^4$$
$$(a + b)^5 = a^5 + 5a^4b + 10a^3b^2 + 10a^2b^3 + 5ab^4 + b^5$$

PATTERNS FOR THE VARIABLE PART

1. The first term is a^n. The exponent on a decreases by 1 for each successive term.
2. The exponent on b increases by 1 for each successive term. The last term is b^n.
3. The degree of each term is n.

Write the variable parts of the terms of the expansion of $(a + b)^6$.

The first term is a^6. For each successive term, the exponent on a decreases by 1, and the exponent on b increases by 1. The last term is b^6.

$a^6, a^5b, a^4b^2, a^3b^3, a^2b^4, ab^5, b^6$

The variable parts of the general expansion of $(a + b)^n$ are

$$a^n, a^{n-1}b, a^{n-2}b^2, \ldots, a^{n-r}b^r, \ldots, ab^{n-1}, b^n$$

A pattern for the coefficients of the terms of the expanded binomial can be found by writing the coefficients in a triangular array known as **Pascal's Triangle.**

Each row begins and ends with the number 1. Any other number in a row is the sum of the two closest numbers above it. For example, $4 + 6 = 10$.

For $(a + b)^1$: 1 1

For $(a + b)^2$: 1 2 1

For $(a + b)^3$: 1 3 3 1

For $(a + b)^4$: 1 4 6 4 1

For $(a + b)^5$: 1 5 10 10 5 1

Write the sixth row of Pascal's Triangle.

To write the sixth row, first write the numbers of the fifth row. The first and last numbers of the sixth row are 1. Each of the other numbers of the sixth row can be obtained by finding the sum of the two closest numbers above it in the fifth row.

1 5 10 10 5 1

1 6 15 20 15 6 1

These numbers will be the coefficients of the terms of the expansion of $(a + b)^6$.

Using the numbers of the sixth row of Pascal's Triangle for the coefficients, and using the pattern for the variable part of each term, we can write the expanded form of $(a + b)^6$ as follows:

$$(a + b)^6 = a^6 + 6a^5b + 15a^4b^2 + 20a^3b^3 + 15a^2b^4 + 6ab^5 + b^6$$

Although Pascal's Triangle can be used to find the coefficients for the expanded form of the power of any binomial, this method is inconvenient when the power of the binomial is large. An alternative method for determining these coefficients is based on the concept of **factorial.**

POINT OF INTEREST

Blaise Pascal (1623–1662) is given credit for this triangle, which he first published in 1654. In that publication, the triangle looked like

	1	2	3	4	5...
1	1	1	1	1	1...
2	1	2	3	4...	
3	1	3	6...		
4	1	4...			
5	1				

Thus the triangle was rotated 45° from the way it is shown today.

The first European publication of the triangle is attributed to Peter Apianus in 1527. However, there are versions of it in a Chinese text by Yang Hui that dates from 1250. In that text, Hui demonstrated how to find the third, fourth, fifth, and sixth roots of a number by using the triangle.

POINT OF INTEREST

The exclamation point was first used as a notation for factorial in 1808 in the book *Elements d'Arithmetique Universelle* for the convenience of the printer of the book. Some English textbooks suggested using the phrase "*n*-admiration" for $n!$

n Factorial

$n!$ (which is read "n factorial") is the product of the first n consecutive natural numbers. $0!$ is defined to be 1.

$$n! = n(n-1)(n-2) \cdot \cdots \cdot 3 \cdot 2 \cdot 1$$

1!, 5!, and 7! are shown at the right.

$1! = 1$

$5! = 5 \cdot 4 \cdot 3 \cdot 2 \cdot 1 = 120$

$7! = 7 \cdot 6 \cdot 5 \cdot 4 \cdot 3 \cdot 2 \cdot 1 = 5040$

Example 1 Evaluate: $\frac{7!}{4!\,3!}$

Solution $\frac{7!}{4!\,3!} = \frac{7 \cdot 6 \cdot 5 \cdot 4 \cdot 3 \cdot 2 \cdot 1}{(4 \cdot 3 \cdot 2 \cdot 1)(3 \cdot 2 \cdot 1)}$ ▶ Write each factorial as a product.

$= 35$ ▶ Simplify.

Problem 1 Evaluate: $\frac{12!}{7!\,5!}$

Solution See page S32. 792

INSTRUCTOR NOTE

Show students that factorials do not operate in standard ways. For instance,

$3! + 4! \neq 7!$ and $\frac{12!}{3!} \neq 4!$

The coefficients in a binomial expansion can be given in terms of factorials. Note that in the expansion of $(a + b)^5$ shown below, the coefficient of a^2b^3 can be given by $\frac{5!}{2!\,3!}$. The numerator is the factorial of the power of the binomial. The denominator is the product of the factorials of the exponents on a and b.

$$(a + b)^5 = a^5 + 5a^4b + 10a^3b^2 + 10a^2b^3 + 5ab^4 + b^5$$

$$\frac{5!}{2!\,3!} = \frac{5 \cdot 4 \cdot 3 \cdot 2 \cdot 1}{(2 \cdot 1)(3 \cdot 2 \cdot 1)} = 10$$

POINT OF INTEREST

Leonard Euler (1707–1783) used the notation $\left(\frac{n}{r}\right)$ and $\left[\frac{n}{r}\right]$ for binomial coefficients around 1784. The notation $\binom{n}{r}$ appeared in the late 1820s.

In general, the coefficients of $(a + b)^n$ are given as the quotients of factorials. The coefficient of $a^{n-r}b^r$ is $\frac{n!}{(n - r)!\,r!}$. The symbol $\binom{n}{r}$ is used to express this quotient of factorials.

$$\binom{n}{r} = \frac{n!}{(n - r)!\,r!}$$

Example 2 Evaluate: $\binom{8}{5}$

Solution $\binom{8}{5} = \frac{8!}{(8 - 5)!\,5!}$ ▶ Write the quotient of the factorials.

$= \frac{8!}{3!\,5!} = \frac{8 \cdot 7 \cdot 6 \cdot 5 \cdot 4 \cdot 3 \cdot 2 \cdot 1}{(3 \cdot 2 \cdot 1)(5 \cdot 4 \cdot 3 \cdot 2 \cdot 1)} = 56$ ▶ Simplify.

Problem 2 Evaluate: $\binom{7}{0}$

Solution See page S32. 1

Using factorials and the pattern for the variable part of each term, we can write a formula for any natural-number power of a binomial.

The Binomial Expansion Formula

INSTRUCTOR NOTE

Students will be exposed to this theorem in courses such as probability and statistics, genetics, and calculus.

$$(a + b)^n = \binom{n}{0}a^n + \binom{n}{1}a^{n-1}b + \binom{n}{2}a^{n-2}b^2 + \cdots + \binom{n}{r}a^{n-r}b^r + \cdots + \binom{n}{n}b^n$$

The Binomial Expansion Formula is used below to expand $(a + b)^7$.

$(a + b)^7$

$$= \binom{7}{0}a^7 + \binom{7}{1}a^6b + \binom{7}{2}a^5b^2 + \binom{7}{3}a^4b^3 + \binom{7}{4}a^3b^4 + \binom{7}{5}a^2b^5 + \binom{7}{6}ab^6 + \binom{7}{7}b^7$$

$$= a^7 + 7a^6b + 21a^5b^2 + 35a^4b^3 + 35a^3b^4 + 21a^2b^5 + 7ab^6 + b^7$$

INSTRUCTOR NOTE

Make sure students observe that there are $n + 1$ terms in a binomial expansion.

Example 3   Write $(4x + 3y)^3$ in expanded form.

Solution $(4x + 3y)^3 = \binom{3}{0}(4x)^3 + \binom{3}{1}(4x)^2(3y) + \binom{3}{2}(4x)(3y)^2 + \binom{3}{3}(3y)^3$

$$= 1(64x^3) + 3(16x^2)(3y) + 3(4x)(9y^2) + 1(27y^3)$$

$$= 64x^3 + 144x^2y + 108xy^2 + 27y^3$$

Problem 3   Write $(3m - n)^4$ in expanded form.

Solution   See page S32. $81m^4 - 108m^3n + 54m^2n^2 - 12mn^3 + n^4$

Example 4   Find the first three terms in the expansion of $(x + 3)^{15}$.

Solution $(x + 3)^{15} = \binom{15}{0}x^{15} + \binom{15}{1}x^{14}(3) + \binom{15}{2}x^{13}(3)^2 + \cdots$

$$= 1x^{15} + 15x^{14}(3) + 105x^{13}(9) + \cdots$$

$$= x^{15} + 45x^{14} + 945x^{13} + \cdots$$

Problem 4   Find the first three terms in the expansion of $(y - 2)^{10}$.

Solution   See page S32. $y^{10} - 20y^9 + 180y^8 + \cdots$

The Binomial Theorem can also be used to write any term of a binomial expansion.

Note below that in the expansion of $(a + b)^5$, the exponent on b is 1 less than the term number.

$$(a + b)^5 = a^5 + 5a^4b + 10a^3b^2 + 10a^2b^3 + 5ab^4 + b^5$$

INSTRUCTOR NOTE

It is confusing to students that the rth term requires $r - 1$ in the formula. Reviewing with students that the binomial expansion begins at $n = 0$ may help them.

The Formula for the rth Term in a Binomial Expansion

The rth term of $(a + b)^n$ is $\binom{n}{r-1}a^{n-r+1}b^{r-1}$.

Example 5 Find the 4th term in the expansion of $(x + 3)^7$.

Solution $\binom{n}{r-1}a^{n-r+1}b^{r-1}$ ▶ Use the Formula for the rth Term in a Binomial Expansion. $r = 4, n = 7, a = x, b = 3$

$$\binom{7}{4-1}x^{7-4+1}(3)^{4-1} = \binom{7}{3}x^4(3)^3$$
$$= 35x^4(27)$$
$$= 945x^4$$

Problem 5 Find the 3rd term in the expansion of $(t - 2s)^7$.

Solution See page S33. $84t^5s^2$

CONCEPT REVIEW 11.4

Determine whether the following statements are always true, sometimes true, or never true.

1. $0! \cdot 4! = 0$ Never true
2. $\frac{4!}{0!}$ is undefined. Never true
3. In the expansion $(a + b)^8$, the exponent on a for the fifth term is 5. Never true
4. By definition, $n!$ is the sum of the first n positive numbers. Never true
5. There are n terms in the expansion of $(a + b)^n$. Never true
6. The sum of the exponents in each term of the expansion of $(a + b)^n$ is n. Always true

EXERCISES 11.4

1 1. What is the factorial of a number?

2. What is a purpose of the Binomial Expansion Formula?

Evaluate.

3. $3!$ 6
4. $4!$ 24
5. $8!$ 40,320
6. $9!$ 362,880
7. $0!$ 1
8. $1!$ 1
9. $\frac{5!}{2!\,3!}$ 10
10. $\frac{8!}{5!\,3!}$ 56
11. $\frac{6!}{6!\,0!}$ 1
12. $\frac{10!}{10!\,0!}$ 1
13. $\frac{9!}{6!\,3!}$ 84
14. $\frac{10!}{2!\,8!}$ 45

Evaluate.

15. $\binom{7}{2}$ 21
16. $\binom{8}{6}$ 28
17. $\binom{10}{2}$ 45
18. $\binom{9}{6}$ 84
19. $\binom{9}{0}$ 1
20. $\binom{10}{10}$ 1
21. $\binom{6}{3}$ 20
22. $\binom{7}{6}$ 7
23. $\binom{11}{1}$ 11
24. $\binom{13}{1}$ 13
25. $\binom{4}{2}$ 6
26. $\binom{8}{4}$ 70

Write in expanded form.

27. $(x + y)^4$ $x^4 + 4x^3y + 6x^2y^2 + 4xy^3 + y^4$

28. $(r - s)^3$ $r^3 - 3r^2s + 3rs^2 - s^3$

29. $(x - y)^5$ $x^5 - 5x^4y + 10x^3y^2 - 10x^2y^3 + 5xy^4 - y^5$

30. $(y - 3)^4$ $y^4 - 12y^3 + 54y^2 - 108y + 81$

31. $(2m + 1)^4$ $16m^4 + 32m^3 + 24m^2 + 8m + 1$

32. $(2x + 3y)^3$ $8x^3 + 36x^2y + 54xy^2 + 27y^3$

33. $(2r - 3)^5$
$32r^5 - 240r^4 + 720r^3 - 1080r^2 + 810r - 243$

34. $(x + 3y)^4$ $x^4 + 12x^3y + 54x^2y^2 + 108xy^3 + 81y^4$

Find the first three terms in the expansion.

35. $(a + b)^{10}$ $a^{10} + 10a^9b + 45a^8b^2$

36. $(a + b)^9$ $a^9 + 9a^8b + 36a^7b^2$

37. $(a - b)^{11}$ $a^{11} - 11a^{10}b + 55a^9b^2$

38. $(a - b)^{12}$ $a^{12} - 12a^{11}b + 66a^{10}b^2$

39. $(2x + y)^8$ $256x^8 + 1024x^7y + 1792x^6y^2$

40. $(x + 3y)^9$ $x^9 + 27x^8y + 324x^7y^2$

41. $(4x - 3y)^8$ $65{,}536x^8 - 393{,}216x^7y + 1{,}032{,}192x^6y^2$

42. $(2x - 5)^7$ $128x^7 - 2240x^6 + 16{,}800x^5$

43. $\left(x + \frac{1}{x}\right)^7$ $x^7 + 7x^5 + 21x^3$

44. $\left(x - \frac{1}{x}\right)^8$ $x^8 - 8x^6 + 28x^4$

45. $(x^2 + 3)^5$ $x^{10} + 15x^8 + 90x^6$

46. $(x^2 - 2)^6$ $x^{12} - 12x^{10} + 60x^8$

Find the indicated term in the expansion.

47. $(2x - 1)^7$; 4th term $-560x^4$

48. $(x + 4)^5$; 3rd term $160x^3$

49. $(x^2 - y^2)^6$; 2nd term $-6x^{10}y^2$

50. $(x^2 + y^2)^7$; 6th term $21x^4y^{10}$

51. $(y - 1)^9$; 5th term $126y^5$

52. $(x - 2)^8$; 8th term $-1024x$

53. $\left(n + \frac{1}{n}\right)^5$; 2nd term $5n^3$

54. $\left(x + \frac{1}{2}\right)^6$; 3rd term $\frac{15}{4}x^4$

55. $\left(\frac{x}{2} + 2\right)^5$; 1st term $\frac{x^5}{32}$

56. $\left(y - \frac{2}{3}\right)^6$; 3rd term $\frac{20}{3}y^4$

APPLYING CONCEPTS 11.4

Solve.

57. Write the seventh row of Pascal's Triangle.
1 7 21 35 35 21 7 1

58. Evaluate $\frac{n!}{(n-2)!}$ for $n = 50$. 2450

59. Simplify $\frac{n!}{(n-1)!}$. n

60. Write the term that contains an x^3 in the expansion of $(x + a)^7$. $35x^3a^4$

Expand the binomial.

61. $\left(x^{\frac{1}{2}} + 2\right)^4$ $x^2 + 8x^{\frac{3}{2}} + 24x + 32x^{\frac{1}{2}} + 16$ **62.** $(x^{-1} + y^{-1})^3$ $\frac{1}{x^3} + \frac{3}{x^2y} + \frac{3}{xy^2} + \frac{1}{y^3}$ **63.** $(1 + i)^6$ $-8i$

[C] **64.** For $0 \leq r \leq n$, show that $\binom{n}{r} = \binom{n}{n-r}$.

Complete solution available in the Solutions Manual.

[C] **65.** For $n \geq 1$, evaluate $\frac{2 \cdot 4 \cdot 6 \cdot 8 \cdots (2n)}{2^n n!}$. 1

A mathematical theorem is used to expand $(a + b + c)^n$. According to this theorem, the coefficient of $a^r b^k c^{n-r-k}$ in the expansion of $(a + b + c)^n$ is $\frac{n!}{r!\,k!(n - r - k)!}$.

[C] **66.** Use the mathematical theorem to find the coefficient of a^2b^3c in the expansion of $(a + b + c)^6$. 60

[C] **67.** Use the mathematical theorem to find the coefficient of $a^4b^2c^3$ in the expansion of $(a + b + c)^9$. 1260

68. Write a summary of the history of the development of Pascal's Triangle. Include some of the properties of the triangle.

Focus on Problem Solving

Forming Negations

Problem solving sometimes requires that you be able to form the **negation** of a statement. If a statement is true, its negation must be false. If a statement is false, its negation must be true. For instance,

"Mark McGwire broke Babe Ruth's single-season home run record" is a true sentence. Accordingly, the sentence "Mark McGwire did not break Babe Ruth's home run record" is false.

"Dogs have twelve paws" is a false sentence, so the sentence "Dogs do not have twelve paws" is a true sentence.

To form the negation of a statement requires a precise understanding of the statement being negated. Here are some phrases that demand special attention.

Statement	Meaning
At least seven	Greater than or equal to seven
More than seven	More than seven (does not include 7)
At most seven	Less than or equal to seven
Less than (or *fewer than*) seven	Less than seven (does not include 7)

Consider the sentence "At least 3 people scored over 90 on the last test." The negation of that sentence is "Fewer than 3 people scored over 90 on the last test."

For each of the following, write the negation of the given sentence.

1. There was at least one error on the test.

2. The temperature was less than 30 degrees.
3. The mountain is more than 5000 feet tall.
4. There are at most 5 vacancies for a field trip to New York.

When the word *all*, *no* (or *none*), or *some* occurs in a sentence, forming the negation requires careful thought. Consider the sentence "All roads are paved with cement." This sentence is not true because there are some roads—dirt roads, for example—that are not paved with cement. Because the sentence is false, its negation must be true. You might be tempted to write "All roads are not paved with cement" as the negation, but that sentence is not true because some roads are paved with cement. The correct negation is "*Some* roads are *not* paved with cement."

Now consider the sentence "Some fish live in aquariums." Because this sentence is true, the negation must be false. Writing "Some fish do not live in aquariums" as the negation is not correct because that sentence is true. The correct negation is "*All* fish do *not* live in aquariums."

The sentence "No houses have basements" is false because there is at least one house with a basement. Because this sentence is false, its negation must be true. The negation is "Some houses do have basements."

Statement	Negation
All *A* are *B*.	Some *A* are not *B*.
No *A* are *B*.	Some *A* are *B*.
Some *A* are *B*.	All *A* are not *B*.
Some *A* are not *B*.	All *A* are *B*.

Write the negation of each sentence.

5. All trees are tall.
6. All cats chase mice.
7. Some flowers have red blooms.
8. No golfers like tennis.
9. Some students do not like math.
10. No honest people are politicians.
11. All cars have power steering.
12. Some televisions are black and white.

Projects and Group Activities

ISBN and UPC Numbers

Every book that is cataloged for the Library of Congress must have an ISBN (International Standard Book Number). An ISBN is a 10-digit number of the form a_1-$a_2a_3a_4$-$a_5a_6a_7a_8a_9$-c. For instance, the ISBN for the Windows version of the CD-ROM containing the *American Heritage Children's Dictionary* is 0-395-73580-7. The first number, 0, indicates that the book is written in English. The next three numbers, 395, indicate the publisher (Houghton Mifflin Company). The next five numbers, 73580, identify the book (*American Heritage Children's Dictionary*), and the last digit, c, is called a *check digit*. This digit is chosen such that the following sum is divisible by 11.

$$a_1(10) + a_2(9) + a_3(8) + a_4(7) + a_5(6) + a_6(5) + a_7(4) + a_8(3) + a_9(2) + c$$

For the *American Heritage Children's Dictionary,*

$$\begin{aligned} 0(10) + 3(9) + 9(8) + 5(7) + 7(6) + 3(5) + 5(4) + 8(3) + 0(2) + c \\ = 235 + c \end{aligned}$$

The last digit of the ISBN is chosen as 7 because $235 + 7 = 242$ and $242 \div 11 = 22$. The value of c could be any number from 0 to 10. The number 10 is coded as an X.

One purpose of the ISBN method of coding books is to ensure that orders placed for books are accurately filled. For instance, suppose a clerk sends an order for the *American Heritage Children's Dictionary* and inadvertently enters the number 0-395-75380-7 (the 3 and 5 have been transposed). Now

$$0(10) + 3(9) + 9(8) + 5(7) + 7(6) + 5(5) + 3(4) + 8(3) + 0(2) + 7 = 244$$

and 244 is not divisible by 11. Thus an error in the ISBN has been made.

1. Determine the check digit for the book *Reader's Digest Book of Facts.* The first nine digits of the ISBN are 0-895-77692-?

2. Is 0-395-12370-4 a possible ISBN?

The UPC (Universal Product Code) is another coding scheme. This number is particularly useful in grocery stores. A checkout clerk passes the number by a scanner that reads the number and records the price on the cash register.

The UPC is a 12-digit number of the form $a_1\text{-}a_2a_3a_4a_5\text{-}a_6a_7a_8a_9a_{10}a_{11}\text{-}c$. The last digit is the check digit and is chosen such that the following sum ends in zero.

$$a_1(3) + a_2 + a_3(3) + a_4 + a_5(3) + a_6 + a_7(3) + a_8 + a_9(3) + a_{10} + a_{11}(3) + c$$

The UPC for the *Consumer Guide, 1997 Cars* is 0-71162-00699-5. Using the expression above,

$$\begin{aligned} 0(3) + 7 + 1(3) + 1 + 6(3) + 2 + 0(3) + 0 + 6(3) + 9 + 9(3) + c \\ = 85 + c \end{aligned}$$

Because $85 + 5 = 90$, which ends in zero, the check digit is 5.

Unlike the ISBN, the UPC coding scheme does not detect all transpositions. If the 6 and 1 are transposed for the *Consumer Guide, 1997 Cars,* then

$$0(3) + 7 + 1(3) + 6 + 1(3) + 2 + 0(3) + 0 + 6(3) + 9 + 9(3) + 5 = 80$$

The number still ends in zero, but the UPC is incorrect.

3. Check the UPC code of a product in your home.

4. Check the ISBN of the intermediate algebra text that you are now using.

Finding the Proper Dosage

The elimination of a drug from the body usually occurs at a rate that is proportional to the amount of the drug in the body. This relationship as a function of time can be given by the formula

$$f(t) = Ae^{kt}$$

where $k = -\frac{\ln 2}{H}$, A is the original dosage of the drug, and H is the time for one-half of the drug to be eliminated.

The three graphs show one administration of a drug, repeated dosages in which the amount of drug in the body remains relatively constant, and dosages in which the amount of drug will increase with time.

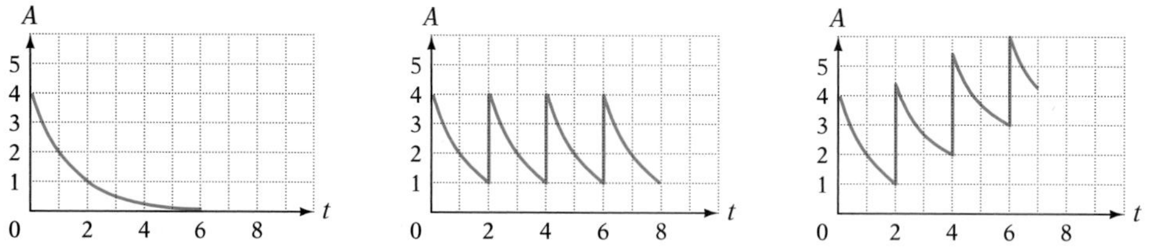

The amount of a drug in the body after n time periods is given by

$$A + Ae^{kt} + Ae^{2kt} + \cdots + Ae^{nkt}$$

where n is the number of dosages and t is the time between doses.

The amount of a drug in the body at time t can be represented by

$$S = \frac{Ae^{kt}(1 - e^{nkt})}{1 - e^{kt}}$$

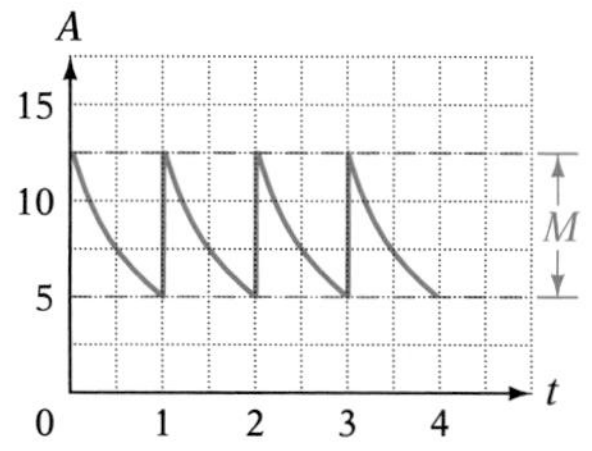

Over a period of time, with continuing dosages at periods of time t, the amount S may increase, decrease, or reach an equilibrium state, as shown in the diagram at the left.

To reach this equilibrium state, the original dosage A is reduced to some maintenance dosage M, given by the equation

$$M = A(1 - e^{nkt})$$

Solve.

1. Cancer cells have been exposed to x-rays. The number of surviving cells depends on the strength of the x-rays applied. Find the strength of the x-rays (in roentgens) for 40% of the cancer cells to be destroyed. Use the equation $A(r) = A_0e^{-0.2r}$.
2. Sodium pentobarbital is to be given to a patient for surgical anesthesia. The operation is assumed to last one-half hour, and the half-life of the sodium pentobarbital is 2 h. The amount of anesthesia in the patient's system should not go below 500 mg. If only one dosage is to be given, find the initial dose of the sodium pentobarbital.
3. One dose of a drug increases the blood level of the drug by 0.5 mg/ml. The half-life of the drug is 8 h, and the dose is given every 4 h. Find the concentration of the drug just before the fourth dose.
4. A dosage of 50 mg of medication every 2 h for 8 h will achieve a desired level of medication.

   a. Find the level of medication after 8 h if the half-life of the medication is 5 h.

   b. Find the maintenance dose after 8 h.

Chapter Summary

Key Words

A *sequence* is an ordered list of numbers. Each of the numbers of a sequence is called a *term* of the sequence. (Objective 11.1.1)

$1, \frac{1}{3}, \frac{1}{9}, \frac{1}{27}$ is a sequence.
The terms are $1, \frac{1}{3}, \frac{1}{9}$, and $\frac{1}{27}$.

A *finite sequence* contains a finite number of terms. (Objective 11.1.1)

2, 4, 6, 8 is a finite sequence.

An *infinite sequence* contains an infinite number of terms. (Objective 11.1.1)

1, 3, 5, 7, ... is an infinite sequence.

The indicated sum of the terms of a sequence is a *series*. (Objective 11.1.2)

1, 5, 9, 13 is a sequence.
The series is $1 + 5 + 9 + 13$.

An *arithmetic sequence*, or *arithmetic progression*, is one in which the difference between any two consecutive terms is constant. The difference between consecutive terms is called the *common difference* of the sequence. (Objective 11.2.1)

3, 9, 15, 21, ... is an arithmetic sequence.
$9 - 3 = 6$; 6 is the common difference.

A *geometric sequence*, or *geometric progression*, is one in which each successive term of the sequence is the same nonzero constant multiple of the preceding term. The common multiple is called the *common ratio* of the sequence. (Objective 11.3.1)

$9, 3, 1, \frac{1}{3} \ldots$ is a geometric sequence.
$\frac{3}{9} = \frac{1}{3}; \frac{1}{3}$ is the common ratio.

An *arithmetic series* is the indicated sum of the terms of an arithmetic sequence. (Objective 11.2.2)

1, 2, 3, 4, 5 is an arithmetic sequence.
The indicated sum is $1 + 2 + 3 + 4 + 5$.

A *geometric series* is the indicated sum of the terms of a geometric series. (Objective 11.3.2)

5, 10, 20, 40 is a geometric series.
The indicated sum is $5 + 10 + 20 + 40$.

n factorial, written $n!$, is the product of the first n natural numbers. 0! is defined to be 1. (Objective 11.4.1)

$5! = 5 \cdot 4 \cdot 3 \cdot 2 \cdot 1 = 120$

Essential Rules and Procedures

Formula for the nth Term of an Arithmetic Sequence
The nth term of an arithmetic sequence with a common difference of d is given by $a_n = a_1 + (n - 1)d$. (Objective 11.2.1)

4, 7, 10, 13, ... $d = 7 - 4 = 3, a_1 = 4$
$a_{10} = 4 + (10 - 1)3 = 4 + 27 = 31$

Formula for the Sum of n Terms of an Arithmetic Series
Let a_1 be the first term of a finite arithmetic sequence, let n be the number of terms, and let a_n be the last term of the sequence. Then the sum of the series is given by $S_n = \frac{n}{2}(a_1 + a_n)$. (Objective 11.2.2)

5, 8, 11, 14, ... $d = 3, a_{12} = 38$
$S_{12} = \frac{12}{2}(5 + 38) = 258$

Formula for the nth Term of a Geometric Sequence
The nth term of a geometric sequence with first term a_1 and common ratio r is given by $a_n = a_1 r^{n-1}$. (Objective 11.3.1)

2, 6, 18, 54, ... $a_1 = 2, r = 3$
$a_{10} = 2(3)^{10-1} = 2(3)^9 = 39{,}366$

Formula for the Sum of n Terms of a Finite Geometric Series
Let a_1 be the first term of a finite geometric sequence, let n be the number of terms, and let r be the common ratio. Then the sum of the series S_n is given by $S_n = \frac{a_1(1 - r^n)}{1 - r}$. (Objective 11.3.2)

$1, 4, 16, 64, \ldots \quad a_1 = 1, r = 4$

$S_8 = \frac{1(1 - 4^8)}{1 - 4} = \frac{1 - 65{,}536}{-3} = 21{,}845$

Formula for the Sum of an Infinite Geometric Series
The sum of an infinite geometric series in which $|r| < 1$ and a_1 is the first term is given by $S = \frac{a_1}{1 - r}$. (Objective 11.3.3)

$2, 1, \frac{1}{2}, \frac{1}{4}, \ldots \quad a_1 = 2, r = \frac{1}{2}$

$S = \frac{2}{1 - \frac{1}{2}} = \frac{2}{\frac{1}{2}} = 4$

The Binomial Expansion Formula
$(a + b)^n =$

$\binom{n}{0}a^n + \binom{n}{1}a^{n-1}b + \binom{n}{2}a^{n-2}b^2 + \cdots + \binom{n}{r}a^{n-r}b^r + \cdots + \binom{n}{n}b^n$

(Objective 11.4.1)

$(x + y)^4$

$= \binom{4}{0}x^4 + \binom{4}{1}x^3y + \binom{4}{2}x^2y^2 + \binom{4}{3}xy^3 + \binom{4}{4}y^4$

$= x^4 + 4x^3y + 6x^2y^2 + 4xy^3 + y^4$

Formula for the rth Term in a Binomial Expansion
The rth term of $(a + b)^n$ is $\binom{n}{r-1}a^{n-r+1}b^{r-1}$. (Objective 11.4.1)

The sixth term of $(2x - y)^9$ is

$\binom{9}{5}(2x)^{9-6+1}(-y)^{6-1} = -2016x^4y^5$

Chapter Review Exercises

1. Write $\sum_{i=1}^{4} 3x^i$ in expanded form.
$3x + 3x^2 + 3x^3 + 3x^4$

2. Find the number of terms in the finite arithmetic sequence $-5, -8, -11, \ldots, -50$. 16

3. Find the 7th term of the geometric sequence $4, 4\sqrt{2}, 8, \ldots$. 32

4. Find the sum of the infinite geometric sequence, $4, 3, \frac{9}{4}, \ldots$. 16

5. Evaluate: $\binom{9}{3}$ 84

6. Write the 14th term of the sequence whose nth term is given by the formula $a_n = \frac{8}{n+2}$. $\frac{1}{2}$

7. Find the 10th term of the arithmetic sequence $-10, -4, 2, \ldots$. 44

8. Find the sum of the first 18 terms of the arithmetic sequence $-25, -19, -13, \ldots$. 468

9. Find the sum of the first five terms of the geometric sequence $-6, 12, -24, \ldots$. -66

10. Evaluate: $\frac{8!}{4!\,4!}$ 70

11. Find the 7th term in the expansion of $(3x + y)^9$.
$2268x^3y^6$

12. Find the sum of the series $\sum_{n=1}^{4}(3n + 1)$. 34

13. Write the 6th term of the sequence whose nth term is given by the formula $a_n = \frac{n+1}{n}$. $\frac{7}{6}$

14. Find the formula for the nth term of the arithmetic sequence $12, 9, 6, \ldots$. $a_n = -3n + 15$

15. Find the 5th term of the geometric sequence $6, 2, \frac{2}{3}, \ldots$. $\frac{2}{27}$

16. Find an equivalent fraction for $0.2\overline{3}$. $\frac{7}{30}$

17. Find the 35th term of the arithmetic sequence $-13, -16, -19, \ldots$. -115

18. Find the sum of the first six terms of the geometric sequence $1, \frac{3}{2}, \frac{9}{4}, \ldots$. $\frac{665}{32}$

19. Find the sum of the first 21 terms of the arithmetic sequence $5, 12, 19, \ldots$. 1575

20. Find the 4th term in the expansion of $(x - 2y)^7$. $-280x^4y^3$

21. Find the number of terms in the finite arithmetic sequence $1, 7, 13, \ldots, 121$. 21

22. Find the 8th term of the geometric sequence $\frac{3}{8}, \frac{3}{4}, \frac{3}{2}, \ldots$. 48

23. Find the sum of the series $\sum_{i=1}^{5} 2i$. 30

24. Find the sum of the first five terms of the geometric sequence $1, 4, 16, \ldots$. 341

25. Evaluate: $5!$ 120

26. Find the 3rd term in the expansion of $(x - 4)^6$. $240x^4$

27. Find the 30th term of the arithmetic sequence $-2, 3, 8, \ldots$. 143

28. Find the sum of the first 25 terms of the arithmetic sequence $25, 21, 17, \ldots$. -575

29. Write the 5th term of the sequence whose nth term is given by the formula $a_n = \frac{(-1)^{2n-1}n}{n^2 + 2}$. $-\frac{5}{27}$

30. Write $\sum_{i=1}^{4} 2x^{i-1}$ in expanded form. $2 + 2x + 2x^2 + 2x^3$

31. Find an equivalent fraction for $0.\overline{23}$. $\frac{23}{99}$

32. Find the sum of the infinite geometric series $4 - 1 + \frac{1}{4} - \cdots$. $\frac{16}{5}$

33. Find the sum of the geometric series $\sum_{n=1}^{5} 2(3)^n$. 726

34. Find the 8th term in the expansion of $(x - 2y)^{11}$. $-42{,}240x^4y^7$

35. Find the sum of the geometric series $\sum_{n=1}^{8} \left(\frac{1}{2}\right)^n$. Round to the nearest thousandth. 0.996

36. Find the sum of the infinite geometric series $2 + \frac{4}{3} + \frac{8}{9} + \cdots$. 6

37. Find an equivalent fraction for $0.6\overline{3}$. $\frac{19}{30}$

38. Write $(x - 3y^2)^5$ in expanded form. $x^5 - 15x^4y^2 + 90x^3y^4 - 270x^2y^6 + 405xy^8 - 243y^{10}$

39. Find the number of terms in the finite arithmetic sequence $8, 2, -4, \ldots, -118$. 22

40. Evaluate: $\frac{12!}{5!\,8!}$ 99

41. Write $\sum_{i=1}^{5} \frac{(2x)^i}{i}$ in expanded form. $2x + 2x^2 + \frac{8x^3}{3} + 4x^4 + \frac{32x^5}{5}$

42. Find the sum of the series $\sum_{n=1}^{4} \frac{(-1)^{n-1}n}{n+1}$. $-\frac{13}{60}$

43. The salary schedule for an apprentice electrician is \$1200 for the first month and a \$40-per-month salary increase for the next 9 months. Find the total salary for the first 9 months. \$12,240

44. The temperature of a hot-water spa is 102°F. Each hour, the temperature is 5% lower than during the previous hour. Find the temperature of the spa after 8 h. Round to the nearest tenth. 67.7°F

Chapter Test

1. Write the 14th term of the sequence whose nth term is given by the formula $a_n = \frac{6}{n+4}$. $\frac{1}{3}$

2. Write the 9th and 10th terms of the sequence whose nth term is given by the formula $a_n = \frac{n-1}{n}$. $\frac{8}{9}, \frac{9}{10}$

3. Find the sum of the series $\sum_{n=1}^{4}(2n+3)$. 32

4. Write $\sum_{i=1}^{4} 2x^{2i}$ in expanded form.
$2x^2 + 2x^4 + 2x^6 + 2x^8$

5. Find the 28th term of the arithmetic sequence $-12, -16, -20, \ldots$. -120

6. Find the formula for the nth term of the arithmetic sequence $-3, -1, 1, \ldots$. $a_n = 2n - 5$

7. Find the number of terms in the finite arithmetic sequence $7, 3, -1, \ldots, -77$. 22

8. Find the sum of the first 15 terms of the arithmetic sequence $-42, -33, -24, \ldots$. 315

9. Find the sum of the first 24 terms of the arithmetic sequence $-4, 2, 8, \ldots$. 1560

10. Evaluate: $\frac{10!}{5!5!}$ 252

11. Find the 10th term of the geometric sequence $4, -4\sqrt{2}, 8, \ldots$. $-64\sqrt{2}$

12. Find the 5th term of the geometric sequence $5, 3, \frac{9}{5}, \ldots$. $\frac{81}{125}$

13. Find the sum of the first five terms of the geometric sequence $1, \frac{3}{4}, \frac{9}{16}, \ldots$. $\frac{781}{256}$

14. Find the sum of the first five terms of the geometric sequence $-5, 10, -20, \ldots$. -55

15. Find the sum of the infinite geometric sequence $2, 1, \frac{1}{2}, \ldots$. 4

16. Find an equivalent fraction for $0.2\overline{3}$. $\frac{7}{30}$

17. Evaluate: $\binom{11}{4}$ 330

18. Find the 5th term in the expansion of $(3x - y)^8$.
$5670x^4y^4$

19. An inventory of supplies for a fabric manufacturer indicated that 7500 yd of material were in stock on January 1. On February 1, and on the first of the month for each successive month, the manufacturer sent 550 yd of material to retail outlets. How much material was in stock after the shipment on October 1? 2550 yd

20. An ore sample contains 320 mg of a radioactive substance with a half-life of 1 day. Find the amount of radioactive material in the sample at the beginning of the fifth day. 20 mg

Cumulative Review Exercises

1. Subtract: $\frac{4x^2}{x^2+x-2} - \frac{3x-2}{x+2}$ $\frac{x^2+5x-2}{(x+2)(x-1)}$

2. Factor: $2x^6 + 16$ $2(x^2+2)(x^4-2x^2+4)$

3. Multiply: $\sqrt{2y}(\sqrt{8xy} - \sqrt{y})$ $4y\sqrt{x} - y\sqrt{2}$

4. Simplify: $\left(\frac{x^{-\frac{3}{4}} \cdot x^{\frac{3}{2}}}{x^{-\frac{5}{2}}}\right)^{-8}$ $\frac{1}{x^{26}}$

5. Solve: $5 - \sqrt{x} = \sqrt{x+5}$ 4

6. Solve: $2x^2 - x + 7 = 0$ $\frac{1}{4} \pm \frac{\sqrt{55}}{4}i$

7. Solve by the addition method: $3x - 3y = 2$
$6x - 4y = 5$
$\left(\frac{7}{6}, \frac{1}{2}\right)$

8. Solve: $2x - 1 > 3$ or $1 - 3x > 7$
$\{x \mid x < -2 \text{ or } x > 2\}$

9. Evaluate the determinant: $\begin{vmatrix} -3 & 1 \\ 4 & 2 \end{vmatrix}$ -10

10. Write $\log_5 \sqrt{\frac{x}{y}}$ in expanded form.
$\frac{1}{2}\log_5 x - \frac{1}{2}\log_5 y$

11. Solve for x: $4^x = 8^{x-1}$ 3

12. Write the 5th and 6th terms of the sequence whose nth term is given by the formula $a_n = n(n-1)$. $a_5 = 20$, $a_6 = 30$

13. Find the sum of the series $\sum_{n=1}^{7} (-1)^{n-1}(n+2)$. 6

14. Solve by the addition method:
$x + 2y + z = 3$ (2, 0, 1)
$2x - y + 2z = 6$
$3x + y - z = 5$

15. Solve for x: $\log_6 x = 3$ 216

16. Divide: $(4x^3 - 3x + 5) \div (2x + 1)$
$2x^2 - x - 1 + \frac{6}{2x+1}$

17. For $g(x) = -3x + 4$, find $g(1 + h)$. $-3h + 1$

18. Find the range of $f(a) = \frac{a^3 - 1}{2a + 1}$ if the domain is $\{0, 1, 2\}$. $\left\{-1, 0, \frac{7}{5}\right\}$

19. Graph: $3x - 2y = -4$

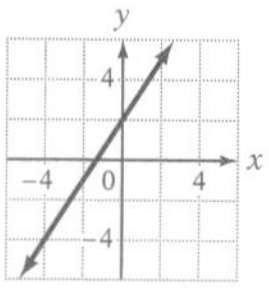

20. Graph the solution set of $2x - 3y < 9$.

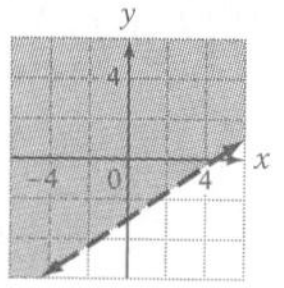

21. A new computer can complete a payroll in 16 min less time than it takes an older computer to complete the same payroll. Working together, both computers can complete the payroll in 15 min. How long would it take each computer, working alone, to complete the payroll?
New computer: 24 min; older computer: 40 min

22. A boat traveling with the current went 15 mi in 2 h. Against the current, it took 3 h to travel the same distance. Find the rate of the boat in calm water and the rate of the current. Boat: 6.25 mph; current: 1.25 mph

23. An 80-milligram sample of a radioactive material decays to 55 mg in 30 days. Use the exponential decay equation $A = A_0\left(\frac{1}{2}\right)^{\frac{t}{k}}$, where A is the amount of radioactive material present after time t, k is the half-life, and A_0 is the original amount of radioactive material, to find the half-life of the 80-milligram sample. Round to the nearest whole number. 55 days

24. A "theater in the round" has 62 seats in the first row, 74 seats in the second row, 86 seats in the third row, and so on in an arithmetic sequence. Find the total number of seats in the theater if there are 12 rows of seats. 1536 seats

25. To test the "bounce" of a ball, the ball is dropped from a height of 10 ft. The ball bounces to 80% of its previous height with each bounce. How high does the ball bounce on the fifth bounce? Round to the nearest tenth. 3.3 ft

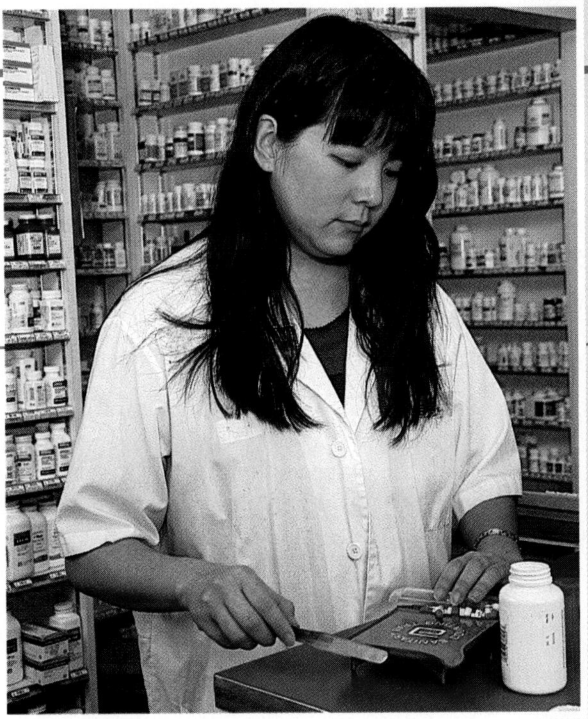

The principal responsibility of a pharmacist is to prepare and dispense medication ordered by physicians. A pharmacist must be knowledgeable about the effects of drugs on people and must have a thorough understanding of procedures for testing drug purity and strength.

12

Conic Sections

OBJECTIVES

Conic Sections

The graphs of four curves—the parabola, the circle, the ellipse, and the hyperbola—are discussed in this chapter. These curves were studied by the Greeks and were known before 400 B.C. The names of these curves were first used by Apollonius around 250 B.C. in *Conic Sections,* the most authoritative Greek discussion of these curves. Apollonius borrowed the names from a school founded by Pythagoras.

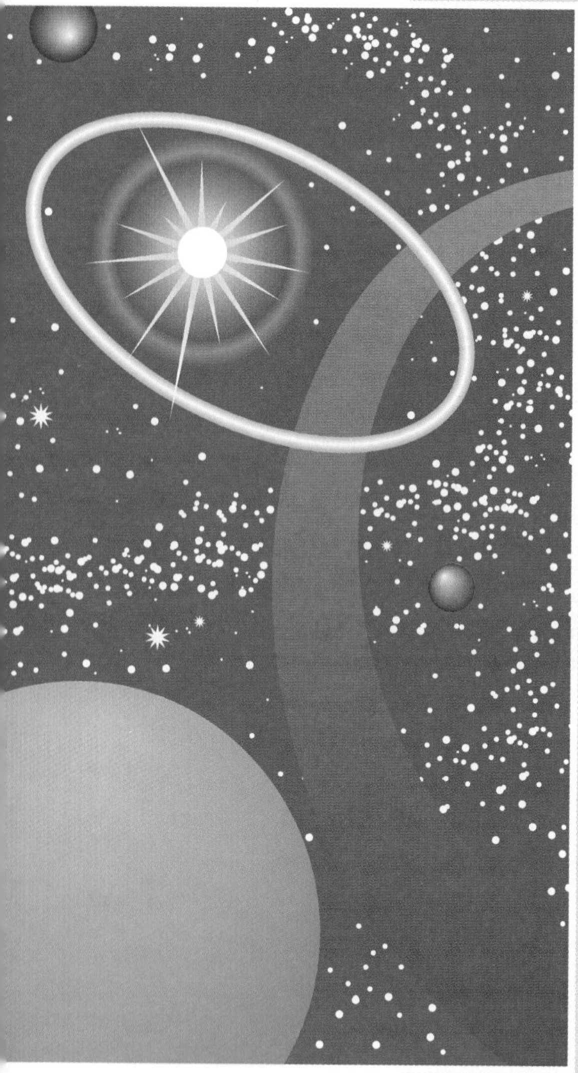

The diagram below shows the path of a planet around the sun. The curve traced out by the planet is an ellipse. The **aphelion** is the position of the planet when it is farthest from the sun. The **perihelion** is the position of the planet when it is nearest to the sun.

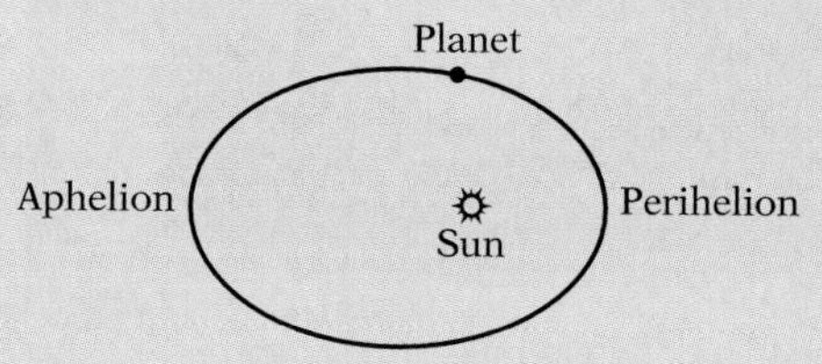

A telescope, like the one at the Palomar Observatory, has a cross section that is in the shape of a parabola. A parabolic mirror has the unusual property that all light rays parallel to the axis of symmetry that hit the mirror are reflected to the same point. This point is called the **focus of the parabola.**

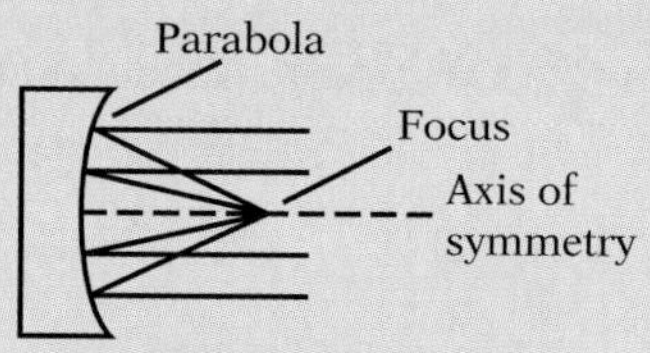

Some comets, unlike Halley's comet, travel with such speed that they are not captured by the sun's gravitational field. The path of the comet as it comes around the sun is in the shape of a hyperbola.

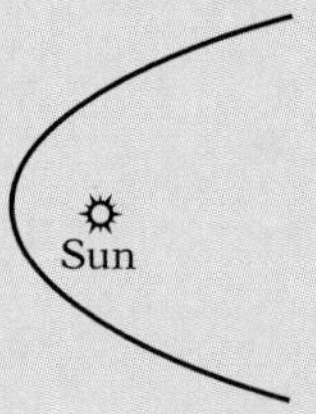

SECTION 12.1
The Parabola

1 Graph parabolas

POINT OF INTEREST

Hypatia (ca. 355–415) is considered the first prominent woman mathematician. She lectured in mathematics and philosophy at the Museum in Alexandria, the most distinguished place of learning in the world at that time. One of the topics on which Hypatia lectured was conic sections. One historian has claimed that with the death (actually the murder) of Hypatia, "the long and glorious history of Greek mathematics was at an end."

The **conic sections** are curves that can be constructed from the intersection of a plane and a right circular cone. The four conic sections are the parabola, circle, ellipse, and hyperbola. The **parabola** was introduced earlier. Here we will review some of that previous discussion and look at equations of parabolas that were not discussed before.

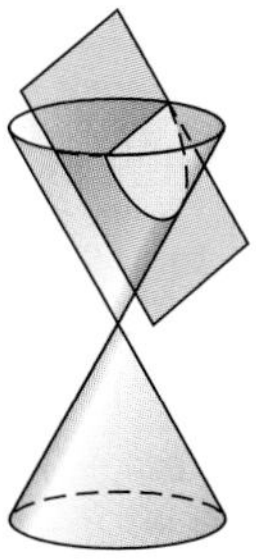

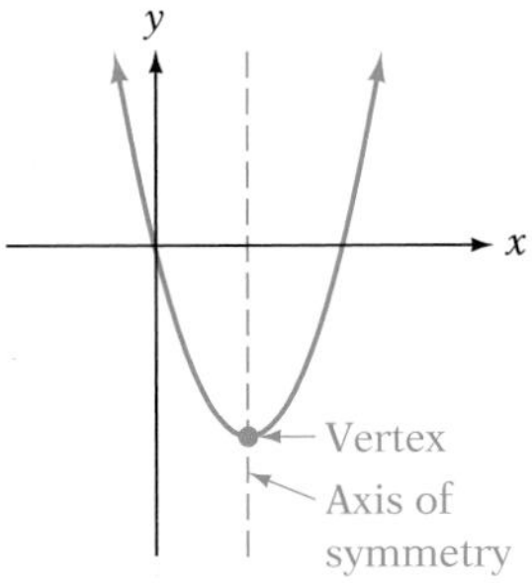

A parabola is a conic section formed by the intersection of a right circular cone and a plane parallel to the side of the cone. Every parabola has an **axis of symmetry** and a **vertex** that is on the axis of symmetry. To understand the axis of symmetry, think of folding the paper along that axis. The two halves of the curve will match up.

INSTRUCTOR NOTE

Some properties of the parabola were covered earlier. Those properties are repeated here for completeness.

The graph of the equation $y = ax^2 + bx + c$, $a \neq 0$, is a parabola with the axis of symmetry parallel to the y-axis. The parabola opens up when $a > 0$ and opens down when $a < 0$. When the parabola opens up, the vertex is the lowest point on the parabola. When the parabola opens down, the vertex is the highest point on the parabola.

The coordinates of the vertex can be found by completing the square.

Find the vertex of the parabola whose equation is $y = x^2 - 4x + 5$.

$$y = x^2 - 4x + 5$$

Group the terms involving x. $\quad y = (x^2 - 4x) + 5$

Complete the square of $x^2 - 4x$. Note that 4 is added and subtracted. Because $4 - 4 = 0$, the equation is not changed. $\quad y = (x^2 - 4x + 4) - 4 + 5$

Factor the trinomial and combine like terms. $\quad y = (x - 2)^2 + 1$

The coefficient of x^2 is positive, so the parabola opens up. The vertex is the lowest point on the parabola, or the point that has the least y-coordinate.

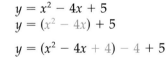

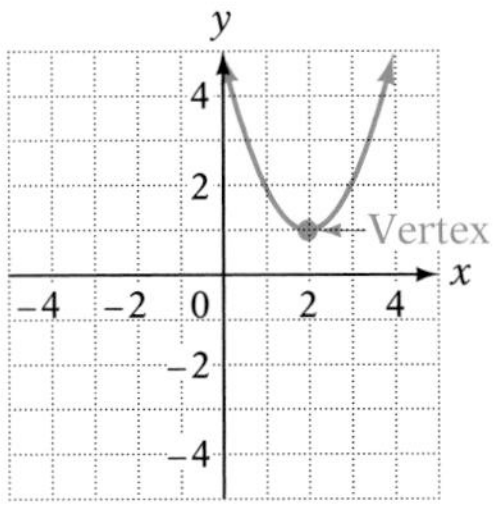

INSTRUCTOR NOTE

This argument is particularly difficult for students. You might have students evaluate the expression $(x - 2)^2$ for various values of x to illustrate that the quantity is always nonnegative and is zero when $x = 2$.

Because $(x - 2)^2 \geq 0$ for all x, the least y-coordinate occurs when $(x - 2)^2 = 0$, which occurs when $x = 2$. This means the x-coordinate of the vertex is 2.

To find the y-coordinate of the vertex, replace x in $y = (x - 2)^2 + 1$ by 2 and solve for y.

$$y = (x - 2)^2 + 1$$
$$= (2 - 2)^2 + 1 = 1$$

The vertex is (2, 1).

POINT OF INTEREST

The suspension cables for some bridges, such as the Golden Gate bridge, hang in the shape of a parabola. Parabolic shapes are also used for mirrors in telescopes and in certain antenna designs.

By following the procedure of the last example and completing the square on the equation $y = ax^2 + bx + c$, we find that the **x-coordinate of the vertex** is $-\frac{b}{2a}$. The y-coordinate of the vertex can then be determined by substituting this value of x into $y = ax^2 + bx + c$ and solving for y.

Because the axis of symmetry is parallel to the y-axis and passes through the vertex, the equation of the **axis of symmetry** is $x = -\frac{b}{2a}$.

Example 1 Find the vertex and the axis of symmetry of the parabola given by the equation $y = x^2 + 2x - 3$. Then sketch the graph of the parabola.

Solution $-\frac{b}{2a} = -\frac{2}{2(1)} = -1$ ▶ The x-coordinate of the vertex is $-\frac{b}{2a}$.

$y = x^2 + 2x - 3$
$y = (-1)^2 + 2(-1) - 3$
$y = -4$

▶ Find the y-coordinate of the vertex by replacing x by -1 and solving for y.

The vertex is $(-1, -4)$.

The axis of symmetry is the line $x = -1$. ▶ The axis of symmetry is the line $x = -\frac{b}{2a}$.

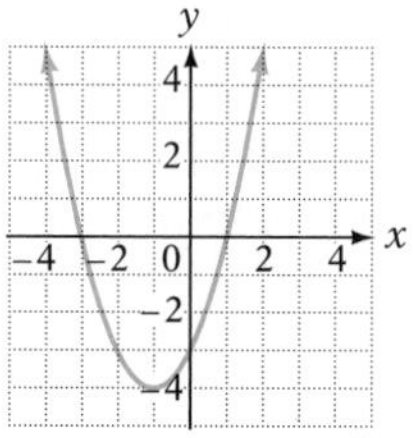

▶ Because a is positive, the parabola opens up. Use the vertex and axis of symmetry to sketch the graph.

Problem 1 Find the vertex and the axis of symmetry of the parabola given by the equation $y = -x^2 + x + 3$. Then sketch the graph of the parabola.

Solution See page S33. Vertex: $\left(\frac{1}{2}, \frac{13}{4}\right)$
Axis of symmetry: $x = \frac{1}{2}$

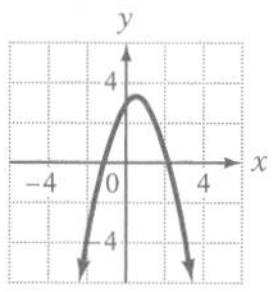

INSTRUCTOR NOTE

Point out that the analysis of $x = ay^2 + by + c$ is essentially the same as that for $y = ax^2 + bx + c$. The difference is that the graph opens left or right instead of up or down.

The graph of an equation of the form $x = ay^2 + by + c$, $a \neq 0$, is also a parabola. In this case, the parabola opens to the right when a is positive and opens to the left when a is negative.

For a parabola of this form, the **y-coordinate of the vertex** is $-\frac{b}{2a}$. The **axis of symmetry** is the line $y = -\frac{b}{2a}$.

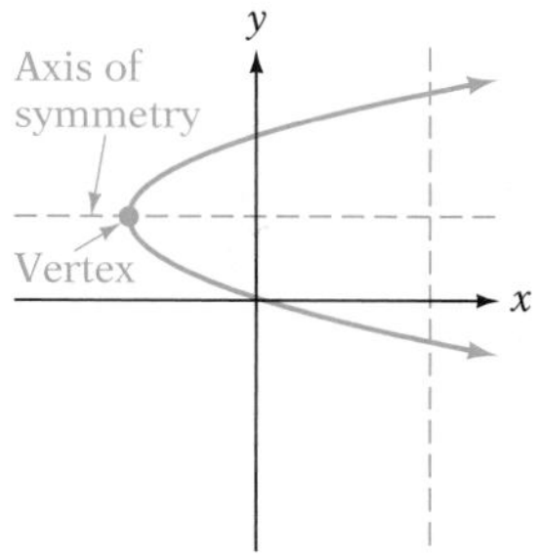

Using the vertical-line test reveals that the graph of a parabola of this form is not the graph of a function. The graph of $x = ay^2 + by + c$ is the graph of a relation.

Example 2 Find the vertex and the axis of symmetry of the parabola whose equation is $x = 2y^2 - 8y + 5$. Then sketch its graph.

Solution $-\frac{b}{2a} = -\frac{-8}{2(2)} = 2$ ▶ Find the y-coordinate of the vertex. $a = 2, b = -8$.

$x = 2y^2 - 8y + 5$
$x = 2(2)^2 - 8(2) + 5$
$x = -3$

▶ Find the x-coordinate of the vertex by replacing y by 2 and solving for x.

The vertex is $(-3, 2)$.

The axis of symmetry is the line $y = 2$. ▶ The axis of symmetry is the line $y = -\frac{b}{2a}$.

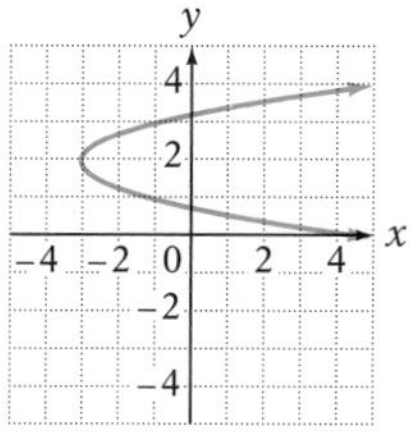

▶ Because a is positive, the parabola opens to the right. Use the vertex and axis of symmetry to sketch the graph.

Problem 2 Find the vertex and the axis of symmetry of the parabola whose equation is $x = -2y^2 - 4y - 3$. Then sketch its graph.

Solution See page S33. Vertex: $(-1, -1)$
Axis of symmetry: $y = -1$

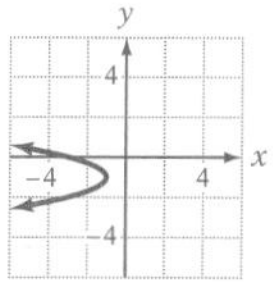

CONCEPT REVIEW 12.1

Determine whether the following statements are always true, sometimes true, or never true.

1. The graph of a parabola is the graph of a function. Sometimes true

2. The axis of symmetry of a parabola passes through the vertex. Always true

3. The graph of a parabola has two x-intercepts. Sometimes true

4. The graph of a parabola has a minimum value. Sometimes true

5. The axis of symmetry of a parabola is the x-axis or the y-axis. Sometimes true

6. The equation of the axis of symmetry of a parabola is $x = -\frac{b}{2a}$.
Sometimes true

EXERCISES 12.1

1 **1.** Describe the vertex and axis of symmetry of a parabola.

2. Explain how, by looking at the equation of a parabola, you can tell whether the parabola opens up or down or opens left or right.

Find the vertex and the axis of symmetry of the parabola given by the equation. Then sketch its graph.

3. $y = x^2 - 2x - 4$
Vertex: $(1, -5)$
Axis of symmetry: $x = 1$

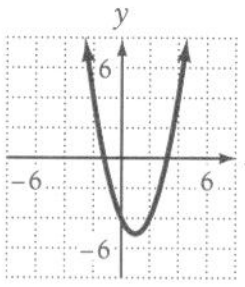

4. $y = x^2 + 4x - 4$
Vertex: $(-2, -8)$
Axis of symmetry: $x = -2$

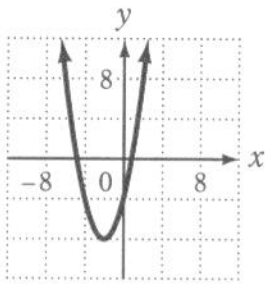

5. $y = -x^2 + 2x - 3$
Vertex: $(1, -2)$
Axis of symmetry: $x = 1$

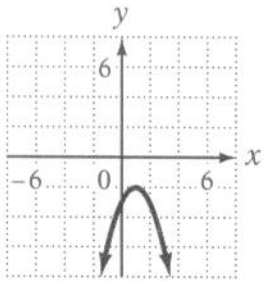

6. $y = -x^2 + 4x - 5$
Vertex: $(2, -1)$
Axis of symmetry: $x = 2$

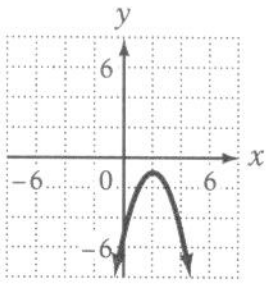

7. $x = y^2 + 6y + 5$
Vertex: $(-4, -3)$
Axis of symmetry: $y = -3$

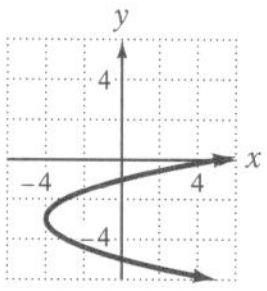

8. $x = y^2 - y - 6$
Vertex: $\left(-\frac{25}{4}, \frac{1}{2}\right)$
Axis of symmetry: $y = \frac{1}{2}$

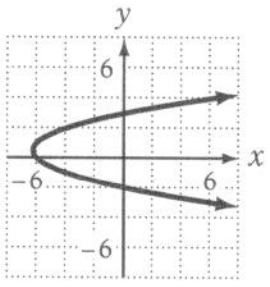

9. $y = 2x^2 - 4x + 1$
Vertex: $(1, -1)$
Axis of symmetry: $x = 1$

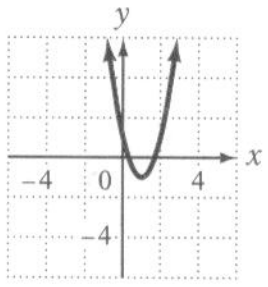

10. $y = 2x^2 + 4x - 5$
Vertex: $(-1, -7)$
Axis of symmetry: $x = -1$

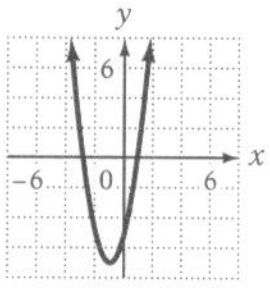

11. $y = x^2 - 5x + 4$
Vertex: $\left(\frac{5}{2}, -\frac{9}{4}\right)$
Axis of symmetry: $x = \frac{5}{2}$

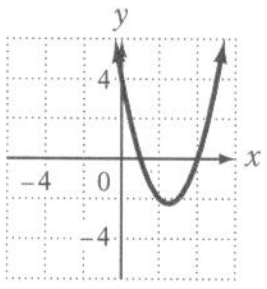

12. $y = x^2 + 5x + 6$
Vertex: $\left(-\frac{5}{2}, -\frac{1}{4}\right)$
Axis of symmetry: $x = -\frac{5}{2}$

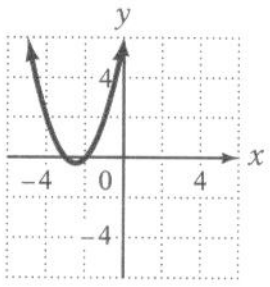

13. $x = y^2 - 2y - 5$
Vertex: $(-6, 1)$
Axis of symmetry: $y = 1$

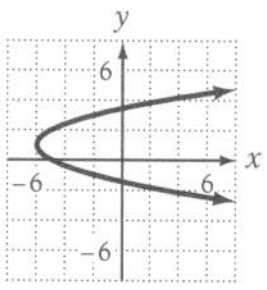

14. $x = y^2 - 3y - 4$
Vertex: $\left(-\frac{25}{4}, \frac{3}{2}\right)$
Axis of symmetry: $y = \frac{3}{2}$

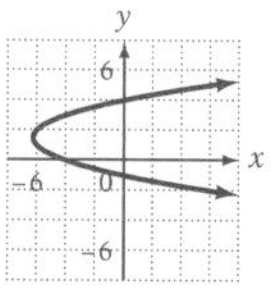

15. $y = -3x^2 - 9x$
Vertex: $\left(-\frac{3}{2}, \frac{27}{4}\right)$
Axis of symmetry: $x = -\frac{3}{2}$

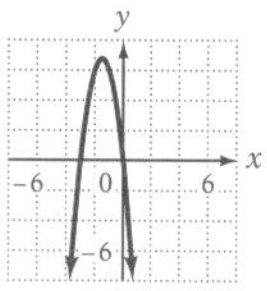

16. $y = -2x^2 + 6x$
Vertex: $\left(\frac{3}{2}, \frac{9}{2}\right)$
Axis of symmetry: $x = \frac{3}{2}$

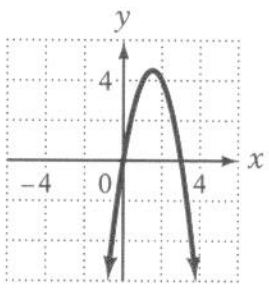

17. $x = -\frac{1}{2}y^2 + 4$
Vertex: $(4, 0)$
Axis of symmetry: $y = 0$

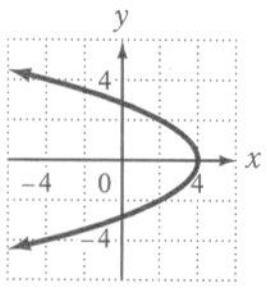

18. $x = -\frac{1}{4}y^2 - 1$
Vertex: $(-1, 0)$
Axis of symmetry: $y = 0$

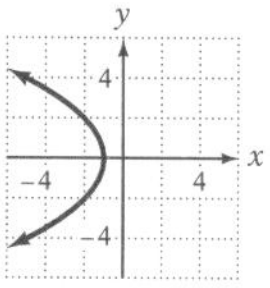

19. $x = \frac{1}{2}y^2 - y + 1$

Vertex: $\left(\frac{1}{2}, 1\right)$

Axis of symmetry: $y = 1$

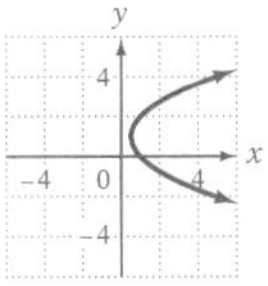

20. $x = -\frac{1}{2}y^2 + 2y - 3$

Vertex: $(-1, 2)$

Axis of symmetry: $y = 2$

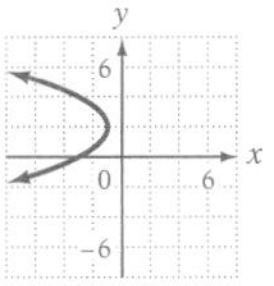

21. $y = \frac{1}{2}x^2 + 2x - 6$

Vertex: $(-2, -8)$

Axis of symmetry: $x = -2$

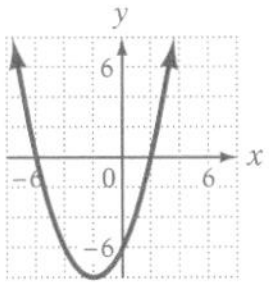

22. $y = -\frac{1}{2}x^2 + x - 3$

Vertex: $\left(1, -\frac{5}{2}\right)$

Axis of symmetry: $x = 1$

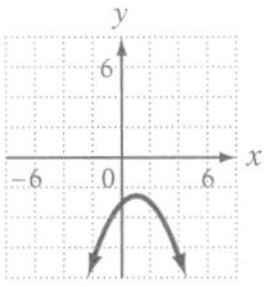

APPLYING CONCEPTS 12.1

Use the vertex and the direction in which the parabola opens to determine the domain and range of the relation.

23. $y = x^2 - 4x - 2$ D: $\{x \mid x \in \text{real numbers}\}$ R: $\{y \mid y \geq -6\}$

24. $y = x^2 - 6x + 1$ D: $\{x \mid x \in \text{real numbers}\}$ R: $\{y \mid y \geq -8\}$

25. $y = -x^2 + 2x - 3$ D: $\{x \mid x \in \text{real numbers}\}$ R: $\{y \mid y \leq -2\}$

26. $y = -x^2 - 2x + 4$ D: $\{x \mid x \in \text{real numbers}\}$ R: $\{y \mid y \leq 5\}$

27. $x = y^2 + 6y - 5$ D: $\{x \mid x \geq -14\}$ R: $\{y \mid y \in \text{real numbers}\}$

28. $x = y^2 + 4y - 3$ D: $\{x \mid x \geq -7\}$ R: $\{y \mid y \in \text{real numbers}\}$

29. $x = -y^2 - 2y + 6$ D: $\{x \mid x \leq 7\}$ R: $\{y \mid y \in \text{real numbers}\}$

30. $x = -y^2 - 6y + 2$ D: $\{x \mid x \leq 11\}$ R: $\{y \mid y \in \text{real numbers}\}$

Recall from the historical feature at the beginning of this chapter that an application of parabolas as mirrors for telescopes was mentioned. The light from a source strikes the mirror and is reflected to a point called the **focus** of the parabola. The focus is $\frac{1}{4a}$ units from the vertex of the parabola on the axis of symmetry in the direction the parabola opens. In the expression $\frac{1}{4a}$, a is the coefficient of the second-degree term. In each of the following, find the coordinates of the focus of the parabola.

[C] **31.** $y = 2x^2 - 4x + 1$ $\left(1, -\frac{7}{8}\right)$

[C] **32.** $y = -\frac{1}{4}x^2 + 2$ $(0, 1)$

[C] **33.** $x = \frac{1}{2}y^2 + y - 2$ $(-2, -1)$

[C] **34.** $x = -y^2 - 4y + 1$ $\left(\frac{19}{4}, -2\right)$

35. *Astronomy* Mirrors used in reflecting telescopes have a cross section that is a parabola. The 200-inch mirror at the Palomar Observatory in California is made from Pyrex, is 2 ft thick at the ends, and weighs 14.75 tons. The cross section of the mirror has been ground to a true parabola within 0.0000015 in. No matter where light strikes the parabolic surface, the light is reflected to a point called the focus of the parabola, as shown in the figure at the right.

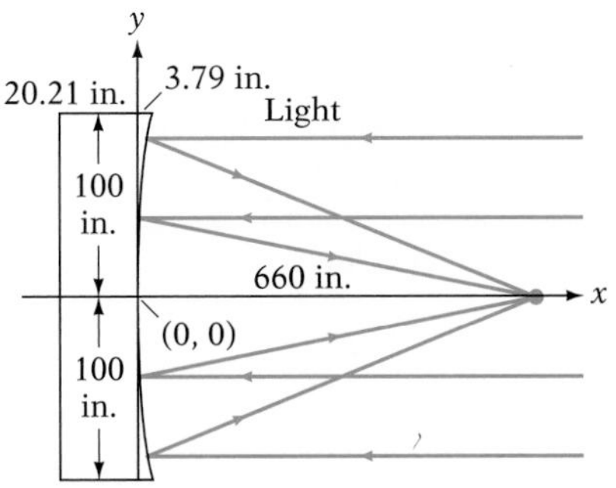

a. Determine an equation of the mirror. Round to the nearest whole number. $x = \frac{1}{2639}y^2$

b. Over what interval for x is the equation valid? $\{0 \leq x \leq 3.79\}$

36. *Meteorology* A radar dish used in the Cassegrain radar system has a cross section that is a parabola. The radar dish, used in weather forecasting, has a diameter of 84 ft. It is made of structural steel and has a depth of 17.7 ft. Signals from the radar system are reflected off clouds, collected by the radar system, and then analyzed.

   a. Determine an equation of the radar dish. Round to the nearest whole number. $x = \frac{1}{100}y^2$

   b. Over what interval for x is the equation valid? $[0, 17.7]$

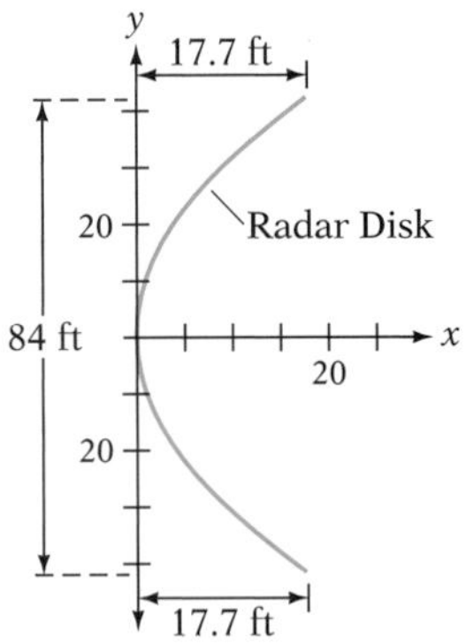

Cassegrain Radar Dish

37. Explain how the concepts of congruence and symmetry are related.

38. Explain how the graph of $f(x) = ax^2$ changes, depending on the value of a.

39. Explain why $x = y^2 + 4$ isn't an equation that defines y as a function of x.

40. When $y = 0$, the x values of the equation $y = x^2 + 2x + 3$ are imaginary numbers. What does this mean for the graph of the equation?

SECTION 12.2 The Circle

1 Find the equation of a circle and then graph the circle

A **circle** is a conic section formed by the intersection of a cone and a plane that is parallel to the base of the cone.

LOOK CLOSELY

As the angle of the plane that intersects the cone changes, different conic sections are formed. For a parabola, the plane is *parallel to the side* of the cone. For a circle, the plane is *parallel to the base* of the cone.

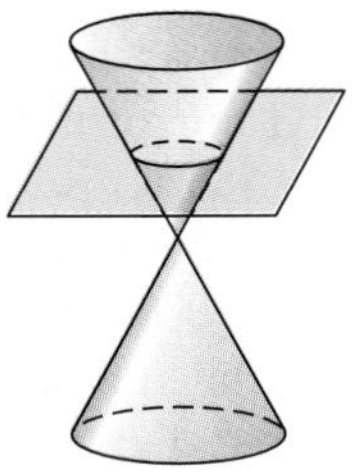

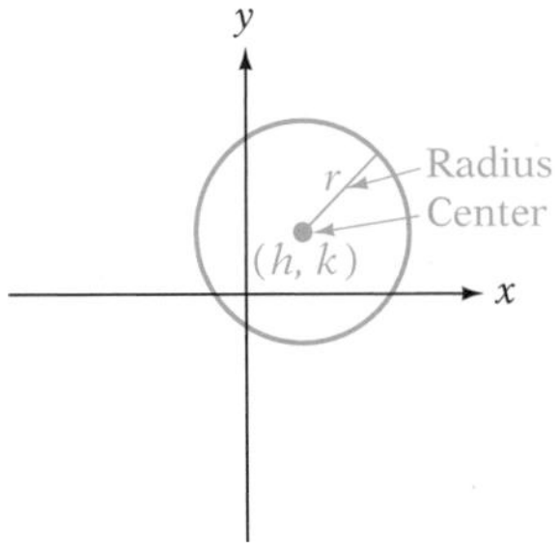

A **circle** can be defined as all the points (x, y) in the plane that are a fixed distance from a given point (h, k) called the **center.** The fixed distance is the **radius** of the circle.

INSTRUCTOR NOTE

The distance formula was discussed earlier in the text. It may be necessary to review it before doing this example.

The equation of a circle can be determined by using the distance formula.

Let (h, k) be the coordinates of the center of the circle, let r be the radius, and let (x, y) be any point on the circle. Then, by the distance formula,

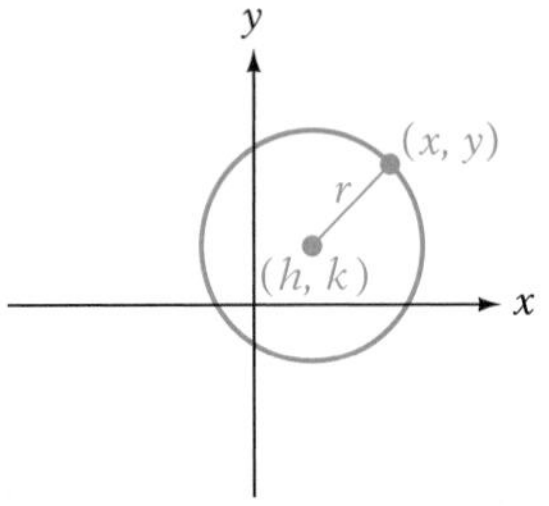

$$r = \sqrt{(x - h)^2 + (y - k)^2}$$

Squaring each side of the equation gives the equation of a circle.

$$r^2 = [\sqrt{(x - h)^2 + (y - k)^2}]^2$$
$$r^2 = (x - h)^2 + (y - k)^2$$

The Standard Form of the Equation of a Circle

Let r be the radius of a circle and let (h, k) be the coordinates of the center of the circle. Then the equation of the circle is given by

$$(x - h)^2 + (y - k)^2 = r^2$$

Recall that the graph of a circle is not the graph of a function. The graph of a circle is the graph of a relation.

Find the equation of the circle with radius 4 and center $C(-1, 2)$.

Use the standard form of the equation of a circle. $(x - h)^2 + (y - k)^2 = r^2$

Replace r by 4, h by -1, and k by 2. $[x - (-1)]^2 + (y - 2)^2 = 4^2$

$(x + 1)^2 + (y - 2)^2 = 16$

To sketch the graph of this circle, draw a circle with center $C(-1, 2)$ and radius 4.

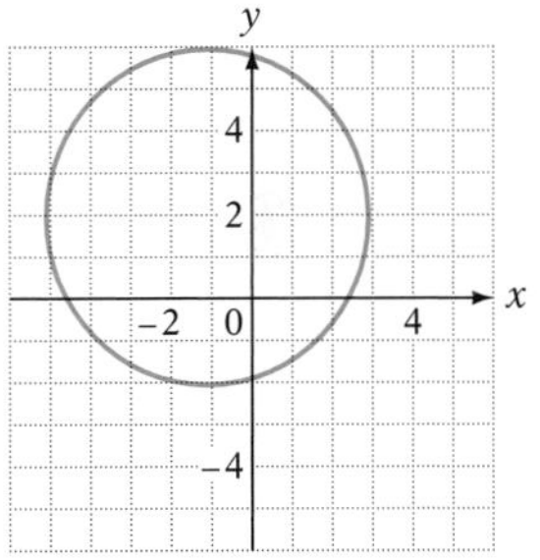

Example 1 Find the equation of the circle with radius 5 and center $C(-1, 3)$. Then sketch its graph.

Solution

$$(x - h)^2 + (y - k)^2 = r^2$$
$$[x - (-1)]^2 + (y - 3)^2 = 5^2$$
$$(x + 1)^2 + (y - 3)^2 = 25$$

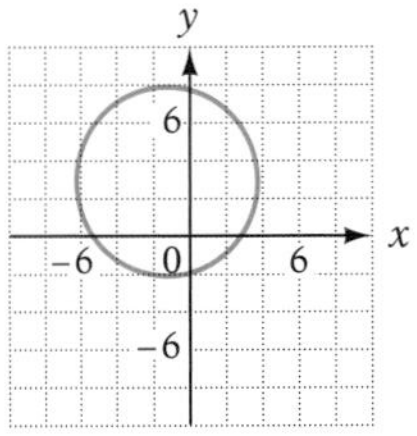

Problem 1 Find the equation of the circle with radius 4 and center $C(2, -3)$. Then sketch its graph.

Solution See page S33. $(x - 2)^2 + (y + 3)^2 = 16$

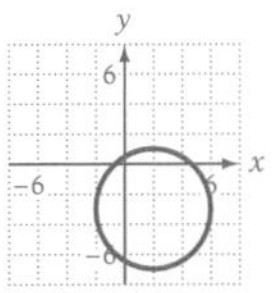

⇒ Find the radius of a circle that passes through the point $P(2, 1)$ and has as its center the point $C(3, -4)$.

Because the center of the circle and a point on the circle are known, use the distance formula to find the radius of the circle.	$\sqrt{(x_2 - x_1)^2 + (y_2 - y_1)^2} = r$
	$\sqrt{(3 - 2)^2 + (-4 - 1)^2} = r$
	$\sqrt{1^2 + (-5)^2} = r$
	$\sqrt{1 + 25} = r$
	$\sqrt{26} = r$

The equation of the circle is $(x - 3)^2 + (y + 4)^2 = 26$. The radius is $\sqrt{26}$. ◄

Example 2   Find the equation of the circle for which a diameter has endpoints $P_1(-4, -1)$ and $P_2(2, 3)$.

Solution
$$x_m = \frac{x_1 + x_2}{2} \qquad y_m = \frac{y_1 + y_2}{2}$$
$$x_m = \frac{-4 + 2}{2} \qquad y_m = \frac{-1 + 3}{2}$$
$$x_m = -1 \qquad y_m = 1$$

▶ Let $(x_1, y_1) = (-4, -1)$ and $(x_2, y_2) = (2, 3)$. Find the center of the circle by finding the midpoint of the diameter.

$$(x_m, y_m) = (-1, 1)$$

$$r = \sqrt{(x_1 - x_m)^2 + (y_1 - y_m)^2}$$
$$r = \sqrt{[-4 - (-1)]^2 + (-1 - 1)^2}$$
$$r = \sqrt{9 + 4}$$
$$r = \sqrt{13}$$

▶ Find the radius of the circle. Use either point on the circle and the coordinates of the center of the circle. P_1 is used here.

$$(x + 1)^2 + (y - 1)^2 = 13$$

▶ Write the equation of the circle with center $C(-1, 1)$ and radius $\sqrt{13}$.

Problem 2   Find the equation of the circle for which a diameter has endpoints $P_1(-2, 1)$ and $P_2(4, -1)$.

Solution   See page S33. $(x - 1)^2 + y^2 = 10$

2 ▪ Write the equation of a circle in standard form and then graph the circle

The equation of a circle can also be expressed as the equation

$$x^2 + y^2 + ax + by + c = 0$$

To rewrite this equation in standard form, it is necessary to complete the square on the x and y terms.

⇒ Write the equation of the circle $x^2 + y^2 + 4x + 2y + 1 = 0$ in standard form.

Subtract the constant term from each side of the equation.	$x^2 + y^2 + 4x + 2y + 1 = 0$
	$x^2 + y^2 + 4x + 2y = -1$

Rewrite the equation by grouping terms involving x and terms involving y.

$$(x^2 + 4x) + (y^2 + 2y) = -1$$

Complete the square on $x^2 + 4x$ and $y^2 + 2y$.

$$(x^2 + 4x + 4) + (y^2 + 2y + 1) = -1 + 4 + 1$$
$$(x^2 + 4x + 4) + (y^2 + 2y + 1) = 4$$

Factor each trinomial.

$$(x + 2)^2 + (y + 1)^2 = 4$$

Example 3 Write the equation of the circle $x^2 + y^2 + 3x - 2y = 1$ in standard form. Then sketch its graph.

Solution

$$x^2 + y^2 + 3x - 2y = 1$$
$$(x^2 + 3x) + (y^2 - 2y) = 1$$

▸ Group terms involving x and terms involving y.

$$\left(x^2 + 3x + \frac{9}{4}\right) + (y^2 - 2y + 1) = 1 + \frac{9}{4} + 1$$

▸ Complete the square on $x^2 + 3x$ and $y^2 - 2y$.

$$\left(x + \frac{3}{2}\right)^2 + (y - 1)^2 = \frac{17}{4}$$

▸ Factor each trinomial.

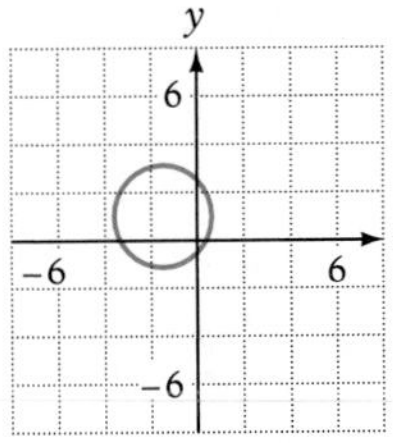

▸ Draw a circle with center $\left(-\frac{3}{2}, 1\right)$ and radius $\sqrt{\frac{17}{4}} = \frac{\sqrt{17}}{2}$. $\left(\frac{\sqrt{17}}{2} \approx 2.1\right)$

Problem 3 Write the equation of the circle $x^2 + y^2 - 4x + 8y + 15 = 0$ in standard form. Then sketch its graph.

Solution See page S33. $(x - 2)^2 + (y + 4)^2 = 5$

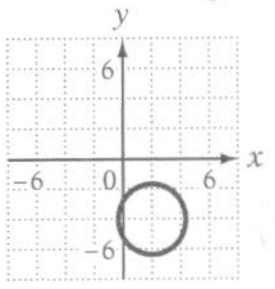

CONCEPT REVIEW 12.2

Determine whether the following statements are always true, sometimes true, or never true.

1. The center of a circle is at the origin of the coordinate axes. Sometimes true

2. The center of the circle given by the equation $(x + 2)^2 + (y + 1)^2 = 16$ is $(2, 1)$. Never true

3. The equation $x^2 + y^2 - 2x + 3y = 12$ is the equation of a circle. Always true

4. The graph of a circle is the graph of a function. Never true

5. $(x - 2)^2 + (y + 4)^2 = -4$ is the equation of a circle. Never true

EXERCISES 12.2

1

1. Describe how the points on the circumference of a circle are related to the center of the circle.

2. What do the values of h, k, and r represent in the equation of a circle in standard form?

Sketch a graph of the equation of the circle.

3. $(x - 2)^2 + (y + 2)^2 = 9$

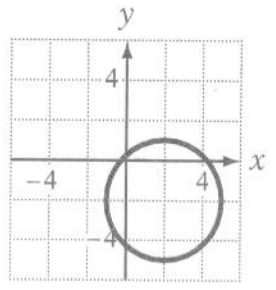

4. $(x + 2)^2 + (y - 3)^2 = 16$

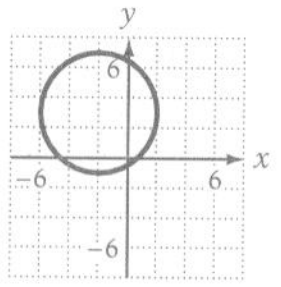

5. $(x + 3)^2 + (y - 1)^2 = 25$

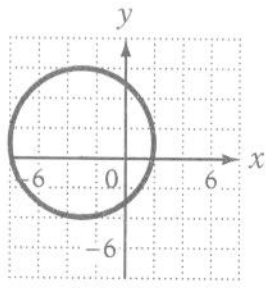

6. $(x - 2)^2 + (y + 3)^2 = 4$

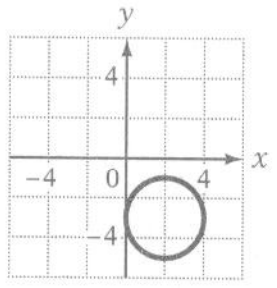

7. $(x - 4)^2 + (y + 2)^2 = 1$

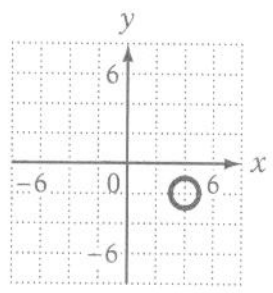

8. $(x - 3)^2 + (y - 2)^2 = 16$

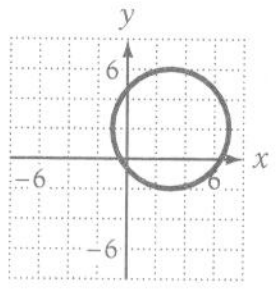

9. $(x + 5)^2 + (y + 2)^2 = 4$

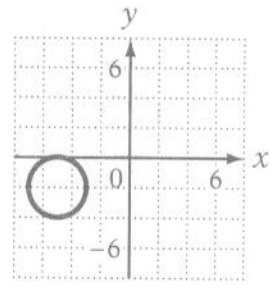

10. $(x + 1)^2 + (y - 1)^2 = 9$

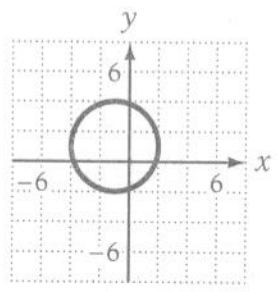

11. Find the equation of the circle with radius 2 and center $C(2, -1)$. Then sketch its graph.
$(x - 2)^2 + (y + 1)^2 = 4$

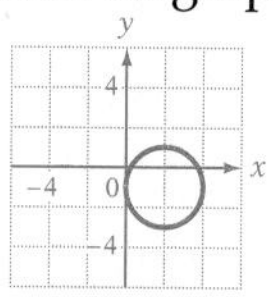

12. Find the equation of the circle with radius 3 and center $C(-1, -2)$. Then sketch its graph.
$(x + 1)^2 + (y + 2)^2 = 9$

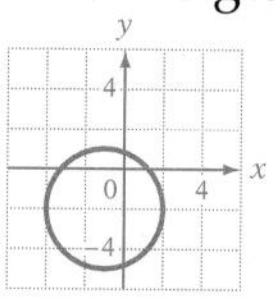

13. Find the equation of the circle that passes through the point $P(1, 2)$ and whose center is the point $C(-1, 1)$. Then sketch its graph.
$(x + 1)^2 + (y - 1)^2 = 5$

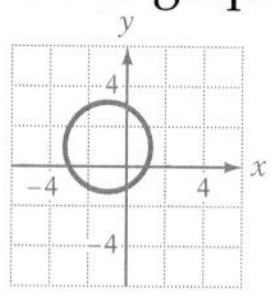

14. Find the equation of the circle that passes through the point $P(-1, 3)$ and whose center is the point $C(-2, 1)$. Then sketch its graph.
$(x + 2)^2 + (y - 1)^2 = 5$

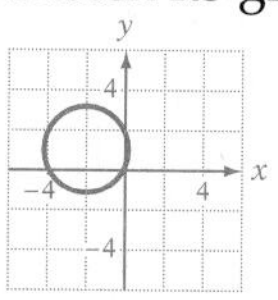

15. Find the equation of the circle for which a diameter has endpoints $P_1(-1, 4)$ and $P_2(-5, 8)$.
$(x + 3)^2 + (y - 6)^2 = 8$

16. Find the equation of the circle for which a diameter has endpoints $P_1(2, 3)$ and $P_2(5, -2)$.
$\left(x - \frac{7}{2}\right)^2 + \left(y - \frac{1}{2}\right)^2 = \frac{17}{2}$

17. Find the equation of the circle for which a diameter has endpoints $P_1(-4, 2)$ and $P_2(0, 0)$.
$(x + 2)^2 + (y - 1)^2 = 5$

18. Find the equation of the circle for which a diameter has endpoints $P_1(-8, -3)$ and $P_2(0, -4)$.
$(x + 4)^2 + \left(y + \frac{7}{2}\right)^2 = \frac{65}{4}$

2 Write the equation of the circle in standard form. Then sketch its graph.

19. $x^2 + y^2 - 2x + 4y - 20 = 0$
$(x - 1)^2 + (y + 2)^2 = 25$

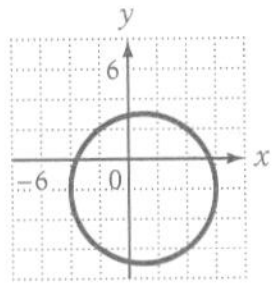

20. $x^2 + y^2 - 4x + 8y + 4 = 0$
$(x - 2)^2 + (y + 4)^2 = 16$

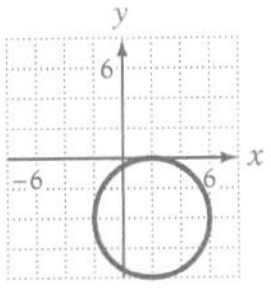

21. $x^2 + y^2 + 6x + 8y + 9 = 0$
$(x + 3)^2 + (y + 4)^2 = 16$

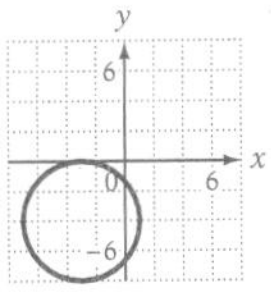

22. $x^2 + y^2 - 6x + 10y + 25 = 0$
$(x - 3)^2 + (y + 5)^2 = 9$

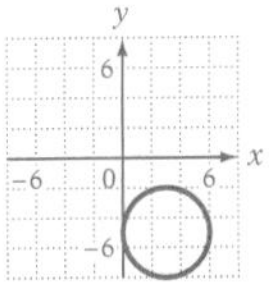

23. $x^2 + y^2 - x + 4y + \frac{13}{4} = 0$
$\left(x - \frac{1}{2}\right)^2 + (y + 2)^2 = 1$

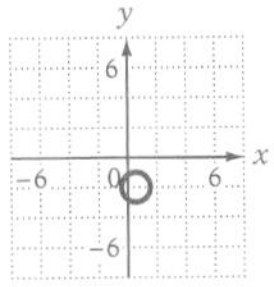

24. $x^2 + y^2 + 4x + y + \frac{1}{4} = 0$
$(x + 2)^2 + \left(y + \frac{1}{2}\right)^2 = 4$

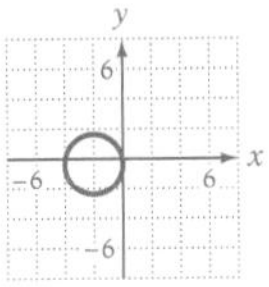

25. $x^2 + y^2 - 6x + 4y + 4 = 0$
$(x - 3)^2 + (y + 2)^2 = 9$

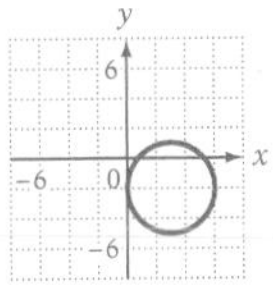

26. $x^2 + y^2 - 10x + 8y + 40 = 0$
$(x - 5)^2 + (y + 4)^2 = 1$

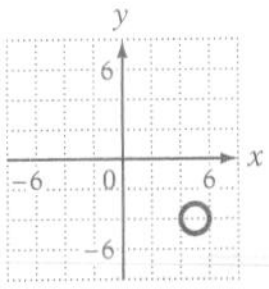

APPLYING CONCEPTS 12.2

In Exercises 27–29, write the equation of the circle in standard form.

27. The circle has its center at the point $C(3, 0)$ and passes through the origin.
$(x - 3)^2 + y^2 = 9$

[C] 28. A diameter of the circle has endpoints $P_1(-2, 4)$ and $P_2(2, -2)$.
$x^2 + (y - 1)^2 = 13$

[C] 29. The circle has radius 1, is tangent to both the x- and y-axes, and lies in quadrant II. $(x + 1)^2 + (y - 1)^2 = 1$

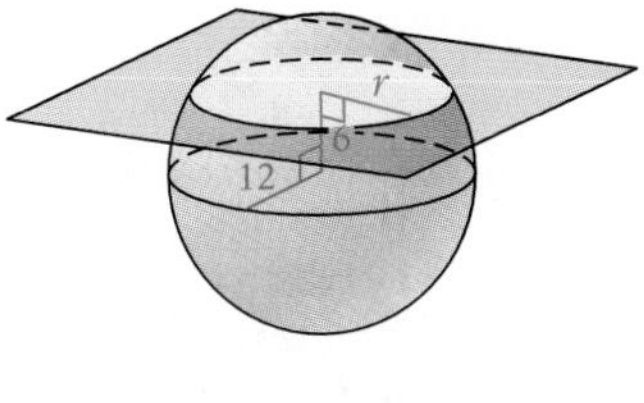

30. *Geometry* The radius of a sphere is 12 in. What is the radius of the circle that is formed by the intersection of a plane and the sphere at a point 6 in. from the center of the sphere? $6\sqrt{3}$ in.

31. Is $x^2 + y^2 = 16$ the equation of a function? Explain how to graph this relation on a graphing utility.

32. Is $x^2 + y^2 + 4x + 8y + 24 = 0$ the equation of a circle? If not, explain why not. If so, find the radius and the coordinates of the center.

33. Explain why the graph of the equation $\frac{x^2}{9} + \frac{y^2}{9} = 1$ is or is not a circle.

34. Explain the relationship between the distance formula and the standard form of the equation of a circle.

35. Write a report on the circles of longitude and latitude. Include a description of the prime meridian, the international date line, and time zones.

SECTION 12.3 The Ellipse and the Hyperbola

1 Graph an ellipse with center at the origin

INSTRUCTOR NOTE

You might point out to students that if β is the angle at which the plane intersects the axis of the cone and α is the angle shown in the figure below, then an ellipse is formed when $\alpha < \beta < 90°$.

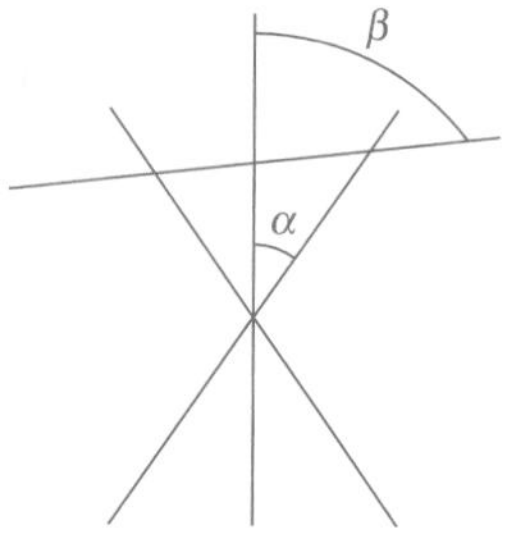

The orbits of the planets around the sun are "oval" shaped. This oval shape can be described as an **ellipse,** which is another of the conic sections.

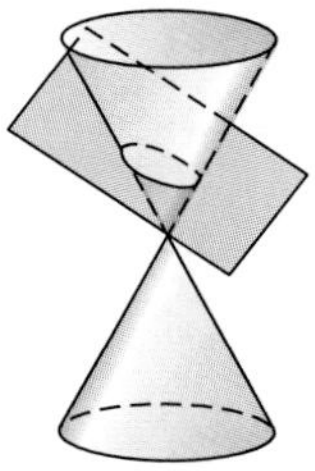

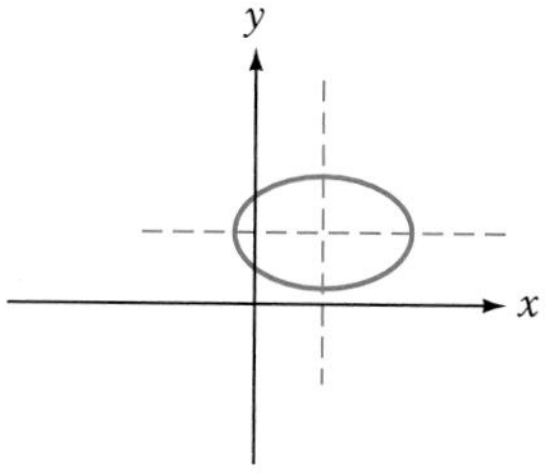

There are two **axes of symmetry** for an ellipse. The intersection of these two axes is the **center** of the ellipse.

An ellipse with center at the origin is shown at the right. Note that there are two x-intercepts and two y-intercepts.

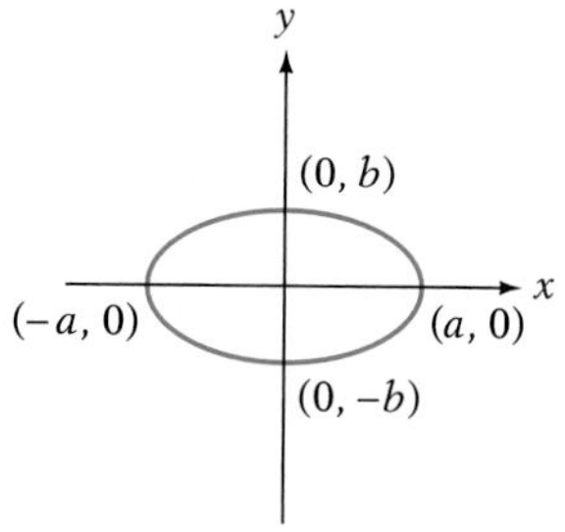

POINT OF INTEREST

The word *ellipse* comes from the Greek word *ellipsis,* which means "deficient." The method by which the early Greeks analyzed the conics caused a certain area in the construction of the ellipse to be less than another area (deficient). The word *ellipsis* in English, which means "omission," has the same Greek root as the word *ellipse.*

The Standard Form of the Equation of an Ellipse with Center at the Origin

The equation of an ellipse with center at the origin is $\frac{x^2}{a^2} + \frac{y^2}{b^2} = 1$. The x-intercepts are $(a, 0)$ and $(-a, 0)$. The y-intercepts are $(0, b)$ and $(0, -b)$.

By finding the x- and y-intercepts for an ellipse and using the fact that the ellipse is "oval" shaped, we can sketch a graph of an ellipse.

Example 1 Sketch a graph of the ellipse given by the equation.

A. $\dfrac{x^2}{9} + \dfrac{y^2}{4} = 1$ B. $\dfrac{x^2}{16} + \dfrac{y^2}{12} = 1$

Solution A. $\dfrac{x^2}{9} + \dfrac{y^2}{4} = 1$ ▶ $a^2 = 9, b^2 = 4$

x-intercepts:
$(3, 0)$ and $(-3, 0)$

▶ The x-intercepts are $(a, 0)$ and $(-a, 0)$.

y-intercepts:
$(0, 2)$ and $(0, -2)$

▶ The y-intercepts are $(0, b)$ and $(0, -b)$.

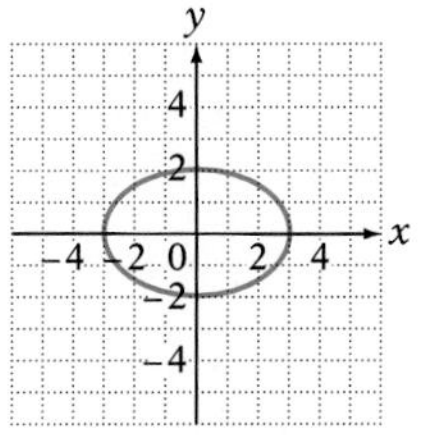

▶ Use the intercepts and symmetry to sketch the graph of the ellipse.

B. $\dfrac{x^2}{16} + \dfrac{y^2}{12} = 1$ ▶ $a^2 = 16, b^2 = 12$

x-intercepts:
$(4, 0)$ and $(-4, 0)$

▶ The x-intercepts are $(a, 0)$ and $(-a, 0)$.

y-intercepts:
$(0, 2\sqrt{3})$ and $(0, -2\sqrt{3})$

▶ The y-intercepts are $(0, b)$ and $(0, -b)$.

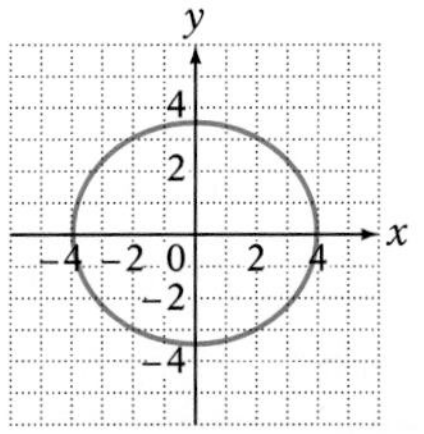

▶ Use the intercepts and symmetry to sketch the graph of the ellipse. $2\sqrt{3} \approx 3.5$

Problem 1 Sketch a graph of the ellipse given by the equation.

A. $\dfrac{x^2}{4} + \dfrac{y^2}{25} = 1$ B. $\dfrac{x^2}{18} + \dfrac{y^2}{9} = 1$

Solution See page S33.

A.

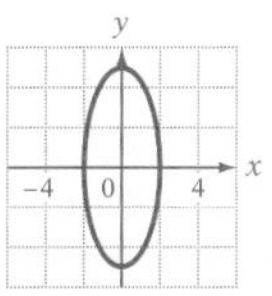

B.

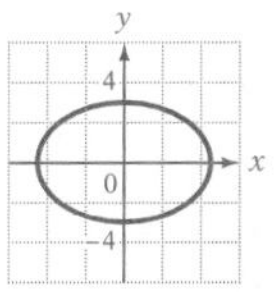

2 Graph a hyperbola with center at the origin

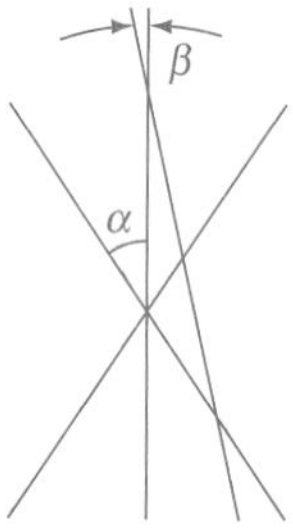

A **hyperbola** is a conic section that is formed by the intersection of a right circular cone and a plane perpendicular to the base of the cone.

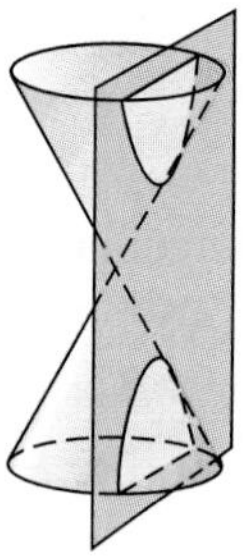

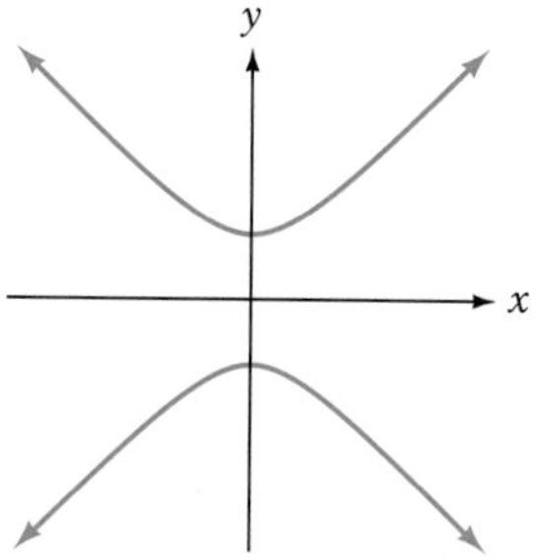

The hyperbola has two **vertices** and an **axis of symmetry** that passes through the vertices. The **center** of a hyperbola is the point halfway between the vertices.

The graphs below show two graphs of a hyperbola with center at the origin.

In the first graph, an axis of symmetry is the x-axis and the vertices are x-intercepts.

In the second graph, an axis of symmetry is the y-axis and the vertices are y-intercepts.

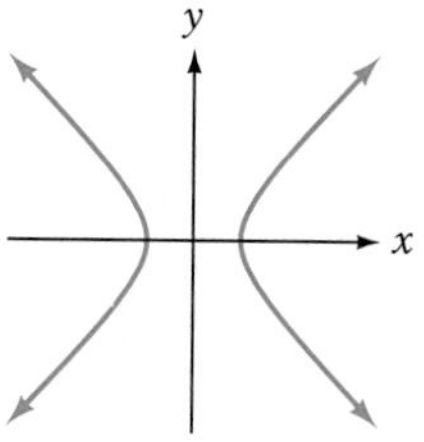

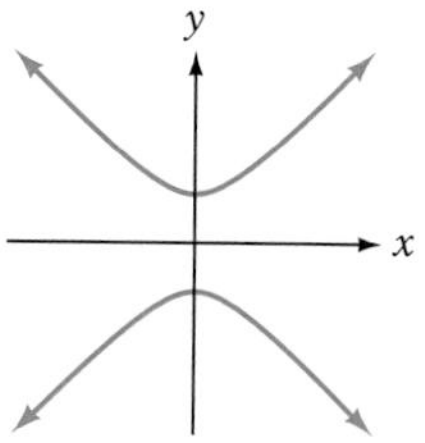

Note that in either case, the graph of a hyperbola is not the graph of a function. The graph of a hyperbola is the graph of a relation.

The Standard Form of the Equation of a Hyperbola with Center at the Origin

The equation of a hyperbola for which an axis of symmetry is the x-axis is $\frac{x^2}{a^2} - \frac{y^2}{b^2} = 1$. The vertices are $(a, 0)$ and $(-a, 0)$.

The equation of a hyperbola for which an axis of symmetry is the y-axis is $\frac{y^2}{b^2} - \frac{x^2}{a^2} = 1$. The vertices are $(0, b)$ and $(0, -b)$.

To sketch a hyperbola, it is helpful to draw two lines that are "approached" by the hyperbola. These two lines are called **asymptotes**. As the hyperbola gets farther from the origin, the hyperbola "gets closer to" the asymptotes.

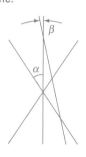

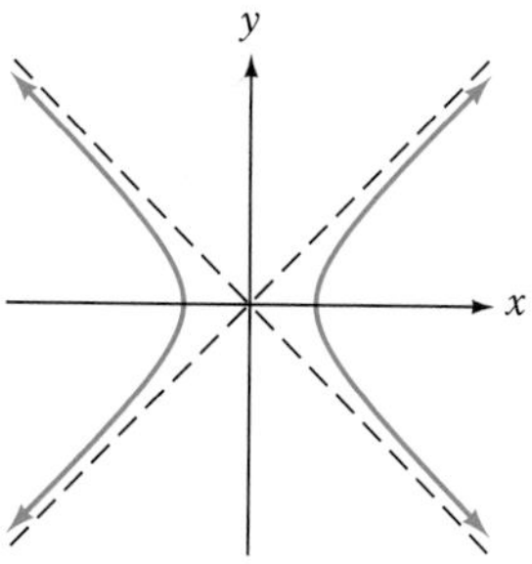

Because the asymptotes are straight lines, their equations are linear equations. The equations of the asymptotes for a hyperbola with center at the origin are $y = \frac{b}{a}x$ and $y = -\frac{b}{a}x$.

Example 2 Sketch a graph of the hyperbola given by the equation.

A. $\frac{x^2}{16} - \frac{y^2}{4} = 1$ B. $\frac{y^2}{16} - \frac{x^2}{25} = 1$

Solution A. $\frac{x^2}{16} - \frac{y^2}{4} = 1$ ▶ $a^2 = 16, b^2 = 4$

Axis of symmetry:
x-axis

Vertices:
$(4, 0)$ and $(-4, 0)$ ▶ The vertices are $(a, 0)$ and $(-a, 0)$.

Asymptotes:
$y = \frac{1}{2}x$ and $y = -\frac{1}{2}x$ ▶ The asymptotes are $y = \frac{b}{a}x$ and $y = -\frac{b}{a}x$.

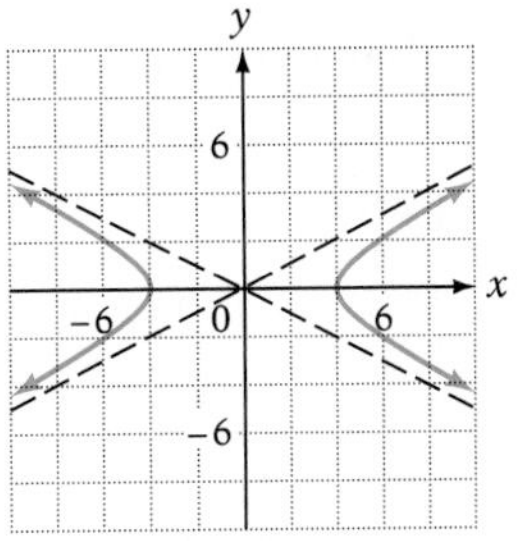

▶ Sketch the asymptotes. Use symmetry and the fact that the hyperbola will approach the asymptotes to sketch its graph.

B. $\frac{y^2}{16} - \frac{x^2}{25} = 1$ ▶ $a^2 = 25, b^2 = 16$

Axis of symmetry:
y-axis

Vertices:
$(0, 4)$ and $(0, -4)$ ▶ The vertices are $(0, b)$ and $(0, -b)$.

Asymptotes:
$y = \frac{4}{5}x$ and $y = -\frac{4}{5}x$ ▶ The asymptotes are $y = \frac{b}{a}x$ and $y = -\frac{b}{a}x$.

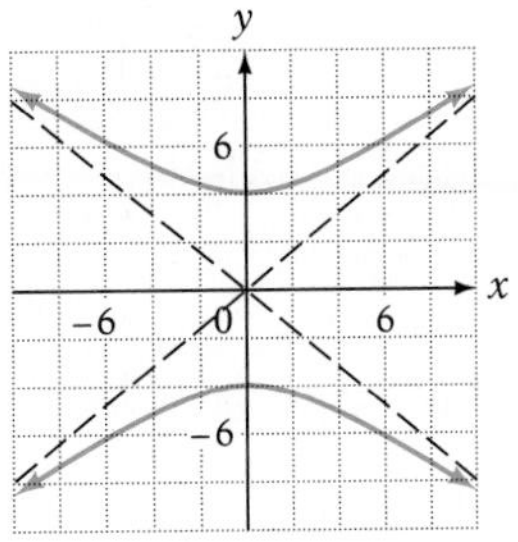

▶ Sketch the asymptotes. Use symmetry and the fact that the hyperbola will approach the asymptotes to sketch its graph.

Problem 2 Sketch a graph of the hyperbola given by the equation.

A. $\frac{x^2}{9} - \frac{y^2}{25} = 1$ B. $\frac{y^2}{9} - \frac{x^2}{9} = 1$

Solution See page S33–S34.

A.

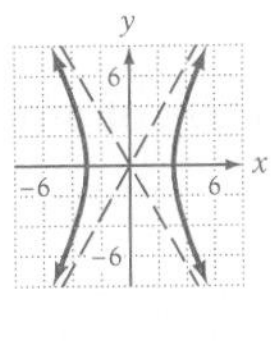

B.

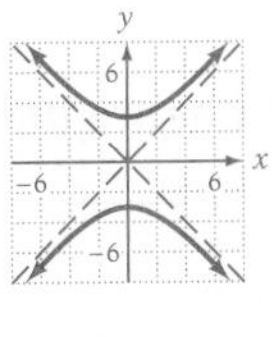

CONCEPT REVIEW 12.3

Determine whether the following statements are always true, sometimes true, or never true.

1. The graph of an ellipse is the graph of a function. Never true

2. The graph of a hyperbola is the graph of a function. Never true

3. An ellipse has two axes of symmetry. Always true

4. An ellipse with center at the origin has two x-intercepts. Always true

5. A hyperbola with center at the origin has two x-intercepts. Sometimes true

6. $4x^2 - y^2 = 16$ is the equation of a hyperbola. Always true

EXERCISES 12.3

1 Sketch the graph of the ellipse given by the equation.

1. $\frac{x^2}{4} + \frac{y^2}{9} = 1$

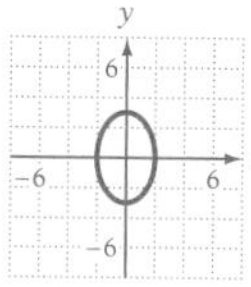

2. $\frac{x^2}{25} + \frac{y^2}{16} = 1$

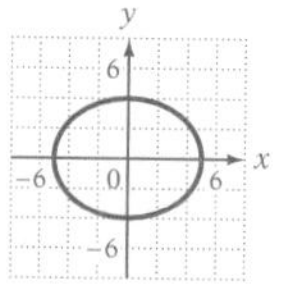

3. $\frac{x^2}{25} + \frac{y^2}{9} = 1$

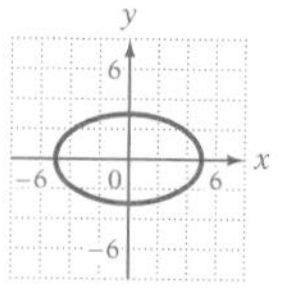

4. $\frac{x^2}{16} + \frac{y^2}{9} = 1$

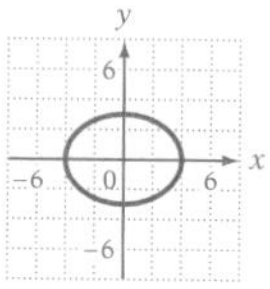

5. $\frac{x^2}{36} + \frac{y^2}{16} = 1$

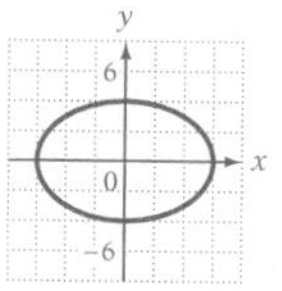

6. $\frac{x^2}{49} + \frac{y^2}{64} = 1$

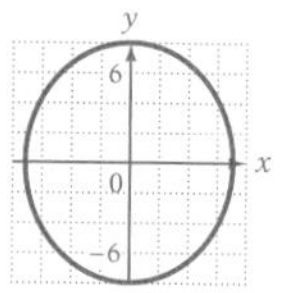

7. $\frac{x^2}{9} + \frac{y^2}{25} = 1$

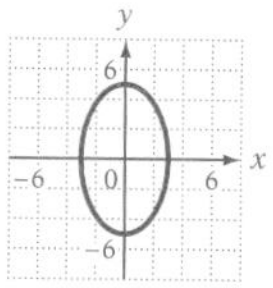

8. $\frac{x^2}{8} + \frac{y^2}{25} = 1$

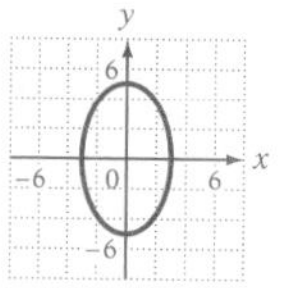

9. $\frac{x^2}{12} + \frac{y^2}{4} = 1$

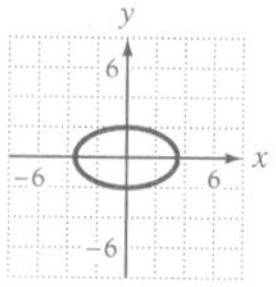

10. $\frac{x^2}{16} + \frac{y^2}{36} = 1$

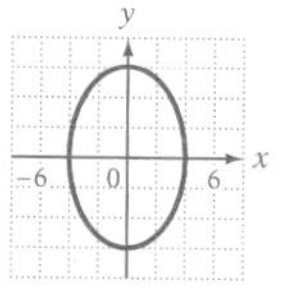

11. $\frac{x^2}{36} + \frac{y^2}{9} = 1$

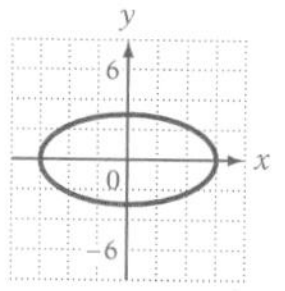

12. $\frac{x^2}{4} + \frac{y^2}{16} = 1$

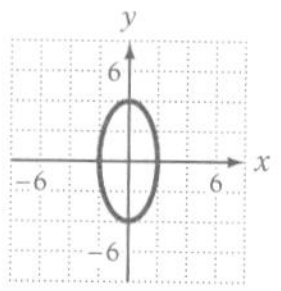

2 13. How can you tell from an equation whether the graph will be that of an ellipse or the graph of a hyperbola?

14. What are the asymptotes of a hyperbola?

Sketch a graph of the hyperbola given by the equation.

15. $\frac{x^2}{9} - \frac{y^2}{16} = 1$

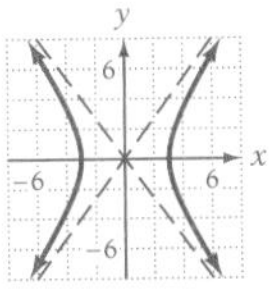

16. $\frac{x^2}{25} - \frac{y^2}{4} = 1$

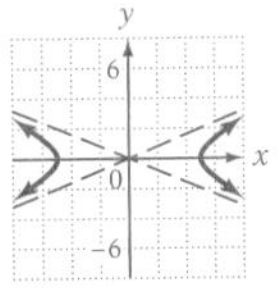

17. $\frac{y^2}{16} - \frac{x^2}{9} = 1$

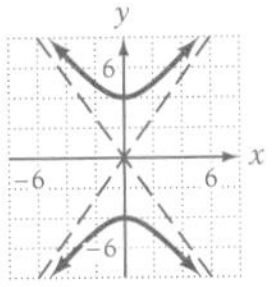

18. $\frac{y^2}{4} - \frac{x^2}{9} = 1$

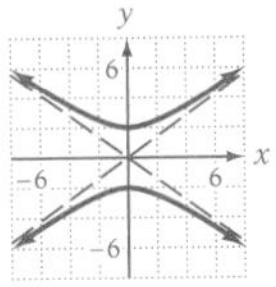

19. $\frac{x^2}{4} - \frac{y^2}{25} = 1$

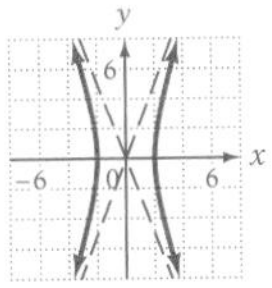

20. $\frac{x^2}{9} - \frac{y^2}{49} = 1$

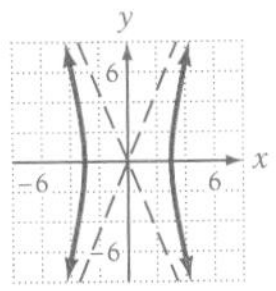

21. $\frac{y^2}{25} - \frac{x^2}{9} = 1$

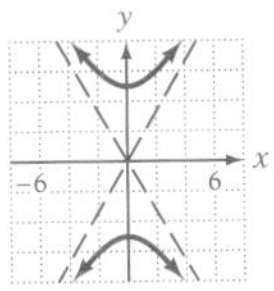

22. $\frac{y^2}{4} - \frac{x^2}{16} = 1$

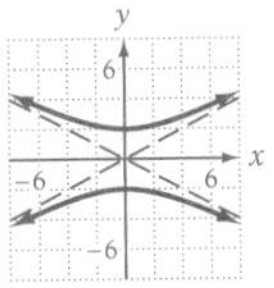

23. $\frac{x^2}{25} - \frac{y^2}{16} = 1$

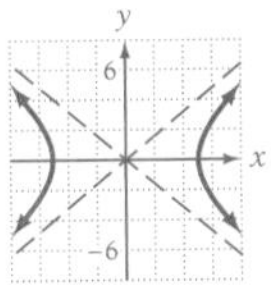

24. $\frac{x^2}{9} - \frac{y^2}{9} = 1$

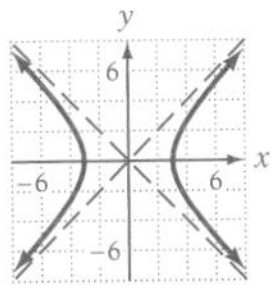

25. $\frac{y^2}{16} - \frac{x^2}{4} = 1$

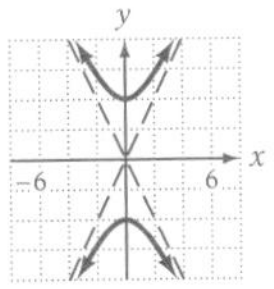

26. $\frac{y^2}{9} - \frac{x^2}{36} = 1$

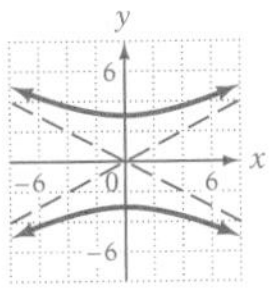

27. $\frac{x^2}{25} - \frac{y^2}{9} = 1$

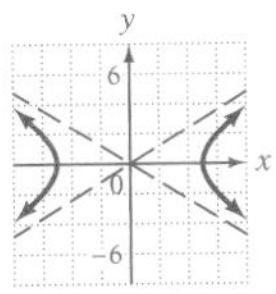

28. $\frac{x^2}{16} - \frac{y^2}{25} = 1$

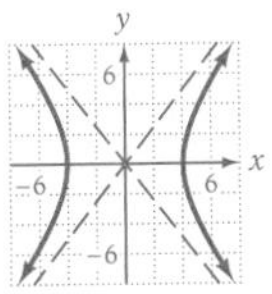

APPLYING CONCEPTS 12.3

Sketch a graph of the conic section given by the equation. (*Hint:* Divide each term by the number on the right side of the equation.)

29. $4x^2 + y^2 = 16$

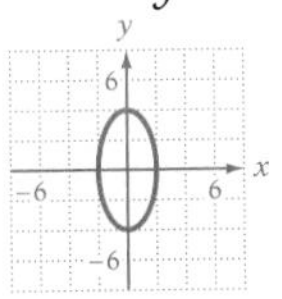

30. $x^2 - y^2 = 9$

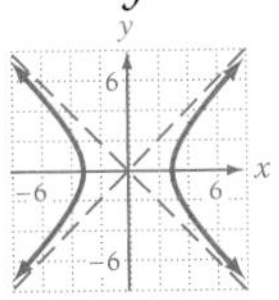

31. $y^2 - 4x^2 = 16$

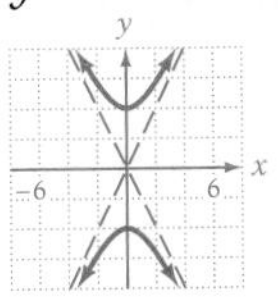

32. $9x^2 + 4y^2 = 144$

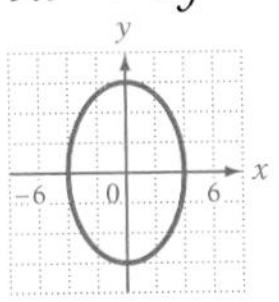

33. $9x^2 - 25y^2 = 225$

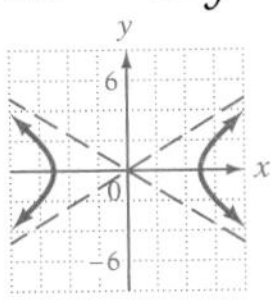

34. $4y^2 - x^2 = 36$

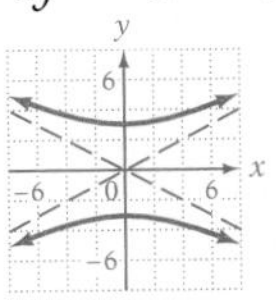

35. *Astronomy* The orbit of Halley's comet is an ellipse with a major axis of approximately 36 AU and a minor axis of approximately 9 AU. (1 AU is 1 astronomical unit and is approximately 92,960,000 mi, the average distance of Earth from the sun.)

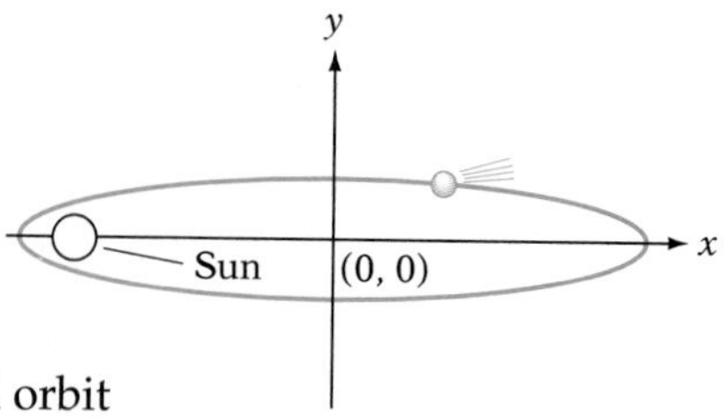

a. Determine an equation for the orbit of Halley's comet in terms of astronomical units. See the diagram at the right.

b. The distance of the sun from the center of Halley's comet's elliptical orbit is $\sqrt{a^2 - b^2}$. The aphelion of the orbit (the point at which the comet is farthest from the sun) is a vertex on the major axis. Determine the distance, to the nearest hundred-thousand miles, from the sun to the point at the aphelion of Halley's comet.

c. The perihelion of the orbit (the point at which the comet is closest to the sun) is a vertex on the major axis. Determine the distance, to the nearest hundred-thousand miles, from the sun to the point at the aphelion of Halley's comet.

a. $\frac{x^2}{324} + \frac{y^2}{20.25} = 1$
b. 3,293,400,000 mi
c. 53,100,000 mi

36. *Astronomy* The orbit of the comet Hale–Bopp is an ellipse as shown at the right. The units are astronomical units (abbreviated AU). 1 AU ≈ 92,960,000 mi.

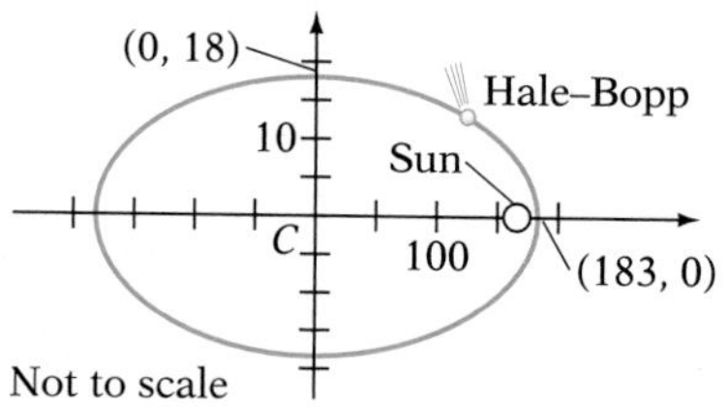

a. Find the equation of the orbit of the comet. $\frac{x^2}{33{,}489} + \frac{y^2}{324} = 1$

b. The distance from the center, *C*, of the orbit to the sun is approximately 182.085 AU. Find the aphelion (the point at which the comet is farthest from the sun) in miles. Round to the nearest million miles. 33,938,000,000 mi

c. Find the perihelion (the point at which the comet is closest to the sun) in miles. Round to the nearest hundred-thousand miles. 85,100,000 mi

37. *Astronomy* As mentioned in the historical feature at the beginning of this chapter, the orbits of the planets are ellipses. The length of the major axis of Mars' orbit is 3.04 AU (see Exercise 35), and the length of the minor axis is 2.99 AU.

a. Determine an equation for the orbit of Mars. $\frac{x^2}{2.310} + \frac{y^2}{2.235} = 1$

b. Determine the aphelion to the nearest hundred-thousand miles. 166,800,000 mi

c. Determine the perihelion to the nearest hundred-thousand miles. 115,800,000 mi

38. Prepare a report on the system of navigation called loran (long-range navigation).

39. Besides the curves presented in this chapter, how else might the intersection of a plane and a cone be represented?

40. Explain why neither the equation of an ellipse nor the equation of a hyperbola represents a function.

S E C T I O N **12.4**

Solving Nonlinear Systems of Equations

1 Solve nonlinear systems of equations

A **nonlinear system of equations** is one in which one or more equations of the system are not linear equations. Some examples of nonlinear systems and their graphs are shown here.

$$x^2 + y^2 = 4$$
$$y = x^2 + 2$$

The graphs intersect at one point.
The system of equations has one solution.

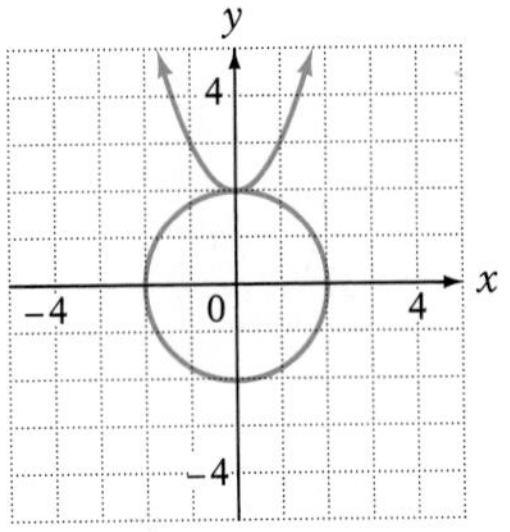

$$y = x^2$$
$$y = -x + 2$$

The graphs intersect at two points.
The system of equations has two solutions.

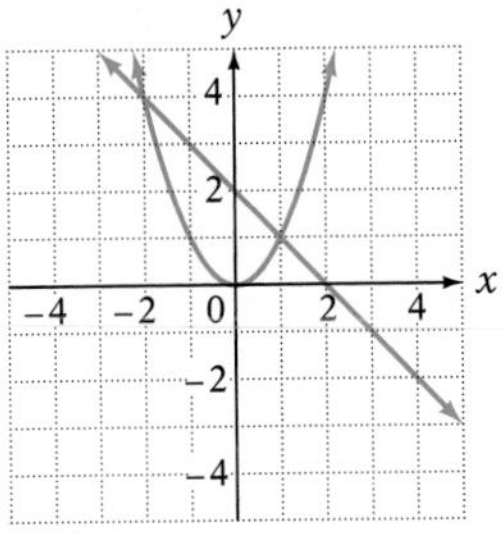

$$(x + 2)^2 + (y - 2)^2 = 4$$
$$x = y^2$$

The graphs do not intersect.
The system of equations has no solutions.

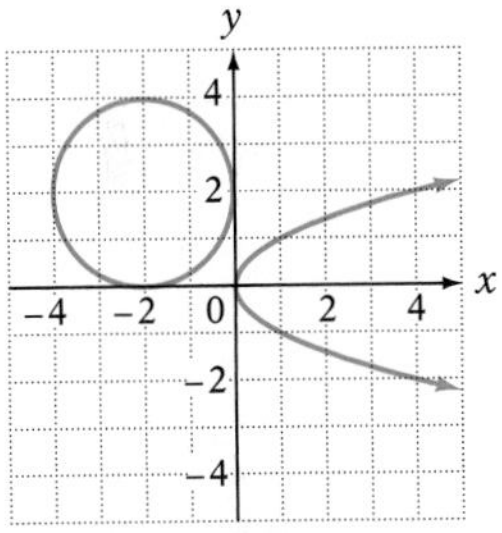

Nonlinear systems of equations can be solved by using either a substitution method or an addition method.

⇒ Solve: $2x - y = 4$ (1)
$\quad\quad\quad y^2 = 4x$ (2)

When a system contains both a linear and a quadratic equation, the substitution method is used.

Solve equation (1) for y.

$$2x - y = 4$$
$$-y = -2x + 4$$
$$y = 2x - 4$$

INSTRUCTOR NOTE

As a general rule, a system of equations that contains both a quadratic equation and a linear equation is solved by the substitution method. If both equations of the system are quadratic, then the addition method is used.

Substitute $2x - 4$ for y into equation (2).

$$y^2 = 4x$$
$$(2x - 4)^2 = 4x$$

Write the quadratic equation in standard form.

$$4x^2 - 16x + 16 = 4x$$
$$4x^2 - 20x + 16 = 0$$

Solve for x by factoring.

$$4(x^2 - 5x + 4) = 0$$
$$4(x - 4)(x - 1) = 0$$

$$x - 4 = 0 \qquad x - 1 = 0$$
$$x = 4 \qquad x = 1$$

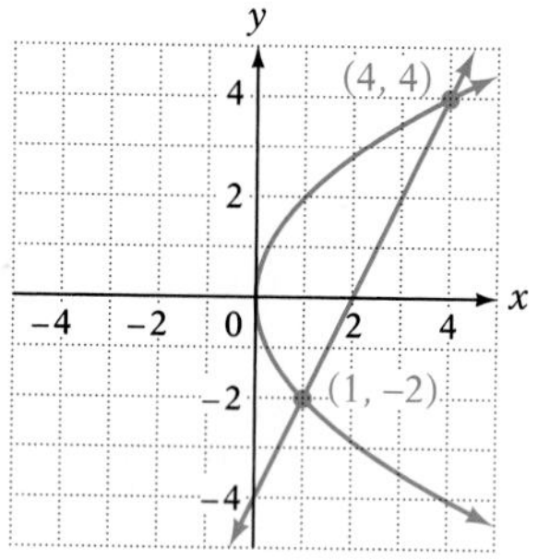

Substitute the values of x into the equation $y = 2x - 4$, and solve for y.

$$y = 2x - 4 \qquad y = 2x - 4$$
$$y = 2(4) - 4 \qquad y = 2(1) - 4$$
$$y = 4 \qquad y = -2$$

The solutions are $(4, 4)$ and $(1, -2)$.

The graph of the system that was solved above is shown at the left. Note that the line intersects the parabola at two points. These points correspond to the solutions.

Solve: $x^2 + y^2 = 4$ (1)
$y = x + 4$ (2)

The system of equations contains a linear equation. The substitution method is used to solve the system.

Substitute the expression for y into equation (1).

$$x^2 + y^2 = 4$$
$$x^2 + (x + 4)^2 = 4$$

Write the equation in standard form.

$$x^2 + x^2 + 8x + 16 = 4$$
$$2x^2 + 8x + 16 = 4$$
$$2x^2 + 8x + 12 = 0$$

Because the discriminant of the quadratic equation is less than zero, the equation has two complex number solutions. Therefore, the system of equations has no real number solutions.

$$b^2 - 4ac = 8^2 - 4(2)(12)$$
$$= 64 - 96$$
$$= -32$$

The graph of the system of equations that was solved above is shown at the right. Note that the two graphs do not intersect.

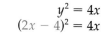

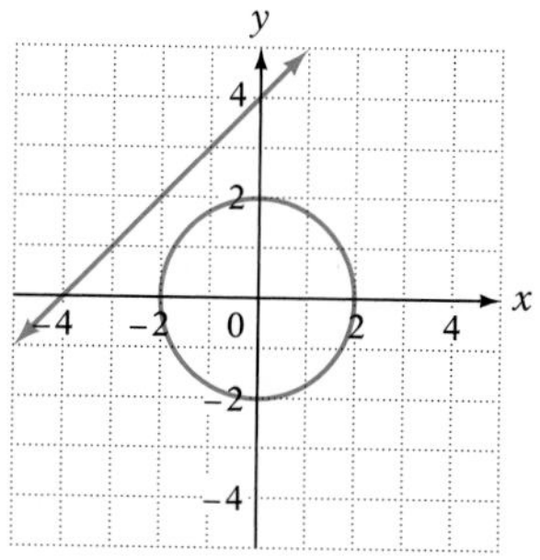

⟹ Solve: $4x^2 + y^2 = 16$ (1)
$\qquad\quad x^2 + y^2 = 4$ (2)

Use the addition method to solve this system of equations.

Multiply equation (2) by -1 and add it to equation (1).

$$\begin{aligned} 4x^2 + y^2 &= 16 \\ -x^2 - y^2 &= -4 \\ \hline 3x^2 &= 12 \end{aligned}$$

Solve for x.

$$x^2 = 4$$
$$x = \pm 2$$

Substitute the values of x into equation (2), and solve for y.

$$\begin{aligned} x^2 + y^2 &= 4 & x^2 + y^2 &= 4 \\ 2^2 + y^2 &= 4 & (-2)^2 + y^2 &= 4 \\ y^2 &= 0 & y^2 &= 0 \\ y &= 0 & y &= 0 \end{aligned}$$

The solutions are $(2, 0)$ and $(-2, 0)$. ⟸

LOOK CLOSELY

Note from the examples in this section that the number of points at which the graphs of the equations of the system intersect is the same as the number of real number solutions of the system of equations.

The graph of the system that was solved above is shown at the right. Note that the graphs intersect at two points.

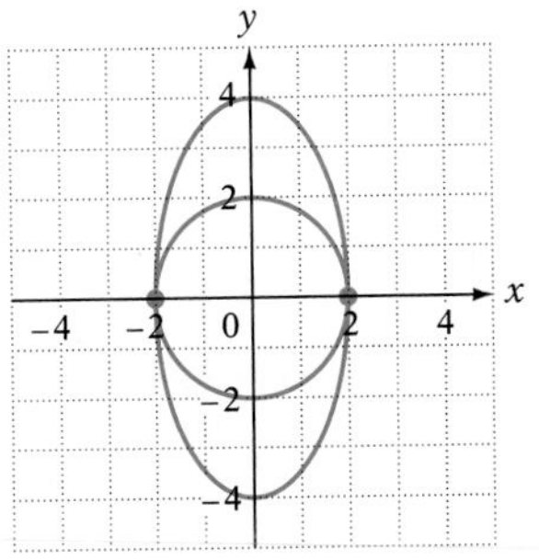

Example 1   Solve.   A. $y = 2x^2 - 3x - 1$ B. $3x^2 - 2y^2 = 26$
$\qquad\qquad\qquad\qquad\quad y = x^2 - 2x + 5$ $\qquad\quad x^2 - y^2 = 5$

Solution   A. (1) $\qquad\qquad\quad y = 2x^2 - 3x - 1$
$\qquad\qquad$ (2) $\qquad\qquad\quad y = x^2 - 2x + 5$

$$\begin{aligned} 2x^2 - 3x - 1 &= x^2 - 2x + 5 \\ x^2 - x - 6 &= 0 \\ (x - 3)(x + 2) &= 0 \end{aligned}$$ ▶ Use the substitution method.

$$\begin{aligned} x - 3 &= 0 & x + 2 &= 0 \\ x &= 3 & x &= -2 \end{aligned}$$

$$\begin{aligned} y &= 2x^2 - 3x - 1 \\ y &= 2(3)^2 - 3(3) - 1 \\ y &= 18 - 9 - 1 \\ y &= 8 \end{aligned}$$ ▶ Substitute the values of x into equation (1).

$$\begin{aligned} y &= 2x^2 - 3x - 1 \\ y &= 2(-2)^2 - 3(-2) - 1 \\ y &= 8 + 6 - 1 \\ y &= 13 \end{aligned}$$

The solutions are $(3, 8)$ and $(-2, 13)$.

B. (1) $3x^2 - 2y^2 = 26$
    (2) $x^2 - y^2 = 5$

$$3x^2 - 2y^2 = 26$$
$$-2x^2 + 2y^2 = -10$$
$$x^2 = 16$$
$$x = \pm 4$$

▶ Use the addition method. Multiply equation (2) by -2.

$$x^2 - y^2 = 5 \qquad\qquad x^2 - y^2 = 5$$
$$4^2 - y^2 = 5 \qquad\qquad (-4)^2 - y^2 = 5$$
$$16 - y^2 = 5 \qquad\qquad 16 - y^2 = 5$$
$$-y^2 = -11 \qquad\qquad -y^2 = -11$$
$$y^2 = 11 \qquad\qquad y^2 = 11$$
$$y = \pm\sqrt{11} \qquad\qquad y = \pm\sqrt{11}$$

▶ Substitute the values of x into equation (2).

The solutions are $(4, \sqrt{11})$, $(4, -\sqrt{11})$, $(-4, \sqrt{11})$, and $(-4, -\sqrt{11})$.

Problem 1   Solve.   A. $y = 2x^2 + x - 3$ B. $x^2 - y^2 = 10$
                              $y = 2x^2 - 2x + 9$ $x^2 + y^2 = 8$

Solution   See page S34.   A. $(4, 33)$ B. No solution

CONCEPT REVIEW 12.4

Determine whether the following statements are always true, sometimes true, or never true.

1. It is possible for two ellipses with centers at the origin to intersect in three points.   Never true

2. Two circles will intersect in two points.   Sometimes true

3. A straight line will intersect a parabola in two points.   Sometimes true

4. An ellipse and a circle will intersect in four points.   Sometimes true

5. When the graphs of the equations of a system of equations do not intersect, the system has no solution.   Always true

6. Two circles will intersect in four points.   Never true

EXERCISES 12.4

1. How do nonlinear systems of equations differ from linear systems of equations?

2. What types of methods are used to solve nonlinear systems of equations?

Solve.

3. $y = x^2 - x - 1$
$y = 2x + 9$
$(-2, 5), (5, 19)$

4. $y = x^2 - 3x + 1$
$y = x + 6$
$(5, 11), (-1, 5)$

5. $y^2 = -x + 3$
$x - y = 1$
$(-1, -2), (2, 1)$

6. $y^2 = 4x$
$x - y = -1$
$(1, 2)$

7. $y^2 = 2x$
$x + 2y = -2$
$(2, -2)$

8. $y^2 = 2x$
$x - y = 4$
$(2, -2), (8, 4)$

9. $x^2 + 2y^2 = 12$
$2x - y = 2$
$(2, 2), \left(-\frac{2}{9}, -\frac{22}{9}\right)$

10. $x^2 + 4y^2 = 37$
$x - y = -4$
$\left(-\frac{27}{5}, -\frac{7}{5}\right), (-1, 3)$

11. $x^2 + y^2 = 13$
$x + y = 5$
$(3, 2), (2, 3)$

12. $x^2 + y^2 = 16$
$x - 2y = -4$
$(-4, 0), \left(\frac{12}{5}, \frac{16}{5}\right)$

13. $4x^2 + y^2 = 12$
$y = 4x^2$
$\left(\frac{\sqrt{3}}{2}, 3\right), \left(-\frac{\sqrt{3}}{2}, 3\right)$

14. $2x^2 + y^2 = 6$
$y = 2x^2$
$(1, 2), (-1, 2)$

15. $y = x^2 - 2x - 3$
$y = x - 6$
No solution

16. $y = x^2 + 4x + 5$
$y = -x - 3$
No solution

17. $3x^2 - y^2 = -1$
$x^2 + 4y^2 = 17$
$(1, 2), (1, -2), (-1, 2), (-1, -2)$

18. $x^2 + y^2 = 10$
$x^2 + 9y^2 = 18$
$(3, 1), (-3, 1), (3, -1), (-3, -1)$

19. $2x^2 + 3y^2 = 30$
$x^2 + y^2 = 13$
$(3, 2), (3, -2), (-3, 2), (-3, -2)$

20. $x^2 + y^2 = 61$
$x^2 - y^2 = 11$
$(6, 5), (6, -5), (-6, 5), (-6, -5)$

21. $y = 2x^2 - x + 1$
$y = x^2 - x + 5$
$(2, 7), (-2, 11)$

22. $y = -x^2 + x - 1$
$y = x^2 + 2x - 2$
$\left(\frac{1}{2}, -\frac{3}{4}\right), (-1, -3)$

23. $2x^2 + 3y^2 = 24$
$x^2 - y^2 = 7$
$(3, \sqrt{2}), (3, -\sqrt{2}),$
$(-3, \sqrt{2}), (-3, -\sqrt{2})$

24. $2x^2 + 3y^2 = 21$
$x^2 + 2y^2 = 12$
$(\sqrt{6}, \sqrt{3}), (\sqrt{6}, -\sqrt{3}),$
$(-\sqrt{6}, \sqrt{3}), (-\sqrt{6}, -\sqrt{3})$

25. $x^2 + y^2 = 36$
$4x^2 + 9y^2 = 36$
No solution

26. $2x^2 + 3y^2 = 12$
$x^2 - y^2 = 25$
No solution

27. $11x^2 - 2y^2 = 4$
$3x^2 + y^2 = 15$
$(\sqrt{2}, 3), (\sqrt{2}, -3),$
$(-\sqrt{2}, 3), (-\sqrt{2}, -3)$

28. $x^2 + 4y^2 = 25$
$x^2 - y^2 = 5$
$(3, 2), (3, -2), (-3, 2), (-3, -2)$

29. $2x^2 - y^2 = 7$
$2x - y = 5$
$(2, -1), (8, 11)$

30. $3x^2 + 4y^2 = 7$
$x - 2y = -3$
$(-1, 1), \left(-\frac{1}{2}, \frac{5}{4}\right)$

31. $y = 3x^2 + x - 4$
$y = 3x^2 - 8x + 5$
$(1, 0)$

32. $y = 2x^2 + 3x + 1$
$y = 2x^2 + 9x + 7$
$(-1, 0)$

APPLYING CONCEPTS 12.4

Solve by graphing. Approximate the solutions of the systems to the nearest thousandth.

33. $y = 2^x$
$x + y = 3$
$(1.000, 2.000)$

34. $y = 3^{-x}$
$x^2 + y^2 = 9$
$(3.000, 0.037), (-0.952, 2.845)$

35. $y = \log_2 x$
$\frac{x^2}{9} + \frac{y^2}{1} = 1$
$(1.755, 0.811), (0.505, -0.986)$

36. $y = \log_3 x$
$x^2 + y^2 = 4$
(1.911, 0.590), (0.111, −1.997)

37. $y = -\log_3 x$
$x + y = 4$
(5.562, −1.562), (0.013, 3.987)

38. $y = \left(\frac{1}{2}\right)^x$
$\frac{x^2}{9} + \frac{y^2}{4} = 1$
(−0.928, 1.902), (2.994, 0.126)

39. Is it possible for two circles with centers at the origin to intersect at exactly two points?

40. Graph $xy > 1$ and $y > \frac{1}{x}$ on different coordinate grids. Dividing each side of $xy > 1$ by x yields $y > \frac{1}{x}$, but the graphs are not the same. Explain.

SECTION 12.5

Quadratic Inequalities and Systems of Inequalities

1 Graph the solution set of a quadratic inequality in two variables

The **graph of a quadratic inequality in two variables** is a region of the plane that is bounded by one of the conic sections (parabola, circle, ellipse, or hyperbola). When graphing an inequality of this type, first replace the inequality symbols with an equals sign. Graph the resulting conic using a dashed curve when the original inequality is less than (<) or greater than (>). Use a solid curve when the original inequality is ≤ or ≥. Use the point (0, 0) to determine which region of the plane to shade. If (0, 0) is a solution of the inequality, then shade the region of the plane containing (0, 0). If not, shade the other portion of the plane.

Graph the solution set of $x^2 + y^2 > 9$.

Change the inequality to an equality.

$x^2 + y^2 > 9$
$x^2 + y^2 = 9$

This is the equation of a circle with center (0, 0) and radius 3.

Because the inequality is >, the graph is drawn as a dashed circle.

Substitute the point (0, 0) into the inequality. Because $0^2 + 0^2 > 9$ is not true, the point (0, 0) should not be in the shaded region.

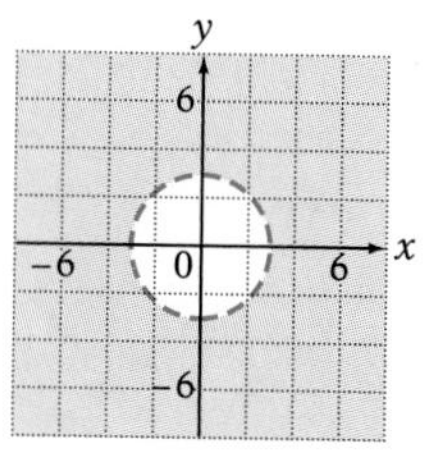

Example 1 Graph the solution set.

A. $y \le x^2 + 2x + 2$ B. $\frac{y^2}{9} - \frac{x^2}{4} \ge 1$

Solution A. $y \le x^2 + 2x + 2$

$y = x^2 + 2x + 2$

▶ Change the inequality to an equality. This is the equation of a parabola that opens up. The vertex is $(-1, 1)$. The axis of symmetry is the line $x = -1$.

▶ Because the inequality is $\le$, the graph is drawn as a solid line.

▶ Substitute the point $(0, 0)$ into the inequality. Because $0 < 0^2 + 2(0) + 2$ is true, the point $(0, 0)$ should be in the shaded region.

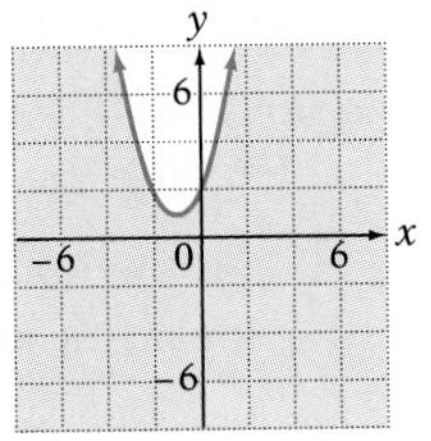

B. $\frac{y^2}{9} - \frac{x^2}{4} \ge 1$

$\frac{y^2}{9} - \frac{x^2}{4} = 1$

▶ Change the inequality to an equality. This is the equation of a hyperbola. The vertices are $(0, -3)$ and $(0, 3)$. The equations of the asymptotes are $y = \frac{3}{2}x$ and $y = -\frac{3}{2}x$.

▶ Because the inequality is $\ge$, the graph is drawn as a solid line.

▶ Substitute the point $(0, 0)$ into the inequality. Because $\frac{0^2}{9} - \frac{0^2}{4} \ge 1$ is not true, the point $(0, 0)$ should not be in the shaded region.

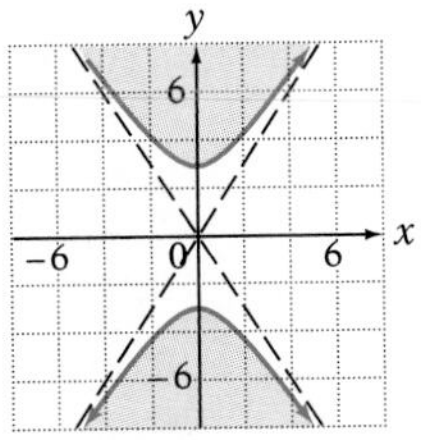

Problem 1 Graph the solution set.

A. $\frac{x^2}{9} + \frac{y^2}{16} \le 1$ B. $\frac{x^2}{9} - \frac{y^2}{4} \le 1$

Solution See page S34. A.

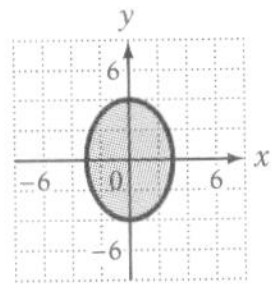

B.

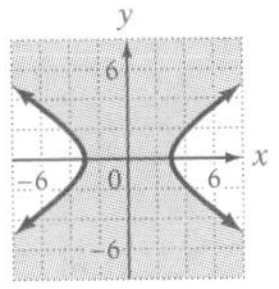

2 Graph the solution set of a nonlinear system of inequalities

INSTRUCTOR NOTE

Have students use parallel lines oriented like this //// to shade the solution set of one inequality and parallel lines oriented like this \\\\ to shade the solution set of the other inequality. The solution set of the system of inequalities is then the region defined by XXXX.

Recall that the solution set of a system of inequalities is the intersection of the solution sets of the individual inequalities. To graph the solution set of a system of inequalities, first graph the solution set for each inequality. The solution set of the system of inequalities is the region of the plane represented by the intersection of the two shaded regions.

Graph the solution set: $x^2 + y^2 \le 16$
$y \ge x^2$

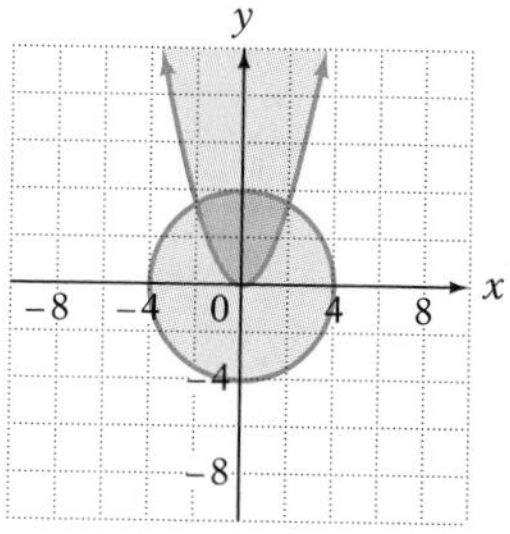

Graph the solution set of each inequality.

$x^2 + y^2 = 16$ is the equation of a circle with a radius of 4. Because the inequality is $\le$, use a solid line.

$y = x^2$ is the equation of a parabola that opens upward. Because the inequality is $\ge$, use a solid line.

Substitute the point (0, 0) into the inequality $x^2 + y^2 \le 16$. $0^2 + 0^2 \le 16$ is a true statement. Shade inside the circle.

Substitute (2, 0) into the inequality $y \ge x^2$. (We cannot use (0, 0), a point on the parabola.) $0 \ge 4$ is a false statement. Shade inside the parabola.

The solution is the intersection of the two shaded regions.

Example 2 Graph the solution set. A. $y > x^2$, $y < x + 2$ B. $\frac{x^2}{9} - \frac{y^2}{16} \ge 1$, $x^2 + y^2 \le 4$

Solution A.

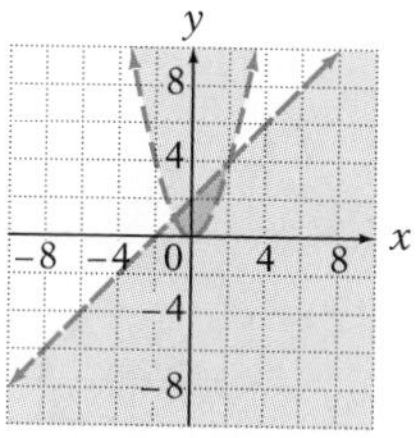

- Graph the solution set of each inequality.
- $y = x^2$ is the equation of a parabola. Use a dashed line. Shade inside the parabola.
- $y = x + 2$ is the equation of a line. Use a dashed line. Shade below the line.

The solution set is the region of the plane represented by the intersection of the solution sets of each inequality.

B.

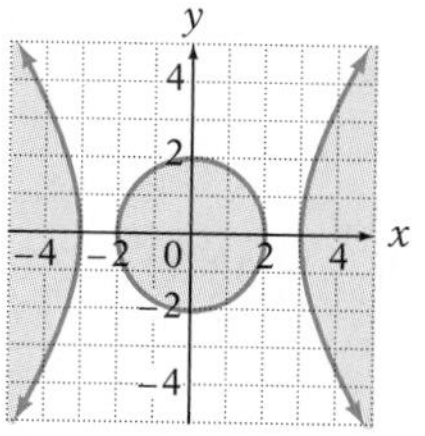

- Graph the solution set of each inequality.
- $\frac{x^2}{9} - \frac{y^2}{16} = 1$ is the equation of a hyperbola. Use a solid line. The point (0, 0) should not be in the shaded region.
- $x^2 + y^2 = 4$ is the equation of a circle. Use a solid line. Shade inside the circle.

The solution sets of the two inequalities do not intersect. This system of inequalities has no real number solution.

Problem 2 Graph the solution set. A. $\frac{x^2}{4} + \frac{y^2}{9} \le 1$, $x > y^2 - 2$ B. $\frac{x^2}{16} + \frac{y^2}{25} \ge 1$, $x^2 + y^2 < 9$

Solution See page S34.

A.

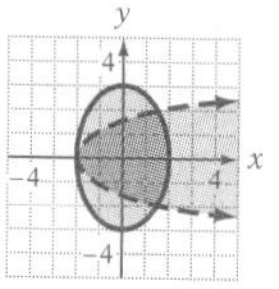

B.

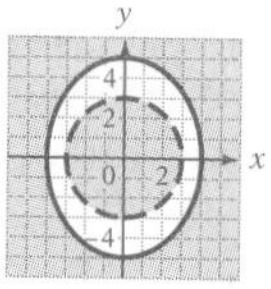

The solution sets of the two inequalities do not intersect. This system of inequalities has no real number solution.

CONCEPT REVIEW 12.5

Determine whether the following statements are always true, sometimes true, or never true.

1. The graph of a quadratic inequality in two variables is a region of the plane that is bounded by one of the conic sections. Always true

2. The solution set of a nonlinear system of inequalities is the union of the solution sets of the individual inequalities. Never true

3. The point (0, 0) is a solution of the inequality $x^2 + y^2 > 4$. Never true

4. The solution set of a quadratic inequality in two variables that is bounded by a hyperbola includes the asymptotes of the hyperbola. Sometimes true

5. The solution set of a quadratic inequality in two variables that is bounded by a conic section includes the boundary. Sometimes true

6. The solution set of a nonlinear system of inequalities is the empty set. Sometimes true

EXERCISES 12.5

1 Graph the solution set.

1. $y \le x^2 - 4x + 3$

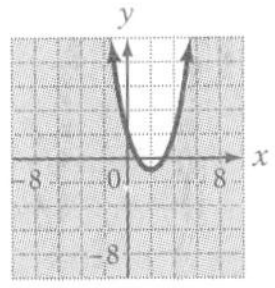

2. $y < x^2 - 2x - 3$

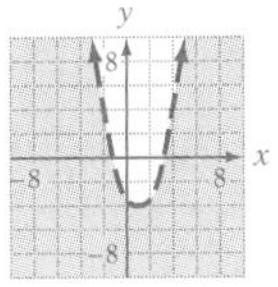

3. $(x - 1)^2 + (y + 2)^2 \le 9$

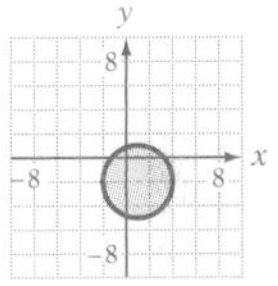

4. $(x + 2)^2 + (y - 3)^2 > 4$

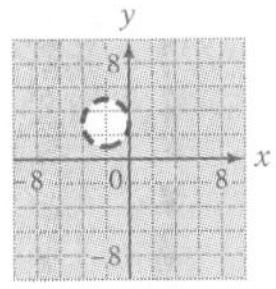

5. $(x + 3)^2 + (y - 2)^2 \ge 9$

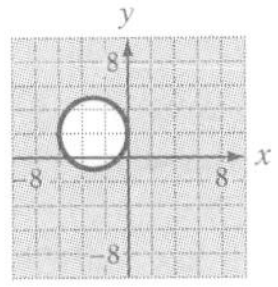

6. $(x - 2)^2 + (y + 1)^2 \le 16$

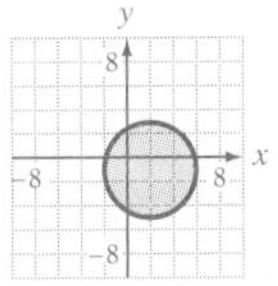

7. $\dfrac{x^2}{16} + \dfrac{y^2}{25} < 1$

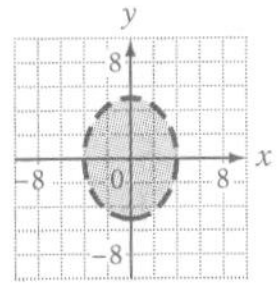

8. $\dfrac{x^2}{9} + \dfrac{y^2}{4} \ge 1$

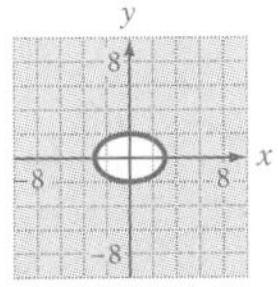

9. $\dfrac{x^2}{25} - \dfrac{y^2}{9} \le 1$

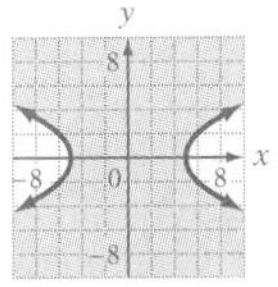

10. $\dfrac{y^2}{25} - \dfrac{x^2}{36} > 1$

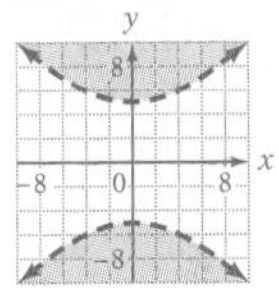

11. $\dfrac{x^2}{4} + \dfrac{y^2}{16} \ge 1$

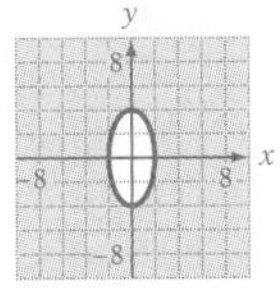

12. $\dfrac{x^2}{4} - \dfrac{y^2}{16} \le 1$

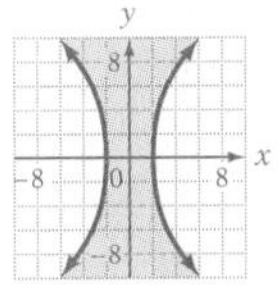

13. $y \le x^2 - 2x + 3$

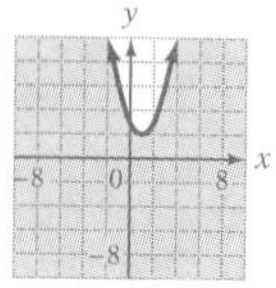

14. $x \le y^2 + 2y + 1$

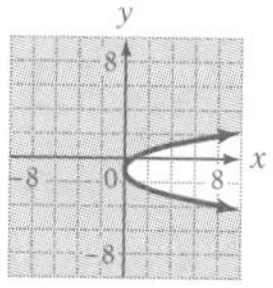

15. $\frac{y^2}{9} - \frac{x^2}{16} \le 1$

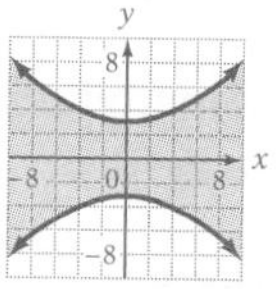

16. $\frac{x^2}{16} - \frac{y^2}{4} < 1$

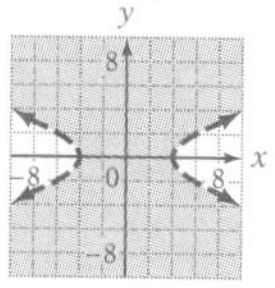

17. $\frac{x^2}{9} + \frac{y^2}{1} \le 1$

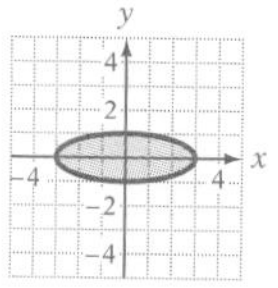

18. $\frac{x^2}{16} + \frac{y^2}{4} > 1$

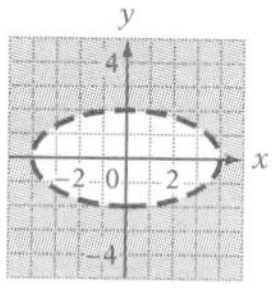

19. $(x - 1)^2 + (y + 3)^2 \le 25$

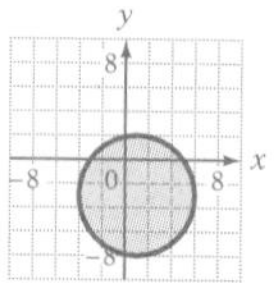

20. $(x + 1)^2 + (y - 2)^2 \ge 16$

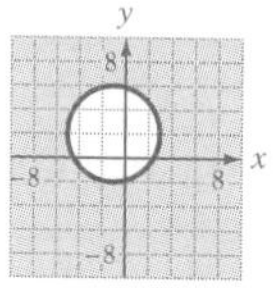

21. $\frac{y^2}{25} - \frac{x^2}{4} \le 1$

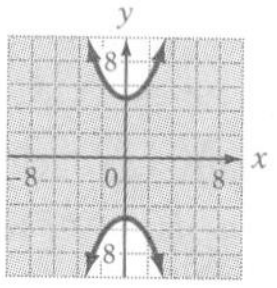

22. $\frac{x^2}{9} - \frac{y^2}{25} \ge 1$

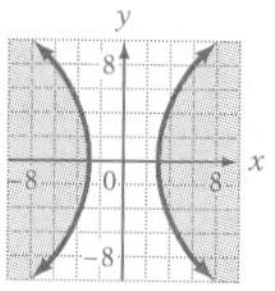

23. $\frac{x^2}{25} + \frac{y^2}{9} \le 1$

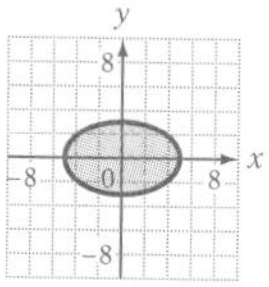

24. $\frac{x^2}{36} + \frac{y^2}{4} \le 1$

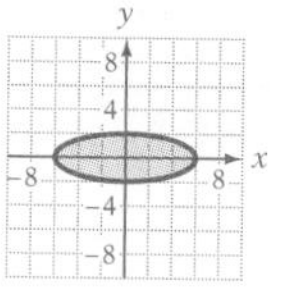

2 **25.** How can you determine which region to shade for the solution set of a nonlinear inequality?

26. How is the solution set drawn for a nonlinear inequality that uses $\le$ or $\ge$ different from the solution set drawn for a nonlinear inequality that uses $<$ or $>$?

Graph the solution set.

27. $y \le x^2 - 4x + 4$
$y + x > 4$

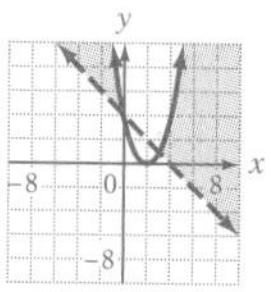

28. $x^2 + y^2 < 1$
$x + y \ge 4$

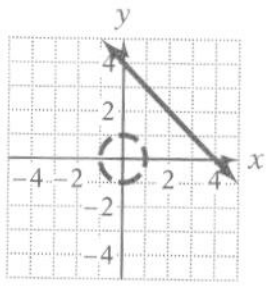

29. $x^2 + y^2 < 16$
$y > x + 1$

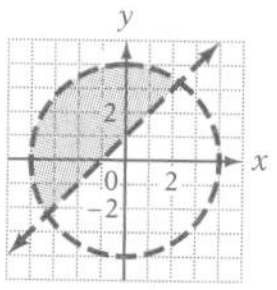

30. $y > x^2 - 4$
$y < x - 2$

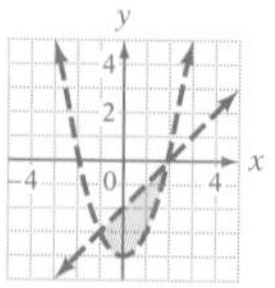

31. $\frac{x^2}{4} + \frac{y^2}{16} \le 1$
$y \le -\frac{1}{2}x + 2$

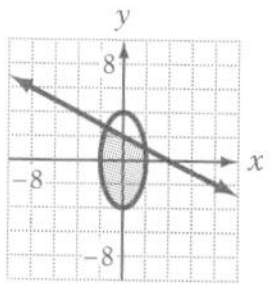

32. $\frac{y^2}{4} - \frac{x^2}{25} \ge 1$
$y \le \frac{2}{3}x + 4$

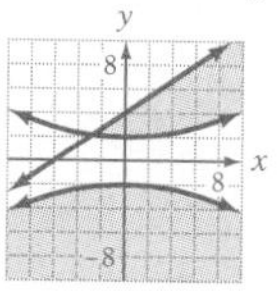

33. $x \geq y^2 - 3y + 2$
$y \geq 2x - 2$

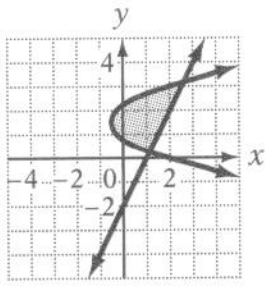

34. $x^2 + y^2 \leq 25$
$y \leq -\frac{1}{3}x + 2$

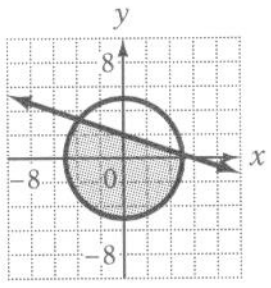

35. $x^2 + y^2 < 25$
$\frac{x^2}{9} + \frac{y^2}{36} < 1$

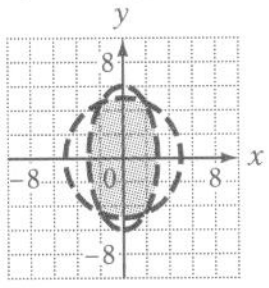

36. $\frac{x^2}{9} - \frac{y^2}{4} < 1$
$\frac{x^2}{25} + \frac{y^2}{9} < 1$

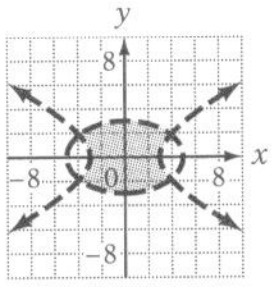

37. $x^2 + y^2 > 4$
$x^2 + y^2 < 25$

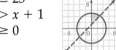

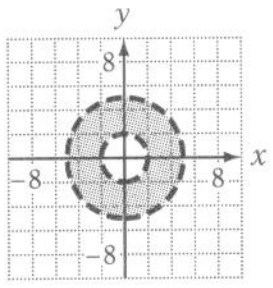

38. $\frac{x^2}{25} + \frac{y^2}{16} \leq 1$
$\frac{x^2}{4} + \frac{y^2}{4} \geq 1$

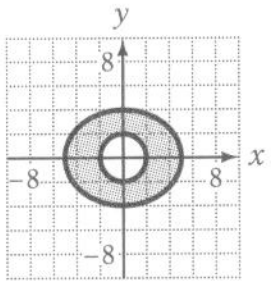

APPLYING CONCEPTS 12.5

Graph the solution set.

39. $y > x^2 - 3$
$y < x + 3$
$x \leq 0$

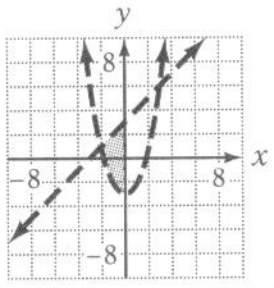

40. $x^2 + y^2 \leq 25$
$y > x + 1$
$x \geq 0$

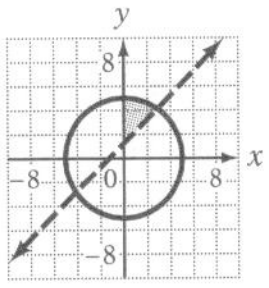

41. $x^2 + y^2 < 3$
$x > y^2 - 1$
$y \geq 0$

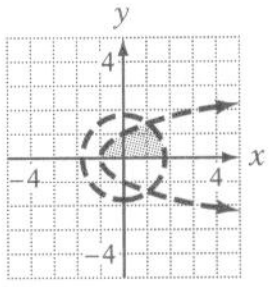

42. $\frac{x^2}{4} - \frac{y^2}{25} \leq 1$
$\frac{x^2}{4} + \frac{y^2}{4} \leq 1$
$y \geq 0$

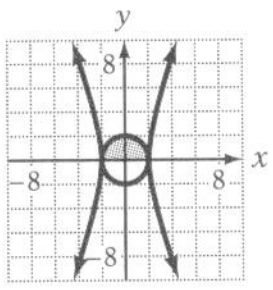

43. $\frac{x^2}{4} + \frac{y^2}{1} \leq 4$
$x^2 + y^2 \leq 4$
$x \geq 0$
$y \leq 0$

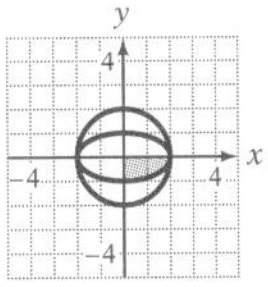

44. $\frac{x^2}{4} + \frac{y^2}{25} \leq 1$
$x > y^2 - 4$
$x \leq 0$
$y \geq 0$

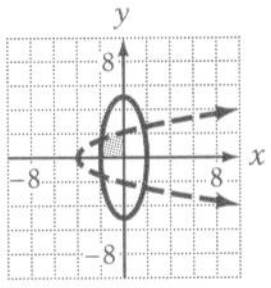

45. $y > 2^x$
$x + y < 4$

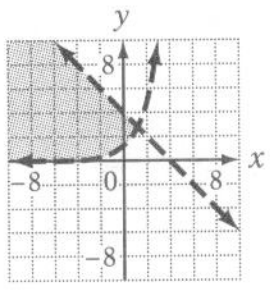

46. $y < \left(\frac{1}{2}\right)^x$
$2x - y \geq 2$

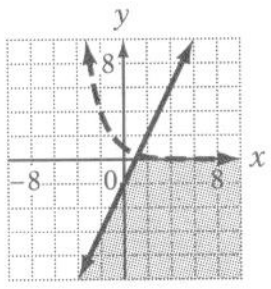

[C] **47.** $y \geq \log_2 x$
$x^2 + y^2 < 9$

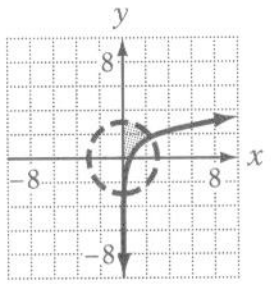

[C] **48.** $y \leq -\log_3 x$
$\frac{x^2}{9} + \frac{y^2}{4} < 1$

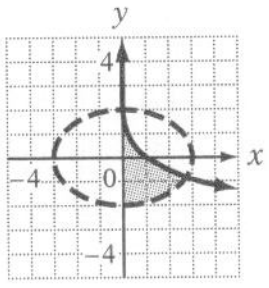

[C] **49.** $y < 3^{-x}$
$\frac{x^2}{4} - \frac{y^2}{1} \geq 1$

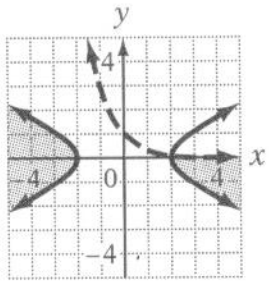

[C] **50.** $y \geq 2^{x-1}$
$2x + 3y > 6$

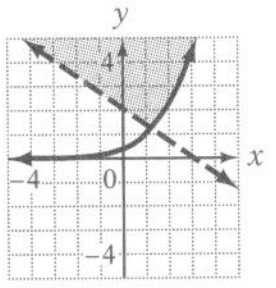

Focus on Problem Solving

Use a Variety of Problem-Solving Strategies

We have examined several problem-solving strategies throughout the text. See if you can apply those techniques to the following problems.

1. Eight coins look exactly alike, but one is lighter than the others. Explain how the different coin can be found in two weighings on a balance scale.
2. For the sequence of numbers 1, 1, 2, 3, 5, 8, 13, . . . , identify a possible pattern and then use that pattern to determine the next number in the sequence.
3. Arrange the numbers 1, 2, 3, 4, 5, 6, 7, 8, and 9 in the squares at the right so that the sum of any row or diagonal is 15. (*Suggestion:* Note that 1, 5, 9; 2, 5, 8; 3, 5, 7; and 4, 5, 6 all add to 15. Because 5 is part of each sum, this suggests that 5 be placed in the center of the squares.)

	5	

4. A restaurant charges \$5.00 for a pizza that has a diameter of 9 in. Determine the selling price of a pizza with a diameter of 18 in. so that the selling price per square inch is the same as for the 9-inch pizza.
5. You have a balance scale and weights of 1 gram, 2 grams, 4 grams, 8 grams, and 16 grams. Using only these weights, can you weigh something that weighs 7 grams? 12 grams?
6. Determine the number of possible paths from A to B for the grid at the right. A path consists of moves right or up along one of the grid lines.

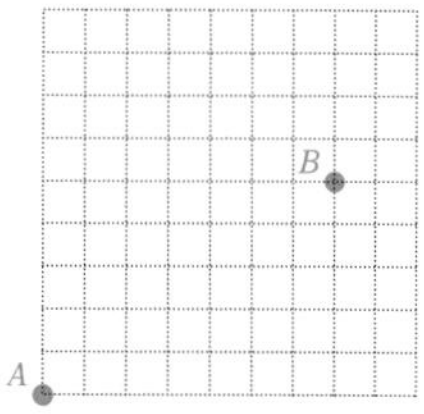

7. Can the checkerboard at the right be covered with dominos (which look like ◻■) so that every square on the board is covered by a domino? Why or why not? (*Note:* The dominos cannot overlap.)

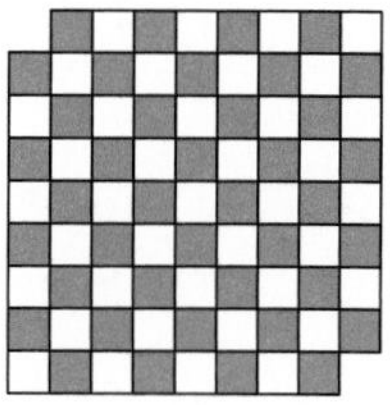

Projects and Group Activities

Graphing Conic Sections Using a Graphing Utility

Consider the graph of the ellipse $\frac{x^2}{16} + \frac{y^2}{9} = 1$ shown on the next page. Because a vertical line can intersect the graph at more than one point, the graph is not the graph of a function. And because the graph is not the graph of a function, the equation does not represent a function. Consequently, the equation cannot be

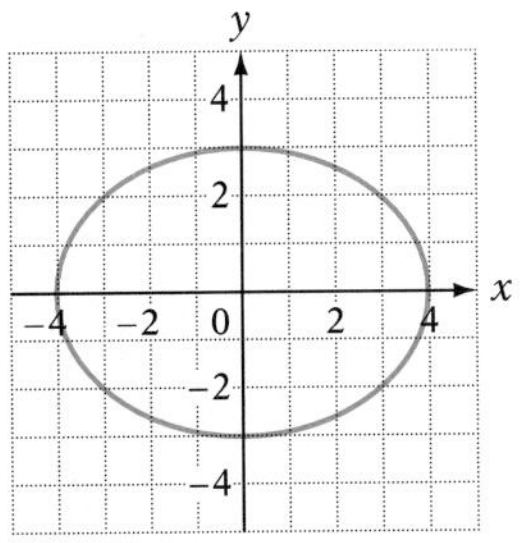

entered into a graphing utility to be graphed. However, by solving the equation for y, we have

$$\frac{y^2}{9} = 1 - \frac{x^2}{16}$$

$$y^2 = 9\left(1 - \frac{x^2}{16}\right)$$

$$y = \pm 3\sqrt{1 - \frac{x^2}{16}}$$

There are two solutions for y, which can be written

$$y_1 = 3\sqrt{1 - \frac{x^2}{16}} \quad \text{and} \quad y_2 = -3\sqrt{1 - \frac{x^2}{16}}$$

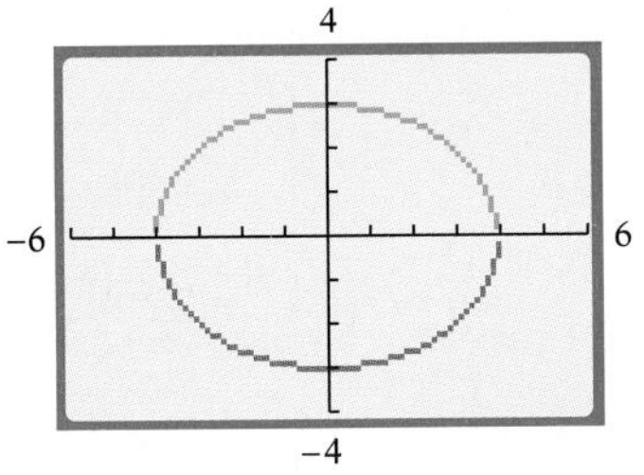

Each of these equations is the equation of a function and therefore can be entered into a graphing utility. The graph of $y_1 = 3\sqrt{1 - \frac{x^2}{16}}$ is shown in blue at the left, and the graph of $y_2 = -3\sqrt{1 - \frac{x^2}{16}}$ is shown in red. Note that, together, the graphs are the graph of an ellipse.

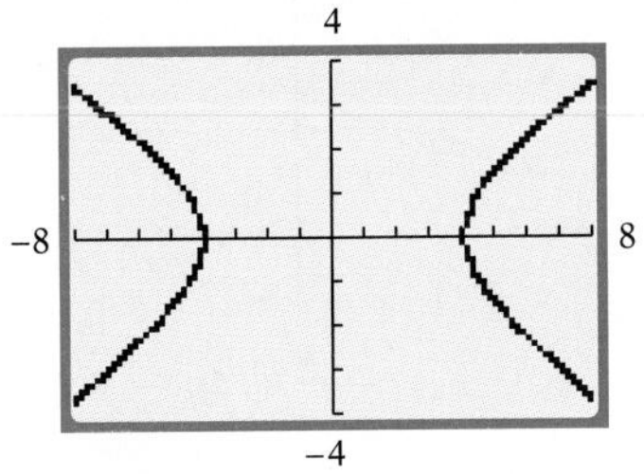

A similar technique can be used to graph a hyperbola. To graph $\frac{x^2}{16} - \frac{y^2}{4} = 1$, solve for y. Then graph each equation, $y_1 = 2\sqrt{\frac{x^2}{16} - 1}$ and $y_2 = -2\sqrt{\frac{x^2}{16} - 1}$.

Guidelines for using graphing calculators are found in the Appendix. In working through these examples or the exercises below, consult the Appendix or the user's manual for your particular calculator.

Solve each equation for y. Then graph using a graphing utility.

1. $\frac{x^2}{25} + \frac{y^2}{49} = 1$ **2.** $\frac{x^2}{4} + \frac{y^2}{64} = 1$ **3.** $\frac{x^2}{16} - \frac{y^2}{4} = 1$ **4.** $\frac{x^2}{9} - \frac{y^2}{36} = 1$

Chapter Summary

Key Words

Conic sections are curves that can be constructed from the intersection of a plane and a cone. The four conic sections are the *parabola, circle, ellipse,* and *hyperbola.* (Objective 12.1.1)

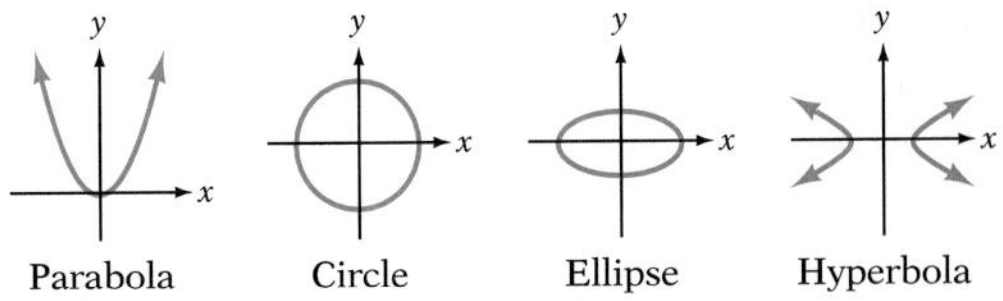

The *asymptotes* of a hyperbola are the two straight lines that are "approached" by the hyperbola. As the graph of the hyperbola gets farther from the origin, the hyperbola gets "closer to" the asymptotes. (Objective 12.3.2)

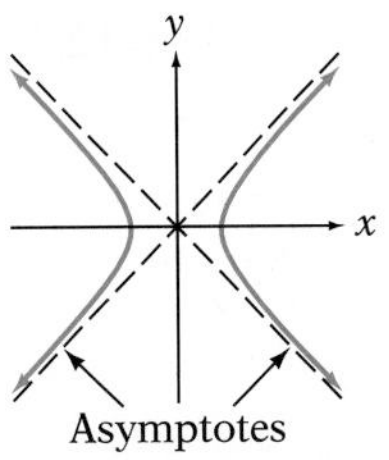

A *nonlinear system of equations* is a system in which one or more of the equations are not linear equations. (Objective 12.4.1)

$x^2 + y^2 = 16$
$y = x^2 - 2$

The *graph of a quadratic inequality in two variables* is a region of the plane that is bounded by one of the conic sections. (Objective 12.5.1)

$y \geq x^2 - 2x - 4$

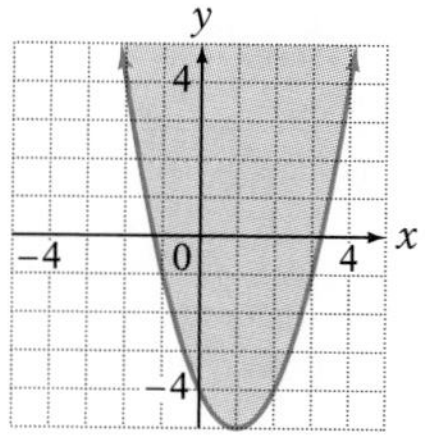

Essential Rules and Procedures

Equation of a parabola (Objective 12.1.1)
$y = ax^2 + bx + c$
When $a > 0$, the parabola opens up.
When $a < 0$, the parabola opens down.

The x-coordinate of the vertex is $-\frac{b}{2a}$.

The axis of symmetry is the line $x = -\frac{b}{2a}$.

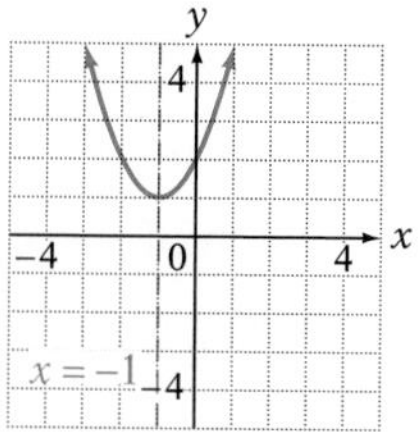

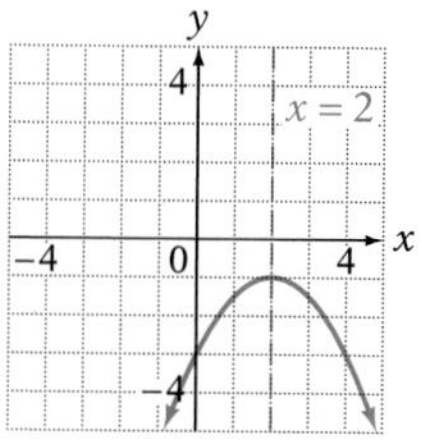

$y = x^2 + 2x + 2$ $\quad$ $y = -\frac{1}{2}x^2 + 2x - 3$

$x = ay^2 + by + c$
When $a > 0$, the parabola opens to the right.
When $a < 0$, the parabola opens to the left.

The y-coordinate of the vertex is $-\frac{b}{2a}$.

The axis of symmetry is the line $y = -\frac{b}{2a}$.

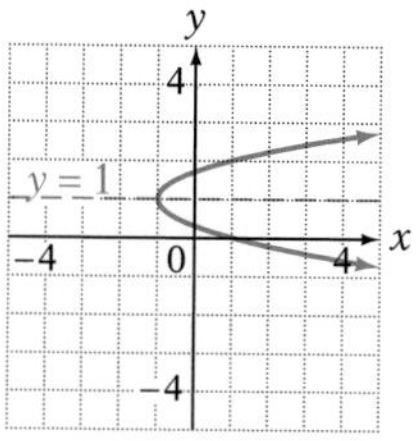

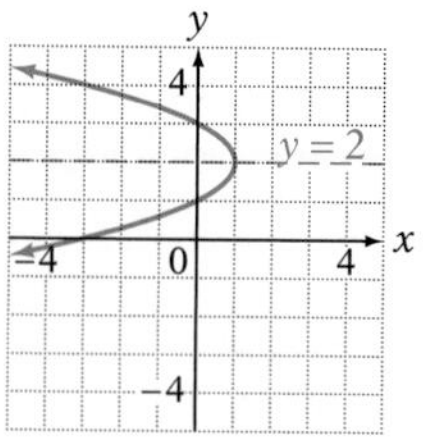

$x = 2y^2 - 4y + 1$ $\quad$ $x = -y^2 + 4y - 3$

Equation of a circle (Objective 12.2.1)
$(x - h)^2 + (y - k)^2 = r^2$
The center of the circle is (h, k), and the radius is r.

$(x - 2)^2 + (y + 1)^2 = 9$
$(h, k) = (2, -1)$
$r = 3$

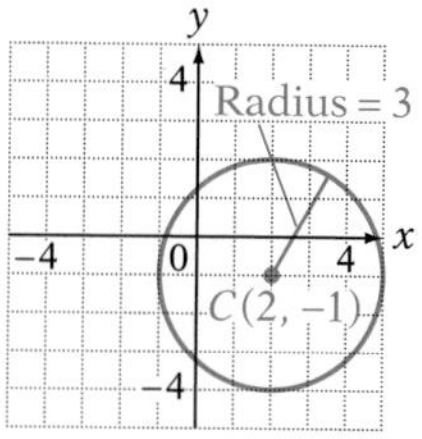

Equation of an ellipse (Objective 12.3.1)
$\frac{x^2}{a^2} + \frac{y^2}{b^2} = 1$
The x-intercepts are $(a, 0)$ and $(-a, 0)$.
The y-intercepts are $(0, b)$ and $(0, -b)$.

$\frac{x^2}{9} + \frac{y^2}{1} = 1$

x-intercepts:
$(3, 0)$ and $(-3, 0)$

y-intercepts:
$(0, 1)$ and $(0, -1)$

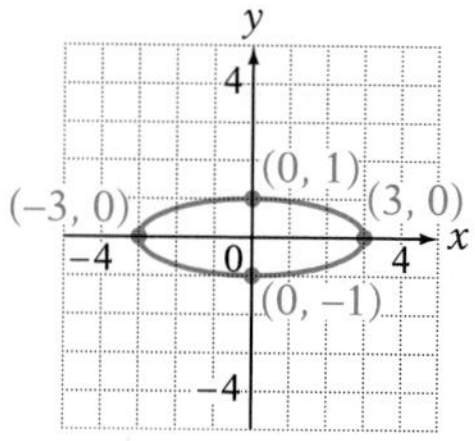

Equation of a hyperbola (Objective 12.3.2)
$\frac{x^2}{a^2} - \frac{y^2}{b^2} = 1$
An axis of symmetry is the x-axis.
The vertices are $(a, 0)$ and $(-a, 0)$.

$\frac{x^2}{9} - \frac{y^2}{4} = 1$

Vertices:
$(3, 0)$ and $(-3, 0)$

Asymptotes:
$y = \frac{2}{3}x$ and $y = -\frac{2}{3}x$

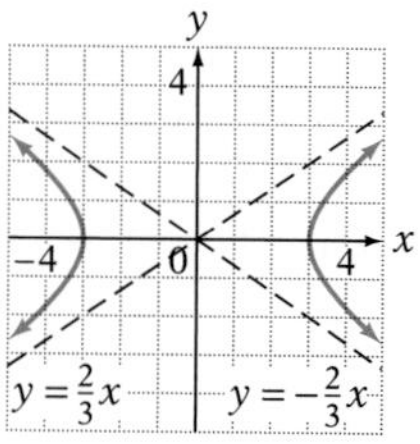

$\frac{y^2}{b^2} - \frac{x^2}{a^2} = 1$
An axis of symmetry is the y-axis.
The vertices are $(0, b)$ and $(0, -b)$.

The equations of the asymptotes are $y = \frac{b}{a}x$ and $y = -\frac{b}{a}x$.

$\frac{y^2}{1} - \frac{x^2}{4} = 1$
Vertices:
$(0, 1)$ and $(0, -1)$
Asymptotes:
$y = \frac{1}{2}x$ and $y = -\frac{1}{2}x$

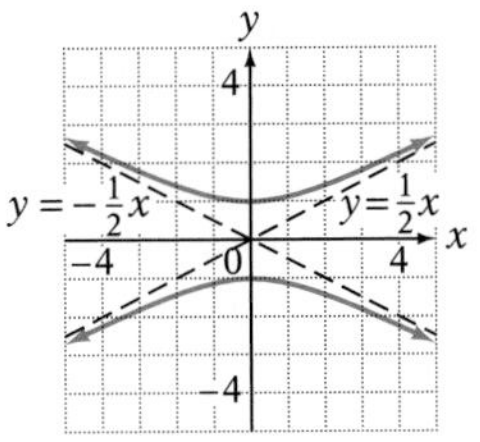

Chapter Review Exercises

1. Find the vertex and axis of symmetry of the parabola $y = x^2 - 4x + 8$.
Vertex: (2, 4); axis of symmetry: $x = 2$

2. Find the vertex and axis of symmetry of the parabola $y = -x^2 + 7x - 8$.
Vertex: $\left(\frac{7}{2}, \frac{17}{4}\right)$; axis of symmetry: $x = \frac{7}{2}$

3. Sketch a graph of $y = -2x^2 + x - 2$.

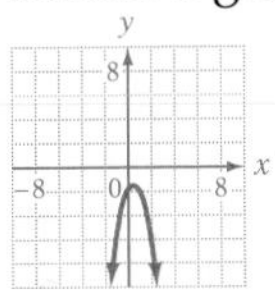

4. Sketch a graph of $x = 2y^2 - 6y + 5$.

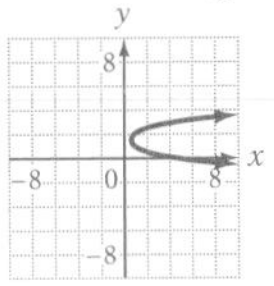

5. Find the equation of the circle that passes through the point $P(2, -1)$ and whose center is the point $C(-1, 2)$. $(x + 1)^2 + (y - 2)^2 = 18$

6. Find the equation of the circle with radius 6 and center $C(-1, 5)$. $(x + 1)^2 + (y - 5)^2 = 36$

7. Sketch a graph of $(x + 3)^2 + (y + 1)^2 = 1$.

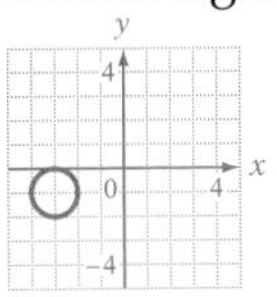

8. Sketch a graph of $x^2 + (y - 2)^2 = 9$.

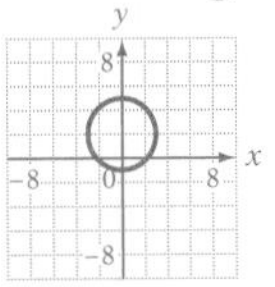

9. Find the equation of the circle that passes through the point $P(4, 6)$ and whose center is the point $C(0, -3)$. $x^2 + (y + 3)^2 = 97$

10. Write the equation $x^2 + y^2 + 4x - 2y = 4$ in standard form. $(x + 2)^2 + (y - 1)^2 = 9$

11. Sketch a graph of $\frac{x^2}{1} + \frac{y^2}{9} = 1$.

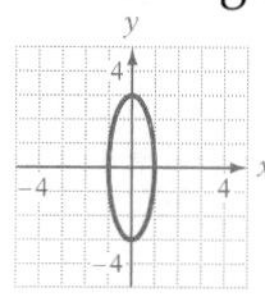

12. Sketch a graph of $\frac{x^2}{25} + \frac{y^2}{9} = 1$.

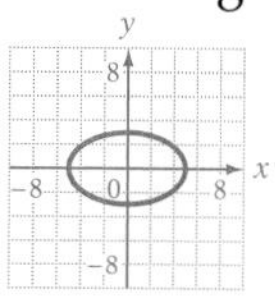

13. Sketch a graph of $\frac{x^2}{25} - \frac{y^2}{1} = 1$.

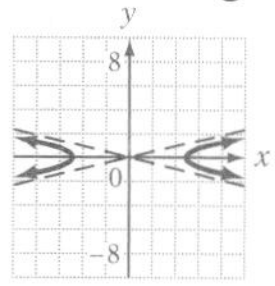

14. Sketch a graph of $\frac{y^2}{16} - \frac{x^2}{9} = 1$.

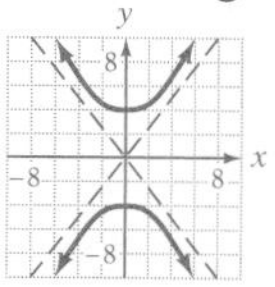

15. Solve: $y = x^2 + 5x - 6$ $(-2, -12)$
$y = x - 10$

16. Solve: $2x^2 + y^2 = 19$
$3x^2 - y^2 = 6$
$(\sqrt{5}, 3), (-\sqrt{5}, 3), (\sqrt{5}, -3), (-\sqrt{5}, -3)$

17. Solve: $x = 2y^2 - 3y + 1$
$3x - 2y = 0$
$\left(\frac{2}{9}, \frac{1}{3}\right), \left(1, \frac{3}{2}\right)$

18. Solve: $y^2 = 2x^2 - 3x + 6$ $(1, \sqrt{5}), (1, -\sqrt{5})$
$y^2 = 2x^2 + 5x - 2$

19. Graph the solution set: $(x - 2)^2 + (y + 1)^2 \leq 16$

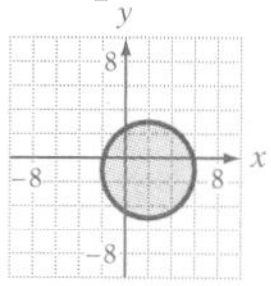

20. Graph the solution set: $\frac{x^2}{9} - \frac{y^2}{16} < 1$

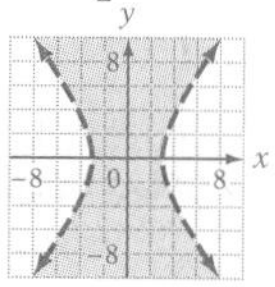

21. Graph the solution set: $y \geq -x^2 - 2x + 3$

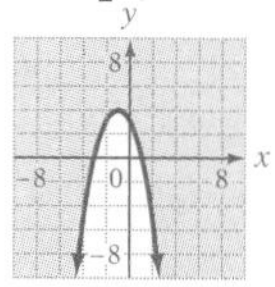

22. Graph the solution set: $\frac{x^2}{16} + \frac{y^2}{4} > 1$

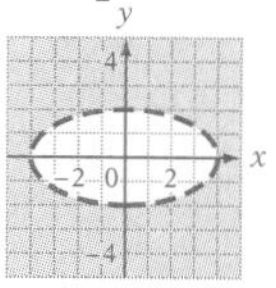

23. Graph the solution set: $y \geq x^2 - 4x + 2$
$y \leq \frac{1}{3}x - 1$

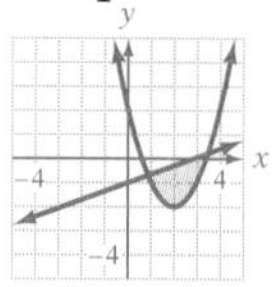

24. Graph the solution set: $\frac{x^2}{25} + \frac{y^2}{16} \leq 1$
$\frac{y^2}{4} - \frac{x^2}{4} \geq 1$

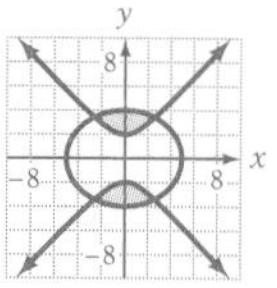

25. Graph the solution set: $\frac{x^2}{9} + \frac{y^2}{1} \geq 1$
$\frac{x^2}{4} - \frac{y^2}{1} \leq 1$

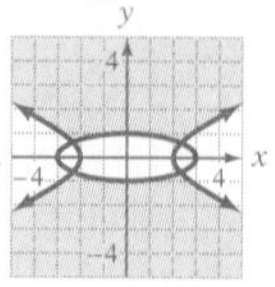

26. Graph the solution set: $\frac{x^2}{16} + \frac{y^2}{4} < 1$
$x^2 + y^2 > 9$

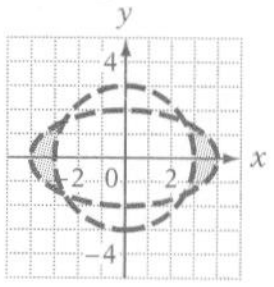

Chapter Test

1. Find the axis of symmetry of the parabola $y = -x^2 + 6x - 5$. $x = 3$

2. Find the vertex of the parabola $y = -x^2 + 3x - 2$. $\left(\frac{3}{2}, \frac{1}{4}\right)$

3. Sketch a graph of $y = \frac{1}{2}x^2 + x - 4$.

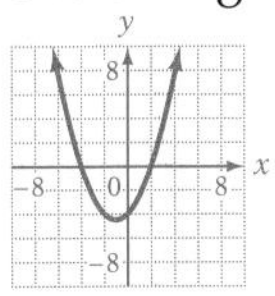

4. Sketch a graph of $x = y^2 - y - 2$.

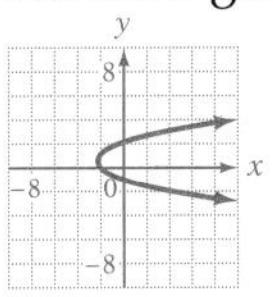

5. Find the equation of the circle with radius 4 and center $C(-3, -3)$. $(x + 3)^2 + (y + 3)^2 = 16$

6. Solve: $x^2 + 2y^2 = 4$
$x + y = 2$
$(2, 0), \left(\frac{2}{3}, \frac{4}{3}\right)$

7. Solve: $x = 3y^2 + 2y - 4$
$x = y^2 - 5y$
$(36, -4), \left(-\frac{9}{4}, \frac{1}{2}\right)$

8. Solve: $x^2 - y^2 = 24$
$2x^2 + 5y^2 = 55$
$(5, 1), (-5, 1), (5, -1), (-5, -1)$

9. Find the equation of the circle that passes through the point $P(2, 4)$ and whose center is the point $C(-1, -3)$. $(x + 1)^2 + (y + 3)^2 = 58$

10. Sketch a graph of $(x - 2)^2 + (y + 1)^2 = 9$.

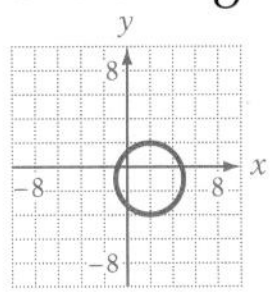

11. Find the equation of the circle with radius 3 and center $C(-2, 4)$. $(x + 2)^2 + (y - 4)^2 = 9$

12. Find the equation of the circle that passes through the point $P(2, 5)$ and whose center is the point $C(-2, 1)$. $(x + 2)^2 + (y - 1)^2 = 32$

13. Write the equation $x^2 + y^2 - 4x + 2y + 1 = 0$ in standard form, and then sketch its graph.
$(x - 2)^2 + (y + 1)^2 = 4$

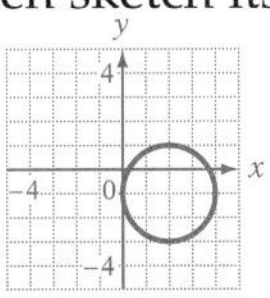

14. Sketch a graph of $\frac{y^2}{25} - \frac{x^2}{16} = 1$.

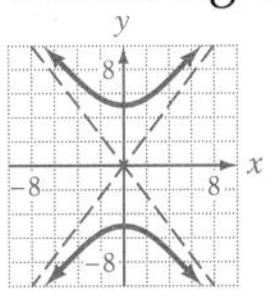

15. Sketch a graph of $\frac{x^2}{9} - \frac{y^2}{4} = 1$.

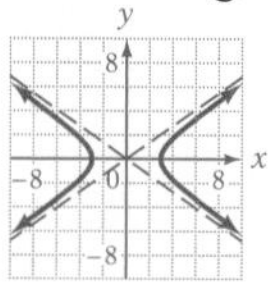

16. Sketch a graph of $\frac{x^2}{16} + \frac{y^2}{4} = 1$.

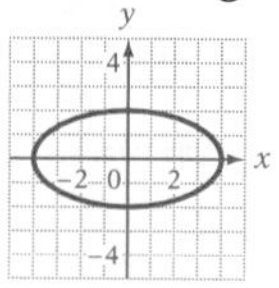

17. Graph the solution set: $\frac{x^2}{16} - \frac{y^2}{25} < 1$

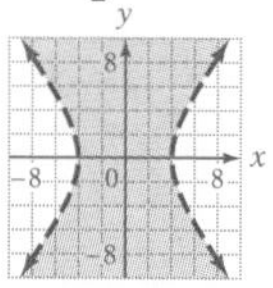

18. Graph the solution set: $x^2 + y^2 < 36$
$x + y > 4$

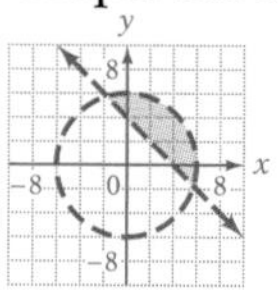

19. Graph the solution set: $\frac{x^2}{25} + \frac{y^2}{4} \leq 1$

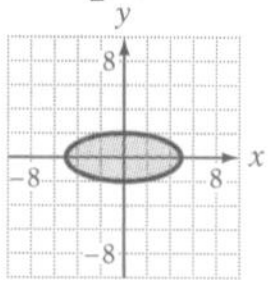

20. Graph the solution set: $\frac{x^2}{25} - \frac{y^2}{16} \geq 1$
$x^2 + y^2 \leq 9$

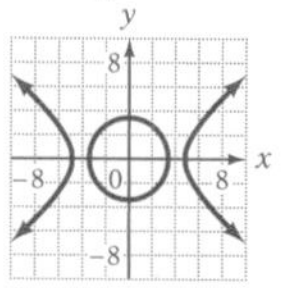

FINAL EXAM

1. Simplify: $12 - 8[3 - (-2)]^2 \div 5 - 3$ $\quad -31$

2. Evaluate $\dfrac{a^2 - b^2}{a - b}$ when $a = 3$ and $b = -4$. $\quad -1$

3. Simplify: $5 - 2[3x - 7(2 - x) - 5x]$ $\quad 33 - 10x$

4. Solve: $\frac{3}{4}x - 2 = 4$ $\quad 8$

5. Solve: $\dfrac{2 - 4x}{3} - \dfrac{x - 6}{12} = \dfrac{5x - 2}{6}$ $\quad \dfrac{2}{3}$

6. Solve: $8 - |5 - 3x| = 1$ $\quad 4, -\dfrac{2}{3}$

7. Solve: $|2x + 5| < 3$ $\quad \{x \mid -4 < x < -1\}$

8. Solve: $2 - 3x < 6$ and $2x + 1 > 4$ $\quad \left\{x \mid x > \dfrac{3}{2}\right\}$

9. Find the equation of the line that contains the point $(-2, 1)$ and is perpendicular to the line $3x - 2y = 6$. $\quad y = -\dfrac{2}{3}x - \dfrac{1}{3}$

10. Simplify: $2a[5 - a(2 - 3a) - 2a] + 3a^2$
$6a^3 - 5a^2 + 10a$

11. Simplify: $\dfrac{3}{2 + i}$ $\quad \dfrac{6}{5} - \dfrac{3}{5}i$

12. Write a quadratic equation that has integer coefficients and has solutions $-\dfrac{1}{2}$ and 2.
$2x^2 - 3x - 2 = 0$

13. Factor: $8 - x^3y^3$ $\quad (2 - xy)(4 + 2xy + x^2y^2)$

14. Factor: $x - y - x^3 + x^2y$
$(x - y)(1 + x)(1 - x)$

15. Divide: $(2x^3 - 7x^2 + 4) \div (2x - 3)$
$x^2 - 2x - 3 - \dfrac{5}{2x - 3}$

16. Divide: $\dfrac{x^2 - 3x}{2x^2 - 3x - 5} \div \dfrac{4x - 12}{4x^2 - 4}$ $\quad \dfrac{x(x - 1)}{2x - 5}$

17. Subtract: $\dfrac{x - 2}{x + 2} - \dfrac{x + 3}{x - 3}$ $\quad -\dfrac{10x}{(x + 2)(x - 3)}$

18. Simplify: $\dfrac{\dfrac{3}{x} + \dfrac{1}{x + 4}}{\dfrac{1}{x} + \dfrac{3}{x + 4}}$ $\quad \dfrac{x + 3}{x + 1}$

19. Solve: $\dfrac{5}{x - 2} - \dfrac{5}{x^2 - 4} = \dfrac{1}{x + 2}$ $\quad -\dfrac{7}{4}$

20. Solve $a_n = a_1 + (n - 1)d$ for d. $\quad d = \dfrac{a_n - a_1}{n - 1}$

21. Simplify: $\left(\dfrac{4x^2y^{-1}}{3x^{-1}y}\right)^{-2}\left(\dfrac{2x^{-1}y^2}{9x^{-2}y^2}\right)^3$ $\quad \dfrac{y^4}{162x^3}$

22. Simplify: $\left(\dfrac{3x^{\frac{2}{3}}y^{\frac{1}{2}}}{6x^2y^{\frac{4}{3}}}\right)^6$ $\quad \dfrac{1}{64x^8y^5}$

23. Subtract: $x\sqrt{18x^2y^3} - y\sqrt{50x^4y}$ $\quad -2x^2y\sqrt{2y}$

24. Simplify: $\dfrac{\sqrt{16x^5y^4}}{\sqrt{32xy^7}}$ $\quad \dfrac{x^2\sqrt{2y}}{2y^2}$

25. Solve by using the quadratic formula:
$2x^2 - 3x - 1 = 0$
$\dfrac{3 \pm \sqrt{17}}{4}$

26. Solve: $x^{\frac{2}{3}} - x^{\frac{1}{3}} - 6 = 0$ $\quad 27, -8$

27. Find the equation of the line containing the points $(3, -2)$ and $(1, 4)$. $\quad y = -3x + 7$

28. Solve: $\dfrac{2}{x} - \dfrac{2}{2x + 3} = 1$ $\quad \dfrac{3}{2}, -2$

29. Solve by the addition method: $3x - 2y = 1$
$5x - 3y = 3$
$(3, 4)$

30. Evaluate the determinant: $\begin{vmatrix} 3 & 4 \\ -1 & 2 \end{vmatrix}$ 10

31. Solve for x: $\log_3 x - \log_3(x - 3) = \log_3 2$ 6

32. Write $\sum_{i=1}^{5} 2y^i$ in expanded form.
$2y + 2y^2 + 2y^3 + 2y^4 + 2y^5$

33. Find an equivalent fraction for $0.5\overline{1}$. $\frac{23}{45}$

34. Find the 3rd term in the expansion of $(x - 2y)^9$.
$144x^7y^2$

35. Solve: $x^2 - y^2 = 4$
$x + y = 1$
$\left(\frac{5}{2}, -\frac{3}{2}\right)$

36. Find the inverse function of $f(x) = \frac{2}{3}x - 4$.
$f^{-1}(x) = \frac{3}{2}x + 6$

37. Write $2(\log_2 a - \log_2 b)$ as a single logarithm with a coefficient of 1. $\log_2 \frac{a^2}{b^2}$

38. Graph $2x - 3y = 9$ by using the x- and y-intercepts.
x-intercept: $\left(\frac{9}{2}, 0\right)$
y-intercept: $(0, -3)$

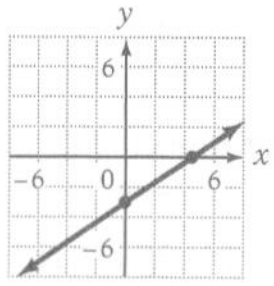

39. Graph the solution set of $3x + 2y > 6$.

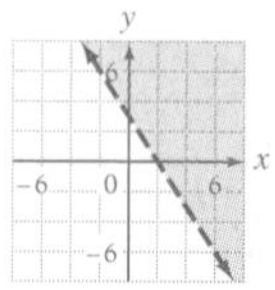

40. Graph: $f(x) = -x^2 + 4$

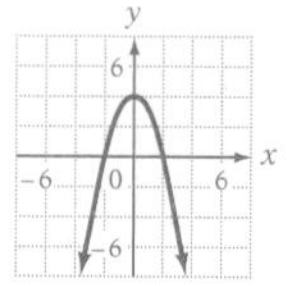

41. Graph: $\frac{x^2}{16} + \frac{y^2}{4} = 1$

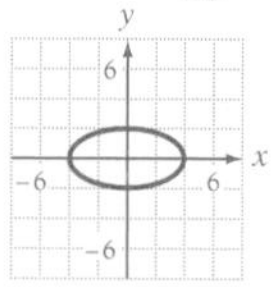

42. Graph: $f(x) = \log_2(x + 1)$

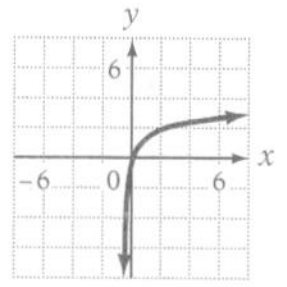

43. Graph $f(x) = x + 2^{-x}$ and approximate to the nearest tenth the values of x for which $f(x) = 2$. $-2, 1.7$

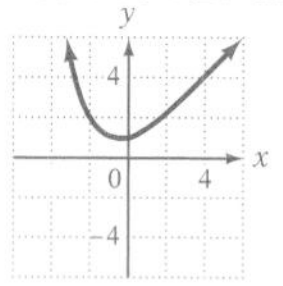

44. Given the graph of $f(x) = \ln x$ shown below, graph $g(x) = \ln(x + 3)$.

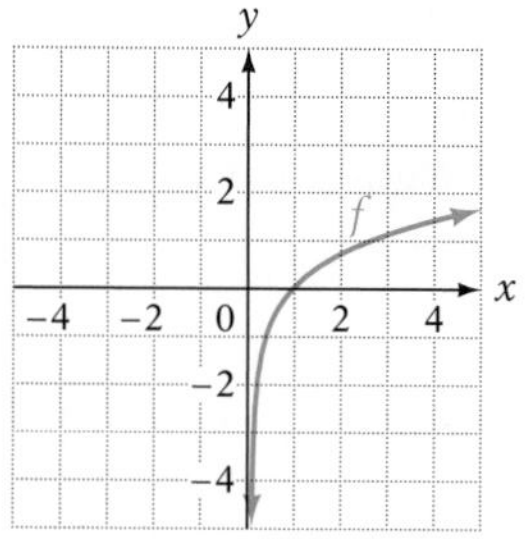

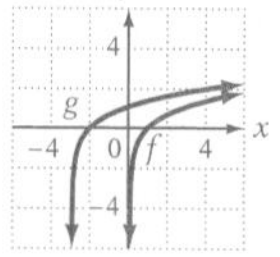

45. An average score of 70–79 in a history class receives a C grade. A student has grades of 64, 58, 82, and 77 on four history tests. Find the range of scores on the fifth test that will give the student a C grade for the course. 69 or better

46. A jogger and a cyclist set out at 8 A.M. from the same point headed in the same direction. The average speed of the cyclist is two and a half times the average speed of the jogger. In 2 h, the cyclist is 24 mi ahead of the jogger. How far did the cyclist ride? 40 mi

47. You have a total of $12,000 invested in two simple interest accounts. On one account, a money market fund, the annual simple interest rate is 8.5%. On the other account, a tax-free bond fund, the annual simple interest rate is 6.4%. The total annual interest earned by the two accounts is $936. How much do you have invested in each account? $8000 at 8.5%; $4000 at 6.4%

48. The length of a rectangle is 1 ft less than three times the width. The area of the rectangle is 140 ft^2. Find the length and width of the rectangle.
Length: 20 ft; width: 7 ft

49. Three hundred shares of a utility stock earn a yearly dividend of $486. How many additional shares of the utility stock would give a total dividend income of $810? 200 additional shares

50. An account executive traveled 45 mi by car and then an additional 1050 mi by plane. The rate of the plane was seven times the rate of the car. The total time of the trip was $3\frac{1}{4}$ h. Find the rate of the plane. 420 mph

51. An object is dropped from the top of a building. Find the distance the object has fallen when the speed reaches 75 ft/s. Use the equation $v = \sqrt{64d}$, where v is the speed of the object and d is the distance. Round to the nearest whole number. 88 ft

52. A small plane made a trip of 660 mi in 5 h. The plane traveled the first 360 mi at a constant rate before increasing its speed by 30 mph. Another 300 mi was traveled at the increased speed. Find the rate of the plane for the first 360 mi.
120 mph

53. The intensity (L) of a light source is inversely proportional to the square of the distance (d) from the source. If the intensity is 8 lumens at a distance of 20 ft, what is the intensity when the distance is 4 ft? 200 lumens

54. A motorboat traveling with the current can go 30 mi in 2 h. Against the current, it takes 3 h to go the same distance. Find the rate of the motorboat in calm water and the rate of the current. Boat: 12.5 mph; current: 2.5 mph

55. An investor deposits $4000 into an account that earns 9% annual interest compounded monthly. Use the compound interest formula $P = A(1 + i)^n$, where A is the original value of the investment, i is the interest rate per compounding period, n is the total number of compounding periods, and P is the value of the investment after n periods, to find the value of the investment after 2 years. Round to the nearest cent. $4785.65

56. Assume the average value of a home increases 6% per year. How much would a house costing $80,000 be worth in 20 years? Round to the nearest dollar. $256,571

APPENDIX:

Guidelines for Using Graphing Calculators

Texas Instruments TI-83

To evaluate an expression

a. Press the [Y=] key. A menu showing \Y1 = through \Y7 = will be displayed vertically with a blinking cursor to the right of \Y1 =. Press [CLEAR], if necessary, to delete an unwanted expression.

b. Input the expression to be evaluated. For example, to input the expression $-3a^2b - 4c$, use the following keystrokes:

[(−)] 3 [ALPHA] A [∧] 2 [ALPHA] B [−] 4 [ALPHA] C [2nd] QUIT

Note the difference between the key for a negative sign [(−)] and the key for a *minus* sign [−].

c. Store the value of each variable that will be used in the expression. For example, to evaluate the expression above when $a = 3$, $b = -2$, and $c = -4$, use the following keystrokes:

3 [STO▷] [ALPHA] A [ENTER] [(−)] 2 [STO▷] [ALPHA] B [ENTER]
[(−)] 4 [STO▷] [ALPHA] C [ENTER]

These steps store the value of each variable.

d. Press [2nd] [VARS] [▷] [1] [1] [ENTER]. The value for the expression, Y1, for the given values is displayed; in this case, Y1 = 70.

To graph a function

a. Press the [Y=] key. A menu showing \Y1 = through \Y7 = will be displayed vertically with a blinking cursor to the right of \Y1 =. Press [CLEAR], if necessary, to delete an unwanted expression.

b. Input the expression for each function that is to be graphed. Press [X,T,θ,n] to input x. For example, to input $y = x^3 + 2x^2 - 5x - 6$, use the following keystrokes:

[X,T,θ,n] [∧] 3 [+] 2 [X,T,θ,n] [∧] 2 [−] 5 [X,T,θ,n] [−] 6

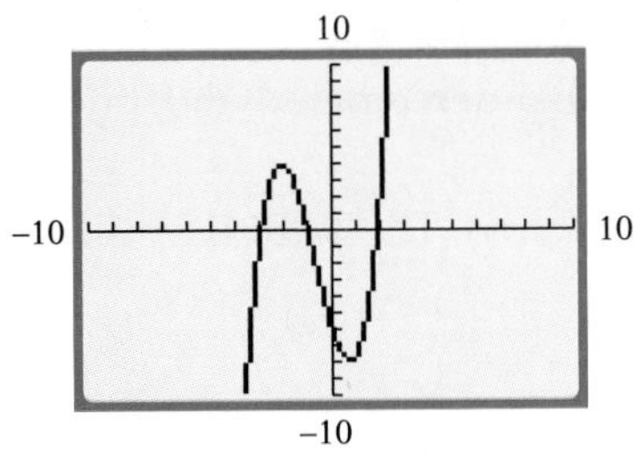

c. Set the domain and range by pressing [WINDOW]. Enter the values for the minimum x value (Xmin), the maximum x value (Xmax), the distance between tick marks on the x-axis (Xscl), the minimum y value (Ymin), the maximum y value (Ymax), and the distance between tick marks on the y-axis (Yscl). Now press [GRAPH]. For the graph shown at the left, Xmin = −10, Xmax = 10, Xscl = 1, Ymin = −10, Ymax = 10, and Yscl = 1. This is called the standard viewing rectangle. Pressing [ZOOM] [6] is a quick way to set the calculator to the standard viewing rectangle. *Note:* This will also immediately graph the function in that window.

d. Press the [Y=] key. The equals sign has a black rectangle around it. This indicates that the function is active and will be graphed when the [GRAPH] key is pressed. A function is deactivated by using the arrow keys. Move the cursor over the equals sign and press ENTER. When the cursor is moved to the right, the black rectangle will not be present, and that equation will not be active.

e. Graphing some radical equations requires special care. To graph the function $y = \sqrt{2x + 3}$, enter the following keystrokes:

Y= 2nd $\sqrt{\ }$ (2 X,T,θ,n + 3)

The graph is shown below.

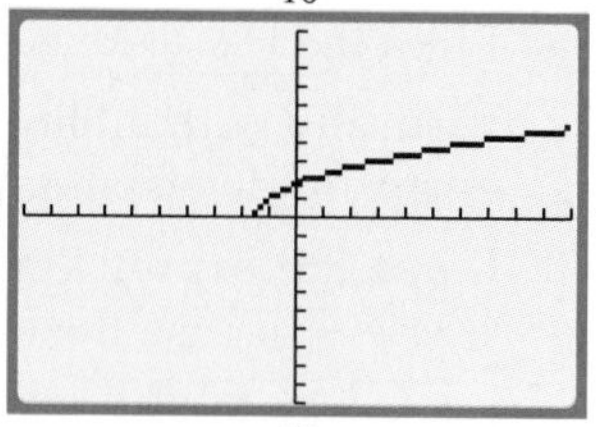

To display the *x*-coordinates of rectangular coordinates as integers

a. Set the viewing window as follows: Xmin = −47, Xmax = 47, Xscl = 10, Ymin = −31, Ymax = 31, Yscl = 10.

b. Graph the function and use the TRACE feature. Press TRACE and then move the cursor with the ◁ and ▷ keys. The values of x and $y = f(x)$ displayed on the bottom of the screen are the coordinates of a point on the graph.

To display the *x*-coordinates of rectangular coordinates in tenths

a. Set the viewing window as follows: ZOOM 4

b. Graph the function and use the TRACE feature. Press TRACE and then move the cursor with the ◁ and ▷ keys. The values of x and $y = f(x)$ displayed on the bottom of the screen are the coordinates of a point on the graph.

To evaluate a function for a given value of *x*, or to produce ordered pairs of a function

a. Input the equation; for example, input $Y_1 = 2x^3 - 3x + 2$.

b. Press 2nd QUIT.

c. Input a value for x; for example, to input 3, press 3 STO▷ X,T,θ,n ENTER.

d. Press VARS ▷ 1 1 ENTER. The value for the expression, Y_1, for the given x value is displayed, in this case, $Y_1 = 47$. An ordered pair of the function is (3, 47).

e. Repeat steps **c.** and **d.** to produce as many pairs as desired. The TABLE feature of the *TI-83* can also be used to determine pairs.

Zoom Features

To zoom in or out on a graph

a. Here are two methods of using ZOOM. The first method uses the built-in features of the calculator. Move the cursor to a point on the graph that is of interest. Press [ZOOM]. The ZOOM menu will appear. Press [2] [ENTER] to zoom in on the graph by the amount shown under the SET FACTORS menu. The center of the new graph is the location at which you placed the cursor. Press [ZOOM] [3] [ENTER] to zoom out on the graph by the amount under the SET FACTORS menu. (The SET FACTORS menu is accessed by pressing [ZOOM] [▷] [4].)

b. The second method uses the ZBOX option under the ZOOM menu. To use this method, press [ZOOM] [1]. A cursor will appear on the graph. Use the arrow keys to move the cursor to a portion of the graph that is of interest. Press [ENTER]. Now use the arrow keys to draw a box around the portion of the graph you wish to see. Press [ENTER]. The portion of the graph defined by the box will be drawn.

c. Pressing [ZOOM] [6] resets the window to the standard 10×10 viewing window.

Solving Equations

This discussion is based on the fact that the solution of an equation can be related to the x-intercepts of a graph. For instance, the real solutions of the equation $x^2 = x + 1$ are the x-intercepts of the graph of $f(x) = x^2 - x - 1$, which are the zeros of f.

To solve $x^2 = x + 1$, rewrite the equation with all terms on one side. The equation is now $x^2 - x - 1 = 0$. Think of this equation as $Y_1 = x^2 - x - 1$. The x-intercepts of the graph of Y_1 are the solutions of the equation $x^2 = x + 1$.

a. Enter $x^2 - x - 1$ into Y_1.

b. Graph the equation. You may need to adjust the viewing window so that the x-intercepts are visible.

c. Press [2nd] CALC [2].

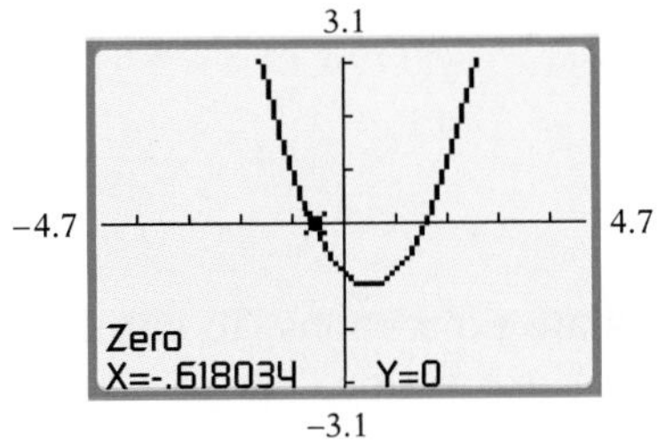

d. Move the cursor to a point on the curve that is to the left of an x-intercept. Press [ENTER].

e. Move the cursor to a point on the curve that is to the right of an x-intercept. Press [ENTER].

f. Press [ENTER].

g. The root is shown as the x-coordinate on the bottom of the screen; in this case, the root is approximately -0.618034. To find the next intercept, repeat steps **c.** through **f.** The SOLVER feature under the MATH menu can also be used to find solutions of equations.

Solving Systems of Equations in Two Variables

To solve a system of equations

The system of equations $\begin{aligned} y &= x^2 - 1 \\ \frac{1}{2}x + y &= 1 \end{aligned}$ will be solved.

a. Solve each equation for y.

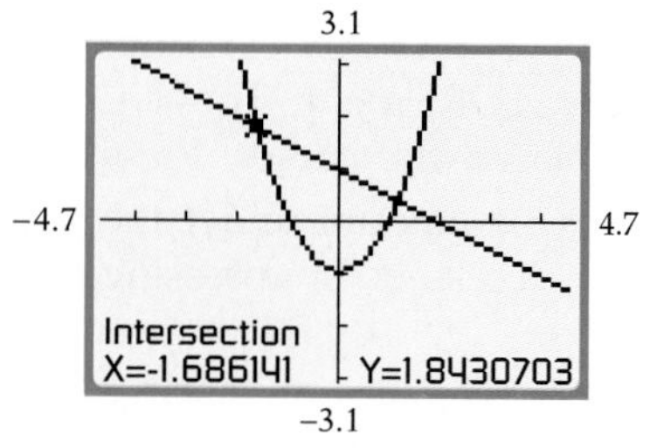

b. Enter the first equation as Y1. For instance, $Y1 = x^2 - 1$.

c. Enter the second equation as Y2. For instance, $Y2 = 1 - \frac{1}{2}x$.

d. Graph both equations. (*Note:* The point(s) of intersection must appear on the screen. It may be necessary to adjust the viewing window so that the point(s) of intersection are displayed.)

e. Press [2nd] CALC [5].

f. Move the cursor to the left of the first point of intersection. Press [ENTER].

g. Move the cursor to the right of the first point of intersection. Press [ENTER].

h. Press [ENTER].

i. The first point of intersection is $(-1.686141, 1.8430703)$.

j. Repeat steps **e.** through **h.** for each point of intersection.

Finding the Minimum or Maximum Value of a Function

a. Enter the function into Y1. The equation $y = x^2 - x - 1$ is used here.

b. Graph the equation. You may need to adjust the viewing window so that the maximum or minimum point is visible.

c. Press [2nd] CALC [3] to determine a minimum value or press [2nd] CALC [4] to determine a maximum value.

d. Move the cursor to a point on the curve that is to the left of the minimum (maximum). Press [ENTER].

e. Move the cursor to a point on the curve that is to the right of the minimum (maximum). Press [ENTER].

f. Press [ENTER].

g. The minimum (maximum) is shown as the y-coordinate on the bottom of the screen; in this case, the minimum value is -1.25.

Sharp EL-9600

To evaluate an expression

a. The SOLVER mode of the calculator is used to evaluate expressions. To enter SOLVER mode, press [2ndF] SOLVER [CL]. The expression $-3a^2b - 4c$ must be entered as the equation $-3a^2b - 4c = t$. The letter t can be any letter other than one used in the expression. Use the following keystrokes to input $-3a^2b - 4c = t$:

[(−)] 3 [ALPHA] A [a^b] 2 [▷] [ALPHA] B [−] 4 [ALPHA] C [ALPHA] [=] [ALPHA] T [ENTER]

Note the difference between the key for a *negative* sign [(−)] and the key for a *minus* sign [−].

b. After you press [ENTER], variables used in the equation will be displayed on the screen. To evaluate the expression for $a = 3$, $b = -2$, and $c = -4$, input each value, pressing [ENTER] after each number. When the cursor moves to T, press [2ndF] EXE. T = 70 will appear on the screen. This is the value of the expression. To evaluate the expression again for different values of a, b, and c, press [2ndF] QUIT and then [2ndF] SOLVER.

c. Press [+ − × ÷] to return to normal operation.

To graph a function

a. Press the [Y=] key. The screen will show Y1 through Y8.

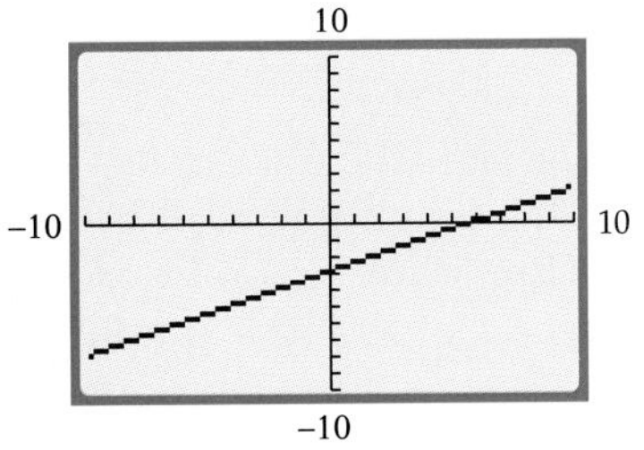

b. Input the expression for a function that is to be graphed. Press [X/θ/T/n] to enter an x. For example, to input $y = \frac{1}{2}x - 3$, use the following keystrokes:

[Y=] [CL] [(] 1 [÷] 2 [)] [X/θ/T/n] [−] 3 [ENTER]

c. Set the viewing window by pressing [WINDOW]. Enter the values for the minimum x value (Xmin), the maximum x value (Xmax), the distance between tick marks on the x-axis (Xscl), the minimum y value (Ymin), the maximum y value (Ymax), and the distance between tick marks on the y-axis (Yscl). Press [ENTER] after each entry. Press [GRAPH]. For the graph shown at the left, enter Xmin = −10, Xmax = 10, Xscl = 1, Ymin = −10, Ymax = 10, and Yscl = 1. Press [GRAPH].

d. Press [Y=] to return to the equation. The equals sign has a black rectangle around it. This indicates that the function is active and will be graphed when the [GRAPH] key is pressed. A function is deactivated by using the arrow keys. Move the cursor over the equals sign and press [ENTER]. When the cursor is moved to the right, the black rectangle will not be present, and that equation will not be active.

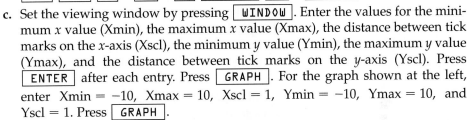

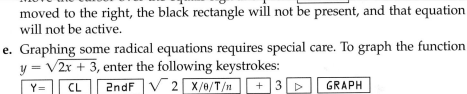

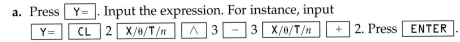

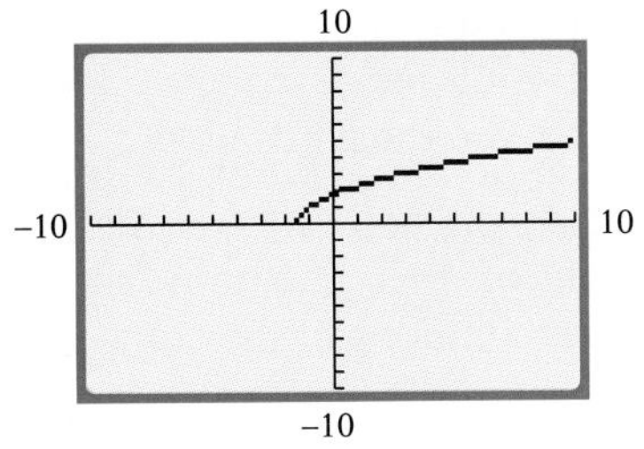

e. Graphing some radical equations requires special care. To graph the function $y = \sqrt{2x + 3}$, enter the following keystrokes:

[Y=] [CL] [2ndF] √ 2 [X/θ/T/n] [+] 3 [▷] [GRAPH]

The graph is shown at the left.

To display the xy-coordinates as integers

a. Press [ZOOM] [▷] 8.

b. Graph the function. Press [TRACE]. Use the left arrow and right arrow keys to trace along the graph of the function. The x- and y-coordinates of the function are shown on the bottom of the screen.

To display the xy-coordinates in tenths

a. Press [ZOOM] [▷] 7.

b. Graph the function. Press [TRACE]. Use the left arrow and right arrow keys to trace along the graph of the function. The x- and y-coordinates of the function are shown on the bottom of the screen.

To evaluate a function for a given value of x, or to produce ordered pairs of the function

a. Press [Y=]. Input the expression. For instance, input

[Y=] [CL] 2 [X/θ/T/n] [∧] 3 [−] 3 [X/θ/T/n] [+] 2. Press [ENTER].

b. Press [+] [−] [×] [÷]. Store the x-coordinate of the ordered pair you want in [X/θ/T/n].

For instance, enter 3 [STO] [X/θ/T/n] [ENTER].

c. Press [VARS] [ENTER] 1 [ENTER]. The value of y, 47, will be displayed on the screen. The ordered pair is (3, 47). The TABLE feature of the calculator can also be used to find many ordered pairs for a function.

Zoom Features

To zoom in or out on a graph

a. Here are two methods of using ZOOM. The first method uses the built-in features of the calculator. Move the cursor to a point on the graph that is of interest. Press [ZOOM]. The ZOOM menu will appear. Press 3 to zoom in on the graph by the amount shown by FACTOR. The center of the new graph is the location at which you placed the cursor. Press [ZOOM] 4 to zoom out on the graph by the amount shown in FACTOR.

b. The second method uses the BOX option under the ZOOM menu. To use this method, press [ZOOM] 2. A cursor will appear on the screen. Use the arrow keys to move the cursor to a portion of the graph that is of interest. Press [ENTER]. Use the arrow keys to draw a box around the portion of the graph you wish to see. Press [ENTER].

Solving Equations

This discussion is based on the fact that the real solutions of an equation can be related to the x-intercepts of a graph. For instance, the real solutions of $x^2 = x + 1$ are the x-intercepts of the graph of $f(x) = x^2 - x - 1$, which are the zeros of f.

To solve $x^2 = x + 1$, rewrite the equation with all terms on one side of the equation. The equation is now $x^2 - x - 1 = 0$. Think of this equation as $Y_1 = x^2 - x - 1$. The x-intercepts of the graph of Y_1 are the solutions of the equation $x^2 = x + 1$.

a. Enter $x^2 - x - 1$ into Y_1.

b. Graph the equation. You may need to adjust the viewing window so that the x-intercepts are visible.

c. Press [2ndF] CALC 5.

d. A solution is shown as the x-coordinate at the bottom of the screen. To find another intercept, move the cursor to the right of the next x-intercept. Then press [2ndF] CALC 5.

Solving Systems of Equations in Two Variables

To solve a system of equations

a. Solve each equation for y.

b. Press [Y=] and then enter both equations.

c. Graph the equations. You may need to adjust the viewing window so that the point of intersection is visible.

d. Press [2ndF] CALC 2 to find the point of intersection. Press [2ndF] CALC 2 again to find another point of intersection.

e. The x- and y-coordinates at the bottom of the screen are the coordinates for the point of intersection.

Finding Maximum and Minimum Values of a Function

a. Press [Y=] and then enter the function.

b. Graph the equation. You may need to adjust the viewing window so that the maximum (minimum) is visible.

c. Press [2ndF] CALC 3 for the minimum value of the function or [2ndF] CALC 4 for the maximum value of the function.

d. The y-coordinate at the bottom of the screen is the maximum (minimum).

Casio *CFX-9850G*

To evaluate an expression

a. Press [MENU] [5]. Use the arrow keys to highlight Y1.

b. Input the expression to be evaluated. For example, to input the expression $-3A^2B - 4C$, use the following keystrokes:

[(−)] 3 [ALPHA] A [x^2] [ALPHA] B [−] 4 [ALPHA] C [EXE]

Note the difference between the key for a *negative* sign [(−)] and the key for a *minus* sign [−].

c. Press [MENU] 1. Store the value of each variable that will be used in the expression. For example, to evaluate the expression above when $A = 3$, $B = -2$, and $C = -4$, use the following keystrokes:

3 → [ALPHA] A [EXE] [(−)] 2 → [ALPHA] B [EXE] [(−)] 4 → [ALPHA] C [EXE]

These steps store the value of each variable.

d. Press [VARS] [F4] [F1] 1 [EXE].

The value of the expression, Y1, for the given values is displayed; in this case, Y1 = 70.

To graph a function

a. Press Menu [5] to obtain the GRAPH FUNCTION Menu.

b. Input the function that you want to graph. Press [X,θ,T] to input the variable x. For example, to input $Y1 = x^3 + 2x^2 - 5x - 6$, use the following keystrokes:

[X,θ,T] [∧] 3 [+] 2 [X,θ,T] [x^2] [−] 5 [X,θ,T] [−] 6 [EXE]

c. Set the viewing window by pressing [SHIFT] [F3] and the Range Parameter Menu will appear. Enter the values for the minimum x value (Xmin), maximum x value (Xmax), the units between tick marks on the x-axis (Xscl), minimum y value (Ymin), maximum y value (Ymax), and the units between tick marks on the y-axis (Yscl). Press [EXE] after each of these six entries.

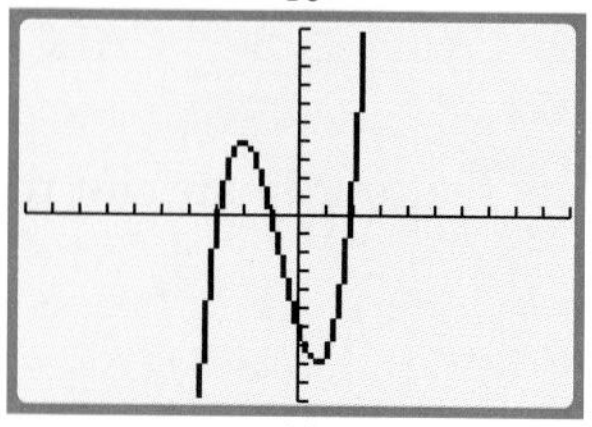

Press EXIT, or SHIFT QUIT, to leave the Range Parameter Menu.

d. Press F6 to draw the graph. For the graph shown at the left, Xmin = −10, Xmax = 10, Xscl = 1, Ymin = −10, Ymax = 10, and Yscl = 1.

e. In the equation for Y_1, there is a rectangle around the equals sign. This indicates that this function is *active* and will be graphed when the F6 key is pressed. A function is deactivated by using the F1 key. After this key is used once, the rectangle around the equals sign will not be present, and that function will not be graphed.

To display the x-coordinates of rectangular coordinates as integers

a. Set the Range as follows: Xmin = −63, Xmax = 63, Xscl = 10, Ymin = −32, Ymax = 32, and Yscl = 10.

b. Graph a function and use the Trace feature. Press F1 and then move the cursor with the ◁ and the ▷ keys. The values of x and $y = f(x)$ displayed on the bottom of the screen are the coordinates of a point on the graph. Observe that the x value is given as an integer.

To display the x-coordinates of rectangular coordinates in tenths

a. Set the Range as follows: Xmin = −6.3 and Xmax = 6.3. A quick way to choose these range parameter settings is to press F1 from the V-Window Menu.

b. Graph a function and use the Trace feature. Press F1 and then move the cursor with the ◁ and the ▷ keys. The values of x and $y = f(x)$ displayed on the bottom of the screen are the coordinates of a point on the graph. Observe that the x value is given in tenths.

To evaluate a function for a given value of x, or to produce ordered pairs of the function

a. Press MENU 5.

b. Input the function to be evaluated. For example, input $2x^3 - 3x + 2$ into Y_1.

c. Press MENU 1.

d. Input a value for x; for example, to input 3, press
3 → X,θ,T EXE

e. Press VARS F4 F1 1 EXE.
The value of Y_1 for the given value $x = 3$ is displayed. In this case, $Y_1 = 47$.

Zoom Features

To zoom in or out on a graph

a. After drawing a graph, press SHIFT Zoom to display the Zoom/Auto Range menu. To zoom in on a graph by a factor of 2 on the x-axis and a factor of 1.5 on the y-axis:
Press F2 to display the Factor Input Screen. Input the zoom factors for each axis: 2 EXE 1 · 5 EXE EXIT. Press F3 to redraw the graph according to the factors specified above. To specify the center point of the

enlarged (reduced) display after pressing SHIFT Zoom, use the arrow keys to move the pointer to the position you wish to become the center of the next display. You can repeat the zoom procedures as needed. If you wish to see the original graph, press F6 F1. This procedure resets the range parameters to their original values and redraws the graph.

b. A second method of zooming makes use of the Box Zoom Function. To use this method, first draw a graph. Then press SHIFT Zoom F1. Now use the arrow (cursor) keys to move the pointer. Once the pointer is located at a portion of the graph that is of interest, press EXE. Now use the arrow keys to draw a box around the portion of the graph you wish to see. Press EXE. The portion of the graph defined by the box will be drawn.

Solving Equations

This discussion is based on the fact that the real solutions of an equation can be related to the x-intercepts of a graph. For instance, the real solutions of $x^2 = x + 1$ are the x-intercepts of the graph of $f(x) = x^2 - x - 1$, which are the zeros of f.

To solve $x^2 = x + 1$, rewrite the equation with all terms on one side. The equation is now $x^2 - x - 1 = 0$. Think of this equation as $Y_1 = x^2 - x - 1$. The x-intercepts of the graph of Y_1 are the solutions of the equation $x^2 = x + 1$.

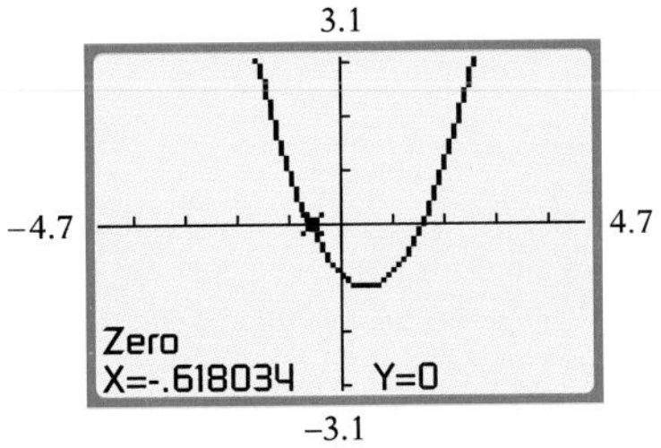

a. Enter $x^2 - x - 1$ into Y_1.

b. Graph the equation. You may need to adjust the viewing window so that the x-intercept is visible.

c. Press SHIFT G-SOLV F1.

d. The root is shown as the x-coordinate on the bottom of the screen; in this case, the root is approximately -0.618034. To find another x-intercept, press the right arrow key.

The EQUA Mode (Press MENU ALPHA A) can also be used to find solutions of linear, quadratic, and cubic equations.

Solving Systems of Two Equations in Two Variables

The following discussion is based on the concept that the solutions of a system of two equations are represented by the point(s) of intersection of the graphs.

The system of equations $\begin{aligned} y &= x^2 - 1 \\ \tfrac{1}{2}x + y &= 1 \end{aligned}$ will be solved.

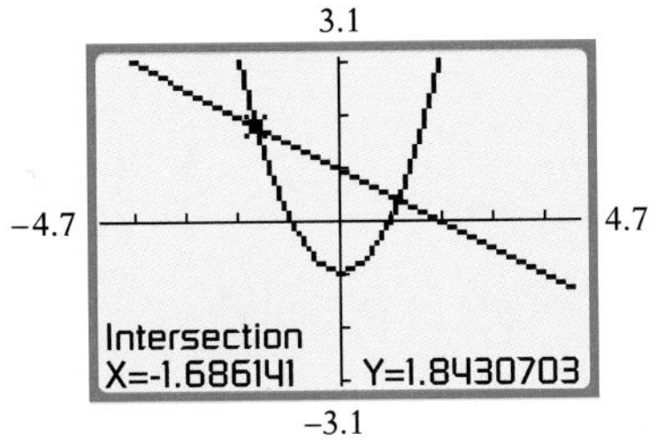

a. Solve each equation for y.

b. Enter the first equation in the Graph Menu as Y_1. For instance, let $Y_1 = x^2 - 1$.

c. Enter the second equation as Y_2. For instance, let $Y_2 = 1 - \frac{1}{2}x$.

d. Graph both equations. (*Note:* The point of intersection must appear on the screen. It may be necessary to adjust the viewing window so that the point of intersection that is of interest is the only intersection point that is displayed.)

e. Press SHIFT G-SOL F5 EXE.

f. The display will show that the graphs intersect at $(-1.686141, 1.8430703)$. To find another point of intersection, repeat step **e.**

Finding the Minimum or Maximum Value of a Function

a. Enter the function into the graphing menu. For this example we have used $y = x^2 - x - 1$.

b. Graph the function. Adjust the viewing window so that the maximum or minimum is visible.

c. Press SHIFT G-SOL F2 EXE for a maximum and F3 EXE for a minimum.

d. The local maximum (minimum) is shown as the y-coordinate on the bottom of the screen; in this case, the minimum value is -1.25.

SOLUTIONS to Chapter 1 Problems

SECTION 1.1

Problem 1 Replace z by each of the elements of the set and determine whether the inequality is true.

$z > -5$

$-10 > -5$ False

$-5 > -5$ False

$6 > -5$ True

The inequality is true for 6.

Problem 2 Replace z by each element of the set and determine the value of the expression.

$|z|$

$|-11| = 11$

$|0| = 0$

$|8| = 8$

Problem 3 $\{1, 3, 5, 7, 9, 11\}$

Problem 4 $\{x \mid x < 7, x \in \text{real numbers}\}$

Problem 5 $A \cup C = \{-5, -2, -1, 0, 1, 2, 5\}$

Problem 6 $E \cap F = \varnothing$

Problem 7

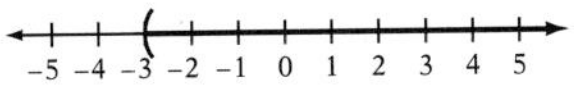

Problem 8

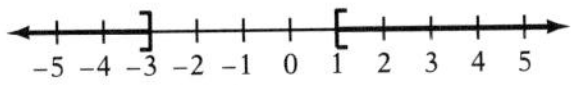

Problem 9 $[-8, -1)$

Problem 10 $\{x \mid x > -12\}$

Problem 11

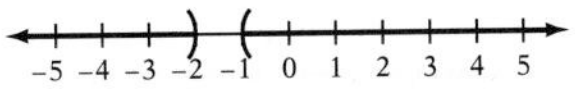

SECTION 1.2

Problem 1
$$\begin{aligned} 6 - (-8) - 10 &= 6 + 8 + (-10) \\ &= 14 + (-10) \\ &= 4 \end{aligned}$$

Problem 2 **A.**
$$\begin{aligned} 2|-7|(-8) &= 2(7)(-8) \\ &= 14(-8) = -112 \end{aligned}$$

B. $$-\frac{-36}{-3} = -\left(\frac{-36}{-3}\right) = -12$$

Problem 3 $$\frac{5}{8} \div \left(-\frac{15}{16}\right) = -\frac{5}{8} \cdot \frac{16}{15} = -\frac{5 \cdot 16}{8 \cdot 15} = -\frac{2}{3}$$

Problem 4 **A.**
$$\begin{aligned} -\frac{5}{12} \div \frac{10}{27} &= -\frac{5}{12} \cdot \frac{27}{10} \\ &= -\frac{5 \cdot 27}{12 \cdot 10} \\ &= -\frac{9}{8} \end{aligned}$$

B.
$$\begin{array}{r} 12.094 \\ -\ 8.729 \\ \hline 3.365 \end{array}$$

$-8.729 + 12.094 = 3.365$

C.
$$\begin{aligned} -8 - |5 - 12| &= -8 - |-7| \\ &= -8 - 7 \\ &= -15 \end{aligned}$$

Problem 5
$$\begin{array}{r} 4.027 \\ \times\ \ 0.49 \\ \hline 36243 \\ 16108 \\ \hline 1.97323 \approx 1.97 \end{array}$$

$-4.027(0.49) \approx -1.97$

Problem 6 $(-4)^2 = (-4)(-4) = 16$

$-4^2 = -(4)(4) = -16$

Problem 7
$$\begin{aligned} -\left(\frac{2}{5}\right)^3 \cdot 5^2 &= -\left(\frac{2}{5}\right)\left(\frac{2}{5}\right)\left(\frac{2}{5}\right) \cdot (5)(5) \\ &= -\frac{8}{5} \end{aligned}$$

Problem 8
$$\begin{aligned} &(3.81 - 1.41)^2 \div 0.036 - 1.89 \\ &= (2.40)^2 \div 0.036 - 1.89 \\ &= 5.76 \div 0.036 - 1.89 \\ &= 160 - 1.89 \\ &= 158.11 \end{aligned}$$

Problem 9
$$\begin{aligned} &\frac{1}{3} + \frac{5}{8} \div \frac{15}{16} - \frac{7}{12} \\ &= \frac{1}{3} + \frac{5}{8} \cdot \frac{16}{15} - \frac{7}{12} \\ &= \frac{1}{3} + \frac{2}{3} - \frac{7}{12} \\ &= 1 - \frac{7}{12} \\ &= \frac{5}{12} \end{aligned}$$

Problem 10 $\frac{11}{12} - \frac{\frac{5}{4}}{2 - \frac{7}{2}} \cdot \frac{3}{4}$

$= \frac{11}{12} - \frac{\frac{5}{4}}{-\frac{3}{2}} \cdot \frac{3}{4}$

$= \frac{11}{12} - \left[\frac{5}{4} \cdot \left(-\frac{2}{3}\right)\right] \cdot \frac{3}{4}$

$= \frac{11}{12} - \left(-\frac{5}{6}\right) \cdot \frac{3}{4}$

$= \frac{11}{12} - \left(-\frac{5}{8}\right)$

$= \frac{37}{24}$

SECTION 1.3

Problem 1 $(x)\left(\frac{1}{4}\right) = \left(\frac{1}{4}\right)(x)$

Problem 2 The Associative Property of Addition

Problem 3 $(b - c)^2 \div ab$

$[2 - (-4)]^2 \div (-3)(2)$

$= [6]^2 \div (-3)(2)$

$= 36 \div (-3)(2)$

$= -12(2)$

$= -24$

Problem 4 $SA = \pi r^2 + \pi rl$

$SA = \pi(5)^2 + \pi(5)(12)$

$SA = 25\pi + 60\pi$

$SA = 85\pi$

$SA \approx 267.04$

The exact area is 85π cm^2.
The approximate area is 267.04 cm^2.

Problem 5 $(2x + xy - y) - (5x - 7xy + y)$

$= 2x + xy - y - 5x + 7xy - y$

$= -3x + 8xy - 2y$

Problem 6 $2x - 3[y - 3(x - 2y + 4)]$

$= 2x - 3[y - 3x + 6y - 12]$

$= 2x - 3[7y - 3x - 12]$

$= 2x - 21y + 9x + 36$

$= 11x - 21y + 36$

SECTION 1.4

Problem 1 the unknown number: n
the difference between 8 and twice the unknown number: $8 - 2n$

$n - (8 - 2n)$

$= n - 8 + 2n$

$= 3n - 8$

Problem 2 the unknown number: n

three-eighths of the number: $\frac{3}{8}n$

five-twelfths of the number: $\frac{5}{12}n$

$\frac{3}{8}n + \frac{5}{12}n = \frac{9}{24}n + \frac{10}{24}n = \frac{19}{24}n$

Problem 3 the amount of caramel: x
the amount of milk chocolate: $x + 3$

Problem 4 the depth of the shallow end: x
the depth of the deep end: $2x + 2$

SOLUTIONS to Chapter 2 Problems

SECTION 2.1

Problem 1 $-3x = 18$

$\frac{-3x}{-3} = \frac{18}{-3}$

$x = -6$

The solution is -6.

Problem 2 $6x - 5 - 3x = 14 - 5x$

$3x - 5 = 14 - 5x$

$3x + 5x - 5 = 14 - 5x + 5x$

$8x - 5 = 14$

$8x - 5 + 5 = 14 + 5$

$8x = 19$

$\frac{8x}{8} = \frac{19}{8}$

$x = \frac{19}{8}$

The solution is $\frac{19}{8}$.

Problem 3

$$
\begin{aligned}
6(5 - x) - 12 &= 2x - 3(4 + x)\\
30 - 6x - 12 &= 2x - 12 - 3x\\
18 - 6x &= -x - 12\\
18 - 5x &= -12\\
-5x &= -30\\
x &= 6
\end{aligned}
$$

The solution is 6.

Problem 4 The LCM of 3, 5, and 30 is 30.

$$
\begin{aligned}
\frac{2x-7}{3} - \frac{5x+4}{5} &= \frac{-x-4}{30}\\
30\left(\frac{2x-7}{3} - \frac{5x+4}{5}\right) &= 30\left(\frac{-x-4}{30}\right)\\
\frac{30(2x-7)}{3} - \frac{30(5x+4)}{5} &= \frac{30(-x-4)}{30}\\
10(2x-7) - 6(5x+4) &= -x - 4\\
20x - 70 - 30x - 24 &= -x - 4\\
-10x - 94 &= -x - 4\\
-9x - 94 &= -4\\
-9x &= 90\\
x &= -10
\end{aligned}
$$

The solution is -10.

Problem 5

Strategy Next year's salary: s
Next year's salary is the sum of this year's salary and the raise.

Solution

$$
\begin{aligned}
s &= 14{,}500 + 0.08(14{,}500)\\
&= 14{,}500 + 1160\\
&= 15{,}660
\end{aligned}
$$

Next year's salary is \$15,660.

SECTION 2.2

Problem 1

Strategy

- Number of 3¢ stamps: x
  Number of 10¢ stamps: $2x + 2$
  Number of 15¢ stamps: $3x$

	Number	Value	Total value
3¢ stamp	x	3	$3x$
10¢ stamp	$2x + 2$	10	$10(2x + 2)$
15¢ stamp	$3x$	15	$45x$

- The sum of the total values of the individual types of stamps equals the total value of all the stamps (156 cents).

Solution

$$
\begin{aligned}
3x + 10(2x + 2) + 45x &= 156\\
3x + 20x + 20 + 45x &= 156\\
68x + 20 &= 156\\
68x &= 136\\
x &= 2
\end{aligned}
$$

$3x = 3(2) = 6$

There are six 15¢ stamps in the collection.

Problem 2

Strategy

- The first number: n
  The second number: $2n$
  The third number: $4n - 3$
- The sum of the numbers is 81.

Solution

$$
\begin{aligned}
n + 2n + (4n - 3) &= 81\\
7n - 3 &= 81\\
7n &= 84\\
n &= 12
\end{aligned}
$$

$2n = 2(12) = 24$

$4n - 3 = 4(12) - 3 = 48 - 3 = 45$

The numbers are 12, 24, and 45.

SECTION 2.3

Problem 1

Strategy

- Pounds of \$3.00 hamburger: x
  Pounds of \$1.80 hamburger: $75 - x$

	Amount	Cost	Value
\$3.00 hamburger	x	3.00	$3.00x$
\$1.80 hamburger	$75 - x$	1.80	$1.80(75 - x)$
Mixture	75	2.20	75(2.20)

- The sum of the values before mixing equals the value after mixing.

Solution

$$
\begin{aligned}
3.00x + 1.80(75 - x) &= 75(2.20)\\
3x + 135 - 1.80x &= 165\\
1.2x + 135 &= 165\\
1.2x &= 30\\
x &= 25
\end{aligned}
$$

$75 - x = 75 - 25 = 50$

The mixture must contain 25 lb of the \$3.00 hamburger and 50 lb of the \$1.80 hamburger.

Problem 2

Strategy ▪ Rate of the second plane: r
Rate of the first plane: $r + 30$

	Rate	Time	Distance
First plane	$r + 30$	4	$4(r + 30)$
Second plane	r	4	$4r$

▪ The total distance traveled by the two planes is 1160 mi.

Solution
$$4(r + 30) + 4r = 1160$$
$$4r + 120 + 4r = 1160$$
$$8r + 120 = 1160$$
$$8r = 1040$$
$$r = 130$$

$r + 30 = 130 + 30 = 160$

The first plane is traveling 160 mph.
The second plane is traveling 130 mph.

SECTION 2.4

Problem 1

Strategy ▪ Amount invested at 7.5%: x

	Principal	Rate	Interest
Amount at 5.2%	3500	0.052	0.052(3500)
Amount at 7.5%	x	0.075	$0.075x$

▪ The sum of the interest earned by the two investments equals the total annual interest earned ($575).

Solution
$$0.052(3500) + 0.075x = 575$$
$$182 + 0.075x = 575$$
$$0.075x = 393$$
$$x = 5240$$

The amount invested at 7.5% is $5240.

Problem 2

Strategy ▪ Pounds of 22% fat hamburger: x
Pounds of 12% fat hamburger: $80 - x$

	Amount	Percent	Quantity
22%	x	0.22	$0.22x$
12%	$80 - x$	0.12	$0.12(80 - x)$
18%	80	0.18	0.18(80)

▪ The sum of the quantities before mixing is equal to the quantity after mixing.

Solution
$$0.22x + 0.12(80 - x) = 0.18(80)$$
$$0.22x + 9.6 - 0.12x = 14.4$$
$$0.10x + 9.6 = 14.4$$
$$0.10x = 4.8$$
$$x = 48$$

$80 - x = 80 - 48 = 32$

The butcher needs 48 lb of the hamburger that is 22% fat and 32 lb of the hamburger that is 12% fat.

SECTION 2.5

Problem 1
$$2x - 1 < 6x + 7$$
$$-4x - 1 < 7$$
$$-4x < 8$$
$$\frac{-4x}{-4} > \frac{8}{-4}$$
$$x > -2$$

$\{x \mid x > -2\}$

Problem 2
$$5x - 2 \le 4 - 3(x - 2)$$
$$5x - 2 \le 4 - 3x + 6$$
$$5x - 2 \le 10 - 3x$$
$$8x - 2 \le 10$$
$$8x \le 12$$
$$\frac{8x}{8} \le \frac{12}{8}$$
$$x \le \frac{3}{2}$$

$\left(-\infty, \frac{3}{2}\right]$

Problem 3
$$-2 \le 5x + 3 \le 13$$
$$-2 - 3 \le 5x + 3 - 3 \le 13 - 3$$
$$-5 \le 5x \le 10$$
$$\frac{-5}{5} \le \frac{5x}{5} \le \frac{10}{5}$$
$$-1 \le x \le 2$$

$[-1, 2]$

Problem 4
$5 - 4x > 1$ and $6 - 5x < 11$
$-4x > -4$ $\quad$ $-5x < 5$
$x < 1$ $\quad$ $x > -1$
$\{x \mid x < 1\}$ $\quad$ $\{x \mid x > -1\}$

$\{x \mid x < 1\} \cap \{x \mid x > -1\} = (-1, 1)$

Problem 5
$2 - 3x > 11$ or $5 + 2x > 7$
$-3x > 9$ $\quad$ $2x > 2$
$x < -3$ $\quad$ $x > 1$
$\{x \mid x < -3\}$ $\quad$ $\{x \mid x > 1\}$

$\{x \mid x < -3\} \cup \{x \mid x > 1\}$
$= \{x \mid x < -3 \text{ or } x > 1\}$

Problem 6

Strategy To find the maximum height, substitute the given values in the inequality $\frac{1}{2}bh < A$ and solve for x.

Solution

$$\frac{1}{2}bh < A$$
$$\frac{1}{2}(12)(x + 2) < 50$$
$$6(x + 2) < 50$$
$$6x + 12 < 50$$
$$6x < 38$$
$$x < \frac{19}{3}$$

The largest integer less than $\frac{19}{3}$ is 6.

$x + 2 = 6 + 2 = 8$

The maximum height of the triangle is 8 in.

Problem 7

Strategy To find the range of scores, write and solve an inequality using N to represent the score on the fifth exam.

Solution

$$80 \le \frac{72 + 94 + 83 + 70 + N}{5} \le 89$$
$$80 \le \frac{319 + N}{5} \le 89$$
$$5(80) \le 5\left(\frac{319 + N}{5}\right) \le 5(89)$$
$$400 \le 319 + N \le 445$$
$$400 - 319 \le 319 - 319 + N \le 445 - 319$$
$$81 \le N \le 126$$

Because 100 is a maximum score, the range of scores that will give the student a B for the course is $81 \le N \le 100$.

SECTION 2.6

Problem 1 **A.** $|x| = 25$

$x = 25 \qquad x = -25$

The solutions are 25 and -25.

B. $|2x - 3| = 5$

$2x - 3 = 5 \qquad 2x - 3 = -5$

$2x = 8 \qquad 2x = -2$

$x = 4 \qquad x = -1$

The solutions are 4 and -1.

C. $|x - 3| = -2$

The absolute value of a number must be nonnegative.

There is no solution to this equation.

D. $5 - |3x + 5| = 3$

$-|3x + 5| = -2$

$|3x + 5| = 2$

$3x + 5 = 2 \qquad 3x + 5 = -2$

$3x = -3 \qquad 3x = -7$

$x = -1 \qquad x = -\frac{7}{3}$

The solutions are -1 and $-\frac{7}{3}$.

Problem 2 $|3x + 2| < 8$

$$-8 < 3x + 2 < 8$$
$$-8 - 2 < 3x + 2 - 2 < 8 - 2$$
$$-10 < 3x < 6$$
$$\frac{-10}{3} < \frac{3x}{3} < \frac{6}{3}$$
$$-\frac{10}{3} < x < 2$$
$$\left\{x \middle| -\frac{10}{3} < x < 2\right\}$$

Problem 3 $|3x - 7| < 0$

The absolute value of a number must be nonnegative. The solution set is the empty set.

$\varnothing$

Problem 4 $|5x + 3| > 8$

$5x + 3 < -8 \quad$ or $\quad 5x + 3 > 8$

$5x < -11 \qquad 5x > 5$

$x < -\frac{11}{5} \qquad x > 1$

$\left\{x \middle| x < -\frac{11}{5}\right\} \qquad \{x | x > 1\}$

$$\left\{x \middle| x < -\frac{11}{5}\right\} \cup \{x | x > 1\}$$
$$= \left\{x \middle| x < -\frac{11}{5} \text{ or } x > 1\right\}$$

Problem 5

Strategy Let b represent the desired diameter of the bushing, T the tolerance, and d the actual diameter of the bushing. Solve the absolute value inequality $|d - b| \le T$ for d.

Solution

$$|d - b| \le T$$
$$|d - 2.55| \le 0.003$$
$$-0.003 \le d - 2.55 \le 0.003$$
$$-0.003 + 2.55 \le d - 2.55 + 2.55 \le 0.003 + 2.55$$
$$2.547 \le d \le 2.553$$

The lower and upper limits of the diameter of the bushing are 2.547 in. and 2.553 in.

SOLUTIONS to Chapter 3 Problems

SECTION 3.1

Problem 1 Replace x by -2 and solve for y.

$$y = \frac{3(-2)}{-2+1} = \frac{-6}{-1} = 6$$

The ordered-pair solution is $(-2, 6)$.

Problem 2

x	y
-3	2
-2	1
-1	0
0	1
1	2

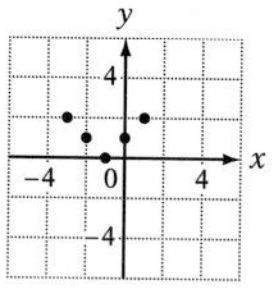

Problem 3 $(x_1, y_1) = (5, -2)$ $(x_2, y_2) = (-4, 3)$

$$\begin{aligned} d &= \sqrt{(x_2 - x_1)^2 + (y_2 - y_1)^2} \\ &= \sqrt{(-4 - 5)^2 + (3 - (-2))^2} \\ &= \sqrt{(-9)^2 + (5)^2} \\ &= \sqrt{81 + 25} \\ &= \sqrt{106} \end{aligned}$$

Problem 4 $(x_1, y_1) = (-3, -5)$ $(x_2, y_2) = (-2, 3)$

$$x_m = \frac{x_1 + x_2}{2} = \frac{-3 + (-2)}{2} = -\frac{5}{2} \qquad y_m = \frac{y_1 + y_2}{2} = \frac{-5 + 3}{2} = -1$$

The midpoint is $\left(-\frac{5}{2}, -1\right)$.

Problem 5

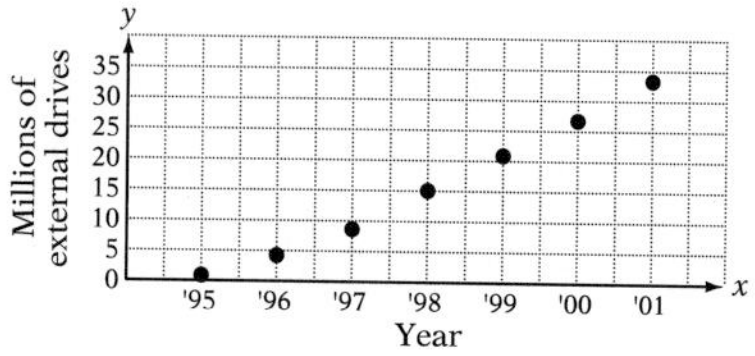

SECTION 3.2

Problem 1 The domain is $\{-1, 3, 4, 6\}$.
The range is $\{5\}$.

Problem 2 $G(x) = \dfrac{2x}{x+4}$

$$G(-2) = \frac{2(-2)}{-2+4} = \frac{-4}{2} = -2$$

Problem 3

$$\begin{aligned} f(x) &= x^2 - x - 2 \\ f(-3) &= (-3)^2 - (-3) - 2 = 10 \\ f(-2) &= (-2)^2 - (-2) - 2 = 4 \\ f(-1) &= (-1)^2 - (-1) - 2 = 0 \\ f(0) &= 0^2 - 0 - 2 = -2 \\ f(1) &= 1^2 - 1 - 2 = -2 \\ f(2) &= 2^2 - 2 - 2 = 0 \end{aligned}$$

The range is $\{-2, 0, 4, 10\}$.

Problem 4 $f(x) = \dfrac{2}{x-5}$

For $x = 5$, $5 - 5 = 0$.

$f(5) = \dfrac{2}{5-5} = \dfrac{2}{0}$, which is not a real number.

5 is excluded from the domain of $f(x) = \dfrac{2}{x-5}$.

SECTION 3.3

Problem 1

x	y
-5	-7
0	-4
5	-1

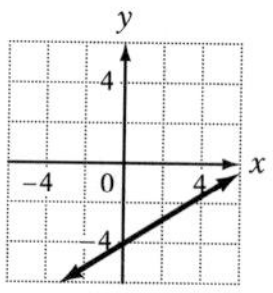

Problem 2

$$\begin{aligned} -3x + 2y &= 4 \\ 2y &= 3x + 4 \\ y &= \frac{3}{2}x + 2 \end{aligned}$$

x	y
-2	-1
0	2
2	5

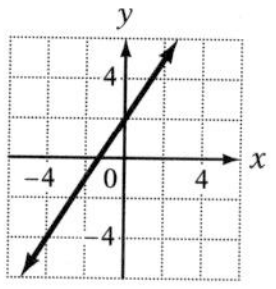

Problem 3

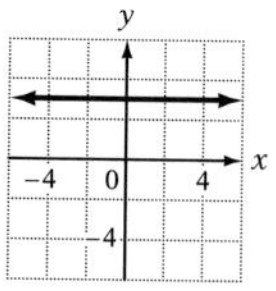

Problem 4 x-intercept:

$$3x - y = 2$$
$$3x - 0 = 2$$
$$3x = 2$$
$$x = \frac{2}{3}$$
$$\left(\frac{2}{3}, 0\right)$$

y-intercept:

$$3x - y = 2$$
$$3(0) - y = 2$$
$$-y = 2$$
$$y = -2$$
$$(0, -2)$$

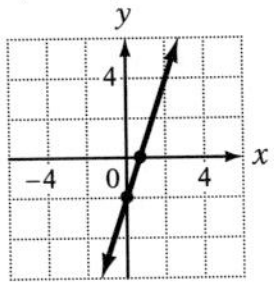

Problem 5

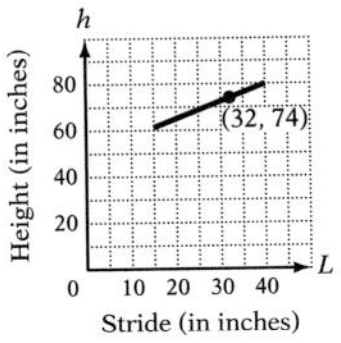

The ordered pair (32, 74) means that a person with a stride of 32 in. is 74 in. tall.

SECTION 3.4

Problem 1 Let $P_1 = (4, -3)$ and $P_2 = (2, 7)$.

$$m = \frac{y_2 - y_1}{x_2 - x_1} = \frac{7 - (-3)}{2 - 4} = \frac{10}{-2} = -5$$

The slope is -5.

Problem 2 $$m = \frac{55{,}000 - 25{,}000}{2 - 5} = \frac{30{,}000}{-3} = -10{,}000$$

A slope of $-10{,}000$ means that the value of the printing press is decreasing by \$10,000 per year.

Problem 3

$$2x + 3y = 6$$
$$3y = -2x + 6$$
$$y = -\frac{2}{3}x + 2$$

$$m = -\frac{2}{3} = \frac{-2}{3}$$

y-intercept $= (0, 2)$

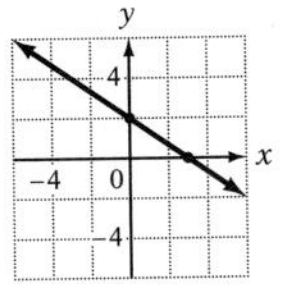

Problem 4 $(x_1, y_1) = (-3, -2)$

$$m = 3 = \frac{3}{1} = \frac{\text{change in } y}{\text{change in } x}$$

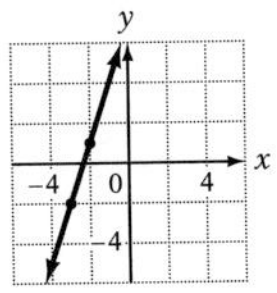

SECTION 3.5

Problem 1 $m = -3 \quad (x_1, y_1) = (4, -3)$

$$y - y_1 = m(x - x_1)$$
$$y - (-3) = -3(x - 4)$$
$$y + 3 = -3x + 12$$
$$y = -3x + 9$$

The equation of the line is $y = -3x + 9$.

Problem 2 Let $(x_1, y_1) = (2, 0)$ and $(x_2, y_2) = (5, 3)$.

$$m = \frac{y_2 - y_1}{x_2 - x_1} = \frac{3 - 0}{5 - 2} = \frac{3}{3} = 1$$
$$y - y_1 = m(x - x_1)$$
$$y - 0 = 1(x - 2)$$
$$y = x - 2$$

The equation of the line is $y = x - 2$.

Problem 3

Strategy

- Select the independent and dependent variables. Because the function will predict the Celsius temperature, that quantity is the dependent variable, y. The Fahrenheit temperature is the independent variable.
- From the given data, two ordered pairs are (212, 100) and (32, 0). Use these ordered pairs to determine the linear function.

Solution Let $(x_1, y_1) = (32, 0)$ and $(x_2, y_2) = (212, 100)$.

$$m = \frac{y_2 - y_1}{x_2 - x_1} = \frac{100 - 0}{212 - 32} = \frac{100}{180} = \frac{5}{9}$$
$$y - y_1 = m(x - x_1)$$
$$y - 0 = \frac{5}{9}(x - 32)$$
$$y = \frac{5}{9}(x - 32), \text{ or } C = \frac{5}{9}(F - 32).$$

The linear function is $f(F) = \frac{5}{9}(F - 32)$.

SECTION 3.6

Problem 1 $(x_1, y_1) = (-2, -3), (x_2, y_2) = (7, 1)$

$$m_1 = \frac{1 - (-3)}{7 - (-2)} = \frac{4}{9}$$

$(x_1, y_1) = (4, 1), (x_2, y_2) = (6, -5)$

$$m_2 = \frac{-5 - 1}{6 - 4} = \frac{-6}{2} = -3$$
$$m_1 \cdot m_2 = \frac{4}{9} \cdot -3 = -\frac{4}{3}$$

No, the lines are not perpendicular.

Problem 2 $5x + 2y = 2$

$$2y = -5x + 2$$
$$y = -\frac{5}{2}x + 1$$
$$m_1 = -\frac{5}{2}$$
$$5x + 2y = -6$$
$$2y = -5x - 6$$
$$y = -\frac{5}{2}x - 3$$
$$m_2 = -\frac{5}{2}$$
$$m_1 = m_2 = -\frac{5}{2}$$

Yes, the lines are parallel.

Problem 3 $x - 4y = 3$

$$-4y = -x + 3$$
$$y = \frac{1}{4}x - \frac{3}{4}$$
$$m_1 = \frac{1}{4}$$
$$m_1 \cdot m_2 = -1$$
$$\frac{1}{4} \cdot m_2 = -1$$
$$m_2 = -4$$
$$y - y_1 = m(x - x_1)$$
$$y - 3 = -4[x - (-2)]$$
$$y - 3 = -4(x + 2)$$
$$y - 3 = -4x - 8$$
$$y = -4x - 5$$

The equation of the line is $y = -4x - 5$.

SECTION 3.7

Problem 1 $x + 3y > 6$

$$3y > -x + 6$$
$$y > -\frac{1}{3}x + 2$$

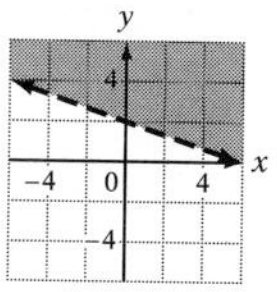

Problem 2 $y < 2$

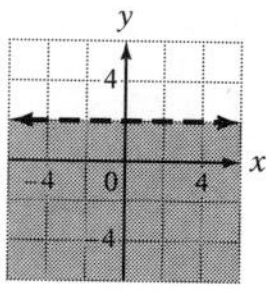

SOLUTIONS to Chapter 4 Problems

SECTION 4.1

Problem 1 **A.**

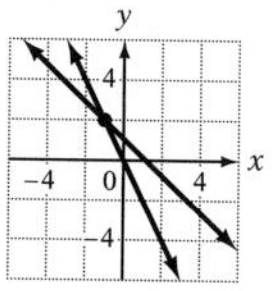

The solution is $(-1, 2)$.

B.

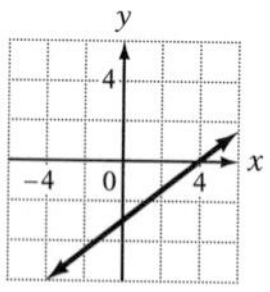

The system of equations is dependent. The solutions are the ordered pairs $\left(x, \frac{3}{4}x - 3\right)$.

Problem 2 **A.** (1) $3x - y = 3$

(2) $6x + 3y = -4$

Solve equation (1) for y.

$$3x - y = 3$$
$$-y = -3x + 3$$
$$y = 3x - 3$$

Substitute into equation (2).

$$6x + 3y = -4$$
$$6x + 3(3x - 3) = -4$$
$$6x + 9x - 9 = -4$$
$$15x - 9 = -4$$
$$15x = 5$$
$$x = \frac{5}{15} = \frac{1}{3}$$

Substitute into equation (1).

$$3x - y = 3$$
$$3\left(\frac{1}{3}\right) - y = 3$$
$$1 - y = 3$$
$$-y = 2$$
$$y = -2$$

The solution is $\left(\frac{1}{3}, -2\right)$.

B. (1) $6x - 3y = 6$
(2) $2x - y = 2$

Solve equation (2) for y.

$$2x - y = 2$$
$$-y = -2x + 2$$
$$y = 2x - 2$$

Substitute into equation (1).

$$6x - 3y = 6$$
$$6x - 3(2x - 2) = 6$$
$$6x - 6x + 6 = 6$$
$$6 = 6$$

$6 = 6$ is a true equation. The system of equations is dependent. The solutions are the ordered pairs $(x, 2x - 2)$.

SECTION 4.2

Problem 1 **A.** (1) $2x + 5y = 6$
(2) $3x - 2y = 6x + 2$

Write equation (2) in the form $Ax + By = C$.

$$3x - 2y = 6x + 2$$
$$-3x - 2y = 2$$

Solve the system: $2x + 5y = 6$
$-3x - 2y = 2$

Eliminate y.

$$2(2x + 5y) = 2(6)$$
$$5(-3x - 2y) = 5(2)$$

$$4x + 10y = 12$$
$$-15x - 10y = 10$$

Add the equations.

$$-11x = 22$$
$$x = -2$$

Replace x in equation (1).

$$2x + 5y = 6$$
$$2(-2) + 5y = 6$$
$$-4 + 5y = 6$$
$$5y = 10$$
$$y = 2$$

The solution is $(-2, 2)$.

B. $2x + y = 5$
$4x + 2y = 6$

Eliminate y.

$$-2(2x + y) = -2(5)$$
$$4x + 2y = 6$$

$$-4x - 2y = -10$$
$$4x + 2y = 6$$

Add the equations.

$$0x + 0y = -4$$
$$0 = -4$$

$0 = -4$ is not a true equation. The system is inconsistent and therefore has no solution.

Problem 2 (1) $x - y + z = 6$
(2) $2x + 3y - z = 1$
(3) $x + 2y + 2z = 5$

Eliminate z. Add equations (1) and (2).

$$x - y + z = 6$$
$$2x + 3y - z = 1$$
(4) $3x + 2y = 7$

Eliminate z. Multiply equation (2) by 2 and add to equation (3).

$$4x + 6y - 2z = 2$$
$$x + 2y + 2z = 5$$
(5) $5x + 8y = 7$

Solve the system of two equations.

(4) $3x + 2y = 7$
(5) $5x + 8y = 7$

Multiply equation (4) by -4 and add to equation (5).

$$-12x - 8y = -28$$
$$5x + 8y = 7$$
$$-7x = -21$$
$$x = 3$$

Replace x by 3 in equation (4).

$$3x + 2y = 7$$
$$3(3) + 2y = 7$$
$$9 + 2y = 7$$
$$2y = -2$$
$$y = -1$$

Replace x by 3 and y by -1 in equation (1).

$$x - y + z = 6$$
$$3 - (-1) + z = 6$$
$$4 + z = 6$$
$$z = 2$$

The solution is $(3, -1, 2)$.

SECTION 4.3

Problem 1 **A.** $\begin{vmatrix} -1 & -4 \\ 3 & -5 \end{vmatrix} = -1(-5) - 3(-4)$

$= 5 + 12 = 17$

The value of the determinant is 17.

B. Expand by cofactors of the first column.

$$\begin{vmatrix} 1 & 4 & -2 \\ 3 & 1 & 1 \\ 0 & -2 & 2 \end{vmatrix}$$

$$= 1\begin{vmatrix} 1 & 1 \\ -2 & 2 \end{vmatrix} - 3\begin{vmatrix} 4 & -2 \\ -2 & 2 \end{vmatrix} + 0$$
$$= 1(2 + 2) - 3(8 - 4)$$
$$= 4 - 12$$
$$= -8$$

The value of the determinant is −8.

Problem 2 $D = \begin{vmatrix} 6 & -6 \\ 2 & -10 \end{vmatrix} = -48$

$D_x = \begin{vmatrix} 5 & -6 \\ -1 & -10 \end{vmatrix} = -56$

$D_y = \begin{vmatrix} 6 & 5 \\ 2 & -1 \end{vmatrix} = -16$

$x = \frac{D_x}{D} = \frac{-56}{-48} = \frac{7}{6}$ $\quad y = \frac{D_y}{D} = \frac{-16}{-48} = \frac{1}{3}$

The solution is $\left(\frac{7}{6}, \frac{1}{3}\right)$.

Problem 3 $D = \begin{vmatrix} 2 & -1 & 1 \\ 3 & 2 & -1 \\ 1 & 3 & 1 \end{vmatrix} = 21$

$D_x = \begin{vmatrix} -1 & -1 & 1 \\ 3 & 2 & -1 \\ -2 & 3 & 1 \end{vmatrix} = 9$

$D_y = \begin{vmatrix} 2 & -1 & 1 \\ 3 & 3 & -1 \\ 1 & -2 & 1 \end{vmatrix} = -3$

$D_z = \begin{vmatrix} 2 & -1 & -1 \\ 3 & 2 & 3 \\ 1 & 3 & -2 \end{vmatrix} = -42$

$x = \frac{D_x}{D} = \frac{9}{21} = \frac{3}{7}$ $\quad y = \frac{D_y}{D} = \frac{-3}{21} = -\frac{1}{7}$

$z = \frac{D_z}{D} = \frac{-42}{21} = -2$

The solution is $\left(\frac{3}{7}, -\frac{1}{7}, -2\right)$.

Problem 4 $\begin{bmatrix} 3 & -5 & -12 \\ 4 & -3 & -5 \end{bmatrix}$

$\begin{bmatrix} 1 & -\frac{5}{3} & -4 \\ 4 & -3 & -5 \end{bmatrix}$ ► Multiply row 1 by $\frac{1}{3}$.

$\begin{bmatrix} 1 & -\frac{5}{3} & -4 \\ 0 & \frac{11}{3} & 11 \end{bmatrix}$ ► Multiply row 1 by −4 and add it to row 2.

$\begin{bmatrix} 1 & -\frac{5}{3} & -4 \\ 0 & 1 & 3 \end{bmatrix}$ ► Multiply row 2 by $\frac{3}{11}$.

$x - \frac{5}{3}y = -4$ $\qquad x - \frac{5}{3}(3) = -4$

$y = 3$ $\qquad x - 5 = -4$

$x = 1$

The solution is (1, 3).

Problem 5 $\begin{bmatrix} 3 & -2 & -3 & 5 \\ 1 & 3 & -2 & -4 \\ 2 & 6 & 3 & 6 \end{bmatrix}$

$\begin{bmatrix} 1 & 3 & -2 & -4 \\ 3 & -2 & -3 & 5 \\ 2 & 6 & 3 & 6 \end{bmatrix}$ ► Interchange rows 1 and 2.

$\begin{bmatrix} 1 & 3 & -2 & -4 \\ 0 & -11 & 3 & 17 \\ 0 & 0 & 7 & 14 \end{bmatrix}$ ► Multiply row 1 by −3 and add it to row 2. ► Multiply row 1 by −2 and add it to row 3.

$\begin{bmatrix} 1 & 3 & -2 & -4 \\ 0 & 1 & -\frac{3}{11} & -\frac{17}{11} \\ 0 & 0 & 7 & 14 \end{bmatrix}$ ► Multiply row 2 by $-\frac{1}{11}$.

$\begin{bmatrix} 1 & 3 & -2 & -4 \\ 0 & 1 & -\frac{3}{11} & -\frac{17}{11} \\ 0 & 0 & 1 & 2 \end{bmatrix}$ ► Multiply row 3 by $\frac{1}{7}$.

$x + 3y - 2z = -4$ $\qquad y - \frac{3}{11}(2) = -\frac{17}{11}$

$y - \frac{3}{11}z = -\frac{17}{11}$ $\qquad y - \frac{6}{11} = -\frac{17}{11}$

$z = 2$ $\qquad y = -1$

$x + 3(-1) - 2(2) = -4$
$x - 3 - 4 = -4$
$x - 7 = -4$
$x = 3$

The solution is (3, −1, 2).

SECTION 4.4

Problem 1

Strategy ▪ Rate of the rowing team in calm water: t
Rate of the current: c

	Rate	Time	Distance
With current	$t + c$	2	$2(t + c)$
Against current	$t - c$	2	$2(t - c)$

▪ The distance traveled with the current is 18 mi.
The distance traveled against the current is 10 mi.

Solution

$$2(t + c) = 18$$
$$\frac{1}{2} \cdot 2(t + c) = \frac{1}{2} \cdot 18 \rightarrow t + c = 9$$
$$2(t - c) = 10$$
$$\frac{1}{2} \cdot 2(t - c) = \frac{1}{2} \cdot 10 \rightarrow t - c = 5$$
$$2t = 14$$
$$t = 7$$

$$t + c = 9$$
$$7 + c = 9$$
$$c = 2$$

The rate of the rowing team in calm water is 7 mph.
The rate of the current is 2 mph.

Problem 2

Strategy ▪ Cost of an orange tree: x
Cost of a grapefruit tree: y

First purchase:

	Amount	Unit cost	Value
Orange trees	25	x	$25x$
Grapefruit trees	20	y	$20y$

Second purchase:

	Amount	Unit cost	Value
Orange trees	20	x	$20x$
Grapefruit trees	30	y	$30y$

▪ The total of the first purchase was \$290.
The total of the second purchase was \$330.

Solution

$$25x + 20y = 290$$
$$20x + 30y = 330$$

$$4(25x + 20y) = 4 \cdot 290$$
$$-5(20x + 30y) = -5 \cdot 330$$

$$100x + 80y = 1160$$
$$-100x - 150y = -1650$$
$$-70y = -490$$
$$y = 7$$

$$25x + 20y = 290$$
$$25x + 20(7) = 290$$
$$25x + 140 = 290$$
$$25x = 150$$
$$x = 6$$

The cost of an orange tree is \$6.
The cost of a grapefruit tree is \$7.

Problem 3

Strategy Number of full-price tickets sold: x
Number of member admission tickets sold: y
Number of student tickets sold: z
There were 750 tickets sold:
$x + y + z = 750$
The income was \$5400:
$10x + 7y + 5z = 5400$
20 more student tickets were sold than full-price tickets: $z = x + 20$

Solution

(1) $x + y + z = 750$
(2) $10x + 7y + 5z = 5400$
(3) $z = x + 20$

Solve the system of equations by substitution. Replace z in equation (1) and equation (2) by $x + 20$.

$$x + y + (x + 20) = 750$$
$$10x + 7y + 5(x + 20) = 5400$$

(4) $2x + y = 730$
(5) $15x + 7y = 5300$

▶ Multiply equation (4) by -7 and add to equation (5).

$$-14x - 7y = -5110$$
$$15x + 7y = 5300$$
$$x = 190$$

Substitute the value of x into equation (4) and solve for y.

$$2x + y = 730$$
$$2(190) + y = 730$$
$$380 + y = 730$$
$$y = 350$$

Substitute the values of x and y into equation (1) and solve for z.

$$x + y + z = 750$$
$$190 + 350 + z = 750$$
$$540 + z = 750$$
$$z = 210$$

The museum sold 190 full-priced tickets, 350 member tickets, and 210 student tickets.

SECTION 4.5

Problem 1 **A.** Shade above the solid line $y = 2x - 3$.
Shade above the dashed line $y = -3x$.

The solution set of the system is the intersection of the solution sets of the individual inequalities.

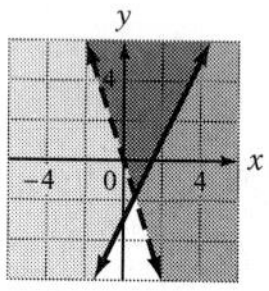

B. $3x + 4y > 12$

$$4y > -3x + 12$$
$$y > -\frac{3}{4}x + 3$$

Shade above the dashed line $y = -\frac{3}{4}x + 3$.

Shade below the dashed line $y = \frac{3}{4}x - 1$.

The solution set of the system is the intersection of the solution sets of the individual inequalities.

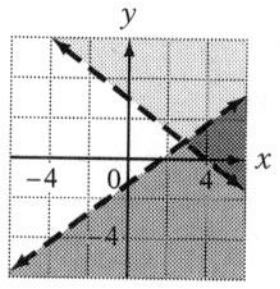

SOLUTIONS to Chapter 5 Problems

SECTION 5.1

Problem 1 $(7xy^3)(-5x^2y^2)(-xy^2) = 35x^4y^7$

Problem 2 **A.** $(y^3)^6 = y^{18}$ **B.** $(x^n)^3 = x^{3n}$

C. $(-2ab^3)^4 = (-2)^4a^{1\cdot 4}b^{3\cdot 4} = 16a^4b^{12}$

D. $6a(2a)^2 + 3a(2a^2) = 6a(2^2a^2) + 6a^3$
$= 6a(4a^2) + 6a^3$
$= 24a^3 + 6a^3 = 30a^3$

Problem 3 **A.** $(2x^{-5}y)(5x^4y^{-3}) = 10x^{-5+4}y^{1+(-3)}$
$= 10x^{-1}y^{-2} = \dfrac{10}{xy^2}$

B. $\dfrac{a^{-1}b^4}{a^{-2}b^{-2}} = a^{-1-(-2)}b^{4-(-2)} = ab^6$

C. $\left(\dfrac{2^{-1}x^2y^{-3}}{4x^{-2}y^{-5}}\right)^{-2} = \left(\dfrac{x^4y^2}{8}\right)^{-2} = \dfrac{x^{-8}y^{-4}}{8^{-2}} = \dfrac{8^2}{x^8y^4}$
$= \dfrac{64}{x^8y^4}$

Problem 4 $942{,}000{,}000 = 9.42 \times 10^8$

Problem 5 $2.7 \times 10^{-5} = 0.000027$

Problem 6 $\dfrac{5{,}600{,}000 \times 0.000000081}{900 \times 0.000000028}$
$= \dfrac{5.6 \times 10^6 \times 8.1 \times 10^{-8}}{9 \times 10^2 \times 2.8 \times 10^{-8}}$
$= \dfrac{(5.6)(8.1) \times 10^{6+(-8)-2-(-8)}}{(9)(2.8)} = 1.8 \times 10^4$

Problem 7

Strategy To find the number of arithmetic operations:
- Find the reciprocal of 1×10^{-7}, which is the number of operations performed in 1 s.
- Write the number of seconds in 1 min (60) in scientific notation.
- Multiply the number of arithmetic operations per second by the number of seconds in 1 min.

Solution $\dfrac{1}{1 \times 10^{-7}} = 10^7$
$60 = 6 \times 10$
$6 \times 10 \times 10^7 = 6 \times 10^8$

The computer can perform 6×10^8 operations in 1 min.

SECTION 5.2

Problem 1

Strategy Evaluate the function when $r = 0.4$.

Solution
$$\begin{aligned} v(r) &= 600r^2 - 1000r^3 \\ &= 600(0.4)^2 - 1000(0.4)^3 \\ &= 32 \end{aligned}$$

The velocity will be 32 cm/s.

Problem 2 The leading coefficient is -3, the constant term is -12, and the degree is 4.

Problem 3

A. This is a polynomial function.

B. This is not a polynomial function. A polynomial function does not have a variable expression raised to a negative power.

C. This is not a polynomial function. A polynomial function does not have a variable expression within a radical.

Problem 4

x	y
-1	3
0	0
1	-1
2	0
3	3

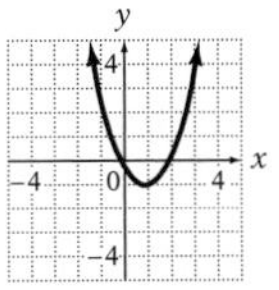

Problem 5

x	y
-3	28
-2	9
-1	2
0	1
1	0
2	-7
3	-26

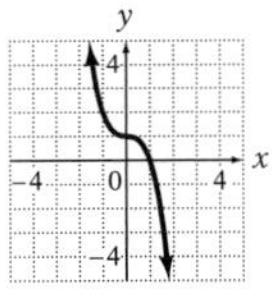

Problem 6

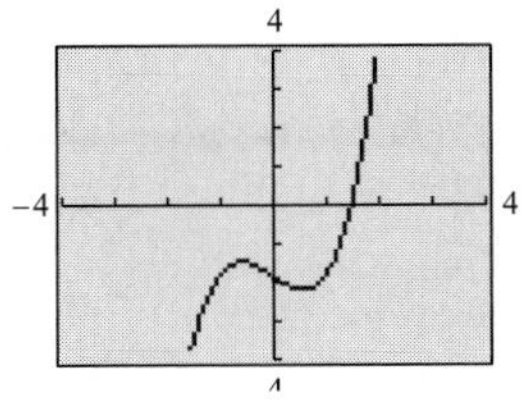

The value of x for which $S(x) = 2$ is approximately 1.8.

Problem 7

$$\begin{array}{r} 5x^2 + 3x - 1 \\ 2x^2 + 4x - 6 \\ \underline{x^2 - 7x + 8} \\ 8x^2 \qquad\quad + 1 \end{array}$$

Problem 8 Rewrite the subtraction as the addition of the opposite.

$$(-5x^2 + 2x - 3) + (-6x^2 - 3x + 7)$$

$$\begin{array}{r} -5x^2 + 2x - 3 \\ \underline{-6x^2 - 3x + 7} \\ -11x^2 - \ \ x + 4 \end{array}$$

Problem 9
$$\begin{aligned} D(x) &= (4x^3 - 3x^2 + 2) - (-2x^2 + 2x - 3) \\ &= (4x^3 - 3x^2 + 2) + (2x^2 - 2x + 3) \\ &= 4x^3 - x^2 - 2x + 5 \end{aligned}$$

SECTION 5.3

Problem 1

A.
$$\begin{aligned} &-4y(y^2 - 3y + 2) \\ &= -4y(y^2) - (-4y)(3y) + (-4y)(2) \\ &= -4y^3 + 12y^2 - 8y \end{aligned}$$

B.
$$\begin{aligned} &x^2 - 2x[x - x(4x - 5) + x^2] \\ &= x^2 - 2x[x - 4x^2 + 5x + x^2] \\ &= x^2 - 2x[6x - 3x^2] \\ &= x^2 - 12x^2 + 6x^3 \\ &= 6x^3 - 11x^2 \end{aligned}$$

C.
$$\begin{aligned} &y^{n+3}(y^{n-2} - 3y^2 + 2) \\ &= y^{n+3}(y^{n-2}) - (y^{n+3})(3y^2) + (y^{n+3})(2) \\ &= y^{n+3+(n-2)} - 3y^{n+3+2} + 2y^{n+3} \\ &= y^{2n+1} - 3y^{n+5} + 2y^{n+3} \end{aligned}$$

Problem 2

$$\begin{array}{r} -2b^2 + 5b - 4 \\ \underline{-3b + 2} \\ -4b^2 + 10b - 8 \\ \underline{6b^3 - 15b^2 + 12b \qquad} \\ 6b^3 - 19b^2 + 22b - 8 \end{array}$$

Problem 3

A.
$$\begin{aligned} &(5a - 3b)(2a + 7b) \\ &= 10a^2 + 35ab - 6ab - 21b^2 \\ &= 10a^2 + 29ab - 21b^2 \end{aligned}$$

B.
$$\begin{aligned} &(2x^n + y^n)(x^n - 4y^n) \\ &= 2x^{2n} - 8x^ny^n + x^ny^n - 4y^{2n} \\ &= 2x^{2n} - 7x^ny^n - 4y^{2n} \end{aligned}$$

Problem 4

A. $(3x - 7)(3x + 7) = 9x^2 - 49$

B. $(3x - 4y)^2 = 9x^2 - 24xy + 16y^2$

C. $(2x^n + 3)(2x^n - 3) = 4x^{2n} - 9$

D. $(2x^n - 8)^2 = 4x^{2n} - 32x^n + 64$

Problem 5

Strategy To find the area, replace the variables b and h in the equation $A = \frac{1}{2}bh$ with the given values and solve for A.

Solution $A = \frac{1}{2}bh$

$$A = \frac{1}{2}(2x + 6)(x - 4)$$
$$A = (x + 3)(x - 4)$$
$$A = x^2 - 4x + 3x - 12$$
$$A = x^2 - x - 12$$

The area is $(x^2 - x - 12)$ ft^2.

Problem 6

Strategy To find the volume, subtract the volume of the small rectangular solid from the volume of the large rectangular solid.

Large rectangular solid: Length $= L_1 = 12x$
Width $= W_1 = 7x + 2$
Height $= H_1 = 5x - 4$

Small rectangular solid: Length $= L_2 = 12x$
Width $= W_2 = x$
Height $= H_2 = 2x$

Solution

V = Volume of large rectangular solid − volume of small rectangular solid

$$V = (L_1 \cdot W_1 \cdot H_1) - (L_2 \cdot W_2 \cdot H_2)$$
$$V = (12x)(7x + 2)(5x - 4) - (12x)(x)(2x)$$
$$V = (84x^2 + 24x)(5x - 4) - (12x^2)(2x)$$
$$V = (420x^3 - 336x^2 + 120x^2 - 96x) - (24x^3)$$
$$V = 396x^3 - 216x^2 - 96x$$

The volume is $(396x^3 - 216x^2 - 96x)$ ft^3.

SECTION 5.4

Problem 1

A.

$$\begin{array}{r}
5x - 1 \\
3x + 4\overline{)15x^2 + 17x - 20} \\
\underline{15x^2 + 20x} \\
-3x - 20 \\
\underline{-3x - 4} \\
-16
\end{array}$$

$$\frac{15x^2 + 17x - 20}{3x + 4} = 5x - 1 - \frac{16}{3x + 4}$$

B.

$$\begin{array}{r}
x^2 + 3x - 1 \\
3x - 1\overline{)3x^3 + 8x^2 - 6x + 2} \\
\underline{3x^3 - x^2} \\
9x^2 - 6x \\
\underline{9x^2 - 3x} \\
-3x + 2 \\
\underline{-3x + 1} \\
1
\end{array}$$

$$\frac{3x^3 + 8x^2 - 6x + 2}{3x - 1} = x^2 + 3x - 1 + \frac{1}{3x - 1}$$

Problem 2

A.

$$\begin{array}{r|rrr}
-2 & 6 & 8 & -5 \\
 & & -12 & 8 \\
\hline
 & 6 & -4 & 3
\end{array}$$

$$(6x^2 + 8x - 5) \div (x + 2)$$
$$= 6x - 4 + \frac{3}{x + 2}$$

B.

$$\begin{array}{r|rrrrr}
3 & 2 & -3 & -8 & 0 & -2 \\
 & & 6 & 9 & 3 & 9 \\
\hline
 & 2 & 3 & 1 & 3 & 7
\end{array}$$

$$(2x^4 - 3x^3 - 8x^2 - 2) \div (x - 3)$$
$$= 2x^3 + 3x^2 + x + 3 + \frac{7}{x - 3}$$

Problem 3

$$\begin{array}{r|rrrr}
3 & 2 & 0 & -4 & -5 \\
 & & 6 & 18 & 42 \\
\hline
 & 2 & 6 & 14 & 37
\end{array}$$

By the Remainder Theorem, $P(3) = 37$.

SECTION 5.5

Problem 1

A. The GCF of $3x^3y$, $6x^2y^2$, and $3xy^3$ is $3xy$.

$$3x^3y - 6x^2y^2 - 3xy^3$$
$$= 3xy(x^2 - 2xy - y^2)$$

B. The GCF of $6t^{2n}$ and $9t^n$ is $3t^n$.

$$6t^{2n} - 9t^n = 3t^n(2t^n - 3)$$

Problem 2

A. $6a(2b - 5) + 7(5 - 2b)$
$$= 6a(2b - 5) - 7(2b - 5)$$
$$= (2b - 5)(6a - 7)$$

B. $3rs - 2r - 3s + 2$
$$= (3rs - 2r) - (3s - 2)$$
$$= r(3s - 2) - (3s - 2)$$
$$= (3s - 2)(r - 1)$$

Problem 3

A. $x^2 + 13x + 42 = (x + 6)(x + 7)$

B. $x^2 - x - 20 = (x + 4)(x - 5)$

C. $x^2 + 5xy + 6y^2 = (x + 2y)(x + 3y)$

Problem 4

$x^2 + 5x - 1$

There are no factors of -1 whose sum is 5.

The trinomial is nonfactorable over the integers.

Problem 5

A. $4x^2 + 15x - 4$

Factors of 4	Factors of −4
1, 4	1, −4
2, 2	−1, 4
	2, −2

Trial Factors	Middle Term
$(4x+1)(x-4)$	$-16x+x=-15x$
$(4x-1)(x+4)$	$16x-x=15x$

$4x^2+15x-4=(4x-1)(x+4)$

B. $10x^2+39x+14$

Factors of 10	Factors of 14
1, 10	1, 14
2, 5	2, 7

Trial Factors	Middle Term
$(x+2)(10x+7)$	$7x+20x=27x$
$(2x+1)(5x+14)$	$28x+5x=33x$
$(10x+1)(x+14)$	$140x+x=141x$
$(5x+2)(2x+7)$	$35x+4x=39x$

$10x^2+39x+14=(5x+2)(2x+7)$

Problem 6 **A.** $6x^2+7x-20$

$a\cdot c=-120$

Factors of −120	Sum
−1, 120	119
−2, 60	58
−3, 40	37
−4, 30	26
−5, 24	19
−6, 20	14
−8, 15	7

$$6x^2+7x-20=6x^2+15x-8x-20$$
$$=(6x^2+15x)-(8x+20)$$
$$=3x(2x+5)-4(2x+5)$$
$$=(2x+5)(3x-4)$$

B. $2-x-6x^2$

$a\cdot c=-12$

Factors of −12	Sum
1, −12	−11
2, −6	−4
3, −4	−1

$$2-x-6x^2=2-4x+3x-6x^2$$
$$=(2-4x)+(3x-6x^2)$$
$$=2(1-2x)+3x(1-2x)$$
$$=(1-2x)(2+3x)$$

Problem 7 **A.** $3a^3b^3+3a^2b^2-60ab$
$$=3ab(a^2b^2+ab-20)$$
$$=3ab(ab+5)(ab-4)$$

B. $40a-10a^2-15a^3=5a(8-2a-3a^2)$
$$=5a(2+a)(4-3a)$$

SECTION 5.6

Problem 1 $x^2-36y^4=x^2-(6y^2)^2$
$$=(x+6y^2)(x-6y^2)$$

Problem 2 $9x^2+12x+4=(3x+2)^2$

Problem 3

A. $8x^3+y^3z^3=(2x)^3+(yz)^3$
$$=(2x+yz)(4x^2-2xyz+y^2z^2)$$

B. $(x-y)^3+(x+y)^3$
$$=[(x-y)+(x+y)]$$
$$\cdot[(x-y)^2-(x-y)(x+y)+(x+y)^2]$$
$$=2x[x^2-2xy+y^2-(x^2-y^2)+x^2+2xy+y^2]$$
$$=2x[x^2-2xy+y^2-x^2+y^2+x^2+2xy+y^2]$$
$$=2x(x^2+3y^2)$$

Problem 4 **A.** $6x^2y^2-19xy+10$
$$=(3xy-2)(2xy-5)$$

B. $3x^4+4x^2-4=(x^2+2)(3x^2-2)$

Problem 5 **A.** $4x-4y-x^3+x^2y$
$$=(4x-4y)-(x^3-x^2y)$$
$$=4(x-y)-x^2(x-y)$$
$$=(x-y)(4-x^2)$$
$$=(x-y)(2+x)(2-x)$$

B. $x^{4n}-x^{2n}y^{2n}=x^{2n+2n}-x^{2n}y^{2n}$
$$=x^{2n}(x^{2n}-y^{2n})=x^{2n}[(x^n)^2-(y^n)^2]$$
$$=x^{2n}(x^n+y^n)(x^n-y^n)$$

SECTION 5.7

Problem 1 **A.**
$$4x^2+11x=3$$
$$4x^2+11x-3=0$$
$$(4x-1)(x+3)=0$$

$4x-1=0 \quad x+3=0$

$4x=1 \quad x=-3$

$x=\frac{1}{4}$

The solutions are $\frac{1}{4}$ and -3.

B. $(x-2)(x+5)=8$
$$x^2+3x-10=8$$
$$x^2+3x-18=0$$
$$(x+6)(x-3)=0$$

$x+6=0 \quad x-3=0$

$x=-6 \quad x=3$

The solutions are -6 and 3.

C.
$$x^3 + 4x^2 - 9x - 36 = 0$$
$$x^2(x + 4) - 9(x + 4) = 0$$
$$(x + 4)(x^2 - 9) = 0$$
$$(x + 4)(x + 3)(x - 3) = 0$$

$$x + 4 = 0 \qquad x + 3 = 0 \qquad x - 3 = 0$$
$$x = -4 \qquad x = -3 \qquad x = 3$$

The solutions are -4, -3, and 3.

Problem 2
$$s(c) = 4$$
$$c^2 - c - 2 = 4$$
$$c^2 - c - 6 = 0$$
$$(c + 2)(c - 3) = 0$$

$$c + 2 = 0 \qquad c - 3 = 0$$
$$c = -2 \qquad c = 3$$

The values of c are -2 and 3.

Problem 3

Strategy Replace D by 20 in the equation $D = \dfrac{n(n - 3)}{2}$ and then solve for n.

Solution
$$D = \frac{n(n - 3)}{2}$$

$$20 = \frac{n(n - 3)}{2}$$

$$40 = n^2 - 3n$$
$$0 = n^2 - 3n - 40$$
$$0 = (n - 8)(n + 5)$$

$$n - 8 = 0 \qquad n + 5 = 0$$
$$n = 8 \qquad n = -5$$

The answer -5 does not make sense in the context of this problem.

The polygon has 8 sides.

SOLUTIONS to Chapter 6 Problems

SECTION 6.1

Problem 1
$$f(x) = \frac{3 - 5x}{x^2 + 5x + 6}$$

$$f(2) = \frac{3 - 5(2)}{2^2 + 5(2) + 6}$$

$$f(2) = \frac{3 - 10}{4 + 10 + 6}$$

$$f(2) = \frac{-7}{20}$$

$$f(2) = -\frac{7}{20}$$

Problem 2
$$x^2 - 4 = 0$$
$$(x - 2)(x + 2) = 0$$

$$x - 2 = 0 \qquad x + 2 = 0$$
$$x = 2 \qquad x = -2$$

The domain is $\{x \,|\, x \neq -2, 2\}$.

Problem 3 $x^2 + 4$ is a positive number.

The domain is $\{x \,|\, x \in \text{real numbers}\}$.

Problem 4 **A.**
$$\frac{6x^4 - 24x^3}{12x^3 - 48x^2} = \frac{6x^3(x - 4)}{12x^2(x - 4)}$$

$$= \frac{6x^3 \cancel{(x - 4)}}{12x^2 \cancel{(x - 4)}} = \frac{x}{2}$$

B.
$$\frac{20x - 15x^2}{15x^3 - 5x^2 - 20x} = \frac{5x(4 - 3x)}{5x(3x^2 - x - 4)}$$

$$= \frac{5x(4 - 3x)}{5x(3x - 4)(x + 1)}$$

$$= \frac{\overset{-1}{\cancel{5x(4 - 3x)}}}{\underset{1}{\cancel{5x(3x - 4)}}(x + 1)}$$

$$= -\frac{1}{x + 1}$$

C.
$$\frac{x^{2n} + x^n - 12}{x^{2n} - 3x^n} = \frac{(x^n + 4)(x^n - 3)}{x^n(x^n - 3)}$$

$$= \frac{(x^n + 4)\cancel{(x^n - 3)}}{x^n\cancel{(x^n - 3)}} = \frac{x^n + 4}{x^n}$$

SECTION 6.2

Problem 1 **A.**
$$\frac{12 + 5x - 3x^2}{x^2 + 2x - 15} \cdot \frac{2x^2 + x - 45}{3x^2 + 4x}$$

$$= \frac{(4 + 3x)(3 - x)}{(x + 5)(x - 3)} \cdot \frac{(2x - 9)(x + 5)}{x(3x + 4)}$$

$$= \frac{(4 + 3x)(3 - x)(2x - 9)(x + 5)}{(x + 5)(x - 3) \cdot x(3x + 4)}$$

$$= \frac{\overset{1}{\cancel{(4 + 3x)}}\overset{-1}{\cancel{(3 - x)}}(2x - 9)\overset{1}{\cancel{(x + 5)}}}{\underset{1}{\cancel{(x + 5)}}\underset{1}{\cancel{(x - 3)}} \cdot x\underset{1}{\cancel{(3x + 4)}}}$$

$$= -\frac{2x - 9}{x}$$

B. $$\frac{2x^2-13x+20}{x^2-16}\cdot\frac{2x^2+9x+4}{6x^2-7x-5}$$

$$=\frac{(2x-5)(x-4)}{(x-4)(x+4)}\cdot\frac{(2x+1)(x+4)}{(3x-5)(2x+1)}$$

$$=\frac{(2x-5)(x-4)(2x+1)(x+4)}{(x-4)(x+4)(3x-5)(2x+1)}$$

$$=\frac{(2x-5)\overset{1}{\cancel{(x-4)}}\overset{1}{\cancel{(2x+1)}}\overset{1}{\cancel{(x+4)}}}{\underset{1}{\cancel{(x-4)}}\underset{1}{\cancel{(x+4)}}(3x-5)\underset{1}{\cancel{(2x+1)}}}$$

$$=\frac{2x-5}{3x-5}$$

Problem 2 **A.** $$\frac{6x^2-3xy}{10ab^4}\div\frac{16x^2y^2-8xy^3}{15a^2b^2}$$

$$=\frac{6x^2-3xy}{10ab^4}\cdot\frac{15a^2b^2}{16x^2y^2-8xy^3}$$

$$=\frac{3x(2x-y)}{10ab^4}\cdot\frac{15a^2b^2}{8xy^2(2x-y)}$$

$$=\frac{45a^2b^2x(2x-y)}{80ab^4xy^2(2x-y)}=\frac{45a^2b^2x\overset{1}{\cancel{(2x-y)}}}{80ab^4xy^2\underset{1}{\cancel{(2x-y)}}}$$

$$=\frac{9a}{16b^2y^2}$$

B. $$\frac{6x^2-7x+2}{3x^2+x-2}\div\frac{4x^2-8x+3}{5x^2+x-4}$$

$$=\frac{6x^2-7x+2}{3x^2+x-2}\cdot\frac{5x^2+x-4}{4x^2-8x+3}$$

$$=\frac{(2x-1)(3x-2)}{(x+1)(3x-2)}\cdot\frac{(x+1)(5x-4)}{(2x-1)(2x-3)}$$

$$=\frac{(2x-1)(3x-2)(x+1)(5x-4)}{(x+1)(3x-2)(2x-1)(2x-3)}$$

$$=\frac{\overset{1}{\cancel{(2x-1)}}\overset{1}{\cancel{(3x-2)}}\overset{1}{\cancel{(x+1)}}(5x-4)}{\underset{1}{\cancel{(x+1)}}\underset{1}{\cancel{(3x-2)}}\underset{1}{\cancel{(2x-1)}}(2x-3)}$$

$$=\frac{5x-4}{2x-3}$$

Problem 3 The LCM is $a(a-5)(a+5)$.

$$\frac{a-3}{a^2-5a}+\frac{a-9}{a^2-25}$$

$$=\frac{a-3}{a(a-5)}\cdot\frac{a+5}{a+5}+\frac{a-9}{(a-5)(a+5)}\cdot\frac{a}{a}$$

$$=\frac{a^2+2a-15}{a(a-5)(a+5)}+\frac{a^2-9a}{a(a-5)(a+5)}$$

$$=\frac{(a^2+2a-15)+(a^2-9a)}{a(a-5)(a+5)}$$

$$=\frac{a^2+2a-15+a^2-9a}{a(a-5)(a+5)}$$

$$=\frac{2a^2-7a-15}{a(a-5)(a+5)}=\frac{(2a+3)(a-5)}{a(a-5)(a+5)}$$

$$=\frac{(2a+3)\overset{1}{\cancel{(a-5)}}}{a\underset{1}{\cancel{(a-5)}}(a+5)}=\frac{2a+3}{a(a+5)}$$

Problem 4 The LCM is $(x-2)(2x-3)$.

$$\frac{x-1}{x-2}-\frac{7-6x}{2x^2-7x+6}+\frac{4}{2x-3}$$

$$=\frac{x-1}{x-2}\cdot\frac{2x-3}{2x-3}-\frac{7-6x}{(x-2)(2x-3)}+\frac{4}{2x-3}\cdot\frac{x-2}{x-2}$$

$$=\frac{2x^2-5x+3}{(x-2)(2x-3)}-\frac{7-6x}{(x-2)(2x-3)}+\frac{4x-8}{(x-2)(2x-3)}$$

$$=\frac{(2x^2-5x+3)-(7-6x)+(4x-8)}{(x-2)(2x-3)}$$

$$=\frac{2x^2-5x+3-7+6x+4x-8}{(x-2)(2x-3)}$$

$$=\frac{2x^2+5x-12}{(x-2)(2x-3)}=\frac{(x+4)(2x-3)}{(x-2)(2x-3)}$$

$$=\frac{(x+4)\overset{1}{\cancel{(2x-3)}}}{(x-2)\underset{1}{\cancel{(2x-3)}}}=\frac{x+4}{x-2}$$

SECTION 6.3

Problem 1 **A.** The LCM of x and x^2 is x^2.

$$\frac{3+\frac{16}{x}+\frac{16}{x^2}}{6+\frac{5}{x}-\frac{4}{x^2}}=\frac{3+\frac{16}{x}+\frac{16}{x^2}}{6+\frac{5}{x}-\frac{4}{x^2}}\cdot\frac{x^2}{x^2}$$

$$=\frac{3\cdot x^2+\frac{16}{x}\cdot x^2+\frac{16}{x^2}\cdot x^2}{6\cdot x^2+\frac{5}{x}\cdot x^2-\frac{4}{x^2}\cdot x^2}$$

$$=\frac{3x^2+16x+16}{6x^2+5x-4}$$

$$=\frac{(3x+4)(x+4)}{(2x-1)(3x+4)}$$

$$=\frac{\overset{1}{\cancel{(3x+4)}}(x+4)}{(2x-1)\underset{1}{\cancel{(3x+4)}}}=\frac{x+4}{2x-1}$$

B. The LCM is $x-3$.

$$\frac{2x+5+\frac{14}{x-3}}{4x+16+\frac{49}{x-3}}$$

$$=\frac{2x+5+\frac{14}{x-3}}{4x+16+\frac{49}{x-3}}\cdot\frac{x-3}{x-3}$$

$$=\frac{(2x+5)(x-3)+\frac{14}{x-3}(x-3)}{(4x+16)(x-3)+\frac{49}{x-3}(x-3)}$$

(Continued on next page)

$$= \frac{2x^2 - x - 15 + 14}{4x^2 + 4x - 48 + 49} = \frac{2x^2 - x - 1}{4x^2 + 4x + 1}$$

$$= \frac{(2x + 1)(x - 1)}{(2x + 1)(2x + 1)} = \frac{\overset{1}{\cancel{(2x + 1)}}(x - 1)}{\underset{1}{\cancel{(2x + 1)}}(2x + 1)}$$

$$= \frac{x - 1}{2x + 1}$$

SECTION 6.4

Problem 1

A.
$$\frac{5}{2x - 3} = \frac{-2}{x + 1}$$

$$(x + 1)(2x - 3)\frac{5}{2x - 3} = (x + 1)(2x - 3)\frac{-2}{x + 1}$$

$$5(x + 1) = -2(2x - 3)$$
$$5x + 5 = -4x + 6$$
$$9x + 5 = 6$$
$$9x = 1$$
$$x = \frac{1}{9}$$

The solution is $\frac{1}{9}$.

B.
$$\frac{4x + 1}{2x - 1} = 2 + \frac{3}{x - 3}$$

$$(2x - 1)(x - 3)\frac{4x + 1}{2x - 1} = (2x - 1)(x - 3)\left(2 + \frac{3}{x - 3}\right)$$

$$(x - 3)(4x + 1) = (2x - 1)(x - 3)2 + (2x - 1)3$$
$$4x^2 - 11x - 3 = 4x^2 - 14x + 6 + 6x - 3$$
$$-11x - 3 = -8x + 3$$
$$-3x = 6$$
$$x = -2$$

The solution is -2.

Problem 2

Strategy   ▪ Time required for the small pipe to fill the tank: x

	Rate	Time	Part
Large pipe	$\frac{1}{9}$	6	$\frac{6}{9}$
Small pipe	$\frac{1}{x}$	6	$\frac{6}{x}$

  ▪ The sum of the part of the task completed by the large pipe and the part of the task completed by the small pipe is 1.

Solution
$$\frac{6}{9} + \frac{6}{x} = 1$$

$$\frac{2}{3} + \frac{6}{x} = 1$$

$$3x\left(\frac{2}{3} + \frac{6}{x}\right) = 3x \cdot 1$$

$$2x + 18 = 3x$$
$$18 = x$$

The small pipe working alone will fill the tank in 18 h.

Problem 3

Strategy   ▪ Rate of the wind: r

	Distance	Rate	Time
With wind	700	$150 + r$	$\frac{700}{150 + r}$
Against wind	500	$150 - r$	$\frac{500}{150 - r}$

  ▪ The time flying with the wind equals the time flying against the wind.

Solution
$$\frac{700}{150 + r} = \frac{500}{150 - r}$$

$$(150 + r)(150 - r)\left(\frac{700}{150 + r}\right) = (150 + r)(150 - r)\left(\frac{500}{150 - r}\right)$$

$$(150 - r)700 = (150 + r)500$$
$$105{,}000 - 700r = 75{,}000 + 500r$$
$$30{,}000 = 1200r$$
$$25 = r$$

The rate of the wind is 25 mph.

SECTION 6.5

Problem 1

Strategy   To find the cost, write and solve a proportion using x to represent the cost.

Solution
$$\frac{2}{3.10} = \frac{15}{x}$$

$$x(3.10)\frac{2}{3.10} = x(3.10)\frac{15}{x}$$

$$2x = 15(3.10)$$
$$2x = 46.50$$
$$x = 23.25$$

The cost of 15 lb of cashews is \$23.25.

Problem 2

Strategy To find the distance:
- Write the basic direct variation equation, replace the variables by the given values, and solve for *k*.
- Write the direct variation equation, replacing *k* by its value. Substitute 5 for *t*, and solve for *s*.

Solution

$$s = kt^2$$
$$64 = k(2)^2$$
$$64 = k \cdot 4$$
$$16 = k$$

$$s = 16t^2$$
$$s = 16(5)^2 = 400$$

The object will fall 400 ft in 5 s.

Problem 3

Strategy To find the resistance:
- Write the basic inverse variation equation, replace the variables by the given values, and solve for *k*.
- Write the inverse variation equation, replacing *k* by its value. Substitute 0.02 for *d*, and solve for *R*.

Solution

$$R = \frac{k}{d^2}$$
$$0.5 = \frac{k}{(0.01)^2}$$
$$0.5 = \frac{k}{0.0001}$$
$$0.00005 = k$$

$$R = \frac{0.00005}{d^2}$$
$$R = \frac{0.00005}{(0.02)^2} = 0.125$$

The resistance is 0.125 ohm.

Problem 4

Strategy To find the strength of the beam:
- Write the basic combined variation equation, replace the variables by the given values, and solve for *k*.
- Write the basic combined variation equation, replacing *k* by its value and substituting 4 for *W*; 8 for *d*, and 16 for *L*. Solve for *s*.

Solution

$$s = \frac{kWd^2}{L} \qquad s = \frac{50Wd^2}{L}$$
$$1200 = \frac{k(2)(12)^2}{12} \qquad s = \frac{50(4)8^2}{16}$$
$$1200 = \frac{k(288)}{12} \qquad s = \frac{12{,}800}{16}$$
$$14{,}400 = 288k \qquad s = 800$$
$$50 = k$$

The strength of the beam is 800 lb.

SECTION 6.6

Problem 1 **A.**

$$\frac{1}{R_1} + \frac{1}{R_2} = \frac{1}{R}$$
$$RR_1R_2\left(\frac{1}{R_1} + \frac{1}{R_2}\right) = RR_1R_2\left(\frac{1}{R}\right)$$
$$RR_1R_2\left(\frac{1}{R_1}\right) + RR_1R_2\left(\frac{1}{R_2}\right) = R_1R_2$$
$$RR_2 + RR_1 = R_1R_2$$
$$R(R_2 + R_1) = R_1R_2$$
$$R = \frac{R_1R_2}{R_2 + R_1}$$

B.

$$t = \frac{r}{r+1}$$
$$t(r+1) = (r+1)\left(\frac{r}{r+1}\right)$$
$$tr + t = r$$
$$t = r - tr$$
$$t = r(1 - t)$$
$$\frac{t}{1-t} = r$$

SOLUTIONS to Chapter 7 Problems

SECTION 7.1

Problem 1 **A.** $64^{\frac{2}{3}} = (2^6)^{\frac{2}{3}} = 2^4 = 16$

B. $16^{-\frac{3}{4}} = (2^4)^{-\frac{3}{4}} = 2^{-3} = \frac{1}{2^3} = \frac{1}{8}$

C. $(-81)^{\frac{3}{4}}$

The base of the exponential expression is negative, and the denominator of the exponent is a positive even number.

Therefore, $(-81)^{\frac{3}{4}}$ is not a real number.

Problem 2 **A.** $\dfrac{x^{\frac{1}{2}}y^{-\frac{5}{4}}}{x^{-\frac{4}{3}}y^{\frac{1}{3}}} = \dfrac{x^{\frac{1}{2}+\frac{4}{3}}}{y^{\frac{1}{3}+\frac{5}{4}}} = \dfrac{x^{\frac{11}{6}}}{y^{\frac{19}{12}}}$

B. $(x^{\frac{3}{4}}y^{\frac{1}{2}}z^{-\frac{2}{3}})^{-\frac{4}{3}} = x^{-1}y^{-\frac{2}{3}}z^{\frac{8}{9}} = \dfrac{z^{\frac{8}{9}}}{xy^{\frac{2}{3}}}$

C. $\left(\dfrac{16a^{-2}b^{\frac{4}{3}}}{9a^4b^{-\frac{2}{3}}}\right)^{-\frac{1}{2}} = \left(\dfrac{2^4a^{-6}b^2}{3^2}\right)^{-\frac{1}{2}} = \dfrac{2^{-2}a^3b^{-1}}{3^{-1}}$

$= \dfrac{3a^3}{2^2b} = \dfrac{3a^3}{4b}$

Problem 3 **A.** $(2x^3)^{\frac{3}{4}} = \sqrt[4]{(2x^3)^3} = \sqrt[4]{8x^9}$

B. $-5a^{\frac{5}{6}} = -5(a^5)^{\frac{1}{6}} = -5\sqrt[6]{a^5}$

Problem 4 **A.** $\sqrt[3]{3ab} = (3ab)^{\frac{1}{3}}$

B. $\sqrt[4]{x^4 + y^4} = (x^4 + y^4)^{\frac{1}{4}}$

Problem 5 **A.** $-\sqrt[4]{x^{12}} = -(x^{12})^{\frac{1}{4}} = -x^3$

B. $\sqrt{121x^{10}y^4} = \sqrt{11^2x^{10}y^4} = 11x^5y^2$

C. $\sqrt[3]{-125a^6b^9} = \sqrt[3]{(-5)^3a^6b^9} = -5a^2b^3$

SECTION 7.2

Problem 1 $\sqrt[5]{128x^7} = \sqrt[5]{2^7x^7} = \sqrt[5]{2^5x^5(2^2x^2)}$

$= \sqrt[5]{2^5x^5}\sqrt[5]{2^2x^2} = 2x\sqrt[5]{4x^2}$

Problem 2 **A.** $3xy\sqrt[3]{81x^5y} - \sqrt[3]{192x^8y^4}$

$= 3xy\sqrt[3]{3^4x^5y} - \sqrt[3]{2^6 \cdot 3x^8y^4}$

$= 3xy\sqrt[3]{3^3x^3}\sqrt[3]{3x^2y}$

$- \sqrt[3]{2^6x^6y^3}\sqrt[3]{3x^2y}$

$= 3xy \cdot 3x\sqrt[3]{3x^2y} - 2^2x^2y\sqrt[3]{3x^2y}$

$= 9x^2y\sqrt[3]{3x^2y} - 4x^2y\sqrt[3]{3x^2y}$

$= 5x^2y\sqrt[3]{3x^2y}$

B. $4a\sqrt[3]{54a^7b^9} + a^2b\sqrt[3]{128a^4b^6}$

$= 4a\sqrt[3]{3^3 \cdot 2a^7b^9} + a^2b\sqrt[3]{2^6 \cdot 2a^4b^6}$

$= 4a\sqrt[3]{3^3a^6b^9}\sqrt[3]{2a} + a^2b\sqrt[3]{2^6a^3b^6}\sqrt[3]{2a}$

$= 4a \cdot 3a^2b^3\sqrt[3]{2a} + a^2b \cdot 2^2ab^2\sqrt[3]{2a}$

$= 12a^3b^3\sqrt[3]{2a} + 4a^3b^3\sqrt[3]{2a}$

$= 16a^3b^3\sqrt[3]{2a}$

Problem 3 $\sqrt{5b}(\sqrt{3b} - \sqrt{10})$

$= \sqrt{15b^2} - \sqrt{50b}$

$= \sqrt{3 \cdot 5b^2} - \sqrt{2 \cdot 5^2b}$

$= \sqrt{b^2}\sqrt{3 \cdot 5} - \sqrt{5^2}\sqrt{2b}$

$= b\sqrt{15} - 5\sqrt{2b}$

Problem 4 **A.** $(2\sqrt[3]{2x} - 3)(\sqrt[3]{2x} - 5)$

$= 2\sqrt[3]{4x^2} - 10\sqrt[3]{2x} - 3\sqrt[3]{2x} + 15$

$= 2\sqrt[3]{4x^2} - 13\sqrt[3]{2x} + 15$

B. $(2\sqrt{x} - 3)(2\sqrt{x} + 3)$

$= 2^2\sqrt{x^2} - 9 = 4x - 9$

Problem 5 **A.** $\dfrac{y}{\sqrt{3y}} = \dfrac{y}{\sqrt{3y}} \cdot \dfrac{\sqrt{3y}}{\sqrt{3y}} = \dfrac{y\sqrt{3y}}{\sqrt{3^2y^2}} = \dfrac{y\sqrt{3y}}{3y}$

$= \dfrac{\sqrt{3y}}{3}$

B. $\dfrac{3}{\sqrt[3]{3x^2}} = \dfrac{3}{\sqrt[3]{3x^2}} \cdot \dfrac{\sqrt[3]{3^2x}}{\sqrt[3]{3^2x}} = \dfrac{3\sqrt[3]{9x}}{\sqrt[3]{3^3x^3}} = \dfrac{3\sqrt[3]{9x}}{3x}$

$= \dfrac{\sqrt[3]{9x}}{x}$

Problem 6 **A.** $\dfrac{4 + \sqrt{2}}{3 - \sqrt{3}} = \dfrac{4 + \sqrt{2}}{3 - \sqrt{3}} \cdot \dfrac{3 + \sqrt{3}}{3 + \sqrt{3}}$

$= \dfrac{12 + 4\sqrt{3} + 3\sqrt{2} + \sqrt{6}}{9 - (\sqrt{3})^2}$

$= \dfrac{12 + 4\sqrt{3} + 3\sqrt{2} + \sqrt{6}}{6}$

B. $\dfrac{\sqrt{2} + \sqrt{x}}{\sqrt{2} - \sqrt{x}} = \dfrac{\sqrt{2} + \sqrt{x}}{\sqrt{2} - \sqrt{x}} \cdot \dfrac{\sqrt{2} + \sqrt{x}}{\sqrt{2} + \sqrt{x}}$

$= \dfrac{\sqrt{2^2} + \sqrt{2x} + \sqrt{2x} + \sqrt{x^2}}{(\sqrt{2})^2 - (\sqrt{x})^2}$

$= \dfrac{2 + 2\sqrt{2x} + x}{2 - x}$

SECTION 7.3

Problem 1 $\sqrt{-45} = i\sqrt{45} = i\sqrt{3^2 \cdot 5} = 3i\sqrt{5}$

Problem 2 $\sqrt{98} - \sqrt{-60} = \sqrt{98} - i\sqrt{60}$

$= \sqrt{2 \cdot 7^2} - i\sqrt{2^2 \cdot 3 \cdot 5}$

$= 7\sqrt{2} - 2i\sqrt{15}$

Problem 3 $(-4 + 2i) - (6 - 8i) = -10 + 10i$

Problem 4 $(16 - \sqrt{-45}) - (3 + \sqrt{-20})$

$= (16 - i\sqrt{45}) - (3 + i\sqrt{20})$

$= (16 - i\sqrt{3^2 \cdot 5}) - (3 + i\sqrt{2^2 \cdot 5})$

$= (16 - 3i\sqrt{5}) - (3 + 2i\sqrt{5})$

$= 13 - 5i\sqrt{5}$

Problem 5 $\sqrt{-3}(\sqrt{27} - \sqrt{-6})$

$= i\sqrt{3}(\sqrt{27} - i\sqrt{6})$

$= i\sqrt{81} - i^2\sqrt{18}$

$= i\sqrt{3^4} - (-1)\sqrt{2 \cdot 3^2}$

$= 9i + 3\sqrt{2}$

$= 3\sqrt{2} + 9i$

Problem 6 **A.** $(4 - 3i)(2 - i) = 8 - 4i - 6i + 3i^2$

$= 8 - 10i + 3i^2 = 8 - 10i + 3(-1)$

$= 5 - 10i$

B. $(3 - i)\left(\dfrac{3}{10} + \dfrac{1}{10}i\right)$

$= \dfrac{9}{10} + \dfrac{3}{10}i - \dfrac{3}{10}i - \dfrac{1}{10}i^2 = \dfrac{9}{10} - \dfrac{1}{10}i^2$

$= \dfrac{9}{10} - \dfrac{1}{10}(-1) = \dfrac{9}{10} + \dfrac{1}{10} = 1$

C. $(3 + 6i)(3 - 6i) = 3^2 + 6^2 = 9 + 36$
$= 45$

Problem 7 $\dfrac{2 - 3i}{4i} = \dfrac{2 - 3i}{4i} \cdot \dfrac{i}{i} = \dfrac{2i - 3i^2}{4i^2} = \dfrac{2i - 3(-1)}{4(-1)}$

$= \dfrac{3 + 2i}{-4} = -\dfrac{3}{4} - \dfrac{1}{2}i$

Problem 8 $\dfrac{2 + 5i}{3 - 2i} = \dfrac{(2 + 5i)}{(3 - 2i)} \cdot \dfrac{(3 + 2i)}{(3 + 2i)}$

$= \dfrac{6 + 4i + 15i + 10i^2}{3^2 + 2^2} = \dfrac{6 + 19i + 10(-1)}{13}$

$= \dfrac{-4 + 19i}{13} = -\dfrac{4}{13} + \dfrac{19}{13}i$

SECTION 7.4

Problem 1 **A.** Because Q contains an odd root, there are no restrictions on the radicand.

The domain is $(-\infty, \infty)$.

B. $3x + 9 \geq 0$
$3x \geq -9$
$x \geq -3$

The domain is $[-3, \infty)$.

Problem 2

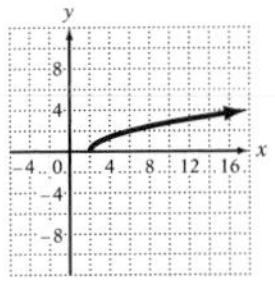

Problem 3

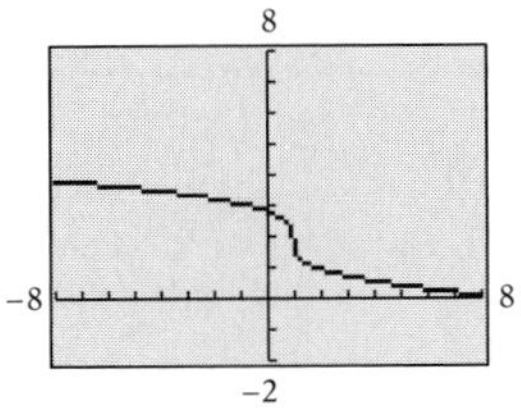

SECTION 7.5

Problem 1 **A.** $\sqrt{4x + 5} - 12 = -5$
$\sqrt{4x + 5} = 7$
$(\sqrt{4x + 5})^2 = 7^2$
$4x + 5 = 49$
$4x = 44$
$x = 11$

Check: $\sqrt{4x + 5} - 12 = -5$

$\sqrt{4x + 5} - 12$	-5
$\sqrt{4 \cdot 11 + 5} - 12$	-5
$\sqrt{44 + 5} - 12$	
$\sqrt{49} - 12$	
$7 - 12$	

$-5 = -5$

The solution is 11.

B. $\sqrt[4]{x - 8} = 3$
$(\sqrt[4]{x - 8})^4 = 3^4$
$x - 8 = 81$
$x = 89$

Check: $\sqrt[4]{x - 8} = 3$

$\sqrt[4]{89 - 8}$	3
$\sqrt[4]{81}$	

$3 = 3$

The solution is 89.

Problem 2 **A.** $x + 3\sqrt{x + 2} = 8$
$3\sqrt{x + 2} = 8 - x$
$(3\sqrt{x + 2})^2 = (8 - x)^2$
$9(x + 2) = 64 - 16x + x^2$
$9x + 18 = 64 - 16x + x^2$
$0 = x^2 - 25x + 46$
$0 = (x - 2)(x - 23)$

$x - 2 = 0 \qquad x - 23 = 0$
$x = 2 \qquad x = 23$

Check: $x + 3\sqrt{x + 2} = 8$

$2 + 3\sqrt{2 + 2}$	8
$2 + 3\sqrt{4}$	
$2 + 3 \cdot 2$	
$2 + 6$	

$8 = 8$

$x + 3\sqrt{x + 2} = 8$

$23 + 3\sqrt{23 + 2}$	8
$23 + 3\sqrt{25}$	
$23 + 3 \cdot 5$	
$23 + 15$	

$38 \neq 8$

23 does not check as a solution. The solution is 2.

B. $\sqrt{x + 5} = 5 - \sqrt{x}$
$(\sqrt{x + 5})^2 = (5 - \sqrt{x})^2$
$x + 5 = 25 - 10\sqrt{x} + x$
$-20 = -10\sqrt{x}$
$2 = \sqrt{x}$
$2^2 = (\sqrt{x})^2$
$4 = x$

Check: $\sqrt{x + 5} = 5 - \sqrt{x}$

$\sqrt{4 + 5}$	$5 - \sqrt{4}$
$\sqrt{9}$	$5 - 2$

$3 = 3$

The solution is 4.

Problem 3

Strategy To find the diagonal, use the Pythagorean Theorem. One leg is the length of the rectangle. The second leg is the width of the rectangle. The hypotenuse is the diagonal of the rectangle.

Solution

$$\begin{aligned} c^2 &= a^2 + b^2 \\ c^2 &= (6)^2 + (3)^2 \\ c^2 &= 36 + 9 \\ c^2 &= 45 \\ (c^2)^{\frac{1}{2}} &= (45)^{\frac{1}{2}} \\ c &= \sqrt{45} \\ c &\approx 6.7 \end{aligned}$$

The diagonal is 6.7 cm.

SOLUTIONS to Chapter 8 Problems

SECTION 8.1

Problem 1

$$\begin{aligned} x^2 - 3ax - 4a^2 &= 0 \\ (x + a)(x - 4a) &= 0 \end{aligned}$$

$$x + a = 0 \qquad x - 4a = 0$$
$$x = -a \qquad x = 4a$$

The solutions are $-a$ and $4a$.

Problem 2

$$\begin{aligned} (x - r_1)(x - r_2) &= 0 \\ \left[x - \left(-\frac{2}{3}\right)\right]\left(x - \frac{1}{6}\right) &= 0 \\ \left(x + \frac{2}{3}\right)\left(x - \frac{1}{6}\right) &= 0 \\ x^2 + \frac{3}{6}x - \frac{2}{18} &= 0 \\ 18\left(x^2 + \frac{3}{6}x - \frac{2}{18}\right) &= 0 \\ 18x^2 + 9x - 2 &= 0 \end{aligned}$$

Problem 3

$$\begin{aligned} 2(x + 1)^2 + 24 &= 0 \\ 2(x + 1)^2 &= -24 \\ (x + 1)^2 &= -12 \\ \sqrt{(x + 1)^2} &= \sqrt{-12} \\ x + 1 &= \pm\sqrt{-12} \\ x + 1 &= \pm 2i\sqrt{3} \end{aligned}$$

$$x + 1 = 2i\sqrt{3} \qquad x + 1 = -2i\sqrt{3}$$
$$x = -1 + 2i\sqrt{3} \qquad x = -1 - 2i\sqrt{3}$$

The solutions are $-1 + 2i\sqrt{3}$ and $-1 - 2i\sqrt{3}$.

SECTION 8.2

Problem 1

A.

$$\begin{aligned} 4x^2 - 4x - 1 &= 0 \\ 4x^2 - 4x &= 1 \\ \frac{4x^2 - 4x}{4} &= \frac{1}{4} \\ x^2 - x &= \frac{1}{4} \end{aligned}$$

Complete the square.

$$\begin{aligned} x^2 - x + \frac{1}{4} &= \frac{1}{4} + \frac{1}{4} \\ \left(x - \frac{1}{2}\right)^2 &= \frac{2}{4} \\ \sqrt{\left(x - \frac{1}{2}\right)^2} &= \sqrt{\frac{2}{4}} \\ x - \frac{1}{2} &= \pm\sqrt{\frac{2}{4}} \\ x - \frac{1}{2} &= \pm\frac{\sqrt{2}}{2} \end{aligned}$$

$$x - \frac{1}{2} = \frac{\sqrt{2}}{2} \qquad x - \frac{1}{2} = -\frac{\sqrt{2}}{2}$$
$$x = \frac{1}{2} + \frac{\sqrt{2}}{2} \qquad x = \frac{1}{2} - \frac{\sqrt{2}}{2}$$

The solutions are $\frac{1 + \sqrt{2}}{2}$ and $\frac{1 - \sqrt{2}}{2}$.

B.

$$\begin{aligned} 2x^2 + x - 5 &= 0 \\ 2x^2 + x &= 5 \\ \frac{2x^2 + x}{2} &= \frac{5}{2} \\ x^2 + \frac{1}{2}x &= \frac{5}{2} \end{aligned}$$

Complete the square.

$$\begin{aligned} x^2 + \frac{1}{2}x + \frac{1}{16} &= \frac{5}{2} + \frac{1}{16} \\ \left(x + \frac{1}{4}\right)^2 &= \frac{41}{16} \\ \sqrt{\left(x + \frac{1}{4}\right)^2} &= \sqrt{\frac{41}{16}} \\ x + \frac{1}{4} &= \pm\sqrt{\frac{41}{16}} \\ x + \frac{1}{4} &= \pm\frac{\sqrt{41}}{4} \end{aligned}$$

$$x + \frac{1}{4} = \frac{\sqrt{41}}{4} \qquad x + \frac{1}{4} = -\frac{\sqrt{41}}{4}$$
$$x = -\frac{1}{4} + \frac{\sqrt{41}}{4} \qquad x = -\frac{1}{4} - \frac{\sqrt{41}}{4}$$

The solutions are $\frac{-1 + \sqrt{41}}{4}$ and $\frac{-1 - \sqrt{41}}{4}$.

Problem 2 **A.** $x^2 + 6x - 9 = 0$

$a = 1, b = 6, c = -9$

$$\begin{aligned} x &= \frac{-b \pm \sqrt{b^2 - 4ac}}{2a} \\ &= \frac{-6 \pm \sqrt{6^2 - 4(1)(-9)}}{2 \cdot 1} \\ &= \frac{-6 \pm \sqrt{36 + 36}}{2} \\ &= \frac{-6 \pm \sqrt{72}}{2} \\ &= \frac{-6 \pm 6\sqrt{2}}{2} = -3 \pm 3\sqrt{2} \end{aligned}$$

The solutions are $-3 + 3\sqrt{2}$ and $-3 - 3\sqrt{2}$.

B.
$$\begin{aligned} 4x^2 &= 4x - 1 \\ 4x^2 - 4x + 1 &= 0 \end{aligned}$$

$a = 4, b = -4, c = 1$

$$\begin{aligned} x &= \frac{-b \pm \sqrt{b^2 - 4ac}}{2a} \\ &= \frac{-(-4) \pm \sqrt{(-4)^2 - 4(4)(1)}}{2 \cdot 4} \\ &= \frac{4 \pm \sqrt{16 - 16}}{8} \\ &= \frac{4 \pm \sqrt{0}}{8} = \frac{4}{8} = \frac{1}{2} \end{aligned}$$

The solution is $\frac{1}{2}$.

Problem 3 $3x^2 - x - 1 = 0$

$a = 3, b = -1, c = -1$

$$\begin{aligned} b^2 - 4ac &= (-1)^2 - 4(3)(-1) \\ &= 1 + 12 = 13 \end{aligned}$$

$13 > 0$

Because the discriminant is greater than zero, the equation has two real number solutions.

SECTION 8.3

Problem 1 **A.**
$$\begin{aligned} x - 5x^{\frac{1}{2}} + 6 &= 0 \\ \left(x^{\frac{1}{2}}\right)^2 - 5\left(x^{\frac{1}{2}}\right) + 6 &= 0 \\ u^2 - 5u + 6 &= 0 \\ (u - 2)(u - 3) &= 0 \end{aligned}$$

$$u - 2 = 0 \qquad u - 3 = 0$$
$$u = 2 \qquad u = 3$$

Replace u with $x^{\frac{1}{2}}$.

$$x^{\frac{1}{2}} = 2 \qquad x^{\frac{1}{2}} = 3$$
$$\left(x^{\frac{1}{2}}\right)^2 = 2^2 \qquad \left(x^{\frac{1}{2}}\right)^2 = 3^2$$
$$x = 4 \qquad x = 9$$

4 and 9 check as solutions.

The solutions are 4 and 9.

B.
$$\begin{aligned} 4x^4 + 35x^2 - 9 &= 0 \\ 4(x^2)^2 + 35(x^2) - 9 &= 0 \\ 4u^2 + 35u - 9 &= 0 \\ (4u - 1)(u + 9) &= 0 \end{aligned}$$

$$4u - 1 = 0 \qquad u + 9 = 0$$
$$4u = 1 \qquad u = -9$$
$$u = \frac{1}{4}$$

Replace u with x^2.

$$x^2 = \frac{1}{4} \qquad x^2 = -9$$
$$\sqrt{x^2} = \sqrt{\frac{1}{4}} \qquad \sqrt{x^2} = \sqrt{-9}$$
$$x = \pm\sqrt{\frac{1}{4}} \qquad x = \pm\sqrt{-9}$$
$$x = \pm\frac{1}{2} \qquad x = \pm 3i$$

The solutions are $\frac{1}{2}$, $-\frac{1}{2}$, $3i$, and $-3i$.

Problem 2
$$\begin{aligned} \sqrt{2x + 1} + x &= 7 \\ \sqrt{2x + 1} &= 7 - x \\ (\sqrt{2x + 1})^2 &= (7 - x)^2 \\ 2x + 1 &= 49 - 14x + x^2 \\ 0 &= x^2 - 16x + 48 \\ 0 &= (x - 4)(x - 12) \end{aligned}$$

$$x - 4 = 0 \qquad x - 12 = 0$$
$$x = 4 \qquad x = 12$$

4 checks as a solution, but 12 does not check as a solution.

The solution is 4.

Problem 3 $\sqrt{2x - 1} + \sqrt{x} = 2$

Solve for one of the radical expressions.
$$\sqrt{2x - 1} = 2 - \sqrt{x}$$
$$(\sqrt{2x - 1})^2 = (2 - \sqrt{x})^2$$
$$2x - 1 = 4 - 4\sqrt{x} + x$$
$$x - 5 = -4\sqrt{x}$$

Square each side of the equation.
$$(x - 5)^2 = (-4\sqrt{x})^2$$
$$x^2 - 10x + 25 = 16x$$
$$x^2 - 26x + 25 = 0$$
$$(x - 1)(x - 25) = 0$$

$$x - 1 = 0 \qquad x - 25 = 0$$
$$x = 1 \qquad\qquad x = 25$$

1 checks as a solution, but 25 does not check as a solution.

The solution is 1.

Problem 4

$$3y + \frac{25}{3y - 2} = -8$$

$$(3y - 2)\left(3y + \frac{25}{3y - 2}\right) = (3y - 2)(-8)$$

$$(3y - 2)(3y) + (3y - 2)\left(\frac{25}{3y - 2}\right) = (3y - 2)(-8)$$

$$9y^2 - 6y + 25 = -24y + 16$$
$$9y^2 + 18y + 9 = 0$$
$$9(y^2 + 2y + 1) = 0$$
$$9(y + 1)(y + 1) = 0$$

$$y + 1 = 0 \qquad y + 1 = 0$$
$$y = -1 \qquad\quad y = -1$$

The solution is -1.

SECTION 8.4

Problem 1

Strategy ▪ This is a geometry problem.
▪ Width of the rectangle: W
Length of the rectangle: $W + 3$
▪ Use the equation $A = L \cdot W$.

Solution $A = L \cdot W$
$$54 = (W + 3)(W)$$
$$54 = W^2 + 3W$$
$$0 = W^2 + 3W - 54$$
$$0 = (W + 9)(W - 6)$$

$$W + 9 = 0 \qquad W - 6 = 0$$
$$W = -9 \qquad\quad W = 6$$

The solution -9 is not possible.

Length $= W + 3 = 6 + 3 = 9$

The length is 9 m.

SECTION 8.5

Problem 1 $2x^2 - x - 10 \le 0$
$$(2x - 5)(x + 2) \le 0$$

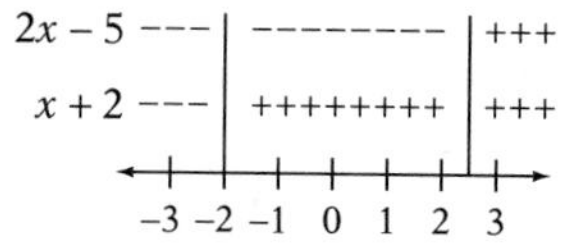

$$\left\{x\,\middle|\,-2 \le x \le \frac{5}{2}\right\}$$

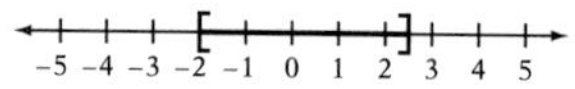

Problem 2 $\dfrac{x}{x - 2} \le 0$

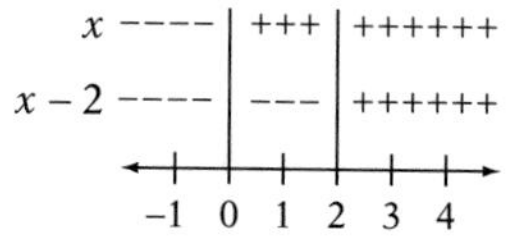

$$\{x \,|\, 0 \le x < 2\}$$

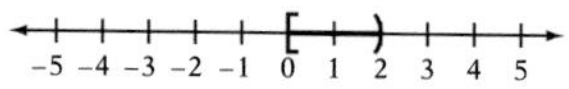

SECTION 8.6

Problem 1 x-coordinate: $-\dfrac{b}{2a} = -\dfrac{0}{2(1)} = 0$

$$y = x^2 - 2$$
$$= 0^2 - 2$$
$$= -2$$

The vertex is $(0, -2)$.

The axis of symmetry is the line $x = 0$.

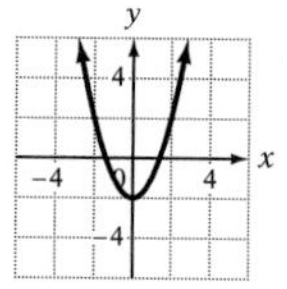

Problem 2

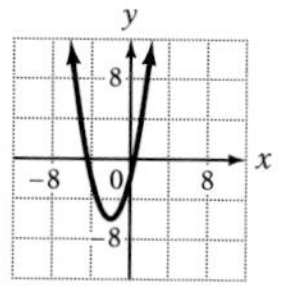

The domain is $\{x \,|\, x \in \text{real numbers}\}$.
The range is $\{y \,|\, y \ge -6\}$.

Problem 3 **A.** $y = 2x^2 - 5x + 2$

$0 = 2x^2 - 5x + 2$

$0 = (2x - 1)(x - 2)$

$2x - 1 = 0 \qquad x - 2 = 0$

$2x = 1 \qquad x = 2$

$x = \frac{1}{2}$

The x-intercepts are $\left(\frac{1}{2}, 0\right)$ and $(2, 0)$.

B. $y = x^2 + 4x + 4$

$0 = x^2 + 4x + 4$

$0 = (x + 2)(x + 2)$

$x + 2 = 0 \qquad x + 2 = 0$

$x = -2 \qquad x = -2$

The x-intercept is $(-2, 0)$.

Problem 4

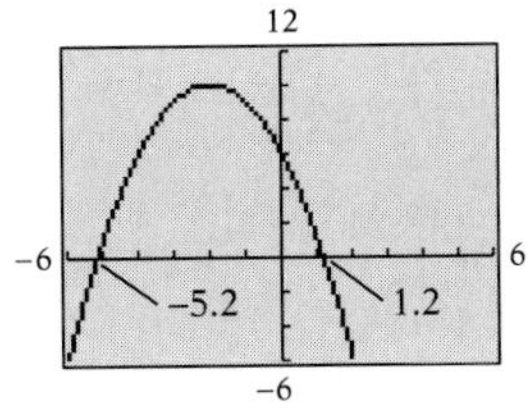

The approximate values of the zeros are -5.2 and 1.2.

Problem 5 $y = x^2 - x - 6$

$a = 1, b = -1, c = -6$

$b^2 - 4ac = (-1)^2 - 4(1)(-6)$

$= 1 + 24 = 25$

Because the discriminant is greater than zero, the parabola has two x-intercepts.

SECTION 8.7

Problem 1 $x = -\frac{b}{2a} = -\frac{-3}{2(2)} = \frac{3}{4}$

$f(x) = 2x^2 - 3x + 1$

$f\left(\frac{3}{4}\right) = 2\left(\frac{3}{4}\right)^2 - 3\left(\frac{3}{4}\right) + 1 = \frac{9}{8} - \frac{9}{4} + 1$

$= -\frac{1}{8}$

Because a is positive, the function has a minimum value.

The minimum value of the function is $-\frac{1}{8}$.

Problem 2

Strategy
- To find the time it takes the ball to reach its maximum height, find the t-coordinate of the vertex.
- To find the maximum height, evaluate the function at the t-coordinate of the vertex.

Solution $t = -\frac{b}{2a} = -\frac{64}{2(-16)} = 2$

The ball reaches its maximum height in 2 s.

$s(t) = -16t^2 + 64t$

$s(2) = -16(2)^2 + 64(2) = -64 + 128 = 64$

The maximum height is 64 ft.

Problem 3

Strategy The perimeter is 44 ft.

$44 = 2L + 2W$

$22 = L + W$

$22 - L = W$

The area is $L \cdot W = L(22 - L) = 22L - L^2$.

- To find the length, find the L-coordinate of the vertex of the function $f(L) = -L^2 + 22L$.
- To find the width, replace L in $22 - L$ by the L-coordinate of the vertex and evaluate.

Solution $L = -\frac{b}{2a} = -\frac{22}{2(-1)} = 11$

The length is 11 ft.

$22 - L = 22 - 11 = 11$

The width is 11 ft.

SOLUTIONS to Chapter 9 Problems

SECTION 9.1

Problem 1   Because any vertical line would intersect the graph at no more than one point, the graph is the graph of a function.

Problem 2   **A.**

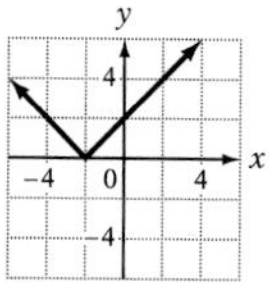

The domain is $\{x \mid x \in \text{real numbers}\}$
The range is $\{y \mid y \geq 0\}$.

B.

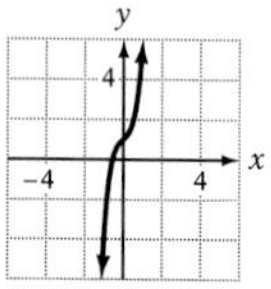

The domain is $\{x \mid x \in \text{real numbers}\}$.
The range is $\{y \mid y \in \text{real numbers}\}$.

C.

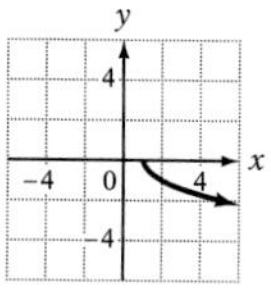

▶ $x - 1 \geq 0$
$x \geq 1$

The domain is $\{x \mid x \geq 1\}$.
The range is $\{y \mid y \leq 0\}$.

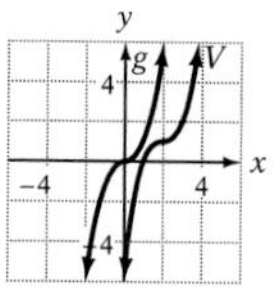

SECTION 9.2

Problem 1   The graph of $g(x) = x^2 - 2x - 3$ is the graph of $f(x) = x^2 - 2x$ shifted 3 units down.

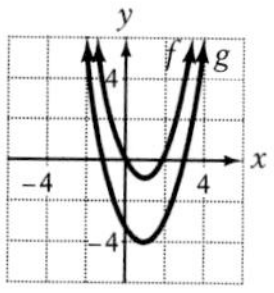

Problem 2   The graph of $g(x) = \sqrt{x - 3}$ is the graph of $f(x) = \sqrt{x}$ shifted horizontally 3 units to the right.

Problem 3   The graph of $V(x) = (x - 2)^3 + 1$ is the graph of $g(x) = x^3$ shifted vertically up 1 unit and shifted horizontally 2 units to the right.

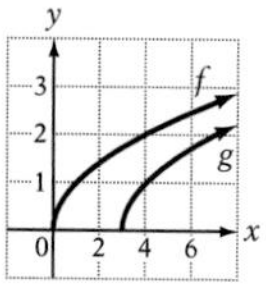

SECTION 9.3

Problem 1
$$(f + g)(-2) = f(-2) + g(-2)$$
$$= [(-2)^2 + 2(-2)]$$
$$+ [5(-2) - 2]$$
$$= (4 - 4) + (-10 - 2)$$
$$= -12$$

$$(f + g)(-2) = -12$$

Problem 2
$$(f \cdot g)(3) = f(3) \cdot g(3)$$
$$= (4 - 3^2) \cdot [3(3) - 4]$$
$$= (4 - 9) \cdot (9 - 4)$$
$$= (-5)(5)$$
$$= -25$$

$$(f \cdot g)(3) = -25$$

Problem 3
$$\left(\frac{f}{g}\right)(4) = \frac{f(4)}{g(4)}$$
$$= \frac{4^2 - 4}{4^2 + 2 \cdot 4 + 1}$$
$$= \frac{16 - 4}{16 + 8 + 1}$$
$$= \frac{12}{25}$$

$$\left(\frac{f}{g}\right)(4) = \frac{12}{25}$$

Problem 4   **A.** $h(x) = x^2 + 1$
$h(0) = 0 + 1 = 1$

$g(x) = 3x - 2$
$g(1) = 3(1) - 2 = 1$

$g[h(0)] = 1$

B. $h[g(x)] = h(3x - 2)$
$$= (3x - 2)^2 + 1$$
$$= 9x^2 - 12x + 4 + 1$$
$$= 9x^2 - 12x + 5$$

SECTION 9.4

Problem 1 **A.** Because any vertical line will intersect the graph at no more than one point, and any horizontal line will intersect the graph at no more than one point, the graph is the graph of a 1–1 function.

B. Because any vertical line will intersect the graph at no more than one point, and any horizontal line will intersect the graph at no more than one point, the graph is the graph of a 1–1 function.

Problem 2

$$f(x) = 4x + 2$$
$$y = 4x + 2$$
$$x = 4y + 2$$
$$4y = x - 2$$
$$y = \frac{1}{4}x - \frac{1}{2}$$
$$f^{-1}(x) = \frac{1}{4}x - \frac{1}{2}$$

Problem 3

$$h[g(x)] = 4\left(\frac{1}{4}x - \frac{1}{2}\right) + 2$$
$$= x - 2 + 2 = x$$
$$g[h(x)] = \frac{1}{4}(4x + 2) - \frac{1}{2}$$
$$= x + \frac{1}{2} - \frac{1}{2} = x$$

The functions are inverses of each other.

SOLUTIONS to Chapter 10 Problems

SECTION 10.1

Problem 1

$$f(x) = \left(\frac{2}{3}\right)^x$$
$$f(3) = \left(\frac{2}{3}\right)^3 = \frac{8}{27}$$
$$f(-2) = \left(\frac{2}{3}\right)^{-2} = \left(\frac{3}{2}\right)^2 = \frac{9}{4}$$

Problem 2

$$f(x) = 2^{2x+1}$$
$$f(0) = 2^{2(0)+1} = 2^1 = 2$$
$$f(-2) = 2^{2(-2)+1} = 2^{-3} = \frac{1}{2^3} = \frac{1}{8}$$

Problem 3

$$f(x) = \pi^x$$
$$f(3) = \pi^3 \approx 31.0063 \qquad f(-2) = \pi^{-2} \approx 0.1013$$
$$f(\pi) = \pi^\pi \approx 36.4622$$

Problem 4

$$f(x) = e^x$$
$$f(1.2) = e^{1.2} \approx 3.3201 \qquad f(-2.5) = e^{-2.5} \approx 0.0821$$
$$f(e) = e^e \approx 15.1543$$

Problem 5 **A.**

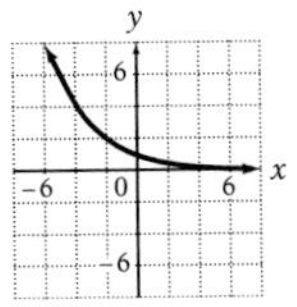

B.

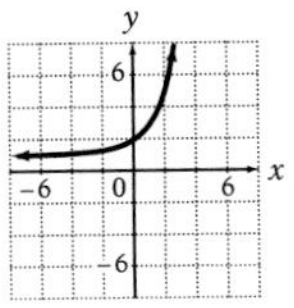

Problem 6 **A.**

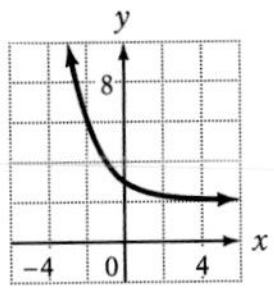

B.

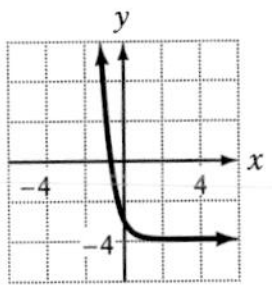

Problem 7

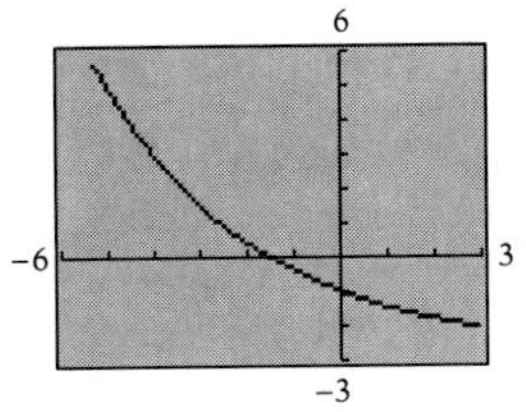

−2.4

SECTION 10.2

Problem 1 $3^{-4} = \frac{1}{81}$ is equivalent to $\log_3 \frac{1}{81} = -4$.

Problem 2 $\log_{10} 0.0001 = -4$ is equivalent to $10^{-4} = 0.0001$.

Problem 3

$$\log_4 64 = x$$
$$64 = 4^x$$
$$4^3 = 4^x$$
$$3 = x$$
$$\log_4 64 = 3$$

Problem 4 $\log_2 x = -4$

$$2^{-4} = x$$
$$\frac{1}{2^4} = x$$
$$\frac{1}{16} = x$$

The solution is $\frac{1}{16}$.

Problem 5

$$\log x = 1.5$$
$$10^{1.5} = x$$
$$31.6228 \approx x$$

The solution is 31.6228.

Problem 6 **A.** $\log_b \frac{x^2}{y} = \log_b x^2 - \log_b y$

$$= 2\log_b x - \log_b y$$

B. $\ln y^{\frac{1}{3}}z^3 = \ln y^{\frac{1}{3}} + \ln z^3$

$$= \frac{1}{3}\ln y + 3\ln z$$

C. $\log_8 \sqrt[3]{xy^2} = \log_8(xy^2)^{\frac{1}{3}}$

$$= \frac{1}{3}\log_8 xy^2$$
$$= \frac{1}{3}(\log_8 x + \log_8 y^2)$$
$$= \frac{1}{3}(\log_8 x + 2\log_8 y)$$
$$= \frac{1}{3}\log_8 x + \frac{2}{3}\log_8 y$$

Problem 7 **A.** $2\log_b x - 3\log_b y - \log_b z$

$$= \log_b x^2 - \log_b y^3 - \log_b z$$
$$= \log_b \frac{x^2}{y^3} - \log_b z$$
$$= \log_b \frac{x^2}{y^3 z}$$

B. $\frac{1}{3}(\log_4 x - 2\log_4 y + \log_4 z)$

$$= \frac{1}{3}(\log_4 x - \log_4 y^2 + \log_4 z)$$
$$= \frac{1}{3}\left(\log_4 \frac{x}{y^2} + \log_4 z\right)$$
$$= \frac{1}{3}\left(\log_4 \frac{xz}{y^2}\right)$$
$$= \log_4\left(\frac{xz}{y^2}\right)^{\frac{1}{3}} = \log_4 \sqrt[3]{\frac{xz}{y^2}}$$

C. $\frac{1}{2}(2\ln x - 5\ln y)$

$$= \frac{1}{2}(\ln x^2 - \ln y^5)$$
$$= \frac{1}{2}\left(\ln \frac{x^2}{y^5}\right)$$
$$= \ln\left(\frac{x^2}{y^5}\right)^{\frac{1}{2}}$$
$$= \ln\sqrt{\frac{x^2}{y^5}}$$

Problem 8 **A.** $\log_{16} 1 = 0$

B. $12\log_3 3 = \log_3 3^{12} = 12$

Problem 9 $\log_4 2.4 = \frac{\ln 2.4}{\ln 4} \approx 0.6315$

Problem 10 $4\log_8(3x + 4) = 4\frac{\log(3x + 4)}{\log 8}$

$$= \frac{4}{\log 8}\log(3x + 4)$$

SECTION 10.3

Problem 1 **A.** $f(x) = \log_2(x - 1)$

$$y = \log_2(x - 1)$$

$y = \log_2(x - 1)$ is equivalent to $2^y = x - 1$.

$$2^y + 1 = x$$

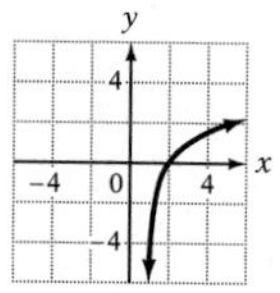

B. $f(x) = \log_3 2x$

$$y = \log_3 2x$$

$y = \log_3 2x$ is equivalent to $3^y = 2x$.

$$\frac{3^y}{2} = x$$

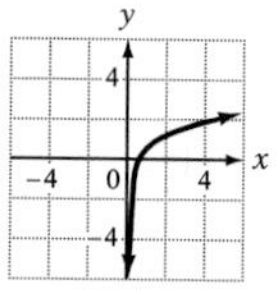

Problem 2

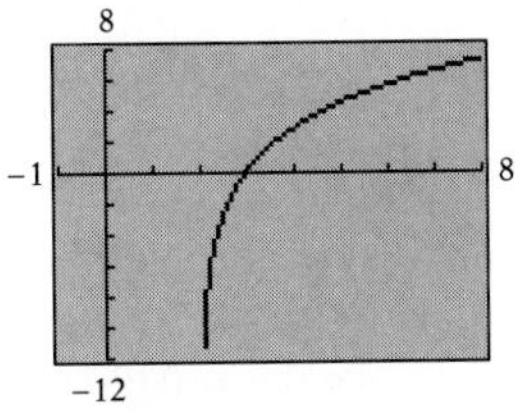

Problem 3

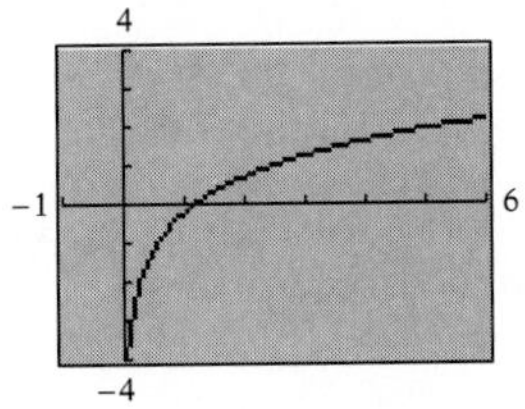

Problem 4

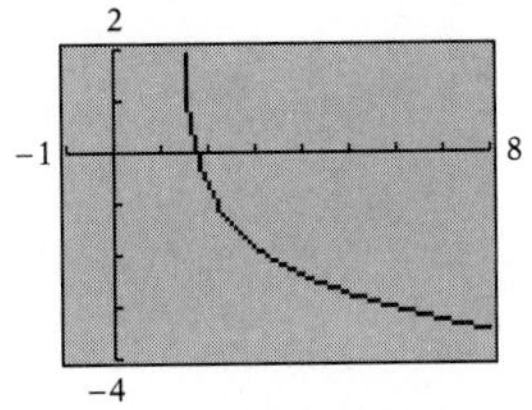

$f(x) = 1$ when $x \approx 1.5$

SECTION 10.4

Problem 1 $10^{3x+5} = 10^{x-3}$

$$\begin{aligned} 3x + 5 &= x - 3 \\ 2x + 5 &= -3 \\ 2x &= -8 \\ x &= -4 \end{aligned}$$

Check: $10^{3x+5} = 10^{x-3}$

$10^{3(-4)+5}$	10^{-4-3}
10^{-12+5}	10^{-7}

$10^{-7} = 10^{-7}$

The solution is -4.

Problem 2 **A.**

$$\begin{aligned} 4^{3x} &= 25 \\ \log 4^{3x} &= \log 25 \\ 3x \log 4 &= \log 25 \\ 3x &= \frac{\log 25}{\log 4} \\ x &= \frac{\log 25}{3 \log 4} \\ x &\approx 0.7740 \end{aligned}$$

The solution is 0.7740.

B.

$$\begin{aligned} (1.06)^x &= 1.5 \\ \log(1.06)^x &= \log 1.5 \\ x \log 1.06 &= \log 1.5 \\ x &= \frac{\log 1.5}{\log 1.06} \\ x &\approx 6.9585 \end{aligned}$$

The solution is 6.9585.

Problem 3

$$\begin{aligned} e^x &= x \\ e^x - x &= 0 \end{aligned}$$

Graph $f(x) = e^x - x$.

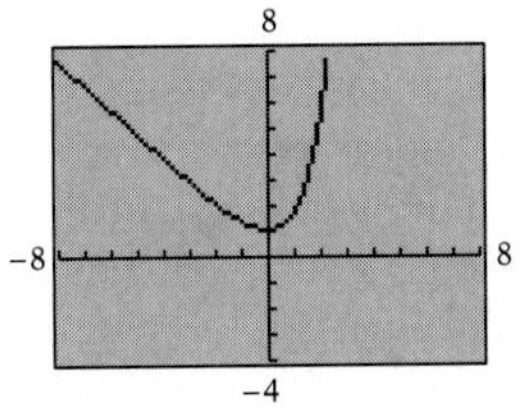

The equation has no real number solutions.

Problem 4 **A.** $\log_4(x^2 - 3x) = 1$

Rewrite in exponential form.

$$\begin{aligned} 4^1 &= x^2 - 3x \\ 4 &= x^2 - 3x \\ 0 &= x^2 - 3x - 4 \\ 0 &= (x + 1)(x - 4) \end{aligned}$$

$$\begin{aligned} x + 1 &= 0 \qquad & x - 4 &= 0 \\ x &= -1 & x &= 4 \end{aligned}$$

The solutions are -1 and 4.

B.

$$\begin{aligned} \log_3 x + \log_3(x + 3) &= \log_3 4 \\ \log_3[x(x + 3)] &= \log_3 4 \end{aligned}$$

Use the One-to-One Property of Logarithms.

$$\begin{aligned} x(x + 3) &= 4 \\ x^2 + 3x &= 4 \\ x^2 + 3x - 4 &= 0 \\ (x + 4)(x - 1) &= 0 \end{aligned}$$

$$\begin{aligned} x + 4 &= 0 \qquad & x - 1 &= 0 \\ x &= -4 & x &= 1 \end{aligned}$$

-4 does not check as a solution. The solution is 1.

Problem 5 $\log(3x - 2) = -2x$

$\log(3x - 2) + 2x = 0$

Graph $f(x) = \log(3x - 2) + 2x$.

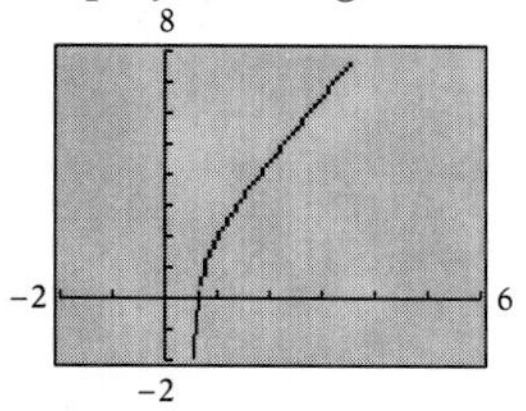

The solution is 0.68.

SECTION 10.5

Problem 1

Strategy To find the year in which a first-class stamp cost \$.22:

- ▶ Replace C by 0.22 and solve for t.
- ▶ Add 1962 to the value of t.

Solution

$$\begin{aligned} C &= 0.04e^{0.057t} \\ 0.22 &= 0.04e^{0.057t} \\ 5.5 &= e^{0.057t} \\ \ln 5.5 &= \ln e^{0.057t} \\ \ln 5.5 &= 0.057t \ln e \\ \ln 5.5 &= 0.057t(1) \\ \ln 5.5 &= 0.057t \\ \frac{\ln 5.5}{0.057} &= t \\ 30 &\approx t \end{aligned}$$

$1962 + 30 = 1992$

According to this model, a first-class stamp cost \$.22 in 1992.

Problem 2

Strategy To find how much oil is needed to last 25 years, replace T by 25 and solve for r.

Solution

$$\begin{aligned} T &= 14.29 \ln(0.00411r + 1) \\ 25 &= 14.29 \ln(0.00411r + 1) \\ \frac{25}{14.29} &= \ln(0.00411r + 1) \\ e^{\frac{25}{14.29}} &= 0.00411r + 1 \\ e^{\frac{25}{14.29}} - 1 &= 0.00411r \\ \frac{e^{\frac{25}{14.29}} - 1}{0.00411} &= r \\ 1156 &\approx r \end{aligned}$$

According to this model, 1156 billion barrels of oil are needed to last 25 years.

SOLUTIONS to Chapter 11 Problems

SECTION 11.1

Problem 1 $a_n = n(n + 1)$

$a_1 = 1(1 + 1) = 2$ The first term is 2.

$a_2 = 2(2 + 1) = 6$ The second term is 6.

$a_3 = 3(3 + 1) = 12$ The third term is 12.

$a_4 = 4(4 + 1) = 20$ The fourth term is 20.

Problem 2 $a_n = \dfrac{1}{n(n + 2)}$

$a_6 = \dfrac{1}{6(6 + 2)} = \dfrac{1}{48}$ The sixth term is $\frac{1}{48}$.

$a_9 = \dfrac{1}{9(9 + 2)} = \dfrac{1}{99}$ The ninth term is $\frac{1}{99}$.

Problem 3

A. $\displaystyle\sum_{n=1}^{4} (7 - n)$

$= (7 - 1) + (7 - 2) + (7 - 3) + (7 - 4)$

$= 6 + 5 + 4 + 3 = 18$

B. $\displaystyle\sum_{i=3}^{6} (i^2 - 2)$

$= (3^2 - 2) + (4^2 - 2) + (5^2 - 2) + (6^2 - 2)$

$= 7 + 14 + 23 + 34 = 78$

Problem 4 $\displaystyle\sum_{n=1}^{5} nx = x + 2x + 3x + 4x + 5x$

SECTION 11.2

Problem 1 $9, 3, -3, -9, \ldots$

$$d = a_2 - a_1 = 3 - 9 = -6$$

$$\begin{aligned} a_n &= a_1 + (n-1)d \\ a_{15} &= 9 + (15-1)(-6) = 9 + (14)(-6) \\ &= 9 - 84 = -75 \end{aligned}$$

Problem 2 $-3, 1, 5, 9, \ldots$

$$d = a_2 - a_1 = 1 - (-3) = 4$$

$$\begin{aligned} a_n &= a_1 + (n-1)d \\ a_n &= -3 + (n-1)4 \\ a_n &= -3 + 4n - 4 \\ a_n &= 4n - 7 \end{aligned}$$

Problem 3 $7, 9, 11, \ldots, 59$

$$d = a_2 - a_1 = 9 - 7 = 2$$

$$\begin{aligned} a_n &= a_1 + (n-1)d \\ 59 &= 7 + (n-1)2 \\ 59 &= 7 + 2n - 2 \\ 59 &= 5 + 2n \\ 54 &= 2n \\ 27 &= n \end{aligned}$$

There are 27 terms in the sequence.

Problem 4 $-4, -2, 0, 2, 4, \ldots$

$$d = a_2 - a_1 = -2 - (-4) = 2$$

$$\begin{aligned} a_n &= a_1 + (n-1)d \\ a_{25} &= -4 + (25-1)2 \\ &= -4 + (24)2 = -4 + 48 = 44 \end{aligned}$$

$$S_n = \frac{n}{2}(a_1 + a_n)$$

$$\begin{aligned} S_{25} &= \frac{25}{2}(-4 + 44) \\ &= \frac{25}{2}(40) = 25(20) = 500 \end{aligned}$$

Problem 5 $\sum_{n=1}^{18} (3n - 2)$

$$\begin{aligned} a_n &= 3n - 2 \\ a_1 &= 3(1) - 2 = 1 \\ a_{18} &= 3(18) - 2 = 52 \end{aligned}$$

$$S_n = \frac{n}{2}(a_1 + a_n)$$

$$S_{18} = \frac{18}{2}(1 + 52) = 9(53) = 477$$

Problem 6

Strategy To find the value of the 20th-place prize:
- Write the equation for the *n*th-place prize.
- Find the 20th term of the sequence.

To find the total amount of prize money being awarded, use the Formula for the Sum of *n* Terms of an Arithmetic Sequence.

Solution 10,000, 9700, . . .

$$d = a_2 - a_1 = 9700 - 10{,}000 = -300$$

$$\begin{aligned} a_n &= a_1 + (n-1)d \\ &= 10{,}000 + (n-1)(-300) \\ &= 10{,}000 - 300n + 300 \\ &= -300n + 10{,}300 \\ a_{20} &= -300(20) + 10{,}300 \\ &= -6000 + 10{,}300 = 4300 \end{aligned}$$

$$S_n = \frac{n}{2}(a_1 + a_n)$$

$$\begin{aligned} S_{20} &= \frac{20}{2}(10{,}000 + 4300) \\ &= 10(14{,}300) = 143{,}000 \end{aligned}$$

The value of the 20th-place prize is $4300.

The total amount of prize money being awarded is $143,000.

SECTION 11.3

Problem 1 $5, 2, \frac{4}{5}, \ldots$

$$r = \frac{a_2}{a_1} = \frac{2}{5}$$

$$a_n = a_1 r^{n-1}$$

$$a_5 = 5\left(\frac{2}{5}\right)^{5-1} = 5\left(\frac{2}{5}\right)^4 = 5\left(\frac{16}{625}\right) = \frac{16}{125}$$

Problem 2 $3, a_2, a_3, -192, \ldots$

$$\begin{aligned} a_n &= a_1 r^{n-1} \\ a_4 &= 3r^{4-1} \\ -192 &= 3r^{4-1} \\ -192 &= 3r^3 \\ -64 &= r^3 \\ -4 &= r \end{aligned}$$

$$\begin{aligned} a_n &= a_1 r^{n-1} \\ a_3 &= 3(-4)^{3-1} = 3(-4)^2 = 3(16) \\ &= 48 \end{aligned}$$

Problem 3 $1, -\frac{1}{3}, \frac{1}{9}, -\frac{1}{27}$

$$r = \frac{a_2}{a_1} = \frac{-\frac{1}{3}}{1} = -\frac{1}{3}$$

$$S_n = \frac{a_1(1 - r^n)}{1 - r}$$

$$S_4 = \frac{1\left[1 - \left(-\frac{1}{3}\right)^4\right]}{1 - \left(-\frac{1}{3}\right)}$$

$$= \frac{1 - \frac{1}{81}}{\frac{4}{3}} = \frac{\frac{80}{81}}{\frac{4}{3}} = \frac{80}{81} \cdot \frac{3}{4} = \frac{20}{27}$$

Problem 4 $\sum_{n=1}^{5} \left(\frac{1}{2}\right)^n$

$$a_n = \left(\frac{1}{2}\right)^n$$

$$a_1 = \left(\frac{1}{2}\right)^1 = \frac{1}{2}$$

$$a_2 = \left(\frac{1}{2}\right)^2 = \frac{1}{4}$$

$$r = \frac{a_2}{a_1} = \frac{\frac{1}{4}}{\frac{1}{2}} = \frac{1}{4} \cdot \frac{2}{1} = \frac{1}{2}$$

$$S_n = \frac{a_1(1 - r^n)}{1 - r}$$

$$S_5 = \frac{\frac{1}{2}\left[1 - \left(\frac{1}{2}\right)^5\right]}{1 - \frac{1}{2}}$$

$$= \frac{\frac{1}{2}\left(1 - \frac{1}{32}\right)}{\frac{1}{2}} = \frac{\frac{1}{2}\left(\frac{31}{32}\right)}{\frac{1}{2}} = \frac{\frac{31}{64}}{\frac{1}{2}}$$

$$= \frac{31}{64} \cdot \frac{2}{1} = \frac{31}{32}$$

Problem 5 $3, -2, \frac{4}{3}, -\frac{8}{9}, \ldots$

$$r = \frac{a_2}{a_1} = -\frac{2}{3}$$

$$S = \frac{a_1}{1 - r} = \frac{3}{1 - \left(-\frac{2}{3}\right)}$$

$$= \frac{3}{1 + \frac{2}{3}} = \frac{3}{\frac{5}{3}} = \frac{9}{5}$$

Problem 6 $0.\overline{36} = 0.36 + 0.0036 + 0.000036 + \cdots$

$$S = \frac{a_1}{1 - r} = \frac{\frac{36}{100}}{1 - \frac{1}{100}} = \frac{\frac{36}{100}}{\frac{99}{100}} = \frac{36}{99} = \frac{4}{11}$$

An equivalent fraction is $\frac{4}{11}$.

Problem 7

Strategy To find the total number of letters mailed, use the Formula for the Sum of n Terms of a Finite Geometric Series.

Solution $n = 6, a_1 = 3, r = 3$

$$S_n = \frac{a_1(1 - r^n)}{1 - r}$$

$$S_6 = \frac{3(1 - 3^6)}{1 - 3} = \frac{3(1 - 729)}{1 - 3} = \frac{3(-728)}{-2}$$

$$= \frac{-2184}{-2} = 1092$$

From the first through the sixth mailings, 1092 letters will have been mailed.

SECTION 11.4

Problem 1 $\frac{12!}{7!5!} = \frac{12 \cdot 11 \cdot 10 \cdot 9 \cdot 8 \cdot 7 \cdot 6 \cdot 5 \cdot 4 \cdot 3 \cdot 2 \cdot 1}{(7 \cdot 6 \cdot 5 \cdot 4 \cdot 3 \cdot 2 \cdot 1)(5 \cdot 4 \cdot 3 \cdot 2 \cdot 1)}$

$$= 792$$

Problem 2 $\binom{7}{0} = \frac{7!}{(7 - 0)!0!} = \frac{7!}{7!0!}$

$$= \frac{7 \cdot 6 \cdot 5 \cdot 4 \cdot 3 \cdot 2 \cdot 1}{(7 \cdot 6 \cdot 5 \cdot 4 \cdot 3 \cdot 2 \cdot 1)(1)} = 1$$

Problem 3 $(3m - n)^4$

$$= \binom{4}{0}(3m)^4 + \binom{4}{1}(3m)^3(-n) + \binom{4}{2}(3m)^2(-n)^2 + \binom{4}{3}(3m)(-n)^3 + \binom{4}{4}(-n)^4$$

$$= 1(81m^4) + 4(27m^3)(-n) + 6(9m^2)(n^2) + 4(3m)(-n^3) + 1(n^4)$$

$$= 81m^4 - 108m^3n + 54m^2n^2 - 12mn^3 + n^4$$

Problem 4

$(y - 2)^{10}$

$$= \binom{10}{0}y^{10} + \binom{10}{1}y^9(-2) + \binom{10}{2}y^8(-2)^2 + \cdots$$

$$= 1(y^{10}) + 10y^9(-2) + 45y^8(4) + \cdots$$

$$= y^{10} - 20y^9 + 180y^8 + \cdots$$

Problem 5 $(t - 2s)^7$

$n = 7, a = t, b = -2s, r = 3$

$$\binom{7}{3-1}(t)^{7-3+1}(-2s)^{3-1} = \binom{7}{2}(t)^5(-2s)^2$$
$$= 21t^5(4s^2) = 84t^5s^2$$

SOLUTIONS to Chapter 12 Problems

SECTION 12.1

Problem 1 $-\frac{b}{2a} = -\frac{1}{2(-1)} = \frac{1}{2}$

$y = -x^2 + x + 3$

$y = -\left(\frac{1}{2}\right)^2 + \frac{1}{2} + 3 = \frac{13}{4}$

The vertex is $\left(\frac{1}{2}, \frac{13}{4}\right)$.

The axis of symmetry is the line $x = \frac{1}{2}$.

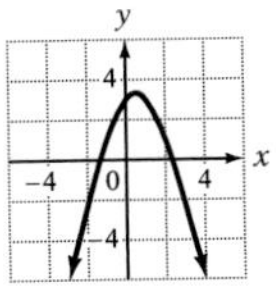

Problem 2 y-coordinate: $-\frac{b}{2a} = -\frac{-4}{2(-2)} = -1$

$x = -2y^2 - 4y - 3$
$x = -2(-1)^2 - 4(-1) - 3$
$x = -1$

The vertex is $(-1, -1)$. The axis of symmetry is the line $y = -1$.

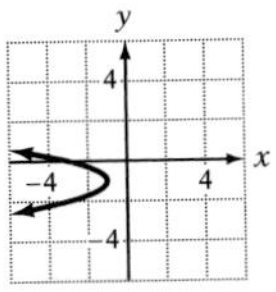

SECTION 12.2

Problem 1

$$(x - h)^2 + (y - k)^2 = r^2$$
$$(x - 2)^2 + [y - (-3)]^2 = 4^2$$
$$(x - 2)^2 + (y + 3)^2 = 16$$

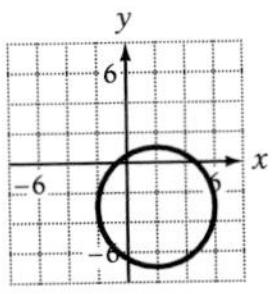

Problem 2 $x_m = \frac{x_1 + x_2}{2}$ $\quad y_m = \frac{y_1 + y_2}{2}$

$x_m = \frac{-2 + 4}{2} = 1$ $\quad y_m = \frac{1 + (-1)}{2} = 0$

Center: $(1, 0)$

$r = \sqrt{(x_1 - x_m)^2 + (y_1 - y_m)^2}$
$r = \sqrt{(-2 - 1)^2 + (1 - 0)^2} = \sqrt{9 + 1}$
$= \sqrt{10}$

Radius: $r = \sqrt{10}$

$$(x - h)^2 + (y - k)^2 = r^2$$
$$(x - 1)^2 + (y - 0)^2 = 10$$
$$(x - 1)^2 + y^2 = 10$$

Problem 3

$$x^2 + y^2 - 4x + 8y + 15 = 0$$
$$(x^2 - 4x) + (y^2 + 8y) = -15$$
$$(x^2 - 4x + 4) + (y^2 + 8y + 16) = -15 + 4 + 16$$
$$(x - 2)^2 + (y + 4)^2 = 5$$

Center: $(2, -4)$
Radius: $\sqrt{5}$

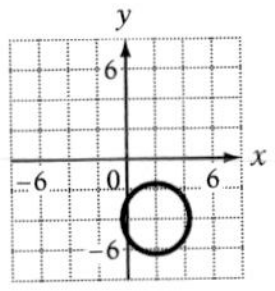

SECTION 12.3

Problem 1 **A.** x-intercepts: $(2, 0)$ and $(-2, 0)$

y-intercepts: $(0, 5)$ and $(0, -5)$

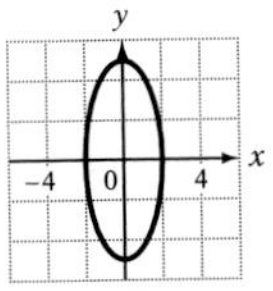

B. x-intercepts: $(3\sqrt{2}, 0)$ and $(-3\sqrt{2}, 0)$

y-intercepts: $(0, 3)$ and $(0, -3)$

$\left(3\sqrt{2} \approx 4\frac{1}{4}\right)$

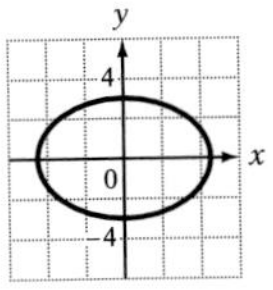

Problem 2 **A.** Axis of symmetry: x-axis

Vertices: $(3, 0)$ and $(-3, 0)$

Asymptotes: $y = \frac{5}{3}x$ and $y = -\frac{5}{3}x$

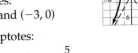

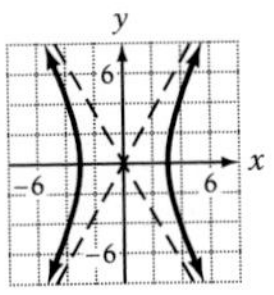

B. Axis of symmetry:
y-axis

Vertices:
(0, 3) and (0, −3)

Asymptotes:
$y = x$ and $y = -x$

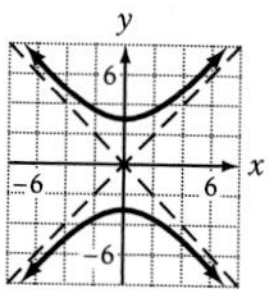

SECTION 12.4

Problem 1 **A.** (1) $y = 2x^2 + x - 3$
(2) $y = 2x^2 - 2x + 9$

Use the substitution method.

$$2x^2 - 2x + 9 = 2x^2 + x - 3$$
$$-3x + 9 = -3$$
$$-3x = -12$$
$$x = 4$$

Substitute into equation (1).

$$y = 2x^2 + x - 3$$
$$y = 2(4)^2 + 4 - 3$$
$$y = 32 + 4 - 3$$
$$y = 33$$

The solution is (4, 33).

B. (1) $x^2 - y^2 = 10$
(2) $x^2 + y^2 = 8$

Use the addition method.

$$2x^2 = 18$$
$$x^2 = 9$$
$$x = \pm\sqrt{9} = \pm 3$$

Substitute into equation (2).

$x^2 + y^2 = 8$	$x^2 + y^2 = 8$
$3^2 + y^2 = 8$	$(-3)^2 + y^2 = 8$
$9 + y^2 = 8$	$9 + y^2 = 8$
$y^2 = -1$	$y^2 = -1$
$y = \pm\sqrt{-1}$	$y = \pm\sqrt{-1}$

y is not a real number. Therefore, the system of equations has no real number solution.

SECTION 12.5

Problem 1 **A.** Graph the ellipse $\frac{x^2}{9} + \frac{y^2}{16} = 1$ as a solid line.

Shade the region of the plane that includes (0, 0).

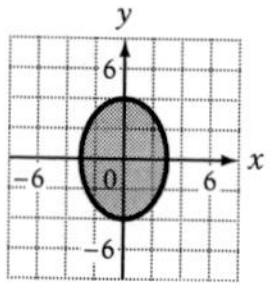

B. Graph the hyperbola $\frac{x^2}{9} - \frac{y^2}{4} = 1$ as a solid line.

Shade the region that includes (0, 0).

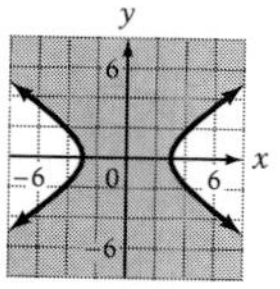

Problem 2 **A.** Graph the ellipse $\frac{x^2}{4} + \frac{y^2}{9} = 1$ as a solid line.

Shade inside the ellipse.

Graph the parabola $x = y^2 - 2$ as a dashed line.

Shade inside the parabola.

The solution set is the region of the plane represented by the intersection of the solution sets of the two inequalities.

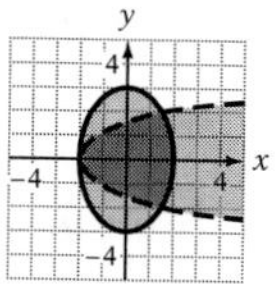

B. Graph the ellipse $\frac{x^2}{16} - \frac{y^2}{25} = 1$ as a solid line.

Shade outside the ellipse.

Graph the circle $x^2 + y^2 = 9$ as a dashed line.

Shade inside the circle.

The solution sets of the two inequalities do not intersect. This system of inequalities has no real number solution.

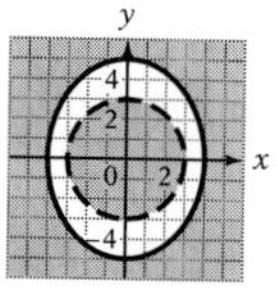

ANSWERS to Chapter 1 Odd-Numbered Exercises

CONCEPT REVIEW 1.1

1. Sometimes true **3.** Sometimes true **5.** Never true **7.** Always true

SECTION 1.1

1. a. Natural numbers: 9, 53 **b.** Whole numbers: 0, 9, 53 **c.** Integers: $-14, 9, 0, 53, -626$ **d.** Positive integers: 9, 53 **e.** Negative integers: $-14, -626$ **3. a.** Integers: $0, -3$ **b.** Rational: $-\frac{15}{2}, 0, -3, 2.\overline{33}$ **c.** Irrational: π, 4.232232223…, $\frac{\sqrt{5}}{4}, \sqrt{7}$ **d.** Real: all **9.** -27 **11.** $-\frac{3}{4}$ **13.** 0 **15.** $\sqrt{33}$ **17.** 91 **19.** $-3, 0$ **21.** 7 **23.** $-2, -1$ **25.** $9, 0, -9$ **27.** $4, 0, 4$ **29.** $-6, -2, 0, -1, -4$ **33.** $\{-2, -1, 0, 1, 2, 3, 4\}$ **35.** $\{2, 4, 6, 8, 10, 12\}$ **37.** $\{3, 6, 9, 12, 15, 18, 21, 24, 27, 30\}$ **39.** $\{-35, -30, -25, -20, -15, -10, -5\}$ **41.** $\{x \mid x > 4, x \in \text{integers}\}$ **43.** $\{x \mid x \geq -2\}$ **45.** $\{x \mid 0 < x < 1\}$ **47.** $\{x \mid 1 \leq x \leq 4\}$ **49.** $\{1, 2, 4, 6, 9\}$ **51.** $\{2, 3, 5, 8, 9, 10\}$ **53.** $\{-4, -2, 0, 2, 4, 8\}$ **55.** $\{1, 2, 3, 4, 5\}$ **57.** $\{6\}$ **59.** $\{5, 10, 20\}$ **61.** $\varnothing$ **63.** $\{4, 6\}$

65. 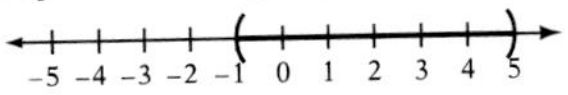

67. 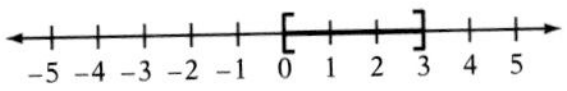

69. 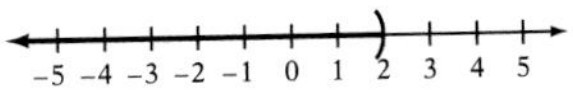

71. 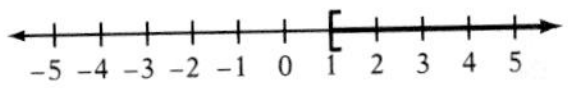

73. 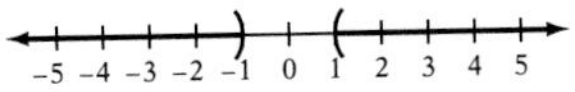

75. 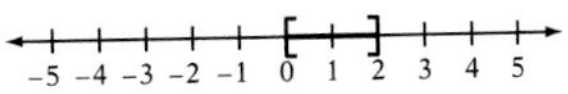

77. 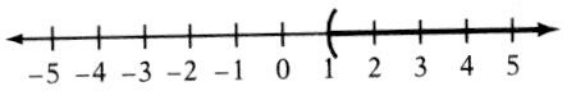

79. 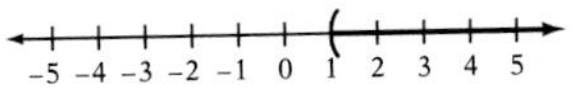

81. $\{x \mid 0 < x < 8\}$ **83.** $\{x \mid -5 \leq x \leq 7\}$ **85.** $\{x \mid -3 \leq x < 6\}$ **87.** $\{x \mid x \leq 4\}$ **89.** $\{x \mid x > 5\}$ **91.** $(-2, 4)$ **93.** $[-1, 5]$ **95.** $(-\infty, 1)$ **97.** $[-2, 6)$ **99.** $(-\infty, \infty)$

101. 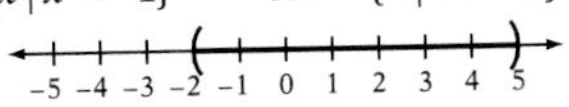

103. 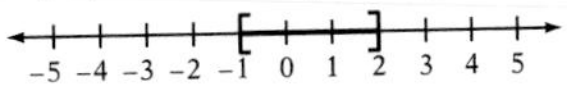

105. 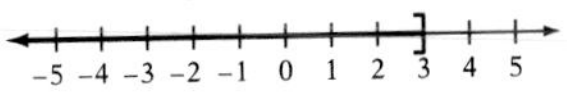

107. 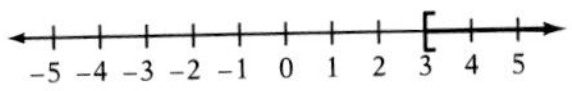

109. 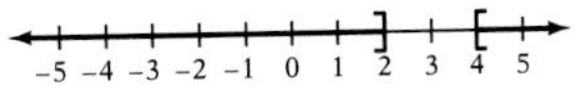

111. 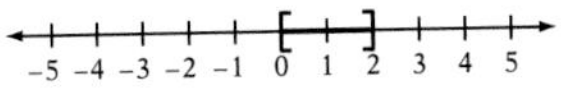

113. 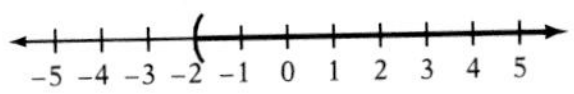

115. A **117.** B **119.** A **121.** R **123.** R **125.** 0

127. 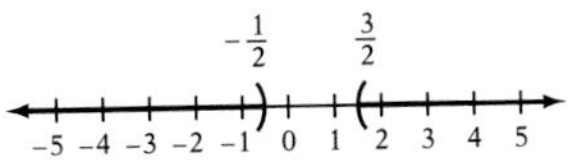

129. 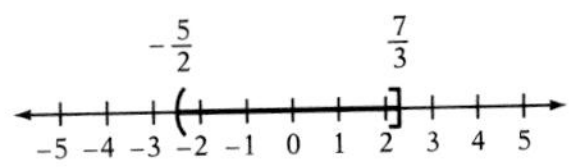

131. 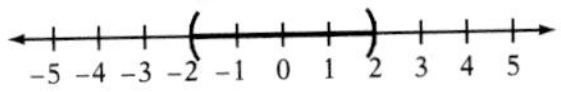

133. 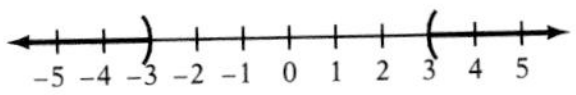

135. $\{x \mid x > 0, x \text{ is an integer}\}$ **137.** $\{x \mid x \geq 15, x \text{ is an odd integer}\}$ **139.** b and c

CONCEPT REVIEW 1.2

1. Sometimes true **3.** Never true **5.** Always true **7.** Never true **9.** Always true

SECTION 1.2

5. -30 **7.** -17 **9.** -96 **11.** -6 **13.** 5 **15.** 11,200 **17.** 98,915 **19.** 96 **21.** 23 **23.** 7 **25.** 17 **27.** -4 **29.** -62 **31.** -25 **33.** 7 **35.** 587 **37.** 26 **39.** -83 **43.** $\frac{43}{48}$ **45.** $-\frac{67}{45}$ **47.** $-\frac{13}{36}$ **49.** $\frac{11}{24}$ **51.** $\frac{13}{24}$ **53.** $-\frac{3}{56}$ **55.** $-\frac{2}{3}$ **57.** $-\frac{11}{14}$ **59.** $-\frac{1}{24}$ **61.** -12.974 **63.** -6.008 **65.** 1.9215 **67.** -6.02 **69.** -6.7 **71.** -1.11 **73.** -2030.44 **75.** 125 **77.** -8 **79.** -125 **81.** 324 **83.** -36 **85.** -72 **87.** 864 **89.** -160 **91.** -6480 **93.** 2,654,208

97. -7 **99.** $\frac{177}{11}$ **101.** $\frac{16}{3}$ **103.** -40 **105.** 32 **107.** 24 **109.** $\frac{2}{15}$ **111.** $\frac{1}{2}$ **113.** $-\frac{97}{72}$
115. 6.284 **117.** 5.4375 **119.** 1.891878 **121.** 0 **123.** No, zero does not have a multiplicative inverse.
125. 9 **127.** 625 **129.** Find b^c; then find $a^{(b^c)}$.

CONCEPT REVIEW 1.3

1. Sometimes true **3.** Sometimes true **5.** Always true

SECTION 1.3

1. 3 **3.** 3 **5.** 0 **7.** 6 **9.** 0 **11.** 1 **13.** $(2 \cdot 3)$ **15.** Division Property of Zero
17. Inverse Property of Multiplication **19.** Addition Property of Zero **21.** Division Property of Zero
23. Distributive Property **25.** Associative Property of Multiplication **29.** 10 **31.** 4 **33.** 0 **35.** $-\frac{1}{7}$
37. $\frac{9}{2}$ **39.** $\frac{1}{2}$ **41.** 2 **43.** 3 **45.** -12 **47.** 2 **49.** 6 **51.** -2 **53.** -14 **55.** 56
57. 256 **59.** The volume of the rectangular solid is 840 in^3. **61.** The volume of the pyramid is 15 ft^3.
63. The volume of the sphere is exactly 4.5π cm^3. The volume of the sphere is approximately 14.14 cm^3.
65. The surface area of the rectangular solid is 94 m^2. **67.** The surface area of the pyramid is 56 m^2.
69. The surface area of the cylinder is exactly 96π in^2. The surface area of the cylinder is approximately 301.59 in^2.
73. $13x$ **75.** $-4x$ **77.** $7a + 7b$ **79.** x **81.** $-3x + 6$ **83.** $5x + 10$ **85.** $x + y$ **87.** $3x - 6y - 5$
89. $-11a + 21$ **91.** $-x + 6y$ **93.** $140 - 30a$ **95.** $-10y + 30x$ **97.** $-10a + 2b$ **99.** $-12a + b$
101. $-2x - 144y - 96$ **103.** $5x - 32 + 3y$ **105.** $x + 6$ **107.** Distributive Property **109.** Incorrect use of the Distributive Property; $2 + 3x = 2 + 3x$ **111.** Incorrect use of the Associative Property of Multiplication; $2(3y) = (2 \cdot 3)y = 6y$ **113.** Commutative Property of Addition **115. a.** The Distributive Property **b.** The Commutative Property of Addition **c.** The Associative Property of Addition **d.** The Distributive Property **117. a.** The Distributive Property **b.** The Associative Property of Multiplication **c.** The Multiplication Property of One

CONCEPT REVIEW 1.4

1. Never true **3.** Never true **5.** Sometimes true

SECTION 1.4

1. $n - (n + 2); -2$ **3.** $\frac{1}{3}n + \frac{4}{5}n; \frac{17}{15}n$ **5.** $5(8n); 40n$ **7.** $17n - 2n; 15n$ **9.** $n^2 - (12 + n^2); -12$
11. $15n + (5n + 12); 20n + 12$ **13.** $2x + (15 - x + 2); x + 17$ **15.** $(34 - x + 2) - 2x; -3x + 36$
17. The population of Milan: P; the population of San Paolo: $4P$ **19.** The amount that Arnold Palmer earned from endorsements: A; the amount that Dennis Rodman earned from endorsements: $\frac{2}{3}A$ **21.** The measure of angle B: x; the measure of angle A: $2x$; the measure of angle C: $4x$ **23.** Flying time between Los Angeles and New York: t; flying time between New York and Los Angeles: $12 - t$ **25.** The sum of twice a number and three **27.** Twice the sum of a number and three **29.** $\frac{1}{2}gt^2$ **31.** Av^2

CHAPTER REVIEW EXERCISES

1. $\frac{3}{4}$ (Obj. 1.3.1) **2.** 0, 2 (Obj. 1.1.1) **3.** $-4, 0, -7$ (Obj. 1.1.1) **4.** $\{-2, -1, 0, 1, 2, 3\}$ (Obj. 1.1.2)
5. $\{x \mid x < -3\}$ (Obj. 1.1.2) **6.** $\{x \mid -2 \le x \le 3\}$ (Obj. 1.1.2) **7.** $\{1, 2, 3, 4, 5, 6, 7, 8\}$ (Obj. 1.1.2)

8. $\{2, 3\}$ (Obj. 1.1.2) **9.** 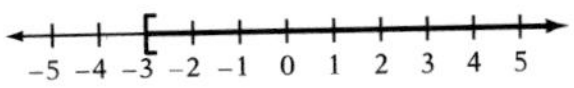 (Obj. 1.1.2) **10.** 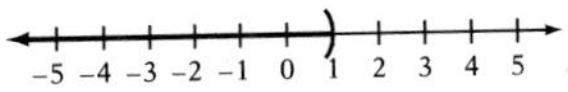 (Obj. 1.1.2)
11. 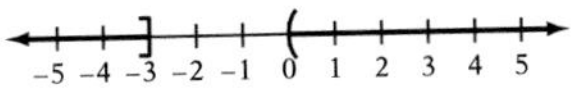 (Obj. 1.1.2) **12.** 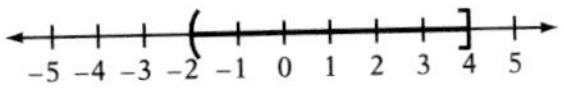 (Obj. 1.1.2) **13.** -15 (Obj. 1.2.1)
14. 12 (Obj. 1.2.1) **15.** 14 (Obj. 1.2.1) **16.** -288 (Obj. 1.2.1) **17.** $\frac{7}{120}$ (Obj. 1.2.2)
18. $\frac{2}{15}$ (Obj. 1.2.2) **19.** $-\frac{5}{8}$ (Obj. 1.2.2) **20.** -2.84 (Obj. 1.2.2) **21.** 3.1 (Obj. 1.2.2)
22. 52 (Obj. 1.2.4) **23.** 20 (Obj. 1.3.2) **24.** $\frac{7}{6}$ (Obj. 1.3.2) **25.** 3 (Obj. 1.3.1) **26.** y (Obj. 1.3.1)
27. (ab) (Obj. 1.3.1) **28.** 4 (Obj. 1.3.1) **29.** The Inverse Property of Addition (Obj. 1.3.1) **30.** The Associative Property of Multiplication (Obj. 1.3.1) **31.** $-6x + 14$ (Obj. 1.3.3) **32.** $16y - 15x + 18$ (Obj. 1.3.3) **33.** $4(x + 4); 4x + 16$ (Obj. 1.4.1) **34.** $2(x - 2) + 8; 2x + 4$ (Obj. 1.4.1) **35.** $2x + (40 - x) + 5; x + 45$ (Obj. 1.4.1) **36.** $[2(9 - x) + 3] - (x + 1); -3x + 20$ (Obj. 1.4.1) **37.** The width of the rectangle: W; the length of the rectangle: $3W - 3$ (Obj. 1.4.2) **38.** The first integer: x; the second integer: $4x + 5$ (Obj. 1.4.2)

CHAPTER TEST

1. 12 (Obj. 1.1.1) **2.** -5 (Obj. 1.1.1) **3.** 12 (Obj. 1.2.1) **4.** -30 (Obj. 1.2.1) **5.** -15 (Obj. 1.2.1)
6. 2 (Obj. 1.2.1) **7.** -100 (Obj. 1.2.3) **8.** -72 (Obj. 1.2.3) **9.** $\frac{25}{36}$ (Obj. 1.2.2) **10.** $-\frac{4}{27}$ (Obj. 1.2.2)
11. -1.41 (Obj. 1.2.2) **12.** -4.9 (Obj. 1.2.2) **13.** 10 (Obj. 1.2.4) **14.** 6 (Obj. 1.2.4) **15.** -5 (Obj. 1.3.2) **16.** 2 (Obj. 1.3.2) **17.** 4 (Obj. 1.3.1) **18.** The Distributive Property (Obj. 1.3.1) **19.** $13x - y$ (Obj. 1.3.3) **20.** $14x + 48y$ (Obj. 1.3.3) **21.** $13 - (n - 3)(9); 40 - 9n$ (Obj. 1.4.1)
22. $\frac{1}{3}(12n + 27); 4n + 9$ (Obj. 1.4.1) **23.** $\{1, 2, 3, 4, 5, 7\}$ (Obj. 1.1.2) **24.** $\{-2, -1, 0, 1, 2, 3\}$ (Obj. 1.1.2)
25. $\{5, 7\}$ (Obj. 1.1.2) **26.** $\{-1, 0, 1\}$ (Obj. 1.1.2) **27.** 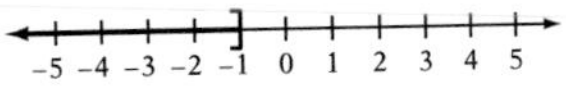 (Obj. 1.1.2)
28. 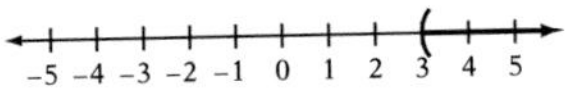 (Obj. 1.1.2) **29.** 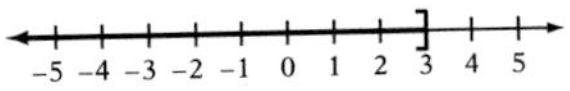 (Obj. 1.1.2)
30. 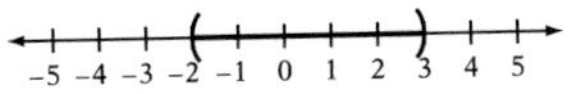 (Obj. 1.1.2)

ANSWERS to Chapter 2 Odd-Numbered Exercises

CONCEPT REVIEW 2.1

1. Never true **3.** Sometimes true **5.** Never true **7.** Always true

SECTION 2.1

5. 9 **7.** -10 **9.** 4 **11.** $\frac{11}{21}$ **13.** $-\frac{1}{8}$ **15.** 20 **17.** -49 **19.** $-\frac{21}{20}$ **21.** -3.73 **23.** $\frac{3}{2}$
25. 8 **27.** No solution **29.** -3 **31.** 1 **33.** $\frac{7}{4}$ **35.** 24 **37.** $\frac{5}{3}$ **39.** $-\frac{3}{2}$ **41.** No solution
43. $\frac{11}{2}$ **45.** -1.25 **47.** 6 **49.** $\frac{1}{2}$ **51.** -6 **53.** $-\frac{4}{3}$ **55.** $\frac{6}{7}$ **57.** $\frac{35}{12}$ **59.** -1 **61.** -33
63. 6 **65.** 6 **67.** $\frac{25}{14}$ **69.** $\frac{3}{4}$ **71.** $-\frac{4}{29}$ **73.** 3 **75.** -10 **77.** $\frac{5}{11}$ **79.** 9.4
81. The customer bought 8 bags of feed. **83.** The charge per hour for labor was \$42. **85.** Her regular hourly rate is \$7. **87.** There were 7 grandstand tickets and 5 upper field box tickets purchased. **89.** There were 2 mezzanine tickets purchased. **91.** -9 **93.** $-\frac{15}{2}$ **95.** No solution **97.** -1 **99.** All real numbers

CONCEPT REVIEW 2.2

1. Always true **3.** Always true **5.** Sometimes true

SECTION 2.2

3. There are 24 dimes in the collection. **5.** The cashier has 26 twenty-dollar bills. **7.** There are eight 20¢ stamps and sixteen 15¢ stamps in the collection. **9.** There are 8 dimes in the bank. **11.** There are five 3¢ stamps in the collection. **13.** There are seven 18¢ stamps in the collection. **17.** The integers are 3 and 7. **19.** The integers are 21 and 29. **21.** The three numbers are 21, 44, and 58. **23.** The three integers are −20, −19, and −18. **25.** There is no solution. **27.** The three integers are 5, 7, and 9. **29.** There is $4.30 in the bank. **31.** There are fourteen 3¢ stamps in the collection. **33.** The sum of the smallest and largest integers is −32. **35.** The number is 312.

CONCEPT REVIEW 2.3

1. Sometimes true **3.** Never true **5.** Never true

SECTION 2.3

3. The vegetable medley costs $1.66 per pound. **5.** There were 320 adult tickets sold. **7.** The mixture must contain 225 L of imitation maple syrup. **9.** The instructor used 8 lb of peanuts. **11.** The cost of the mixture is $4.04 per pound. **13.** The mixture must contain 37.5 gal of cranberry juice. **15.** The bicyclist overtakes the in-line skater 11.25 mi from the starting point. **17.** The speed of the first plane is 420 mph. The speed of the second plane is 500 mph. **19.** The distance to the island is 43.2 mi. **21.** The rate of the first plane is 255 mph. The rate of the second plane is 305 mph. **23.** The distance between the student's home and the bicycle shop is 2.8 mi. **25.** The two trains will pass each other in 2 h. **27.** It takes the runner $\frac{2}{9}$ h to travel from the front of the parade to the end of the parade. **29.** The cars are $3\frac{1}{3}$ mi apart 2 min before impact. **31.** No

CONCEPT REVIEW 2.4

1. Never true **3.** Never true

SECTION 2.4

3. The amount invested in the CD is $15,000. **5.** An additional $2000 must be invested at an annual simple interest rate of 10.5%. **7.** The amount invested at 8.5% is $3000. The amount invested at 6.4% is $5000. **9.** She must invest an additional $3000 at an annual interest rate of 10%. **11.** The amount that should be invested at 4.2% is $8000. The amount that should be invested at 6% is $5600. **15.** The resulting alloy is 30% silver. **17.** The resulting alloy is $33\frac{1}{3}$% silver. **19.** 500 lb of the 12% aluminum alloy is needed. **21.** 25 L of the 65% solution and 25 L of the 15% solution were used. **23.** 3 qt of water must be added. **25.** 30 oz of pure water must be added. **27.** The result is 6% real fruit juice. **29.** The resulting alloy is 31.1% copper. **31.** The amount invested at 9% was $5000. The amount invested at 8% was $6000. The amount invested at 9.5% was $9000. **33.** The cost per pound of the tea mixture is $4.84. **35.** 60 g of pure water were in the beaker before the acid was added. **37.** **a.** 1996 **b.** 1997

CONCEPT REVIEW 2.5

1. Always true **3.** Sometimes true **5.** Sometimes true

SECTION 2.5

3. $\{x|x < 5\}$ **5.** $\{x|x \le 2\}$ **7.** $\{x|x < -4\}$ **9.** $\{x|x > 3\}$ **11.** $\{x|x > 4\}$ **13.** $\{x|x > -2\}$ **15.** $\{x|x \ge 2\}$ **17.** $\{x|x \le 3\}$ **19.** $\{x|x < -3\}$ **21.** $(-\infty, 5]$ **23.** $[1, \infty)$ **25.** $(-\infty, -5)$ **27.** $\left(-\infty, \frac{23}{16}\right)$ **29.** $\left[\frac{8}{3}, \infty\right)$ **31.** $(-\infty, 1)$ **33.** $(-\infty, 3)$ **35.** $(-1, 2)$ **37.** $(-\infty, 1] \cup [3, \infty)$ **39.** $(-2, 4)$ **41.** $(-\infty, -3) \cup (0, \infty)$ **43.** $[3, \infty)$ **45.** $\varnothing$ **47.** $\varnothing$ **49.** $(-\infty, 1) \cup (3, \infty)$ **51.** $\{x|-3 < x < 4\}$ **53.** $\{x|3 < x < 5\}$ **55.** $\left\{x|x > 3 \text{ or } x \le -\frac{5}{2}\right\}$ **57.** $\{x|x > 4\}$ **59.** $\varnothing$ **61.** $\{x|x \text{ is a real number}\}$ **63.** $\{x|-4 < x < -2\}$ **65.** $\{x|x < -4 \text{ or } x > 3\}$ **67.** $\{x|-4 \le x < 1\}$ **69.** $\left\{x|x > \frac{27}{2} \text{ or } x < \frac{5}{2}\right\}$ **71.** $\left\{x|-10 \le x \le \frac{11}{2}\right\}$ **73.** The smallest integer is 3. **75.** The maximum width of the rectangle is 2 ft. **77.** The four integers are 15, 16, 17, 18; or 16, 17, 18, 19; or 17, 18, 19, 20. **79.** The length of the second side could be 5 in., 6 in., or 7 in. **81.** A customer can use a cellular phone 160 min before the charges exceed the first option. **83.** The AirTouch Plan is less expensive for more than 460 pages. **85.** A call must be 7 min or less if it is to be cheaper to pay with coins. **87.** If a business chooses the Glendale Federal Bank, then the business writes more than 200 checks per month. **89.** The car can travel between 429 mi and 536.25 mi on a full tank of gasoline. **91.** $\{1, 2\}$ **93.** $\{1, 2, 3, 4, 5, 6\}$ **95.** $\{1, 2, 3, 4\}$ **97.** $\{1, 2\}$ **99.** The temperature is between 25°C and 30°C. **101.** The largest whole number of minutes the call would last is 10 min.

CONCEPT REVIEW 2.6

1. Always true **3.** Never true **5.** Never true

SECTION 2.6

3. -7 and 7 **5.** -3 and 3 **7.** No solution **9.** -5 and 1 **11.** 8 and 2 **13.** 2 **15.** No solution **17.** $\frac{1}{2}$ and $\frac{9}{2}$ **19.** 0 and $\frac{4}{5}$ **21.** -1 **23.** No solution **25.** 4 and 14 **27.** 4 and 12 **29.** $\frac{7}{4}$ **31.** No solution **33.** -3 and $\frac{3}{2}$ **35.** $\frac{4}{5}$ and 0 **37.** No solution **39.** No solution **41.** -2 and 1 **43.** $-\frac{3}{5}$ **47.** $\{x|x > 3 \text{ or } x < -3\}$ **49.** $\{x|x > 1 \text{ or } x < -3\}$ **51.** $\{x|4 \le x \le 6\}$ **53.** $\{x|x \le -1 \text{ or } x \ge 5\}$ **55.** $\{x|-3 < x < 2\}$ **57.** $\left\{x|x > 2 \text{ or } x < -\frac{14}{5}\right\}$ **59.** $\varnothing$ **61.** $\{x|x \text{ is a real number}\}$ **63.** $\left\{x|x \le -\frac{1}{3} \text{ or } x \ge 3\right\}$ **65.** $\left\{x|-2 \le x \le \frac{9}{2}\right\}$ **67.** $\{x|x = 2\}$ **69.** $\left\{x|x < -2 \text{ or } x > \frac{22}{9}\right\}$ **71.** The lower and upper limits of the diameter of the bushing are 1.742 in. and 1.758 in. **73.** The lower and upper limits of the amount of medicine to be given to the patient are 2.3 cc and 2.7 cc. **75.** The lower and upper limits of the amount of voltage on which the motor will run are 195 volts and 245 volts. **77.** The lower and upper limits of the diameter of the piston are $3\frac{19}{64}$ in. and $3\frac{21}{64}$ in. **79.** The lower and upper limits of the resistor are 13,500 ohms and 16,500 ohms. **81.** The lower and upper limits of the resistor are 53.2 ohms and 58.8 ohms. **83.** $2, -\frac{2}{3}$ **85.** $\{x|-7 \le x \le 8\}$ **87.** $\{y|y \ge -6\}$ **89.** $\{b|b \le 7\}$ **91.** $|x - 2| < 5$ **93.** **a.** $\le$ **b.** $\ge$ **c.** $\ge$ **d.** $=$ **e.** $=$

CHAPTER REVIEW EXERCISES

1. -9 (Obj. 2.1.1) **2.** $-\frac{1}{12}$ (Obj. 2.1.1) **3.** 7 (Obj. 2.1.1) **4.** $\frac{2}{3}$ (Obj. 2.1.1) **5.** $\frac{8}{5}$ (Obj. 2.1.1) **6.** 6 (Obj. 2.1.1) **7.** $\frac{26}{17}$ (Obj. 2.1.2) **8.** $\frac{5}{2}$ (Obj. 2.1.2) **9.** $-\frac{17}{2}$ (Obj. 2.1.2)

10. $-\frac{9}{19}$ (Obj. 2.1.2) **11.** $\left(\frac{5}{3}, \infty\right)$ (Obj. 2.5.1) **12.** $(-4, \infty)$ (Obj. 2.5.1) **13.** $\left\{x \mid x \le -\frac{87}{14}\right\}$ (Obj. 2.5.1) **14.** $\left\{x \mid x \ge \frac{16}{13}\right\}$ (Obj. 2.5.1) **15.** $(-1, 2)$ (Obj. 2.5.2) **16.** $(-\infty, -2) \cup (2, \infty)$ (Obj. 2.5.2) **17.** $\left\{x \mid -3 < x < \frac{4}{3}\right\}$ (Obj. 2.5.2) **18.** $\{x \mid x$ is a real number$\}$ (Obj. 2.5.2) **19.** $-\frac{5}{2}$ and $\frac{11}{2}$ (Obj. 2.6.1) **20.** $-\frac{8}{5}$ (Obj. 2.6.1) **21.** No solution (Obj. 2.6.1) **22.** $\{x \mid 1 \le x \le 4\}$ (Obj. 2.6.2) **23.** $\left\{x \mid x \le \frac{1}{2} \text{ or } x \ge 2\right\}$ (Obj. 2.6.2) **24.** $\varnothing$ (Obj. 2.6.2) **25.** The lower and upper limits of the diameter of the bushing are 2.747 in. and 2.753 in. (Obj. 2.6.3) **26.** The lower and upper limits of the amount of medicine to be given are 1.75 cc and 2.25 cc. (Obj. 2.6.3) **27.** The integers are 6 and 14. (Obj. 2.2.2) **28.** The integers are −1, 0, and 1. (Obj. 2.2.2) **29.** There are 7 quarters in the collection. (Obj. 2.2.1) **30.** The mixture costs $4.25 per ounce. (Obj. 2.3.1) **31.** The mixture must contain 52 gal of apple juice. (Obj. 2.3.1) **32.** The speed of the first plane is 440 mph. The speed of the second plane is 520 mph. (Obj. 2.3.2) **33.** The amount invested at 10.5% was $3000. The amount invested at 6.4% was $5000. (Obj. 2.4.1) **34.** 375 lb of the 30% tin alloy and 125 lb of the 70% tin alloy were used. (Obj. 2.4.2) **35.** The executive's amount of sales must be $55,000 or more. (Obj. 2.5.3) **36.** The range of scores to receive a B grade is $82 \le x \le 100$. (Obj. 2.5.3)

CHAPTER TEST

1. −2 (Obj. 2.1.1) **2.** $-\frac{1}{8}$ (Obj. 2.1.1) **3.** $\frac{5}{6}$ (Obj. 2.1.1) **4.** 4 (Obj. 2.1.1) **5.** $\frac{32}{3}$ (Obj. 2.1.1) **6.** $-\frac{1}{5}$ (Obj. 2.1.1) **7.** 1 (Obj. 2.1.2) **8.** −24 (Obj. 2.1.2) **9.** $\frac{12}{7}$ (Obj. 2.1.2) **10.** $(-\infty, -3]$ (Obj. 2.5.1) **11.** $(-1, \infty)$ (Obj. 2.5.1) **12.** $\{x \mid x > -2\}$ (Obj. 2.5.2) **13.** $\varnothing$ (Obj. 2.5.2) **14.** 3 and $-\frac{9}{5}$ (Obj. 2.6.1) **15.** 7 and −2 (Obj. 2.6.1) **16.** $\left\{x \mid -\frac{1}{3} \le x \le 1\right\}$ (Obj. 2.6.2) **17.** $\{x \mid x > 2 \text{ or } x < -1\}$ (Obj. 2.6.2) **18.** No solution (Obj. 2.6.1) **19.** It costs less to rent from agency A if the car is driven less than 120 mi. (Obj. 2.5.3) **20.** The lower and upper limits of the amount of medication to be given are 2.9 cc and 3.1 cc. (Obj. 2.6.3) **21.** There are six 24¢ stamps. (Obj. 2.2.1) **22.** The price of the hamburger mixture is $2.20/lb (Obj. 2.3.1) **23.** The jogger ran a total distance of 12 mi. (Obj. 2.3.2) **24.** The amount invested at 7.8% was $5000. The amount invested at 9% was $7000. (Obj. 2.4.1) **25.** 100 oz of pure water must be added. (Obj. 2.4.2)

CUMULATIVE REVIEW EXERCISES

1. −108 (Obj. 1.2.3) **2.** 3 (Obj. 1.2.4) **3.** −64 (Obj. 1.2.4) **4.** −8 (Obj. 1.3.2) **5.** The Commutative Property of Addition (Obj. 1.3.1) **6.** $\{3, 9\}$ (Obj. 1.1.2) **7.** $-17x + 2$ (Obj. 1.3.3) **8.** $25y$ (Obj. 1.3.3) **9.** 2 (Obj. 2.1.1) **10.** $\frac{1}{2}$ (Obj. 2.1.1) **11.** 1 (Obj. 2.1.1) **12.** 24 (Obj. 2.1.1) **13.** 2 (Obj. 2.1.2) **14.** 2 (Obj. 2.1.2) **15.** $-\frac{13}{5}$ (Obj. 2.1.2) **16.** $\{x \mid x \le -3\}$ (Obj. 2.5.1) **17.** $\varnothing$ (Obj. 2.5.2) **18.** $\{x \mid x > -2\}$ (Obj. 2.5.2) **19.** −1 and 4 (Obj. 2.6.1) **20.** −4 and 7 (Obj. 2.6.1) **21.** $\left\{x \mid \frac{1}{3} \le x \le 3\right\}$ (Obj. 2.6.2) **22.** $\left\{x \mid x > 2 \text{ or } x < -\frac{1}{2}\right\}$ (Obj. 2.6.2) **23.**

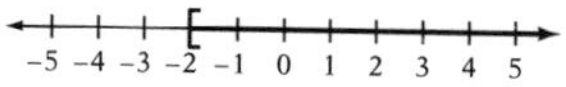

(Obj. 1.1.2)

24.

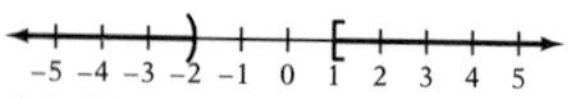

(Obj. 1.1.2) **25.** $(3n + 6) + 3n$; $6n + 6$ (Obj. 1.4.1) **26.** The first integer is 1. (Obj. 2.2.2) **27.** There are eleven 9¢ stamps. (Obj. 2.2.1) **28.** 48 adult tickets were sold. (Obj. 2.3.1) **29.** The speed of the faster plane is 340 mph. (Obj. 2.3.2) **30.** 3 L of 12% acid solution must be in the mixture. (Obj. 2.4.2) **31.** $6500 was invested at 9.8%. (Obj. 2.4.1)

ANSWERS to Chapter 3 Odd-Numbered Exercises

CONCEPT REVIEW 3.1

1. Always true **3.** Never true **5.** Never true

SECTION 3.1

3.

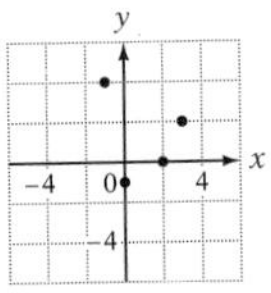

5. A is (0, 3); B is (1, 1); C is (3, −4); D is (−4, 4). **7.**

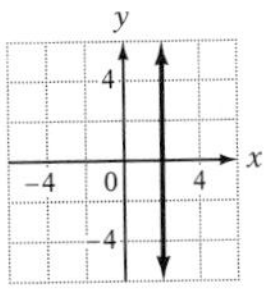

9.

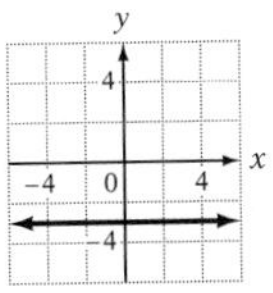

11.

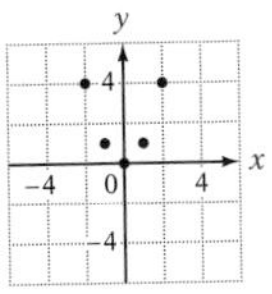

13.

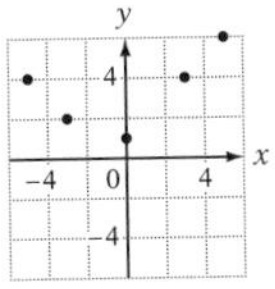

15.

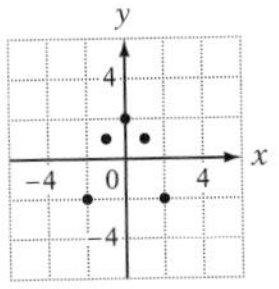

17.

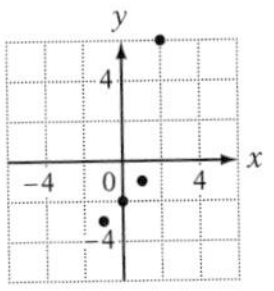

21. Length: $\sqrt{17}$; midpoint: $\left(\frac{7}{2}, 3\right)$ **23.** Length: $\sqrt{5}$; midpoint: $\left(-1, \frac{7}{2}\right)$ **25.** Length: $\sqrt{26}$; midpoint: $\left(-\frac{1}{2}, -\frac{9}{2}\right)$ **27.** Length: $\sqrt{85}$; midpoint: $\left(2, \frac{3}{2}\right)$ **29.** Length: 3; midpoint: $\left(\frac{7}{2}, -5\right)$ **31. a.** There are 275 calories in a hamburger. **b.** There are 1100 mg of sodium in a Big Mac. **33. a.** In 1994 the number of people using cell phones first exceeded 200 million. **b.** There were 573 million people using cell phones in 1997.

35.

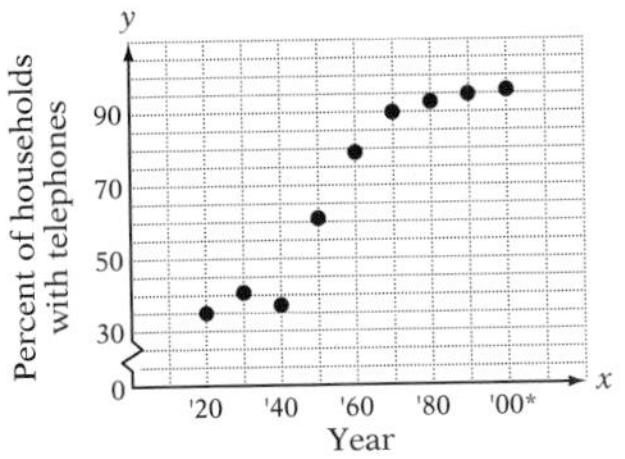

*Projected

37.

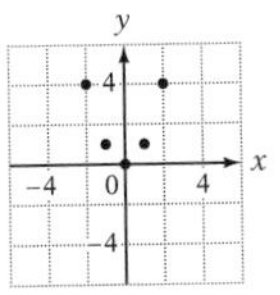

39.

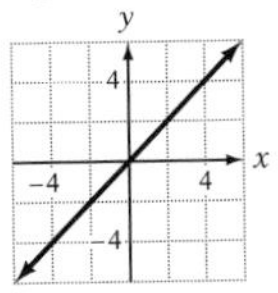

CONCEPT REVIEW 3.2

1. Sometimes true **3.** Sometimes true **5.** Always true

SECTION 3.2

3. Yes. No element of the domain is paired with more than one element of the range. **5.** Yes. No element of the domain is paired with more than one element of the range. **7.** No. The number 6 in the domain is paired with 2 and 3. Therefore, an element of the domain is paired with more than one element of the range. **9.** Function **11.** Function **13.** Function **15.** Not a function **17.** Yes **21.** 11 **23.** −4 **25.** 4 **27.** 10 **29.** 5 **31.** 0 **33.** 24 **35.** −4 **37.** 1 **39.** $\frac{3t}{t+2}$ **41.** 6 **43.** $a^3 - 3a + 4$ **45. a.** A person with an annual income of \$40,135 pays \$2387 in state income taxes. **b.** A person with an annual income of \$40,249 pays \$2393 in state income taxes. **47. a.** The appraiser receives a fee of \$3000. **b.** The appraiser receives a fee of \$950. **49.** Domain: {1, 2, 3, 4, 5}; range: {1, 4, 7, 10, 13} **51.** Domain: {0, 2, 4, 6}; range: {1, 2, 3, 4}

53. Domain: {1, 3, 5, 7, 9}; range: {0}   **55.** Domain: {−2, −1, 0, 1, 2}; range: {0, 1, 2}   **57.** 1   **59.** −8
61. None   **63.** None   **65.** 0   **67.** None   **69.** None   **71.** −2   **73.** None   **75.** Range: {−3, 1, 5, 9}
77. Range: {−23, −13, −8, −3, 7}   **79.** Range: {0, 1, 4}   **81.** Range: {2, 14, 26, 42}   **83.** Range: $\left\{-5, \frac{5}{3}, 5\right\}$
85. Range: $\left\{-1, -\frac{1}{2}, -\frac{1}{3}, 1\right\}$ **87.** Range: {−38, −8, 2}   **89.** $4h$ **91.** $-3h$ **93.** The car will skid for 61.2 ft.
95. a. The parachutist will be falling at a rate of 20 ft/s after 5 s.   **b.** The parachutist will be falling at a rate of
28 ft/s after 15 s.   **97. a.** The temperature of the cola will be 60°F after 10 h.   **b.** The temperature of the cola will
be 50°F after 20 h.   **99. a.**   **b.** The 435 representatives in the House of Representatives
are apportioned among the states according to their
populations.

CONCEPT REVIEW 3.3

1. Never true   **3.** Always true   **5.** Sometimes true

SECTION 3.3

3.   **5.**   **7.**   **9.**   **11.**

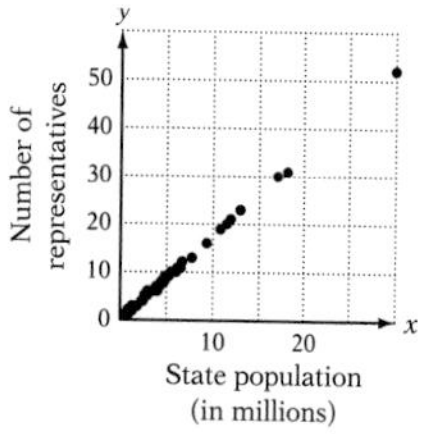

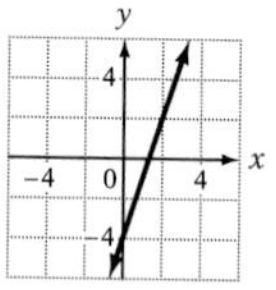

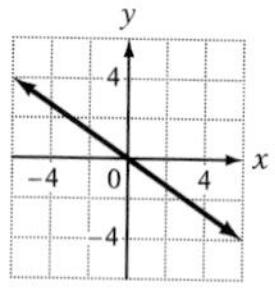

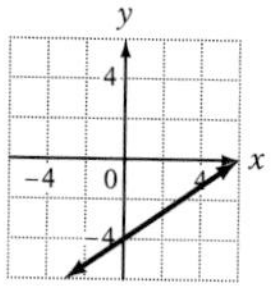

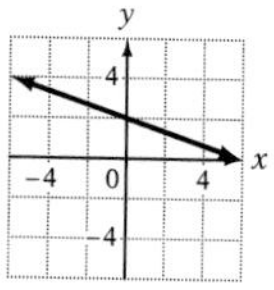

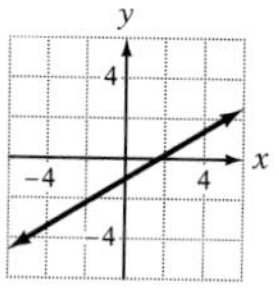

15.   **17.**   **19.**   **21.**

23. x-intercept: (−4, 0)   **25.** x-intercept: $\left(\frac{9}{2}, 0\right)$ **27.** x-intercept: (2, 0)   **29.** x-intercept: $\left(\frac{5}{3}, 0\right)$
y-intercept: (0, 2) y-intercept: (0, −3) y-intercept: (0, −4) y-intercept: $\left(0, \frac{5}{2}\right)$

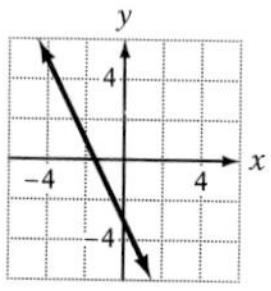

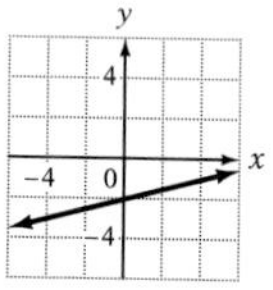

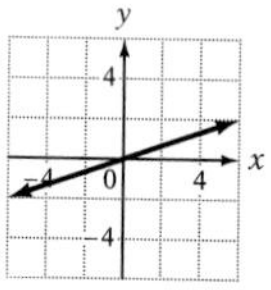

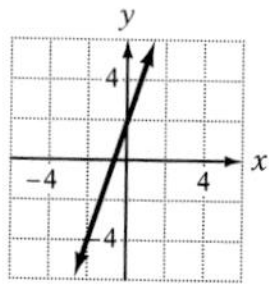

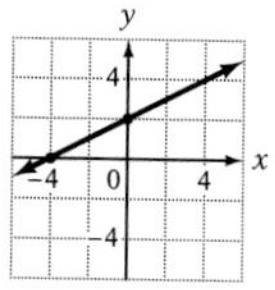

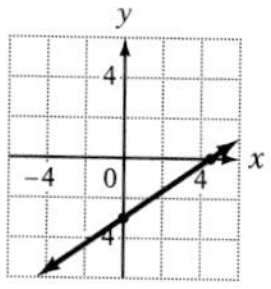

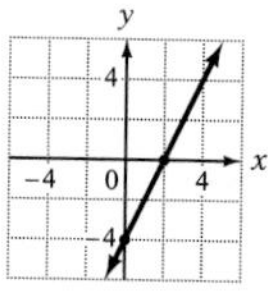

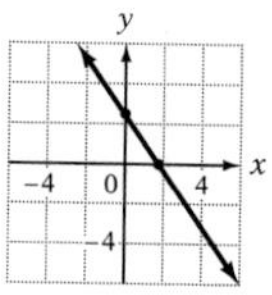

31. x-intercept: $\left(\frac{4}{3}, 0\right)$ **33.** x-intercept: (3, 0)   **35.**   The laborer will earn $169.50
y-intercept: (0, 2) y-intercept: $\left(0, -\frac{9}{5}\right)$ for working 30 h.

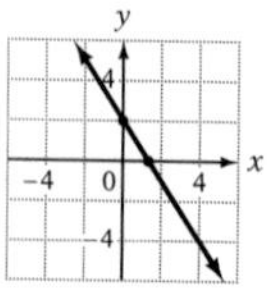

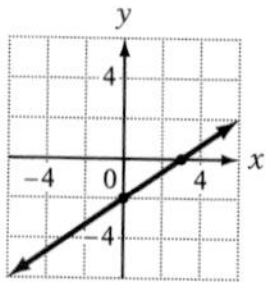

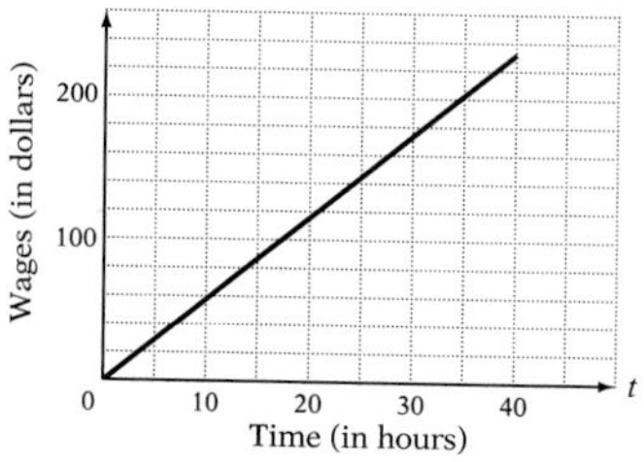

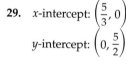

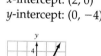

37. 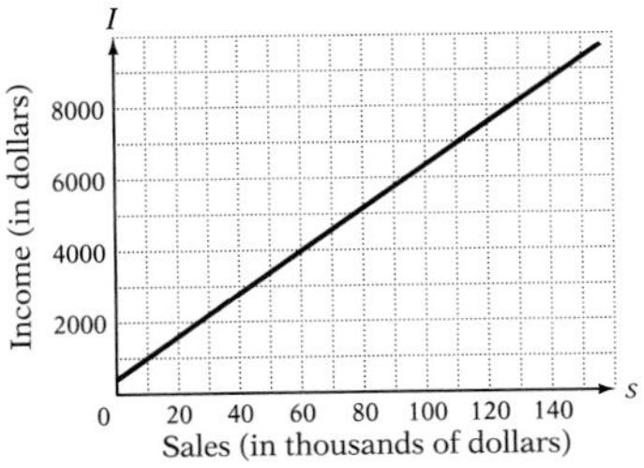

The realtor will earn \$4000 for selling \$60,000 worth of property.

39. 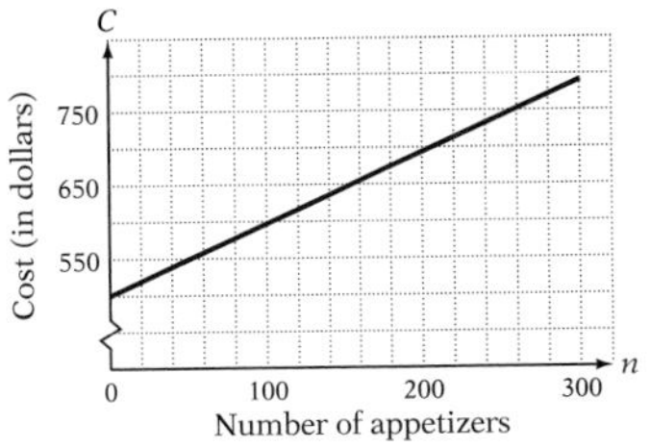

The caterer will charge \$614 for 120 hot appetizers.

41. The sale price is \$160. **43.** 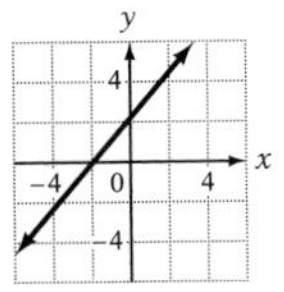 **45.** 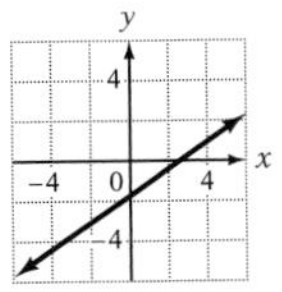 **47.** 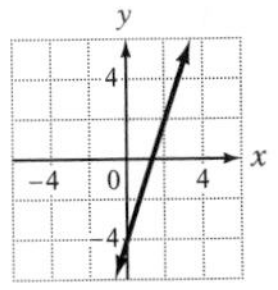

49. a. 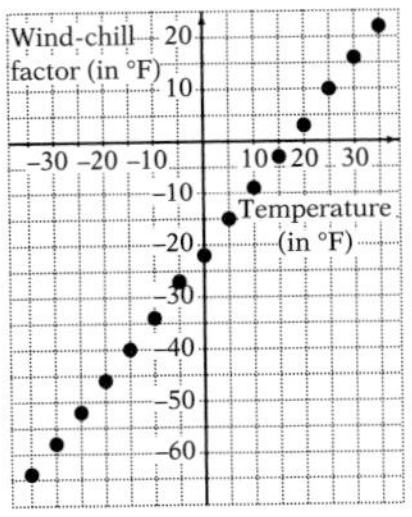

b. Yes
c. Domain: $\{-35, -30, -25, -20, -15, -10, -5, 0, 5, 10, 15, 20, 25, 30, 35\}$; range: $\{-64, -58, -52, -46, -40, -34, -27, -22, -15, -9, -3, 3, 10, 16, 22\}$
d. The ordered pair $(0, -22)$ does not satisfy the function.
e. The x-intercept is $(17, 0)$. Given a wind speed of 10 mph, the wind-chill factor is 0°F when the air temperature is 17°F. The y-intercept is $(0, -21.3)$. Given a wind speed of 10 mph, the wind-chill factor is -21.3°F when the air temperature is 0°F.

CONCEPT REVIEW 3.4

1. Always true **3.** Never true

SECTION 3.4

3. -1 **5.** $\frac{1}{3}$ **7.** $-\frac{2}{3}$ **9.** $-\frac{3}{4}$ **11.** Undefined **13.** $\frac{7}{5}$ **15.** 0 **17.** $-\frac{1}{2}$ **19.** Undefined
21. The slope is 90. The price of the "Flutter" Beanie Baby was increasing at a rate of \$90 per month. **23.** The slope is -0.05. For each mile the car is driven, approximately 0.05 gal of fuel is used. **25.** Line A represents Lois's distance; line B, Tanya's distance; and line C, the distance between. **27. a.** The ramp does not meet the ANSI requirements. **b.** The ramp does meet the ANSI requirements. **29.** 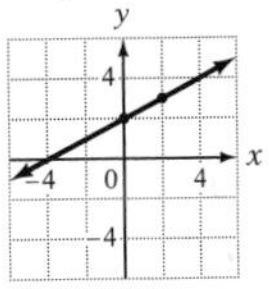 **31.** 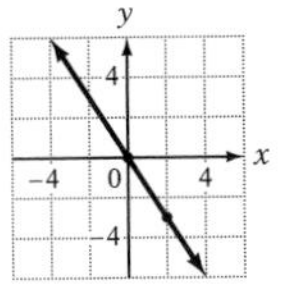

33. 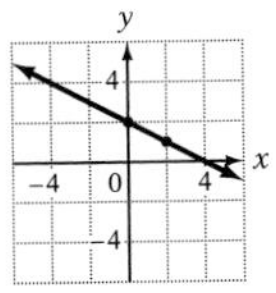 **35.** 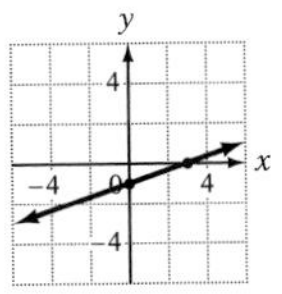 **37.** 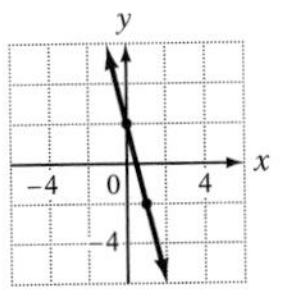 **39.** 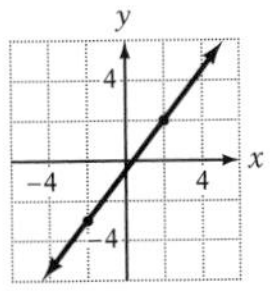 **41.** 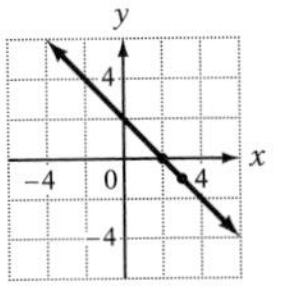

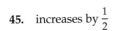

43. increases by 2 **45.** increases by $\frac{1}{2}$ **47.** 10

49. a. and b.

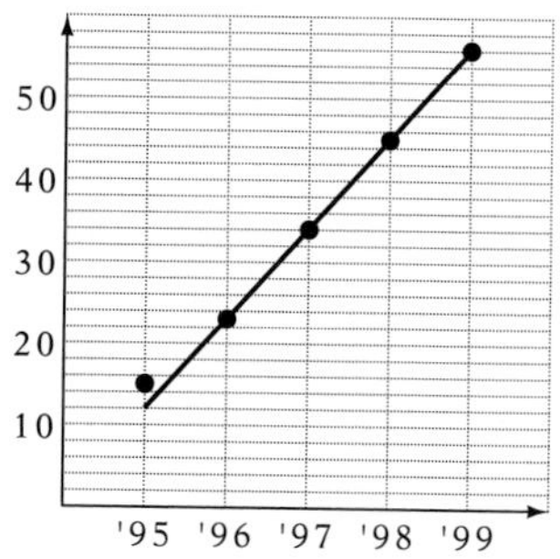

c. There would be 67 million on-line households in 2000.
d. The number of households on-line increased by 11 million between 1997 and 1998.
e. The number of on-line households is increasing at a rate of 11 million per year.
f. The model would estimate that there were 1 million households on-line in 1994.

CONCEPT REVIEW 3.5

1. Always true 3. Never true 5. Never true 7. Never true

SECTION 3.5

3. $y = 2x + 5$ 5. $y = \frac{1}{2}x + 2$ 7. $y = -\frac{5}{3}x + 5$ 9. $y = -3x + 4$ 11. $y = \frac{1}{2}x$ 13. $y = 3x - 9$
15. $y = -\frac{2}{3}x + 7$ 17. $y = -x - 3$ 19. $x = 3$ 21. $y = -3$ 23. $y = -2x + 3$ 25. $x = -5$
27. $y = x + 2$ 29. $y = -2x - 3$ 31. $y = \frac{1}{3}x + \frac{10}{3}$ 33. $y = -\frac{3}{2}x + 3$ 35. $y = x - 1$
37. $y = \frac{2}{3}x + \frac{5}{3}$ 39. $y = \frac{1}{2}x - 1$ 41. $y = -4$ 43. $y = \frac{3}{4}x$ 45. $x = -2$ 47. $y = x - 1$
49. $y = -x + 3$ 51. $y = 1200x$; the model predicts the plane will be 13,200 ft above sea level in 11 min.
53. $y = 0.59x + 4.95$; the model predicts that using a cell phone for 13 min in one month will cost $12.62.
55. $y = 63x$; the model predicts that there are 315 calories in a 5-ounce serving of lean hamburger.
57. $y = -0.431x + 864$; the model predicts that in 1967 a plane could travel from New York to Paris in 16.2 h.
59. $y = -0.032x + 16$; the model predicts that this car would use 11.2 gal of gas for a trip of 150 m.
61. $\left(-\frac{b}{m}, 0\right)$ 63. 0 65. Changing b moves the graph of the line up or down.

CONCEPT REVIEW 3.6

1. Sometimes true 3. Never true 5. Always true 7. Always true

SECTION 3.6

3. Yes 5. No 7. No 9. Yes 11. Yes 13. Yes 15. No 17. Yes 19. Yes
21. $y = \frac{2}{3}x - \frac{8}{3}$ 23. $y = \frac{1}{3}x - \frac{1}{3}$ 25. $y = -\frac{5}{3}x - \frac{14}{3}$ 27. $\frac{A_1}{B_1} = -\frac{B_2}{A_2}$ 29. Any equation of the form $y = 2x + b$, where $b \neq -13$, or of the form $y = -\frac{3}{2}x + c$, where $c \neq 8$. 31. x-intercept: (9, 0); y-intercept (0, 9)

CONCEPT REVIEW 3.7

1. Never true 3. Always true 5. Always true

SECTION 3.7

3. 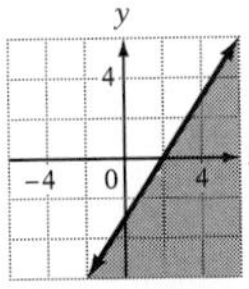

5. 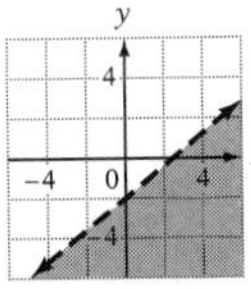

7. 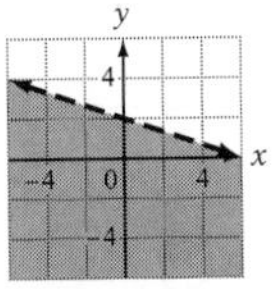

9. 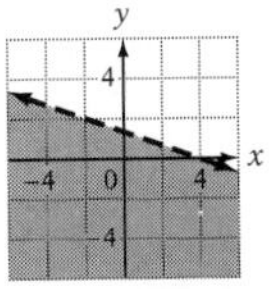

11. 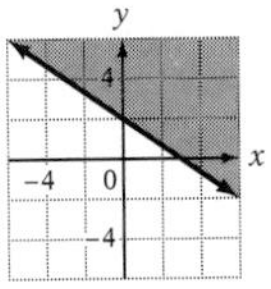

13. 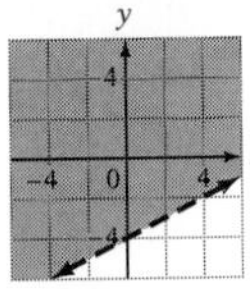

15. 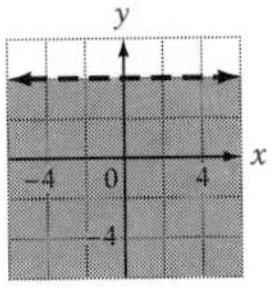

17. 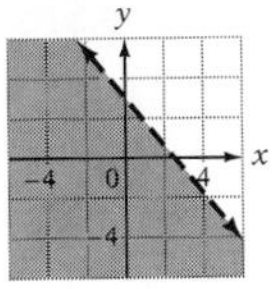

19. 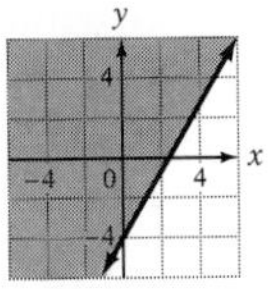

CHAPTER REVIEW EXERCISES

1. (4, 2) (Obj. 3.2.1)

2. Midpoint: $\left(\frac{1}{2}, \frac{9}{2}\right)$; length: $\sqrt{26}$ (Obj. 3.1.2)

3. 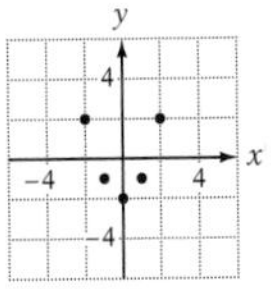 (Obj. 3.1.1)

4. 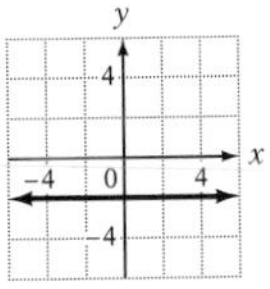 (Obj. 3.1.1)

5. $P(-2) = -2$; $P(a) = 3a + 4$ (Obj. 3.2.1)

6. Domain: $\{-1, 0, 1, 2, 5\}$; range: $\{0, 2, 3\}$ (Obj. 3.2.1)

7. Range: $\{-2, -1, 2\}$ (Obj. 3.2.1)

8. -4 (Obj. 3.2.1)

9. x-intercept: (3, 0)
y-intercept: (0, -2) 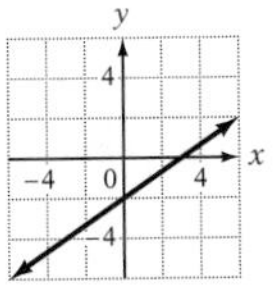 (Obj. 3.3.2)

10. 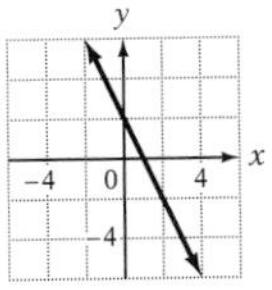 (Obj. 3.3.1)

11. 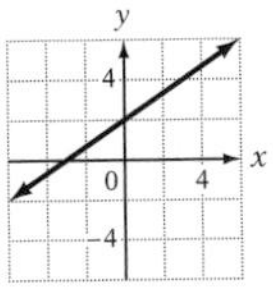 (Obj. 3.3.2)

12. -1 (Obj. 3.4.1)

13. x-intercept: $\left(-\frac{4}{3}, 0\right)$
y-intercept: (0, -2) 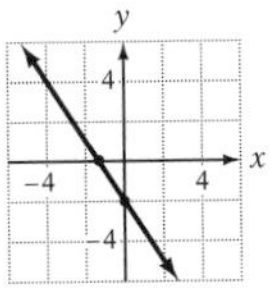 (Obj. 3.3.2)

14. 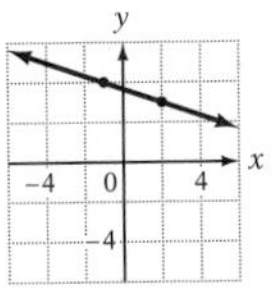 (Obj. 3.4.2)

15. $y = \frac{5}{2}x + \frac{23}{2}$ (Obj. 3.5.1)

16. $y = -\frac{7}{6}x + \frac{5}{3}$ (Obj. 3.5.2)

17. $y = -3x + 7$ (Obj. 3.6.1)

18. $y = \frac{2}{3}x - \frac{8}{3}$ (Obj. 3.6.1)

19. $y = \frac{3}{2}x + 2$ (Obj. 3.6.1)

20. $y = -\frac{1}{2}x - \frac{5}{2}$ (Obj. 3.6.1)

21. 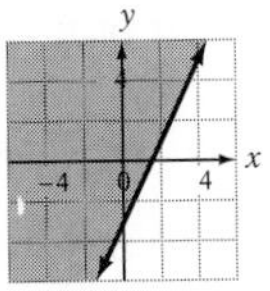 (Obj. 3.7.1)

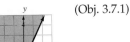

22. 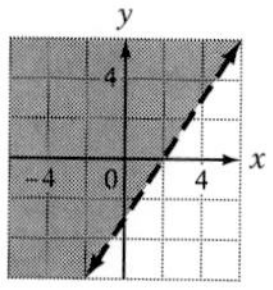 (Obj. 3.7.1) **23.** 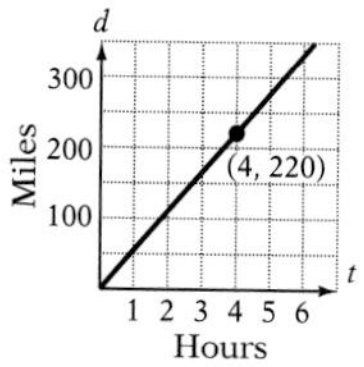

After 4 h the car has traveled 220 mi. (Obj. 3.3.3)

24. The slope is 20. The cost of manufacturing one calculator is \$20. (Obj. 3.4.1) **25.** $y = 80x + 25{,}000$; this model predicts that the cost to build a house that contains 2000 ft^2 will be \$185,000. (Obj. 3.5.3)

CHAPTER TEST

1. 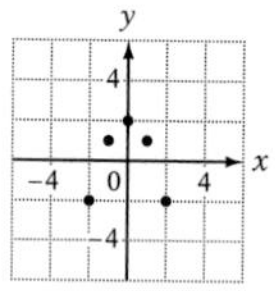 (Obj. 3.1.1) **2.** $(-3, 0)$ (Obj. 3.1.1) **3.** 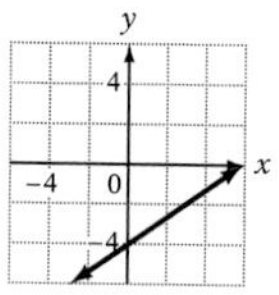 (Obj. 3.3.1)

4. 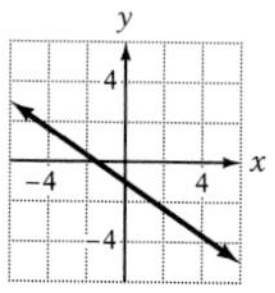 (Obj. 3.3.2) **5.** $x = -2$ (Obj. 3.5.1) **6.** Midpoint: $\left(-\frac{1}{2}, 5\right)$; length: $\sqrt{117}$ (Obj. 3.1.2)

7. $-\frac{1}{6}$ (Obj. 3.4.1) **8.** 9 (Obj. 3.2.1) **9.** 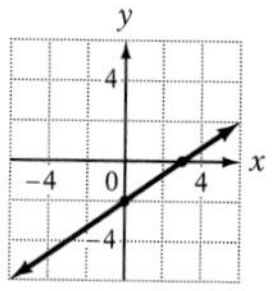 (Obj. 3.3.2) **10.** 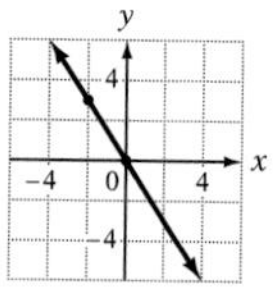 (Obj. 3.4.2)

11. $y = \frac{2}{5}x + 4$ (Obj. 3.5.1) **12.** 0 (Obj. 3.2.1) **13.** $y = -\frac{7}{5}x + \frac{1}{5}$ (Obj. 3.5.2) **14.** $y = -3$ (Obj. 3.5.1)
15. Domain: $\{-4, -2, 0, 3\}$; range: $\{0, 2, 5\}$ (Obj. 3.2.1) **16.** $y = -\frac{3}{2}x + \frac{7}{2}$ (Obj. 3.6.1) **17.** $y = 2x + 1$ (Obj. 3.6.1) **18.** 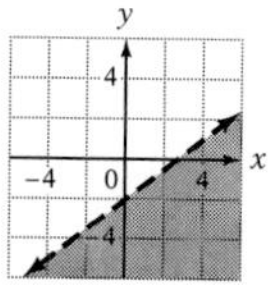 (Obj. 3.7.1) **19.** $y = -\frac{3}{10}x + 175$; this model predicts that 85 students will enroll when the tuition is \$300. (Obj. 3.5.3)

20. The slope is $-\frac{10{,}000}{3}$. The value of the house decreases by \$3333.33 each year. (Obj. 3.4.1)

CUMULATIVE REVIEW EXERCISES

1. Commutative Property of Multiplication (Obj. 1.3.1) **2.** $\frac{9}{2}$ (Obj. 2.1.2) **3.** $\frac{8}{9}$ (Obj. 2.1.2)
4. $-\frac{1}{14}$ (Obj. 2.1.2) **5.** $\left\{x \mid x < -1 \text{ or } x > \frac{1}{2}\right\}$ (Obj. 2.5.2) **6.** $\frac{5}{2}$ and $-\frac{3}{2}$ (Obj. 2.6.1)
7. $\left\{x \mid 0 < x < \frac{10}{3}\right\}$ (Obj. 2.6.2) **8.** 8 (Obj. 1.2.4) **9.** -18 (Obj. 1.3.2)
10. 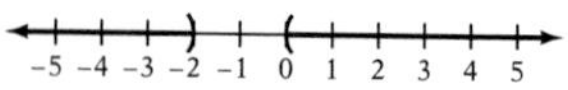

(Obj. 1.1.2) **11.** $\varnothing$ (Obj. 2.5.2) **12.** 14 (Obj. 3.2.1)

13. $(-8, 13)$ (Obj. 3.1.1) **14.** $-\frac{7}{4}$ (Obj. 3.4.1) **15.** $y = \frac{3}{2}x + \frac{13}{2}$ (Obj. 3.5.1)

16. $y = -\frac{5}{4}x + 3$ (Obj. 3.5.2) **17.** $y = -\frac{3}{2}x + 7$ (Obj. 3.6.1) **18.** $y = -\frac{2}{3}x + \frac{8}{3}$ (Obj. 3.6.1)

19.

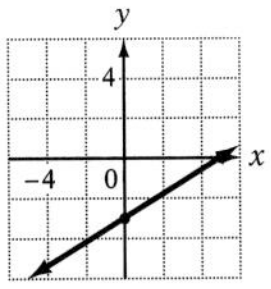

(Obj. 3.3.2)

20.

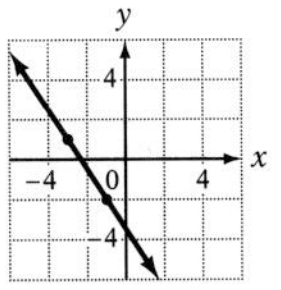

(Obj. 3.4.2)

21.

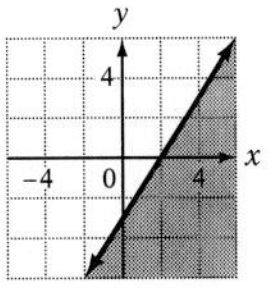

(Obj. 3.7.1)

22. There are 12 nickels in the coin purse. (Obj. 2.2.1) **23.** The rate of the first plane is 200 mph, and the rate of the second plane is 400 mph. (Obj. 2.3.2) **24.** The mixture consists of 48 lb of \$3 coffee and 32 lb of \$8 coffee. (Obj. 2.3.1) **25.** $y = -2500x + 15{,}000$; the value of the truck decreases by \$2500 each year. (Obj. 3.4.1)

ANSWERS to Chapter 4 Odd-Numbered Exercises

CONCEPT REVIEW 4.1

1. Always true **3.** Sometimes true **5.** Always true

SECTION 4.1

3.

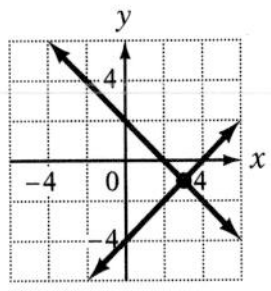

$(3, -1)$

5.

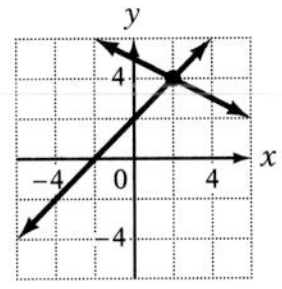

$(2, 4)$

7.

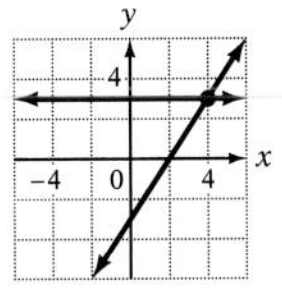

$(4, 3)$

9.

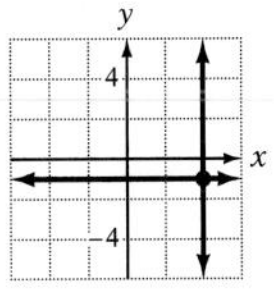

$(4, -1)$

11.

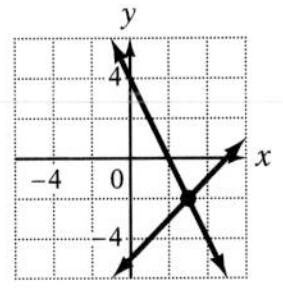

$(3, -2)$

13.

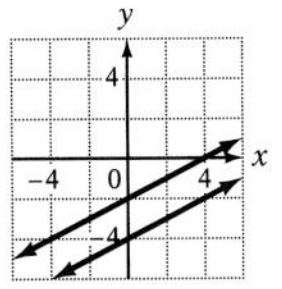

Inconsistent

15.

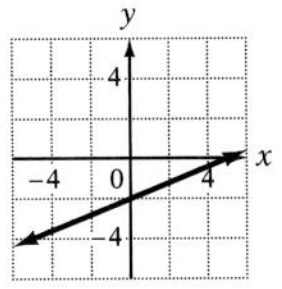

$\left(x, \frac{2}{5}x - 2\right)$

17.

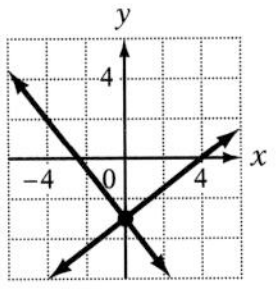

$(0, -3)$

21. $(2, 1)$ **23.** $(1, 1)$ **25.** $(2, 1)$ **27.** $(-2, -3)$ **29.** $(3, -4)$ **31.** $(0, -1)$ **33.** $\left(\frac{1}{2}, 3\right)$ **35.** $(1, 2)$ **37.** $(-1, 2)$ **39.** $(-2, 5)$ **41.** $(0, 0)$ **43.** $(-4, -6)$ **45.** $(1, 5)$ **47.** $(5, 2)$ **49.** $(1, 4)$ **51.** 2 **53.** $\frac{1}{2}$ **55.** The two numbers are 26 and 18. **57.** The two numbers are 8 and 11. **59.** $(2, 1)$ **61.** $\left(\frac{13}{11}, \frac{13}{5}\right)$ **63.** $(1.20, 1.40)$ **65.** $(0.64, -0.10)$ **67.** **a.** No **b.** 2009 **c.** The population is decreasing at a rate of 17,500 people per year. **d.** The population is increasing at a rate of 14,600 people per year.

CONCEPT REVIEW 4.2

1. Always true **3.** Sometimes true **5.** Never true

SECTION 4.2

3. (6, 1) **5.** (1, 1) **7.** (2, 1) **9.** $(x, 3x - 4)$ **11.** $\left(-\frac{1}{2}, 2\right)$ **13.** Inconsistent **15.** (−1, −2) **17.** (2, 5) **19.** $\left(\frac{1}{2}, \frac{3}{4}\right)$ **21.** (0, 0) **23.** $\left(\frac{2}{3}, -\frac{2}{3}\right)$ **25.** (1, −1) **27.** $\left(x, \frac{4}{3}x - 6\right)$ **29.** (5, 3) **31.** $\left(\frac{1}{3}, -1\right)$ **33.** $\left(\frac{5}{3}, \frac{1}{3}\right)$ **35.** Inconsistent **41.** (−1, 2, 1) **43.** (6, 2, 4) **45.** (4, 1, 5) **47.** (3, 1, 0) **49.** (−1, −2, 2) **51.** Inconsistent **53.** (2, 1, −3) **55.** (2, −1, 3) **57.** (6, −2, 2) **59.** (0, −2, 0) **61.** (2, 3, 1) **63.** (1, 1, 3) **65.** (2, 1) **67.** (2, −1) **69.** (2, 0, −1) **71.** $A = 2, B = 3, C = -3$ **73.** The number of nickels is $3z - 5$, the number of dimes is $-4z + 35$, and the number of quarters is z, when $z = 2, 3, 4, 5, 6 7$, or 8. **75.** (1, 1) **77.** (3, 2)

CONCEPT REVIEW 4.3

1. Always true **3.** Sometimes true **5.** Always true

SECTION 4.3

3. 11 **5.** 18 **7.** 0 **9.** 15 **11.** −30 **13.** 0 **17.** (3, −4) **19.** (4, −1) **21.** $\left(\frac{11}{14}, \frac{17}{21}\right)$ **23.** $\left(\frac{1}{2}, 1\right)$ **25.** Not possible by Cramer's Rule **27.** (−1, 0) **29.** (1, −1, 2) **31.** (2, −2, 3) **33.** Not possible by Cramer's Rule **35.** $\left(\frac{68}{25}, \frac{56}{25}, -\frac{8}{25}\right)$ **39.** (1, 3) **41.** (−1, −3) **43.** (3, 2) **45.** Inconsistent **47.** (−2, 2) **49.** (0, 0, −3) **51.** (1, −1, −1) **53.** Inconsistent **55.** $\left(\frac{1}{3}, \frac{1}{2}, 0\right)$ **57.** $\left(\frac{1}{5}, \frac{2}{5}, -\frac{3}{5}\right)$ **59.** $\left(\frac{1}{4}, 0, -\frac{2}{3}\right)$ **61.** −14 **63.** 0 **65.** **a.** 0 **b.** 0 **67.** The area of the polygon is 239 ft^2.

CONCEPT REVIEW 4.4

1. Never true **3.** Always true

SECTION 4.4

1. The rate of the plane in calm air is 150 mph. The rate of the wind is 10 mph. **3.** The rate of the cabin cruiser in calm water is 14 mph. The rate of the current is 2 mph. **5.** The rate of the plane in calm air is 165 mph. The rate of the wind is 15 mph. **7.** The rate of the boat in calm water is 19 km/h. The rate of the current is 3 km/h. **9.** The rate of the plane in calm air is 105 mph. The rate of the wind is 15 mph. **11.** The rate of the boat in calm water is 16.5 mph. The rate of the current is 1.5 mph. **13.** The cost of the pine is $.18 per foot. The cost of the redwood is $.30 per foot. **15.** The cost per unit of gas is $.08. **17.** There are 18 quarters in the bank. **19.** The company plans to manufacture 60 color TVs during the week. **21.** The pharmacist should use 200 mg of the first powder and 450 mg of the second powder. **23.** The model IV costs $4000. **25.** $8000 was invested at 9%, $6000 at 7%, and $4000 at 5%. **27.** The distances are $d_1 = 6$ in., $d_2 = 3$ in., and $d_3 = 9$ in. **29.** There were 10 nickels, 5 dimes, and 4 quarters. **31.** There was $14,000 invested at 12%, $10,000 invested at 8%, and $9000 invested at 9%. **33.** The endorsements for Michael Jordan were $38 million, for Shaquille O'Neal $23 million, and for Arnold Palmer $16 million. **35.** The measures of the two angles are 9° and 81°. **37.** There are 25 nickels, 10 dimes, and 5 quarters in the coin bank.

CONCEPT REVIEW 4.5

1. Sometimes true **3.** Always true

SECTION 4.5

3.   **5.**   **7.**   **9.**   **11.**

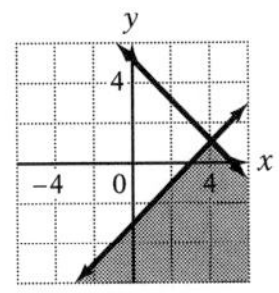

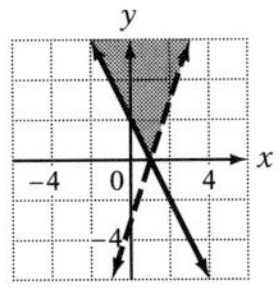

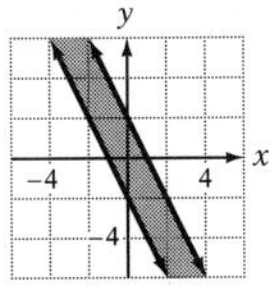

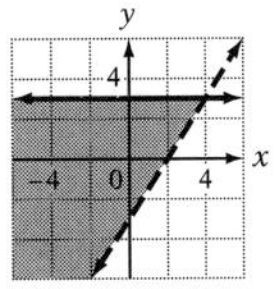

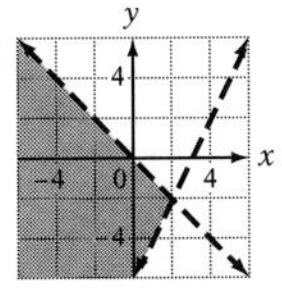

13.   **15.**   **17.**   **19.**   **21.**

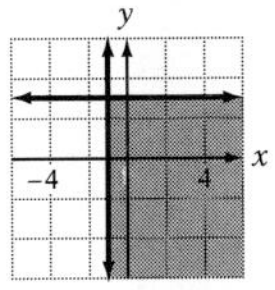

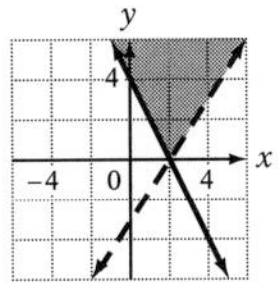

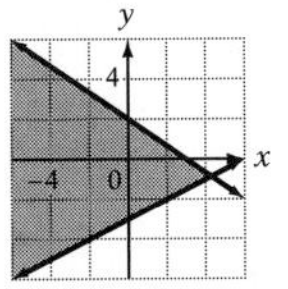

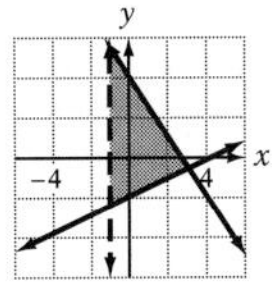

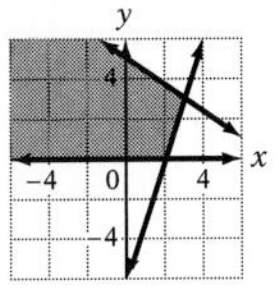

23.   **25.**

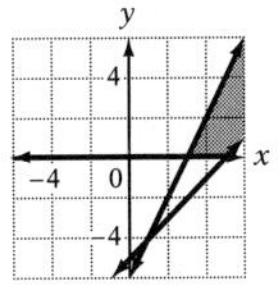

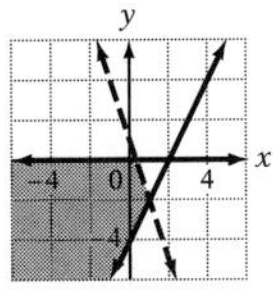

CHAPTER REVIEW EXERCISES

1. Inconsistent (Obj. 4.1.2) **2.** $\left(x, -\dfrac{1}{4}x + \dfrac{3}{2}\right)$ (Obj. 4.1.2) **3.** $(-4, 7)$ (Obj. 4.2.1)

4. $\left(x, \dfrac{1}{3}x - 2\right)$ (Obj. 4.2.1) **5.** $(5, -2, 3)$ (Obj. 4.2.2) **6.** $(3, -1, -2)$ (Obj. 4.2.2) **7.** 28 (Obj. 4.3.1)

8. 0 (Obj. 4.3.1) **9.** $(3, -1)$ (Obj. 4.3.2) **10.** $\left(\dfrac{110}{23}, \dfrac{25}{23}\right)$ (Obj. 4.3.2) **11.** $(-1, -3, 4)$ (Obj. 4.3.2)

12. $(2, 3, -5)$ (Obj. 4.3.2) **13.** $(1, -1, 4)$ (Obj. 4.2.2) **14.** $\left(\dfrac{8}{5}, \dfrac{7}{5}\right)$ (Obj. 4.3.2) **15.** $\left(\dfrac{1}{2}, -1, \dfrac{1}{3}\right)$

(Obj. 4.3.3) **16.** 12 (Obj. 4.3.1) **17.** $(2, -3)$ (Obj. 4.3.2) **18.** $(2, -3, 1)$ (Obj. 4.3.3)

19.   (Obj. 4.1.1) **20.**   (Obj. 4.1.1) **21.**   (Obj. 4.5.1)

$(0, 3)$ $(x, 2x - 4)$

22.   (Obj. 4.5.1) **23.** The rate of the cabin cruiser in calm water is 16 mph. The rate of the current is 4 mph. (Obj. 4.4.1)

24. The rate of the plane in calm air is 175 mph. The rate of the wind is 25 mph. (Obj. 4.4.1)

25. The number of children attending on Friday was 100. (Obj. 4.4.2)

26. The chef must use 3 oz of meat, 5 oz potatoes, and 4 oz of green beans. (Obj. 4.2.2)

CHAPTER TEST

1. $\left(\dfrac{3}{4}, \dfrac{7}{8}\right)$ (Obj. 4.1.2) **2.** $(-3, -4)$ (Obj. 4.1.2) **3.** $(2, -1)$ (Obj. 4.1.2) **4.** $(-2, 1)$ (Obj. 4.3.3)

5. Inconsistent (Obj. 4.2.1) **6.** $(1, 1)$ (Obj. 4.2.1) **7.** Inconsistent (Obj. 4.2.2) **8.** $(2, -1, -2)$

(Obj. 4.3.3) **9.** 10 (Obj. 4.3.1) **10.** -32 (Obj. 4.3.1) **11.** $\left(-\dfrac{1}{3}, -\dfrac{10}{3}\right)$ (Obj. 4.3.2) **12.** $\left(\dfrac{1}{5}, -\dfrac{6}{5}, \dfrac{3}{5}\right)$

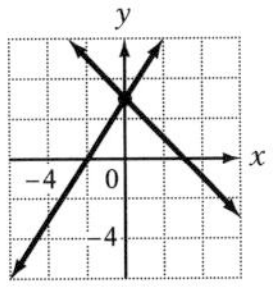

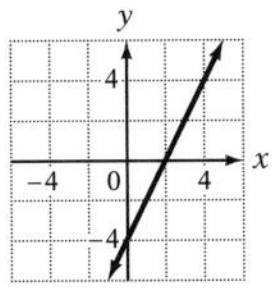

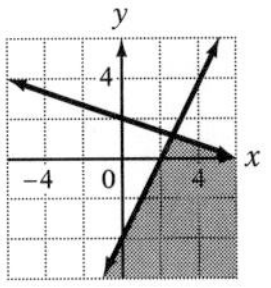

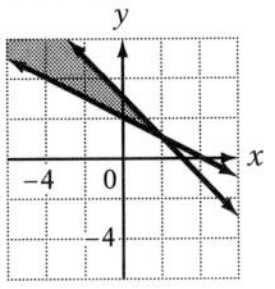

(Obj. 4.3.2) **13.** $(0, -2, 3)$ (Obj. 4.3.2) **14.**   (Obj. 4.1.1) **15.**   (Obj. 4.1.1)

$(3, 4)$ $(-5, 0)$

16.   (Obj. 4.5.1) **17.**   (Obj. 4.5.1) **18.** $(-0.14, 2.43)$ (Obj. 4.1.1)

19. The rate of the plane in calm air is 150 mph. The rate of the wind is 25 mph. (Obj. 4.4.1) **20.** The cost per yard of cotton is \$9. The cost per yard of wool is \$14. (Obj. 4.4.2)

CUMULATIVE REVIEW EXERCISES

1. $-\dfrac{11}{28}$ (Obj. 2.1.2) **2.** $y = 5x - 11$ (Obj. 3.5.2) **3.** $3x - 24$ (Obj. 1.3.3) **4.** -4 (Obj. 1.3.2)

5. $\{x \mid x < 6\}$ (Obj. 2.5.2) **6.** $\{x \mid -4 < x < 8\}$ (Obj. 2.6.2) **7.** $\{x \mid x > 4 \text{ or } x < -1\}$ (Obj. 2.6.2)

8. -98 (Obj. 3.2.1) **9.** $\{0, 1, 5, 8, 16\}$ (Obj. 3.2.1) **10.** 1 (Obj. 3.2.1) **11.** $3h$ (Obj. 3.2.1)

12.   (Obj. 1.1.2) **13.** $y = -\dfrac{2}{3}x + \dfrac{5}{3}$ (Obj. 3.5.1) **14.** $y = -\dfrac{3}{2}x + \dfrac{1}{2}$ (Obj. 3.6.1)

15. $2\sqrt{10}$ (Obj. 3.1.2) **16.** $\left(-\dfrac{1}{2}, 4\right)$ (Obj. 3.1.2) **17.**   (Obj. 3.4.2)

18.   (Obj. 3.7.1) **19.** $(-5, -11)$ (Obj. 4.1.2) **20.** $(1, 0, -1)$ (Obj. 4.2.2) **21.** 3 (Obj. 4.3.1)

22.   (Obj. 4.1.1) **23.** $(2, -3)$ (Obj. 4.3.2) **24.** (Obj. 4.5.1)

$(2, 0)$

25. There are 16 nickels in the purse. (Obj. 2.2.1) **26.** The amount of water that should be added is 60 ml. (Obj. 2.4.2) **27.** The rate of the wind is 12.5 mph. (Obj. 4.4.1) **28.** The cost per pound of steak is \$5. (Obj. 4.4.2) **29.** The lower and upper limits of the resistor are 10,200 ohms and 13,800 ohms. (Obj. 2.6.3) **30.** The slope is 40. The account executive earns \$40 for each \$1000 of sales. (Obj. 3.4.1)

ANSWERS to Chapter 5 Odd-Numbered Exercises

CONCEPT REVIEW 5.1

1. Never true **3.** Never true **5.** Never true

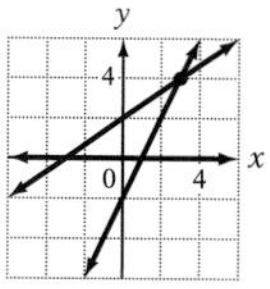

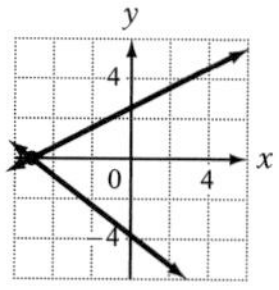

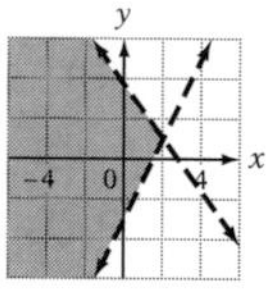

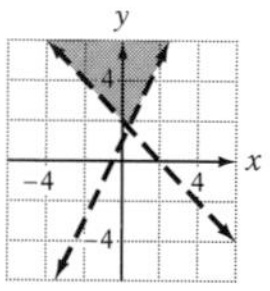

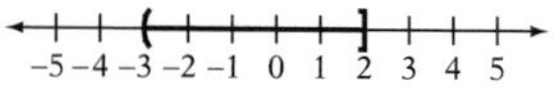

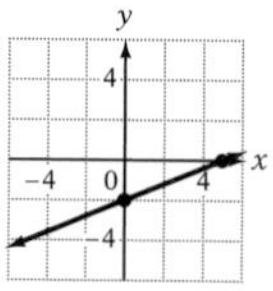

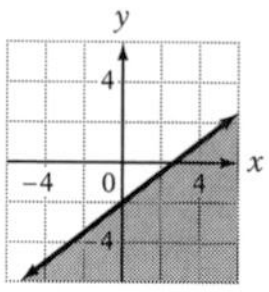

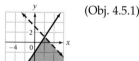

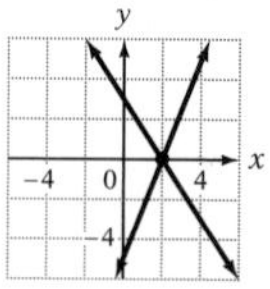

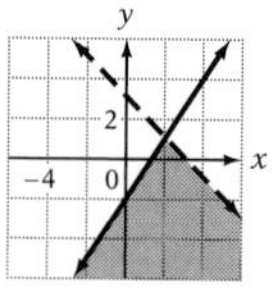

SECTION 5.1

3. a^4b^4 **5.** $-18x^3y^4$ **7.** x^8y^{16} **9.** $81x^8y^{12}$ **11.** $729a^{10}b^6$ **13.** x^5y^{11} **15.** $729x^6$ **17.** $a^{18}b^{18}$ **19.** $4096x^{12}y^{12}$ **21.** $64a^{24}b^{18}$ **23.** x^{2n+1} **25.** y^{6n-2} **27.** a^{2n^2-6n} **29.** x^{15n+10} **31.** $-6x^5y^5z^4$ **33.** $-12a^2b^9c^2$ **35.** $-6x^4y^4z^5$ **37.** $-432a^7b^{11}$ **39.** $54a^{13}b^{17}$ **43.** a^3 **45.** $\frac{a^5}{b^3}$ **47.** 243 **49.** y^3 **51.** $\frac{a^3b^2}{4}$ **53.** $\frac{1}{x^8}$ **55.** $\frac{1}{125x^6}$ **57.** x^9 **59.** a^2 **61.** $\frac{1}{x^6y^{10}}$ **63.** $\frac{1}{2187a}$ **65.** $\frac{y^6}{x^3}$ **67.** $\frac{y^8}{x^4}$ **69.** $\frac{b^2c^2}{a^4}$ **71.** $-\frac{9a}{8b^6}$ **73.** $\frac{b^{10}}{a^{10}}$ **75.** $-\frac{1}{y^{6n}}$ **77.** $\frac{x^{n-5}}{y^6}$ **79.** $\frac{4x^5}{9y^4}$ **81.** $\frac{8b^{15}}{3a^{18}}$ **83.** $x^{12}y^6$ **85.** $\frac{b^2}{a^4}$ **87.** 5×10^{-8} **89.** 4.3×10^6 **91.** 9.8×10^9 **93.** 0.0000000000062 **95.** 634,000 **97.** 4,350,000,000 **99.** 2.848×10^{-10} **101.** 4.608×10^{-3} **103.** 9×10^9 **105.** 2×10^{10} **107.** 6.92×10^{-1} **109.** 1.38 **111.** 6×10^{-15} **113.** The computer can perform 3×10^{10} operations in 1 min. **115.** Light travels 2.592×10^{13} m in 1 day. **117.** The proton is 1.83664508×10^3 times heavier than the electron. **119.** Light travels 5.865696×10^{12} mi in 1 year. **121.** The weight of one orchid seed is 3.2258065×10^{-8} oz. **123.** It will take the satellite 8.86×10^3 h to reach Saturn. **125.** The volume of the cell is $1.41371669 \times 10^{-11}$ mm³ **127.** The space ship travels across the galaxy in 2.4×10^{13} h. **129.** No **131.** Yes **133.** Yes **135.** -4 **137.** The perimeter is $10x^n$. **139.** The area is $20x^2y^2$. **141.** $6x^3y^6$ **143.** $\frac{4}{m^4}$ **145.** No. Let $a = -1$ and $b = 1$.

CONCEPT REVIEW 5.2

1. Always true **3.** Always true **5.** Sometimes true

SECTION 5.2

3. 13 **5.** 10 **7.** -11 **9.** The length of the wave is 23.1 m. **11.** There must be 56 games scheduled. **13.** The baked Alaska needs 260 in³ of meringue. **15. a.** The leading coefficient is -1. **b.** The constant term is 8. **c.** The degree is 2. **17.** The expression is not a polynomial. **19.** The expression is not a polynomial. **21. a.** The leading coefficient is 3. **b.** The constant term is π. **c.** The degree is 5. **23. a.** The leading coefficient is -5. **b.** The constant term is 2. **c.** The degree is 3. **25. a.** The leading coefficient is 14. **b.** The constant term is 14. **c.** The degree is 0. **27.**

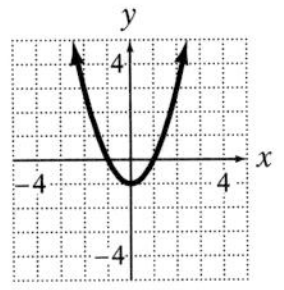

29.

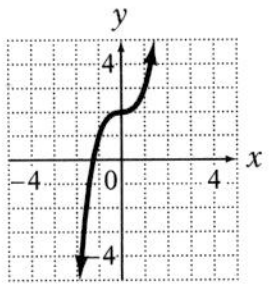

31.

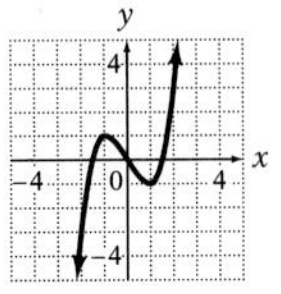

35. $6x^2 - 6x + 5$ **37.** $-x^2 + 1$ **39.** $5y^2 - 15y + 2$ **41.** $7a^2 - a + 2$ **43.** $5x^3 - 4x^2 + 4x - 7$ **45.** $9x^n + 5$ **47.** $3x^4 - 8x^2 + 2x$ **49.** $-2x^3 + 4x^2 + 8$ **51.** 1.5 **53.** 3.4, 0.6 **55.** $-1.9, 0.35, 1.5$ **57.** $1.6, -0.6$ **59.** -2 **61.** $P(x) + Q(x)$ is a fourth-degree polynomial **63.** The maximum deflection of the beam is 3.125 in. **65.** 2 **67.**

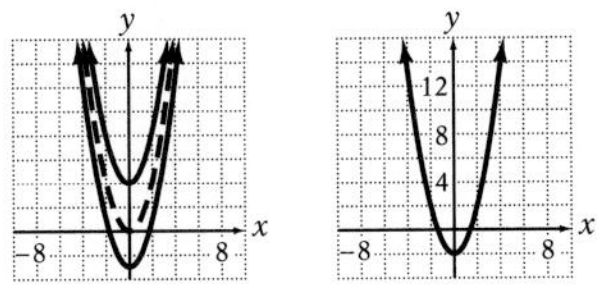

The graph of k is the graph of f moved 2 units down.

CONCEPT REVIEW 5.3

1. Always true **3.** Always true **5.** Always true

SECTION 5.3

3. $2x^2 - 6x$ **5.** $6x^4 - 3x^3$ **7.** $6x^2y - 9xy^2$ **9.** $x^{n+1} + x^n$ **11.** $x^{2n} + x^ny^n$ **13.** $-4b^2 + 10b$ **15.** $-6a^4 + 4a^3 - 6a^2$ **17.** $9b^5 - 9b^3 + 24b$ **19.** $-20x^5 - 15x^4 + 15x^3 - 20x^2$ **21.** $-2x^4y + 6x^3y^2 - 4x^2y^3$ **23.** $x^{3n} + x^{2n} + x^{n+1}$ **25.** $a^{2n+1} - 3a^{n+2} + 2a^{n+1}$ **27.** $5y^2 - 11y$ **29.** $6y^2 - 31y$ **31.** $15x^2 - 61x + 56$ **33.** $14x^2 - 69xy + 27y^2$ **35.** $3a^2 + 16ab - 35b^2$ **37.** $15x^2 + 37xy + 18y^2$ **39.** $30x^2 - 79xy + 45y^2$ **41.** $2x^2y^2 - 3xy - 35$ **43.** $x^4 - 10x^2 + 24$ **45.** $x^4 + 2x^2y^2 - 8y^4$ **47.** $x^{2n} - 9x^n + 20$ **49.** $10b^{2n} + 18b^n - 4$ **51.** $3x^{2n} + 7x^nb^n + 2b^{2n}$ **53.** $x^3 + 8x^2 + 7x - 24$ **55.** $a^4 - a^3 - 6a^2 + 7a + 14$ **57.** $6a^3 - 13a^2b - 14ab^2 - 3b^3$ **59.** $6b^4 - 6b^3 + 3b^2 + 9b - 18$ **61.** $6a^5 - 15a^4 - 6a^3 + 19a^2 - 20a + 25$ **63.** $x^4 - 5x^3 + 14x^2 - 23x + 7$ **65.** $3b^3 - 14b^2 + 17b - 6$ **67.** $x^{3n} - 4x^{2n}y^n + 2x^ny^{2n} + y^{3n}$ **69.** $b^2 - 49$ **71.** $b^2 - 121$ **73.** $25x^2 - 40xy + 16y^2$ **75.** $x^4 + 2x^2y^2 + y^4$ **77.** $4a^2 - 9b^2$ **79.** $x^4 - 1$ **81.** $4x^{2n} + 4x^ny^n + y^{2n}$ **83.** $25a^2 - 81b^2$ **85.** $4x^{2n} - 25$ **87.** $36 - x^2$ **89.** $9a^2 - 24ab + 16b^2$ **91.** $9x^{2n} + 12x^n + 4$ **93.** $x^{2n} - 9$ **95.** $x^{2n} - 2x^n + 1$ **97.** $4x^{2n} + 20x^ny^n + 25y^{2n}$ **99.** The area is $\left(x^2 + \frac{x}{2} - 3\right)$ ft^2. **101.** The area is $(x^2 + 12x + 16)$ ft^2. **103.** The volume is $(4x^3 - 72x^2 + 324x)$ in^3. **105.** The volume is $(2x^3 - 7x^2 - 15x)$ cm^3. **107.** The volume is $(4x^3 + 32x^2 + 48x)$ cm^3. **109.** $9x^2 - 30x + 25$ **111.** $2x^2 + 4xy$ **113.** $6x^5 - 5x^4 + 8x^3 + 13x^2$ **115.** $5x^2 - 5y^2$ **117.** $4x^4y^2 - 4x^4y + x^4$ **119.** -1 **121.** a^2 **123.** $2x^2 + 11x - 21$ **125.** $x^2 - 2xy + y^2$ **127.** 256

CONCEPT REVIEW 5.4

1. Sometimes true **3.** Always true

SECTION 5.4

3. $x + 8$ **5.** $x^2 + \frac{2}{x-3}$ **7.** $3x + 5 + \frac{3}{2x+1}$ **9.** $5x + 7 + \frac{2}{2x-1}$ **11.** $4x^2 + 6x + 9 + \frac{18}{2x-3}$ **13.** $3x^2 + 1 + \frac{1}{2x^2-5}$ **15.** $x^2 - 3x - 10$ **17.** $-x^2 + 2x - 1 + \frac{1}{x-3}$ **19.** $2x^3 - 3x^2 + x - 4$ **21.** $x - 4 + \frac{x+3}{x^2+1}$ **23.** $2x - 3 + \frac{1}{x-1}$ **25.** $3x + 1 + \frac{10x+7}{2x^2-3}$ **27.** $2x - 8$ **29.** $3x + 3 - \frac{1}{x-1}$ **31.** $x - 4 + \frac{7}{x+4}$ **33.** $x - 2 + \frac{16}{x+2}$ **35.** $2x^2 - 3x + 9$ **37.** $x^2 - 3x + 2$ **39.** $x^2 - 5x + 16 - \frac{41}{x+2}$ **41.** $x^2 - x + 2 - \frac{4}{x+1}$ **43.** $4x^2 + 8x + 15 + \frac{12}{x-2}$ **45.** $2x^3 - 3x^2 + x - 4$ **47.** $3x^3 + 2x^2 + 12x + 19 + \frac{33}{x-2}$ **49.** $3x^3 - x + 4 - \frac{2}{x+1}$ **51.** $2x^3 + 6x^2 + 17x + 51 + \frac{155}{x-3}$ **53.** $x^2 - 5x + 25$ **55.** $P(3) = 8$ **57.** $R(4) = 43$ **59.** $P(-2) = -39$ **61.** $Q(2) = 31$ **63.** $F(-3) = 190$ **65.** $P(5) = 122$ **67.** $R(-3) = 302$ **69.** $Q(2) = 0$ **71.** $x - y$ **73.** $a^2 + ab + b^2$ **75.** $x^4 - x^3y + x^2y^2 - xy^3 + y^4$ **77.** 3 **79.** -1 **81.** $x - 3$

CONCEPT REVIEW 5.5

1. Always true **3.** Sometimes true

SECTION 5.5

3. $3a(2a - 5)$ **5.** $x^2(4x - 3)$ **7.** Nonfactorable **9.** $x(x^4 - x^2 - 1)$ **11.** $4(4x^2 - 3x + 6)$ **13.** $5b^2(1 - 2b + 5b^2)$ **15.** $x^n(x^n - 1)$ **17.** $x^{2n}(x^n - 1)$ **19.** $a^2(a^{2n} + 1)$ **21.** $6x^2y(2y - 3x + 4)$ **23.** $4a^2b^2(-4b^2 - 1 + 6a)$ **25.** $y^2(y^{2n} + y^n - 1)$ **27.** $(a + 2)(x - 2)$ **29.** $(x - 2)(a + b)$ **31.** $(x + 3)(x + 2)$ **33.** $(x + 4)(y - 2)$ **35.** $(a + b)(x - y)$ **37.** $(y - 3)(x^2 - 2)$ **39.** $(3 + y)(2 + x^2)$ **41.** $(2a + b)(x^2 - 2y)$ **43.** $(y - 5)(x^n + 1)$ **45.** $(x + 1)(x^2 + 2)$ **47.** $(2x - 1)(x^2 + 2)$ **51.** $(x - 5)(x - 3)$ **53.** $(a + 11)(a + 1)$ **55.** $(b + 7)(b - 5)$ **57.** $(y - 3)(y - 13)$ **59.** $(b + 8)(b - 4)$

61. $(a - 7)(a - 8)$ **63.** $(y + 12)(y + 1)$ **65.** $(x + 5)(x - 1)$ **67.** $(a + 6b)(a + 5b)$ **69.** $(x - 12y)(x - 2y)$ **71.** $(y + 9x)(y - 7x)$ **73.** $(x - 36)(x + 1)$ **75.** $(a + 9)(a + 4)$ **77.** Nonfactorable **79.** $(2x + 5)(x - 8)$ **81.** $(4y - 3)(y - 3)$ **83.** $(2a + 1)(a + 6)$ **85.** Nonfactorable **87.** $(5x + 1)(x + 5)$ **89.** $(11x - 1)(x - 11)$ **91.** $(12x - 5)(x - 1)$ **93.** $(4y - 3)(2y - 3)$ **95.** Nonfactorable **97.** Nonfactorable **99.** $(2x - 3y)(3x + 7y)$ **101.** $(4a + 7b)(a + 9b)$ **103.** $(5x - 4y)(2x - 3y)$ **105.** $(8 - x)(3 + 2x)$ **107.** Nonfactorable **109.** $(3 - 4a)(5 + 2a)$ **111.** $y^2(5y - 4)(y - 5)$ **113.** $2x(2x - 3y)(x + 4y)$ **115.** $5(5 + x)(4 - x)$ **117.** $4x(10 + x)(8 - x)$ **119.** $2x^2(5 + 3x)(2 - 5x)$ **121.** $a^2b^2(ab - 5)(ab + 2)$ **123.** $5(18a^2b^2 + 9ab + 2)$ **125.** Nonfactorable **127.** $2a(a^2 + 5)(a^2 + 2)$ **129.** $3y^2(x^2 - 5)(x^2 - 8)$ **131.** $3b^2(3a + 4)(5a - 6)$ **133.** $3xy(6x - 5y)(2x + 3y)$ **135.** $6b^2(2a - 3b)(4a + 3b)$ **137.** $5(2x^n - 3)(x^n + 4)$ **139.** $xy(3x + y)(x - y)$ **141.** $3b(b^2 + 5)(b^2 - 2)$ **143.** $5, -5, 1, -1$ **145.** $3, -3, 9, -9$ **147.** $1, -1, 5, -5$

CONCEPT REVIEW 5.6

1. Never true **3.** Never true **5.** Never true

SECTION 5.6

3. $(x + 4)(x - 4)$ **5.** $(2x + 1)(2x - 1)$ **7.** $(b - 1)^2$ **9.** $(4x - 5)^2$ **11.** $(xy + 10)(xy - 10)$ **13.** Nonfactorable **15.** $(x + 3y)^2$ **17.** $(2x + y)(2x - y)$ **19.** $(a^n + 1)(a^n - 1)$ **21.** $(a + 2)^2$ **23.** $(x - 6)^2$ **25.** $(4x + 11)(4x - 11)$ **27.** $(1 + 3a)(1 - 3a)$ **29.** Nonfactorable **31.** Nonfactorable **33.** $(5 + ab)(5 - ab)$ **35.** $(5a - 4b)^2$ **37.** $(x^n + 3)^2$ **39.** $(x - 3)(x^2 + 3x + 9)$ **41.** $(2x - 1)(4x^2 + 2x + 1)$ **43.** $(x - y)(x^2 + xy + y^2)$ **45.** $(m + n)(m^2 - mn + n^2)$ **47.** $(4x + 1)(16x^2 - 4x + 1)$ **49.** $(3x - 2y)(9x^2 + 6xy + 4y^2)$ **51.** $(xy + 4)(x^2y^2 - 4xy + 16)$ **53.** Nonfactorable **55.** Nonfactorable **57.** $(a - 2b)(a^2 - ab + b^2)$ **59.** $(x^{2n} + y^n)(x^{4n} - x^{2n}y^n + y^{2n})$ **63.** $(xy - 3)(xy - 5)$ **65.** $(xy - 5)(xy - 12)$ **67.** $(x^2 - 6)(x^2 - 3)$ **69.** $(b^2 - 18)(b^2 + 5)$ **71.** $(x^2y^2 - 6)(x^2y^2 - 2)$ **73.** $(x^n + 2)(x^n + 1)$ **75.** $(3xy - 5)(xy - 3)$ **77.** $(2ab - 3)(3ab - 7)$ **79.** $(2x^2 - 15)(x^2 + 1)$ **81.** $(2x^n - 1)(x^n - 3)$ **83.** $(2a^n + 5)(3a^n + 2)$ **85.** $3(2x - 3)^2$ **87.** $a(3a - 1)(9a^2 + 3a + 1)$ **89.** $5(2x + 1)(2x - 1)$ **91.** $y^3(y + 11)(y - 5)$ **93.** $(4x^2 + 9)(2x + 3)(2x - 3)$ **95.** $2a(2 - a)(4 + 2a + a^2)$ **97.** $(x + 2)(x + 1)(x - 1)$ **99.** $(x + 2)(2x^2 - 3)$ **101.** $(x + 1)(x + 4)(x - 4)$ **103.** $b^3(ab - 1)(a^2b^2 + ab + 1)$ **105.** $x^2(x - 7)(x + 5)$ **107.** Nonfactorable **109.** $2x^3(3x + 1)(x + 12)$ **111.** $(4a^2 + b^2)(2a + b)(2a - b)$ **113.** Nonfactorable **115.** $3b^2(b - 2)(b^2 + 2b + 4)$ **117.** $x^2y^2(x - 3y)(x - 2y)$ **119.** $2xy(4x + 7y)(2x - 3y)$ **121.** $(x - 2)(x + 1)(x - 1)$ **123.** $4(b - 1)(2x - 1)$ **125.** $(y + 1)(y - 1)(2x + 3)(2x - 3)$ **127.** $(x + 2)(x - 2)^2(x^2 + 2x + 4)$ **129.** $a^2(a^n - 3)^2$ **131.** $x^n(2x - 1)(x - 3)$ **133.** $20, -20$ **135.** $8, -8$ **137.** $(x - 1)(x^2 + x + 1)(a - b)$ **139.** $(y^{2n} + 1)^2(y^n - 1)^2(y^n + 1)^2$ **141.** The sum of the numbers is 16. **143.** The palindromic perfect squares less than 500 are 1, 4, 9, 121, and 484. **145.** $(x^2 - 4x + 8)(x^2 + 4x + 8)$

CONCEPT REVIEW 5.7

1. Always true **3.** Sometimes true **5.** Sometimes true

SECTION 5.7

3. -4 and -6 **5.** 0 and 7 **7.** 0 and $-\frac{5}{2}$ **9.** $-\frac{3}{2}$ and 7 **11.** 7 and -7 **13.** $\frac{4}{3}$ and $-\frac{4}{3}$ **15.** -5 and 1 **17.** $-\frac{3}{2}$ and 4 **19.** 0 and 9 **21.** 0 and 4 **23.** 7 and -4 **25.** $\frac{2}{3}$ and -5 **27.** $\frac{2}{5}$ and 3 **29.** $\frac{3}{2}$ and $-\frac{1}{4}$ **31.** 7 and -5 **33.** 9 and 3 **35.** $-\frac{4}{3}$ and 2 **37.** $\frac{4}{3}$ and -5 **39.** 5 and -3 **41.** -11 and -4 **43.** 4 and 2 **45.** 2 and 1 **47.** 2 and 3 **49.** -4, -1, and 1 **51.** 1 and 2 **53.** $-\frac{1}{2}$ and 1 **55.** $\frac{1}{2}$ **57.** -9, -1, and 1 **59.** The number is -10 or 9.

61. The integers are 8 and 9. **63.** The number is 0, 3, or 4. **65.** The width is 8 in. The length is 21 in. **67.** The value of x is 2 m or 3 m. **69.** The velocity of the rocket is 100 m/s. **71.** The height is 4 cm. The base is 12 cm. **73.** The object will hit the ground in 5 s. **75.** The length of the larger rectangle is 11 cm. The width is 8 cm. **77.** $2a$ and $-5a$ **79.** $3a$ and $-3a$ **81.** $5a$ and $-3a$ **83.** 175 or 15 **85.** The length is 18 in. The width is 8 in. **87.** For example, $x^3 - 4x^2 + x + 6 = 0$ **89.** **a.** The profit for U.S. airlines for 1997 was \$6.83 billion. **b.** The percent error is 11.4%.

CHAPTER REVIEW EXERCISES

1. $2x^2 - 5x - 2$ (Obj. 5.2.2) **2.** $4x^2 - 8xy + 5y^2$ (Obj. 5.2.2) **3.** $70xy^2z^6$ (Obj. 5.1.2) **4.** $-\dfrac{48y^9z^{17}}{x}$ (Obj. 5.1.2) **5.** $-\dfrac{x^3}{4y^2z^3}$ (Obj. 5.1.2) **6.** $\dfrac{c^{10}}{2b^{17}}$ (Obj. 5.1.2) **7.** 9.3×10^7 (Obj. 5.1.3) **8.** 0.00254 (Obj. 5.1.3) **9.** 2×10^{-6} (Obj. 5.1.3) **10.** -7 (Obj. 5.2.1) **11.**

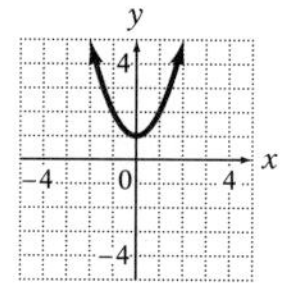

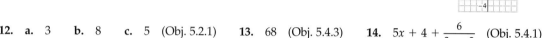

(Obj. 5.2.1)

12. **a.** 3 **b.** 8 **c.** 5 (Obj. 5.2.1) **13.** 68 (Obj. 5.4.3) **14.** $5x + 4 + \dfrac{6}{3x - 2}$ (Obj. 5.4.1) **15.** $2x - 3 - \dfrac{4}{6x + 1}$ (Obj. 5.4.1) **16.** $4x^2 + 3x - 8 + \dfrac{50}{x + 6}$ (Obj. 5.4.1/5.4.2) **17.** $x^3 + 4x^2 + 16x + 64 + \dfrac{252}{x - 4}$ (Obj. 5.4.1/5.4.2) **18.** $12x^5y^3 + 8x^3y^2 - 28x^2y^4$ (Obj. 5.3.1) **19.** $a^{3n+3} - 5a^{2n+4} + 2a^{2n+3}$ (Obj. 5.3.1) **20.** $9x^2 + 8x$ (Obj. 5.3.1) **21.** $x^{3n+1} - 3x^{2n} - x^{n+2} + 3x$ (Obj. 5.3.2) **22.** $x^4 + 3x^3 - 23x^2 - 29x + 6$ (Obj. 5.3.2) **23.** $6x^3 - 29x^2 + 14x + 24$ (Obj. 5.3.2) **24.** $25a^2 - 4b^2$ (Obj. 5.3.3) **25.** $16x^2 - 24xy + 9y^2$ (Obj. 5.3.3) **26.** $6a^2b(3a^3b - 2ab^2 + 5)$ (Obj. 5.5.1) **27.** $x^2(5x^{n+3} + x^{n+1} + 4)$ (Obj. 5.5.1) **28.** $(y - 3)(x - 4)$ (Obj. 5.5.2) **29.** $(a + 2b)(2x - 3y)$ (Obj. 5.5.2) **30.** $(x + 5)(x + 7)$ (Obj. 5.5.3) **31.** $(3 + x)(4 - x)$ (Obj. 5.5.3) **32.** $(x - 7)(x - 9)$ (Obj. 5.5.3) **33.** $(3x - 2)(2x - 9)$ (Obj. 5.5.4) **34.** $(8x - 1)(3x + 8)$ (Obj. 5.5.4) **35.** $(xy + 3)(xy - 3)$ (Obj. 5.6.1) **36.** $(2x + 3y)^2$ (Obj. 5.6.1) **37.** $(x^n - 6)^2$ (Obj. 5.6.1) **38.** $(6 + a^n)(6 - a^n)$ (Obj. 5.6.1) **39.** $(4a - 3b)(16a^2 + 12ab + 9b^2)$ (Obj. 5.6.2) **40.** $(2 - y^n)(4 + 2y^n + y^{2n})$ (Obj. 5.6.2) **41.** $(3x^2 + 2)(5x^2 - 3)$ (Obj. 5.6.3) **42.** $(6x^4 - 5)(6x^4 - 1)$ (Obj. 5.6.3) **43.** $(3x^2y^2 + 2)(7x^2y^2 + 3)$ (Obj. 5.6.3) **44.** $3a^2(a^2 - 6)(a^2 + 1)$ (Obj. 5.6.4) **45.** $(x^n + 2)^2(x^n - 2)^2$ (Obj. 5.6.4) **46.** $3ab(a - b)(a^2 + ab + b^2)$ (Obj. 5.6.4) **47.** -2, 0, and 3 (Obj. 5.7.1) **48.** $\dfrac{5}{2}$ and 4 (Obj. 5.7.1) **49.** -4, 0, and 4 (Obj. 5.7.1) **50.** -1, -6, and 6 (Obj. 5.7.1) **51.** The distance from Earth to the Great Galaxy of Andromeda is 1.137048×10^{23} mi. (Obj. 5.1.4) **52.** The sun generates 1.09×10^{21} horsepower. (Obj. 5.1.4) **53.** The area is $(10x^2 - 29x - 21)$ cm^2. (Obj. 5.3.4) **54.** The volume is $(27x^3 - 27x^2 + 9x - 1)$ ft^3. (Obj. 5.3.4) **55.** The area is $(5x^2 + 8x - 8)$ in^2. (Obj. 5.3.4) **56.** The two integers are -6 and -4, or 4 and 6. (Obj. 5.7.2) **57.** The number is -8 or 7. (Obj. 5.7.2) **58.** The length of the rectangle is 12 m. (Obj. 5.7.2)

CHAPTER TEST

1. $2x^3 - 4x^2 + 6x - 14$ (Obj. 5.2.2) **2.** $64a^7b^7$ (Obj. 5.1.1) **3.** $\dfrac{2b^7}{a^{10}}$ (Obj. 5.1.2) **4.** 5.01×10^{-7} (Obj. 5.1.3) **5.** 6.048×10^5 (Obj. 5.1.3) **6.** $\dfrac{x^{12}}{16y^4}$ (Obj. 5.1.2) **7.** $35x^2 - 55x$ (Obj. 5.3.1) **8.** $6a^2 - 13ab - 28b^2$ (Obj. 5.3.2) **9.** $6t^5 - 8t^4 - 15t^3 + 22t^2 - 5$ (Obj. 5.3.2) **10.** $9z^2 - 30z + 25$ (Obj. 5.3.3) **11.** $2x^2 + 3x + 5$ (Obj. 5.4.1) **12.** $x^2 - 2x - 1 + \dfrac{2}{x - 3}$ (Obj. 5.4.1/5.4.2) **13.** -3 (Obj. 5.2.1/5.4.3) **14.** -8 (Obj. 5.4.3) **15.** $(2a^2 - 5)(3a^2 + 1)$ (Obj. 5.6.3) **16.** $3x(2x - 3)(2x + 5)$ (Obj. 5.6.4) **17.** $(4x - 5)(4x + 5)$ (Obj. 5.6.1) **18.** $(4t + 3)^2$ (Obj. 5.6.1) **19.** $(3x - 2)(9x^2 + 6x + 4)$

(Obj. 5.6.2) **20.** $(3x - 2)(2x - a)$ (Obj. 5.5.2) **21.** $-\frac{1}{3}$ and $\frac{1}{2}$ (Obj. 5.7.1) **22.** $-1, -\frac{1}{6}$, and 1 (Obj. 5.7.1) **23.** The area of the rectangle is $(10x^2 - 3x - 1)$ ft^2. (Obj. 5.3.4) **24.** It takes 12 h for the space vehicle to reach the moon. (Obj. 5.1.4)

CUMULATIVE REVIEW EXERCISES

1. 4 (Obj. 1.2.4) **2.** $-\frac{5}{4}$ (Obj. 1.3.2) **3.** Inverse Property of Addition (Obj. 1.3.1) **4.** $-18x + 8$ (Obj. 1.3.3) **5.** $-\frac{1}{6}$ (Obj. 2.1.1) **6.** $-\frac{11}{4}$ (Obj. 2.1.1) **7.** $x^2 + 3x + 9 + \frac{24}{x - 3}$ (Obj. 5.4.1/5.4.2) **8.** -1 and $\frac{7}{3}$ (Obj. 2.6.1) **9.** 18 (Obj. 3.2.1) **10.** -2 (Obj. 3.2.1) **11.** $\{-4, -1, 8\}$ (Obj. 3.2.1) **12.** $-\frac{1}{6}$ (Obj. 3.4.1) **13.** $y = -\frac{3}{2}x + \frac{1}{2}$ (Obj. 3.5.1) **14.** $y = \frac{2}{3}x + \frac{16}{3}$ (Obj. 3.6.1) **15.** $\left(-\frac{7}{5}, -\frac{8}{5}\right)$ (Obj. 4.3.2) **16.** $\left(-\frac{9}{7}, \frac{2}{7}, \frac{11}{7}\right)$ (Obj. 4.2.2) **17.**

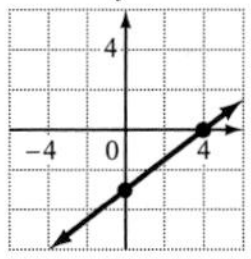

(Obj. 3.3.2) **18.**

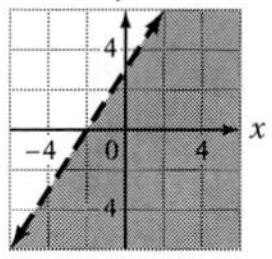

(Obj. 3.7.1)

19.

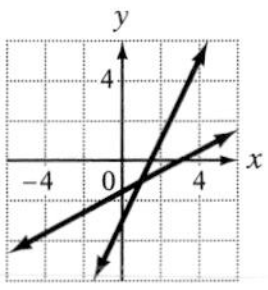

(1, −1) (Obj. 4.1.1) **20.**

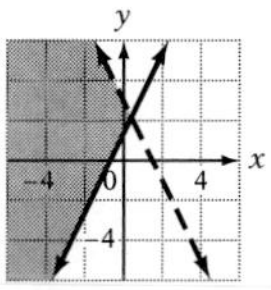

(Obj. 4.5.1) **21.** $\frac{b^5}{a^8}$ (Obj. 5.1.2)

22. $\frac{y^2}{25x^6}$ (Obj. 5.1.2) **23.** $\frac{21}{8}$ (Obj. 5.1.2) **24.** $4x^3 - 7x + 3$ (Obj. 5.3.2) **25.** $-2x(2x - 3)(x - 2)$ (Obj. 5.5.4) **26.** $(x - y)(a + b)$ (Obj. 5.5.2) **27.** $(x - 2)(x + 2)(x^2 + 4)$ (Obj. 5.6.4) **28.** $2(x - 2)(x^2 + 2x + 4)$ (Obj. 5.6.4) **29.** The integers are 9 and 15. (Obj. 2.2.2) **30.** 40 oz of pure gold must be mixed with the alloy. (Obj. 2.3.1) **31.** The slower cyclist travels at 5 mph, the faster cyclist at 7.5 mph. (Obj. 2.3.2) **32.** The additional investment is \$4500. (Obj. 2.4.1) **33.** $m = 50$; a slope of 50 means the average speed was 50 mph. (Obj. 3.4.1)

ANSWERS to Chapter 6 Odd-Numbered Exercises

CONCEPT REVIEW 6.1

1. Never true **3.** Always true **5.** Never true

SECTION 6.1

3. $f(4) = 2$ **5.** $f(-2) = -2$ **7.** $f(-2) = \frac{1}{9}$ **9.** $f(3) = \frac{1}{35}$ **11.** $f(-1) = \frac{3}{4}$ **13.** $\{x \mid x \neq 3\}$ **15.** $\{x \mid x \neq -4\}$ **17.** $\{x \mid x \neq -3\}$ **19.** $\{x \mid x \neq -2, 4\}$ **21.** $\left\{x \mid x \neq -\frac{5}{2}, 2\right\}$ **23.** $\{x \mid x \neq 0\}$ **25.** $\{x \mid x \in \text{real numbers}\}$ **27.** $\{x \mid x \neq -3, 2\}$ **29.** $\{x \mid x \neq -6, 4\}$ **31.** $\{x \mid x \in \text{real numbers}\}$ **33.** $\left\{x \mid x \neq \frac{2}{3}, \frac{3}{2}\right\}$ **35.** $\left\{x \mid x \neq 0, \frac{1}{2}, -5\right\}$ **39.** $1 - 2x$ **41.** $3x - 1$ **43.** $2x$ **45.** $-\frac{2}{x}$ **47.** $\frac{x}{2}$

49. The expression is in simplest form. 51. $4x^2 - 2x + 3$ 53. $a^2 + 2a - 3$ 55. $\frac{x^n - 3}{4}$ 57. $\frac{x - 3}{x - 5}$

59. $\frac{x + y}{x - y}$ 61. $-\frac{x + 3}{3x - 4}$ 63. $-\frac{x + 7}{x - 7}$ 65. The expression is in simplest form. 67. $\frac{a - b}{a^2 - ab + b^2}$

69. $\frac{x + y}{3x}$ 71. $\frac{x + 2y}{3x + 4y}$ 73. $-\frac{2x - 3}{2(2x + 3)}$ 75. $\frac{x - 2}{a - b}$ 77. $\frac{x^2 + 2}{(x + 1)(x - 1)}$ 79. $\frac{xy + 7}{xy - 7}$ 81. $\frac{a^n - 2}{a^n + 2}$

83. decrease 85. a.

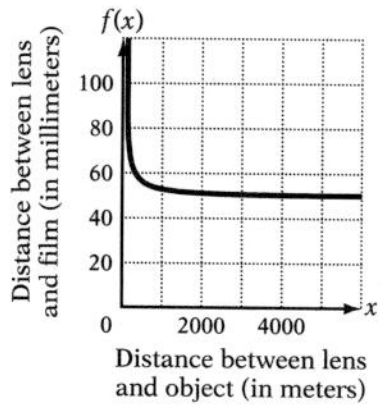

b. The ordered pair (2000, 51) means that when the distance between the object and the lens is 2000 m, the distance between the lens and the film is 51 mm.

c. For $x = 50$, the expression $\frac{50x}{x - 50}$ is undefined. For $0 < x < 50$, $f(x)$ is negative, and distance cannot be negative. Therefore, the domain is $x > 50$.

d. For $x > 1000$, $f(x)$ changes very little for large changes in x.

CONCEPT REVIEW 6.2

1. Always true 3. Never true

SECTION 6.2

3. $\frac{15b^4xy}{4}$ 5. 1 7. $\frac{2x - 3}{x^2(x + 1)}$ 9. $-\frac{x - 1}{x - 5}$ 11. $\frac{2(x + 2)}{x - 1}$ 13. $-\frac{x - 3}{2x + 3}$ 15. $\frac{x^n - 6}{x^n - 1}$ 17. $x + y$

19. $\frac{a^2y}{10x}$ 21. $\frac{3}{2}$ 23. $\frac{5(x - y)}{xy(x + y)}$ 25. $\frac{(x - 3)^2}{(x + 3)(x - 4)}$ 27. $-\frac{x + 2}{x + 5}$ 29. $\frac{x^{2n}}{8}$ 31. $\frac{(x - 3)(3x - 5)}{(2x + 5)(x + 4)}$

33. -1 35. 1 37. $-\frac{2(x^2 + x + 1)}{(x + 1)^2}$ 39. $-\frac{13}{2xy}$ 41. $\frac{1}{x - 1}$ 43. $\frac{15 - 16xy - 9x}{10x^2y}$ 45. $-\frac{5y + 21}{30xy}$

47. $-\frac{6x + 19}{36x}$ 49. $\frac{10x + 6y + 7xy}{12x^2y^2}$ 51. $-\frac{x^2 + x}{(x - 3)(x - 5)}$ 53. -1 55. $\frac{5x - 8}{(x + 5)(x - 5)}$ 57. $-\frac{3x^2 - 4x + 8}{x(x - 4)}$

59. $\frac{4x^2 - 14x + 15}{2x(2x - 3)}$ 61. $\frac{2x^2 + x + 3}{(x + 1)^2(x - 1)}$ 63. $\frac{x + 1}{x + 3}$ 65. $\frac{x + 2}{x + 3}$ 67. $-\frac{x^n - 2}{(x^n + 1)(x^n - 1)}$ 69. $\frac{12x + 13}{(2x + 3)(2x - 3)}$

71. $\frac{2x + 3}{2x - 3}$ 73. $\frac{2x + 5}{(x + 3)(x - 2)(x + 1)}$ 75. $\frac{3x^2 + 4x + 3}{(2x + 3)(x + 4)(x - 3)}$ 77. $\frac{x - 2}{x - 3}$ 79. $-\frac{2x^2 - 9x - 9}{(x + 6)(x - 3)}$

81. $\frac{3x^2 - 14x + 24}{(3x - 2)(2x + 5)}$ 83. $\frac{2}{x^2 + 1}$ 85. $-x - 1$ 87. $\frac{y - 2}{x^4}$ 89. $\frac{4y}{(y - 1)^2}$ 91. $\frac{5(x + 4)(x - 4)}{2(x + 2)(x - 2)(x - 3)}$

93. $-\frac{(a + 3)(a - 2)(5a - 2)}{(3a + 1)(a + 2)(2a - 3)}$ 95. $\frac{6(x - 1)}{x(2x + 1)}$ 97. $\frac{8}{15} \neq \frac{1}{8}$ 99. $\frac{3}{y} + \frac{6}{x}$ 101. $\frac{4}{b^2} + \frac{3}{ab}$

103. a. $f(x) = \frac{2x^3 + 528}{x}$ b.

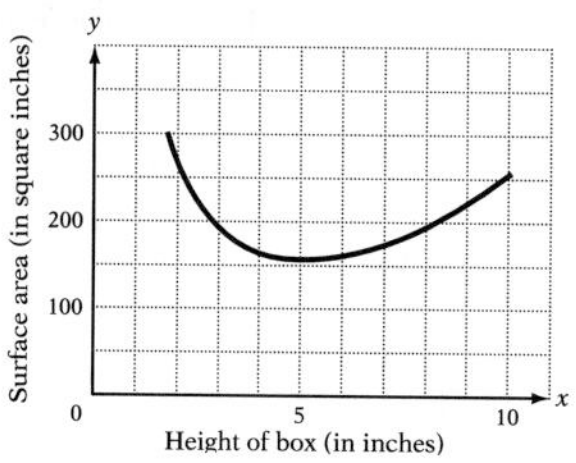

c. When the height of the box is 4 in., 164 in^2 of cardboard is needed.

d. The height of the box is 5.1 in.

e. The minimum amount of cardboard is 155.5 in^2.

CONCEPT REVIEW 6.3

1. Sometimes true 3. Always true

SECTION 6.3

3. $\frac{5}{23}$ 5. $\frac{2}{5}$ 7. $\frac{1-y}{y}$ 9. $\frac{5-a}{a}$ 11. $\frac{b}{2(2-b)}$ 13. $\frac{4}{5}$ 15. $-\frac{1}{2}$ 17. $\frac{2a^2-3a+3}{3a-2}$ 19. $3x+2$
21. $\frac{x+2}{x-1}$ 23. $-\frac{(x-3)(x+5)}{4x^2-5x+4}$ 25. $\frac{1}{x+4}$ 27. $\frac{x-7}{x+5}$ 29. $-\frac{2(a+1)}{7a-4}$ 31. $\frac{x+y}{x-y}$ 33. $-\frac{2x}{x^2+1}$
35. $\frac{6(x-2)}{2x-3}$ 37. $\frac{2y}{x}$ 39. $x-1$ 41. $\frac{c-3}{c-2}$ 43. $\frac{2a(2a-3)}{3a-4}$ 45. $-\frac{1}{x(x+h)}$ 47. $\frac{3n^2+12n+8}{n(n+2)(n+4)}$

49. a. $P(x) = \frac{Cx(x+1)^{60}}{(x+1)^{60}-1}$ b.

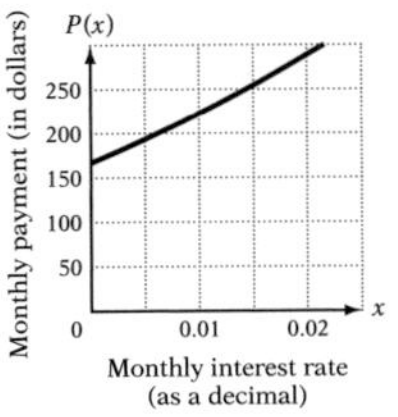

c. The interval of annual interest rates is 0% to 22.8%.
d. The ordered pair (0.006, 198.96) means that when the monthly interest rate on a car loan is 0.6%, the monthly payment on the loan is $198.96.
e. The monthly payment with a loan amount of $10,000 and an annual interest rate of 8% is $203.

CONCEPT REVIEW 6.4

1. Never true 3. Always true 5. Always true

SECTION 6.4

3. -5 5. -1 7. $\frac{5}{2}$ 9. 0 11. -3 and 3 13. 8 15. 5 17. -3 19. No solution
21. 4 and 10 23. -1 25. -3 and 4 27. 2 29. $\frac{7}{2}$ 31. No solution 35. It would take 1.2 h to process the data with both computers working. 37. With both panels working, it would take 18 min to raise the temperature 1 degree. 39. With both members working together, it would take 3 h to wire the telephone lines. 41. Working alone, the new machine would take 10 h to package the transistors. 43. Working alone, the smaller printer would take 80 min to print the payroll. 45. It would take the apprentice 20 h to shingle the roof working alone. 47. It would take approximately 1 min to make the pizza with all three employees working. 49. In 30 min, the bathtub will start to overflow. 51. With all three pipes open, it would take 30 h to fill the tank. 53. The rate of the first skater is 15 mph; the rate of the second skater is 12 mph. 55. The rate of the freight train is 45 mph. The rate of the passenger train is 59 mph. 57. The cyclist was riding at the rate of 12 mph. 59. The motorist walks at the rate of 4 mph. 61. The business executive can walk at the rate of 6 ft/s. 63. The rate of the single-engine plane is 180 mph. The rate of the jet is 720 mph. 65. The rate of the Gulf Stream is 6 mph. 67. The rate of the current is 3 mph. 69. The original fraction is $\frac{17}{21}$. 71. With all three printers operating, it would take $5\frac{1}{3}$ min to print the checks. 73. The bus usually travels at 60 mph. 75. The total time is $\frac{5}{8}$ h.

CONCEPT REVIEW 6.5

1. Never true 3. Never true 5. Never true

SECTION 6.5

3. There are 4000 ducks in the preserve. 5. The caterer will need an additional 3 gal. 7. The dimensions of the room are 17 ft by 22 ft. 9. In a shipment of 3200, 4 diodes would be defective. 11. An additional 0.5 oz of

medicine is required. **13.** An additional 34 oz of insecticide is required. **15.** The height of the person is 5.3 ft. **17.** The length of the whale is 96 ft. **21.** When the company sells 300 products, the profit is \$37,500. **23.** The pressure is 13.5 lb/in^2. **25.** The horizon is 24.03 mi away from a point that is 800 ft high. **27.** In 4 s, the ball will roll 80 ft. **29.** The person has a face length of 10.2 in. **31.** When the volume is 150 ft^3, the pressure is $66\frac{2}{3}$ lb/in^2. **33.** The pressure is 22.5 lb/in^2. **35.** The repulsive force is 112.5 lb when the distance is 1.2 in. **37.** The resistance is 56.25 ohms. **39.** The wind force is 180 lb. **41. a.**

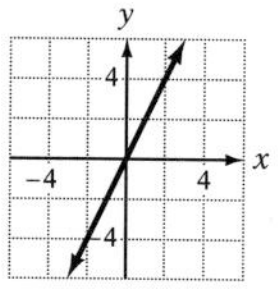

b. A linear function

43. a.

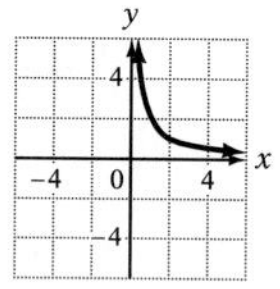

b. Yes **45.** x is halved. **47.** directly, inversely **49.** directly

51. a. Dollars per share **b.** 87.11, 94.56, 113.41, 95.87, 93.10 **c.** 3.9%

CONCEPT REVIEW 6.6

1. Always true **3.** Never true

SECTION 6.6

1. $W = \frac{P - 2L}{2}$ **3.** $C = \frac{S}{1 - r}$ **5.** $R = \frac{PV}{nT}$ **7.** $m_2 = \frac{Fr^2}{Gm_1}$ **9.** $R = \frac{E - Ir}{I}$ **11.** $b_2 = \frac{2A - hb_1}{h}$ **13.** $R_2 = \frac{RR_1}{R_1 - R}$ **15.** $d = \frac{a_n - a_1}{n - 1}$ **17.** $H = \frac{S - 2WL}{2W + 2L}$ **19.** $x = \frac{-by - c}{a}$ **21.** $x = \frac{d - b}{a - c}$ **23.** $x = \frac{ac}{b}$ **25.** $x = \frac{ab}{a + b}$ **27.** $5 + 2y$ **29.** $0, y$ **31.** $\frac{y}{2}, 5y$ **33.** $f = \frac{w_2 f_2 + w_1 f_1}{w_1 + w_2}$

CHAPTER REVIEW EXERCISES

1. 4 (Obj. 6.1.1) **2.** $\frac{2}{21}$ (Obj. 6.1.1) **3.** $\{x \mid x \neq 3\}$ (Obj. 6.1.1) **4.** $\{x \mid x \neq -3, 2\}$ (Obj. 6.1.1) **5.** $\{x \mid x \in \text{real numbers}\}$ (Obj. 6.1.1) **6.** $3a^{2n} + 2a^n - 1$ (Obj. 6.1.2) **7.** $-\frac{x + 4}{x(x + 2)}$ (Obj. 6.1.2) **8.** $\frac{x^2 + 3x + 9}{x + 3}$ (Obj. 6.1.2) **9.** $\frac{1}{a - 1}$ (Obj. 6.2.1) **10.** x (Obj. 6.2.1) **11.** $-\frac{(x + 4)(x + 3)(x + 2)}{6(x - 1)(x - 4)}$ (Obj. 6.2.1) **12.** $\frac{x^n + 6}{x^n - 3}$ (Obj. 6.2.1) **13.** $\frac{3x + 2}{x}$ (Obj. 6.2.1) **14.** $-\frac{x - 3}{x + 3}$ (Obj. 6.2.1) **15.** $\frac{21a + 40b}{24a^2b^4}$ (Obj. 6.2.2) **16.** $\frac{4x - 1}{x + 2}$ (Obj. 6.2.2) **17.** $\frac{3x + 26}{(3x - 2)(3x + 2)}$ (Obj. 6.2.2) **18.** $\frac{9x^2 + x + 4}{(3x - 1)(x - 2)}$ (Obj. 6.2.2) **19.** $-\frac{5x^2 - 17x - 9}{(x - 3)(x + 2)}$ (Obj. 6.2.2) **20.** $\frac{x - 4}{x + 5}$ (Obj. 6.3.1) **21.** $\frac{x - 1}{4}$ (Obj. 6.3.1) **22.** $\frac{15}{13}$ (Obj. 6.4.1) **23.** −3 and 5 (Obj. 6.4.1) **24.** 6 (Obj. 6.4.1) **25.** No solution (Obj. 6.4.1) **26.** $R = \frac{V}{I}$ (Obj. 6.6.1) **27.** $N = \frac{S}{1 - Q}$ (Obj. 6.6.1) **28.** $r = \frac{S - a}{S}$ (Obj. 6.6.1) **29.** The number of tanks of fuel is $6\frac{2}{3}$. (Obj. 6.5.1) **30.** The number of miles is 48. (Obj. 6.5.1) **31.** Working alone, the apprentice would take 104 min to install the ceiling fan. (Obj. 6.4.2) **32.** With both pipes open, it would take 40 min to empty the tub. (Obj. 6.4.2)

33. It would take 4 h for the three painters to paint the dormitory together. (Obj. 6.4.2) **34.** The rate of the current is 2 mph. (Obj. 6.4.3) **35.** The rate of the bus is 45 mph. (Obj. 6.4.3) **36.** The rate of the tractor is 30 mph. (Obj. 6.4.3) **37.** The pressure is 40 lb. (Obj. 6.5.2) **38.** The illumination is 300 lumens 2 ft from the light. (Obj. 6.5.2) **39.** The resistance of the cable is 0.4 ohm. (Obj. 6.5.2)

CHAPTER TEST

1. $\frac{v(v+2)}{2v-1}$ (Obj. 6.1.2) **2.** $-\frac{2(a-2)}{3a+2}$ (Obj. 6.1.2) **3.** $\frac{6(x-2)}{5}$ (Obj. 6.2.1) **4.** 2 (Obj. 6.1.1)
5. $\frac{x+1}{3x-4}$ (Obj. 6.2.1) **6.** $\frac{x^n-1}{x^n+2}$ (Obj. 6.2.1) **7.** $\frac{4y^2+6x^2-5xy}{2x^2y^2}$ (Obj. 6.2.2) **8.** $\frac{2(5x+2)}{(x-2)(x+2)}$ (Obj. 6.2.2)
9. $\{x \mid x \neq -3, 3\}$ (Obj. 6.1.1) **10.** $-\frac{x^2+5x-2}{(x+1)(x+4)(x-1)}$ (Obj. 6.2.2) **11.** $\frac{x-4}{x+3}$ (Obj. 6.3.1)
12. $\frac{x+4}{x+2}$ (Obj. 6.3.1) **13.** 2 (Obj. 6.4.1) **14.** No solution (Obj. 6.4.1) **15.** $x = \frac{c}{a-b}$ (Obj. 6.6.1)
16. It would take 80 min to empty the full tank with both pipes open. (Obj. 6.4.2) **17.** The office requires 14 rolls of wallpaper. (Obj. 6.5.1) **18.** Working together, the landscapers would complete the task in 10 min. (Obj. 6.4.2) **19.** The rate of the cyclist is 10 mph. (Obj. 6.4.3) **20.** The stopping distance of the car is 61.2 ft. (Obj. 6.5.2)

CUMULATIVE REVIEW EXERCISES

1. $\frac{36}{5}$ (Obj. 1.2.4) **2.** $\frac{15}{2}$ (Obj. 2.1.2) **3.** 7 and 1 (Obj. 2.6.1) **4.** $\{x \mid x \neq 3\}$ (Obj. 6.1.1) **5.** $\frac{3}{7}$ (Obj. 6.1.1) **6.** 3.5×10^{-8} (Obj. 5.1.3) **7.** No solution (Obj. 6.4.1) **8.** $\frac{1}{9}$ and 4 (Obj. 5.7.1) **9.** $\frac{8b^3}{a}$ (Obj. 5.1.2) **10.** $\{x \mid x \leq 4\}$ (Obj. 2.5.1) **11.** $-4a^4 + 6a^3 - 2a^2$ (Obj. 5.3.1) **12.** $(2x^n - 1)(x^n + 2)$ (Obj. 5.6.3)
13. $(xy - 3)(x^2y^2 + 3xy + 9)$ (Obj. 5.6.2) **14.** $x + 3y$ (Obj. 6.1.2) **15.** $y = \frac{3}{2}x + 2$ (Obj. 3.6.1)

16. $\frac{x}{x^2-1}$ (Obj. 5.1.2) **17.** 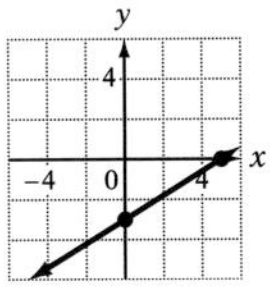

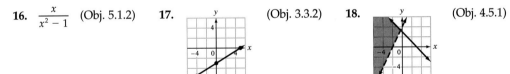

(Obj. 3.3.2) **18.** 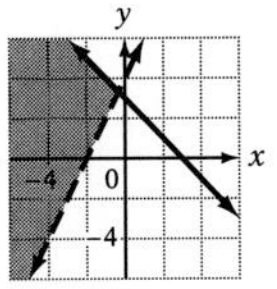

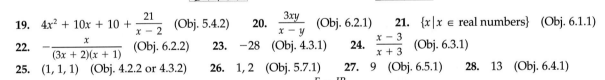

(Obj. 4.5.1)

19. $4x^2 + 10x + 10 + \frac{21}{x-2}$ (Obj. 5.4.2) **20.** $\frac{3xy}{x-y}$ (Obj. 6.2.1) **21.** $\{x \mid x \in \text{real numbers}\}$ (Obj. 6.1.1)
22. $-\frac{x}{(3x+2)(x+1)}$ (Obj. 6.2.2) **23.** −28 (Obj. 4.3.1) **24.** $\frac{x-3}{x+3}$ (Obj. 6.3.1)
25. (1, 1, 1) (Obj. 4.2.2 or 4.3.2) **26.** 1, 2 (Obj. 5.7.1) **27.** 9 (Obj. 6.5.1) **28.** 13 (Obj. 6.4.1)
29. $a^4 + 5a^3 - 3a^2 - 11a + 20$ (Obj. 5.3.2) **30.** $r = \frac{E - IR}{I}$ (Obj. 6.6.1) **31.** No (Obj. 3.6.1)
32. 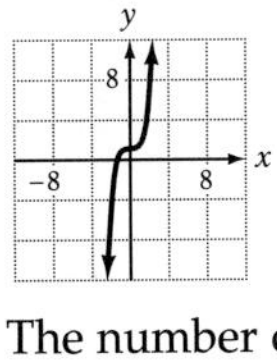

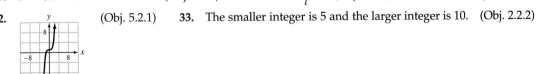

(Obj. 5.2.1) **33.** The smaller integer is 5 and the larger integer is 10. (Obj. 2.2.2)

34. The number of pounds of almonds is 50. (Obj. 2.3.1) **35.** The number of people expected to vote is 75,000. (Obj. 6.5.1) **36.** It would take the new computer 14 min to do the job working alone. (Obj. 6.4.2) **37.** The rate of the wind is 60 mph. (Obj. 6.4.3) **38.** The frequency of vibration of the open pipe organ is 80 vibrations/min. (Obj. 6.5.2)

ANSWERS to Chapter 7 Odd-Numbered Exercises

CONCEPT REVIEW 7.1

1. Sometimes true **3.** Always true **5.** Always true

SECTION 7.1

3. 2 **5.** 27 **7.** $\frac{1}{9}$ **9.** 4 **11.** Not a real number **13.** $\frac{343}{125}$ **15.** x **17.** $y^{\frac{1}{2}}$ **19.** $x^{\frac{1}{12}}$ **21.** $a^{\frac{7}{12}}$
23. $\frac{1}{a}$ **25.** $\frac{1}{y}$ **27.** $y^{\frac{3}{2}}$ **29.** $\frac{1}{x}$ **31.** $\frac{1}{x^4}$ **33.** a **35.** $x^{\frac{3}{10}}$ **37.** a^3 **39.** $\frac{1}{x^{\frac{1}{2}}}$ **41.** $y^{\frac{2}{3}}$ **43.** x^4y
45. $x^6y^3z^9$ **47.** $\frac{x}{y^2}$ **49.** $\frac{x^{\frac{3}{2}}}{y^{\frac{1}{4}}}$ **51.** x^2y^8 **53.** $\frac{1}{x^{\frac{11}{12}}}$ **55.** $\frac{1}{y^{\frac{5}{2}}}$ **57.** $\frac{1}{b^{\frac{7}{8}}}$ **59.** a^5b^{13} **61.** $\frac{m^2}{4n^{\frac{3}{2}}}$
63. $\frac{y^{\frac{17}{2}}}{x^3}$ **65.** $\frac{16b^2}{a^{\frac{1}{3}}}$ **67.** $\frac{y^2}{x^3}$ **69.** $a - a^2$ **71.** $y + 1$ **73.** $a - \frac{1}{a}$ **75.** $\frac{1}{a^{10}}$ **77.** $a^{\frac{n}{6}}$ **79.** $\frac{1}{b^{\frac{2m}{3}}}$
81. x^{10n^2} **83.** $x^{3n}y^{2n}$ **85.** $\sqrt{5}$ **87.** $\sqrt[3]{b^4}$ **89.** $\sqrt[3]{9x^2}$ **91.** $-3\sqrt[5]{a^2}$ **93.** $\sqrt[4]{x^6y^9}$ **95.** $\sqrt{a^9b^{21}}$
97. $\sqrt[3]{3x - 2}$ **99.** $14^{\frac{1}{2}}$ **101.** $x^{\frac{1}{3}}$ **103.** $x^{\frac{4}{3}}$ **105.** $b^{\frac{3}{5}}$ **107.** $(2x^2)^{\frac{1}{3}}$ **109.** $-(3x^5)^{\frac{1}{2}}$ **111.** $3xy^{\frac{2}{3}}$
113. $(a^2 + 2)^{\frac{1}{2}}$ **115.** y^7 **117.** $-a^3$ **119.** a^7b^3 **121.** $11y^6$ **123.** a^2b^4 **125.** $-a^3b^3$ **127.** $5b^5$
129. $-a^2b^3$ **131.** $5x^4y$ **133.** Not a real number **135.** $2a^7b^2$ **137.** $-3ab^5$ **139.** y^3 **141.** $3a^5$
143. $-a^4b$ **145.** ab^5 **147.** $2a^2b^5$ **149.** $-2x^3y^4$ **151.** $\frac{4x}{y^7}$ **153.** $\frac{3b}{a^3}$ **155.** $2x + 3$ **157.** $x + 1$
159. c **161.** a **163.** x **165.** b **167.** ab^2 **169.** $\frac{3}{5}$ **171.** $\frac{3}{4}$

CONCEPT REVIEW 7.2

1. Sometimes true **3.** Never true **5.** Always true

SECTION 7.2

3. $3\sqrt{2}$ **5.** $7\sqrt{2}$ **7.** $2\sqrt[3]{9}$ **9.** $2\sqrt[3]{2}$ **11.** $x^2yz^2\sqrt{yz}$ **13.** $2ab^4\sqrt{2a}$ **15.** $3xyz^2\sqrt{5yz}$
17. $-5y\sqrt[3]{x^2y}$ **19.** $-6xy^3\sqrt[3]{x^2}$ **21.** $ab^2\sqrt[3]{a^2b^2}$ **23.** $2\sqrt{2}$ **25.** $3\sqrt[3]{7}$ **27.** $-6\sqrt{x}$ **29.** $-2\sqrt{2}$
31. $\sqrt{2x}$ **33.** $3\sqrt{3a} - 2\sqrt{2a}$ **35.** $10x\sqrt{2x}$ **37.** $2x\sqrt{3xy}$ **39.** $xy\sqrt{2y}$ **41.** $5a^2b^2\sqrt{ab}$ **43.** $9\sqrt[3]{2}$
45. $-7a\sqrt[3]{3a}$ **47.** $-5xy^2\sqrt[3]{x^2y}$ **49.** $16ab\sqrt[4]{ab}$ **51.** $6\sqrt{3} - 6\sqrt{2}$ **53.** $6x^3y^2\sqrt{xy}$ **55.** $-9a^2\sqrt{3ab}$
57. 16 **59.** $2\sqrt[3]{4}$ **61.** $xy^3\sqrt{x}$ **63.** $8xy\sqrt{x}$ **65.** $2x^2y\sqrt[3]{2}$ **67.** $2ab\sqrt[4]{3a^2b}$ **69.** 6 **71.** $x - \sqrt{2x}$
73. $4x - 8\sqrt{x}$ **75.** $x - 6\sqrt{x} + 9$ **77.** $84 + 16\sqrt{5}$ **79.** $672x^2y^2$ **81.** $4a^3b^3\sqrt[3]{a}$ **83.** $-8\sqrt{5}$
85. $x - y^2$ **87.** $12x - y$ **89.** $x - 3\sqrt{x} - 28$ **91.** $\sqrt[3]{x^2} + \sqrt[3]{x} - 20$ **93.** $4\sqrt{x}$ **95.** $b^2\sqrt{3a}$ **97.** $\frac{\sqrt{5}}{5}$
99. $\frac{\sqrt{2x}}{2x}$ **101.** $\frac{\sqrt{5x}}{x}$ **103.** $\frac{\sqrt{5x}}{5}$ **105.** $\frac{3\sqrt[3]{4}}{2}$ **107.** $\frac{3\sqrt[3]{2x}}{2x}$ **109.** $\frac{\sqrt{2xy}}{2y}$ **111.** $\frac{2\sqrt{3ab}}{3b^2}$ **113.** $2\sqrt{5} - 4$
115. $\frac{3\sqrt{y} + 6}{y - 4}$ **117.** $-5 + 2\sqrt{6}$ **119.** $8 + 4\sqrt{3} - 2\sqrt{2} - \sqrt{6}$ **121.** $\frac{\sqrt{15} + \sqrt{10} - \sqrt{6} - 5}{3}$ **123.** $\frac{3\sqrt[4]{2x}}{2x}$
125. $\frac{2\sqrt[5]{2a^3}}{a}$ **127.** $\frac{\sqrt[5]{16x^2}}{2}$ **129.** $\frac{1 + 2\sqrt{ab} + ab}{1 - ab}$ **131.** $\frac{5x\sqrt{y} + 5y\sqrt{x}}{x - y}$ **133.** $2\sqrt{2}$ **135.** $29\sqrt{2} - 45$
137. $\frac{3\sqrt{y + 1} - 3}{y}$ **139.** $\sqrt[6]{x + y}$ **141.** $\sqrt[4]{2y(x + 3)^2}$ **143.** $\sqrt[6]{a^3(a + 3)^2}$ **145.** $16^{\frac{1}{4}}, 2$ **147.** $32^{-\frac{1}{5}}, \frac{1}{2}$
149. **a.** $D(x) = 2.79\sqrt[10]{x^3}$ **b.** The model predicts a national debt of \$4.52 trillion in 1993. **c.** The national debt will be \$6.41 trillion.

CONCEPT REVIEW 7.3

1. Always true   **3.** Always true

SECTION 7.3

3. $2i$ **5.** $7i\sqrt{2}$ **7.** $3i\sqrt{3}$ **9.** $4 + 2i$ **11.** $2\sqrt{3} - 3i\sqrt{2}$ **13.** $4\sqrt{10} - 7i\sqrt{3}$ **15.** $2ai$ **17.** $7x^6i$
19. $12ab^2i\sqrt{ab}$ **21.** $2\sqrt{a} + 2ai\sqrt{3}$ **23.** $3b^2\sqrt{2b} - 3bi\sqrt{3b}$ **25.** $5x^3y\sqrt{y} + 5xyi\sqrt{2xy}$ **27.** $2a^2bi\sqrt{a}$
29. $2a\sqrt{3a} + 3bi\sqrt{3b}$ **31.** $10 - 7i$ **33.** $11 - 7i$ **35.** $-6 + i$ **37.** $-4 - 8i\sqrt{3}$ **39.** $-\sqrt{10} - 11i\sqrt{2}$
41. $6 - 4i$ **43.** 0 **45.** $11 + 6i$ **47.** -24 **49.** -15 **51.** $-5\sqrt{2}$ **53.** $-15 - 12i$ **55.** $3\sqrt{2} + 6i$
57. $-10i$ **59.** $29 - 2i$ **61.** 1 **63.** 1 **65.** 89 **67.** 50 **69.** $-\dfrac{4}{5}i$ **71.** $-\dfrac{5}{3} + \dfrac{16}{3}i$
73. $\dfrac{30}{29} - \dfrac{12}{29}i$ **75.** $\dfrac{20}{17} + \dfrac{5}{17}i$ **77.** $\dfrac{11}{13} + \dfrac{29}{13}i$ **79.** $-\dfrac{1}{5} + \dfrac{\sqrt{6}}{10}i$ **81.** $-1 + 4i$ **83.** -1 **85.** i
87. -1 **89.** -1 **91.** $(y + i)(y - i)$ **93.** $(x + 5i)(x - 5i)$ **95. a.** No   **b.** No   **97.** $-\dfrac{\sqrt{2}}{2} + i\dfrac{\sqrt{2}}{2}$

CONCEPT REVIEW 7.4

1. Always true   **3.** Never true

SECTION 7.4

3. $\{x \mid x \in \text{real numbers}\}$ **5.** $\{x \mid x \ge -1\}$ **7.** $\{x \mid x \ge 0\}$ **9.** $\{x \mid x \ge 0\}$ **11.** $\{x \mid x \ge 2\}$
13. $(-\infty, \infty)$ **15.** $(-\infty, 3]$ **17.** $[2, \infty)$ **19.** $\left(-\infty, \dfrac{2}{3}\right]$ **21.**

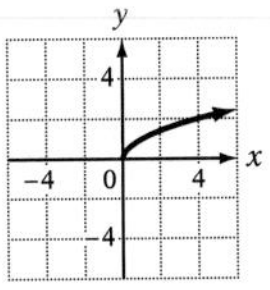

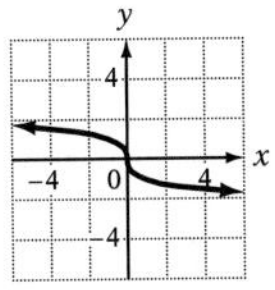

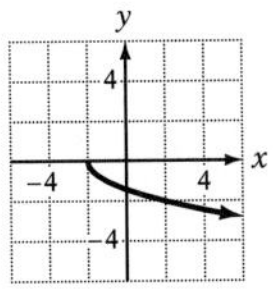

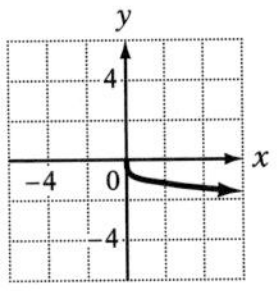

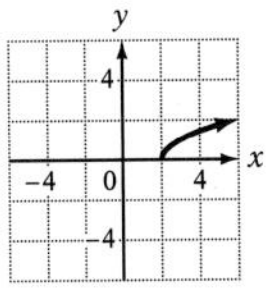

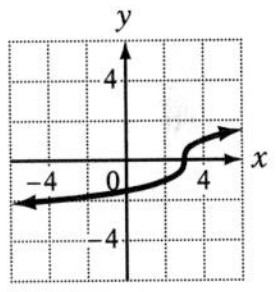

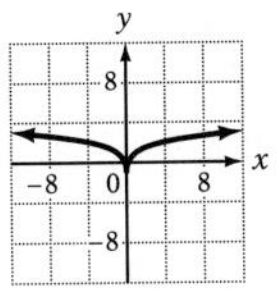

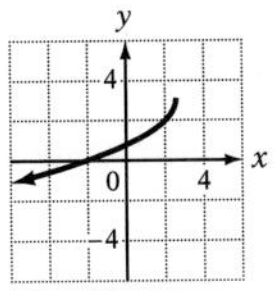

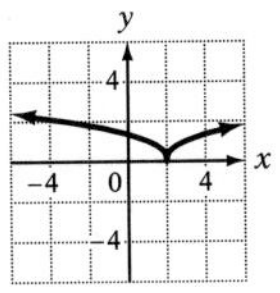

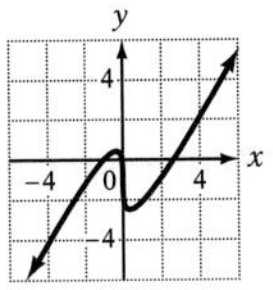

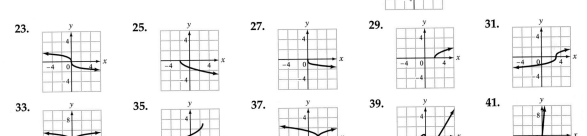

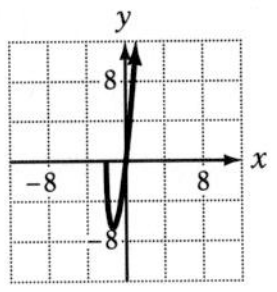

CONCEPT REVIEW 7.5

1. Sometimes true   **3.** Sometimes true

SECTION 7.5

3. 25 **5.** 27 **7.** 48 **9.** -2 **11.** No solution   **13.** 9 **15.** -23 **17.** -16 **19.** $\dfrac{9}{4}$ **21.** 14
23. 7 **25.** 7 **27.** -122 **29.** 35 **31.** 45 **33.** 23 **35.** -4 **37.** 10 **39.** 1 **41.** 3

43. 21 **45.** $\frac{25}{72}$ **47.** 5 **49.** $-2, 4$ **51.** 6 **53.** 0, 1 **55.** $-2, 1$ **57.** 2, 11 **59.** 3
61. The air pressure is 909.8 mb. The wind speed increases. **63.** The mean distance is 295,000 km. **65.** The elephant weighs 3700 lb. **67.** The distance is 10 ft. **69.** The periscope must be 8.17 ft above the water. **71.** The distance is 225 ft. **73.** The distance is 96 ft. **75.** The length of the pendulum is 4.67 ft.
77. 27 **79.** $d = \frac{v^2}{64}$ **81.** $r = \frac{\sqrt{V\pi h}}{\pi h}$ **83.** 2 **85.** The distance between the corners is 7.81 in.
87. **a.** 2, −2; integer, rational number, real number **b.** $i, -i$; imaginary **c.** $\frac{16}{15}$; rational number, real number **d.** $2\sqrt{2}, -2\sqrt{2}$; irrational number, real number **e.** 16; integer, rational number, real number **f.** −19,683; integer, rational number, real number **89.** $\sqrt{6}$

CHAPTER REVIEW EXERCISES

1. $\frac{1}{3}$ (Obj. 7.1.1) **2.** $\frac{1}{x^5}$ (Obj. 7.1.1) **3.** $\frac{1}{a^{10}}$ (Obj. 7.1.1) **4.** $\frac{160y^{11}}{x}$ (Obj. 7.1.1) **5.** $3\sqrt[4]{x^3}$ (Obj. 7.1.2) **6.** $7x^{\frac{2}{3}}y$ (Obj. 7.1.2) **7.** $3a^2b^3$ (Obj. 7.1.3) **8.** $-7x^3y^8$ (Obj. 7.1.3) **9.** $-2a^2b^4$ (Obj. 7.1.3) **10.** $3ab^3\sqrt{2a}$ (Obj. 7.2.1) **11.** $-2ab^2\sqrt[5]{2a^3b^2}$ (Obj. 7.2.1) **12.** $xy^2z^2\sqrt{xz}$ (Obj. 7.2.1) **13.** $5\sqrt{6}$ (Obj. 7.2.2) **14.** $4x^2\sqrt{3xy} - 4x^2\sqrt{5xy}$ (Obj. 7.2.2) **15.** $2a^2b\sqrt{2b}$ (Obj. 7.2.2) **16.** $6x^2\sqrt{3y}$ (Obj. 7.2.2) **17.** 40 (Obj. 7.2.3) **18.** $4xy^2\sqrt[3]{x^2}$ (Obj. 7.2.3) **19.** $3x + 3\sqrt{3x}$ (Obj. 7.2.3) **20.** $31 - 10\sqrt{6}$ (Obj. 7.2.3) **21.** $-13 + 6\sqrt{3}$ (Obj. 7.2.3) **22.** $5x\sqrt{x}$ (Obj. 7.2.4) **23.** $\frac{8\sqrt{3y}}{3y}$ (Obj. 7.2.4) **24.** $\frac{x\sqrt{x} - x\sqrt{2} + 2\sqrt{x} - 2\sqrt{2}}{x - 2}$ (Obj. 7.2.4) **25.** $\frac{x + 2\sqrt{xy} + y}{x - y}$ (Obj. 7.2.4) **26.** $6i$ (Obj. 7.3.1) **27.** $5i\sqrt{2}$ (Obj. 7.3.1) **28.** $7 - 4i$ (Obj. 7.3.1) **29.** $10\sqrt{2} + 2i\sqrt{3}$ (Obj. 7.3.1) **30.** $9 - i$ (Obj. 7.3.2) **31.** $-12 + 10i$ (Obj. 7.3.2) **32.** $14 + 2i$ (Obj. 7.3.2) **33.** $-4\sqrt{2} + 8i\sqrt{2}$ (Obj. 7.3.2) **34.** $10 - 9i$ (Obj. 7.3.2) **35.** −16 (Obj. 7.3.3) **36.** $7 + 3i$ (Obj. 7.3.3) **37.** $-6\sqrt{2}$ (Obj. 7.3.3) **38.** $39 - 2i$ (Obj. 7.3.3) **39.** $6i$ (Obj. 7.3.4) **40.** $\frac{2}{3} - \frac{5}{3}i$ (Obj. 7.3.4) **41.** $\frac{14}{5} + \frac{7}{5}i$ (Obj. 7.3.4) **42.** $1 + i$ (Obj. 7.3.4) **43.** $-2 + 7i$ (Obj. 7.3.4)
44. −24 (Obj. 7.5.1) **45.** $\left\{x \mid x \geq \frac{2}{3}\right\}$ (Obj. 7.4.1) **46.** $\{x \mid x \in \text{real numbers}\}$ (Obj. 7.4.1)
47.

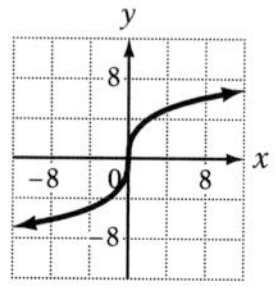

(Obj. 7.4.2) **48.**

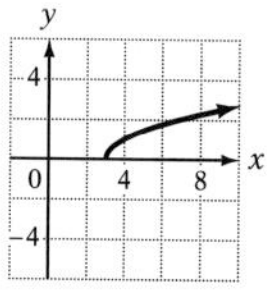

(Obj. 7.4.2) **49.** $\frac{13}{3}$ (Obj. 7.5.1) **50.** −2 (Obj. 7.5.1)

51. The width is 5 in. (Obj. 7.5.2) **52.** The amount of power is 120 watts. (Obj. 7.5.2) **53.** The distance required is 242 ft. (Obj. 7.5.2) **54.** The distance is 6.63 ft. (Obj. 7.5.2)

CHAPTER TEST

1. $r^{\frac{1}{6}}$ (Obj. 7.1.1) **2.** $\frac{64x^3}{y^6}$ (Obj. 7.1.1) **3.** $\frac{b^3}{8a^6}$ (Obj. 7.1.1) **4.** $3\sqrt[5]{y^2}$ (Obj. 7.1.2) **5.** $\frac{1}{2}x^{\frac{3}{4}}$ (Obj. 7.1.2) **6.** $\{x \mid x \leq 4\}$ (Obj. 7.4.1) **7.** $(-\infty, \infty)$ (Obj. 7.4.1) **8.** $3abc^2\sqrt[3]{ac}$ (Obj. 7.2.1) **9.** $8a\sqrt{2a}$ (Obj. 7.2.2) **10.** $-2x^2y\sqrt[3]{2x}$ (Obj. 7.2.2) **11.** $-4x\sqrt{3}$ (Obj. 7.2.3) **12.** $14 + 10\sqrt{3}$ (Obj. 7.2.3) **13.** $2a - \sqrt{ab} - 15b$ (Obj. 7.2.3) **14.** $4x + 4\sqrt{xy} + y$ (Obj. 7.2.3) **15.** $\frac{4x^2}{y}$ (Obj. 7.2.4) **16.** 2 (Obj. 7.2.4) **17.** $\frac{x + \sqrt{xy}}{x - y}$ (Obj. 7.2.4) **18.** −4 (Obj. 7.3.3) **19.** $-3 + 2i$ (Obj. 7.3.2) **20.** $18 + 16i$ (Obj. 7.3.3) **21.** $-\frac{4}{5} + \frac{7}{5}i$ (Obj. 7.3.4) **22.** 4 (Obj. 7.3.2) **23.** 4 (Obj. 7.5.1) **24.** −3 (Obj. 7.5.1) **25.** The distance is 576 ft. (Obj. 7.5.2)

CUMULATIVE REVIEW EXERCISES

1. The Distributive Property   (Obj. 1.3.1)   **2.** $-x - 24$ (Obj. 1.3.3)   **3.** $\varnothing$ (Obj. 1.1.2)   **4.** 16   (Obj. 7.5.1)

5. $\dfrac{3}{2}$ (Obj. 2.1.1)   **6.** $\dfrac{2}{3}$ (Obj. 2.1.2)   **7.** $\{x \mid x \geq -1\}$ (Obj. 2.5.1)   **8.** $\{x \mid 4 < x < 5\}$ (Obj. 2.5.2)

9. $\left\{x \mid x < 2 \text{ or } x > \dfrac{8}{3}\right\}$ (Obj. 2.6.2)   **10.** $(8a + b)(8a - b)$ (Obj. 5.6.1)   **11.** $x(x^2 + 3)(x + 1)(x - 1)$

(Obj. 5.6.4)   **12.** $\dfrac{2}{3}$ and -5 (Obj. 5.7.1)   **13.**   (Obj. 7.4.2)   **14.** Yes   (Obj. 3.6.1)

15. $\dfrac{3x^3}{y}$ (Obj. 5.1.2)   **16.** $\dfrac{y^8}{x^2}$ (Obj. 7.1.1)   **17.** $-x\sqrt{5x}$ (Obj. 7.2.2)   **18.** $11 - 5\sqrt{5}$ (Obj. 7.2.3)

19. $\dfrac{x\sqrt[3]{4y^2}}{2y}$ (Obj. 7.2.4)   **20.** $-\dfrac{3}{5} + \dfrac{6}{5}i$ (Obj. 7.3.4)   **21.** [number line from -5 to 5, open interval from -1 to 3]   (Obj. 1.1.2)

22. $\{x \mid x \in \text{real numbers}\}$ (Obj. 7.4.1)   **23.** 34   (Obj. 3.2.1)   **24.** $y = \dfrac{1}{3}x + \dfrac{7}{3}$ (Obj. 3.5.2)

25. 3   (Obj. 4.3.1)   **26.** $\left(\dfrac{17}{4}, \dfrac{9}{2}\right)$ (Obj. 4.3.2)   **27.** [graph of line] $m = \dfrac{3}{2}, b = 3$ (Obj. 3.3.2)

28.   (Obj. 3.7.1)   **29.** There are nineteen 18¢ stamps.   (Obj. 2.2.1)

30. The additional investment must be $4000.   (Obj. 2.4.1)   **31.** The length is 12 ft. The width is 6 ft.   (Obj. 5.7.2)
32. The rate of the plane is 250 mph.   (Obj. 6.4.3)   **33.** The time is 1.25 s.   (Obj. 5.1.4)   **34.** The height of the periscope is 32.7 ft.   (Obj. 7.5.2)   **35.** The slope is 0.08. A slope of 0.08 means that the annual interest income is 8% of the investment.   (Obj. 3.4.1)

ANSWERS to Chapter 8 Odd-Numbered Exercises

CONCEPT REVIEW 8.1

1. Sometimes true   **3.** Never true

SECTION 8.1

3. 0 and 4   **5.** -5 and 5   **7.** -2 and 3   **9.** 3   **11.** 0 and 2   **13.** -2 and 5   **15.** 2 and 5   **17.** 6

and $-\dfrac{3}{2}$ **19.** $\dfrac{1}{4}$ and 2   **21.** -4 and $\dfrac{1}{3}$ **23.** $-\dfrac{2}{3}$ and $\dfrac{9}{2}$ **25.** -4 and $\dfrac{1}{4}$ **27.** -2 and 9   **29.** -2

and $-\dfrac{3}{4}$ **31.** 2 and -5 **33.** $-\dfrac{3}{2}$ and -4 **35.** $2b$ and $7b$ **37.** $-c$ and $7c$ **39.** $-\dfrac{b}{2}$ and $-b$ **41.** $\dfrac{2a}{3}$

and $4a$ **43.** $x^2 - 7x + 10 = 0$ **45.** $x^2 + 6x + 8 = 0$ **47.** $x^2 - 5x - 6 = 0$ **49.** $x^2 - 9 = 0$
51. $x^2 - 8x + 16 = 0$ **53.** $x^2 - 5x = 0$ **55.** $x^2 - 3x = 0$ **57.** $2x^2 - 7x + 3 = 0$ **59.** $4x^2 - 5x - 6 = 0$

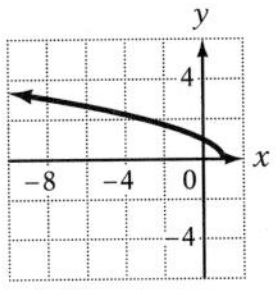

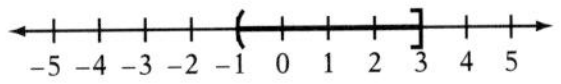

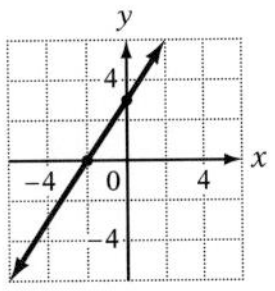

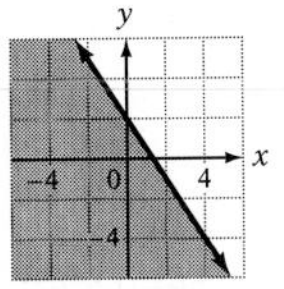

61. $3x^2 + 11x + 10 = 0$ **63.** $9x^2 - 4 = 0$ **65.** $6x^2 - 5x + 1 = 0$ **67.** $10x^2 - 7x - 6 = 0$ **69.** $8x^2 + 6x + 1 = 0$ **71.** $50x^2 - 25x - 3 = 0$ **73.** 7 and −7 **75.** $2i$ and $-2i$ **77.** 2 and −2 **79.** $\frac{9}{2}$ and $-\frac{9}{2}$ **81.** $7i$ and $-7i$ **83.** $4\sqrt{3}$ and $-4\sqrt{3}$ **85.** $5\sqrt{3}$ and $-5\sqrt{3}$ **87.** $3i\sqrt{2}$ and $-3i\sqrt{2}$ **89.** 7 and −5 **91.** 0 and −6 **93.** −7 and 3 **95.** $-5 + 5i$ and $-5 - 5i$ **97.** $4 + 2i$ and $4 - 2i$ **99.** $9 + 3i$ and $9 - 3i$ **101.** 0 and 1 **103.** $-\frac{1}{5}$ and 1 **105.** $-\frac{3}{2}$ and 0 **107.** $\frac{7}{3}$ and 1 **109.** $-5 + \sqrt{6}$ and $-5 - \sqrt{6}$ **111.** $2 + 2\sqrt{6}$ and $2 - 2\sqrt{6}$ **113.** $-1 + 2i\sqrt{3}$ and $-1 - 2i\sqrt{3}$ **115.** $3 + 3i\sqrt{5}$ and $3 - 3i\sqrt{5}$ **117.** $\frac{-2 + 9\sqrt{2}}{3}$ and $\frac{-2 - 9\sqrt{2}}{3}$ **119.** $-\frac{1}{2} + 2i\sqrt{10}$ and $-\frac{1}{2} - 2i\sqrt{10}$ **121.** $x^2 - 2 = 0$ **123.** $x^2 + 1 = 0$ **125.** $x^2 - 8 = 0$ **127.** $x^2 - 12 = 0$ **129.** $x^2 + 12 = 0$ **131.** $-\frac{5z}{y}$ and $\frac{5z}{y}$ **133.** $y + 2$ and $y - 2$ **135.** 1

CONCEPT REVIEW 8.2

1. Always true **3.** Never true **5.** Never true

SECTION 8.2

3. 5 and −1 **5.** −9 and 1 **7.** 3 **9.** $-2 + \sqrt{11}$ and $-2 - \sqrt{11}$ **11.** $3 + \sqrt{2}$ and $3 - \sqrt{2}$ **13.** $1 + i$ and $1 - i$ **15.** $\frac{1 + \sqrt{5}}{2}$ and $\frac{1 - \sqrt{5}}{2}$ **17.** $3 + \sqrt{13}$ and $3 - \sqrt{13}$ **19.** 3 and 5 **21.** $2 + 3i$ and $2 - 3i$ **23.** $-3 + 2i$ and $-3 - 2i$ **25.** $1 + 3\sqrt{2}$ and $1 - 3\sqrt{2}$ **27.** $\frac{1 + \sqrt{17}}{2}$ and $\frac{1 - \sqrt{17}}{2}$ **29.** $1 + 2i\sqrt{3}$ and $1 - 2i\sqrt{3}$ **31.** $-\frac{1}{2}$ and −1 **33.** $\frac{1}{2}$ and $\frac{3}{2}$ **35.** $-\frac{1}{2}$ and $\frac{4}{3}$ **37.** $\frac{1}{2} + i$ and $\frac{1}{2} - i$ **39.** $\frac{1}{3} + \frac{1}{3}i$ and $\frac{1}{3} - \frac{1}{3}i$ **41.** $\frac{2 + \sqrt{14}}{2}$ and $\frac{2 - \sqrt{14}}{2}$ **43.** $-\frac{3}{2}$ and 1 **45.** $1 + \sqrt{5}$ and $1 - \sqrt{5}$ **47.** $\frac{1}{2}$ and 5 **49.** $2 + \sqrt{5}$ and $2 - \sqrt{5}$ **51.** −3.236 and 1.236 **53.** 1.707 and 0.293 **55.** 0.309 and −0.809 **59.** 5 and −2 **61.** 4 and −9 **63.** $4 + 2\sqrt{22}$ and $4 - 2\sqrt{22}$ **65.** 3 and −8 **67.** $\frac{1}{2}$ and −3 **69.** $-\frac{1}{4}$ and $\frac{3}{2}$ **71.** $1 + 2\sqrt{2}$ and $1 - 2\sqrt{2}$ **73.** 10 and −2 **75.** $6 + 2\sqrt{3}$ and $6 - 2\sqrt{3}$ **77.** $\frac{1 + 2\sqrt{2}}{2}$ and $\frac{1 - 2\sqrt{2}}{2}$ **79.** $\frac{1}{2}$ and 1 **81.** $\frac{-5 + \sqrt{7}}{3}$ and $\frac{-5 - \sqrt{7}}{3}$ **83.** $\frac{2}{3}$ and $\frac{5}{2}$ **85.** $2 + i$ and $2 - i$ **87.** $-3 + 2i$ and $-3 - 2i$ **89.** $3 + i$ and $3 - i$ **91.** $-\frac{3}{2}$ and $-\frac{1}{2}$ **93.** $-\frac{1}{2} + \frac{5}{2}i$ and $-\frac{1}{2} - \frac{5}{2}i$ **95.** $\frac{-3 + \sqrt{6}}{3}$ and $\frac{-3 - \sqrt{6}}{3}$ **97.** $1 + \frac{\sqrt{6}}{3}i$ and $1 - \frac{\sqrt{6}}{3}i$ **99.** $-\frac{3}{2}$ and −1 **101.** $-\frac{3}{2}$ and 4 **103.** Two complex **105.** One real **107.** Two real **109.** Two real **111.** 0.873 and −6.873 **113.** 3.236 and −1.236 **115.** 2.468 and −0.135 **117.** $\frac{\sqrt{2}}{2}$ and $-2\sqrt{2}$ **119.** $-3\sqrt{2}$ and $\frac{\sqrt{2}}{2}$ **121.** $\frac{\sqrt{3}}{2} + \frac{1}{2}i$ and $\frac{\sqrt{3}}{2} - \frac{1}{2}i$ **123.** $2a$ and $-a$ **125.** $\frac{a}{2}$ and $-2a$ **127.** $1 + \sqrt{y + 1}$ and $1 - \sqrt{y + 1}$ **129.** $\{p \mid p < 9, p \in \text{real numbers}\}$ **131.** $\{p \mid p > 1, p \in \text{real numbers}\}$ **133.** $b^2 + 4 > 0$ for any real number b. **135.** The ball takes 4.43 s to hit the ground. **137. a.** 1.05: The model predicts that the population will increase. **b.** 0.96: The model predicts that the population will decrease.

CONCEPT REVIEW 8.3

1. Always true **3.** Always true

SECTION 8.3

3. 3, −3, 2, −2 **5.** 2, −2, $\sqrt{2}$, $-\sqrt{2}$ **7.** 1 and 4 **9.** 16 **11.** $2i, -2i, 1, -1$ **13.** $4i, -4i, 2, -2$ **15.** 16 **17.** 512 and 1 **19.** $2, 1, -1 + i\sqrt{3}, -1 - i\sqrt{3}, -\frac{1}{2} + \frac{\sqrt{3}}{2}i, -\frac{1}{2} - \frac{\sqrt{3}}{2}i$ **21.** $1, -1, 2, -2, i, -i, 2i, -2i$ **23.** −64 and 8 **25.** $\frac{1}{4}$ and 1 **27.** 3 **29.** 9 **31.** 2 and −1 **33.** 0 and 2 **35.** $-\frac{1}{2}$ and 2 **37.** −2 **39.** 1 **41.** 1 **43.** −3 **45.** 10 and −1 **47.** $-\frac{1}{2} + \frac{\sqrt{7}}{2}i$ and $-\frac{1}{2} - \frac{\sqrt{7}}{2}i$ **49.** 1 and −3 **51.** 0 and −1 **53.** $\frac{1}{2}$ and $-\frac{1}{3}$ **55.** 6 and $-\frac{2}{3}$ **57.** $\frac{4}{3}$ and 3 **59.** $-\frac{1}{4}$ and 3 **61.** $\frac{-5 + \sqrt{33}}{2}$ and $\frac{-5 - \sqrt{33}}{2}$ **63.** 4 and −6 **65.** 5 and 4 **67.** $3, -3, i, -i$ **69.** $i\sqrt{6}, -i\sqrt{6}, 2, -2$ **71.** $\sqrt{2}$ **73.** 4

CONCEPT REVIEW 8.4

1. Sometimes true **3.** Always true

SECTION 8.4

1. The length is 40 ft. The width is 16 ft. **3.** The length is 13 ft. The width is 5 ft. **5.** The rate returning was 55 mph. **7.** The projectile takes 12.5 s to return to Earth. **9.** The maximum speed is 72.5 km/h. **11.** The rocket will be 300 ft above the ground after 1.74 s and after 10.76 s. **13.** It would take the smaller pipe 12 min. It would take the larger pipe 6 min. **13.** Working alone, it would take the smaller pipe 12 min to fill the tank. It would take the larger pipe 6 min. **15.** The rate of the cruise ship for the first 40 mi was 10 mph. **17.** The rate of the wind is 20 mph. **19.** The rate of the crew in calm water is 6 mph. **21.** The number is $\frac{1}{4}$ or 8. **23.** The integers are −7 and −6 or 6 and 7. **25.** The width is 8 cm. The length is 16 cm. The height is 2 cm. **27.** The sales will first exceed 150 million units in 2001. **29.** The bottom of the scoop of ice cream is 2.3 in. from the bottom of the cone.

CONCEPT REVIEW 8.5

1. Sometimes true **3.** Sometimes true

SECTION 8.5

3. $\{x \mid x < -2 \text{ or } x > 4\}$ 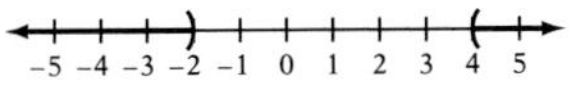

5. $\{x \mid x \le 1 \text{ or } x \ge 2\}$ 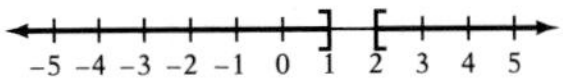

7. $\{x \mid -3 < x < 4\}$ 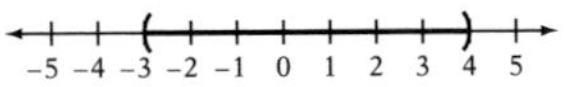

9. $\{x \mid x < -2 \text{ or } 1 < x < 3\}$ 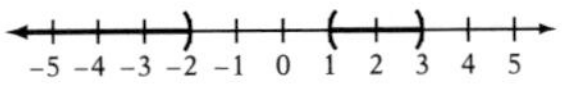

11. $\{x \mid x < -4 \text{ or } x > 4\}$ **13.** $\{x \mid x < 2 \text{ or } x > 2\}$ **15.** $\{x \mid -3 \le x \le 12\}$ **17.** $\left\{x \mid x \le \frac{1}{2} \text{ or } x \ge 2\right\}$ **19.** $\left\{x \,\middle|\, \frac{1}{2} < x < \frac{3}{2}\right\}$ **21.** $\{x \mid x \le -3 \text{ or } 2 \le x \le 6\}$ **23.** $\left\{x \,\middle|\, -\frac{3}{2} < x < \frac{1}{2} \text{ or } x > 4\right\}$ **25.** $\{x \mid x \le -3 \text{ or } -1 \le x \le 1\}$ **27.** $\{x \mid -2 \le x \le 1 \text{ or } x \ge 2\}$

29. $\{x \mid x < -2 \text{ or } x > 4\}$ 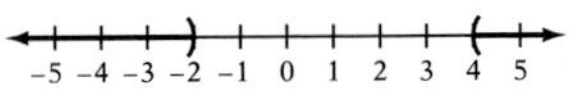

31. $\{x \mid -1 < x \le 3\}$ 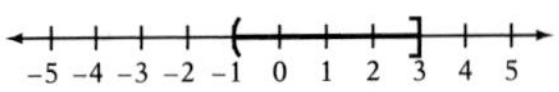

33. $\{x \mid x \le -2 \text{ or } 1 \le x < 3\}$ 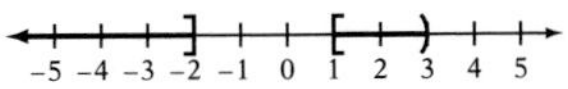

35. $\{x \mid x < -1 \text{ or } x > 2\}$ **37.** $\{x \mid -1 < x \le 0\}$ **39.** $\{x \mid -2 < x \le 0 \text{ or } x > 1\}$ **41.** $\left\{x \mid x < 0 \text{ or } x > \frac{1}{2}\right\}$

43. 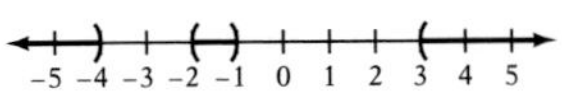

45. 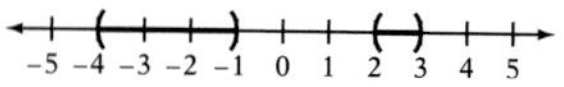

47. 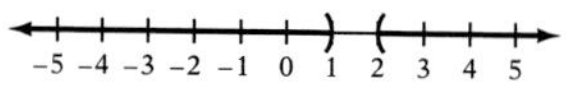

49. 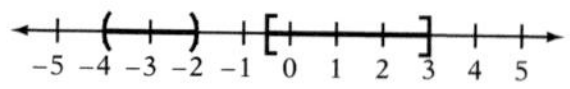

51. 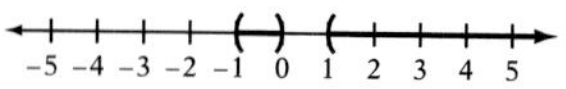

53. 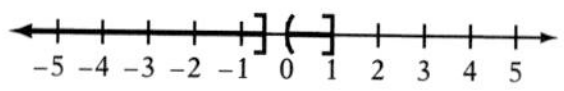

CONCEPT REVIEW 8.6

1. Sometimes true 3. Sometimes true 5. Always true

SECTION 8.6

3. Vertex: (0, 0)
Axis of symmetry: $x = 0$

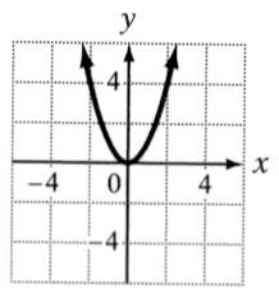

5. Vertex: (0, −2)
Axis of symmetry: $x = 0$

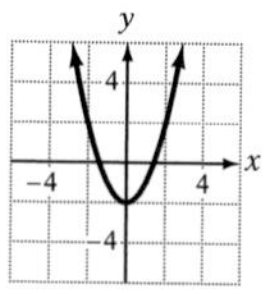

7. Vertex: (0, 3)
Axis of symmetry: $x = 0$

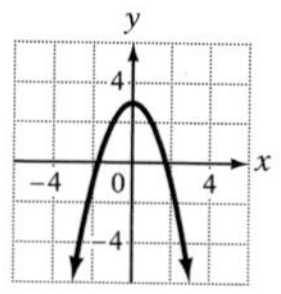

9. Vertex: (0, 0)
Axis of symmetry: $x = 0$

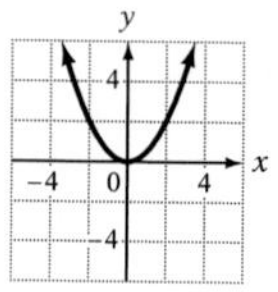

11. Vertex: (0, −1)
Axis of symmetry: $x = 0$

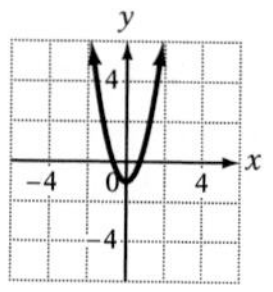

13. Vertex: (1, −1)
Axis of symmetry: $x = 1$

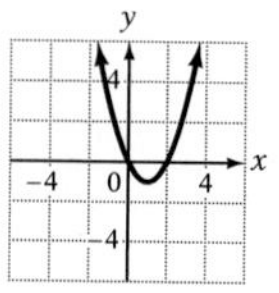

15. Vertex: (1, 2)

Axis of symmetry: $x = 1$

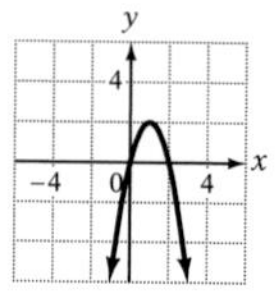

17. Vertex: $\left(\frac{1}{2}, -\frac{9}{4}\right)$

Axis of symmetry: $x = \frac{1}{2}$

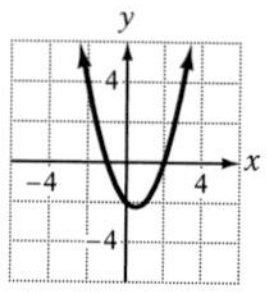

19. Vertex: $\left(\frac{1}{4}, -\frac{41}{8}\right)$

Axis of symmetry: $x = \frac{1}{4}$

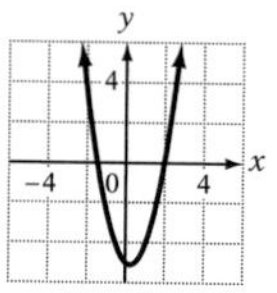

21. 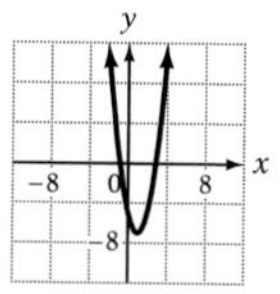

Domain: $\{x \mid x \in \text{real numbers}\}$
Range: $\{y \mid y \geq -7\}$

23. 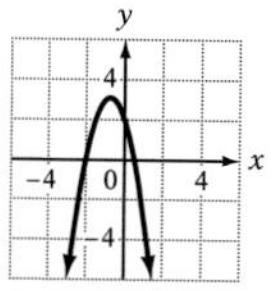

Domain: $\{x \mid x \in \text{real numbers}\}$
Range: $\left\{y \mid y \leq \frac{25}{8}\right\}$

25. 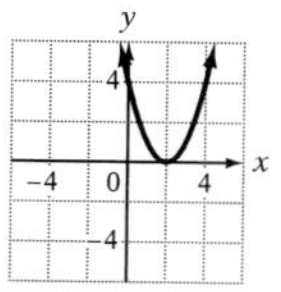

Domain: $\{x \mid x \in \text{real numbers}\}$
Range: $\{y \mid y \geq 0\}$

27. 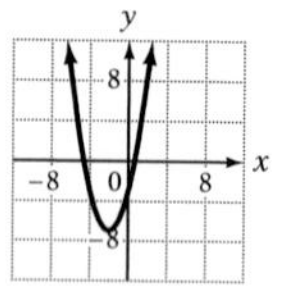

Domain: $\{x \mid x \in \text{real numbers}\}$
Range: $\{y \mid y \geq -7\}$

29. 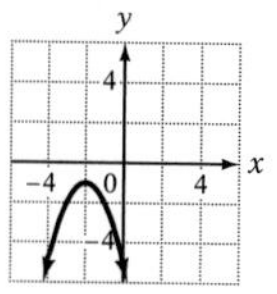

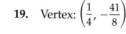

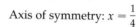

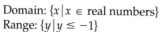

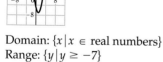

Domain: $\{x \mid x \in \text{real numbers}\}$
Range: $\{y \mid y \leq -1\}$

33. (2, 0) and (−2, 0)
35. (0, 0) and (2, 0)
37. (2, 0) and (−1, 0)

39. (3, 0) and $\left(-\frac{1}{2}, 0\right)$ 41. $\left(-\frac{2}{3}, 0\right)$ and (7, 0) 43. (5, 0) and $\left(\frac{4}{3}, 0\right)$ 45. $\left(\frac{2}{3}, 0\right)$ 47. $\left(\frac{\sqrt{2}}{3}, 0\right)$ and $\left(-\frac{\sqrt{2}}{3}, 0\right)$ 49. $\left(-\frac{1}{2}, 0\right)$ and (3, 0) 51. $(-2 + \sqrt{7}, 0)$ and $(-2 - \sqrt{7}, 0)$ 53. No x-intercepts

55. $(-1 + \sqrt{2}, 0)$ and $(-1 - \sqrt{2}, 0)$ **57.**

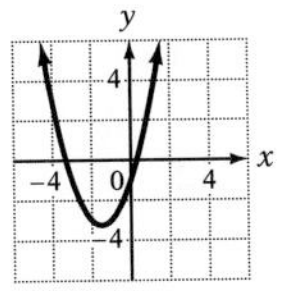

-3.3 and 0.3

59.

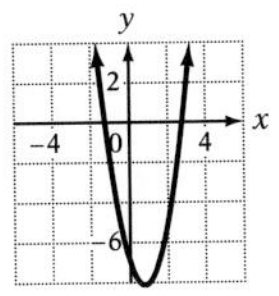

-1.3 and 2.8

61.

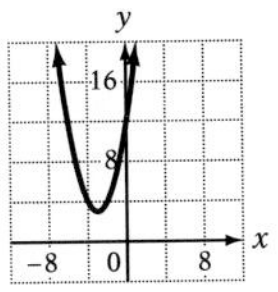

No x-intercepts

63. No x-intercepts **65.** Two **67.** One **69.** No x-intercepts **71.** Two **73.** No x-intercepts

75. One **77.** Two **79.** -2 **81.** 1 **83.** $\frac{1 + \sqrt{19}}{3}$ **85.** The graph becomes thinner.

87. The graph is higher on the rectangular coordinate system. **89.** $y = (x - 2)^2 + 3$; vertex: $(2, 3)$

91. $y = \left(x + \frac{1}{2}\right)^2 + \frac{7}{4}$; vertex: $\left(-\frac{1}{2}, \frac{7}{4}\right)$ **93.** $y = 3x^2 - 6x + 5$

CONCEPT REVIEW 8.7

1. Sometimes true **3.** Always true **5.** Sometimes true

SECTION 8.7

3. Minimum: 2 **5.** Maximum: -1 **7.** Minimum: -2 **9.** Maximum: -3 **11.** Minimum: $-\frac{73}{8}$

13. Minimum: $-\frac{11}{4}$ **15.** Maximum: $-\frac{2}{3}$ **17.** Minimum: $-\frac{1}{12}$ **19.** The maximum height is 150 ft.

21. To minimize the cost, 100 lenses should be produced. **23.** The minimum height is 24.36 ft. **25.** Yes.

27. The maximum height is 141.6 ft. **29.** The speed of 41 mph will yield the maximum fuel efficiency of 33.658 mi/gal. **31.** 3.0 **33.** 0.5

35. a.

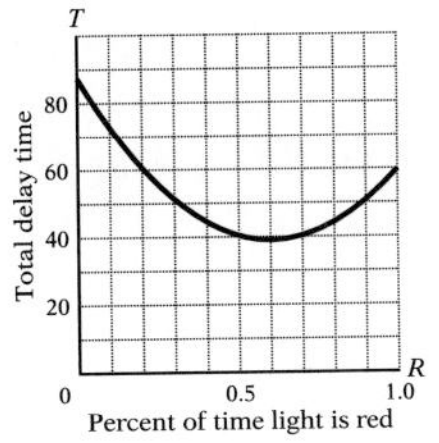

b. R is the percent of time the red light is in the horizontal direction. The percent cannot be less than 0% or more than 100%. **c.** The traffic light should remain red in the horizontal direction approximately 62% of the time.

CHAPTER REVIEW EXERCISES

1. 0 and $\frac{3}{2}$ (Obj. 8.1.1) **2.** $-2c$ and $\frac{c}{2}$ (Obj. 8.1.1) **3.** $-4\sqrt{3}$ and $4\sqrt{3}$ (Obj. 8.1.3) **4.** $-\frac{1}{2} - 2i$ and $-\frac{1}{2} + 2i$ (Obj. 8.1.3) **5.** $-\frac{17}{4}$ (Obj. 8.7.1) **6.** 3 (Obj. 8.7.1) **7.** $3x^2 + 8x - 3 = 0$ (Obj. 8.1.2) **8.** -5 and $\frac{1}{2}$ (Obj. 8.1.1) **9.** $-1 - 3\sqrt{2}$ and $-1 + 3\sqrt{2}$ (Obj. 8.1.3) **10.** $-3 - i$ and $-3 + i$ (Obj. 8.2.1) **11.** 2 and 3 (Obj. 8.3.3) **12.** $-\sqrt{2}, \sqrt{2}, -2, 2$ (Obj. 8.3.1) **13.** $\frac{25}{18}$ (Obj. 8.3.2) **14.** -8 and $\frac{1}{8}$ (Obj. 8.3.1) **15.** 2 (Obj. 8.3.2) **16.** $3 - \sqrt{11}$ and $3 + \sqrt{11}$ (Obj. 8.2.2) **17.** 5 and $-\frac{4}{9}$ (Obj. 8.3.3) **18.** $\frac{1 + \sqrt{3}}{2}$ and $\frac{1 - \sqrt{3}}{2}$ (Obj. 8.2.2) **19.** $\frac{5}{4}$ (Obj. 8.3.2) **20.** 3 and -1 (Obj. 8.3.3) **21.** -1 (Obj. 8.3.3) **22.** $\frac{-3 + \sqrt{249}}{10}$ and $\frac{-3 - \sqrt{249}}{10}$ (Obj. 8.3.3) **23.** $\frac{-11 - \sqrt{129}}{2}$ and $\frac{-11 + \sqrt{129}}{2}$ (Obj. 8.3.3) **24.** $x = 3$ (Obj. 8.6.1) **25.** $\left(\frac{3}{2}, \frac{1}{4}\right)$ (Obj. 8.6.1) **26.** No x-intercepts (Obj. 8.6.2) **27.** Two (Obj. 8.6.2)

28. $\left(\frac{-3-\sqrt{5}}{2}, 0\right)$ and $\left(\frac{-3+\sqrt{5}}{2}, 0\right)$ (Obj. 8.6.2) **29.** $(-2, 0)$ and $\left(\frac{1}{2}, 0\right)$ (Obj. 8.6.2) **30.** $-\frac{1}{3}+\frac{\sqrt{5}}{3}i$ and $-\frac{1}{3}-\frac{\sqrt{5}}{3}i$ (Obj. 8.6.2) **31.** $\left\{x \mid -3 < x < \frac{5}{2}\right\}$ (Obj. 8.5.1) **32.** $\left\{x \mid x \le -4 \text{ or } -\frac{3}{2} \le x \le 2\right\}$ (Obj. 8.5.1)

33. $\left\{x \mid x < \frac{3}{2} \text{ or } x \ge 2\right\}$ 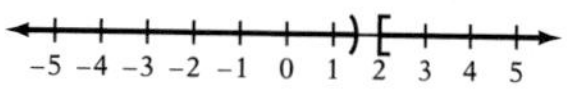

(Obj. 8.5.2)

34. $\left\{x \mid x \le -3 \text{ or } \frac{1}{2} \le x < 4\right\}$ 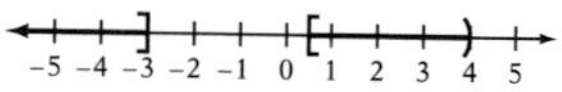

(Obj. 8.5.2)

35. 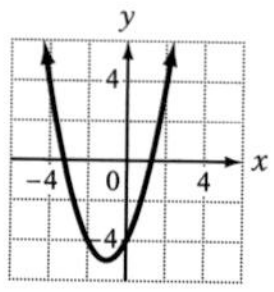

(Obj. 8.6.1)

Domain: $\{x \mid x \in \text{real numbers}\}$
Range: $\{y \mid y \ge -5\}$

36. 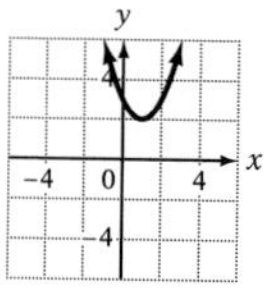

(Obj. 8.6.1)

Vertex: $(1, 2)$
Axis of symmetry: $x = 1$

37. The width of the rectangle is 5 cm. The length of the rectangle is 12 cm. (Obj. 8.4.1) **38.** The integers are 2, 4, and 6 or −6, −4, −2. (Obj. 8.4.1) **39.** Working alone, the new computer can print the payroll in 12 min. (Obj. 8.4.1) **40.** The rate of the first car is 40 mph. The rate of the second car is 50 mph. (Obj. 8.4.1)

CHAPTER TEST

1. $\frac{3}{2}$ and -2 (Obj. 8.1.1) **2.** $\frac{3}{4}$ and $-\frac{4}{3}$ (Obj. 8.1.1) **3.** 9 (Obj. 8.7.1) **4.** $3x^2 - 8x - 3 = 0$ (Obj. 8.1.2) **5.** $-3 + 3\sqrt{2}$ and $-3 - 3\sqrt{2}$ (Obj. 8.1.3) **6.** $-2 + \sqrt{5}$ and $-2 - \sqrt{5}$ (Obj. 8.2.2) **7.** $\frac{-3+\sqrt{41}}{2}$ and $\frac{-3-\sqrt{41}}{2}$ (Obj. 8.6.2) **8.** $\frac{1}{6}-\frac{\sqrt{95}}{6}i$ and $\frac{1}{6}+\frac{\sqrt{95}}{6}i$ (Obj. 8.2.2) **9.** $-3 - \sqrt{10}$ and $-3 + \sqrt{10}$ (Obj. 8.3.3) **10.** $\frac{1}{4}$ (Obj. 8.3.1) **11.** $-3, 3, -\sqrt{2}, \sqrt{2}$ (Obj. 8.3.1) **12.** 4 (Obj. 8.3.2) **13.** No solution (Obj. 8.3.2) **14.** Two (Obj. 8.6.2) **15.** $(-4, 0)$ and $\left(\frac{3}{2}, 0\right)$ (Obj. 8.6.2) **16.** $x = -\frac{3}{2}$ (Obj. 8.6.1)

17. 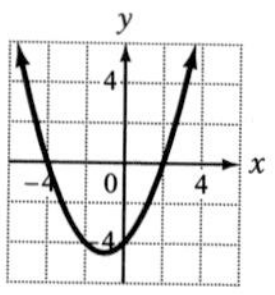

(Obj. 8.6.1)

Domain: $\{x \mid x \in \text{real numbers}\}$
Range: $\{y \mid y \ge -4.5\}$

18. $\left\{x \mid -4 < x \le \frac{3}{2}\right\}$ 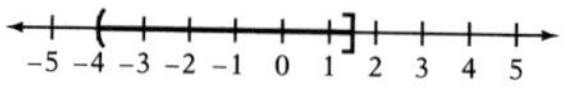

(Obj. 8.5.2)

19. The height of the triangle is 4 ft. The base of the triangle is 15 ft. (Obj. 8.4.1) **20.** The rate of the canoe in calm water is 4 mph. (Obj. 8.4.1)

CUMULATIVE REVIEW EXERCISES

1. 14 (Obj. 1.3.2) **2.** −28 (Obj. 2.1.2) **3.** $-\frac{3}{2}$ (Obj. 3.4.1) **4.** $y = x + 1$ (Obj. 3.6.1) **5.** $-3xy(x^2 - 2xy + 3y^2)$ (Obj. 5.5.1) **6.** $(2x - 5)(3x + 4)$ (Obj. 5.5.4) **7.** $(x + y)(a^n - 2)$ (Obj. 5.5.2) **8.** $x^2 - 3x - 4 - \frac{6}{3x - 4}$ (Obj. 5.4.1) **9.** $\frac{x}{2}$ (Obj. 6.2.1) **10.** $2\sqrt{5}$ (Obj. 3.1.2) **11.** $b = \frac{2S - an}{n}$ (Obj. 6.6.1) **12.** $-8 - 14i$ (Obj. 7.3.3) **13.** $1 - a$ (Obj. 7.1.1) **14.** $\frac{x\sqrt[3]{4y^2}}{2y}$ (Obj. 7.2.4) **15.** $-\frac{3}{2}$ and -1 (Obj. 8.3.3) **16.** $-\frac{1}{2}$ (Obj. 8.3.3) **17.** $2, -2, \sqrt{2}$, and $-\sqrt{2}$ (Obj. 8.3.1) **18.** 0 and 1 (Obj. 8.3.2)

19. $\left\{x \mid -2 < x < \frac{10}{3}\right\}$ (Obj. 2.6.2) **20.** $\left(\frac{5}{2}, 0\right)$ and $(0, -3)$ (Obj. 3.3.2) **21.**

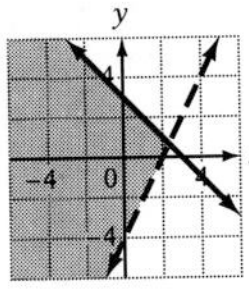

(Obj. 4.5.1)

22. $(1, -1, 2)$ (Obj. 4.2.2) **23.** $-\frac{7}{3}$ (Obj. 6.1.1) **24.** $\{x \mid x \neq 5, -3\}$ (Obj. 6.1.1)

25. $\{x \mid x < -3 \text{ or } 0 < x < 2\}$

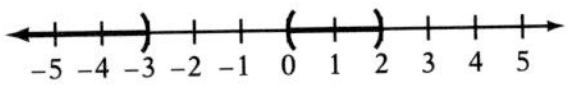

(Obj. 8.5.1)

26. $\{x \mid -3 < x \leq 1 \text{ or } x \geq 5\}$

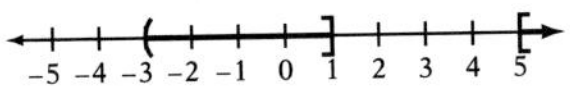

(Obj. 8.5.2) **27.** The lower and upper limits of the length of the piston rod are $9\frac{23}{64}$ in. and $9\frac{25}{64}$ in. (Obj. 2.6.3) **28.** The area of the triangle is $(x^2 + 6x - 16)$ ft^2. (Obj. 5.3.4) **29.** Two complex number solutions (Obj. 8.2.2) **30.** $m = -\frac{25{,}000}{3}$; a slope of $-\frac{25{,}000}{3}$ means that the value of the building decreases $\frac{\$25{,}000}{3}$ each year. (Obj. 3.4.1)

ANSWERS to Chapter 9 Odd-Numbered Exercises

CONCEPT REVIEW 9.1

1. Sometimes true **3.** Sometimes true **5.** Always true

SECTION 9.1

3. Yes **5.** No **7.** Yes **9.**

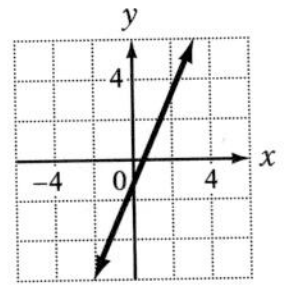

Domain: $\{x \mid x \in \text{real numbers}\}$
Range: $\{y \mid y \in \text{real numbers}\}$

11.

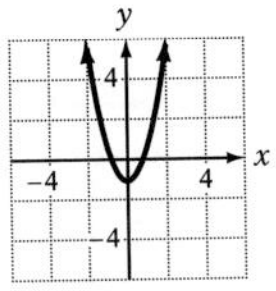

Domain: $\{x \mid x \in \text{real numbers}\}$
Range: $\{y \mid y \geq -1\}$

13.

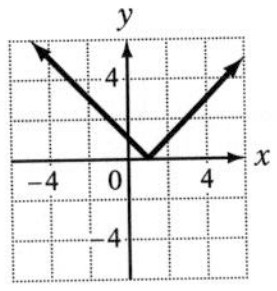

Domain: $\{x \mid x \in \text{real numbers}\}$
Range: $\{y \mid y \geq 0\}$

15.

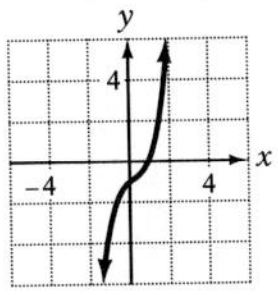

Domain: $\{x \mid x \in \text{real numbers}\}$
Range: $\{y \mid y \in \text{real numbers}\}$

17.

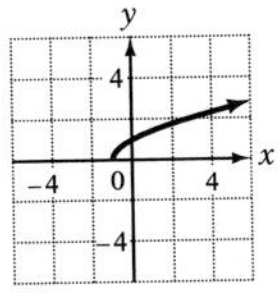

Domain: $\{x \mid x \geq -1\}$
Range: $\{y \mid y \geq 0\}$

19.

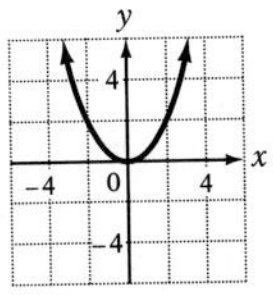

Domain: $\{x \mid x \in \text{real numbers}\}$
Range: $\{y \mid y \geq 0\}$

21.
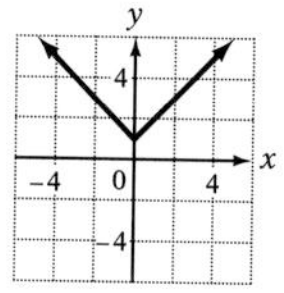

Domain: $\{x|x \in \text{real numbers}\}$
Range: $\{y|y \geq 1\}$

23.
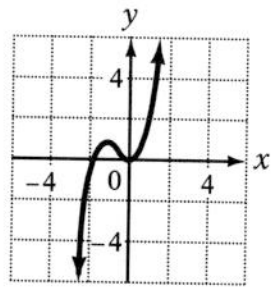

Domain: $\{x|x \in \text{real numbers}\}$
Range: $\{y|y \in \text{real numbers}\}$

25.
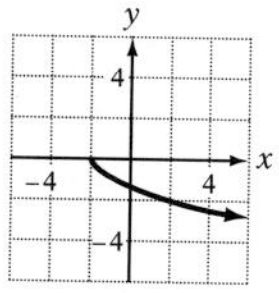

Domain: $\{x|x \geq -2\}$
Range: $\{y|y \leq 0\}$

27.
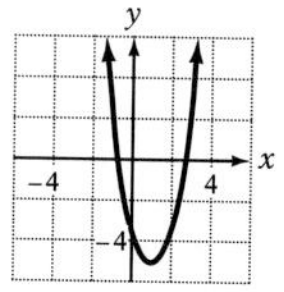

Domain: $\{x|x \in \text{real numbers}\}$
Range: $\{y|y \geq -5\}$

29.
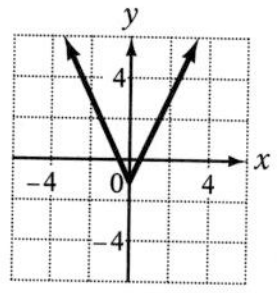

Domain: $\{x|x \in \text{real numbers}\}$
Range: $\{y|y \geq -1\}$

31.
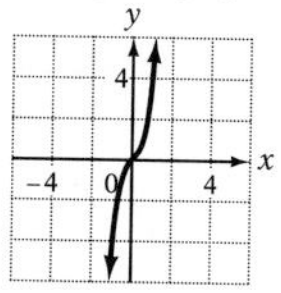

Domain: $\{x|x \in \text{real numbers}\}$
Range: $\{y|y \in \text{real numbers}\}$

33.
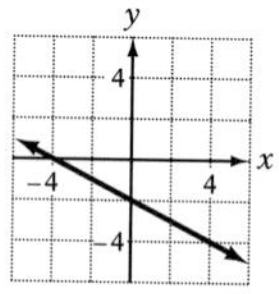

Domain: $\{x|x \in \text{real numbers}\}$
Range: $\{y|y \in \text{real numbers}\}$

35.
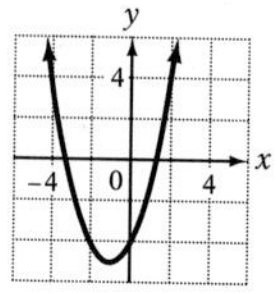

Domain: $\{x|x \in \text{real numbers}\}$
Range: $\{y|y \geq -5\}$

37.
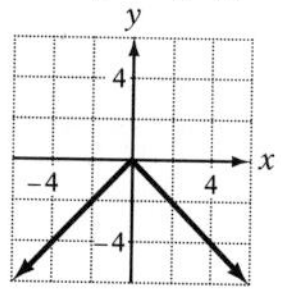

Domain: $\{x|x \in \text{real numbers}\}$
Range: $\{y|y \leq 0\}$

39.
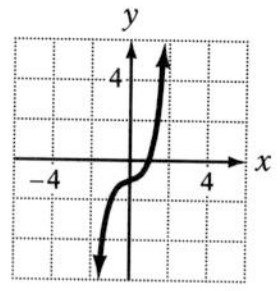

Domain: $\{x|x \in \text{real numbers}\}$
Range: $\{y|y \in \text{real numbers}\}$

41. c **43.** $a = 18$ **45.** 77 **47.** -3
49. **a.** 0 **b.** $(0, 0)$ **51.** 1

CONCEPT REVIEW 9.2

1. Always true **3.** Always true **5.** Never true

SECTION 9.2

3.
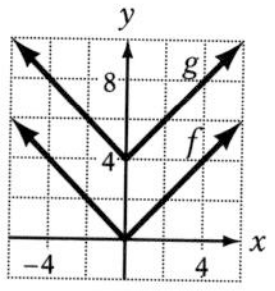

5.
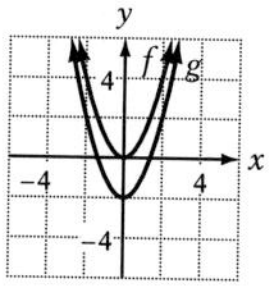

7.
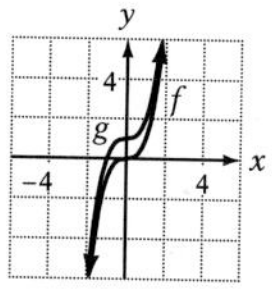

9.
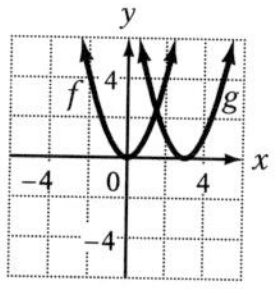

11.
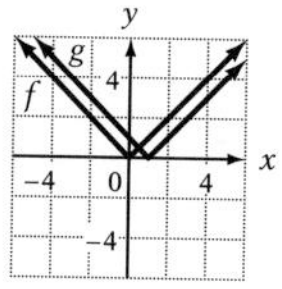

13.
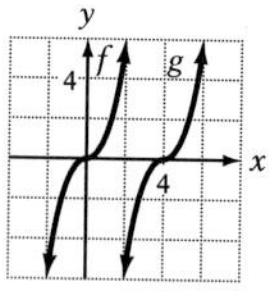

15.
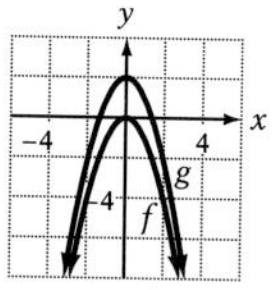

17.
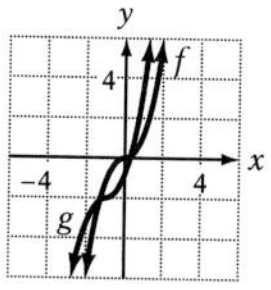

19.
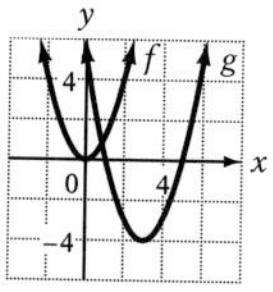

21.
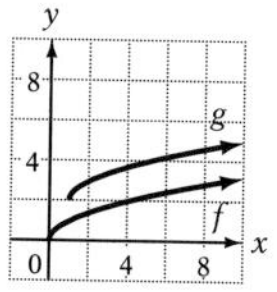

23.
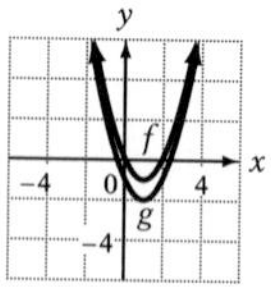

25.
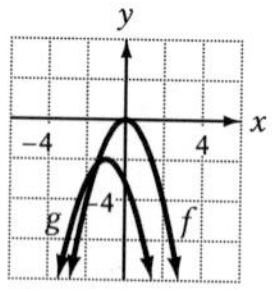

27.
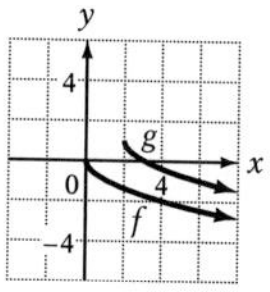

29.
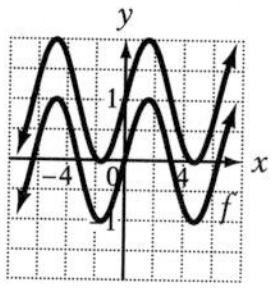

31.
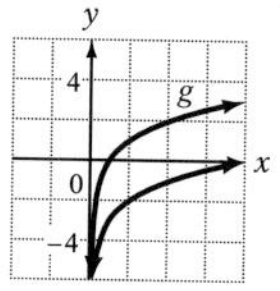

33.
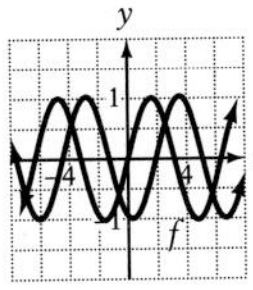

35.
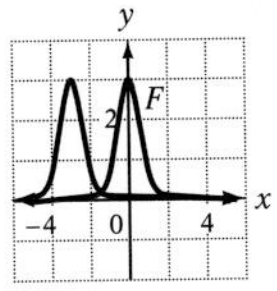

37.
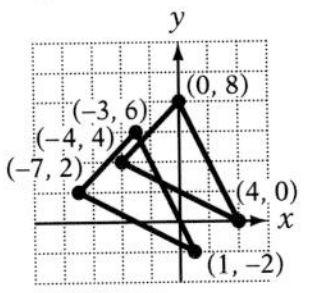

CONCEPT REVIEW 9.3

1. Always true **3.** Never true

SECTION 9.3

1. 5 **3.** 1 **5.** 0 **7.** $-\frac{29}{4}$ **9.** $\frac{2}{3}$ **11.** 2 **13.** 39 **15.** −8 **17.** $-\frac{4}{5}$ **19.** −2 **21.** 7
25. −13 **27.** −29 **29.** $8x - 13$ **31.** 1 **33.** 5 **35.** $x^2 + 1$ **37.** 5 **39.** 11
41. $3x^2 + 3x + 5$ **43.** −3 **45.** −27 **47.** $x^3 - 6x^2 + 12x - 8$ **49.** $6h + h^2$ **51.** $2 + h$
53. $h + 2a$ **55.** −1 **57.** −6 **59.** $6x - 13$ **61.** −2 **63.** −2 **65.** 0 **67.** 6

CONCEPT REVIEW 9.4

1. Always true **3.** Never true **5.** Sometimes true

SECTION 9.4

3. Yes **5.** No **7.** Yes **9.** No **11.** No **13.** No **17.** {(0, 1), (3, 2), (8, 3), (15, 4)} **19.** No inverse **21.** $f^{-1}(x) = \frac{1}{4}x + 2$ **23.** No inverse **25.** $f^{-1}(x) = x + 5$ **27.** $f^{-1}(x) = 3x - 6$
29. $f^{-1}(x) = -\frac{1}{3}x - 3$ **31.** $f^{-1}(x) = \frac{3}{2}x - 6$ **33.** $f^{-1}(x) = -3x + 3$ **35.** $f^{-1}(x) = \frac{1}{2}x + \frac{5}{2}$ **37.** No inverse **39.** Yes **41.** No **43.** Yes **45.** Yes **47.** range **49.** 3 **51.** 0 **53.** 2 **55.** $\frac{7}{3}$

57.
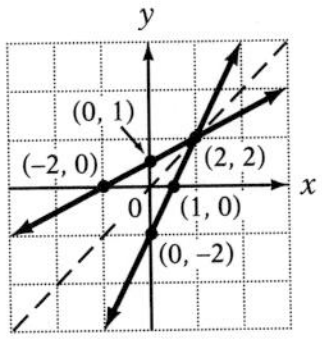

59.
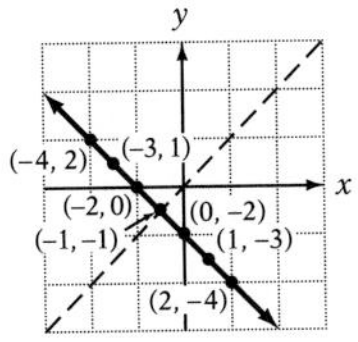

The function and its inverse have the same graph.

61.
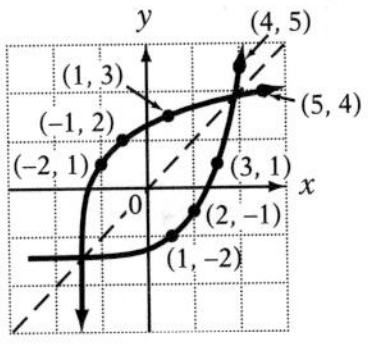

CHAPTER REVIEW EXERCISES

1. Yes (Obj. 9.1.1)

2.
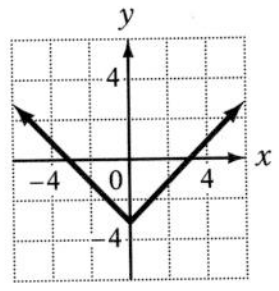

(Obj. 9.1.1)

Domain: $\{x \mid x \in \text{real numbers}\}$
Range: $\{y \mid y \geq -3\}$

3.
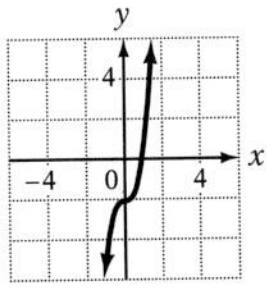

(Obj. 9.1.1)

Domain: $\{x \mid x \in \text{real numbers}\}$
Range: $\{y \mid y \in \text{real numbers}\}$

4. (Obj. 9.1.1) **5.** (Obj. 9.2.1) **6.** (Obj. 9.2.1)

Domain: $[-4, \infty)$
Range: $[0, \infty)$

7. (Obj. 9.2.1) **8.** (Obj. 9.2.1) **9.** 7 (Obj. 9.3.1) **10.** -9 (Obj. 9.3.1)

11. 70 (Obj. 9.3.1) **12.** $\frac{12}{7}$ (Obj. 9.3.1) **13.** $12x^2 + 12x - 1$ (Obj. 9.3.2) **14.** 5 (Obj. 9.3.2) **15.** 10
(Obj. 9.3.2) **16.** $6x^2 + 3x - 16$ (Obj. 9.3.2) **17.** Yes (Obj. 9.4.1) **18.** $\{(1, -2), (3, 2), (-4, 5), (9, 7)\}$
(Obj. 9.4.2) **19.** No (Obj. 9.4.1) **20.** Yes (Obj. 9.4.1) **21.** $f^{-1}(x) = 2x - 16$ (Obj. 9.4.2)
22. $f^{-1}(x) = -\frac{1}{6}x + \frac{2}{3}$ (Obj. 9.4.2) **23.** $f^{-1}(x) = \frac{3}{2}x + 18$ (Obj. 9.4.2) **24.** Yes (Obj. 9.4.2)

CHAPTER TEST

1. No (Obj. 9.1.1) **2.** -2 (Obj. 9.3.1) **3.** 234 (Obj. 9.3.1) **4.** $-\frac{13}{2}$ (Obj. 9.3.1) **5.** -5 (Obj. 9.3.1)

6. 5 (Obj. 9.3.2) **7.** (Obj. 9.2.1) **8.** (Obj. 9.2.1) **9.** (Obj. 9.1.1)

Domain: $(-\infty, 3]$
Range: $(-\infty, 0]$
(Obj. 9.2.1)

10. (Obj. 9.1.1) **11.** (Obj. 9.2.1) **12.** (Obj. 9.2.1)

Domain: $\{x \mid x \in \text{real numbers}\}$
Range: $\{y \mid y \geq -2\}$

13. $f^{-1}(x) = 4x + 16$ (Obj. 9.4.2) **14.** $\{(6, 2), (5, 3), (4, 4), (3, 5)\}$ (Obj. 9.4.2) **15.** Yes (Obj. 9.4.2)
16. $2x^2 - 4x - 5$ (Obj. 9.3.2) **17.** $f^{-1}(x) = 2x + 6$ (Obj. 9.4.2) **18.** No (Obj. 9.4.2)
19. (Obj. 9.1.1) **20.** No (Obj. 9.4.1)

Domain: $\{x \mid x \in \text{real numbers}\}$
Range: $\{y \mid y \in \text{real numbers}\}$

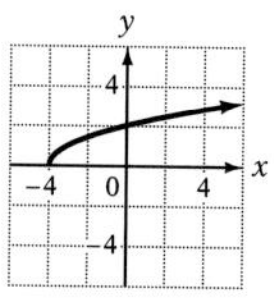

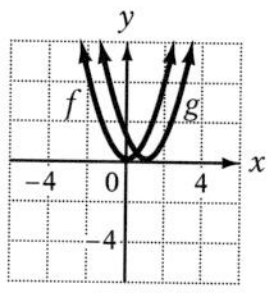

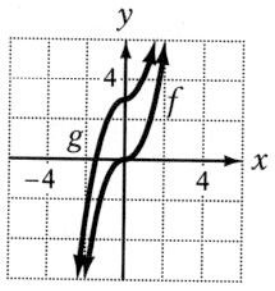

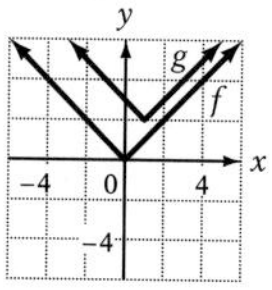

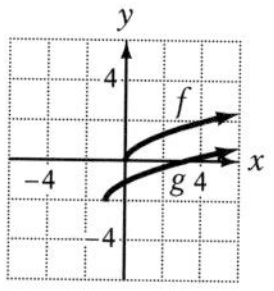

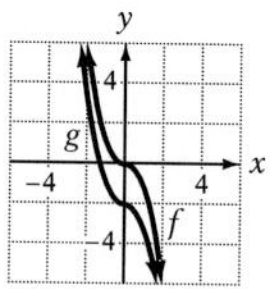

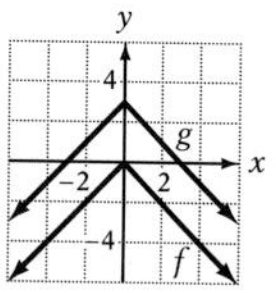

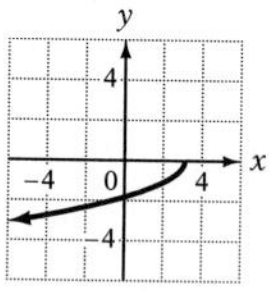

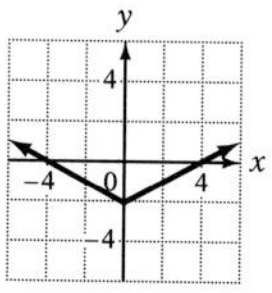

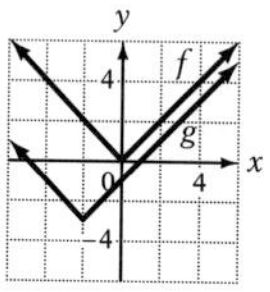

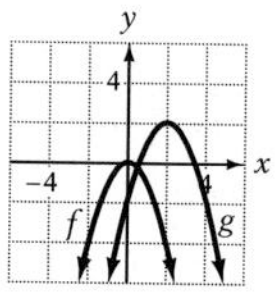

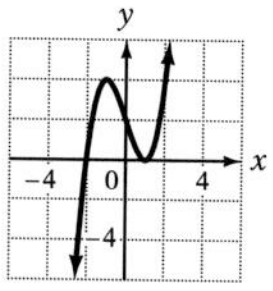

CUMULATIVE REVIEW EXERCISES

1. $-\frac{23}{4}$ (Obj. 1.3.2) **2.** 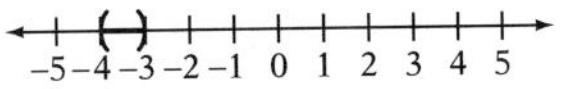

(Obj. 1.1.2) **3.** 3 (Obj. 2.1.2) **4.** $\{x \mid x < -2 \text{ or } x > 3\}$ (Obj. 2.5.2) **5.** 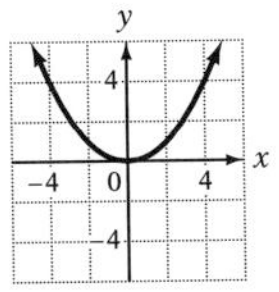

(Obj. 8.6.1) **6.** 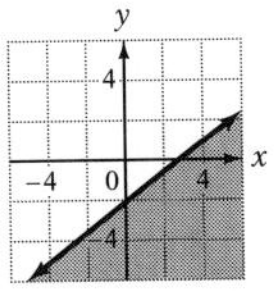

(Obj. 3.7.1)

Vertex: (0, 0)
Axis of symmetry: $x = 0$

7. $\{x \mid x \in \text{real numbers}\}$ (Obj. 2.6.2) **8.** $-\frac{a^{10}}{12b^4}$ (Obj. 5.1.2) **9.** $2x^3 - 4x^2 - 17x + 4$ (Obj. 5.3.2) **10.** $(a + 2)(a - 2)(a^2 + 2)$ (Obj. 5.6.4) **11.** $xy(x - 2y)(x + 3y)$ (Obj. 5.6.4) **12.** -3 and 8 (Obj. 5.7.1) **13.** $\{x \mid x < -3 \text{ or } x > 5\}$ (Obj. 8.5.1) **14.** $\frac{5}{2x - 1}$ (Obj. 6.2.2) **15.** -2 (Obj. 6.4.1) **16.** $-3 - 2i$ (Obj. 7.3.4) **17.** $y = -2x - 2$ (Obj. 3.5.2) **18.** $y = -\frac{3}{2}x - \frac{7}{2}$ (Obj. 3.6.1) **19.** $\frac{1}{2} + \frac{\sqrt{3}}{6}i$ and $\frac{1}{2} - \frac{\sqrt{3}}{6}i$ (Obj. 8.2.2) **20.** 3 (Obj. 7.5.1) **21.** -3 (Obj. 8.7.1) **22.** $\{1, 2, 4, 5\}$ (Obj. 3.2.1) **23.** Yes (Obj. 3.2.1) **24.** 2 (Obj. 7.5.1) **25.** $g[h(2)] = 10$ (Obj. 9.3.2) **26.** $f^{-1}(x) = -\frac{1}{3}x + 3$ (Obj. 9.4.2) **27.** The cost per pound of the mixture is \$3.96. (Obj. 2.3.1) **28.** 25 lb of the 80% copper alloy must be used. (Obj. 2.4.2) **29.** An additional 4.5 oz of insecticide is required. (Obj. 6.5.1) **30.** It would take the larger pipe 4 min to fill the tank. (Obj. 6.4.2) **31.** A force of 40 lb will stretch the spring 24 in. (Obj. 6.5.2) **32.** The frequency is 80 vibrations per minute. (Obj. 6.5.2)

ANSWERS to Chapter 10 Odd-Numbered Exercises

CONCEPT REVIEW 10.1

1. Never true **3.** Never true **5.** Never true

SECTION 10.1

3. **a.** 9 **b.** 1 **c.** $\frac{1}{9}$ **5.** **a.** 16 **b.** 4 **c.** $\frac{1}{4}$ **7.** **a.** 1 **b.** $\frac{1}{8}$ **c.** 16 **9.** **a.** 7.3891 **b.** 0.3679 **c.** 1.2840 **11.** **a.** 54.5982 **b.** 1 **c.** 0.1353 **13.** **a.** 16 **b.** 16 **c.** 1.4768 **15.** **a.** 0.1353 **b.** 0.1353 **c.** 0.0111 **17.** 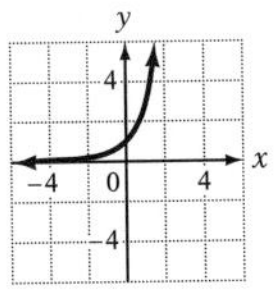
19. 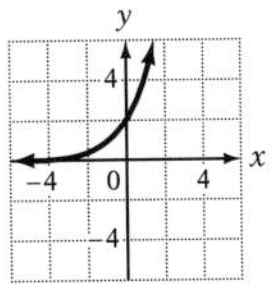
21. 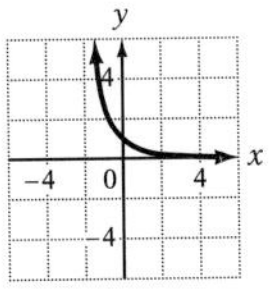

23. 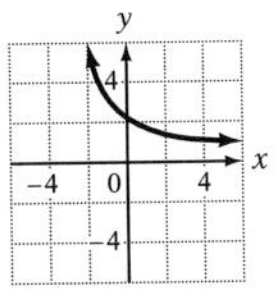
25. 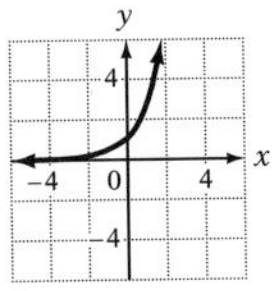
27. 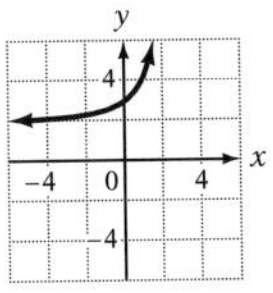
29. 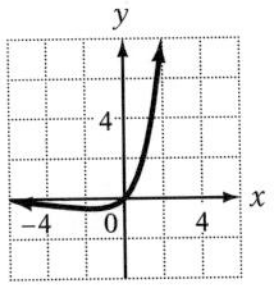
31. 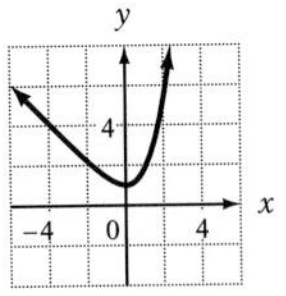

33. 1.6

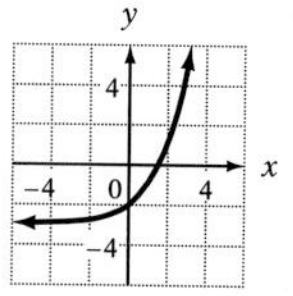

35. 1.1

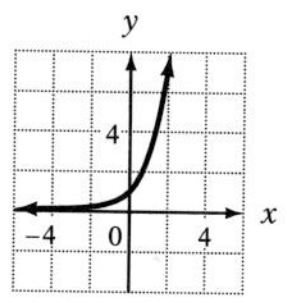

37.

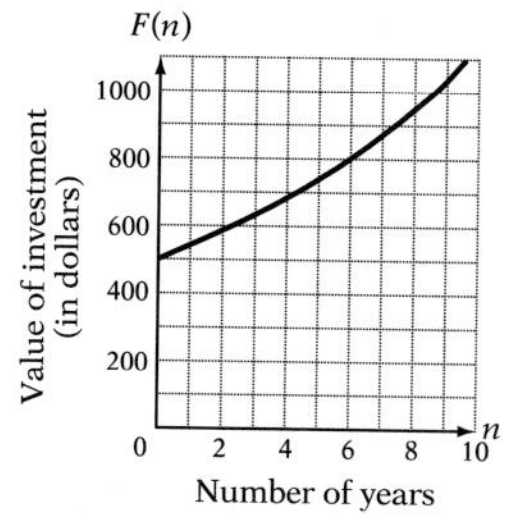

The investment will be worth \$1000 in 9 years.

39.

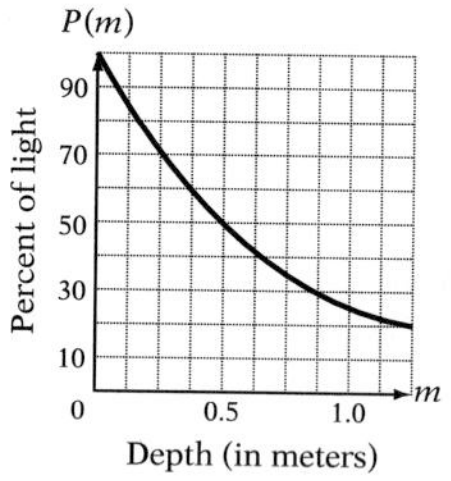

50% of the light will reach to a depth of 0.5 m.

41. 0

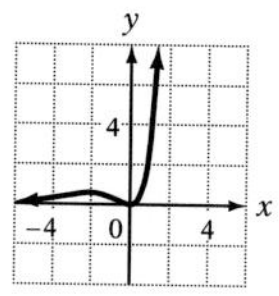

43. 0.80

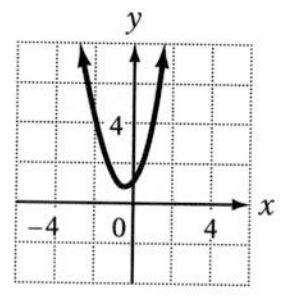

45. As n increases, $\left(1 + \frac{1}{n}\right)^n$ becomes closer to e.

47. **a.**

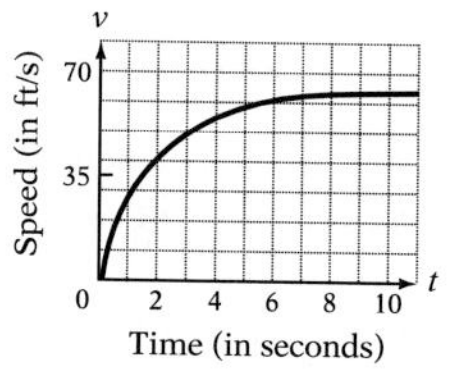

b. The point (4, 55.3) means that after 4 s, the object is moving at a speed of 55.3 ft/s.

CONCEPT REVIEW 10.2

1. Always true **3.** Never true **5.** Always true **7.** Always true

SECTION 10.2

3. $\log_5 25 = 2$ **5.** $\log_4 \frac{1}{16} = -2$ **7.** $\log_{10} x = y$ **9.** $\log_a w = x$ **11.** $3^2 = 9$ **13.** $10^{-2} = 0.01$
15. $e^y = x$ **17.** $b^v = u$ **19.** $\log_3 81 = 4$ **21.** $\log_2 128 = 7$ **23.** $\log 100 = 2$ **25.** $\ln e^3 = 3$
27. $\log_8 1 = 0$ **29.** $\log_5 625 = 4$ **31.** 9 **33.** 64 **35.** $\frac{1}{7}$ **37.** 1 **39.** 316.23 **41.** 0.02
43. 7.39 **45.** 0.61 **49.** $\log_8 x + \log_8 z$ **51.** $5 \log_3 x$ **53.** $\ln r - \ln s$ **55.** $2 \log_3 x + 6 \log_3 y$
57. $3 \log_7 u - 4 \log_7 v$ **59.** $2 \log_2 r + 2 \log_2 s$ **61.** $2 \log_9 x + \log_9 y + \log_9 z$ **63.** $\ln x + 2 \ln y - 4 \ln z$
65. $2 \log_8 x - \log_8 y - 2 \log_8 z$ **67.** $\frac{1}{2} \log_7 x + \frac{1}{2} \log_7 y$ **69.** $\frac{1}{2} \log_2 x - \frac{1}{2} \log_2 y$ **71.** $\frac{3}{2} \ln x + \frac{1}{2} \ln y$
73. $\frac{3}{2} \log_7 x - \frac{1}{2} \log_7 y$ **75.** $\log_3 \frac{x^3}{y}$ **77.** $\log_8 x^4 y^2$ **79.** $\ln x^3$ **81.** $\log_5 x^3 y^4$ **83.** $\log_4 \frac{1}{x^2}$
85. $\log_3 \frac{x^2 z^2}{y}$ **87.** $\log_b \frac{x}{y^2 z}$ **89.** $\ln x^2 y^2$ **91.** $\log_6 \sqrt{\frac{x}{y}}$ **93.** $\log_4 \frac{s^2 r^2}{t^4}$ **95.** $\log_5 \frac{x}{y^2 z^2}$ **97.** $\ln \frac{t^3 v^2}{r^2}$
99. $\log_4 \sqrt{\frac{x^3 z}{y^2}}$ **101.** 1.3863 **103.** 1.0415 **105.** 0.8617 **107.** 2.1133 **109.** -0.6309 **111.** 0.2727
113. 1.6826 **115.** 1.9266 **117.** $\frac{\log(3x - 2)}{\log 3}$ **119.** $\frac{\log(4 - 9x)}{\log 8}$ **121.** $\frac{5}{\log 9} \log(6x + 7)$ **123.** $\frac{\ln(x + 5)}{\ln 2}$
125. $\frac{\ln(x^2 + 9)}{\ln 3}$ **127.** $\frac{7}{\ln 8} \ln(10x - 7)$ **129.** 8 **131.** 6 **133.** 2 **135.** 3 **139.** **a.** b **b.** b
141. 13.12 **143.** -6.59 **145.** 15.1543

CONCEPT REVIEW 10.3

1. Never true **3.** Always true

SECTION 10.3

3. 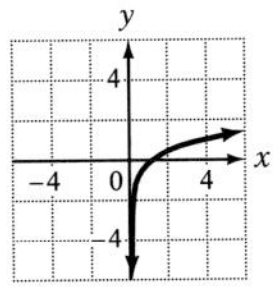
5. 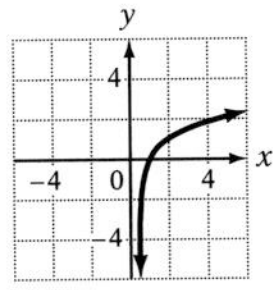
7. 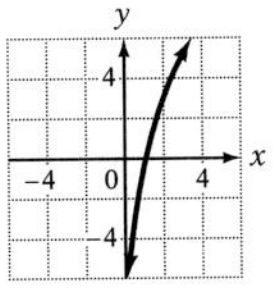
9. 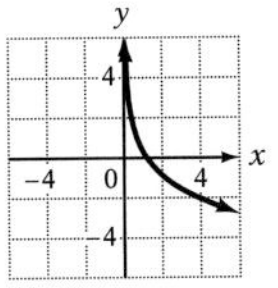
11. 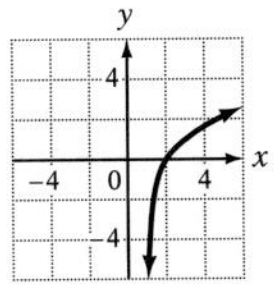
13. 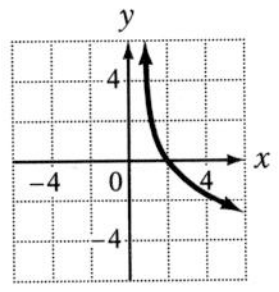
15. 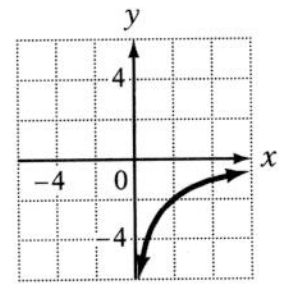
17. 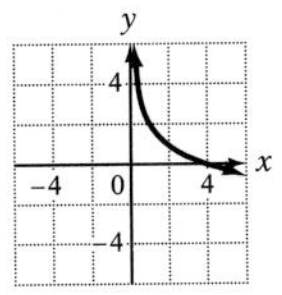
19. 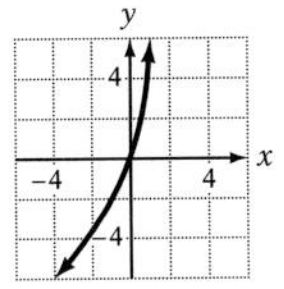
21. 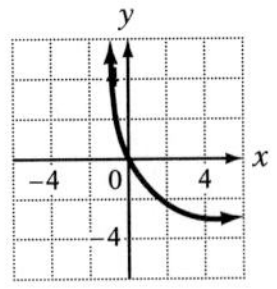
23. 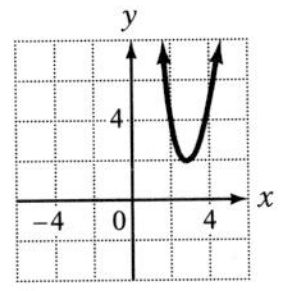
25. 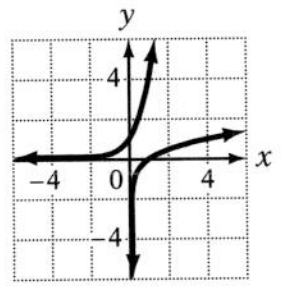
27. 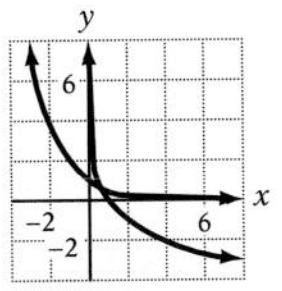
29. $\{x \mid x > -2\}$ **31.** $\{x \mid x \text{ is a real number}\}$

33. $\{x \mid x < -2 \text{ or } x > 0\}$ **35.** $-\ln x + 2$ **37.** $\dfrac{e^{x-3}}{2}$ **39.** 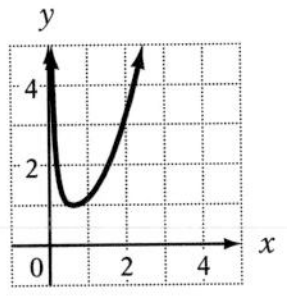
0.8

41. **a.** 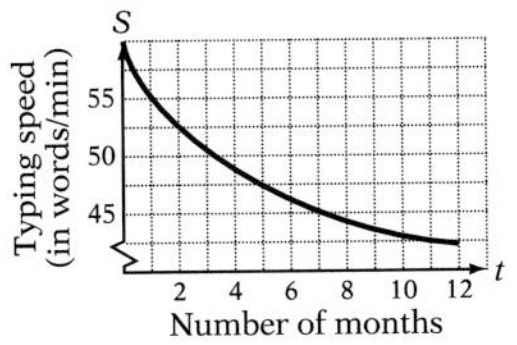

b. The point (4, 49) means that after 4 months, the typist's proficiency has dropped to 49 words per minute.

43. **a.** 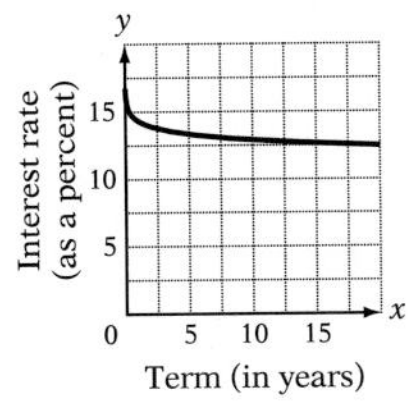

b. A security that has a yield of 13% has a term of 8.5 years.
c. The model predicts an interest rate of 12.2%.

CONCEPT REVIEW 10.4

1. Always true **3.** Never true **5.** Always true

SECTION 10.4

3. 1 **5.** -3 **7.** 1.1133 **9.** 0.7211 **11.** -1.5850 **13.** 1.7095 **15.** 1.3222 **17.** -2.8074 **19.** 3.5850 **21.** 1.1309 **23.** 1 **25.** 6 **27.** 1.3754 and -1.3754 **29.** 1.5805 **31.** -0.6309 **33.** 1.6610 **35.** -1.5129 **37.** 1.5 **39.** 0.63 **41.** -1.86 and 3.44 **43.** -1.16 **47.** 8 **49.** $\frac{11}{2}$

51. 2 and -4 **53.** $\frac{5}{3}$ **55.** $\frac{1}{2}$ **57.** 1,000,000,000 **59.** 4 **61.** 3 **63.** 3 **65.** 2
67. No solution **69.** 4 **71.** $\frac{5 \pm \sqrt{33}}{2}$ **73.** 1.76 **75.** 2.42 **77.** -1.51 and 2.10 **79.** 1.7233
81. 0.8060 **83.** 5.6910 **85.** (3, 1) **87.** (502, 498) **89.** $(-4, -9)$
91. **a.**

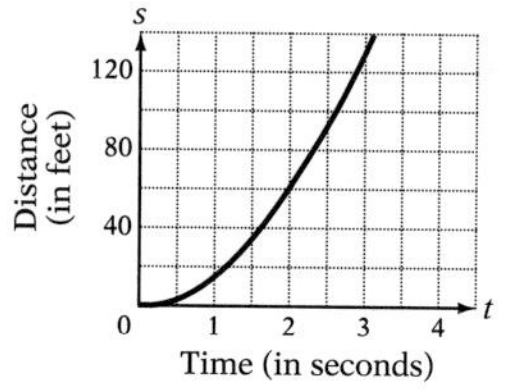

b. It will take the object approximately 2.64 s to fall 100 ft.

CONCEPT REVIEW 10.5

1. Always true **3.** Always true

SECTION 10.5

3. The value of the investment is $11,202.50 after 20 years. **5.** The plant manager must deposit $480,495.83. **7.** It will take 3.5 h to decay to 20 mg. **9.** The half-life of the promethium-147 is 2.5 years. **11.** It will take 6 weeks of practice to make 80% of the welds correctly. **13.** The pH of the solution is 1.35. **15.** 79% of the light will pass through the glass. **17.** The aluminum must be 0.4 cm thick. **19.** The magnitude of the earthquake is 5.2. **21.** The Tyrannosaurus weighed approximately 25,190 kg. **23.** The earthquake in Japan was 63.1 times as intense as the San Francisco earthquake. **25.** There are 65 decibels in normal conversation. **27.** The star is 150.7 parsecs from Earth. **29.** No, Antares is 3.5 times farther from Earth than Pollux. **31.** 742.9 billion barrels of oil will last 20 years. **33.** The population will be 9×10^6 after 5 P.M. **35.** The annual interest rate is 8.5%.
37. $r = \frac{1}{0.00411}(e^{T/14.29} - 1)$ **39.** $y = Ae^{(k \ln 2)t}$ **41.** **a.** The rocket requires 14 s to reach a height of 1 mi.
b. The velocity of the rocket is 813 ft/s. **c.** The domain of the velocity function is [0, 96).

CHAPTER REVIEW EXERCISES

1. 2 (Obj. 10.2.1) **2.** $\log_3 \sqrt{\frac{x}{y}}$ (Obj. 10.2.2) **3.** 1 (Obj. 10.1.1) **4.** -3 (Obj. 10.4.1) **5.** 1 (Obj. 10.1.1) **6.** $\frac{1}{9}$ (Obj. 10.2.1) **7.** $\log_2 32 = 5$ (Obj. 10.2.1) **8.** 6 (Obj. 10.4.2)
9. $\frac{1}{2}\log_6 x + \frac{3}{2}\log_6 y$ (Obj. 10.2.2) **10.** 2 (Obj. 10.4.1) **11.** -2 (Obj. 10.4.1) **12.** $\frac{1}{3}$ (Obj. 10.1.1)
13. 4 (Obj. 10.2.1) **14.** 2 (Obj. 10.4.2) **15.** 2.3219 (Obj. 10.2.2) **16.** 1.7251 (Obj. 10.2.2) **17.** 3 (Obj. 10.4.1) **18.** 4 (Obj. 10.1.1) **19.** $\frac{1}{2}\log_5 x - \frac{1}{2}\log_5 y$ (Obj. 10.2.2) **20.** $\frac{1}{4}$ (Obj. 10.4.2) **21.** 125 (Obj. 10.2.1) **22.** $\frac{1}{2}$ (Obj. 10.4.2) **23.** $\log_b \frac{x^3}{y^5}$ (Obj. 10.2.2) **24.** 4 (Obj. 10.1.1) **25.** 2.6801 (Obj. 10.2.2) **26.** -0.5350 (Obj. 10.4.1) **27.**

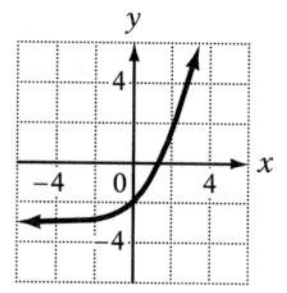

(Obj. 10.1.2) **28.**

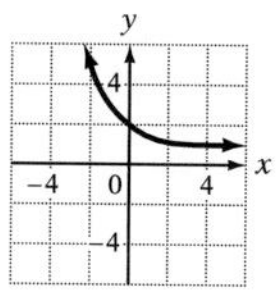

(Obj. 10.1.2)

29. 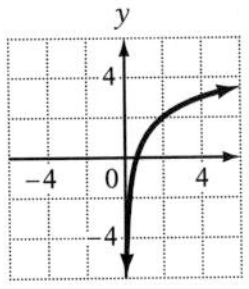 (Obj. 10.3.1) **30.** 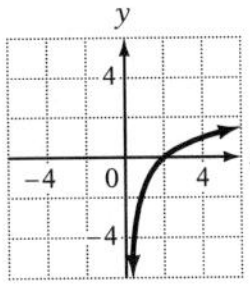 (Obj. 10.3.1) **31.** 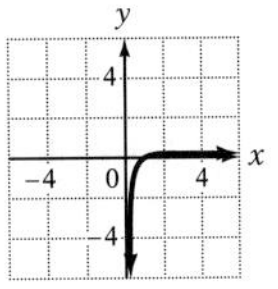 1.0 (Obj. 10.3.1)

32. The half-life is 33 h. (Obj. 10.5.1) **33.** The material must be 0.602 cm thick. (Obj. 10.5.1)

CHAPTER TEST

1. 1 (Obj. 10.1.1) **2.** $\frac{1}{64}$ (Obj. 10.1.1) **3.** 3 (Obj. 10.2.1) **4.** $\frac{1}{16}$ (Obj. 10.2.1)
5. $\frac{2}{3}\log_5 x + \frac{5}{3}\log_5 y$ (Obj. 10.2.2) **6.** $\log_5 \sqrt{\frac{x}{y}}$ (Obj. 10.2.2) **7.** 3 (Obj. 10.4.2) **8.** $\frac{1}{3}$ (Obj. 10.1.1)
9. 2.5789 (Obj. 10.4.1) **10.** 10 (Obj. 10.4.2) **11.** 3 (Obj. 10.4.1) **12.** −4 (Obj. 10.4.1) **13.** $\frac{3}{2}$ (Obj. 10.4.2) **14.** 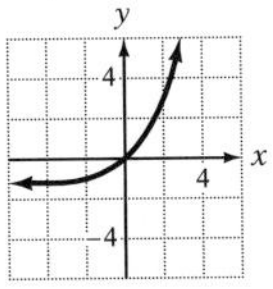 (Obj. 10.1.2) **15.** 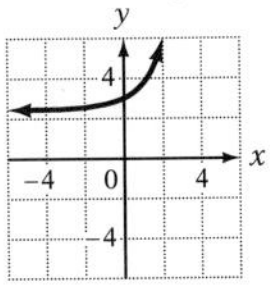 (Obj. 10.1.2)

16. 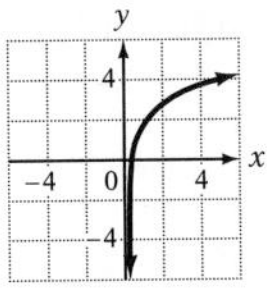 (Obj. 10.3.1) **17.** 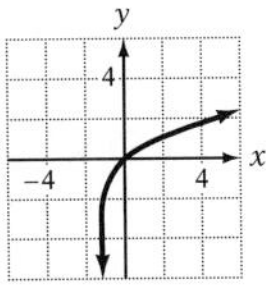 (Obj. 10.3.1) **18.** 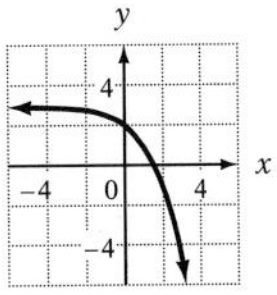 1.6 (Obj. 10.1.2)

19. The value of the investment after 6 years is $15,661. (Obj. 10.5.1) **20.** The half-life is 24 h. (Obj. 10.5.1)

CUMULATIVE REVIEW EXERCISES

1. $\frac{8}{7}$ (Obj. 2.1.2) **2.** $L = \frac{S - 2WH}{2W + 2H}$ (Obj. 6.6.1) **3.** $\{x \mid 1 \le x \le 4\}$ (Obj. 2.6.2) **4.** $(4x^n + 3)(x^n + 1)$ (Obj. 5.6.3) **5.** $\{x \mid -5 \le x \le 1\}$ (Obj. 8.5.1) **6.** $\frac{x - 3}{x + 3}$ (Obj. 6.3.1) **7.** $\frac{x\sqrt{y} + y\sqrt{x}}{x - y}$ (Obj. 7.2.4)
8. $-4x^2y^3\sqrt{2x}$ (Obj. 7.2.2) **9.** $-\frac{1}{5} + \frac{2}{5}i$ (Obj. 7.3.4) **10.** $y = 2x - 6$ (Obj. 3.6.1) **11.** $3x^2 + 8x - 3 = 0$ (Obj. 8.1.2) **12.** $2 + \sqrt{10}$ and $2 - \sqrt{10}$ (Obj. 8.2.1, 8.2.2) **13.** $\{-6, -4, 0\}$ (Obj. 3.2.1) **14.** 4 (Obj. 9.3.2)
15. (0, −1, 2) (Obj. 4.2.2) **16.** $\left(-\frac{1}{2}, -2\right)$ (Obj. 4.1.2) **17.** 243 (Obj. 10.1.1) **18.** 64 (Obj. 10.2.1)
19. 8 (Obj. 10.4.1) **20.** 1 (Obj. 10.4.2) **21.** 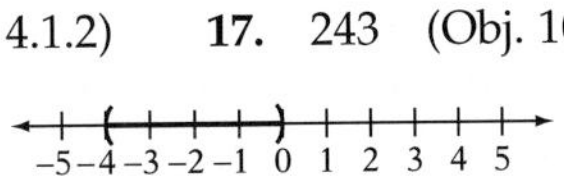 (Obj. 1.1.2)

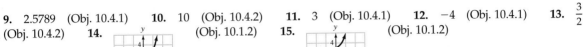

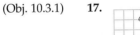

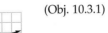

22. 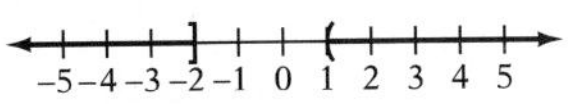 (Obj. 8.5.2) **23.** 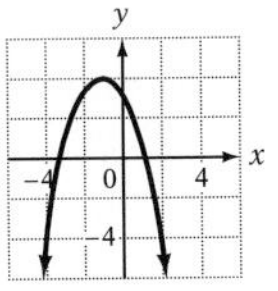 (Obj. 8.6.1) **24.** 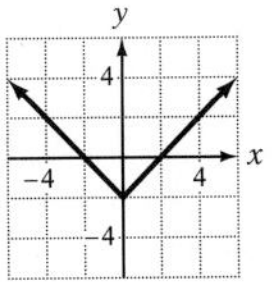 (Obj. 9.1.1)

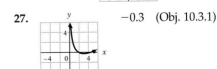

25. 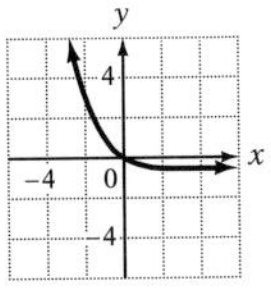 (Obj. 10.1.2) **26.** 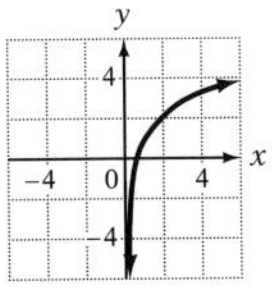 (Obj. 10.3.1) **27.** 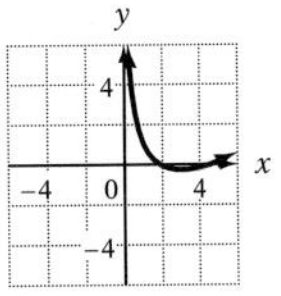 −0.3 (Obj. 10.3.1)

28. 800 lb of the alloy containing 25% tin and 1200 lb of the alloy containing 50% tin were used. (Obj. 2.4.2)

29. To earn \$3000 or more a month, the sales executive must make sales amounting to \$31,250 or more. (Obj. 2.5.3) **30.** With both printers operating, it would take 7.5 min to print the checks. (Obj. 6.4.2) **31.** When the volume is 25 ft^3, the pressure is 500 lb/in^2. (Obj. 6.5.2) **32.** The cost per yard of the wool carpet is \$12. (Obj. 4.4.2) **33.** The value of the investment after 5 years is \$15,657. (Obj. 10.5.1)

ANSWERS to Chapter 11 Odd-Numbered Exercises

CONCEPT REVIEW 11.1

1. Always true **3.** Always true

SECTION 11.1

3. 2, 3, 4, 5 **5.** 3, 5, 7, 9 **7.** 0, −2, −4, −6 **9.** 2, 4, 8, 16 **11.** 2, 5, 10, 17 **13.** $\frac{1}{2}, \frac{2}{5}, \frac{3}{10}, \frac{4}{17}$ **15.** $0, \frac{3}{2}, \frac{8}{3}, \frac{15}{4}$ **17.** 1, −2, 3, −4 **19.** $\frac{1}{2}, -\frac{1}{5}, \frac{1}{10}, -\frac{1}{17}$ **21.** −2, 4, −8, 16 **23.** $\frac{2}{9}, \frac{2}{27}, \frac{2}{81}, \frac{2}{243}$ **25.** 15 **27.** $\frac{12}{13}$ **29.** 24 **31.** $\frac{32}{243}$ **33.** 88 **35.** $\frac{1}{20}$ **37.** 38 **41.** 42 **43.** 28 **45.** 60 **47.** $\frac{25}{24}$ **49.** 28 **51.** $\frac{99}{20}$ **53.** $\frac{137}{120}$ **55.** −2 **57.** $-\frac{7}{12}$ **59.** $\frac{2}{x} + \frac{4}{x} + \frac{6}{x} + \frac{8}{x}$ **61.** $\frac{x}{2} + \frac{x^2}{3} + \frac{x^3}{4} + \frac{x^4}{5}$ **63.** $\frac{x^2}{3} + \frac{x^3}{5} + \frac{x^4}{7}$ **65.** $x + x^3 + x^5 + x^7$ **67.** $a_n = 2n - 1$ **69.** $a_n = -2n + 1$ **71.** $a_n = 4n$ **73.** log 384 **75.** 3 **77.** 2, 3, 5 **79.** $\sum_{i=1}^{n} \frac{1}{i}$

CONCEPT REVIEW 11.2

1. Sometimes true **3.** Sometimes true

SECTION 11.2

3. 141 **5.** 50 **7.** 71 **9.** $\frac{27}{4}$ **11.** 17 **13.** 3.75 **15.** $a_n = n$ **17.** $a_n = -4n + 10$ **19.** $a_n = \frac{3n + 1}{2}$ **21.** $a_n = -5n - 3$ **23.** $a_n = -10n + 36$ **25.** 42 **27.** 16 **29.** 20 **31.** 20 **33.** 13 **35.** 20 **37.** 650 **39.** −605 **41.** $\frac{215}{4}$ **43.** 420 **45.** −210 **47.** −5 **49.** In 9 weeks the person will walk 60 min per day. **51.** There are 2180 seats in the theater. **53.** The salary for the tenth month is \$3150. The total salary for the 10-month period is \$24,750. **55.** The nth distance is $a_n = 7.66n + 392.19$. **57.** 8 **59.** −3 **61.** The sum of the angles of a dodecagon is 1800°. The formula for the sum of the angles of an n-sided polygon is $180(n - 2)$. **63.** $a_n = 2n - 1$

CONCEPT REVIEW 11.3

1. Always true **3.** Sometimes true **5.** Never true

SECTION 11.3

3. 131,072 **5.** $\frac{128}{243}$ **7.** 16 **9.** 6, 4 **11.** $-2, \frac{4}{3}$ **13.** 12, −48 **15.** 2186 **17.** $\frac{2343}{64}$ **19.** 62 **21.** $\frac{121}{243}$ **23.** 1364 **25.** 2800 **27.** $\frac{2343}{1024}$ **29.** $\frac{1360}{81}$ **31.** 9 **33.** $\frac{18}{5}$ **35.** $\frac{7}{9}$ **37.** $\frac{8}{9}$ **39.** $\frac{2}{9}$ **41.** $\frac{5}{11}$ **43.** $\frac{1}{6}$ **45.** The nth term is $a_n = 3^{n-1}$. **47.** There will be 7.8125 mg of radioactive material in the sample at the beginning of the seventh day. **49.** The ball bounces to a height of 2.6 ft on the fifth bounce. **51.** The value of the land in 15 years will be \$82,103.49. **53.** The value of the house in 30 years will be \$432,194.24. **55.** G; $-\frac{1}{2}$ **57.** A; 9.5 **59.** N; 25 **61.** G; x^2 **63.** A; 4 log x **65.** 27 **67.** The common difference is log 2. **69.** The common ratio is 8. **71.** **a.** \$80.50 of the loan is repaid in the 27th payment. **b.** The total amount repaid after 20 payments is \$1424.65. **c.** The unpaid balance on the loan after 20 payments is \$3575.35.

CONCEPT REVIEW 11.4

1. Never true **3.** Never true **5.** Never true

SECTION 11.4

3. 6 **5.** 40,320 **7.** 1 **9.** 10 **11.** 1 **13.** 84 **15.** 21 **17.** 45 **19.** 1 **21.** 20 **23.** 11 **25.** 6 **27.** $x^4 + 4x^3y + 6x^2y^2 + 4xy^3 + y^4$ **29.** $x^5 - 5x^4y + 10x^3y^2 - 10x^2y^3 + 5xy^4 - y^5$ **31.** $16m^4 + 32m^3 + 24m^2 + 8m + 1$ **33.** $32r^5 - 240r^4 + 720r^3 - 1080r^2 + 810r - 243$ **35.** $a^{10} + 10a^9b + 45a^8b^2$ **37.** $a^{11} - 11a^{10}b + 55a^9b^2$ **39.** $256x^8 + 1024x^7y + 1792x^6y^2$ **41.** $65{,}536x^8 - 393{,}216x^7y + 1{,}032{,}192x^6y^2$ **43.** $x^7 + 7x^5 + 21x^3$ **45.** $x^{10} + 15x^8 + 90x^6$ **47.** $-560x^4$ **49.** $-6x^{10}y^2$ **51.** $126y^5$ **53.** $5n^3$ **55.** $\frac{x^5}{32}$ **57.** 1 7 21 35 35 21 7 1 **59.** n **61.** $x^2 + 8x^{\frac{3}{2}} + 24x + 32x^{\frac{1}{2}} + 16$ **63.** $-8i$ **65.** 1 **67.** 1260

CHAPTER REVIEW EXERCISES

1. $3x + 3x^2 + 3x^3 + 3x^4$ (Obj. 11.1.2) **2.** 16 (Obj. 11.2.1) **3.** 32 (Obj. 11.3.1) **4.** 16 (Obj. 11.3.3) **5.** 84 (Obj. 11.4.1) **6.** $\frac{1}{2}$ (Obj. 11.1.1) **7.** 44 (Obj. 11.2.1) **8.** 468 (Obj. 11.2.2) **9.** −66 (Obj. 11.3.2) **10.** 70 (Obj. 11.4.1) **11.** $2268x^3y^6$ (Obj. 11.4.1) **12.** 34 (Obj. 11.2.2) **13.** $\frac{7}{6}$ (Obj. 11.1.1) **14.** $a_n = -3n + 15$ (Obj. 11.2.1) **15.** $\frac{2}{27}$ (Obj. 11.3.1) **16.** $\frac{7}{30}$ (Obj. 11.3.3) **17.** −115 (Obj. 11.2.1) **18.** $\frac{665}{32}$ (Obj. 11.3.2) **19.** 1575 (Obj. 11.2.2) **20.** $-280x^4y^3$ (Obj. 11.4.1) **21.** 21 (Obj. 11.2.1) **22.** 48 (Obj. 11.3.1) **23.** 30 (Obj. 11.2.2) **24.** 341 (Obj. 11.3.2) **25.** 120 (Obj. 11.4.1) **26.** $240x^4$ (Obj. 11.4.1) **27.** 143 (Obj. 11.2.1) **28.** −575 (Obj. 11.2.2) **29.** $-\frac{5}{27}$ (Obj. 11.1.1) **30.** $2 + 2x + 2x^2 + 2x^3$ (Obj. 11.1.2) **31.** $\frac{23}{99}$ (Obj. 11.3.3) **32.** $\frac{16}{5}$ (Obj. 11.3.3) **33.** 726 (Obj. 11.3.2) **34.** $-42{,}240x^4y^7$ (Obj. 11.4.1) **35.** 0.996 (Obj. 11.3.2) **36.** 6 (Obj. 11.3.3) **37.** $\frac{19}{30}$ (Obj. 11.3.3) **38.** $x^5 - 15x^4y^2 + 90x^3y^4 - 270x^2y^6 + 405xy^8 - 243y^{10}$ (Obj. 11.4.1) **39.** 22 (Obj. 11.2.1) **40.** 99 (Obj. 11.4.1) **41.** $2x + 2x^2 + \frac{8x^3}{3} + 4x^4 + \frac{32x^5}{5}$ (Obj. 11.1.2) **42.** $-\frac{13}{60}$ (Obj. 11.1.2) **43.** The total salary for the 9-month period is \$12,240. (Obj. 11.2.3) **44.** The temperature is 67.7°F. (Obj. 11.3.4)

CHAPTER TEST

1. $\frac{1}{3}$ (Obj. 11.1.1) **2.** $\frac{8}{9}, \frac{9}{10}$ (Obj. 11.1.1) **3.** 32 (Obj. 11.1.2) **4.** $2x^2 + 2x^4 + 2x^6 + 2x^8$ (Obj. 11.1.2) **5.** -120 (Obj. 11.2.1) **6.** $a_n = 2n - 5$ (Obj. 11.2.1) **7.** 22 (Obj. 11.2.1) **8.** 315 (Obj. 11.2.2) **9.** 1560 (Obj. 11.2.2) **10.** 252 (Obj. 11.4.1) **11.** $-64\sqrt{2}$ (Obj. 11.3.1) **12.** $\frac{81}{125}$ (Obj. 11.3.1) **13.** $\frac{781}{256}$ (Obj. 11.3.2) **14.** -55 (Obj. 11.3.2) **15.** 4 (Obj. 11.3.3) **16.** $\frac{7}{30}$ (Obj. 11.3.3) **17.** 330 (Obj. 11.4.1) **18.** $5670x^4y^4$ (Obj. 11.4.1) **19.** The inventory after the October 1 shipment was 2550 yd. (Obj. 11.2.3) **20.** There will be 20 mg of radioactive material in the sample at the beginning of the fifth day. (Obj. 11.3.4)

CUMULATIVE REVIEW EXERCISES

1. $\frac{x^2 + 5x - 2}{(x + 2)(x - 1)}$ (Obj. 6.2.2) **2.** $2(x^2 + 2)(x^4 - 2x^2 + 4)$ (Obj. 5.6.4) **3.** $4y\sqrt{x} - y\sqrt{2}$ (Obj. 7.2.3) **4.** $\frac{1}{x^{26}}$ (Obj. 7.1.1) **5.** 4 (Obj. 7.5.1) **6.** $\frac{1}{4} - \frac{\sqrt{55}}{4}i$ and $\frac{1}{4} + \frac{\sqrt{55}}{4}i$ (Obj. 8.2.2) **7.** $\left(\frac{7}{6}, \frac{1}{2}\right)$ (Obj. 4.2.1) **8.** $\{x \mid x < -2 \text{ or } x > 2\}$ (Obj. 2.5.2) **9.** -10 (Obj. 4.3.1) **10.** $\frac{1}{2}\log_5 x - \frac{1}{2}\log_5 y$ (Obj. 10.2.2) **11.** 3 (Obj. 10.4.1) **12.** $a_5 = 20; a_6 = 30$ (Obj. 11.1.1) **13.** 6 (Obj. 11.1.2) **14.** (2, 0, 1) (Obj. 4.2.2) **15.** 216 (Obj. 10.4.2) **16.** $2x^2 - x - 1 + \frac{6}{2x + 1}$ (Obj. 5.4.1) **17.** $-3h + 1$ (Obj. 3.2.1) **18.** $\left\{-1, 0, \frac{7}{5}\right\}$ (Obj. 3.2.1)

19. 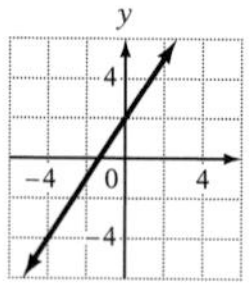

(Obj. 3.3.2) **20.** 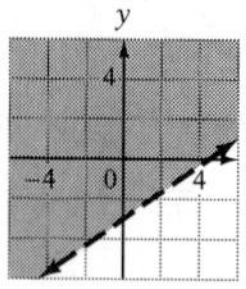

(Obj. 3.7.1)

21. The new computer would take 24 min to complete the payroll. The older computer would take 40 min to complete the payroll. (Obj. 6.4.2) **22.** The rate of the boat in calm water is 6.25 mph. The rate of the current is 1.25 mph. (Obj. 4.4.1) **23.** The half-life is 55 days. (Obj. 10.5.1) **24.** The total number of seats in the theater is 1536. (Obj. 11.2.3) **25.** The height the ball reaches on the fifth bounce is 3.3 ft. (Obj. 11.3.4)

ANSWERS to Chapter 12 Odd-Numbered Exercises

CONCEPT REVIEW 12.1

1. Sometimes true **3.** Sometimes true **5.** Sometimes true

SECTION 12.1

3. Vertex: (1, −5)
Axis of symmetry: $x = 1$

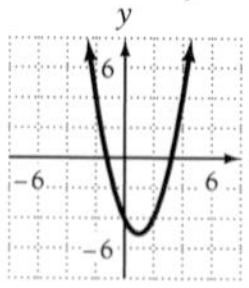

5. Vertex: (1, −2)
Axis of symmetry: $x = 1$

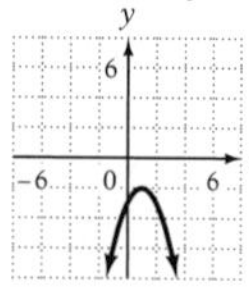

7. Vertex: (−4, −3)
Axis of symmetry: $y = -3$

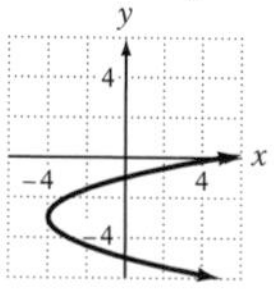

9. Vertex: $(1, -1)$
Axis of symmetry: $x = 1$

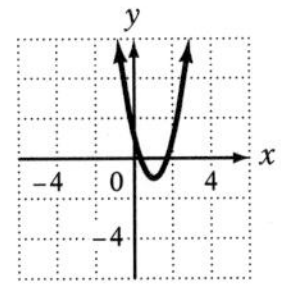

11. Vertex: $\left(\frac{5}{2}, -\frac{9}{4}\right)$
Axis of symmetry: $x = \frac{5}{2}$

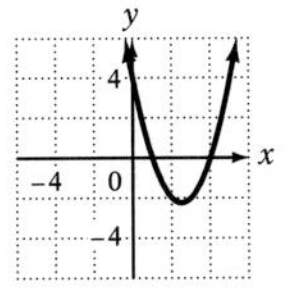

13. Vertex: $(-6, 1)$
Axis of symmetry: $y = 1$

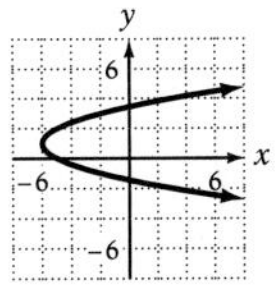

15. Vertex: $\left(-\frac{3}{2}, \frac{27}{4}\right)$
Axis of symmetry: $x = -\frac{3}{2}$

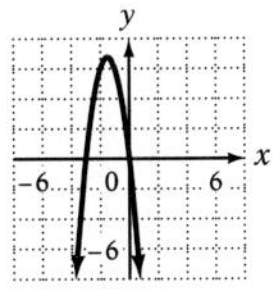

17. Vertex: $(4, 0)$
Axis of symmetry: $y = 0$

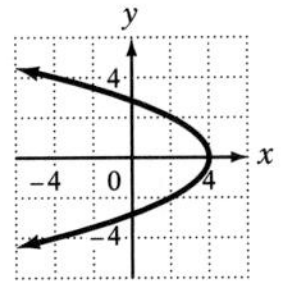

19. Vertex: $\left(\frac{1}{2}, 1\right)$
Axis of symmetry: $y = 1$

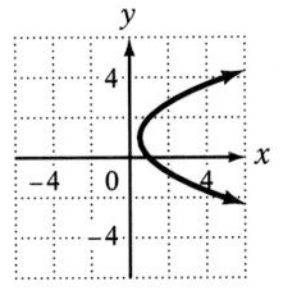

21. Vertex: $(-2, -8)$
Axis of symmetry: $x = -2$

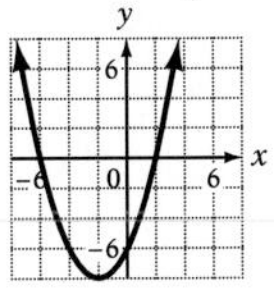

23. Domain: $\{x \mid x \in \text{real numbers}\}$
Range: $\{y \mid y \geq -6\}$

25. Domain: $\{x \mid x \in \text{real numbers}\}$
Range: $\{y \mid y \leq -2\}$

27. Domain: $\{x \mid x \geq -14\}$
Range: $\{y \mid y \in \text{real numbers}\}$

29. Domain: $\{x \mid x \leq 7\}$
Range: $\{y \mid y \in \text{real numbers}\}$

31. $\left(1, -\frac{7}{8}\right)$

33. $(-2, -1)$

35. **a.** The equation of the mirror is $x = \frac{1}{2639}y^2$. **b.** The equation is valid over the interval $\{0 \leq x \leq 3.79\}$.

CONCEPT REVIEW 12.2

1. Sometimes true **3.** Always true **5.** Never true

SECTION 12.2

3.

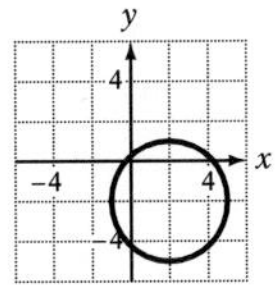

5.

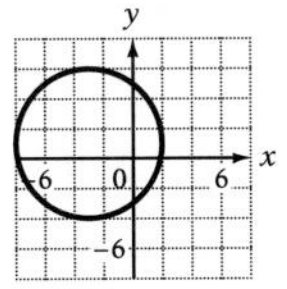

7.

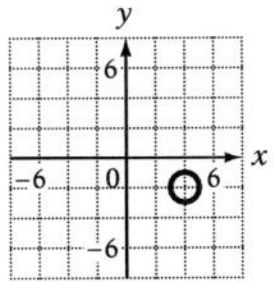

9.

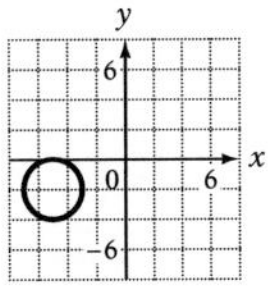

11. $(x - 2)^2 + (y + 1)^2 = 4$

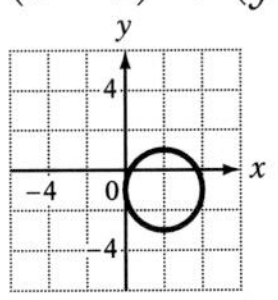

13. $(x + 1)^2 + (y - 1)^2 = 5$

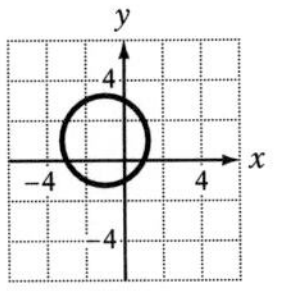

15. $(x + 3)^2 + (y - 6)^2 = 8$

17. $(x + 2)^2 + (y - 1)^2 = 5$

19. $(x-1)^2+(y+2)^2=25$

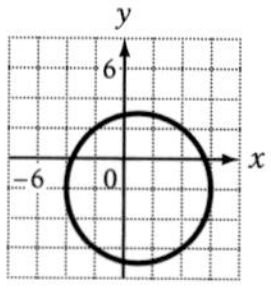

21. $(x+3)^2+(y+4)^2=16$

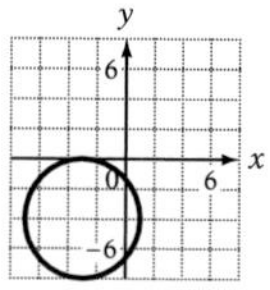

23. $\left(x-\frac{1}{2}\right)^2+(y+2)^2=1$

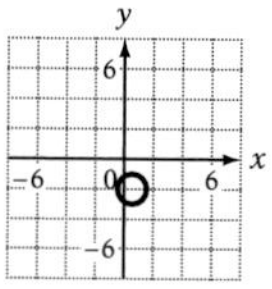

25. $(x-3)^2+(y+2)^2=9$

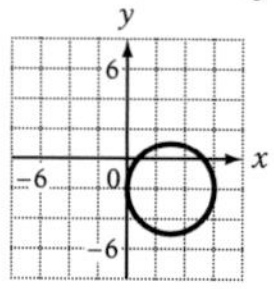

27. $(x-3)^2+y^2=9$ **29.** $(x+1)^2+(y-1)^2=1$

CONCEPT REVIEW 12.3

1. Never true **3.** Always true **5.** Sometimes true

SECTION 12.3

1.

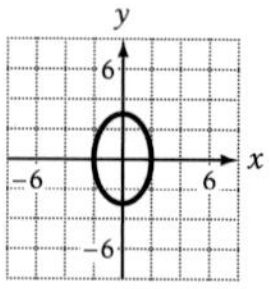

3.

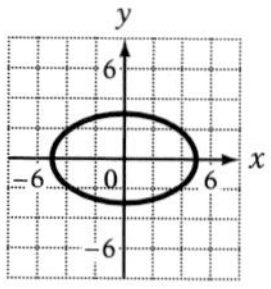

5.

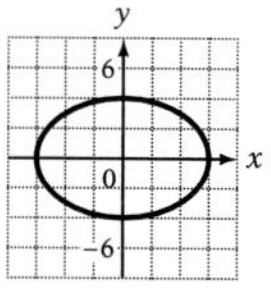

7.

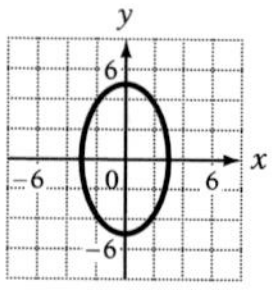

9.

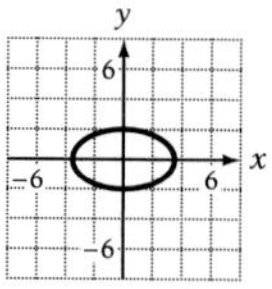

11.

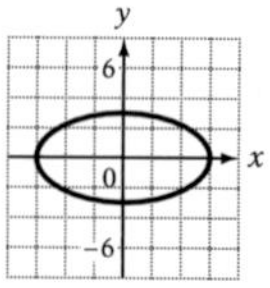

15.

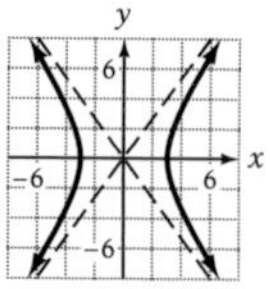

17.

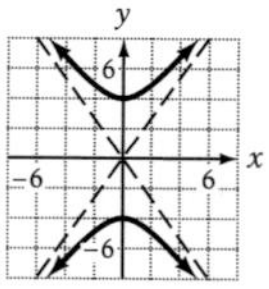

19.

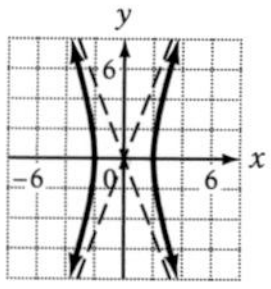

21.

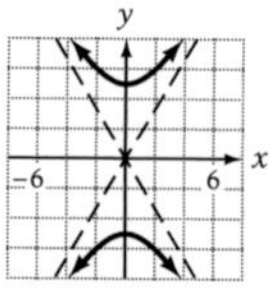

23.

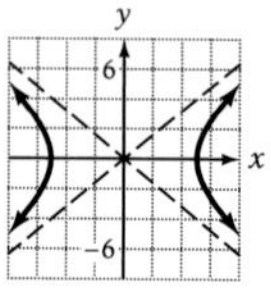

25.

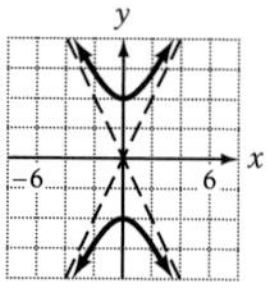

27.

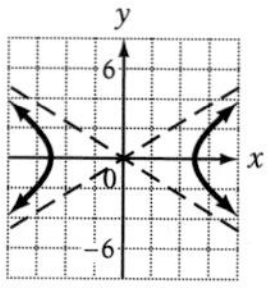

29.

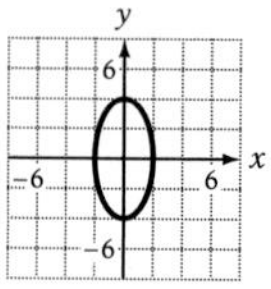

31.

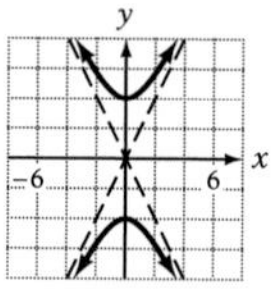

33.

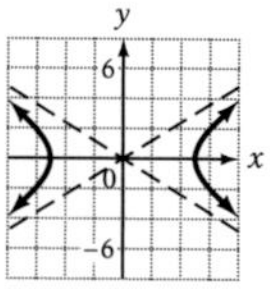

35. **a.** $\frac{x^2}{324}+\frac{y^2}{20.25}=1$ **b.** The aphelion is 3,293,400,000 mi. **c.** The perihelion is 53,100,000 mi.

37. **a.** $\frac{x^2}{2.310}+\frac{x^2}{2.235}=1$ **b.** The aphelion is 166,800,000 mi. **c.** The perihelion is 115,800,000 mi.

CONCEPT REVIEW 12.4

1. Never true **3.** Sometimes true **5.** Always true

SECTION 12.4

3. $(-2, 5)$ and $(5, 19)$ **5.** $(-1, -2)$ and $(2, 1)$ **7.** $(2, -2)$ **9.** $(2, 2)$ and $\left(-\frac{2}{9}, -\frac{22}{9}\right)$ **11.** $(3, 2)$ and $(2, 3)$ **13.** $\left(\frac{\sqrt{3}}{2}, 3\right)$ and $\left(-\frac{\sqrt{3}}{2}, 3\right)$ **15.** No solution **17.** $(1, 2)$, $(1, -2)$, $(-1, 2)$, and $(-1, -2)$ **19.** $(3, 2)$, $(3, -2)$, $(-3, 2)$, and $(-3, -2)$ **21.** $(2, 7)$ and $(-2, 11)$ **23.** $(3, \sqrt{2})$, $(3, -\sqrt{2})$, $(-3, \sqrt{2})$, and $(-3, -\sqrt{2})$ **25.** No solution **27.** $(\sqrt{2}, 3)$, $(\sqrt{2}, -3)$, $(-\sqrt{2}, 3)$, and $(-\sqrt{2}, -3)$ **29.** $(2, -1)$ and $(8, 11)$ **31.** $(1, 0)$ **33.** $(1.000, 2.000)$ **35.** $(1.755, 0.811)$, $(0.505, -0.986)$ **37.** $(5.562, -1.562)$, $(0.013, 3.987)$

CONCEPT REVIEW 12.5

1. Always true **3.** Never true **5.** Sometimes true

SECTION 12.5

1. 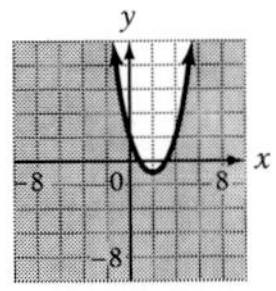
3. 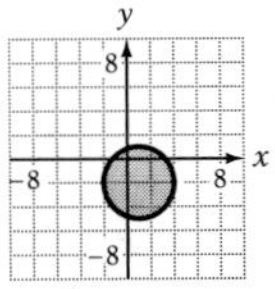
5. 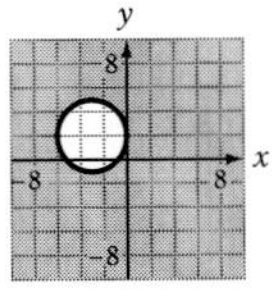
7. 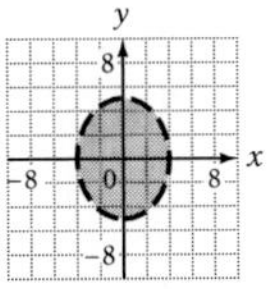
9. 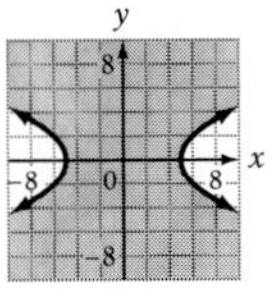
11. 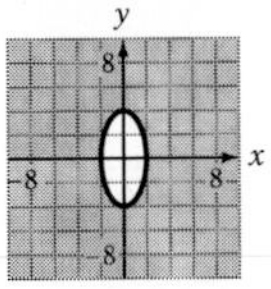
13. 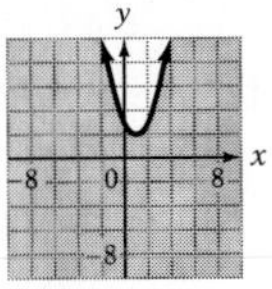
15. 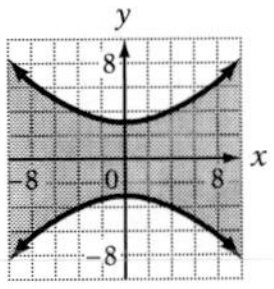
17. 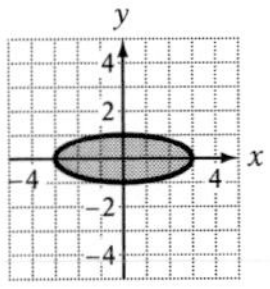
19. 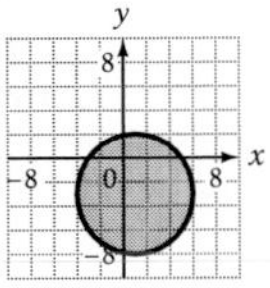
21. 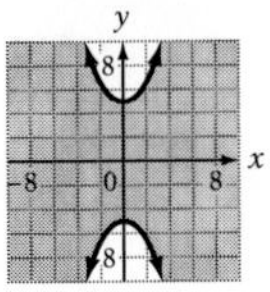
23. 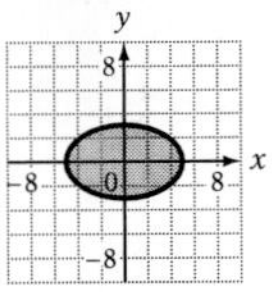
27. 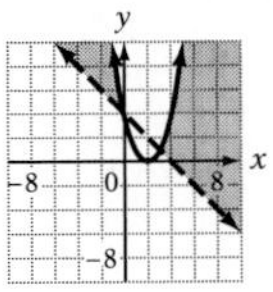
29. 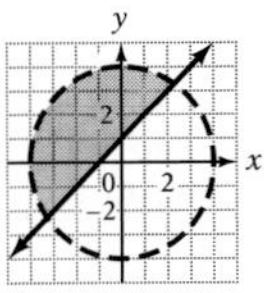
31. 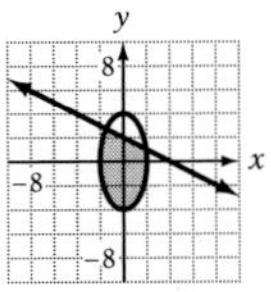
33. 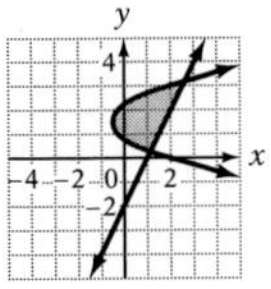
35. 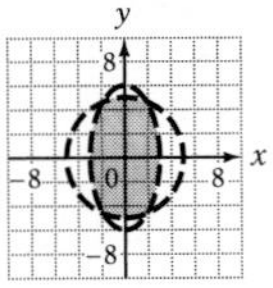
37. 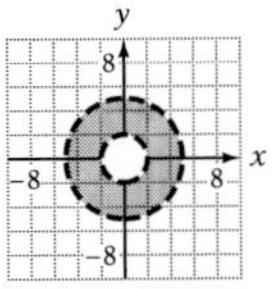
39. 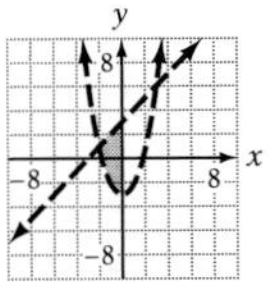
41. 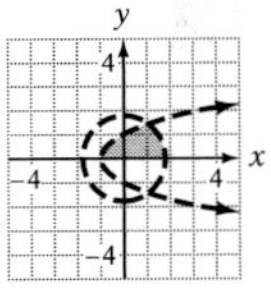
43. 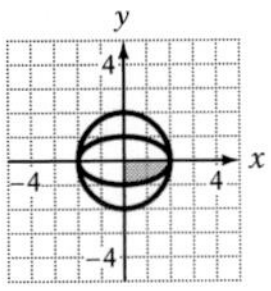
45. 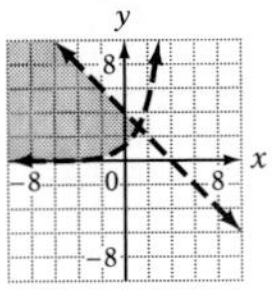
47. 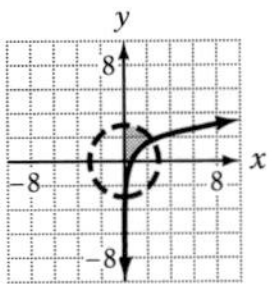
49. 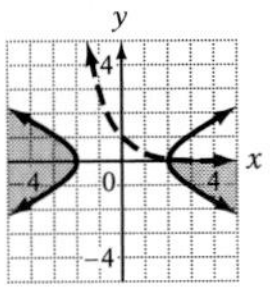

CHAPTER REVIEW EXERCISES

1. Vertex: $(2, 4)$; axis of symmetry: $x = 2$ (Obj. 12.1.1)

2. Vertex: $\left(\frac{7}{2}, \frac{17}{4}\right)$; axis of symmetry: $x = \frac{7}{2}$ (Obj. 12.1.1)

3. 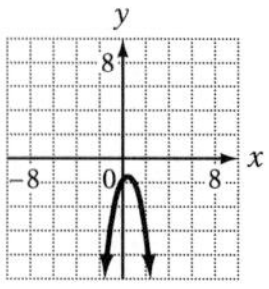

(Obj. 12.1.1)

4. 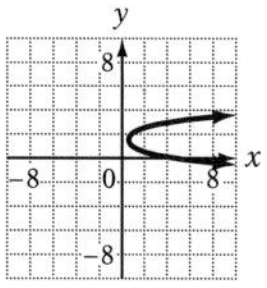

(Obj. 12.1.1)

5. $(x + 1)^2 + (y - 2)^2 = 18$ (Obj. 12.2.1)

6. $(x + 1)^2 + (y - 5)^2 = 36$ (Obj. 12.2.1)

7. 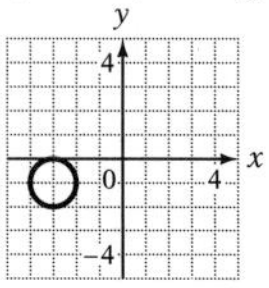

(Obj. 12.2.1)

8. 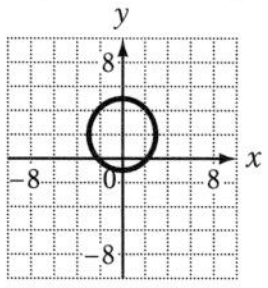

(Obj. 12.2.1)

9. $x^2 + (y + 3)^2 = 97$ (Obj. 12.2.1)

10. $(x + 2)^2 + (y - 1)^2 = 9$ (Obj. 12.2.2)

11. 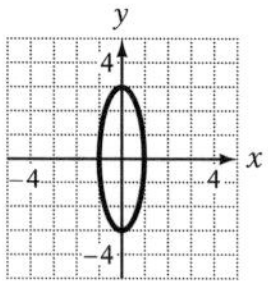

(Obj. 12.3.1)

12. 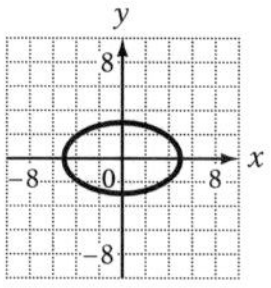

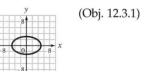

(Obj. 12.3.1)

13. 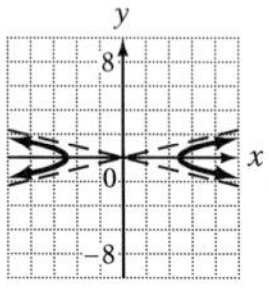

(Obj. 12.3.2)

14. 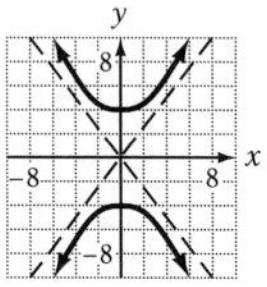

(Obj. 12.3.2)

15. $(-2, -12)$ (Obj. 12.4.1)

16. $(\sqrt{5}, 3), (-\sqrt{5}, 3), (\sqrt{5}, -3)$, and $(-\sqrt{5}, -3)$ (Obj. 12.4.1)

17. $\left(\frac{2}{9}, \frac{1}{3}\right)$ and $\left(1, \frac{3}{2}\right)$ (Obj. 12.4.1)

18. $(1, \sqrt{5})$ and $(1, -\sqrt{5})$ (Obj. 12.4.1)

19. 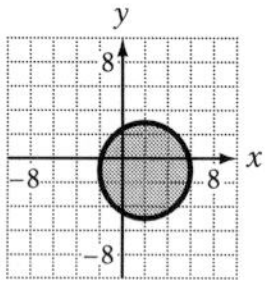

(Obj. 12.5.1)

20. 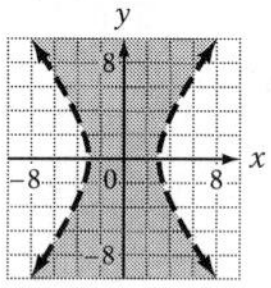

(Obj. 12.5.1)

21. 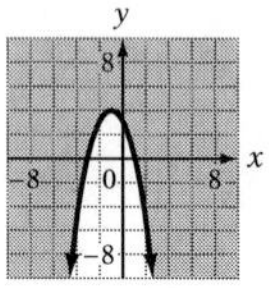

(Obj. 12.5.1)

22. 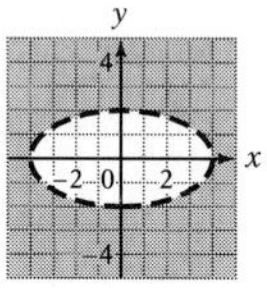

(Obj. 12.5.1)

23. 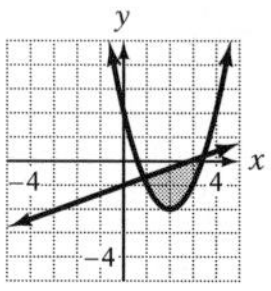

(Obj. 12.5.2)

24. 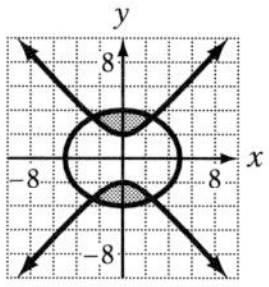

(Obj. 12.5.2)

25. 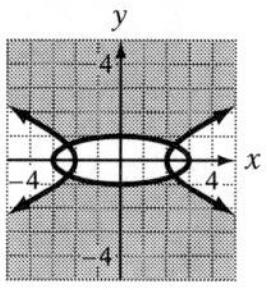

(Obj. 12.5.2)

26. 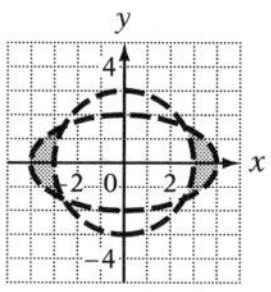

(Obj. 12.5.2)

CHAPTER TEST

1. $x = 3$ (Obj. 12.1.1)

2. $\left(\frac{3}{2}, \frac{1}{4}\right)$ (Obj. 12.1.1)

3. 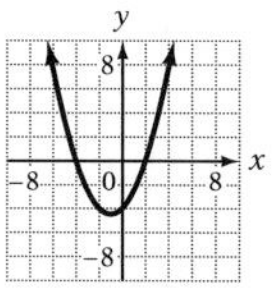

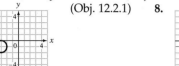

(Obj. 12.1.1)

4. 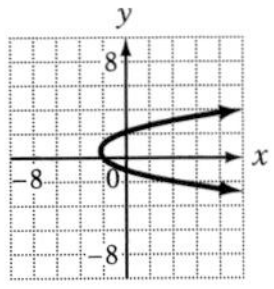 (Obj. 12.1.1) **5.** $(x+3)^2+(y+3)^2=16$ (Obj. 12.2.1) **6.** (2, 0) and $\left(\frac{2}{3}, \frac{4}{3}\right)$ (Obj. 12.4.1)

7. (36, −4) and $\left(-\frac{9}{4}, \frac{1}{2}\right)$ (Obj. 12.4.1) **8.** (5, 1), (−5, 1), (5, −1), and (−5, −1) (Obj. 12.4.1)
9. $(x+1)^2+(y+3)^2=58$ (Obj. 12.2.1) **10.** 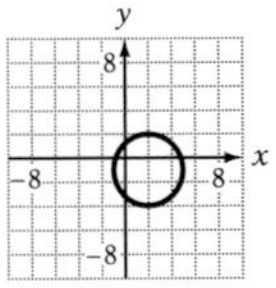 (Obj. 12.2.1)

11. $(x+2)^2+(y-4)^2=9$ (Obj. 12.2.1) **12.** $(x+2)^2+(y-1)^2=32$ (Obj. 12.2.1)
13. $(x-2)^2+(y+1)^2=4$ 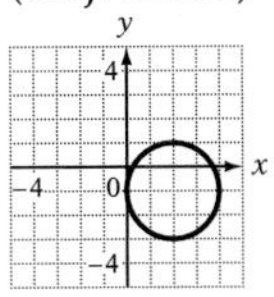 (Obj. 12.2.2) **14.** 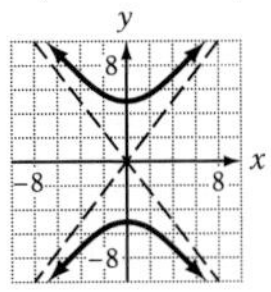 (Obj. 12.3.2)

y-intercept: (0, −3)

15. 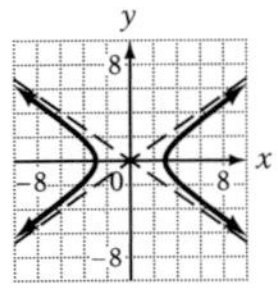 (Obj. 12.3.2) **16.** 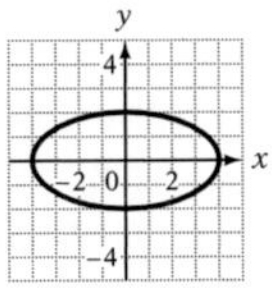 (Obj. 12.3.1) **17.** 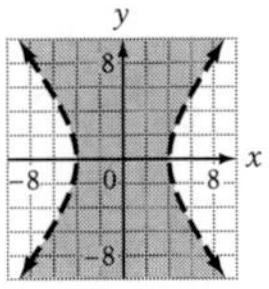 (Obj. 12.5.1)

18. 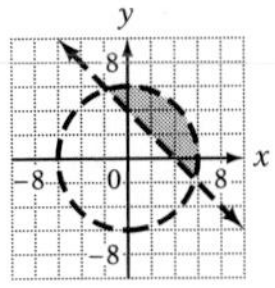 (Obj. 12.5.2) **19.** 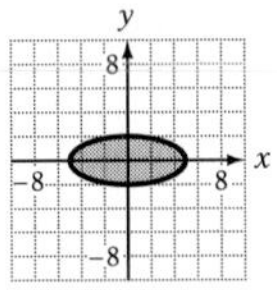 (Obj. 12.5.1) **20.** 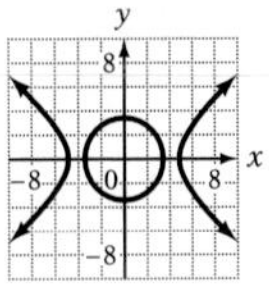 (Obj. 12.5.2)

FINAL EXAM

1. −31 (Obj. 1.2.4) **2.** −1 (Obj. 1.3.2) **3.** $33-10x$ (Obj. 1.3.3) **4.** 8 (Obj. 2.1.1) **5.** $\frac{2}{3}$ (Obj. 2.1.2) **6.** 4 and $-\frac{2}{3}$ (Obj. 2.6.1) **7.** $\{x \mid -4 < x < -1\}$ (Obj. 2.6.2) **8.** $\left\{x \mid x > \frac{3}{2}\right\}$ (Obj. 2.5.2) **9.** $y=-\frac{2}{3}x-\frac{1}{3}$ (Obj. 3.6.1) **10.** $6a^3-5a^2+10a$ (Obj. 5.3.1) **11.** $\frac{6}{5}-\frac{3}{5}i$ (Obj. 7.3.4) **12.** $2x^2-3x-2=0$ (Obj. 8.1.2) **13.** $(2-xy)(4+2xy+x^2y^2)$ (Obj. 5.6.2) **14.** $(x-y)(1+x)(1-x)$ (Obj. 5.6.4) **15.** $x^2-2x-3-\frac{5}{2x-3}$ (Obj. 5.4.1) **16.** $\frac{x(x-1)}{2x-5}$ (Obj. 6.2.1) **17.** $-\frac{10x}{(x+2)(x-3)}$ (Obj. 6.2.2) **18.** $\frac{x+3}{x+1}$ (Obj. 6.3.1) **19.** $-\frac{7}{4}$ (Obj. 6.4.1) **20.** $d=\frac{a_n-a_1}{n-1}$ (Obj. 6.6.1) **21.** $\frac{y^4}{162x^3}$ (Obj. 5.1.2) **22.** $\frac{1}{64x^8y^5}$ (Obj. 7.1.1) **23.** $-2x^2y\sqrt{2y}$ (Obj. 7.2.2) **24.** $\frac{x^2\sqrt{2y}}{2y^2}$ (Obj. 7.2.4) **25.** $\frac{3+\sqrt{17}}{4}$ and $\frac{3-\sqrt{17}}{4}$ (Obj. 8.2.2) **26.** 27 and −8 (Obj. 8.3.1) **27.** $y=-3x+7$ (Obj. 3.5.2) **28.** $\frac{3}{2}$ and −2 (Obj. 8.3.3) **29.** (3, 4) (Obj. 4.2.1) **30.** 10 (Obj. 4.3.1) **31.** 6 (Obj. 10.4.2) **32.** $2y+2y^2+2y^3+2y^4+2y^5$ (Obj. 11.1.2) **33.** $\frac{23}{45}$ (Obj. 11.3.3) **34.** $144x^7y^2$ (Obj. 11.4.1)

35. $\left(\frac{5}{2}, -\frac{3}{2}\right)$ (Obj. 12.4.1) **36.** $f^{-1}(x) = \frac{3}{2}x + 6$ (Obj. 9.4.2) **37.** $\log_2 \frac{a^2}{b^2}$ (Obj. 10.2.2)

38. x-intercept: $\left(\frac{9}{2}, 0\right)$
y-intercept: $(0, -3)$

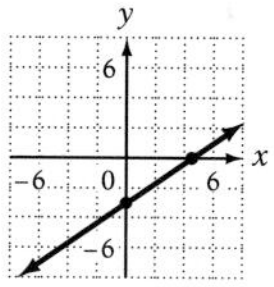

(Obj. 3.3.2) **39.**

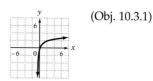

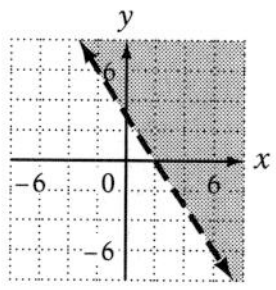

(Obj. 3.7.1)

40.

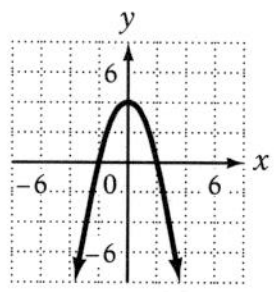

(Obj. 8.6.1/12.1.1) **41.**

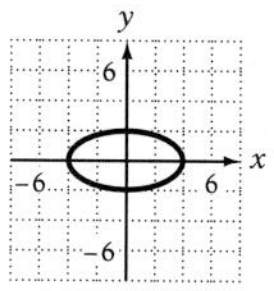

(Obj. 12.3.1) **42.**

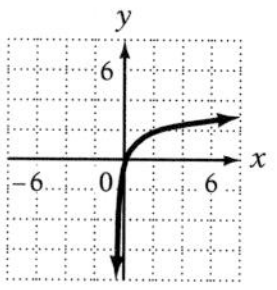

(Obj. 10.3.1)

43.

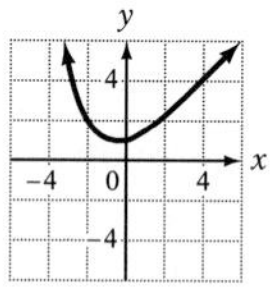

−2, 1.7 (Obj. 10.1.2) **44.**

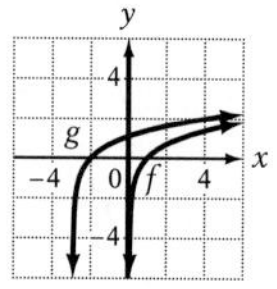

(Obj. 10.3.1)

45. The range of scores is 69 or better. (Obj. 2.5.3) **46.** The cyclist traveled 40 mi. (Obj. 2.3.2) **47.** The amount invested at 8.5% is $8000. The amount invested at 6.4% is $4000. (Obj. 2.4.1) **48.** The width is 7 ft and the length is 20 ft. (Obj. 8.4.1) **49.** The number of additional shares to be purchased is 200. (Obj. 6.5.1) **50.** The rate of the plane is 420 mph. (Obj. 6.4.3) **51.** The distance the object has fallen is 88 ft. (Obj. 7.5.2) **52.** The rate of the plane for the first 360 mi is 120 mph. (Obj. 8.4.1) **53.** The intensity is 200 lumens. (Obj. 6.5.2) **54.** The rate of the boat in calm water is 12.5 mph. The rate of the current is 2.5 mph. (Obj. 4.4.1) **55.** The value of the investment is $4785.65. (Obj. 10.5.1) **56.** The value of the house will be $256,571. (Obj. 11.3.4)

GLOSSARY

abscissa The first number in an ordered pair; it measures a horizontal distance and is also called the first coordinate of an ordered pair. (Section 3.1)

absolute value of a number The distance of the number from zero on the number line. (Sections 1.1/2.6)

absolute value equation An equation that contains the absolute value symbol. (Section 2.6)

addition method Method of finding an exact solution of a system of linear equations. (Section 4.2)

additive inverses Numbers that are the same distance from zero on the number line but lie on different sides of zero. (Sections 1.1/1.3)

analytic geometry Geometry in which a coordinate system is used to study relationships between variables. (Section 3.1)

antilogarithm If $\log_b M = N$, then the antilogarithm, base b, of N is M. (Section 10.2)

arithmetic progression A sequence in which the difference between any two consecutive terms is constant; also called an arithmetic sequence. (Section 11.2)

arithmetic sequence A sequence in which the difference between any two consecutive terms is constant; also called an arithmetic progression. (Section 11.2)

arithmetic series The indicated sum of the terms of an arithmetic sequence. (Section 11.2)

asymptotes The two straight lines that a hyperbola "approaches." (Section 12.3)

axes The two number lines that form a rectangular coordinate system; also called coordinate axes. (Section 3.1)

axis of symmetry of a parabola A line of symmetry that passes through the vertex of the parabola. (Sections 8.6/12.1)

base In an exponential expression, the number that is taken as a factor as many times as indicated by the exponent. (Section 1.2)

binomial A polynomial of two terms. (Section 5.2)

binomial expansion A method of finding a natural number power of a binomial. (Section 11.4)

center of a circle The central point that is equidistant from all the points that make up a circle. (Section 12.2)

center of a hyperbola The point halfway between the two vertices. (Section 12.3)

center of an ellipse The intersection of the two axes of symmetry of the ellipse. (Section 12.3)

characteristic The integer part of a common logarithm. (Section 10.2)

circle The set of all points (x, y) in the plane that are a fixed distance from a given point (h, k) called the center. (Section 12.2)

clearing denominators Multiplying each side of an equation by the least common multiple of the denominators. (Section 7.4)

closed interval An interval that contains its endpoints. (Section 1.1)

cofactor of an element of a matrix $(-1)^{i+j}$ times the minor of that element, where i is the row number of the element and j is its column number. (Section 4.3)

combined variation A variation in which two or more types of variation occur at the same time. (Section 6.5)

common difference of a sequence The difference between any two consecutive terms in an arithmetic progression. (Section 11.2)

common logarithms Logarithms to the base 10. (Section 10.2)

common ratio of a sequence In a geometric sequence, each successive term of the sequence is the same nonzero constant multiple of the preceding term. This common multiple is called the common ratio of the sequence. (Section 11.3)

completing the square Adding to a binomial the constant term that makes it a perfect-square trinomial. (Section 8.2)

complex fraction A fraction whose numerator or denominator contains one or more fractions (Sections 1.2/6.3)

complex number A number of the form $a + bi$, where a and b are real numbers and $i = \sqrt{-1}$. (Section 7.3)

composite number A natural number that is not a prime number. (Section 1.1)

composition of functions The operation on two functions f and g denoted by $f \circ g$. The value of the composition of f and g is given by $(f \circ g)(x) = f[g(x)]$. (Section 9.3)

compound inequality Two inequalities joined with a connective word such as "and" or "or." (Section 2.5)

compound interest Interest that is computed not only on the original principal but also on the interest already earned. (Section 10.5)

conditional equation An equation that is true if the variable it contains is replaced by the proper value. $x + 2 = 5$ is a conditional equation. (Section 2.1)

conic section A curve that can be constructed from the intersection of a plane and a right circular cone. The four conic sections are the parabola, hyperbola, ellipse, and circle. (Section 12.1)

conjugates Binomial expressions that differ only in the sign of a term. The expressions $a + b$ and $a - b$ are conjugates. (Section 7.2)

consecutive even integers Even integers that follow one another in order. (Section 2.2)

consecutive integers Integers that follow one another in order. (Section 2.2)

consecutive odd integers Odd integers that follow one another in order. (Section 2.2)

constant function A function given by $f(x) = b$, where b is a constant. (Section 3.3)

constant of proportionality k in the direct variation equation $y = kx$; also called the constant of variation. (Section 6.5)

constant of variation k in the direct variation equation $y = kx$; also called the constant of proportionality. (Section 6.5)

constant term A term that contains no variable part. (Section 1.3)

coordinate axes The two number lines that form a rectangular coordinate system; also called simply axes. (Section 3.1)

coordinates of a point The numbers in the ordered pair that is associated with the point. (Section 3.1)

Cramer's Rule A method of solving a system of equations. (Section 4.3)

cube root of a perfect cube One of the three equal factors of the perfect cube. (Section 5.6)

cubic function A third-degree polynomial function. (Section 5.2)

degree of a monomial The sum of the exponents of the variables. (Section 5.1)

degree of a polynomial The greatest of the degrees of any of its terms. (Section 5.2)

dependent system of equations A system of equations whose graphs coincide. (Section 4.1)

dependent variable A variable whose value depends on that of another variable, which is known as the independent variable. (Section 3.2)

descending order The terms of a polynomial in one variable are arranged in descending order when the exponents of the variable decrease from left to right. (Section 5.2)

determinant The number associated with a square matrix. (Section 4.3)

direct variation A special function that can be expressed as the equation $y = kx$, where k is a constant called the constant of variation or the constant of proportionality. (Section 6.5)

discriminant For an equation of the form $ax^2 + bx + c = 0$, the quantity $b^2 - 4ac$ is called the discriminant. (Sections 4.3/8.2)

distance between two points on the number line The absolute value of the difference between the coordinates of the two points. (Section 2.6)

domain The set of the first coordinates of all the ordered pairs of a function. (Section 1.1/3.2)

double root When a quadratic equation has two solutions that are the same number, the solution is called a double root of the equation. (Section 8.1)

element of a matrix A number in a matrix. (Section 4.3)

element of a set The elements in the set. (Section 1.1)

ellipse An oval shape that is one of the conic sections. (Section 12.3)

empty set The set that contains no elements. (Section 1.1)

equation A statement of the equality of two mathematical expressions. (Section 2.1)

equivalent equations Equations that have the same solution. (Section 2.1)

evaluating the function Determining $f(x)$ for a given value of x. (Section 3.2)

evaluating a variable expression Replacing the variable in a variable expression by a numerical value and then simplifying. (Section 1.3)

even integer An integer that is divisible by 2. (Section 2.2)

expanding by cofactors A technique for finding the value of a 3×3 or larger determinant. (Section 4.3)

exponent In an exponential expression, the raised number that indicates how many times the factor, or base, occurs in the multiplication. (Section 1.2)

exponential equation An equation in which the variable occurs in the exponent. (Section 10.4)

exponential form The expression 2^6 is in exponential form. Compare factored form. (Section 1.2)

exponential function The exponential function with base b is defined by $f(x) = b^x$, where b is a positive real number not equal to 1. (Section 10.1)

factored form The multiplication $2 \cdot 2 \cdot 2 \cdot 2 \cdot 2$ is in factored form. Compare exponential form. (Section 1.2)

factoring a polynomial Writing the polynomial as a product of other polynomials. (Section 5.5)

factoring a quadratic trinomial Expressing the trinomial as the product of two binomials. (Section 5.5)

finite sequence A sequence that contains a finite number of terms. (Section 11.1)

finite set A set in which all the elements can be listed. (Section 1.1)

first coordinate of an ordered pair The first number of the ordered pair; it measures a horizontal distance and is also called the abscissa. (Section 3.1)

first-degree equation An equation of the form $ax + b = 0$, where a is not equal to zero. (Section 2.1)

FOIL A method of finding the product of two binomials. The letters stand for First, Outer, Inner, and Last. (Section 5.3)

function A relation in which no two ordered pairs that have the same first coordinate have different second coordinates. (Section 3.2)

functional notation Notation used for those equations that define functions. The letter f is commonly used to name a function. (Section 3.2)

general term of a sequence In the sequence $a_1, a_2, a_3, \ldots, a_n, \ldots$, the general term of the sequence is a_n. (Section 11.1)

geometric progression A sequence in which each successive term of the sequence is the same nonzero constant multiple of the preceding term; also called a geometric sequence. (Section 11.3)

geometric sequence A sequence in which each successive term of the sequence is the same nonzero constant multiple of the preceding term; also called a geometric progression. (Section 11.3)

geometric series The indicated sum of the terms of a geometric sequence. (Section 11.3)

graph of a function A graph of the ordered pairs that belong to the function. (Section 3.3)

graph of a quadratic inequality in two variables A region of the plane that is bounded by one of the conic sections. (Section 12.5)

graph of a real number A heavy dot placed directly above the number on the number line. (Section 1.1)

graph of an ordered pair The dot drawn at the coordinates of the point in the plane. (Section 3.1)

graphing a point in the plane Placing a dot at the location given by the ordered pair; also called plotting a point in the plane. (Section 3.1)

greatest common factor (GCF) The GCF of two or more numbers is the largest integer that divides all of them. (Section 1.2)

half-plane The solution set of a linear inequality in two variables. (Section 3.7)

horizontal shift The displacement of a graph to the right or left on the coordinate axes; also known as a horizontal translation. (Section 9.2)

horizontal translation The displacement of a graph to the right or left on the coordinate axes; also known as a horizontal shift. (Section 9.2)

hyperbola A conic section formed by the intersection of a right circular cone and a plane perpendicular to the base of the cone. (Section 12.3)

hypotenuse In a right triangle, the side opposite the 90° angle. (Section 7.5)

identity An equation in which any replacement for the variable will result in a true equation. $x + 2 = x + 2$ is an identity. (Section 2.1)

imaginary number A number of the form ai, where a is a real number and $i = \sqrt{-1}$. (Section 7.3)

imaginary part of a complex number For the complex number $a + bi$, b is the imaginary part. (Section 7.3)

inconsistent system of equations A system of equations that has no solution. (Section 4.1)

independent system of equations A system of equations that has exactly one solution. (Section 4.1)

independent variable A variable whose value determines that of another variable, which is known as the dependent variable. (Section 3.2)

index In the expression $\sqrt[n]{a}$, n is the index of the radical. (Section 7.1)

infinite geometric series The indicated sum of the terms of an infinite geometric sequence. (Section 11.3)

infinite sequence A sequence that contains an infinite number of terms. (Section 11.1)

infinite set A set in which all the elements cannot be listed. (Section 1.1)

integers The numbers $\ldots, -3, -2, -1, 0, 1, 2, 3, \ldots$. (Section 1.1)

intersection of two sets The set that contains all elements that are common to both of the sets. (Section 1.1)

interval notation A type of set-builder notation in which the property that distinguishes the elements of the set is their location within a specified interval. (Section 1.1)

inverse of a function The set of ordered pairs formed by reversing the coordinates of each ordered pair of the function. (Section 9.4)

inverse variation A function that can be expressed as the equation $y = k/x$, where k is a constant. (Section 6.5)

irrational number The decimal representation of an irrational number never terminates or repeats. (Sections 1.1/7.2)

joint variation A variation in which a variable varies directly as the product of two or more variables. A joint variation can be expressed as the equation $z = kxy$, where k is a constant. (Section 6.5)

least common multiple (LCM) The LCM of two or more numbers is the smallest number that is a multiple of each of those numbers. (Section 1.2)

least common multiple of two or more polynomials The simplest polynomial of least degree that contains the factors of each polynomial. (Section 6.2)

leg In a right triangle, one of the two sides that are not opposite the 90° angle. (Section 7.5)

like terms Terms of a variable expression that have the same variable part. Having no variable part, constant terms are like terms. (Section 1.3)

linear equation in three variables An equation of the form $Ax + By + Cz = D$, where A, B, C, and D are constants. (Section 4.2)

linear equation in two variables An equation of the form $y = mx + b$ or $Ax + By = C$. (Section 3.3)

linear function A function that can be expressed in the form $f(x) = mx + b$. Its graph is a straight line. (Section 3.3)

linear inequality in two variables An inequality of the form $y > mx + b$ or $Ax + By > C$. The symbol $>$ could be replaced by $\geq$, $<$, or $\leq$. (Section 3.7)

literal equation An equation that contains more than one variable. (Section 6.6)

logarithm For b greater than zero and not equal to 1, the statement y = the logarithm of x, base b, is equivalent to $x = b^y$. (Section 10.2)

main diagonal The elements $a_{11}, a_{22}, a_{33}, \ldots, a_{nm}$ of a matrix. (Section 4.3)

mantissa The decimal part of a common logarithm. (Section 10.2)

matrix A rectangular array of numbers. (Section 4.3)

maximum The value of a function at its highest point. (Section 8.7)

minimum The value of a function at its lowest point. (Section 8.7)

minor of an element The minor of an element in a 3×3 determinant is the 2×2 determinant obtained by eliminating the row and column that contain the element. (Section 4.3)

monomial A number, a variable, or a product of a number and variables. (Section 5.1)

monomial factor The GCF of all the terms in a polynomial. (Section 5.5)

multiplicative inverse The multiplicative inverse of a nonzero real number a is $1/a$; is also called the reciprocal of a. (Sections 1.2/1.3)

***n* factorial** The product of the first n natural numbers; n factorial is written $n!$. (Section 11.4)

natural exponential function The function defined by $f(x) = e^x$, where $e \approx 2.71828$. (Section 10.1)

natural logarithm When e (the base of the natural exponential function) is used as the base of a logarithm, the logarithm is referred to as the natural logarithm and is abbreviated $\ln x$. (Section 10.2)

natural numbers The numbers 1, 2, 3,...; also called the positive integers. (Section 1.1)

negative integers The numbers ..., −3, −2, −1. (Section 1.1)

negative slope The slope of a line that slants downward to the right. (Section 3.4)

nonfactorable over the integers A polynomial is nonfactorable over the integers if it does not factor using only integers. (Section 5.5)

nonlinear system of equations A system of equations in which one or more of the equations are not linear equations. (Section 12.4)

n*th root of *a A number b such that $b^n = a$. The nth root of a can be written $a^{\frac{1}{n}}$ or $\sqrt[n]{a}$. (Section 7.1)

null set A set that contains no elements; also called an empty set. (Section 1.1)

numerical coefficient The number part of a variable term. (Section 1.3)

odd integer An integer that is not divisible by 2. (Section 2.2)

one-to-one function In a one-to-one function, given any y, there is only one x that can be paired with the given y. (Section 9.4)

open interval An interval that does not contain its endpoints. (Section 1.1)

opposites Numbers that are the same distance from zero on the number line but lie on different sides of zero; also called additive inverses. (Section 1.1)

order $m \times n$ A matrix of m rows and n columns is of order $m \times n$. (Section 4.3)

Order of Operations Agreement Rules that specify the order in which operations are performed in simplifying numerical expressions. (Section 1.2)

ordered pair A pair of numbers expressed in the form (a, b) and used to locate a point in a rectangular coordinate system. (Section 3.1)

ordinate The second number of an ordered pair; it measures a vertical distance and is also called the second coordinate of an ordered pair. (Section 3.1)

origin The point of intersection of the two number lines that form a rectangular coordinate system. (Section 3.1)

parabola The graph of a quadratic function is called a parabola. (Sections 8.6/12.1)

parallel lines Lines that have the same slope and thus do not intersect. (Section 3.6)

Pascal's Triangle A pattern for the coefficients of the terms of the expansion of the binomial $(a + b)^n$ can be formed by writing the coefficients in a triangular array known as Pascal's Triangle. (Section 11.4)

percent mixture problems A problem that involves combining two ingredients with different percent concentrations into a single mixture. (Section 2.4)

perfect cube The product of the same three factors. (Section 5.6)

perfect square The product of a term and itself. (Section 5.6)

perpendicular lines Lines that intersect at right angles. (Section 3.6)

plotting a point in the plane Placing a dot at the location given by the ordered pair; also called graphing a point in the plane. (Section 3.1)

point-slope formula The equation $y - y_1 = m(x - x_1)$, where m is the slope of a line and (x_1, y_1) is a point on the line. (Section 3.5)

polynomial A variable expression in which the terms are monomials. (Section 5.2)

positive integers The numbers 1, 2, 3,...; also called the natural numbers. (Section 1.1)

positive slope The slope of a line that slants upward to the right. (Section 3.4)

prime number A number that is divisible only by itself and 1. (Section 1.1)

principal square root The positive square root. (Section 7.1)

proportion An equation that states the equality of two ratios or rates. (Section 6.5)

Pythagorean Theorem The square of the hypotenuse of a right triangle is equal to the sum of the squares of the two legs. (Section 7.5)

quadrant One of the four regions into which a rectangular coordinate system divides the plane. (Section 3.1)

quadratic equation An equation of the form $ax^2 + bx + c = 0$, where a, b, and c are constants and a is not equal to zero; also called a second-degree equation. (Sections 7.3/8.1)

quadratic formula A general formula derived by applying the method of completing the square to the standard form of a quadratic equation. (Section 8.2)

quadratic function A function that can be expressed by the equation $f(x) = ax^2 + bx + c$, where a is not equal to zero. (Sections 5.2/8.6)

quadratic in form An equation that can be written in the form $au^2 + bu + c = 0$. (Section 8.3)

quadratic inequality in one variable An inequality that can be written in the form $ax^2 + bx + c < 0$ or $ax^2 + bx + c > 0$, where a is not equal to zero. The symbols $\leq$ and $\geq$ can also be used. (Section 8.5)

quadratic trinomial A trinomial of the form $ax^2 + bx + c$, where a, b, and c are nonzero constants. (Section 5.5)

radical The symbol $\sqrt{\ }$. (Section 7.1)

radical equation An equation that contains a variable expression in a radical. (Section 7.5)

radical function A function that contains a variable expression in a radical. (Section 7.4)

radicand In a radical expression, the expression within the radical. (Section 7.1)

radius The fixed distance, from the center of a circle, of all points that make up the circle. (Section 12.2)

range The set of the second coordinates of all the ordered pairs of a function. (Section 3.3)

rate The quotient of two quantities that have different units. (Section 6.5)

rate of work The part of a task that is completed in one unit of time. (Section 6.4)

ratio The quotient of two quantities that have the same unit. (Section 6.5)

rational expression A fraction in which the numerator and denominator are polynomials. (Section 6.1)

rational function A function that is written in terms of a rational expression. (Section 6.1)

rational number A number of the form a/b, where a and b are integers and b is not equal to zero. (Section 1.1)

rationalizing the denominator The procedure used to remove a radical from the denominator. (Section 7.2)

real numbers The rational numbers and the irrational numbers taken together. (Section 1.1)

real part of a complex number For the complex number $a + bi$, a is the real part. (Section 7.3)

reciprocal The reciprocal of a nonzero number a is $1/a$; also called the multiplicative inverse. (Section 1.2)

reciprocal of a rational expression The rational expression with the numerator and denominator interchanged. (Section 6.2)

rectangular coordinate system A coordinate system formed by two number lines, one horizontal and one vertical, that intersect at the zero point of each line. (Section 3.1)

relation A set of ordered pairs. (Section 3.2)

repeating decimal A decimal formed when dividing the numerator of a fraction by its denominator, in which a digit or a sequence of digits in the decimal repeat infinitely. (Section 1.1)

root(s) of an equation The replacement value(s) of the variable that will make the equation true; also called the solution(s) of the equation. (Section 2.1)

roster method A method of designating a set by enclosing a list of its elements in braces. (Section 1.1)

scatter diagram A graph of ordered-pair data. (Section 3.1)

scientific notation Notation in which each number is expressed as the product of a number between 1 and 10 and a power of 10. (Section 5.1)

second coordinate of an ordered pair The second number of the ordered pair; it measures a vertical distance and is also called the ordinate. (Section 3.1)

second-degree equation An equation of the form $ax^2 + bx + c = 0$, where a, b, and c are constants and a is not equal to zero; also called a quadratic equation. (Section 8.1)

sequence An ordered list of numbers. (Section 11.1)

series The indicated sum of the terms of a sequence. (Section 11.1)

set A collection of objects. (Section 1.1)

set-builder notation A method of designating a set that makes use of a variable and a certain property that only elements of that set possess. (Section 1.1)

sigma notation Notation used to represent a series in a compact form; also called summation notation. (Section 11.1)

simplest form A rational expression is in simplest form when the numerator and denominator have no common factors. (Section 7.1) For radical expressions, see Section 7.2.

slope A measure of the slant, or tilt, of a line. (Section 3.4)

slope-intercept form of a straight line The equation $y = mx + b$, where m is the slope of the line and $(0, b)$ is the y-intercept. (Section 3.4)

solution(s) of an equation The replacement value(s) of the variable that will make the equation true; also called the root(s) of the equation. (Section 2.1)

solution of an equation in two variables An ordered pair whose coordinates make the equation a true statement. (Section 3.1)

solution of a system of equations in three variables An ordered triple that is a solution of each equation of the system. (Section 4.2)

solution of a system of equations in two variables An ordered pair that is a solution of each equation of the system. (Section 4.1)

solution set of an inequality A set of numbers, each element of which, when substituted for the variable, results in a true inequality. (Section 2.5)

solution set of a system of inequalities The intersection of the solution sets of the individual inequalities. (Section 4.5)

solving an equation Finding a root, or solution, of the equation. (Section 2.1)

square matrix A matrix that has the same number of rows as columns. (Section 4.3)

square root of a perfect square One of the two equal factors of the perfect square. (Section 5.6)

standard form of a quadratic equation A quadratic equation is in standard form when the polynomial is in descending order and is equal to zero. (Section 8.1)

substitution method Method of finding a solution of a system of linear equations. (Section 4.1)

summation notation Notation used to represent a series in a compact form; also called sigma notation. (Section 11.1)

synthetic division A shorter method of dividing a polynomial by a binomial of the form $x - a$. This method uses only the coefficients of the variable terms. (Section 5.4)

system of equations Two or more equations considered together. (Section 4.1)

system of inequalities Two or more inequalities considered together. (Section 4.5)

term of a sequence Each of the numbers in the sequence. (Section 11.1)

terminating decimal A decimal formed when dividing the numerator of a fraction by its denominator and the remainder is zero. (Section 1.1)

terms of a variable expression The addends of the expression. (Section 1.3)

tolerance of a component The acceptable amount by which the component may vary from a given measurement. (Section 2.6)

trinomial A polynomial of three terms. (Section 5.2)

undefined slope The slope of a vertical line is undefined. (Section 3.4)

uniform motion The motion of an object whose speed and direction do not change. (Section 2.3)

union of two sets The set that contains all elements that belong to either of the sets. (Section 1.1)

value mixture problem A problem that involves combining two ingredients that have different prices into a single blend. (Section 2.3)

value of a function The value of the dependent variable for a given value of the independent variable. (Section 3.2)

variable A letter of the alphabet used to stand for some number. (Section 1.1)

variable expression An expression that contains one or more variables. (Section 1.3)

variable term A term composed of a numerical coefficient and a variable part. When the numerical coefficient is 1 or -1, the 1 is not usually written. (Section 1.3)

vertex of a parabola The point on the parabola with the smallest y-coordinate or the largest y-coordinate. (Sections 8.6/12.1)

vertical shift The displacement of a graph up or down on the coordinate axes; also known as a vertical translation. (Section 9.2)

vertical translation The displacement of a graph up or down on the coordinate axes; also known as a vertical shift. (Section 9.2)

whole numbers The numbers 0, 1, 2, 3,.... The whole numbers include the natural numbers and zero. (Section 1.1)

***x*-coordinate** The abscissa in an ordered pair. (Section 3.1)

***x*-intercept** The point at which a graph crosses the x-axis. (Section 3.3)

***xy*-coordinate system** A rectangular coordinate system in which the horizontal axis is labeled x and the vertical axis is labeled y. (Section 3.1)

***y*-coordinate** The ordinate in an ordered pair. (Section 3.1)

***y*-intercept** The point at which a graph crosses the y-axis. (Section 3.3)

zero slope The slope of a horizontal line. (Section 3.4)

INDEX

Table of Properties

Properties of Real Numbers

The Associative Property of Addition
If a, b, and c are real numbers, then $(a + b) + c = a + (b + c)$.

The Commutative Property of Addition
If a and b are real numbers, then $a + b = b + a$.

The Addition Property of Zero
If a is a real number, then $a + 0 = 0 + a = a$.

The Multiplication Property of Zero
If a is a real number, then $a \cdot 0 = 0 \cdot a = 0$.

The Inverse Property of Addition
If a is a real number, then $a + (-a) = (-a) + a = 0$.

The Associative Property of Multiplication
If a, b, and c are real numbers, then $(a \cdot b) \cdot c = a \cdot (b \cdot c)$.

The Commutative Property of Multiplication
If a and b are real numbers, then $a \cdot b = b \cdot a$.

The Multiplication Property of One
If a is a real number, then $a \cdot 1 = 1 \cdot a = a$.

The Inverse Property of Multiplication
If a is a real number and $a \neq 0$, then $a \cdot \frac{1}{a} = \frac{1}{a} \cdot a = 1$.

Distributive Property
If a, b, and c are real numbers, then $a(b + c) = ab + ac$ or $(b + c)a = ba + ca$.

Properties of Equations

Addition Property of Equations
The same number or variable term can be added to each side of an equation without changing the solution of the equation.

Multiplication Property of Equations
Each side of an equation can be multiplied by the same nonzero number without changing the solution of the equation.

Properties of Inequalities

Addition Property of Inequalities
If $a > b$, then $a + c > b + c$.
If $a < b$, then $a + c < b + c$.

Multiplication Property of Inequalities
If $a > b$ and $c > 0$, then $ac > bc$.
If $a < b$ and $c > 0$, then $ac < bc$.
If $a > b$ and $c < 0$, then $ac < bc$.
If $a < b$ and $c < 0$, then $ac > bc$.

Properties of Exponents

If m and n are integers, then $x^m \cdot x^n = x^{m+n}$.

If m and n are integers, then $(x^m)^n = x^{mn}$.

If $x \neq 0$, then $x^0 = 1$.

If m and n are integers and $x \neq 0$, then $\frac{x^m}{x^n} = x^{m-n}$.

If m, n, and p are integers, then $(x^m \cdot y^n)^p = x^{mp}y^{np}$.

If n is a positive integer and $x \neq 0$, then $x^{-n} = \frac{1}{x^n}$ and $\frac{1}{x^{-n}} = x^n$.

If m, n, and p are integers and $y \neq 0$, then $\left(\frac{x^m}{y^n}\right)^p = \frac{x^{mp}}{y^{np}}$.